AF616154

Organic Peroxides

Organic Peroxides

Edited by
W. ANDO
University of Tsukuba, Tsukuba, Japan

JOHN WILEY & SONS
Chichester · New York · Brisbane · Toronto · Singapore

Other Wiley Editorial Offices

John Wiley & Sons, Inc., 605 Third Avenue,
New York, NY 10158-0012, USA

Jacaranda Wiley Ltd, G.P.O. Box 859, Brisbane,
Queensland 4001, Australia

John Wiley & Sons (Canada) Ltd, 22 Worcester Road,
Rexdale, Ontario M9W 1L1, Canada

John Wiley & Sons (SEA) Pte Ltd, 37 Jalan Pemimpin #05-04,
Block B, Union Industrial Building, Singapore 2057

Library of Congress Cataloging-in-Publication Data:

Organic Peroxides / edited by W. Ando.
p. cm.
Includes bibliographical references and index.
ISBN 0 471 93438 0 (cloth)
1. Peroxides. I. Andō, Wataru, 1934-
QD305.E707 1992 92-1320
547′.23—dc20 CIP

British Library Cataloguing in Publication Data

A catalogue record for this book is available from the British Library
ISBN 0 471 93438 0

Typeset in Times 10/12pt by APS, Salisbury, Wiltshire
Printed and bound in Great Britain by Biddles Ltd, Guildford, Surrey

Contents

List of Contributors

W. Adam
Institut für Organische Chemie der Universität Würzburg, Am Hubland, W-8700 Würzburg, Germany

T. Akasaka
Department of Chemistry, University of Tsukuba, Tsukuba, Ibaraki 305, Japan

M. Akita
Research Laboratory of Resources Utilization, Tokyo Institute of Technology, 4259 Nagatsuda, Midori-ku, Yokohama 227, Japan

W. Ando
Department of Chemistry, University of Tsukuba, Tsukuba, Ibaraki 305, Japan

E. L. Clennan
Department of Chemistry, University of Wyoming, Laramie, WY 82071, USA

V. Conte
Centro di Studio sui Meccanismi di Reazioni Organiche-CNR, Istituto Chimica Organica dell'Università, Via Marzolo 1, I-35131 Padova, Italy

R. Curci
CNR 'Center MISO,' Department of Chemistry, University of Bari, 70126 Bari, Italy

F. Di Furia
Centro di Studio sui Meccanismi di Reazioni Organiche-CNR, Istituto Chimica Organica dell'Università, Via Marzolo 1, I-35131 Padova, Italy

C. S. Foote
Department of Chemistry and Biochemistry, University of California at Los Angeles, 405 Hilgard Avenue, Los Angeles, CA 90024–1569, USA

K. Fujimori
Department of Chemistry, University of Tsukuba, Tsukuba, Ibaraki 305, Japan

L. Hadjiarapoglou
Institut für Organische Chemie der Universität Würzburg, Am Hubland, W-8700 Würzburg, Germany

M. Heil
Institut für Organische Chemie der Universität Würzburg, Am Hubland, W-8700 Würzburg, Germany

M. Hirobe
Faculty of Pharmaceutical Sciences, University of Tokyo, Hongo, Bunkyo-ku, Tokyo 113, Japan

N. Kitajima
Research Laboratory of Resources Utilization, Tokyo Institute of Technology, 4259 Nagatsuda, Midori-ku, Yokohama 227, Japan

Y. H. Kim
Department of Chemistry, Korea Advanced Institute of Science and Technology, 373-1 Kusung-dong, Yusung-gu, Taejon 305-701

S. Matsugo
Department of Chemical and Biochemical Engineering, Faculty of Engineering, Toyama University, Gofuku 3190, Toyama 930, Japan

R. Mello
CNR Center 'MISO,' Department of Chemistry, University of Bari, 70126 Bari, Italy

K. J. McCullough
Department of Chemistry, Heriot-Watt University, Edinburgh EH14 4AS, UK

K. Mizuno
Department of Applied Chemistry, College of Engineering, University of Osaka Prefecture, Sakai, Osaka 591, Japan

G. Modena
Centro di Studio sui Meccanismi di Reazioni Organische-CNR, Istituto Chimica Organica dell'Università, Via Marzolo 1, I-35131 Padova, Italy

Y. Moro-oka
Research Laboratory of Resources Utilization, Tokyo Institute of Technology, 4259 Nagatsuda, Midori-ku, Yokohama 227, Japan

T. Mosandl
Institut für Organische Chemie der Universität Würzburg, Am Hubland, W-8700 Würzburg, Germany

T. Nagano
Faculty of Pharmaceutical Sciences, University of Tokyo, Hongo, Bunkyo-ku, Tokyo 113, Japan

E. Niki
Research Center for Advanced Science and Technology, University of Tokyo, 4–6–1 Komaba, Meguro-ku, Tokyo 153, Japan

M. Nojima
Department of Process Engineering, Osaka University, Yamadaoka, Suita, Osaka 565, Japan

Y. Otsuji
Department of Applied Chemistry, College of Engineering, University of Osaka Prefecture, Sakai, Osaka 591, Japan

B. Plesničar
Department of Chemistry, Edvard Kardelij University, PO Box 537, YU-641001 Ljubljana, Yugoslavia

N. A. Porter
Paul M. Gross Chemical Laboratory, Department of Chemistry, Duke University, Durham, NC 27706, USA

C. R. Saha-Möller
Institut für Organische Chemie der Universität Würzburg, Am Hubland, W-8700 Würzburg, Germany

I. Saito
Department of Synthetic Chemistry, Faculty of Engineering, Kyoto University, Kyoto 606, Japan

Y. Sawaki
Department of Applied Chemistry, Faculty of Engineering, Nagoya University, Chikusa-ku, Nagoya 464-01, Japan

K. Takada
Department of Chemistry, Faculty of Science, Hokkaido University, Sapporo 060, Japan

H. Takizawa
Faculty of Pharmaceutical Sciences, University of Tokyo, Hongo, Bunkyo-ku, Tokyo 113, Japan

K. Yamaguchi
Department of Chemistry, Faculty of Science, Hokkaido University, Sapporo 060, Japan

Preface

In the 30 years that have elapsed since the appearance of the first comprehensive edition of Davies's book in 1961, substantial changes have taken place in research into the chemistry of peroxides. The high reactivity of many organic peroxides is appealing, for it suggests that they may undergo a wide variety of reactions. During the past decade, these reactions have been receiving especially intensive study and extremely interesting results are continually appearing in the literature. Some important reactions with numerous new oxidants have been developed that can bring about certain selective oxidations. These include not only new types of stable peroxide, but also unstable peroxides which may play a role in biological systems. Synthetic and theoretical interest in the organic peroxides is also widespread and increasing rapidly, but no comprehensive account of the chemistry of these compounds is available. These treatises are devoted to detailed discussions of specific oxidants or topics involving oxidation of organic compounds.

The level of all chapters in this book is such that experts in these areas of research and students and researchers who require a thorough and rigorous discussion of these topics should find them useful. In general, emphasis is on the scope and preparative use as well as the mechanistic aspects of the various oxidations.

We hope that this volume will prove valuable to the active researchers in this field or those contemplating entering the field including organic, physical organic chemists, and biochemists. The editor will be very grateful to readers who inform him of mistakes and omissions relating to this book. In closing, let me acknowledge some of those who have contributed indirectly, though not insubstantially to the success of this project.

Tsukuba, January 9, 1992 Wataru Ando

Introduction

W. ADAM
Institute of Organic Chemistry, University of Würzburg, Am Hubland, W-8700 Würzburg, Germany

The chemistry of organic peroxides, which entails the synthesis, characterization and transformation of derivatives of hydrogen peroxide, has a long history and strong tradition. Unquestionably, it constitutes a prominent field of organic chemistry of pronounced multidisciplinary character, which ranges from biological aspects (involvement of peroxides in the autoxidative metabolism) to industrial applications (initiators for polymerization, cross-linking agents, bleaches, disinfectants, etc). The large volume of hydrogen peroxide (*ca* 1.5×10^6 tons per year) and derivatives that are industrially produced world-wide clearly expresses the commercial importance of the chemistry of organic peroxides.

The seminal work by Baeyers' School at the end of the last century matured to prominence through the intensive efforts of Bartlett, Criegee, Hock, Kharash, Milas, Prileshajew, Rieche, Rüchardt, Schenck, Walling, and numerous others. The Baeyer–Villiger oxidation of aldehydes and ketones to carboxylic acids and esters, the Hock cleavage of hydroperoxides (the industrial manufacture of acetone and phenol from cumene hydroperoxide), and the Prileshajew epoxidation of olefins to epoxides by peroxy acids, constitute monumental examples portraying the significance and value of organic peroxides for the synthetic chemist. As a culmination, the success of the enantioselective Sharpless epoxidation of allylic alcohols to afford optically active epoxy alcohols through the $Ti(OR)_4$-catalysed action of hydroperoxides in the presence of tartrate as chiral auxiliary, discovered only a little over a decade ago, must be mentioned. It is in no way inferior to what enzymatic oxyfunctionalizations are capable of.

Numerous monographs and treatises have been written on organic peroxides over the years, a 'classic' in my mind being A. G. Davies' book *Organic Peroxides*, which appeared about 30 years ago. In a compact and handy form (215 pages), it summarized, at the time, most of the important facets of the chemistry of organic peroxides, both on the preparative and mechanistic fronts. Unquestionably, it kindled much of the present-day flurry of activity in this field. For myself, I am happy to admit that this stimulating book enticed me to become active in the chemistry of peroxides, although fertile ground had been laid already in my doctoral thesis with F. D. Greene (MIT) and my postdoctoral interlude with R. Criegee (Karlsruhe). Three decades ago, Davies' monograph brought the novice up to date, while the expert found it to be a rich source for interesting projects. To update this book thirty years later is, because of the enormous developments in the field, a difficult task and requires an ambitious colleague like Professor Ando to confront this challenge.

In the meantime, others have not been idle! Besides the many reviews on the subject, the treatise (three volumes) by D. Swern, *Organic Peroxides*; the volume *Organic Peroxide Chemistry* in S. Patai's series; *Advances in Oxygenated Processes*, by A. L. Baumstark (Ed.); the volume of the proceedings *The Role of Oxygen in Chemistry and Biochemistry*, by W. Ando and Y. Moro-oka (Eds.); and the encyclopaedia (two volumes) by H. Kropf *Organische Peroxoverbindungen* in the Houben–Weyl collection (in German) must be mentioned. In particular, the latter, most recent, publication represents an invaluable source of information on which classes of organic peroxides are known and what one can do with them chemically. Yet, there is a need for a modern, streamlined coverage of the pertinent developments since Davies' monograph, to put the interested individual (layman and specialist alike) at the frontier of important happenings. Such an arduous undertaking is no longer manageable for one single author and Professor Ando was very astute in engaging the help of a variety of experts in the field. On one hand one bargains for the heterogeneity in the presentation of the personalized chapters, but on the other hand, one gains the necessary authoritative diversity and scope.

Much exciting work has been reported during the last thirty years and only a few of the most salient highlights can be addressed in these introductory remarks. For example, the prominence that photooxygenation (singlet oxygen) has attained as a valuable preparative method of organic peroxides is truly impressive and would no doubt please H. Kautsky, the discoverer of singlet oxygen. The fact that polyoxides $R—(O)_n—R'$ can be prepared, handled and used for oxidations is a remarkable breakthrough for peroxide chemistry. Although mentioned in the old peroxide literature (Baeyer, Staudinger), who would have thought three decades ago that dioxetanes and dioxiranes were sufficiently stable for isolation. The seminal work by Kopecky on dioxetanes and by Edwards on dioxiranes has unleashed intensive activity on these most strained cyclic peroxides. Through their chemiluminescent decomposition, the dioxetanes have been developed into valuable photochemical and photobiological probes, while dioxiranes have assumed the role of 'wonder oxidant' for the selective oxyfunctionalization of organic substrates. The latter development has received its impetus through the incisive work by R. W. Murray, who first succeeded in isolating dimethyldioxirane (as acetone solution).

The intervention of organic peroxides in biological transformations has particularly come into focus during the last thirty years. It should be quite obvious that in the cellular degradation of organic matter through enzymatic autoxidation (lipoxygenases, cytochrome P-450 monoxygenases, prostaglandin cycloxygenases, luciferases, etc.), peroxide intermediates must intervene. In fact, whenever free radical chemistry and redox processes (single-electron transfer) are involved in the presence of molecular oxygen, organic peroxides play an eminent role. However, the problem, of course, is to detect and isolate such labile peroxide intermediates. Notable examples are the α-peroxy lactone derived from firefly luciferin, which serves as the active ingredient in firefly bioluminescence and the endoperoxide in prostaglandin chemistry. In the case of 1,2,4-trioxane *qinghaosu*, a natural peroxide with potent antimalarial activity, nature has been more generous in that this material is sufficiently stable for isolation from the Chinese medicinal plant *Artemesia annua*. A number of other stable, biologically derived, peroxides could have been mentioned and undoubtedly many are yet waiting for discovery. Be this as it may, such natural products have kindled interesting peroxide chemistry over the last decades and much of it is covered in this monograph.

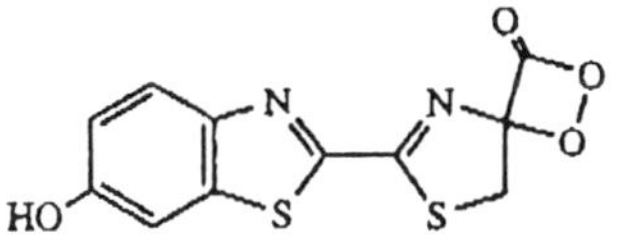

α-peroxy lactone from firefly luciferin

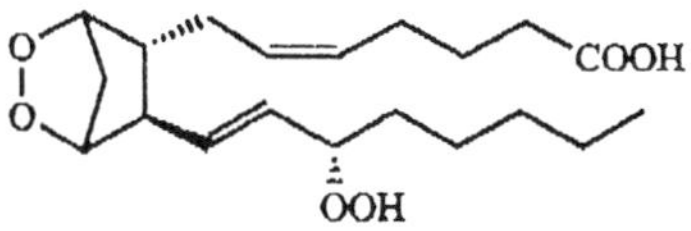

prostaglandin endoperoxide

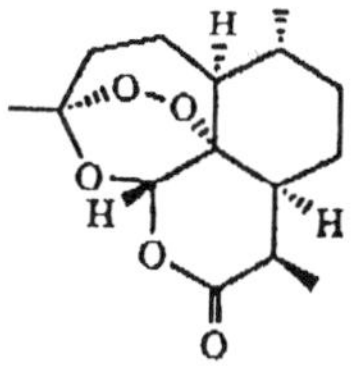

1,2,4-trioxane (*qinghaosu*)

The enzymatic generation of organic peroxides in the living cell necessarily implies pathological consequences. Among the most prominent disorders to be mentioned are ageing, ischaemia, diabetes and cancer. Although the biological cell is well equipped to combat peroxide-promoted cellular damage through inactivation of the peroxidic oxidants by the glutathione protection system and antioxidants such as ascorbic acid, tocopherol, etc., the oxidative destruction of biomolecules such as nucleic acids, lipids, sugars, proteins, hormones, etc., is inevitable. Much timely progress has been made in this area of biological chemistry over recent years, but the real challenges lie still ahead for the curious peroxide chemist.

As expected from a follow-up of Davies' monograph, the present sequel covers a broad scope of interesting topics of current peroxide chemistry. This ranges from a sound theoretical foundation (Chapter 1) to important biological and medical aspects (Chapters 14 and 15). Of course, all important classes of organic peroxides, e.g. alkyl hydroperoxides (Chapter 2), dialkyl peroxides (Chapter 3), dioxiranes (Chapter 4), dioxetanes (Chapter 5), endoperoxides (Chapter 6), diacyl peroxides and variations (Chapter 7 and 8), peroxy acids and esters (Chapter 9), polyoxides (Chapter 10), organometallic peroxides (Chapter 11), heteroatom peroxides (Chapter 12), and ozonides (Chapter 13) have been reviewed. This 'potpourri' of interesting peroxide chemistry, written by experts in the field, is hoped to stimulate further intensive activity in this fascinating area of multidisciplinary chemistry. Organic peroxide chemistry must for the next thirty years meet the challenge of developing oxidants which operate efficiently, selectively, catalytically, and environmentally safely, to provide mechanistic insight into oxygen transfer processes and to unravel the medical consequences of organic peroxides involved in enzymatic autoxidation. The present monograph is expected to serve this ambitious role and Professor Ando is to be congratulated for having had the courage to take on such an immense job and follow through with it to fruition. He and his contributors have done us all an important service!

Würzburg, February 28, 1992 Waldemar Adam

1 Theoretical and General Aspects of Organic Peroxides

KIZASHI YAMAGUCHI and KAZUHIRO TAKADA
Department of Chemistry, Faculty of Science, Hokkaido University, Sapporo 060, Japan

YOSHIO OTSUJI and KAZUHIKO MIZUNO
Department of Applied Chemistry, College of Engineering, University of Osaka Prefecture, Sakai, Osaka 591, Japan

Organic Peroxides. Edited by W. Ando

1 INTRODUCTION

Peroxy compounds play important roles in oxidation reactions.[1–3] In most cases, they are produced by the reaction with molecular dioxygen or its activated form, singlet dioxygen, superoxide anion radical, etc. An understanding of the chemical properties of these species is essential for the elucidation of the chemistry of oxidation of organic compounds. These species also play important roles in combustion, polymerization, biological metabolism, cancer, aging, pollution and various aspects of environmental science and technology.

In the past decade remarkable progress has been made in theoretical chemistry, particularly in computational procedures for the elucidation of the structures of unstable molecules and transition states and those of stable molecules.[4] These advanced theoretical methods have been applied to the determination of fully optimized geometries of unstable intermediates and also to the location of transition structures in oxidation reactions of organic compounds with various peroxy compounds. The mechanisms of many newly developed oxidation reactions have been elucidated by the use of such modern techniques.[5]

Among theoretical approaches, molecular orbital theory is now a useful technique for organic chemists and several efficient program packages for calculations are available.[6,7] The *ab initio* closed-shell Hartree–Fock molecular orbital method has been widely applied in determining equilibrium geometries, ionization potentials and other properties of organic peroxides. Cremer[8] has summarized the experimental and theoretical results for these species, particularly hydrogen peroxide, peracids and cyclic peroxy compounds such as ozonides. In preparing this chapter, attempts were made to avoid duplication of the material covered in Cremer's review, and instead attention has been focused mainly on results obtained after 1980.

As the development[4,5] of theoretical methods treating peroxy compounds has been remarkable, the computational methods adopted for this class of compounds are outlined in Section 2 and abbreviations and notations used in this review are summarized in Section 3. The electronic structures of selected peroxy compounds, including transition metal peroxo complexes and transition metal hydroperoxides, are described in Section 4. In Sections 5–8, some theoretical aspects of chemical reactions involving peroxy compounds are described. The topics described in this review are limited to those of chemical importance, and experimental details are omitted. Recent developments in the experimental chemistry of peroxy compounds are described in other chapters of this book.

2 THEORETICAL BACKGROUND

2.1 MO Theory for Heterolysis and Homolysis of Peroxides

The *ab initio* restricted Hartree–Fock (RHF) molecular orbital (MO) method is now routinely utilized for the elucidation of electronic structures of closed-shell organic molecules. The RHF wavefunction, $\Phi(\text{RHF})$, is defined by a single Slater determinant as

$$\Phi(\text{RHF}) = |\phi_1\bar{\phi}_1\phi_2\bar{\phi}_2\ldots\phi_N\bar{\phi}_N| \tag{1}$$

where the RHF MOs ϕ are determined by solving the so-called Hartree–Fock (HF) equation under the condition that two electrons with up and down spins occupy the same MO.[5]

In order to elucidate geometrical parameters such as bond lengths and angles, a variety of *ab initio* RHF calculations have been carried out for peroxy compounds of the type XOOY, where X and Y denote a variety of functional groups.[8] The results calculated for hydroperoxides, peracids, cyclic peroxides and other peroxy compounds at their equilibrium geometries are qualitatively consistent with available experimental data. Many of the theoretical results have already been summarized in a previous book.[2]

However, the intrinsic deficiency of the RHF MO method is that it cannot describe a dissociation process of a covalent bond even in a qualitative sense. An approach to this problem within the Hartree–Fock MO framework is to use the unrestricted HF (UHF) theory.[9] The UHF wavefunction is defined also by a single Slater determinant:[5,10]

$$\Psi(\mathrm{UHF}) = |\psi_1^+ \overline{\psi_1^-} \psi_2^+ \overline{\psi_2^-} \ldots \psi_N^+ \overline{\psi_N^-}| \tag{2}$$

The UHF MOs ψ_i^+ and ψ_i^- for up(+) and down(−) spins are different from each other and lead to the concept of the different orbitals for different spins (DODS). The bifurcation of the RHF MOs into the DODS MOs:

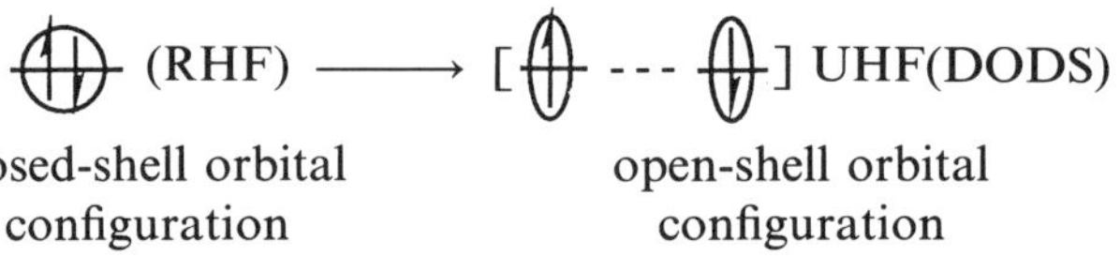

$$\text{(RHF)} \longrightarrow [\;\cdots\;]\ \text{UHF(DODS)}$$

closed-shell orbital configuration → open-shell orbital configuration

occurs at the triplet instability threshold. The triplet instability for the RHF solution can be examined by the stability condition implemented in the GAUSSIAN program package.[6]

For example, consider a dissociation process of the oxygen—oxygen (O—O) bond of H_2O_2

$$\mathrm{HO{\div}OH} \longrightarrow \mathrm{HO}\cdot\uparrow + \downarrow\cdot\mathrm{OH} \tag{3}$$

where $\div$ and $\cdot\uparrow(\downarrow)$ denote a covalent single bond and a localized electron with up (down) spin, respectively. Figure 1 illustrates the orbital correlation diagram obtained by using the 6–31G* basis set (see also Figure 5). The RHF solution provides the unreasonable orbital correlation for the O—O bond dissociation because of the closed-shell constraint in equation 1, which is chemically described as

$$\mathrm{HO{\div}OH} \longrightarrow [\mathrm{HO}\!:^- + {}^+\mathrm{OH}] \quad \text{or} \quad [\mathrm{HO}^+ + {}^-\!:\mathrm{OH}] \tag{4}$$

where − and + denote anionic and cationic sites, respectively. On the other hand, the UHF solution provides the proper bond dissociation process to give two hydroxyl radicals through a single diradical state, as can be seen from Figure 1A

$$\mathrm{HO{\div}OH} \longrightarrow [\mathrm{HO}\cdot\uparrow + \downarrow\cdot\mathrm{OH}] \longrightarrow 2\cdot\mathrm{OH} \tag{5}$$

The UHF orbital energies become identical with those of singly occupied MOs of two hydroxyl radicals at the dissociation limit.

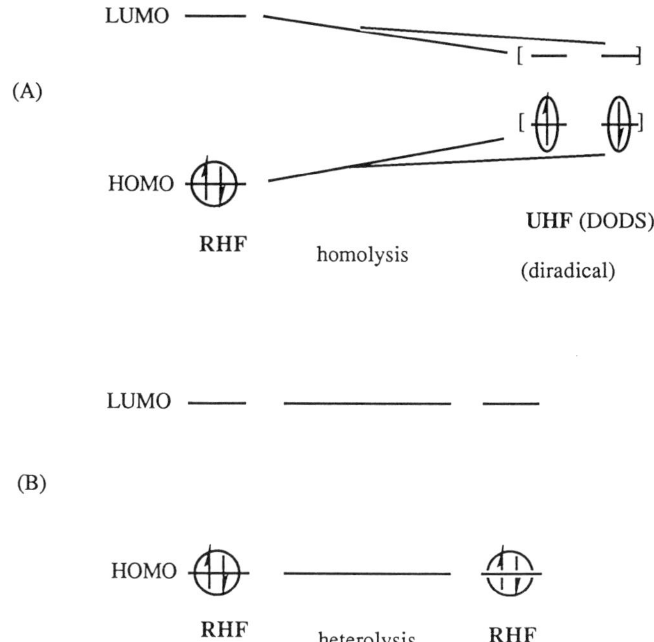

Figure 1. Orbital correlation diagrams (A) for homolysis of an oxygen—oxygen single bond by the unrestricted Hartree–Fock (UHF) method and (B) for its heterolysis by the restricted Hartree–Fock (RHF) method

Another example of a bond dissociation process is the heterolysis of the Li—O bond of LiOOR in water:

$$\text{LiOOR} + (\text{H}_2\text{O})_{m+n} \longrightarrow \text{Li}^+(\text{H}_2\text{O})_m + (\text{H}_2\text{O})_n{}^-\text{OOR} \qquad (6)$$

In this case, the closed-shell character remains unchanged throughout the heterolysis reaction as illustrated in Figure 1B. Therefore, the RHF solution is reliable enough for tracing this process by the MO theory.

Generally, the UHF MO method offers a qualitatively correct picture of the homolytic dissociation of a covalent bond. However, the UHF solution for the resulting singlet diradical suffers from the spin contaminations arising from high-spin states.[11] Procedures[11,12] for eliminating this deficiency will be discussed later.

A basic problem of *ab initio* approaches is that the methods cannot reproduce binding energies of chemical bonds within the Hartree–Fock framework (single Slater determinant) in general, because of the lack of electron correlation. The exact computation of this correlation energy is one of the important problems in quantum chemistry. A conventional approach to this problem is to use the multi-configurations (many Slater determinants) described below.

2.2 MCSCF Theory for Oxygen, Oxyradicals and Biradicals

A more reliable method for the elucidation of electronic structures of molecules, particularly of open-shell species such as oxy and peroxy radicals, is the multi-configuration self-consistent-field (MCSCF) method.[4] The MCSCF wavefunction is generally constructed from the many Slater determinants. In this section, characteristics of this theory are explained, in relation to the valence-bond theory.

2.2.1 Active electrons (spins) and active orbitals: a VB model

Besides the MO approach, valence-bond (VB)-type theories are often utilized for the description of open-shell species. In a simple VB approach, each electron (spin) enters into one-atomic orbital or one-localized orbital. For example, one, two and three unpaired electrons are considered for H, O and N atoms, respectively, and the directions of their s and p atomic orbitals are defined as follows:

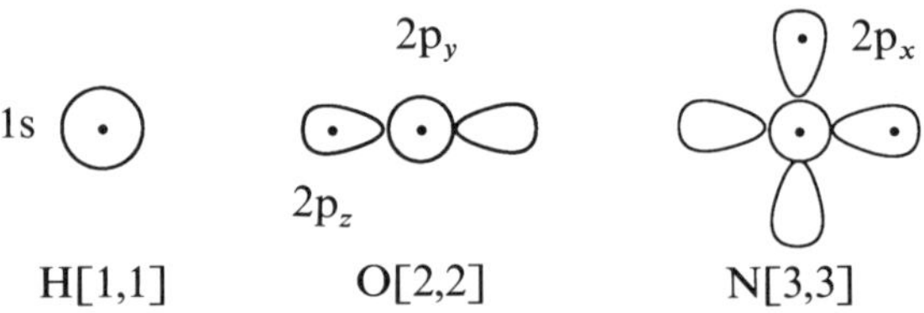

The H, O and N atoms are regarded as one orbital, one-electron [1,1], two-orbital, two-electron [2,2] and three-orbital, three-electron [3,3] systems, respectively.

The Heisenberg model (a simple VB model)[13] can be applied to the spin coupling processes of these atoms and is represented by

$$\mathbf{H} = -\Sigma 2 J_{ij} \mathbf{S}_i \cdot \mathbf{S}_j \tag{7}$$

where J_{ij} is the effective exchange integral and $\mathbf{S}_\mathrm{p}$ ($\mathrm{p} = i, j$) is the spin at the orbital p. According to the extended Kanamori–Goodenough (KG) rule for the radical coupling process,[13] the J_{ij} values for overlapping AOs are usually negative in sign and the overlapping AOs lead to the singlet-type spin coupling which is described schematically as (↑ ↓). On the other hand, the orthogonal AOs lead to the triplet-type spin coupling (↑ ↑).

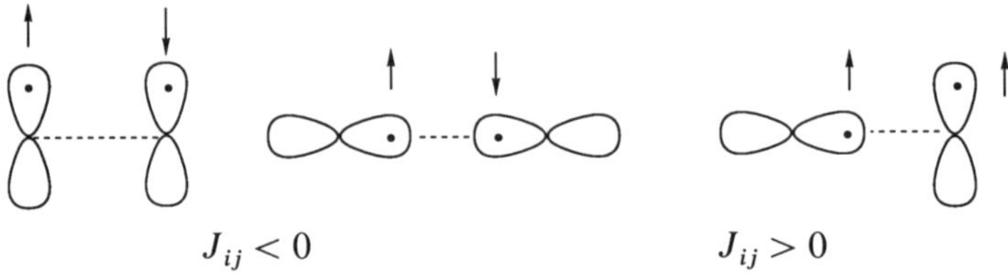

The extended KG rule[13] also indicates that in the ground state, atomic and molecular oxygens are triplet and atomic nitrogen is quartet.

The spin coupling processes for radical reactions, which satisfy the extended KG rule, can easily be described. Some examples are shown in Figure 2. The J_{ij} value for the recombination of H and O (process 1 in Figure 2) is negative in sign because of the orbital overlap between 1s and $2p_z$ AOs, and this process leads to formation of a singlet pair. The resulting hydroxyl radical is doublet and constitutes a three-orbital, three-electron [3,3] system. Similarly, the

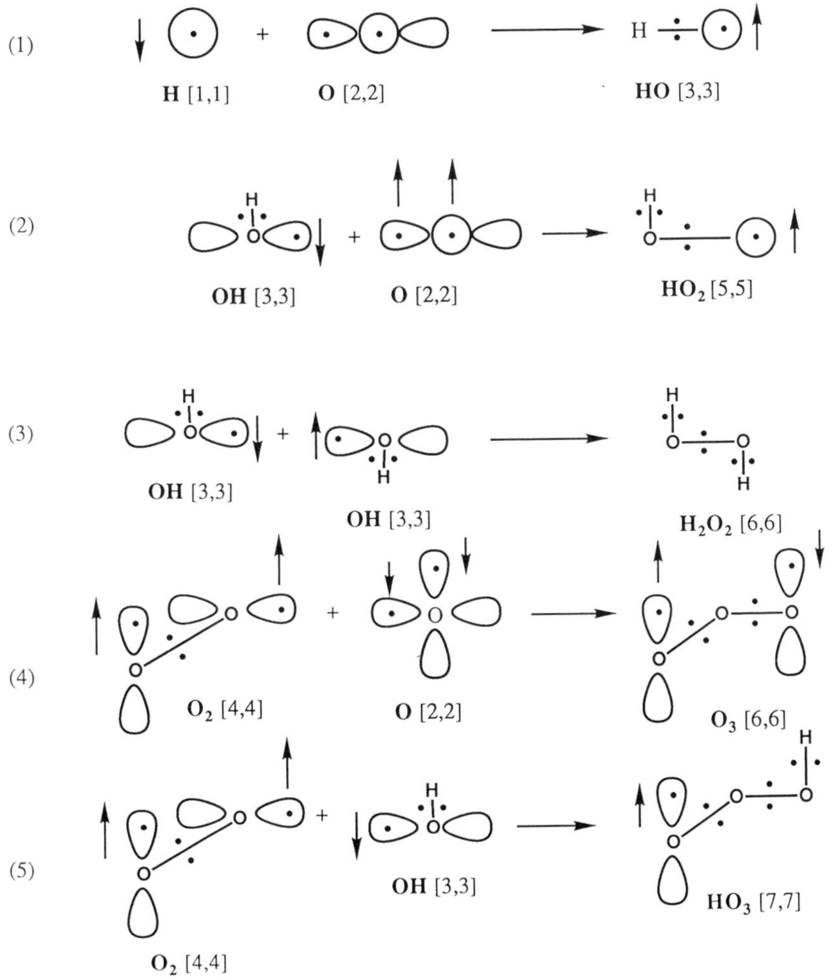

(1) **H** [1,1] + **O** [2,2] → **HO** [3,3]

(2) **OH** [3,3] + **O** [2,2] → **HO_2** [5,5]

(3) **OH** [3,3] + **OH** [3,3] → **H_2O_2** [6,6]

(4) **O_2** [4,4] + **O** [2,2] → **O_3** [6,6]

(5) **O_2** [4,4] + **OH** [3,3] → **HO_3** [7,7]

Figure 2. Spin coupling schemes for association reactions of oxygen and oxyradical on the basis of the Heisenberg model, where [*m*, *n*] denotes the *m*-orbital, *n*-electron system

recombination between doublet hydroxyl radical and triplet atomic oxygen (process 2) provides a low-spin (doublet) hydroperoxyl radical which is a five-orbital, five-electron [5,5] system. The recombination of two OH radicals (process 3) provides singlet hydrogen peroxide which is a six-orbital, six-electron [6,6] system. Ozone is regarded as a [6,6] system arising from the recombination between triplet molecular oxygen and triplet atomic oxygen (process 4)

Hydrogen peroxide can also be regarded as a two-orbital, two-electron system [2,2] if OH radical is regarded as a one-orbital, one-electron [1,1] system since the O—H single bond is not relevant in the recombination process (3). Ozone can also be regarded as a [2,2] system for a description of its singlet diradical state since the O—O σ-bond is stable near the equilibrium geometry. Hence the numbers of active orbitals and active electrons[14–16] participating in the recombination processes under consideration must be carefully selected; note that the concept of active orbital and active electron is an extended form of the concept of frontier orbital and frontier electron.

2.2.2 Selection of active orbitals and active electrons

The selection of active orbitals and active electrons is an important problem in the MCSCF calculations since it determines their reliability and accuracy. The determination of approximate natural orbitals is also of practical importance as trial orbitals for rapid convergence of the solution. These problems have been studied extensively in the past decade,[16] and the conclusions and implications of such studies have been summarized in a book.[4]

Yamaguchi[16] and Pulay and Hamilton[17] have shown that the UHF solutions can be used for selecting active orbitals and electrons and determining approximate natural orbitals that are starting trial orbitals for effective MCSCF calculations. For this purpose, the UHF solution of equation 2, which is represented by canonical MOs, is transformed into the corresponding MOs representation:[16]

$$\Psi(\text{UHF}) = |\chi_1\overline{\eta_1}\chi_2\overline{\eta_2}\cdots\chi_N\overline{\eta_N}| \tag{8}$$

where χ_i and η_i are the corresponding MOs which are given by approximate natural orbitals φ_j:

$$\chi_i = \cos\omega\ \varphi_{h-i} + \sin\omega\ \varphi_{l+i}\ (h = \text{HO}, l = \text{LU}) \tag{9a}$$

$$\eta_i = \cos\omega\ \varphi_{h-i} - \sin\omega\ \varphi_{l+i}\ (h = \text{HO}, l = \text{LU}) \tag{9b}$$

where ω is the orbital mixing parameter.

The natural orbital (NO) pairs (φ_{h-i}, φ_{l+i}) are usually the bonding and antibonding MOs for a chemical bond i. The occupation numbers (n) of these NOs are given by the orbital overlap between the corresponding orbital pairs (χ_i

and η_i) as

$$n_{h-i}(\text{UHF}) = 1 + T_i;\; n_{l+i}\,(\text{UHF}) = 1 - T_i \tag{10a}$$

where

$$T_{ij} = \int \chi_i \eta_j \, dS = \delta_{ij} \tag{11}$$

This means that the corresponding MO (CMO) χ_i for up-spin is orthogonal to all the other corresponding MOs η_j for down-spin except for its own partner MO η_i ($j = i$). Therefore, CMO expression enables us to select the diradical-type electron pairs in unstable molecules.

Figure 3 shows the corresponding π-MOs for ozone, which are left and right split orbitals such as those of the generalized valence-bond (GVB) orbitals. These orbitals are responsible for the singlet diradical character of ozone. On the

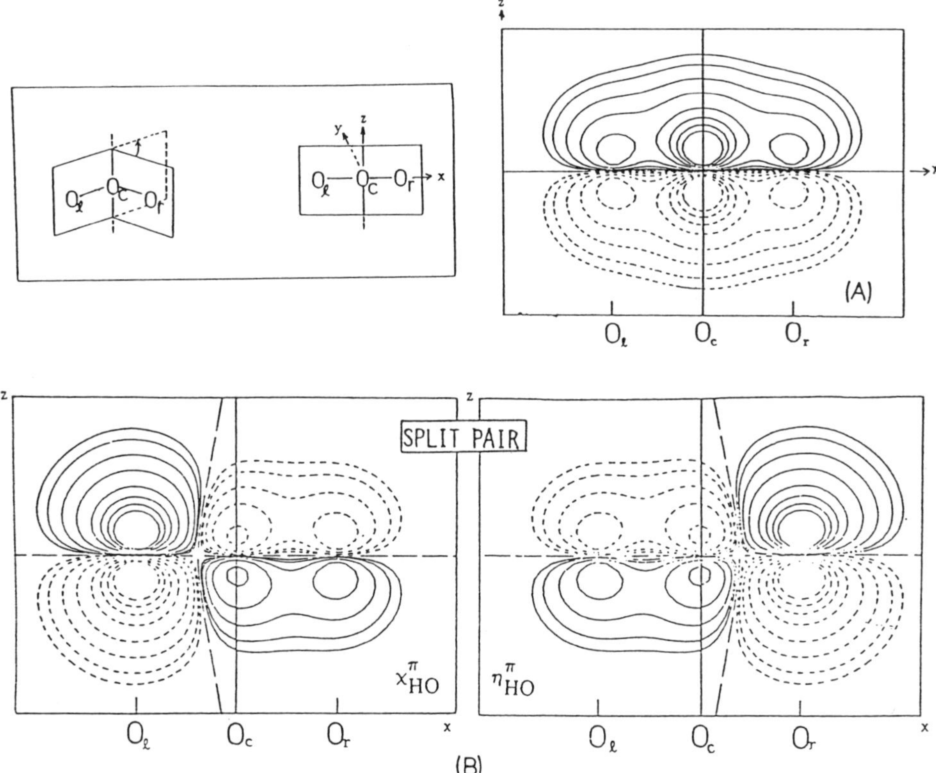

Figure 3. Contour maps of (A) the symmetric π-orbital (NHOMO) of ozone and (B) the split corresponding π-orbitals (HOMOs) (χ_{HO} and η_{HO})

other hand, the lower π-MO and σ-bonding MOs by UHF are symmetry-adapted closed-shell orbitals like RHF orbitals, in contrast to the bond orbitals in the GVB theory. Figure 4 illustrates schematically the shapes and occupation numbers of the UHF natural orbitals (UNO) for singlet ozone at the experimental geometry.[18] The occupation numbers of the highest occupied (HO) and lowest unoccupied (LU) NOs are about 1.24 and 0.76, respectively. The orbital overlap (T_{HO}) between χ_{HO} and η_{HO} is 0.24, which is close to the value ($T_{HO} = 0.28$) in the GVB theory.[19] The occupation numbers of the oxygen—oxygen antibonding orbitals are larger than 0.0025, showing that these are

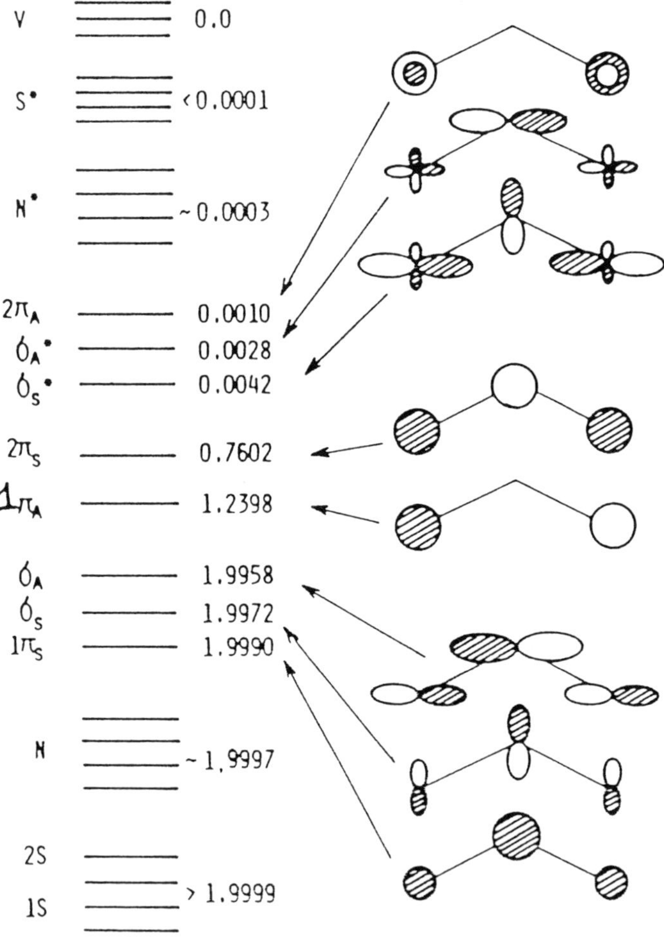

Figure 4. Shapes and occupation numbers (n) of natural orbitals of ozone for the experimental geometry by the UHF 4-31G method

active NOs for extensive MCSCF calculations. Hence the six NOs and six electrons [6,6] are selected from the occupation number of the UHF NOs without any artificial assumption or presumption for ozone. The [6,6] system thus selected is also consistent with the simple VB (Heisenberg) model for ozone. When the oxygen lone pairs of ozone are taken into consideration for more accurate MCSCF calculations, the π-type lone pair at the central oxygen should be primarily considered as an active orbital, as can be seen from the occupation numbers of UHF NOs in Figure 4. In this case, the eight orbitals and eight electrons [8,8] are considered as an active reaction subsystem (ARS).[14–16]

2.2.3 MCSCF calculations

There are several types of MCSCF wavefunctions. These wavefunctions are generally given by the linear combination of multiple Slater determinants as

$$\Phi(\text{MCSCF}) = \Sigma D_j \Phi_i \tag{12}$$

where both the expansion coefficients D_j and MOs ϕ_p involved in Φ_i are determined by the self-consistent-field (SCF) procedure like the RHF and UHF SCF calculations. The MCSCF wavefunction of the GVB perfect pairing (PP) type[19] includes special types of excited configurations in equation 12, which are responsible for a generalization of a simple VB model. In fact, the natural orbitals (NO) by GVB are bond orbitals.

bonding O—O bond NO

antibonding O—O bond NO

On the other hand, the wavefunction of the complete active space SCF(CASSCF) type[15] includes all of the possible Slater determinants constructed of the m active orbitals and n active electrons. The CASSCF wavefunctions are invariant under the orbital transformations within active orbitals, and therefore involve all the radical-type VB electron configurations in the Heisenberg model, together with all the ionic configurations. This corresponds to the configuration interaction (CI) in the VB theory. Hence the CASSCF solution can provide a well-balanced description of radical and ionic configurations in a chemical sense. When the UHF NOs (UNO) are utilized as trial orbitals for CASSCF calculations, this is called the UNO-CAS procedure.[16,17] The UNO-CAS procedure can be applied even to MCSCF calculations of complex radicals. For practical calculations by these procedures, readers can consult the GAUSSIAN 90 series of program packages.[6]

The interrelationship between the UHF and UNO–CAS methods is clear in the case of diradical species in which the corresponding orbitals are essentially equivalent to the closed-shell MOs ϕ except for a diradical pair:[20]

$$\chi_i = \eta_i \approx \phi_i \quad (i = 1, 2, \ldots, h-1) \tag{13}$$

The singlet UHF orbital configuration $\chi_{HO}\eta_{HO}$ for diradicals does not express the pure singlet state. The spin projection of the configuration into the singlet state (PUHF) is necessary for quantitative purposes:

$$\Psi(\text{PUHF}) = |\phi_1\overline{\phi_1}\phi_2\overline{\phi_2}\ldots\{\chi_{HO}\overline{\eta_{HO}} + \eta_{HO}\overline{\chi_{HO}}\}| \tag{14a}$$

$$= \cos^2\omega|\phi_1\overline{\phi_1}\phi_2\overline{\phi_2}\ldots\varphi_{HO}\overline{\varphi_{HO}}| - \sin^2\omega|\phi_1\overline{\phi_1}\phi_2\overline{\phi_2}\ldots\varphi_{LU}\overline{\varphi_{LU}}\}| \tag{14b}$$

where ω is defined by equation 9. The PUHF wavefunction is just the 2×2 multi-configuration wavefunction constructed from the ground- and doubly excited-state configurations constructed of UNO.

The diradical character y is defined by the weight (W_d) of the doubly excited configuration in the MCSCF theory and it is formally expressed by the orbital overlap T_{HO} in the PUHF and GVB theories:[20]

$$y = 2\,W_d = 1 - 2T_{HO}/(1 + T_{HO}^2) \tag{15a}$$

The diradical characters by the PUHF method are 0 and 100% for closed-shell and pure diradical states, respectively. The occupation numbers of UNO by the PUHF method are generally given by

$$n_{h-i}(\text{PUHF}) = [n_{h-i}(\text{UHF})]^2/(1 + T_{HO}^2) \tag{10b}$$

$$n_{l+i}(\text{PUHF}) = [n_{l+i}(\text{UHF})]^2/(1 + T_{HO}^2) \tag{10c}$$

Therefore, the occupation number of LUMO is just the diradical character for chemical bond i:

$$n_{LU}(\text{PUHF}) = y \tag{15b}$$

Bofill and Pulay[21] carried out UNO–CAS calculations of an O—O dissociated hydrogen peroxide and singlet diradical state of ozone, setting the [2,2] system as CAS. Their results are summarized in Tables 1 and 2. The occupation number of the antibonding NO for the former species is 0.71 and therefore the orbital overlap T_{HO} is 0.29. The diradical property of the O—O dissociated hydrogen peroxide at $R = 2.052$ Å is calculated to be about 40%. The convergence of the UNO–CAS calculations is rapid, indicating that UNO is a very good trial MO for the MCSCF calculations.[21] The UNO–CAS calculations for ozone were also carried out by assuming the [6,6]-type CAS. The results are given in Table 2. Judging from the magnitude of D_j in equation 12, the diradical characters for the O—O σ-bonds of ozone are weak at the equilibrium geometry.

Table 1. Total energies and occupation numbers for the O—O dissociated hydrogen peroxide[a]

Method[b]	E_{total} (kcal mol^{-1})	n_{HO}	n_{LU}
UHF	−150.819006	1.2904	0.7096
APUHF	−150.914217	1.5356	0.4644
UNO-CAS[2,2]	−150.831105	1.5470	0.4530
PUHF	−150.841324		
PUMP2	−151.174082		
PUMP4	−151.205557		
APUM2	−151.169306		
APUM4	−151.196250		

[a] R(O—O) = 2.052 Å.
[b] 6-311G**.

Table 2. Total energies and occupation numbers of ozone[a]

Method[b,c]	E_{total} (kcal mol^{-1})	n_{HO}	n_{LU}
RHF	−223.907148	2.0	0.0
UHF	−224.008019	1.2398	0.7602
APUHF	−224.016613	1.4535	0.5645
UNO-CAS[2,2]	−224.014973	1.5005	0.4995
UNO-CAS[6,6]	−224.104310	1.6154	0.3846
UNO-CAS[8,8]	−224.148721		
PUHF	−224.022268		
PUMP2	−224.342192		
PUMP4	−224.366855		
APUMP2	−224.348036		
APUMP4	−224.375970		
MRCI//CAS[2,2]	−224.293912		
UHF QCISD(T)	−224.375930		

[a] R(O—O) = 1.29 Å; ∠O—O—O = 116.0°.
[b] 4-31G.
[c] $l = s + 1 - s + 3$ in equation 17, see text.

2.3 MP Perturbation Theory for Total Energy Calculations

The Hartree–Fock wavefunctions do not involve the so-called dynamic electron correlation effect. Møller–Plesset (MP) perturbation calculations have been used successfully as a method for simple and efficient computations of electron correlation energies.[22] The RHF MP calculations have been carried out for closed-shell molecules and singlet species with weak diradical characters. For open-shell molecules, the reference wavefunction is of the open-shell RHF (ORHF), UHF, or MCSCF type:

$$\Psi(\mathrm{XMP}n) = \Psi(\mathrm{X}) + \Psi(\mathrm{MP}_n \text{ correction}) \; (\mathrm{X} = \mathrm{ORHF, UHF, MCSCF}) \quad (16)$$

The second term involves the correction arising from the single (S), double (D), triple (T) and quadruple (Q) excitations from the ground state. For example, the fourth-order ($n = 4$) MP procedure UMP4 involves usually all these excitations, namely UMP4(SDTQ). The UMP wavefunction is an acceptable wavefunction for open-shell species such as radicals and biradicals at, or close to, the equilibrium geometries. For singlet diradicals, however, the UMP wavefunction is often heavily contaminated by higher spin states, especially by the triplet state. The ordinary fourth-order UMP (UMP4) theory usually does not remove this contamination. Schlegel[12] proposed a spin annihilation procedure for the UMP_n ($n = 2$–4) wavefunctions based on Lowdin's projection operator $\hat{\mathbf{O}}$:[23]

$$\hat{\mathbf{O}} = \prod_{S=J+1}^{J+l} [\mathbf{S}^2 - J(J+1)]/[S(S+1) - J(J+1)] \qquad (17)$$

where S is given by the difference between up (N_α) and down (N_β) spins, i.e. $S = \frac{1}{2}(N_\alpha - N_\beta)$ and where $l = 1$ and $l = 2$ correspond, respectively, to Schlegel[12] and Knowles and Handy[24] projections, which often suffer from the so-called size-inconsistent problem; the size inconsistency means that the total energy of a cluster consisting of n molecules is not equal to n times the energy of one molecule even at the dissociation limit.

A size-consistent spin projection procedure has also been proposed.[11,25] In the case of the singlet diradical, if the contamination of the singlet wavefunction is assumed to have arisen only from the triplet state, then the singlet nth order UMP wavefunction can be written as

$$^1\Psi(\text{UMP}n) = C_\text{s}\, {}^1\Phi(\text{PUMP}n) + C_\text{t}\, {}^3\Phi(\text{PUMP}n) \qquad (18)$$

where ${}^q\Phi(\text{PUMP}n)$ denotes the pure q spin state UMP wavefunction, and C_s and C_t are the coefficients of the pure singlet and triplet components, respectively. Because of the second term in equation 18, the ${}^1\Phi(\text{UMP}n)$ wavefunction will have an expectation value ${}^1\langle \mathbf{S}^2 \rangle$ of the total spin operator different from zero. Therefore, the coefficient of the triplet component can be calculated by

$$C_\text{t}^2 = {}^1\langle \mathbf{S}^2 \rangle / {}^3\langle \mathbf{S}^2 \rangle \qquad (19)$$

The total energy of the pure singlet state is approximately given by

$$^1E(\text{APUMP}) = {}^1E(\text{UMP}) + f[{}^1E(\text{UMP}) - {}^3E(\text{UMP})] \qquad (20)$$

where $f = C_\text{t}^2/(1 - C_\text{t}^2)$ and APUMP denotes the approximate PUMP. The effective exchange integral J_{ij} in equation 7 is also easily calculated under the assumption in equation 19 by

$$J_{ij} = [{}^1E(\text{UMP}) - {}^3E(\text{UMP})]/[{}^3\langle \mathbf{S}^2 \rangle - {}^1\langle \mathbf{S}^2 \rangle] \qquad (21)$$

Figure 5 shows the potential energy curve of the O—O dissociation process of hydrogen peroxide calculated by the UMP2 and APUMP2 methods. A small

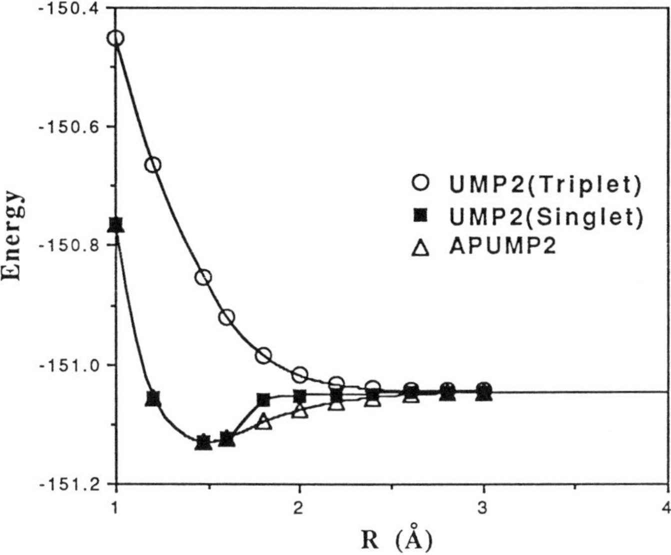

Figure 5. Potential curves of the O—O dissociation process of H_2O_2 obtained by UMP and APUMP methods

hump is seen on the UMP2 potential energy curve, but this defect is removed by the spin projection as described above. The shape of the APUMP2 curve is reasonable. The O—O bond energies calculated by the APUMP2 STO-3G, 4-31G and 6-31G* methods are 38.3, 51.3 and 54.3 kcal mol^{-1}, respectively, and these are in reasonable agreement with the experimental value of 51 kcal mol^{-1}.[8] Since the bond energy calculated by the APUHF method is only 3.8 kcal mol^{-1}, the dynamic electron correlation correction is crucial even for semi-quantitative purposes. The equilibrium bond length by the APUMP2 method is 1.48 Å, which is compatible with the experimental value (1.46 Å).

As shown in Table 2, the total energies for ozone calculated by the CASSCF method using [2,2] and [6,6] systems are lowered significantly by employing the APUMP2 and APUMP4 methods, indicating that the dynamic correlation corrections are crucial for the calculation of electronic energies. The MR MPn method is desirable for the purpose as shown in equation 16. Such a correction is also important for a quantitative estimation of activation barriers of chemical reactions (see Sections 5 and 6).

2.4 MR CI Theory for Ground and Excited States of Peroxides

In principle, the electron correlation energy can be exactly calculated by the CI method based on the MCSSCF wavefunction; this is referred to as the multi-reference (MR) CI method:[4]

$$\Phi(\text{MR CI}) = \Phi_0(\text{MCSCF}) + (\text{many excited configuration from } \Phi_0) \quad (22)$$

The MR CI method provides quantitative wavefunctions for both the ground and excited states, where the latter states are often characterized as ionic and zwitterionic states.[8] The theoretical aspects of the MR CI calculation have already been given in an excellent review.[27]

Many MR CI calculations of peroxy compounds have been carried out on the basis of approximate natural orbitals determined by the GVB[19] and UHF calculations.[16] The results have been summarized in other reviews.[19,20] The results of the MR CI calculation for the ground state of ozone are given in Table 2, in comparison with those of the CASSCF and APUMP calculations. Since ozone is essentially a diradical species, the 2×2 MCSCF calculation was performed before the MR CI calculations leading to the two-reference (2R) single and double (SD) CI: MR CI/CAS [2,2]. However, judging from the total energies, the MR CI wavefunction setting the main configurations from CASSCF [6,6] as the reference is compatible with those of exact (PUMP4) and approximate (APUMP4) projected MP calculations. This result is consistent with the necessity of the multiple spin projection ($l = 3$ in equation 17) in the PUMP approach.

2.5 Full Geometry Optimization by Energy Gradient Methods

The total electronic energy E of a molecule is a function of nuclear coordinates $\mathbf{q}$ under the Born–Oppenheimer approximation. Analytical derivatives[28] of total energies E obtained from the RHF, UHF and MCSCF wavefunctions with respect to nuclear coordinates $\mathbf{q}$ give useful information about detailed paths of chemical reactions.The first and second derivatives are usually referred to as energy gradient $\mathbf{G}$ and Hessian $\mathbf{H}$, respectively:[28]

$$\mathbf{G} = \delta E(\mathbf{q})/\delta\mathbf{q} \quad \text{and} \quad \mathbf{H} = \delta E^2(\mathbf{q})/\delta\mathbf{q}\delta\mathbf{q} \quad (23)$$

where $\mathbf{q} = (q_1, q_2, \ldots, q_N)$.

These derivatives can be used to determine optimized geometries of unstable intermediates and to locate transition structures for chemical reactions.[29]

The energy gradient should disappear at a stationary point on the energy hypersurface, and it is a true minimum only if all the eigenvalues of the Hessian matrix are positive in sign. On the other hand, the transition state (TS) is defined as a saddle point with only one negative eigenvalue of the Hessian matrix. This is called the first-order TS. If the Hessian matrix has two and three negative eigenvalues, the saddle points are referred to as second- and third-order TSs, respectively, although these points have no chemical significance. Figure 6 illustrates characteristics of a potential energy hypersurface.

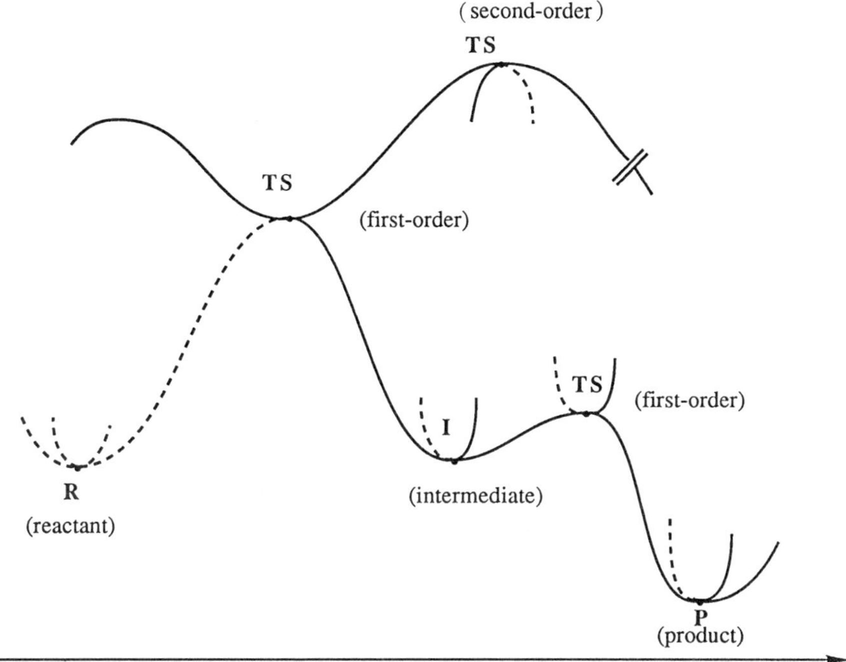

Figure 6. Characteristics of potential energy surfaces with stationary points corresponding to first- and second-order transition states (TS), intermediates (I), etc.

The energy gradient techniques have been successfully applied to full geometry optimizations of organic molecules. Figure 7 shows fully optimized geometries of hydroperoxy radical by the open-shelf RHF (ORHF), UHF and MCSCF CI techniques, together with the experimental geometry. The geometries optimized by the ORHF and UHF methods are similar, whereas those by the MCSCF CI and GVB CI methods are close to each other. The O—O bond lengths by the former two methods are shorter by about 0.02 Å than the experimental value, whereas those by the latter two methods are longer by about 0.03 Å. The UMP2 (6–31G*) treatment is sufficient to remove these small errors.

Figure 8 shows the open-chain dimers of the hydroperoxy radical.[30] The Hessian matrix **H** for the planar C_{2h} structure has three negative eigenvalues (imaginary frequencies), indicating that the C_{2h} structure is not a minimum but a third-order TS. The optimization based on the C_i symmetry yields one negative eigenvalue, indicating that the C_i structure is not only a true minimum but also a true TS. If the structural constraints are removed, a true minimum can be found on the energy hypersurface. Thus, the analytical Hessian matrix is a

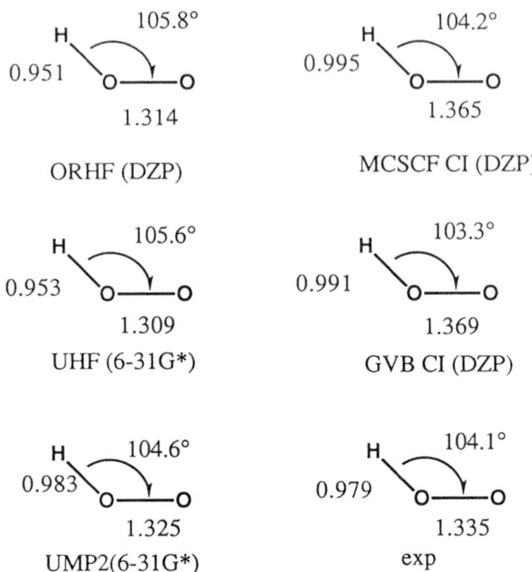

Figure 7. Optimized geometries of hydroperoxy radical by several theoretical methods, together with experimental geometry

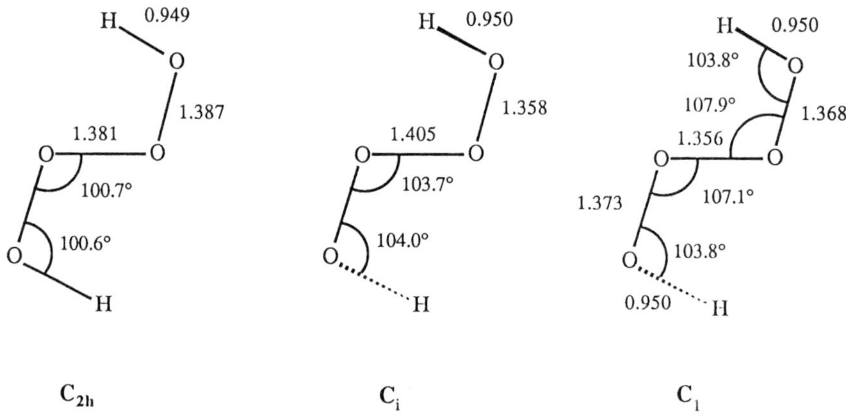

Figure 8. Optimized geometries of the dimer of HO_2. The C_{2h}, C_i and C_1 symmetries correspond to second-order TS, first-order TS and local minimum, respectively

powerful tool for a theoretical search for a local minimum on the energy hypersurface.

The dimerization energy of HO_2 is endothermic ($D_e = -18.9$ kcal mol^{-1}) at the RHF (DZP) level, in accord with thermochemical estimation. However, it becomes exothermic ($D_e = 9.9$ kcal mol^{-1}) after the RHF single and double

(SD) CI plus Davidson correction.[30] Hence full geometry optimization and correlation correction are generally crucial for semiquantitative discussions on geometries and energies of peroxy compounds.[8]

2.6 Location of Transition Structures

The key characteristics of potential energy surfaces for chemical reactions can be elucidated from *ab initio* quantum mechanical calculations. Location of transition structures (TSs) for chemical reactions has been elucidated by the use of the energy gradient technique. As mentioned in the preceding sections, the transition state is a saddle point of the energy hypersurface, where the Hessian matrix has only one negative eigenvalue (imaginary frequency) and all the remaining eigenvalues are positive in sign. The intrinsic reaction coordinate (IRC)[31] is defined as a steepest descent path which connects a reactant with product(s) through a transition state. The activation barrier for chemical reactions can be calculated as the difference in energies between total energy for TS and reactants.

As an example, consider the addition reaction of hydrogen atom toward ozone: $H + O_3 \rightarrow OH + O_2$. Several paths are conceivable for the reaction:[32]

$$H + O_3 \rightarrow \mathbf{TS1} \rightarrow HO_3 \tag{24a}$$

$$HO_3 \rightarrow \mathbf{TS2} \rightarrow OH + O_2 \tag{24b}$$

$$HO_3 \rightarrow \text{no TS} \rightarrow O + HO_2 \tag{24c}$$

$$O + HO_2 \rightarrow \mathbf{TS3} \rightarrow OH + O_2 \tag{24d}$$

$$H + O_3 \rightarrow \text{no TS} \rightarrow H + O + O_2 \tag{24e}$$

In order to describe these paths in a well balanced manner, the CASSCF calculations at least for the seven-orbital, seven-electron [7,7] system need to be performed, since both the O—H and O—O dissociation processes are involved. The CASSCF calculations for the [7,7] system have been carried out to locate transition structures of the above elementary steps.[32]

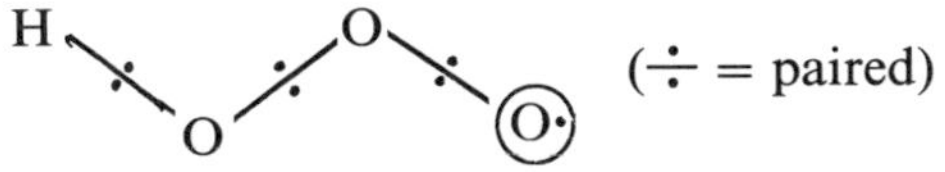

H O O O (÷ = paired)

Figure 9 illustrates the profiles of potential energy curves and the transition structures **TS1**–**TS3**.[32] The HO_3 system was calculated to be 79 kcal mol^{-1} more stable than the $H + O_3$ system; the enthalpy of formation of HO_3 was estimated to be 62 kcal mol^{-1} from the thermochemical data. The activation barrier for the addition step, namely transition structure **TS1**, was only 2.7 kcal mol^{-1}. The same barrier was obtained for **TS2** that was brought about by the

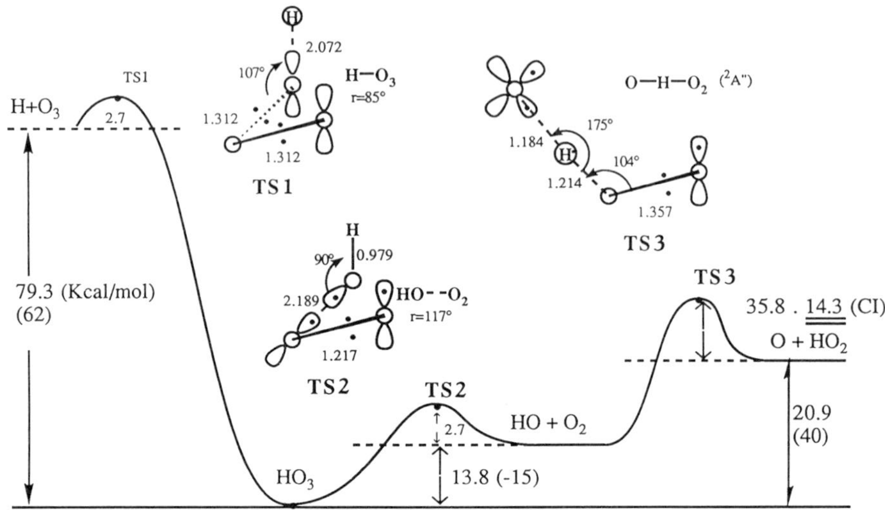

Figure 9. Potential energy profiles and transition structures obtained for the addition reaction of a hydrogen atom to ozone by the CASSCF method. The thermochemical data are given in parentheses

recombination step between OH and O_2. However, the OH + O_2 system was less stable by 13.8 kcal mol^{-1} than HO_3, although the thermochemical estimation suggested that the former is more stable than the latter by 15 kcal mol^{-1}. The activation barriers for the hydrogen abstraction reaction (**TS3**) were calculated to be 35.8 and 14.3 kcal mol^{-1} by the CASSCF and single reference (SR) SD CI methods, respectively. This may indicate that the extensive MR CI method is necessary to reproduce thermochemical relative energies for all the channels involved in the H + O_3 reaction; this reaction is one of the important elementary steps occurring in the atmosphere.

3. NOTATIONS AND ABBREVIATIONS

The notations and abbreviations used in this chapter are as follows: MO = molecular orbital; NO = natural orbitals; CMO = corresponding MO; UNO = UHF NO; VB = valence bond; GVB = generalized VB; DODS = different orbitals for different spins; RHF = restricted Hartree–Fock; UHF = unrestricted Hartree–Fock; MP = Møller–Plesset perturbation method; RMP = RHF MP; UMP = UHF MP; PUMP = spin-projected UMP; APUMP = approximate PUMP; MR CI = multi-reference configuration interaction; MR MP = multi-reference Møller–Plesset perturbation; CASSCF = complete active space self-consistent field; MCSCF = multi-configuration self-consistent field.

4 ELECTRONIC STRUCTURES OF PEROXY COMPOUNDS

4.1 Classification of Oxygen-containing Reactive Species

Oxygen-containing reactive species can be classified into two groups according to their electronic and chemical properties.[1–3] Net charges on oxygen atoms of the species are useful indices for classification. Table 3 summarizes net charges on oxygen atoms of typical oxygen-containing reactive species with and without formal charges, whose geometries are fully optimized by the energy gradient technique.[28]

The cationic species, $O^{+\bullet}$, HO^+ and $HO_2{}^+$, act usually as electrophiles toward organic substrates such as alkenes and aromatic compounds. The anionic species HO^- and $O^{-\bullet}$ act as nucleophiles toward electron-deficient substrates. The superoxide radical anion has both anionic and radical centers, but it exhibits nucleophilic properties toward electron-deficient substrates. Neutral oxygen radicals usually act as electrophiles toward electron-rich substrates such as alkenes. The net charges on oxygen atoms of these species are formally negative, as shown in Table 3. Therefore, electrophilic reactions of these species are frontier-orbital controlled reactions,[33] rather than charge density-controlled reactions.

4.2 Oxygenated Dipoles and Diradicals

Carbonyl oxides exhibit electrophilic, nucleophilic and diradical characters in their chemical reactions, depending on the substituents attached and the nature of substrates, and also on the reaction conditions.[34] The diverse reactivities of carbonyl oxides are ascribed to their electronic nature, which can be described

Table 3. Net charges on the terminal (inner) oxygen atom of oxygen-containing compounds by the Hartree–Fock method[a]

System	Basis set	Cation	Neutral	Anion
HO	STO-3G	0.510	−0.146	−0.742
	6-31G*	0.361	−0.444	−1.206
CH_3O	STO-3G	0.413	−0.108	−0.640
	6-31G*	0.235	−0.329	−0.893
HO_2	STO-3G	0.414	−0.014	−0.583
		(0.143)	(−0.182)	(−0.391)
	6-31G*	0.345	−0.076	−0.676
		(0.054)	(−0.391)	(−0.675)
CH_3O_2	STO-3G	0.326	−0.015	−0.544
		(0.151)	(−0.192)	(−0.332)
	6-31G*	0.266	−0.080	−0.648
		(0.053)	(−0.296)	(−0.484)

[a]Geometries were optimized by each basis set.

as 1,3-diradical and 1,3-dipole. A dual electronic nature of this type is also observed for a variety of peroxy compounds such as those shown in Figure 10. The interpretation of the electronic features of these species is described in this section on the basis of the theories outlined in the preceding sections.

4.2.1 Dissociation of a C—O bond of dioxirane and orbital bifurcation

Consider the least-motion dissociation of a C—O bond of dioxirane.[35] The RHF solution provides a reasonable description of this molecule at a near-equilibrium geometry. Two electrons with different spins occupy the C—O σ-bonding MO, forming a familiar closed-shell orbital configuration at the equilibrium geometry:

At a C—O dissociated geometry, the closed-shell RHF solution converges to a spurious excited state consisting of the ionic structure $^{+}$C—O—O^{-}, because of

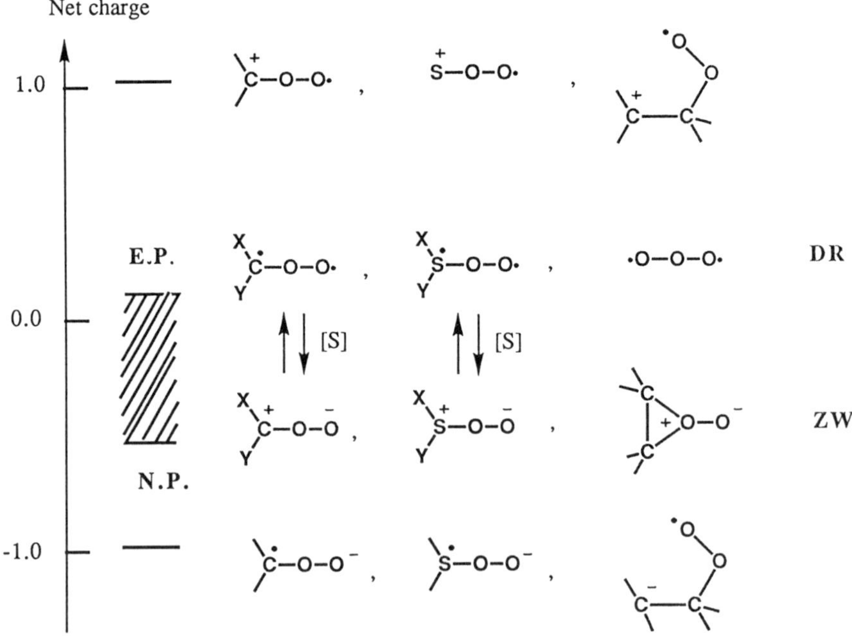

Figure 10. Electronic properties of peroxy compounds. E.P. and N.P. denote electrophilic and nucleophilic, respectively. The hatched region implies that the electronic structures are variable with substituents and environmental effects

the double occupancy constraint. It clearly provides an unreasonable description of a C—O dissociated dioxirane.

This breakdown of the closed-shell approximation is closely related to the occurrence of the triplet instability for the RHF solution.[35] It predicts that the RHF solution should be reorganized into a more stable UHF solution which allows the DODS for a singlet diradical with no constraint of the double occupancy.[10] This guarantees that the occupied σ-MO of dioxirane bifurcates into the pσ-MOs which are mainly localized on the terminal carbon and oxygen atoms. Since each electron with different spin occupies each pσ-type orbital, the DODS orbital configuration corresponds to a singlet (σ,σ) diradical (DR) configuration as illustrated in Figure 11D. By 90° rotation of the terminal

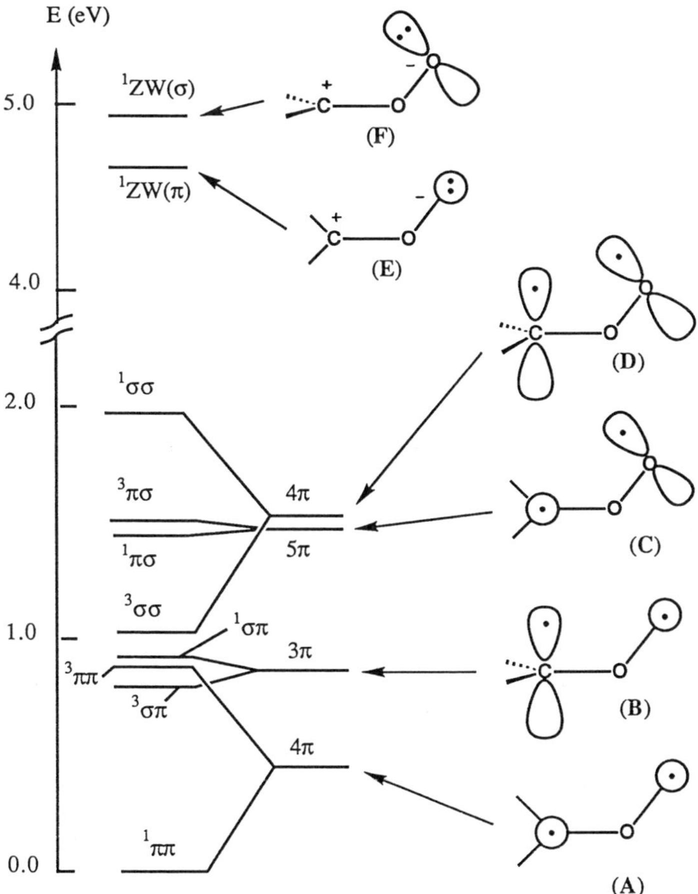

Figure 11. Energy levels of methylene peroxide by the UNO CI method. (A), (B), (C) and (D) denote $\pi\pi$, $\sigma\pi$, $\pi\sigma$ and $\sigma\sigma$ diradical configurations, respectively, which are also characterized by the number of π-electrons; (E) and (F) denote the zwitterionic (ZW) structures for planar and perpendicular conformations, respectively

methylene group, the pσ-orbital on the carbon atom rotates by the same amount without a significant change of the orbital shape, becoming a π-orbital. The resulting DODS orbital configuration expresses the (π,σ) DR configuration of the C—O ring-opened dioxirane. Similarly, the (σ,π) DR configuration is obtained by 90° rotation of the pσ-orbital on the oxygen atom. The (π,π) DR configuration can be obtained by the double rotations. The relaxation of the double occupancy constraint is thus crucial for a reasonable MO description of 1,3-DR states of methylene peroxide (carbonyl oxide). The MR CI calculations by the use of UHF natural orbitals (UNO) were carried out for these four diradical states, together with the two zwitterionic states. The energy levels obtained by the UNO CI method[36] are shown in Figure 11. The (π,π) DR state becomes the most stable state among 1,3-DR states at $\angle$COO = 120°.

In general, the closed-shell HOMO bifurcates into the DODS HOMOs for 1,3-dipolar species with DR characters. The orbital bifurcation does not occur for stable 1,3-dipoles such as allyl anion.[20] The mathematical (triplet instability) condition for the orbital bifurcation is approximately given by

$$\lambda_0 = (\varepsilon_j - \varepsilon_i) - (ii|jj) - (ij|ij) \tag{25}$$

where the first term is the HOMO–LUMO energy gap, and the second and third terms are the Coulombic repulsion energy and the exchange energy, respectively. From equation 25, the triplet instability ($\lambda_0 < 0$) occurs when the electron repulsion exceeds the orbital energy gap, which is generally small for unstable species.

4.2.2 Diradical characters of 1,3-dipolar species

The semi-empirical (INDO) and *ab initio* calculations have been performed for 1,3-dipolar species that are discussed by Huisgen. The results are listed in Table 4. These results[35] indicate that 1,3-dipolar species can be classified into four groups: (I) closed-shell 1,3-dipoles or zwitterions (ZW); (II) 1,3-dipoles with weak DR character; (III) 1,3-dipoles with moderate DR character; and (IV) 1,3-dipoles with strong DR character. Interestingly, this MO-based classification[35,37] is parallel to that proposed by Huisgen.[34] Type A and C species in Huisgen's classification are regarded as group I, whereas type B species such as azomethinium betaines and oxygenated dipoles are classified into group II and group III, respectively. 1,3-Dipoles without octet stabilization belong to group IV. A detailed discussion of this classification has been given in a previous review.[37]

It can be seen from Table 4 that the diradical character of oxygenated dipoles is in the range 41–53%. These values are independent of the theoretical methods employed. In fact, the DR characters of ozone are 48%, 55% and 53% by the GVB (spd basis set), PUHF (4–31G) and PUHF (INDO) methods, respectively.

Table 4. Net charges (P_X), bond orders (P_{XY}), orbital overlaps (T_{HO}) and diradical characters (y) of 1,3-dipolar species (A—B—C)

No.	Compound[a]	I_p (eV)[b]	P_A	P_B	P_C	P_{AB}	P_{BC}	T_{HO}	y
1	$HCNCH_2$	9.69	−0.15	0.24	−0.26	0.72	0.69	1.00	0.00
2	HCNNH	10.9	−0.15	0.32	−0.37	0.83	0.56	1.00	0.00
3	HCNO	12.9	−0.15	0.37	−0.36	0.91	0.40	1.00	0.00
4	$NNCH_2$	11.0	−0.14	0.32	−0.26	0.71	0.71	1.00	0.00
5	NNNH	12.4	−0.13	0.40	−0.37	0.82	0.57	1.00	0.00
6	NNO	14.5	−0.13	0.46	−0.34	0.90	0.43	1.00	0.00
7	$HCCHCH_2$	9.52	−0.04	0.11	−0.25	0.62	0.77	1.00	0.00
8	HCCHNH (*t*)	10.6	−0.01	0.18	−0.37	0.69	0.70	1.00	0.00
8	HCCHNH (*c*)	10.8	−0.03	0.19	−0.36	0.72	0.66	1.00	0.00
9	HCCHO	12.2	0.01	0.24	−0.39	0.77	0.60	1.00	0.00
10	$NCHCH_2$	10.9	−0.05	0.15	−0.23	0.57	0.81	1.00	0.00
11	NCHNH (*t*)	12.0	−0.03	0.22	−0.34	0.64	0.74	1.00	0.00
11	NCHNH (*c*)	12.2	−0.04	0.24	−0.33	0.67	0.71	1.00	0.00
12	NCHO	13.8	−0.01	0.29	−0.36	0.71	0.64	1.00	0.00
13	H_2CNHCH_2	9.59	−0.16	0.13	−0.16	0.60	0.60	0.74	0.04
14	H_2CNHNH (*t*)	10.1(11.3)	−0.13	0.20	−0.26	0.64	0.57	0.75	0.04
14	H_2CNHNH (*c*)	10.0(11.3)	−0.14	0.18	−0.27	0.63	0.57	0.74	0.05
15	H_2CNHO	12.5	−0.12	0.28	−0.32	0.72	0.65	0.83	0.02
16	HN^2NHNH (*t*/*t*)	11.9	−0.22	0.23	−0.22	0.59	0.59	0.69	0.06
16	HNNHNH (*t*/*c*)	11.8(11.8)	−0.23	0.22	−0.24	0.58	0.58	0.68	0.07
16	HNNHNH (*c*/*c*)	11.7	−0.24	0.21	−0.24	0.58	0.58	0.67	0.08
17	HNNHO (*t*)	13.5	−0.20	0.29	−0.24	0.62	0.57	0.71	0.07
17	HNNHO (*c*)	13.4	−0.23	0.28	−0.24	0.61	0.57	0.70	0.06
18	ONHO	14.4	−0.21	0.33	−0.21	0.69	0.69	0.69	0.07
19	H_2COCH_2	11.1	0.00	−0.08	0.00	0.40	0.40	0.33	0.41
20	H_2CONH (*t*)	13.3	0.00	−0.05	−0.09	0.40	0.39	0.32	0.42
20	H_2CONH (*c*)	13.4	0.00	−0.04	−0.08	0.40	0.39	0.32	0.42
21	H_2COO	11.8(15.5)	0.00	−0.04	−0.08	0.39	0.38	0.32	0.42

Table 4. *(continued)*

No.	Compound[a]	I_p (eV)[b]	P_A	P_B	P_C	P_{AB}	P_{BC}	T_{HO}	y
22	HNONH (*t*/*t*)	13.6	−0.08	−0.01	−0.08	0.38	0.38	0.30	0.45
22	HNONH (*t*/*c*)	13.6(13.7)	−0.09	−0.01	−0.08	0.38	0.38	0.30	0.46
22	HNONH (*c*/*c*)	13.7	−0.08	−0.00	−0.08	0.37	0.37	0.29	0.46
23	HNOO (*t*)	14.2(15.5)	−0.08	0.03	−0.06	0.37	0.37	0.28	0.48
23	HNOO (*c*)	14.3(16.0)	−0.07	0.04	−0.07	0.36	0.36	0.28	0.48
24	OOO	16.8	−0.04	0.09	−0.04	0.35	0.35	0.25	0.53
25	$H_2CCH_2CH_2$	10.4	−0.05	0.07	−0.05	0.39	0.39	0.07	0.85
26	H_2CCH_2NH (*t*)	10.5(12.6)	−0.08	0.15	−0.16	0.40	0.35	0.11	0.79
26	H_2CCH_2NH (*c*)	10.5(12.5)	−0.08	0.14	−0.18	0.40	0.35	0.11	0.79
27	H_2CCH_2O	10.7(14.9)	−0.11	0.21	−0.17	0.41	0.32	0.12	0.76
28	$HNCH_2NH$ (*t*/*t*)	12.7	−0.18	0.24	−0.18	0.36	0.36	0.13	0.74
28	$HNCH_2NH$ (*t*/*c*)	12.6(12.6)	−0.19	0.23	−0.20	0.36	0.36	0.13	0.75
28	$HNCH_2NH$ (*c*/*c*)	12.5	−0.21	0.22	−0.21	0.36	0.36	0.13	0.75
29	$HNCH_2O$ (*t*)	12.9(15.0)	−0.20	0.28	−0.18	0.37	0.33	0.14	0.73
29	$HNCH_2O$ (*c*)	12.7(14.9)	−0.22	0.28	−0.20	0.37	0.33	0.13	0.74
30	OCH_2O	15.2	−0.19	0.33	−0.19	0.33	0.33	0.15	0.72

[a] *c* and *t* denote the *cis* and *trans* conformations, respectively.
[b] I_p for η_π is given in parentheses if $I_p(\chi_\pi) \neq I_p(\eta_\pi)$.

Those of methylene peroxide (carbonyl oxide) are 38% and 42% by the GVB (spd) and PUHF (INDO) methods, respectively. The HOMOs of oxygenated dipoles are significantly localized on terminal heavy atoms, but are still delocalized on the whole molecular skeletons. As an example, the HOMOs expressed by corresponding MOs (CMOs) for the ring-opened ethylene oxide (carbonyl ylide) are shown in Figure 12.

Since oxygenated dipoles have both non-radical and diradical characters, the chemical behaviour of the species may be complex. Harding and Goddard[38] emphasized the latter property for ozone and carbonyl oxide and explained ozonolysis reactions of alkenes in terms of a diradical mechanism. However, note that the UHF and CASSCF calculations of these species cannot rule out the possibility of a concerted mechanism for the reaction, particularly for solution-phase ozonolyses (see Section 6).

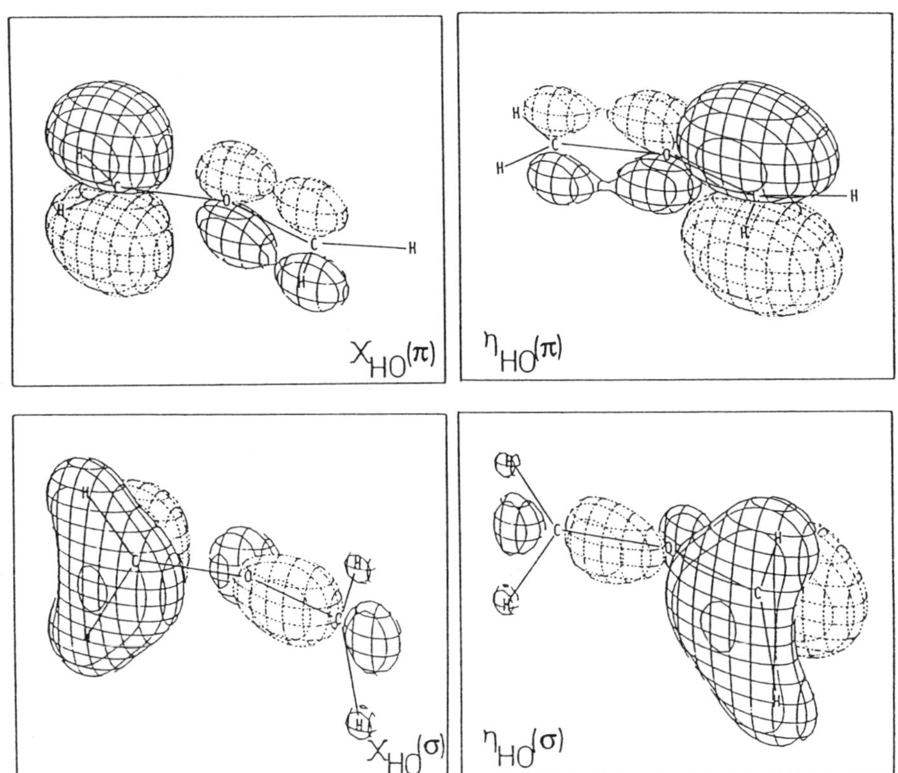

Figure 12. Graphic illustrations of corresponding molecular orbitals (CMO) of the (π, π) and (σ, σ) diradical states of carbonyl ylide, i.e. ring-opened ethylene oxide

4.2.3 Geometries and stabilities of ozone and carbonyl oxide

There are three important problems for reliable *ab initio* calculations of molecular systems:[4] selection of basis sets, geometry optimizations, and correction for electron correlation in HF calculations. These problems are examined by taking ozone as an example of a symmetrical oxygenated dipole. Table 5 summarizes the geometrical parameters optimized by several methods. The O—O bond length of ozone by the RHF 6-31G* method is shorter by 0.074 Å than the experimental value. The UHF 6-31G* and GVB DZP methods provide similar O—O bond lengths, which are longer by about 0.02 Å than the experimental value. This implies that the experimental geometry of ozone is a mixture of the closed-shell structure (O=O=O) by the RHF method and the diradical structure (·O—O—O·) by the GVB and UHF methods. The full geometry optimizations by the RMP2 method that takes both the configurations into consideration reproduces the experimental geometry.[39]

The UHF 4-31G and 6-31G* methods provide smaller O—O bond lengths than the experimental value, indicating that the DZ basis set is not flexible enough for ozone. The UHF 4-31G and 6-31G* methods provide smaller O—O—O angles than the experimental value. This may be attributable to the spin contamination effect.

The energy calculations were carried out by the use of the UMP and APUMP 4-31G* methods for the geometries given in Table 5. Table 6 summarizes the calculated results. The UMP*n* calculations show that the UHF (6-31G*), GVB and MCSCF geometries are more stable than the experimental structure. This is due to the overestimation of the diradical stabilization by UMP. On the other hand, the APUMP*n* method predicts that the experimental structure is the most stable; for example, it is more stable by 0.6 kcal mol^{-1} than the GVB DZP structure and by 1.7 kcal mol^{-1} than the UHF 6-31G* structure. On the other hand, the RHF 6-31G* closed-shell structure by the RHF MP (RMP) 6-31G*

Table 5. Optimized geometries of ozone calculated by several methods

Method	*R*(O—O)(Å)	∠O—O—O(°)
RHF 4-31G	1.255	119.2
RHF 6-31G*	1.204	119.0
UHF 4-31G	1.402	109.2
UHF 6-31G*	1.295	111.6
GVB DZP[a]	1.299	116.0
MCSCF DZ[b]	1.327	117.3
RMP2	1.277	117.0
Exp.[b]	1.278	116.8

[a]From ref. 38
[b]From ref. 8

Table 6. Relative energies for several optimized geometries of ozone calculated by the APUMP 4-31G* method

Method	RHF 4-31G	RHF 6-31G*[a]	UHF 4-31G	UHF 6-31G*	GVB DZ	MCSCF DZ	Exp.
UHF	0.0	7.9(50.7)	5.3	−3.9	−2.9	−1.0	−2.5
UMP2	0.0	8.8(−21.4)	4.0	−3.2	−3.2	−2.3	−2.5
UMP3	0.0	9.0(0.7)	2.6	−3.8	−3.6	−2.7	−2.7
UMP4	0.0	10.2(−8.6)	0.2	−4.1	−4.3	−4.1	−3.0
UMP∞	0.0	10.4(−7.8)	−0.3	−4.1	−4.4	−4.4	−3.1
APUHF	0.0	4.7(59.3)	11.8	−0.5	−0.1	2.8	−0.8
APUMP2	0.0	4.5(−3.7)	16.5	2.3	1.4	4.3	0.2
APUMP3	0.0	4.8(15.8)	13.7	1.2	0.7	3.3	−0.3
APUMP4	0.0	6.2(10.5)	12.4	1.2	0.1	2.0	−0.5
APUMP∞	0.0	6.5(11.6)	11.9	1.1	−0.1	1.7	−0.6

[a]RMP*n* values are given in parentheses.

method was less stable by 10 kcal mol^{-1} than the diradical states by the UHF 6-31G* and GVB DZP methods. Hence ozone is a diradical species as described by the APUMP method.

Table 7 summarizes the geometric parameters optimized for methylene peroxide, with an unsymmetrical oxygenated dipole. Compared with the optimized geometry by the CASSCF DZP method, the RHF 6-31G* method underestimates the C—O bond length, but it overestimates the O—O bond length. The RHF structure is given by the closed-shell structure C:[35]

$$\rangle C{=}O^{+}{-}O^{-} \qquad \cdot\rangle C{-}O{-}O\cdot \qquad \overset{\delta\dot{+}}{C}\,\cdots\,O\,\cdots\,\overset{\delta\dot{-}}{O}$$

C DR MIX

Table 7. Optimized geometries of methylene peroxide calculated by several methods

Method	*R*(C—O)(Å)	*R*(O—O)(Å)	∠C—O—O(°)
RHF 4-31G	1.213	1.660	114.2
RHF 6-31G*	1.200	1.482	114.5
UHF 4-31G	1.385	1.366	114.3
UHF 6-31G*	1.354	1.286	114.8
Av. 6-31G*[a]	1.277	1.384	114.7
GVB CI DZP	1.343	1.362	116.6
RMP2 DZP	1.297	1.295	120.3
CAS SCF DZP	1.275	1.313	119.5

[a]Av. = average geometry of RHF and UHF 6-31G* results.

On the other hand, the C—O and O—O bond lengths are close to those of the DR structure given by the UHF and GVB DZP methods. The CASSCF bond lengths give the average (MIX) of the RHF and UHF 6-31G* values. The RMP2 geometry is also close to the CASSCF geometry. Hence it may be concluded that the geometries of oxygenated dipoles with moderate diradical chartacters can be reproduced by the RMP2 method.[40]

The APUMP 4-31G* calculations were carried out for several optimized geometries of methylene peroxide. Table 8 summarizes the calculated results. The APUMP4 method predicts that the CASSCF geometry is the most stable. The UHF 6-31G* DR and RHF 6-31G* closed-shell structures are less stable than the CASSCF structure by about 5.4 and 17.0 kcal mol^{-1}, respectively. Thus, methylene peroxide is a DR species described by the APUMP4//UHF 6-31G* method. However, the ionic character C is introduced to some extent after the reoptimization by the CASSCF method. The energy gain after the refinement was about 5.4 kcal mol^{-1}.

4.2.4 Substituent effects on carbonyl oxides

Diradical properties of oxygenated dipoles are weakened by introducing polar substituents. In particular, the push-pull stabilization by polar substituents is known to be effective for unsymmetrical oxygenated dipoles such as substituted carbonyl oxides of the type of XYCOO. The *ab initio* MO calculations were carried out for XYCOO in order to elucidate the nature of the push-pull effect.[39,41] For this purpose, the diradical-type geometries of these species were constructed by using the optimized geometry of the $\pi\pi$ DR state of methylene peroxide and standard CX and CY bond lengths, and the UHF and RHF 4-31G calculations were performed for these assumed geometries. Table 9 summarizes the net charges on the parent skeletons COO of XYCOO species. The net

Table 8. Relative energies (kcal mol^{-1}) for several optimized geometries of methylene peroxides calculated by the APUMP 4-31G* method

Method	UHF 4-31G	RHF 6-31G*	UHF 6-31G*	RMP2 DZP	CASSCF DZP	Av. 6-31G*
UHF	0.0	8.0	−3.4	−0.9	1.0	4.6
UMP2	0.0	−2.7	−3.8	−4.7	−4.2	1.9
UMP3	0.0	−1.2	−3.4	−3.5	−2.6	2.6
UMP4	0.0	−3.6	−2.0	−2.7	−2.2	1.6
UMP∞	0.0	−3.9	−1.8	−2.5	−2.0	1.4
APUHF	0.0	11.3	−5.6	−5.0	−2.9	3.3
APUMP2	0.0	7.1	−7.8	−12.6	−12.6	−2.5
APUMP3	0.0	6.6	−7.1	−10.5	−9.9	−0.9
APUMP4	0.0	7.3	−4.3	−9.6	−9.7	−3.1
APUMP∞	0.0	7.5	−4.8	−9.2	−9.3	−3.3

Table 9. Relative energies between DR or ZW (kcal mol^{-1}) and net charges for these states

					Net charges								
			E(DR–ZW)[a]		DR[b]			ZW[b]			ZW + H_2O[c]		
No.	X	Y	1	2	C	O	O	C	O	O	C	O	O
1	H	H	32.76	−4.99	0.00	−0.40	−0.01	0.23	−0.40	−0.26	0.24	−0.40	−0.27
2	CH_3	H	27.73	−11.31	0.16	−0.42	−0.02	0.38	−0.44	−0.27	0.22	−0.40	−0.27
3	H	CH_3	25.67	−12.07	0.17	−0.42	−0.02	0.41	−0.44	−0.28	0.25	−0.40	−0.28
4	CH_3	CH_3	19.96	−16.13	0.32	−0.44	−0.02	0.53	−0.48	−0.28	0.25	−0.39	−0.28
5	F	H	30.38	−16.98	0.54	−0.43	0.03	0.75	−0.44	−0.23	0.39	−0.44	−0.28
6	H	F	33.48	−19.30	0.55	−0.41	0.01	0.75	−0.43	−0.24	0.41	−0.44	−0.27
7	F	F	30.92	−27.33	1.08	−0.42	0.05	1.28	−0.43	−0.21	0.54	−0.48	−0.29
8	OH	H	27.74	−29.58	0.47	−0.43	0.02	0.71	−0.47	−0.27	0.76	−0.43	−0.23
9	H	OH	19.60	−35.72	0.55	−0.44	−0.01	0.71	−0.49	−0.26	0.76	−0.42	−0.24
10	OH	OH	12.49	−54.64	0.98	−0.47	0.04	1.20	−0.48	−0.27	1.29	−0.43	−0.21
11	OCH_3	H	27.20	−23.45	0.49	−0.45	0.02	0.74	−0.49	−0.25	0.72	−0.47	−0.28
12	H	OCH_3	15.99	−37.02	0.49	−0.46	0.01	0.73	−0.51	−0.26	0.72	−0.49	−0.26
13	NH_2	H	5.16	−39.85	0.43	−0.42	−0.03	0.67	−0.52	−0.30	1.22	−0.51	−0.27
14	H	NH_2	8.05	−38.76	0.43	−0.42	−0.02	0.69	−0.53	−0.29	0.74	−0.48	−0.25
15	HO	CH_3			0.67	−0.45	0.01	0.87	−0.51	−0.29	0.74	−0.50	−0.26
16	H	CHO			0.18	−0.44	0.02	0.34	−0.43	−0.24	0.68	−0.52	−0.31
17	OH	F			1.01	−0.43	0.05	1.25	−0.47	−0.25	0.69	−0.52	−0.29

[a] 1, Without H_2O; 2, with H_2O.
[b] DR-type geometry is used.
[c] ZW-type geometry is used.

charges on the terminal carbon and oxygen atoms of the CH_2OO species calculated by the UHF method are essentially zero, showing that this species is in the neutral DR state, $\cdot COO\cdot$. On the other hand, the net charges on the carbon atom of the same species calculated by the RHF method are in the range 0.24–1.20, whereas those of the terminal oxygen are in the range -0.21 to -0.30. The RHF solution is therefore responsible for the zwitterionic state, $^{+}C—O—O^{-}$. Consequently, the energy difference (equation 26a) between total energies of the singlet (projected) UHF (APUHF) and those of RHF appears to correspond approximately to the DR–ZW energy gap:

$$\Delta E(\text{DR–ZW}) = E(\text{Z}) - E(\text{RHF}) \tag{26a}$$

where Z denotes the APUHF or UHF method.

The calculated energy gaps are summarized in Table 9. Electron-donating substituents such as OH and NH_2reduce the energy gap significantly, whereas electron-withdrawing substituents such as F and CHO increase the energy gap. The calculated results clearly indicate the push–pull effect for XYCOO, namely, when substituents XY are electron donors the O_2 part acts as an electron acceptor. The reverse is true for electron-withdrawing substituents. This is ensured by the fact that the energy gaps are well correlated with Hammett's σ values of polar substituents, as illustrated in Figure 13:

$$\Delta E(\text{DR–ZW}) = -40.1\sigma_t + 34.5 + \Delta E_s \tag{26b}$$

where σ_t is the sum of the σ values of the substituents X and Y, and ΔE_s denotes the solvation energy.

In order to investigate the solvation effect, the full geometry optimizations of the methylene peroxide plus water system were also carried out by the UHF and RHF (4–31G) energy gradient techniques.[35] The optimized geometries are shown in Figure 14. The optimized geometry of the $\pi\pi$ DR state by the UHF method does not change appreciably even if water is added as a model polar solvent. On the other hand, the optimized geometry by the RHF method is different from the DR geometry, which is regarded as a non-radical structure

$$>C{=}O^{+}—O^{-}\cdots H—O—H.$$

The DR–ZW energy gap of the XYCOO plus H_2O system was also estimated by assuming that their geometries are identical with those of the $CH_2OO + H_2O$ system. The energy gaps calculated by the UHF and RHF 4–31G method for these systems are also given in Table 9. Interestingly, the ZW states are more stable than the DR states in all the XYCOO species. This indicates that carbonyl oxides become zwitterions in polar solution, and then significant geometric changes occur accordingly. In fact, zwitterionic carbonyl oxides can be trapped by alcohols. The calculated results are consistent with the available experimental data given in other chapters in this book.

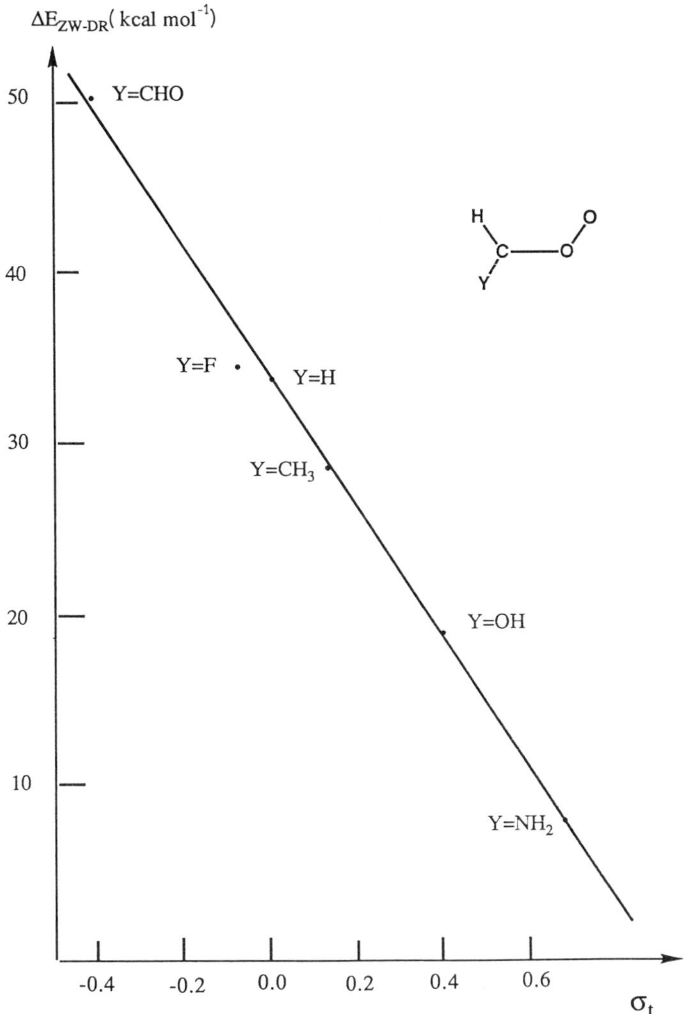

Figure 13. Correlations of the ZW – DR enerlgy gap for substituted carbonyl oxides by the APUHF and RHF 4-31G methods with Hammett's sigma (σ_t) constant

4.3 Transition Metal Peroxo Complexes

Transition metal-catalyzed oxygenations are useful reactions in organic synthesis and also have received considerable attention in the field of biological oxidation. Key intermediates of these reactions are transition metal peroxo complexes. In this section, the electronic structures and reactivity features of the complex of the type $LnMOO^{m+}$ (M = Cr, Fe, Ni; Ln = ligands) are discussed on the basis of the *ab initio* MO calculations.[11,42–45] A proposed mechanism of the transition metal-catalyzed oxygenation is illustrated in Figure 15.

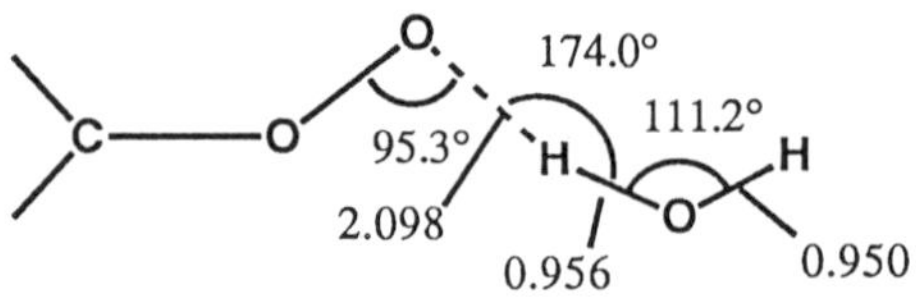

174.0°
111.2°
95.3°
2.098
0.956
0.950

ZW-H_2O

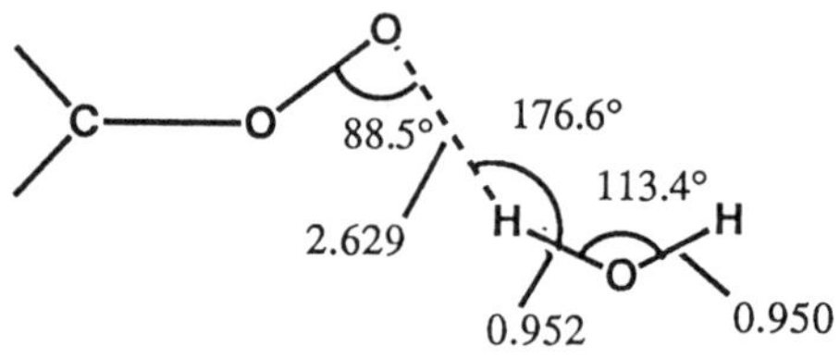

88.5°
176.6°
113.4°
2.629
0.952
0.950

ΠΠ–DR plus H_2O

Figure 14. Optimized geometries for DR and ZW states with one clustering water

M-O-O
5
M(+3)
e⁻
M(+2)
O2
M(+3)-O-O
1
e⁻
ROH
RH
M(+5)=O
4
H⁺
M(+3)-O-OH
3
H⁺
M(+3)-O-O
δ−
2
H2O

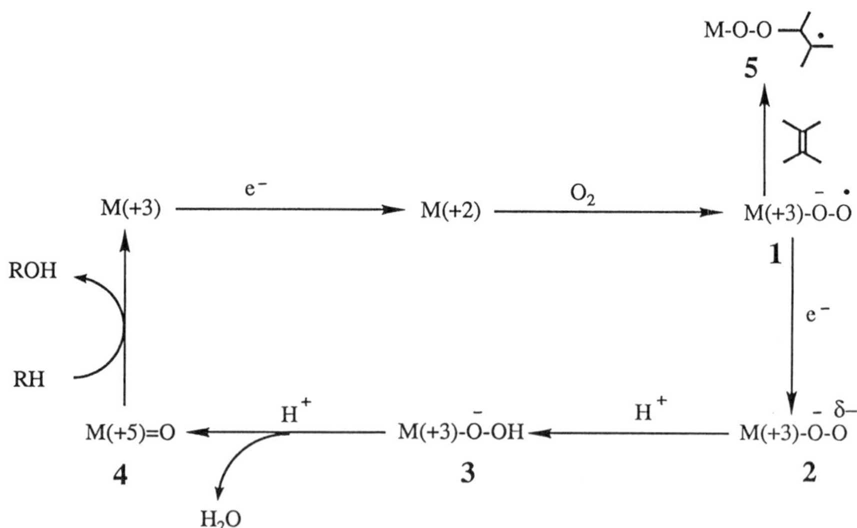

Figure 15. Possible mechanism for activations of molecular oxygen catalyzed by transition metal ions M(+*m*) [≡ M^{m+}] in artificial and biological systems. The key intermediates **1–5** have been proposed on theoretical and experimental grounds

4.3.1 Stable and unstable metal–molecular oxygen bonds

Consider the chemical bonding between a transition metal ion $M(+m)$ $[\equiv M^{m+}]$ and molecular oxygen (O_2). The π_z^*-orbital of O_2 interacts with the d_{z^2}-orbital of $M(+m)$ in the case of the side-on conformation of the complex $(MO_2)^{m+}$, giving the bonding (σ) and antibonding (σ^*) MOs as illustrated in Figure 16A.[45] Similarly, the orbital interaction between the $\pi_y^*(O_2)$- and $d_{yz}[M(+m)]$-orbitals provides the π- and π^*-MOs as shown in Figure 16B.

The RHF solution for the ground state of $(MO_2)^{m+}$ gives the $(\sigma)^2(\pi)^2$ configuration as shown in Figure 17A. However, the RHF solution often suffers from triplet instability and is reorganized into the more stable UHF solution. As shown in Figure 17B, the UHF MOs are spin-polarized so as to incorporate electron correlations in quasi-degenerate RHF MOs:[45]

$$\sigma^+ = \cos\theta\sigma + \sin\theta\sigma^* \tag{27a}$$

$$\pi^+ = \cos\omega\pi + \sin\omega\pi^* \tag{27b}$$

where θ and ω are the orbital mixing parameters. The UHF MOs, (σ^+, π^+) and (σ^-, π^-) for $(MO_2)^{m+}$ are more or less localized on the $M(+m)$ and the terminal oxygen atom, respectively, as in the case of 1,3-diradicals which are described in Section 4.2.2. These are reduced respectively, to the fragment orbitals, (d_{z^2}, d_{yz}) of $M(+m)$ and $(\pi_z^*, \pi_y^*)$of O_2 at the dissociation limit, $M(+m) + O_2$. On the other hand, the UHF MOs become equivalent to the RHF MOs if the electron correlation is weak near the equilibrium geometry of $(MO_2)^{m+}$. Hence the qualitative characteristics of transition metal–oxygen bonds may be understood in terms of the MO concepts obtained by the most stable HF solution.

4.3.2 *Ab initio* MO calculations

Ab initio UHF calculations[45] have been carried out for superoxo transition metal complexes with and without ligands in order to confirm the above qualitative MO picture and to elucidate their electronic structures. The $(NH_2)_4^{2-}$ group was considered as a model for porphyrin and macrocyclic polyamines, whereas NH_3, imidazole (Im) and ^-SH were examined as axial ligands. Figure 18 illustrates the geometries assumed for the complexes plus water system in which water was regarded as a model for the hydrogen bonding. The basis sets used for transition metals are the Tatewaki–Huzinaga MINI[46] supplemented by the 4p AO with the same exponent as that for the 4s AO. The MINI plus diffuse basis sets were used for second-row atoms. Table 10 summarizes the calculated total energies, spin multiplicities, orbital configurations and net charges.

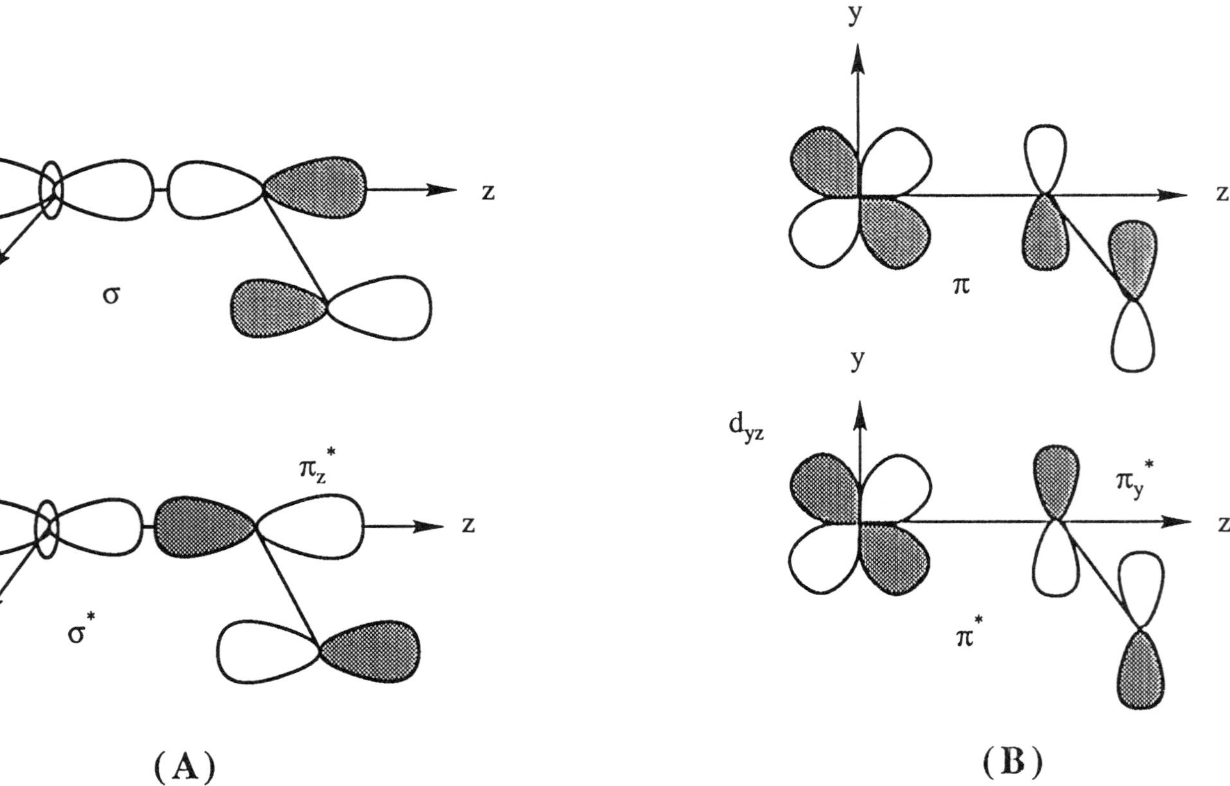

Figure 16. (A) Bonding (σ) and antibonding (σ^*) sigma molecular orbitals of transition metal peroxide (MO_2) constructed from d_{z^2} of M and π_z^* of O_2 and (B) the bonding π and antibonding π^* molecular orbitals constructed from d_{yz} of M and π_y^* of O_2

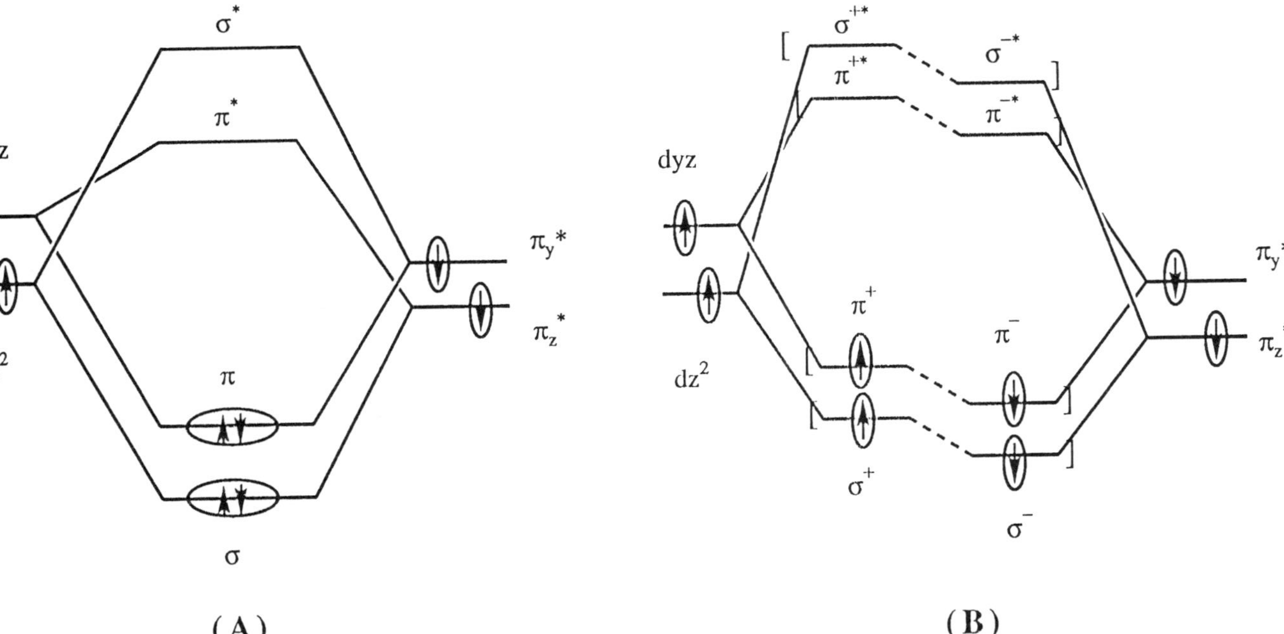

Figure 17. Molecular orbital correlation diagrams for transition metal peroxo compounds with closed-shell (stable) singlet pairs (A) and open-shell (unstable) singlet pairs (B)

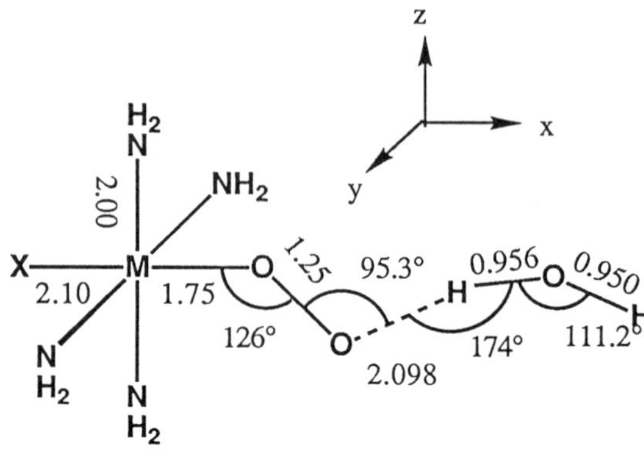

X = SH$^-$, NH$_3$. M = Cr, Fe, Ni.

Figure 18. Coordinate axes and geometries for transition metal peroxo compounds. The porphyrin and other electron-donating ligands are represented by $(NH_2)_4{}^{2-}$ and X. The hydrogen bonding stabilization is examined by addition of water

Table 10 shows that the net charges $\Delta Q(O_2)$ on O_2 in superoxo transition metal complexes are parallel to the electron-donating abilities ΔQ(ED) of ligands, which are defined as the net electron transfer from ligands to the naked core $(MO_2)^{m+}$. Figure 19 shows the linear relationship between $\Delta Q(O_2)$ and ΔQ(ED):

$$\Delta Q(O_2) = -0.726\, \Delta Q(\text{ED}) + 0.292 \tag{28}$$

The $\Delta Q(O_2)$ values are responsible for the superoxide (SOD) character, which is also related to the charge-transfer (CT) interaction between O_2 and LnM^{m+}. On the basis of ΔQ(ED) values, the superoxo transition metal complexes can be classified into three types:[45] type I with weak SOD character [$0.0 < \Delta Q(\text{ED}) < 0.5$], type II with intermediate SOD character [$0.5 < \Delta Q(\text{ED}) < 1.5$] and type III with strong SOD character [$1.5 < \Delta Q(\text{ED})$].

The net charge and spin density populations indicate that naked cores $(MO_2)^{m+}$ and $(MO_2)^{m+}$ with weak electron-donating ligands (type I) are regarded as electrophilic oxygen-transfer agents and may give, for example, radical-addition intermediates such as **5** in Figure 15 in the oxygenation reaction of alkenes. The $FeO_2(NH_2)_4$ complexes with and without an axial N-ligand exhibit the intermediate SOD character (type II), although the isolobal Cr and Ni analogs indicate the large SOD character (type III group). Therefore, the electronic structures of these iron–O_2 complexes are labile and sensitive to environmental effects such as the hydrogen bond, gegen ion and solvation. Thus, hemoglobin (HbO_2) and myoglobin (MbO_2) act as oxygen carriers under certain biological conditions since the Fe—O_2 bonds are weak. The J_{ij} values in

Table 10. Total energy (E_t), spin multiplicity ($2S + 1$), orbital configuration and net charge for superoxo transition metal complexes with and without ligands

No.	System	$2S+1$	E_t(kcal mol^{-1})	Orbital configuration[d, e]							Net charge		
				$d\pi_\perp$	$d\delta_1$	$d\delta_2$	$d\pi$	$d\sigma$	π_z^*	π_y^*	M	O	O
1	CrO_2^+	3	−1190.8562	↑	↑		↑	↑	↓	↓	1.86	−0.37	0.51
2		7	−1190.8458	↑	↑		↑	↑	↑	↑	1.86	−0.43	0.57
3	FeO_2^+	3	−1409.8678	↑	↑	↑↓	↑	↑	↓	↓	1.82	−0.41	0.59
4		7	−1409.8597	↑	↑	↑↓	↑	↑	↑	↑	1.84	−0.44	0.60
5		3	−1409.7120	↑↓	↑↓		↑↓		↑	↑	1.74	−0.29	0.55
6		3	−1409.6341		↑↓	↑↓		↑↓	↑	↑	1.82	−0.49	0.67
7		3	−1409.7633	↑↓	↑	↓	↓	↑	↑	↑	1.78	−0.36	0.59
8		5	−1409.8326	↑	↑	↑↓	↑	↑		↑↓	1.84	−0.30	0.46
9		5	−1409.7551	↑	↑↓	↑	↑↓		↑	↑	1.73	−0.31	0.58
10	NiO_2^{2+}	3	−1654.1217	↑↓	↑↓	↑↓	↑↓		↑	↑	1.73	−0.36	0.62
11	$CrO_2(NH_2)_4$	3	−1412.5185	↑	↑		↑	↑	↓	↓	1.40	−0.51	−0.26
12		7	−1412.5049	↑	↑		↑	↑	↑	↑	1.43	−0.69	−0.38
13	$CrO_2(NH_2)_4NH_3$	3	−1468.3490	↑	↑		↑	↑	↓	↓	1.70	−0.53	−0.36
14	$FeO_2(NH_2)_4$	3	−1631.4674	↑↓	↑	↑	↑	↑	↓	↓	1.24	−0.35	0.08
15		1	−1631.4290	↑	↑↓		↑↓	↑	↓	↓	1.14	−0.33	0.04
16		5	−1631.4275	↑	↑↓		↑↓	↑	↑	↑	1.13	−0.33	0.07
17		3	−1631.4281	↑↓	↑↓		↑↓		↑	↑	1.09	−0.35	0.10
18	$FeO_2(NH_2)_4NH_3$	3	−1687.2787	↑↓	↑	↑	↑	↑	↓	↓	1.42	−0.31	−0.01
19		1	−1687.2317	↑	↑↓		↑↓	↑	↓	↓	1.28	−0.33	0.04
20		5	−1687.2309	↑	↑↓		↑↓	↑	↑	↑	1.27	−0.32	0.07
21		3	−1687.2523	↑↓	↑↓		↑↓		↑	↑	1.20	−0.33	0.09
22	$FeO_2(NH_2)_4NH_3H_2O$	3	−1762.8357	↑↓	↑	↑	↑	↑	↓	↓	1.44	−0.31	−0.07
23	[a]$FeO_2(NH_2)_4IM$	3	−1854.1634	↑↓	↑	↑	↑	↑	↓	↓	1.15	−0.26	−0.10
24	$NiO_2(NH_2)_4$	3	−1875.8652	↑↓	↑↓	↑↓	↑↓		↑	↑	1.26	−0.59	−0.34
25		1	−1875.8254	↑↓	↑↓	↑↓	↑↓		↑	↓	1.14	−0.37	−0.11

Table 10. *(continued)*

No.	System	$2S+1$	E_t(kcal mol^{-1})	Orbital configuration[d, e]							Net charge		
				$d\pi_\perp$	$d\delta_1$	$d\delta_2$	$d\pi$	$d\sigma$	π_z^*	π_y^*	M	O	O
26	$FeO_2(NH_2)_4SH$[b]	3	−2027.5515	↑↓	↑	↑	↑	↑	↑	↑↓	1.40	−0.62	−0.44
27		5	−2027.5642	↑	↑	↑	↑↓	↑	↑↓	↓	1.52	−0.61	−0.60
28		7	−2027.5512	↑	↑↓	↑	↑	↑	↑	↑↓	1.41	−0.84	−0.24
29	$FeO_2(NH_2)_4SHH_2O$[b]	3	−2103.1253	↑↓	↑	↑	↑	↑	↓	↑↓	1.41	−0.56	−0.51
30		5	−2103.1413	↑	↑	↑	↑↓	↑	↑↓	↓	1.53	−0.60	−0.63
31		7	−2103.1253	↑	↑↓	↑	↑	↑	↑	↑↓	1.44	−0.86	−0.26
32	$[FeO_2(NH_2)_4SH]$[c]	2	−2027.3030	↑	↑↓		↑↓	↑	↓	↑↓	1.16	−0.53	−0.59
33		4	−2027.2910	↑	↑↓		↑↓	↑	↑	↑↓	1.15	−0.65	−0.44

[a] Im = imidazole.
[b] There is one hole in the ligand part and the formal charge is −1.
[c] The formal charge is −2.
[d] $d\pi_\perp = d\pi_{xz}$, $d\delta_1 = d_{xy}$, $d\delta_2 = d_{x^2-y^2}$, $d\sigma = d_{z^2}$, $d\pi = d_{yz}$.
[e] The ligand part ($d\pi_\perp d\delta_1 d\delta_2$) is the lowest spin state in the case of porphyrin.

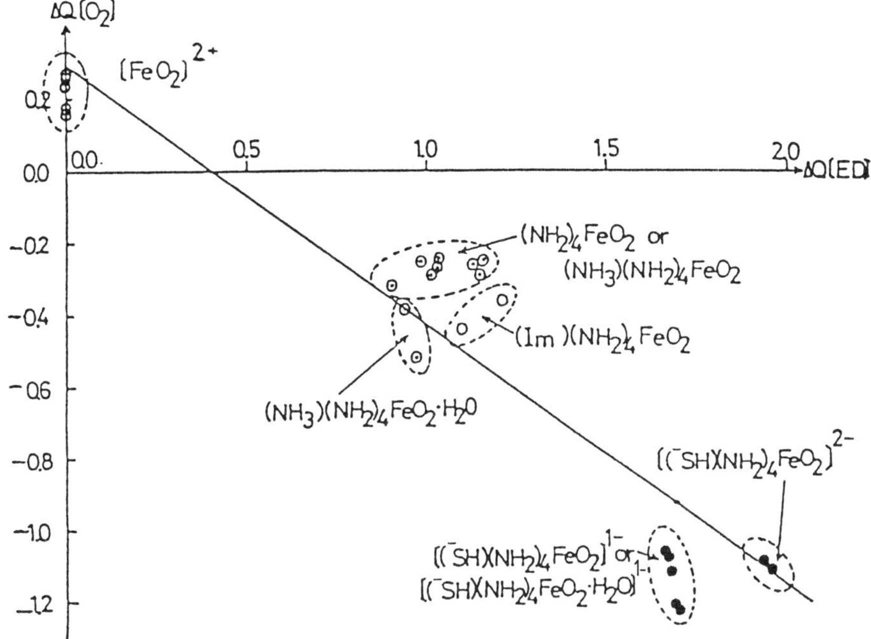

Figure 19. Variation of the superoxide (SOD) character for superoxo transition metal complexes. The net charge of the O_2 part is given by $\Delta Q[O_2]$, whereas the net electron transfer from ligands and environments to naked iron–peroxo core $[MO_2]^{2+}$ is given by ΔQ(ED). The superscript is the formal charge of the iron–peroxo species. Im denotes imidazole as an axial ligand

the Heisenberg model are negative in sign for these complexes, being responsible for the low-spin ground state of the complex.[42–45]

On the other hand, for activation of O_2 with iron complexes having N-axial ligands, protic solvents (S) and strong reducing agents (B) may be necessary under artificial conditions:[47]

$$\begin{aligned}\text{PorFe(II)} + O_2 &\xrightarrow{\text{S}} [\text{Por Fe(III)—O}^-\text{—O}]_s \\ &\longrightarrow [\text{PorFe(III)—O}^-\text{—O}^-\text{—H}^+]_s \qquad (3)\end{aligned}$$

This is compatible with the fact that the shunt path is favorable for artificial N-axial Fe systems with an intermediate SOD character as in the case of HRP:[48]

$$\begin{aligned}\text{PorFe(III)} + H_2O_2 &\xrightarrow{\text{S}} \text{PorFe(III)—O—O—H(H}^+) \xrightarrow{-H_2O} \\ &\text{Por}\overset{+}{\cdot}\,\text{Fe(IV)=O(HRP-I)}\end{aligned}$$

$$\begin{aligned}&\text{PorFe(III)} + \text{O} \longrightarrow \text{PorFe(V)=O} \longrightarrow \text{Por}\overset{+}{\cdot}\,\text{Fe(IV)=O} \\ &\text{(artificial P-450)} \qquad (4)\end{aligned}$$

The iron–O_2 complexes exhibit large SOD character if the thiolate ligand is used as an axial ligand (Nos 26–33 in Table 10). This demonstrates the crucial role of cys-S^- for activation of O_2 that facilitates the generation of an active P-450 intermediate such as **4** (Figure 15) in biological systems:[49]

$$\text{cys-S}^-\text{Fe(III)—O—O}^- \xrightarrow{\text{H}^+,\text{e}} \mathbf{3} \xrightarrow{-\text{H}_2\text{O}}$$

$$\text{cys-S}^-\text{Fe(V)=O or cys-S=Fe(IV)=O}$$

The type III group is a nucleophilic oxygen-transfer reagent of the type M—O—O like the superoxide radical anion.

4.4 Transition Metal Hydroperoxide Complexes

The electronic properties of oxygen compounds are modified by their coordination to transition metal ions and metal complexes. For example, transition metal peroxo compounds exhibit electrophilic and nucleophilic characteristics, depending on the nature of the transition metal ions and ligands, and also on the reaction conditions as shown in the preceding section. Situations are similar in the cases of transition metal hydroperoxide complexes.[50] The electronic properties of these compounds are sensitive to the formal oxidation number of transition metal ions and ligands.

4.4.1 Intermolecular interaction between hydroperoxyl radical and transition metal ions

Consider the intermolecular interaction between hydroperoxyl radical (·OOH) and a transition metal ion M(II). Figure 20 illustrates the σ- and π-radical sites of the ·OOH radical and a metal ion M(II). In order to elucidate the relative stability between these states, *ab initio* UHF MO(MINI) calculations were carried out for these complexes, assuming the molecular structure shown in Figure 21. Table 11 summarizes the results of the calculations. The σ-radical state of ·OOH is more stable than the π-radical state; note that the notations σ and π are exchanged in Figure 20 compared with the normal ones. On the other hand, the π-type complex is more stable than the σ-type complex because of the intermolecular interaction between M(II) and OOH. The net charges (ΔQ) on the hydroperoxy group in the σ- and π-type MnOOH complexes are about 0.24, indicating the electron-donating property of ·OOH to the manganese ion Mn(II), which is a high-spin five d-electron (d^5) system. The corresponding values for the naked cobalt Co(II) complexes are 0.4–0.5, where Co(II) is a low-spin seven-electron (d^7) system. These results indicate that the divalent metal ions effectively enhance the electrophilicity of the hydroperoxy group. If the

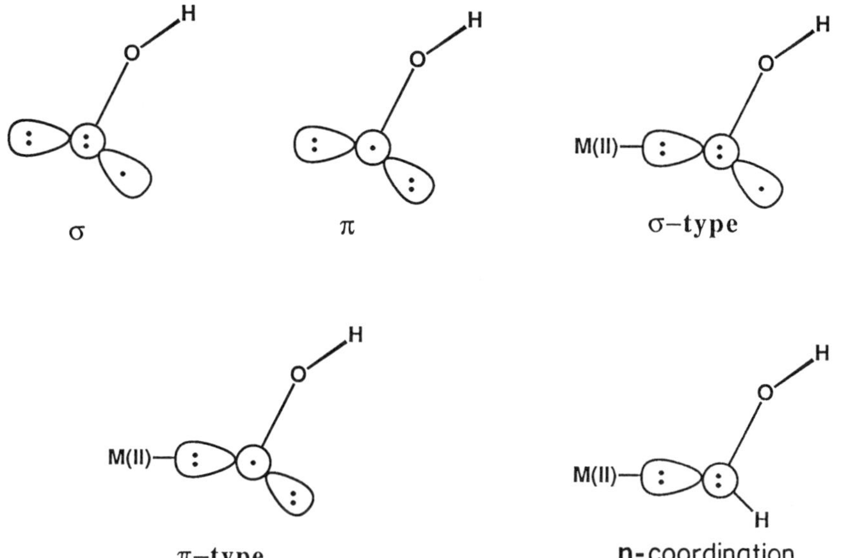

Figure 20. Electronic structures of the hydroperoxy radical and the transition metal hydroperoxide. σ and π denote the σ- and π-radical states, respectively. M(II) means a transition metal ion

partial back-donation of the $d\pi_x$ electron to the O—O antibonding orbital $\sigma^*[OO]$ of the OOH group occurs thermally, the O—O cleavage reaction may be induced in the following way:

$$\mathrm{Mn(II)\dot{O}OH} \rightarrow \mathrm{Mn(II)}\cdots\mathrm{\dot{O}}\cdots\mathrm{OH} \rightarrow \mathrm{Mn(III)}—\mathrm{O}^- + \cdot\mathrm{OH}$$

The hydroxyl radical (·OH) thus produced undergoes radical reactions.

The back-donating ability from the metal ions to the O—O σ^*-bonds is negligible for d^0 transition metal complexes such as LnTi(IV)OOR and LnV(V)OOR where Ln represents ligand. Therefore, these species do not undergo the O—O cleavage reaction. This is also responsible for the non-radical oxygen transfer reactions that are observed when these complexes react with organic substrates as discussed in Section 5.

4.4.2 Electronic structures of transition metal hydroperoxides

The $(NH_2)_4$ group is regarded as a model for ligands (Ln) in transition metal complexes such as bleomycin.[51] *Ab initio* MO calculations have been carried out for several transition metal complexes with this ligand[50] and the results are summarized in Table 11.

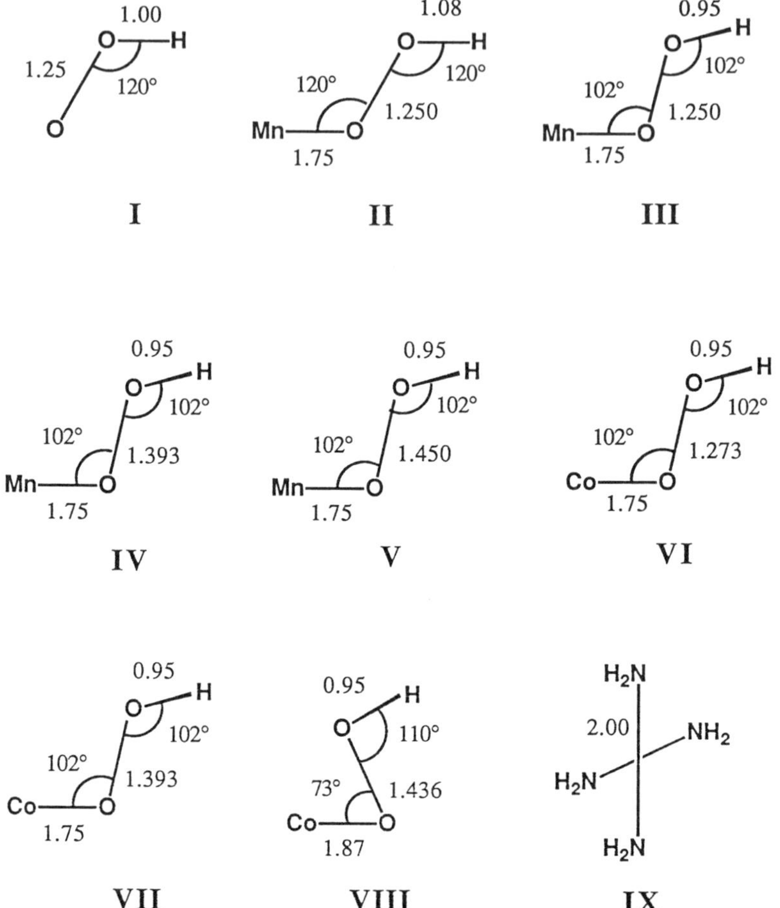

Figure 21. Molecular structures of manganese and cobalt hydroperoxides with short and long O—O bonds

The total net charge on the hydroperoxy group is -0.89 for the manganese complexes $(NH_2)_4Mn(III)OOH$. It becomes -0.95 when ammonia is taken as the sixth ligand (L′) of the complex, indicating that the hydroperoxy group in these complexes is essentially the monoanion (^{-}OOH). When the manganese intermediates become trivalent, they exhibit the high-spin (HS: $S = 4/2$) or intermediate-spin (IS: $S = 2/2$) state as illustrated in Figure 22A.

The net charge on the hydroperoxy group is -0.24 for the singlet cobalt complex $(NH_2)_4Co(III)OOH$ with the assumed O—O bond length $R(O—O) =$ 1.274 Å, while it becomes -0.75 for the same complex with the elongated O—O

Table 11. Total energies, spin multiplicities, net charges and total net charge (ΔQ) of the hydroperoxy group[a] for the manganese and cobalt hydroperoxide complexes by the *ab initio* unrestricted Hartree–Fock (UHF) molecular orbital (MO) method

System	Structure	$2S+1$	E_t (a.u.)	Net charge				ΔQ[a]
				M	O_1	O_2	H	
$\cdot$OOH (σ)	I	2	−149.2148		−0.32	−0.07	0.38	0.0
(π)	I	2	−149.1874		−0.28	−0.13	0.40	0.0
$^{-}$OOH	I	1	−149.2528		−0.96	−0.38	0.34	−1.0
HOOH		1	−150.5581		−0.44	−0.44	0.44	0.0
MnOOH (σ)	II	5	−1297.8794	1.78	−0.54	0.24	0.53	0.22
(π)	II	5	−1297.9176	1.78	−0.64	0.34	0.52	0.22
(σ)	III	5	−1297.9107	1.76	−0.26	0.05	0.46	0.24
(π)	III	5	−1297.9171	1.76	−0.20	0.00	0.44	0.24
	IV	7	−1297.8989	1.76	−0.25	0.03	0.46	0.24
	V	5	−1298.9149	1.76	−0.23	0.02	0.45	0.24
CoOOH	VI	1	−1529.2856	1.54	−0.27	0.25	0.49	0.47
	VI	3	−1529.2820	1.63	−0.39	0.29	0.48	0.38
	VII	1	−1529.3029	1.50	0.03	0.00	0.47	0.50
	VII	3	−1529.2970	1.61	−0.07	0.00	0.46	0.39
	VIII	1	−1529.3127	1.47	0.14	−0.08	0.48	0.53
	VIII	3	−1529.3077	1.54	0.07	−0.08	0.43	0.46
$(NH_2)_4$MnOOH	III · L	5	−1519.4987	1.34	−1.12	−0.07	0.30	−0.89
	III · L	7	−1519.4980	1.32	−1.11	−0.09	0.31	−0.89
$(NH_3)(NH_2)_4$MnOOH	III · L	5	−1575.2846	1.56	−1.14	−0.12	0.30	−0.95
	III · L	7	−1575.2845	1.55	−1.12	−0.14	0.30	−0.95
$(NH_2)_4$CoOOH	V · L	1	−1750.9257	0.93	−0.52	−0.10	0.38	−0.24
	V · L	3	−1750.9361	0.98	−0.80	0.26	0.38	−0.16
$(NH_2)_4$CoOOH	VI · L	1	−1750.9882	1.05	−1.00	−0.05	0.30	−0.75
	VI · L	3	−1750.9491	0.94	−0.68	0.20	0.36	−0.12

[a] See text.
[b] a.u. = atomic unit

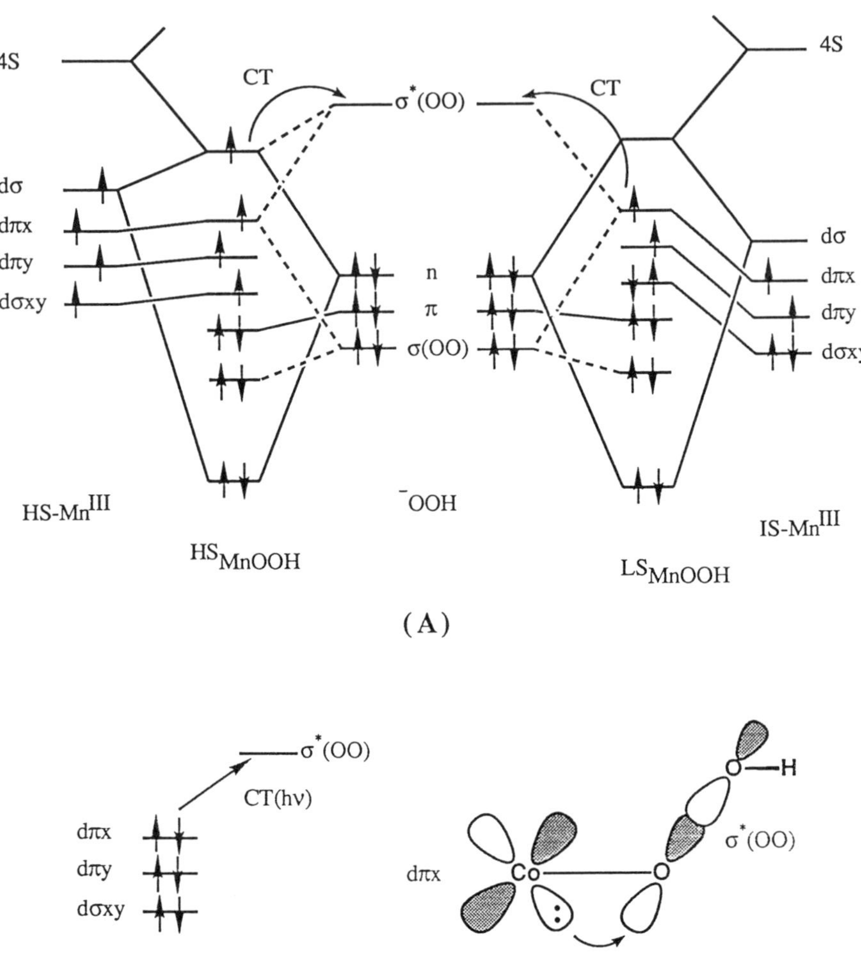

Figure 22. (A) Electronic structures of high-spin (HS) and low-spin (LS) manganese hydroperoxides and (B) photo-induced O—O cleavage of cobalt hydroperoxide

bond, $R(\mathrm{O{-}O}) = 1.393$ Å. The latter complex is largely stabilized compared with the former because of the charge-transfer (CT) and other interactions. The hydroperoxy group of the latter complex is essentially regarded as the monoanion and the cobalt ion as trivalent [Co(III)] in the low-spin (LS: $S = 0/2$) state. The *ab initio* calculations revealed that the spin state is different for Co(III) and Mn(III) ions in the hydroperoxido complexes.

The net charge on the OOH group is about -0.12 for the cobalt complexes at the triplet excited state. The back electron donation from the O-anion site to Co(III) occurs in this excited state.

Experiments have demonstrated[52] that the manganese and iron hydroperoxido complexes undergo the thermal O—O cleavage reaction, affording the metal oxo species. However, the cobalt hydroperoxide complexes do not show such reactivity, but undergo the O—O cleavage on electronic excitation. The difference in the spin states of Mn(III) and Co(III) in the complexes may be responsible for the difference in their reactivities. The unpaired $d\pi_x$ spin in the intermediate spin ($S = 2/2$) Mn(III) complex interacts with the antibonding O—O orbital if the electron-deficient group (e.g. H^+) assists the lowering of the σ^*(OO)-orbital energy as illustrated in Figure 20. Similarly, the non-bonding σ_n-electron interacts with the σ^*(OO) orbital in the HS ($S = 4/2$) Mn(III) complex. The partial electron transfer from Mn(III) to the σ^*(OO)-orbital induces O—O cleavage, affording the Mn(V)=O and H_2O (HO^- and H^+). The activation energy ΔE‡ for the O—O cleavage reaction can be estimated from the O—O bond energy E(O—O) and the $d\pi$–$p\pi$ bonding energy $E[\pi(\text{Mn—O})]$ between Mn(V) and O^{2-}:

$$\Delta E^{\ddagger} = k\{E(\text{O—O}) - E[\pi(\text{Mn—O})]\} \tag{29}$$

where k is a constant. The activation energy could be small or negative since the latter term contributes at the transition state. On the other hand, the $d\pi_x$ lone pair is largely stabilized in the case of low-spin cobalt complexes, hence thermal hopping of the $d\pi_x$-electron into the σ^*(OO) orbital does not occur. The $\Delta E^{\ddagger}$ value should be large because of the lacking of $d\pi$–$p\pi$ bonding energy. The photo-induced CT excitation from $d\pi_x$ to the σ^*(OO) orbital is crucial in inducing O—O cleavage, as illustrated in Figure 22B, in accord with experiments.[52]

4.4.3 Isolobal analogies

The discussion given in the preceding section suggests that isolobal analogies exist between transition metal peroxides and organic peroxides. For example, carbonyl oxides XYCOO could be regarded as a model for metal (M) molecular oxygen complexes such as oxygenated myoglobin (MbO_2) and hemoglobin (HbO_2). The simple VB consideration provides the spin-paired structure $^1(p\pi)^1(d_{yz})$ with 100% diradical character for the MOO complex, whereas the carbonyl oxide model (CH_2OO) suggests a moderate DR character ($y = 40\%$) for the same complex. The singlet–triplet splitting of an MbO_2 model is calculated to be 10.4 kcal mol^{-1} by the GVB CI method.[43] This value is very close to the corresponding values for oxygenated organic 1,3-dipoles such as carbonyl oxides (Table 4). This indicates that the GVB method may predict a similar y value (*ca* 40%) for the iron–peroxo complex:[43]

$$\begin{array}{ccc} \mathrm{O} & \mathrm{O}\cdot & \mathrm{O}\,\delta^{-}\cdot \\ / & / & /\!\!/ \\ \mathrm{O} & \mathrm{O} & \mathrm{O} \\ | & | & | \\ \mathrm{M} & \mathrm{M}\cdot & \mathrm{M}\,\delta^{+}\cdot \\ (y = 0 \sim 100\%) & y = 100\% & y = 40\% \end{array}$$

Neutron diffraction analysis[53] has revealed hydrogen bonding stabilization of the bound molecular oxygen in MbO_2 bearing a histidine ligand. The observed O—H bond distance in this oxygenated complex is 1.98 Å, which is close to that of the ZW-H_2O structure of the carbonyl oxide in Figure 14. The carbonyl oxide model strongly suggests the formation of an MbO_2-histidine complex described by a mixture of DR and ZW structures. The hydrogen bonding stabilization as represented by Fe—O—O—H—N is therefore assumed to serve as a means of controlling oxygen affinity in Hb and Mb systems, as in the case of the model system in Figure 14. This experiment and a theoretical study on solvation show that the carbonyl oxide model (unsymmetrical oxygenated dipolar model) is more appropriate than the ozone model (symmetrical oxygenated dipolar model) for MbO_2 and HbO_2. The former model further suggests that the proton transfer occurs from an amino acid to FeO_2, giving the iron–oxo species plus hydroxy anion or iron–hydroperoxide complex under biological conditions as discussed in the preceding section:

$$FeO_2 \longrightarrow Fe{=}O + {}^{-}OH + H^{+} \longrightarrow Fe{=}O + H_2O \qquad (30a)$$

$$FeO_2 \longrightarrow FeOOH \qquad (30b)$$

5 EPOXIDATION OF ALKENES

The epoxidation of alkenes is an important reaction in organic synthesis.[54] The reaction can be achieved by using a variety of peroxy compounds such as dioxiranes, peroxy acids and transition metal hydroperoxides as oxidizing agents. The structures of the peroxy compounds are shown in Figure 23. In the epoxidation reactions, the electrophilic oxygen is transferred from the peroxy compounds to alkenes. In this section, the electronic features of peroxy compounds and the transition structures in the epoxidation reactions are discussed on the basis of the results of *ab initio* semi-empirical MO calculations.

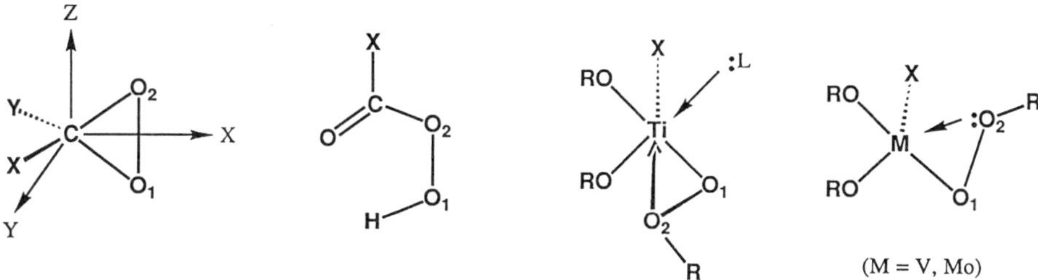

Figure 23. Molecular structures of typical epoxidation reagents: dioxirane, peracids and transition metal hydroperoxides

5.1 Dioxiranes

Dioxiranes are often used as reagents for the epoxidation of alkenes. Table 12 summarizes total electronic energies, dipole moments, gross orbital charges and net charges calculated for parent and substituted dioxiranes. Substituted dioxiranes are formally regarded as adducts between dications of carbenes XYC(2+) and oxygen dianion O_2^{2-}. Since the electron transfer from oxygen dianion to XYC(2+) is about 1.20–1.35, the net charge on an oxygen atom of dioxiranes is about −0.40 to −0.35, where the net charge on the carbon atom varies in the range 0.3–1.3, depending on the nature of substituents X and Y. The situation is similar with ring-opened dioxiranes, i.e. carbonyl oxides. The electrophilic nature of dioxiranes cannot be explained by the electrostatic interactions between dioxiranes and alkenes because the oxygen atoms bear a considerable negative net charge. This implies that the orbital interactions between dioxiranes and alkenes become important in the epoxidation of alkenes.

In order to elucidate the nature of the electrophilicity of dioxiranes, the MO energy levels of the parent dioxirane were calculated by the *ab initio* 4–31G MO method. The results are shown in Figure 24. The HOMO and next HOMO (NHOMO) of dioxiranes are the π-type O—O antibonding and bonding MOs, respectively, like the peroxy anion $^{-}$OOH, and these MOs are occupied by four electrons. The gross orbital charges on $2\pi_y$ AO orbitals (lone pairs) of oxygens are about 1.95 as shown in Table 12. On the other hand, the gross orbital charge on the $2p_z$ AO is about 1.0, which is responsible for the O—O covalent bonding. The LUMO of dioxirane corresponds to the σ-type O—O antibonding MO. The LUMO energy levels are lowered by introducing electron-withdrawing substituents. Thus, the calculations indicate that the LUMO serves effectively as an electron-accepting orbital for electron-donating alkenes, particularly in the case of difluorodioxirane.

The relative stability between the ring form of dioxirane and its ring-opened form (methylene peroxide) was calculated by the Schlegel PUMP2/6–31G*

Table 12. Total energy (E_t), LUMO energy level (LU), gross orbital charges (GC), dipole moments and net charges (NC) for substituted oxiranes (XYOO) by the RHF 4–31G method

No.	X	Y	E_t (kcal mol^{-1})	LU(σ_A^*)	μ(D)	GC(O)				NC	
						s	x	y	z	C	O
1	H	H	−188.3124	0.156	3.73	2.033	1.380	1.967	0.993	0.319	−0.375
2	CH_3	H	−227.3050	0.163	3.93	2.043	1.388	1.957	1.003	0.449	−0.388
3	CH_3	CH_3	−266.2959	0.170	4.06	2.056	1.395	1.947	1.006	0.568	−0.400
4	OH	H	−263.0774	0.143	1.74	2.044	1.401	1.947	1.002	0.772	−0.391
5	OH	OH	−337.8343	0.131	0.28	2.054	1.408	1.930	1.003	1.222	−0.391
6	OCH_3	H	−302.0428	0.150	2.01	2.045	1.407	1.946	1.002	0.799	−0.396
7	NH_2	H	−243.2696	0.178	4.75	2.048	1.411	1.949	1.008	0.723	−0.412
8	F	H	−287.0475	0.124	3.12	2.044	1.373	1.950	0.999	0.843	−0.363
9	F	F	−385.7731	0.093	1.36	2.058	1.344	1.939	1.001	1.345	−0.338
10	CHO	H	−300.8560	0.062	0.57	2.054	1.354	1.954	1.001	0.346	−0.359

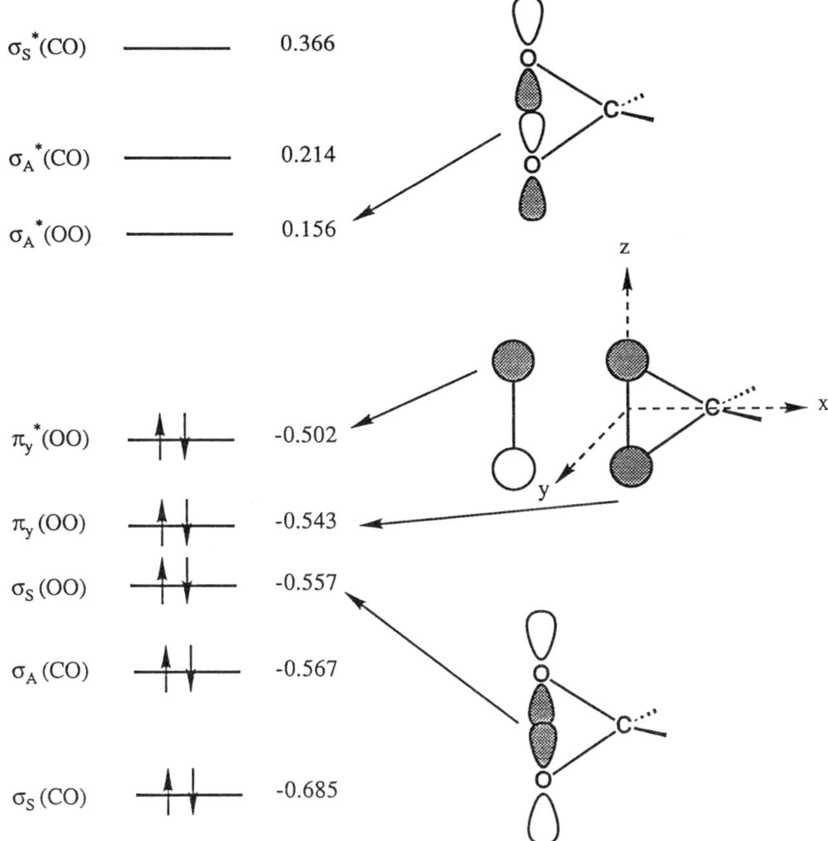

Figure 24. Orbital energy diagrams calculated for parent dioxirane by the RHF 4–31G method

method.[55] The ring form was more stable by 41.0 kcal mol^{-1} than the singlet diradical (DR) state of the ring-opened methylene peroxide. The energy gap between the DR and ring forms for difluorodioxirane was 67 kcal mol^{-1}.[55]

The locations of transition structures for the ring-closure reactions of the ring-opened DR to dioxiranes were also determined by the use of the HF 3–21G energy gradient procedures and the MP2/6–31G* energy calculations. Figure 25 illustrates schematically the calculated results for difluorodioxirane. The activation barrier for the closure of difluoromethylene peroxide (difluorocarbonyl oxide) to difluorodioxirane via the transition structure **TS4** of C_1 symmetry was only 10.4 kcal mol^{-1}. On the other hand, the activation barrier from unsubstituted methylene peroxide to the **TS4** state was 33.6 kcal mol^{-1}. This value was reduced to 22.8 kcal mol^{-1} by the MP4SDTQ/6–31G* method. Hence the

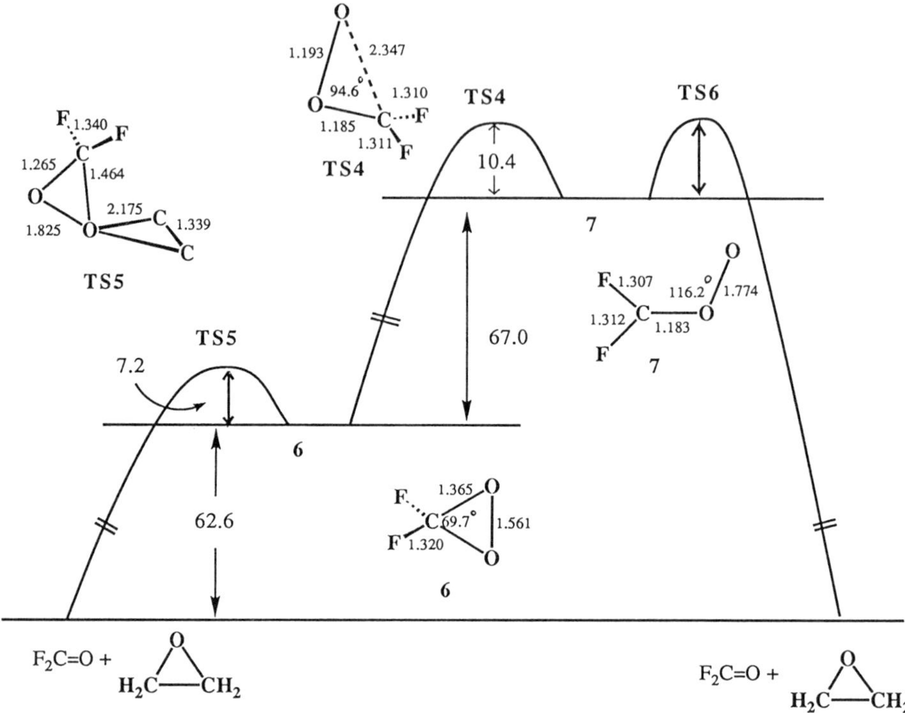

Figure 25. Potential energy profiles for the difluorodioxirane plus ethylene system and the optimized geometries of the transition structures (**TS4**–**TS6**) and the intermediates (**6** and **7**) by the RHF 3-21G method

calculations indicate that difluoromethylene peroxide undergoes ring closure to the dioxirane more readily than does methylene peroxide. The barrier for the conversion of acetaldehyde oxide to methyldioxirane was experimentally estimated to be about 16 kcal mol^{-1}.[56] The calculated barrier heights are in reasonable agreement with the experimental values.[55]

The activation energy for the oxygen transfer reaction from difluorodioxirane (**6**) to ethylene via the transition structure **TS5** was calculated to be 7.2 kcal mol^{-1} after the zero-point energy correction, while the corresponding barrier was 51.1 kcal mol^{-1} for parent dioxirane. Hence the computations also indicated that the activation barriers for oxygen transfer reactions are reduced by introducing electron-withdrawing groups into dioxirane. This is compatible with the fact that the antibonding O—O energy levels are lowered by attaching electron-withdrawing substituents on the carbon atom of dioxiranes as shown in Table 12. The *ab initio* 4-31G calculation was also carried out for the **TS5** transition structure for the oxygen transfer reaction from dioxirane to tetramethylethylene (TME). The net electron transfer from dioxirane to TME in this reaction was found to be 0.12, indicating the electrophilic nature of methylene peroxide. Unfortunately, the activation barrier was not reported for **TS6** of the oxygen transfer from the ring-opened form **7** toward ethylene.

It is known that methyl(trifluoromethyl)dioxirane reacts with polycyclic hydrocarbons to give alcohol derivatives.[57] The oxidation reaction of this sort was thought to proceed via an 'oxenoid' mechanism that involves a concerted O-atom insertion by dioxiranes into the C—H bonds of the hydrocarbons. This result is also compatible with the fact that dioxiranes are strongly electrophilic.

5.2 Carbonyl Oxides

The ring-opened dioxiranes, carbonyl oxides, also serve as oxygen transfer reagents to alkenes. The geometries of transition structures (TSs) for oxygen transfer reactions of carbonyl oxides to alkenes were determined by the energy gradient techniques using the Hartree–Fock STO-3G method.[58] The five different attacking modes were examined and three of them were found to be genuine TSs with one imaginary frequency. Figure 26 illustrates such genuine TSs for the epoxidation of ethylene with methylene peroxide. The activation barriers for the epoxidation calculated by the HF STO-3G//HF STO-3G method are 8.2, 8.3 and 13.5 kcal mol^{-1} for **TS7a**, **TS7b** and **TS7c**, respectively. The first two values are reduced to 4.8 and 4.7 kcal mol^{-1}, respectively, by the HF 6–31G*//HF STO-3G method. For substituted ethylenes, **TS7b** seems more favorable than **TS7a** because of steric repulsion.

In **TS7b**, the HOMO of ethylene interacts with the O—O σ-antibonding MO of carbonyl oxide as in the case of dioxirane, while the π-HOMO of carbonyl oxide can overlap with the LUMO of ethylene. These orbital interactions stabilize **TS7b**. Figure 27 illustrates the orbital interaction schemes between the $\pi\pi$ diradical (DR) state of carbonyl oxide and ethylene and between its zwitterion (ZW) state and ethylene. Figure 27A suggests that O—O homolysis may occur in the $\pi\pi$ DR state to give the nπ* DR excited state of formaldehyde and the ground state of dioxirane.[58] Indeed, the electronically excited aldehydes were detected in the course of the ozonolysis reaction of alkenes in the gas phase;[19] note that in this reaction, carbonyl oxides are produced as active reaction intermediates.

On the other hand, the O—O heterolysis is expected to occur in the ZW state of carbonyl oxides in oxygen transfer reactions in the solution phase, giving the ground-state products. In this case, the reaction can be described as an S_N2 reaction on the terminal oxygen atom of carbonyl oxides as illustrated in Figure 27B. Thus, the computations show that **TS7b (TS7a)** is in line with an S_N2 reaction which involves O-atom or oxene transfer as a key step and leads directly to oxiranes and the ground state of carbonyl compounds.

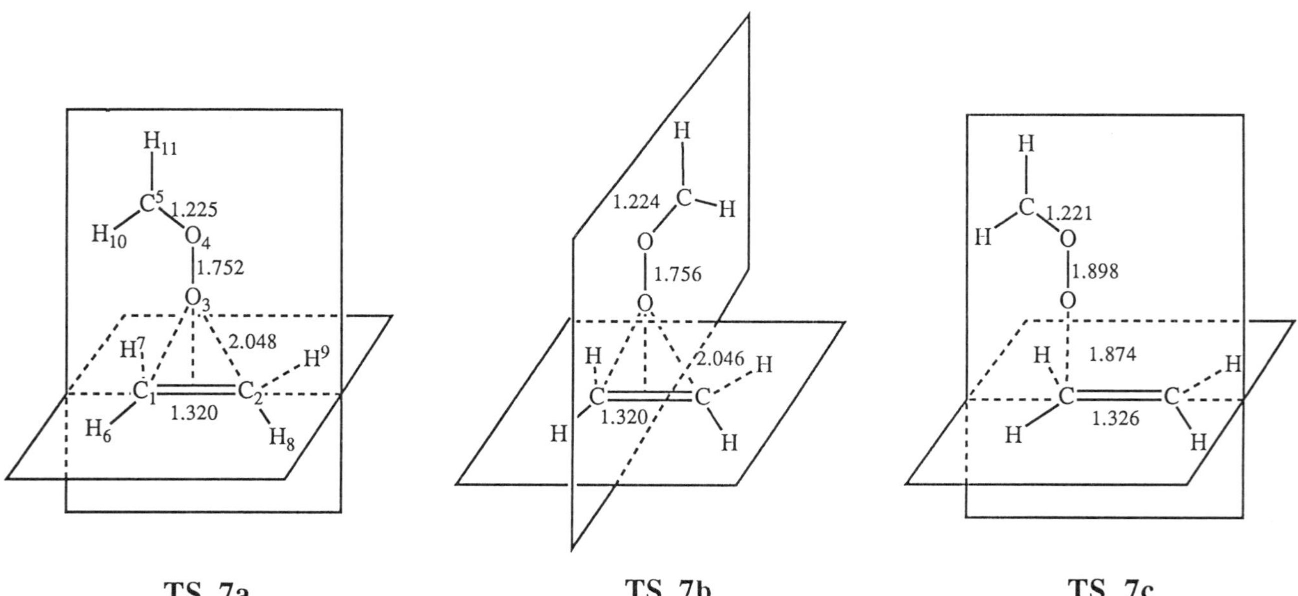

Figure 26. Optimized geometries of the transition structures (**TS7a–c**) for different attacking modes of carbonyl oxide at the ethylene double bond

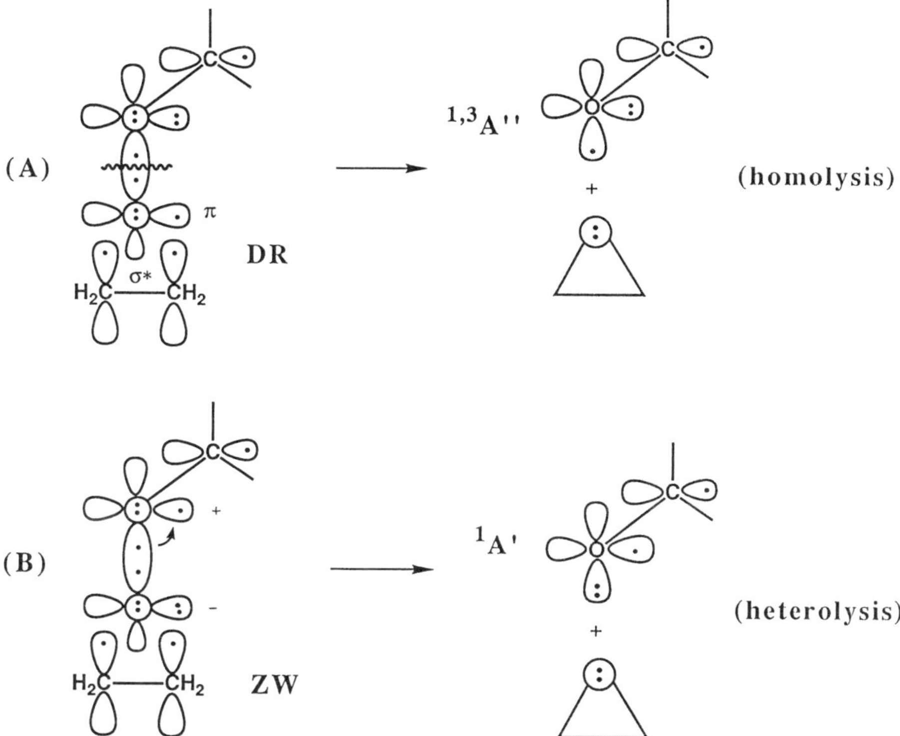

Figure 27. Schematic representation of the electronic configurations for (A) homolysis and (B) heterolysis of the O—O bond of the carbonyl oxide plus ethylene system

5.3 Peroxy Acids

Peroxy acids are powerful agents for the epoxidation of alkenes. The total electronic energies, gross orbital charges and net charges for alkyl peroxides and peroxy acids are given in Table 13. The gross orbital charges on the $2p_z$ AO of the terminal oxygen atom (O_1) and the net charges on the same atom of peroxy acids decrease slightly on introducing electron-withdrawing groups on the carbon atom of peroxy acids. However, the negative net charges on O_1 indicate that the epoxidation of alkenes by peroxy acids should be frontier-orbital controlled.

Five extreme models of plausible transition structures for the epoxidation of ethylene with peroxyformic acid were investigated by the HF STO-2G(4G) method.[60] The fully optimized geometry of a symmetrical TS (**TS8**) is illustrated in Figure 28. This geometry indicates that little reorganization is recognized in either of the reactant molecules at the transition state. The net charge

Table 13. Total energy (E_t), gross orbital charges (GC) and net charges (NC) for alkyl hydroperoxides (XOOH) by *ab initio* RHF MO (4-31G) method

Structure	E_t(a.u.)	R(O—O) (Å)	GC(O_1)			NC	
			x	y	z	O_1	O_2
H—OOH	−150.5581	1.464	1.44	1.98	1.07	−0.44	−0.44
CH_3—OOH	−189.5215	1.460	1.44	1.98	1.07	−0.42	−0.43
t-Bu—OOH	−306.4101	1.480	1.44	1.99	1.09	−0.44	−0.43
HO—OOH	−225.2168	1.442	1.44	1.98	1.04	−0.39	−0.11
F—OOH	−249.1721	1.419	1.44	1.96	1.03	−0.34	0.07
CF_3—OOH	−485.7215	1.447	1.45	1.99	0.94	−0.32	−0.44
HCO—OOH	−263.1059	1.402	1.55	1.99	0.89	−0.38	−0.40
CH_3CO—OOH	−302.0979	1.400	1.56	1.99	0.90	−0.39	−0.42
CF_3CO—OOH	−598.2590	1.400	1.56	1.99	0.85	−0.34	−0.41

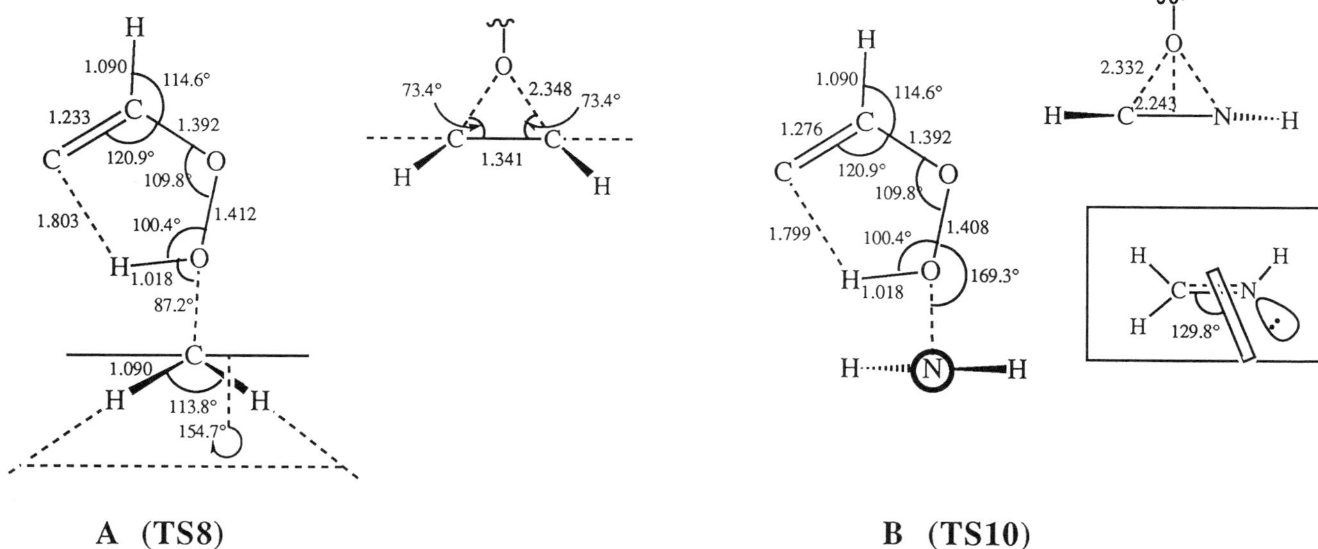

Figure 28. Optimized transition structures (**TS8** and **TS10**) of the oxygen-transfer reactions of peracid toward (A) ethylene and (B) methanimine

population on **TS8** shows that electrons are transformed from ethylene to peroxyformic acid. This is consistent with experimental observations that peroxy acids act as electrophiles in the epoxidation of alkenes.

The activation barrier for the symmetrical **TS8** is 27 kcal mol^{-1}, whereas the lowest activation energy for the unsymmetrical **TS9** that represents a terminal carbon attack is approximately 16 kcal mol^{-1}. The latter value is in good agreement with experimental values (15–18 kcal mol^{-1}).[61] The secondary deuterium isotope effect on the epoxidation of *p*-phenylstyrene[62] has revealed that the oxygen atom of the peroxy acid attacks preferentially the α-carbon of the C—C double bond of the styrene, indicating that the transition structure of this reaction is expected to be unsymmetrical **TS9**, rather than the symmetrical **TS8**.

Six plausible transition structures were examined for the oxidation of imines with peroxyformic acids.[63] Figure 28B illustrates the fully optimized transition structure for a central attack on the C=N bond. **TS10** is analogous to the transition structure in the epoxidation of alkenes and involves nucleophilic attack of π-bonding electrons of the imine on the electrophilic oxygen atom of peroxy acids. However, an unsymmetrical TS (**TS11**) which involves attack of the terminal oxygen atom (O_1) of HCOOOH on the nitrogen atom of methanimine is more stable than the symmetrical **TS10**. The oxidation of the imine with HCOOOH proceeds, in fact, via this terminal attack and the formation of the oxaziridine is an exothermic reaction of 18.8 kcal mol^{-1} with an activation barrier ranging from 1.5 to 4 kcal mol^{-1}.[63]

5.4 Transition Metal Alkyl Peroxides

Alkyl peroxides do not have sufficient reactivity toward alkenes under ordinary conditions. The transfer of oxygen from alkyl hydroperoxides to alkenes is effectively catalyzed by high-valent d^0 transition metal complexes [such as Mo(VI), V(V) and Ti(IV) complexes].[64] The X-ray crystal structure of (dipic)VO(OO-*t*-Bu)H_2O (**8**) in Figure 29 revealed that the complex has a deformed pentagonal bipyramidal structure with the peroxo oxygens in the equatorial plane.[65] A striking structural feature of this complex is that the proximal oxygen—metal bond is only slightly longer than that of the coordinated distal oxygen—metal bond as illustrated in Figure 29A. *Ab initio* UHF calculations by the use of the MINI basis set were carried out for the model complex as shown in Figure 29B, in which environmental effects of ligands and solvents are expressed by changes in the formal charges and spin states of the vanadium cation. The calculated results are summarized in Table 14. The gross atomic charges show that the methylperoxido group of the model complex is essentially regarded as a monoanion in the cases of $[VOOCH_3]^{m+}$ ($m = 0, 1$) (**9**), irrespective of spin states. In contrast, the methylperoxido group of $[VOOCH_3]^{2+}$ corresponds to the π-radical state $\cdot$OOR. Interestingly, the

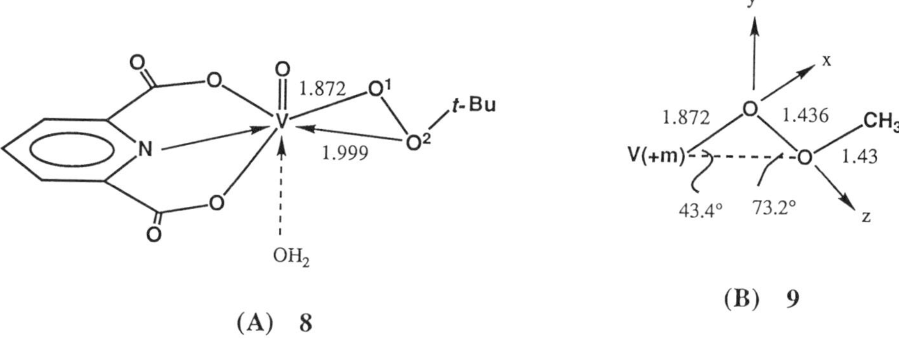

Figure 29. (A) X-ray structure (**8**) of the vanadium *tert*-butyl hydroperoxide and (B) the assumed geometry (**9**) of a model compound

gross orbital charge on the $2p_z$ AO of $[VOOCH_3]^+$ is close to that of peroxytrifluoroacetic acid (see Table 13). This implies that a peroxy acid constitutes a formal model of the complex **8**.

The orbital interaction scheme between OOR and V($+m$) is illustrated in Figure 30.[66] The π^*-OO HOMO of OOR interacts with the vacant d_{yz} AO of vanadium ion, while the OO antibonding σ^*-LUMO of OOR is stabilized by the interaction with the vacant d_{xz} AO of vanadium ion. The former stabilization is important for non-radical oxygen transfer reactions of high-valent d^0 transition metal complexes. On the other hand, the latter stabilization is essential for the electrophilic epoxidation of alkenes by these species. In addition, the computations show that the radical-type addition of OOR, which is facilitated by the stabilization from the π^*-OO–d_{yz} interaction, may occur under the specific condition that the vanadium cation is extremely electron deficient.

The asymmetrical epoxidation of allylic alcohols is a useful and important reaction in organic synthesis.[54] This reaction can be achieved by utilizing transition metal hydroperoxides. The most familiar is the titanium tetraisopropoxide-*tert*-butyl hydroperoxide system (**10**), which is called the Sharpless catalyst.[67] The molecular structure of **10** is illustrated in Figure 31. It has been observed that electron-withdrawing groups on the substrates accelerate the rate of the epoxidation, whereas electron-donating substituents retard the reaction.

Extended Hückel calculations[68] were carried out in order to elucidate the transition structure for the epoxidation of allylic alcohols by **10**. The transition structure **TS12** from these calculations is shown in Figure 31B. The electrophilic atom in the hydroperoxide group attacks the carbon—carbon double bonds of allylic alcohols from below the olefinic plane.[68] This is in accord with the

Table 14. Total energy (E_t), gross orbital charges (GC) and net charges (NC) for methylperoxovanadium complexes ($VO_1O_2CH_3$) by *ab initio* UHF (MINI) method

Structure	$2S+1$	E_T(a.u.)	$GC(O_1)$			$GC(O_2)$			NC	
			x	y	z	x	y	z	O_1	O_2
$[VOOCH_3]^0$	1	−1130.5659	1.74	1.83	0.82	1.44	1.89	1.25	−0.70	−0.01
	3	−1130.6446	1.72	1.84	0.83	1.46	1.89	1.23	−0.60	0.04
	5	−1130.6971	1.73	1.81	0.85	1.47	1.89	1.21	−0.63	0.03
$[H_2VOOCH_3]^0$	1	−1131.6301	1.75	1.94	0.82	1.40	1.92	0.98	−0.59	0.04
$[VOOCH_3]^+$	2	−1130.4257	1.65	1.85	0.87	1.49	1.89	1.19	−0.56	−0.04
	4	−1130.4728	1.67	1.80	0.89	1.49	1.90	1.17	−0.59	−0.06
$[VOOCH_3]^{2+}$	1	−1129.9188	1.61	1.07	1.15	1.61	1.86	0.96	0.01	0.07
	3	−1129.9085	1.83	1.08	1.15	1.61	1.85	0.97	0.01	0.06

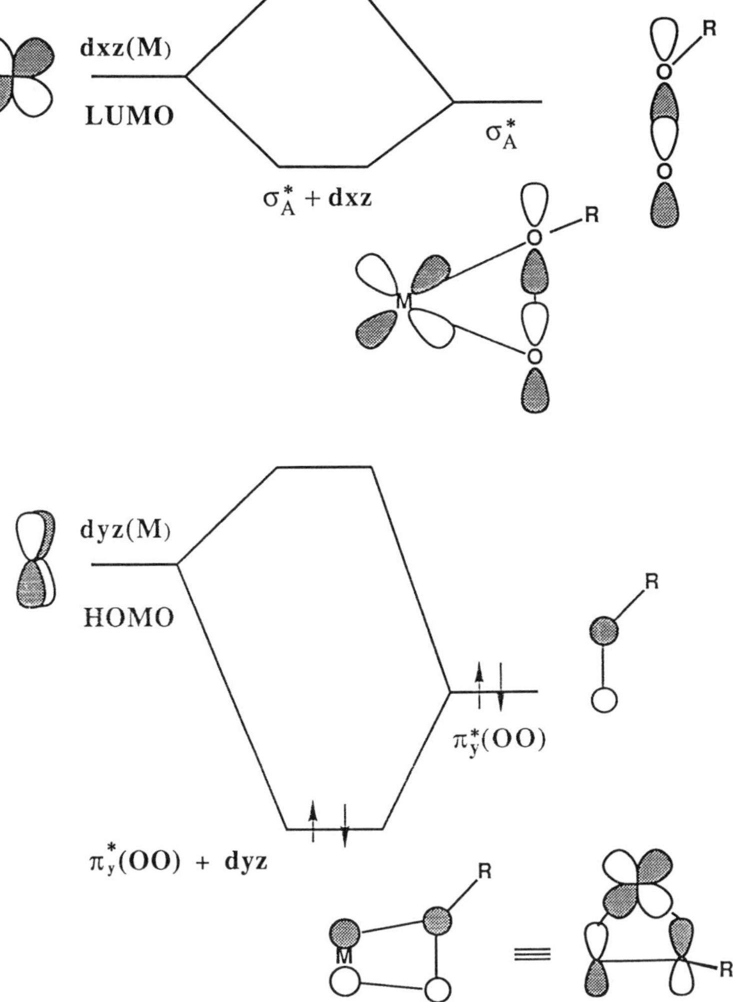

Figure 30. The σ- and π-type orbital interaction schemes between a transition metal ion (M) and alkylhydroperoxy group

mechanism proposed by Finn and Sharpless on experimental grounds.[54] The computations revealed the following characteristics:[68]

1. A spiro orientation of the alkene around the olefinic moiety and Ti—O—O—H centered on the O_1 atom (see Figure 31) of the hydroperoxo group is preferred because of a favorable interaction between the π^*-orbital of the alkene and the lone pair of electrons on the O_1 atom, which is oriented in the direction perpendicular to the titanium—peroxy bond.

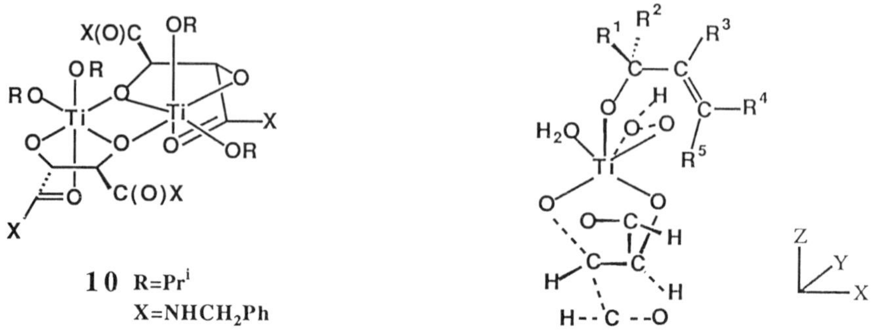

10 R=Pr[i]
X=$NHCH_2Ph$

(A)

(B) TS12

Figure 31. (A) Molecular structure of the titanium–tartrate complex (**10**) and (B) a model of the transition structure (**TS12**) of its epoxidation reaction with alkenes

2. The olefinic π-orbital of allylic alcohols can interact with the antibonding Ti(d)—O_1(p) π^*-orbital. The electron-donating groups attached to the olefinic part of allylic alcohols push up the energy level of the olefinic π-orbital. This intensifies the interaction between the hyperoxide group of the complex and the olefinic orbital and enhances the reaction rate.
3. The asymmetric structure of the titanium complex reflects its electronic structure. Both electronic and steric factors determine the stereochemical course of the Finn and Sharpless epoxidation.[54]

Hence the *ab initio*[66] and extended Hückel[68] calculations for the model and true complexes are complementary for the theoretical understanding of the Finn and Sharpless epoxidation.[54]

6 ELECTROPHILIC REACTIONS INVOLVING PEROXY COMPOUNDS

6.1 Transition Structures of Electrophilic Reactions of Oxygen Radicals

Thermal and photo-induced O—O cleavage of transition metal hydroperoxides generate hydroxyl radical and/or transition metal oxo species.[64] The chemical properties of the latter species resemble those of "oxenes" such as the oxygen atom O. These reactive species play important roles in the chemistry of combustion[69] and pollution[70] and in the oxidation of organic compounds and polymerization. In this section, the transition structures[71] for addition reactions of ·OH and O toward ethylene are discussed.

The addition of •OH to ethylene produces the 2-hydroxyethyl radical, which is converted into a variety of products. Kinetic studies have revealed that the reaction of •OH with ethylene in the gas phase gives a small negative Arrhenius activation energy, *ca.* -0.9 to -0.3 kcal mol^{-1}.[72]

The transition structure for the electrophilic addition of •OH to C_2H_4 was elucidated by calculation at the UHF 3-21G and 6-31G* levels.[71] Figure 32A shows the optimized geometries of the transition structure **TS13**. Prior to the transition state, •OH forms a hydrogen-bonded complex with the ethylene π-orbital.[71] **TS13** has only one imaginary frequency (5561 cm^{-1}). At the HF and MP levels, the barrier height for the •OH addition reaction was overestimated by 7–10 kcal mol^{-1}.[71] After spin projection by the PUMP4 6-31G* method,[12] the barrier is reduced to -0.9 kcal mol^{-1} (ref. 71) and is shifted 0.3 Å closer to the reactants, and the calculated activation barrier was found to be compatible with the observed data.[72]

The initial step of the addition of O to C_2H_4 was also examined by the UHF 6-31G* method.[73] Figure 32B illustrates the calculated transition-state structure **TS14**. The activation barrier height was calculated by the UMP*n*/6-31G* and the APUMP*n*/6-31G* methods by assuming the UHF/6-31G* geometry. The barrier heights obtained by the former method were 10.2 (UMP2), 8.7 (UMP4) and 8.4 (UMP) kcal mol^{-1}, and those obtained by the latter method were 4.2 (APUMP2), 2.6 (APUMP4) and 2.3 (APUMP) kcal mol^{-1}. The latter values were compatible with the observed rates of reaction. These results indicate that the spin projection effect on the calculated energy is appreciable for radical-type transition structures, since the chemical bonds responsible for the reaction are inevitably more or less spin-polarized.[11,12,71,73]

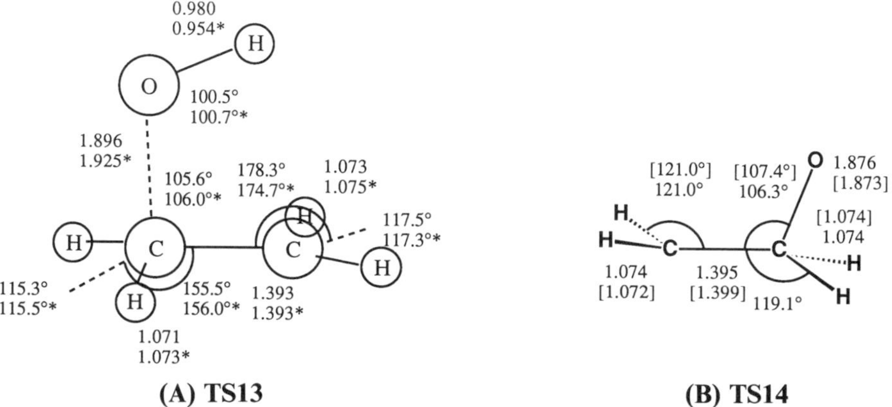

Figure 32. (A) Optimized transition structure (**TS13**) of the addition reaction between •OH and ethylene by the UHF 3-21G and 6-31G* (asterisks) method and (B) that (6-31G*) of the O plus ethylene system. CASSCF results are given in parentheses

6.2 Regioselectivities in Oxyradical Addition Reactions

A protonated form of $O_2^{\cdot-}$, the hydroperoxyl radical (HOO·), is known to be an initiator of radical chain reactions of alkenes although its concentration is nearly 1% at pH 7 in aqueous solutions. The regioselectivity of this species in radical reactions has been studied by the *ab initio* UHF STO-3G energy gradient technique.[74] Figure 33 illustrates the structures of the adducts obtained by the addition of $HO_2\cdot$ to the 2- and 3-positions of propene and vinyl alcohol. The relative stabilities of these radical adducts are shown in Table 15.[74] Attack on the 3-position is preferred to attack on the 2-position, indicating the electrophilic nature of $HO_2\cdot$, in contrast to the nucleophilicity of $O_2^{\cdot-}$.

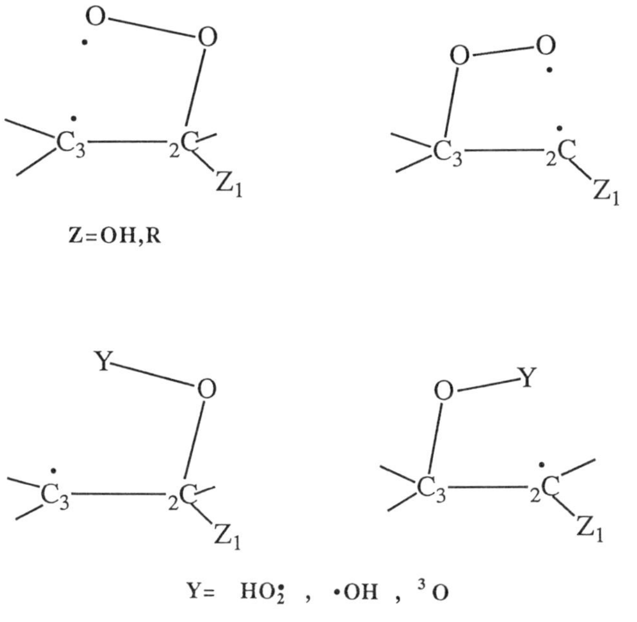

Figure 33. Schematic illustrations of regioselectivities for addition reactions of active oxygens toward electron-rich unsymmetrical alkenes

Table 15. Relative stabilities between 2- and 3-attacks of oxygen radicals

	$\Delta E_{rel.}$				
	$HO_2\cdot$	·OH	^3O	$^1O_2$	$^3O_2$
Propene	2.82	1.84	2.07	1.93	1.93
Vinyl alcohol	5.39	1.63	3.45	4.18	4.14

Ab initio geometry optimization calculations were also performed for the addition products of singlet and triplet molecular oxygens, hydroxyl radical and oxygen atom toward propene and vinyl alcohol. The results of the *ab initio* calculations shown in Table 15 confirm the electrophilic nature of these oxygen radicals. Note that the regioselectivity in the oxygen radical addition to electron-rich alkenes is opposite to that in the oxyanion addition to radical cations of the same compounds.[74]

6.3 Reactions of Singlet Oxygen with Alkenes

Singlet oxygen is one of the most important reactive species in oxidation reactions.[41] The addition of singlet oxygen ($^1O_2$) to carbon—carbon double bonds has been extensively investigated from both the theoretical[41,75–78] and experimental points of view. The mechanisms so far proposed for the reactions of $^1O_2$ with alkenes include[41] (1) a concerted supra-supra attack, (2) a concerted supra-antara attack, (3) a non-concerted attack on a double bond to generate a 1,4-diradical (DR) intermediate, (4) a non-concerted attack on a double bond to generate a 1,4-zwitterion (ZW) intermediate and (5) formation of a perepoxide intermediate. The structures of these intermediates are illustrated schematically in Figure 34.

The locations of transition structures and unstable intermediates for the reaction of $^1O_2$ with ethylene via the above five reaction pathways have been carried out by the CASSCF 4-31G (STO-3G) energy gradient techniques by the use of the six-active orbital, six-electron system.[78] The analytical Hessian matrix on the saddle point for the C_{2v} supra-supra concerted pathway [path (A) in Figure 34] has three negative eigenvalues. This saddle point is therefore classified as a third-order saddle point, which is of no chemical interest (see Figure 6). The saddle point for the C_2 supra-antara concerted pathway [path (B) in Figure 34] is classified as a second-order saddle point with two negative eigenvalues, and this point is also of no chemical interest. The situation is the same even for the C_s peroxirane-like saddle point and the C_s *syn* diradicaloid saddle point. Hence the concerted pathways are ruled out in the case of the addition reaction of $^1O_2$ to ethylene on the basis of the CASSCF computations.[78]

For the non-concerted, broken-symmetry pathway of C_1 mode, a first-order saddle point was found. The transition structure for this pathway is shown as **TS15** in Figure 35A. This structure is located at 29 kcal mol^{-1} above the dissociation limit [$^1O_2(\Delta) + C_2H_4$] and the CCOO dihedral angle is 102°. **TS15** has a *gauche* conformation[40] around the C—O bond and is responsible for the diradical path (C) in Figure 34. The other *gauche* diradical intermediate (**11**) with a true local minimum is shown in Figure 35B. The transition structure **TS15** has a CCOO dihedral angle of 74° and is located at 23 kcal mol^{-1} above the dissociation limit. This is converted into dioxetane (**12**), whose optimized geometry is given in Figure 35C.

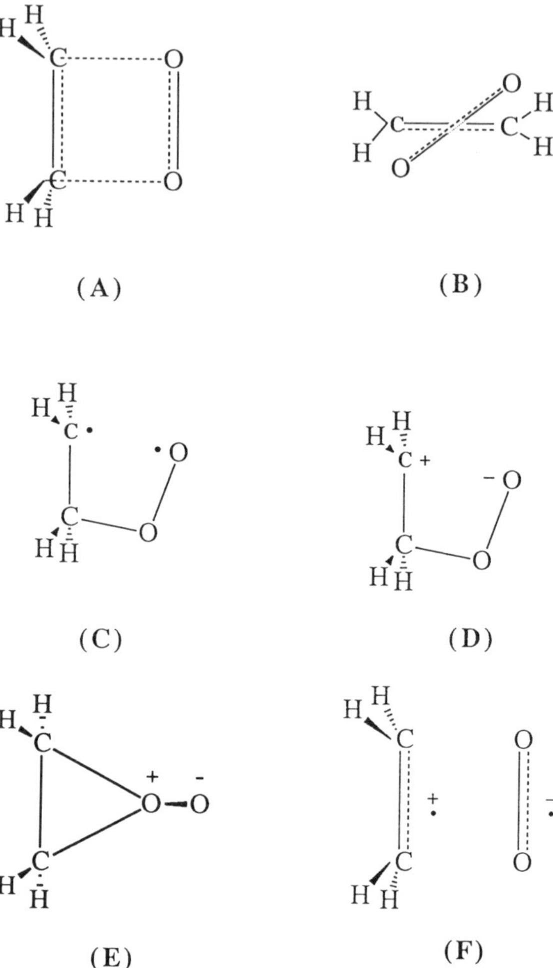

Figure 34. Possible modes (A–F) of transition structures and unstable intermediates for singlet oxygen reaction with ethylene

A perepoxidic species has frequently been proposed as a possible intermediate for singlet oxygen reactions.[41] The CASSCF computations showed that peroxirane (**13**) has a true minimum on the energy hypersurface, although the peroxirane-like saddle point is not a true transition state. The optimized geometry of **13** is shown in Figure 35D. Perepoxide is assumed to come from *gauche* **TS15**.

At the level of the CASSCF computations, the dioxetane-forming reaction, which is typical of singlet oxygen reactions, is described as a two-step reaction that proceeds through a *gauche* diradical intermediate.[78] However, it is

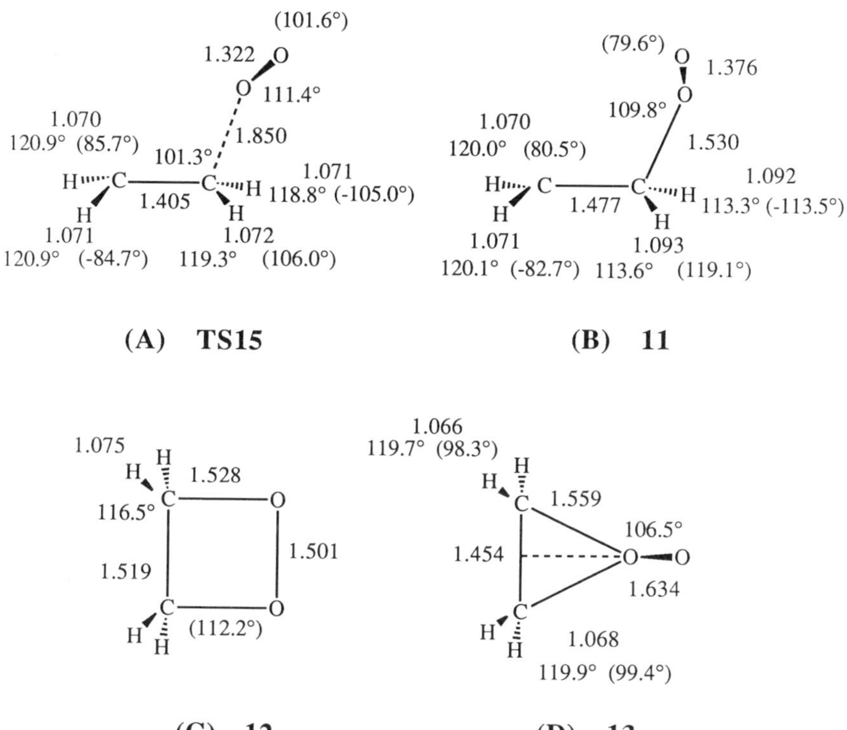

Figure 35. Optimized structures of the *gauche* transition structure (**TS15**) and intermediate (**11**), dioxetane (**12**) and perepoxide intermediate (**13**) by the CASSCF 4-31G method

noteworthy that these theoretical results may not be conclusive, because the CASSCF computations do not involve the correction of the dynamic electron correlation. In fact, the heat of dioxetane formation is too small (only 4 kcal mol^{-1}) at the CASSCF 4-31G level compared with the value from thermochemical estimation (30 kcal mol^{-1}).

Table 16 summarizes the relative energies for several key intermediates obtained after the dynamic correlation corrections by the APUMP, contracted CI (CCI) method,[77] etc. Judging from the activation barriers by these methods, singlet oxygen is not so reactive toward ethylene, in accord with experiment. The APUMP2 calculations showed that the activation barrier for the *gauche* **TS15** is lower by about 6 kcal mol^{-1} than the second-order perepoxide-like TS. The CCI method predicts that the latter is more stable by about 4 kcal mol^{-1} than the second-order *syn* diradicaloid TS. Note that the stability of dioxetane is underestimated in the calculation at the CASSCF level,[78] whereas the MINDO/3 method overestimates the stability of dioxetane.[79]

Table 16. Relative energies (kcal mol^{-1}) for singlet oxygen + ethylene system by several theoretical methods

System	APUMP4[a]	CASSCF[b]	CASSCF[c]	CCI[c]	GVB[d]	MINDO/3[e]
$C_2H_4 + {}^1O_2$	0.0(0.0)	0.0	0.0	0.0	0.0	0.0
PE(TS)	37.6(29.7)	(13.6)	43.2	35.6		11.4
${}^1\sigma\sigma$-DRTS	(32.5)	33.2	49.3	40.3		
${}^1\sigma\pi$-DRTS[f]	31.0	29.5				
${}^1\sigma\pi$-DR	16.7(21.8)	22 ~ 23			9	−13
^{1}-$\pi\pi$-DR	(19.7)				9	
PE	21.1(23.3)	(0.0)	27.4	31.8	16.5	−16.1
DO	−29.4(−35.0)	−4.0	−12.8	−30.5	−36.9	−65.5

[a] APUMP4 (6-31G*) at CASSCF optimized geometry in Ref. 78 (APUMP2 values by assuming the geometry in Ref. 77 are given in parentheses).
[b] From Ref. 78.
[c] From Ref. 77.
[d] From Ref. 76.
[e] From Ref. 79.
[f] *gauche* TS in Ref. 78. is assumed.

6.4 Reactions of Ozone with Alkenes

Reactions of ozone with alkenes have received a great deal of attention as a useful tool for organic synthesis and the determination of the structures of organic compounds.[34,80] The chemical properties of ozonides have also been the subject of considerable interest. Many theoretical and experimental studies have been carried out to elucidate the mechanism of the reactions. The Criegee mechanism[81] is now generally accepted for most reactions of ozone with alkenes in solution. On the other hand, side-reactions occur in the gas phase because the reactions[82] are highly exothermic. In this section, recent advances in this field are described.

6.4.1 Concerted cycloaddition of ozone to ethylene

Ab initio computations and microwave investigations have elucidated the detailed mechanism of the reaction of ozone with ethylene.[83] Interaction betweeen ozone and ethylene forms a van der Waals complex (**14**). This complex lies in a shallow minimum on the reaction coordinate (see Figure 38). Figure 36A illustrates the geometry of the complex. The C—O distance was found to be about 3.2 Å by microwave spectroscopy, and was calculated to be 3.5, 2.4 and 3.05 Å by the HF 6-31G(2d,p), MP2 6-31G(d,p) and MP4(SDTQ) 6-31G(d,p) methods, respectively. The stability of the complex is dependent on the level of computations. A very large stabilization energy (5.8 kcal mol^{-1} after zero-point correction) was predicted by the MP2 6-31G(d,p) method. However, the best *ab initio* calculation predicts that the stabilization energy is in the range 0.7-1.0 kcal mol^{-1}.[83]

The complex proceeds to the transition structure **TS16**. Figure 36B illustrates the fully optimized geometries of **TS16** by the RHF and RMP2 6-31G(d,p) methods.[84] The C—O bond length at the RHF level is shortened after the correlation correction by the MP2 method, and was concluded to be 2.0 Å. Judging from the shape of **TS16**, the ozone-ethylene addition reaction is assumed to occur in a concerted fashion, compatible with the Woodward-Hoffmann rule.[85] The calculated barrier heights after the zero-point correction are 16.3, 5.5 and 2.5 kcal mol^{-1} by the RHF, RMP2 and RMP4(SDTQ) 6-31G(d,p) methods, respectively.[84] The semi-empirical AM1 method[86] predicts that the barrier is about 11 kcal mol^{-1}. The theoretical value of 5.5 kcal mol^{-1} by the RMP2 method is in good agreement with the experimental estimate of 5 kcal mol^{-1}.

TS16 is converted into the first ozonide, 1,2,3-trioxolane (**15**). Figure 36C shows the fully optimized geometry of the first ozonide by the RMP2-6-31G(d,p) method.[84] The RMP2 geometry has the O-envelope form, in accord with the results of microwave spectroscopy. The calculated bond lengths and angles are also in excellent agreement with the experimental values. The calculated heats of reaction for the conversion from $O_3 + C_2H_4$ to the first

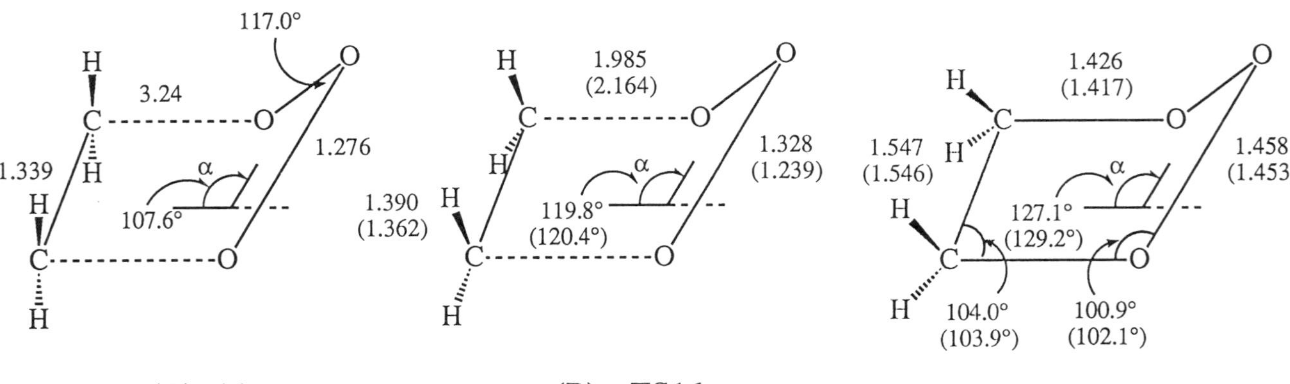

Figure 36. (A) Experimental geometry of the ozone plus ethylene complex (**14**); (B) optimized transition structure (**TS16**) for 1,3-dipolar cycloaddition of ozone to ethylene by the RMP2 6–31G(d,p) (RHF values in parentheses); (C) optimized structure of first ozonide (**15**) by RMP2 6–31G(d,p) method (experimental values are given in parentheses)

ozonide are -49.2 (MP2), -51.5 (MP4) and -53 (GVB) kcal mol^{-1}. These values also agree with the thermochemically estimated values (-45 ± 6 kcal mol^{-1}).[87]

6.4.2 Decomposition of ozonide and addition of carbonyl oxide to formaldehyde and ethylene

Ab initio RHF 4-31G calculations were carried out to locate the transition structure (**TS17**) for the thermal decomposition of the first ozonide (FO) **15** to the carbonyl oxide (CH_2OO) and formaldehyde (HCHO). **TS17** is also the transition structure for the recombination of CH_2OO and HCHO. The calculated optimized structure of **TS17** is illustrated in Figure 37A. The shape of **TS17** indicates that the transition state is located at a later part of the reaction coordinate for the decomposition of FO to CH_2OO and HCHO and at an early part for the recombination of CH_2OO and HCHO. The activation barriers for these processes were calculated to be 29.5 and 20.1 kcal mol^{-1}, respectively, by the APUMP4 4-31G*//RHF 4-31G method. The thermochemically estimated values of the activation barriers for these processes are 20 and 10 kcal mol^{-1}, respectively. On the other hand, the AM1 method[86] predicts that the forward and backward barriers are 19.5 and 30 kcal mol^{-1}, respectively. Since **TS17** is more stable than **TS16** by 25 kcal mol^{-1}, the decomposition reaction of FO via **TS17** is facile in the gas phase, and even in solution.

Figure 37B shows the transition structure (**TS18**) for the other recombination process from CH_2OO and HCHO to give the secondary ozonide (SO). Judging from the geometric structure, **TS18** is also located at an early stage of this recombination reaction. Interestingly, the activation barrier for this process is formally negative (-2.6 kcal mol^{-1}) by the APUMP4 method, suggesting that a van der Waals complex (**16**) is formed between CH_2OO and HCHO. The depth of the minimum corresponding to the formation of the complex is estimated to be over 3.0 kcal mol^{-1}. The energy barrier from the complex to **TS18** is 4.0 kcal mol^{-1} by the GVB plus thermochemical estimation[88] and 2.1 kcal mol^{-1} by the AM1 method.[86] From these data, it is concluded that the recombination to give SO **17** via **TS18** is facile.

During the course of the formation of SO, the conformational change of carbonyl oxides is expected to occur because of their diradical character. The rotational energy barrier of methylene peroxide was calculated to be about 10 kcal mol^{-1},[35] in accord with the experimental values of 8–16 kcal mol^{-1}.[81b] The low energy barriers for carbonyl oxides are compatible with the stereoselectivity in the ozonolysis of alkenes.

There is the possibility that in the course of the ozonolysis of ethylene, carbonyl oxide may react with ethylene, leading to the formation of dioxirane. Figure 37C illustrates the transition structure (**TS19**) for the addition reaction of CH_2OO to C_2H_4 calculated by the RHF STO-3G energy gradient technique.[58]

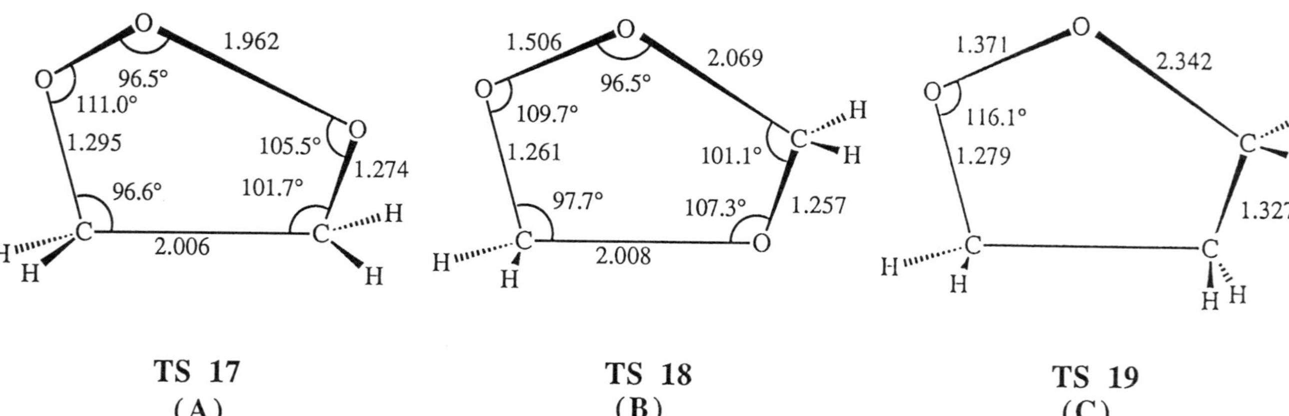

Figure 37. (A) Optimized transition structure (**TS17**) for the decomposition of first ozonide and (B) that (**TS18**) of 1,3-dipolar cycloaddition between carbonyl oxide and formaldehyde by the RHF 4–31G method, and (C) the optimized transition structure (**TS19**) for the reaction between carbonyl oxide and ethylene

Although **TS19** is also at an early stage of the reaction coordinate, the activation barrier is calculated to be 11.1 kcal mol^{-1} by the RHF 6-31G* method. This indicates that the cycloaddition between CH_2OO and HCHO is more favorable than that between CH_2OO and C_2H_4. Hence the theoretical results support the concerted mechanism for the ozonolysis reactions of alkenes in solution.

Figure 38 illustrates the characteristic of the potential energy hypersurface for the ozonolysis reaction of ethylene revealed by APUMP4 4-31G* calculations. The calculated entrance barrier with the zero-point correction is extremely small (2.0 kcal mol^{-1}), showing that the addition step is very rapid. The calculated heats of reaction are compatible with the thermochemical values (given in parentheses).

6.4.3 Diradical reactions in gas-phase ozonolysis

Several side-reactions take place in the gas-phase ozonolysis of alkenes, since the reaction is highly exothermic. For example, the 1,5-diradical intermediate in Figure 38 may be produced from the first ozonide (FO).[82] This is an energetically accessible process because the heat of this reaction is −78

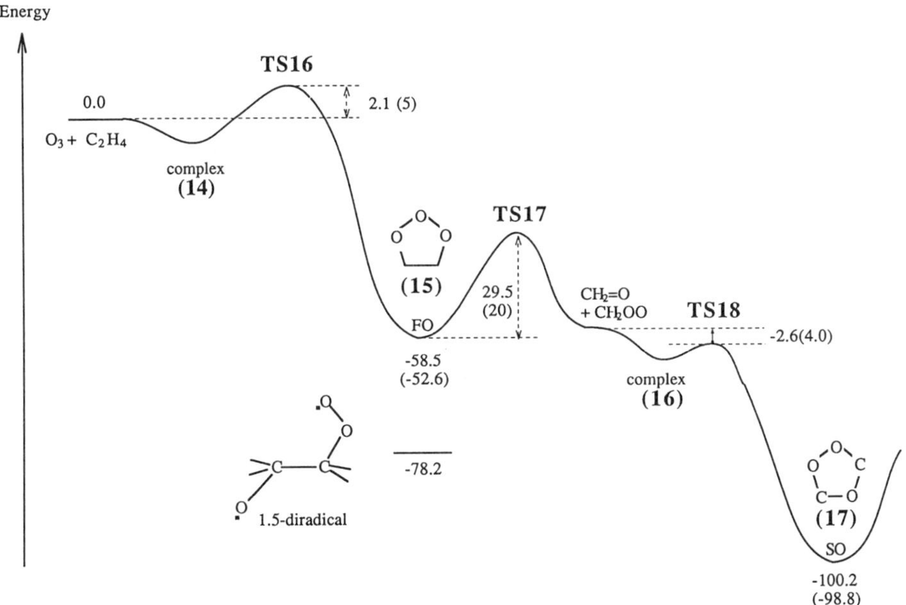

Figure 38. Potential energy profiles and relative energies calculated for ozonolysis of ethylene by the *ab initio* APUMP4 4-31G* calculation

kcal mol^{-1}. The following three different decomposition reactions are expected to occur in the 1,5-diradical:[35,88,89]

$$\text{1,5-DR} \longrightarrow \begin{cases} O + CH_2CH_2OO & (31a) \\ HCHO + CH_2OO & (31b) \\ CH_2CH_2O + O_2 & (31c) \end{cases}$$

where the spin and space symmetries of the products are determined by their conservation rules. The 1,4-peroxy diradical thus produced decomposes to the ground-state and excited-state formaldehydes. The ring-opened oxirane (CH_2CH_2O) rearranges and decomposes to acetoxyl and vinoxyl radicals:

$$CH_2CH_2O \longrightarrow \begin{cases} CH_3CHO^* \longrightarrow \cdot CH_3 + \cdot H & (32a) \\ CH_2{-}CH{=}O + \cdot H & (32b) \end{cases}$$

Hydrogen atom reacts with ozone to produce the excited state of the hydroxyl radical ($\cdot OH$) and the Meinel band emission is observed (see equation 23).

The 1,5-diradical has a large excess energy at high temperatures and therefore undergoes several successive reactions: 1,2-hydrogen migration, α-hydrogen abstraction and hydrogen migration:[82]

$$\text{1,5-DR} \longrightarrow \begin{cases} (HO)CH{-}CH_2OO & (33a) \\ OHC{-}CH_2OOH & (33b) \\ OHC{-}CH_2OO + \cdot H & (33c) \end{cases}$$

The 1,4-DR in equation 33a can cyclize to form dioxetane, which decomposes into formic acid and excited formaldehyde. The peroxide species in equations 33b and 33c are precursors for excited glyoxal.

For carbonyl oxides, diradical-type reactions such as intermolecular hydrogen abstraction are expected to occur in the presence of good hydrogen donors such as α-methylstyrene. The details are given in other chapters.

7 RING-OPENING REACTIONS OF ORGANIC PEROXIDES

7.1 Dioxetene and Related Species

The Woodward–Hoffmann rules are now providing useful guidelines for understanding and predicting the course of chemical reactions.[84] These rules were derived from the symmetry of MOs primarily participating in the chemical transformations. Recently, *ab initio* MO calculations have been applied to the location of the transition structures (TSs) of Woodward–Hoffmann symmetry-allowed and symmetry-forbidden reactions. The calculations have shown that the symmetry-allowed TSs are generally more favorable and stable than the corresponding symmetry-forbidden TSs.[40] In this section, the reactivity features of the ring-opening reactions of dioxetene and related species are discussed on the basis of the results of *ab initio* MO calculations.

Dioxetene has been proposed as an intermediate in the oxygenation of acetylene that leads to glyoxal. Figure 39 illustrates the optimized geometry of **TS20** for the conrotatory opening of dioxetene by the CASSCF energy gradient technique.[90] Before the CASSCF calculations, the four active orbitals, four active electrons [4,4] system was selected on the basis of the occupation numbers of UHF-NO (see Section 2). The orbital overlap (T_{HOMO}) for the up- and down-spin HOMOs (χ_{HO} and η_{HO}) was 0.5438 and that (T_{NHOMO}) for the next HOMOs was 0.8130. This indicates that UHF MOs are considerably spin-polarized even for the concerted TS, and the electron correlation correction is crucial. The situation is similar with isoelectronic species such as oxetene and dithietene.[90]

The activation barrier height for the TS was 10.2, 10.4 and 10.9 kcal mol^{-1} by the CASSCF 6-31G*, RMP2 and RMP4 (SDTQ) methods, respectively.[90] These small values for the activation barrier are in accord with the symmetry-allowed character of the conrotatory ring-opening reaction of dioxetene.

The barrier height for the conrotatory ring opening (**TS21**) of cyclobutane to butadiene is about 37 kcal mol^{-1} by the RMP2 and CASSCF 6-31G* methods, and this was reduced to about 35 kcal mol^{-1} after the zero-point correction. The latter value is compatible with the experimental value of 32.9 kcal mol^{-1}.[91] The activation energy for the conrotatory ring opening of oxetane is 26.6 kcal mol^{-1} by the RMP4 (SDTQ) 6-31G* method,[90] which is also in good agreement with the experimental value of 25.09 kcal mol^{-1}.[92]

The *ab initio* computations involving electron correlation effects generally reproduce the activation barriers for the Woodward-Hoffmann concerted reactions.[40]

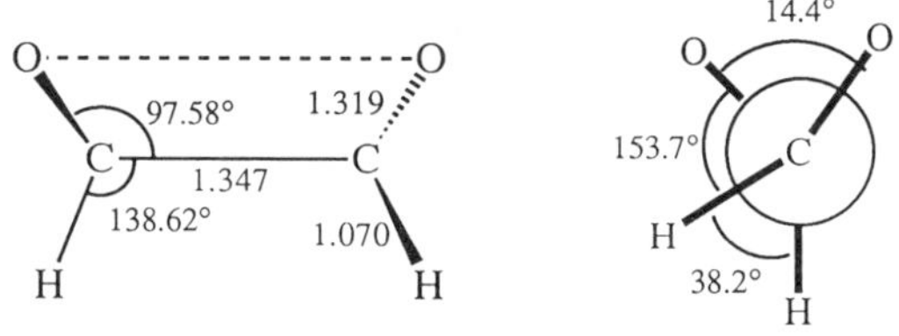

Figure 39. Optimized transition structure (**TS20**) for the conrotatory opening of dioxetene by the CASSCF 3-21G method

7.2 Dioxetane and Related Species

The thermal ring opening reaction of 1,2-dioxetanes is often accompanied by the emission of light, usually phosphorescence. Light-emitting chemical reactions are called chemiluminescent reactions.[41]

The chemiluminescence of 1,2-dioxetanes is a spin-forbidden non-adiabatic reaction, i.e. a chemical change from the lowest singlet potential surface (S_0) to the lowest triplet surface (T_1). *Ab initio* CASSCF 4–31G calculations for four-orbital, four-electron and six-orbital, eight-electron systems were carried out in order to locate the transition structures for this reaction.[93] Multi-reference (MR) second-order Møller–Plesset (MP) 6–31G* calculations were also performed to elucidate the energetics of this reaction. Figure 40 illustrates the optimized geometries calculated by the CASSCF energy gradient technique for the singlet (σ,σ) diradical (DR) intermediate (**18**) and the triplet (σ,π) DR intermediate (**19**) which are generated from dioxetane (DO). Figure 41 shows the optimized geometries of the key transition structures **TS22**, **TS23** and **TS24**. ^{1}**TS22** is an early TS which is responsible for the O—O dissociation of DO. ^{1}**TS23** is a late TS which is responsible for the C—C cleavage of the singlet (σ,σ) DR intermediate. The activation barrier leading to ^{1}**TS22** was found to be extremely low (only 2.0 kcal mol^{-1}) by the MR MP2 6–31G* method.[93] The

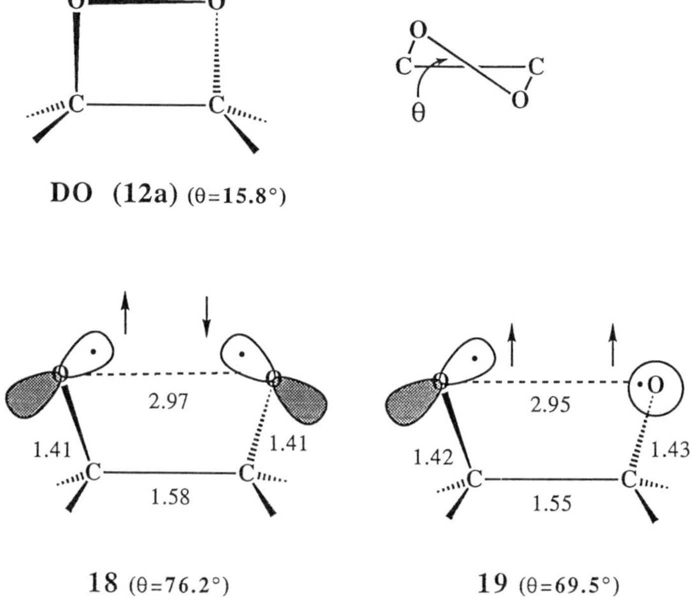

Figure 40. Optimized geometries of dioxetane (**12a**), singlet $\sigma\sigma$-diradical (**18**) and triplet $\sigma\pi$-diradical (**19**) intermediates by the CASSCF 4–31G method

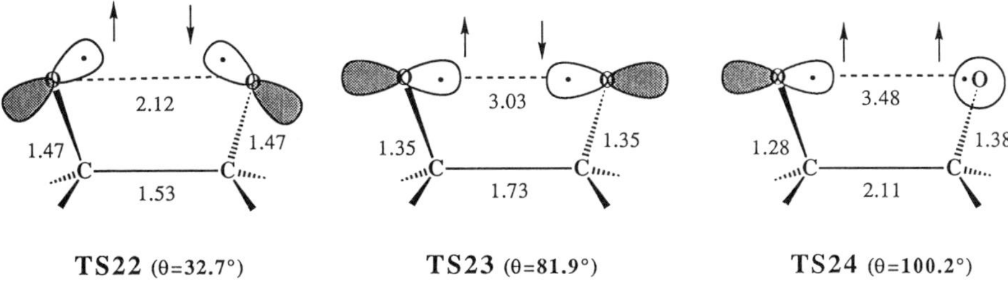

Figure 41. Optimized geometries of early (**TS22**) and late (**TS23**) transition structures for the decomposition of dioxetane in the lowest singlet state and the transition structure (**TS24**) on the lowest triplet surface

singlet (σ,σ) DR intermediate (**18**) and ^{1}**TS23** on the S_0 surface were less stable by 1.0 and 5.0 kcal mol^{-1}, respectively, than the triplet (σ,σ) DR intermediate (**19**). The heat of decomposition of DO to two $H_2C{=}O$ is about -57 kcal mol^{-1}, in accord with a thermochemical estimate of -60 kcal mol^{-1}.

Interestingly, the lowest triplet (T_1) surface and the S_0 surface cross each other twice in the region of the biradical minimum, in accord with the diradical mechanism. The first crossing point is very close to the ^{1}**TS22** for the O—O cleavage, where the spin–orbit interaction should be effective because of the orthogonality between the π- and σ-localized orbitals on the oxygen. On the other hand, there is no spin–orbit coupling near the second crossing point because the radical orbitals are delocalized significantly over the C—C bond region which is cleaved. The transition structure ^{3}**TS24** on the T_1 surface was located above 21 kcal mol^{-1}, indicating that ^{3}**TS24** is an intrinsic transition structure for the chemiluminescence reaction of DO into the triplet $CH_2C{=}O$ and the ground-state $CH_2C{=}O$. This activation barrier is indeed in acceptable agreement with the experimental activation energy of 22 kcal mol^{-1}.[93] Hence the CASSCF calculations followed by the MR MP2 calculation are likely to elucidate key pathways of the chemiluminescent reaction of dioxetanes. Figure 42 illustrates the overall reaction pathways of dioxetanes revealed by the *ab initio* computations.[93] The remaining theoretical problem is to estimate quantitatively the rate of the non-adiabatic transition from the S_0 to T_1 surface near the surface crossing points.

Schmidt *et al.*[94] proposed a chemically initiated electron exchange luminescence (CIEEL) mechanism for the decomposition of dimethyldioxetanone. The basic assumptions in the CIEEL mechanism are (1) the activated transfer of an electron from an electron donor to peroxide and (2) that the O—O bond of the peroxide cleaves either simultaneously with the transfer of the electron or varies rapidly following its arrival. The details of the theoretical treatments of CIEEL are given elsewhere[41] and therefore only the oxygenation reactions via radical ions are considered here.

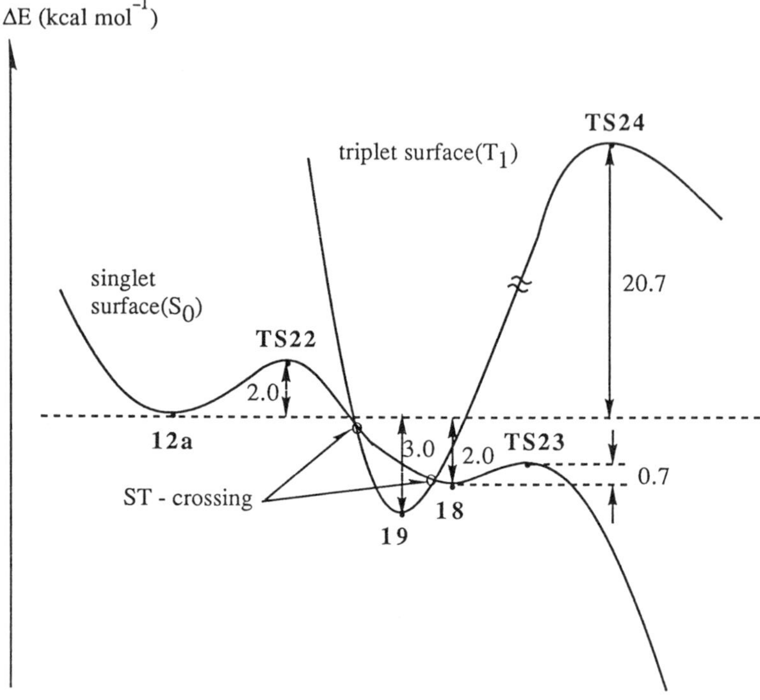

Figure 42. Potential energy profiles and relative energies calculated by the MR MP2 6–31G* method

8 OXYGENATION INVOLVING RADICAL ION SPECIES

Radical ion species play an important role as key intermediates in the oxygenation of organic compounds. They can be generated by photo-induced electron transfer from electron-donating to electron-accepting compounds or by electrochemical redox reactions. In the photo-oxygenation of electron-rich organic compounds (D) via photo-induced electron transfer, cyanoaromatic compounds such as 1,4-dicyanonaphthalene (DCN) and 9,10-dicyanoanthracene (DCA) are frequently used as efficient electron-transfer sensitizers (Sens), and pairs of radical cations and radical anions are generated on photoexcitation of Sens as illustrated in equation 34a.[95] In the electrochemically promoted oxygenation, radical cations $D^{+\bullet}$ of electron-rich compounds are produced by a one-electron anodic oxidation of D. Radical anions of electron-poor compounds $A^{-\bullet}$ can also be generated by a one-electron cathodic reduction of A (equations 34b–c).[96]

$$D + Sens \rightarrow D^{+\bullet} + Sens^{-\bullet} \quad (34a)$$

$$D - e^{-} \rightarrow D^{+\bullet} \quad (34b)$$

$$A + e^{-} \rightarrow A^{-\bullet} \quad (34c)$$

Several elementary processes have been proposed for the oxygenation of organic compounds involving these radical ion species. Such processes are shown in equations 35.

$$D^{+\bullet} + O_2^{-\bullet} \rightarrow DO_2 \quad (35a)$$

$$A + O_2^{-\bullet} \rightarrow AO_2^{-\bullet} \quad (35b)$$

$$D^{+\bullet} + {}^3O_2 \rightarrow DO_2^{+\bullet} \quad (35c)$$

$$A^{-\bullet} + {}^3O_2 \rightarrow AO_2^{-\bullet} \quad (35d)$$

The process in equation 35a occurs in the photo-oxygenation of organic compounds in which superoxide ion $O_2^{-\bullet}$ is produced by electron transfer from $Sens^{-\bullet}$ to molecular dioxygen ${}^3O_2$ (equation 36).

$$Sens^{-\bullet} + {}^3O_2 \rightarrow Sens + O_2^{-\bullet} \quad (36)$$

The process in equation 35b is recognized mostly in the electrochemical oxygenation and less commonly in the photo-oxygenation. Recently, this process has also been applied chemically by the reaction of A with KO_2 and $Me_4N^+O_2^{\bar{\bullet}}$. The process in equation 35c occurs in both the photo-oxygenation and the electrochemical oxygenation. The process in equation 35d is known to occur in photo-oxygenation, although less commonly.

The other important process in the photo-oxygenation of organic compounds is the sensitized photo-oxygenation in which singlet oxygen is involved as a key intermediate as illustrated in equations 37. In this reaction, the formation of $O_2^{-\bullet}$ is observed when D has a higher electron-donating ability (equation 37d and F in Figure 34).[41,97] In this section, the reactivity features of the reactions of the types in equations 37 are discussed and compared.

$$Sens \rightarrow {}^3Sens^* \quad (37a)$$

$${}^3Sens^* + {}^3O_2 \rightarrow Sens + {}^1O_2 \quad (37b)$$

$$D + {}^1O_2 \rightarrow DO_2 \quad (37c)$$

$$D + {}^1O_2 \rightarrow D^{+\bullet} + O_2^{\bar{\bullet}} \quad (37d)$$

The concept of frontier orbital interactions is utilized as a principal tool for explanation of their reaction modes. Obviously, the MR CI-type calculations described in Section 2.4 are required for rigorous, quantitative treatments of electronic processes of the reactions in equations 35 which involve open-shell species. However, quantitative calculations are almost impossible for large molecules associated with polar solvents such as those involved in this section. Therefore, the conclusions derived from the orbital interaction theory[41,74] in this section should be taken as qualitative.

8.1 Reactions of Radical Cations with $O_2^{-\bullet}$

The reaction of a radical cation ($D^{+\bullet}$) with $O_2^{-\bullet}$ is one of the important elementary steps in electron-transfer sensitized photo-oxygenations. The MO interactions between $D^{+\bullet}$ and $O_2^{-\bullet}$ are schematically depicted by **20** in Figure 43. In contrast, the MO interactions between D and singlet molecular oxygen ($^1O_2$) are described by **21**.[41] Since $O_2^{-\bullet}$ acts as an electron donor to $D^{+\bullet}$, the charge-transfer (CT) interaction between the singly occupied π_2^* of $O_2^{-\bullet}$ and the vacant LUMO of $D^{+\bullet}$, and also the CT interaction between the doubly occupied π_1^* of $O_2^{-\bullet}$ and the singly occupied HOMO (SOMO) of $D^{+\bullet}$, are particularly important. These are referred to as the LUMO-π_2^* and HOMO-π_1^* interactions, respectively, as illustrated in **22**. The former interaction predicts that the (2 + 2) cycloaddition is symmetry-allowed in the coupling of $O_2^{-\bullet}$ with radical cations of dienes and aromatic compounds as illustrated in **22a** and **22b**, whereas the latter interaction induces a back-electron transfer to produce a singlet (or triplet) molecular oxygen and the ground state of the substrate (equation 38).

$$D^{+\bullet} + O_2^{-\bullet} \rightarrow D + {}^{1,3}O_2 \tag{38}$$

On the other hand, in oxygenation with the use of $^1O_2$, the HOMO-π_2^* interaction between organic substrates and $^1O_2$ is of primary importance. This interaction predicts that the (4 + 2) cycloadditions between dienes and $^1O_2$ and between aromatic compounds and $^1O_2$ are symmetry-allowed, as illustrated in **23a** and **23b**; note that in these cases, $^1O_2$ acts as an electron acceptor. These selection rules are consistent with the reaction schemes in Figure 44 proposed for the DCA-sensitized oxygenation of dienes and aromatic compounds.[98]

When an alkene has an electron-donating substituent, its MO is largely unsymmetrical, as shown in **24**. The LCAO coefficients (C_i) of the HOMO are greater at the 3-position than at the 2-position; note that $|C_3| > |C_2|$ for the HOMO, whereas $|C_2| > |C_3|$ for the LUMO. The situation is totally reversed for an alkene with an electron-withdrawing group W, as illustrated in **25**. In the DCA-sensitized oxygenation of alkenes having electron-donating groups such as alkyl, alkoxy and phenyl, the LUMO-π_2^* interaction is of primary importance. This interaction predicts that an initial attack of $O_2^{-\bullet}$ occurs at the 2-position of radical cations of these alkenes. On the other hand, the HOMO-π_2^* interaction predicts that the attack of $^1O_2$ occurs at the 3-position of electron-rich alkenes. Hence the orbital interactions indicate the reverse regioselectivity for initial attacks of $O_2^{-\bullet}$ and $^1O_2$ in the oxygenation of unsymmetrical alkenes.[74,99,100]

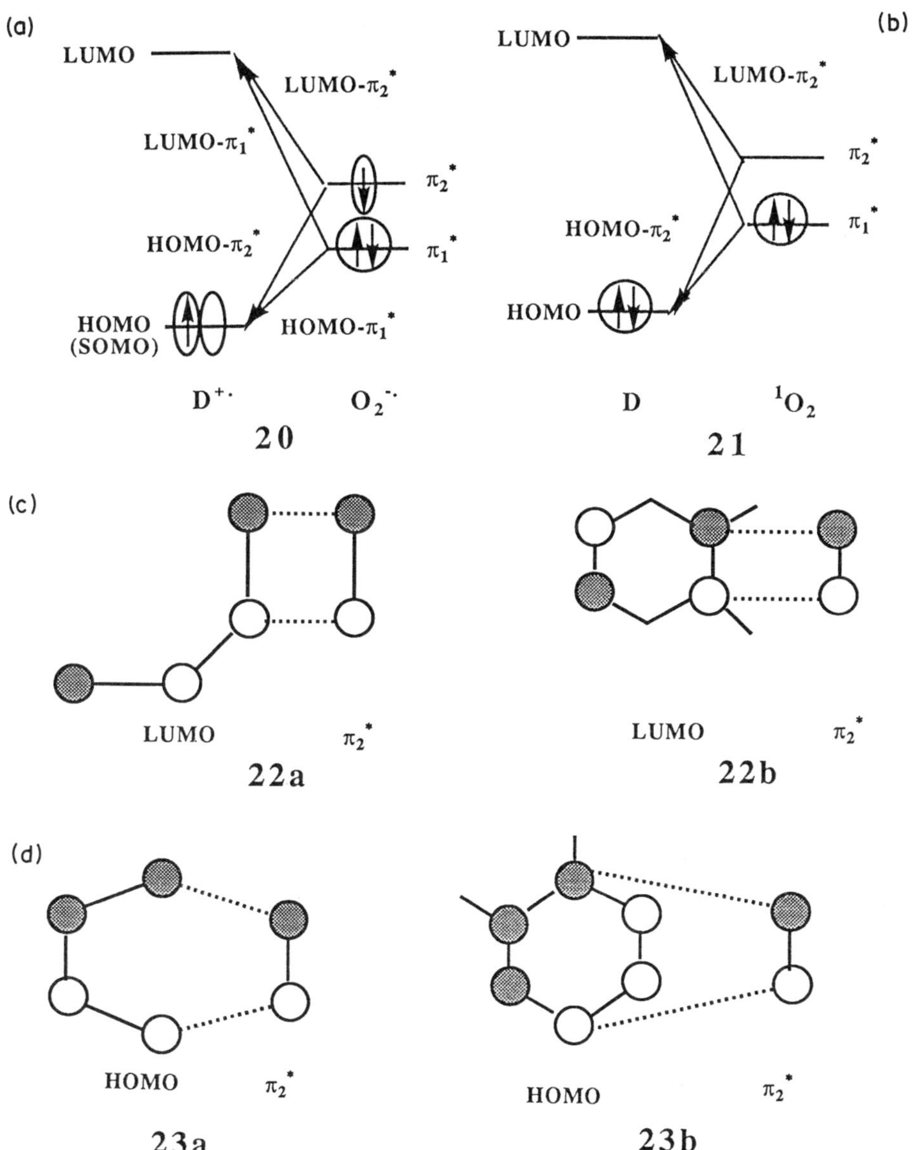

Figure 43. (a) Orbital interactions (**20**) between radical cation ($D^{+\bullet}$) and superoxide radical anion ($O_2^{-\bullet}$); (b) orbital interaction (**21**) between substrate (D) and singlet molecular oxygen ($^1O_2$); (c) LUMO and π_2^* interactions (**22a**, **22b**) between radical cation ($D^{+\bullet}$) and superoxide radical anion ($O_2^{-\bullet}$); (d) HOMO and π_2* interaction (**23a**, **23b**) between substrate (D) and singlet molecular oxygen ($^1O_2$)

(continued)

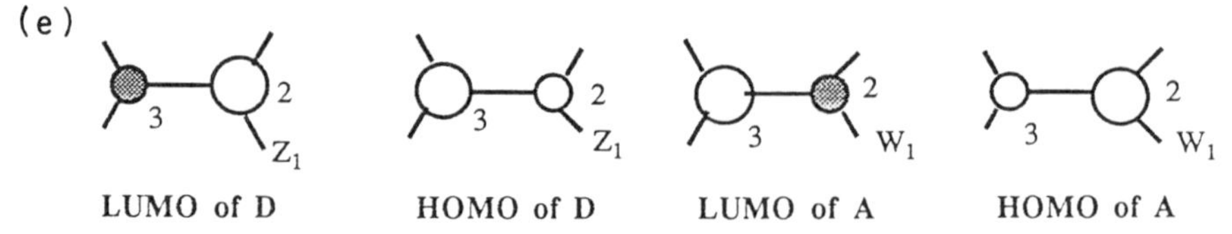

(e)

LUMO of D HOMO of D LUMO of A HOMO of A

24: Z=R, OR, NRR' 25: W=CN, CHO, CO_2CH_3

Figure 43. (e) HOMO and LUMO (**24**, **25**) of unsymmetrical alkenes

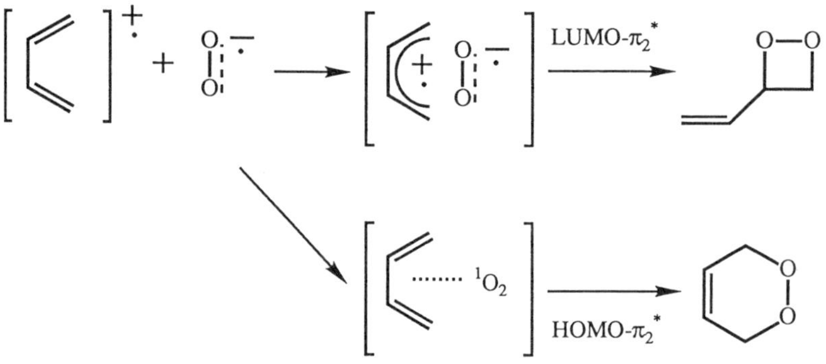

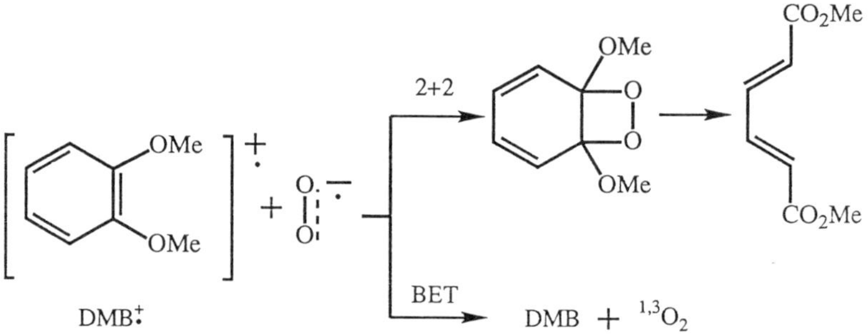

LUMO-π_2^*

HOMO-π_2^*

CO_2Me

OMe

2+2

OMe

OMe

OMe

CO_2Me

DMB$^{+\cdot}$

BET

DMB + $^{1,3}O_2$

Figure 44. Reactions of radical cations of butadiene and 1,2-dimethoxybenzene with $O_2^{-\cdot}$

The stabilization by orbital interactions of the types **20–23** becomes negligible at large intermolecular distances. However, since the Coulombic attraction between positive and negative charges is long-range, the Coulombic interaction may play an important role in some sorts of recombination reactions between radical cations and $O_2^{-\cdot}$. The Coulombic interaction between the radical cation of vinyl alcohol (Z = OH) and $O_2^{-\cdot}$ was therefore calculated by using the net charges.[74] As illustrated in **26** and **27** in Figure 45, two different modes of attack

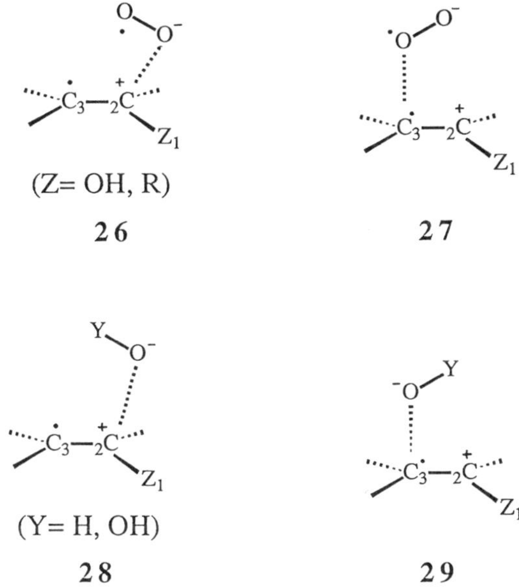

Figure 45. Coulombic interactions (**26–29**) between radical cation of vinyl alcohol (electron-rich alkene) and superoxide radical anion

were examined. The differences in the Coulombic energies between **26** and **27**, $\Delta E_Q = E_c\,(\mathbf{26}) - E_c\,(\mathbf{27})$, are summarized in Table 17. Attack on the 2-position (**26**) is preferred to attack on the 3-position (**27**) even at a relatively large intermolecular distance ($R = 4.0$ Å). Hence both the orbital and Coulombic interactions predict that the initial attack of $O_2^{-\bullet}$ occurs at the 2-position of the vinyl alcohol radical cation.

The Coulombic interactions between the radical cation of vinyl alcohol and oxyanions, HO^-, HO_2^- and $O^{-\bullet}$, were also calculated as illustrated in **28** and **29** in Figure 45.[74] The calculated energy differences $\Delta E_Q = E_c\,(\mathbf{28}) - E_c\,(\mathbf{29})$ in Table 17 confirm that the attack of oxyanions occurs preferentially at the 2-position of the radical cations.

Table 17. Difference[a] (kcal mol^{-1}) of the Coulombic energies upon the attack of oxyanions on the 2- and 3-positions of vinyl alcohol radical cation

	$O_2^{-\bullet}$	HO^-	HO_2^-	$O^{-\bullet}$
ΔE_Q	−1.4	−1.9	−3.3	−2.8

[a] $\Delta E_Q = E_c$ (**26** or **28**) − E_c (**27** or **29**); see Figure 45.

Mattes and Farid[100] have clearly demonstrated that $O_2^{-\bullet}$ attacks the 2-position of the radical cation of 1,1-dimethylindene [**30**, Z = $C(CH_3)_2$], whereas $^1O_2$ adds at the 3-position of this compound. Sawyer and Valentine[101] have shown that the reaction of $O_2^{-\bullet}$ with the radical cation of methylviologen (**31**) occurs at the 2-position. *Ab initio* (STO-3G) calculations were carried out for radical cations of 1,1-dimethylindene (**30**), methylviologen (**31**), *N*-methylpyridine (**32**), indole (**33**), phenol (**34**) and 4-methylanisole (**35**).[74] The calculated net charges are given in Table 18 and Figure 46. The LCAO

Table 18. Net charges on 2- and 3-positions of radical cations of electron-donating compounds

Position	**30**	**31**[a]	**32**	**33**	**34**	**35**
2	0.05	0.12	0.12	0.05	0.22	0.17
3	−0.01	−0.01	−0.04	0.01	−0.04	−0.03

[a]INDO method was used for this molecule; see Figure 46.

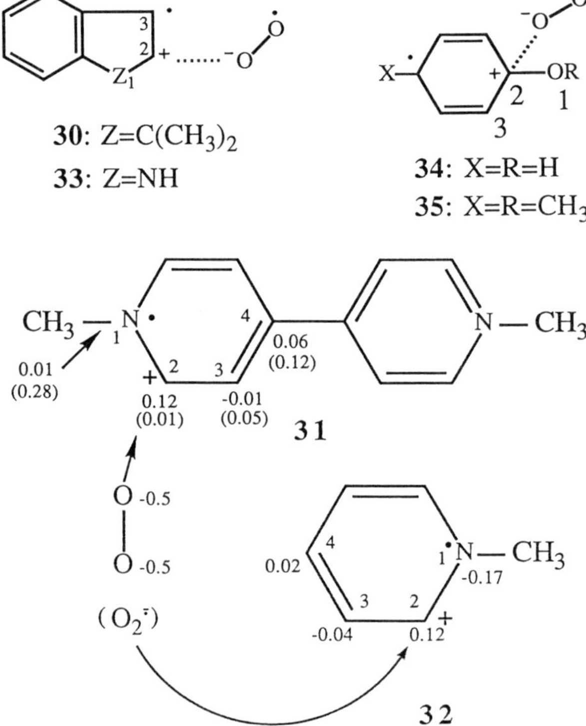

Figure 46. Coulombic interactions (**30–35**) between aromatic radical cations and $O_2^{-\bullet}$. Net charges on carbon atoms of methylviologen (**31**) and *N*-methylpyridine (**32**) are given in this figure, and spin densities are given in parentheses

coefficients of the LUMO of **30** are shown in **36** in Figure 47. The net positive charges are the largest at the 2-position for all the species examined. Hence the orbital and Coulombic interactions predict that nucleophilic attack of $O_2^{-\bullet}$ on the 2-position is preferred for all the radical cations, in agreement with experimental results. It is interesting that the reaction of pyrrole derivatives **37** with $^1O_2$ gives the hydroperoxide **38** in a highly regioselective manner.[102] The interaction between the HOMO of **37** and the π_2* of $^1O_2$ indicates that the concerted (2 + 2) cycloaddition (a) is symmetry-forbidden, whereas the concerted (2 + 4) cycloaddition (b) is symmetry-allowed, but sterically forbidden. As a result, an ene-type reaction occurs between **37** and $^1O_2$ as illustrated in Figure 48.

8.2 Nucleophilic Addition of $O_2^{-\bullet}$ to Electron-deficient Alkenes[103]

The LUMO-π_2* and the HOMO-π_1* interactions between an electron-deficient alkene A and $O_2^{-\bullet}$ are schematically represented by **39** and **40**, respectively, in Figure 47. The latter interaction between the doubly occupied pairs induces a symmetry-imposed exchange repulsion. Since the latter repulsive

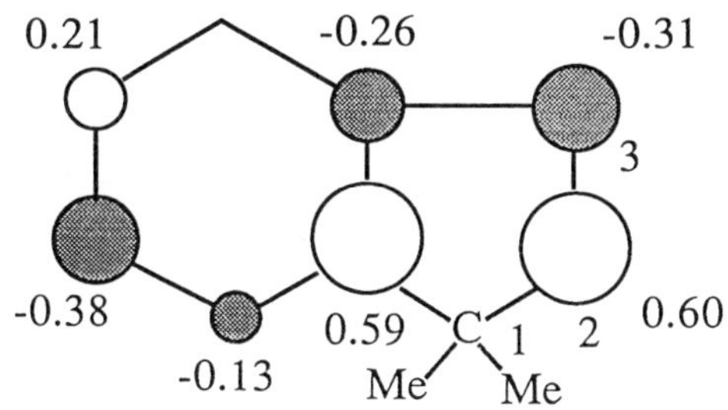

36: LUMO of **30**

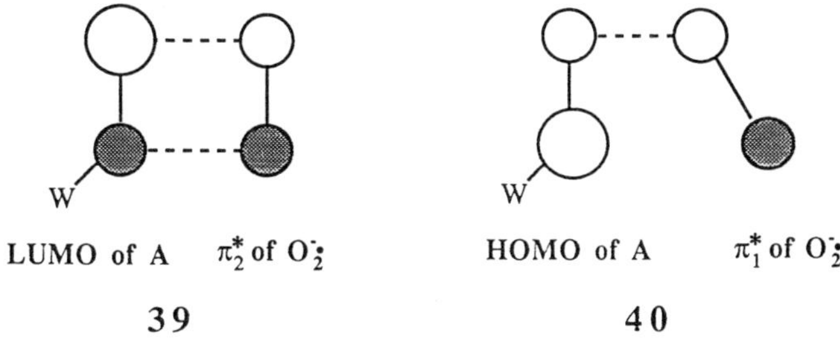

Figure 47. LCAO coefficients (**36**) of the LUMO for dimethylindene (**30**) and orbital interactions (**39**, **40**) between the frontier orbitals of electron-deficient alkenes (A) and superoxide radical anion

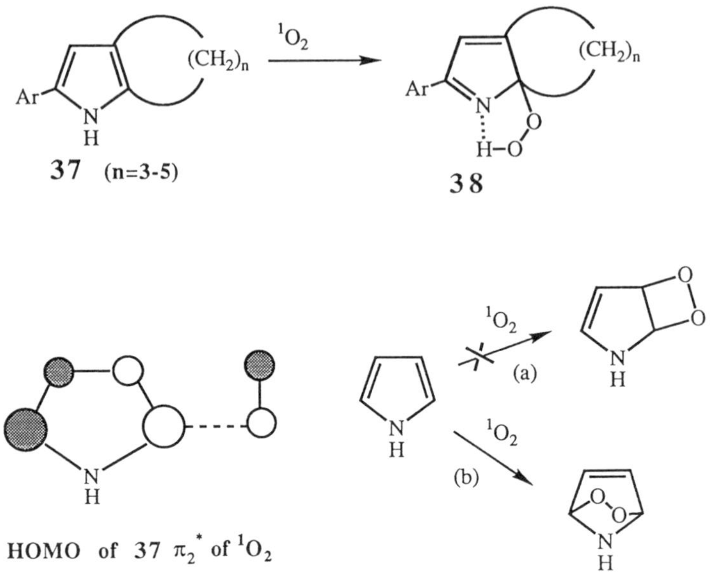

Figure 48. Reaction scheme and orbital interaction for substituted pyrroles and singlet oxygen

interaction is larger than the former attractive interaction, the concerted cycloaddition between the A and $O_2^{-\bullet}$ species is unfavorable. A non-concerted cycloaddition involving an alkeneperoxy radical anion intermediate is conceivable.

Full geometry optimizations about the ethyleneperoxy radical anion were carried out by using the UHF 4-31G energy gradient technique.[41] The calculations show that the unpaired electron and negative charges are localized on the terminal oxygen and carbon atoms, respectively. This indicates that the addition of anionic oxygen of $O_2^{-\bullet}$ to the C—C double bond is nucleophilic in character. Therefore, the introduction of an electron-withdrawing group at the terminal carbon atom is effective for the stabilization of peroxy radical anion intermediates. It has been pointed out that the nucleophilic addition of $O_2^{-\bullet}$ to electron-rich alkenes hardly occurs under the usual reaction conditions.[103]

Ab initio MO calculations have been carried out for the intermediates formed by the addition of $O_2^{-\bullet}$, HO_2^-, HO^- and $O^{-\bullet}$ to acrylonitrile.[74] The net charges are shown in Figure 49. The calculated net charges on the terminal carbon atoms are close to each other among the intermediates, indicating that the nucleophilicities of these oxygen anions resemble each other. *Ab initio* MO calculations (4-31G) have also been performed for the nucleophilic adducts of $O_2^{-\bullet}$ to the carbonyl groups of HCHO, HCOOH and $HCOOCH_3$.[74] The results are shown in Figure 50. The net charges on the carbonyl oxygen are not

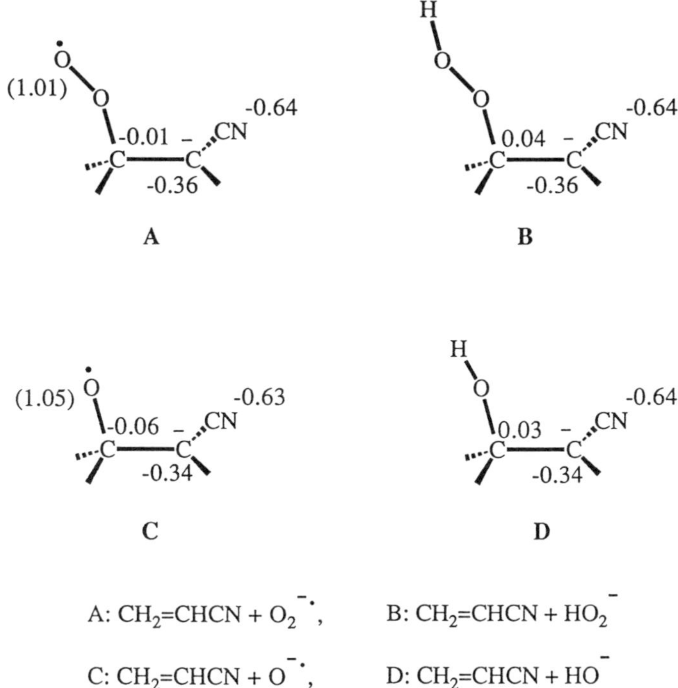

Figure 49. Charge distributions (spin densities in parentheses) in radical anion intermediates obtained by the nucleophilic addition of oxyanions ($O_2^{-\bullet}$, HO_2^-, $O^{\bar{\bullet}}$ and HO^-) to acrylonitrile

very different among the intermediates, indicating that the nucleophilic addition step may be similar. *Ab initio* MO calculations also showed that methoxide anion becomes a leaving group in the elimination step from the radical anion generated from methyl formate, as illustrated in equation 39.

$$\text{CH}_3\text{O}(\text{H})\text{C}{=}\text{O} + \text{O}_2^{\bar{\bullet}} \longrightarrow \dot{\text{O}}{-}\text{O}{-}\text{C}(\text{H})(\text{OCH}_3){-}\bar{\text{O}} \longrightarrow \dot{\text{O}}{-}\text{O}{-}\text{C}(\text{H}){=}\text{O} + \bar{\text{O}}\text{CH}_3 \quad (39)$$

If an alkene with an electron-withdrawing group has a larger electron-accepting ability, a complete one-electron transfer (ET) would occur from $O_2^{-\bullet}$ to the LUMO of the alkene to produce the alkene radical anion and $^3O_2$. This ET reaction may be followed by the attack of $^3O_2$. In this case, the LUMO–π_2^* interaction is of particular importance, showing a possible attack of $^3O_2$ on the 2-position of the radical anion (equation 40). Therefore, the regioselectivity

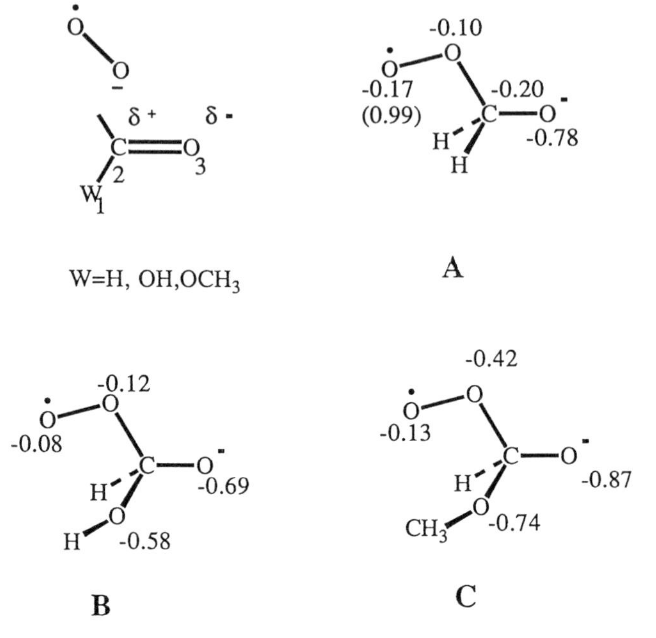

A: $HCHO + O_2^{-\bullet}$, B: $HCOOH + O_2^{-\bullet}$, C: $HCOOCH_3 + O_2^{-\bullet}$

Figure 50. Charge distributions (spin densities in parentheses) in radical anion intermediates obtained by the nucleophilic addition of $O_2^{-\bullet}$ to carbonyl compounds (**A**–**C**)

should be the same for both the ET reaction and the Michael addition toward electron-deficient double bonds with $O_2^{-\bullet}$.

$$\text{>C=C<(W)} + O_2^{-\bullet} \xrightarrow{\text{ET}} \text{>}\dot{C}\text{–}\bar{C}\text{<(W)} + {}^3O_2 \qquad (40)$$

$$\text{>}\dot{C}\text{–}\bar{C}\text{<(W)} + {}^3O_2 \xrightarrow{\text{RC}} \cdot O\text{–}O\text{–}C\text{–}\bar{C}\text{<(W)}$$

8.3 Reactions of Radical Cations with $^3O_2$

The other important mode of reaction involving dioxygen species is the coupling of a substrate radical cation with triplet molecular dioxygen:

$$D^{+\bullet} + {}^3O_2 \longrightarrow DO_2\cdot \xrightarrow{+e^-} \text{product} \qquad (41)$$

The HOMO-π_1^* interaction between $D^{+\bullet}$ and $^3O_2$ indicates that the (4 + 2) reaction between the radical cation of a diene and $^3O_2$ is formally symmetry-allowed, as illustrated in **41**, whereas the (2 + 2) cycloaddition is symmetry-forbidden in the case of the alkene radical cation and $^3O_2$ system, as illustrated in **42** in Figure 51.[98,104] The peroxy radical cation intermediate formed via **42** may receive an electron from a radical anion of sensitizer ($Sens^{-\bullet}$) or from a substrate itself. If the peroxy radical cation receives an electron from the substrate, then a radical chain reaction takes place as in the cases of electrochemical and Lewis acid-catalyzed oxygenation of unsaturated compounds with molecular dioxygen. Such a radical chain mechanism is also observed in the photosensitized oxygenation. The typical pathways for the oxygenation of dienes via their radical cations are shown in Figure 52.[74,104]

Since the orbital interaction consideration predicts that the reaction between a radical cation of an alkene and $^3O_2$ occurs in a stepwise rather than a concerted manner, an alkeneperoxy radical cation intermediate may be formed via the pathway shown in Figure 53. In Figure 53, the pathways of the reaction between the radical cation of an alkene and $O_2^{-\bullet}$ are also shown. The

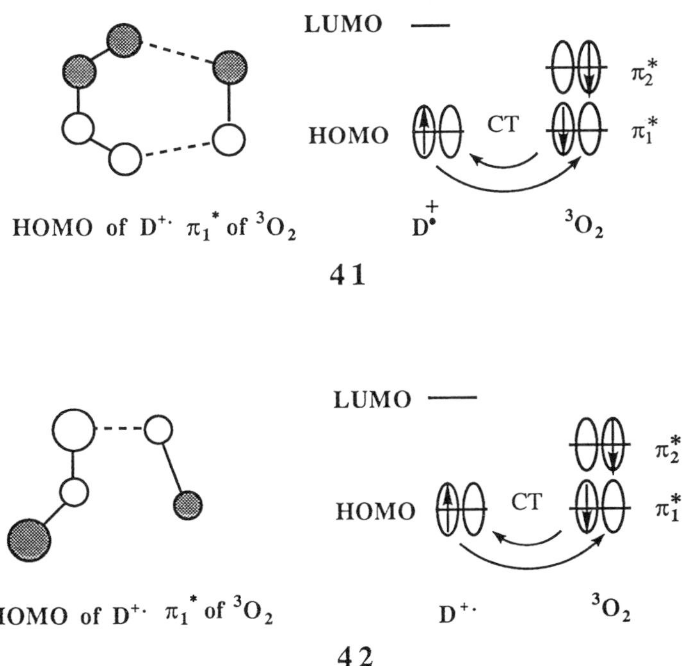

Figure 51. Orbital interactions between radical cations of alkadiene (**41**) (and alkene **42**) and triplet molecular oxygen ($^3O_2$)

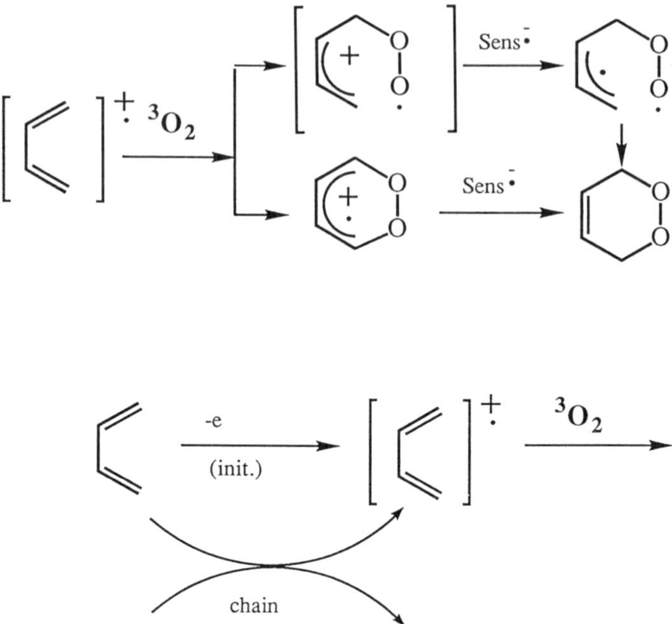

Figure 52. Radical coupling between the radical cation of butadiene and $^3O_2$ in the electron transfer-sensitized photo-oxygenation and the electrochemical oxygenation

intermediate 1,4-radical thus formed has four different electronic states which are expressed by the occupation numbers of π- and σ-type molecular orbitals as illustrated in Figure 54. Full geometry optimizations were carried out for these states in the case of ethyleneperoxy radical cation by using the UHF 4–31G energy gradient technique.[41] Assuming the optimized geometries obtained, more reliable UHF (6–31G**) calculations were performed to elucidate the relative stabilities among the states examined. It has been shown that (1) the $\sigma^0\pi^1$ radical, which is an initial addition product, is less stable than the $\pi^0\pi^1$

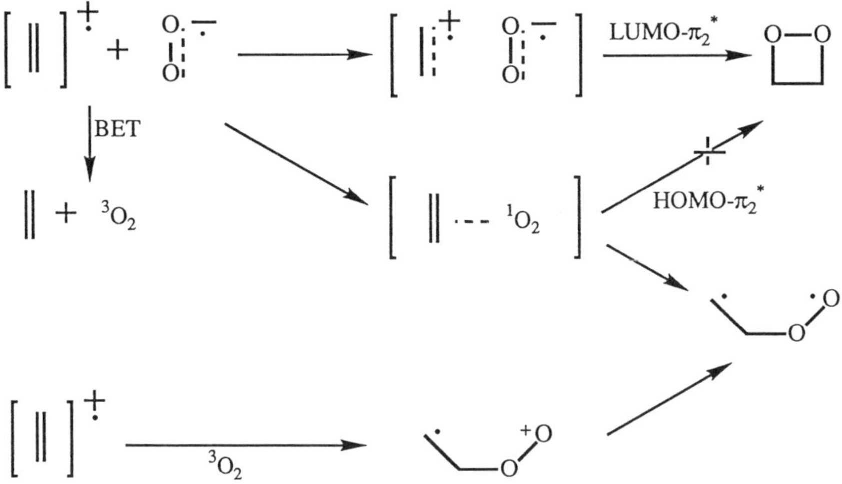

Figure 53. Reactions of the radical cation of ethylene with $O_2^{-\cdot}$, $^1O_2$, and $^3O_2$

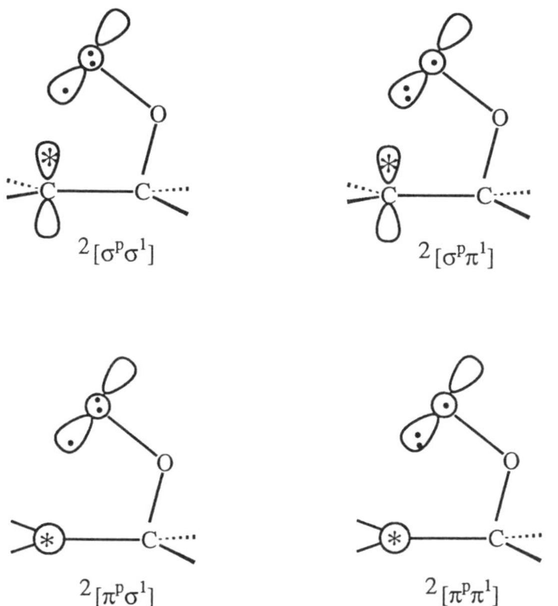

Figure 54. Electronic structures of four different states of ethyleneperoxy radical cation ($p = 0$, $* = +$) and ethyleneperoxy radical anion ($p = 2$, $* = -$)

radical, indicating that a rotation of the terminal methylene group is likely in the course of the addition step, and that (2) since the $\sigma^0\sigma^1$ radical is less stable than the $\pi^0\pi^1$ radical, the cyclization of the ethyleneperoxy radical cation to give the radical cation of dioxetane is energetically unfavorable. Then, the *ab initio* MO calculations predict that (3) dioxetane formation should occur after the electron transfer from Sens$^{-\cdot}$, $O_2^{-\cdot}$ or the substrate alkene to the alkeneperoxy radical cation, and the oxygenation reaction is non-stereospecific. If the electron transfer occurs from the substrate alkene, the dioxetane-forming reaction becomes a radical chain reaction. The alkeneperoxy radical cation is sometimes converted into epoxide via unclarified pathways. Such situations are summarized in Figure 55.[41]

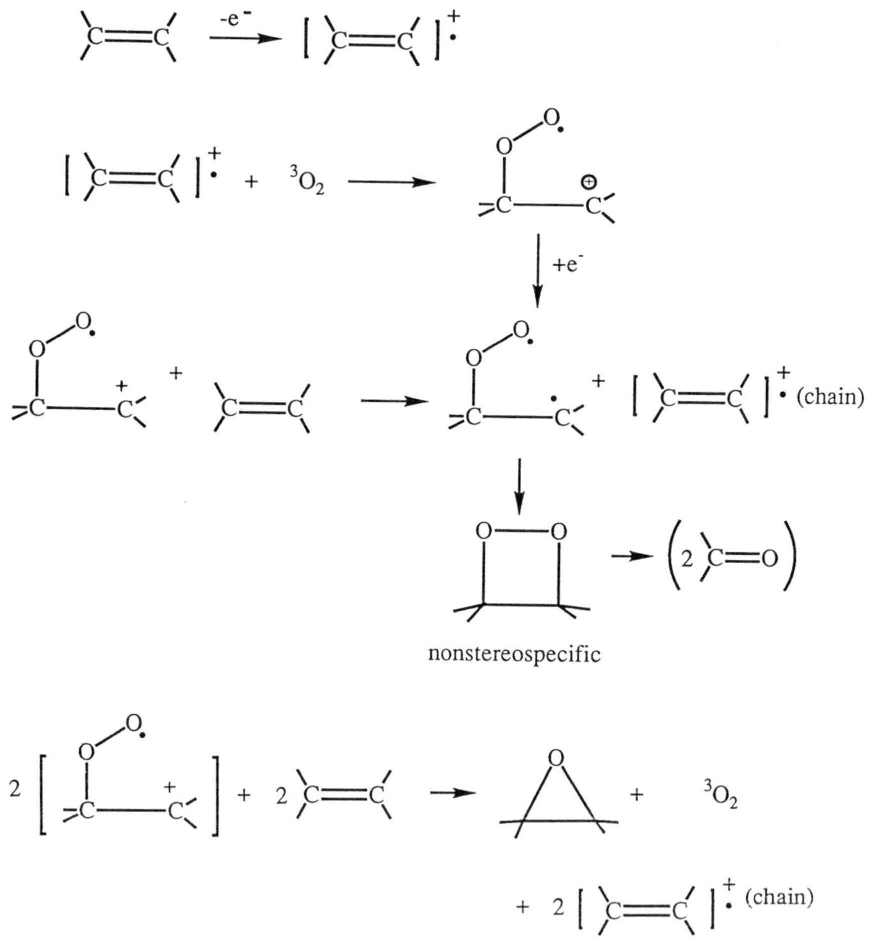

-e⁻ ; + $^3O_2$; +e⁻ ; (chain) ; nonstereospecific ; $^3O_2$; (chain)

Figure 55. Radical chain reactions in the oxygenation of alkene via radical cation

The HOMO-π_1^* interaction in **42** in Figure 51 shows that $^3O_2$ attacks the 3-position of radical cations of unsymmetrical electron-rich alkenes. The regioselectivity in the radical coupling reaction between $D^{+\bullet}$ and $^3O_2$ is different from that in the coupling reaction between $D^{+\bullet}$ and $O_2^{-\bullet}$.

The other mode of reaction involving $D^{+\bullet}$ and $^3O_2$ is a free radical coupling between $^3O_2$ and a free radical that is derived from $D^{+\bullet}$ by deprotonation. For example, the DCA-sensitized oxygenation of toluene gives benzaldehyde and benzoic acid.[105,106] This reaction proceeds via the pathway shown in Figure 56.

The electron transfer-mediated oxygenation of small-ring compounds provides interesting examples. The DCA-sensitized photo-oxygenation of cyclopropane derivatives (**43** or **44**) affords dioxolane derivatives **48** and **49** via the pathway illustrated in Figure 57.[107] The key intermediate in this photo-

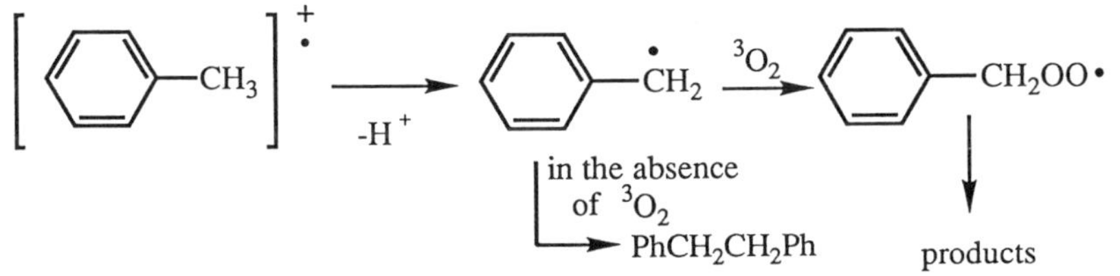

-H$^+$
$^3O_2$
$CH_2OO\bullet$
in the absence of $^3O_2$
$PhCH_2CH_2Ph$
products

Figure 56. Free-radical oxygenation of toluene via radical cation

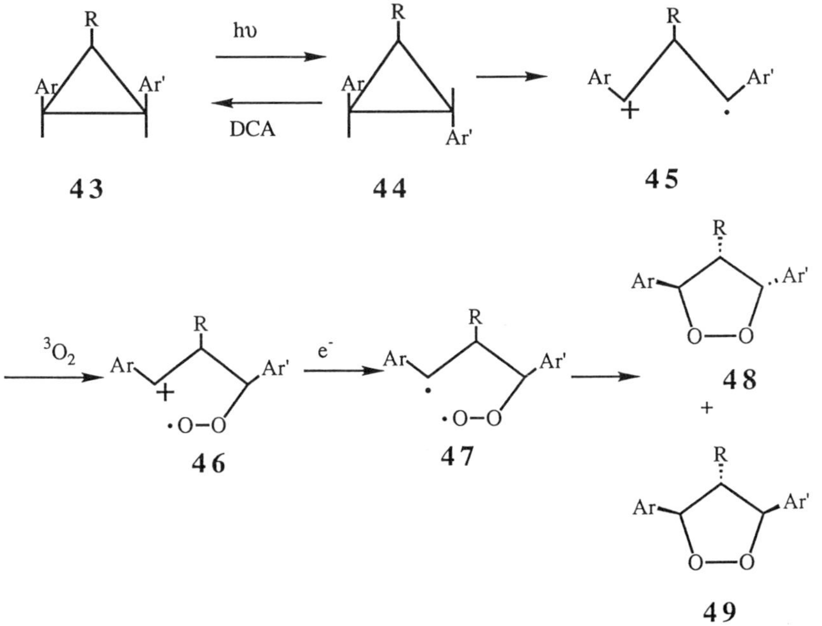

hυ
DCA
43 **44** **45** **46** **47** **48** **49**

Figure 57. DCA-sensitized photo-oxygenations of arylcyclopropanes; **44–49** are key reaction intermediates and products

oxygenation is the open-chain radical cation **45**, in which the cation center is localized on the carbon bearing a more powerful electron-donating substituent, while the radical center is localized on the other carbon. The attack of $^3O_2$ occurs on **45** to give **46**. A back-electron transfer from $DCA^{-\bullet}$ or the substrates, followed by cyclization, affords a stereoisomeric mixture of dioxolanes **48** and **49**. This reaction involves a radical chain mechanism, indicating that the back-electron transfer occurs to some extent from the substrates to **46**. The 1,5-biradical species **47** is stereochemically non-rigid. Accordingly, the **49**:**50** isomer ratio is determined by their thermodynamic stabilities; if cyclopropanes have alkyl substituents on the 3-position, **49** and **50** are produced in about a 9:1 ratio.

The DCA-sensitized photo-oxygenation of electron-rich methylenecyclopropanes and cyclobutanes also affords the cyclic peroxides via a similar mechanism.[108–110] Ando *et al.*[108] have demonstrated that the DCA-sensitized photo-oxygenation of the adamantylidenecyclopropane **50** gives the 4-alkylidenedioxolane derivative **51** (Figure 58) in a highly *exo*-selective manner.

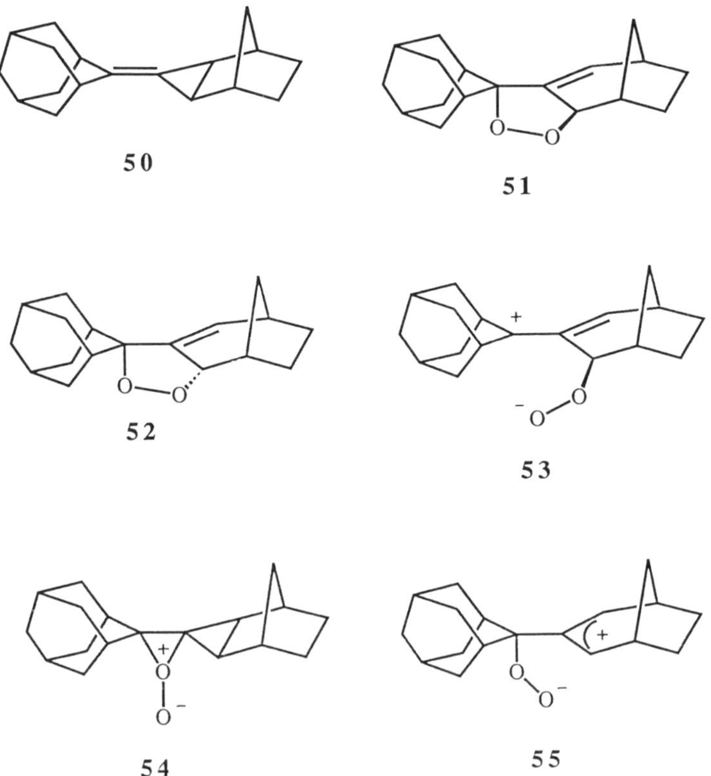

Figure 58. Key intermediates for oxygenation reactions of adamantylidene compounds

In contrast, the oxygenation of **50** by $^1O_2$ gives stereoselectively the *endo*-adduct **52**. They proposed that the key step in the former reaction is the recombination of the radical cation of **50** with $O_2^{-\bullet}$, which is produced by an electron transfer from $DCA^{-\bullet}$ to $^3O_2$. In this process, $O_2^{-\bullet}$ attacks the radical cation stereospecifically from the less hindered *exo* face to give the zwitterionic peroxide **53**. The cyclization of **53** gives **51**. It is noteworthy that the radical cation of **50**, which can be produced by an electrochemical oxidation, does not react with $^3O_2$. On the other hand, $^1O_2$ attacks **50** stereospecifically from the less hindered *endo* face to give the peroxides **54** and/or **55** that are converted to **51**. These examples suggest that the stereochemical courses and the modes of reactions may be varied by changing the reaction conditions and substrates.

The photo-oxygenation of hetero-three-membered ring compounds via electron transfer has also been investigated.[111–113] Schaap and co-workers reported that the DCA-sensitized photo-oxygenation of stilbene oxide gives *cis*-2,4-diphenyl-1,2,4-trioxolane with high stereoselectivity. They proposed a mechanism involving the reaction of an opened ylide structure of stilbene oxide with $^1O_2$. In the case of aziridines, the stereoselectivity was dependent on the bulkiness of the *N*-substituents.

8.4 Reactions of Radical Anions with $^3O_2$

Since molecular dioxygen has a strong electron-accepting ability, the reaction between radical anions and $^3O_2$ causes electron transfer to generate $O_2^{-\bullet}$ (equation 42).[114]

$$A^{-\bullet} + {}^3O_2 \rightarrow A + O_2^{-\bullet} \tag{42}$$

However, the addition of $^3O_2$ to $A^{-\bullet}$ is observed under special conditions. The photoreaction of DCA under a dioxygen atmosphere in the presence of an electron donor such as biphenyl affords 9,10-anthraquinone.[106] A key intermediate in this reaction is the radical anion of DCA. The SOMO–π_2^* or SOMO–π_1^* interaction between $DCA^{-\bullet}$ and $^3O_2$ predicts that the cycloaddition in a concerted manner is symmetry-forbidden. The most probable mode of attack may be an attack of $^3O_2$ at the carbons of the radical and anion centers of $DCA^{-\bullet}$.

Frimer *et al.*[103a] proposed an electron-transfer mechanism for the reaction of electron-deficient alkenes A with KO_2; the first step of this reaction is electron transfer from $O_2^{-\bullet}$ to A to generate the radical anion $A^{-\bullet}$. This radical anion reacts with $^3O_2$, to give regioselectively the radical anion of type A in Figure 49.

The photo-excitation of a CT band between adamantylideneadamantane and tetracyanoethylene (TCNE) in the presence of $^3O_2$ gives epoxides of both adamantylideneadamantane and TCNE.[115] This reaction proceeds via the itnermediate illustrated in equation 43

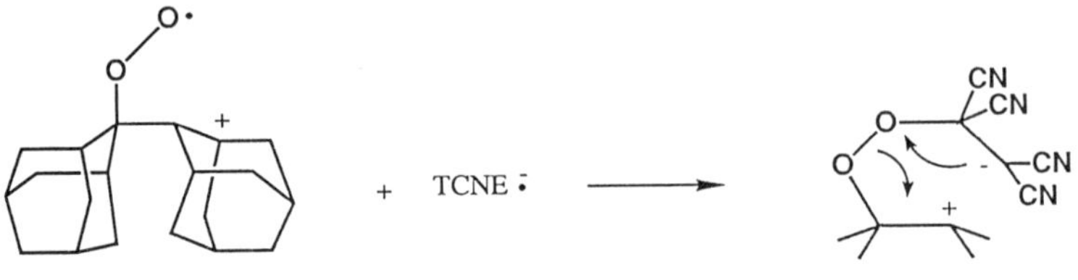

(43)

In conclusion, the orbital interaction arguments are wholly compatible with the experimental results of oxygenation reactions involving radical ions, although they are qualitative and even speculative in some cases. The more reliable *ab initio* MR CI calculations are necessary for further quantitative discussions of the reactions examined in this section.

9 CONCLUSION

Several important conclusions can be drawn from the theoretical results, although the topics reviewed here are limited. First, we can recognize that the *ab initio* energy gradient method developed in the field of quantum chemistry can be successfully applied to locate transition structures for various types of oxygenation reactions. The molecular structures of the transition states can hardly be determined except for the so-called recent 'computer experiments' described in this chapter. In fact, the reliable transition structures shown in the figures have been determined in the past decade. Second, to this end, accurate wavefunctions involving electron correlation effects are necessary because of the radical and biradical characters of oxygen and peroxide species, leading to the beyond Hartree-Fock approaches such as the HF plus Møller–Plesset (MP) method followed by the spin projection (PUMP, APUMP), multi-configuration SCF (CASSCF) method, MR MP and the CI methods. Third, non-adiabatic and dynamic theoretical treatments are apparently necessary for chemiluminescence reactions and excited reactions involving electron-transfer processes. *Ab initio* calculations of the reaction rates are still difficult even for gas-phase oxygenation reactions. These are largely future problems.

The primary emphasis in this chapter has been placed on the theoretical approaches to the elucidation of the structures and chemical properties of reactive species generated from peroxy compounds. We believe that the theoretical results summarized may provide general insights into oxygenation reactions, in combination with various experimental results given in other chapters of this book. The mechanisms of a variety of oxidation reactions have been increasingly clarified from the cooperative efforts of theory and experiment. In fact, we are confident from the content of this chapter that the cooperative work between theoretical and experimental chemists will open up new horizons in the chemistry of oxygenation reactions.

As mentioned in the Introduction, peroxy compounds and their reactions play important roles in a wide range of fields including organic synthesis, industrial chemistry, atmospheric chemistry, biology and medicine. *Ab initio* simulations of oxygenation reactions in various environments are desirable to elucidate the mechanisms of complex oxygenation reactions in these systems. For example, the location of transition structures of transition metal-catalyzed oxygenations and oxygenations catalyzed by enzymes are very limited at present. Great progress can be expected in these fields with future experimental and theoretical efforts.

10 REFERENCES

1. A. G. Davies, *Organic Peroxides*, Butterworths, London (1961).
2. S. Patai (Ed.), *The Chemistry of Peroxides*, Wiley, Chichester (1983).
3. H. Kropf (Ed.), *Organische Peroxo-Verbindungen*, Georg Thieme, New York (1988).
4. K. P. Lawley (Ed.), *Advances in Chemical Physics*, Vol. 69 I and II, Wiley, New York (1987).
5. W. J. Hehre, L. Radom, P. v. R. Schleyer and J. A. Pople, *Ab initio Molecular Orbital Theory*, Wiley, New York (1986).
6. M. Frish, M. Head-Gordon, G. W. Trucks *et al.*, *GAUSSIAN 90 Series*, Carnegie–Mellon Quantum Chemistry Publishing Unit, Carnegie–Mellon University, Pittsburgh, PA (1990).
7. (a) E. Clementi (Ed.), *Modern Techniques in Computational Chemistry: MOTECC-90*, IBM, Kingston, NY (1990); (b) e.g. *AMPAC*, QCPE Publ. 506, Indiana University, Bloomington, IN, (1990).
8. D. Cremer, in ref. 2, Chapt. 1.
9. K. Yamaguchi, T. Fueno and H. Fukutome, *Chem. Phys. Lett.*, **22**, 466 (1973).
10. K. Yamaguchi, *Chem. Phys. Lett.*, **33**, 330 (1975).
11. K. Yamaguchi, Y. Takahara and T. Fueno, in *Applied Quantum Chemistry* (V. H. Smith, H. F. Schaefer III and K. Morokuma, Eds), Reidel, Dordrecht (1986), p. 155.
12. H. B. Schlegel, *J. Chem. Phys.*, **84**, 4530 (1986).
13. K. Yamaguchi, Y. Yoshioka and T. Fueno, *Chem. Phys.*, **20**, 171 (1977).
14. K. Rudenberg, L. M. Cheung and S. T. Elbert, *Int. J. Quantum Chem.*, **16**, 1069 (1979).
15. B. Roos, *Int. J. Quantum Chem.*, **Symp. 14**, 175 (1980).
16. K. Yamaguchi, *Int. J. Quantum Chem.*, **Symp. 14**, 269 (1980).
17. P. Pulay and T. P. Hamilton, *J. Chem. Phys.*, **88**, 4926 (1988).
18. K. Yamaguchi, *Int. J. Quant. Chem.*, **18**, 101 (1980).
19. W. A. Goddard, III, T. H. Dunning, Jr, W. J. Hunt and P. J. Hay, *Acc. Chem. Res.*, **6**, 368 (1973).
20. K. Yamaguchi, in *SCF Theory and Application* (R. Carbo and K. Klobukowski, Eds), Elsevier, New York (1990), p. 727.
21. J. M. Bofill and P. Pulay, *J. Chem. Phys.*, **90**, 3657 (1989).
22. C. Møller and M. S. Plesset, *Phys. Rev.*, **46**, 618 (1934).
23. P. O. Lowdin, *Phys. Rev.*, **97**, 1509 (1955).

24. P. J. Knowles and N. C. Handy, *J. Chem. Phys.*, **88**, 6991 (1988).
25. K. Yamaguchi, F. Jensen, A. Dorigo and K. N. Houk, *Chem. Phys. Lett.*, **149**, 537 (1988).
26. Y. Takahara, K. Yamaguchi and T. Fueno, *Chem. Phys. Lett.*, **157**, 211 (1989).
27. I. Shavitt, in *Applications of Electronic Structure Theory* (H. F. Schaefer, III, Ed.), Plenum Press, New York (1977), p. 189.
28. P. Pulay, in *Applications of Electronic Structure Theory* (H. F. Schaefer, III, Ed.), Plenum Press, New York (1970), p. 153.
29. F. Bernardi and M. A. Robb, in ref. 4, Vol. I, p. 155.
30. G. Fitzgerald, T. J. Lee, H. F. Schaefer, III, and R. J. Bartlett, *J. Chem. Phys.*, **83**, 6275 (1985).
31. K. Fukui, *J. Phys. Chem.*, **74**, 4161 (1970).
32. M. Dupuis, G. Fitzgerald, B. Hammond, W. A. Lester, Jr, and H. F. Schaefer, III, *J. Chem. Phys.*, **84**, 2691 (1986).
33. (a) G. Klopmann, *J. Am. Chem. Soc.*, **90** 223 (1968); (b) L. Salem, *J. Am. Chem. Soc.*, **90**, 545 (1968).
34. R. Huisgen, in *1,3-Dipolar Cycloaddition Chemistry*, Vols. I and II, A. Padwa (Ed.), Wiley–Interscience, New York (1984).
35. K. Yamaguchi, *J. Mol. Struct. (Theochem)*, **103**, 101 (1983).
36. K. Yamaguchi, S. Yabushita, T. Fueno, S. Kato, K. Morokuma and S. Iwata, *Chem. Phys. Lett.*, **71**, 583 (1980).
37. K. N. Houk and K. Yamaguchi, in ref. 34, Chapt. 13.
38. L. B. Harding and W. A. Goddard, III, *J. Am. Chem. Soc.*, **102**, 439 (1980).
39. *Ab initio* data without references are results newly calculated for this chapter by the authors.
40. K. Yamaguchi, T. Takahara, T. Fueno and K. N. Houk, *Theor. Chim. Acta*, **73**, 337 (1988).
41. K. Yamaguchi, in *Singlet Oxygen*, Vol. III (A. A. Frimer, Ed.), CRC Press, Boca Raton, FL (1985), p. 119.
42. A. Dedieu, M.-M. Rohmer and A. Veillard, *Adv. Quantum Chem.*, **16**, 43 (1982).
43. B. D. Olafson and W. A. Goddard, III, *Proc. Natl. Acad. Sci. USA*, **74**, 1315 (1977).
44. T. Nozawa, M. Hatano, U. Nagashima, S. Obara and H. Kashiwagi, *Bull. Chem. Soc. Jpn.*, **56**, 1721 (1983).
45. K. Yamaguchi, Y. Takahara and T. Fueno, in *The Role of Oxygen in Chemistry and Biochemistry* (W. Ando and Y. Moro-oka, Eds.), Elsevier, Amsterdam (1988), p. 263.
46. H. Tatewaki and S. Huzinaga, *J. Chem. Phys.*, **71**, 4339 (1971).
47. (a) J. T. Groves and T. E. Nemo, *J. Am. Chem. Soc.*, **105**, 5786 (1983); (b) I. Tabushi, M. Kodera and M. Yokoyama, *J. Am. Chem. Soc.*, **107**, 4466 (1985); (c) M. F. Powell, E. F. Pai and T. C. Bruice, *J. Am. Chem. Soc.*, **106**, 3277 (1984); (d) B. Meunier, E. Guilment, M.-E. De Carvalho and R. Poibblanc, *J. Am. Chem. Soc.*, **106**, 6668 (1984).
48. (a) R. D. Jones, D. A. Summerville and F. Basolo, *Chem. Rev.*, **79**, 139 (1979); (b) S. Kimura, I. Yamazaki and T. Kitagawa, *Biochemistry*, **20**, 4632 (1981); (c) I. Morishima, Y. Takamuki and Y. Shirto, *J. Am. Chem. Soc.*, **106**, 7666 (1984).
49. R. E. White and M. J. Coon, *Annu. Rev. Biochem.*, **49**, 315 (1980).
50. K. Yamaguchi, Y. Takahara, T. Fueno, I. Saito and T. Matsuura, in *Medical, Biochemical and Chemical Aspects of Free Radicals* (O. Hayaishi, E. Niki, M. Kondo and T. Yoshikawa, Eds), Elsevier, Amsterdam (1989), p. 993.
51. Y. Sugiura, T. Takita and H. Umezawa, in *Metal Ions in Biological Systems*, Vol. 19 (H. Sigel, Ed.), Marcel Dekker, New York (1985), p. 81.
52. T. Morii, T. Matsuura, I. Saito, T. Suzuki, J. Kuwahara and Y. Sugiura, *J. Am. Chem. Soc.*, **108**, 7089 (1986).

53. S. E. Phillips and B. P. Schoenborn, *Nature (London)*, **292**, 81 (1981).
54. M. G. Finn and K. B. Sharpless, in *Asymmetric Synthesis*, Vol. 5 (J. D. Morrison, Ed.), Academic Press, New York (1986), p. 247.
55. M. Rahman, M. L. McKee, P. B. Shevlin and R. Sztyrbicka, *J. Am. Chem. Soc.*, **110**, 4002 (1988).
56. J. T. Herron, R. I. Martinez and R. E. Huie, *Int. J. Chem. Kinet.*, **14**, 201 (1982).
57. R. Mello, M. Fiorentino, C. Fusco and R. Curci, *J. Am. Chem. Soc.*, **111**, 6749 (1989).
58. D. Cremer and C. W. Bock, *J. Am. Chem, Soc.*, **108**, 3375 (1986).
59. R. Atkinson, B. J. Finlayson and J. N. Pitts, *J. Am. Chem. Soc.*, **95**, 7592 (1973).
60. B. Plesnicar, M. Taservski and A. Azman, *J. Am. Chem. Soc.*, **100**, 744 (1978).
61. D. Swern, in *Organic Peroxides*, Vol. 2 (D. Swern, Ed.), Wiley, New York (1971), p. 355.
62. R. P. Hanzlik and G. O. Shearer, *J. Am. Chem. Soc.*, **97**, 5231 (1975).
63. A. Azman, J. Koller and B. Plesnicar, *J. Am. Chem. Soc.*, **101**, 1107 (1979).
64. R. A. Sheldon and J. Kochi, *Metal-catalyzed Oxidations of Organic Compounds*, Academic Press, New York (1981).
65. H. Mimoun, *Angew. Chem., Int. Ed. Engl.*, **21**, 734 (1982).
66. R. D. Bach, G. J. Wolber and B. A. Coddens, *J. Am. Chem. Soc.*, **106**, 6098 (1984).
67. I. D. Williams, S. F. Pedersen, K. B. Sharpless and S. J. Lippard, *J. Am. Chem. Soc.*, **106**, 6430 (1984).
68. K. A. Jørgensen, R. A. Wheeler and R. Hoffmann, *J. Am. Chem. Soc.*, **109**, 3240 (1987).
69. W. C. Gardiner (Ed.), *Combustion Chemistry*, Springer, New York (1984).
70. W. Jaeschke (Ed.), *Chemistry of Multiphase Atmospheric Systems*, Springer (1986).
71. C. Sosa and H. B. Schlegel, *J. Am. Chem. Soc.*, **109**, 7007 (1987).
72. R. Atkinson, R. A. Perry and J. N. Pitts, Jr, *J. Chem. Phys.*, **66**, 1197 (1977).
73. T. Fueno, Y. Takahara and K. Yamaguchi, *Chem. Phys. Lett.*, **167**, 291 (1990).
74. K. Yamaguchi, in *Oxygen Radicals in Chemistry and Biology* (W. Bors, M. Saran and D. Tait, Eds.), Walter de Gruyter, New York (1984), p. 65.
75. K. Yamaguchi, S. Yabushita, T. Fueno and K. N. Houk, *J. Am. Chem. Soc.*, **103**, 5043 (1981).
76. L. B. Harding and W. A. Goddard, III, *J. Am. Chem. Soc.*, **99**, 4520 (1977).
77. M. Hottokka, B. Roos and P. J. Siegbahn, *J. Am. Chem. Soc.*, **105**, 5263 (1983).
78. G. Tonachini, H. B. Schlegel, F. Bernardi and M. A. Robb, *J. Am. Chem. Soc.*, **112**, 483 (1990).
79. M. J. S. Dewar and W. Thiel, *J. Am. Chem. Soc.*, **97**, 3978 (1975).
80. P. S. Bailey, *Ozonolysis in Organic Chemistry*, Academic Press, New York (1978).
81. (a) R. Criegee, *Angew. Chem., Int. Ed. Engl.*, **14**, 745, 765 (1975); (b) R. Huisgen, *Angew. Chem., Int. Ed. Engl.*, **16**, 572 (1977).
82. H. E. O'Neal and C. Blumstein, *Int. J. Chem. Kinet.*, **5**, 397 (1973).
83. C. W. Gillies, J. Z. Gillies, R. D. Suenram, F. J. Lovcas, E. Kraka and D. Cremer, *J. Am. Chem. Soc.*, **113**, 2412 (1991).
84. M. L. McKee and C. M. Rohlfing, *J. Am. Chem. Soc.*, **111**, 2497 (1989).
85. R. B. Woodward and R. Hoffmann, *The Conservation of Orbital Symmetry*, Verlag Chemie, Weinheim (1970).
86. M. J. S. Dewar, J. C. Hwang and D. R. Kuhn, *J. Am. Chem. Soc.*, **113**, 735 (1991).
87. P. S. Nangia and S. W. Benson, *J. Am. Chem. Soc.*, **102**, 3105 (1980).
88. W. R. Wadt and W. A. Goddard, III, *J. Am. Chem. Soc.*, **97**, 3004 (1975).
89. D. Cremer, *J. Am. Chem. Soc.*, **103**, 3627 (1981) and his series of papers for ozonolysis reactions.
90. H. Yu, W.-T. Chan and J. D. Goddard, *J. Am. Chem. Soc.*, **112**, 7529 (1990).

91. W. Cooper and W. D. Walters, *J. Am. Chem. Soc.*, **80**, 4220 (1958).
92. L. E. Friedrich and G. B. Schuster, *J. Am. Chem. Soc.*, **93**, 4602 (1971).
93. M. Reguero, F. Bernardi, A. Bottoni, M. Olivucci and M. A. Robb, *J. Am. Chem. Soc.*, **113**, 1566 (1991).
94. S. P. Schmidt, M. A. Vincent, C. E. Dykstra and G. B. Schuster, *J. Am. Chem. Soc.*, **103**, 1292 (1981).
95. M. A. Fox and M. Chanon (Eds.), *Photoinduced Electron Transfer, Parts A–D*, Elsevier, Amsterdam (1988).
96. (a) S. Torii, *Recent Advances in Electroorganic Synthesis*, Kodansha, Tokyo (1987); (b) S. F. Nelsen, *Acc. Chem. Res.*, **20**, 269 (1987).
97. I. Saito, T. Matsuura and K. Inoue, *J. Am. Chem. Soc.*, **103**, 188 (1981).
98. (a) J. Eriksen, C. S. Foote and T. L. Parker, *J. Am. Chem. Soc.*, **99**, 6455 (1977); (b) J. Eriksen and C. S. Foote, *J. Am. Chem. Soc.*, **102**, 6083 (1980).
99. L. Lopez, in *Photoinduced Electron Transfer I, Topics in Current Chemistry* 156 (J. Mattay, Ed.), Springer-Verlag, Berlin (1990), p. 117.
100 S. L. Mattes and S. Farid, *J. Am. Chem. Soc.*, **104**, 1454 (1982).
101. D. T. Sawyer and J. S. Valentine, *Acc. Chem. Res.*, **14**, 393 (1981).
102. K. Mizuno, N. Ohmura, S. Nakanishi and Y. Otsuji, *J. Chem. Soc., Chem. Commun.*, 355 (1983).
103. (a) A. A. Frimer, I. Rosenthal and S. Hot, *Tetrahedron Lett.*, 4631 (1977); (b) Y. Sawaki and Y. Ogata, *Bull. Chem. Soc. Jpn.*, **54** 793 (1981); (c) T. S. Calderwood, R. C. Neuman, Jr, and D. T. Sawyer, *J. Am. Chem. Soc.*, **105**, 3337 (1983); (d) J. L. Roberts, Jr, T. S. Calderwood and D. T. Sawyer, *J. Am. Chem. Soc.*, **105**, 7691 (1983).
104. (a) D. H. R. Barton, R. K. Haynes, G. Leclarc, P. D. Magnus and I. D. Menzies, *J. Chem. Soc., Perkin Trans. 1*, 2055 (1983).
105. (a) I. Saito, K. Tamoto and T. Matsuura, *Tetrahedron Lett.*, 2889 (1979); (b) J. Santamaria, P. Gabillet and L. Bokobza, *Tetrahedron Lett.*, **25**, 2139 (1984); (c) J. Santamaria and R. Ouchabane, *Tetrahedron*, **42**, 5559 (1986).
106. K. Mizuno, N. Ichinose, T. Tamai and Y. Otsuji, *Tetrahedron Lett.*, **26**, 5823 (1985).
107. K. Mizuno, N. Kamiyama, N. Ichinose and Y. Otsuji, *Tetrahedron*, **41**, 2207 (1985); *Chem. Lett.*, 477 (1983); *J. Org. Chem.*, **55**, 4079 (1990).
108. T. Akasaka and W. Ando, *J. Am. Chem. Soc.*, **109**, 1260 (1987).
109. (a) Y. Takahashi, T. Miyashi and T. Makai, *J. Am. Chem. Soc.*, **105**, 6511 (1983); (b) T. Miyashi, Y. Takahashi, T. Makai, H. D. Roth and M. L. M. Schilling, *J. Am. Chem. Soc.*, **107**, 1079 (1985).
110. (a) K. Mizuno, K. Murakami, N. Kamiyama and Y. Otsuji, *J. Chem. Soc., Chem. Commun.*, 462 (1983); (b) K. Gollnick and A. Schnatterrer, *Tetrahedron Lett.*, **25**, 2735 (1984).
111. (a) A. P. Schaap, L. Lopez and S. D. Gagnon, *J. Am. Chem. Soc.*, **105**, 663 (1983); (b) A. P. Schaap, S. Siddiqui, S. D. Gagnon and L. Lopez, *J. Am. Chem Soc.*, **105**, 5149 (1983); (c) A. P. Schaap, S. Siddiqui, G. Prasad, A. F. M. Rahman and J. F. Oliber, *J. Am. Chem. Soc.*, **106**, 6087 (1984).
112. S. Futamura, S. Kusunose, H. Ohta and Y. Kamiya, *J. Chem. Soc., Perkin Trans. 1*, 15 (1984).
113. (a) T. Miyashi, Y. Takahashi, K. Yokogawa and T. Mukai, *J. Chem. Soc., Chem. Commun.*, 175 (1987); (b) T. Miyashi, M. Kamata and T. Mukai, *J. Am. Chem. Soc.*, **109**, 2780 (1987).
114. L. T. Spada and C. S. Foote, *J. Am. Chem. Soc.*, **102**, 391 (1980).
115. R. Akaba, H. Sakuragi and K. Tokumaru, *Chem. Lett.*, 1677 (1984).

2 Alkyl Hydroperoxides

NED A. PORTER
Paul M. Gross Chemical Laboratory, Department of Chemistry, Duke University, Durham, NC 27706, USA

Alkyl hydroperoxides, with the composition RO—OH, are structurally the simplest organic derivatives of hydrogen peroxide.[1,2] Synthetic approaches to hydroperoxides generally involve the construction of C—O bonds and utilize reagents that already have the O—O bond present. Hence reagents that are useful in the preparation of hydroperoxides are hydrogen peroxide, singlet and

Organic Peroxides. Edited by W. Ando

triplet molecular oxygen, the superoxide anion and ozone. The chemistry of hydroperoxides is rich and centers around the weak O—O and O—H bonds in the functional group, the easy reduction of the O—O bond, the basicity of the O—O group and the acidity of the O—H bond. Hydroperoxides of low molecular weight are explosive and difficult to purify and the general rule of thumb is that at least five carbons per peroxide bond are necessary to confer stability to the compound.

1 PREPARATION FROM HYDROGEN PEROXIDE

The hydrogen peroxide molecule and its anion are strong nucleophilic reagents and many of the successful preparations of hydroperoxides rely on this property. Thus, reaction of hydrogen peroxide or its anion with alkyl sulfonates, alcohols, carboxylates or halides leads to alkylation and formation of the alkyl hydroperoxide (equation 1). Dialkylation may occur and alkyl peroxides may also be prepared by this sequence.

$$\text{H-O-O-H} \left(\text{H-O-O}^-\right) + \text{R-X} \longrightarrow \text{R-OO-H} + \text{H-X}\,(\text{X}^-) \qquad (1)$$

$$\text{X} = \text{-OSO}_2\text{R, -OH, -O-CO-R, -OP(OR)}_2\text{, Cl, Br, I....}$$

Several approaches have been developed to provide quantitative information about nucleophiles and the anion of hydrogen peroxide is one of the most nucleophilic reagents in all of the schemes developed. In reactions of nucleophiles with methyl iodide,[3,4] for example, HOO^- has an n_{MeI} value of 7.80 compared with methoxide, $n_{MeI} = 6.29$, azide, $n_{MeI} = 5.78$, and iodide, $n_{MeI} = 7.42$. Ritchie and Sawada[5] developed a measure of nucleophilicity, N^+, based on the reaction of nucleophiles with organic cations and the anion of hydrogen peroxide is also one of the most reactive nucleophiles studied in this scheme. Thus, HOO^- has an N^+ value of 8.08 compared with methoxide, $N^+ = 7.28$, azide, $N^+ = 7.60$, and hydroxylamine, $N^+ = 5.05$. The high nucleophilicity of hydrogen peroxide and its anion has been described in terms of the α- effect (two adjacent atoms bearing lone pairs of electrons) and the theory of this effect has been discussed in some detail.[6,7]

1.1 Reaction of Hydrogen Peroxide with Alkyl Halides

The synthesis of secondary or primary hydroperoxides or diakyl peroxides was historically achieved by the reaction of hydrogen peroxide with halides, mesylates or tosylates under conditions to promote an S_N2 reaction. These hydroperoxides are, however, sensitive to base-catalyzed decomposition and yields are generally low. Cookson *et al.*[8] investigated the reaction of primary, secondary and tertiary alkyl bromides and iodides with hydrogen peroxide and silver trifluoroacetate or silver tetrafluoroborate. The silver salt is soluble in diethyl ether or hexane and the reaction is conveniently carried out by mixing the halide, hydrogen peroxide and the silver salt (equation 2), filtering the precipitated silver halide and purifying the hydroperoxide by chromatography. Yields vary from 40 to 60% by this procedure. Formation of primary hydroperoxides is generally via the iodide while secondary and tertiary hydroperoxides are prepared via the bromide. Frimer[9] reported the preparation of cyclohexenyl hydroperoxide in nearly quantitative yield by the reaction of the allylic chloride with silver nitrate, sodium hydrogencarbonate, and 98% hydrogen peroxide in diethyl ether. Also reported was the conversion of 3-chloro-2-methoxycyclohexene to the allylic hydroperoxide by reaction with silver triflate, pyridine and 98% hydrogen peroxide in tetrahydrofuran (THF). Hydrogen peroxide with such a low water content is extremely reactive, however, and extreme caution should be used whenever procedures calling for its use are followed. The use of 98% hydrogen peroxide in conjunction with pyridine as described[9] has resulted in uncontrolled exothermic reactions.[10]

$$\text{H-O-O-H} + \text{R-X} + \text{AgO-CO-CF}_3 \longrightarrow \text{R-OO-H} + \text{Ag-X} + \text{HO-CO-CF}_3 \quad (2)$$

Dialkyl peroxides are also conveniently prepared by this procedure if excess of alkyl halide and silver salt is used or if a hydroperoxide is used in place of hydrogen peroxide in the reaction described in equation 2. In fact, Kopecky *et al.*[44] pioneered the use of the silver salt procedure for the preparation of dioxetanes and the use of this procedure for the preparation of other peroxides and endoperoxides derives from this important work in the late 1960s.

The reaction of alkyl halides with hydrogen peroxide and silver salts, following the procedure of Cookson *et al.*,[8] has been put to extensive use. Thus, Porter and co-workers[12,13] described the reaction of the two dibromides *trans*-**1** and *cis*-**1** with hydrogen peroxide–silver trifluoroacetate. The course of the reaction is dramatically different for the two stereoisomeric dibromides and suggests that the reaction occurs with inversion of configuration at the carbon center undergoing substitution. The *cis*-dibromide presumably is converted into the intermediate bromohydroperoxide, *trans*-**2**, which is then converted into the endoperoxide **3**. The *trans*-dibromide is presumably converted into a *cis*-bromohydroperoxide intermediate which cannot be converted to the endoperoxide **3** because of the backside displacement geometric requirement. In the presence of silver salt and hydrogen peroxide, *cis*-**2** is therefore converted into the dihydroperoxide product. While the bromohydroperoxides *cis*- and *trans*-**2** are not isolable intermediates in the chemistry observed for the dibromides **1**, these compounds have been independently synthesized and studied. The *trans*-bromohydroperoxide is converted cleanly and efficiently into the endoperoxide **3** on reaction with silver trifluoroacetate whereas the *cis*-bromohydroperoxide reacts slowly with silver trifluoroacetate and gives a mixture of products.

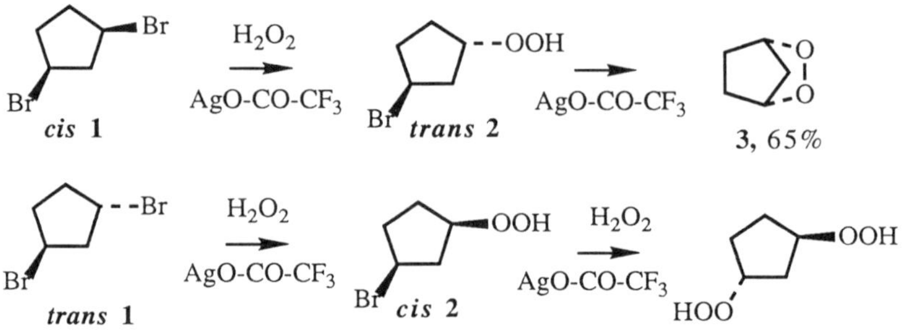

The stimulus for much of the work on hydroperoxide and endoperoxide chemistry in the past 20 years was the isolation and characterization of the prostaglandin endoperoxides PGG_2 and PGH_2 by Hamberg and co-workers in 1973.[14,15] The silver salt–hydrogen peroxide sequence has been successfully used to prepare several hydroperoxides, dialkyl peroxides and endoperoxides,[16–22] including the prostaglandin endoperoxides and their analogs.

Comment should be made on the reactions of the bromides **4** and **5** with hydrogen peroxide and silver trifluoroacetate (Table 1). The reaction of **4** with silver ion apparently leads to an orbital symmetry-controlled ring opening of the cyclopropyl bromide to provide a pentadienyl cation of specific geometry. Groups on the cyclopropane *trans* to the bromide leaving group open outward and if the alkyl groups on **4** are *trans* to the bromide, the cation formed, **6**, gives products that reflect its geometry.

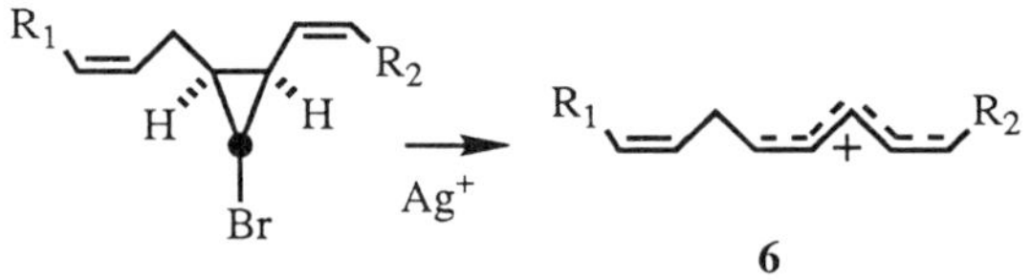

The reaction of the bromide **5** with silver ion–hydrogen peroxide gives a product mixture that contains cyclic peroxides and cyclopropylcarbinyl hydroperoxides and none of the simple substitution products. The formation of the cyclopropylcarbinyl products can be understood based on the rearrangement of an intermediate homoallylic cation to the cyclopropyl carbinyl cation and entrapment of this species with hydrogen peroxide. The cyclic peroxide–hydroperoxides observed in the reaction apparently form by a free-radical process (see Section 3.2) and it seems likely that these products might be eliminated if antioxidants were included in the reaction mixture or if the reaction was carried out in the absence of oxygen.

Table 1. Alkyl hydroperoxides, dialkyl peroxides and endoperoxides prepared from alkyl halides, silver salts and hydrogen peroxide or hydroperoxides

Reactants	Peroxide	Ref.
n-$C_6H_{13}I + H_2O_2$	n-$C_6H_{13}OOH$	8
s-$BuBr + H_2O_2$	s-$BuOOH$	8
$(n$-$Pr)_3CBr + H_2O_2$	$(n$-$Pr)_3COOH$	8
s-$BuBr + t$-$BuOOH$	s-$BuOO$-t-Bu	8
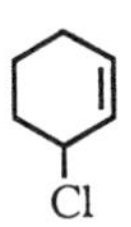 Cl + H_2O_2	 OOH	9
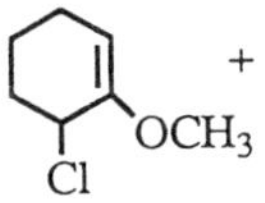 Cl, OCH_3 + H_2O_2	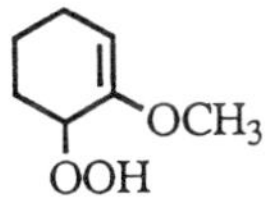 OOH, OCH_3	9
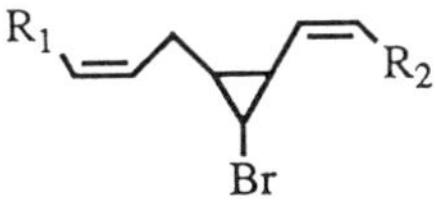 R_1, R_2, Br **4**	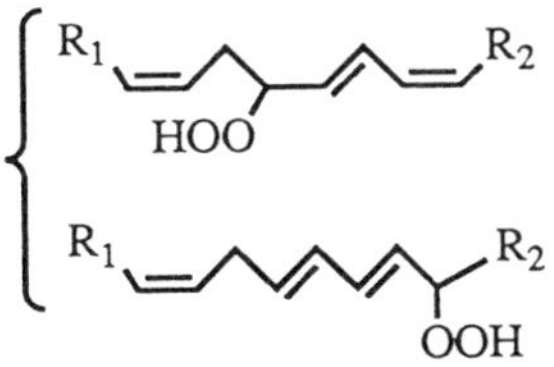 R_1, R_2, HOO; R_1, R_2, OOH	16, 17
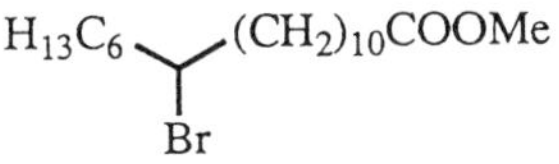 $H_{13}C_6$, $(CH_2)_{10}COOMe$, Br + H_2O_2	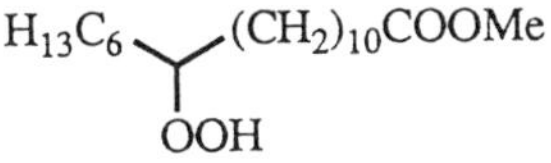 $H_{13}C_6$, $(CH_2)_{10}COOMe$, OOH	18
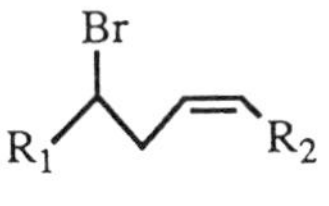 Br, R_1, R_2 **5** + H_2O_2	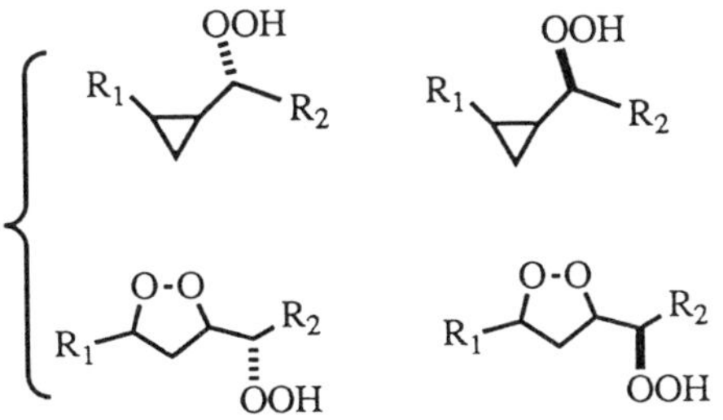 OOH, R_1, R_2; OOH, R_1, R_2; O-O, R_1, R_2, OOH; O-O, R_1, R_2, OOH	18
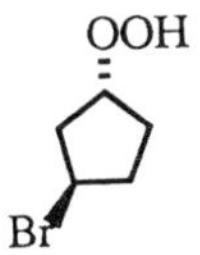 OOH, Br	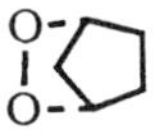 O, O	12, 13

(continued)

Table 1. (*continued*)

Reactants	Peroxide	Reference
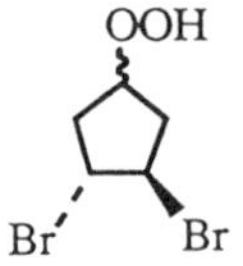	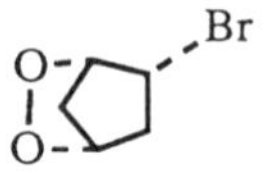	19
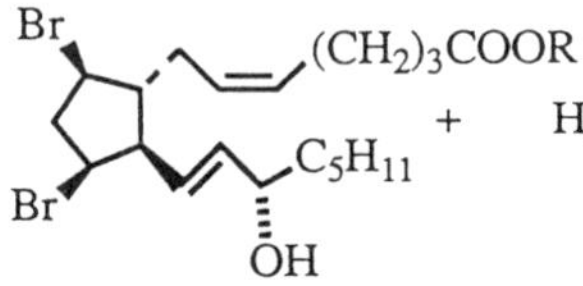 + H_2O_2	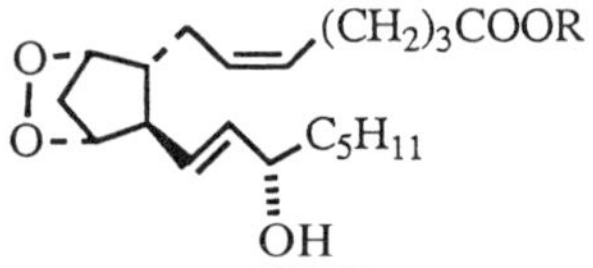 PGH_2	20, 21
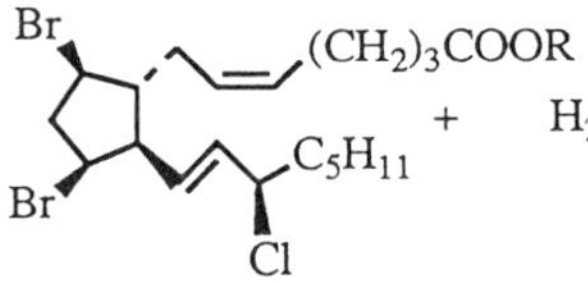 + H_2O_2	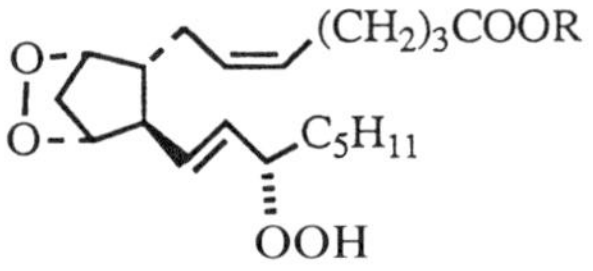 PGG_2	22

In addition to providing an efficient route to endoperoxides such as PGG_2 and PGH_2, the reaction of bromohydroperoxides or bromoperoxides with silver ion may lead to peroxide or hydroperoxide transfer reactions.[23–25] Thus, reaction of the bromohydroperoxide **7a** or the bromoperoxide **7b** with silver ion gives the product **8a** or **8b** and the cyclic peroxide **9**. A peroxonium intermediate has been proposed to account for these products. Other reactions proposed to proceed by peroxonium intermediates are the conversions **10a** into **11a** and **10b** into **11b**.

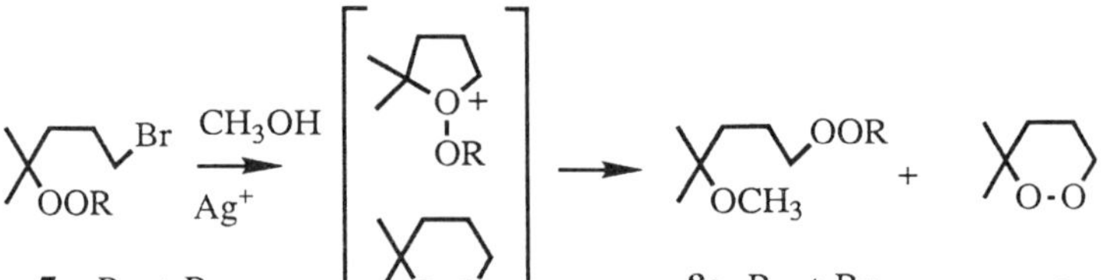

CH_3OH, Ag^+

7a, R= t-Bu
7b, R=H

8a, R= t-Bu
8b, R=H

9

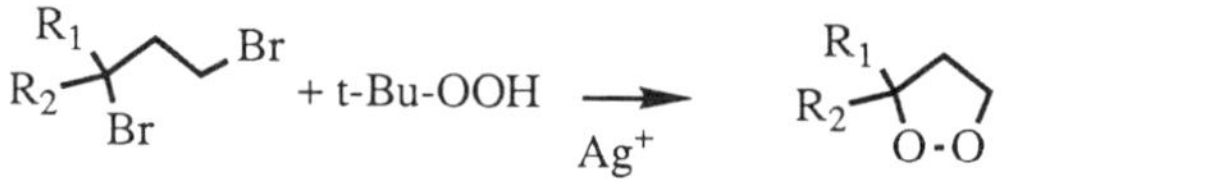

+ t-Bu-OOH, Ag^+

10 a, $R_1 = R_2 = CH_3$
b, R_1 = H, R_2 =Ph

11 a, $R_1 = R_2 = CH_3$
b, R_1 = H, R_2 =Ph

1.2 Reaction of Hydrogen Peroxide with Alkyl Phosphates and Sulfonates

The reaction of alkyl sulfonates with H_2O_2 is one of the oldest methods for the preparation of hydroperoxides and it continues to be an important procedure for hydroperoxide synthesis.[1,2] Primary hydroperoxides are suitable for preparation by displacement of mesylates and sulfonates with basic hydrogen peroxide, for example, and the method has been used extensively. Elimination is usually competitive with substitution for secondary sulfonates. The yields of product hydroperoxides are moderate in the case of primary sulfonates and poor for secondary substrates. Tertiary hydroperoxides can be prepared by reaction of 30% H_2O_2 with tertiary alkyl sulfates prepared from alkenes or alcohols with sulfuric acid. Examples of conversions utilizing sulfonates and basic hydrogen peroxide are shown in Table 2.[26–32]

The preparation of fatty acid hydroperoxides has been attempted by the nucleophilic substitution procedure and some success has been achieved by this method. These compounds, many of which are derived from arachidonic acid, have diverse and important biological properties. The synthesis of 5-HPETE was reported to proceed via the mesylate by reaction at −100°C in dichloromethane.[29]

The synthesis of 12-HPETE was also reported by this method and it was suggested that the displacement reaction proceeds by inversion of configuration at the reaction center.[30] Subsequently, it was shown that the reaction results in a complete racemization.[31] An alternative method for the conversion of a fatty acid alcohol into the corresponding hydroperoxide was reported by Nagata *et al.*[33] Reaction of the alcohol with chlorodiethyl phosphite and triethylamine in hexane gave the phosphite which was allowed to react with anhydrous hydrogen peroxide in dichloromethane at 0°C for 1.5 h.[32] The mixture of product hydroperoxides was formed in good yield by this procedure.

1.3 Reaction of Hydrogen Peroxide with Alkenes, Alcohols and Ethers

Alkenes bearing electron-donating groups react with acidic hydrogen peroxide to give hydroperoxides. Tertiary hydroperoxides are readily prepared by this procedure, 2-methylbut-2-ene being converted into *tert*-pentyl hydroperoxide with hydrogen peroxide and catalytic sulfuric acid.[1,2] Vinyl ethers are also particularly reactive with acidic hydrogen peroxide and this reactivity has been used to prepare α-alkoxyhydroperoxides. The chemistry proceeds through intermediate carbenium ions, and reaction of alcohols that provide the same cations on reaction with acid can serve as precursors to the same hydroperoxide products. Other electrophiles can be used to initiate the reaction with alkenes and a general reaction scheme that applies for these conversions is outlined in Scheme 1. Among the other electrophiles that have been used to initiate addition to alkenes are mercurinium ions derived from mercury(II) salts such as

Table 2. Alkyl hydroperoxides prepared from alkyl sulfonates and phosphites

Reactants	Peroxide	Ref.
$CH_3CH{=}CH(CH_2)_3OMs + H_2O_2$	$CH_3CH{=}CH(CH_2)_3OOH$	26, 27
$CH_2{=}CH(CH_3)(CH_2)_2OMs + H_2O_2$	$CH_2{=}CH(CH_3)(CH_2)_2OOH$	26, 27
$CH_2{=}CHCH{=}CH(CH_2)_2OMs + H_2O_2$	$CH_2{=}CHCH{=}CH(CH_2)_2OOH$	28
$(CH_3)_2C{=}CH(CH_2)_2OMs + H_2O_2$	$(CH_3)_2C{=}CH(CH_2)_2OOH$	27
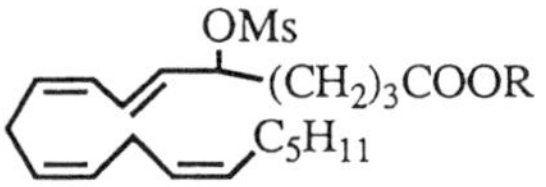 OMs, $(CH_2)_3COOR$, C_5H_{11} + H_2O_2	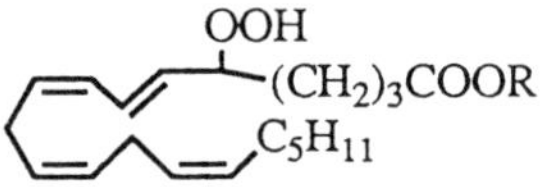 OOH, $(CH_2)_3COOR$, C_5H_{11} **5-HPETE**	29
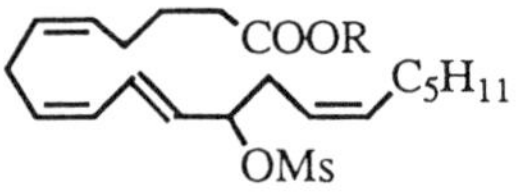 COOR, C_5H_{11}, OMs + H_2O_2	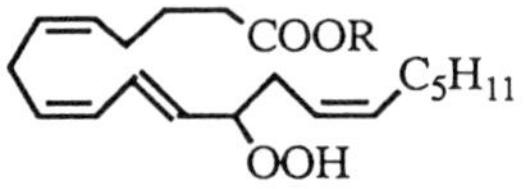 COOR, C_5H_{11}, OOH **12-HPETE**; COOR, C_5H_{11}, HOO 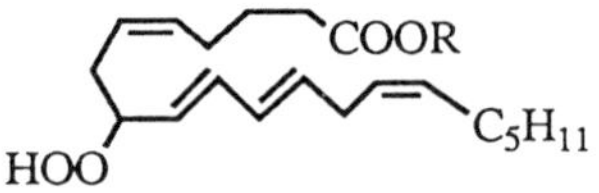	30–32
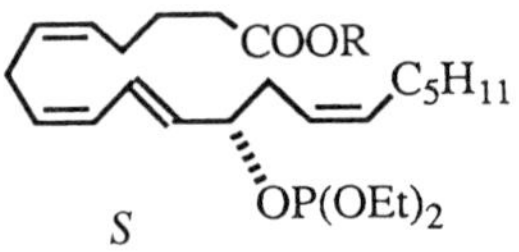 COOR, C_5H_{11}, $OP(OEt)_2$ *S* + H_2O_2	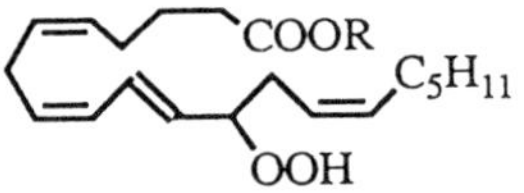 COOR, C_5H_{11}, OOH **11-HPETE**, *S*:*R* = *65*:*35*; 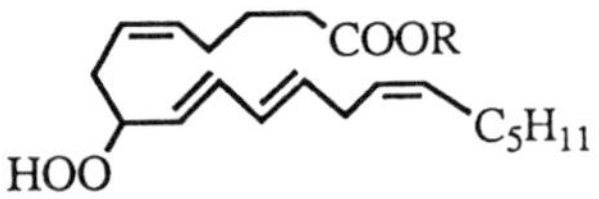 COOR, C_5H_{11}, HOO	33

the acetate, trifluoroacetate or nitrate. Halogen electrophiles can be added to alkenes along with hydrogen peroxide and useful sources of electrophilic halogen are *tert*-butyl hypochlorite, 1,3-dibromo-5,5-dimethylhydantoin, *N*-bromosuccinimide, *N*-iodosuccinimide and 1,3-diiodo-5,5-dimethylhydantoin. Alkyl hydroperoxides can substitute for hydrogen peroxide in these reactions

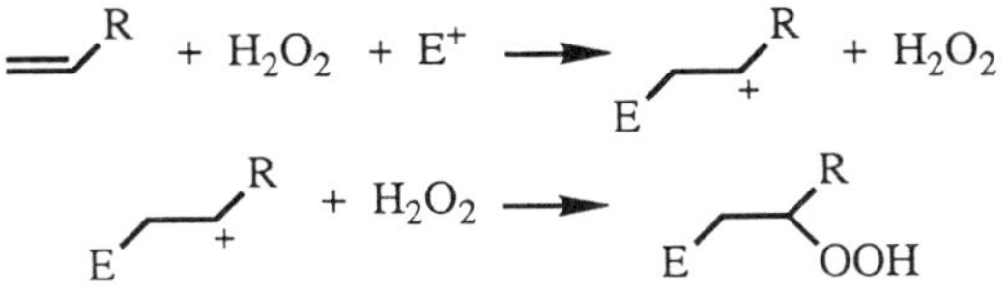

R + H_2O_2 + E^+ → E, R, + + H_2O_2 (1)

E, R, + + H_2O_2 → E, R, OOH (2)

R = alkyl, aryl, alkoxyl
E^+= H^+, HgX^+, Br^+, Cl^+..

Scheme 1

and dialkyl peroxides result from this reagent substitution. Table 3 shows several examples of electrophile-catalyzed addition of hydrogen peroxide or alkyl hydroperoxide to alkenes. Note that dialkyl peroxides are sometimes formed in the peroxymercuration reaction, and cyclic peroxides may be formed in a one-step reaction from dienes, hydrogen peroxide and mercury (II) salts. Cyclopropanes may also serve as substrates for reactions of electrophiles and hydrogen peroxide. Thus, bicyclopentane reacts with *N*-bromosuccinimide and hydrogen peroxide to provide the stereoisomeric cyclopentane bromohydroperoxides[43] and an analogous addition occurs with substituted cyclopropanes.[44] The ring strain in these substrates presumably provides the driving force for the reaction. The addition of Br^+–H_2O_2 to alkenes has been established to be *anti* in those cases which have been examined in detail.

The conversion of alcohols or ethers to hydroperoxides is particularly efficient if the intermediate carbocation is stabilized by electron-donating groups. Hemiacetals or acetals, for example,[45,46] are readily converted into α-alkoxyhydroperoxides in the presence of acid and hydrogen peroxide. Epoxides can also be converted to β-hydroxyhydroperoxides by reaction with acid and hydrogen peroxide.[47]

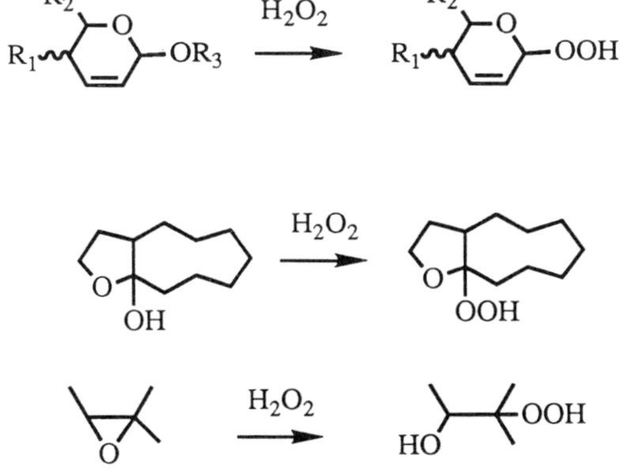

R_2, O, R_1, OR_3 —H_2O_2→ R_2, O, R_1, OOH

R_1= H, OAc
R_2= CH_2OAc
R_3= H, Me, Et

Ref. 45

O, OH —H_2O_2→ O, OOH *Ref. 46*

O —H_2O_2→ HO, OOH *Ref. 47*

Table 3. Alkyl hydroperoxides prepared from alkenes, electrophiles and hydrogen peroxide

Reactants	Peroxide	Reference
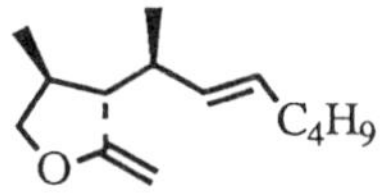 (structure, C_4H_9) + H_2O_2	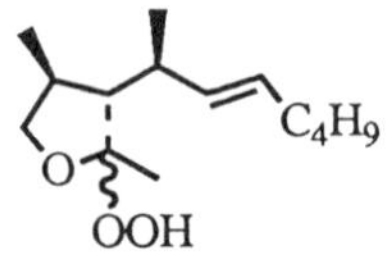 (structure, C_4H_9, OOH)	34
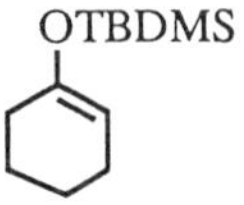 OTBDMS (structure) + H_2O_2	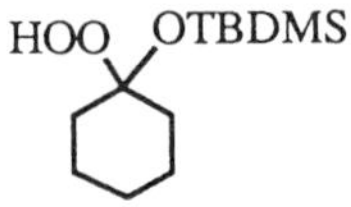 HOO OTBDMS (structure)	35
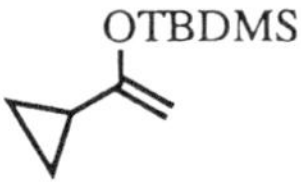 OTBDMS (structure) + H_2O_2	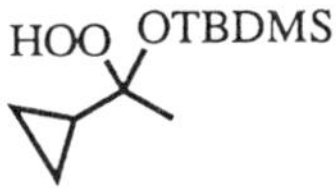 HOO OTBDMS (structure)	35
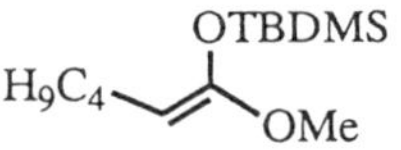 H_9C_4 OTBDMS OMe (structure) + H_2O_2	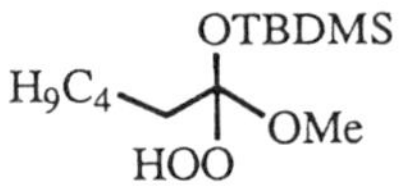 H_9C_4 OTBDMS OMe HOO (structure)	35
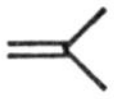 (structure) + H_2O_2 + $Hg(O_2CCF_3)_2$	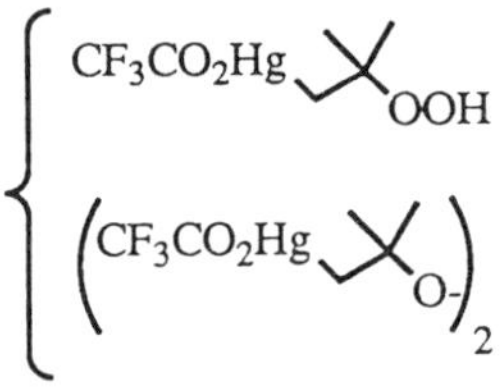 CF_3CO_2Hg OOH (structure); $(CF_3CO_2Hg$ O-$)_2$ (structure)	36
 (structure) + H_2O_2 + $Hg(OAc)_2$	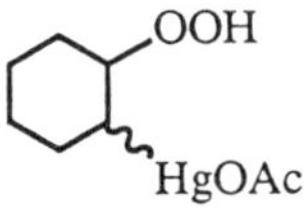 OOH HgOAc (structure)	37
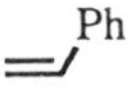 Ph (structure) + H_2O_2 + $Hg(O_2CCF_3)_2$	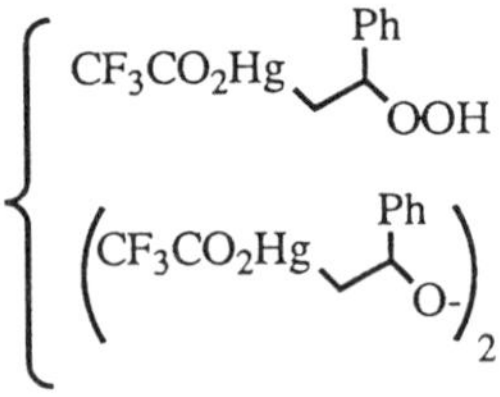 CF_3CO_2Hg Ph OOH (structure); $(CF_3CO_2Hg$ Ph O-$)_2$ (structure)	36
 (structure) + H_2O_2 + $Hg(O_2CCF_3)_2$	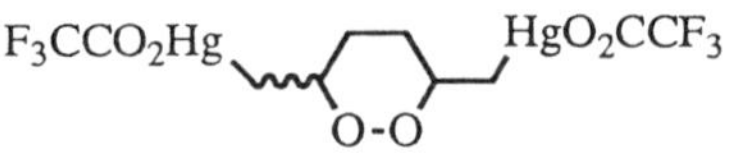 F_3CCO_2Hg O-O HgO_2CCF_3 (structure)	38

Table 3. (*continued*)

Reactants	Peroxide	Reference
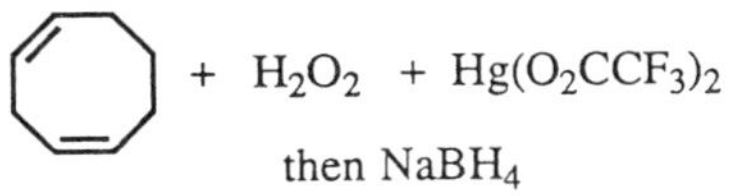 + H_2O_2 + $Hg(O_2CCF_3)_2$ then $NaBH_4$	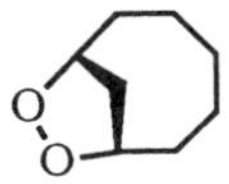 O O	39
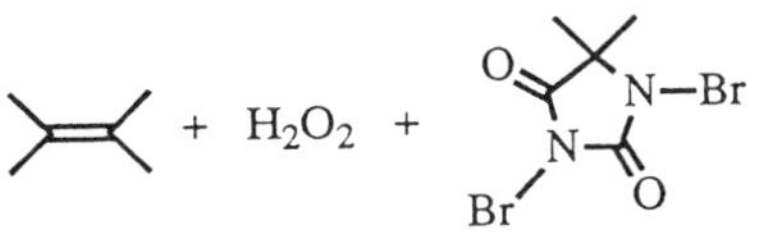 + H_2O_2 + O N–Br N Br O	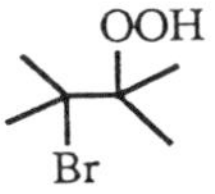 OOH Br	40
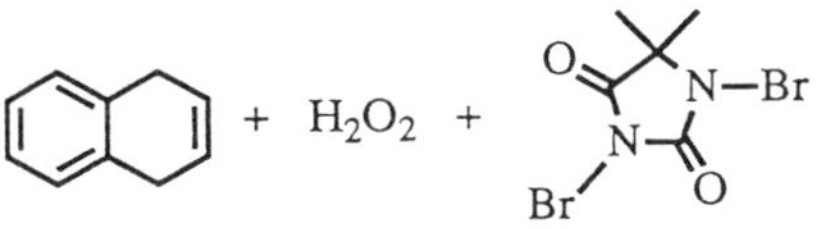 + H_2O_2 + O N–Br N Br O	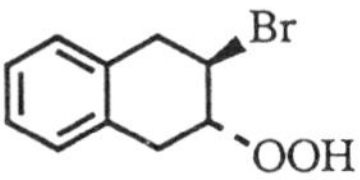 Br OOH	41
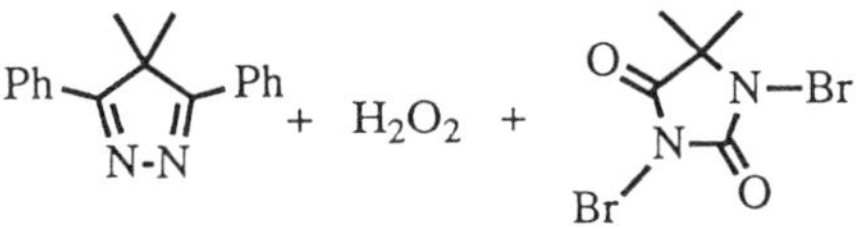 Ph Ph N-N + H_2O_2 + O N–Br N Br O	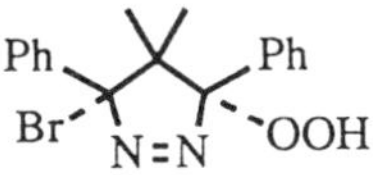 Ph Ph Br N=N OOH	42
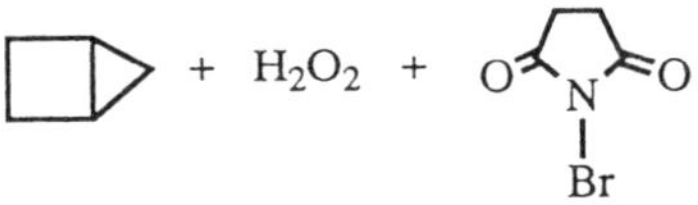 + H_2O_2 + O N O Br	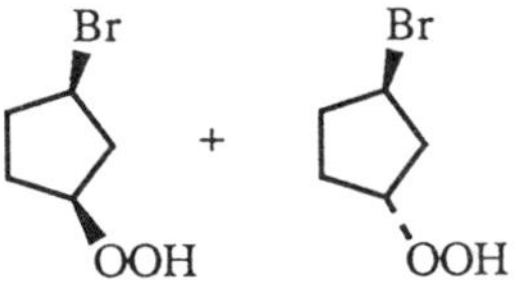 Br OOH + Br OOH	43
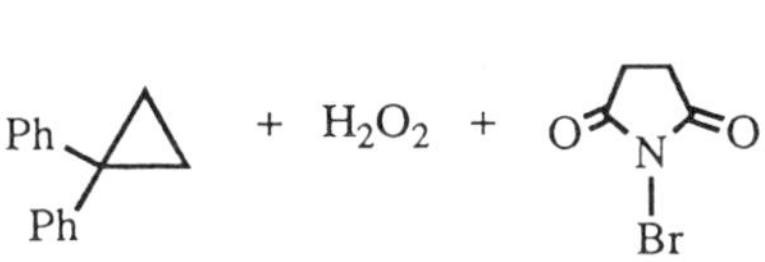 Ph Ph + H_2O_2 + O N O Br	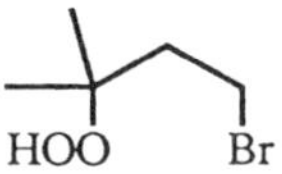 HOO Br	44

2 PREPARATION FROM SUPEROXIDE ANION

Synthetic applications of the superoxide anion, O_2^{-}, were not extensively reported before the late 1970s, when crown ethers were shown to solubilize superoxide alkali metal salts and the more soluble tetramethylammonium salt became available. In 1970, a notable synthetic application of electrochemically generated superoxide was made.[48,49] Superoxide was found to react with alkyl halides or sulfonates to form dialkyl peroxides or alkyl hydroperoxides. The

$$\text{R-Br} + O_2^{-} \longrightarrow \text{R-OO}\bullet + \text{Br}^{-} \quad (1)$$

$$\text{R-OO}\bullet + O_2^{-} \longrightarrow \text{R-OO}^{-} + O_2 \quad (2)$$

$$\text{R-OO}^{-} + H^{+} \longrightarrow \text{R-OOH} \quad (3)$$

$$\text{R-OO}^{-} + \text{R-Br} \longrightarrow \text{R-OO-R} + \text{Br}^{-} \quad (4)$$

$$\text{R-OO-R} + O_2^{-} \longrightarrow \text{R-O}^{-}\ \bullet\text{O-R} + O_2 \quad (5)$$

Scheme 2

scheme for this reaction, shown in Scheme 2, was proposed by Dietz *et al.*[48] and involves an initial S_N2 displacement of the alkyl bromide by superoxide followed by reduction of the intermediate peroxyl radical by superoxide and a subsequent S_N2 displacement by the peroxide anion to give the dialkyl peroxide. The yield of hydroperoxide is sometimes low compared with dialkyl peroxide and alcohol formed in the reaction.[50–52] For example, in dimethyl sulfoxide (DMSO) (20 ml), 1-octyl bromide (3.3 mmol) and potassium superoxide (10 mmol), 68% of the dialkyl peroxide is formed together with 20% of the alcohol, whereas a similar reaction in dimethylformamide (DMF) gives 30% of the dialkyl peroxide and 30% of the hydroperoxide. Dimethyl sulfone is observed in the reactions utilizing DMSO as solvent and reaction of the hydroperoxide anion with DMSO has been proposed to account for the formation of this product. This reaction also indicates that DMF may be the solvent of choice for the preparation of hydroperoxides.

Tosylates and mesylates can also be used as substrates for reaction with superoxide, octadecyl tosylate giving 11% ROOR, 40% ROOH and 20% ROH. Cyclic and secondary alkyl bromides and tosylates are susceptible to elimination in reactions with potassium superoxide. The carbon—oxygen bond-forming reaction at secondary carbon occurs with nearly complete inversion of configuration.

A synthesis of a hydroperoxide analog to the AIDS drug AZT was recently reported that utilizes the superoxide method in DMF as solvent.[53] Inversion of configuration was reported and the yield of the secondary cyclic hydroperoxide

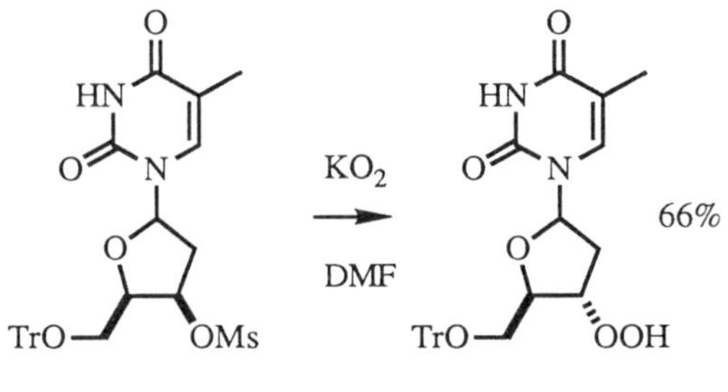

O HN O N O TrO OMs —KO_2, DMF→ O HN O N O TrO OOH 66%

was excellent. Conditions reported for the conversion were 3 equiv. KO_2 and 2.5 equiv. crown ether in DMF at 0°C for 30 min.

3 PREPARATION BY AUTOXIDATION

The spontaneous reaction of molecular oxygen with organic molecules is commonly referred to as autoxidation. This reaction proceeds via carbon free radicals which react by addition with molecular oxygen. Ground-state molecular oxygen is a triplet molecule having two unpaired electrons and the reaction of oxygen with carbon radicals occurs very rapidly, at or near the diffusion controlled limit.[54,56] Fragmentation of the peroxyl radical is the reverse of radical addition to molecular oxygen and this process is known to occur in several systems where there is a stabilized radical intermediate. Benson and Nangia[57] estimated that fragmentation of a benzyl peroxyl radical would be endothermic by about 14 kcal mol^{-1} and energy barriers of this magnitude allow this unimolecular process to be competitive with other bimolecular radical processes.

3.1 Mechanism of Hydrocarbon Autoxidation

The autoxidative process is commonly represented as consisting of chain initiation propagation and termination steps as shown in Scheme 3.

Initiation: $$\text{In}\bullet + \text{RH} \longrightarrow \text{R}\bullet + \text{InH} \quad (1)$$

Propagation: $$\text{R}\bullet + \text{O}_2 \longrightarrow \text{ROO}\bullet \quad (2)$$

$$\text{ROO}\bullet + \text{RH} \xrightarrow{k_p} \text{R}\bullet + \text{ROOH} \quad (3)$$

Termination: $$2\text{ROO}\bullet \longrightarrow [\text{ROOOOR}] \dashrightarrow \overset{\text{non radical}}{\text{products}}\ \text{O}_2 \quad (4)$$

Scheme 3

In the initial propagation step of autoxidation, molecular oxygen adds to R·. At partial oxygen pressures above 100 mmHg, this addition approaches the diffusion-controlled limit (*ca* $10^9\,l\,mol^{-1}\,s^{-1}$). This means that the major radical in solution is the peroxy radical ROO· and not R·. As a result, it is unlikely that any termination reactions involving R· will take place, provided that oxygen is present in sufficient concentration. In the second propagation reaction (equation 2 in Scheme 3), ROO· abstracts hydrogen atom from RH at rate k_p to generate more R· in what is the rate-determining step in the propagation sequence. For each hydroperoxide product formed, another radical R· is generated. This process could proceed *ad infinitum*, save for termination reactions that interrupt propagation.

The rate constant k_p for hydrogen atom abstraction depends primarily on the strength of the C—H bond being broken. Since ROO· is strongly resonance stabilized and is a comparatively unreactive radical, it is fairly selective in abstraction from hydrocarbons and prefers the most weakly bound hydrogen atom.[58] Products of autoxidation of hydrocarbons having many different types of carbon—hydrogen bonds can be understood based on the stability of intermediate radicals.

Termination of autoxidation reactions generally involves the coupling of two peroxyl radicals giving rise to unstable tetraoxides. Tertiary tetraoxides decompose to free radicals at low temperatures giving peroxyl radicals at temperatures above −80°C and alkoxyl radicals above −30°C. For this reason, the termination reactions of tertiary peroxyl radicals are inefficient. That is, the primary products of termination coupling are unstable and give rise to radicals that can continue the chain process. Secondary and primary peroxyl radicals terminate efficiently by a mechanism in which the tetraoxide decomposes to give molecular oxygen, an alcohol and a carbonyl compound. This mechanism is known as the Russell mechanism after Glenn Russell of Iowa State University, who discovered this sequence of reactions by an elegant series of kinetic and oxygen-labeling experiments.[59]

Russell termination:

$$2\ H_9C_4\text{-}OO\bullet \longrightarrow H_7C_3\text{-}CH(H)\text{-}O\text{-}O\text{-}O\text{-}O\text{-}C_4H_9 \longrightarrow H_7C_3\text{-}CH{=}O + HO\text{-}C_4H_9 + O_2$$

The reaction of carbon radicals with molecular oxygen occurs at or near the diffusion-controlled limit while the hydrogen atom transfer step occurs much more slowly. Thus, at or near atmospheric oxygen pressures, the hydrogen atom transfer step is rate determining. This fact, along with the knowledge that in reactions having long chains the rate of initiation must be equal to the rate of termination, gives the rate expression

$$-\mathrm{d}O_2/\mathrm{d}t = k_p[\mathrm{R{-}H}]R_i^{1/2}/2k_t^{1/2}$$

In this equation, the rate of oxygen (or R—H) consumption is shown to be proportional to the rate constant k_p and the concentration of the substrate (R—H), proportional to the square root of the rate of radical initiation (R_i) and inversely proportional to the square root of twice the termination rate, k_t.[60–62]

The fact that the rate of autoxidation is dependent on the concentration of substrate is frequently overlooked, and this fact has important practical consequences. Autoxidation of any organic substrates is concentration dependent and even reactive compounds do not autoxidize readily at millimolar concentrations or below. The substrate should therefore be present in high concentrations in order to promote autoxidation. The rate of initiation is, of course, also critical in promoting autoxidation.

Experimentally, it is possible to measure the rates of oxygen consumption and radical initiation at a known substrate concentration and this gives the quantity known as the substrate oxidizability, $k_p/2k_t^{1/2}$. The termination rate constant k_t can also be determined by absolute methods and in this way the rate constant for propagation can be calculated.

3.2 Hydroperoxide Products from Autoxidation

Products of autoxidation of representive molecules are shown in Table 4. For low molecular weight alkanes with activating groups such as phenyl or ether oxygens, hydroperoxide products can be extracted into aqueous base and purified by reacidification and extraction with organic solvents. Autoxidation occurs with substrates that have allylic OH or N—H bonds. Thus, enols or phenols in addition to hydrazones and imines undergo autoxidation readily. Enolates and phenolates also undergo autoxidation and representative conversions of these substrates are presented in Table 4.

The mechanism of autoxidation of simple alkenes, dienes and polyenes is complex. A reasonable mechanism for the formation of the products derived from simple alkenes involves hydrogen atom abstraction at both allylic positions of the chain to give the resonance-stabilized allylic radicals as shown in Scheme 4. These allylic radicals exist as stereochemical geometric isomers and have partial double bond character between the carbons in the allylic system,

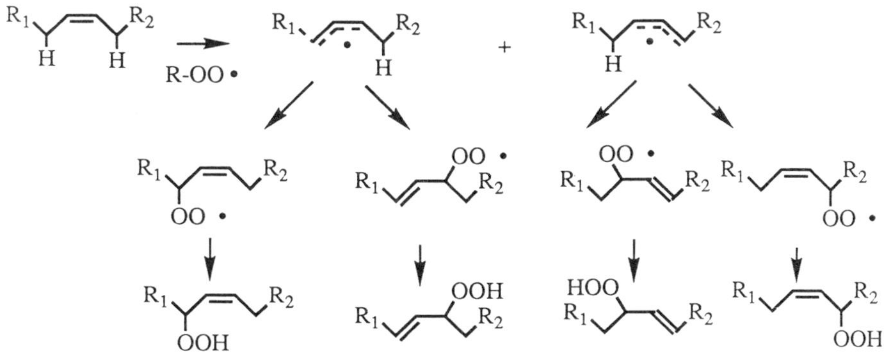

Scheme 4

Table 4. Alkyl hydroperoxides prepared by autoxidation

Reactants	Peroxide	Reference
	OOH	63
	OOH	64
O	HOO, O	65
O, O, H	O, O, OOH	66
	OOH	67
R_1, R_2	R_1, R_2, OOH; R_1, R_2, OOH; HOO, R_1, R_2; OOH, R_1, R_2	68, 69
R_1, R_2	HOO, R_1, R_2; HOO, R_1, R_2; OOH, R_2, R_1; OOH, R_1, R_2	70–73

Table 4. (*continued*)

Reactants	Peroxide	Reference
	HOO	74
O, N	HOO, N, O	75
O, HN, R_1, R_2	O, HN, R_1, HOO, R_2	76
OH, CN	HOO, O, CN	77
OH	HOO, O	78
OH	O, OOH	79
OH, *t*-Bu, *t*-Bu, Ar, base catalyzed	O, *t*-Bu, *t*-Bu, Ar, OOH + O, *t*-Bu, *t*-Bu, OOH, Ar	80

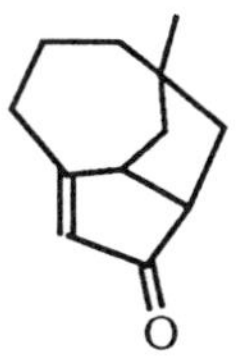

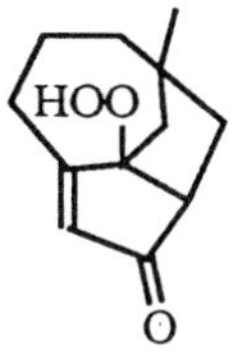

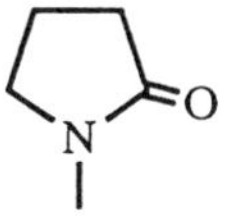

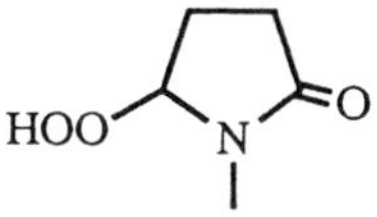

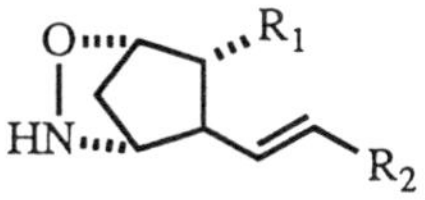

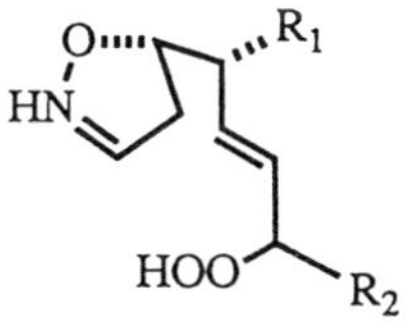

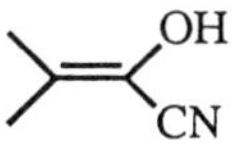

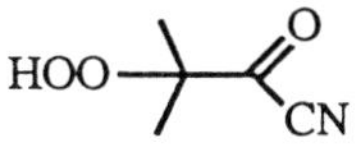

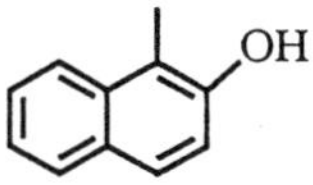

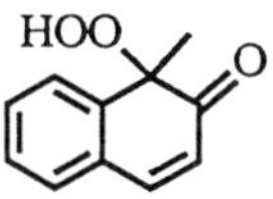

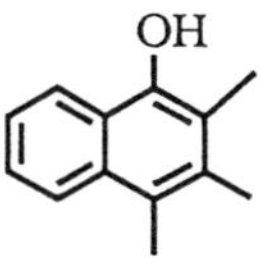

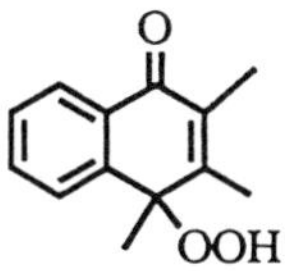

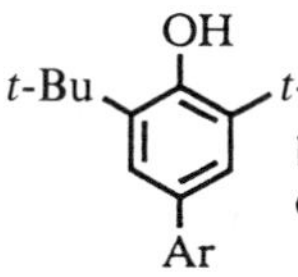

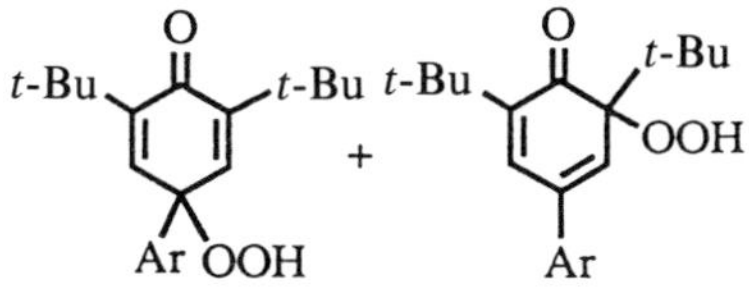

and these radicals can exist as *trans–trans*, *trans–cis*, *cis–trans* and *cis–cis* geometric isomers. It seems most reasonable to write the structure of intermediate radicals having the *cis* configuration at what was formerly the *cis* double bond of oleate and having the more stable *trans* configuration at the other partial double bond. Addition of oxygen to the allylic radicals would give rise to four different radicals formed from addition at either end of the two radicals and the major hydroperoxide products from autoxidation of simple alkenes derive from hydrogen atom transfer to these four major peroxyl radicals.

While monoalkenes only undergo autoxidation at reasonable rates at elevated temperatures, homoconjugated dienes autoxidize readily at room temperature. The major products formed from autoxidation of homoconjugated dienes, $RCH{=}CHCH_2CH{=}CHR$, such as linoleic acid, at room temperature and under atmospheric oxygen are hydroperoxides formed by oxygen addition at the end carbons of the diene unit.[70–73] The products formed have two different stereochemistries of the conjugated diene unit present: one of the products has *trans–cis* and the other *trans–trans* diene stereochemistry. Other minor products have been isolated from autoxidation of linoleate. Non-conjugated hydroperoxides have been isolated but these are formed in very small amounts. While the major products all result from abstraction of an H atom from the carbon linking the two alkenes, the minor products result from abstraction of H atoms that are only singly allylic.[81] At higher temperatures or with low oxygen concentrations present, a more complex product mixture that includes carbon radical dimers is formed.[82]

The mechanism for the formation of the four major linoleate products has been discussed in detail elsewhere[83,84] and will be described here in outline form. Briefly, the mechanism proceeds by H-atom removal from the center carbon of the system to generate a pentadienyl radical with stereochemistry as shown in Scheme 5. The geometry of the pentadienyl radical is written as *cis–trans–trans–cis*. The second step in autoxidation is addition of oxygen to the intermediate radical. In other pentadienyl systems, products are observed from reaction of oxygen at the center carbon of the pentadienyl system but in those systems the terminally substituted hydroperoxides are sterically crowded.[85]

To simplify the discussion, consider only oxygen coupling at one end of the radical (recognizing that similar chemistry occurs at the other end), which gives a peroxyl radical with *trans–cis* conjugated diene geometry. Entrapment of this radical before β-fragmentation leads to the kinetic product, the *trans–cis* compound. If good H-atom donors are not present (R–H or antioxidants, such as tocopherol), then β-fragmentation occurs and oxygen coupling at the opposite end of the carbon radical leads ultimately to the thermodynamically more stable *trans–trans* product. The competition between H-atom transfer to give kinetic product and β-fragmentation to give thermodynamic product thus accounts for the four major products in this autoxidation.

The mechanism outlined in Scheme 5 is consistent with the experimental

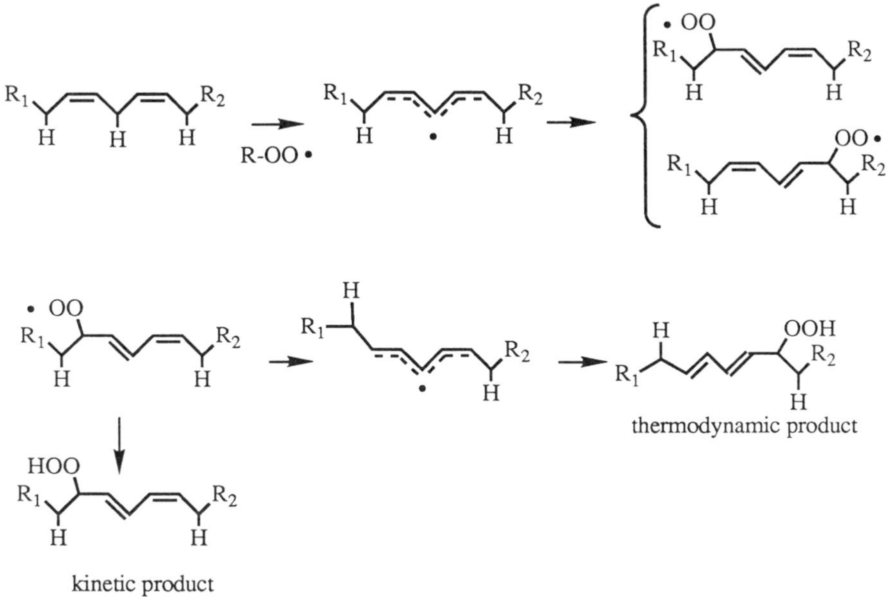

kinetic product

thermodynamic product

Scheme 5

observation that high concentrations of diene lead to more *trans–cis* products. This is because at higher [R—H], the reaction is driven toward kinetic control, since the diene is an H-atom donor. Second, addition of other good H-atom donors, such as 1,4-cyclohexadiene[84] and tocopherol,[86,87] causes the product mixture to consist of primarily *trans–cis* products (kinetic control).

The product mixtures obtained in polyene autoxidation may be extremely complex. Cyclic peroxide products have been suggested[88] to be important components of such autoxidation mixtures and a reasonable understanding of

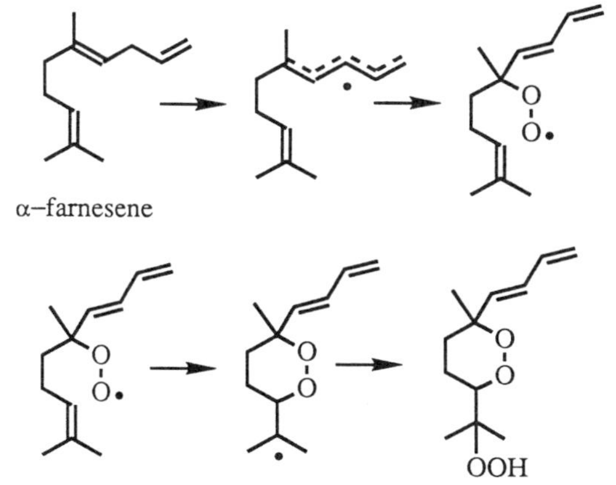

α–farnesene

peroxyl radical cyclization has emerged. The autoxidation of α-farnesene was one of the earliest studies to conclude that cyclic peroxides were formed in polyene autoxidation.[89]

Cyclic peroxides are also obtained in the autoxidation of the diene **12**, the compounds **13**, **14** and **15** being isolated from the reaction.[90] The alcohol **14** and the ketone **15** apparently result from termination reactions of intermediate peroxyl radicals while the hydroperoxide is the major product of autoxidation at 60°C initiated by an azo compound.

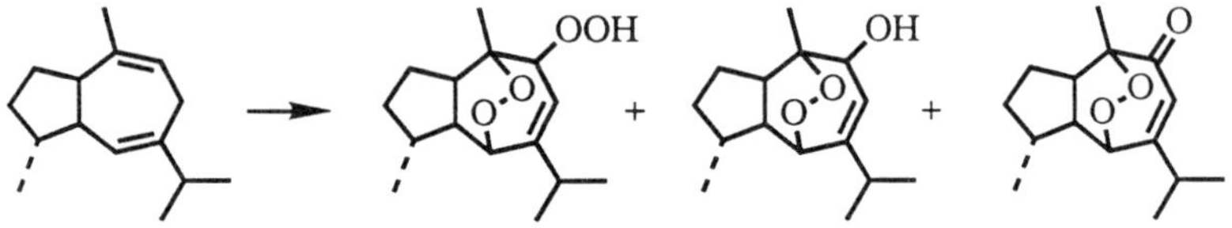

OOH OH O

O O O O O O

+ +

12 **13** **14** **15**

Cyclic peroxide products have been isolated in the autoxidation of arachidonate and other triene or tetraene fatty acids. To simplify the product mixture of cyclic peroxide products, specific hydroperoxides have been used as precursors to particular peroxyl radicals and this chemistry will be discussed in detail in Section 7.1.

3.3 Hydroperoxide Products by Decarboxylation or Decarbonylation Reactions

Esters of *N*-hydroxypyridine-2-thione are very convenient sources of carbon radicals[91] and these radical precursors have been used as a means for hydroperoxide preparation.[92,93] Thus, esters such as **16** react with *tert*-butanethiol under oxygen to give hydroperoxide products with yields in excess of 70%. The mechanism presented in Scheme 6 has been proposed for the conversions.

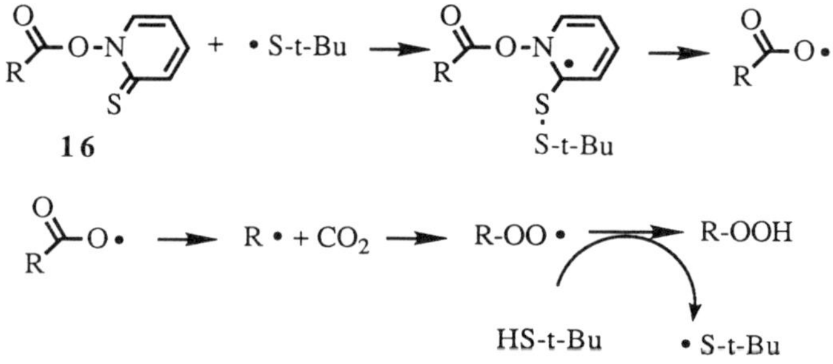

O O O

R O-N S + • S-t-Bu → R O-N S S-t-Bu → R O•

16

O

R O• → R • + CO_2 → R-OO • → R-OOH

HS-t-Bu • S-t-Bu

Scheme 6

Autoxidation of aldehydes usually leads to the corresponding carboxylic acids. Deformylation of aldehydes to give hydrocarbons via a radical process involving the loss of carbon monoxide is also well known and there have been some examples reported of replacement of tertiary aldehyde groups with hydroperoxide groups by a presumed decarbonylation–oxygen addition sequence.[94,95]

4 PREPARATION BY REACTION WITH SINGLET MOLECULAR OXYGEN

The reaction of molecular oxygen with organic molecules can be stimulated by photolysis of the reactants in the presence of triplet sensitizers such as rose bengal, methylene blue or hematoporphyrins. The reactive intermediate in these reactions is singlet oxygen, $^1\Delta_g$, and the reaction of this species with alkyl-substituted alkenes leads to allylic hydroperoxides. This reaction, which is an ene-type reaction, invariably results in migration of a double bond together with formation of the hydroperoxide. Other possible pathways that may compete with the ene reaction are (4 + 2) and (2 + 2) cycloaddition chemistry that gives rise to cyclic peroxide products. The mechanism of the ene reaction has been the focus of much debate and proposals that suggest concerted mechanisms in addition to biradical, zwitterionic or perepoxide intermediates have been made.[96,97] Consideration of the details of the mechanism of this reaction is beyond the scope of this chapter. Representative applications of the singlet oxygen ene reaction that give rise to allylic hydroperoxides are presented in Table 5.

H + $^1O_2$ → OOH

5 PREPARATION BY OZONOLYSIS

The reaction of alkenes with ozone leads to carbonyl oxide intermediates (Criegee zwitterions) that can be trapped with alcohols to give alkoxyhydroperoxides. The preparation of hydroperoxides by ozonization of alkenes is the focus of Chapter 13.

R, R —O_3→ R, O, O• ⟷ R, +, =O, O– —R'OH→ OOH, R, OR'

Table 5. Alkyl hydroperoxides prepared from alkenes by reaction with singlet oxygen

Reactants	Peroxide	Reference
	 OOH	98
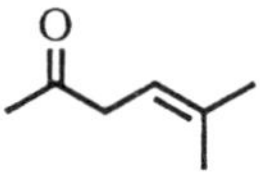 O	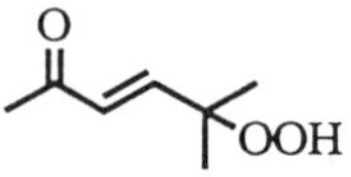 O OOH	99
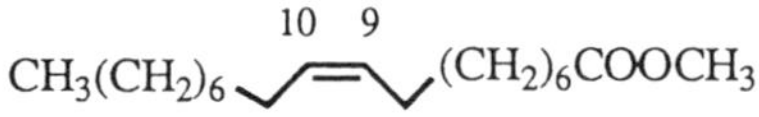 10 9 $CH_3(CH_2)_6$ (CH$_2$)$_6$COOCH$_3$	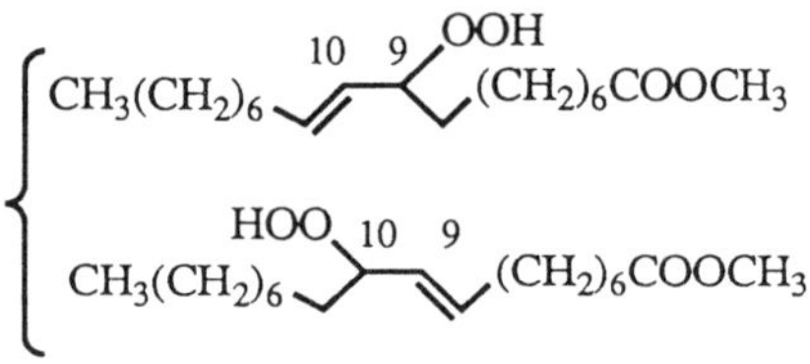 10 9 OOH $CH_3(CH_2)_6$ $(CH_2)_6COOCH_3$ HOO 10 9 $CH_3(CH_2)_6$ $(CH_2)_6COOCH_3$	63, 68, 100–103
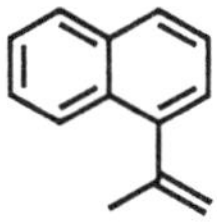	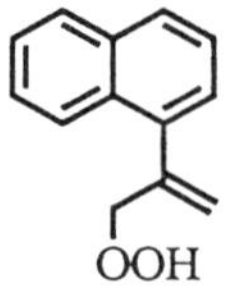 OOH	104
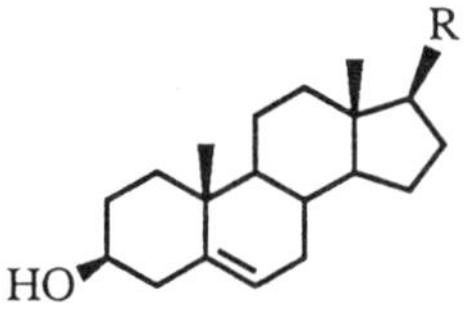 R HO	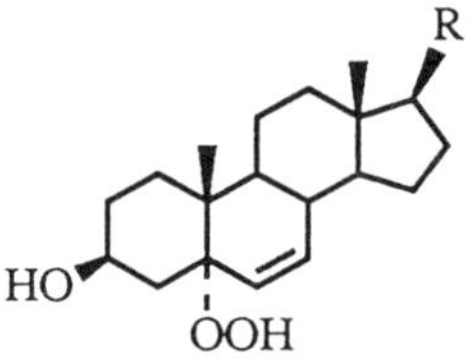 R HO OOH	105–107
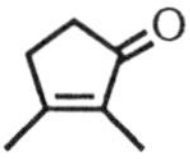 O	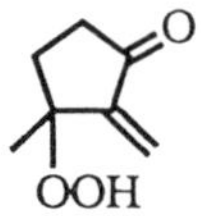 O OOH	108
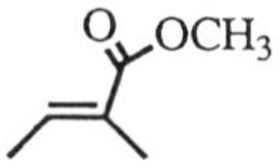 O OCH$_3$	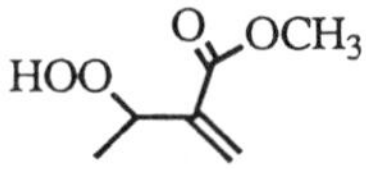 O OCH$_3$ HOO	109
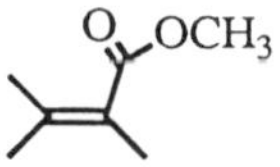 O OCH$_3$	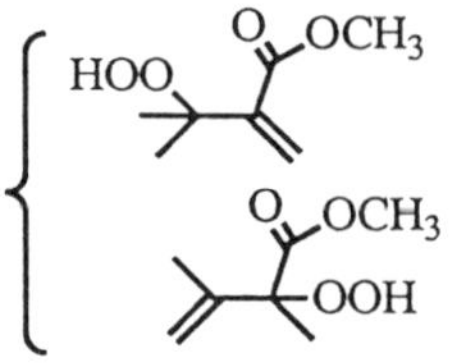 O OCH$_3$ HOO (88) O OCH$_3$ OOH (12)	109

Table 5. (*continued*)

Reactants	Peroxide	Reference
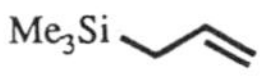 Me_3Si (allyl)	1 : 4 { Me_3Si…OOH ; Me_3Si…OOH } 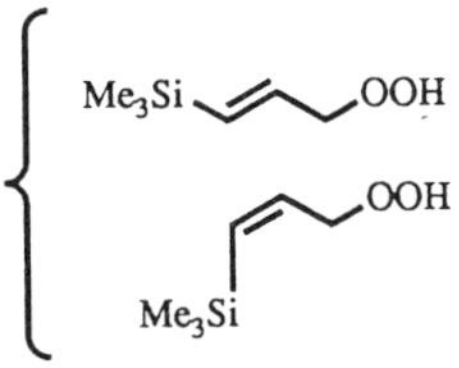	110
	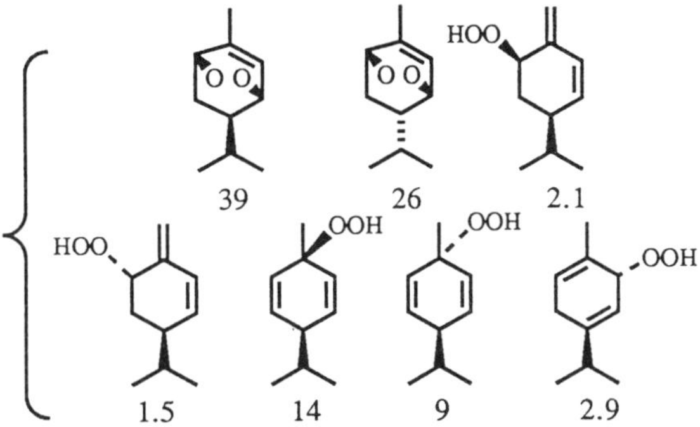 39 ; 26 ; 2.1 ; 1.5 ; 14 ; 9 ; 2.9 (HOO, OOH)	111, 112
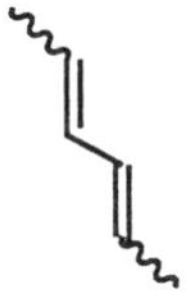	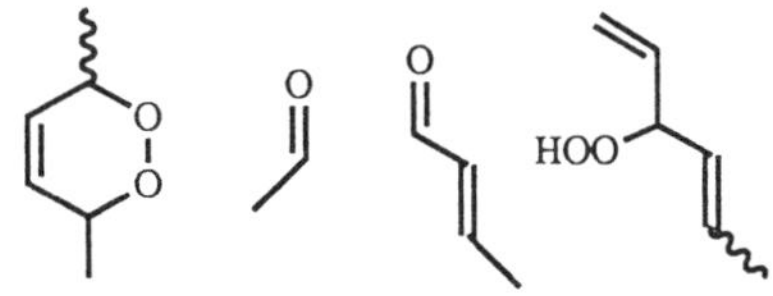 O–O ; O ; O ; HOO	113
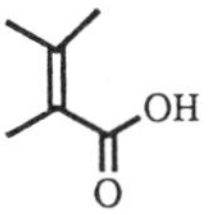 OH, O	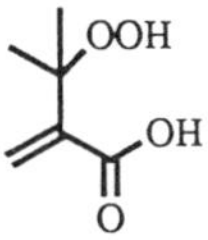 OOH, OH, O	114
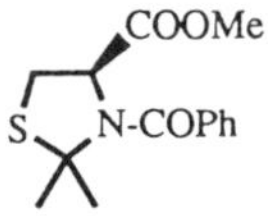 COOMe, S, N-COPh	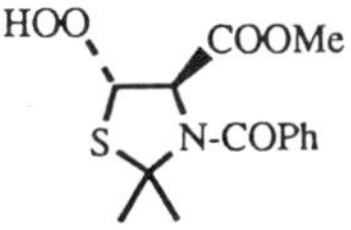 HOO, COOMe, S, N-COPh	115, 116

6 PREPARATION FROM PEROXIDE PRECURSORS

Peroxide derivatives, R—OO—X, in which one substituent attached to the peroxide functional group is an alkyl group and the other substituent is modified such that it is removable under mild conditions, may serve as hydroperoxide progenitors. Among the X groups that have been used as hydroperoxide precursors are hydroperoxide addition compounds to aldehydes, ketones and amides. Thus, peracetals, perketals and peraminals can be converted into the hydroperoxide by a mild acid catalyst. The preparation of such hydroperoxide derivatives was pioneered by Rieche and co-workers in the 1950s and 1960s, although the reconversion of such derivatives to hydroperoxides was not examined in great detail.[117–121]

$$R{-}O{-}O{-}X \longrightarrow R{-}O{-}O{-}H$$

X= C(OR')R''$_2$, SiR$_3$, SnR$_3$

Hydroperoxide derivatives can be prepared from acetals, ketals, or vinyl ethers as shown in Scheme 7, and peracetals ROOCR′HOR″, perorthoesters ROOC(OR′)$_3$, peraminals ROOCR′NR$_2$″OR″ and perketals ROOCR$_2$OR″ have been made by this approach. Perorthoesters and peraminals are unstable to most chromatographic conditions whereas peracetals and perketals are reasonably stable to silica gel and normal- or reversed-phase solvents.

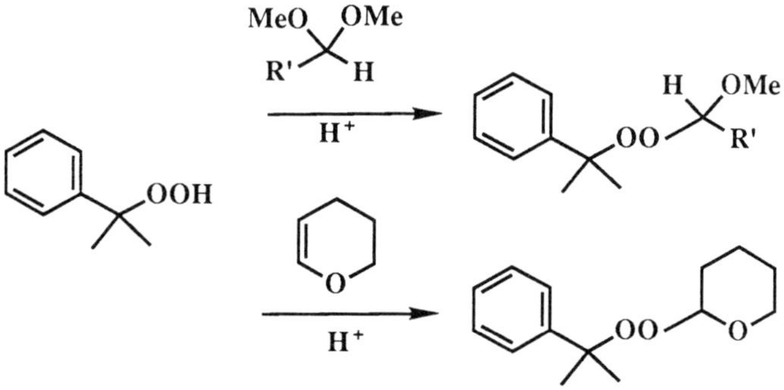

MeO OMe
R' H
H⁺
OOH
OO
H OMe
R'
O
H⁺
OO
O

Scheme 7

Generally, sulfonic acids are necessary as catalysts for the conversion of peracetals into hydroperoxides and although some hydroperoxides tolerate these conditions of deprotection, sensitive hydroperoxides, such as those with unsaturation present, give decomposition under these harsh conditions of deprotection. Perketals can be deprotected under very mild conditions (THF–water–acetic acid).

The hydroperoxide protection–deprotection sequence involving peracetals and perketals has been used as a means of resolution of racemic hydroperoxides.[63,122] The strategy is to derivatize a racemic hydroperoxide with a homochiral peracetal or perketal, separate the resulting diastereomeric compounds by chromatography and regenerate the optically pure hydroperoxides by hydrolysis of the hydroperoxide derivative. While several hydroperoxide derivatives were examined in preliminary experiments, more detailed studies were carried out with the peracetal **17** and the perketals **18** and **19**. These detailed studies involved assessment of chromatographic characteristics of the derivatives from assorted hydroperoxides, determination of conditions for deprotection and evaluation of hydroperoxide stability under deprotection conditions.

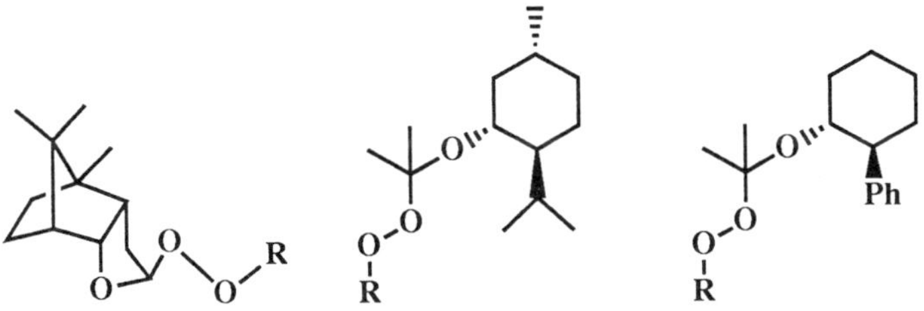

17 **18** **19**

Perketals **18** and **19** were readily formed, were stable to chromatography under appropriate conditions and the perketal protecting group was removed without the need to use strong acid catalysts. Although perketals are completely stable to normal-phase chromatography, some decomposition occurs in reversed-phase solvents, such as acetonitrile–water or methanol–water unless 0.01% triethylamine is added. Diastereomers could be isolated with isomeric purities of 99% or better from perketals **18** and **19** derived from α-phenethyl hydroperoxide. The chromatographic resolution of the diastereomers of a phenethyl hydroperoxide was more efficient as the perketal **19** than as **18**. The phenylcyclohexanol derivative was therefore chosen for a broader screen of hydroperoxide resolution. It should be pointed out that the exploration of perketal structure as it relates to resolution has not been exhaustive, since it was possible to resolve every hydroperoxide examined as the perketal **19**. In Figure 1 are shown a few of the hydroperoxides that have been resolved by this method.

Perketal hydroperoxide protecting groups have been used in organic synthesis. Dussault and Sahli[123] protected allylic hydroperoxides as perketals and followed this by ozonolysis of the alkene to give protected hydroperoxyaldehydes (**20**). Horner–Emmons olefination or Wittig chemistry with stabilized ylides and **20** subsequently provided new allylic peroxides, which were deprotected to afford the allylic hydroperoxides. This new approach to hydroperoxide synthesis may find a wide variety of applications.

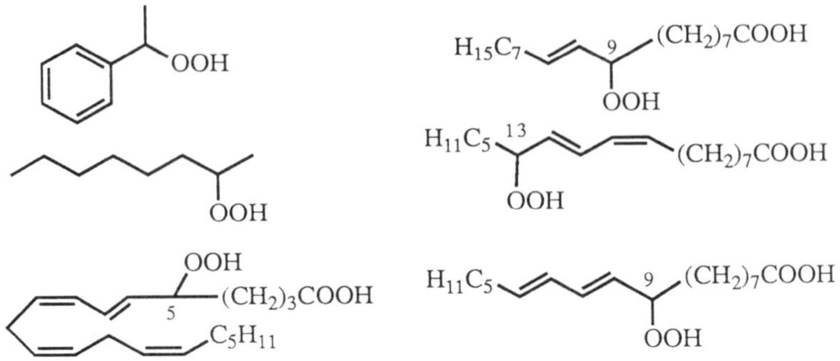

Figure 1. Hydroperoxides resolved as perketals

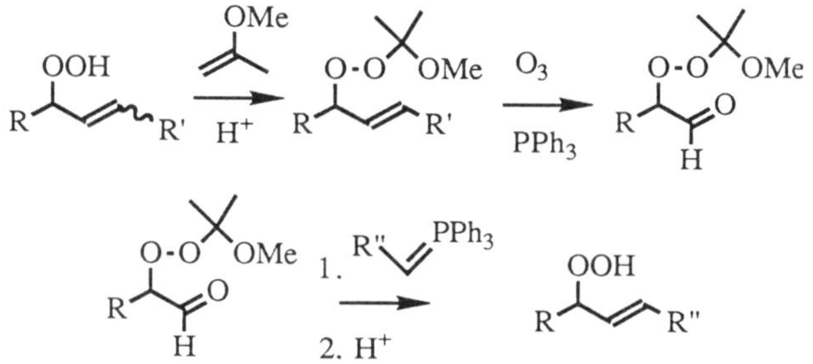

Other workers have utilized perketals as peroxide-protecting groups. Thus, synthesis of a diacylglycerophosphocholine hydroperoxide (**21**) has been reported[124] that utilizes the same perketal reported by Dussault and Sahli[123] as a protecting group. The perketal is stable to the conditions required for hydrolysis of an ester and DCC coupling of an acid and an alcohol.

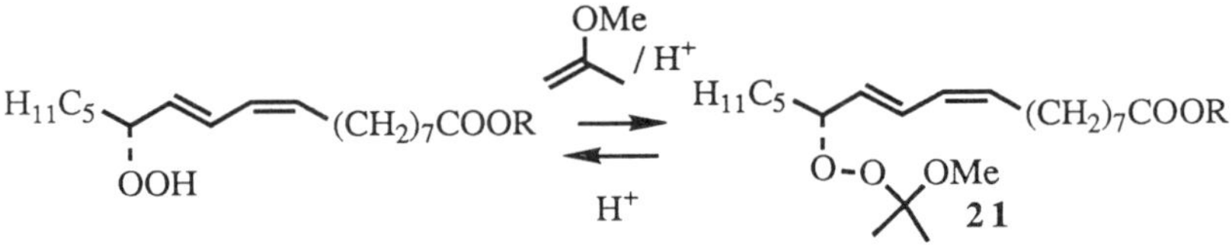

Hydroperoxides can be prepared from the corresponding silyl peroxides by reaction of the silyl peroxide with methanol. The silyl peroxides can be prepared from the corresponding trialkylsilyl chloride and the hydroperoxide[125] or by singlet oxygenation of silylenol ethers and disilylketene acetals.[126–129] Examples of such conversions involving silyl peroxides are shown.

$$\text{OSiMe}_3 \xrightarrow{^1O_2} \text{OOSiMe}_3,\ \text{O} \xrightarrow{\text{MeOH}} \text{OOH},\ \text{O}$$

$$\text{Me}_3\text{SiO},\ \text{OSiMe}_3 \xrightarrow{^1O_2} \text{Me}_3\text{SiO},\ \text{O},\ \text{OOSiMe}_3 \xrightarrow{\text{MeOH}} \text{HO},\ \text{O},\ \text{OOH}$$

Recent studies by Haynes and Vonwiller[130] have shown that stannyl peroxides are versatile compounds providing direct access to a variety of peroxide derivatives and serving as protecting groups for hydroperoxides. Treatment of hydroperoxides with tributyltin methoxide in diethyl ether quantitatively converts the hydroperoxide into stannyl peroxides. Such stannyl peroxides are unstable but they react with trityl chloride to give the trityl peroxide (67%) and chloromethyl methyl ether to give the methoxymethyl peroxide (80%). *tert*-Butyldimethylsilyl peroxides may be prepared from the stannyl peroxide by reaction with *tert*-butyldimethylsilyl triflate. Primary and secondary hydroperoxides can be stannylated and the product stannyl peroxides converted in good yield into the derivatives described above, whereas tertiary peroxides can be converted to stannyl peroxides but in some cases, particularly allylic systems, these peroxides undergo rearrangement reactions.

$$\text{ROOH} + \text{Bu}_3\text{Sn-OMe} \longrightarrow \text{ROO-SnBu}_3 \longrightarrow \begin{cases} \text{ROO-CPh}_3 \\ \text{ROO-Si(Me)}_2\text{t-Bu} \\ \text{ROO-CH}_2\text{OCH}_3 \end{cases}$$

7 RADICAL DECOMPOSITION OF HYDROPEROXIDES—FORMATION OF PEROXYL RADICALS

7.1 Peroxyl Radical Cyclization

The hydroperoxide O—H bond is relatively weak and reactions of hydroperoxides with oxyl radicals lead to the formation of the peroxyl radical. Thus, Howard and Ingold[131] and Walling and Padwa[132] first showed that hydroperoxides compete efficiently with hydrocarbons for *tert*-butoxyl radicals. On a per hydrogen basis *tert*-butoxyl radicals are over 1300 times more reactive towards hydroperoxide hydrogens than with cyclohexane hydrogens and it is therefore possible to generate peroxyls from hydroperoxide precursors.

$$R{-}O{-}OH + R'{-}O\bullet \longrightarrow R{-}O{-}O\bullet + R'{-}OH$$

Transfer reactions of hydroperoxide hydrogens between peroxyl radicals are also relatively rapid. For example, transfer of the hydrogen of cumyl hydroperoxide to a tetralin-1-peroxyl radical has been reported[133] to have a rate constant of $2.5 \times 10^3\ l\,mol^{-1}s^{-1}$ whereas abstraction of hydrogens from hydrocarbons by peroxyls generally occurs with much lower rate constants. The fact that hydroperoxide hydrogens are readily abstracted provides a strategy for the generation of specific peroxyl radicals from the corresponding hydroperoxide precursors and the chemistry of specific peroxyls may be investigated without the need to proceed from a hydrocarbon precursor.

$$R{-}O{-}OH + R'{-}O{-}O\bullet \longrightarrow R{-}O{-}O\bullet + R'{-}O{-}O{-}H$$

Porter and co-workers[134–137] first utilized this strategy to investigate systematically the cyclization reactions of peroxyl radicals. Unsaturated primary hydroperoxides were reacted under oxygen with radical initiators [di-*tert*-butyl peroxyoxalate (DBPO) or di-*tert*-butyl hyponitrite (DTBN)] that serve as a source of *tert*-butoxyl radicals, and radical cyclization reactions were observed from the peroxyls generated. As an example, the hydroperoxide **22** (0.01 M) was reacted in benzene under oxygen at 30°C with DBPO (0.001 M) and the cyclized peroxide **23** was obtained in nearly pure form by high-performance liquid chromatography. *Threo* and *erythro* diastereomers result from addition of oxygen at both faces of the prochiral radical formed in the cyclization. Hydroperoxides such as **23** tend to be unstable and it is frequently convenient to reduce the hydroperoxide to the corresponding stable alcohol (i.e. **24**) for ease of isolation. Several hydroperoxides were examined by this method and there is a strong tendency for peroxyl radicals to cyclize in an *exo* mode. Even in cases where there is a thermodynamic bias for *endo* cyclization, the *exo* mode is preferred, **25–26**.

Other strategies have been developed for initiation of sequences involving peroxyl radical cyclization. Thus, Beckwith and co-workers[138–141] pioneered the use of a thiol–oxygen–co-oxidation (TOCO) sequence in systems where peroxyl cyclization can intervene in the chain sequence. This scheme, shown here for penta-1,4-diene, involves thiol radical addition, oxygen entrapment, radical cyclization, oxygen entrapment and hydrogen atom transfer. Peroxyl radical cyclization leads preferentially to the *cis*-substituted dioxolane.

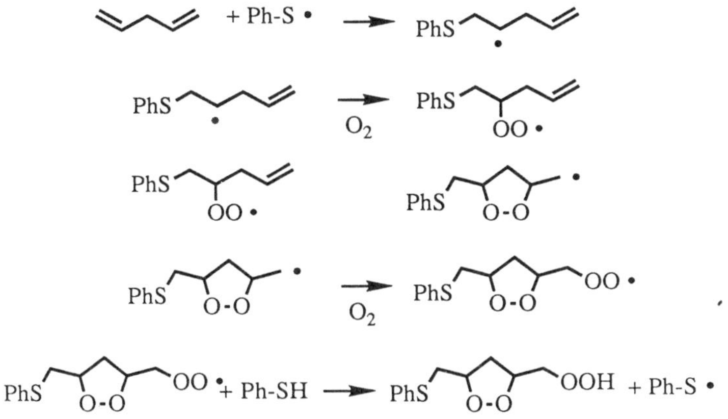

A modification of the TOCO sequence has been reported by Feldman and co-workers[142,143] that incorporates a cyclopropylcarbinyl–homoallyl rearrangement into the TOCO sequence. Thus, diphenyl disulfide–azobisisobutyronitrile (AIBN) and oxygen convert the vinyl cyclopropane **27** to the dioxolane **28**.

27 **28**

Hydroperoxides have also been used as precursors to peroxyl radicals that lead to polycyclic peroxide compounds.[144–146] Thus, the hydroperoxide **29** is converted into the polyperoxide compound **30** on reaction with DBPO–oxygen. The mechanism for the sequence is similar to that described above for **22** and there is stereoselectivity in the 6-*exo* cyclization that gives preference to the *trans*-substituted 1,2-dioxane. Such polycyclization sequences are apparently important in the degradation of rubber, polybutadiene and isoprenoid natural products such as squalene, ubiquinones and other prenyl derivatives.[147–152] Further, the role of the autoxidation of lipids and isoprenoids in the development of flavors and aromas has been well established.[153] Such polycyclic peroxides may therefore be key intermediates in the degradation of commercially and biologically important polyalkenes.

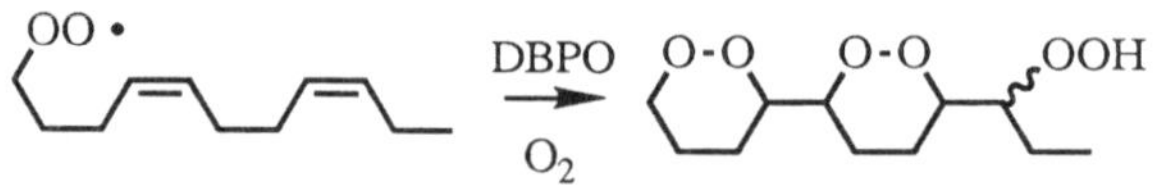

29 **30**

Cyclic peroxide products have also been isolated in the autoxidation of arachidonate and other triene or tetraene fatty acids. To simplify the product mixture of cyclic peroxide products, specific hydroperoxides have been used as precursors to particular peroxyl radicals, much like the simple systems described earlier.

For autoxidation of arachidonic acid, there are three centers at C-7, C-10 and C-13 that have carbons flanked by two double bonds. These carbons are analogous to C-11 of linoleate and are readily attacked by chain-carrying peroxyl radicals. Thus, three pentadienyl radicals are formed as intermediates in arachidonate autoxidation and these three pentadienyl radicals can give rise to six different peroxyl radicals by addition of oxygen at the ends of the carbon radical. For example, addition of oxygen to both ends of the radical shown leads to the 11- and 15-peroxyl radicals. These same peroxyls can be independently

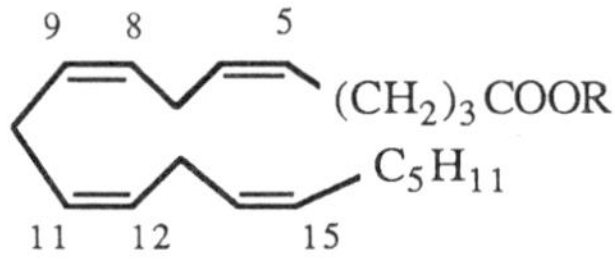

Arachidonic Acid

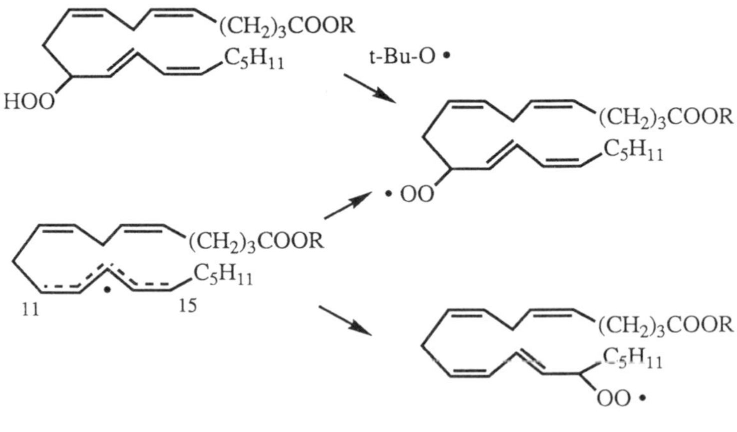

generated by reaction of the corresponding hydroperoxide with *tert*-butoxyl radicals.

Of the two radicals shown above, only the radical at C-11 has the possibility of cyclization. The C-15-substituted peroxyl radical is flanked by only the conjugated diene unsaturation and the proximal *trans* double bond separates the peroxyl radical from potential cyclization sites of unsaturation. On the other hand, the C-11 peroxyl radical has an isolated double bond positioned such that a favorable 5-*exo* cyclization can occur and products are observed that result from this relatively facile cyclization ($k_c \approx 10^3\ s^{-1}$). Four of the six peroxyl radicals have isolated double bonds placed such that a 5-*exo* cyclization can occur. These four peroxyl radicals have oxygen substitution at positions C-8, C-9, C-11 and C-12 on the carbon chain. The peroxyl radicals substituted at positions C-5 and C-15 have no such double bond and no cyclization products are observed from these radicals.

Since four different peroxyl radicals can give cyclic products and since multiple products are possible after cyclization (see below), methods have been developed to study the cyclization reactions of specific peroxyl radicals. Thus, hydroperoxides serve as precursors to specific peroxyl radicals and the chemistry of peroxyl radicals generated in this manner is easier to study than is the case when random autoxidation is used. The hydroperoxide from γ-linolenate can be formed by the enzyme soybean lipoxygenase and on reaction with free-radical initiators several cyclic peroxide products are observed. Thus, for example, O'Connor *et al.*[154] investigated the cyclization reactions of the 9-hydroperoxide from γ-linolenate and found several cyclic peroxide compounds.

Monocyclic peroxides, bicyclic peroxides and epoxy alcohols, all formed as several stereoisomers, give rise to hundreds of different products in the autoxidation of arachidonates. Other polyene structures may give more or less complicated product mixtures depending on the number of double bonds in the system. At least three double bonds are required for radical cyclization pathways to exist and the more double bonds present, the more complex will be the mixture of products.

In a process catalyzed by cyclooxygenase enzymes, arachidonic acid is stereospecifically converted into the bicyclic endoperoxide–hydroperoxide PGG_2. Some of the products of autoxidation of arachidonic acid have structures analogous to PGG_2 and the formation of these compounds by radical cyclization routes has been the focus of many studies.[154–157]

$(CH_2)_3COOH$, C_5H_{11} —cyclooxygenase→ $(CH_2)_3COOH$, C_5H_{11}, OOH

PGG_2

Cyclooxygenase enzymes are heme-iron containing enzymes and free-radical mechanisms have been written for the conversion of arachidonic acid to PGG_2. In fact, products formed in polyene autoxidation have similar structures to the prostaglandin endoperoxides but the stereochemistry of the products formed without the enzyme catalyst is uncontrolled whereas the enzyme gives only the stereochemistry shown above. The evidence for the enzyme mechanism is circumstantial with the exception that some spin-labeling experiments have been carried out that suggest the intermediacy of free radicals in the enzymatic conversion.

In Table 6 are shown the structures of cyclic peroxide products formed from hydroperoxide precursors. The reaction of the hydroperoxide **31** is particularly noteworthy. The sequence involved in the conversion of **31**—**32** includes a fragmentation of the peroxyl to give a pentadienyl radical, addition of oxygen at the opposite end of the pentadienyl radical, and peroxyl radical serial cyclization.

7.2 Peroxyl Radical Rearrangements

Peroxyl radicals generated from hydroperoxides undergo chemistry other than radical cyclization. Whereas cyclization or fragmentation reactions such as those described in Section 7.1 give rise to rearranged structures, these examples are intramolecular cases of the general reactions of addition and fragmentation. There are a few other examples of radical rearrangements that apparently do not fit under the other general categories of radical propagation reactions. One of these rearrangements, the acetate 3,2 rearrangement, will be mentioned here since analogous rearrangements occur in peroxyl radicals generated in the autoxidation of lipids such as cholesterol and oleic acid. The acetate 3,2 rearrangement is shown in Scheme 8 and the rearrangement involves the transfer of acetate from a β-position on the radical to the radical site.[161–163] The evidence suggests that the rearrangement occurs without the formation of an intermediate and that the transition state, shown in Scheme 8 in brackets, has considerable polar character. The rate of rearrangement increases in polar solvents and the reaction has been written with a transition state that involves considerable transfer of charge to the acetate unit from the two-carbon fragment. The rearrangement is classified as a 3,2 rearrangement with the

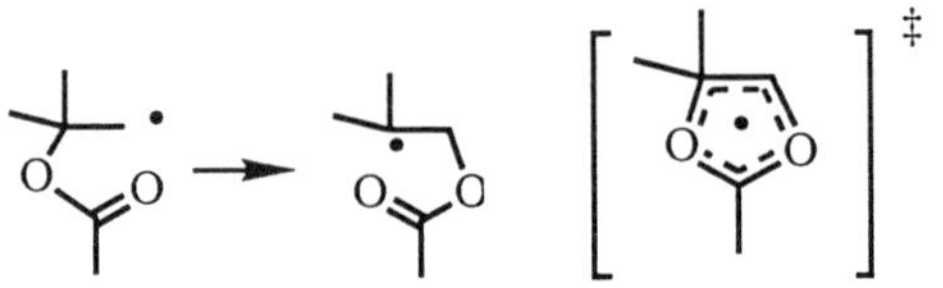

Scheme 8

Table 6. Cyclic peroxides prepared from hydroperoxide precursors via intermediate peroxyl radicals

Hydroperoxide	Cyclic peroxide	Reference
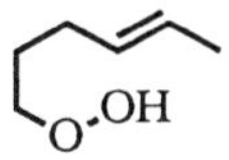	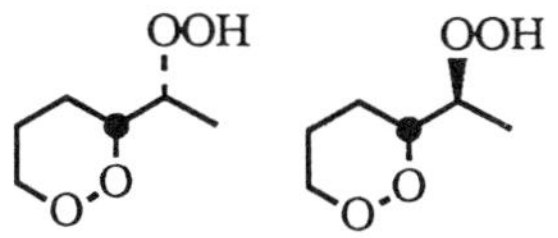	134, 135
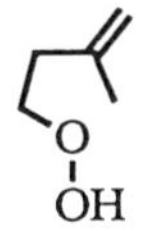	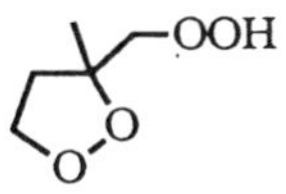	134, 135
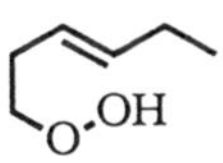	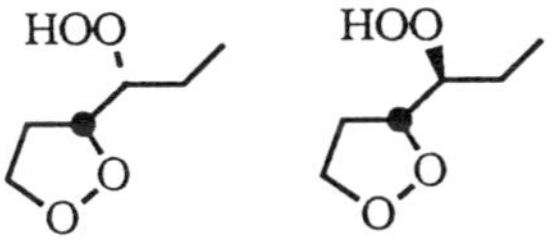	134, 135
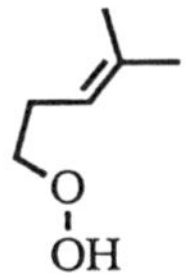	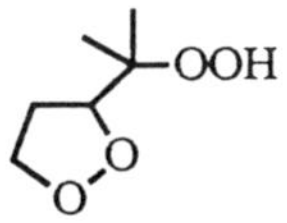	134, 135
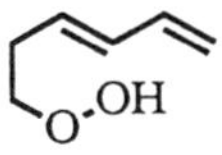	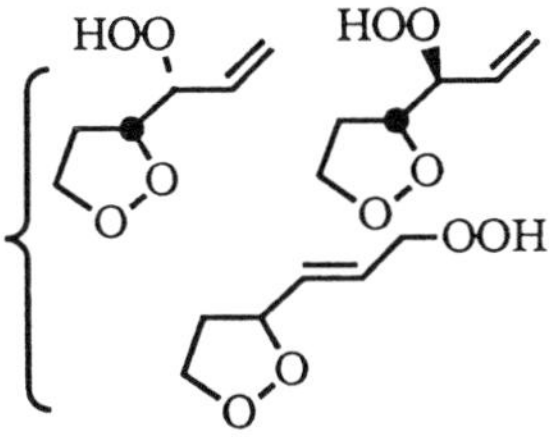	136
	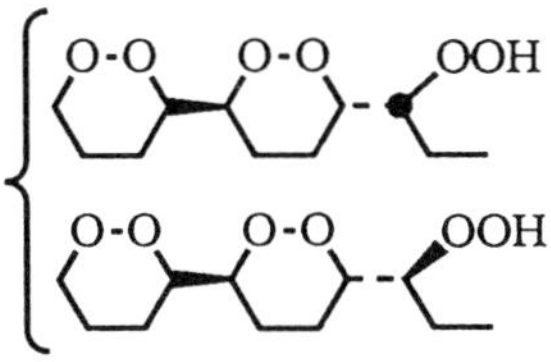	144, 145
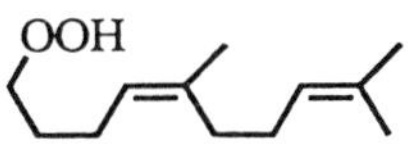	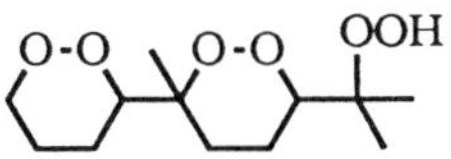	144, 145

Table 6. (*continued*)

Hydroperoxide	Cyclic peroxide	Reference
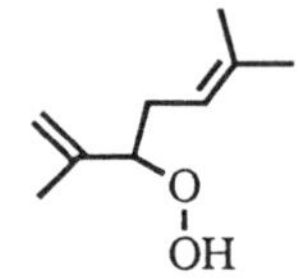	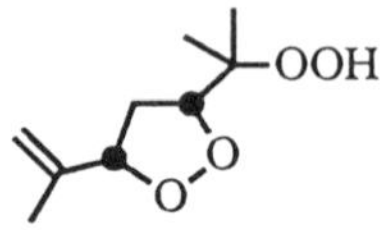	158
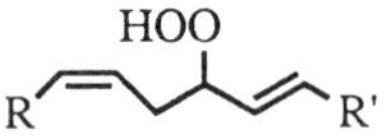	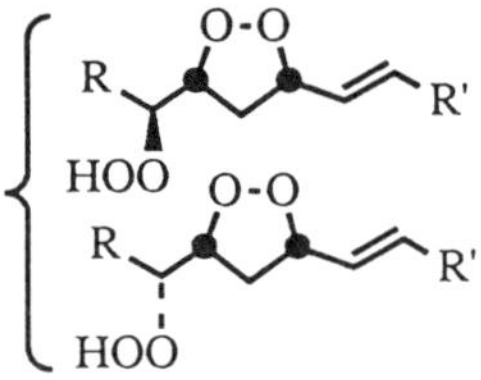	159
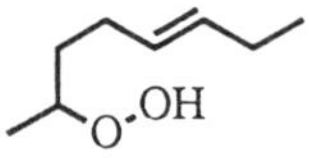	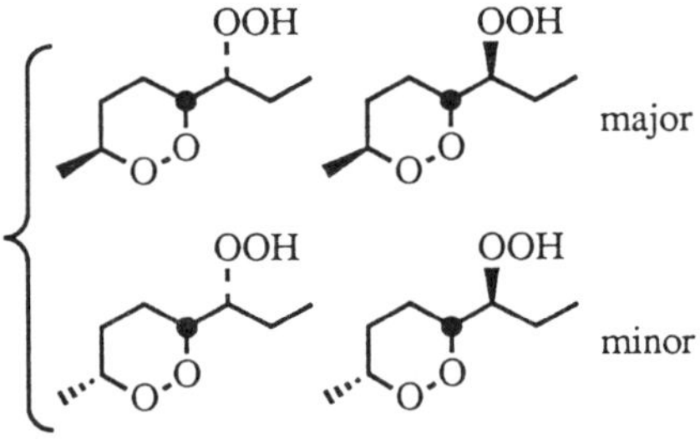 major minor	160
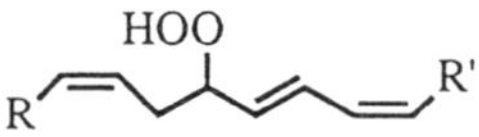	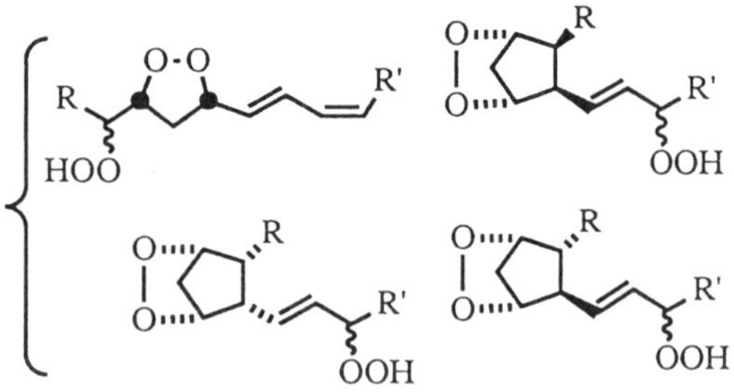	154–157
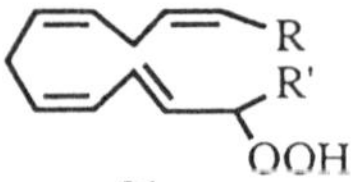 **31**	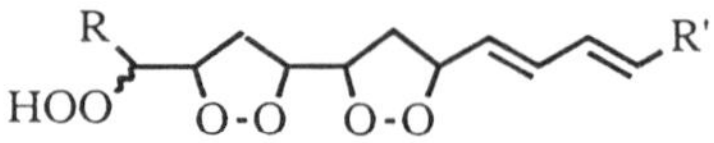 **32**	146

acetate group being the 3 unit and the carbon framework being the 2 atom component of the 3,2 rearrangement.

Allyl peroxyl radicals undergo similar 3,2 rearrangements and the evidence here also suggests that the process is a concerted reaction without intermediates being formed. Allyl peroxyl radicals are important intermediates in processes as diverse as the drying of paint and the peroxidative destruction of cell membranes. The rearrangement was first observed by Schenck *et al.*[164] in cholesterol hydroperoxides and examples in simple systems were first reported by Brill.[165,166] The general rearrangement is shown in Scheme 9. The

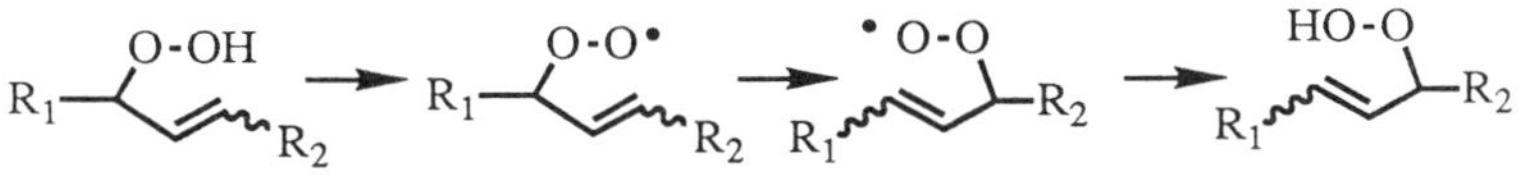

Scheme 9

rearrangement is catalyzed by free-radical initiators or by light and it is inhibited by phenolic antioxidants. In these 3,2 rearrangements of peroxyls, the oxygens are the 2 component with the 3 component being the vinylic and allylic framework. The initial mechanism proposed for the rearrangement suggested that it proceeded through an intermediate dioxolanyl radical **33**. Subsequent work by Brill[166] and Porter and co-workers[167,168] provided evidence against the intermediate **33** since authentic radicals similar to **33** undergo reactions (such as oxygen entrapment) to give products not observed in the rearrangement product mixture. Brill[166] subsequently suggested that the open peroxyl radicals were not intermediates in the rearrangement but that the hydroperoxide gave a common cyclic intermediate in which the unpaired electron was localized in an antibonding orbital on oxygen. An EPR investigation of allyl peroxyl radicals did not support this suggestion however.[169]

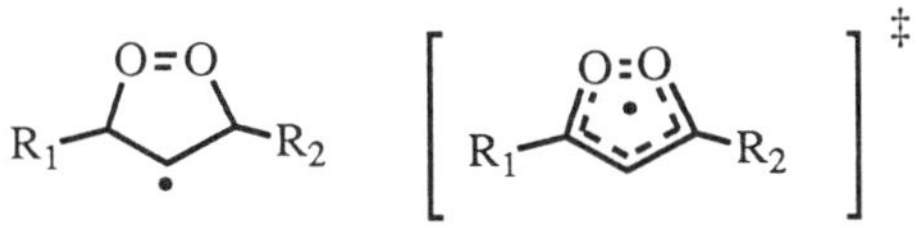

33 **34**

Porter and Wujek[168] suggested that the rearrangement proceeds from an open peroxyl radical through a transition state, **34**, with no intermediate like **33** being formed. Evidence provided to support this view was the fact that rearrangement proceeded without incorporation of atmospheric oxygen and that the rearrangement is highly stereoselective in acyclic systems. Thus, rearrangement of the oleate hydroperoxide **35** under a $^{36}O_2$ atmosphere gives an equilibrium mixture of **35** and **36** without any incorporation of $^{36}O_2$ into the hydroperoxide. This rules out a fragmentation–readdition process analogous to the mechanism proposed for diene hydroperoxides.

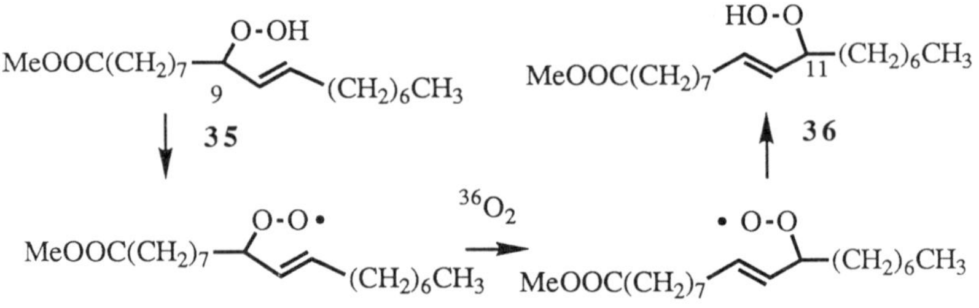

The rearrangement is highly stereoselective, the rearrangement product is of the opposite configuration from the starting material and, even after extensive rearrangement when the 9- and 11-hydroperoxides are present in nearly equal amounts, both hydroperoxides are still significantly enriched in one enantiomer.[170] There is a temperature dependence on the observed selectivity with product hydroperoxide being isolated with optical purities approaching 98% when the rearrangement is carried out at 20°C, whereas rearrangements at 50°C lead to products with a greater loss of stereochemical integrity. As an example of the stereoselectivity of the rearrangement, (9*R*)-**35** with enantiomeric excess >99% rearranges at 22°C to give a mixture of (9*R*)-**35** and (11*S*)-**36** with enantiomeric excess of 97.2 ± 0.5% if the product mixture is analyzed at the time when **35**/**36** = 1.7.

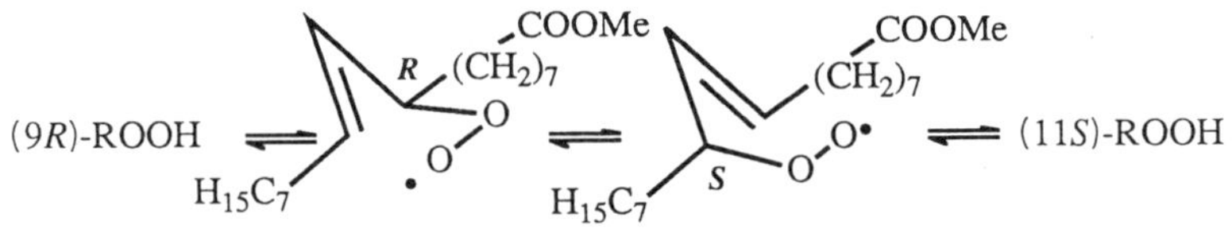

The stereochemical course of the rearrangement of acyclic allylic hydroperoxides supports the proposal that the rearrangement proceeds by a concerted 3,2 peroxyl radical mechanism. Furthermore, the transfer of chirality across the allyl system during rearrangement suggests a model for the rearrangement in acyclic peroxyls in which substituents on the allyl system occupy pseudo-equatorial positions in a envelope-like transition-state structure **37**. Such a transition state model suggests that the stereochemistry of the

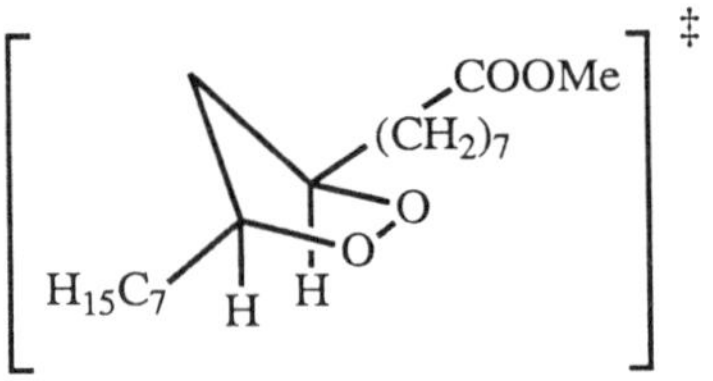

37

product stereogenic center should depend on the geometry of the double bonds in the reactant and product hydroperoxides.

Beckwith, Davies and co-workers[169,171–175] have examined the Schenck cholesterol hydroperoxide and related rearrangements in detail. Initial rearrangement of the 5α-hydroperoxide gives the 7α-hydroperoxide as shown in Scheme 10. Subsequent to this rearrangement, the 7α hydroperoxide rearranges to the 7β compound. The data show unequivocally that the 5α to 7α rearrangement occurs without incorporation of labeled oxygen (the reactions are carried out under $^{36}O_2$), indicating that this rearrangement occurs without fragmentation–readdition while the rearrangement of 7α to 7β occurs with incorporation of atmospheric oxygen. The cholesterol 7α- to 7β-hydroperoxide rearrangement must, however, be a fragmentation–readdition process. This amounts to nothing more than the reversible addition of molecular oxygen to the radical formed by fragmentation of the 7α-peroxyl radical.

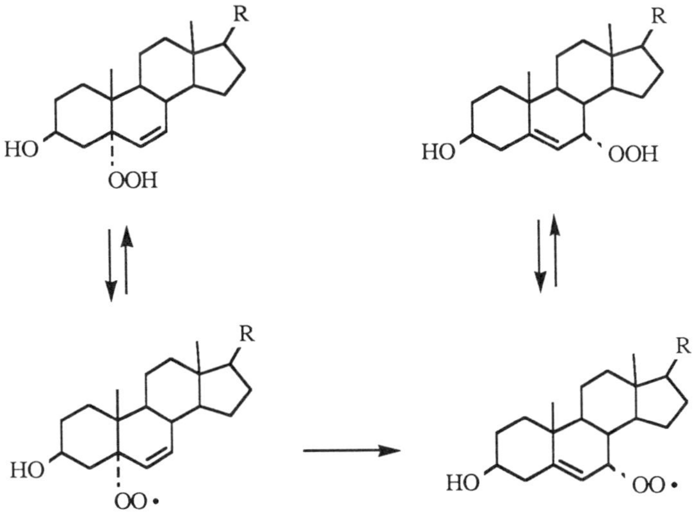

Scheme 10

One model for the 3,2 peroxyl radical transition state is that of an allyl radical–molecular oxygen complex as shown in **38**. Analysis of the symmetry of

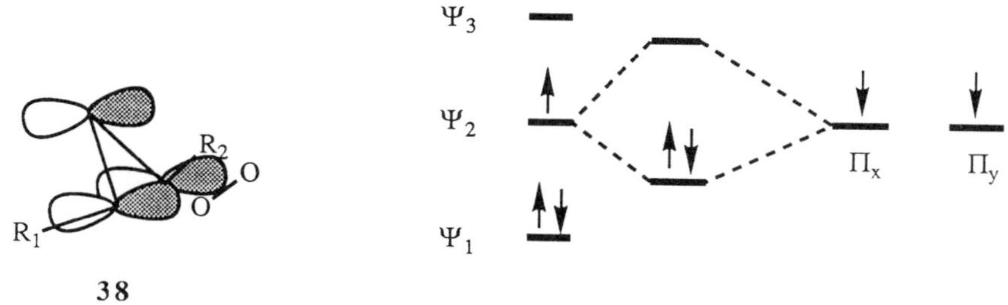

38

this complex suggests a favorable interaction between Ψ_2 of the allyl radical and Π_x^* of molecular oxygen. This leads to a bonding orbital filled with paired electrons and leaves the unpaired electron in Π_y^* of molecular oxygen, orthogonal to the allyl Π system.

7.3 Metal-catalyzed Formation of Peroxyl Radicals

Haynes and Vonwiller[176–178] have recently suggested that peroxyl radicals can be efficiently generated from allylic hydroperoxides by reaction with iron(III) and copper(II) catalysts. The products depend dramatically on the particularly catalyst employed and the general structure of the allylic hydroperoxide. The most successful reagent for initiating peroxyl radical reactions was copper(II) triflate–octanoic acid in acetonitrile at −20 to 5°C. Cyclic peroxide products analogous to those reported in Table 6[46,154,157–159] were reported in good yield starting from acyclic hydroperoxides after relatively short reaction times (*ca.* 5 h).

HOO R R' Fe(III) or Cu(II) O-O R R' HOO O-O R R' HOO O O R R' OOH O O R R' OOH

Simple allylic hydroperoxides with no double bonds present to allow for peroxyl radical cyclization undergo fragmentation reactions on reaction with copper(II) triflate or an iron(III)–phenanthroline complex. Thus, pinene hydroperoxide fragments to a ketoaldehyde on reaction with one of these reagents.

OOH Pinene hydroperoxide cupric triflate O O H

The preparation of dehydroqinghaosu and qinghaosu is an elegant application of the use of this catalyst. Thus, treatment under oxygen of **39**, one of the hydroperoxides formed from singlet oxidation of qinghao acid, with iron(III) and/or copper(II) catalysts gave 30–50% of the antimalarial compound dehydroqinghaosu. A reasonable mechanism for this transformation has been suggested that involves a 4-*exo* cyclization of the peroxyl radical derived from **39** followed by oxygen entrapment to give the hydroperoxide **40**. The dioxetane **40** is suggested to fragment to give a dicarbonyl hydroperoxide that cyclizes under acid catalysis to the hemiacetal precursor of qinghaosu.

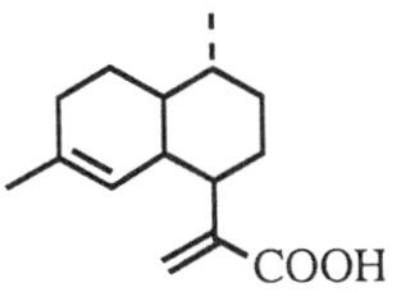

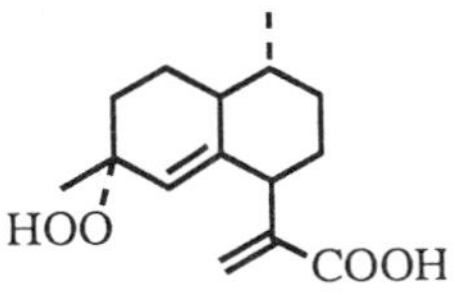

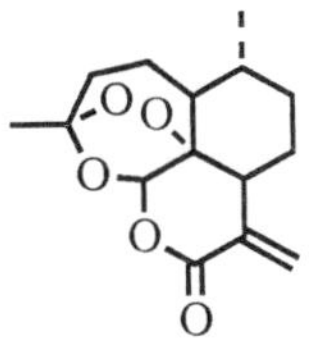

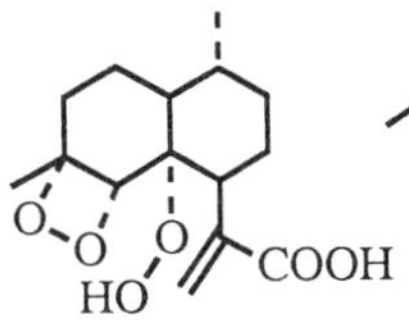

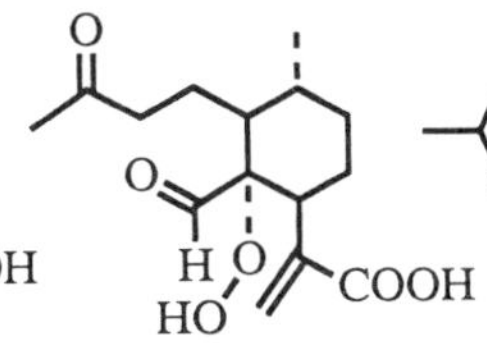

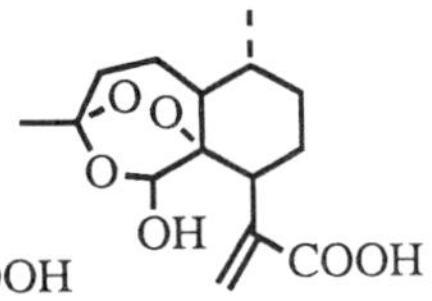

8 PEROXIDE BOND-CLEAVING REACTIONS

Although the hydroperoxide O–O bond is weak (40–43 kcal mol^{-1}), thermolysis of alkyl hydroperoxides is usually complicated by induced decomposition. Alkenes, alcohols, amines, carboxylic acids, aldehydes and ketones induce decomposition of hydroperoxides. Thus, clean homolytic thermolysis chemistry of hydroperoxides is usually not observed. The chemistry of O–O scission of hydroperoxides is rich, however, with metal- and acid-catalyzed processes being particularly important. Metal-catalyzed decomposition of hydroperoxides is discussed in detail elsewhere in this volume and we present here a discussion of metal- and acid-catalyzed hydroperoxide reactions with emphasis on the synthetic and biological applications of these reactions.

8.1 Generation of Alkoxyl or Hydroxyl Radicals

The homolytic cleavage of the O–O bond by low-valent transition metals provides an effective route to alkoxyl radicals.[179–181] In chemistry pioneered by Kochi, iron(II) ion-promoted reactions of hydroperoxides proceed by reactions indicative of alkoxyl radical intermediates. Reactions observed from the putative alkoxyl radical intermediates are δ-hydrogen atom abstraction and fragmentation, reactions typical of alkoxyls.

Hydrogen abstraction reactions provide intramolecular functionalization of carbon frameworks. For example, reaction of 2-hexyl hydroperoxide with iron(II) sulfate–copper(II) acetate gives a mixture of hex-3- and 4-hexen-1-ol whereas 3-methyl-1-butyl hydroperoxide gives 3-methyl-3-buten-l-ol.[182] The mechanism proposed to account for the formation of the products observed is presented in Scheme 11. If copper(II) halides or pseudohalides replace the copper(II) acetate in the reaction, the product formed in moderate to good yield is the δ-carbon halide or pseudohalide substituted product.[183]

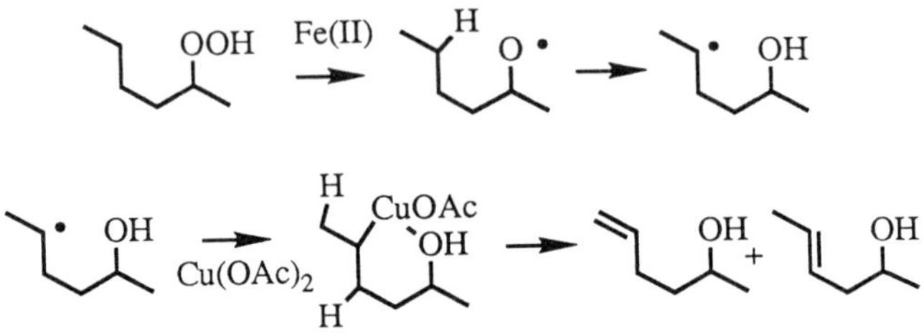

Scheme 11

The adjacent C—C bond to the alkoxyl radical is exceptionally weak and β-fragmentation processes are typical of alkoxyls because of this.[184] The iron(II) ion-promoted fragmentation reaction of hydroperoxides therefore frequently results in the formation of carbon radicals by scission of the hydroperoxide O–O bond and subsequent β-scission of an alkoxyl C–C bond.[184] In fact, metal ion-promoted ring opening of hydroperoxides has been known since the 1950s as a synthetically useful reaction.

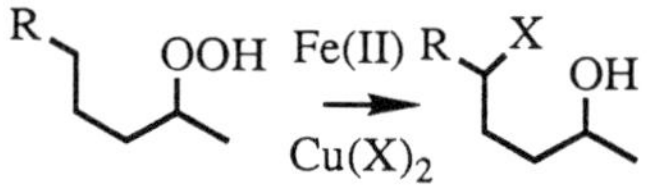

X= SCN, N_3, Cl, Br, I

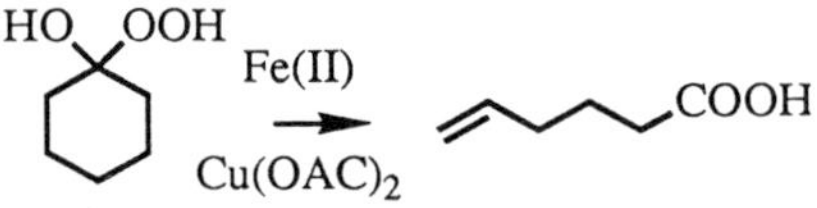

A useful modification of the ring-opening reaction is in the generation of dimer dicarboxylic acids from α-silyloxy hydroperoxides.[185] In this reaction, which is run in the absence of copper(II) salts, the intermediate radical dimerizes to give the dicarboxylic acids in 50–90% yield. For example, the α-silyloxy hydroperoxide prepared from cycloheptanone reacts in methanol under nitrogen at 0°C with 1.2 equiv. of $FeSO_4$ to give 72% of the 16-carbon 1,ω-dicarboxylic acid.

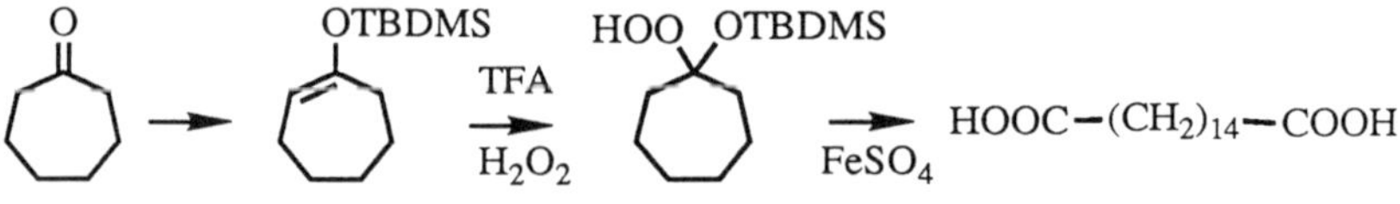

HOOC—$(CH_2)_{14}$—COOH

Other examples of ring fragmentation strategies using hydroperoxide–Fe(II)–Cu(II) are the conversions **41** → **42**[186] and **43** → **44** + **45**.[187] It has been suggested that the product **45** is formed by a fragmentation of the alkoxyl derived from **43** to give the primary radical followed by δH-atom transfer to give the radical **46**.

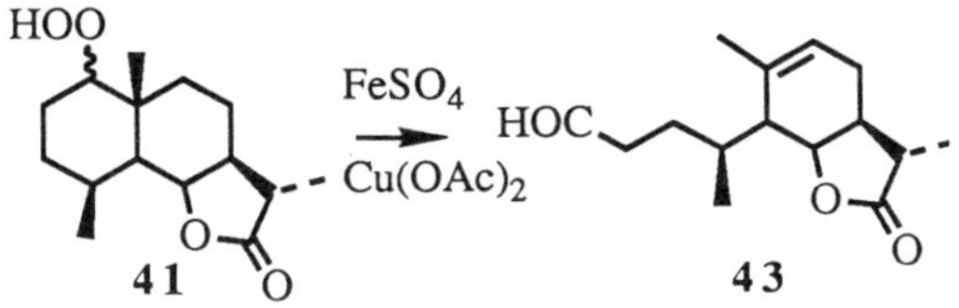

HOO FeSO$_4$ Cu(OAc)$_2$ HOC

41 43

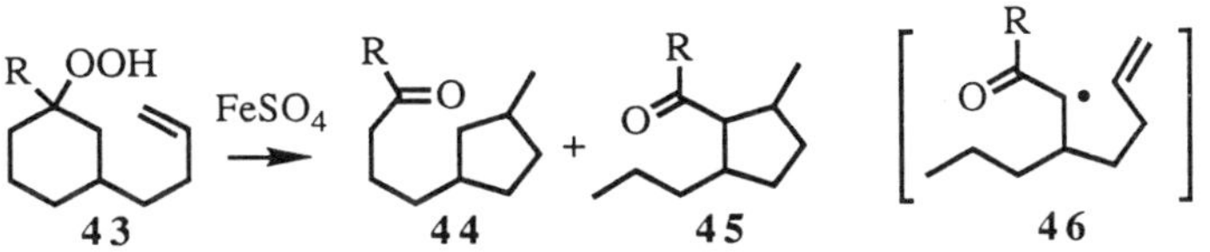

R OOH FeSO$_4$ R O R O R

43 44 45 46

A fragmentation strategy that involves the use of ozonolysis derived α-alkoxy-hydroperoxides[188,189] has been used extensively in organic synthesis. This sequence, in which alkenes serve as precursors of the hydroperoxides, has been used in both cyclic and acyclic systems. Schreiber and co-workers[184,190–195] have made particularly good use of this fragmentation to construct macrocyclic systems from hydroperoxides. Examples of the

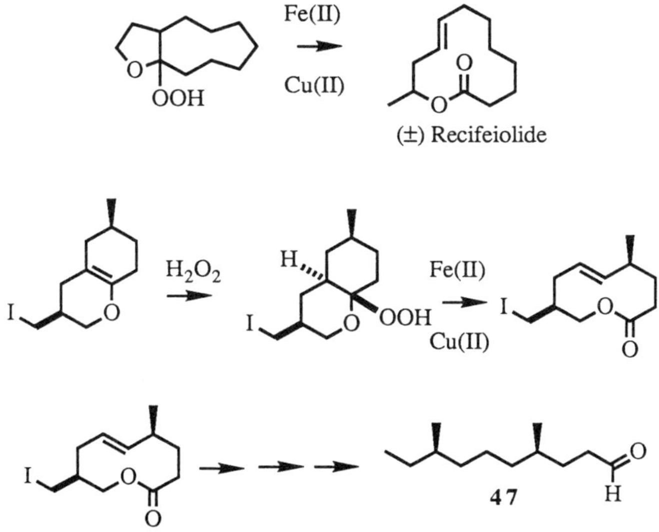

O OOH Fe(II) Cu(II) O O

(±) Recifeiolide

I O H$_2$O$_2$ H I O OOH Fe(II) Cu(II) I O O

I O O → → → O H 47

Scheme 12

application of this strategy, presented in Scheme 12, are the preparation of (±)-recifeiolide and **47**, the aggregation pheromone of the red flour beetle.

$FeCl_3$ → $R\text{-}CH_2CH_2\text{-}Cl$

$FeSO_4$, $Cu(OAc)_2$ → $R\text{-}CH{=}CH_2$

$FeCl_3$ → $R\text{-}CH_2CH_2\text{-}H$

$FeSO_4$ → $R(CH_2CH_2)_2R$

(starting material: R–CH₂CH₂–CH(OOH)OCH₃)

8.2 Generation of Alkoxyl Radicals from Lipid Hydroperoxides

Lipid hydroperoxides such as those formed from autoxidation of polyunsaturated fatty acids react with transition metals in their low oxidation states to give lipid alkoxyl radicals.[196] The alkoxyl radical generated is very reactive

$$\text{L-OOH} + Fe^{2+} \text{-------->} \text{LO}\bullet + Fe^{3+}$$

and fragmentation, H-atom abstraction and cyclization reactions are observed from these intermediates. Fragmentation of oleate hydroperoxides gives 20% cleavage whereas conjugated diene hydroperoxides usually give less than 10% of the total product mixture as fragmentation products. The unconjugated diene hydroperoxide, 10-hydroperoxy-8,12-octadecadienoic acid, gives in excess of 80% of the fragmentation product on reaction with $FeSO_4$ or $FeCl_3$-cysteine.[197] Fragmentation of the unconjugated diene alkoxyl radical is enhanced relative to the other radicals by virtue of the fact that a stabilized allyl is a product in this case.

$H_{11}C_5$–CH=CH–CH₂–CH(OOH)–CH=CH–$(CH_2)_6COOH$ —Fe(II), Cu(II)→ $H_{11}C_5$–CH=CH–CH₂–CH(O•)–CH=CH–$(CH_2)_6COOH$

$H_{11}C_5$–CH=CH–CH₂–CH(O•)–CH=CH–$(CH_2)_6COOH$ → $H_{11}C_5$–CH=CH–CH₂• + OHC–CH=CH–$(CH_2)_6COOH$

Epoxide formation by alkoxyl radical cyclization is a major reaction pathway for conjugated diene alkoxyl radicals. Thus, linoleate conjugated diene hydroperoxides rearrange to give several epoxide products that result from apparent cyclization of an intermediate alkoxyl.[198–201] The allyl radical formed from cyclization reacts with oxygen to give allylic hydroperoxides and these hydroperoxides can react further with iron(II) to give alcohol and ketone products. Reaction of conjugated dienyl hydroperoxides under anaerobic conditions with iron(II) in the presence of thiols[202] gives products where R_1 = SR in the epoxide products. With hematin as a catalyst, a portion of the epoxyallylic radicals diffuse from the catalyst to combine with O_2 and the remainder react by an 'oxygen rebound' mechanism[203] whereby an iron oxo species is formed from the hydroperoxide and transfers OH back to the radical. Alkoxyl radicals generated from lipid hydroperoxides by photolysis lead to similar epoxide products.[204,205]

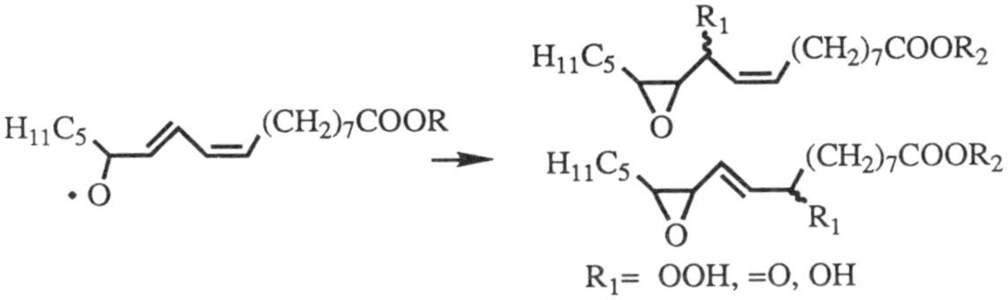

R_1 = OOH, =O, OH

9 ACID-CATALYZED DECOMPOSITION

Acid-catalyzed rearrangement reactions of hydroperoxides have been known for more than 50 years. Attack of a proton at the hydroperoxide oxygen bonded to carbon may result in loss of hydrogen peroxide and formation of a carbocation. Treatment of 9-hydroperoxy-9-phenylxanthene with acid, for example, gives hydrogen peroxide and the xanthylium perchlorate.[206] An excess of strong acid is required for optimum xanthenyl cation and hydrogen peroxide production. Treatment of the hydroperoxide with weak acids has been shown to produce the corresponding dialkylperoxides.

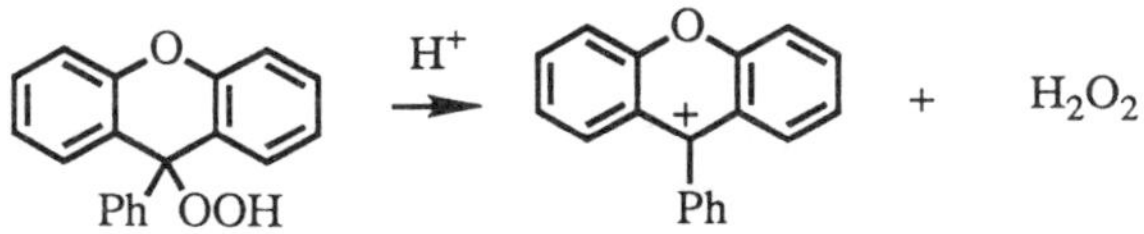

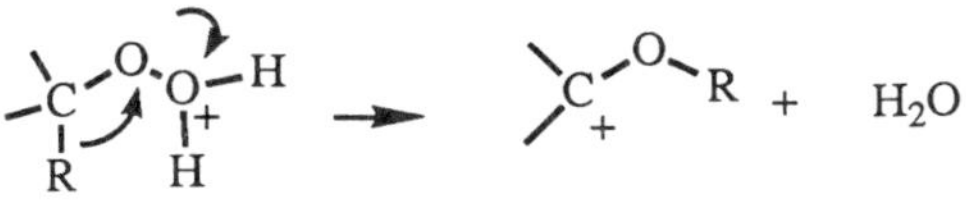

Attack of acid on the hydroperoxide oxygen remote from the carbon leads to O—O heterolysis. Like the similar reaction of peroxides and peroxy esters, the reaction depends on the propensity of the alkyl or aryl group to undergo a 1,2-shift from carbon to oxygen. The mechanism of this rearrangement, first observed by Hock,[207] has been a focus of extensive research and has been reviewed elsewhere.[208] The rearrangement is similar to the Criegee perester rearrangement.

9.1 Use of the Hock Cleavage in Synthesis

Hock cleavage of hydroperoxides has been used extensively in synthesis and is also frequently observed with hydroperoxides derived from lipid peroxidation. Under normal Hock cleavage conditions (H^+–H_2O), the oxocarbenium ion generated reacts with water to give the hemiacetal which leads to a carbonyl and an alcohol or phenol. Cumyl hydroperoxide gives acetone and phenol in quantitative yield, for example, and two carbonyls are formed from Hock cleavage of allylic or dienylic hydroperoxides where vinyl migration and hydration of the carbocation give one of the products as an enol.[209] A variation of the Hock cleavage is one in which the oxycarbenium ion undergoes elimination of a β-proton to give a divinyl ether. The benzoxepine **49** is generated from the hydroperoxide **48** under Hock conditions, presumably by the β-proton elimination route.[210]

H_2O

OH H

OOH H^+

48 49

Ronald and co-workers[211–215] have made extensive use of the Hock cleavage in the construction of bicyclic peroxyketals. Reaction of the brosylate **50**, for example, gives 80% of the product **51**. The mechanism for this interesting

OBs OOH OOH OH

H_2O_2

50

H^+ H_2O_2

OOH

51

conversion is obviously complex, and the formation and rearrangement of the intermediate cyclobutyl hydroperoxide has been suggested to be a key step in the rearrangement. This compound, if formed as an intermediate, is never isolated. The cyclopropylcarbinyl hydroperoxide was isolated under neutral conditions but it is also converted to product **51** with acid catalysts. Other examples of the use of the Hock cleavage in synthesis are the conversion of **52** to **53**[216] and the analogous Criegee rearrangement of the hydroperoxide **54** to **55** in the presence of acetic anhydride.[217]

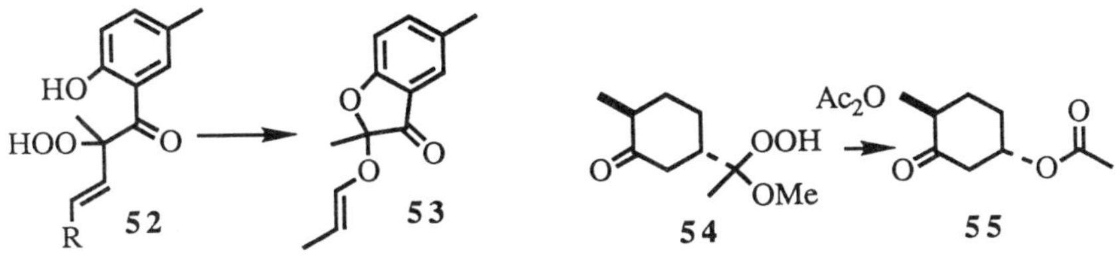

9.2 Hock Cleavage of Lipid Hydroperoxides

Fatty acid hydroperoxides react under Hock rearrangement conditions to give aldehydes. The linoleate-13-hydroperoxide **56** gives hexanal and methyl-12-oxo-9-dodecanoate on reaction with Lewis acids in aprotic solvents.[218] The role of such fragmentation has been suggested to be important in the production of volatile products in foods. The mechanism proposed for this fragmentation involves the carbon to oxygen rearrangement of a vinyl group of the conjugated system.

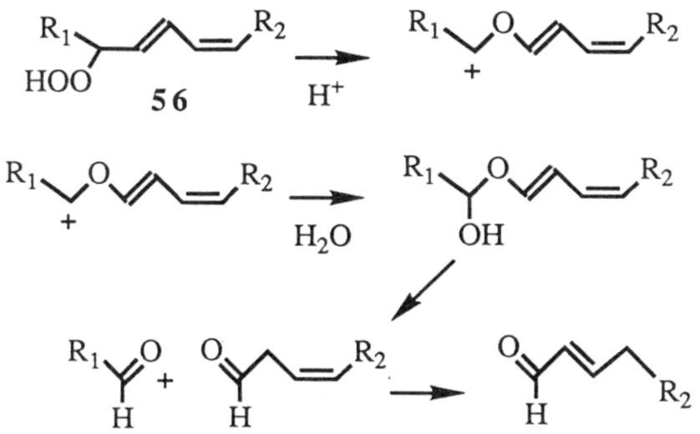

Protic acids catalyze the conversion of dienyl hydroperoxides to epoxy alcohols analogous to the products observed in iron(II)-catalyzed rearrangements. Thus, sulfuric acid in methanol–water at 100°C catalyzes the conversion of the linoleate-13-hydroperoxide **56** to a mixture of epoxy alcohols of which the (11*R*, 12*S*, 13*S*)-12,13-epoxy-11-hydroxyoctadec-9-enoic acid **57** is the major stereoisomeric product formed.[219] The mechanism for the formation presumably proceeds via an intermediate allyl cation. This mechanism is substantially different from the standard Hock cleavage mechanism that involves a 1,2 carbon to oxygen migration and raises the possibility that the oxycarbenium ion formed in the Hock cleavage of allyl and dienyl hydroperoxides arises from an analogous cyclization–scission pathway. This mechanism is presented in Scheme 13 and shows a pathway for formation of epoxy alcohols and the Hock oxycarbenium ion.

R_1= HOOC$(CH_2)_6$
R_2=C_5H_{11}

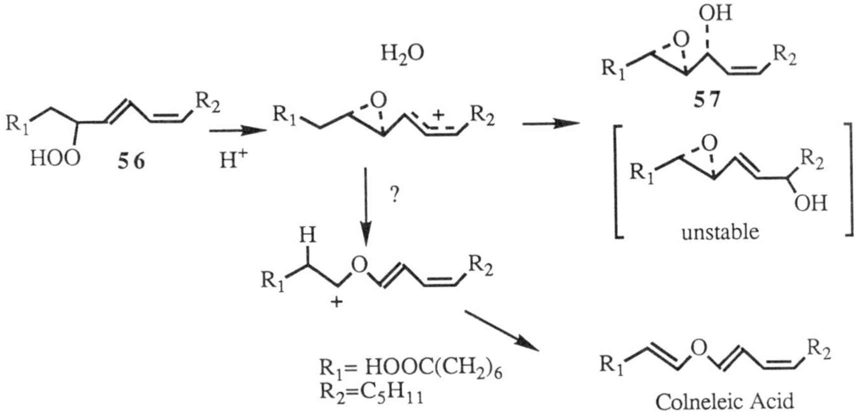

Scheme 13

The enzymatic equivalent of Hock rearrangement with β-proton elimination also takes place in the case of lipid hydroperoxides. Colneic acid and colneleic acids are divinyl ethers derived from linoleic and linolenic acids that have structures related to fatty acid hydroperoxides by a putative Hock rearrangement with β-proton elimination. Galliard and Philips[220] made the interesting discovery that an enzyme from potato was capable of converting the linoleate and linolenate-9-hydroperoxides to these compounds[220] and Crombie *et al.*[22] showed that the vinyl ether oxygens of colneic acid and colneleic acids are derived from molecular oxygen. Crombie *et al.* proposed a mechanism for colneleic acid biosynthesis that involves the intermediacy of the epoxyallyl cation on the pathway to the oxycarbenium ion. This is also not unreasonable for the general Hock cleavage mechanism. Hence the Hock cleavage of allyl or dienyl hydroperoxides may not be a 1,2 carbon to oxygen migration concerted with O—O cleavage but may rather be epoxide formation simultaneous to O—O cleavage followed by fragmentation of the allyl cation to give the Hock oxycarbenium ion.

10 EPOXIDATION OF ALKENES

Hydroperoxides are important reagents used in the conversion of alkenes to epoxides. Epoxides are sometimes observed as products in the autoxidation of alkenes and an addition–fragmentation mechanism has been written to account for their formation. It has been suggested that in the autoxidation of styrene, intramolecular radical attack on the peroxide bond is the critical step in the unzipping of the styrene–oxygen copolymer.[222] The radical intermediate in this

sequence results from the addition of a peroxyl to styrene, and hydroperoxide chemistry that proceeds through intermediate peroxyl radicals may therefore also give rise to epoxide products. Carbon radicals generated from peroxyl radical cyclization lead to epoxide products by intramolecular homolytic substitution (s_Hi) on the peroxide bond and there is a backside stereochemical preference for attack of a carbon radical on the peroxide bond.[223,224] The radical **58**, with fixed equatorial stereochemistry, attacks the peroxide bond with a rate of *ca* $2 \times 10^4\ s^{-1}$ whereas the rate of radical attack on the peroxide of the axial radical **59** is $<2 \times 10\ s^{-1}$.

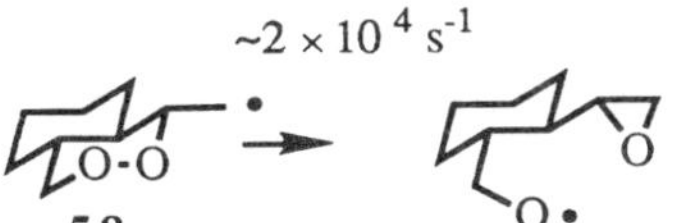

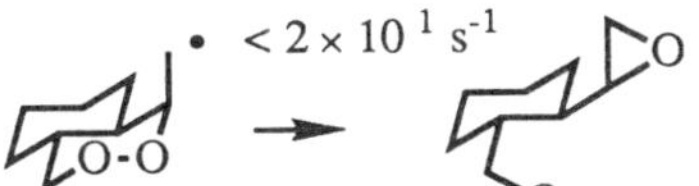

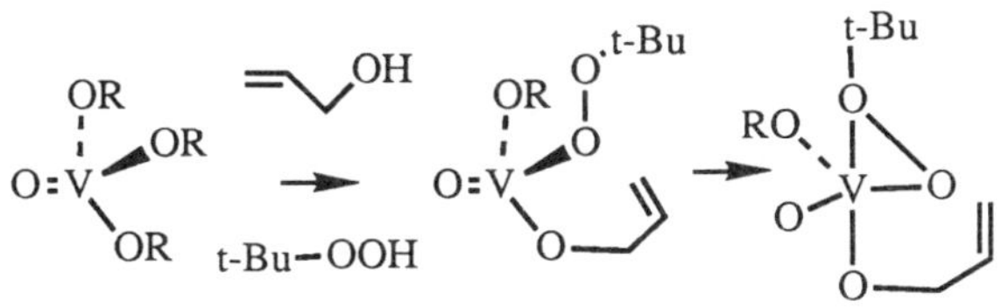

The study of metal-catalyzed oxygenations of alkenes and alkynes by hydroperoxides has led to important advances in synthetic chemistry.[225] Vanadium, molybdenum and titanium compounds[226–228] catalyze the epoxidation of allylic alcohols and remarkable selectivities have been achieved with the use of these reagents. A mechanistic picture for the vanadium-catalyzed epoxidations has been proposed that involves exchange of the allylic alcohol and the hydroperoxide onto the vanadium followed by an oxygen transfer from peroxide to alkene in this intermediate. A similar mechanism has been proposed for titanium-catalyzed epoxidations. A single-crystal X-ray analysis of some titanium–tartrate derivatives showed that these compounds, which are analogous to the catalysts used in Sharpless epoxidation, are actually binuclear titanium species.[229]

Following the work of Sharpless and co-workers on allylic epoxidations utilizing *tert*-butyl hydroperoxide, Adam and co-workers[230–232] explored the use of titanium catalysts in the decomposition of allylic hydroperoxides. Treatment of the allylic alcohol **60** with titanium isopropoxide gives rise to a 95:5 diastereomeric mixture (*erythro* presumably favored) of epoxy alcohols **61** with yields in excess of 80%. It was concluded from these studies that oxygen transfer does not take place intramolecularly and that the chain carrier is the allylic alcohol. A convenient 'one-pot' procedure has been developed so that epoxy alcohols derived from cholesterol can be prepared by singlet oxygen oxidation in the presence of titanium isopropoxide.

OOH, t-Bu **60** → O, OOH, *, *, t-Bu **61**

Epoxides are also formed from alkenes and hydroperoxides by mechanisms not involving free-radical intermediates or metal catalysis. Hydroperoxide–alkene complex formation and carbon displacement on the O—O bond give rise to epoxides by a mechanism that is apparently analogous to peracid epoxidation.[233,234] The non-catalyzed (metal ion-free) oxygen atom transfer reactions for many hydroperoxides are slow reactions unless the hydroperoxide is substituted at the α-position with heteroatoms. Hydroperoxides with α-substituents have been examined for their ability to epoxidize alkenes in a stereospecific manner. Hydroperoxy ethers, amines, carbonyl compounds, nitriles[234] and diazenes[235–241] showed this capability. Intramolecular epoxidation of homoallylic, bishomoallylic and more remote alkenes was observed, an example being the conversion of the orthoester of oleic acid to oleate epoxide on treatment with hydrogen peroxide. Hydroperoxides substituted α to carbonyls are remarkably effective epoxidation reagents and give epoxides that are cleanly stereospecific. The α-hydroperoxide of ethyl diphenylacetate gives, for example, an 85% yield of the epoxide from β-methylstyrene.[234]

$CH_3(CH_2)_6CH{=}CH(CH_2)_7C(OMe)_3 \xrightarrow{H_2O_2} CH_3(CH_2)_6CH{=}CH(CH_2)_7C(OOH)(OMe)_2 \rightarrow CH_3(CH_2)_6\text{-epoxide-}(CH_2)_7C(O)OMe$

Several chiral α-substituted hydroperoxides were examined to test whether these reagents would give epoxides with significant enantiomeric excess but in all cases the results were disappointing.

Cyclic α-azo hydroperoxides are extremely reactive in epoxidation reactions whereas acyclic analogs are two orders of magnitude less reactive. These hydroperoxides readily oxidize amines, sulfides and alkenes in protic or aprotic media. Diverse data including ρ values, relative selectivities, solvent effects and deuterium isotope effects support a mechanism of oxygen atom transfer from these hydroperoxides similar to that of peracids. Intramolecular hydrogen bonding of the hydroperoxide hydrogen and the azo linkage is apparently important in the transition state for epoxidation.

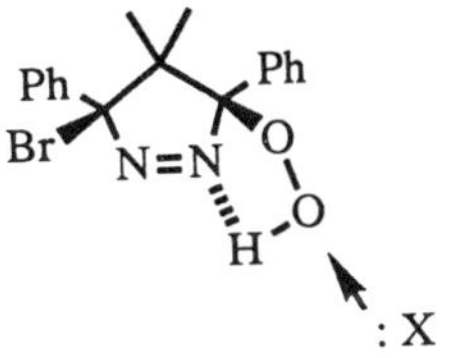

Ph Ph Br N=N O O H :X

Hydroperoxides have been implicated as important intermediates in biological oxidations. Many of the α-substituted hydroperoxides that are particularly reactive in reactions with amines, sulfides and alkenes have a resemblance to flavin hydroperoxides which are putative intermediates in flavin-dependent oxygenases.[242] Chemical models for these oxygenases have been reported by Bruice and co-workers[243–245], Rastetter and co-workers[246,247] and others,[248–250] and an enantioselective oxidation of methyl aryl sulfides with the chiral flavinopohane **62** has been reported. This compound acted as an asymmetric, autorecycling catalyst for mono-oxygenation in the presence of excess of H_2O_2.[250]

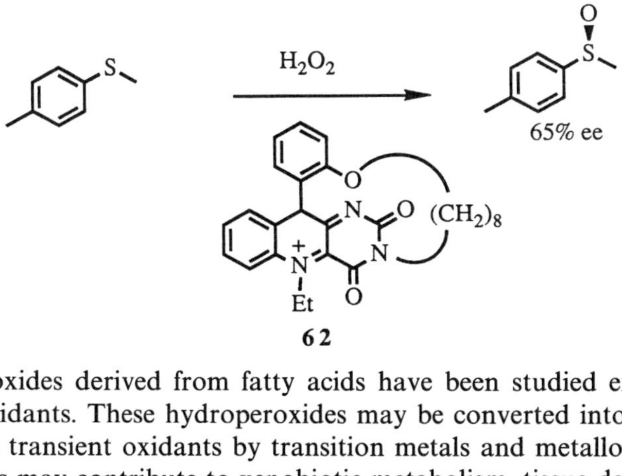

S H_2O_2 O S 65% ee

O N O $(CH_2)_8$ N + N Et O

62

Hydroperoxides derived from fatty acids have been studied extensively as biological oxidants. These hydroperoxides may be converted into a variety of powerful but transient oxidants by transition metals and metalloproteins and such oxidants may contribute to xenobiotic metabolism, tissue degradation or tumor promotion, among other things.[251–253] Metal–oxo complexes formed from heme and hydroperoxides are extremely reactive, as are alkoxyls generated by one-electron reduction of hydroperoxides. Lipid hydroperoxides epoxidize dihydroaromatic compounds to give mutagenic and carcinogenic epoxides, and a mechanism has been proposed that suggests that peroxyl radicals derived from an intermediate alkoxyl are responsible for epoxidation of the dihydroaromatic compounds.[254–256] The intermediate alkoxyls formed by reaction of the hydroperoxide with heme presumably cyclize to give carbon radicals which trap oxygen to give peroxyls. Peroxyls generated in this way or from carbon radicals formed by β-scission are proposed to epoxidize dihydroaromatics such as 7,8-dihydroxy-7,8-dihydrobenzo[*a*]pyrene (BP-7,8-diol), a proximate carcinogenic metabolite of benzo[*a*]pyrene.

HO OH + R_1 OOH R_2 → hematin HO OH O

BP-7,8-diol

11 REFERENCES

1. A. G. Davies, *Organic Peroxides*, Butterworths, London (1961), pp. 1–40.
2. R. Hiatt, *Organic Peroxides*, Vol. 1 (D. Swern, Ed.), Wiley–Interscience, New York (1971), pp. 1–153.
3. C. G. Swain and C. B. Scott, *J. Am. Chem. Soc.*, **75**, 141 (1953).
4. R. G. Pearson, H. Sobel and J. Songstad, *J. Am. Chem. Soc.*, **90**, 319 (1968).
5. C. D. Ritchie and M. Sawada, *J. Am. Chem. Soc.*, **99**, 3754 (1977).
6. S. Wolfe, D. J. Mitchell, H. B. Schlegel, C. Minot and O. Eisenstein, *Tetrahedron Lett.*, **23**, 615 (1982).
7. M. M. Heaton, *J. Am. Chem. Soc.*, **100**, 2004 (1978).
8. P. G. Cookson, A. G. Davies and B. P. Roberts, *J. Chem. Soc., Chem. Commun.*, 1022 (1976).
9. A. A. Frimer, *J. Org. Chem.*, **42**, 3194 (1977).
10. N. A. Porter and T. Capps, unpublished work.
11. K. R. Kopecky, J. H. van de Sande and C. Mumford, *Can. J. Chem.*, **46**, 25 (1968).
12. N. A. Porter and D. W. Gilmore, *J. Am. Chem. Soc.*, **99**, 3503 (1977).
13. N. A. Porter, J. R. Nixon and D. W. Gilmore, in *Organic Free Radicals* (W. A. Pryor, Ed.), American Chemical Society, Washington, DC (1978), pp. 89–101.
14. M. Hamberg and B. Samuelsson, *Proc. Natl. Acad. Sci. USA*, **70**, 899 (1973).
15. M. Hamberg, J. Svensson, T. Wakabayashi and B. Samuelsson, *Proc. Natl. Acad. Sci. USA*, **71**, 345 (1974).
16. N. A. Porter, D. H. Roberts and C. B. Ziegler, Jr, *J. Am. Chem. Soc.*, **102**, 5912 (1980).

17. N. A. Porter, C. B. Ziegler, Jr, F. F. Khouri and D. H. Roberts, *J. Org. Chem.*, **50**, 2252 (1985).
18. E. Bascetta and F. D. Gunstone, *J. Chem. Soc., Perkin Trans. 1*, 2217 (1984).
19. A. J. Bloodworth and H. J. Eggelte, *J. Chem. Soc., Perkin Trans. 1*, 327 (1981).
20. N. A. Porter, J. D. Byers, R. C. Mebane, W. Gilmore and J. R. Nixon, *J. Org. Chem.*, **43**, 2088 (1978).
21. N. A. Porter, J. D. Byers, K. M. Holden and D. B. Menzel, *J. Am. Chem. Soc.*, **101**, 4319 (1979).
22. N. A. Porter, J. D. Byers, A. E. Ali and T. E. Eling, *J. Am Chem. Soc.*, **102**, 1181 (1980).
23. N. A. Porter and J. C. Mitchell, *Tetrahedron Lett.*, **24**, 543 (1983).
24. J. C. Mitchell, S. Heaton and N. A. Porter, *Tetrahedron Lett.*, **25**, 3769 (1984).
25. J. C. Mitchell, unpublished work.
26. M. O. Funk, R. Isaac and N. A. Porter, *J. Am. Chem. Soc.*, **97**, 1281 (1975).
27. N. A. Porter, M. O. Funk, D. Gilmore, R. Isaac and J. Nixon, *J. Am. Chem. Soc.*, **98**, 6000 (1976).
28. J. R. Nixon, M. A. Cudd and N. A. Porter, *J. Org. Chem.*, **43**, 4048 (1978).
29. E. J. Corey, A. Marfat, J. R. Falck and J. O. Albright, *J. Am. Chem. Soc.*, **102**, 1433 (1980).
30. E. J. Corey, J. O. Albright, A. E. Barton and S. Hashimoto, *J. Am. Chem. Soc.*, **102**, 1435 (1980).
31. R. Zomboni and J. Rokach, *Tetrahedron Lett.*, **24**, 999 (1983).
32. G. Just, C. Luthe and M. T. P. Viet, *Can. J. Chem.*, **61**, (1983).
33. R. Nagata, M. Kwakami, T. Matsuura and I. Saito, *Tetrahedron Lett.*, **30**, 2817 (1989).
34. F. E. Ziegler and R. T. Wester, *Tetrahedron Lett.*, **25**, 617 (1984).
35. I. Saito, R. Nagata, K. Yuba and T. Matsuura, *Tetrahedron Lett.*, **24**, 1737 (1983).
36. A. J. Bloodworth and M. E. Loveitt, *J. Chem. Soc., Perkin Trans. 1*, 1031 (1977).
37. E. Schmitz, A. Rieche and O. Brede, *J. Prakt. Chem.*, **312**, 30 (1970).
38. A. J. Bloodworth and M. E. Loveitt, *J. Chem. Soc., Perkin Trans. 1*, 522 (1978).
39. A. J. Bloodworth and B. P. Leddy, *Tetrahedron Lett.*, **8**, 729 (1979).
40. K. R. Kopecky, J. E. Filby, C. Mumford, P. A. Lockwood and J. Din, *Can. J. Chem.*, **53**, 1103 (1975).
41. A. L. Baumstark and P. C. Vasquez, *J. Org. Chem.*, **51**, 5213 (1986).
42. M. E. Landis, R. L. Lindsey, W. H. Watson and V. Zabel, *J. Org. Chem.*, **45**, 525 (1980).
43. N. A. Porter and D. W. Gilmore, *J. Am. Chem. Soc.*, **99**, 3503 (1977).
44. W. Adam, A. Birke, C. Cadiz, S. Diaz and A. Rodriguiez, *J. Org. Chem.*, **43**, 1154 (1978).
45. M. Chimielewski, J. Jurczak and S. Maciejewski, *Carbohydr. Res.*, **165**, 111 (1987).
46. S. L. Schreiber, *J. Am. Chem. Soc.*, **102**, 6163 (1980).
47. V. Subramanyam, C. L. Brizuela and A. H. Soloway, *J. Chem. Soc., Chem. Commun.*, 508 (1976).
48. R. Dietz, A. E. J. Forno, B. E. Larcombe and M. E. Peover, *J. Chem. Soc. B*, 816 (1970).
49. M. V. Merritt and D. T. Sawyer, *J. Org. Chem.*, **35**, 2157 (1970).
50. R. A. Johnson, E. G. Nidy and M. V. Merrit, *J. Am. Chem. Soc.*, **100**, 7960 (1978).
51. E. J. Corey, K. C. Nicolaou and M. Shibasaki, *J. Chem. Soc., Chem. Commun.*, 658 (1975).
52. J. S. Filippo, Jr, C. I. Chern and J. S. Valentine, *J. Org. Chem.*, **40**, 1678 (1975).
53. S. L. Schreiber and N. Ikemoto, *Tetrahedron Lett.*, **29**, 3211 (1988).

54. K. U. Ingold, *Pure Appl. Chem.*, **56**, 1767 (1984).
55. B. Maillard, K. U. Ingold and J. C. Scaiano, *J. Am. Chem. Soc.*, **105**, 5095 (1983).
56. J. A. Howard, *Rev. Chem. Intermed.*, **5**, 1 (1984).
57. S. W. Benson and P. S. Nangia, *Acc. Chem. Res.*, **12**, 223 (1979).
58. J. A. Howard, *Adv. Free-Radical Chem.*, **4**, 49 (1972).
59. G. A. Russell, *J. Am. Chem. Soc.*, **79**, 3871 (1957).
60. L. R. C. Barclay, S. J. Locke, J. M. MacNeil, J. Van Kessel, G. W. Burton and K. U. Ingold, *J. Am. Chem Soc.*, **106**, 2479 (1984).
61. L. R. C. Barclay, S. J. Locke, J. M. MacNeil and J. Van Kessel, *Can. J. Chem.*, **103**, 2633 (1985).
62. L. R. C. Barclay, D. Kong and J. Van Kessel, *Can. J. Chem.*, **64**, 2103 (1986).
63. N. A. Porter, P. Dussault, R. A. Breyer, J. Kaplan and J. Morelli, *Chem. Res. Toxicol.*, **3**, 236 (1990).
64. K. Tanaka and J. Imamura, *Chem. Lett.*, 1347 (1974).
65. U. Anthoni, C. Larsen, P. H. Neilsen and C. Christophersen, *Acta Chem. Scand., Ser. B*, **41**, 216 (1987).
66. J. Wolpers and W. Ziegenbein, *Tetrahedron Lett.*, **42**, 3889 (1971).
67. J. A. Howard and K. U. Ingold, *Can. J. Chem.*, **42**, 1250 (1964).
68. E. N. Frankel, R. F. Garwood, J. R. Vinson and B. C. L. Weedon, *J. Chem. Soc., Perkin Trans. 1*, 2707 (1982).
69. E. N. Frankel, R. F. Garwood, B. P. S. Khambay, G. P. O. Moss and B. C. L. Weedon, *J. Chem. Soc., Perkin Trans. 1*, 2233 (1984).
70. N. A. Porter, L. S. Lehman, B. A. Weber and K. J. Smith, *J. Am. Chem. Soc.*, **103**, 6447 (1981).
71. H. W.-S. Chan, G. Levett and J. A. Matthew, *Chem. Phys. Lipids*, **24**, 245 (1979).
72. H. Weenen and N. A. Porter, *J. Am. Chem. Soc.*, **104**, 5216 (1982).
73. Y. Yamamoto, E. Niki and Y. Kamiya, *Lipids*, **17**, 870 (1982).
74. R. V. Venkateswaran and P. Goswami, *Tetrahedron Lett.*, **22**, 1943 (1977).
75. R. S. Drago and R. Riley, *J. Am. Chem. Soc.*, **112**, 215 (1990).
76. G. L. Bundy, J. M. Baldwin and D. C. Peterson, *J. Org. Chem.*, **48**, 976 (1983).
77. R. J. Kazlauskas, *J. Org. Chem.*, **53**, 4635 (1988).
78. J. Carnduff and D. G. Leppard, *J. Chem. Soc., Perkin Trans. 1*, 2570 (1976).
79. H. Greenland, J. T. Pinhey and S. Sternhell, *Aust. J. Chem.*, **40**, 325 (1987).
80. A. Nishinaga, T. Shimizu, T. Fujii and T. Matsuura, *J. Org. Chem.*, **45**, 4997 (1980).
81. F. Haslbeck, W. Grosch and J. Firl, *Biochim. Biophys. Acta*, **750**, 185 (1983).
82. K. Miyashita, K. Fujimoto and T. Kaneda, *Agric. Biol. Chem.*, **48**, 2511 (1984).
83. N. A. Porter, *Acc. Chem. Res.*, **19**, 262 (1986).
84. N. A. Porter and D. G. Wujek, *J. Am. Chem. Soc.*, **106**, 2626 (1984).
85. A. L. J. Beckwith, D. M. O'Shea and D. H. Roberts, *J. Am. Chem. Soc.*, **108**, 6408 (1986).
86. H. Weenen and N. A. Porter, *J. Am. Chem. Soc.*, **104**, 5216 (1982).
87. K. Peers and D. T. Coxon, *Chem. Phys. Lipids*, **32**, 49 (1983).
88. P. Haverkamp-Begeman, W. J. Woestenberg and S. Leer, *J. Agric. Food Chem.*, **16**, 679 (1968).
89. E. F. L. J. Anet, *Aust, J. Chem.*, **22**, 2403 (1969).
90. H. Morita, N. Tomioka, Y. Iitaka and H. Itokawa, *Chem. Pharm. Bull.*, **36**, 2984 (1988).
91. D. H. R. Barton, D. Crich and W. B. Motherwell, *Tetrahedron Lett.*, **41**, 3901 (1985).
92. D. H. R. Barton, D. Crich and W. B. Motherwell, *J. Chem. Soc., Chem. Commun.*, 242 (1984).
93. A. J. Bloodworth, D. Crich and T. Melvin, *J. Chem. Soc., Chem. Commun.*, 786 (1987).

94. R. Caput, L. Mangoni, L. Previtera and R. Iaccarino, *Tetrahedron Lett.*, **9**, 2047 (1973).
95. R. Caput, L. Previtera, P. Monaca and L. Mangoni, *Tetrahedron Lett.*, **30**, 963 (1974).
96. H. H. Wasserman and R. W. Murray (Eds), *Singlet Oxygen*, Academic Press, New York (1979).
97. A. A. Frimer and L. M. Stephenson, in *Singlet Oxygen*, Vol. II (A. A. Frimer, Ed.), CRC Press, Boca Raton, FL (1985), pp. 67–92.
98. C. S. Foote and G. Uhde, *Org. Photochem. Synth.*, **1**, 60 (1971).
99. N. Furutachi, Y. Nakadaira and K. Nakanishi, *J. Chem. Soc., Chem. Commun.*, 1625 (1968).
100. J. Terao and S. Matsushita, *J. Am. Oil Chem. Soc.*, **54**, 234 (1977).
101. H. W.-S. Chan, *J. Am. Oil Chem. Soc.*, **54**, 100 (1977).
102. J. Terao and S. Matsushita, *Agric. Biol. Chem.*, **41**, 2467 (1977).
103. N. A. Khan, *Chem. Ind. (London)*, 1000 (1973).
104. M. Matsumoto and K. Kondo, *Tetrahedron Lett.*, 3935 (1975).
105. M. J. Kulig and L. L. Smith, *J. Org. Chem.*, **38**, 3639 (1973).
106. M. J. Kulig and L. L. Smith, *J. Org. Chem.*, **39**, 3398 (1974).
107. L. L. Smith, *Cholesterol Autoxidation*, Plenum Press, New York (1981).
108. B.-M. K. Kwon, R. C. Kanner and C. S. Foote, *Tetrahedron Lett.*, **30**, 903 (1989).
109. M. Orfanopoulos and C. S. Foote, *Tetrahedron Lett.*, **27**, 5991 (1985).
110. N. Shimizu, F. Shibata, S. Imazu and Y. Tsuno, *Chem. Lett.*, 1071 (1987).
111. R. Matusch and G. Schmidt, *Angew. Chem., Int. Ed. Engl.*, **27**, 717 (1988).
112. R. Matusch and G. Schmidt, *Helv. Chim. Acta*, **72**, 51 (1989).
113. K. E. O'Shea and C. S. Foote, *J. Am. Chem. Soc.*, **110**, 7167 (1988).
114. W. Adam, A. Griesbeck and D. Kappes, *J. Org. Chem.*, **51**, 4479 (1986).
115. T. Takata, K. Hoshino and E. Takeuchi, *Tetrahedron Lett.*, **25**, 4767 (1984).
116. T. Takata and W. Ando, *Tetrahedron Lett.*, **27**, 1591 (1986).
117. E. Schmitz, A. Rieche and E. Beyer, *Chem. Ber.*, **94**, 2921 (1961).
118. A. Rieche, C. Bischoff and P. Dietrich, *Chem. Ber.*, **94**, 2932 (1961).
119. A. Rieche and C. Bischoff, *Chem. Ber.*, **94**, 2772 (1961).
120. A. Rieche, E. Schmitz and E. Beyer, *Chem. Ber.*, **91**, 1935 (1958).
121. J. Rigaudy and G. Izoret, *C. R. Acad. Sci.*, **236**, 2086 (1953).
122. P. H. Dussault and N. A. Porter, *J. Am. Chem. Soc.*, **110**, 6276 (1988).
123. P. Dussault and A. Sahli, *Tetrahedron Lett.*, **31**, 5117 (1990).
124. N. Baba, K. Yoneda, S. Tahara, J. Iwasa, T. Kaneko and M. Matsuo, *J. Chem. Soc., Chem. Commun.*, 1281 (1990).
125. A. J. Bloodworth and H. J. Eggelte, *J. Chem. Soc., Perkin Trans. 1*, 1375 (1981).
126. C. Jefford and C. G. Rimbault, *Tetrahedron Lett.*, 2375 (1977).
127. W. Adam and J. Fierro, *J. Org. Chem.*, **43**, 1159 (1978).
128. G. M. Rubottom and M. L. Nieves, *Tetrahedron Lett.*, 2423 (1972).
129. W. Adam, O. Cueto and V. Ehrig, *J. Org. Chem.*, **41**, 370 (1976).
130. R. K. Haynes, S. C. Vonwiller, *J. Chem. Soc., Chem. Commun.*, 448 (1990).
131. J. A. Howard and K. U. Ingold, *Can. J. Chem.*, **47**, 3797 (1969).
132. C. Walling and A. Padwa, *J. Am. Chem. Soc.*, **85**, 1593 (1963).
133. J. A. Howard, K. U. Ingold and W. J. Schwalm, *Adv. Chem. Ser.*, **75**, 6 (1968).
134. M. O. Funk, R. Isaac and N. A. Porter, *J. Am. Chem. Soc.*, **97**, 1281 (1975).
135. N. A. Porter, M. O. Funk, D. W. Gilmore, R. Isaac and J. R. Nixon, *J. Am. Chem. Soc.*, **98**, 6000 (1976).
136. N. A. Porter, J. R. Nixon and D. W. Gilmore, In *Organic Free Radicals* (W. A. Pryor, Ed.), American Chemical Society, Washington, DC (1978), pp. 89–101.
137. N. A. Porter, A. N. Roe and A. T. McPhail, *J. Am. Chem. Soc.*, **102**, 7574 (1980).

138. A. L. J. Beckwith and R. D. Wagner, *J. Am. Chem. Soc.*, **101**, 7099 (1979).
139. A. L. J. Beckwith and R. D. Wagner, *J. Chem. Soc., Chem. Commun.*, 485 (1980).
140. P. J. Barker, A. L. J. Beckwith and Y. Fung, *Tetrahedron Lett.*, **24**, 97 (1983).
141. A. L. J. Beckwith and R. D. Wagner, *J. Org. Chem.*, **46**, 3638 (1981).
142. K. S. Feldman, R. E. Simpson and M. Pravez, *J. Am. Chem. Soc.*, **108**, 1328 (1986).
143. K. S. Feldman and R. E. Simpson, *J. Am. Chem. Soc.*, **111**, 4878 (1989).
144. A. N. Roe, A. T. McPhail and N. A. Porter, *J. Am. Chem. Soc.*, **105**, 1199 (1983).
145. N. A. Porter, A. N. Roe and A. T. McPhail, *J. Am. Chem. Soc.*, **102**, 7574 (1980).
146. J. A. Khan and N. A. Porter, *Angew. Chem., Int. Ed. Engl.*, **21**, 217, (1982).
147. E. M. Bevilacqua, *J. Am. Chem. Soc.*, **77**, 5394 (1955).
148. F. R. Mayo, *Ind. Eng. Chem.*, **52**, 614 (1960).
149. M. A. Golub and M. S. Hsu, *Rubber Chem. Technol.*, **48**, 953 (1975).
150. R. L. Pecsok, P. C. Painter, J. R. Shelton and J. L. Koenig, *Rubber Chem. Technol.*, **49**, 1010 (1976).
151. J. R. Shelton, R. L. Pecsok and J. L. Koenig, *ACS Symp. Ser.*, No. 95, 75 (1979).
152. E. M. Bevilacqua, *Rubber Plant Age*, **80**, 271 (1956).
153. M. K. Supran, *ACS Symp. Ser.*, No. 75, 48 (1978).
154. D. E. O'Connor, E. D. Mihelich and M. C. Coleman, *J. Am. Chem. Soc.*, **106**, 3577 (1984).
155. N. A. Porter, M. O. Funk, D. W. Gilmore, S. R. Issac, D. B. Menzel, J. R. Nixon and J. H. Roycroft, in *Biochemical Aspects of Prostaglandins and Thromboxanes* (N. Kharasch and J. Fried, Eds), Academic Press, New York (1977), pp. 39–53.
156. W. A. Pryor, J. P. Stanley and E. Blair, *Lipids*, **11**, 370 (1976).
157. E. J. Corey, J. O. Albright, A. E. Barton and S. Hashimoto, *J. Am. Chem. Soc.*, **102**, 1435 (1980).
158. H. A. J. Carless and R. J. Batten, *Tetrahedron Lett.*, **23**, 4735 (1982).
159. E. D. Mihelich, *J. Am. Chem. Soc.*, **102**, 7143 (1980).
160. N. A. Porter and P. J. Zuraw, *J. Org. Chem.*, **49**, 1345 (1984).
161. S. Saebo, A. L. J. Beckwith and L. Radom, *J. Am. Chem. Soc.*, **106**, 5119 (1984).
162. D. A. Lindsay, J. Lusztyk and K. U. Ingold, *J. Am. Chem. Soc.*, **106**, 7087 (1984).
163. L. R. C. Barclay, J. Lusztyk and K. U. Ingold, *J. Am. Chem. Soc.*, **106**, 1793 (1984).
164. G. O. Schenck, O.-A. Neumuller and W. Eisfeld, *Justus Liebigs Ann. Chem.*, **618**, 202 (1958).
165. W. E. Brill, *J. Am. Chem. Soc.*, **87**, 3286 (1965).
166. W. E. Brill, *J. Chem. Soc., Perkin Trans. 2*, 62 (1984).
167. N. Porter and P. Zuraw, *J. Chem. Soc., Chem. Commun.*, 1472 (1985).
168. N. Porter and J. S. Wujek, *J. Org. Chem.*, **52**, 5085 (1987).
169. A. L. J. Beckwith, A. G. Davies, I. G. E. Davison, A. Maccoll and M. H. Mruzek *J. Chem. Soc., Perkin Trans. 2*, 815 (1989).
170. N. A. Porter, J. K. Kaplan and P. H. Dussault, *J. Am. Chem. Soc.*, **112**, 1266 (1990).
171. A. L. J. Beckwith, A. G. Davies, I. G. E. Davison, A. Maccoll and M. H. Mruzek, *Chem. Soc., Chem. Commun.*, 475 (1988).
172. D. V. Avila, A. G. Davies and I. G. E. Davison, *J. Chem. Soc., Perkin Trans. 2*, 1847 (1988).
173. A. G. Davies and I. G. E. Davison, *J. Chem. Soc., Perkin Trans. 2*, 825 (1989).
174. A. G. Davies and W. J. Kinart, *J. Chem. Res.*, 22 (1989).
175. A. G. Davies, I. G. E. Davison and C. H. Schiesser, *J. Chem. Soc., Chem. Commun.*, 742 (1989).
176. R. K. Haynes and S. C. Vonwiller, *J. Chem. Soc., Chem. Commun.*, 449 (1990).
177. R. K. Haynes and S. C. Vonwiller, *J. Chem. Soc., Chem. Commun.*, 451 (1990).
178. R. K. Haynes and S. C. Vonwiller, *J. Chem. Soc., Chem. Commun.*, 1102 (1990).

179. G. Sosnovsky and D. J. Rawlinson, in *Organic Peroxides* Vol. 2, (D. Swern, Ed.), Wiley–Interscience, New York (1971), Chapt. 2.
180. J. K. Kochi, in *Free Radicals*, Vol. 1, Wiley–Interscience, New York (1973), Chapt. 11.
181. R. A. Sheldon, in *The Chemistry of Functional Groups, Peroxides* (S. Patai, Ed.), Wiley–Interscience, New York (1983), Chapt. 6.
182. Z. Cekovic, L. Dimitruevic, G. Diokic and T. Srinic, *Tetrahedron*, **35**, 2021 (1979).
183. Z. Cekovic and M. Cvetkovic, *Tetrahedron Lett.*, **23**, 3791 (1982).
184. S. L. Schreiber, B. Hulin and W.-F. Liew, *Tetrahedron*, **42**, 2945 (1986).
185. I. Saito, R. Nagata, K. Yuba and T. Matsuura, *Tetrahedron Lett.*, **24**, 4439 (1983).
186. A. Gonzalez, A. Galindo, H. Mansilla and A. Trigos, *Tetrahedron Lett.*, **28**, 4203 (1982).
187. Z. Cekovic and R. Saicic, *Tetrahedron Lett.*, **27**, 5981 (1986).
188. G. Cardinale, J. A. M. Laan and J. P. Ward, *Tetrahedron*, **41**, 2899 (1985).
189. G. Cardinale, J. C. Grimmelikhuysen, J. A. M. Laan, F. P. van Lier, D. van der Steen and J. P. Ward, *Tetrahedron*, **45**, 2899 (1989).
190. S. L. Schreiber, *J. Am. Chem. Soc.*, **102**, 6163 (1980).
191. S. L. Schreiber, R. E. Claus and J. Reagan, *Tetrahedron Lett.*, **23**, 3867 (1982).
192. S. L. Schreiber and W.-F. Liew, *Tetrahedron Lett.*, **24**, 2363 (1983).
193. S. L. Schreiber and W.-F. Liew, *J. Am. Chem. Soc.*, **107**, 2980 (1985).
194. S. L. Schrieber and B. Hulin, *Tetrahedron Lett.*, **27**, 4561 (1986).
195. S. L. Schreiber, T. Sammakia, B. Hulin and G. Schulte, *J. Am. Chem. Soc.*, **108**, 2106 (1986).
196. H. W. Gardner, *Free Radical Biol. Med.*, **7**, 65 (1989).
197. R. Labeque and L. J. Marnett, *J. Am. Chem. Soc.*, **109**, 2828 (1987).
198. H. W. Gardner, K. Eskins, G. Grams and G. E. Inglett, *Lipids*, **7**, 324 (1972).
199. H. W. Gardner, R. Kleiman and D. Weisleder, *Lipids*, **9**, 696 (1974).
200. M. Hamberg, *Lipids*, **10**, 87 (1975).
201. H. W. Gardner and R. Kleiman, *Biochim. Biophys. Acta*, **665**, 113 (1981).
202. H. W. Gardner, R. D. Plattner and D. Weisleder, *Biochim. Biophys. Acta*, **834**, 65 (1985).
203. T. A. Dix and L. J. Marnett, *J. Biol. Chem.*, **260**, 5351 (1985).
204. P. Schieberle, Y. Trebert, J. Firl and W. Grosch, *Chem. Phys. Lipids*, **37**, 99 (1985).
205. P. Schieberle, Y. Trebert, J. Firl and W. Grosch, *Chem. Phys. Lipids*, **41**, 101 (1986).
206. B. Taljaard, A. Goosen and C. W. McCleland, *S. Afr. J. Chem.*, **40**, 139 (1987).
207. H. Hock, *Angew. Chem.*, **49**, 595 (1936).
208. A. A. Frimer, *Chem. Rev.*, **79**, 359 (1979).
209. A. A. Frimer, *J. Photochem.*, **25**, 211 (1984).
210. A. M. Jeffrey and D. M. Jerina, *J. Am. Chem. Soc.*, **94**, 4048 (1972).
211. R. C. Ronald and T. S. Lillie, *J. Am. Chem. Soc.*, **105**, 5709 (1983).
212. T. S. Lillie and R. C. Ronald, *J. Org. Chem.*, **50**, 5084 (1985).
213. R. C. Ronald and T. S. Lillie, *Tetrahedron Lett.*, **27**, 5787 (1986).
214. R. C. Ronald, S. M. Ruder and T. S. Lillie, *Tetrahedron Lett.*, **28**, 131 (1987).
215. S. M. Ruder and R. C. Ronald, *Tetrahedron Lett.*, **28**, 135 (1987).
216. H. Iio, H. Nagaoka and Y. Kishi, *Tetrahedron Lett.*, **22**, 2451 (1981).
217. S. L. Schreiber and W.-F. Liew, *Tetrahedron Lett.*, **24**, 2363 (1983).
218. H. W. Gardner and R. D. Planter, *Lipids*, **19**, 294 (1984).
219. H. W. Gardner, D. Weisleder and E. C. Nelson, *J. Org. Chem.*, **49**, 508 (1984).
220. T. Galliard and D. R. Philips, *Biochem. J.*, **129**, 743 (1972).
221. L. Crombie, D. O. Morgan and E. H. Smith, *J. Chem. Soc., Chem. Commun.*, 502 (1986).

222. K. U. Ingold, *Acc. Chem. Res.*, **2**, 1 (1969).
223. N. A. Porter, M. A. Cudd, R. W. Miller and A. T. McPhail, *J. Am. Chem. Soc.*, **102**, 415 (1980).
224. N. A. Porter and J. R. Nixon, *J. Am. Chem. Soc.*, **100**, 7116 (1978).
225. K. B. Sharpless and T. R. Verhoeven, *Aldrichim. Acta*, **12**(4), 63 (1979).
226. K. B. Sharpless, in *Proceedings of the Robert A. Welch Foundation Conferences on Chemical Research, XXVII, Stereospecificity in Chemistry and Biochemistry*, (1984), Chapt. 3. p. 59.
227. B. E. Rossiter, T. Katsuki and K. B. Sharpless, *J. Am. Chem. Soc.*, **103**, 464 (1981).
228. V. S. Martin, S. S. Woodard, T. Katsuki, Y. Yamada, M. Ikeda and K. B. Sharpless, *J. Am. Chem. Soc.*, **103**, 6237 (1981).
229. I. D. Williams, S. F. Pedersen, K. B. Sharpless and S. J. Lippard, *J. Am. Chem. Soc.*, **106**, 6431 (1984).
230. W. Adam, A. Griesbeck and E. Staab, *Angew. Chem., Int. Ed. Engl.*, **25**, 269 (1986).
231. W. Adam and E. Staab, *Liebigs Ann. Chem.*, 757 (1988).
232. W. Adam, M. Braun, A. Griesbeck, V. Lucchini, E. Staab and B. Will, *J. Am. Chem. Soc.*, **111**, 203 (1989).
233. C. Walling and L. Heaton, *J. Am. Chem. Soc.*, **87**, 38 (1965).
234. J. Rebek, Jr, and R. McCready, *J. Am. Chem. Soc.*, **102**, 5602 (1980).
235. T. Tezuka and M. Iwaki, *J. Chem. Soc., Perkin Trans. 1*, 2507 (1984).
236. D. W. Dixon and M. Barbush, *J. Org. Chem.*, **50**, 3194 (1985).
237. J. H. Liu and R. G. Weiss, *J. Org. Chem.*, **50**, 3657 (1985).
238. T. Tezuka and S. Ando, *Chem. Lett.*, 1621 (1985).
239. E. Y. Osei-Twum, D. McCallion, A. S. Nazran, R. Panicucci, P. A. Risbood and J. Warkentin, *J. Org. Chem.*, **49**, 336 (1984).
240. A. L. Baumstark and P. C. Vasquez, *J. Org. Chem.*, **52**, 1939 (1987).
241. A. L. Baumstark, M. Dotrong and P. C. Vasquez, *Tetrahedron Lett.*, **28**, 1963 (1987).
242. T. C. Bruice, *Acc. Chem. Res.*, **13**, 256 (1980).
243. C. Kemal, T. W. Chan and T. C. Bruice, *J. Am. Chem. Soc.*, **99**, 7272 (1977).
244. S. Ball and T. W. Bruice, *J. Am. Chem. Soc.*, **102**, 6498 (1980).
245. S. Muto and T. W. Bruice, *J. Am. Chem. Soc.*, **104**, 2284 (1982).
246. W. H. Rastetter, T. R. Gadek, J. P. Tane and J. W. Frost, *J. Am. Chem. Soc.*, **108**, 2228 (1979).
247. J. W. Frost and W. H. Rastetter, *J. Am. Chem. Soc.*, **103**, 5242 (1981).
248. A. Miller, *Tetrahedron Lett.*, **23**, 753 (1982).
249. A. E. Miller, J. J. Bischoff, C. Bizub, P. Luminoso and S. Smiley, *J. Am. Chem. Soc.*, **108**, 7773 (1986).
250. S. Shinkai, T. Yamaguchi, O. Manabe and F. Toda, *J. Chem. Soc., Chem. Commun.*, 1399 (1988).
251. L. J. Marnett, *Carcinogenesis*, **8**, 1365 (1981).
252. A. Sevanian and P. Hochstein, *Annu. Rev. Nutr.*, **5**, 375 (1985).
253. T. W. Kensler and B. G. Taffe, *Adv. Free Radical Biol. Med.*, **2**, 347 (1986).
254. T. A. Dix and L. J. Marnett, *J. Biol. Chem.*, **260**, 5351 (1985).
255. R. Labeque and L. J. Marnett, *Biochemistry*, **27**, 7060 (1988).
256. T. A. Dix, R. Fontana, A. Panthani and L. J. Marnett, *J. Biol. Chem.*, **260**, 5358 (1985).

3 Dialkyl Peroxides

SEIICHI MATSUGO
Department of Chemical and Biochemical Engineering, Faculty of Engineering, Toyama University, Gofuku 3190, Toyama 930, Japan

and

ISAO SAITO
Department of Synthetic Chemistry, Faculty of Engineering, Kyoto University, Kyoto 606, Japan

1 INTRODUCTION

Dialkyl peroxides are derivatives of hydrogen peroxides with the general formula ROOR′, where R and R′ are acyclic or cyclic alkyl groups. In this chapter, various aspects of the synthesis and reactions of dialkyl peroxides,

Organic Peroxides. Edited by W. Ando

together with their fundamental physical properties, will be reviewed by selecting only recent examples from among innumerable references. In view of the wide interest in the biological activities of peroxides, we have also incorporated important naturally occurring cyclic dialkyl peroxides. We have additionally attempted to illustrate the use of dialkyl peroxides in organic synthesis.

2 PHYSICAL PROPERTIES OF DIALKYL PEROXIDES

2.1 Structural Properties of Dialkyl Peroxides

Dialkyl peroxides are structurally related to hydrogen peroxide. The C—O—O valence angle is *ca* 105°. ROOR′ is not linear or positioned in the same plane. The two-dimensional angle (ϕ) of the peroxides was determined by X-ray analysis or IR spectrometry. The most extensively studied peroxide is hydrogen peroxide itself. Based on the Raman spectra, Penny and Sutherland[1] reported the salient conformational features of hydrogen peroxide. The eclipsed conformation of the hydroperoxide (*cis*, $\phi = 0°$) is 7 kcal mol^{-1} less stable than that of the skew conformation. The two-dimensional angle of the skew conformation of hydrogen peroxide was calculated to be 120°.[2] In the case where both substituents were CF_3, namely CF_3OOCF_3, E_{cis} became 16.6 ± 6 kcal mol^{-1}, which is much larger than that of HOOH.[3] The stability of the skew conformation can be rationalized by considering the least electron pair repulsion[1] and by LCAO-MO-SCF calculation.[2] Direct evidence for the stability and conformational structure of the peroxides can be obtained from X-ray analyses. There are X-ray data on more than 50 dialkyl peroxides. The characteristics of interest to determine the properties of dialkyl peroxides are the following: (i) the O—O bond length; (ii) the C—O bond length; (iii) the C—O—O angle (θ); and (iv) the two-dimensional angle of the peroxide (ϕ in Figure 1). Table 1 summarizes typical X-ray data for various types of peroxides.

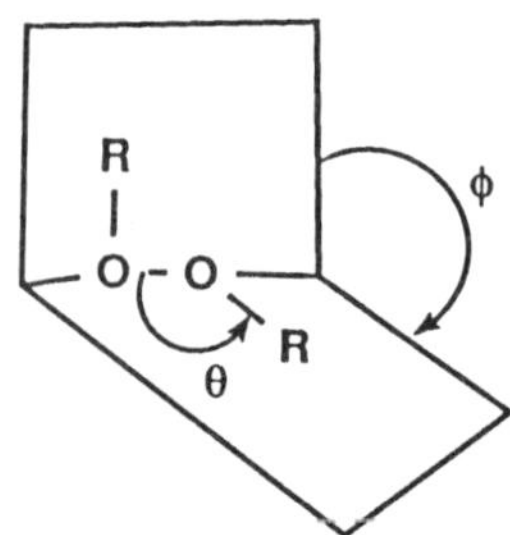

Figure 1. Valence bond angle (θ) and two-dimensional angle (ϕ) of dialkyl peroxides

Table 1. Geometric parameters of dialkyl peroxides

Peroxide	Bond length (Å)		Bond angle (°)	
	O—O	C—O	COO	ϕ
Acyclic:				
$(CF_3O)_2$[4]	1.419	1.399	107	123
Bis(triphenylmethyl) peroxide (**1**)[5]	1.480	1.461	107.5	180
Cyclic:				
5-membered (**2**)[6]	1.483	1.45	107.2	20
6-membered (**3**)[7]	1.45	1.46	110	61
8-membered (**4**)[8]	1.45	1.42	108.5	120
9-membered (**5**)[9]	1.477	1.417	107.6	135.6
4-membered (**6**)[10]	1.491	1.475	89	21

Marsden *et al.*[4] examined the change in the O—O and C—O bond lengths of the $XOOCF_3$ by changing X (H, F, Cl, CF_3). For $HOOCF_3$ and $ClOOCF_3$, the O—O and C—O bond lengths were 1.45 and 1.37 Å, respectively, whereas in CF_3OOCF_3 they were 1.40 and 1.42 Å, respectively. This result shows that the introduction of electron-withdrawing substituents shortens the C—O bond and lengthens the O—O bond. Bis(triphenylmethyl) peroxide (**1**) has a very unique conformation in which the two-dimensional angle (ϕ) is 180°. However, the O—O bond length of this peroxide is nearly the same as those of other cyclic or acyclic peroxides.

The O—O bond of the five-membered peroxide **2** is longer than those of six- and eight-membered peroxides. An X-ray study of the naturally occurring ozonide gilvanol (**7**) also showed the O—O bond length to be 1.485 Å.[11] The repulsion of the lone pair of electrons was the major factor for the long bond length of the five-membered peroxide.

Six-membered peroxides have been studied systematically for many years. This type of peroxide is isomerized thermally and preferentially occupies chair conformations. According to ^{1}HNMR studies, the diperoxide of acetone (**8**) occupies a chair conformation in solution and the energy necessary for the interconversion was calculated to be 12.3 kcal mol^{-1}.[12]

Other six-membered peroxides occupied a similar ideal chair conformation and the deviation from the ideal chair conformation has been thought to be very small. The bond angle (*ca* 4°) around the ring bond and an axial substituent was larger than that of the ideal tetrahedral angle. On the other hand, the bond angle around a ring bond and an equatorial substituent was smaller than the ideal tetrahedral angle. Groth[9] attributed these distortions to the repulsive non-bonded interactions which occurred across the face of the peroxide ring involving the peroxide oxygen and a hydrogen on the axial carbon atom.

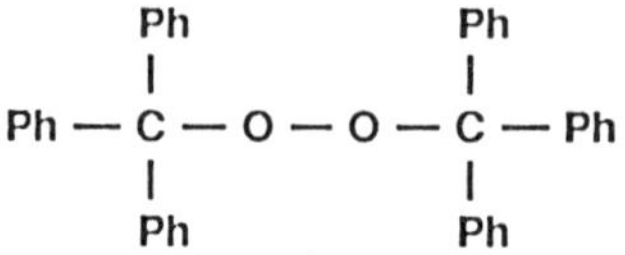

1

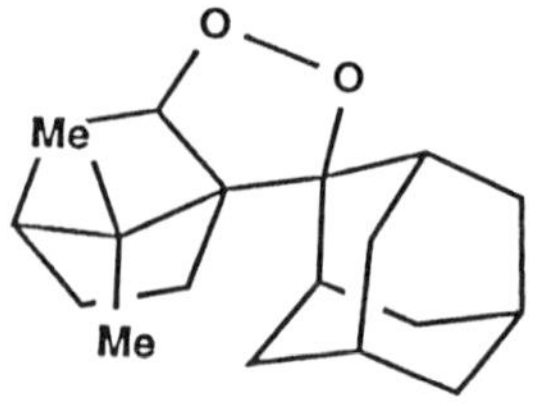

2

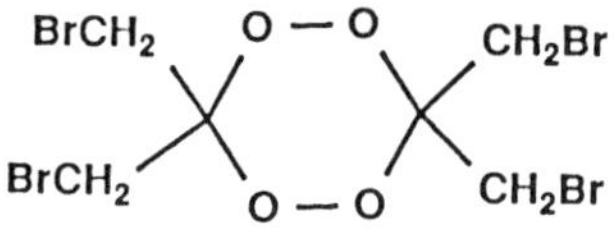

3

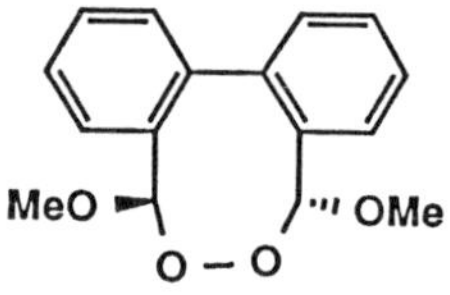

4

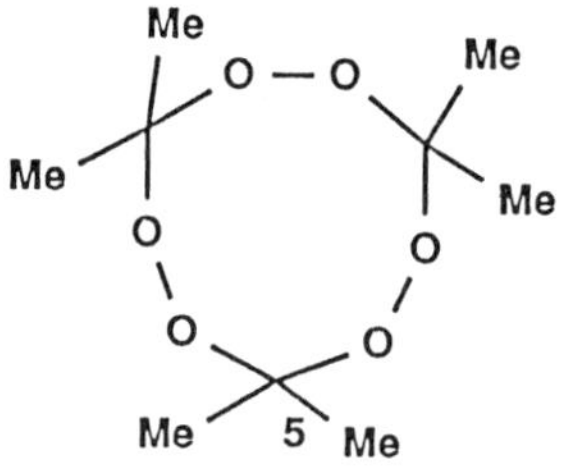

5

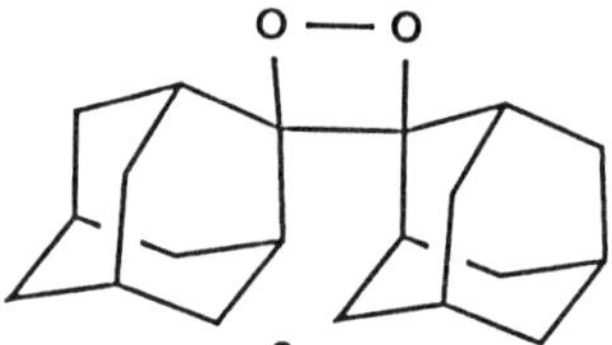

6

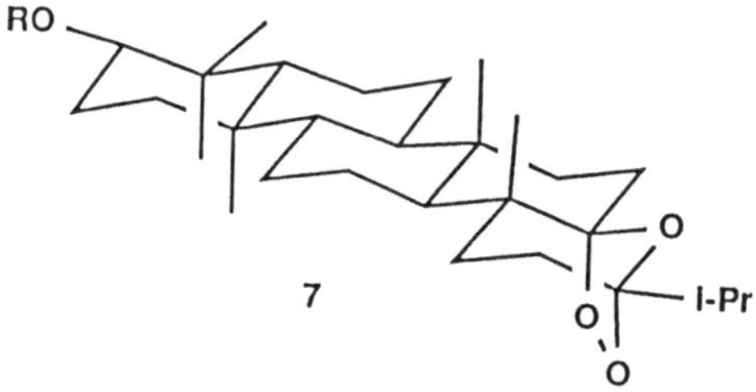

7

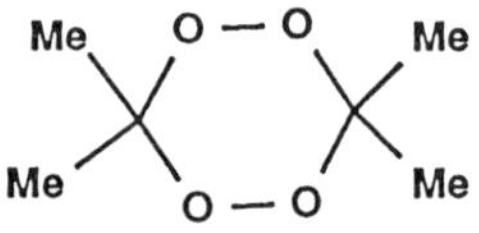

8

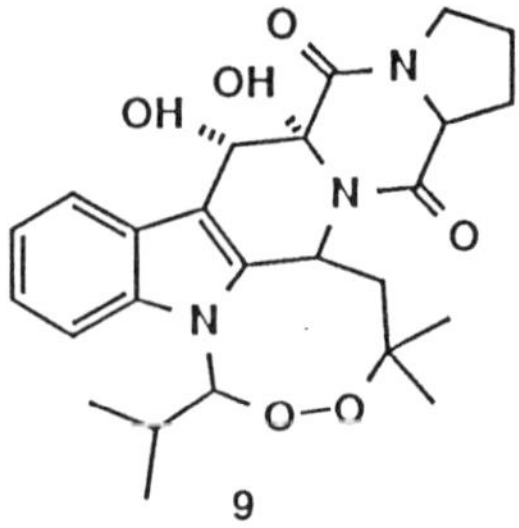

9

The ozonation of phenanthrene in methanol followed by acid-catalyzed methanolysis gives rise to the formation of stable eight-membered cyclic peroxide (**4**). The crystal structure of **4** occupies a C_2 symmetric conformation with the molecular twofold symmetry axis coincident with a crystallographic twofold axis of space group C_2/c ($Z = 4$).[8] The torsional angles around the bonds of the eight-membered ring occupying a symmetry-independent conformation are 120, -31, -52, 3 and 65°, starting with the O—O bond. The dihedral angle between the planes of the phenyl rings was found to be 62°. The reported bond lengths along the chain C—O—C—O—O were 1.434, 1.396, 1.416 and 1.452 Å, respectively (Figure 2).

Another type of eight-membered peroxide of an indole alkaloid (**9**) has also been reported.[13] In this case, the conformation of the peroxide ring is different from the former case. Endocyclic torsional angles and bond distances, starting with the O—O bond and rotating in a clockwise direction around the eight-membered peroxide ring, were as follows: 137° (1.51 Å), -65° (1.43 Å), -26° (1.46 Å), 11° (1.40 Å), 80° (1.51 Å), -90° (1.54 Å), 55° (1.53 Å) and -77° (1.47 Å).

The very explosive 'trimeric ketone peroxide' (**5**) is the only specimen of nine-membered cyclic peroxides. Groth[9] carefully analyzed the monoclinic crystal structure of this peroxide. According to the result, it possesses high internal consistency, and its symmetry is C_3, nearly D_3. Average deviations from the means of the geometric parameters were as follows: O—O 1.477(4) Å, C—O 1.417(3) Å, C—C 1.517(5) Å, ∠C—O—O 107.6(2)°, ∠O—C—O 112.3(3)°, ∠C—C—C 113.8(1)°; endocyclic torsional angles for C_3 symmetry: ∠C—O—O—C 135.6(3)°, ∠O—O—C—O $-56.3(8)$°, ∠O—C—O—O $-59.2(8)$°. The O—O bond length is equal to that in the dimeric ketone peroxides; however, the C—O bond length is significantly shorter in the trimer. The molecular threefold axis was directed along the crystallographic direction $u = -0.016\,a + 0.086\,b + 0.035\,c$, which was inclined by 22° to the b

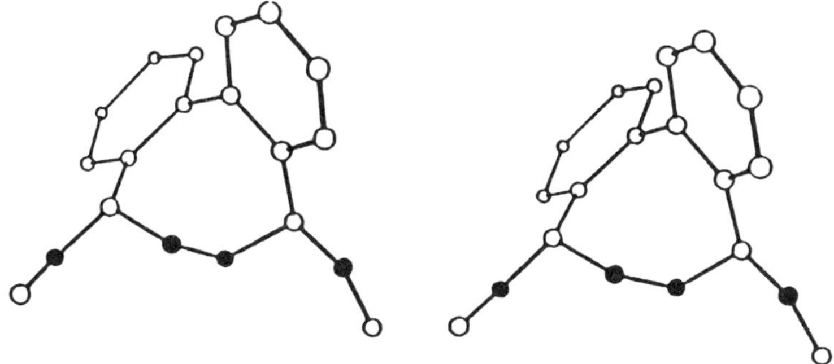

Figure 2. Stereoscopic drawing of the C_2 symmetric conformation of **4**

monoclinic symmetry axis. A stable four-membered peroxide (dioxetane **6**) which melts at 163°C was obtained by the singlet oxygen reaction of adamantylideneadamantane.[14] The X-ray crystallographic analysis showed the C_2 symmetry of **6**. All bonds of the dioxetane were found to be significantly stretched (O—O 1.491, C—O 1.475, C—C 1.549 Å) compared with the corresponding bonds of other peroxides (Table 1).

2.2 Thermochemistry of Dialkyl Peroxides

The thermochemistry of peroxides, including dialkyl peroxides, was comprehensively reviewed by Benson and Shaw[15] in 1970, and they also described a basic group additivity scheme for peroxides in general. In this section we briefly describe the thermochemical data for peroxides such as heats of formation and bond strengths. Experimentally determined heats of formation of dialkyl peroxides are listed in Table 2.

The heat of formation of one peroxy radical has been determined to be 10.5 ± 2.5 kJ mol^{-1} (2.5 ± 0.6 kcal mol^{-1}) from measurements of the equilibrium constants, as shown below.[18] This value showed a marked consistency with the other experimental data or calculations.

$$\cdot OH + NO_2 \rightleftharpoons HOO\cdot + NO \quad (1)$$

The bond strength of peroxides is of considerable interest in view of their explosive characteristics on thermolysis. The O—O bond strength can be derived from the kinetics of thermal decomposition of dialkyl peroxides. The bond strengths of various dialkyl peroxides are listed in Table 3. These bond strengths were similar, 155 ± 5 kJ mol^{-1} (37.1 ± 1 kcal mol^{-1}), independent of the alkyl group. Changing the hydrogen to an electron-withdrawing group such as fluorine strengthens the O—O bond to 193 kJ mol^{-1} (46.2 kcal mol^{-1}), whereas replacement of an acetyl group with a methyl group weakens the O—O bond strength to 125.8 kJ mol^{-1} (30.1 kcal mol^{-1}). A marked strengthening of the O—O bond was observed on replacing one *tert*-butyl group with $SiMe_3$, $GeEt_3$ or $SnEt_3$.

Table 2. Measured heats of formation of dialkyl peroxides

	H_f°	
Compound	kJ mol^{-1}	kcal mol^{-1}
Dimethyl peroxide[16]	−125.8	−30.1
Diethyl peroxide[16]	−192.7	−46.1
Di-*tert*-butyl peroxide[16]	−340.7	−81.5
Hydrogen peroxide[17]	−135.9	−32.5

Table 3. Bond strengths (D°_{298})

Compound	D°_{298} kJ mol^{-1}	D°_{298} kcal mol^{-1}
MeO—OMe[19]	155.0	37.1
EtO—OEt[20]	158.6	37.9
n-PrO—OPr-*n*[20]	155.2	37.1
i-PrO—OPr-*i*[20]	157.7	37.7
s-BuO—OBu-*s*[21]	152.3	36.4
t-BuO—OBu-*t*[22]	152.0	36.4
neo-$C_5H_{11}O—OC_5H_{11}$-*neo*[23]	152.3	36.4
$CF_3O—OCF_3$[24]	193	46.2
$Me_3SiO—OBu$-*t*[25]	197	47
$Et_3GeO—OBu$-*t*[25]	192	46
$Et_3SnO—OBu$-*t*[25]	205	49
MeCOO—OCOMe	125.8	30.1

3 SYNTHESIS OF DIALKYL PEROXIDES

There are several methods for the preparation of dialkyl peroxides, which are classified into three types: (i) acid- or base-catalyzed, (ii) metal-catalyzed; and (iii) photochemical.

3.1 Acid- or Base-catalyzed Synthesis of Dialkyl Peroxides

The most commonly used synthesis of dialkyl peroxide **13** is the reaction of hydrogen peroxide (**10**) or alkyl hydroperoxide (**12**) with alkylating reagents (RX, **11**) in the presence of a catalytic amount of acid or base (equation 2).

$$\underset{\mathbf{10}}{H_2O_2} + \underset{\mathbf{11}}{RX} \xrightarrow[-HX]{} \underset{\mathbf{12}}{ROOH} \xrightarrow[-HX]{RX} \underset{\mathbf{13}}{ROOR} \quad (2)$$

3.1.1 Synthesis of symmetrical dialkyl peroxides

A general synthetic procedure for dialkyl peroxides is the reaction of an alkyl sulfonate (**14**) with 0.5 molar equiv. of hydrogen peroxide.[26] This reaction is heavily dependent on the molar concentration of hydrogen peroxide. For example, an excess of hydrogen peroxide reduces the yield of dialkyl peroxides and increases the yield of alkyl hydroperoxides (equation 3).

$$\underset{\mathbf{14}}{2\,CH_3SO_3R} + \underset{\mathbf{10}}{H_2O_2} \xrightarrow[\text{base}]{} \underset{\mathbf{13}}{ROOR} + 2\,CH_3SO_3H \quad (3)$$

Di(tertiary akyl) peroxides can be prepared easily by the reaction of tertiary alcohols (**15**) with hydrogen peroxide.[27] The alkenes **16** which afford a tertiary carbocation by acid catalysis also produce the di(tertiary alkyl) peroxide (**17**) by the reaction of hydrogen peroxide in the presence of acid.[28] Carbonyl compounds such as benzaldehyde also produce a dialkyl peroxide (**18**) by the action of mineral acid, although by-product formation is usually observed in these cases.[29] α-Dialkylaminomethanol, produced by the condensation of dialkylamine and formaldehyde, reacts with hydrogen peroxide to afford a dialkylaminomethyl peroxide (**19**, equation 4).[30]

$$\text{(i)}\quad \underset{\mathbf{15}}{ROH} + H^+ \longrightarrow R^+ \xrightarrow{H_2O_2} \underset{\mathbf{13}}{ROOR} + H^+$$

$$\text{(ii)}\quad \underset{\mathbf{16}}{(Me)(n\text{-}Pr)C{=}CH_2} + H_2O_2 \xrightarrow{70\%\ H_2SO_4} \underset{\mathbf{17}}{\left(n\text{-}Pr-\overset{Me}{\underset{Me}{C}}-O\right)_2} \qquad (4)$$

$$\text{(iii)}\quad RCHO + H_2O_2\,(\text{anhyd}) \xrightarrow[\text{ether}]{} \underset{\mathbf{18}}{R\underset{OH}{C}HOO\underset{OH}{C}HR}$$

$$\text{(iv)}\quad PhNHMe + HCHO + H_2O_2 \longrightarrow \underset{\mathbf{19}}{(PhMeNCH_2O\text{-})_2}$$

3.1.2 Synthesis of unsymmetrical dialkyl peroxides

Dialkyl peroxides possessing two different alkyl groups can be prepared by nucleophilic substitution of alkyl hydroperoxides with alkylating reagents in the presence of base or acid. The most commonly used procedure is the alkylation of ROOH with dialkyl sulfate.[31] This procedure can be used when R is a primary,[32] secondary[33] or tertiary alkyl group[34] (equation 5).

$$ROOH + R'X \longrightarrow ROOR' \qquad (5)$$

$$(X{=}OSO_2OR'',\ SO_2OR'',\ OH,\ Cl)$$

$$2\,H_2O_2 + n\text{-}BuOSO_2OMe \longrightarrow \underset{\mathbf{20}}{n\text{-}BuOOMe}$$

Reaction of alkenes with alkyl hydroperoxides also affords dialkyl peroxides under acid catalysis.[35] Acetals or ketals are used for the preparation of dialkyl peroxides. A molar excess of hydrogen peroxide gave diperoxyalkyl-substituted compounds.[36] The reaction of *tert*-butyl hypochlorite and alkenes in the presence of alkyl hydroperoxides gave α-chlorodialkyl peroxides (equation 6).[37]

$$\mathrm{ROOH} + \mathrm{R'R''C{=}CHR'''} \xrightarrow{\mathrm{H^+}} \mathrm{R'R''C(OOR)CH_2R'''} \quad \mathbf{21}$$

$$\mathrm{Me_2C(OR')_2} \underset{}{\overset{\mathrm{H^+, ROOH}}{\rightleftharpoons}} \mathrm{Me_2C(OOR)(OR')}\ \mathbf{22} \overset{\mathrm{H^+, ROOH}}{\rightleftharpoons} \mathrm{Me_2C(OOR)_2}\ \mathbf{23} \tag{6}$$

$$\mathrm{ROOH} + \mathrm{t\text{-}BuOCl} + \mathrm{R'R''C{=}CHR'''} \longrightarrow \mathrm{R'R''C(Cl)CH(OOR)R'''}\ \mathbf{24} + \mathrm{t\text{-}BuOH}$$

Work on this type of reaction up to 1970 was summarized in a review by Sosnovsky and Rawlinson.[38] A comprehensive review dealing with the preparation of dialkyl peroxides has also been published.[39] We briefly describe newer synthetic methods reported since 1983.

Kropf *et al.*[40] reported the acid- and base-catalyzed reaction of *t*-BuOOH (**25**) with oxirane (**26**), where they observed a different mode of addition of *t*-BuOOH to the epoxide ring (equation 7). Druliner[41] described the synthesis of a series of di(cycloalkyl) peroxides (**29**) by the reaction of cycloalkyl methanesulfonates with potassium superoxide in the presence of a phase-transfer catalyst (equation 8).

$$\underset{\mathbf{25}}{\mathrm{t\text{-}BuOOH}} + \underset{\mathbf{26}}{\mathrm{Me_2C\text{—}CH_2\ (epoxide)}} \longrightarrow \underset{\mathbf{27}}{\mathrm{t\text{-}BuOO\text{—}CMe_2\text{—}CH_2OH}} + \underset{\mathbf{28}}{\mathrm{t\text{-}BuOO\text{—}CH_2\text{-}CMe_2\text{—}OH}} \tag{7}$$

OSO_3Me — cyclopentyl; Bu_4NBr / KO_2, n-hexane → cyclopentyl–O–O–cyclopentyl **29** (8)

Feldman *et al.*[42] obtained a series of polyoxygenated hydrocarbons (**30**) via a radical-mediated oxygenation of vinylcyclopropanes (equation 9). By the reaction of alk-3-enyl hydroperoxide (**31**) with *tert*-butyl hypochlorite in dichloromethane in the presence of pyridine or silica gel, cyclic 1,2-dioxolanes (**32**, **33**) were prepared by Bloodworth and Tallant[43] (equation 10).

Ph, H, CH_2CH_2Ph; O_2, $(PhSe)_2$, AIBN, 0^oC, hv, CH_3CN → O–O, Ph, CH_2CH_2Ph, H **30** (9)

OOH **31**; t-BuOCl / CH_2Cl_2, pyridine → CH_2Cl, Me, O–O **32** + CH_2Cl, CH_2Cl, O–O **33** (10)

+ CH_2Cl, Cl, OOH **34**

The isolation of pure enantiomers of fatty acid hydroperoxides is very important. Dussault and Porter[44] succeeded in isolating pure enantiomers from racemic fatty acid hydroperoxides (**35**) in very high yields by the use of a peroxyketal (**36**) as a protected hydroperoxide. The enantiomer thus isolated by high-performance liquid chromatography from 9-hydroperoxylinoleic acid ester showed more than a 93% enantiomeric excess (ee).

H, O, R, H + R'OOH; PPTS, CH_2Cl_2 → R'OO, H, O, R, H **36** (11)

R'OOH = $CH_3(CH_2)_6$ … $(CH_2)_7COOMe$, OOH (35)

R = Ph

PPTS = pyridinium p-toluenesulfonate

The perketals of hydroperoxides such as **37** are stable enough to survive reaction conditions such as Wittig olefination (equation 12). Dussault and Sahli[45] prepared various perketals which were smoothly converted into the corresponding hydroperoxides (**38**) under very mild reaction conditions.

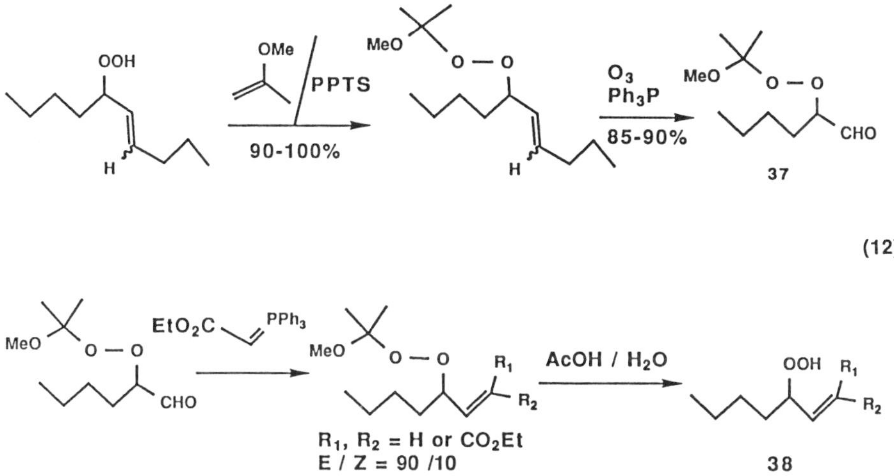

OOH H OMe PPTS 90-100% MeO O—O H O_3 Ph_3P 85-90% MeO O—O CHO **37**

(12)

MeO O—O CHO EtO_2C PPh_3 MeO O—O R_1 R_2 AcOH / H_2O OOH R_1 R_2

R_1, R_2 = H or CO_2Et
E / Z = 90 /10

38

3.2 Metal-catalyzed Synthesis of Dialkyl Peroxides

Metal-catalyzed synthesis of dialkyl peroxides has been widely used. The general synthetic procedure for this type of reaction is shown in equation 13.

$$\begin{aligned} ROOH + M^{n+} &\longrightarrow M^{(n+1)+} + RO^{\bullet} \\ RO^{\cdot} + R'H &\longrightarrow ROH + R'^{\bullet} \\ R'^{\cdot} + M^{(n+1)+} &\longrightarrow R'^{+} + M^{n+} \\ R'^{+} + ROOH &\longrightarrow ROOR' + H^{+} \end{aligned} \quad (13)$$

The most commonly used metal ion in this type of reaction is Cu^+.[46] For example, the reaction of diphenylacetonitrile (**39**) with *tert*-butyl hydroperoxide (**40**) in the presence of a catalytic amount of copper (I) bromide in benzene affords a coupled peroxide (**41**) in 86% yield (equation 14).[46] A similar type of reaction is also observed with other metal ions such as Co^{2+} [47] or Pb^{2+}.[48] However, the yields of peroxides are smaller than those with Cu^+.

$$\underset{\mathbf{39}}{(C_6H_5)_2CHCN} + \underset{\mathbf{40}}{\text{t-BuOOH}} \xrightarrow[C_6H_6]{CuBr} \underset{\mathbf{41}}{(C_6H_5)_2C(CN)\text{-OO-t-Bu}} \quad (14)$$

After the first discovery of the synthesis of dioxetanes (**42**) via hydroperoxy bromination and subsequent silver salt-promoted cyclization (equation 15),[49] several groups have utilized this methodology for the preparation of various types of dialkyl peroxides. Adam *et al.*[50] obtained various types of 1,2-dioxolanes (**43**) from cyclopropane derivatives by this route (equation 16). Using *t*-BuOOH as peroxyl source, Porter and Mitchell[51] demonstrated the facile conversion of 1,3-dibromoalkanes into 1,2-dioxolanes (**44**) in the presence of silver salt (equation 17).

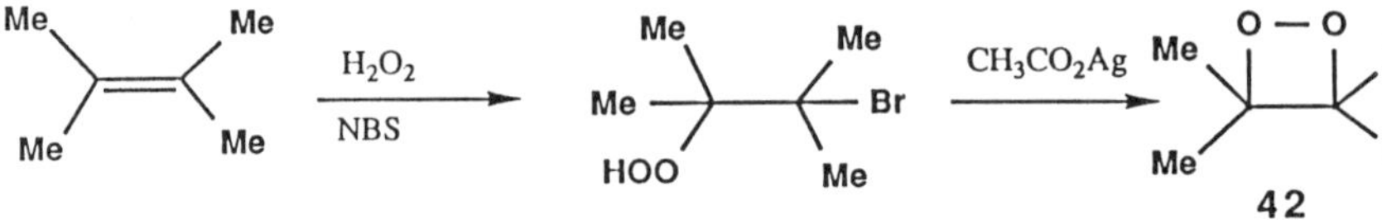

(15)

42

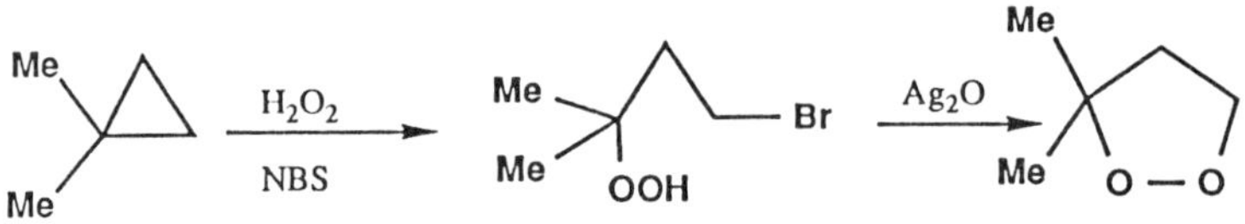

(16)

43

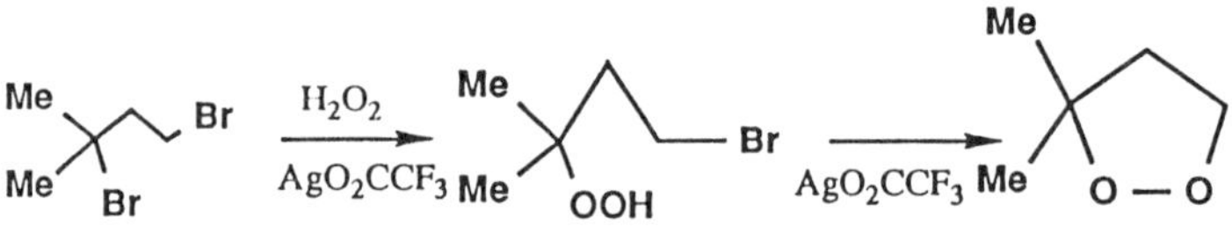

(17)

44

Bloodworth and Courtneidge[52] reported the conversion of phenylcyclopropane to the corresponding γ-bromoalkyl *tert*-butyl peroxide (**45**) via peroxymercuration followed by bromodemercuration (equation 18). Soon after, they demonstrated the conversion of cyclopropanes into 1,2-dioxolanes (**46**) by the sequence of *tert*-butyl peroxymercuration, bromodemercuration and silver salt-induced cyclization (equation 19).[53]

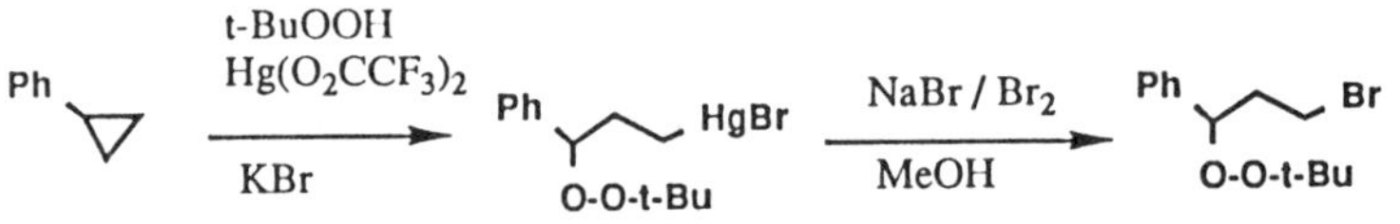

(18)

45

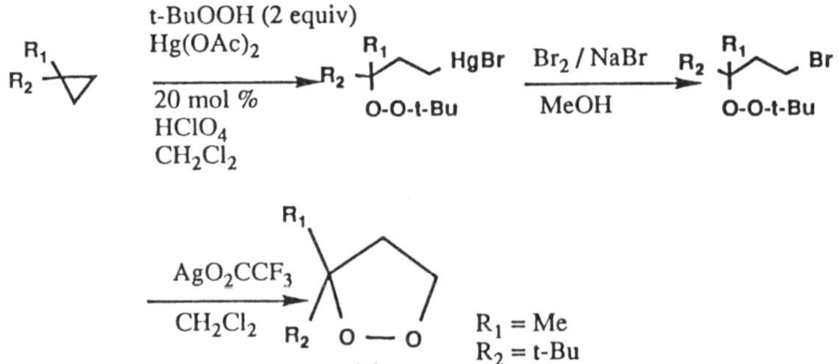

(19)

46

$R_1 = Me$
$R_2 = t\text{-}Bu$

Peroxymetallation is also applicable to the preparation of various types of α-iodo-β-*tert*-butyl peroxyethanes from the corresponding ethene derivatives.[54] In the first stage of the reaction, stereospecific *trans* addition takes place to afford two regioisomers (**47** and **48**), which further react with iodine to give mixtures of **49** and **50** from **47**, and **51** and **52** from **48**, respectively (equation 20).

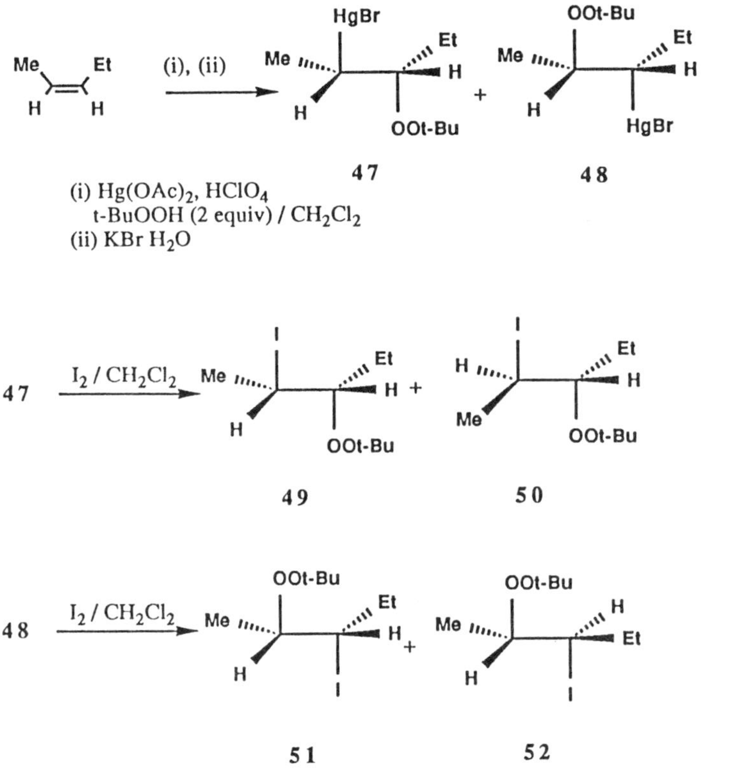
Me
Et
H
H
(i), (ii)
HgBr
Me
Et
H
H
OOt-Bu
+
OOt-Bu
Me
Et
H
H
HgBr
47
48
(i) $Hg(OAc)_2$, $HClO_4$
t-BuOOH (2 equiv) / CH_2Cl_2
(ii) KBr H_2O
47
I_2 / CH_2Cl_2
I
Me
Et
H +
H
OOt-Bu
I
H
Et
H
Me
OOt-Bu
49
50
48
I_2 / CH_2Cl_2
OOt-Bu
Me
Et
H
+
H
I
OOt-Bu
Me
H
Et
H
I
51
52

(20)

The reaction of **49** or **51** with trifluoroacetic acids gives **53** and **54** in a 5:3 ratio. The result is explained in terms of the intermediacy of a perepoxide (**55**) and subsequent attack of the trifluoromethyl group at a less hindered site. In the reactions of **50** and **52** with silver trifluoromethylacetate, **56** and **57** were obtained, respectively, with retention of stereochemistry (equation 21). In these reactions, a cyclic transition state as shown in equation 21 has been proposed. In support of this hypothesis, when methanol was added to the reaction system of **51**, the formation of a methoxy adduct was observed, whereas in the reaction of **52**, no formation of such an adduct was observed. Using a similar methodology, Bloodworth and Hargreaves[55] reported cyclic peroxide formation from bicyclic alkanes and phenylcyclopropanes.[56] Using anhydrous hydrogen peroxide and 1,3-dibromocyclopentane in the presence of silver trifluoroacetate. Takahashi *et* *al.*[57] obtained 1-phenyl-2,3-dioxabicyclo[2.2.1]heptane (**58**) together with the formation of three other peroxides (equation 22).

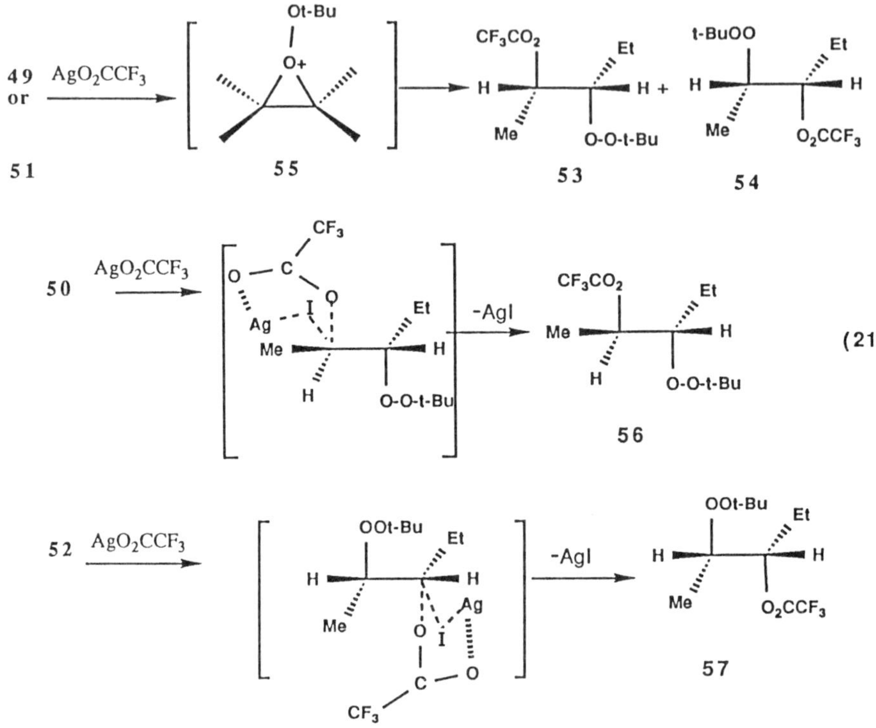

(21)

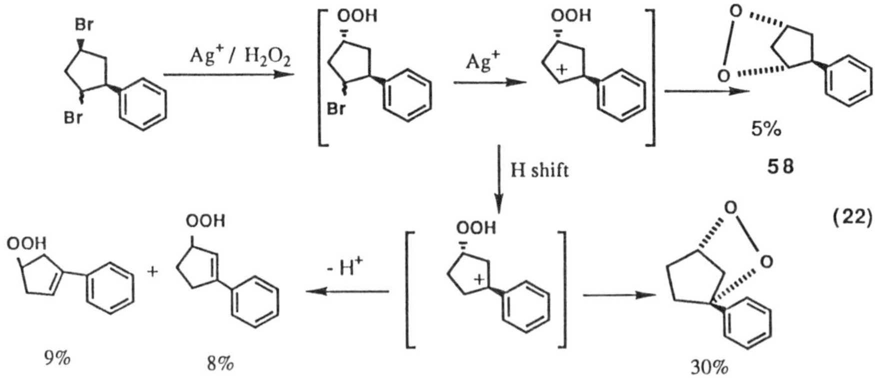

(22)

Isayama and Mukaiyama[58] reported the peroxysilylation of alkenes using triethylsilane and oxygen by the catalysis of bis(trifluoroacetylacetonato)-cobalt(II) in yields of 30–99% (equation 23). Murahashi *et al.*[59] reported a novel ruthenium-catalyzed oxidation of amides and lactams with *t*-BuOOH. In these reactions the α-position of the *N*-functional group was selectively oxidized in very high yields (equation 24). Recently, the formation of 1,2-dioxacyclohexanes (**59**) by the oxidation of alkenes in the presence of $Mn(acac)_3$ (equation 25) has been reported.[60]

R + Et_3SiH + O_2 —$Co(acac)_2$, r. t.→ R–CH(OOSiEt$_3$)–CH$_3$ (23)

R = Ph, $PhCH_2CH_2$, n-C_9H_{19}

N-COR tetrahydroisoquinoline —$RuCl_2(PPh_3)_3$, t-BuOOH→ 1-(OO-t-Bu) N-COR tetrahydroisoquinoline (24)

R = CH_3
R = Ph

$Ph_2C=CH_2$ + O_2 —$Mn(acac)_3$→ **59** (COCH$_3$, OH, CH$_3$, Ph, Ph, O–O) (25)

3.3 Peroxide Formation by Photochemical Methods

Photochemical methods for the preparation of peroxides can be divided into two types. One is the photooxygenation method using singlet oxygen, where electron-rich alkenes are oxidized to peroxidic compounds such as four-membered peroxide dioxetanes or six-membered endoperoxides. The other is the electron transfer type of photooxidations. We only describe here a few examples of the latter type. Eriksen *et al.*[61] reported that the DCA-sensitized (DCA = 1,9-dicyanoanthracene) photooxidation of cyclohexa-1,3-diene gave the corresponding endoperoxide (**60**) (equation 26). Schaap *et al.*[62] reported the formation of a 1,2,4-trioxolane (**61**) from the DCA-sensitized photooxidation of the corresponding epoxides (equation 27). The formation of 1,2-dioxolanes (**62**) has also been reported by Mizuno *et al.* (equation 28).[63]

cyclohexa-1,3-diene —DCA / O_2, hν→ **60** (26)

$$\text{Ph-epoxide-Ph} \xrightarrow[h\nu]{DCA / O_2} \mathbf{61} \quad (27)$$

$$\text{Ar,Ar'-cyclopropane} \xrightarrow[h\nu]{DCA / O_2} \mathbf{62a} + \mathbf{62b} \quad (28)$$

The DCA-sensitized photooxidation of hexa-1,5-diene derivatives gave dioxabicyclo[2.2.2]octane derivatives (**63**) (equation 29).[64] The formation of 1,2-dioxolanes (**64**, **65**) by the quinone-sensitized photooxidation of methylenecyclopropanes has also been reported (equation 30).[65] Irradiation of donor–acceptor complexes of cyclopropanes or methylenecyclopropanes with tetracyanoethylene (TCNE) also gave 1,2-dioxolanes (**66**, **67**) (equation 31).[66] The formation of 1,2,4-trioxolane was also observed in the reaction of TCNE with oxiranes and 1,2-dioxa-4-azapentane from azirizines. Recently, Gollnick *et al.*[67] reported that the DCA-sensitized photooxidation of 1,1,2,2-tetraarylcyclopropanes gave mainly 1,2-dioxolanes with small amounts of allylic hydroperoxides (equation 32). In these reactions, cyclopropanes bearing electron-donating substituents gave the corresponding 1,2-dioxolanes exclusively. However, cyclopropanes bearing less electron-donating substituents gave 1,2-dioxolanes (**68**) accompanied by the formation of allylic hydroperoxides (**69**) and other further oxidized products. They explained their results by considering a chain reaction involving the reaction of the 1,3-radical cation of cyclopropane with triplet oxygen rather than the reaction with 1,3-triplet biradical.

$$\text{Ar-diene-Ar} \xrightarrow{h\nu / DCA} [\text{Ar-radical cation-Ar}] \xrightarrow{O_2} \mathbf{63} \quad (29)$$

$$\text{methylenecyclopropane}(C_6H_4X)_2 \xrightarrow[O_2]{h\nu / sens.} \mathbf{64} + \mathbf{65} \quad (30)$$

$X = H, p\text{-}Cl, p\text{-}CH_3, p\text{-}OCH_3$

sens. = chloranil; anthraquinone; phenanthraquinone; benzoquinone

hν / TCNE, O_2 (31)

66 67

X = Y = p-MeOPh,
X = p-MeOPh, Y = Ph
X = Y = p-MePh
X = Y = p-ClPh
X = Y = C_6H_5

hν / DCA, O_2 (32)

68 69

3.4 Peroxide Formation by Autoxidation

Dialkyl peroxides were produced by the coupling of alkyl peroxyls ($RO_2\cdot$) with an alkyl radical ($R\cdot$) or by the self-termination of alkyl peroxyls (equation 33).[68]

$$R^{\bullet} + RO_2^{\bullet} \longrightarrow ROOR$$
$$2(t\text{-}RO_2^{\bullet}) \longrightarrow t\text{-}ROOR\text{-}t + O_2 \qquad (33)$$

Reaction of alkyl peroxyls with alkyl radicals takes place during the autoxidation of hydrocarbons at low oxygen pressure. The important aspect of this reaction is that it is heavily dependent on the resonance stability of the alkyl radical formed during the reaction. The most popular example of this reaction is the formation of di(triphenylmethyl) peroxide (**70**) from the autooxidation of triphenylmethane (equation 34).[69]

$$Ph_3CH \xrightarrow{O_2} Ph_3COOCPh_3 \qquad (34)$$

70

The self-termination of alkyl peroxyls also affords dialkyl peroxides, although the rate constants for the formation of dialkyl peroxides (equation 33) are small (Table 4), and this is one of the reasons why the parent hydrocarbons undergo autoxidation very easily.

Table 4. Rate constants (k) for the formation of di-tertiary alkyl peroxides from tertiary alkyl peroxyls[70]

Peroxyl	k (l mol^{-1} s^{-1}) $\times 10^{-3a}$	log[A (l mol^{-1} s^{-1})]	E (kJ mol^{-1})
$(CH_3)_3CO_2\cdot$	1.2	9.2	35.5
$C_6H_5C(CH_3)_2O_2\cdot$	6.0	10.7	39.7
$(C_6H_5)_2C(CH_3)O_2\cdot$	64	—	—

[a]at 303 K.

4 DECOMPOSITION OF DIALKYL PEROXIDES

4.1 Photochemical Decomposition of Dialkyl Peroxides

Dialkyl peroxides show continuous and very weak absorption without a peak in the range 200–320 nm, very similar to those of alkyl hydroperoxides or diacyl peroxides.[71–73] UV irradiation of dialkyl peroxides usually gives in the initial stage two alkoxy radicals via O—O homolysis, which are apt to react further with other substrates or decompose to more stable compounds. The radicals thus formed by photoirradiation are in some cases different from those produced by thermolysis because some of the radicals produced are vibrationally excited, i.e. 'hot radicals' are produced by the photoexcitation.[74] Hot radicals are so unstable that they do not react further with other molecules but decompose smoothly.[74] For example, photolysis of diisopropyl peroxide (**71**) gives an excited radical (**72**, *i*-PrO*·) which smoothly decomposes to give mainly acetaldehyde and methyl radical before it reacts with other molecules (equation 35).[75]

$$\underset{\mathbf{71}}{(CH_3)_2CHO{-}OCH(CH_3)_2} \xrightarrow{h\nu} \underset{\mathbf{72}}{(CH_3)_2CHO^{*}\cdot} + (CH_3)_2CHO$$

$$\underset{\mathbf{72}}{(CH_3)_2CHO^{*}\cdot} \longrightarrow CH_3CHO + CH_3\cdot \qquad (35)$$

On photoirradiation in gas phase, dimethyl peroxide (**73**) gives methanol, formaldehyde, CO and a trace of H_2 (equation 36).[76] The reaction mode is considered to be a chain mechanism involving methoxy radical as a chain carrier. The photolysis (313 nm) of diethyl peroxide (**74**) in the gas phase to give CO, methane, formaldehyde, acetaldehyde, ethanol, acetone and biacetyl has also been studied (equation 37). When the photoreaction was carried out in the liquid phase such as in carbon tetrachloride, cyclohexane, or water, the reaction products were ethanol and acetaldehyde.[77] The quantum yields of the decomposition in carbon tetrachloride and in water were 2 and 5, respectively. The difference can be explained by a small amount of metal catalyst present in water. In support of this, Cu^{2+} ion was proved to be a good catalyst for the decomposition of the peroxide (equation 38).

$$\underset{\mathbf{73}}{CH_3O\text{-}OCH_3} \xrightarrow{h\nu} 2\ CH_3O\bullet$$

$$CH_3O\bullet + \underset{\mathbf{73}}{CH_3O\text{-}OCH_3} \longrightarrow \bullet CH_2O\text{-}OCH_3 \longrightarrow HCHO + CH_3O\bullet \quad (36)$$

$$HCHO^{*} \xrightarrow{h\nu} H\bullet + CHO\bullet \longrightarrow H_2 + CO$$

$$\underset{\mathbf{74}}{EtO\text{-}OEt} \xrightarrow{h\nu} 2\ EtO\bullet$$

$$EtO\bullet + \underset{\mathbf{74}}{EtO\text{-}OEt} \longrightarrow EtOH + EtO\text{-}O\dot{C}HCH_3$$

$$EtO\text{-}O\dot{C}HCH_3 \longrightarrow EtO\bullet + CH_3CHO$$

$$CH_3CHO^{*} \xrightarrow{h\nu} CH_3CO\bullet + H\bullet \longrightarrow CH_4 + CO$$

$$2(CH_3C\dot{O}) \longrightarrow CH_3COCOCH_3 \quad (37)$$

$$CH_3CO\bullet \longrightarrow CH_3\bullet + CO$$

$$CH_3\bullet + CH_3CO\bullet \longrightarrow CH_3COCH_3$$

$$2\ EtO\bullet \longrightarrow CH_3CHO + EtOH$$

Cu^{2+} catalyzed

$$EtO\bullet + Cu^{2+} \longrightarrow Cu^{+} + H^{+} + CH_3CHO$$

$$(38)$$

$$Cu^{+} + \underset{\mathbf{74}}{EtO\text{-}OEt} \longrightarrow Cu^{2+} + EtO^{-} + EtO\bullet$$

When the benzoquinone-type peroxide **75** was photolyzed by sunlight it afforded a ring-contracted cyclopentadienone (**76**) exclusively, whereas when the same photolysis was carried out in a protic solvent such as methanol, solvent-participated products (**77**, **78**) were obtained. These products can be explained by considering the cyclopentenone radical (equation 39).[78,79]

(39)

4.2 Thermal Decomposition of Dialkyl Peroxides

On thermolysis, dialkyl peroxides also afford diradicals depending on the substituents attached on the peroxy bridge. Arrhenius parameters for the thermolysis of simple dialkyl peroxides are listed in Table 5,[80] where no relationship between alkyl substituents attached to the O—O bond and the rates of the thermolysis is observed.

The alkoxyl radicals formed by thermolysis undergo a variety of reactions such as (i) hydrogen abstraction, (ii) addition to alkenes, (iii) β-scission and (iv) if they could escape from the cage, disproportionation. The rate constant of the hydrogen abstraction from toluene by *tert*-butoxyl radical was 2.3×10^5 l mol^{-1} s^{-1} at 303 K.[81] Since alkoxyl radical adds rapidly to double bonds, it can be used as a radical initiator of vinyl polymerization.[82] The rate constant of β-scission is highly dependent on the nature of the alkyl radical produced during the reaction and the ketone group produced. As a typical alkoxyl

Table 5. Rate constants and Arrhenius parameters for the thermolysis of dialkyl peroxides[80]

Dialkyl peroxide	10^3k (s^{-1}) at 448 K	log[A (s^{-1})]	E (kJ mol^{-1})
CH_3OOCH_3	4	15.6	154.2
$CH_3CH_2OOCH_2CH_3$	7.9	16.1	156
$(CH_3)_3COOC(CH_3)_3$	2.5	15.9	159

radical, the rate constant of *tert*-butoxy radical in CCl_4 was represented by $\log [k\ (s^{-1})] = 12.5 - 58.1/\theta$, where $\theta = 2.303\ RT$ kJ mol^{-1} and at 303 K, $k = 3 \times 10^2\ s^{-1}$.[83] Rearrangement of alkoxyl radical is observed in the reaction of triphenylmethyloxy radical (**79**), which rearranges to afford diphenylphenoxylmethyl radical (**80**) (equation 40).[84]

$$Ph_3COOCPh_3\ (\mathbf{70}) \longrightarrow 2\ Ph_3CO^\bullet\ (\mathbf{79})$$
$$Ph_3CO^\bullet\ (\mathbf{79}) \longrightarrow Ph_2\dot{C}OPh\ (\mathbf{80}) \qquad (40)$$

The decomposition rates of dialkyl peroxides are usually faster than the predicted values as the induced decomposition of peroxides via hydrogen abstraction takes place during the reaction (equation 41). The induced decomposition rates increase in the order tertiary alkyl < primary alkyl < secondary alkyl peroxides, as might be predicted from the ease of hydrogen abstraction.

$$R_2CHOOCHR_2 + R_2CHO^\bullet \longrightarrow R_2\dot{C}OOCHR_2 + R_2CHOH$$
$$R_2\dot{C}OOCHR_2 \longrightarrow R_2C{=}O + R_2CHO^\bullet \qquad (41)$$
$$R_2CHOOCHR_2 + R^{1\bullet} \longrightarrow R_2CHO^\bullet + R^1OCHR_2$$

The five-membered cyclic peroxide 3,3,5,5-tetramethyl-1,2-dioxolane (**81**) decomposes in benzene in the presence of a radical scavenger to give acetone and diradical (equation 42).[85] The activation parameters for the reaction are $\log [A\ (s^{-1})] = 15.85 \pm 0.42$ and $E = 186 \pm 4$ kJ mol^{-1}. These values suggest a stepwise decomposition mechanism. In the absence of a radical scavenger these values become smaller because of the induced decomposition. The cyclic peroxide **82** also undergoes a thermal decomposition to afford diradicals, which further decompose with C—O or C—C bond scission (equation 45).[86,87] Cyclohexanone diperoxide (**83**), cf. photolysis gives three types of products, cyclohexanone, lactone (23%) and cyclododecane (44%) (equation 44).[88]

$$\mathbf{81} \rightleftharpoons {}^\bullet O\text{–}C(CH_3)_2CH_2C(CH_3)_2\text{–}O^\bullet \longrightarrow (CH_3)_2C{=}O + {}^\bullet CH_2C(CH_3)_2O^\bullet \qquad (42)$$

$$\mathbf{82} \longrightarrow 2\,R_1COR_2 + O_2 \quad ; \quad \longrightarrow 2R_1\cdot + (R_2CO_2)_2 \qquad (43)$$

$$\mathbf{83} \longrightarrow \ldots \qquad (44)$$

2 (cyclohexanone) + O_2 ; (lactone) + CO_2 ; (cycloalkane) + $2CO_2$

4.3 Metal-catalyzed Decomposition of Dialkyl Peroxides

The decomposition of the dialkyl peroxides **84** by metal ions is very similar to that of hydroperoxides, as shown in equation 45. By this procedure alkoxy radical and alkoxide salt were produced. The cleanest reaction of this type is that using Cu^+ ion[89] because of the possibility of using it at low temperature to control the reaction.

$$\underset{\mathbf{84}}{ROOR'} + Cu(I) \longrightarrow RO\cdot + R'OCu(II)$$

$$RO\cdot + R''H \longrightarrow ROH + R''\cdot \qquad (45)$$

$$R''\cdot + R'OCu(II) \longrightarrow R''OR' + Cu(I)$$

$$\longrightarrow R''^{+} + R'OCu(I)$$

The reaction rates of the decomposition are not influenced by the metal ions and the structure of the peroxides. For example, the second-order rate of $S_2O_8^{2-}$, HOOH, *t*-BuOOH (in water, 25°C) and $(PhCOO)_2$ (in EtOH, 25°C) in the presence of Fe^{2+} were 70, 54, 13 and 25 $mol^{-1}\ s^{-1}$, respectively.[90,91] The reaction of dialkyl peroxides with Grignard reagents has been used for the synthesis of unsymmetrical ethers (equation 46). For example, dimethyl peroxide reacts with phenylmagnesium bromide to give anisole in 77% yield; however, di-*tert*-butyl peroxide did not give *tert*-butyl phenyl ether under the same reaction conditions.[20] The preparation of *tert*-butyl phenyl ether can be achieved by the reaction of di-*tert*-butyl peroxide (**85**) with PhLi in 39% yield.[92]

$$CH_3OOCH_3 + PhMgX \longrightarrow PhOCH_3 + CH_3OH + R\text{-}R' \text{ etc.}$$

$$\underset{\mathbf{85}}{t\text{-BuOOBu-}t} + PhLi \longrightarrow t\text{-BuOPh } (39\%) \qquad (46)$$

4.4 Acid- and Base-catalyzed Decomposition of Dialkyl Peroxides

Although acid-catalyzed reactions of hydroperoxides have been widely used, the acid-catalyzed reaction of dialkyl peroxides has only been studied in strongly acidic media. In strongly acidic media, dialkyl peroxides (**86**) equilibrate with hydroperoxides (**88**) and carbonium ions (**87**),[93] which easily undergo rearrangement such as Criegee-type rearrangement,[94] or are trapped by nucleophiles (equation 47.)[95] For example, *tert*-butyl triphenylmethyl peroxide (**89**) gives benzophenone, phenol and *t*-BuOH quantitatively in benzene with the catalysis by benzenesulfonic acid.[94] A trapping reaction has been reported for the decomposition of di(triphenylmethyl) peroxide (**90**) by sulfonic acid. By this procedure di(triphenylmethyl) peroxide (**90**) gives triphenylmethylcarbonium ion, which is trapped by water to afford triphenylmethanol in 92% yield (equation 48).[96]

$$\underset{\mathbf{86}}{R_3C\text{-}O\text{-}O\text{-}R^1} + H^+ \rightleftharpoons \underset{\mathbf{87}}{R_3C^+} + \underset{\mathbf{88}}{R^1OOH} \overset{H^+}{\rightleftharpoons} H_2O_2 + R^{1+} \qquad (47)$$

$R_3C^+ \rightleftharpoons$ alkene + H^+; $R_3C^+ \overset{H_2O}{\rightleftharpoons} H^+ + R_3COH$

$$\underset{\mathbf{89}}{Ph_3COO\text{-}t\text{-}Bu} \xrightarrow[C_6H_6]{PhSO_3H} Ph_2CO + PhOH + t\text{-BuOH} \quad \text{(quantitative)}$$

$$\underset{\mathbf{90}}{Ph_3COOCPh_3} \xrightarrow[AcOH]{H_2SO_4} 2Ph_3C^+ \xrightarrow{H_2O} \underset{92\%}{2Ph_3COH} \qquad (48)$$

The base-catalyzed reaction of dialkyl peroxides has been studied for more than 40 years. Primary of secondary dialkyl peroxides are well known to decompose with the generation of ketone and alkoxy anion in the presence of bases such as potassium hydroxide, sodium ethoxide or piperidine.[97]

Kinetic studies using benzyl *tert*-butyl peroxide in chlorobenzene in the presence of a tertiary amine (equation 49) revealed that the decomposition rate is sensitive to the concentrations of amines and peroxides.[98] The rate of the decomposition is heavily dependent on the amine basicity. On increasing the amine basicity the rate of peroxide decomposition is increased. The Arrhenius *A* factor is small, indicating at least partial separation of charge in the transition state of the rate-determining step of the reaction.

$$\text{Ph-C(Me)(H)-O-O t-Bu} \xrightarrow{:B} \text{PhCOMe} + {}^{-}\text{O-t-Bu} + \text{BH}^{+} \qquad (49)$$

The yields of this type of decomposition are excellent in most cases. For example, di(α-methoxybenzyl) peroxide (**91**) gives the corresponding acetophenone (92%), benzaldehyde (100%) and methanol with catalysis by triethylamine (equation 50).[99] This procedure is useful because it can break C—C bonds selectively.

$$\underset{\mathbf{91}}{\text{Ph-C(H)(OMe)-O-O-C(H)(OMe)-Ph}} \xrightarrow{Et_3N} \text{PhCOOMe (92\%)} + \text{PhCHO (100\%)} + \text{MeOH} \qquad (50)$$

4.5 Utilization of Peroxide Decomposition in Organic Synthesis

The main purpose of using peroxides in organic synthesis is the oxyfuctionalization of organic substrates. Thus, several alkyl μ-peroxo metal complexes have been prepared and the reactivities of these complexes have also been examined. Palladium(II) *tert*-butyl peroxyl (**92**), which is easily obtained by the reaction of palladium dicarboxylate (**93**) and *t*-BuOOH, oxidizes terminal alkenes to methyl ketones with high selectivity (equation 51).[100] The reaction mechanism of this reaction involves the complexation of the alkene to the metal followed by its insertion into the palladium—oxygen bond. The transition state is like a five-membered cyclic intermediate, which further decomposes into methyl ketone and the palladium *tert*-butoxy complex. A similar type of μ-oxo complex of Mo, V and Ti has been postulated in the oxidation of alkenes.[101–103]

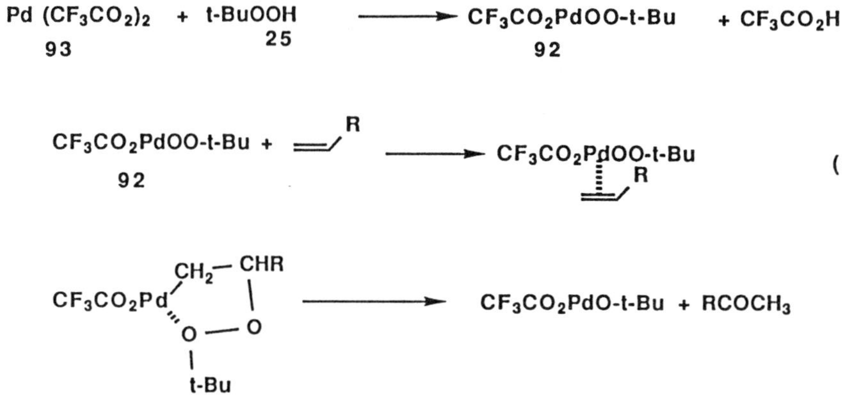

$$Pd(CF_3CO_2)_2 \; (\mathbf{93}) + t\text{-}BuOOH \; (\mathbf{25}) \longrightarrow CF_3CO_2PdOO\text{-}t\text{-}Bu \; (\mathbf{92}) + CF_3CO_2H$$

$$CF_3CO_2PdOO\text{-}t\text{-}Bu \; (\mathbf{92}) + CH_2{=}CHR \longrightarrow CF_3CO_2Pd(OO\text{-}t\text{-}Bu)(CH_2{=}CHR) \qquad (51)$$

$$CF_3CO_2Pd[CH_2\text{-}CHR\text{-}O\text{-}O(t\text{-}Bu)] \longrightarrow CF_3CO_2PdO\text{-}t\text{-}Bu + RCOCH_3$$

Another utilization of dialkyl peroxides is in the selective reduction of the peroxide bond and subsequent inter- or intramolecular reaction. It is a commonly used approach for the reduction of cyclic peroxides, as exemplified by ascaridole (**94**) in equation 52.[104]

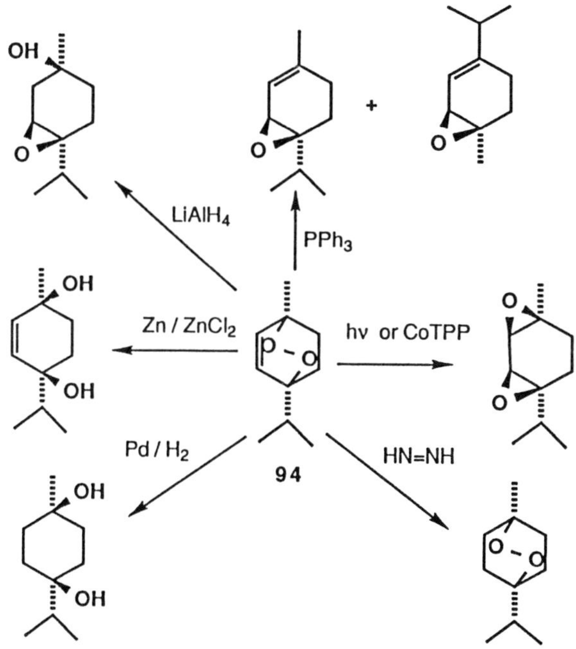

LiAlH$_4$; PPh$_3$; Zn / ZnCl$_2$; hv or CoTPP; Pd / H$_2$; HN=NH; **94** (52)

Jefford *et al.*[105] obtained a 1,2,4-trioxane ring system (**95**) from the reaction of ascaridole (**94**) with cyclopentanone in the presence of trimethylsilyl triflate and

2,6-di-*tert*-butylpyridine, which further reduced to the *cis*-1,2-diol (**96**) (equation 53).

TMSOTf, DTBP Zn / AcOH (53)

94 95 96

TMSOTf = trimethylsilyl triflate
DTBP = 2,6-di-t-butylpyridine

Rearrangement of 1,4-endoperoxides (**97**) has been widely used for the preparation of diepoxides, but side-reactions such as the formation of an epoxy ketone occur in most cases. The utilization of CoTPP for such a reaction significantly improved the formation of diepoxides (equation 54).[106] For the rearrangement of 1,4-peroxides (**98**), the most important in a biological context is that of prostaglandin endoperoxide to thromboxane. Takahashi and Kishi,[107] using a simple model compound (**98**) of prostaglandin endoperoxide, obtained a thromboxane analog (**99**) with catalysis by Fe^{3+} or Cu^{+} ion in *ca* 20% yield together with other products (equation 55).

CoTPP 100% 97 Δ 28% 32% 40% (54)

98 $Fe_2(SO_4)_3$ 99 (22%) (39%) (29%) (55)

By the hydrogenation of 1,2-dioxolanes (**100**) using Pd–charcoal, 1,2-*threo*-2,3-*erythro*-1,3-diols (**101**) were obtained in quantitative yields (equation 56).[108] A similar type of reaction has also been reported by Feldman and Simpson.[109]

In this case they obtained 1,3-diols (**103**, **104**) by the reaction of a 1,2-dioxolane derivative (**102**) in *ca* 33% yield with a *syn*/*anti* ratio of 4.3:1. By the reaction of 1,2-dioxolanes with silica gel at room temperature, *β*-keto alcohols were obtained, which were reduced by $NaBH_4$ to afford 1,3-diols with a *syn*/*anti* ratio of 3:1 (equation 57).

R Ar Ar O—O 100 + H_2 $\xrightarrow[\text{MeOH}]{\text{Pd / C}}$ R Ar Ar OH OH 101 (56)

O—O Ph 102 $\xrightarrow[\text{EtOH}]{NaBH_4}$ [R(C=O)CH₂CH(OH)R'] R, R' = Ph, CH=CH₂ $\xrightarrow{NaBH_4}$ OH OH Ph 103 + OH OH Ph 104 (57)

102 → O OH Ph + OH O Ph $\xrightarrow{NaBH_4}$ 103 + 104

5 BIOLOGICAL ACTIVITIES OF NATURALLY OCCURRING AND SYNTHETIC PEROXIDES

5.1 Peroxides from Land Organisms

Ascaridole (**94**) was the first natural peroxide isolated from chenopodium oil.[110] Many years ago Wieland and Prelog[111] isolated ergosterol peroxide (**105**) from *Asperigillus fumigatus*. The ergosterol peroxide **105** has also been found in the extracts of many organisms such as *Trichophyton schonleini*,[112] *Penicillium scherotignum*,[113] *Trametes kusanoana*,[114] *Piptoporus betulinus*,[115] *Aspergillus niger*,[116] *Ganoderma applanatuna*,[117] *Naematolona fasiculare*[118] and *Cetraria richardsonii*.[117] The biologically active peroxides **106a–c**, which strongly inhibit the growth of some plant roots, were found in *Eucalyptus grandis*.[119] The biological activity of these compounds has been suggested to be connected with the peroxide functionality. In 1979 the medicinal drug quinghaosu (**107**) was isolated from *Artemisia annua* L.[120] Quinghaosu showed a wide range of biological activities toward many malaria strains, including drug-resistant strains. This compound possesses a characteristic trioxane ring and has been a target for many synthetic organic chemists. In 1983 Schmid and Hofheinz[121] reported the first total synthesis and since then many synthetic efforts have been made to prepare this skeleton.[120]

Two indole alkaloids, verruculogen (**108**)[122] and fumitremorgin A (**109**),[123] having an eight-membered peroxide ring have been isolated from *Penicillium verruculasum* and *Aspergillus fumigatus*, respectively. Peroxide **110**, named as oxanthromycin and having an anthraquinone ring, has been isolated from *Nocardia madurae*.[124] The antibacterial spectrum of **110** has been studied for 18 strains, among which **110** showed strong antibacterial activity toward *M. gypseum* ATCC 10215 (MIC value 2 μg ml^{-1}).

From the aerial parts of *Artemisia maritima* L. (Asteraceae), a cyclic five-membered sesquiterpene peroxy semiketal (**111**) was isolated.[125] The structure of **111** was elucidated by spectroscopic methods. Sesquiterpene peroxides (**112**–**114**) were first isolated from the rhizomes of *Alpinia japonica*.[126] Soon afterwards these six sesquiterpene peroxides were isolated from the *n*-hexane- and chloroform-soluble fractions obtained from the rhizomes of *A. intermedia*.[127]

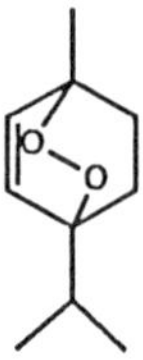

94 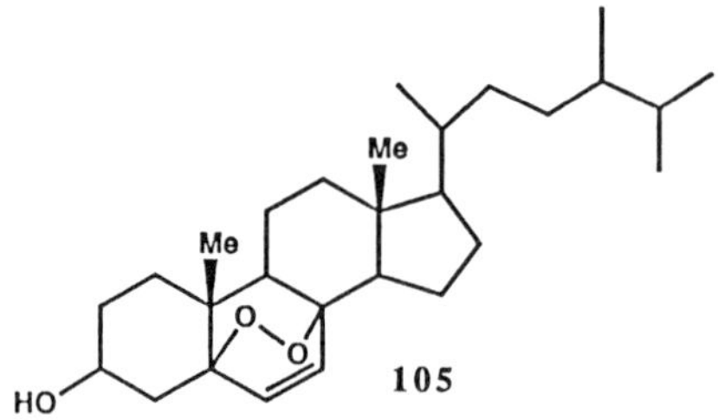 **105**

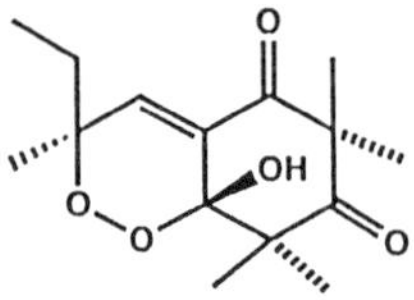

106a 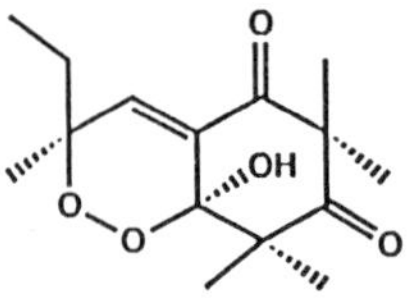 **106b** 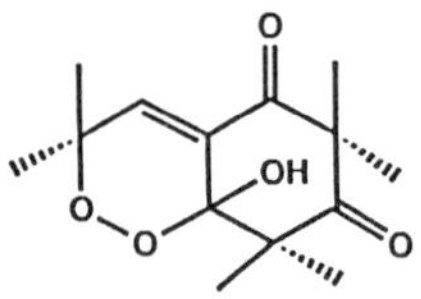 **106c**

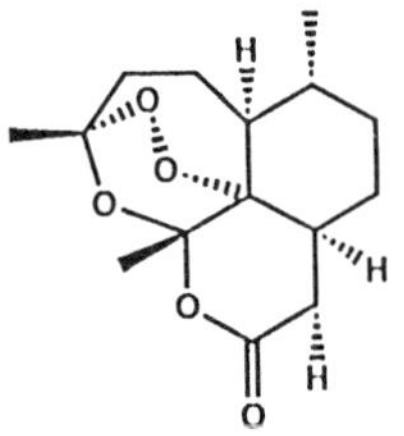

107 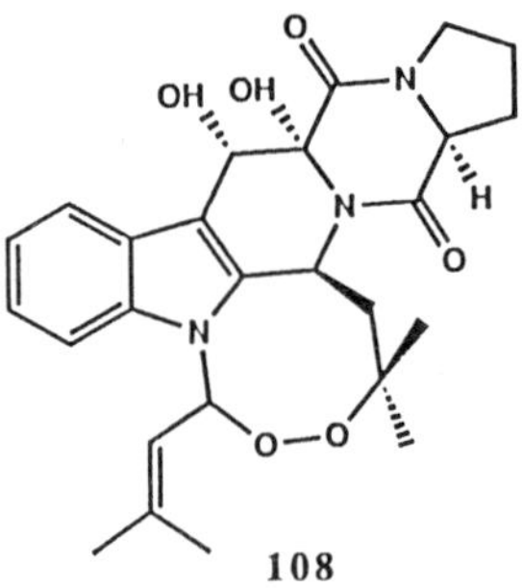 **108** 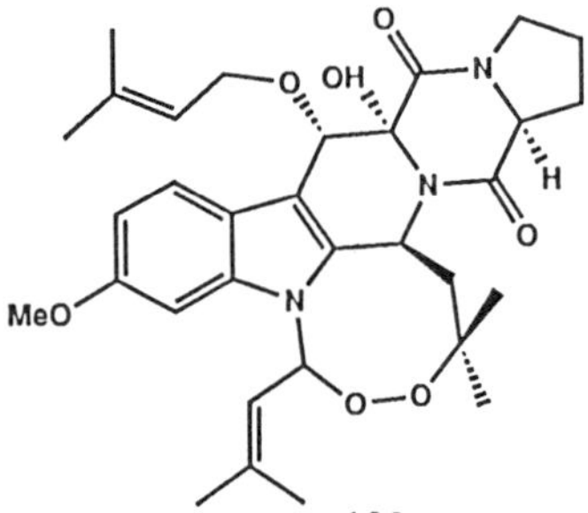 **109**

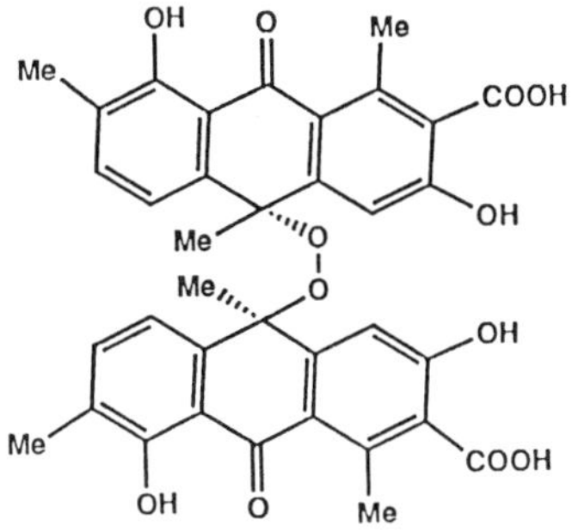

110 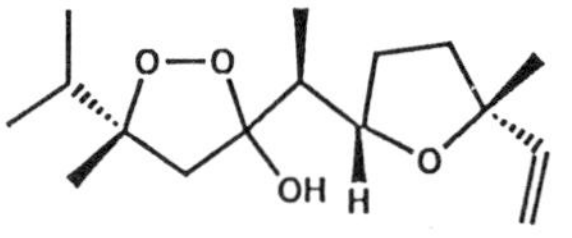 111

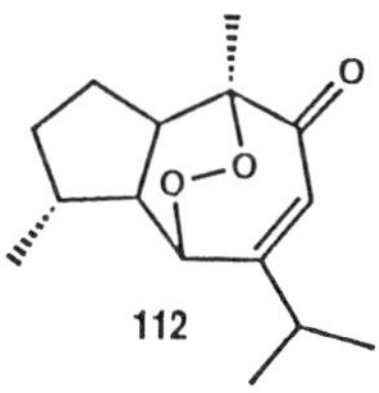

112 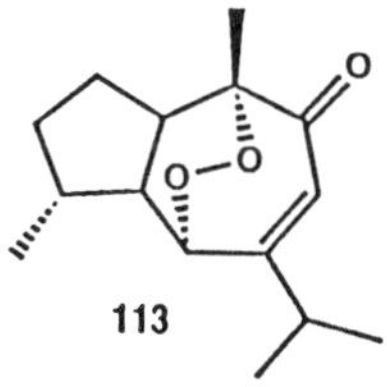 113 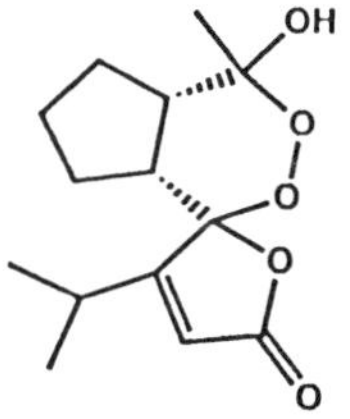 114

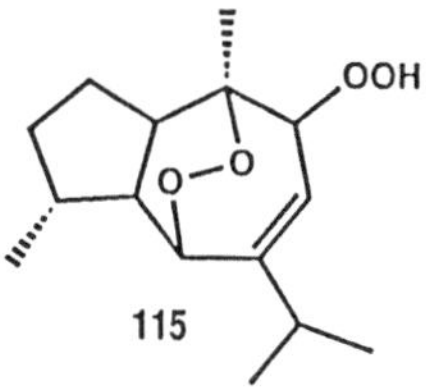

115 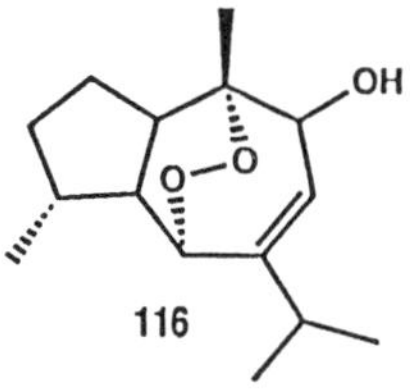 116 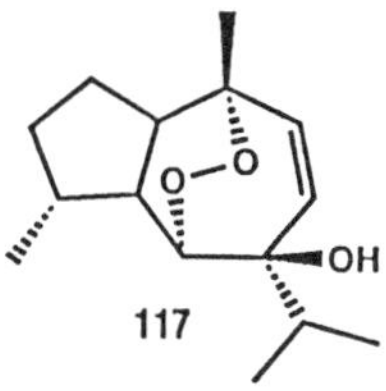 117

5.2 Marine Natural Peroxide Compounds

There are many biologically active peroxides in marine organisms, especially in marine sponges. The first peroxide isolated from marine organisms was rhodophytin **118** from *Laurencia rhodomelacia rhodophta.*[128] Steroidal-type peroxides have also been isolated from marine organisms. For example, Sheith and Djerassi[129] isolated four steroid peroxides (**119a**–**d**) from deep-sea sponges in California. Marine sponges are the treasure house of the peroxidic compounds and many peroxidic compounds have been isolated. Higgs and Faulkner[130] found that ethanol extracts of the sponge *Plakortis halichondriodes* inhibit the growth of *Staphylococcus aureus* and *Escherichia coli* and the biological activity of the extracts was due to the major metabolite plakortin (**120**). Further, they isolated four peroxidic compounds (**121a, 121b, 122a, 122b**) from the dichloromethane extract of *Chondrosia collectrix.* However, they could not isolate each of the pure peroxides. From neutral solution they isolated **121a** and **122a** in a ratio of 4:1, and from acidic solution they isolated **121b** and **122b** in a ratio of 1:1.[131] These peroxides showed weak antibacterial activities, although their activities were lost when they were kept for many hours at room temperature. This suggests the possibility that on standing the peroxide solution O—O bond breakage might occur, resulting in a loss of biological activity.

At nearly the same time, but independently, Kashman and Roten[132] isolated the sesquiterpene peroxide muquibilin (**123**) from *Prianos* sponge from the sea at Eilat. However, they do not discuss the precise stereochemical structure or the biological activity of **123**. Manes *et al.*[133] also isolated muqubilin (**123**) and the related ester (**124**) from methanol extracts of a large soft drab sponge from the Tongan coral reefs and found that **123** at 16 μg ml^{-1} concentration induces 100% inhibition of the cleavage of fertilized sea urchin eggs.

Albericci and co-workers[134–136] isolated five peroxidic compounds (**125a**, **125b**, **126a**, **126b**, **127a**) from the dichloromethane extracts of the sponge *Sigmosceptrella laevis* from Laing Island, New Guinea. They examined the biological activity of these peroxides using *Lebistes reticulatis*. The LD_{50} of the extracts was 25 mg ml^{-1}, but the acid portion of the extracts showed LD_{50} of 5 mg ml^{-1}, which suggests that the carboxylic acid portion of the peroxides contributes considerably to the biological activity found. Phillipson and Rinehart[137] isolated two peroxycarboxylic acids (**128a**, **129a**) and an ester (**128b**) from the toluene extracts of the sponge *Plakidae* sp. from the Caribbean. The crude toluene extract showed various biological activities, e.g. it inhibited the growth of *Saccharomyces cerevisiae* and *Penicillium atrovenetum*. Further, the LD_{50} of the crude extract toward L 1210 cell was 0.14 μg ml^{-1}. The key compounds of these biological activities were determined to be the two peroxy carboxylic acids (**128a**, **129a**) and no biological activity was observed with the methyl esters (**128b**, **129b**). Further, they isolated plakortic acid (**121b**). The methyl ester is known as a plakortin (**120**). The biological activity of **121b** was examined using *S. cerevisiae*, *P. atrovenetum* and *B. subtilis* by a sensitive disk method. Compound **121b** inhibited the growth of *S. cerevisiae* to a 40 mm radius, the growth of *P. atrovenetum* to a 34 mm radius and the growth of *B. subtilius* to a 23 mm radius at a 100 μg per disk concentration. The other peroxidic compounds, **128a** and **129a**, also showed the inhibition of the growth. For example, **128a** at a 100 μg per disk showed the inhibition of the growth of *S. cerevisiae* and *P. atrovenetum* to 24 and 25 mm radii, respectively. The inhibition radii of *S. cerevisiae* and *P. atrovenetum* by **129a** at 100 mg per disk were 20 and 18 mm, respectively. By contrast, the methyl ester of the peroxide plakortin (**120**) showed no inhibition for the strains of *E. coli* and *B. subtilis* even at 100 mg per disk concentration.

Wells[138] isolated the cyclic peroxide chondrillin (**130a**) from *Chondrilla* species from the Great Barrier Reef in Australia. Ten years later Quinoa *et al.*[139] isolated the structurally related xestins A (**130c**) and B (**130b**) from a sponge of the genus *Xestospongia*, and a series of related peroxy ketals (**130d–f**) have been isolated from *Plakortis lita* by Sakemi *et al.* [140] The biological activities of the cyclic peroxides (**130a–g**) have been studied using P 388 mouse leukemia cells. The IC_{50} values of **130c–g** were 0.05–0.3 μg ml^{-1}; however, the cyclic peroxides **130a** and **b** are ten times less active than **130c–g**. Capon and MacLeod[141,142] also isolated a series of structurally related peroxides (**131–135**) from marine sponges.

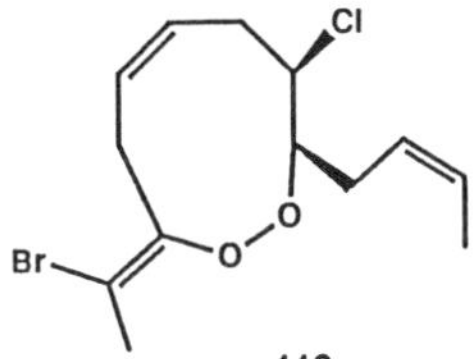

118

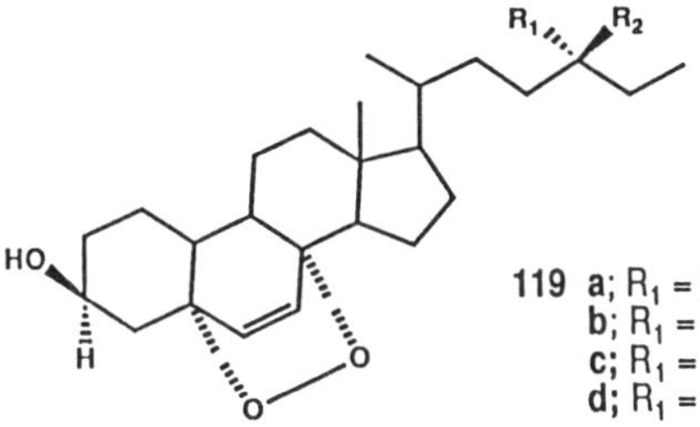

119 a; $R_1 = R_2 = H$
b; $R_1 = H$, $R_2 = Me$
c; $R_1 = R_2 = -CH_2-$
d; $R_1 = R_2 = Et$

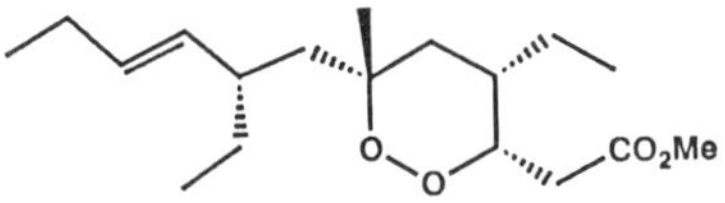

120

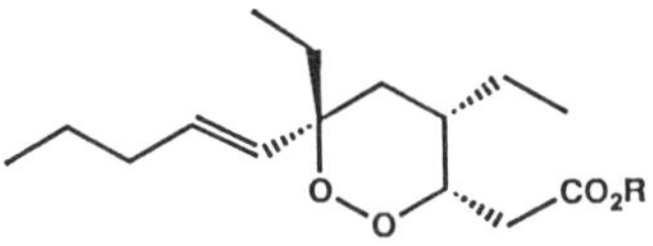

121 a; R = Me
b; R = H

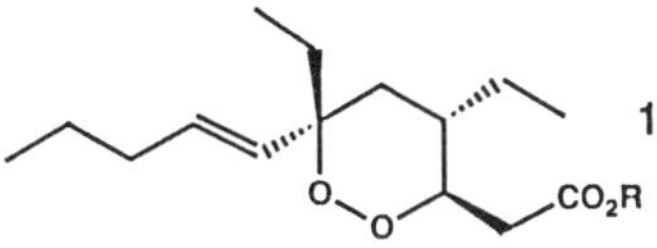

122 a; R = Me
b; R = H

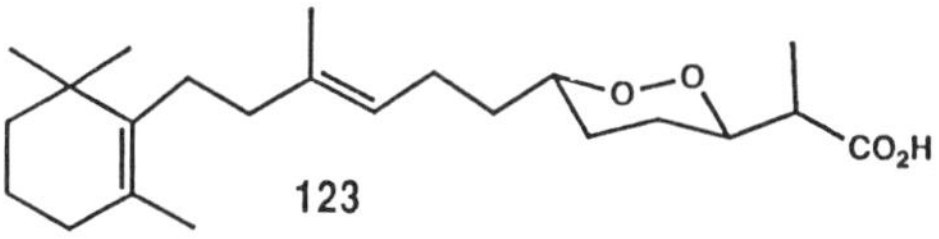

123

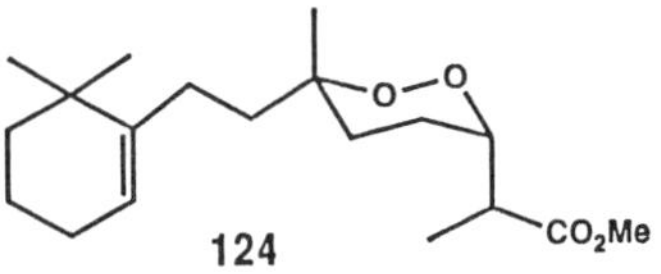

124

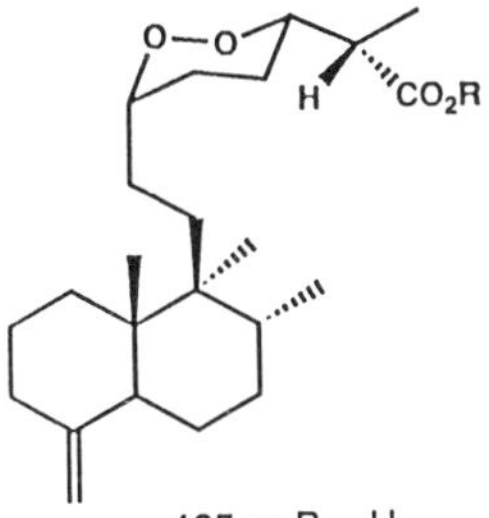

125 a; R = H
b; R = Me

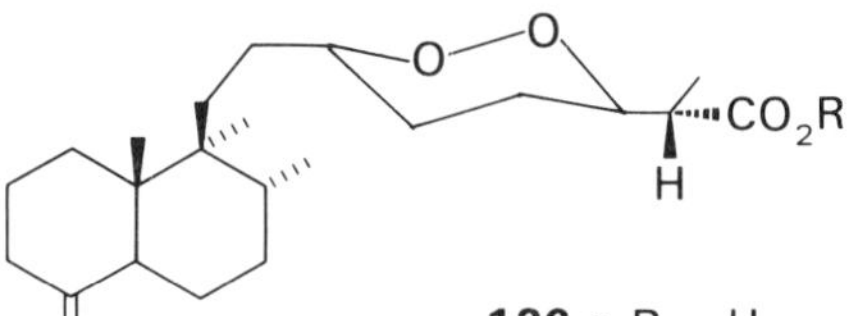

126 a R = H
b R = Me

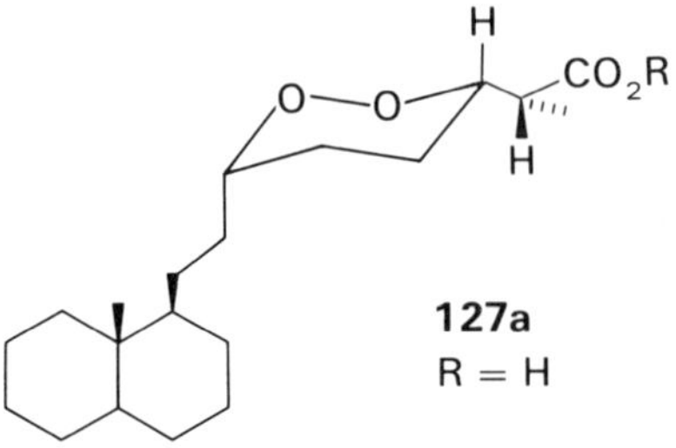

CO_2R

127a
R = H

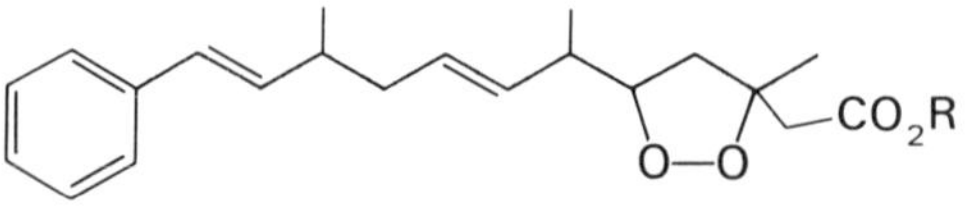

CO_2R

128 a R = H
b R = Me

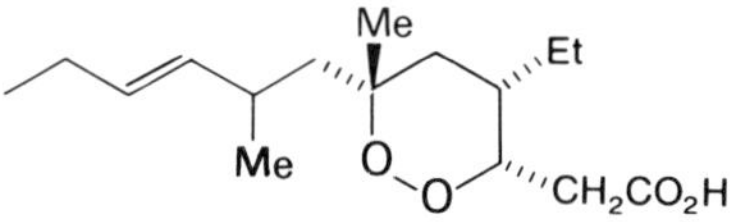

Me Et Me CH_2CO_2H

129

R_2 H R_1 CO_2Me

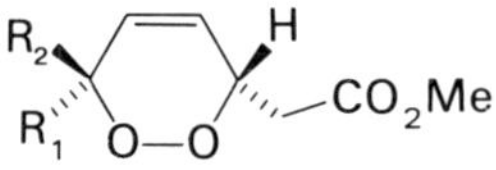

130

a $R_1 = (CH_2)_{15}CH_3$; R_2OMe
b R_1 = OMe; $R_2 = -(CH_2)_{13}$ Me
c $R_1 = -(CH_2)_{13}$ Me; R_2 = OMe
d R_1 = OMe; $R_2 = -(CH_2)_{11}CH_3$
e R_1 = OMe; $R_2 = -(CH_2)_9$ Me
f R_1 = OMe; $R_2 = -(CH_2)_7$ Me
g R_1 = OMe; $R_2 = -(CH_2)_9$ Me

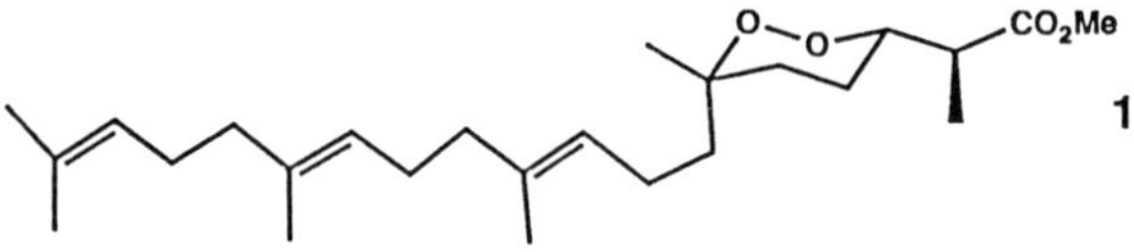

CO_2Me

131

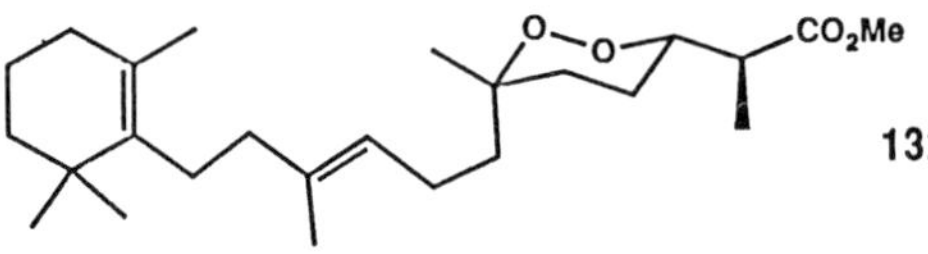

CO_2Me

132

CO_2R CO_2R CO_2R

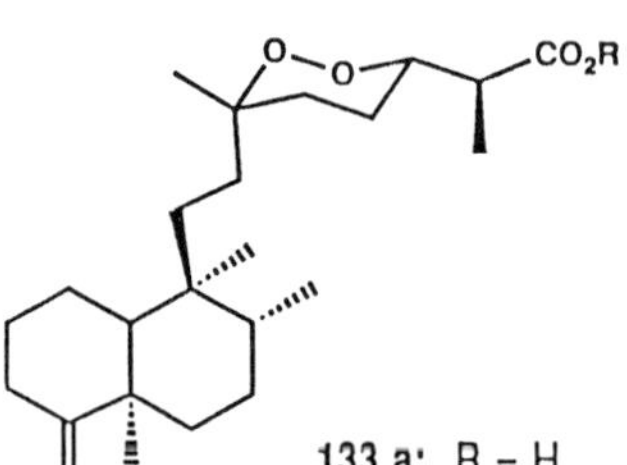

133 a; R = H
b ; R = Me

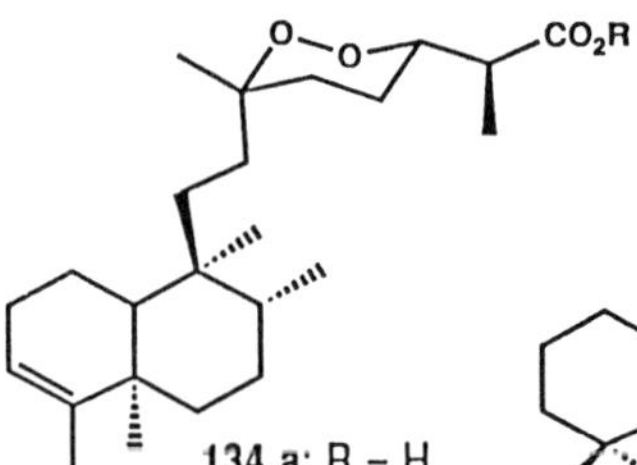

134 a; R = H
b; R = Me

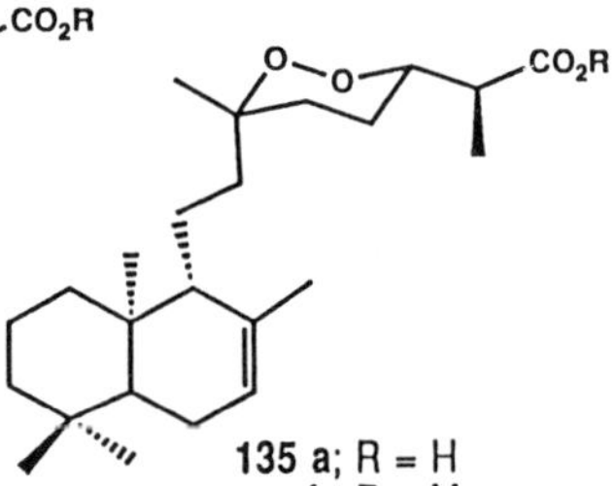

135 a; R = H
b; R = Me

The correlation of peroxide bond with biological activity, especially antibacterial activity has not yet been clarified. If the peroxide bond is changed to —S—S—, the biological activity was reported to be lost. Hence it is likely that the biological activity might be related to the presence of the O—O bond. Another problem is the clarification of the exact reactive species for the biological activity, namely, the peroxide itself or the active oxygen species generated by its decomposition.

5.3 Biological Activity of Synthetic Peroxides

It is well known that certain peroxides such as lipid peroxides, which are formed by the action of active oxygen species with unsaturated lipid membranes, show some cell toxicities that cause various diseases, such as ageing, degradation of membranes and blood diseases. The major causes of this biological damage were considered to be the active oxygen species such as hydroxyl radical, alkoxy radical and superoxide. Based on these preliminary observations, the biological activities of some typical peroxides have been studied. Merka *et al.*[143] studied the biological activities of thirteen peroxides (ROOH and ROOR′). Among them peracetic acid and performic acid showed strong antibacterial activities.

5.3.1 Benzoyl peroxide

Kligman *et al.*[144] found that benzoyl peroxide can improve some skin diseases, which they considered to be due to its antibacterial activity. Fanta *et al.*[145] found that 5–10% ointment preparations of benzoyl peroxide cause inhibition of *propiobacterium acnes.* Higashi *et al.*[146] examined the minimum inhibitory concentration (MIC) of benzoyl peroxide using *Propiobacterium acnes* and obtained a value of 100 μg ml^{-1}. Cunliffe and Holland[147] examined the antibacterial activities of 20% benzoyl peroxide, 0.3% gentamicin and 2% phenoxyethanol using three strains, *Staphylococcus aureus, Proteus* sp. and *Pseudomonas aeruginosa.* In every case benzoyl peroxide showed the highest activity. Bossche *et al.*[148] found that a combination of benzoyl peroxide and misonidazole inhibited the growth of *Staphylococcus* and *Propiobacterium* at low concentrations compared with the use of the individual compounds. Patane and Pistillo[149] examined antibacterial activities against many bacterial strains and found that Gram-negative strains are more sensitive than Gram-positive strains.

5.3.2 *tert*-Butyl Hydroperoxide and Other Peroxides

tert-Butyl hydroperoxide has been used as a potent inhibitor of the growth of bacterial strains of fish since 1969.[150] The antibacterial activity of this dose is strong enough to kill all bacterial strains at least at 50 ppm concentrations. The compound showed strong toxicity toward many living organisms; for example, the LD_{50} to mice is 800 mg kg^{-1} (injection) and to rats 800 mg kg^{-1} (oral).[151] Ivanov and Korotich[152] examined the accurate toxicity of *t*-BuOOH for rats and mice by oral administration. The LD_{50} for rats and mice were 370 and 320 mg kg^{-1}, respectively, and the LC_{50} for mice was 2739 mg m^{-3}. The thresholds for irritation of the respiratory system of humans and rats were 45 and 180 mg m^{-3}, respectively. The root growth inhibitor **136** also has a peroxide bridge in the ring. Recently, the antibacterial activity of simple cyclic peroxides (**137a**, **137b**) has been studied in connection with peroxide decompositions.[153] The antibacterial activities of these peroxides were shown to be due to the active oxygen species generated by their decomposition, and not to their peroxide bonds.

All the results presented here provide new possibilities for using peroxides as potent radical-releasing drugs. However, many further studies are needed because many difficulties such as the stability and solubility of peroxides still exist. An important problem is how to generate active oxygen radicals effectively at specific sites in biological systems.

6 REFERENCES

1. W. G. Penny and G. B. B. Sutherland, *Trans. Faraday Soc.*, **30**, 898 (1934).
2. Y. Amako and P. A. Gighere, *Can. J. Chem.*, **40**, 765 (1962).
3. J. R. Durig and D. W. Wertz, *J. Mol. Spectrosc.*, **25**, 465 (1986).
4. C. J. Marsden, L. S. Bartell and F. P. Diodafi, *J. Mol. Struct.*, **39**, 253 (1977).
5. C. Glidewell, D. C. Liles, D. J. Walton and G. M. Sheldrick, *Acta Crystallog., Sect. B*, **35**, 500 (1979).
6. D. A. Langs, M. G. Erman, G. T. Detitta, D. J. Coughlin and R. G. Saloman, *J. Cryst. Mol. Struct.*, **8**, 239 (1978).
7. M. Schulz, K. Kirshke and E. Hohne, *Chem. Ber.*, **100**, 2242 (1967).
8. J. N. Brown, R. L. R. Towns, M. J. Kovelean and A. H. Andrut, *J. Org. Chem.*, **41**, 3756 (1976).
9. P. Groth, *Acta Chem. Scand.*, **23**, 1311 (1969).
10. J. Hess and A. Vos, *Acta Crystallogr., Sect. B*, **33**, 3527 (1977).
11. H. Itokawa, Y. Tachi, Y. Kamano and Y. Iitaka, *Chem. Pharm. Bull.*, **26**, 331 (1978).
12. R. W. Murray and M. L. Kapla, *Tetrahedron*, **23**, 1575 (1967).
13. J. Fayos, D. Lokensgard, J. Clardy, R. J. Cole and J. W. Kirksey, *J. Am. Chem. Soc.*, **96**, 6785 (1974).
14. J. H. Wieringa, J. Strating, W. Adam and H. Wynberg, *Tetrahedron Lett.*, 169 (1972).

15. S. W. Benson and R. Shaw, in *Organic Peroxides*, Vol. 1 (D. Swern, Ed.), Wiley-Interscience, New York, London (1970).
16. G. Baker, J. H. Littlefair, R. Shaw and J. C. Thynne, *J. Chem. Soc.*, 6970 (1965).
17. *JANAF Thermochemical Tables*, Dow Chemical, Midland, MI (accumulated from 1966).
18. C. J. Howard, *J. Am. Chem. Soc.*, **102**, 6937 (1980).
19. L. Batt and R. D. McCullough, *Int. J. Chem. Kinet.*, **8**, 911 (1976).
20. L. Batt, K. Christie, R. T. Milne and A. J. Summers, *Int. J. Chem. Kinet.*, **6**, 877 (1974).
21. R. F. Walker and L. Phillips, *J. Chem. Soc. A*, 2103 (1968).
22. D. K. Lewis, *Can. J. Chem.*, **54**, 581 (1976).
23. M. J. Perona and D. M. Golden, *Int. J. Chem. Kinet.*, **5**, 55 (1973).
24. J. Czarnowski and H. J. Schumacher, *Int. J. Chem. Kinet.*, **13**, 639 (1981).
25. I. B. Rabinovich, E. G. Kiparisova and Yu. A. Alexksandrov, *Dokl. Akad. Nauk. SSSR*, **20**, 1116 (1971).
26. W. A. Pryor and D. M. Muston, *J. Org. Chem.*, **29**, 512 (1964).
27. N. A. Milas and D. M. Surgenor, *J. Am. Chem. Soc.*, **68**, 205 (1946).
28. A. E. Batog and M. K. Romantssevich, *Zh. Prikl. Khim.*, **41**, 1149 (1968).
29. A. I. Schreibert, N. V. Elsakov, A. P. Khardin, A. A. Popova, V. I. Ermachenko and E. M. Kozlov, *Zh. Org. Khim.*, **4**, 1190 (1968).
30. L. Horner and K. H. Knapp, *Justus Liebigs Ann. Chem.*, **622**, 79 (1959).
31. F. Welch, H. R. Williams and H. S. Mosher, *J. Am. Chem. Soc.*, **77**, 551 (1955).
32. H. Hock and H. Kropf, *Chem. Ber.*, **88**, 1544 (1955).
33. F. F. Rust, F. H. Seubold, Jr, and W. E. Vaughan, *J. Am. Chem. Soc.*, **72**, 338 (1950).
34. N. A. Milas and L. H. Perry, *J. Am. Chem. Soc.*, **68**, 1938 (1946).
35. A. G. Davies, R. V. Foster and A. M. White, *J. Chem. Soc.*, 2200 (1954).
36. A. Rieche and C. Bischoff, *Chem. Ber.*, **94**, 2457 (1961).
37. K. Weissermel and M. Lederer, *Chem. Ber.*, **96**, 77 (1963).
38. G. Sosnovsky and D. J. Rawlinson, in *Organic Peroxides*, Vol. 2 (D. Swern, Ed.), Wiley-Interscience, New York, London (1970) p. 153.
39. R. A. Sheldon, in *The Chemistry of Peroxides* (S. Patai, Ed.) Wiley, Chichester (1983) p. 161.
40. H. Kropf, H. M. Amirabadi, M. Mosebach, A. Torkler and H. von Wallis, *Synthesis* 587 (1983).
41. J. D. Druliner, *Synth. Commun.*, **13**, 115 (1983).
42. K. S. Feldman, R. E. Simpson and M. Parvez, *J. Am. Chem. Soc.*, **108**, 1328 (1986).
43. A. J. Bloodworth and N. A. Tallant, *Tetrahedron Lett.*, **31**, 7077 (1990).
44. P. H. Dussault and N. A. Porter, *J. Am. Chem. Soc.*, **110**, 6276 (1988).
45. P. Dussault and A. Sahli, *Tetrahedron Lett.*, **31**, 5117 (1990).
46. M. S. Kharasch and G. Sosnovsky, *Tetrahedron*, **3**, 105 (1958).
47. W. Triebs and G. Pellmann, *Chem. Ber.*, **87**, 1201 (1954).
48. H. Kropf and D. Goschenhofer, *Tetrahedron Lett.*, 239 (1968).
49. K. R. Kopecky, J. E. Filby, P. A. Lockwood and J.-Y. Ding, *Can. J. Chem.*, **53**, 1103 (1975).
50. W. Adam, A. Birke, C. Cadiz, S. Diaz and A. Rodriguez, *J. Org. Chem.*, **43**, 1154 (1978).
51. N. A. Porter and J. C. Mitchell, *Tetrahedron Lett.*, **24**, 543 (1983).
52. A. J. Bloodworth and J. L. Courtneidge, *J. Chem. Soc., Perkin Trans. 1*, 1807 (1982).
53. A. J. Bloodworth, K. H. Chan and C. J. Cooksey, *J. Org. Chem.*, **51**, 2110 (1986).
54. A. J. Bloodworth, K. J. Bowyer and J. C. Mitchel, *J. Org. Chem.*, **52**, 1124 (1987).
55. A. J. Bloodworth and N. Hargreaves, *Tetrahedron Lett.*, **28**, 2783 (1987).
56. A. J. Bloodworth and G. M. Lampman, *J. Org. Chem.*, **53**, 2668 (1988).

57. K. Takahashi, M. Shiro and M. Kishi, *J. Org. Chem.*, **53**, 3098 (1988).
58. S. Isayama and T. Mukaiyama, *Chem. Lett.*, 573 (1989).
59. S. Murahashi, T. Naota and K. Yonemura, *J. Am. Chem. Soc.*, **110**, 8256 (1988).
60. S. Tategami, T. Yamada, H. Nishion, J. D. Korp and K. Kuorokawa, *Tetrahedron Lett.*, **31**, 6371 (1990).
61. J. Eriksen, C. S. Foote and T. L. Parker, *J. Am. Chem. Soc.*, **99**, 6455 (1977).
62. A. P. Schaap, L. Lopez and S. D. Gagnon, *J. Am. Chem. Soc.*, **105**, 663 (1983).
63. K. Mizuno, N. Ichinose and Y. Otsuji, *Tetrahedron*, **41**, 2207 (1985).
64. T. Miyashi, A. Konno and Y. Takahashi, *J. Am. Chem. Soc.*, **110**, 3676 (1988).
65. Y. Takahashi, T. Miyashi and T. Mukai, *J. Am. Chem. Soc.*, **105**, 6511 (1983).
66. T. Miyashi, M. Kamata and T. Mukai. *J. Am. Chem. Soc.*, **109**, 2780 (1987).
67. K. Gollnick, X. L. Xiao and U. Paulmann, *J. Org. Chem.*, **55**, 5945 (1990).
68. J. A. Howard, in *Free Radicals*, Vol. 2. (J. K. Kochi, Ed.), Wiley–Interscience, New York, London, (1973) Chap. 12.
69. G. A. Russel and A. G. Bemis, *J. Am. Chem. Soc.*, **88**, 5491 (1966).
70. W. A. Howard, in *The Chemistry of Peroxides* (S. Patai, Ed.) Wiley, Chichester (1983) Chap. 8.
71. M. H. Wijnen, *J. Chem. Phys.*, **27**, 710 (1957).
72. M. H. Wijnen, *J. Chem. Phys.*, **28**, 271 (1958).
73. W. H. Wijinen, *J. Chem. Phys.*, **27**, 710 (1957).
74. G. F. Sheats and W. A. Noyes, Jr, *J. Am. Chem. Soc.*, **77**, 1421 (1955).
75. G. R. McMillan, *J. Am. Chem. Soc.*, **83**, 3018 (1961).
76. L. M. Toth and H. S. Johnson, *J. Am. Chem. Soc.*, **91**, 1276 (1969).
77. G. A. Holder, *J. Chem Soc., Perkin Trans. 2*, 1089 (1972).
78. H. Lind and H. Loeliger, *Tetrahedron Lett.*, 2569 (1976).
79. H. Lind, T. Winkler and H. Loelinger, *J. Polym. Sci. Polym. Symp.*, **57**, 225 (1976).
80. S. W. Benson and R. Shaw, in *Organic Peroxides*, Vol. 1 (D. Swern, Ed.), Wiley–Interscience, New York, London (1970) Chap. 2.
81. H. Paul, R. D. Small, Jr, and J. C. Scaiano, *J. Am. Chem. Soc.*, **100**, 4520 (1978).
82. J. K. Allen and J. C. Berington, *Proc. R. Soc. London, Ser. A*, **262**, 271 (1961).
83. K. U. Ingold, in *Free Radicals*, Vol. 1 (J. K. Kochi, Ed.), Wiley–Interscience, New York, London (1973) Chap. 2.
84. H. Wieland, *Chem. Ber.*, **44**, 2553 (1911).
85. W. H. Richardson, R. McGinness and H. E. O'Neal, *J. Org. Chem.*, **46**, 1887 (1981).
86. K. J. McCullough, A. R. Morgan, D. C. Nonhebel and P. L. Pauson, *J. Chem. Res. (S)*, 35 (1980).
87. K. J. McCullough, A. R. Morgan, D. C. Nonhebel and P. L. Pauson, *J. Chem. Res. (S)*, 36 (1980).
88. P. R. Story, D. D. Denson, C. E. Bishop, B. C. Clark, Jr and J. C. Farine, *J. Am. Chem. Soc.*, **90**, 817 (1968).
89. J. K. Kochi and A. Bemis, *Tetrahedron* **24**, 5099 (1968).
90. W. L. Reynolds and R. Lumry, *J. Chem. Phys.*, **23**, 2560 (1955).
91. S. Hasegawa, N. Nishimura, S. Mitsumoto and K. Yokoyama, *Bull. Chem. Soc. Jpn.*, **36**, 522 (1963).
92. G. A. Baramki, H. S. Chang and J. T. Edward, *Can. J. Chem.*, **40**, 441 (1962).
93. R. Hiatt, in *Organic Peroxides*, Vol. 3 (D. Swern, Ed.), Wiley–Interscience, New York, London (1972) Chap. 1.
94. V. P. Maslenikov, V. P. Sergeeva and V. A. Shushunov, *Zh. Obshch. Khim.*, **37**, 1727 (1967).
95. A. G. Davies, R. V. Foster and R. Nery, *J. Chem. Soc.*, 2204 (1954).
96. J. Tanaka, *J. Org. Chem.*, **26**, 4203 (1961).
97. N. Kornblum and H. E. DelaMare, *J. Am. Chem. Soc.*, **73**, 880 (1951).

98. R. P. Bell and A. O. Mcdougall, *J. Am. Chem. Soc.*, 1697 (1958).
99. W. P. Keaveney, M. G. Berger and J. J. Pappas, *J. Org. Chem.*, **32**, 1537 (1967).
100. H. Mimoun, R. Charpentier, A. Mitschler, J. Fischer and R. Weiss, *J. Am. Chem. Soc.*, **102**, 1047 (1980).
101. R. A. Sheldon, in *Aspects of Homogeneous Catalysis*, Vol. 4 (R. Ugo, Ed.), Reidel, Dordrecht (1981) p. 3.
102. K. B. Sharpless and T. R. Verhoeven, *Aldrichim. Acta*, **12**, 63 (1979).
103. J. Sobczak and J. J. Ziolkowski, *J. Mol. Cat.*, **13**, 11 (1981).
104. A. A. Frimer in *The Chemistry of Peroxides* (S. Patai, Ed.) Wiley, Chichester (1983) Chap. 7.
105. C. W. Jefford, A. Jaber and J. Boukouvalas, *J. Chem. Soc., Chem. Commun.*, 1916 (1989).
106. M. Balci and Y. Sutbeyaz, *Tetrahedron Lett.*, **24**, 311 (1983).
107. K. Takahashi and M. Kishi, *Tetrahedron Lett.*, **29**, 4595 (1988).
108. N. Ichinose, K. Mizuno, T. Tamai and Y. Otsuji, *J. Org. Chem.*, **55**, 4079 (1990).
109. K. S. Feldman and R. E. Simpson, *Tetrahedron Lett.*, **30**, 6985 (1989).
110. E. K. Nelson, *J. Am. Chem. Soc.*, **33**, 1404 (1911).
111. P. Wieland and V. Prelog, *Helv. Chim. Acta*, **30**, 1028 (1947).
112. Baumslaugh, G. Just and F. Blank, *Nature (London)*, **202**, 1218 (1964).
113. S. M. Clarke and M. McKenzie, *Nature (London)*, **213**, 505 (1967).
114. Y. Takahashi and T. Takahashi, *Bull. Chem. Soc. Jpn.*, **39**, 848 (1966).
115. H. K. Adam, I. M. Campbell and N. J. McCorkindale, *Nature (London)*, **216**, 397 (1967).
116. H. K. Bhat, G. N. Qazi, C. L. Chopra, K. L. Dhar and C. K. Atal, *J. Indian Chem. Soc.*, **56**, 934 (1979).
117. V. N. Sviridonov and L. I. Strigina, *Khim. Prir. Soedin.*, 669 (1976).
118. J. Diak, *Planta Med.*, **32**, 181 (1977).
119. W. D. Chow, W. Nicolls and M. Sterns, *Tetrahedron Lett.*, 1353 (1971).
120. J.-M. Liu, M.-Y. Ni, Y.-F. Fan, Y.-Y. Tu, Z.-H. Wu, Y.-L. Wu and W.-S. Chou, *Acta Chim. Sin.*, **37**, 129 (1979).
121. G. Schmid and W. Hofheinz, *J. Am. Chem. Soc.*, **105**, 624 (1983).
122. M. Yamazaki, K. Susago and K. Miyashi, *Chem. Commun.*, 408 (1974).
123. M. Yamazaki, H. Fujimoto and T. Kawasaki, *Tetrahedron Lett.*, 1241 (1975).
124. M. Patel, A. C. Horan, V. P. Gullo, D. Loebenberg, J. A. Marguez, G. H. Miller and I. A. Waitz, *J. Antibiot.*, **37**, 413 (1984).
125. G. Rucker, E. Breitmaier, R. Mayer and D. Manns, *Arch. Pharm.*, **320** 437 (1987).
126. H. Itokawa, H. Morita, K. Osawa, K. Watanabe and Y. Iitaka, *Chem. Pharm. Bull.*, **35**, 2849 (1987).
127. H. Itokawa, H. Morita, T. Kobayashi, K. Watanabe and Y. Iitaka, *Chem. Pharm. Bull.*, **35**, 2860 (1987).
128. W. Fenical, *J. Am. Chem. Soc.*, **96**, 5580 (1974).
129. Y. M. Sheith and C. Djerassi, *Tetrahedron*, **30**, 4095 (1974).
130. M. D. Higgs and D. Faulkner, *J. Chem. Org.*, **43**, 3454 (1978).
131. D. B. Stiele and D. Faulkner, *J. Org. Chem.*, **44**, 964 (1979).
132. Y. Kashman and M. Rotem, *Tetrahedron Lett.*, **19**, 1707 (1979).
133. L. V. Manes, G. J. Bakus and P. Crews, *Tetrahedron Lett.*, **25**, 931 (1984).
134. M. Albericci, M. C. Lempereur, J. C. Braekman, D. Dalcze, B. Tursch, J. P. Declereq, G. Germain and M. V. Meerssche, *Tetrahedron Lett.*, **19**, 2687 (1979).
135. M. Albericci, J. C. Braekman, D. Dalcze and B. Tursch, *Tetrahedron*, **38**, 1881 (1982).
136. C. P. Leopardi, G. Germain, M. V. Meerssche, M. Albericci, J. C. Beckman, D. Dalloze and B. Tursch *J. Chem. Soc., Perkin Trans. 2*, 1523 (1982).

137. D. W. Phillipson and K. L. Rinehart Jr, *J. Am. Chem. Soc.*, **105**, 7735 (1983).
138. R. J. Wells, *Tetrahedron Lett.*, **16**, 2637 (1976).
139. E. Quinoa, E. Kho, L. V. Manes, P. Crews and G. J. Bakus, *J. Org. Chem.*, **51**, 4260 (1986).
140. S. Sakemi, T. Higa, U. Anthoni and C. Christophersen, *Tetrahedron*, **43**, 263 (1987).
141. R. J. Capon and J. K. MacLeod, *J. Org. Chem.*, **52**, 339 (1987).
142. R. J. Capon and J. K. MacLeod, *J. Nat. Prod.*, **50**, 225 (1987).
143. V. Merka, V. Zikesh and F. Shita, *Voen.-Med. Zh.*, 46 (1967).
144. A. M. Kligman, J. J. Leyden and R. Stewart, *Int. J. Dermatol.*, **16**, 413 (1977).
145. D. Fanta, H. Bardach and C. Poitscheck, *Arch. Dermatol. Res.*, **264**, 369 (1979).
146. N. Higashi, H. Teranishi and K. Shinohara, *Hifu*, **21**, 15 (1979).
147. W. J. Cunliffe and K. T. Holland, *Acta Derm.-Venereol.*, **59**, 555 (1979).
148. H. V. Bossche, F. Cornelissen and J. V. Cutsem, *Br. J. Dermatol.*, **107**, 343 (1982).
149. A. M. Patane and M. Pistillo, *Ann. Sclavo.*, **24**, 513 (1982).
150. J. J. Cavallo and R. A. Reynolds, *US Pat.* 3 622 351 (1969).
151. A. A. Bukhalovskii, *Tokisikol. Gig. Prod. Neftekhim. Neftekhim. Prozvod.*, 77 (1972).
152. N. G. Ivanov and L. P. Korotich, *Tokisikol. Nov. Prom. Khim. Veshchestv*, **15**, 90 (1979).
153. S. Matsugo, N. Kayamori, T. Ohta and T. Konishi, *Chem. Pharm. Bull.*, **39**, 545 (1991).

4 Dioxiranes, Three-membered Ring Cyclic Peroxides

WALDEMAR ADAM and LAZAROS P. HADJIARAPOGLOU
Institut für Organische Chemie der Universität Würzburg, Am Hubland, W-8700 Würzburg, Germany

RUGGERO CURCI and ROSSELLA MELLO
CNR 'Center MISO,' Department of Chemistry, University of Bari, 70126 Bari, Italy

1 INTRODUCTION AND HISTORICAL PERSPECTIVES

The first reference to dioxiranes **1**, three-membered ring cyclic peroxides, was made as far back as 1899 in the conversion of menthone into the corresponding lactone, which was studied by Bayer and Villiger;[1] in fact, it was suggested that this reaction proceeded through spirodioxirane **1d** as intermediate. Nearly 80 years later, the first isolation[2] of dioxiranes was claimed in a US Patent in 1972, namely the unstable and explosive perfluorodimethyl- and chlorodifluoromethyl(trifluoromethyl)dioxiranes (**1**:$R^1 = R^2 = CF_3$ and $R^1 = CClF_2$, $R^2 = CF_3$); to obtain these, the dilithium salts of the corresponding ketone hydrates were oxidized by fluorine at -80 to -30°C. It was not until 1978 that the parent dioxirane H_2CO_2 (**1a**: $R^1 = R^2 = H$) was detected[3] in the low-

Organic Peroxides. Edited by W. Ando

temperature, low-pressure gas-phase ozonolysis of ethylene, and its structure was rigorously established by microwave spectroscopy.

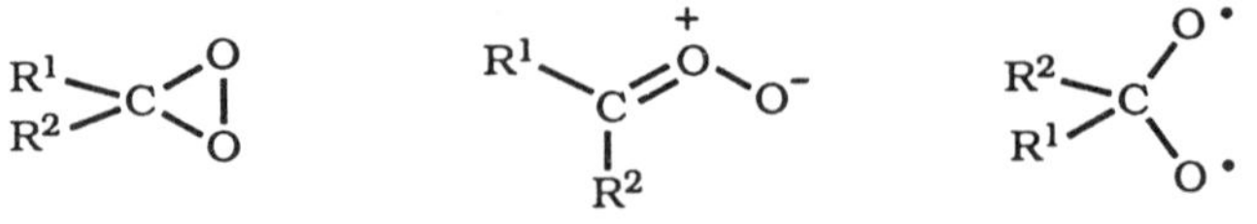

1 **2** **3**

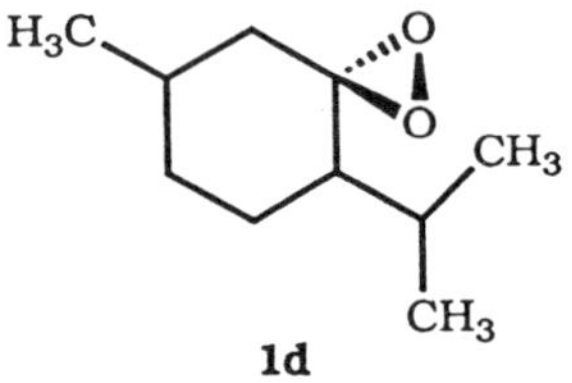

1d

In 1974 it was recognized[4] that simple ketones, e.g. acetone, catalyzed the decomposition of commercial caroate (triple salt, $2KHSO_5 \cdot KHSO_4 \cdot K_2SO_4$) with evolution of oxygen gas in the pH range 6–12, which suggests the intermediacy of dioxiranes.[4,5]

These historically relevant landmarks aside, the scene for the far-reaching developments in dioxirane chemistry was set in the late 1970s through the careful kinetic and labelling studies of Edwards and co-workers,[5] who demonstrated the generation of dioxiranes in the ketone–caroate system. This paved the way for the development of preparative dioxirane chemistry under *in situ* conditions.[5]

A major breakthrough for synthesis came in 1985, when dimethyldioxirane (**1b**) was isolated[6] as an acetone solution by codistillation at reduced pressure from buffered acetone–caroate mixtures. This permitted full spectral characterization[7,8] of these unique organic peroxides. Thus, a new oxidant was made generally available for preparative purposes that is extremely efficient in its ability to transfer oxygen atoms to a variety of organic substrates, yet selective in its reactivity, and conveniently prepared from readily available starting materials.[9–11]

2 GENERATION AND ISOLATION

After it had been recognized that at neutral pH simple ketones catalyze the decomposition of caroate with evolution of oxygen gas,[4] it was demonstrated[12] that caroate in the presence of ketones serves as an effective oxidant for a wide variety of substrates. Kinetic, stereochemical and ^{18}O-labelling studies established dioxiranes unequivocally as the actual oxygen transfer agent (Scheme 1).[5] Dioxiranes, generated *in situ* (caroate–ketone system), were

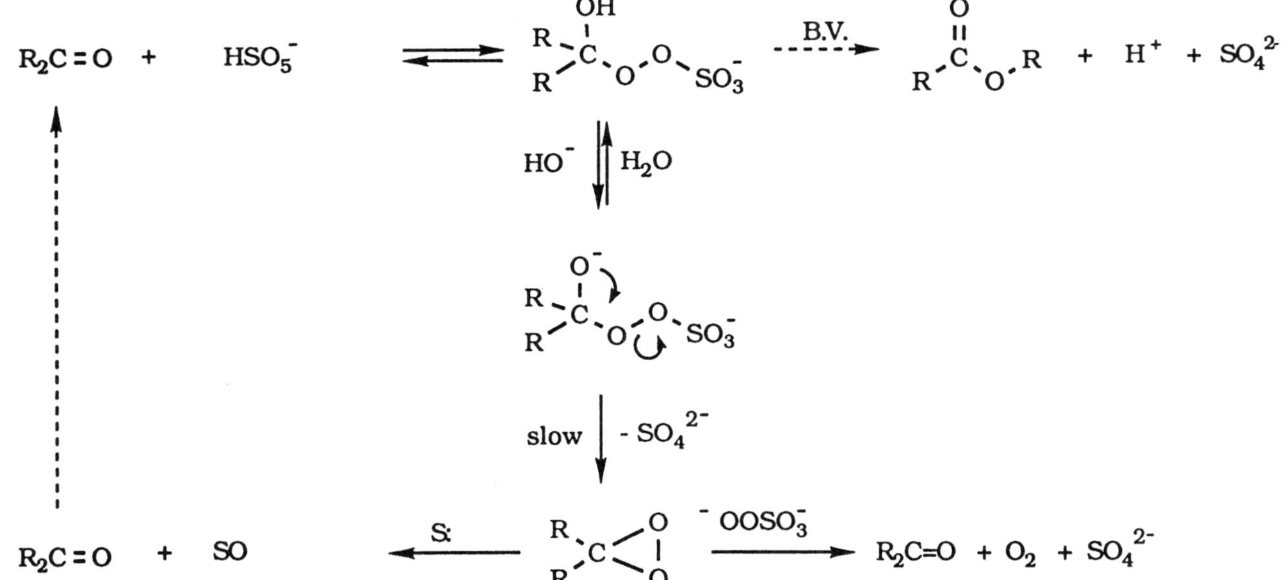

Scheme 1. Mechanism of dioxirane formation. S: and SO represent substrate and oxidized substrate, respectively

quickly adopted as remarkably efficient and stereoselective oxidants for preparative applications, especially for epoxidations and heteroatom oxidations.[12]

A convenient and practical procedure is to use the biphasic butan-2-one–water mixture in the presence of $NaHCO_3$[13a] or Na_2HPO_4–NaH_2PO_4[14] to buffer the pH at *ca* 7.3–7.5. Unless the substrate is not soluble in this medium, a co-solvent and phase-transfer catalyst are not necessary. In this way cyclohexanone also can be used as a dioxirane source.

More recently it has been shown that dioxiranes are also formed in the reaction of ketones with the bis(trimethylsilyl) derivative of Caro's acid, $Me_3SiOOSO_2OSiMe_3$, in anhydrous CH_2Cl_2.[15] There are numerous reports in the literature that dioxiranes are involved as transient species in peroxide chemistry, but it is beyond the scope of this chapter to cover them here.

A major achievement, which kindled the surge of present-day activity in dioxirane chemistry, was the isolation of dimethyldioxirane as an acetone solution.[6] Simple removal of the volatile dioxirane from the generating medium by application of a slight vacuum in a current of inert gas afforded ketone solutions [*ca* 0.05–0.10 M for dimethyldioxirane and 0.5–0.8 M for methyl(trifluoromethyl)dioxirane] of the desired dioxiranes. The availability of isolated dioxiranes (as ketone solutions) permitted these novel oxidants to be employed under non-hydrolytic conditions and thereby opened up a new era of oxidation chemistry.

A simplified procedure[14a] for the preparation of isolated dimethyldioxirane (**1b**) from acetone–caroate mixtures consists of adding solid caroate to a mixture of acetone and water that contains $NaHCO_3$. After complete addition, a slight vacuum (50–100 Torr) is applied to the system and the dioxirane–acetone vapors are condensed into a receiving flask kept at $-78°C$. The resulting dioxirane solutions are assayed for peroxide content (*ca* 0.1 M) by reacting them with phenyl methyl sulfide to yield the corresponding sulfoxide and determination of the latter by ^1H NMR spectroscopy. Alternative analytical methods to determine the dioxirane content include iodimetry and quantitative UV–visible spectrophotometry. Methyl(trifluoromethyl)dioxirane (**1c**) was prepared[16] from trifluoroacetone in this way as a *ca* 0.8 M solution in CF_3COCH_3. Most recently even ketone-free solutions of dioxirane **1c** could be obtained[17] (*ca* 0.8 M in CH_2Cl_2, $CHCl_3$, CCl_4, Freons, etc.) by water extraction of the trifluoroacetone in the form of its hydrate.

The isolation of dioxiranes (as ketone solutions) permitted their complete spectroscopic characterization. The available spectral data of dioxiranes **1b** and **c** are given in Table 1. The dioxirane ring carbon is significant,[7,8] with a ^{13}C resonance of *ca* 100 ppm and a single ^{17}O resonance signal[7] at *ca* 297 ppm. A weak ($\varepsilon \approx 13$) n–σ^* absorption is exhibited in the UV region 320–350 nm, which extends out into the visible region to *ca* 450 nm, hence the pale yellow color of this most strained cyclic peroxide.[6–8,16]

Table 1. Spectral data for dioxiranes **1b** and **1c**

Dioxirane	UV (nm)	IR (cm^{-1})	^{1}H NMR (δ, ppm)	^{13}C NMR (δ, ppm)	^{17}O NMR (δ, ppm)	^{19}F NMR (δ, ppm)
$(CH_3)_2CO_2$ **1b**	335 ($\varepsilon \approx 13$)[a]	3012[a], 3005, 2999, 1209, 1196(w), 1094, 1080(w), 1059(w), 1034(w), 899, 784.	1.65[a]	22.69 (CH_3)[a] 102.0 (ring C)	302[b]	
$(CH_3)(CF_3)CO_2$ **1c**	347 ($\varepsilon \approx 9$)[c]	3020(w)[d], 1447 (m), 1427(m), 1402(s), 1330(s), 755(m), 745(m), 715(s), 615(s).	1.97[c]	14.51 (CH_3)[c] 97.32 (ring C) 122.2 (CF_3)	297[c]	−81.5[c]

[a]All measurements in acetone solution at *ca* 0.1 M.
[b]^{17}O-enriched acetone was used as precursor.[7]
[c]All measurements in trifluoroacetone.
[d]All measurements with ketone-free dioxirane[17] in $CF_2ClCFCl_2$.

Structural parameters such as bond lengths and angles are available[3] for the parent dioxirane **1a** from microwave spectra. It is significant that the O—O bond (1.52 Å) is the longest so far known.[18] This bond length would imply an unstable entity, expected easily to undergo homolysis of the O—O bond to afford the bisoxymethylene diradical **3a**. Theoretical studies on dioxirane **1a**, carbonyl oxide **2a** and bisoxymethylene diradical **3a** confirmed that of these three species the dioxirane **1a** is the lowest energy form, the bisoxymethylene diradical **3a** being about 15 kcal mol^{-1} higher. In view of the low activation barrier (<1 kcal mol^{-1}) estimated for the reclosure,[19,20] the highly strained dioxirane **1a** has sufficient kinetic stability to persist below 40°C in the absence of oxidizable substrates. For comparison, an activation barrier of 22.8 kcal mol^{-1} was calculated[19–21] for the rearrangement **2c** → **1c**, although this process is exothermic by 34.5 kcal mol^{-1}. Indeed, this rearrangement[22] could only be achieved by irradiation of matrix-isolated dioxiranes. Moreover, it was shown recently[17] that the thermolysis (*ca* 60°C) and photolysis (*ca* > 380 nm) of ketone-free dioxirane **1c** in halocarbon solutions involve a complex radical chain decomposition in which $CH_3\cdot$ and $CF_3\cdot$ radicals serve as propagating species (Scheme 2).

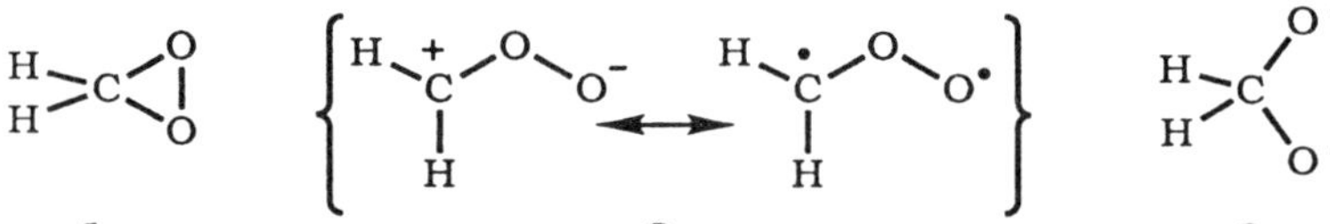

1a **2a** **3a**

3 CHEMICAL REACTIVITY

The hitherto known transformations[9–11] undergone by dioxiranes are summarized in Figure 1. These include the epoxidation of alkenes[5,12a,12c,12d,12f,13,23] (transformation 1), polycyclic aromatic hydrocarbons[12e,24] (transformation 2), allenes[25] (transformation 3), enol derivatives[26–37] (transformation 4), and α,β-unsaturated ketones[38,39] (transformation 5). Even insertions into C—H bonds of alkanes[40,41] (transformation 6) and of aldehydes[6,42] (transformation 7) and into Si—H bonds of silanes[43] (transformation 8) have been reported. Sulfides[8,44] were oxidized to sulfoxides or sulfones (transformation 9), primary amines[45] to nitro compounds (transformation 10), chloride ion[3] to the hypochlorite ion (transformation 11), phosphine sulfides[46] to the phosphine oxides (transformation 12) and imines[47] to nitrones (transformation 13). Further, dimethyldioxirane catalyzed the isomerization of quadricyclane[48] to norbornadiene; the latter was then epoxidized by the dioxirane.[13d] Although the thermal rearrangement of dimethyldioxirane into methyl acetate in solution is still rather circumstantial, methyl acetate was formed when dimethyldioxirane was

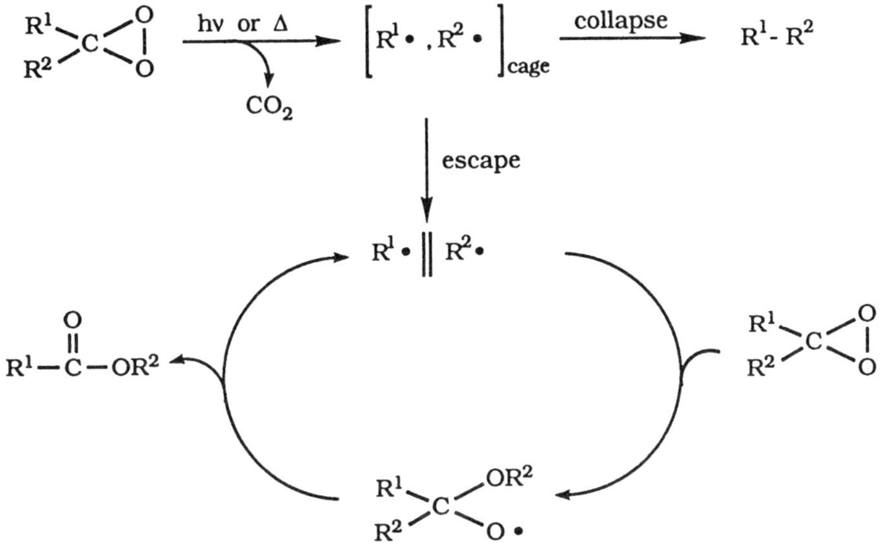

Scheme 2. Proposed free-radical chain mechanism for the photolysis and thermolysis of ketone-free dioxirane **1c** in solution

treated with BF_3[6] or photolyzed under matrix isolation.[22] Also, methyl trifluoroacetate was produced during the thermal or photochemical decomposition of methyl(trifluoromethyl)dioxirane (**1c**).[16,17] A detailed mechanistic study revealed, however, that a complex free-radical chain decomposition is involved (Scheme 2).[17]

On account of all this interesting chemistry, dioxiranes have been recognized as efficient oxygen transfer agents. They are selective (chemo-, regio- and stereo-) despite their pronounced reactivity, mild towards the oxidized product, perform under strictly neutral conditions and possess catalytic activity. We predict that for most preparative applications, dioxiranes will replace peracid oxidations in the future, especially when the *in situ* mode (caroate–ketone system) can be employed or the hydrolytic lability of the substrate and oxidized product demands the use of isolated dioxiranes.

3.1 Epoxidations

The most extensively studied and, for preparative purposes, useful chemical transformation with dioxiranes is their ability to convert a variety of alkenes into the corresponding epoxides. The fact that dioxirane is inert towards the oxidized product and performs under strictly neutral conditions (the precursor ketone is the only by-product of the reaction) permits the isolation of extremely labile epoxides.

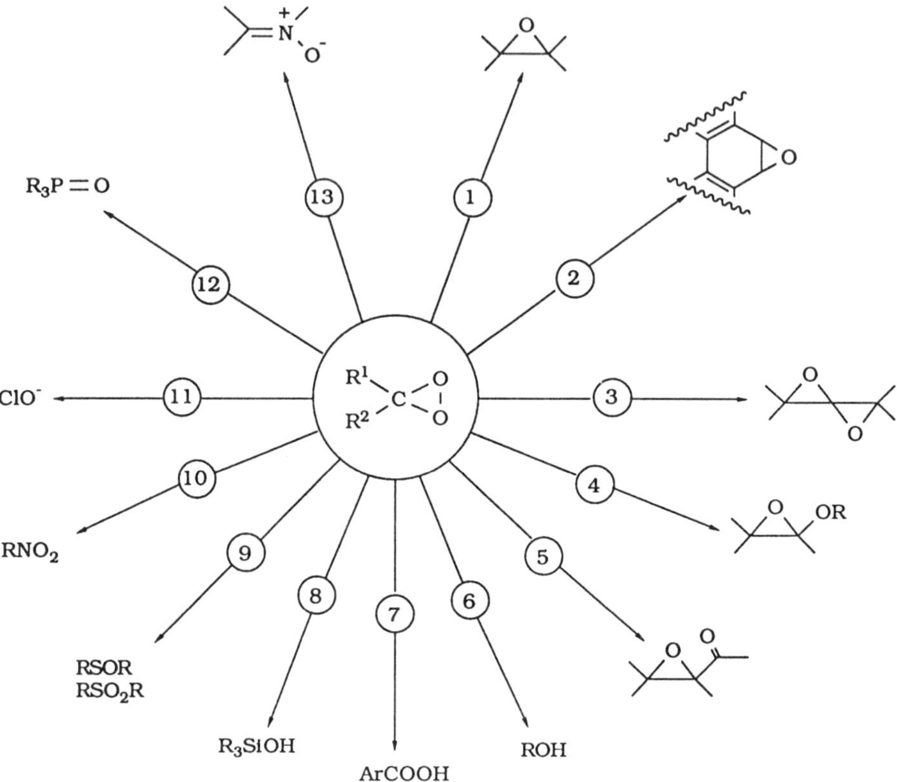

Figure 1. Valuable synthetic transformations with dioxiranes

Dimethyldioxirane is the most extensively used dioxirane[9–11] for synthetic purposes. In the *in situ* mode,[5,12] the epoxidations are carried out in aqueous media for water-soluble substrates. For water-insoluble substrates, biphasic systems of water and organic solvents (dichloromethane, benzene) are used; in this case a phase-transfer catalyst such as Bu_4N^+ HSO_4^- or 18-crown-6 is helpful. Replacement of acetone with butan-2-one[13a,14b] permits the *in situ* mode to be conducted in a biphasic system without the need for any co-solvent or phase-transfer catalyst. Although it has been reported that pH control is unnecessary with hydrogen carbonate buffer,[13b] in our experience it is necessary always to control the pH at 7.3–7.5 by means of base.[5,12,14b] Of course, these problems are obviated when isolated dioxiranes are employed.

The significance of the neutral medium is illustrated in the isolation of acid-sensitive epoxides such as oxaspiropentanes,[12f] which are obtained in the epoxidation of the corresponding methylenecyclopropanes by dimethyldioxirane. Such oxaspirocyclopropanes are not easily obtained without rearrangement to the respective cyclobutanones. Benzvalene epoxide was isolated in over 90% yield within 3 h at −30°C when an etheral solution of benzvalene was treated with isolated dimethyldioxirane (equation 1).

CH_3COCH_3, ether, -30 °C, N_2, 3 h (1)

Employing various *cis*- and *trans*-alkenes, it was demonstrated[5,12a,12d,23] that isolated or *in situ*-generated dioxirane transfers the oxygen atom stereoselectively to the alkene; (*Z*)-alkenes give (*Z*)-epoxides and (*E*)-alkenes give (*E*)-epoxides. Moreover, (*Z*)-alkenes are more reactive than (*E*)-alkenes and the rates of epoxidation by dioxirane are much less sensitive to the degree of alkyl substitution than peracids.[23] In addition to aryl-substituted alkenes,[12a,12d,23] electron-poor substrates[5,12a,13b,49] such as cinnamic esters are also efficiently epoxidized by dimethyldioxirane.

Enantiomeric excesses up to 12% were observed in the epoxidation of simple alkenes[12d] when optically active ketones were employed as precursors of dioxiranes (Scheme 3). These enantioselectivities are more than twice those recorded for optically active peracids.[50] Recently, higher optical yields (up to

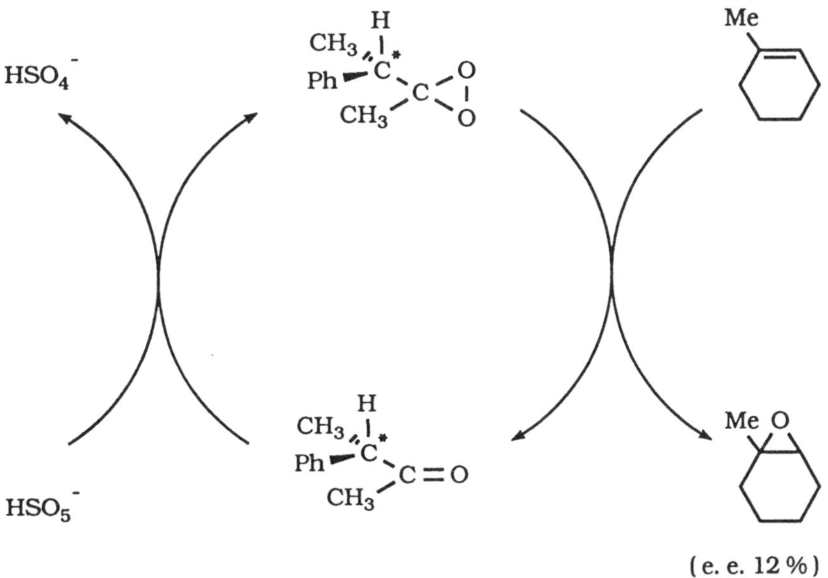

Scheme 3. Enantioselective epoxidation of simple alkenes by *in situ*-generated optically active dioxiranes

24% e.e.) have been recorded[51] by employing 4,4,4-trifluoro-3-phenyl-3-methoxybutan-2-one. The advantage of these epoxidations in the *in situ* mode is that such reactions require only *catalytic* amounts of optically active ketones.

Various polycyclic aromatic hydrocarbons were transformed into arene oxides by dimethyldioxirane (transformation 2 in Figure 1).[6,12e,24a] Phenanthrene[6] was epoxidized by isolated dimethyldioxirane to yield the 9,10-oxide in 83% yield at 25°C. The more reactive methyl-(trifluoromethyl)dioxirane[16] (**1c**) converted (over 80%) phenanthrene at −20°C within 5 min to the 9,10-oxide in 93% yield. Naphthalene was quantitatively converted exclusively into its *anti*-1,2;3,4-dioxide (equation 2). In this context, since it had been demonstrated[3] that gas-phase ozonolysis of ethylenes produces dioxiranes, it was postulated[12e] that arene oxides in polluted atmospheres may be derived from dioxirane epoxidations; urban atmospheres frequently contain relatively high concentrations of ozone, alkenes and polycyclic aromatic hydrocarbons.

Freon A113 (or CH_2Cl_2); Na_2HPO_4
-20 °C; 5 min; 96 % conv.; 98 % yield (2)

Dimethyldioxirane is the reagent of choice for the epoxidation of allenes[25] (transformation 3 in Figure 1) to their dioxides (equation 3).

CH_3COCH_3 ; K_2CO_3 (3)

Various enol ethers (transformation 4 in Figure 1) were transformed into the corresponding epoxides either by dimethyldioxirane[27] or by the more reactive methyl(trifluoromethyl)dioxirane,[28] both being used in isolated form. A highlight is the synthesis of the 8,9-epoxide of aflatoxin B_1,[27] a well-known carcinogen, accomplished by the epoxidation of aflatoxin B_1 with isolated dimethyldioxirane at room temperature (equation 4). On account of its sensitive nature towards hydrolysis, numerous previous efforts to prepare this epoxide

CH_2Cl_2 , CH_3COCH_3
ca 20 °C; 15 min (4)

failed; its availability permitted the direct investigation of its chemistry and mutagenicity.

Other successful epoxidations include 2,7-dioxabicyclo[3.2.0]heptan-3-one,[29] pyranose and furanose glycals,[30] enol phosphates,[26a] enol esters and lactones,[26b] and vinyl formamides.[31] The *O*-tetrabenzyl-2-hydroxyglycal was epoxidized[32] by dimethyldioxirane quantitatively at −30° within 2 h. Removal of the acetone at below −30°C permitted the isolation of this labile epoxide, which is stable at −30°C for several hours but decomposes rapidly within 10 min at room temperature.

Benzofuran epoxides, a hitherto unknown class of epoxides which were postulated as intermediates in the oxidative metabolism of the corresponding furans, were obtained[33,34] in excellent yields in the epoxidation of benzofurans by isolated dimethyldioxirane (equation 5). The epoxides rearrange into the quinone methides[34] and, depending on the substituents of the benzo moiety, the isolation of one or the other valence isomer can be achieved.

R4, R3, R2, R1, Me, Me, O; CH_3, CH_3, C, O, O; CH_2Cl_2, CH_3COCH_3; -78 to -30 °C; N_2; 3-12 h

R4, Me, R3, O, R2, O, Me, R1 ⇄ R4, Me, Me, R3, O, R2, O, R1 (5)

The Rubottom reaction[35] (the peracid oxidation of silyl enol ethers to α-hydroxy ketones) was postulated to proceed through an epoxide intermediate. These epoxides were readily prepared[32,36] by epoxidation with isolated dimethyldioxirane (equation 6) or observed spectroscopically[37] when dimethyldioxirane-d_6 was employed.

$OSiMe_3$; CH_3, CH_3, C, O, O; CH_2Cl_2, CH_3COCH_3; -40 °C; N_2; 3h; $OSiMe_3$, O (6)

It is evident from the previous examples that the epoxidation of electron-rich alkenes by isolated dioxiranes works well. The advantages over peracids are that dioxiranes are more reactive and labile epoxides are isolable. Equally valuable from the preparative point of view is the fact that dimethyldioxirane also performs epoxidations of electron-poor alkenes (transformation 5 in Figure 1). Excellent yields of such epoxides can be obtained, provided that a large excess of dioxirane (up to tenfold) is used at longer reaction times (up to 4 days) and elevated temperatures (up to 35°C). In this manner the relatively sluggish β-oxo enol ethers,[38a] flavones[38b] (equation 7) and isoflavones[38c] were converted into their epoxides in excellent yields.

$$\text{flavone} \xrightarrow[\text{0 °C; } N_2\text{; 23-65 h}]{(CH_3)_2CO_2,\ CH_2Cl_2 \cdot CH_3COCH_3} \text{flavone epoxide} \qquad (7)$$

Similarly, various α,β-unsaturated ketones[14b,39] were converted into their corresponding epoxides either by isolated dimethyldioxirane or by *in situ*-generated ethyl(methyl)dioxirane. Even the novel (*E*)-2′-hydroxychalcone oxides[14b] could be obtained essentially quantitatively by epoxidation with isolated dimethyldioxirane (equation 8). These epoxides were not accessible on treatment of such chalcones with alkaline hydrogen peroxide (the Weitz–Scheffer reaction),[52] the reagent of choice for the epoxidations of electron-poor alkenes.[53]

$$\text{(E)-2'-hydroxychalcone} \xrightarrow[\text{ca. 20° C; } N_2\text{; 23-35 h}]{(CH_3)_2CO_2,\ CH_2Cl_2 \cdot CH_3COCH_3} \text{chalcone oxide} \qquad (8)$$

3.2 Insertions

The efficient oxyfunctionalization of unactivated C—H bonds of alkanes by dioxiranes to give alcohols under extremely mild conditions is undoubtedly the highlight so far of dioxirane chemistry (transformation 6 in Figure 1). One of the first examples[40] concerned the reaction of dimethyldioxirane (**1b**) with *cis*- and *trans*-1,2-dimethylcyclohexanes, which afforded the corresponding 1,2-dimethylcyclohexan-1-ols stereoselectively in *ca* 100 and 45% yield, respectively. For dimethyldioxirane (**1b**) these oxidations require reaction times of several hours and a large excess of oxidant;[40] however, methyl(trifluoromethyl)dioxirane[16] (**1c**) converts *cis*- and *trans*-1,2-dimethylcyclohexanes into the corresponding cyclohexanols in over 90% yields within 4–10 min even at −20°C.[16]

When secondary alcohols are produced, these are further oxidized to the corresponding ketones.[40,41] For instance, an impressive reaction is the virtually complete conversion of cyclohexane into cyclohexanone by **1c** (equation 9) in over 95% yield at −18°C within *ca* 20 min.[41] There is no parallel in oxidation chemistry for this astounding transformation. Equally astounding is the selective hydroxylation of adamantane[54] by dioxirane **1c** to afford the tetrahydroxyadamantane (equation 10). Optically active 2-phenylbutane was converted[55] by **1c** in over 90% yield with complete retention of configuration (equation 11).

$CF_3(CH_3)C(O_2)$; CH_2Cl_2, CF_3COCH_3; -22 °C; 18 min (9)

$CF_3(CH_3)C(O_2)$ excess; CH_2Cl_2, CF_3COCH_3; -20 °C (10)

$CF_3(CH_3)C(O_2)$; CH_2Cl_2; -23 °C; 1h; > 80% conv.; >95% yield (11)

R(-) [o.p. 72%] → *S*(-) [o.p. 72%]

That dioxiranes are useful reagents for the conversion of alcohols into carbonyl products is illustrated[56] in equation 12. Interestingly, secondary hydroperoxides[57] are not affected by dioxiranes; thus allylic hydroperoxides were transformed into epoxy hydroperoxides, whereas allylic alcohols afforded epoxy ketones.

$CF_3(CH_3)C(O_2)$; CH_2Cl_2, CF_3COCH_3; -20 °C; 15 min; 96% conv.; 92% yield (12)

$[\alpha]_D$= +19.6° ($CDCl_3$) *o.p. > 96%* → $[\alpha]_D$= +61.8° ($CDCl_3$) *o.p. > 96%*

Acyclic and cyclic ethers were readily oxidized[58] by dioxirane **1c** with C—O cleavage. For example, *tert*-butyl methyl ether was cleaved into *tert*-butanol and presumably formic acid, whereas tetrahydrofuran afforded γ-butyrolactone in high yield (equation 13).

$$\text{THF} \xrightarrow[CH_2Cl_2,\ 0^\circ C]{CF_3(CH_3)C(O_2)} \left[\text{2-hydroxytetrahydrofuran}\right] \xrightarrow[CH_2Cl_2,\ 0^\circ C]{CF_3(CH_3)C(O_2)} \gamma\text{-butyrolactone} \quad (13)$$

Aldehydes[6] undergo oxygen insertions with dimethyldioxirane to yield the corresponding carboxylic acids (transformation 7 in Figure 1). The fact that no ozonides were formed indicates that dioxiranes do not isomerize to carbonyl oxides, since it is well established that carbonyl oxides[22] react with carbonyl compounds to form ozonides by 1,3-dipolar cycloaddition.

The reaction of dimethyldioxirane with a series of substituted benzaldehydes[42] slowly produced the corresponding benzoic acids. *p*-Methoxybenzoic acid was isolated in 100% yield when the reaction of *p*-anisaldehyde with dimethyldioxirane was carried out under a nitrogen atmosphere. When the reaction was run under an atmosphere of ^{17}O-enriched oxygen gas and an excess of aldehyde, in addition to 71% of the acid also 21% *p*-anisyl formate was isolated. The *p*-methoxybenzoic acid was found to be ^{17}O enriched, which indicates a complex oxidation process with incorporation of molecular oxygen.[42] We suspect that a dioxirane-initiated free-radical chain process is responsible, analogous to the thermal decomposition of dioxirane in solution.[17]

Silanes[43] cleanly undergo oxygen insertions with dioxiranes to yield the silanols (transformation 8 in Figure 1) in excellent yields and with retention of configuration. Methyl(trifluoromethyl)dioxirane (**1c**) was found to be considerably more reactive than dimethyldioxirane (**1c**) and an appreciable primary kinetic isotope effect ($k_H/k_D = 2.6$) was measured when Et_3SiH and Et_3SiD were used.

3.3 Heteroatom Oxidations

It has already been mentioned that dioxirane solutions are assayed for peroxide content by treating them with excess of thioanisole.[8] Indeed, generally sulfides[6,44] are easily oxidized to sulfoxides by dioxiranes and the latter to their sulfones (transformation 9 in Figure 1). The oxidation can be controlled at the sulfoxide stage, by utilizing stoichiometric amounts of the dioxirane.

When a series of *para*-substituted thioanisoles were employed, it was found that both the sulfide to sulfoxide and the sulfoxide to sulfone oxidations by dimethyldioxirane (**1b**) gave negative Hammett ρ values for both oxygen transfers, which indicates the electrophilic nature of dioxiranes. On the other hand, dimethyldioxirane oxidation of thianthrene-5-oxide afforded preferentially the 5,5-dioxide (equation 14), the mole fraction of sulfoxide attack[59] being $X_{SO} = 0.68 \pm 0.03$. For comparison, the X_{SO} values for *m*-CPBA (*meta*-chloroperbenzoic acid) and carbonyl oxides are 0.36 and over 0.80, respectively. Thus, qualitatively, dimethyldioxirane is more nucleophilic than *m*-CPBA and less nucleophilic than carbonyl oxides.

Ox_{Elec} Ox_{Nucl} (14)

High enantiomeric excesses[44b] (e.e.), up to 90% in one case, were achieved in the oxidation of prochiral sulfides to chiral sulfoxides with *in situ*-generated dioxiranes by using bovine serum albumin as a chiral phase-transfer catalyst. Another preparatively useful transformation is the efficient oxidation of thiophenes to the corresponding *S,S*-dioxides, which was recently employed in the oxidation of thiophenophanes.[60]

The recent isolation of α-oxosulfones[61] from the oxidation of α-oxosulfides by excess of dimethyldioxirane at −40°C (equation 15) is another example of the efficiency of dimethyldioxirane under neutral conditions. These labile sulfones, which possess synthetic potential, could previously be obtained only by the ozonolysis of α-diazosulfones.[62]

$$R^1{-}S{-}C(=O){-}R^2 \xrightarrow[CH_3COCH_3\,,\ CH_2Cl_2;\ -40\,^\circ C;\ N_2;\ 3\,h]{(CH_3)_2CO_2} R^1{-}S(=O)_2{-}C(=O){-}R^2 \qquad (15)$$

Regarding *N*-oxidation, the conversion of pyridine into pyridine *N*-oxide by the dioxirane generated *in situ* from cyclohexanone and caroate was among the first examples to be reported.[12b] 2-Aminopyridine also gave quantitatively[63] the

N-oxide with dioxirane **1b**. However, oxidation of primary amines[45] gave nitro compounds rapidly and quantitatively (transformation 10 in Figure 1). A stepwise process was postulated, with hydroxylamines and nitroso compounds as intermediate products, since both were separately oxidized by dimethyldioxirane to their nitro derivatives. This oxidation can be partially controlled at the hydroxylamine stage[64] (equation 16) by performing the reaction at -45°C instead of at room temperature.

AcO, AcO, AcO, O, OCH_3, NH_2 —[$(CH_3)_2C(O_2)$; CH_3COCH_3, -45 to *ca* 20 °C]→ AcO, AcO, AcO, O, OCH_3, NHOH (16)

The *in situ*-generated dimethyldioxirane[65] was employed with success for the oxidation of aromatic amines to nitro compounds on a large scale. Isocyanates[66] were transformed to the corresponding nitro compounds (equation 17) in a stepwise reaction by involving the initial formation of amines, which are further oxidized to nitro compounds.

O=C=N, N=C=O —[$(CH_3)_2C(O_2)$; CH_3COCH_3, water; *ca* 20 °C: dark; 1.5 h]→ O_2N, NO_2 (17)

Secondary amines[67] are cleanly oxidized by dimethyldioxirane (equimolar amounts) to hydroxylamines in nearly quantitative yields. With a second mole of dioxirane, further oxidation yields either nitrones or nitroxides,[68] depending on whether hydrogens are present on the α-C atoms.

An intriguing report[69] concerns the oxidation of DNA by dimethyldioxirane. It was shown that adenine moieties were selectively oxidized and this oxidation was employed for the site-specific cleavage of DNA.

The oxidation of chloride ion to hypochlorite ion (transformation 11 in Figure 1) was in fact the first oxygen transfer reaction performed with *in situ*-generated dimethyldioxirane.[4] The oxidation of phosphine sulfides and phosphorothiolates[46] into their corresponding oxygen analogues (transformation 12 in Figure 1) has been reported to occur also rapidly, cleanly and in high yield.

A wide range of imines[47] (transformation 13 in Figure 1) were converted by dioxirane **1b** into their corresponding nitrones in variable yields, but with no formation of oxaziridine (equation 18). In contrast, *m*-CPBA oxidation of the corresponding imines usually affords oxaziridines as the major product.

$$\underset{R}{\overset{R}{>}}C\overset{O}{\frown}N{-}R \xleftarrow[\substack{CH_2Cl_2\\ 0\,^\circ C:1.2\ h}]{m\text{-CPBA}} \underset{R}{\overset{R}{>}}C{=}N{-}R \xrightarrow[\substack{CH_2Cl_2,\ CH_3COCH_3\\ 0\,^\circ C;\ 2\ h}]{(CH_3)_2C\langle{}^{O}_{O}|} \underset{R}{\overset{R}{>}}C{=}\overset{+}{N}\langle{}^{O^-}_{R} \qquad (18)$$

4 MECHANISTIC ASPECTS

4.1 Epoxidations

From the above, it is apparent that the most characteristic and distinctive feature of dioxiranes concerns their propensity for transferring oxygen atoms to a variety of donor substrates (S:) to yield oxidation products SO and the ketone that pertains to the dioxirane (equation 19). A wealth of experimental facts has been accumulated during the last 15 years on the mechanism of this oxygen transfer. Thus, the kinetics obeyed the second-order rate law, first order each with respect to dioxirane and substrate. The observed alkene reactivity[9–11] follows the order ED—C=C— > —C=C— > EA—C=C—, where ED is an electron donor and EA is an electron acceptor. For example,[23,49] β-methylstyrene is about 10^3 times more reactive toward dimethyldioxirane (**1a**) than ethyl *trans*-cinnamate. Hammett plots for styrenes and ethyl cinnamates substituted in the phenyl ring gave negative ρ values, i.e. -0.90[23] and -1.53.[49] These indicate that positive charge builds up in the alkene substrate during the oxygen atom transfer and demonstrate the electrophilic nature of dioxiranes. The reactivity of alkyl-substituted alkenes follows the epoxidation order $R_2C{=}CR_2 > R_2C{=}CHR > RCH{=}CHR \geqslant R_2C{=}CH_2 > RCH{=}CH_2$, i.e. the more alkylated alkenes react faster. Steric interactions are significant since *trans*-alkenes are *ca* eight times less reactive than their corresponding *cis* isomers. The oxygen atom transfer proceeds rigorously stereoselectively, i.e. *cis*-alkenes give the *cis*-epoxides and *trans*-alkenes the *trans*-epoxides.

$$S{:} \; + \; R_2C\langle{}^{O}_{O}| \longrightarrow SO \; + \; R_2C{=}O \qquad (19)$$

All these experimental facts for dioxiranes parallel those for the epoxidation by peracids. Consequently, the so-called 'butterfly' transition state (Scheme 4) provides a consistent mechanistic picture for the oxygen atom transfer

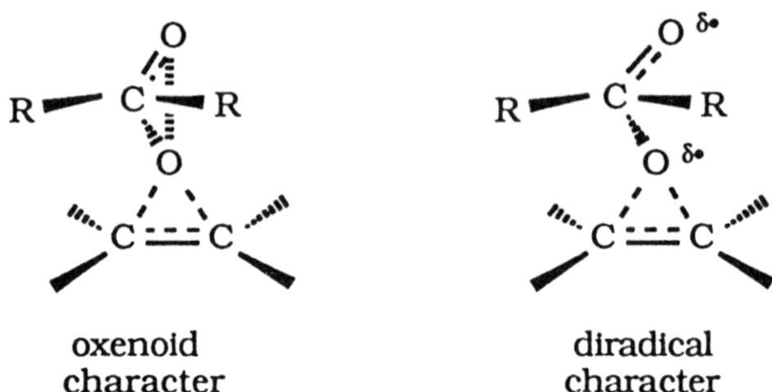

Scheme 4. 'Butterfly' mechanism for the epoxidation of alkenes by dioxirane

processes. In view of the weak peroxide bond in dioxiranes, it is expected that this oxenoid transition state has appreciable diradical character (Scheme 4).

An interesting question pertains to the chemical behavior of the bisoxymethylene diradical **3**. Of the three possible electronic configurations,[18–21] **3i**, **3ii**, and **3iii** (Scheme 5), the [σ, σ]-type 1,3-diradical **3i** is the

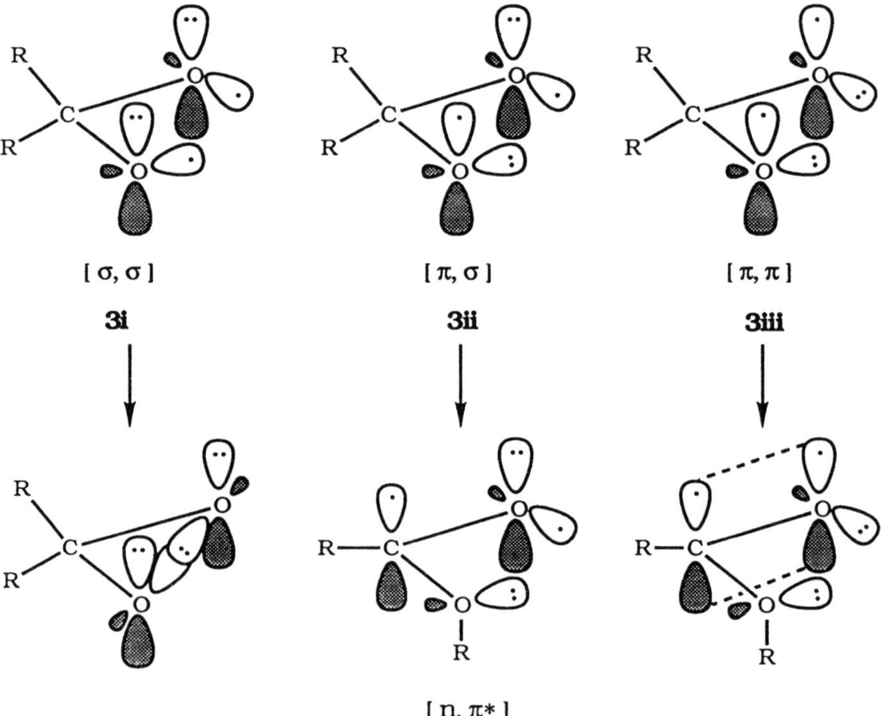

Scheme 5. Electronic configurations of bisoxymethylene diradical **3** and their transformation

one that results from homolysis of the peroxide bond in the dioxirane **1** on thermal activation. One might think that such a diradical should spontaneously rearrange into an ester product. However, only the [π, π]-type species **3iii** would readily rearrange into an ester product in view of the optimum alignment of the radical lobes with the migrating sigma bond. On the other hand, the [σ, π]-type 1,3-diradical **3ii** would generate an electronically excited ester product on rearrangement (a problem of σ/π crossing[19–21]) and, in fact, on photochemical activation (π, σ^* excitation) dioxiranes do rearrange into esters.[22] Thus, the [σ, σ] configuration **3i** is protected by an electronic barrier towards its most obvious transformation, namely rearrangement into the ester product, and instead it cyclizes back to the dioxirane.

4.2 Heteroatom Oxidations

Much less experimental work has been expended on the oxygen atom transfer to sulfur, nitrogen and phosphorus heteroatoms. Sulfides are oxidized faster to sulfoxides than the latter to sulfones, which emphasizes once again the expected electrophilic nature of dioxiranes in their propensity to transfer oxygen atoms. When placed in a competitive situation, however, as in the case with thianthrene-5-oxide as substrate, dimethyldioxirane is more nucleophilic than peracids (*m*-CPBA), but more electrophilic than carbonyl oxides.[59] Nonetheless, Hammett plots[44a] gave negative ρ values for aryl sulfides and sulfoxides, i.e. -0.77 and -0.76 respectively, which demonstrates that positive charge builds up at the sulfur site and thereby the dioxirane acts as an electrophilic oxidant.

Biphilic insertion into the peroxide bond of the dioxirane to give a hypervalent adduct, e.g. sulfurane from sulfides and phosphoranes from phosphines, has not been documented so far, because such intermediates could not be detected. This type of intermediate product has been observed in the oxygen transfer of the homologous 1,2-dioxetanes with sulfur and phosphorus heteroatom substrates.[69] Challenging mechanistic work will be necessary to understand the transfer of oxygen atoms to heteroatoms by dioxiranes.

4.3 Sigma Bond Insertions

In the oxyfunctionalization of alkanes by insertion into C—H sigma bonds, methyl(trifluoromethyl)dioxirane (**1c**) is significantly more reactive than dimethyldioxirane (**1b**), e.g. *ca* 600-fold in the conversion of cumene to cumol and over 7000-fold for adamantane to adamantan-1-ol. These are remarkable reactivity trends.

Concerning alkane reactivities, the data in Figure 2 are mechanistically revealing. The preference for tertiary C—H bonds is particularly evident for adamantane, but k_{rel} diminishes with increasing strain of the cycloalkane that bears the bridgehead C—H bond. For the alkylbenzenes oxygen insertion by dioxirane **1b**[40] shows the relative rates (statistically corrected) $PhCH_3$ (1) <

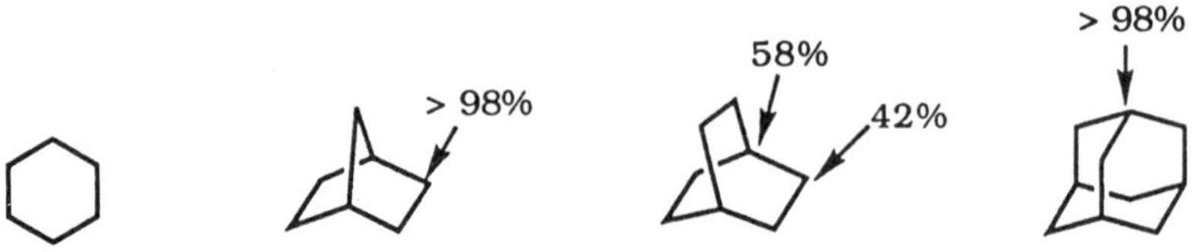

k_{rel} = 0.77 1.00 9.2 146

E_a = 14.3 kcal mol^{-1}
log$A$ = 9.9
(k_H/k_D) = 2.2

Figure 2. Relative rates and regioselectivities for the oxyfunctionalization of cycloalkanes by dioxirane **1c**

$PhCH_2CH_3$(24) < $PhCH(CH_3)_2$ (91). These selectivities are much more pronounced than for hydrogen abstraction by t-BuO·, whose relative rates[70] are $PhCH_3$ (1) < $PhCH_2CH_3$(3.2) < $PhCH(CH_3)_2$ (6.8). Moderate H/D kinetic isotope effects have been measured for oxygen insertion into the C—H bonds of alkanes and into the Si—H bond of silanes by dioxiranes.

More indicative for mechanistic rationalization is the rigorous stereoselectivity. For example, no epimerization takes place at the tertiary C—H reaction center of the stereoisomeric 1,2-dimethylcyclohexanes (Section 3.3) or decalins. Even more striking is the observation[55] that optically active 2-phenylbutane was selectively oxidized by **1b** at the benzylic position with complete retention of configuration.

These product and rate data, kinetic isotope effects and complete stereoselectivity, i.e. retention of configuration, cannot be reconciled in terms of a mechanism that involves free radicals. Rather, analogous to the mechanism of oxygen transfer by dioxiranes in the epoxidation of alkenes, also for the insertion into sigma bonds (alkanes, silanes) a 'butterfly'-like activated complex (oxenoid mechanism) accommodates best the experimental facts accumulated to date (equation 20). Of course, the transition state may have pronounced diradical character.

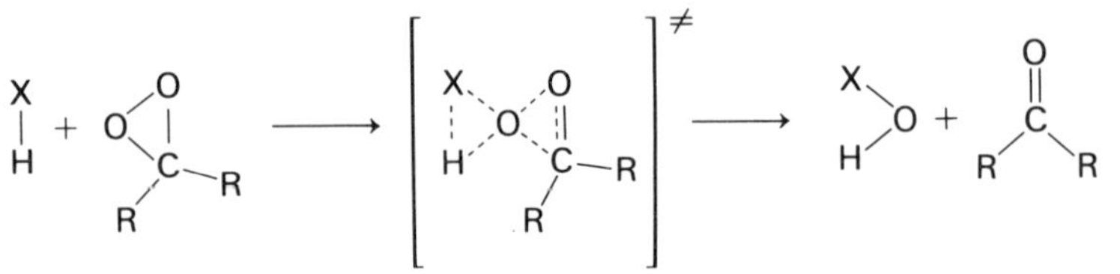

(20)

4.4 Electron Transfer Oxidations

With donor substrates having relatively low oxidation potentials, at least in principle, electron transfer (ET) reactions are possible with dioxiranes. The valence isomerization of quadricyclane into norbornadiene,[13d,48] followed by epoxidation of the latter, may involve an ET process.

Generally such ET processes are well established for all types of peroxides, most recently for the next higher homologues of the dioxiranes, the 1,2-dioxetanes[71] (cf. Chapter 5). For the latter chemiluminescence has been observed with appropriate fluorescent polycyclic arenes, e.g. rubrene, triggered by the so-called CIEEL (chemically induced electron exchange chemiluminescence) process. This possibility was recognized for dioxiranes some time ago[72] and recently reiterated[9] (equation 21).

$R_2C(O_2)$ + S: → $R_2C{=}O$ + SO

→ $\| R_2C(O^-)(O^{\cdot})\ \ S^{\cdot +} \|$ → S^* → $h\nu$

$\| R_2C(O^-)(O^{\cdot})\ \ S^{\cdot +} \|$ ⇸ $R{-}C({=}O){-}OR$ (21)

Indeed, weak chemiluminescence has been demonstrated in the oxidation of rubrene[58] by dioxiranes; however, the efficiency is low in view of the facile oxygen atom transfer to such substrates. The catalytic decomposition of dioxirane **1c** by pyrene,[24b] accompanied by substantial oxygen atom transfer to the latter, has been shown by ESR to involve electron transfer chemistry.

The observation[58] of the phenothiazine radical cation (detected by its UV-visible absorption spectrum) confirms electron transfer (equation 21), but the fate of the dioxirane radical anion is unclear. However, in this context the reaction of ketone-free dioxirane **1b** with the stable nitroxide R_2NO is of interest. In this complex reaction the highly colored oxammonium ion is formed,[73] presumably by electron transfer (Scheme 6). The substantial amounts of methoxyamine and trifluoroacetone as products reflect chemistry derived from the intermediate dioxirane radical anion. β-Scission affords methyl radicals which are scavenged by the nitroxide to give the methoxyamine. The fate of the lost oxygen atom (possibly molecular oxygen) is unclear, but it is mechanistically significant that the dioxirane radical anion does not rearrange

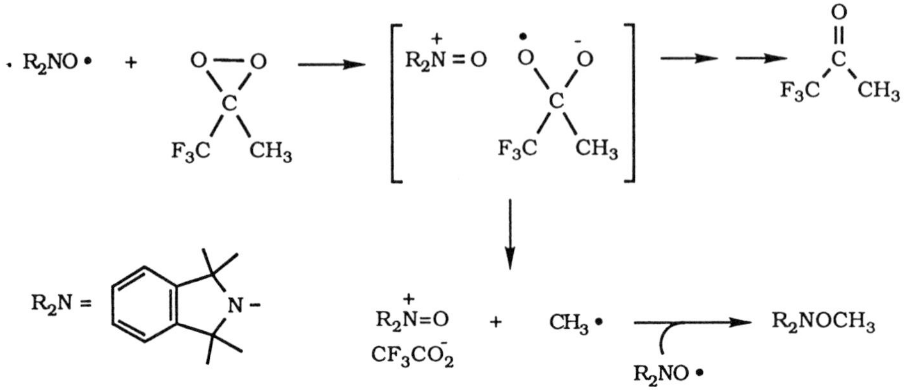

Scheme 6. Postulated mechanism for the reaction of $R_2NO\cdot$ with dioxirane **1c**

into ester products (equation 21). Future work should address the behavior of dioxirane radical anions and also radical cations to understand the complex electron transfer chemistry.

5 ACKNOWLEDGEMENTS

The group at Würzburg thanks the Deutsche Forschungsgemeinschaft (SFB 347 'Selektive Reaktionen Metall-aktivierter Moleküle'), the Fonds der Chemischen Industrie and the Stifterverband for generous funding. The research in Bari was financed by the Italian National Research Council (CNR, Rome) and by the Ministry of University, Science and Technological Research of Italy (MURST 40). Generous gifts of potassium monoperoxysulfate from Degussa (Hanau, Germany) and Peroxide-Chemie (Munich, Germany) are gratefully appreciated.

6 REFERENCES

1. A. Bayer and V. Villiger, *Chem. Ber.*, **32**, 3625 (1899).
2. R. I. Talbott and P. G. Thompson, *US Pat.*, 3 632 606 (1972).
3. R. D. Suenram and F. J. Lovas, *J. Am. Chem. Soc.*, **100**, 5117 (1978).
4. R. E. Montgomery, *J. Am. Chem. Soc.*, **96**, 7820 (1974).
5. (a) J. O. Edwards, R. H. Pater, R. Curci and F. Di Furia, *Photochem. Photobiol.*, **30**, 63 (1979); (b) R. H. Pater, *PhD Thesis*, Brown University (1977).
6. R. W. Murray and R. Jeyaraman, *J. Org. Chem.*, **50**, 2847 (1985).
7. L. Cassidei, M. Fiorentino, R. Mello, O. Sciacovelli and R. Curci, *J. Org. Chem.*, **52**, 699 (1987).
8. W. Adam, Y.-Y. Chan, D. Cremer, D. Scheutzow and M. Schindler, *J. Org. Chem.*, **52**, 2800 (1987).
9. W. Adam, R. Curci and J. O. Edwards, *Acc. Chem. Res.*, **22**, 205 (1989).

10. (a) R. W. Murray, *Chem. Rev.*, **89**, 1187 (1989); (b) R. W. Murray, in *Molecular Structure and Energetics. Unconventional Chemical Bonding*, Vol. 6 (J. F. Liebman and A. Greenberg, Eds), VCH, New York (1988), pp. 311–351.
11. R. Curci, in *Advances in Oxygenated Processes*, Vol. 2 (A. L. Baumstark, Ed.), JAI Press, Greenwich, CT(1990), Chap. 1.
12. (a) R. Curci, M. Fiorentino, L. Troisi, J. O. Edwards and R. H. Pater, *J. Org. Chem.*, **45**, 4758 (1980); (b) A. R. Gallopo and J. O. Edwards, *J. Org. Chem.*, **46**, 1684 (1981); (c) G. Cicala, R. Curci, M. Fiorentino and O. Laricchiuta, *J. Org. Chem.*, **47**, 2670, (1982); (d) R. Curci, M. Fiorentino and M. R. Serio, *J. Chem. Soc., Chem. Commun.*, 155 (1984); (e) R. Jeyaraman and R. W. Murray, *J. Am. Chem. Soc.*, **106**, 2462 (1984); (f) A. Hofland, H. Steinberg and T. J. de Boer, *Recl. Trav. Chim. Pays-Bas*, **104**, 350 (1985).
13. (a) W. Adam, L. Hadjiarapoglou and R. Siegmeier, *Ger. Pat. Appl.* P 40 23 741.9 (1990); (b) P. F. Corey and F. E. Ward, *J. Org. Chem.*, **51**, 1925 (1986); (c) K. C. Nicolaou, C. V. C. Prasad and W. W. Ogilvie, *J. Am. Chem. Soc.*, **112**, 4988 (1990); (d) R. W. Murray, M. K. Pillay and R. Jeyaraman, *J. Org. Chem.*, **53**, 3007 (1988); (e) J. Bujons, F. Camps and A. Messeguer, *Tetrahedron Lett.*, **31**, 5235 (1990); (f) H. Wolleb, *Eur. Pat.*, EP 0 370 946 A2 (1990); (g) D. Schlotter, *Diplomarbeit*, University of Würzburg (1990); (h) M. J. Ford, J. G. Knight, S. V. Ley and S. Vile, *SynLett.*, 331 (1990).
14. (a) W. Adam, J. Bialas and L. Hadjiarapoglou, *Chem. Ber.*, **124**, 2377 (1991); (b) W. Adam, L. Hadjiarapoglou and A. Smerz, *Chem. Ber.*, **224**, 227 (1991).
15. M. Camporeale, T. Fiorani, L. Troisi, W. Adam, R. Curci and J. O. Edwards, *J. Org. Chem.*, **55**, 93 (1990).
16. R. Mello, M. Fiorentino, O. Sciacovelli and R. Curci, *J. Org. Chem.*, **53**, 3890 (1988).
17. W. Adam, R. Curci, M. E. Gonźalez-Nuñez and R. Mello, *J. Am. Chem. Soc.*, **113**, 7654 (1991).
18. D. Cremer, in *The Chemistry of Functional Groups, Peroxides*, (S. Patai, Ed.) Wiley–Interscience, New York (1983), Chap. 1.
19. L. B. Harding and W. A. Goddard, *J. Am. Chem. Soc.*, **100**, 7180 (1978).
20. J. Gaus and D. Cremer, *Chem. Phys. Lett.*, **133**, 420 (1987).
21. (a) D. Cremer, *J. Am. Chem. Soc.*, **101**, 7199 (1979); (b) D. Cremer, T. Schimdt, J. Gauss and T. P. Radhakrishnan, *Angew. Chem.*, **100**, 431 (1988); *Angew. Chem., Int. Ed. Engl.*, **27**, 427, (1988).
22. W. Sander, *Angew. Chem.*, **102**, 362 (1990); *Angew. Chem., Int. Ed. Engl.*, **27**, 344 (1990).
23. (a) A. L. Baumstark and C. J. McCloskey, *Tetrahedron Lett.*, **28**, 3311 (1987); (b) A. L. Baumstark and P. C. Vasquez, *J. Org. Chem.*, **53**, 3437 (1988).
24. (a) S. K. Agarwal, D. R. Boyd, W. B. Jennings, R. M. McGuckin and G. A. O'Kane, *Tetrahedron Lett.*, **30**, 123 (1989); (b) R. Mello, F. Ciminale, M. Fiorentino, C. Fusco, T. Prencipe and R. Curci, *Tetrahedron Lett.*, **31**, 6097 (1990).
25. (a) J. K. Crandall and D. J. Batal, *J. Org. Chem.*, **53**, 1338 (1988); (b) J. K. Crandall and D. J. Batal, *Tetrahedron Lett.*, **29**, 4791 (1988); (c) J. K. Crandall, D. J. Batal, D. P. Sebesta and F. Liu, *J. Org. Chem.*, **56**, 1159 (1991); (d) J. K. Crandall and E. Rambo, *J. Org. Chem.*, **55**, 5929 (1990).
26. (a) W. Adam, L. Hadjiarapoglou and J. Klicic, *Tetrahedron Lett.*, **31**, 6517 (1990); (b) W. Adam, L. Hadjiarapoglou, V. Jäger and B. Seidel, *Tetrahedron Lett.*, **30**, 4223 (1989).
27. S. W. Baertschi, K. D. Raney, M. P. Stone and T. M. Harris, *J. Am. Chem. Soc.*, **110**, 7929 (1988).
28. L. Troisi, L. Cassidei, L. Lopez, R. Mello and R. Curci, *Tetrahedron Lett.*, **30**, 257 (1989).

29. R. Hambalek and G. Just, *Tetrahedron Lett.*, **31**, 4693 (1990).
30. (a) R. L. Halcomb and S. J. Danishefsky, *J. Am. Chem. Soc.*, **111**, 6661 (1989); (b) K. Chow and S. J. Danishefsky, *J. Org. Chem.*, **55**, 4211 (1990).
31. J. E. Baldwin and I. A. O'Neil, *Tetrahedron Lett.*, **31**, 2047 (1990).
32. W. Adam, L. Hadjiarapoglou and X. Wang, *Tetrahedron Lett.*, **32**, 1295 (1991).
33. W. Adam, L. Hadjiarapoglou, T. Mosandl, C. Saha-Möller and D. Wild, *Angew. Chem.*, **103**, 187 (1991); *Angew Chem. Int. Ed. Engl.*, **30**, 200 (1991).
34. M. Sauter, *Diplomarbeit*, University of Würzburg (1991).
35. G. M. Rubottom, M. A. Vasquez and D. R. Pelegrina, *Tetrahedron Lett.*, **15**, 4319 (1974).
36. (a) W. Adam, L. Hadjiarapoglou and X. Wang, *Tetrahedron Lett.*, **30**, 6497 (1989); (b) W. Adam and F. Prechtl, unpublished results.
37. H. K. Chenault and S. J. Danishefsky, *J. Org. Chem.*, **54**, 4249 (1989).
38. (a) W. Adam and L. Hadjiarapoglou, *Chem. Ber.*, **123**, 2077 (1990); (b) W. Adam, D. Golsch, L. Hadjiarapoglou and T. Patonay, *Tetrahedron Lett.*, **32**, 1041 (1991); (c) W. Adam, L. Hadjiarapoglou and A. Levai *Synthesis*, 435, (1992).
39. W. Adam, L. Hadjiarapoglou and B. Nestler, *Tetrahedron Lett.*, **31**, 331 (1989); (b) F. Prechtl, *Diplomarbeit*, University of Würzburg (1990); (c) A. Amstrong and S. V Ley, *SynLett.*, 323 (1990).
40. R. W. Murray, R. Jeyaraman and L. Mohan, *J. Am. Chem. Soc.*, **108**, 2470 (1986).
41. R. Mello, M. Fiorentino, M. Fusco and R. Curci, *J. Am. Chem. Soc.*, **111**, 6749 (1990).
42. A. L. Baumstark, M. Beeson and P. C. Vasquez, *Tetrahedron Lett.*, **30**, 5567 (1989).
43. W. Adam, R. Curci and R. Mello, *Angew. Chem.*, **102**, 916 (1990); *Angew Chem., Int. Ed. Engl.*, **27**, 890 (1990).
44. (a) R. W. Murray, R. Jeyaraman and M. K. Pillay, *J. Org. Chem.*, **52**, 746 (1987); (b) S. Colonna and N. Gaggero, *Tetrahedron Lett.*, **30**, 6233 (1989).
45. (a) R. W. Murray, R. Jeyaraman and L. Mohan, *Tetrahedron Lett.*, **27**, 2355 (1986); (b) R. W. Murray, S. N. Rajadhyaska and L. Mohan, *J. Org. Chem.*, **54**, 5783 (1989); (c) M. D. Coburn, *J. Heterocycl. Chem.*, **26**, 1883 (1989).
46. F. Sanchez-Baeza, F. Durand, D. Barcelo and A. Messeguer, *Tetrahedron Lett.*, **31**, 3359 (1990).
47. D. R. Boyd, P. B. Coulter, M. R. McGuckin, N. D. Sharma, W. B. Jennings and V. E. Wilson, *J. Chem. Soc., Perkin Trans. 1*, 301 (1990).
48. R. W. Murray and M. K. Pillay, *Tetrahedron Lett.*, **29**, 15 (1988).
49. R. W. Murray and D. L. Shiang, *J. Chem. Soc., Perkin Trans. 2*, 349 (1990).
50. (a) W. Pirkle and R. Rinaldi, *J. Org. Chem.*, **42**, 2080 (1977); (b) J. Rebek, Jr, and R. McCready, *J. Am. Chem. Soc.*, **102**, 5602 (1980).
51. R. Curci *et al.*, unpublished results.
52. (a) G. B. Payne and P. A. Williams, *J. Chem. Org.*, **24**, 54 (1959); (b) H. House and R. S. Rho, *J. Am. Chem. Soc.*, **80**, 2428 (1958); (c) G. Berti, *Top Stereochem.*, **7**, 166 (1973), and references cited therein.
53. (a) R. Curci and F. Di Furia, *Tetrahedron Lett.*, 4085 (1974); (b) R. Curci, F. Di Furia and M. Meneghin, *Gazz. Chim. Ital.*, **108**, 123 (1978).
54. R. Mello, L. Cassidei, M. Fiorentino, C. Fusco and R. Curci, *Tetrahedron Lett.*, **31**, 3067 (1990).
55. W. Adam, G. Asenio, R. Curci, M. E. Gonzáles-Nuñez and R. Mello, *J. Org. Chem.*, **57**, 953 (1992).
56. R. Mello, L. Cassidei, M. Fiorentino, C. Fusco, W. Hümmer, V. Jäger and R. Curci, *J. Am. Chem. Soc.*, **113**, in press (1991).
57. W. Adam, M. Abou-El-Zahab and C. R. Saha-Möller, *Liebigs Ann. Chem.*, 967, (1991).

58. R. Curci, L. D'Accolri, M. Fiorentino, C. Fusco, W. Adam, M. E. González-Nuñez, R. Mello, *Tetrahedron Lett.*, submitted for publication.
59. W. Adam, W. Haas and G. Sieker, *J. Am. Chem. Soc.*, **106**, 5020 (1984).
60. Y. Miyahara and T. Inazu, *Tetrahedron Lett.*, **31**, 5955 (1990).
61. W. Adam and L. Hadjiarapoglou, *Tetrahedron Lett.*, **33**, 469 (1992).
62. K. Schank and F. Werner, *Justus Liebigs Ann. Chem.*, 1977 (1979).
63. W. Adam, D. Golsch and L. Hadjiarapoglou, unpublished results.
64. M. D. Wittman, R. L. Halcomb and S. J. Danishefsky, *J. Org. Chem.*, **55**, 1981 (1990).
65. D. L. Zabrowski, A. E. Moorman and K. R. Beck, Jr, *Tetrahedron Lett.*, **29**, 4501 (1988).
66. P. E. Eaton and G. E. Wicks, *J. Org. Chem.*, **53**, 5353 (1988).
67. R. W. Murray and M. Singh, *Synth. Commun.*, **19**, 3509 (1989).
68. (a) R. W. Murray and M. Singh, *Tetrahedron Lett.*, **29**, 4677 (1988); (b) R. W. Murray and M. Singh, *J. Org. Chem.*, **55**, 2954 (1990).
69. R. Davies, H. Jeremy, D. R. Boyd, S. Kumar, N. D. Sharma and C. Stevenson, *Biochem. Biophys. Res. Commun.*, **169**, 87 (1990).
70. A. L. Baumstark, M. E. Landis and P. J. Brooks, *J. Org. Chem.*, **44**, 4251 (1979).
71. C. Walling and W. Thaler, *J. Am. Chem. Soc.*, **83**, 3877 (1961); W. Adam, T. Mosandl and C. Saha-Möller, this book, Chap. 5.
72. W. Adam and R. Curci, *Chim. Ind. (Milan)*, **63**, 20 (1981).
73. W. Adam, S. Bottle, R. Mello, *J. Chem. Soc., Chem. Commun.*, 771 (1991).

5 Dioxetanes and α-Peroxy Lactones, Four-membered Ring Cyclic Peroxides

WALDEMAR ADAM, MARKUS HEIL, THOMAS MOSANDL and CHANTU R. SAHA-MÖLLER

Institut für Organische Chemie der Universität Würzburg, Am Hubland, W-8700 Würzburg, Germany

1 INTRODUCTION

The four-membered ring cyclic peroxides 1,2-dioxetanes (**1**) and α-peroxy lactones (**2**) are 1,2-dioxa analogues of the cyclobutane ring system. Because of the unique property of these *hyperenergetic* molecules to generate electronically

Organic Peroxides. Edited by W. Ando

excited fragments on thermolysis, they are mechanistically relevant in bioluminescent processes and have received a great deal of attention during the last two decades. In this chapter an overview of dioxetane chemistry is given, which includes physical, chemical and biological aspects. A book[1] has been published on this subject and numerous specialized reviews[2] have appeared since.

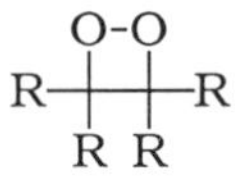

1 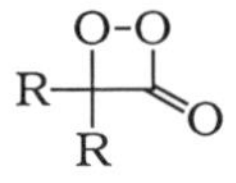 **2**

Although the intermediacy of 1,2-dioxetane was proposed *ca* 85 years ago[3] in the oxidative cleavage of the enol **3** (equation 1), only in 1969 was the first stable 1,2-dioxetane successfully synthesized.[4] Shortly after the preparation of this dioxetane, the synthesis of a stable α-peroxy lactone by dicyclohexylcarbodiimide dehydration of α-peroxy acids was reported.[5] Since that time, intensive efforts have been expended to explore the chemistry of dioxetanes. In recent years a great deal of research has been invested in making use of dioxetane chemiluminescence in molecular biology, particularly for immunoassays.[2c,e]

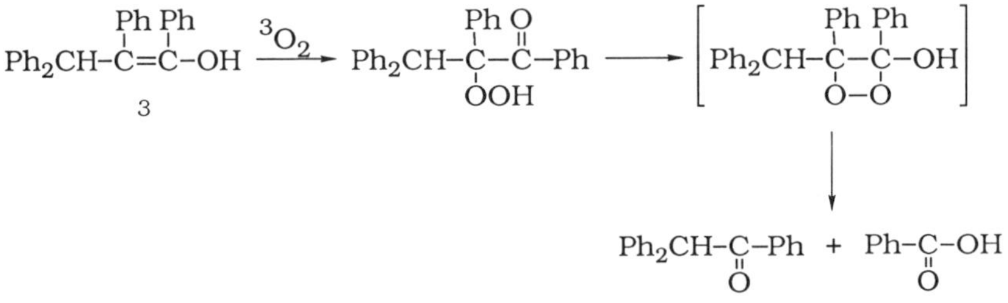

$$Ph_2CH\text{-}C(Ph)\text{=}C(Ph)\text{-}OH \ (\mathbf{3}) \xrightarrow{^3O_2} Ph_2CH\text{-}C(Ph)(OOH)\text{-}C(=O)\text{-}Ph \longrightarrow \left[Ph_2CH\text{-}C(Ph)\text{-}C(Ph)\text{-}OH \ (O\text{-}O) \right] \longrightarrow Ph_2CH\text{-}C(=O)\text{-}Ph + Ph\text{-}C(=O)\text{-}OH \quad (1)$$

2 SYNTHESIS

Although more than two decades have passed since Kopecky and Mumford[4] reported the first isolable 1,2-dioxetane, namely the trimethyl derivative, the methods for the preparation of dioxetanes are still rather limited. Only the *β*-halo hydroperoxide cyclization and singlet oxygen ($^1O_2$) cycloaddition to electron-rich alkenes have general applicability for the synthesis of these four-membered ring cyclic peroxides (Scheme 1).

$$R_2C{=}CR_2 \xrightarrow[H_2O_2]{DDH} R\text{-}\underset{R}{\overset{HOO}{C}}\text{-}\underset{R}{\overset{Br}{C}}\text{-}R \xrightarrow[Ag^+]{HO^- \text{ or}} R\text{-}\underset{R}{\overset{O\text{-}}{C}}\text{-}\underset{R}{\overset{O}{C}}\text{-}R \xleftarrow{^1O_2} R_2C{=}CR_2$$

DDH = 1,3-dibromo-5,5-dimethylhydantoin (Br-N, N-Br, Me, Me, O, O)

Scheme 1. Kopecky and singlet oxygen routes to dioxetanes

2.1 1,2-Dioxetanes

2.1.1 The Kopecky method

The dehydrohalogenation of β-halo hydroperoxides to 1,2-dioxetanes by means of base or silver ion is known as the Kopecky method,[6] which is given in general terms in Scheme 1. β-Bromo hydroperoxides are normally used in this reaction, but in some cases the β-chloro and β-iodo hydroperoxides were employed. The β-bromo hydroperoxides are usually prepared by bromination of suitable alkenes with 1,3-dibromo-5,5-dimethylhydantoin (DDH) in the presence of concentrated (*ca* 85%) hydrogen peroxide. The Kopecky method has been used for the preparation of numerous dioxetanes and some representative derivatives are featured in Table 1.

The primary and secondary β-bromo hydroperoxides can be cyclized to dioxetanes by base (HO^-, RO^-), whereas for the tertiary bromides silver oxide, acetate and particularly succinimide work better.[7] The silver salts must be freshly prepared because the metallic silver present owing to light exposure catalyzes decomposition of the dioxetane.

The cyclization reactions must be carried out at low temperatures (−30 to +10°C), but in the case of the thermally more stable dioxetanes, temperatures up to 30°C can be tolerated. In the silver ion-promoted cyclizations, solvents such as alkanes (pentane, hexane, cyclohexane) and haloalkanes (carbon tetrachloride, dichloromethane, dichlorodifluoromethane) are advantageous. In the reactions with base, protic solvents (water and/or methanol, ethanol) are employed, but two-phase solvent systems such as aqueous methanol and pentane are also useful. Under the latter conditions the base-catalyzed decomposition of the dioxetane can be suppressed.

2.1.2 Singlet oxygen cycloaddition

The (2 + 2) cycloaddition of $^1O_2$ with electron-rich alkenes is a most convenient and versatile method for the preparation of 1,2-dioxetanes. Shortly after the successful synthesis of the first stable dioxetane by Kopecky and Mumford[4] in

Table 1. 1,2-Dioxetanes prepared by dehydrohalogenation of β-halo hydroperoxides[2k]

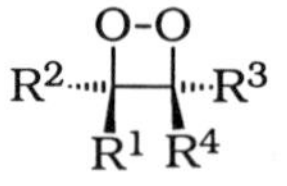

R^1	R^2	R^3	R^4	Solvent	Temperature (°C)	Yield (%)	M.p. (°C)
CH_3	CH_3	CH_3	CH_3	CH_2Cl_2	25	30	76–77
CH_3	CH_3	CH_3	H	CH_3OH–H_2O	0	*ca* 30	5–7
CH_3	CH_3	H	H	CH_2Cl_2–H_2O	0	26	6
CH_3	H	CH_3	H	CH_2Cl_2–H_2O	20	27	—[a]
CH_3	H	H	CH_3	CH_2Cl_2–H_2O	20	39	—[a]
CH_3	H	H	H	CH_2Cl_2–H_2O	0–20	3	—[b]
H	H	H	H	CH_2Cl_2–H_2O	0–20	*ca* 1	—[b]
Bz	Bz	H	H	CH_3OH	−30	38	60–61
Ph	Ph	H	H	CH_3OH	−30	21	—[b]
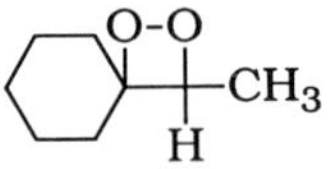 (spirocyclohexane dioxetane, CH_3, H)				CH_3OH–H_2O	0	12.5	−11
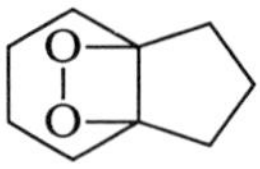 (bicyclic dioxetane)				CH_2Cl_2	0	11	*ca* 20

[a]Liquid.
[b]Isolated as a solution in the stated organic solvent.

1969, the preparation of dioxetanes **3a** and **b**,[8] **4**[9] and **5**[10] by the $^1O_2$ routes was reported. The thermal stability of the spiroadamantane dioxetane **5** is truly impressive, since it melts at 185–186°C with chemiluminescent decomposition. Since that time, a large number of dioxetanes, which are not accessible by the Kopecky method, have been synthesized by $^1O_2$ cycloaddition (Table 2).

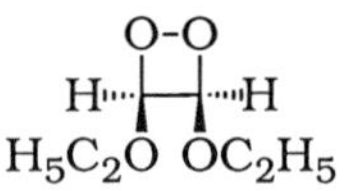

3a 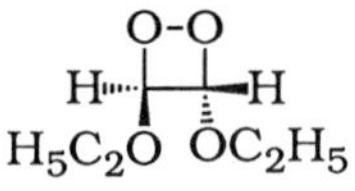 **3b** 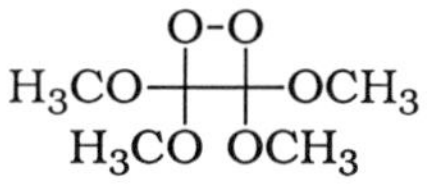 **4** 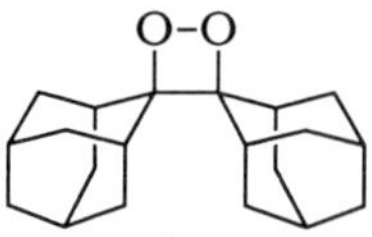 **5**

The $^1O_2$ can be generated either by photosensitization of triplet dioxygen or chemically. The former method is the more common and convenient procedure. The sensitizer–solvent systems which are frequently used are tetraphenylporphine (TPP) in dichloromethane, Rose Bengal in methanol, polymer-bound Rose Bengal (Sensitox)[11] in acetone, and methylene blue in methanol.

Excellent chemical sources of $^1O_2$ are triphenylphosphite ozonide (**6**)[12] and 1,4-dimethylnaphthalene endoperoxide (**7**);[13] recently, triethylsilyl

Table 2. 1,2-Dioxetanes prepared by $^1O_2$ cycloaddition[2k]

Dioxetane	Reaction conditions		Yield (%)	M.p. (°C)
	Solvent	Temperature (°C)		
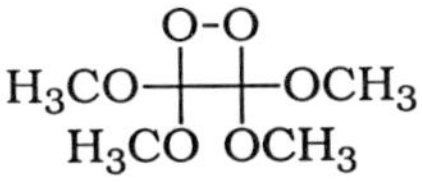 O-O; H_3CO, OCH_3, H_3CO, OCH_3	$(C_2H_5)_2O$	−70	94	−9 to 8
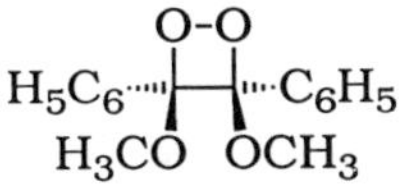 O-O; H_5C_6, C_6H_5, H_3CO, OCH_3	CH_3OH-$(C_2H_5)_2O$	−50	80	67–68
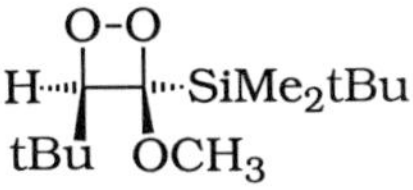 O-O; H, $SiMe_2tBu$, tBu, OCH_3	CH_2Cl_2	−80	97	—[a]
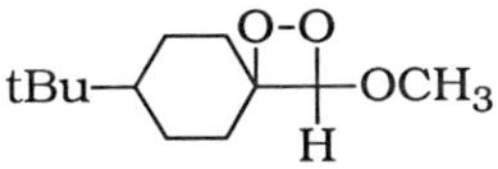 tBu, O-O, OCH_3, H	CH_2Cl_2	−80	60	—[b]
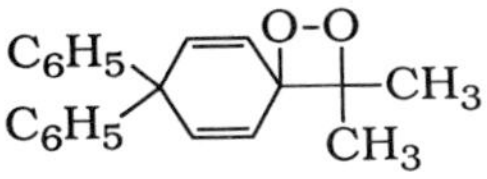 C_6H_5, C_6H_5, O-O, CH_3, CH_3	CH_3OH- CH_3COCH_3	−70	82	72–74
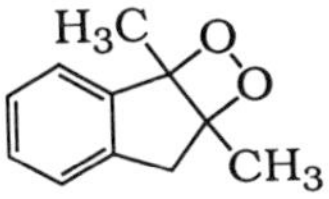 H_3C, O, O, CH_3	CH_3OH- CH_3COCH_3	−78	39	40–41
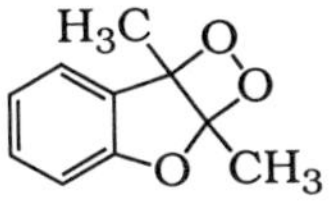 H_3C, O, O, O, CH_3	CH_2Cl_2	−10	53	53–54
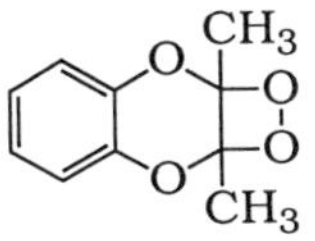 CH_3, O, O, O, O, CH_3	CH_2Cl_2	−78	33	118–122
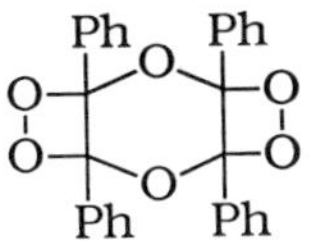 Ph, Ph, O, O, O, O, O, O, Ph, Ph	CH_2Cl_2	−78	22	111
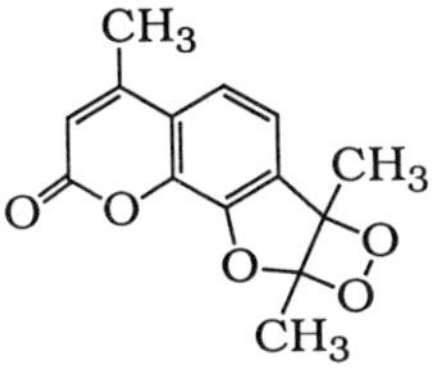 CH_3, O, O, CH_3, O, O, O, CH_3	CH_2Cl_2	−15	44	100–101

[a]Liquid.
[b]Isolated as a solution in the stated organic solvent.

hydrotrioxide (**8**) has been introduced.[14] Interestingly, like triphenylphosphite ozonide, this hydrotrioxide also reacts directly with alkenes to give dioxetanes.[15]

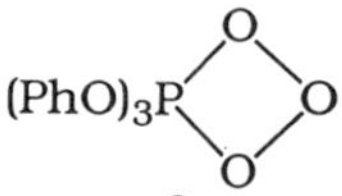

(PhO)$_3$P(O–O–O)

6

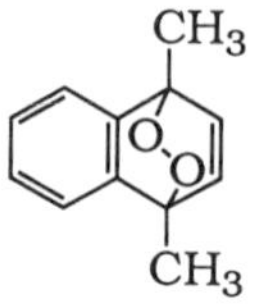

7

$(C_2H_5)_3Si{-}OOOH$

8

The most serious side-reactions in the $^1O_2$ reaction of alkenes are the ene reaction, when allylic hydrogen atoms are present (equation 2), and (4 + 2) cycloaddition (equation 3) when aryl or vinyl substituents are present in the substrate. Although *bis*-adamantylidene, the precursor to dioxetane **5**, possesses allylic hydrogen atoms, they are located at bridgehead positions and do not undergo ene reaction with $^1O_2$. Consequently, the corresponding dioxetane is obtained in high yield.[10] An interesting case to mention here in this context is the example of benzpyrene **9**,[16] which with triphenylphosphite ozonide gave less (4 + 2) cycloaddition and more (2 + 2) cycloaddition than $^1O_2$ generated by photosensitization.

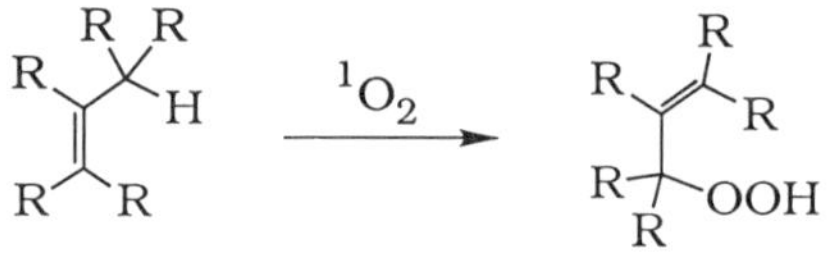

$$\xrightarrow{^1O_2} \quad \text{(2)}$$

$$\xrightarrow{^1O_2} \quad \text{(3)}$$

OCH$_3$

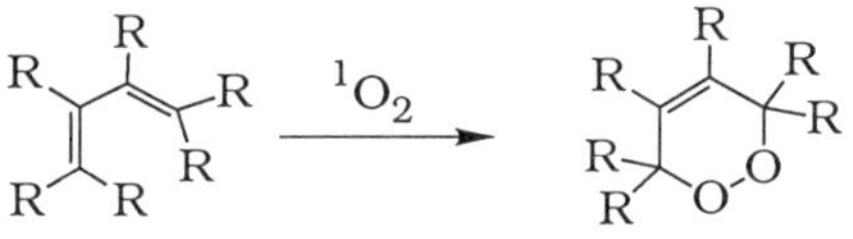

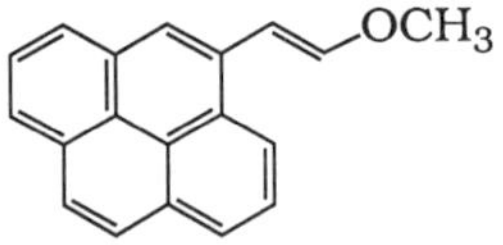

9

2.1.3 Other methods

In addition to the Kopecky method and the $^1O_2$ route, some other preparations of dioxetanes have been reported. Although these methods are not generally applicable, they are the methods of choice for the synthesis of functionalized dioxetanes. Intramolecular peroxymercuration of allylic hydroperoxides[17] with mercury(II) trifluoroacetate and subsequent reaction with molecular bromine at low temperature gave bromomethyl-substituted dioxetanes in acceptable yields (equation 4).

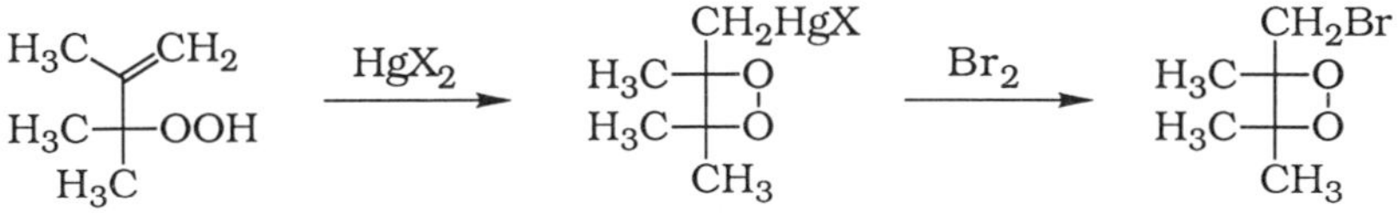

(4)

Allylic hydroperoxides were converted in good yields into hydroxymethyl-substituted dioxetanes by peroxy acid epoxidation and subsequent base-catalyzed ring closure of the epoxy hydroperoxides (Scheme 2).[18] These dioxetanes, particularly the 3-(hydroxymethyl)-3,4,4-trimethyl-1,2-dioxetane, can be transformed with a variety of electrophiles such as carboxylic acids, chlorocarbonates, isocyanates, trialkylsilyl chlorides and trialkyloxonium salts into functionalized derivatives (Scheme 2). After the transformation into its chlorocarbonyl derivative by means of phosgene or its equivalent, this dioxetane can also be functionalized with nucleophiles such as alcohols, amines, amino acids and oximes (Scheme 2).[19]

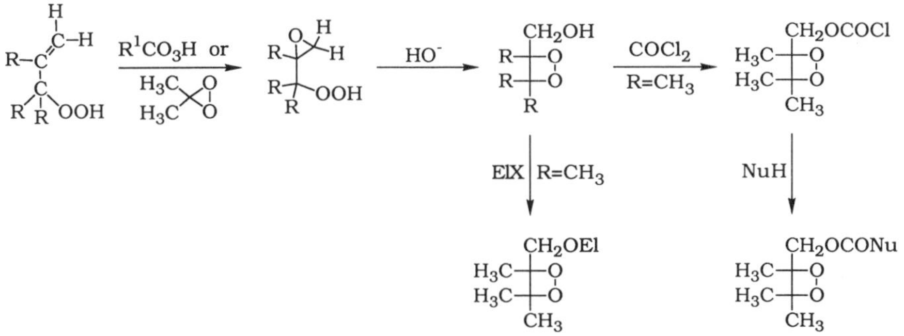

Scheme 2. Synthesis of hydroxymethyl-substituted dioxetanes and their transformations with electrophiles and nucleophiles

The reaction of isolable cyclobutadienes with triplet dioxygen[2k,o,20] gives stable dioxetanes **10**–**12** in excellent yields. Triplet dioxygen also reacts with sterically hindered tetrasubstituted alkenes, e.g. *bis*-adamantylidene, in the presence of one-electron oxidants to give the corresponding dioxetanes.[2f] The ozonolysis of vinyl silanes also affords dioxetanes[21] (equation 5). This potentially useful method has recently been employed in the synthesis of *qinghaosu* analogues.[22]

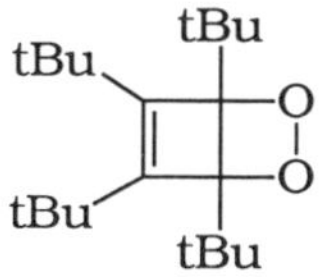

10 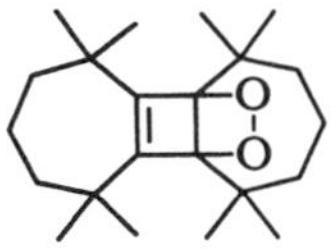 **11** 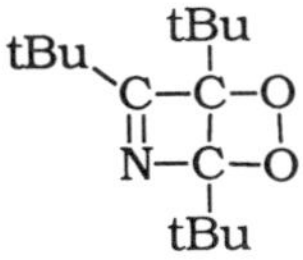 **12**

$$(H_3C)_2C{=}C(CH_3)Si(CH_3)_3 \xrightarrow{O_3} \text{1,2-dioxetane: } H_3C\text{-}C(OSi(CH_3)_3)\text{-O-O-}C(CH_3)\text{-}CH_3 \quad (5)$$

2.2 α-Peroxy Lactones

2.2.1 Cyclization of α-hydroperoxy acids

The cyclization of α-hydroperoxy acids by means of the mild dehydration agent dicyclohexylcarbodiimide (DCC) constitutes a general method for the synthesis of α-peroxy lactones[2k] (equation 6). In Table 3 are summarized the α-peroxy lactones that have been synthesized by this route.

$$R_2C(OOH)COOH \xrightarrow[-50\,^{\circ}C]{DCC} \text{α-peroxy lactone } R_2C\text{-}C(=O)\text{-O-O} \quad (6)$$

Because of their pronounced sensitivity towards acid- and base-catalyzed Grob fragmentation,[23] the preparation of α-hydroperoxy acids, the precursors to the α-peroxy lactones, should be carried out under mild, preferably neutral, conditions. In Scheme 3 are outlined the available methods[2k] for the preparation of α-hydroperoxy acids. Of these, the oxygenation of enolates with triplet dioxygen (route c) is the most widely employed.

Table 3. α-Peroxy lactones prepared by carbodiimide cyclization of α-hydroperoxy acids[2k]

R^1	R^2	Method[a]	Yield (%)	M.p. (°C)
CH_3	CH_3	A	68	10
		B	40	
CF_3	CF_3	B	—	—[b]
$t\text{-}C_4H_9$	H	A,B	50	2
$n\text{-}C_3H_7$	CH_3	B	20	—[b]
$t\text{-}C_4H_9$	$t\text{-}C_4H_9$	A	35	5
C_6H_5	$n\text{-}C_4H_9$	B	14	—[b]
C_6H_5	C_6H_5	B	10	—[b]

[a]Method A, dehydration of α-hydroperoxy acids; method B, photooxygenation of ketenes.
[b]Not isolated.

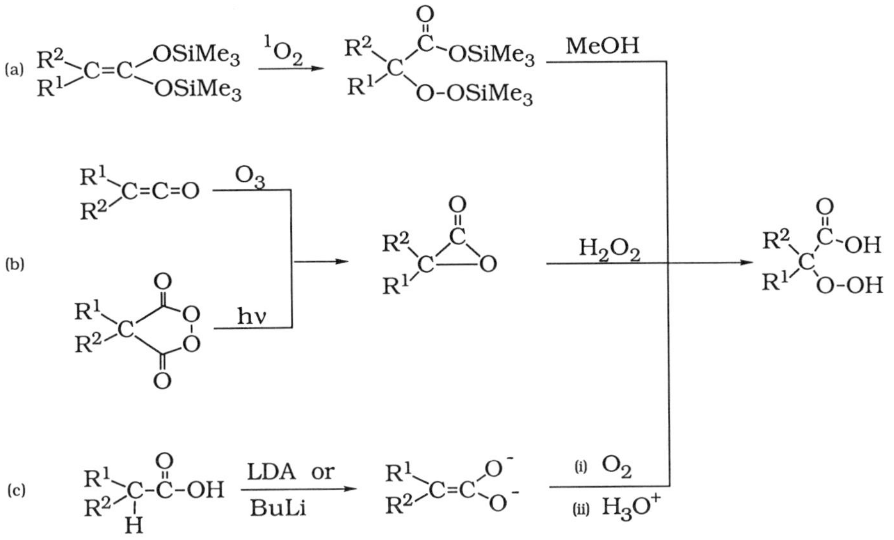

(a) $R^2R^1C{=}C(OSiMe_3)_2 \xrightarrow{^1O_2} R^2R^1C(C(=O)OSiMe_3)(O\text{-}OSiMe_3) \xrightarrow{MeOH}$

(b) $R^1R^2C{=}C{=}O \xrightarrow{O_3}$; cyclic diacyl peroxide $\xrightarrow{h\nu}$ → α-lactone $\xrightarrow{H_2O_2} R^2R^1C(C(=O)OH)(O\text{-}OH)$

(c) $R^1R^2CH\text{-}C(=O)\text{-}OH \xrightarrow{LDA\ or\ BuLi} R^1R^2C{=}C(O^-)_2 \xrightarrow{(i)\ O_2;\ (ii)\ H_3O^+}$

Scheme 3. Alternative routes for α-hydroperoxy acids

2.2.2 Singlet oxygen cycloaddition

The most direct method for the preparation of α-peroxy lactones is the (2 + 2) cycloaddition of $^1O_2$ to ketenes (equation 7). A few α-peroxy lactones have been prepared by this method[24] (Table 3). Apparently ene reaction and (4 + 2) cycloaddition are not serious side reactions since ketenes bearing allylic hydrogen atoms and/or aryl substituents afford the desired α-peroxy lactones in fair yields. Ketenes are prone to autoxidation,[25] which leads to explosive polymeric peroxides; therefore, the utmost care should be exercised in the synthesis and isolation of α-peroxy lactones by this preparative route.

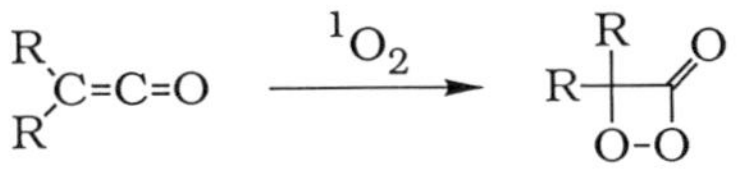

$$R_2C{=}C{=}O \xrightarrow{^1O_2} R_2C\text{—}C(=O)\text{—}O\text{—}O\ (\text{four-membered ring}) \tag{7}$$

3 CHEMICAL TRANSFORMATIONS

The chemical transformations of dioxetanes can be classified in terms of different modes of bond cleavage of the four-membered ring (Scheme 4). Mode A (the most common decomposition route) and mode B (only one example) were observed in the thermolysis and photolysis of dioxetanes, whereas mode C (lateral bond cleavage) also takes place in acid-catalyzed reactions. Mode D (O—O bond rupture) was detected in lithium aluminum hydride reductions, in

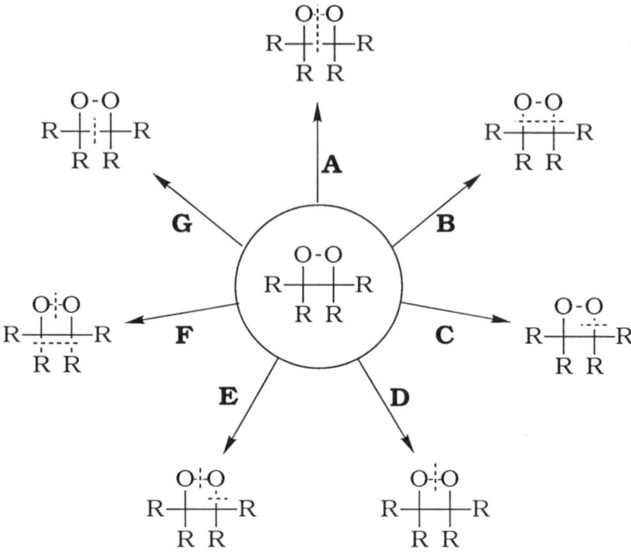

Scheme 4. Bond cleavage modes of dioxetanes

electron-transfer reactions and in the biphilic attack of phosphorus, arsenic, antimony and sulfur nucleophiles. The addition reaction of dioxetanes to alkenes belongs also to the mode D transformation. Base-induced reactions of dioxetanes constitute mode E transformations, and mode F takes place in the thermolysis of a few heteroatom-substituted dioxetanes. Chemistry derived from the scission of the C—C bond of the dioxetane ring (mode G) with preservation of the peroxide bond has to date not been reported.

3.1 Mode A Transformations

The most common transformation of dioxetanes is their thermal or photolytic decomposition into carbonyl fragments by O—O and C—C bond cleavage. For the thermal decomposition of dioxetanes, both diradical and concerted mechanisms have been argued.[2g,26] A merged mechanism, in which the kinetic features reflect diradical character and the excited state products reflect the concerted nature along the complex reaction coordinate, was proposed to reconcile best the available experimental data on the two extreme mechanistic views;[27] however, recently experimental evidence was provided[28] in favor of the diradical mechanism. For example, the thermolysis of 3,3-dimethyl-1,2-dioxetane in cyclohexa-1,4-diene as solvent (cf. mode D) gave the *vic*-diol by trapping of the intermediate 1,4-dioxyl diradical.

Easily oxidized transition metal ions[29] and amines[30] catalyze the decomposition of dioxetanes in the dark; an electron-transfer mechanism has been proposed for this decomposition route. In contrast, univalent rhodium and iridium complexes react with dioxetanes by insertion of the metal into the O—O bond.[31] Subsequent cleavage of the metallocycle releases the carbonyl fragments and the metal complex. Although the insertion step of this reaction belongs to the category of mode D transformations, the overall process represents cleavage of the dioxetane into two carbonyl fragments and thus belongs to the mode A category.

When dioxetanes bear fluorescent electron donor groups, e.g. easily oxidized diaryl substituents, the above electron-transfer process affords the corresponding carbonyl product with strong luminescence[32] (Scheme 5). This phenomenon is known as chemically induced electron exchange luminescence (CIEEL)[33] and is of general scope. Intramolecular electron transfer from the aniline moiety to the dioxetane ring with subsequent cleavage generates the radical ion pair; final electron back-transfer produces the singlet excited state, the emitter of the observed fluorescence (Scheme 5).[32a]

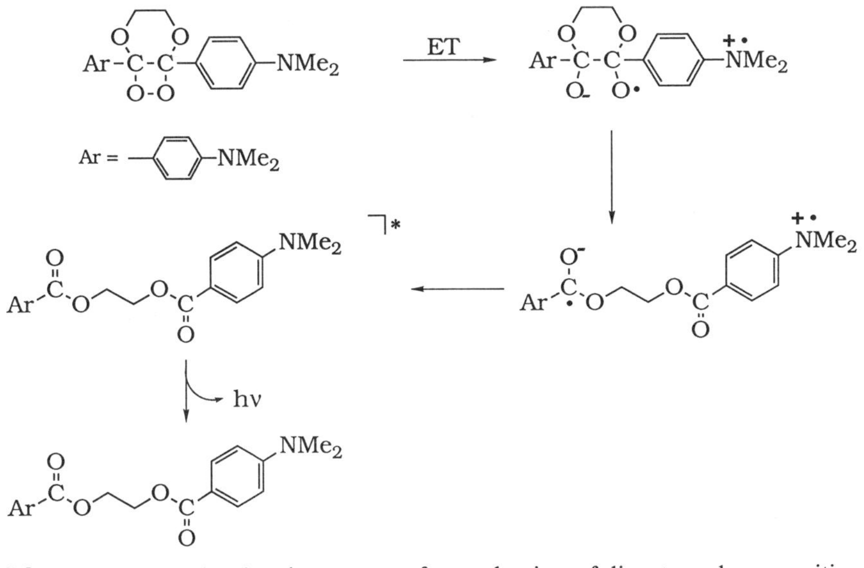

Scheme 5. Intramolecular electron-transfer mechanism of dioxetane decomposition

An intermolecular case of the CIEEL process is the efficient chemiluminescence of α-peroxy lactones with electron-rich arenes, e.g. rubrene.[34] The bioluminescence of fireflies derives its high efficiency from such intramolecular CIEEL mechanisms with α-peroxy lactones as intermediates.[35] Recent applications of the CIEEL effect employed thermally stable dioxetanes by triggering their CIEEL decomposition through chemical and enzymatic[2c] ester hydrolysis (cf. Section 5).

3.2 Mode B Transformations

Experimental evidence for double C—O bond rupture to liberate dioxygen was postulated in the gas-phase pyrolysis of the bridged annulene in equation 8.[36] The dioxetane was assumed as the reaction intermediate in this unusual transformation. Other examples of this type appear not to have been reported.

500-600°C; $-O_2$ (8)

3.3 Mode C Transformations

The simple C—O bond scission with preservation of the dioxetane ring C—C bond (mode C) is well documented in dioxetane transformations. These include thermolysis[20a,37] and catalysis by Brønsted and Lewis acids (equation 9).[20a,38] The cyclobutadiene dioxetane in equation 9a is isomerized on heating into its regioisomer, whereas the alkoxy-substituted dioxetanes in equation 9b rearranges by C—O bond cleavage into the corresponding allylic hydroperoxides. The benzofuran dioxetane in equation 9c required assistance by CF_3CO_2H, whereas the silyl migration in equation 9d occurred at room temperature on its own.

protic solvent, 30°C, few min; n-hexane, HCl, 25°C, 3 min (9a)

OtBu; CCl_4, ca. 20°C; OOH, OtBu (9b)

H_3C, CH_3; CF_3CO_2H, $CHCl_3$, -10°C, 3h; CH_2, OOH, CH_3 (9c)

tBu-C(H)(O-O)-C(OAr)-OSi$(CH_3)_2$tBu; C_6H_6, ca. 20°C; H, tBu-C(C(=O)-OAr)OOSi$(CH_3)_2$tBu (9d)

Under acidic conditions, alcohols add nucleophilically to dioxetanes with α-hetero atom (N, O) substitution to form labile α-alkoxy hydroperoxides (equation 10).[39,40] Similarly, aldehydes add to such dioxetanes to give 1,2,4-trioxanes under Brønsted and Lewis acid catalysis (equation 11).[41] The latter are more expediently prepared by perhydrolysis of epoxides and subsequent cyclization with carbonyl compounds,[42] which allows also the incorporation of the less reactive ketones.

HCl, EtOH, reflux, 1h (10a)

CF_3CO_2H, CH_3OH, 0°C, 6h (10b)

+ RCHO; Amberlyst-15, CH_2Cl_2, 25°C, 12h (11)

In the Rose Bengal (RB)-sensitized photooxygenation of indoles in the presence of trimethylsilyl cyanide dioxetane adducts were also formed (equation 12).[43] Although this transformation can be classified as a mode C process, because the adduct with trimethylsilyl cyanide could have originated from the intermediate dioxetane, a zwitterionic perepoxide was proposed as the precursor.

$^1O_2$, $(CH_3)_3SiCN$, CH_3CN, -30°C (12)

3.4 Mode D Transformations

In the thermolysis of dioxetanes, the 1,4-dioxyl diradicals were proposed as intermediates derived from homolytic O—O bond cleavage.[26] Indeed, as already stated,[28] such a short-lived diradical was trapped in the case of 3,3-dimethyl-1,2-dioxetane when the reaction was performed in the hydrogen donor solvent cyclohexa-1,4-diene (equation 13).

$$\begin{array}{c} O\text{-}O \\ H_3C\text{-}C\text{-}C\text{-}H \\ H_3C\ \ H \end{array} \xrightarrow[50^{\circ}C,\ 5h]{\text{cyclohexadiene} \rightarrow \text{benzene}} \begin{array}{c} HO\ \ OH \\ H_3C\text{-}C\text{-}C\text{-}H \\ H_3C\ \ H \end{array} \quad (13)$$

A prominent case in the category of mode D transformations is the quantitative lithium aluminum hydride reduction of dioxetanes to *vic*-diols (equation 14).[6] This reduction, in which the dioxetane C—C bond is preserved, was used extensively for proof of the structure of dioxetanes.

$$\begin{array}{c} O\text{-}O \\ R\text{-}C\text{-}C\text{-}R \\ R\ \ R \end{array} \xrightarrow[\substack{Et_2O \\ -78^{\circ}C}]{LiAlH_4} \begin{array}{c} HO\ \ OH \\ R\text{-}C\text{-}C\text{-}R \\ R\ \ R \end{array} \quad (14)$$

Dioxetanes were also found to be reduced to *vic*-diols by thiols such as the biologically relevant glutathione, cysteine and penicillamine; the thiols were oxidized to the corresponding disulfides (equation 15).[45] Simple thiols also effect this reduction, but substantial catalyzed cleavage into carbonyl products (mode A) additionally takes place. Such reduction of the peroxide bond in dioxetanes (mode D) has also been reported[44b] for other biological reductants, e.g. dihydronicotinamide adenine dinucleotide (NADH), riboflavin adenosine diphosphate ($FADH_2$), ascorbic acid and tocopherol. Again, appreciable catalyzed decomposition of the dioxetane in the dark into carbonyl products (mode A) has been observed. An electron-transfer mechanism was postulated for these redox processes.[44]

$$\begin{array}{c} O\text{-}O \\ R\text{-}C\text{-}C\text{-}R \\ R\ \ R \end{array} + 2\,R'SH \longrightarrow \begin{array}{c} HO\ \ OH \\ R\text{-}C\text{-}C\text{-}R \\ R\ \ R \end{array} + R'S\text{-}SR' \quad (15)$$

An important case of mode D chemistry is the reaction of dioxetanes with trivalent phosphorus nucleophiles, which has been extensively investigated.[45] The trivalent phosphorus of phosphines, phosphites or phosphinates first inserts into the O—O bond to afford the corresponding phosphorane. On warming, the phosphorane eliminates the phosphorus oxide to yield the epoxide (equation 16). Divalent sulfur compounds[40,46] also react with dioxetanes in this manner. Analogous biphilic reactions have been reported[47] for triphenylarsine and triphenylantimony; the resulting arsenic(V) and antimony(V) adducts were spectroscopically characterized in these reactions (equation 16). As already mentioned under mode A transformations, univalent transition metal complexes of rhodium and iridium perform similar biphilic insertions.[31]

$$\begin{array}{c} O\text{-}O \\ R\text{-}C\text{-}C\text{-}R \\ R\ \ R \end{array} + XR'_3 \longrightarrow \begin{array}{c} R'_3 \\ X \\ O\quad O \\ R\text{-}C\text{-}C\text{-}R \\ R\quad R \end{array} \xrightarrow[-\,R'_3PO]{X\,=\,P} \begin{array}{c} O \\ R\text{-}C\text{-}C\text{-}R \\ R\ \ R \end{array} \quad (16)$$

X = P, As, Sb

Recently it has been found that 3,3-disubstituted dioxetanes **13a–d** show a dramatically higher reactivity with nucleophiles than their tetrasubstituted analogs in terms of mode D transformations.[48] The sterically exposed oxygen atom (proximate to the CH_2 group) of the dioxetane peroxide bond becomes the site of nucleophilic attack to produce a zwitterionic or anionic adduct.

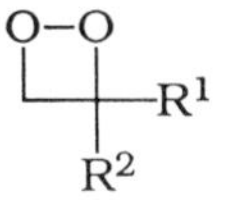

13a-d

a : $R^1 = CH_3$, $R^2 = Ph$
b : $R^1 = CH_2Cl$, $R^2 = Ph$
c : $R^1 = CH_2Br$, $R^2 = Ph$
d : $R^1 = R^2 = CH_2Ph$

Numerous reaction channels become available to the zwitterionic or anionic intermediates, depending upon their chemical nature. These include addition, deoxygenation, reduction and fragmentation. Thus, the reaction of olefins with 3,3-disubstituted dioxetanes involves nucleophilic attack by the olefinic π bond on the dioxetane peroxide bond to form, initially, an intermediary dipolar ion (Scheme 6) by heterolytic O—O bond scission (mode D). Subsequent ring closure leads to the 1,4-dioxane or intramolecular proton abstraction gives the β-hydroxy ether (Scheme 6).[44] Moreover, the dipolar intermediate was trapped by methanol to afford the expected hydroxy ether adduct. In view of these results, the intermediacy of the 1,6-dipole is unquestionable. Analogously the reaction of 3,3-disubstituted dioxetanes with enamines as π-nucleophiles (equation 17) affords β-hydroxy β'-keto ethers after hydrolysis of the primary adducts. Similarly, nucleophilic attack on the dioxetane peroxide bond was observed with n-butyllithium to produce the β-hydroxy ether in equation 18.[48c]

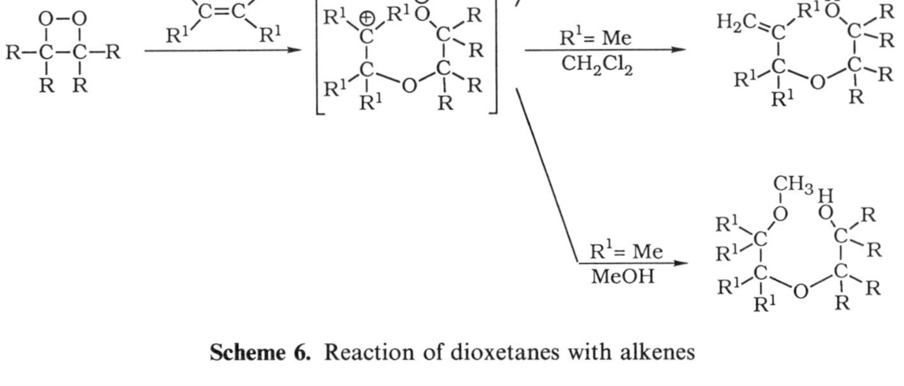

$R^1 = OEt$, CH_2Cl_2

$R^1 = Me$, CH_2Cl_2

$R^1 = Me$, MeOH

Scheme 6. Reaction of dioxetanes with alkenes

13c + (X = O, CH_2) $\xrightarrow[\text{-20 bis 0 °C, 0.5 - 3 h,}]{CH_2Cl_2,\ Ar,}$ $\xrightarrow[\text{40 - 50 °C, 30 min}]{1N\ HCl}$ (17)

$$H_2C\text{-}C(\text{-O-O-})\text{-Ph}(CH_2X) + n\text{-BuLi} \xrightarrow[-78^{\circ}C]{THF} \left[\begin{array}{c} n\text{-BuO}\ \ O^{\ominus}Li^{\oplus} \\ H_2C-C-Ph \\ CH_2X \end{array}\right] \quad \begin{array}{l} \xrightarrow[X=H,\ Cl,\ Br]{H^+} n\text{-BuO-}H_2C\text{-}C(OH)(Ph)\text{-}CH_2X \\ \xrightarrow[X=Br]{H^+} n\text{-BuO-}H_2C\text{-}C(Ph)(\text{-O-}CH_2\text{-}) \end{array} \tag{18}$$

Particularly valuable as a mechanistic tool to substantiate the observed S_N2 chemistry was the bromomethyl-substituted dioxetane **13c**.[48c,d] Debromination through internal nucleophilic displacement by the resulting proximate alkoxide ion constitutes the mechanistic diagnosis for S_N2 reactivity of these electrophilic peroxides. In particular, the reaction of 3,3-disubstituted dioxetanes **13a–d** with a large number of heteroatom nucleophiles establishes the S_N2 reactivity of these strained peroxides (Scheme 7).[48d] Hence, while hydroxylamine derivatives were

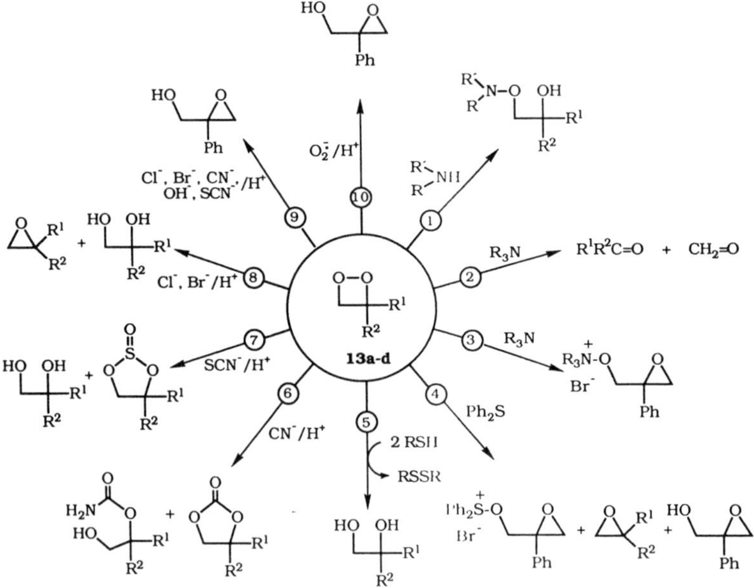

Scheme 7. S_N2 reactions of 3,3-disubstituted dioxetanes with a variety of neutral and anionic heteroatom nucleophiles

isolated from the reaction with secondary amines (transformation 1), tertiary amines, e.g. triethylamine and DABCO, in the case of dioxetanes **13a, d** gave exclusively dioxetane decomposition products (transformation 2). On the other hand, with the same tertiary amines, the bromomethyl-substituted dioxetane **13c** afforded alkoxyammonium epoxides exclusively (transformation 3). The same type of adduct, namely a labile alkoxysulfonium epoxide, was detected as an intermediate in the deoxygenation of **13c** by diphenylsulfide (transformation 4). In the reactions of cyanide (transformation 6) and thiocyanate ions (transformation 7) with **13c, d,** five-membered ring insertion products, namely carbonates and cyclic sulfites, were isolated. The halide ions Cl^- and Br^- (transformation 8) and mercaptans (transformation 5) reduced **13c** to the corresponding diol. Moreover, appreciable amounts of 2,3-epoxy-2-phenyl-1-propanol were obtained in the reactions of **13c** with anionic nucleophiles (transformation 9). Thus, with HO^- and O_2^- (transformation 10) the epoxyalcohol was formed essentially exclusively (Scheme 7).

3.5 Mode E Transformations

Amines[49a,b] and sodium hydroxide[49c] have been observed to convert dioxetanes into α-hydroxy carbonyl products, the so-called Kornblum–De La Mare reaction.[50] This case represents mode E chemistry. For example, the reaction of β-chloro-*tert*-butyl hydroperoxide with sodium hydroxide in aqueous methanol afforded 2-methylpropane-1,2-diol (equation 19). This unusual result was rationalized in terms of proton abstraction by the sodium hydroxide from the intermediate dioxetane with concomittant O—O bond cleavage to form α-hydroxy isobutyraldehyde; the latter is converted into the 1,2-diol by a crossed-Cannizarro reaction with formaldehyde, which is produced *in situ* from dioxetane cleavage. However, recent results[48d] confirmed that nucleophilic attack of the hydroxide ion on the dioxetane peroxide bond is preferred over the Kornblum–De La Mare reaction, at least for 3,3-disubstituted dioxetanes (cf. transformation 9 in Scheme 7).

$$CH_3-C(OOH)(CH_3)-CH_2Cl \xrightarrow[MeOH/H_2O]{NaOH} \left[CH_3-\overset{O-O}{C(H_3C)-CH_2} \xrightarrow{NaOH} CH_3-C(HO)(CH_3)-C(=O)-H \right] \xrightarrow{H_2CO,\ NaOH} CH_3-C(HO)(CH_3)-CH_2OH + HCO_2^- \quad (19)$$

Another unusual example of a mode E transformation was observed in the thermal rearrangement of naphthofuran and quinolinofuran dioxetanes into the corresponding spiro epoxides (equation 20).[51] It was proposed that electron transfer from the benzo ring to the dioxetane moiety with O—O bond cleavage generates the radical ion pair; instead of the usual dioxetane C—C cleavage, radical coupling and C—O bond rupture affords the spiroepoxide.

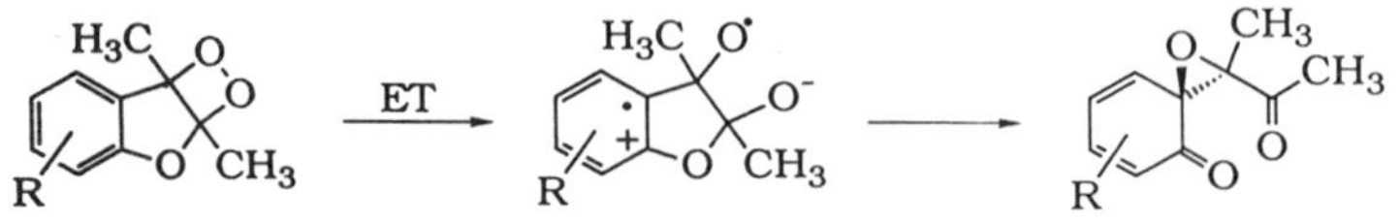

(20)

3.6 Mode F Transformations

Only a few examples of mode F transformations have been observed for dioxetanes, in which breaking of the peroxide and two vicinal substituent bonds takes place with preservation of the dioxetane ring C—C linkage. An early case was the decomposition of the tetrathioketene acetal dioxetane[52] into the cyclic oxalate (equation 21). A more recent report concerns the cleavage of the azacyclobutadiene dioxetane (equation 22) into the α-dione and pivalonitrile.[20b]

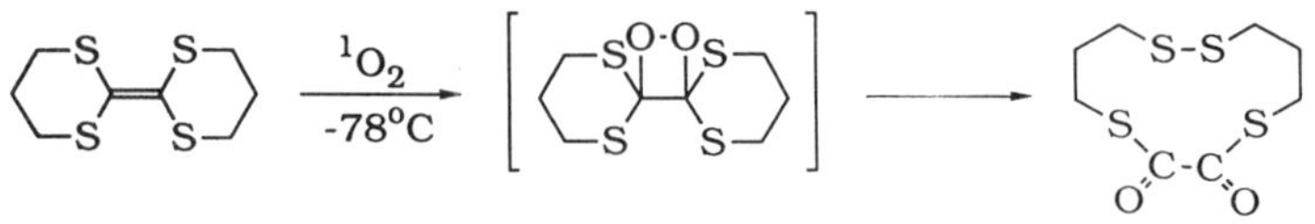

$^1O_2$, -78°C (21)

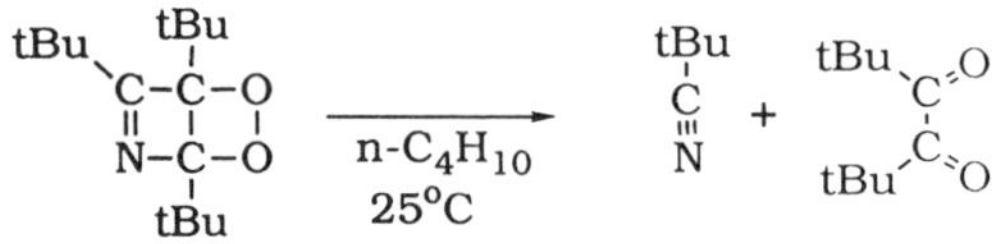

n-C_4H_{10}, 25°C (22)

4 CHEMILUMINESCENCE

The emission of light by four-membered ring peroxides on thermolysis is the most characteristic property of these substances.[2d] The chemical energy stored in these energy-rich molecules is converted mainly into vibrational and to some extent into electronic energy, the latter being released in the form of light. The fundamental steps of this process are detailed in equation 23. Thus, the ground-state reactant R_0 (the four-membered ring cyclic peroxide) acquires sufficient thermal energy on heating to form the activated complex (‡). This activated complex can dissociate into the electronically excited product P* (the carbonyl fragment). The last step is very unusual, since a crossover from a ground-state energy surface to an excited state surface—a non-adiabatic path—has taken place. Subsequently, the electronically excited product rids itself of the excess

$$\begin{aligned} R_o + \Delta &\longrightarrow (\ddagger) \\ (\ddagger) &\longrightarrow P^* \text{ and } P_o \\ P^* &\longrightarrow P_o + h\nu \end{aligned} \tag{23}$$

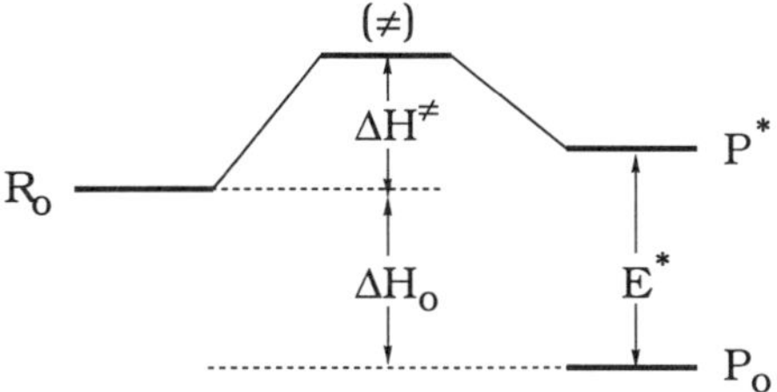

Figure 1. Energy diagram for a chemiluminescent decomposition. ΔH_0 is the reaction enthalpy, $\Delta H^{\neq}$ the activation enthalpy and E^* the excitation energy

energy by exhibiting luminescence, fluorescence in the case of singlet excited product $^{S}P^*$ and phosphorescence in the case of triplet excited product $^{T}P^*$. The reaction profile for a general chemiluminescent reaction is shown in Figure 1.

One of the most important criteria for a chemiluminescent process is that of energy sufficiency. The sum of activation enthalpy ($\Delta H^{\neq}$) and reaction enthalpy (ΔH_0) must be greater than the excitation energy (E^*) of the electronically excited product. A substance that conforms to this energy condition is designated energy sufficient.

Since the lowest excited states, namely n,π* type, of simple carbonyl compounds such as aldehydes and ketones range between *ca* 75 and 85 kcal mol^{-1} for singlet excitation and *ca* 60 and 75 kcal mol^{-1} for triplet excitation, the energy sufficiency criterion demands that at least 70–85 kcal mol^{-1} of energy must be made available in the transition state of decomposition to create one of the carbonyl fragments in its n,π* excited state. This energy sufficiency criterion is usually fulfilled for 1,2-dioxetanes and α-peroxy lactones. Typically, the activation enthalpies ($\Delta H^{\neq}$), determined by isothermal kinetic methods, range between 20 and 30 kcal mol^{-1} and the heats of reaction (ΔH_0), estimated from thermochemical calculations,[53] vary between 60 and 90 kcal mol^{-1}.

4.1 Activation Parameters

The kinetics of the thermal decomposition of 1,2-dioxetanes and α-peroxy lactones are first order. A variety of experimental methods can be used to determine the rates. We distinguish between isothermal and non-isothermal methods and between chemiluminescent and non-chemiluminescent methods. These include[26] direct chemiluminescence of the excited carbonyl product, energy transfer chemiluminescence of the chemienergized excited carbonyl product to an efficient fluorescer, dioxetane consumption or carbonyl product formation by NMR spectroscopy, iodometry of the cyclic peroxide and infrared spectrometry of α-peroxy lactone consumption or appearance of the carbonyl product. Of these kinetic methods, the chemiluminescence techniques are by far the most convenient and sensitive.

Most activation parameters of four-membered ring peroxides have been determined by kinetic analysis under isothermal conditions with chemiluminescence methods. The light emission is followed by photometry. Either the fluorescence or phosphorescence of the excited carbonyl fragment is directly monitored or that of an efficient fluorescer, which is excited by energy transfer of the chemienergized carbonyl product. The rate constants of the thermolysis should be measured over a temperature range as wide as possible (at least 20°C), and the activation parameters are determined by the Arrhenius (equation 24) or the Eyring method (equation 25). Unfortunately, dark catalysis can be a problem in this technique (cf. Section 3).

$$\ln k = \ln A - E_a/RT \tag{24}$$

$$\ln (k/T) = \ln k/h + \Delta S^{\neq}/R - \Delta H^{\neq}/RT \tag{25}$$

In addition to isothermal kinetics, a number of non-isothermal techniques have been developed. The temperature drop method[54] is advantageous since one can determine the activation energy (E_a) exclusively for the chemiluminescent process and thereby obviate problems of dark catalysis.[29,30] In this case, the chemiluminescence intensity I_1 is monitored at a constant temperature T_1, the reaction mixture is abruptly cooled by about 15°C to a lower temperature T_2, maintained at this temperature, and the chemiluminescence intensity I_2 is again recorded. From the temperature dependence the activation energy (E_a) is calculated according to equation 26.

$$E_a = \frac{R \ln(I_1/I_2)}{(1/T_2) - (1/T_1)} \tag{26}$$

The major advantage of this non-isothermal method, apart from convenience, is that in view of the high sensitivity of modern photometric equipment for monitoring chemiluminescence emissions, a negligible fraction of the four-membered ring cyclic peroxide is consumed during the time period of decomposition. Since bulk consumption of the dioxetane, as is necessarily the case in isothermal methods, is thereby avoided, one only registers the light-producing process and thereby obviates the problems of dark catalysis.

This convenient method has been modified in the form of the so-called step function analysis at various temperatures and with continuous temperature variation.[55] In the former,[55a] multiple temperature changes are employed; in the latter,[55b] the temperature is continuously varied in a linear fashion with time. Finally, differential thermal analysis[56] has also been utilized to assess activation parameters. In contrast to the above non-isothermal methods, in the latter the peroxide is completely consumed and therefore, instead of initial rates, one measures total rates. The advantage of minimizing dark catalysis is thereby lost.

When properly carried out, the activation parameters for the thermolysis of four-membered ring cyclic peroxides obtained by the described techniques agree well, within experimental error. Some typical examples are listed in Table 4. The activation enthalpies ($\Delta H^{\neq}$) lie in the range 13–36 kcal mol^{-1}, most values clustering around 26 kcal mol^{-1}. The activation entropies ($\Delta S^{\neq}$) lie within 0 ± 5 e.u., but when they drop much below -5 e.u., catalytic dark decomposition should be suspected.

Unfortunately, only limited information can be extracted from these thermal stability data in terms of structural features and substitution patterns. On the one hand, easily oxidized substituents lower the thermal stability (e.g. CIEEL);[33] on the other, it has been suggested[2g] that the more conformationally rigid the dioxetane ring is towards puckering, the higher is its thermal stability. Moreover, it was observed that increased steric interactions at geminal positions of the four-membered ring lead to high $\Delta H^{\neq}$ values with little effect on the $\Delta S^{\neq}$ terms.

The bulk of the thermal stability data[2g] support the diradical mechanism, but for ring-fused dioxetanes the concerted mechanism provides a more consistent explanation. However, when considering the activation data (rate-determining step) together with the excitation yields (product-determining step), the best compromise between these two mechanistic extremes is the 'merged mechanism', as brought out by the complete series of methylated dioxetane derivatives.[27] However, a recent[57] theoretical study suggests that the cleavage of the peroxide bond in dioxetanes requires negligible energy to produce the 1,4-dioxy diradical. The observed thermal activation[27] derives from cleavage of the dioxetane C—C bond to afford triplet excited ketone. We know of no experimental evidence in support of this most provocative theoretical suggestion.

4.2 Excitation Yields

The thermal decomposition of 1,2-dioxetanes and α-peroxy lactones generates sufficient energy for the formation of one electronically excited product, which manifests itself by light emission. The values of the singlet excitation yields (Φ^{S}), triplet excitation yields (Φ^{T}) and the spin state selectivity (Φ^{T}/Φ^{S}) characterize a particular dioxetane.[2g] Chemiluminescent techniques and photochemical transformations represent the two experimental methods for the determination of chemiexcitation yields.

In the photophysical techniques, the physical properties of the electronically excited state are utilized, specifically its luminescence. If for the excited state product its fluorescence quantum yield (Φ^{fl}) in the case of singlet excited state and its phosphorescence quantum yield (Φ^{ph}) in the case of triplet excited state are known, the determination of the direct chemiluminescence quantum yield (Φ^{DC}) allows us to calculate the desired $\Phi^{S/T}$ parameters (equation 27).[2o] The

Table 4. Activation parameters and excitation yields of some 1,2-dioxetanes[2d]

Entry	Dioxetane	$\Delta H^{\neq}$ (kcal mol^{-1})^a	$\Delta S^{\neq}$ (e.u.)^a	Φ^S (%)^b	Φ^T (%)^b
1	O-O; H–C–C–H; H H	18.9 ± 0.8	-12.6 ± 0.4	0.00031 ± 0.00006	0.24 ± 0.60
2	O-O; H_3C–C–C–H; H_3C H	22.5 ± 0.2	-5.1 ± 0.4	0.018 ± 0.002	7.8 ± 1.9
3	O-O; $PhCH_2$–C–C–H; $PhCH_2$ H	23.6 ± 0.1	-2.0 ± 0.5	0.76	22
4	O-O; H_3C–C–C–CH_3; H_3C CH_3	24.9 ± 0.7	-2.8 ± 2.3	0.25 ± 0.05	35 ± 7
5	O-O; H_3C–C–C=O; H_3C	21.7	0 ± 1	0.1	1.5

6	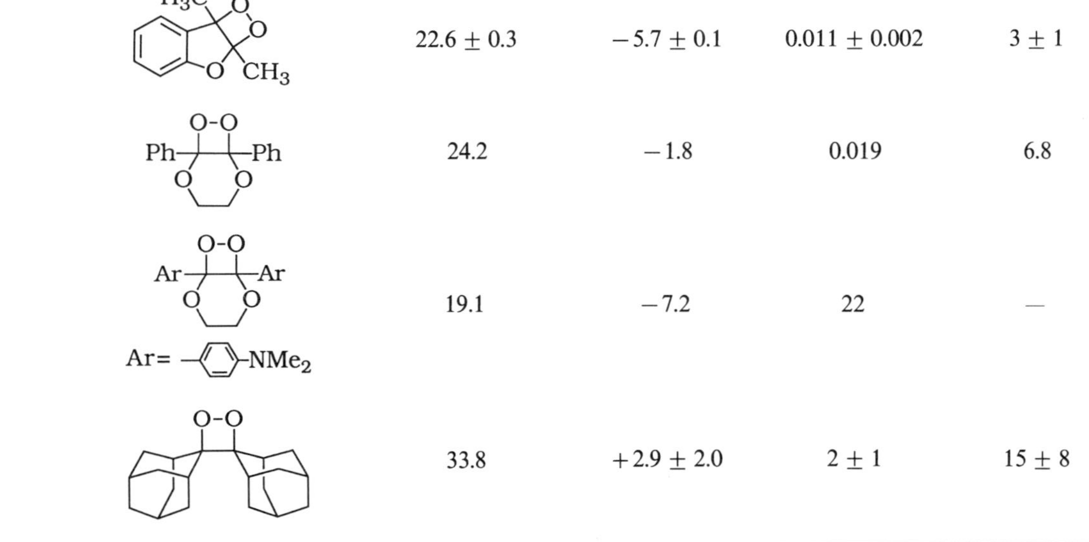	22.6 ± 0.3	−5.7 ± 0.1	0.011 ± 0.002	3 ± 1
7		24.2	−1.8	0.019	6.8
8		19.1	−7.2	22	—
9		33.8	+2.9 ± 2.0	2 ± 1	15 ± 8

[a]Most of the activation parameters were determined by measuring direct or enhanced chemiluminescence with isothermal kinetics.
[b]Most of the excitation yields were determined by using energy transfer chemiluminescence with DBA and DPA as fluorophores.

experimental definition of the direct chemiluminescence quantum yield (Φ^{DC}) is given by equation 28, in which the initial rate of photon production (I_0^{DC}) and the rate of peroxide decomposition ($k[C_0]$) must be measured. With a suitable photometer or photon-counting apparatus these values are readily assessed. Once the direct chemiluminescence quantum yield (Φ^{DC}) has been acquired experimentally, the singlet and triplet excitation yields ($\Phi^{S/T}$) can be calculated from equation 27.

$$\Phi^{DC} = \Phi^{S/T}\Phi^{fl/ph} \tag{27}$$

$$\Phi^{DC} = I_o{}^{DC}/k[C_o] \tag{28}$$

By far the most chemiluminescence yields have been determined by energy transfer to suitable fluorophores and subsequent measurement of the enhanced chemiluminescence quantum yied (Φ^{EC}). One distinguishes between enhanced luminescence chemienergized by singlet–singlet energy transfer (SS) and triplet–singlet energy transfer (TS). The former permits the determination of singlet excitation yields (Φ^S) and the latter triplet excitation yields (Φ^T). The chosen fluorophores are generally polycyclic aromatic compounds whose photophysical properties are known.[58]

The fluorescer of choice in the singlet–singlet energy-transfer process is 9,10-diphenylanthracene (DPA) in organic solvents and its sulfonate derivative (DPAS) in aqueous media. By means of steady-state kinetics, the relationship in equation 29 for the DPA-enhanced chemiluminescence quantum yield (Φ_{DPA}) is obtained in terms of the singlet excitation yield (Φ^S), the efficiency of singlet–singlet energy transfer (Φ_{ET}) and the DPA fluorescence quantum yield (Φ_{DPA}). At infinite concentration of DPA, the Φ_{ET} term is unity. One measures the DPA-enhanced chemiluminescence intensity (I_{DPA}) as a function of DPA concentration and constructs a plot of $1/I_{DPA}$ versus 1/[DPA]. The intercept of such a double reciprocal plot represents the DPA-enhanced chemiluminescence intensity at infinite DPA concentration ($I_{[DPA]}$). The experimental definition of the DPA-enhanced chemiluminescence quantum yield (Φ_{DPA}) is given in equation 30.

$$\Phi_{DPA}^{EC} = \Phi^S\Phi_{ET}^{SS}\Phi_{DPA}^{fl} \tag{29}$$

$$\Phi_{DPA}^{EC} = I_{[DPA]\infty}^{EC}/k[C_o] \tag{30}$$

The determination of the triplet excitation yields requires 9,10-dibromoanthracene (DBA) or its sulfonate derivative (DBAS). Other heavy atom-containing fluorophores, which efficiently undergo triplet–singlet energy transfer with triplet excited carbonyl products,[59] can also be used. Steady-state kinetics afford the expression given in equation 31 for the DBA-enhanced chemiluminescence quantum yield (Φ_{DBA}). As with DPA-enhanced chemiluminescence, to ensure that all chemienergized carbonyl triplets are intercepted by DBA through triplet–singlet energy transfer, the DBA-enhanced chemiluminescence intensity (I_{DBA}) is determined at infinite DBA concentration by constructing a double reciprocal plot of $1/I_{DBA}$ versus 1/[DBA]. Using equation 32, again analogous to the singlet excitation yield (Φ^S), the triplet excitation yield (Φ^T) is calculated. Note that these expressions neglect the contribution of energy transfer by singlet excited product to DBA. This is usually a valid assumption, since for simple dioxetanes and α-peroxy lactones the Φ^S values (DPA method) are generally at least two orders of magnitude smaller than the Φ^T values (DBA method).

$$\Phi_{DBA}^{EC} = \Phi^T \Phi_{ET}^{TS} \Phi_{DBA}^{fl} \quad (31)$$

$$\Phi_{DBA}^{EC} = I_{[DBA]\infty}^{EC}/k[C_o] \quad (32)$$

Non-chemiluminescent methods for the determination of excitation yields involve photochemical transformations.[2o] Since simple 1,2-dioxetanes or α-peroxy lactones produce very low yields of excited singlets, usually this method is not of sufficient sensitivity to allow accurate determinations of $\Phi^{S/T}$. If the quantum yield of photochemical reaction (Φ_{photo}) is known and the relative yield of photochemical product (Φ_{chem}) is determined experimentally, the yield of excited states ($\Phi^{S/T}$) can be calculated according to equation 33. Representative transformations include Norrish type I and type II cleavages and cyclohexadienone and cyclopentyl ketone rearrangements.

$$\Phi_{chem} = \Phi^{S/T} \Phi_{photo} \quad (33)$$

More convenient is intermolecular energy transfer to a photolabile reactant for the determination of chemiexcitation yields in the thermolysis of four-membered ring cyclic peroxides, because a particular 1,2-dioxetane or α-peroxy lactone need not be tailor-made for the chemical transformation in question. Compared with intramolecular reactions (equation 33), one has to know additionally the efficiency of energy transfer (Φ_{ET}) of the excited carbonyl fragment to the reactant (equation 34).

$$\Phi_{chem} = \Phi^{S/T} \Phi_{ET} \Phi_{photo} \quad (34)$$

Again one takes recourse in a double reciprocal plot of the yield of photoproduct *versus* the concentration of the photolabile substrate. A variety of photochemical transformations have been carried out in this way,[2o] which include *cis–trans* isomerization of alkenes, alkene dimerization, oxetane formation, the di-π-methane and dienone rearrangements, hydrogen abstraction, azoalkane denitrogenation and photocyclizations.

For symmetrically substituted 1,2-dioxetanes, only one electronically excited product fragment is possible. However, in unsymmetrically substituted dioxetanes, two different electronically excited carbonyls can be formed and the identification of the excited states is a major problem. Such detailed information about energy partitioning between the two possible excited carbonyl products is rare,[60] although these data would provide valuable mechanistic insight.

In general, it is difficult to draw definitive conclusions about the dependence between the excitation parameters and structural variations, but some interesting empirical trends emerge.[1,2o] For dioxetanes that chemienergize n,π* excited states of carbonyl products, the triplet yields are high ($\Phi^T > 10\%$) and the singlet yields low ($\Phi^S < 1\%$). On the other hand, in the case of dioxetanes that chemienergize π,π* or charge-transfer excited states of carbonyl products, e.g. by the intramolecular CIEEL mechanism (cf. Section 3),[32,33] the singlet excitation yields are fairly high ($\Phi^S > 10\%$) compared with the n,π* chemienergizers. Although the triplet yields are generally not reported for these systems (triplet yields are difficult to measure when the singlet yields are high), it is apparent from the large Φ^S values that the Φ^T values must be fairly low (cf. entry 8 in Table 4).

Provided that the dioxetane is energy sufficient to chemienergize a particular carbonyl product, empirically no definite correlation between the excited state energy of the product and excitation yield can be discerned. The effects are small and are buried within the experimental error. Generally n,π* are preferred over π,π* triplet states, even if the latter are of lower excitation energy. This trend does not apply to π,π* singlet excited states, which are produced by the intramolecular electron exchange mechanism (CIEEL).[32,33] In the case of unsymmetrically substituted 1,2-dioxetanes, a correlation of the partitioning of the available energy between the two possible fragments was attempted.[61] Although some obvious exceptions exist, it has been concluded that n,π* triplet states of lower energy are formed preferentially.

An impressive contrast exists between the 3,3,4,4-tetramethyl-1,2-dioxetane and the α-dimethylperoxy lactone. Although both produce excited acetone, the total efficiency of excited state production drops from 35 to 1.5% (cf. entries 4 and 5 in Table 4) for the α-peroxy lactone; it should be noted that this decrease is exclusively in the triplet yield, since the singlet excitation yields are essentially negligible compared with the triplet yields (Table 4). Although the decomposition of the α-peroxy lactone is about 22 kcal mol^{-1} more exothermic than that of the tetramethyldioxetane, the former is the less efficient system to produce excited acetone. Consequently, the available energy of these reactions does not necessarily parallel the observed chemienergization efficiencies.

5 BIOLOGICAL ASPECTS

The fact that biological systems are capable of generating electronically excited states is amply demonstrated in bioluminescence, in which singlet excited states are produced through enzymatic reactions.[2e] Whereas the phenomenon of bioluminescence is abundantly observed in nature, the question must be raised of whether in the living cell triplet excited states intervene, particularly in metabolic processes with molecular oxygen by the action of dioxygenases. Indeed, there is ample evidence that 1,2-dioxetanes play a central role in the mechanism of enzymatic oxygenations.[62] Thus, in the aerobic metabolism of aromatic compounds, 1,2-dioxetanes were postulated as labile intermediates. Two important examples are the oxidation of benzene to catechol by the microorganism *Pseudomonas putida* (equation 35)[63] and the cleavage of *trans*-7,8-dihydroxy-7,8-dihydrobenzo[*a*]pyrene by P-450 microsomes, in the latter case with light emission (equation 36).[64]

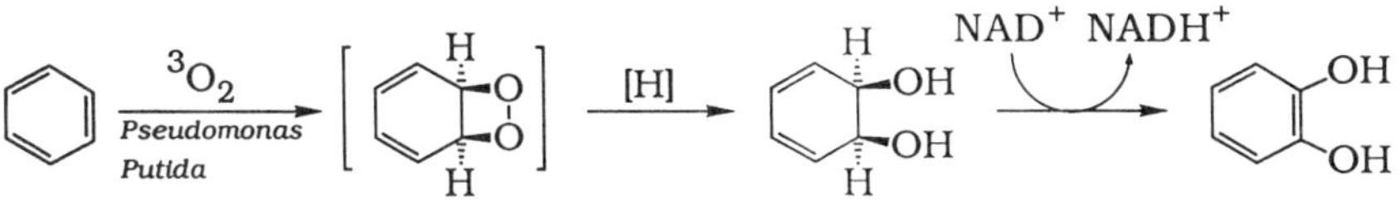

$^3O_2$, *Pseudomonas Putida*; [H]; NAD$^+$ NADH$^+$ (35)

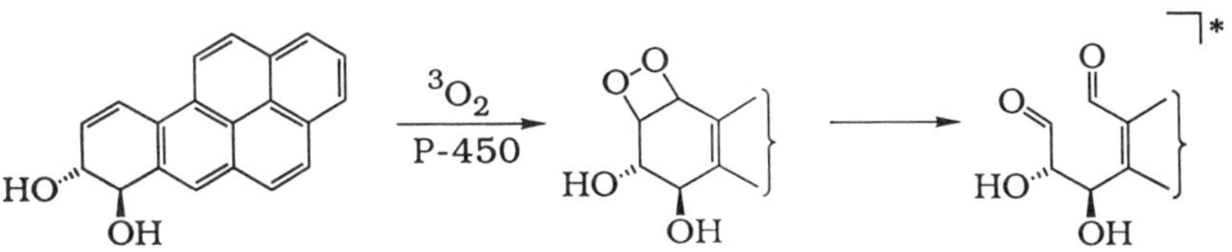

$^3O_2$, P-450 (36)

One of the most efficient enzymatic systems which generates triplet-state acetone is the horseradish peroxidase (HRP)-catalyzed oxygenation of isobutyraldehyde (equation 37).[62] Consequently, the existence of enzymatically generated triplet excited states in the cell is feasible. The provocative question is now raised on what these excited states do in the biological system. In this connection, the isobutyraldehyde–peroxidase–O_2 system has been extensively studied and the induction of interesting and important dark photobiological transformations observed. These include photoreactions of flavin, provitamin D, vitamin K, chloropromazine and arachidonic acid, which have all been performed with enzyme-generated triplet acetone.[62]

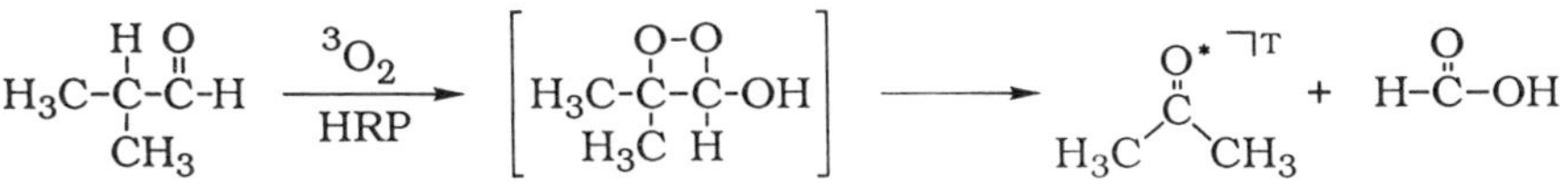

$$\mathrm{H_3C{-}CH(CH_3){-}CHO} \xrightarrow[\text{HRP}]{^3O_2} \left[\mathrm{H_3C{-}C(CH_3)(O{-})(H){-}C(O{-})(H){-}OH}\right] \longrightarrow \left[(\mathrm{H_3C})_2\mathrm{C{=}O}\right]^{T} + \mathrm{H{-}C(=O){-}OH} \quad (37)$$

Extensive work has been carried out with alkyl-substituted 1,2-dioxetanes to induce photochemical damage of the genetic material in biological targets. The investigations on the photogenotoxicity of dioxetanes revealed that, indeed, DNA can be damaged when treated with dioxetanes under physiological conditions. With isolated DNA, pyrimidine dimers were detected (equation 38), whereas with superhelical PM2 DNA and pyrimidine dimer-specific repair endonucleases a correspondence between the amount of dimer formation and the triplet excitation flux of different alkyl-substituted dioxetanes was observed.[38] Moreover, these toxicological results suggested that the damage, e.g. single strand breaks and generation of AP sites, does not correspond to that caused by direct UV (260 nm) irradiation, but appears to be more similar to that caused by singlet oxygen and radical species. Additionally, photobinding in the dark of psoralens to isolated DNA, promoted by the decomposition of 1,2-dioxetanes, was detected recently (equation 39).[65]

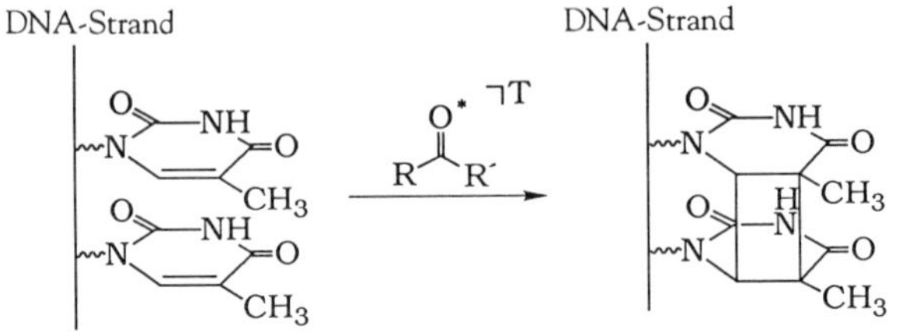

DNA-Strand DNA-Strand (38)

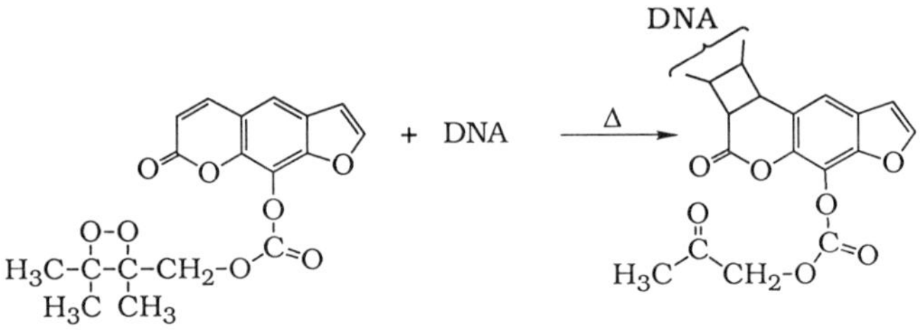

DNA + DNA $\xrightarrow{\Delta}$ (39)

Also in bacteria and mammalian cells the main damage is single strand breakage and induction of micronuclei at the chromosomal level, most probably caused by active oxygen species.[38] For example, in *Escherichia coli* bacteria dioxetanes produce the dose-dependent SOS function sfiA and in mammalian cells (HL-60, SHE) they induce micronuclei. It was therefore surprising that these alkyl-substituted dioxetanes are not mutagenic in several *Salmonella typhimurium* strains (Ames test). Nevertheless, recently it was observed[51,66] that benzofuran dioxetanes are highly mutagenic in the *S. typhimurium* TA100. In fact, these derivatives are the first known dioxetanes which possess potent mutagenicity, presumably not of photochemical origin but rather by alkylation of DNA (Table 5).[40]

In this context, it is of interest and importance to mention that glutathione efficiently reduces dioxetanes to the corresponding inactive vicinal diols (cf. Section 3). Undoubtedly such glutathione reduction of dioxetanes provides an effective deactivation mechanism of their photogenotoxicity.

The genotoxicity of 1,2-dioxetanes cannot only be ascribed to photochemical damage, either promoted directly by triplet states or indirectly by singlet oxygen. Free radicals, produced from triplet states or through redox processes of these four-membered ring cyclic peroxides with cell components, could also be involved; furthermore, alkylating epoxides, formed by deoxygenation of intermediary dioxetanes, may be responsible (Scheme 8). However, overall, the intensity of these DNA-damaging effects is fairly moderate, presumably owing to effective detoxification by the protective mechanisms of the living cell.

Table 5. Mutagenicity of benzofuran dioxetanes in *Salmonella typhimurium* strain TA100[38,51]

Dioxetane	Revertants per μmol[a]
H_3C, O, O, H_3C, O, O, CH_3	370 000
H_3C, O, O, O, CH_3	117 000
H_3CO, CH_3, O, O, O, CH_3	24 000
H_3C, O, O, O, CH_2O-C(=O)-CH_3	5220
CH_3, H_3C, O, O, O, O, O, CH_3	115 000
CH_3, CH_3, O, O, O, O, O, CH_3	50 000

[a]Calculated by linear regression analysis from the linear portion of the dose–response plots; values are within an error of *ca* 10% and the spontaneous revertant frequencies per plate were 150–200.

Some of the biological processes induced by 1,2-dioxetanes and α-peroxy lactones are expected to be benign, whereas others are surely malignant. Since dioxetanes are sources of excited states or active oxygen species and can be generated enzymatically, the toxicological data corroborate the hypothesis that these cyclic peroxides are also involved in spontaneous mutations[67] and oxidative stress[68].

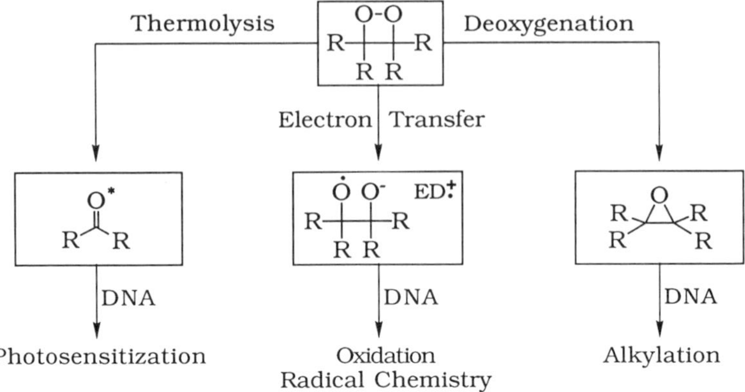

Scheme 8. Possible pathways of DNA damage by dioxetanes

In addition to the above-mentioned toxicological aspects, recently 1,2-dioxetanes have been used as effective chemiluminescence immunoassay probes in clinical analysis.[2c] For example, phosphoryloxy-substituted dioxetanes[69] are commercially available as highly chemiluminescent enzyme substrates for alkaline phosphatase (equation 40). The CIEEL effect (cf. Section 3)[33] serves as an ultra-sensitive detection system for antibodies and/or antigens bound to alkaline phosphatase; levels down to 1×10^{-20} mol can be revealed and such ultra-sensitive chemiluminescent probes can compete effectively with radiochemical techniques. For the same purpose, the functionalized adamantylidene 1,2-dioxetanes (**14**) was used to generate singlet excited xanthone[69c] on enzyme (AP)-triggered decomposition.

Alkaline Phosphatase (40)

14 **15**

Enzyme-promoted immunoassay is not limited to alkaline phosphatase-labeled substrates. Recently[69b] a β-D-galactoside-substituted dioxetane (**15**) was used as a highly luminescent detection method for β-D-galactosidase. It is expected that the use of relatively stable, CIEEL-active dioxetanes will offer new perspectives in clinical, biological and bioorganic chemistry.[12b,70]

6 ACKNOWLEDGEMENTS

We thank the Deutsche Forschungsgemeinschaft (SFB 172 'Molekulare Mechanismen kanzerogener Primärveränderungen'), the Fonds der Chemischen Industrie and the Wilhelm-Sander-Stiftung for generous financial support.

7 REFERENCES

1. W. Adam and G. Cilento, *Chemical and Biological Generation of Electronically Excited States*, Academic Press, New York (1982).
2. (a) W. Adam, in *Ullmann's Encyclopedia of Industrial Chemistry*, Vol. A15, VCH, Weinheim (1990), p. 548; (b) S. Albrecht, H. Brandl and W. Adam, *Chem. Unserer Zeit*, **24**, 227 (1990); (c) S. Beck and H. Köster, *Anal. Chem.*, **62**, 2258 (1990); (d) W. Adam, A. Beinhauer and H. Hauer, in *CRC Handbook of Photochemistry*, Vol. II (J. C. Scaiano, Ed.), CRC Press, Boca Raton, FL (1989), p. 271; (e) A. K. Campbell, *Chemiluminescence*, Ellis Horwood, Chichester (1988); (f) A. Krebs, in *Methoden der Organischen Chemie (Houben-Weyl), Organische Peroxo-Verbindungen*, Vol. E13, Part 1 (H. Krop, Ed.), Georg Thieme, Stuttgart (1988), pp. 401, 620; (g) A. L. Baumstark, in *Advances in Oxygenated Processes*, Vol. 1, (A. L. Baumstark, Ed.), JAI Press, Greenwich, CT (1988), p. 31; (h) E. J. H. Bechara, in *Advances in Oxygenated Processes*, Vol. 1, (A. L. Baumstark, Ed.), JAI Press, Greenwich, CT (1988), p. 123; (i) K.-D. Gundermann and F. McCapra, *Chemiluminescence in Organic Chemistry*, Springer, Berlin, Heidelberg (1987); (j) J. Schölmerich, R. Andreesen, A. Kapp, M. Ernst and W. G. Woods, *Bioluminescence and Chemiluminescence*, Wiley, Chichester (1987); (k) W. Adam and F. Yany, in *The Chemistry of Heterocyclic Compounds, Small Ring Heterocycles*, Vol. 42, Part 3 (A. Hassner, Ed.), Wiley, New York, (1985), p. 351; (l) T. Wilson, in *Singlet O_2*, Vol. II, Part 1 (A. A. Frimer, Ed.), CRC Press, Boca Raton, FL (1985), p. 37; (m) J. G. Burr, *Chemi- and Bioluminescence*, Marcel Dekker, New York (1985); (n) A. L. Baumstark, in *Singlet O_2*, Vol. II, Part 1 (A. A. Frimer, Ed.), CRC Press, Boca Raton, FL (1985), p. 1; (o) W. Adam, in *The Chemistry of Peroxides*, (S. Patai, Ed.), Wiley, Chichester (1983), p. 829; (p) W. Adam, W. J. Baader, C. Babatsikos and E. Schmidt, *Bull. Soc. Chim. Belg.*, **93**, 605 (1984); (q) W. Adam and G. Cilento, *Angew. Chem., Int. Ed. Engl.*, **22**, 529 (1983); (r) G. B. Schuster and S. P. Schmidt, *Adv. Phys. Org. Chem.*, **18**, 187 (1982).
3. E. P. Kohler, *J. Am. Chem. Soc.*, **36**, 177 (1906).
4. R. R. Kopecky and C. Mumford, *Can. J. Chem.*, **47**, 709 (1969).
5. W. Adam and J.-C. Liu, *J. Am. Chem. Soc.*, **94**, 2894 (1972).
6. W. Adam and G. Cilento, *Chemical and Biological Generation of Electronically Excited States*, Academic Press, New York (1982), p. 85.
7. F. Vargas, *PhD Thesis*, University of Würzburg (1989).

8. P. D. Bartlett and A. P. Schaap, *J. Am. Chem. Soc.*, **92**, 3223 (1970).
9. S. Mazur and C. S. Foote, *J. Am. Chem. Soc.*, **92**, 3225 (1970).
10. J. H. Wieringa, J. Strating, H. Wynberg and W. Adam, *Tetrahedron Lett.*, 169 (1972).
11. A. P. Schaap, A. L. Thayer, E. C. Blossey and D. C. Neckers, *J. Am. Chem. Soc.*, **97**, 3741 (1975).
12. R. W. Murray and M. L. Kaplan, *J. Am. Chem. Soc.*, **91**, 5358 (1969).
13. H. H. Wasserman and D. L. Larson, *J. Chem. Soc., Chem. Commun.*, 253 (1972).
14. E. J. Corey, M. M. Mehrotra and A. U. Khan, *J. Am. Chem. Soc.*, **108**, 2472 (1986).
15. G. H. Posner, K. S. Webb, W. M. Nelson, T. Kishimoto and H. H. Seliger, *J. Org. Chem.*, **54**, 3252 (1989).
16. G. H. Posner, M. Weitzberg, W. M. Nelson, B. L. Murr and H. H. Seliger, *J. Am. Chem. Soc.*, **109**, 278 (1987).
17. W. Adam and K. Sakanishi, *J. Am. Chem. Soc.*, **100**, 3935 (1978).
18. D. Leclercq, J.-P. Bats, P. Picard and J. Moulines, *Synthesis*, 778 (1982).
19. (a) W. Adam, V. Bhushan, R. Fuchs and U. Kirchgässner, *J. Org. Chem.*, **52**, 3059 (1987); (b) W. Adam, R. Fuchs and U. Kirchgässner, *Chem. Ber.*, **120**, 1565 (1987).
20. (a) A. Krebs, H. Schmalstieg, O. Jarchow and K.-H. Klaska, *Tetrahedron Lett.*, **21**, 3171 (1980); (b) U. J. Vogelbacher, M. Ledermann, T. Schach, G. Michels, U. Hees and M. Regitz, *Angew. Chem.*, **100**, 304 (1988).
21. G. Büchi and H. Wüest, *J. Am. Chem. Soc.*, **100**, 3935 (1978).
22. M. A. Avery, W. K. M. Chong and G. Detre, *Tetrahedron Lett.*, **31**, 1799 (1990).
23. C. A. Grob and P. W. Schiess, *Angew. Chem., Int. Ed. Engl.*, **6**, 1 (1967).
24. N. J. Turro, Y. Ito, M.-F. Chow, W. Adam, O. Rodriguez and F. Yany, *J. Am. Chem. Soc.*, **102**, 5836 (1977).
25. H. Staudinger, K. Dyckeroff, H. W. Klever and L. Ruzicka, *Chem. Ber.*, **58**, 1079 (1925).
26. W. Adam and G. Cilento, *Chemical and Biological Generation of Electronically Excited States*, Academic Press, New York (1982), p. 115.
27. W. Adam and W. J. Baader, *J. Am. Chem. Soc.*, **107**, 410 (1985).
28. (a) W. H. Richardson, M. B. Lovett and L. Olson, *J. Org. Chem.*, **54**, 3523 (1989); (b) M. Heil, PhD Thesis, University of Würzburg (1992).
29. P. D. Bartlett, A. L. Baumstark and M. E. Landis, *J. Am. Chem. Soc.*, **96**, 5557 (1974).
30. D. C.-S. Lee and T. Wilson, in *Chemiluminescence and Bioluminescence*, (M. J. Cormier, D. M. Hercules and J. Lee, Eds.), Plenum Press, New York (1973), p. 265.
31. P. D. Bartlett and J. S. McKennis, *J. Am. Chem. Soc.*, **99**, 5334 (1977).
32. (a) K. A. Zaklika, T. Kissel, A. L. Thayer, P. A. Burns and A. P. Schaap, *Photochem. Photobiol.*, **30**, 5334 (1979); (b) H. Nakamura and T. Goto, *Photochem. Photobiol.*, **30**, 27 (1979).
33. G. B. Schuster, *Acc. Chem. Res.*, **12**, 366 (1979).
34. (a) G. B. Schuster, *J. Am. Chem. Soc.*, **101**, 5851 (1979); (b) W. Adam, O. Cueto and F. Yan, *J. Am. Chem. Soc.*, **100**, 2587 (1978).
35. W. Adam and G. Cilento, *Chemical and Biological Generation of Electronically Excited States*, Academic Press, New York (1982), p. 229.
36. E. Vogel, G. Markowitz, L. Schmalstieg, S. Ito, R. Breuckmann and W. R. Roth, *Angew. Chem.*, **96**, 719 (1984).
37. (a) M. Balci, Y. Taskesenligil and M. Harmandar, *Tetrahedron Lett.*, **30**, 3339 (1989); (b) W. Adam, J. del Fierro, F. Quiroz and F. Yany, *J. Am. Chem. Soc.*, **102**, 2127 (1980); (c) W. Adam, E. Kades and X. Wang, *Tetrahedron Lett.*, **31**, 2259 (1990).
38. W. Adam, A. Beinhauer, T. Mosandl, C. Saha-Möller, F. Vargas, B. Epe, E. Müller, D. Schiffmann and D. Wild, *Environ. Health Persp.*, **88**, 89 (1990).
39. F. Abelló, J. Boix, J. Gómez, J. Morell and J.-J. Bonet, *Helv. Chim. Acta*, **58**, 2549 (1975).

40. W. Adam, L. Hadjiarapoglou, T. Mosandl, C. R. Saha-Möller and D. Wild, *Angew. Chem.*, **103**, 187 (1991).
41. C. W. Jefford, J. Boukouvalas and S. Kohmoto, *Helv. Chim. Acta*, **66**, 2615 (1983).
42. (a) W. Adam and A. Rios, *Chem. Commun.*, 882 (1971); (b) M. Schultz, in *Organic Peroxides*, Vol. 3 (D. Swern, Ed.), Wiley, New York (1972), p. 109; (c) C. Singh, *Tetrahedron Lett.*, **37**, 6901 (1990).
43. I. Saito, H. Nakagawa, Y.-H. Kuo, K. Obata and T. Matsuura, *J. Am. Chem. Soc.*, **107**, 5279 (1985).
44. (a) W. Adam, B. Epe, D. Schiffmann, F. Vargas and D. Wild, *Angew. Chem., Int. Ed. Engl.*, **27**, 429 (1988); (b) W. Adam, F. Vargas, B. Epe, D. Schiffmann and D. Wild, *Free Rad. Res. Commun.*, **5**, 253 (1989); (c) W. Adam, S. Hückmann and F. Vargas, tetrahedron Lett., **30**, 6315 (1989).
45. P. D. Bartlett, A. L. Baumstark, M. E. Landis and C. L. Lerman, *J. Am. Chem. Soc.*, **96**, 5267 (1974).
46. (a) H. H. Wasserman and I. Saito, *J. Am. Chem. Soc.*, **97**, 905 (1975); (b) B. S. Cambell, D. B. Denney, D. Z. Denney and L. S. Shih, *J. Am. Chem. Soc.*, **97**, 3850 (1975).
47. A. L. Baumstark, M. E. Landis and P. J. Brooks, *J. Org. Chem.*, **44**, 4251 (1979).
48. (a) W. Adam, S. Andler and M. Heil, *Angew. Chem. Int. Ed. Engl.*, **30**, 1365 (1991). (b) W. Adam, M. Heil and V. Voerckel, *J. Org. Chem.*, **57**, 2680 (1992). (c) W. Adam and M. Heil, *Chem. Ber.*, **125**, 235 (1992); (d) W. Adam and M. Heil, *J. Am. Chem. Soc.*, **114** (1992), in press.
49. (a) H. H. Wasserman and S. Terao, *Tetrahedron Lett.*, **21**, 1735 (1975); (b) W. Ando, T. Saiki and T. Migita, *J. Am. Chem. Soc.*, **97**, 5028 (1975); (c) W. H. Richardson and V. F. Hodge, *J. Am. Chem. Soc.*, **93**, 3996 (1971).
50. N. Kornblum and H. E. DeLaMare, *J. Am. Chem. Soc.*, **73**, 880 (1951).
51. W. Adam, H. Hauer, T. Mosandl, C. R. Saha-Möller, W. Wagner and D. Wild, *Liebigs Ann. Chem.*, 1227 (1990).
52. W. Adam and C.-J. Lui, *J. Chem. Soc., Chem. Commun.*, 73 (1972).
53. (a) H. E. O'Neal and W. H. Richardson, *J. Am. Chem. Soc.*, **93**, 1828 (1971). (b) W. H. Richardson and H. E. O'Neal, *J. Am. Chem. Soc.*, **94**, 8665 (1972).
54. T Wilson and A. P. Schaap, *J. Am. Chem. Soc.*, **93**, 4126 (1971).
55. (a) H.-C. Steinmetzer, A. Yekta and N. J. Turro, *J. Am. Chem. Soc.*, **96**, 282 (1974). (b) W. Adam and K. Sakanishi, *Photochem. Photobiol.*, **30**, 45 (1979).
56. (a) P. Lechtken, *Chem. Ber.*, **109**, 2862 (1976); (b) H. Keul, *Chem. Ber.*, **108**, 1198 (1975).
57. M. Reguero, F. Bernardi, A. Bottoni, M. Olivucci and M. Robb, *J. Am. Chem. Soc.*, **113**, 1566 (1991).
58. V. A. Belyakov and R. F. Vasil'ev, *Photochem. Photobiol.*, **11**, 179 (1970).
59. W. Adam, G. Cilento and K. Zinner, *Photochem. Photobiol.*, **32**, 87 (1980).
60. K. A. Horn and G. B. Schuster, *J. Am. Chem. Soc.*, **100**, 6649 (1978).
61. W. H. Richardson, M. B. Lovett, M. E. Price and J. H. Anderegg, *J. Am. Chem. Soc.*, **101**, 4683 (1979).
62. (a) G. Cilento, *Acc. Chem. Res.*, **13**, 225 (1980); (b) G. Cilento, *Photochem. Photobiol. Rev.*, **5**, 199 (1980); (c) G. Cilento and W. Adam, *Photochem. Photobiol.*, **48**, 361 (1988).
63. (a) D. T. Gibson, M. Hensley, H. Yoshioka and T. J. Mabry, *Biochemistry*, **9**, 1616 (1970); (b) D. T. Gibson, G. E. Cardini, F. E. Maseles and R. E. Kallio, *Biochemistry*, **9**, 1631 (1970); (c) F. Lingens, in *Schrift-Reihe Verein für Wasser, Boden und Lufthygiene 80*, Gustav Fischer, Stuttgart (1988), p. 39.
64. A. Thompson, W. H. Biggley, G. H. Posner, J. R. Lever and H. H. Seliger, *Biochim. Biophys. Acta*, **882**, 210 (1986).

65. W. Adam, T. Mosandl, F. Dall'Acqua and D. Vedaldi, *J. Photochem. Photobiol., B: Biol.*, **8**, 431 (1991).
66. W. Adam, O. Albrecht, E. Feineis, I. Reuther, C. R. Saha-Möller, P. Seufert-Baumbach and D. Wild, *Liebigs Ann. Chem.*, 33 (1991).
67. K. C. Smith and N. J. Sargentini, *Photochem. Photobiol.*, **42**, 801 (1985).
68. H. Sies, *Angew. Chem.*, **98**, 1061 (1986).
69. (a) A. P. Schaap, M. D. Sandison and R. S. Handley, *Tetrahedron Lett.*, **28**, 1159 (1987); (b) A. C. Brouwer, J. C. Hummelen, T. M. Luider, F. van Bolhuis and H. Wynberg, *Tetrahedron Lett.*, **29**, 3139 (1988); (c) I. Bronstein, B. Edwards and J. C. Voyta, *J. Biolumin. Chemilumin.*, **4**, 99 (1989).
70. W. Adam, S. Albrecht, H. Brandl and C. Schönfels, *Chem. Unserer Zeit*, **26**, 63 (1992).

6 Endoperoxides

E. L. CLENNAN

Department of Chemistry, University of Wyoming, Laramie, WY 82071, USA

and

C. S. FOOTE

Department of Chemistry and Biochemistry, University of California at Los Angeles, 405 Hilgard Avenue, Los Angeles, CA 90024–1569, USA

The first synthesis of a steroidal endoperoxide was reported in 1928 by Windaus and Brunken.[1] This method was extended to simple 1,3-dienes with the synthesis of ascaridole, **1**, by Schenck and Ziegler in 1944.[2] In the late 1960s and 1970s, a flurry of activity in the synthesis and chemistry of endoperoxides occurred as a result of the suggestion that endoperoxides were key intermediates in the

Organic Peroxides. Edited by W. Ando

bioconversion of polyunsaturated fatty acids into prostaglandins.[3,4] Various aspects of the early endoperoxide literature have been discussed in excellent reviews by Gollnick and Schenck,[5] Balci,[6] Saito and Nittala,[7] Bloodworth and Eggelte[8] and Porter and Wujeck.[9] This chapter focuses for the most part on development in endoperoxide chemistry between 1977 and 1991.

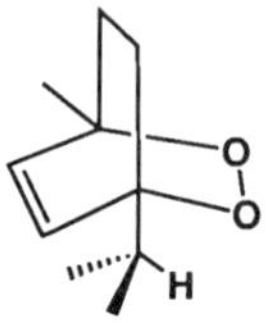

1

1 SYNTHESIS OF ENDOPEROXIDES

1.1 Photo-oxidation

The reaction of 1,3-dienes with singlet oxygen is the most general and frequently used route for the formation of endoperoxides. The formation of endoperoxide **2**[10] and of **3** and its bisendoperoxide homologue[11,12] are recent applications of this very versatile procedure.

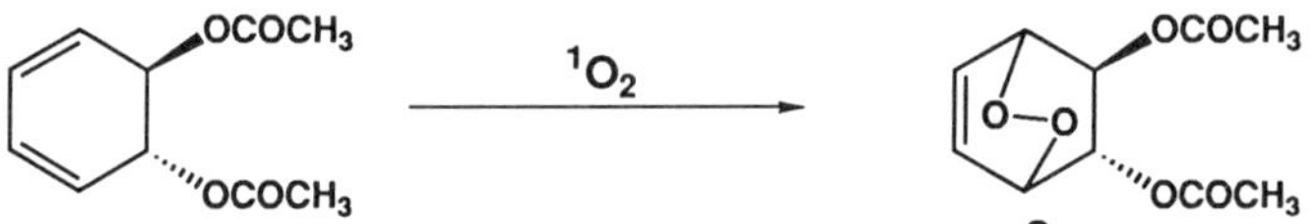

OCOCH$_3$ OCOCH$_3$ $\xrightarrow{^1O_2}$ OCOCH$_3$ OCOCH$_3$ **2**

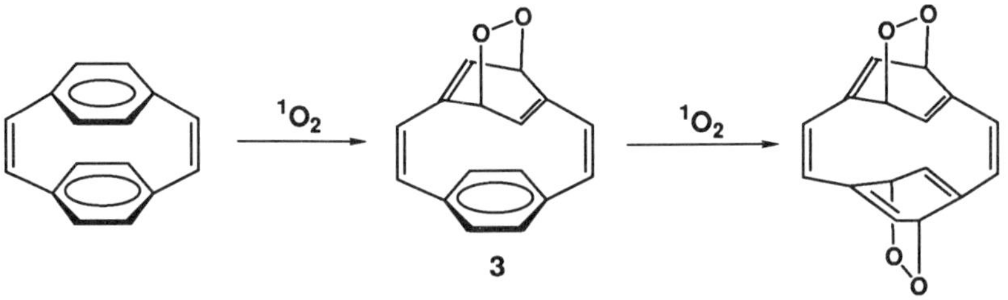

$\xrightarrow{^1O_2}$ **3** $\xrightarrow{^1O_2}$

Singlet oxygen can be produced in synthetically useful amounts either chemically or photochemically. The photochemical process involves population of a sensitizer singlet or triplet excited state which transfers its excitation energy to oxygen (equations 1 and 2).[13] Singlet oxygen formation via quenching of the longer lived triplet state is by far the most commonly encountered process. In either case, the energy difference between the initial and final excited state, $^1S_1$ or

$^3S_1$ and $^3S_1$ or $^1S_0$, respectively, must exceed 22 kcal mol^{-1}, the excitation energy of $^1\Delta_g$ oxygen.

$$^1S_1 + O_2(^3\Sigma_g^-) \rightarrow {}^3S_1 + O_2(^1\Delta_g) \qquad (1)$$

$$^3S_1 + O_2(^3\Sigma_g^-) \rightarrow {}^1S_0 + O_2(^1\Delta_g) \qquad (2)$$

A large number of sensitizing dyes are available for endoperoxide formation, including the commonly utilized Rose Bengal, **4**, methylene blue, **5**, and zinc tetraphenylporphine, **6**. The availability of polymer-bound sensitizers such as Rose Bengal[14] has dramatically improved the synthetic utility of the photosensitized reaction pathway. These polymer-bound reagents act as heterogenous sensitizers and can be conveniently removed from the reaction mixture by filtration. Chemical sources of singlet oxygen such as triphenyl phosphite ozonide, **7**,[15,16] and the anthracene endoperoxide **8**,[17] on the other hand, produce byproducts, triphenyl phosphate and an anthracene, respectively, which can be difficult to separate from the sensitive endoperoxide product.

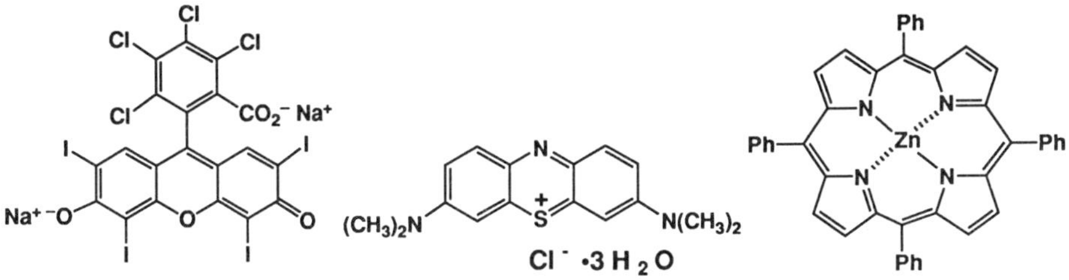

Cl Cl Cl Cl CO_2^- Na^+ I I Na^+ ^-O O O I I

4

N $(CH_3)_2N$ $\overset{+}{S}$ $N(CH_3)_2$ Cl^- $\cdot 3H_2O$

5

Ph Ph Ph Ph N N Zn N N

6

$(C_6H_5O)_3P$ (O–O–O ring) $\xrightarrow{T > -35°C}$ $(C_6H_5O)_3P{=}O$ + $^1O_2$

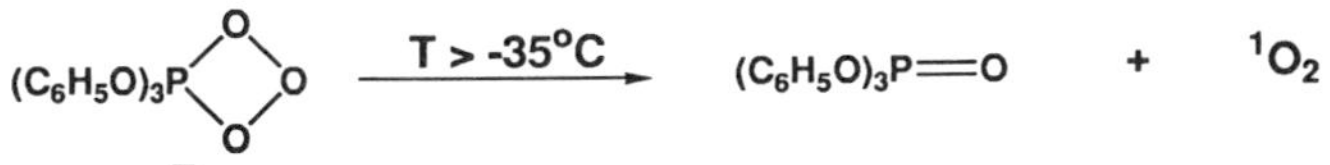

7

Ph Ph CO_2Me CO_2Me $\xrightarrow{\text{heat}}$ Ph Ph CO_2Me CO_2Me + $^1O_2$

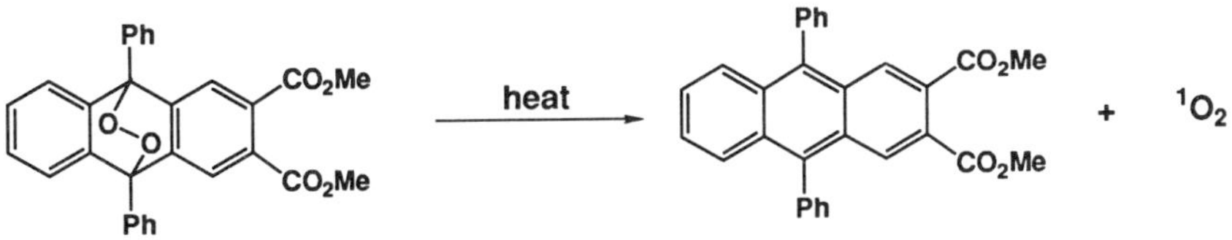

8

Careful control of the photo-oxidation reaction conditions is critical for the efficient formation of endoperoxides. Radical-derived byproducts can often be minimized by running the photo-oxidations under conditions which produce the endoperoxides as rapidly as possible. This can often be accomplished by running the reactions at low temperatures in deuterated or halogenated solvents. Photo-oxidations have very low or even negative enthalpies of activation and, as a result, the reactions are almost as (or even more) rapid at −78°C than at room temperature.[18] Deuterated and halogenated solvents promote endoperoxide formation as a result of the long lifetime and relatively high steady-state concentrations of singlet oxygen attainable in these media.[19] Use of a filter to remove ultraviolet light or a lamp with very low ultraviolet content is also very useful in preventing UV-initiated rearrangements of the peroxides.

Even under optimum conditions, photo-oxidation of acyclic dienes occurs at a much slower rate than addition of singlet oxygen to five-, six- or seven-membered ring dienes (Table 1).[20,21] As a result, hydroperoxides and dioxetanes are often encountered as byproducts during the synthesis of monocyclic endoperoxides, but are rarely encountered during the synthesis of bicyclic endoperoxides.[22] Stereospecific addition of singlet oxygen to acyclic dienes is normally observed, but cannot always be assumed. During photooxidation, (*E*, *E*)-, (*E*, *Z*)- and (*Z*, *Z*)-2,4-hexadienes undergo singlet oxygen-induced isomerizations, and mixtures of isomeric endoperoxides, hydroperoxides and dioxetanes are produced.[23,24] Exceptions to the normally rapid photo-oxidations of cyclic dienes are the very slow endoperoxidations of eight-membered ring dienes which are caused by the lack of planarity of the diene.[25,26] Adam and Klug[27] reported that irradiation for 14 days was necessary to add singlet oxygen to cyclooctatetraene.

1.2 Triplet Oxygen Routes

Endoperoxides can also be synthesized by the reactions of triplet oxygen with radical cations or diradicals. Radical cations usually react with triplet oxygen

Table 1. Rate constants for endoperoxide formation[a]

Compound	k (l mol^{-1} s^{-1})	Compound	k (l mol^{-1} s^{-1})
	2.6×10^5		7.1×10^6
	1.0×10^5		1.1×10^6
	2.4×10^5		6.5×10^4
	1.0×10^8		

[a]Refs 20 and 21.

via a chain reaction mechanism as depicted in equations 3, 4 and 5. These reactions are usually conducted in polar solvents in order to allow ion-pair diffusion and reaction with oxygen to compete with back electron transfer. Under these conditions, the chain reaction will proceed as long as the endoperoxide radical cation ($SO_2^{+\bullet}$) is a stronger oxidant than the substrate radical cation (i.e. reaction 5 is thermodynamically downhill). A variety of different oxidants, Ox, have been used to initiate the chain reaction as illustrated by the reactions of **9**–**14**.

$$S + Ox \rightleftharpoons S^{+\bullet} + Ox^{-\bullet} \tag{3}$$

$$S^{+\bullet} + {}^3O_2 \rightleftharpoons SO_2^{+\bullet} \tag{4}$$

$$SO_2^{+\bullet} + S \rightleftharpoons SO_2 + S^{+\bullet} \tag{5}$$

9 $\xrightarrow[h\nu,\ O_2]{Ph_3C^+\ BF_4^-}$

10 $\xrightarrow[h\nu,\ 2h,\ O_2]{Ph_3C^+\ BF_4^-}$

11 $\xrightarrow[O_2]{-e^-}$

12 $\xrightarrow[CH_3CN]{h\nu,\ O_2,\ DCA}$ Ar = p-MeOPh

13 $\xrightarrow[CH_3CN]{h\nu,\ O_2,\ DCA}$ Ar = p-MeOPh

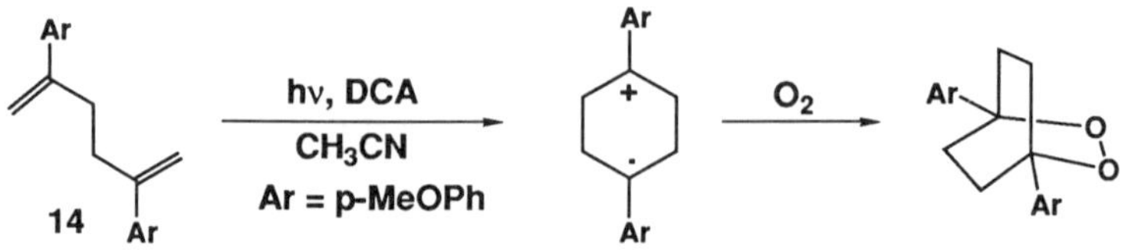

Ergosteryl acetate, **9**, reacts photochemically in the presence of a catalytic amount of trityl fluoroborate to give a quantitative yield of the endoperoxide.[28] Other catalysts, such as $AlCl_3$, BBr_3, BCl_3, I_2, $SbCl_5$, $SnCl_4$ and SnI_4, also function as effective photochemical catalysts for this interconversion.[29] Transition metal catalysts such as $FeBr_3$, $FeCl_3$, $VOCl_3$, WCl_6 and $MoCl_5$, on the other hand, perform the same transformation with only thermal activation. Singlet oxygen was not involved in these reactions, since lumisteryl acetate was unreactive under a variety of Lewis acid-catalyzed conditions but could be readily converted to its endoperoxide with singlet oxygen.[30] Sterically unencumbered dienes such as cyclohexa-1,3-diene preferentially undergo the electron transfer-catalyzed Diels–Alder reaction[31] rather than oxidation. Even the radical cation of 1-*tert*-butyl-1,3-cyclohexadiene, **10**, forms a cation-radical dimer prior to reaction with oxygen.[32] Diene **15**, with both the 1- and 4-positions blocked towards dimerizations, reacts successfully with the sterically small oxygen molecule, giving chain lengths as long as 260.[33]

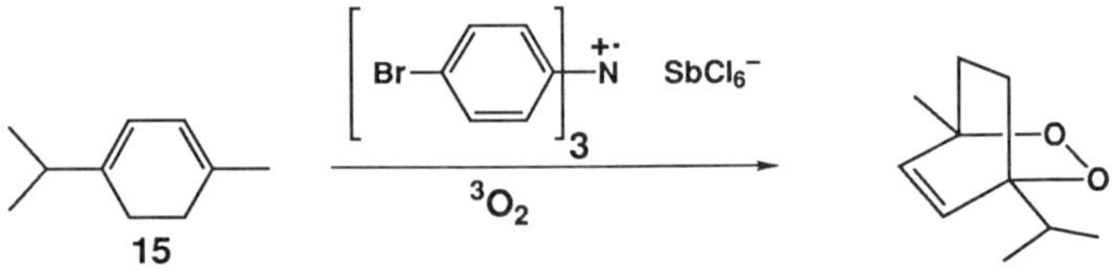

An electrode can also be used to catalyze the oxidation of **11**.[33] Electrochemical catalysis also makes it convenient to calculate the chain length of the reaction by comparison of the number of coulombs passed in the presence and absence of oxygen.[34]

The reactions of **12**, **13** and **14** were photochemically initiated using 9,10-dicyanoanthracene (DCA) as the sensitizer.[35] The photo-oxidation of 1,1-di(*p*-anisyl)ethylene, **12**, produced the endoperoxide with a quantum yield of approximately 10, consistent with addition of $\mathbf{12}^{+\bullet}$ to **12** followed by the chain propagation steps given by equations 4 and 5.[36] The radical cation chain character of the reaction of **13**,[37] but not **14**, was also demonstrated by observation of a quantum yield in excess of 1.

Determination of the quantum yield is an important criterion to establish operation of the mechanism given by equations 3, 4 and 5. Historically, a non-chain DCA-sensitized photo-oxidation process given by equations 6, 7 and 8 was first recognized by Foote and co-workers[35] and experimentally established by the direct observation of both the substrate radical cation[38] and the sensitizer radical anion[38,39] and by inhibiting the reaction by trapping superoxide with benzoquinone.[40] In addition, Foote and co-workers[41] detected an enhanced formation of singlet oxygen from DCA and demonstrated with several alkenes that reaction with singlet oxygen can compete with radical ion formation.[42]

$$^{1}\mathrm{DCA}^{*} + \mathrm{S} \rightarrow \mathrm{DCA}^{-\bullet} + \mathrm{S}^{+\bullet} \quad (6)$$

$$\mathrm{DCA}^{-\bullet} + \mathrm{O_2} \rightarrow \mathrm{DCA} + \mathrm{O_2}^{-\bullet} \quad (7)$$

$$\mathrm{S}^{+\bullet} + \mathrm{O_2}^{-\bullet} \rightarrow \mathrm{SO_2} \quad (8)$$

Endoperoxides can also be generated by trapping of triplet diradicals with oxygen.[43] Singlet diradicals, on the other hand, are generally too short-lived for bimolecular trapping, even at extremely high triplet oxygen pressures.[44] Triplet diradicals which have been successfully trapped have been generated by sensitized photolysis of azoalkanes, in Norrish type II and Paterno–Büchi reactions, and from the thermal cleavage of strained sigma bonds. Examples of these processes are illustrated by the reactions of **16**–**21**.

Triplet diradicals from azoalkanes can be most efficiently generated by sensitized photolysis using a monochromatic laser light source (Figure 1). Even under these conditions, intersystem crossing (k_{isc}) or oxygen-induced intersystem crossing ($k_i[O_2]$) can compete with oxygen trapping (Figure 1). The mole fraction of oxygen-trapped products from the triplet, χ_{TP}, increases as the concentration of oxygen, the rate of trapping (k_t) and the lifetime of the diradical ($^3\tau$) increase, as depicted in equation 9.[45] At infinite oxygen concentration, all of the triplet diradical can be trapped and the endoperoxide produced in high yield if the oxygen-induced intersystem crossing rate (k_i) is zero. In practice, however, the limit is approximately 10 atm of oxygen, since benzophenone, the most commonly used sensitizer, is quenched at higher oxygen concentrations.

$$\frac{1}{\chi_{TP}} = 1 + \frac{k_{isc} + k_i[O_2]}{k_t(^3\tau)[O_2]} \tag{9}$$

Azoalkane **16** undergoes triplet-sensitized formation of the very labile and short-lived (*ca* 0.1 ns) cyclohexane-1,4-diyl. At 10 atm of oxygen, however, only

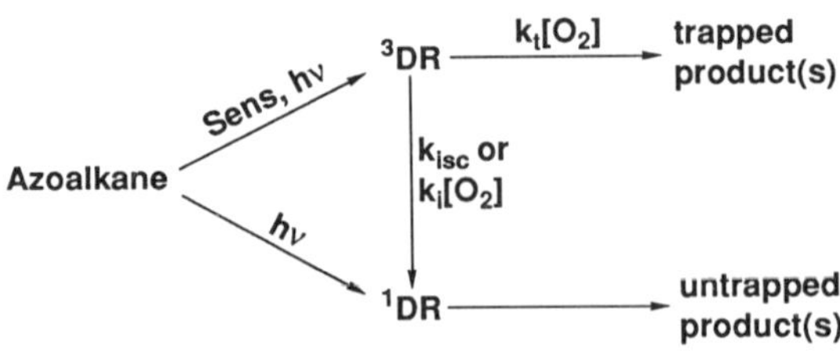

Figure 1. Photochemical reactions of azo compounds

4% of trapped products were observed.[46] On the other hand, trimethylenemethane diradicals usually exist as long-lived ground-state triplets and, as a result, synthetically useful amounts of endoperoxides are produced during the decompositions of **17**.[47] The 1,3-diyl from decomposition of **18** also has an appreciable lifetime and is effectively trapped. Unfortunately, the bicyclo[2.2.1]endoperoxide derived from **18** is susceptible to photodestruction (perhaps by UV) over an extended photolysis time, since the yield is 73% at 21% conversion and only 43% at 54% conversion.[47]

The Norrish type II reaction of **19**,[48] the Paterno–Büchi reaction of **20**,[49] and the thermal cleavage of the strained σ-bond in **21**[50] also form trappable diradicals. The reaction of **20**, in particular, is synthetically attractive since it allows a one-step entry into the 1,2,4-trioxane ring system[51–53] found in artemisinin (qinghaosu).[54,55]

1.3 Nucleophilic Addition and Substitution

Endoperoxides can also be constructed by intramolecular cyclizations of either preformed or *in situ*-generated hydroperoxy groups. This has most often been accomplished by nucleophilic additions to mercurium ions, halonium ions, epoxides or carbonyls, or by nucleophilic substitution of halides, tosylates or triflates. Examples of these processes are illustrated by the reactions of **22**–**34**.

OOH Br **22** $\xrightarrow{Ag^+}$ O–O

23 Br Br or **24** Br Br $\xrightarrow[H_2O_2]{Ag^+}$ O O + O O + OOH + OOH

HOO **25** $\xrightarrow[CH_2Cl_2]{Hg(OAc)_2}$ O O Hg(OAc)

26 $\xrightarrow[HgX_2]{H_2O_2}$ XHg O O HgX

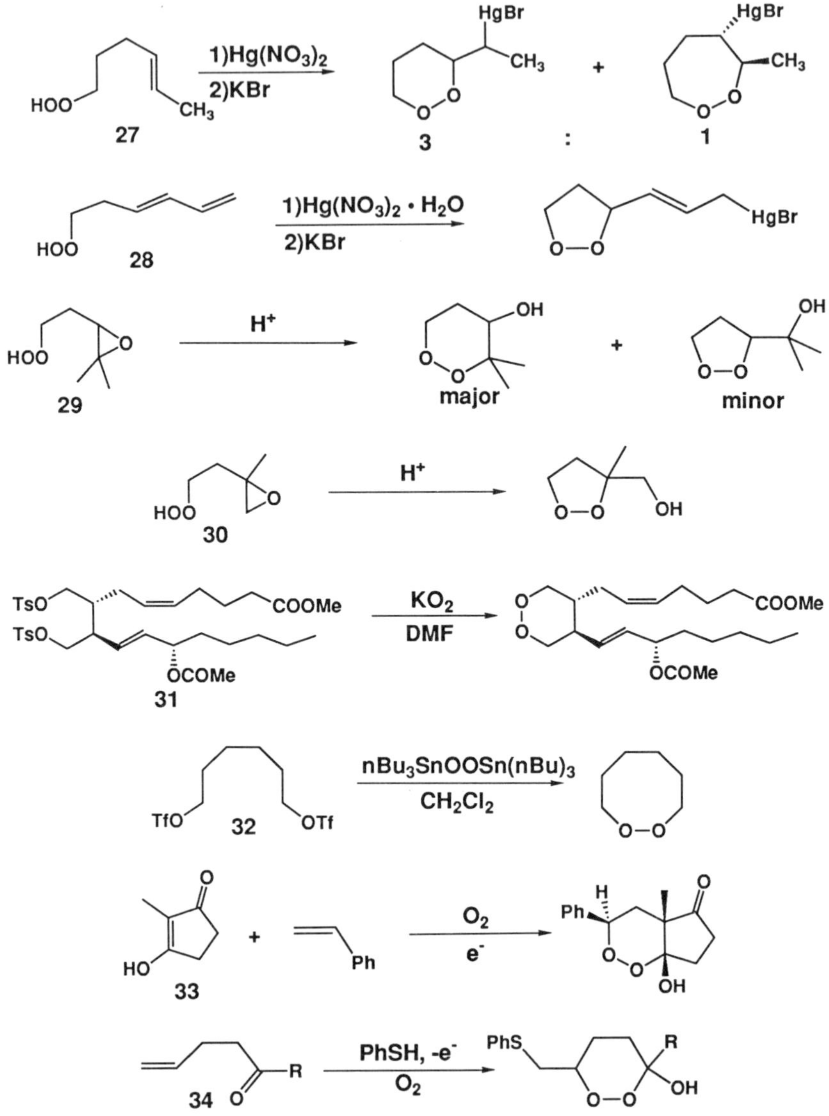

The intramolecular silver-assisted nucleophilic substitution in **22** proceeds quantitatively to give 3,3-dimethyl-1,2-dioxolane.[56] The reaction proceeds via the very novel dialkylperoxonium ion intermediate **35**, which can be totally diverted to 2,2-dimethyltetrahydrofuran in the presence of methylphenyl sulfoxide.[57] The reaction of the primary bromide **22** is undoubtedly a Ag^+-assisted S_N2 displacement. The formation of identical product mixtures during the reactions of isomers **23** and **24**, however, suggests that in some cases

Ag^+-assisted S_N1 reactions also occur.[58] Similar Ag^+-assisted reactions can also be used to synthesize bicyclic endoperoxides.[59–61]

22 → 35 —PhS(O)Me→ + PhS(O_2)Me

The reactions of substrates **25**–**28** involve hydroperoxy attack on a mercurium ion to give an organomercury-substituted endoperoxide. Hydridodemercuration can be subsequently accomplished by treatment of the mercury-substituted endoperoxide with basic sodium tetrahydroborate.[62] Hydroperoxy mercuration of **25** gives exclusively a [5.2.1] bicyclic endoperoxide.[63] Stereochemical constraints in the cyclooctane ring system of its isomer **36**, however, prevent the formation of an endoperoxide, and only a bicyclic ether, presumably formed via a dialkylperoxonium ion, was observed.[63] Sequential nucleophilic additions to mercurium ions formed during the reaction of **26** give a bicyclic endoperoxide directly and circumvent the necessity of isolating very sensitive hydroperoxy alkenes.[64,65]

36 + N–Br → (Br) + N–OH

Monocyclic endoperoxide formation via the peroxymercuration route occurs as predicted by the Baldwin rules.[66] The reaction of **27** forms a single diastereomer of both six- and seven-membered ring endoperoxides in a 3:1 ratio via 6-*exo*-trig- and 7-*endo*-trig-like transition states, respectively.[67] The reaction of **28** gives a five-membered ring endoperoxide via the energetically preferred 5-*exo*-trig transition state.[67]

In contrast to the peroxymercuration route, trichloroacetic acid-catalyzed opening of the hydroperoxy epoxide **29** does not conform to the Baldwin rules.[68] The 6-*endo*-trig rather than the 5-*exo*-trig transition state is preferred. A reversal in this stereochemical preference of the reaction of the structurally similar hydroperoxy epoxide **30** suggests that it is the stability of a carbocation intermediate rather than trajectory control which determines the product outcome.

Endoperoxides can also be synthesized using either superoxide, as illustrated by the reaction of **31**,[69] or bis(tri-*n*-butyltin) peroxide as shown by the reaction of **32**.[70,71] In the tri-*n*-butyltin peroxide route, dialkyl ethers and 2-butanol are also produced as minor byproducts. The formation of these products can be conveniently explained within the framework of the mechanism presented in

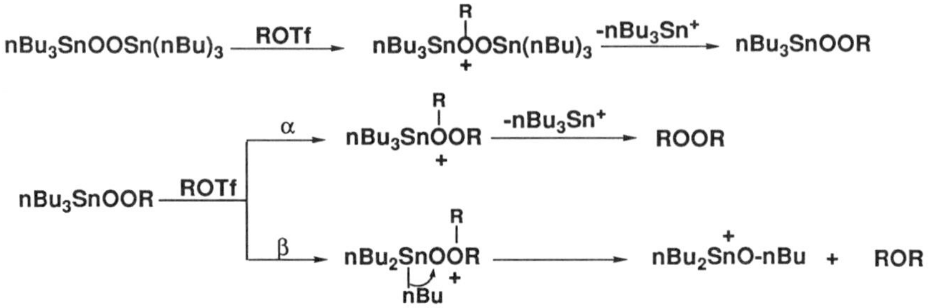

$$nBu_3SnOOSn(nBu)_3 \xrightarrow{ROTf} nBu_3Sn\overset{+}{O}(R)OSn(nBu)_3 \xrightarrow{-nBu_3Sn^+} nBu_3SnOOR$$

$$nBu_3SnOOR \xrightarrow{ROTf} \begin{cases} \alpha: nBu_3Sn\overset{+}{O}(R)OR \xrightarrow{-nBu_3Sn^+} ROOR \\ \beta: nBu_2SnO\overset{+}{O}(R)R(nBu) \longrightarrow nBu_2\overset{+}{S}nO\text{-}nBu + ROR \end{cases}$$

Figure 2. Tin peroxide reactions

Figure 2. Alkylation of the tin peroxide to give a peroxonium ion intermediate followed by loss of a tri-*n*-butyltin cation gives an alkyltin peroxide intermediate. This pivotal intermediate has two chemically distinguishable oxygens, and alkylation at the oxygen α to tin ultimately produces the endoperoxide, whereas alkylation β to tin ultimately produces the ether and butan-1-ol.

The reactions of **33** and **34** both involve nucleophilic additions of a hydroperoxy group to a carbonyl and are electrochemically and chemically initiated short radical-chain reactions. The reactions of **33** occur by loss of an electron and a proton to give a β-diketo radical which adds to 1-substituted or 1,1-disubstituted alkenes to generate a radical which is trapped by oxygen.[72] The resulting hydroperoxy radical acts as the chain carrier and abstracts a hydrogen atom from the substrate and collapses to the product. The radical chain reaction of **34**, on the other hand, is initiated by addition of phenylthio radical to the double bond.[73]

1.4 Peroxy Radical Cyclizations

The importance of peroxy radical cyclizations in biological autoxidation pathways has been recognized for some time. The viability of autoxidation to produce the pivotal endoperoxide intermediate in the arachidonic acid cascade has been convincingly demonstrated.[74–76] Peroxy radical cyclization has also become a synthetically useful procedure as a result of the development of an understanding of the underlying principles which govern these reactions and as a result of the development of methodologies other than autoxidation to initiate the cyclizations.

Autoxidations produce complicated mixtures of hydroperoxides and endoperoxides whose exact compositions are dependent on the type and concentrations of hydrogen atom donors present in the reaction mixture. As a result, they are rarely synthetically useful. The formation of a complicated mixture of hydroperoxy endoperoxides during autoxidation of methyl linolenate, **37**, however, can be circumvented by sequential lipoxygenase-catalyzed oxidation and autoxidation.[77] Under these conditions, only a single hydroperoxy endoperoxide in 31.5% yield and a mixture of polar products in 43% yield were observed. The carbon radical intermediate in these reactions can also be generated by addition of phenylthio radical formed by coautoxidation of thiophenol with diene, **38**,[78] or triene, **39**.[79] The reaction is both regio- and stereospecific. The phenylthio radical adds regiospecifically to the least substituted double bond, followed by capture with oxygen and stereospecific ring closure to the 1,2-dioxolane. The preferential formation of dioxolanes with *cis*-3,5-substituents is consistent with products observed in closely related carbon-centered radical ring closures.[80]

37 $(CH_2)_7CO_2Me$ — 1) Lipoxygenase, 2) O_2, 40°C, 96 h → HOO, O—O, $(CH_2)_7CO_2Me$

38 — PhSH, O_2 → PhS, OOH, O—O

PhS· ; PhS ; O_2 ; PhS, O, O· ; PhS, O—O ; PhSH, O_2

39 — PhSH, O_2 → PhS, OOH, O—O

Peroxy radical cyclizations can also be initiated by either hydrogen atom abstraction or by oxidation of an alkenyl hydroperoxide. The scope of these processes is illustrated by the reactions of **27** and **40**–**44**. Abstraction of a hydrogen atom from the hydroperoxy group in **27** with *tert*-butoxy radicals, generated by decomposition of di-*tert*-butyl peroxyoxalate (DBPO), resulted in 6-*exo*-trig cyclization and nearly quantitative formation of a 1,2-dioxane.[68] The use of secondary rather than primary hydroperoxy alkenes in these reactions preferentially produces the *trans*-3,6-disubstituted-1,2-dioxanes, suggesting a chair-like transition state for the cyclization step.[81] DBPO-initiated cyclization of **40**, which cannot react by the 6-*exo*-trig mode, resulted in a stereospecific 5-*exo*-trig cyclization to give a *cis*-3,5-disubstituted-1,2-dioxolane.[82] Peroxy radical serial cyclizations by both 6-*exo*-trig and 5-*exo*-trig modes to form five diastereomeric bis-1,2-dioxanes, **41**,[83,84] and six dioxolanes, **42**,[85] respectively, have also been reported. The exclusive formation of a 1,2-dioxolane during the DBPO-initiated cyclization of **43** suggests that, when there is a choice, 5-*exo*-trig is preferred to 6-*exo*-trig ring closure.[86] In contrast to peroxy radical cyclization, polar cyclization (Hg^{2+}) of **43** is less regiospecific, and both the 1,2-dioxolane and 1,2-dioxane were observed.

HOO
DBPO
O_2
HOO
27

HOO
DBPO
O_2
HOO
O—O
40

OOH
1) *tert*-BuO•, O_2
2)Ph_3P
OH
O—O O—O
41

COOMe
1) *tert*-BuO•, O_2
2)Ph_3P
O-O O-O
CO_2Me
OOH
OH
42

1) (*tert*-BuOOCO)$_2$
O_2
2) Ph_3P
OH
OOH
O—O
43

1) $Hg(NO_3)_2$
2) KBr
BrHg
O—O
+
HgBr
O
O

Hydrogen abstraction with excited acetophenone[68] and succinimide radical have also been reported. The succinimide radical abstraction is a propagation step in a chain reaction initiated by electron transfer from an alkenyl hydroperoxide such as **44** to *N*-iodosuccinimide, and results in the formation of iodo-substituted endoperoxides.[87]

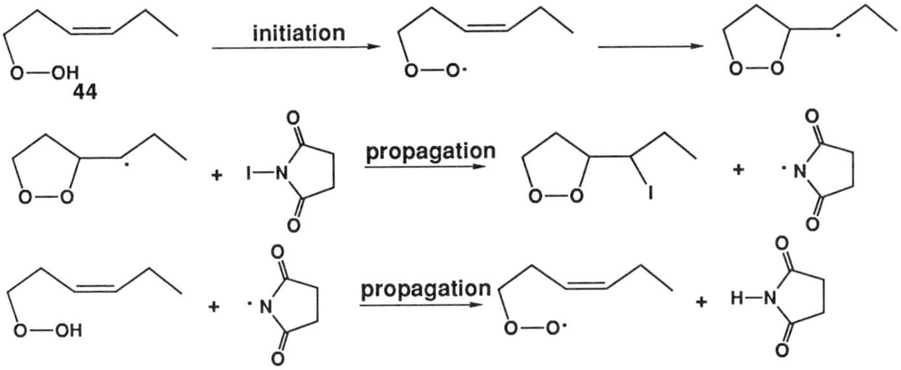

2 PHYSICAL PROPERTIES OF ENDOPEROXIDES

2.1 Gas- and Solution-phase Oxidation Potentials

Endoperoxides have not been examined with the same theoretical vigor as hydrogen peroxide.[88] In 1969, however, Kearns[89] assigned the endoperoxide HOMO as an antibonding combination of the oxygen lone pairs (n_O) and the LUMO as the antibonding O—O bond.[89] This assignment of the HOMO is consistent with theoretical calculations on H_2O_2[90] and with Raman studies[91] which place the O—O stretch at 880 cm^{-1}, and with photoelectron spectroscopic (PES) studies[92] which place the O—O stretch of H_2O_2 at 1080 cm^{-1}. A PES study of **1** verified these suggestions by assigning the first ionization potential ($IP_v = 8.42$ eV, $IP_a = 8.07$ eV) to the antisymmetric lone pair combination (n^-), the second ionization potential ($IP_v = 9.70$ eV, $IP_a = 9.35$ eV) to the C—C π-bond and the third ionization potential ($IP_v = 10.71$ eV, $IP_a = 10.26$ eV) to the antisymmetric combination of the C—O σ bonds (σ^-).[92] Table 2 lists the solution-phase oxidation potentials ($E^{0\prime}$) from cyclic voltammetric (CV) experiments for endoperoxides **1** and **45**–**50**.[93] The lifetimes of the radical cations decrease in the order **46** > **47** > **45** ≫ **49** > **48**, **50**. The loss of a proton from the α-carbon limits the lifetime of the radical cations.

Endoperoxide **46** has 'Bredt's-rule kinetic protection' and is stable under all CV conditions, whereas endoperoxides which have their bridgehead hydrogens in an eight-membered ring are unstable on the CV time scale at room temperature. Bridgehead-substituted endoperoxides are also stable under all CV conditions.[94] These solution-phase oxidation potentials are linearly related to the gas-phase PES oxidation potentials, indicative of a small relaxation energy

Table 2. Gas-phase (PES)[a] and solution-phase (CV)[b] oxidation potentials of endoperoxides

Compound	n^- (eV)	σ^- (eV)	$E^{0\prime}$ (V)	Compound	n^- (eV)	σ^- (eV)	$E^{0\prime}$ (V)[a]
1	8.42	10.71	2.2 (irrev.)[c]	**53**	9.26	10.47	
45	8.99	11.23	2.40	**54**	10.0	10.2	
46	8.82	10.84	2.27	**55**	9.35	9.76	
47	8.97	10.37	2.35	**56**	10.37	10.37	
48	9.05	10.03	2.48 (irrev.)	**49**			2.44 (irrev.)
51	8.50	10.36		**50**			2.57 (irrev.)
52	9.86	11.13					

[a]Ref. 95.
[b]vs. SCE, Ref. 93 except as noted.
[c]Ref. 94.

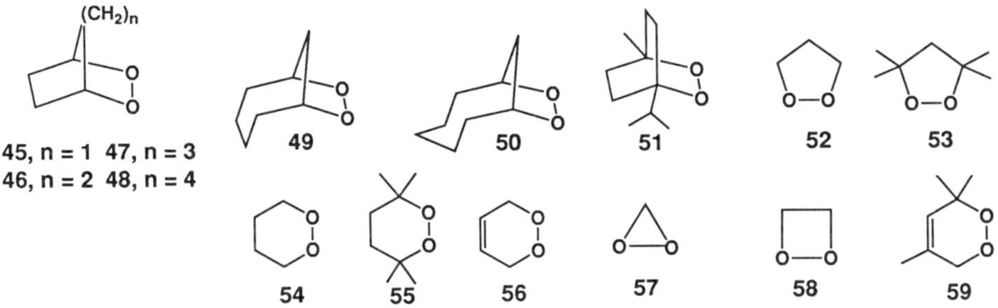

$(CH_2)_n$

45, n = 1 47, n = 3
46, n = 2 48, n = 4

49 50 51 52 53

54 55 56 57 58 59

difference between the vertical and adiabatic endoperoxide radical cations. This is consistent with the MO picture, which suggests that the electron is lost from the molecular orbital formed from the antisymmetric combination of pure p orbitals.

2.2 Conformational Analysis

It has been demonstrated that for compounds **1**, **45–48** and **51–56** the splitting between n^- (the antisymmetric lone pair combination) and σ- (the antisymmetric combinations of the C—O σ-bond) is a sensitive function of the C—O—O—C dihedral angle (Table 2).[95,96] The COOC bond angle in **45** is structurally constrained to 0° and a large n^--σ^- splitting of 2.24 eV is observed. The increased flexibility in **48** allows a larger COOC angle of 60°, which is accompanied by a decreased lone pair–lone pair interaction and a smaller n^--σ^- splitting of 0.98 eV.

Microwave data demonstrate that the lone pair–lone pair interaction in **56** is minimized in the gas phase by adopting a half-chair conformation.[97] The decreased lone pair–lone pair interaction is also accompanied by a shortening of the O—O bond length (1.516 Å in dioxirane **57**, 1.491 Å in dioxetane **58**, 1.483 Å in **52** and 1.463 Å in **56**).[88] A complete line-shape analysis of the variable-temperature ^{1}H NMR spectrum of **56** also indicates that it adopts a half-chair conformation in solution. The activation barriers for the half-chair–half-chair interconversion of **56** are $\Delta G\ddagger = 9.7$ kcal mol^{-1}; $\Delta H\ddagger = 8.6$ kcal mol^{-1} and $\Delta S\ddagger = -5.4$ e.u. The activation barriers for the half-chair–half-chair interconversion in 1,3,3-trimethyl-4,5-dioxyacyclohexene, **59**, are $\Delta G\ddagger = 11$ kcal mol^{-1}, $\Delta H\ddagger = 12.6$ kcal mol^{-1} and $\Delta S\ddagger = +7.4$ e.u.[98] The free energies for both of these interconversions are higher than that reported for cyclohexene ($\Delta G\ddagger = 5.3$ kcal mol^{-1}) and perhaps reflect a dramatic increase in the free energy content of the boat as a result of lone pair–lone pair interactions. The corresponding barrier in *cis*-hexadiene endoperoxide is $\Delta G\ddagger = 10.5$ kcal mol^{-1}.[24] A similar barrier is observed in alkoxy endoperoxides.[99]

2.3 Mass Spectrometry, ^{17}O NMR and UV Spectrometry

Marine organisms have proven to be a rich source of naturally occurring endoperoxides.[100–102] The steroidal endoperoxide 5α,8α-epidioxycholesta-6,9(11),24-triene-3β-ol, **60**,[103] the nor-sesterterpene endoperoxide muqbilin, **61**,[104] the norditerpene endoperoxide, **62**,[105] and endoperoxides apparently derived from fatty acids with ethyl and methyl side-chains, plakortin, **63**,[106–108] and chondrillin, **64**,[109] in addition to artemisinin and related compounds mentioned above, are a few representatives of such endoperoxides that have been isolated and identified.

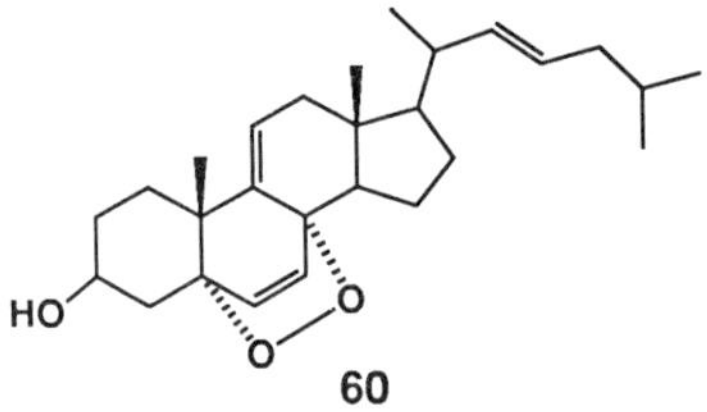

HO O O 60

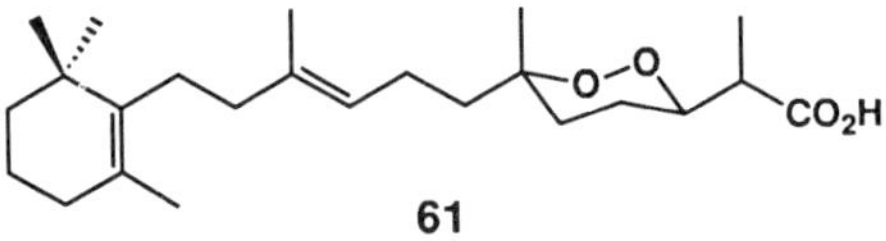

O O CO_2H 61

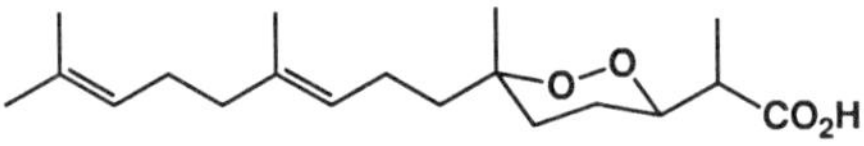

O O CO_2H 62

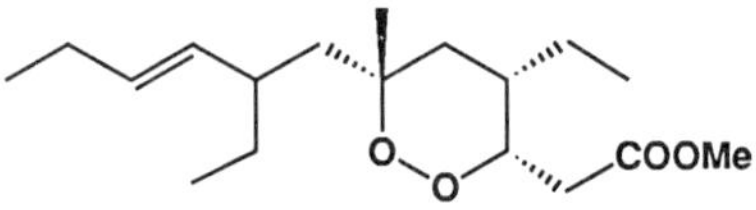

O O COOMe 63

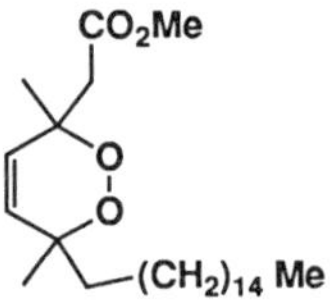

CO_2Me O O $(CH_2)_{14}$ Me 64

Structural identification of these endoperoxides has involved examination of their spectral properties, independent synthesis, chemical modification and X-ray analysis. Electron impact (EI) ionization mass spectrometry has not been extensively utilized, undoubtedly as a result of the perceived thermal instability of endoperoxides.[110] The few studies which have been conducted, however, have demonstrated that useful information is available using both electron impact[111] and soft ionization techniques.[112,113] Steroidal endoperoxides such as **60** exhibit small (1–10%) molecular ions and base peaks derived from retro (4 + 2) loss of oxygen.[103] The EI mass spectrum of 2,3-dioxabicyclo[2.2.2]oct-5-ene, **65**, also exhibits a small molecular ion (12%), a reasonably intense peak for $[M - HCO]^+$ and a base peak for loss of oxygen (Figure 3). Methane and isobutane chemical ionization (CI) of **65** exhibit low-intensity $[M + H]^+$ ions as a result of the rapid rate of dissociative proton transfer. The $[M + NH_4]^+$ ion, however, is the base peak in the ammonia CI mass spectrum (Figure 3) because it is more stable and/or the rate of addition competes with dissociative proton transfer.[114]

^{17}O NMR chemical shifts[115] for several bicyclic and acyclic peroxides are found between 232 and 318 ppm[116] (Table 3), considerably downfield from ether,[117] alcohol[118] or ester oxygen,[117] and diagnostic of the peroxide linkage. The ^{17}O chemical shifts of both the bicyclic and acyclic peroxides are linearly related to the ^{13}C chemical shifts of the corresponding hydrocarbons but with different slopes. The very different linear correlations for these two peroxide series is indicative of at least two factors determining the chemical shifts. The slope (12.98) of the δ_O versus δ_C correlation for the bicyclic endoperoxides shows that the factor which dominates the chemical shifts in these compounds is 13 times more important than in the hydrocarbon analogues. This increased effect is attributed to lone pair–lone pair interactions. The acyclic peroxides, which have very similar geometries, experience downfield shifts as the alkyl groups are varied from primary to tertiary, as expected for a predominant diamagnetic contribution to the screening constant.

66 67 68 69

ROOR
70 R = tBu
71 R = iPr
72 R = Et
73 R = nPr

74

Endoperoxides exhibit weak absorption bands ($\varepsilon = 10$–$100\,l\,mol^{-1}\,cm^{-1}$) which increase in intensity from approximately 340 to 200 nm with no apparent

Table 3. ^{17}O and ^{13}C chemical shifts (ppm) of peroxides and hydrocarbon analogues[a]

Compound	$\delta(^{17}O)$	$\delta(^{13}C)$	Compound	$\delta(^{17}O)$	$\delta(^{13}C)$
66	318 (per) 169 (ether)		**70**	260	38.5
67	310	29.4	**47**	259	25.9
45	303	30.1	**71**	255	37.1
1	283	29.5 34.8	**54**	254	27.3
52	280	26.5	**72**	253	31.8
68	277	26.4	**46**	250	26.0
69	275		**73**	249	29.5
65	265	25.8	**74**	247	
51	263	28.4 33.6	**48**	232	24.5

[a]Ref. 116.

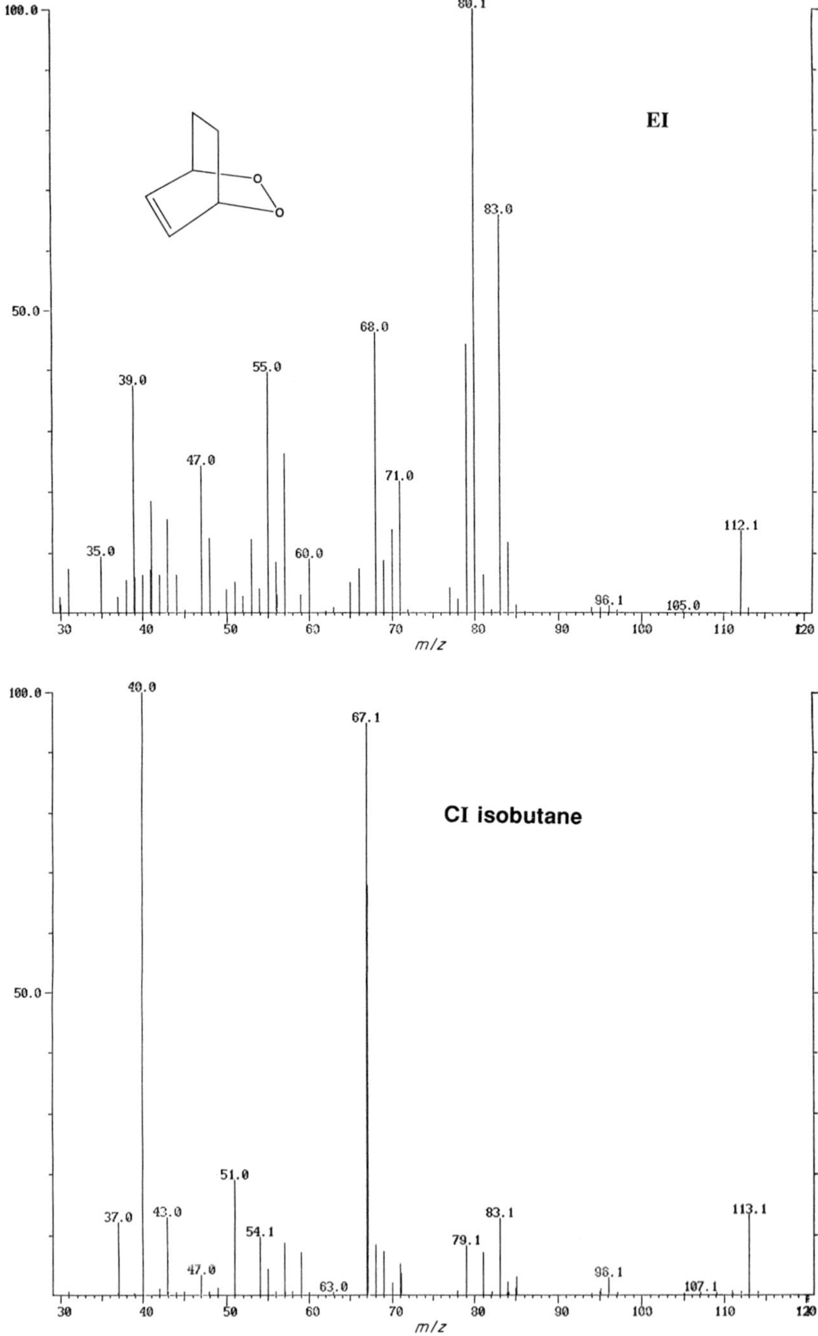

Figure 3. EI and CI mass spectra of cyclohexadiene endoperoxide

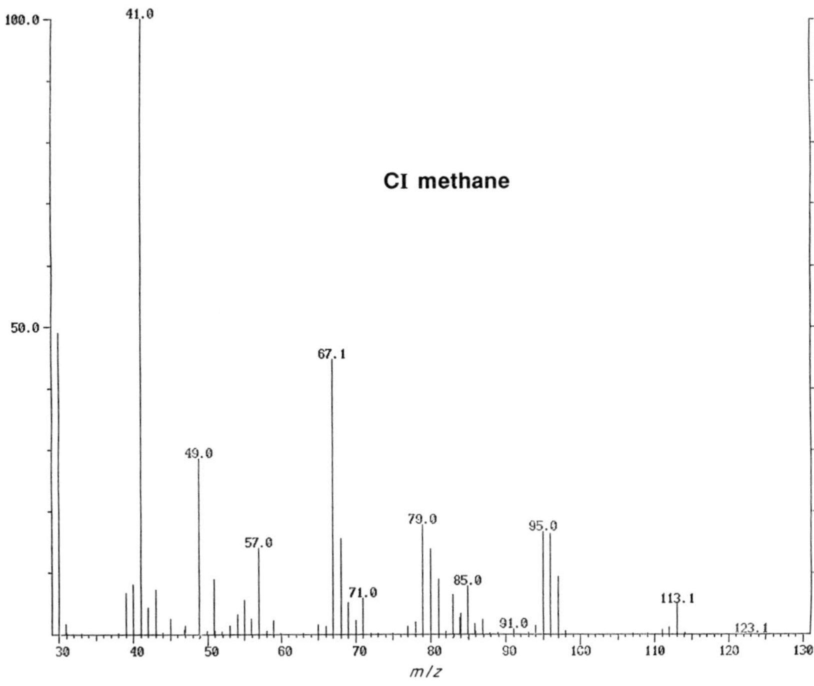
CI methane
100.0
50.0
41.0
49.0
57.0
67.1
71.0
79.0
85.0
91.0
95.0
113.1
123.1
m/z

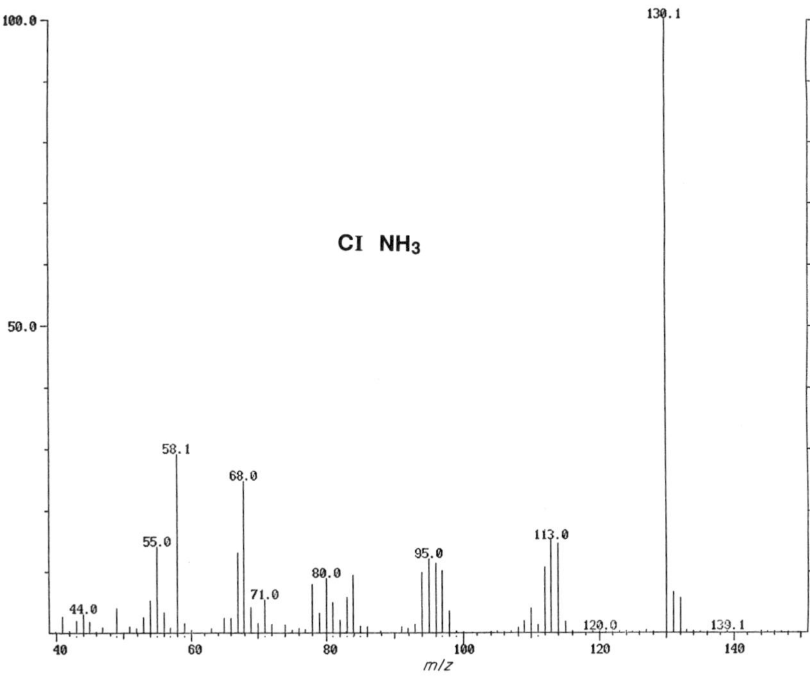
CI NH3
100.0
50.0
44.0
55.0
58.1
68.0
71.0
80.0
95.0
113.0
120.0
130.1
139.1
m/z

maximum.[119] Ascaridole, **1**, is unique in that its UV trace is more intense than those of many other peroxides and is characterized by a low-intensity maximum at 233 nm ($\varepsilon \approx 166$ l mol^{-1} cm^{-1}) and a long tail which extends beyond 360 nm.[120] It has been suggested that the maximum is a result of the interaction of the oxygen lone pairs with the alkenic antibonding orbital.

3 THERMOLYSIS OF ENDOPEROXIDES

The bond-dissociation enthalpies of dialkyl peroxides are 37 ± 1 kcal mol^{-1},[121] substantially smaller than for C—C bonds (e.g. 88 kcal mol^{-1} for ethane and 85 kcal mol^{-1} for propane),[122] and are independent of the identities of the alkyl groups. As a result of this low bond dissociation energy, endoperoxides are susceptible to thermal decomposition. Ascaridole, **1**, for example, decomposes with an activation free energy ($\Delta G\ddagger$) of 31.4 ± 0.2 kcal mol^{-1}, an activation enthalpy ($\Delta H\ddagger$) of 30.6 ± 0.1 kcal mol^{-1} and an activation entropy ($\Delta S\ddagger$) of 2.6 e.u. at 111°C.[123]

3.1 Unsaturated Endoperoxides

In 1944, Schenck and Ziegler[2] demonstrated that bisepoxides and epoxy ketones are general thermolysis products of unsaturated [2.2.2] endoperoxides.[124–129] Thermolysis of unsaturated [3.2.2] endoperoxides[130] such as **68**[131] also produces epoxy ketones and bisepoxides. The thermal stability of these unsaturated [3.2.2] endoperoxides is a subtle function of the conformation of the 3-atom bridge.[132] After 6 h of refluxing in toluene, 71% of **68**, 100% of **75** and only 11% of **76** were converted into product. The [4.2.2] endoperoxide **74**,[131] is significantly more stable than either its [2.2.2] or [3.2.2] homologue. Under these severe thermolysis conditions, only a bisepoxide and unidentified degradation products are isolated. Adam[133] suggested that the epoxy ketone is a secondary product formed by rearrangement of the bisepoxide. Carless *et al.*[131] suggested, however, that the 1,3-diradical formed by closure of the first epoxide can either collapse to the bisepoxide or rearrange by a hydrogen shift to an enol, which then tautomerizes to the ketone.

68 → 90% + 10%

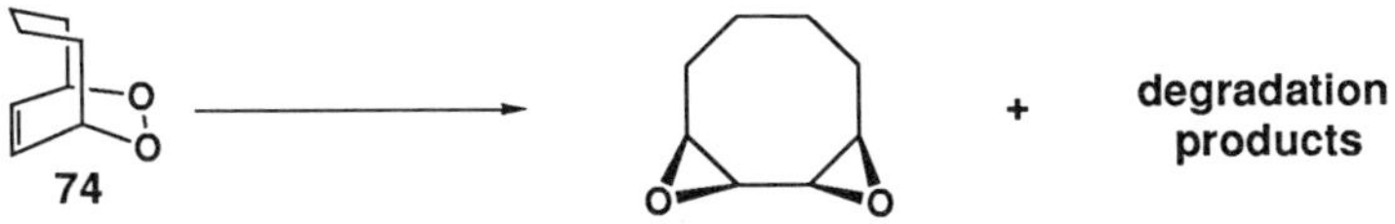

74 → + degradation products

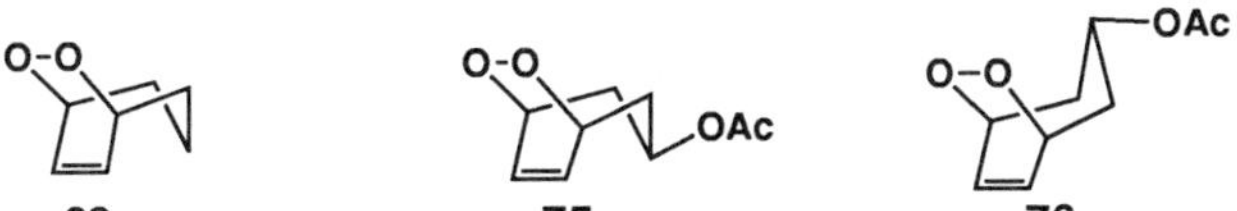

68 75 76

Thermolysis of 2,3-dioxabicyclo[2.2.1]hept-5-ene, **77**, occurs by a novel C—C bond cleavage not observed in its larger bicyclic ring homologues.[134,135] Substituents at the bridgehead in **78**,[95] and at the 7-position in **79**,[136] but not on the double bond in **80** and **81**,[137–139] influence the extent of C—C cleavage. The dimethylsilyl endoperoxide **82** has been suggested to decompose via a similar pathway to give a novel bicyclic product.[140,141]

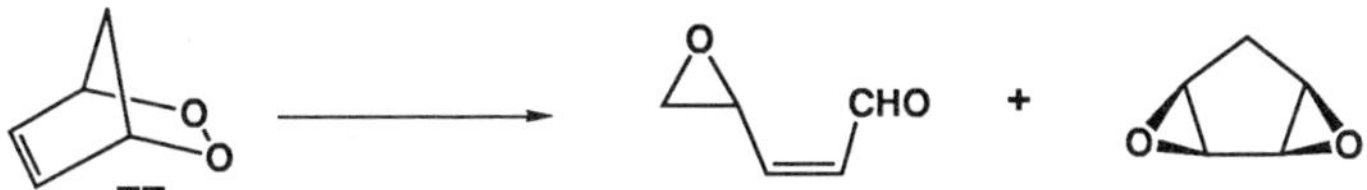

77 → CHO +

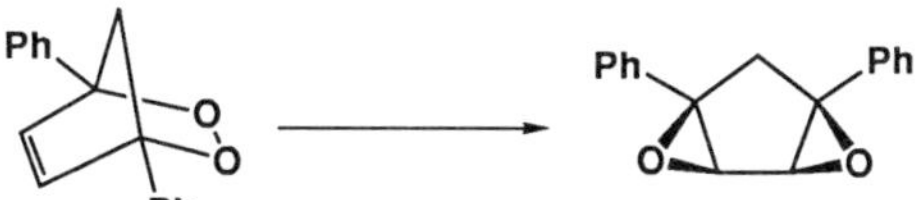

78 →

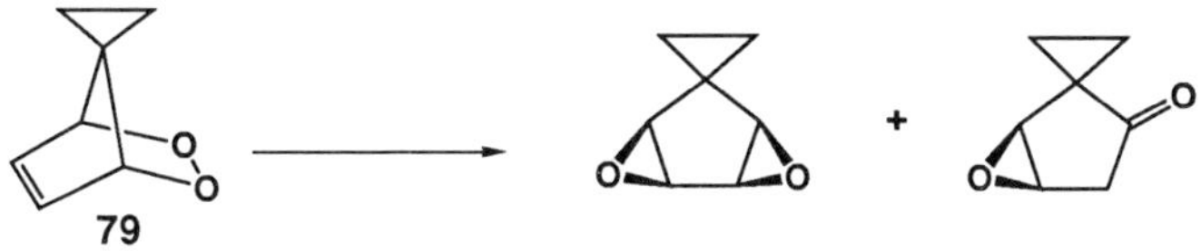

79 → +

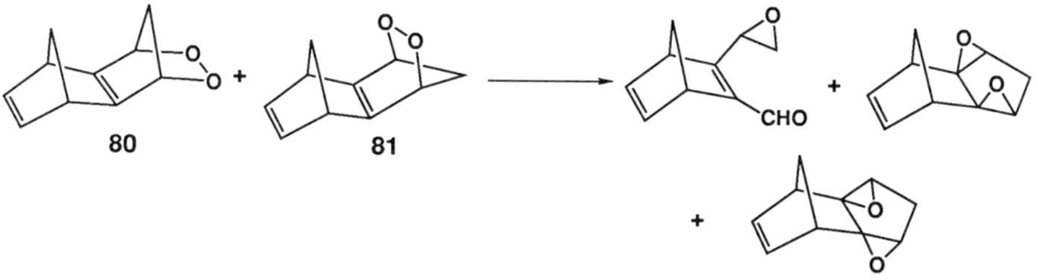

80 + 81 → CHO + +

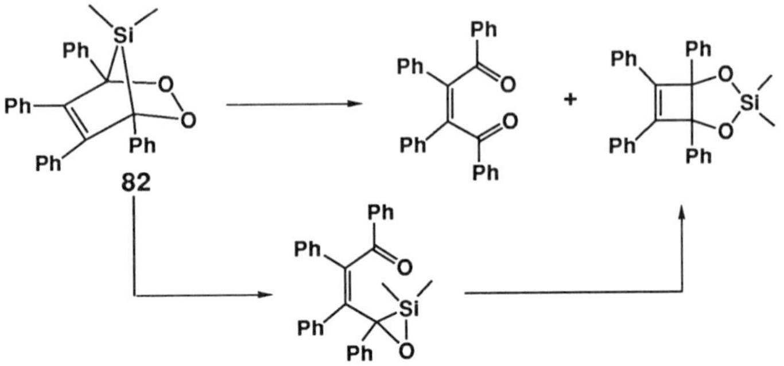

Fulvene endoperoxides decompose thermally by loss of oxygen or by homolytic cleavage of the O—O bond (Figure 4).[142] The diradical generated by homolytic cleavage can subsequently decompose by β-cleavage[143] or by collapse to the bisepoxide. The factors which influence the partitioning between the loss of oxygen and homolytic cleavage and the exact decomposition pathway for the diradical are not completely understood. It is known, however, that the reaction conditions[144] and the substitution pattern on the endoperoxide exert dramatic effects on product composition. Diphenylfulvene, **83**,[142] and pentaphenylfulvene endoperoxide, **84**,[145] decompose in aprotic solvents to regenerate the fulvene and to give the bisepoxide. In isopropanol, **84** also decompose to give 30% of a diketone, presumably by rearrangement of the endoperoxide to the dioxetane followed by decomposition.[145] In contrast, dimethylfulvene endoperoxide, **85**,[146–149] decomposes by homolytic scission of the O—O bond, followed by β-cleavage to give a variety of products instead of bisepoxide formation. The diethoxyfulvene endoperoxide **86**[142] decomposes to give the fulvene as the only identifiable product in 30–50% yield. On the other

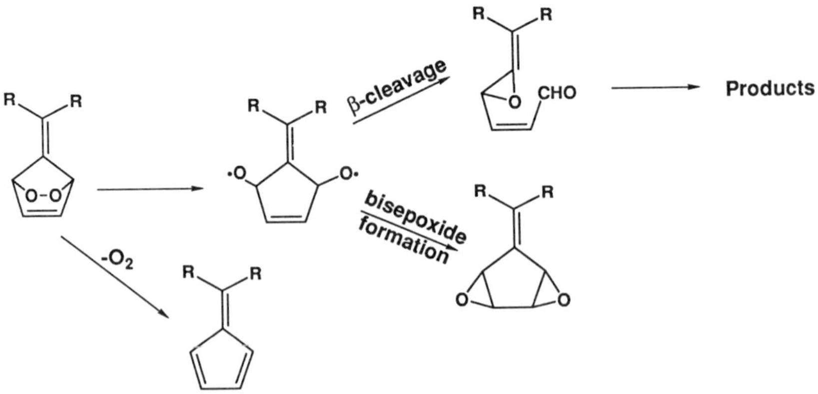

Figure 4. Reactions of fulvene endoperoxides

hand, the ethylenedithiofulvene endoperoxide **87**[142] decomposes via β-cleavage of a diradical intermediate.

Ph Ph Ph Ph Ph Ph
O-O
83
+
O O

H Ph H Ph H Ph
Ph Ph Ph Ph + Ph Ph
xylene
100°C
O-O
O O
Ph 84 Ph Ph Ph Ph Ph
Ph
Ph
O
iPrOH
O
Ph
Ph
H Ph

CH_3 CH_3
CH_3 CH_3
O
O-O
O + O OH + O OH
85

EtO OEt EtO OEt
O-O
86

S S
S S
O S
S
+
O
CHO
O-O
CHO
87

3.2 Saturated Endoperoxides

Adam and Sanabia[150] reported in 1977 that 3,3,6,6-tetramethyl-1,2-dioxane, **88a**, thermally decomposes to acetone and ethylene. The activation parameters for the thermolysis of **88a** (Table 4) are very similar to those of di-*tert*-butyl peroxide, **70**,[151] suggesting an O—O bond homolysis mechanism. The exceedingly small (<1% per D) four-atom deuterium isotope effect [$k_H/k_D = k(\mathbf{88a})/k(\mathbf{88b})$] is 1.03 ± 0.03 and is consistent with a simple O—O bond homolysis without any change of the hybridization at C-4 and C-5 in the rate-determining step. Thermolysis of the *meso*-d_2 compound **88c** gave more than 60% of the *trans*- and the thermolysis of the *dl*-d_2 compound **88d** more than 60% of the *cis*-dideuterioethylene. The mechanism consistent with these observations is depicted in Figure 5. The stereochemistry of the reaction demands that the double β-scission to give ethylene occurs primarily through the more stable *trans*-periplanar conformation of the 1,6-diradical. The reversible O—O bond homolysis ($k_{-OO} > 0$) is consistent with the observation that the cyclic peroxalate **89** thermally decomposes to give **88a** as a minor product. In contrast,

Table 4. Activation barriers for thermolysis of endoperoxides

Compound	$\Delta H‡$ (kcal mol^{-1})	$\Delta S‡$ (e.u.)	$\Delta G‡$ (kcal mol^{-1})
45[a]	20.7 ± 1.8	-19 ± 5	30.2
46[a]	33.1 ± 1.2	3 ± 3	31.4
53[b]	27.0 ± 1.0	-24.0	39.4
70[c]	40.8	21.1 ± 1.4	30.3
88a[d]	32 ± 1	13 ± 2	38

[a]Ref. 153.
[b]Ref. 154.
[c]Ref. 151.
[d]Ref. 150.

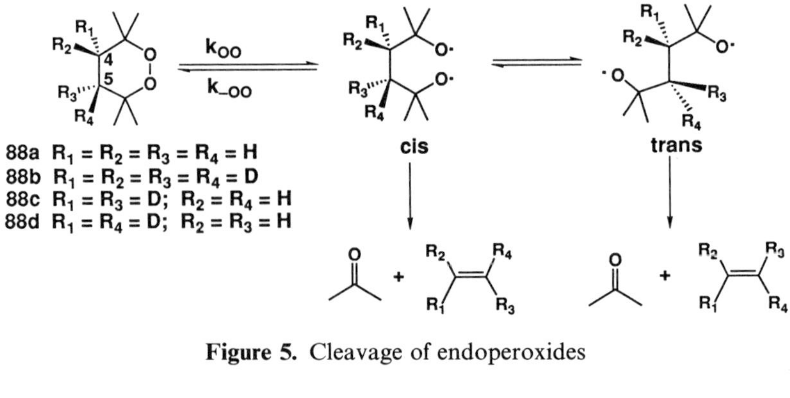
R1
R2
R3
R4
4
5
O
O
k_OO
k_-OO
O·
·O
R3
cis
trans
88a R1 = R2 = R3 = R4 = H
88b R1 = R2 = R3 = R4 = D
88c R1 = R3 = D; R2 = R4 = H
88d R1 = R4 = D; R2 = R3 = H
+

Figure 5. Cleavage of endoperoxides

gas-phase pyrolysis of a 4:1 mixture of *trans*- and *cis*-3,6-dimethyl-1,2-dioxacyclohexane, **90**, resulted in dehydrogenation and a 1,5-hydrogen shift rather than loss of ethylene.[152]

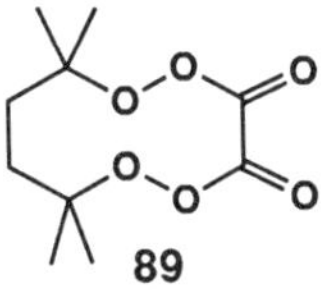

89

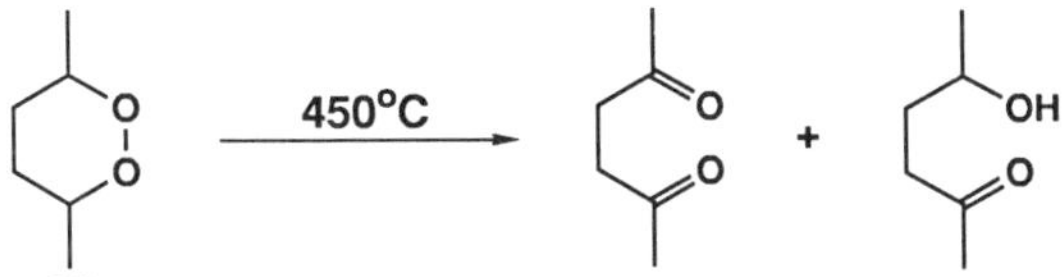

450°C

90

Thermolysis of the prostaglandin endoperoxide analogue 2,3-dioxabicyclo-[2.2.1]heptane, **45**, resulted in the formation of two primary products, **91** and **92**.[155] Bicyclic ether **93** is most likely a rearrangement product derived from the epoxyaldehyde **91** and the hydroxycyclopentanone **94** is only formed in D_2O. The partial rate constant for the formation of **91** is insensitive and the partial rate constant for the formation of **92** is very sensitive to solvent polarity. The activation enthalpy for thermolysis of **45** is much smaller than those reported for other peroxides (Table 4) as a result of the effect of strain on the strength of the O—O bond.[153] Endoperoxide **45** is reasonably stable, however, because of a very negative activation entropy.

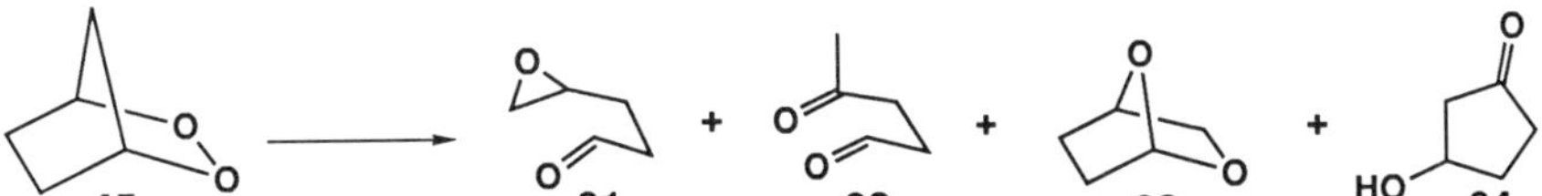

45 **91** **92** **93** **94**

These activation barriers are very similar to those reported for thermolysis of **53**[154] (which has been argued to undergo concerted decomposition) but very different from the activation barriers for **70** and **88a** which decompose by simple O—O bond homolysis. A mechanism consistent with these observations is presented in Figure 6. A concerted pathway with diradical character is responsible for the formation of **91** and a concerted pathway with ionic character for the formation of **92**. The one-carbon bridge is better aligned than the two-carbon bridge with the O—O bond and, as a consequence, concerted cleavage of the smaller bridge is preferred. Consistent with this analysis is the observation that 1,5-dimethyl-6,7-dioxabicyclo[3.2.1]octane, **95**, also prefers one-carbon to three-carbon bridge cleavage.[156,157]

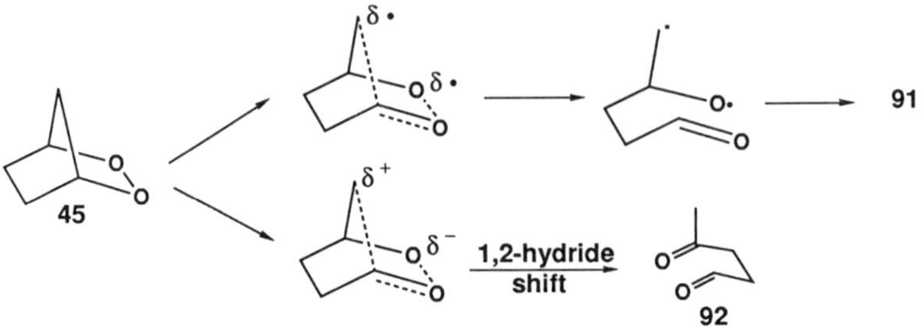

Figure 6. Cleavage of bicyclic endoperoxides

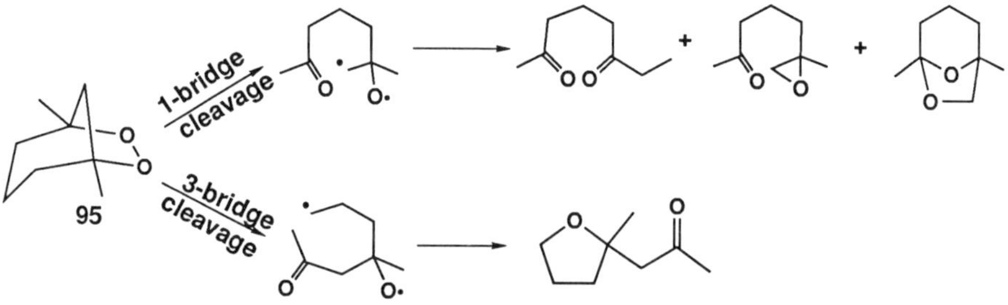

Bloodworth and co-workers[152,158] examined the flash vacuum pyrolysis (FVP) of a series of [*n*.2.1] and [*n*.2.2] bicyclic endoperoxides and observed different reactivity patterns. The [*n*.2.2] bicyclic endoperoxides decompose by loss of hydrogen or ethylene. The lack of products derived from cleavage of the three- and four-carbon bridges is consistent with a concerted double β-scission in the cycloalkanedioxy radical intermediate. The [*n*.2.1] bicyclic endoperoxides, on the other hand, decompose by cleavage of the one-carbon bridge. The absence of ethylene during the decomposition of the [2.2.1] bicyclic endoperoxide **45** can be best explained by a concerted fragmentation, completely bypassing the formation of the cycloalkanedioxy radical.

[n.2.2]

FVP 450°C

46 n = 2	42%	56%
47 n = 3	85%	10%
48 n = 4	61%	6%

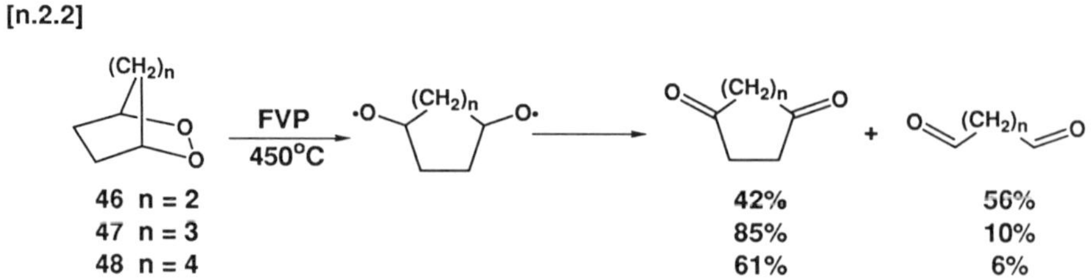

[n.2.1]

45 n = 2
96 n = 3
49 n = 4
50 n = 5

> 90% n = 3,4, and 5

> 90% n = 2

3.3 Aromatic Endoperoxides

Aromatic endoperoxides derived from substituted benzenes,[159,160] substituted and unsubstituted naphthalenes,[161,162] anthracenes,[163,164] naphthacenes[165,166] and polycyclic aromatic hydrocarbons[167] are a few examples of those which have been reported. These endoperoxides decompose thermally by either loss of oxygen and regeneration of the parent aromatic hydrocarbon or by homolytic cleavage of the O—O bond.

The oxygen liberated during decomposition has been shown by trapping[168,169] and by direct spectroscopic detection[170] to be, at least in part, in the singlet ($^1\Delta_g$) state. The use of methyl-substituted polyvinylnaphthalene polymers as reversible singlet oxygen carriers takes advantage of this phenomenon and provides a convenient source of singlet oxygen which can be easily removed from the reaction medium.[171] The yield of singlet oxygen can vary from approximately 30% to nearly 100% of the oxygen liberated during decomposition of the endoperoxide (Table 5).[172,173] The yield of singlet oxygen is high for those endoperoxides which exhibit near-zero or slightly negative entropies of activation but very low for those endoperoxides which exhibit positive entropies of activation (Table 5). These observations suggests that two mechanisms for oxygen production are operational: a concerted mechanism characterized by near-zero or negative entropies of activation and a mechanism characterized by positive activation entropies involving stepwise cleavage of the C—O bonds. The observation of a dramatic effect of external magnetic fields on the yield of singlet oxygen from decomposition of those endoperoxides which have positive but not near-zero or negative entropies of activation is consistent with the suggestion of a diradical intermediate in the stepwise reactions, where efficient spin–orbit coupling can occur.[174]

Homolytic cleavage of these peroxides ultimately leads to a bisepoxide whose existence has been demonstrated in several cases by trapping with maleic anhydride.[166] The products derived from the bisepoxide in the absence of an external trapping agent depend on the identity of the aromatic nucleus and the reaction conditions. A typical pathway is depicted in Fig. 7.

Table 5. Activation barriers and singlet oxygen yields for thermolysis of endoperoxides

Compound	$\Delta H\ddagger$ (kcal mol^{-1})	$\Delta S\ddagger$ (e.u.)	$^1O_2$ yield (%)
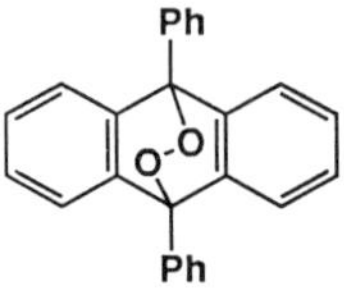 97	32.5 ± 0.2	9.6 ± 0.5	32 ± 1
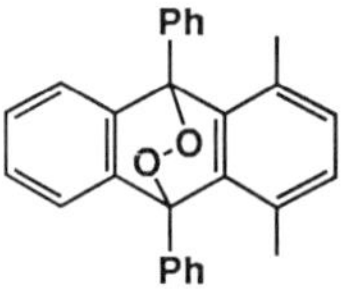 98	31.8 ± 0.3	7.4 ± 0.8	52 ± 4
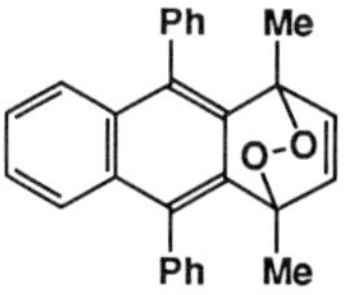 99	29.8 ± 0.3	-1.8 ± 0.8	92 ± 1
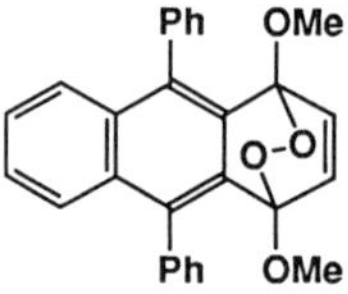 100	24.2 ± 0.2	-0.3 ± 0.7	95 ± 5
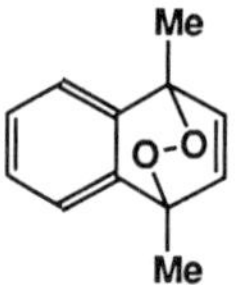 101	24.2 ± 0.2	2 ± 1	76 ± 1
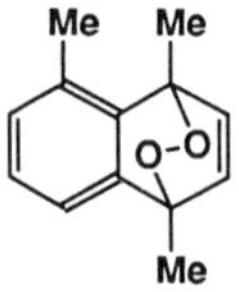 102	26.0 ± 0.2	2.6 ± 0.5	69 ± 1

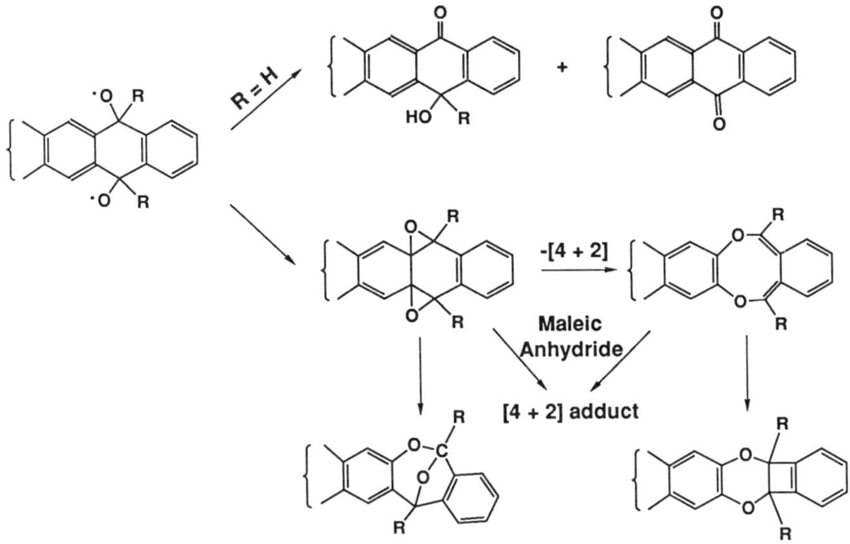

Figure 7. Thermolysis of aromatic endoperoxides

4 PHOTOLYSIS OF ENDOPEROXIDES

4.1 Photophysical Studies

In 1969, Kearns predicted that the first excited singlet and triplet states of endoperoxides with electronic configurations $[\cdots \pi_{OO}{}^{*1}\sigma_{OO}{}^{*1}]$ (Figure 8) would decompose by cleavage of the O—O bond.[89,175] The population of the doubly excited state $[\cdots \pi_{OO}{}^{*1}\sigma_{CO}{}^{*1}]$, on the other hand, was predicted to strengthen the O—O bond and decompose by dissociation into the diene and $^{1}\Sigma_{g}{}^{+}$ oxygen. The Kearns theory predicts an exception to the general axiom which states that organic photoreactions in solution do not occur efficiently from upper electronically excited states owing to the expected rapid decay to the lowest excited state of the same symmetry (Kasha's rule).[176] Despite the stringent constraints placed on reactions occurring from upper excited states, the Kearns theory has been verified by the observation of wavelength-dependent photochemistry for both dialkyl peroxides[177] and endoperoxides.[178]

Rigaudy *et al.*[178] reported the first observation of wavelength-dependent product distribution during a study of the photochemistry of 9,10-diphenylanthracene endoperoxide, **97**. Irradiation at 254 nm resulted primarily in cycloreversion to give 9,10-diphenylanthracene, while irradiation at longer wavelengths resulted in O—O cleavage and rearrangement as the major process. Competition between cycloreversion and O—O cleavage has sub-

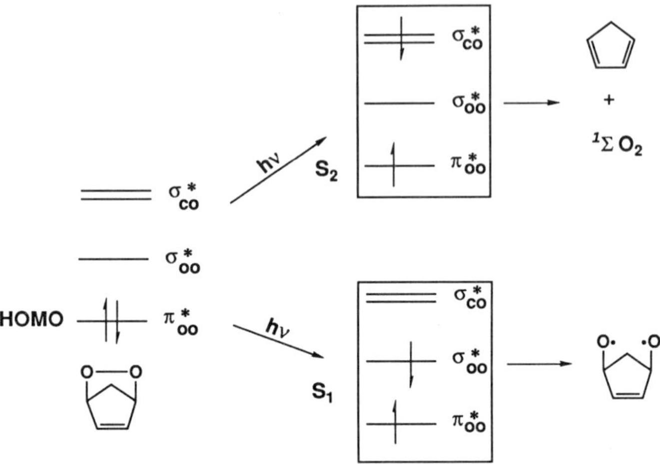

Figure 8. Peroxide orbitals and photolysis

sequently been demonstrated to be a general phenomenon for naphthalene,[179,180] anthracene,[181–184] tetracene[185,186] and polycyclic aromatic[186–189] endoperoxides.

The ratios of quantum yields of cycloreversion to rearrangement via O—O cleavage (Φ_{CR}/Φ_R) for several endoperoxides are listed in Table 6.[167] The predominant reaction for the unsubstituted endoperoxides of anthracene, **103**, and tetracene, **106**, is rearrangement. Addition of substituents to the bridgehead carbons in **103** and **106** to give **104** and **107** produces a dramatic decrease in the rearrangement quantum yields. The buttressing effect of these bridgehead substituents with additional substituents as in **105** and **108** produces an even further decrease in rearrangement quantum yields. Schmidt and Brauer[167] have argued that it is unlikely that these variations are caused by substituent effects on the O—O cleavage, and that the observed decrease in rearrangement reflects the destabilization of the primary photoproduct, the bisepoxide (Figure 9). If this suggestion is correct, and if a 1,6-dioxybiradical precedes bisepoxide formation, the substituent-induced competition between rearrangement and cycloreversion would require reversible formation of the O—O bond (Figure 9). It seems likely that the main substituent effect is on k_{-1}, reclosure of the

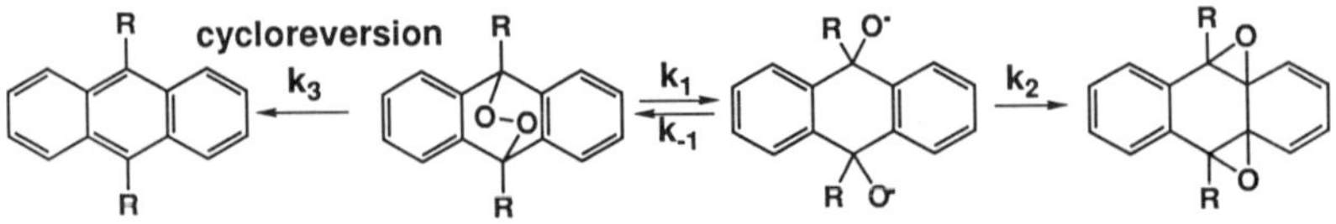

Figure 9. Competing reactions in aromatic endoperoxide photolysis

Table 6. Quantum yields for cycloreversion (Φ_{CR}) and rearrrangement (Φ_R) of endoperoxides[a]

Compound[b]	Φ_{CR}/Φ_R	λ (nm)[c]	Compound[b]	Φ_{CR}/Φ_R	λ (nm)[c]
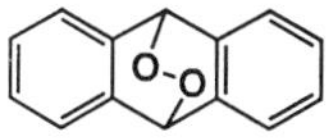 103	0.22/0.96	270	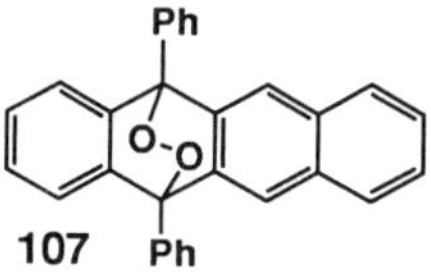 107	0.073/0.26	248
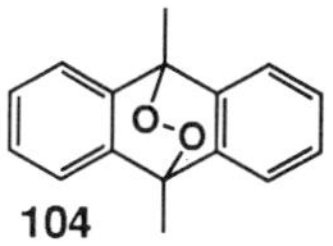 104	0.35/0.34	270	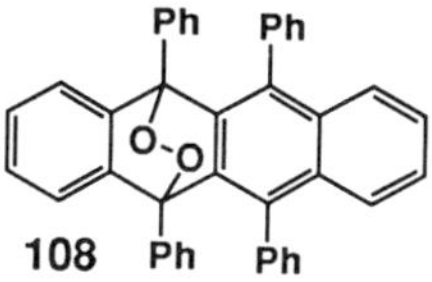 108	0.013/0.20	313
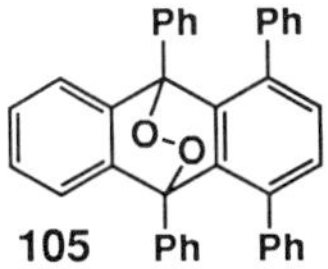 105	0.091/0.13	270	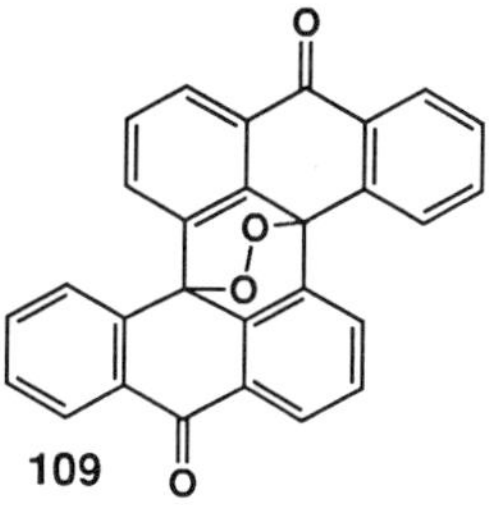 109[d]	0.26/0.0043	313
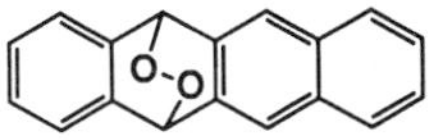 106	0.055/0.95	248			

[a]$\Phi_R(\lambda) = [1 - \Phi_{CR}(\lambda)]\Phi_R$; see ref. 167 for details.
[b]Ref. 167 except where indicated
[c]Wavelength.
[d]Ref. 190.

diradical, which should be promoted by the buttressing effect of bulky substituents.

The very high value of the cycloreversion to rearrangement quantum yield for **109** ($\Phi_{CR}/\Phi_R \approx 60$) has prompted the suggestion of its use as a chemical actinometer in the 248–334 nm range.[190] This high quantum yield ratio also provided Schmidt and co-workers[191,192] with the opportunity to study the cycloreversion process quantitatively.

The quantum yield for cycloreversion of **109**, Φ_{CR}, at wavelengths less than 334 nm is 0.26 but decreases sharply at longer wavelengths to a value of 7×10^{-4} at 387.5 nm.[190] The increase in Φ_{CR} with decreasing wavelength eliminates the possibility that cycloreversion is occurring from the lowest excited state (S_1), which should be populated by irradiation at the longer wavelengths. The decrease in Φ_{CR} when S_1 is irradiated also eliminates the possibility that cycloreversion occurs from a vibrationally hot ground state, since the vibrational levels of S_0 are expected to be populated more rapidly from S_1 than from upper excited states (Figure 8). The lowest triplet state, T_1, is also not involved, since triplet sensitization was ineffective at promoting cycloreversion. Identical values of Φ_{CR} at 290 and 313 nm strongly implicate thermally equilibrated S_2 as the photoreactive state for cycloreversion. The small cycloreversion quantum yield of 0.26 is a result of competing internal conversion to S_1 or S_0. The alternative possibility of competing intersystem crossing to T_1 is unlikely since no solvent or internal heavy-atom effect on the cycloreversion quantum yield was observed.

Schmidt *et al.*[191] also established the adiabaticity of the cycloreversion by successfully trapping singlet oxygen with 1,3-diphenylisobenzofuran and tetramethylethylene. A minor adiabatic channel ($<1\%$) which results in the formation of the excited hydrocarbon was also detected during a picosecond laser study of 1,4-dimethyl-9,10-diphenylanthracene endoperoxide, **98**.[193]

Cycloreversion in 1,4-dimethyl-9,10-diphenylanthracene endoperoxide, **110**, can occur from higher excited states than S_3.[186,194] Endoperoxide **110** does not exhibit a limiting plateau for the fractional anthracene yield as a function of wavelength as expected and observed for **98**, which undergoes cycloreversion exclusively from S_2 (Figure 10).[195,196]

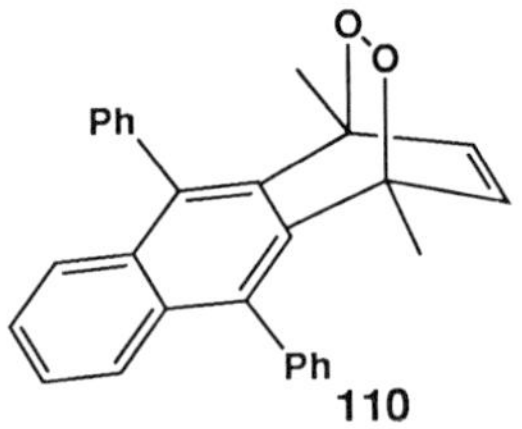

110

The time constants for regeneration of the aromatic hydrocarbon as revealed by picosecond[193,195,197–199] and subpicosecond[200] studies are of the order of tens of picoseconds. In contrast, the absence of any measurable quantum yield for luminescence from the reactive excited state is indicative of its very short lifetime. These results demand that decomposition must be a stepwise process with a rate-determining decomposition of a biradical intermediate (Figure 11).

4.2 Product Studies

Irradiation of the [*n*.2.1] saturated dioxabicycloalkanes **45**, **49**, **50** and **96** in benzene through a Pyrex filter for 20–40 h resulted predominantly in C_1 bridge cleavage to give epoxyaldehydes, **111**, as the major products.[201] Direct

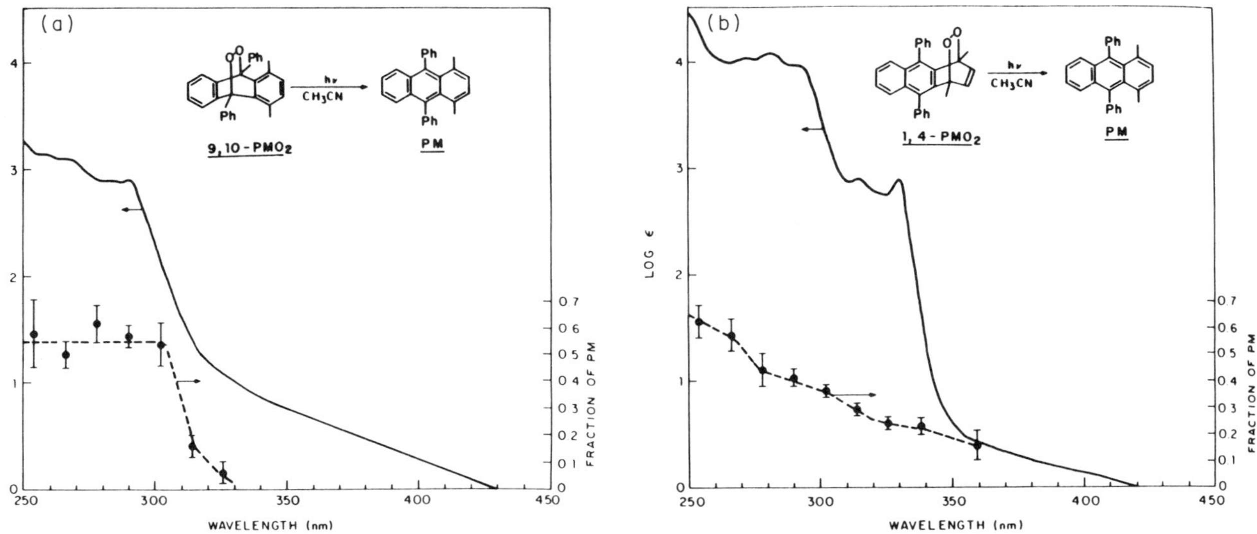

Figure 10. Wavelength dependence of endoperoxide photodissociation. (a) Wavelength dependence of the photodissociation of 9,10-PMO_2. Comparison of the electronic absorption spectrum of 9,10-PMO_2 with the fractional yield of PM produced on photolysis of 9,10-PMO_2 as a function of excitation wavelength, using acetonitrile as solvent at ambient temperature. (b) Wavelength dependence of the photodissociation of 1,4-PMO_2. Comparison of the electronic absorption spectrum of 1,4-PMO_2 with the fractional yield of PM produced on photolysis of 1,4-PMO_2 as a function of excitation wavelength, using acetonitrile as solvent at ambient temperature. Reproduced with permission from K. B. Eisenthal, N. J. Turro, C. G. Dupuy, D. A. Hrovat, J. Langan, T. . Jenny and E. V. Sitzmann, *J. Phys. Chem.*, **90**, 5168 (1986). Copyright 1986, American Chemical Society

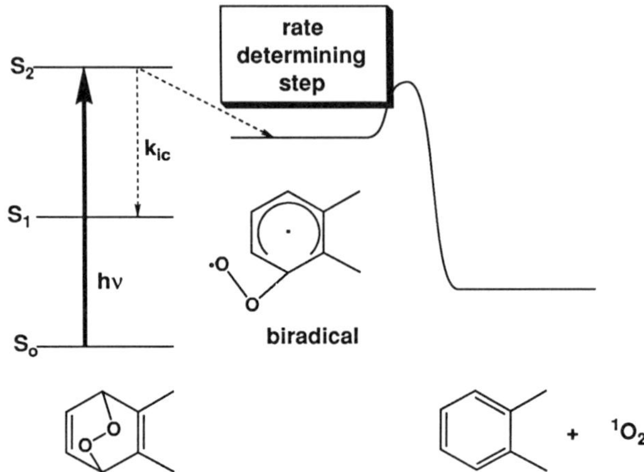

Figure 11. Photodissociation of endoperoxides vs excited state

irradiation of the bridgehead-substituted endoperoxide **95** also resulted in C_1 bridge cleavage and quantitative formation of an epoxy ketone, **113**.[156] Benzophenone-sensitized photolysis of endoperoxide **95**[156] and its unsubstituted analogue, **96**, however, resulted in the formation of the 6,8-dioxabicyclo[3.2.1]octane skeleton, **114**. This abrupt change in product distribution is a result of the population of the longer-lived triplet diradical **115** which, unlike the singlet, has enough time to rotate and add to the aldehyde or ketone group (Figure 12). Ketoaldehydes, **112**, which are formed by C_1 bridge cleavage in greater than 90% yield during gas-phase thermolysis,[152,158] were formed only as minor products in the photolysis of endoperoxides **49**, **50** and **96**. The formation of approximately 10% of 3-hydroxycycloalkanones, **116**, during photolysis of these endoperoxides was attributed to induced decomposition.

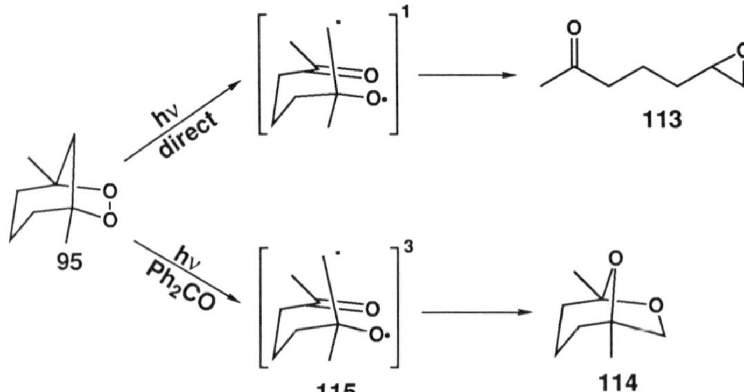

Figure 12. Direct vs sensitized photolysis of endoperoxides

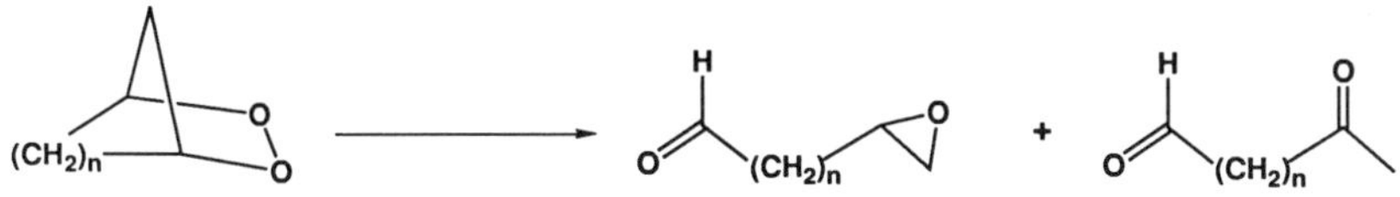

(CH2)n
45 n = 2 49 n = 4
96 n = 3 50 n = 5

111

112

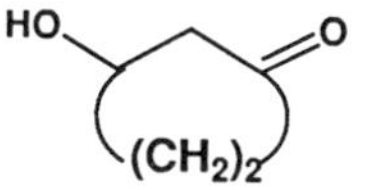

116

Irradiation of the saturated [*n*.2.2] dioxabicycloalkanes **46**–**48** did not produce bridge-cleavage products, but instead resulted in dehydrogenation to form cycloalkane-1,4-diones, **117**. Cyclopentane-1,3-dione was also formed as a dehydrogenation byproduct in 25% yield during the photolysis of the bicyclo[2.2.1]endoperoxide **45**. These dehydrogenations undoubtedly occur via mechanisms similar to that proposed for the pyrolysis reactions.[152,158]

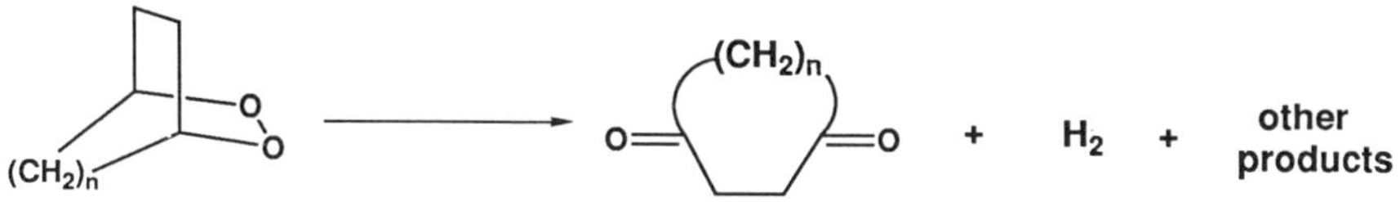

46: n = 2 48: n = 4
47: n = 3

117 + H_2 + other products

In 1970, Maheshwari, De Mayo, and Wiegand[120] demonstrated that irradiation of the unsaturated endoperoxide ascaridole, **1**, at 366 nm resulted in the exclusive formation of isoascaridole, **118**. Photolysis of **1** at 185 nm, however, also led to formation of an α-terpinene, **119**, triene, **120**, and *p*-cymene, **121**, by loss of oxygen.[202]

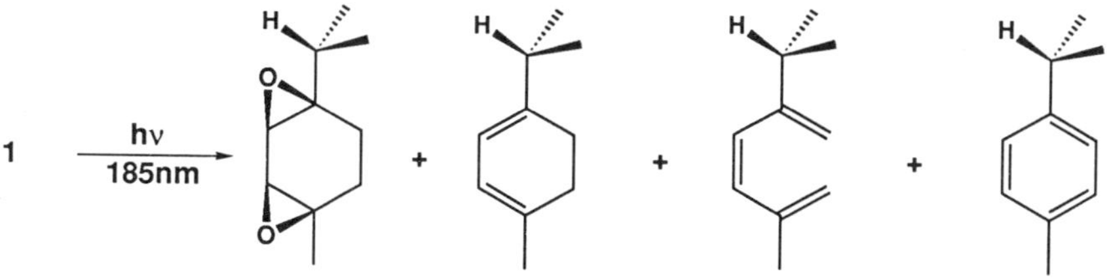

1 —hν, 185nm→ 118 + 119 + 120 + 121

Irradiation of unsaturated bicyclic endoperoxides **65**, **122** and **123**, which, unlike **1**, have one or two bridgehead hydrogens, resulted in the formation of both bisepoxides and epoxy ketones. A very early example of this reaction was reported in 1953 by Conca and Bergmann,[203] who observed the formation of an epoxy ketone during irradiation of Δ^3-2,5-peroxidocholestene. In 1985, Carless *et al.*[131] examined the photochemical behavior of several unsaturated endoperoxides and reported that formation of epoxy ketones also occurred during photolysis of the larger bicyclo[3.2.2] and [4.2.2] ring systems.

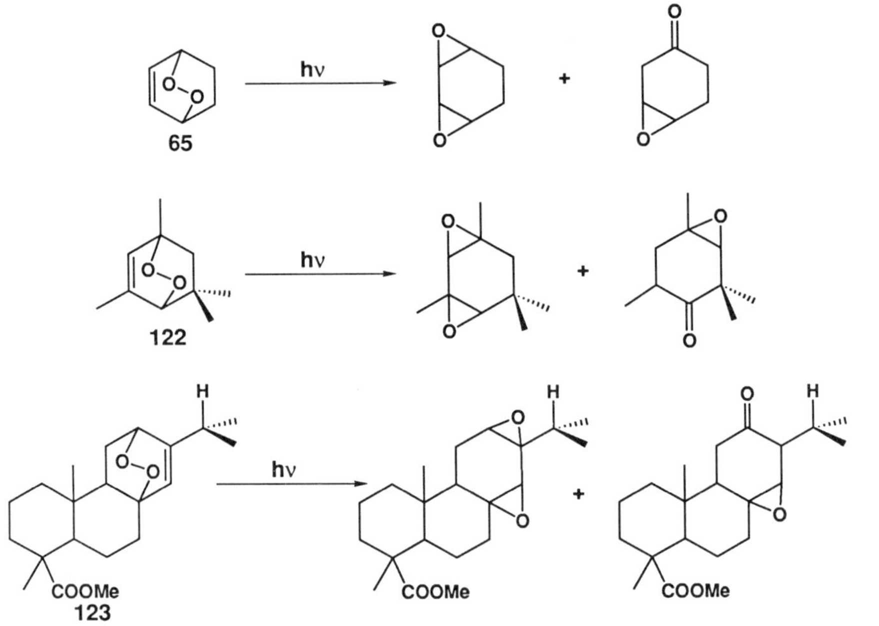

The epoxy ketone/bisepoxide ratio is a sensitive function of substituent stereochemistry, as illustrated by the reactions of the *tert*-butyl-substituted bicyclo[2.2.2]oct-2-enes **124** and **125**. The preference for epoxy ketone during the photolysis of endoperoxide **125** is a result of the near perfect overlap between the spin-bearing orbital and the migrating hydrogen. This overlap is considerably reduced in endoperoxide **124** and, as a result, the bisepoxide becomes the major product (Figure 13). The bisepoxide **126** is susceptible to overphoto-oxidation in methanol to give solvent incorporation products and, as a result, the epoxy ketone/bisepoxide ratio would be expected to change as a function of time.[204]

Rigaudy *et al.*[205] have extensively examined the photochemical rearrangements of naphthalene, anthracene and tetracene endoperoxides. In all the aromatic endoperoxides examined, a bisepoxide was identified as a pivotal intermediate. Photo-oxidation of the 9,10-diphenylanthracene endoperoxide **97** at low temperature resulted in the direct observation of the bisepoxide **127**,

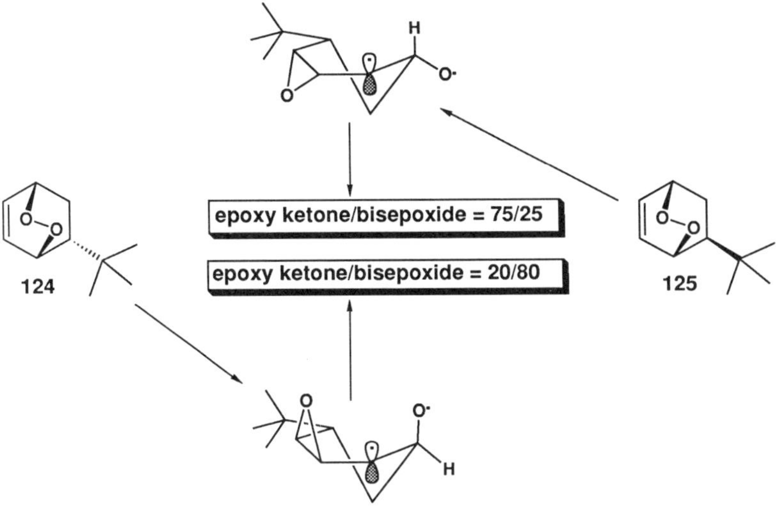

Figure 13. Conformational effects on peroxide photolysis

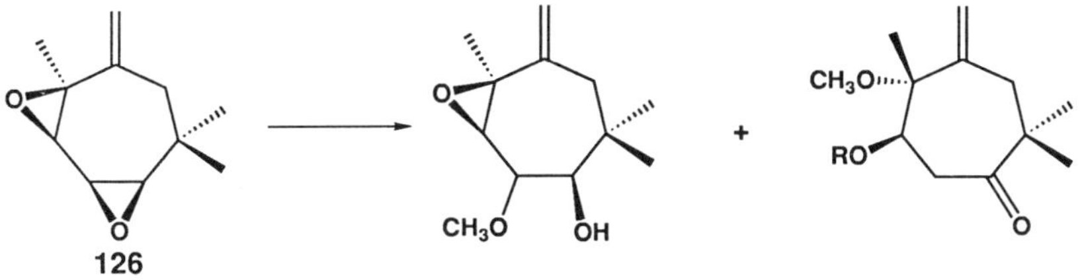

which decomposed either photochemically or thermally to give the rearrangement products.[260] Rubrene endoperoxide, **108**, decomposes photochemically to form two bisepoxides. The one formed in the smallest yield, **128**, which requires loss of the benzene resonance energy, was successfully trapped with *N*-methylmaleimide.[207] In the absence of a trap, bisepoxide **128** undergoes a homolytic cleavage to alkoxy radical **129**. The bisepoxide formed in the largest yield **130**, requires the energetically less costly loss of the naphthalene resonance energy and undergoes ring expansion to form the major product.

Photochemical decompositions of **131** and **132** provide evidence for a stepwise mechanism of bisepoxide formation via a bisalkoxy radical. In these endoperoxides, double β-cleavage of the biradical intermediate competes with bisepoxide formation.[208] The bisalkoxy radical intermediate has also been trapped by the proximal phenyl ring in endoperoxide **98**.[209]

Ph Ph O O Ph Ph

thermal

Ph O O Ph 97 hν Ph O O Ph 127

photochemical

O HO Ph + Ph O O Ph + Ph O O Ph

Ph Ph O O Ph Ph 108 hν Ph Ph O O Ph Ph 128 O NMe O Ph Ph O O Ph Ph O NMe O

Ph Ph O• O Ph Ph 129 Ph OPh O Ph Ph

Ph OPh O Ph Ph

Ph OPh O Ph Ph

Ph Ph O O Ph Ph 130 Ph + O Ph O- Ph Ph Ph Ph O O Ph Ph

131 $\xrightarrow{h\nu}$ + +

132 $\xrightarrow{h\nu}$ + +

98 $\xrightarrow{h\nu}$ + +

In 1983, Buckland and Davidson[210] examined the photochemical decompositions of phosphorus-containing endoperoxides, **133**, which result in expulsion of stabilized phosphorus species. In the case of the phosphonium salts, the ylide was identified by trapping with hexanal. The mechanistic details of these interesting reactions are not entirely clear.

133 $\xrightarrow{h\nu}$ + Phosphorus Species

a. R = $OP(O)(OEt)_2$ d. R = $CH_2P^+Ph_3\ ClO_4^-$
b. R = $CH_2P(O)(OEt)_2$ e. R = $PhCHP^+Ph_3\ Cl^-$
c. R = $CH_2P^+Ph_3\ Br^-$

5 BIMOLECULAR REACTIONS OF ENDOPEROXIDES

5.1 Metal-promoted Decompositions

The reactions of endoperoxides with a wide variety of metals, including cobalt, copper, iron, palladium, rhodium, ruthenium, tin, titanium and zinc, have been examined.[211] The metals can promote the decompositions of the endoperoxides by acting as Lewis acids or as one- or two-electron donors. The electron-donation processes can occur by inner- or outer-sphere mechanisms and the reactions can be catalytic or stoichiometric in the metals.

In 1980, Foote and co-workers[212] reported that cobalt *meso*-tetraphenylporphine (CoTPP) promoted a high-yield catalytic rearrangement of endoperoxides to bisepoxides. These reactions, as represented by **134** and **135**, are rapid at low temperature and occur readily with 5 mol% CoTPP. Similar catalyzed rearrangements have since been reported for a large number of unsaturated bicyclic endoperoxides.[129,130,213–216]

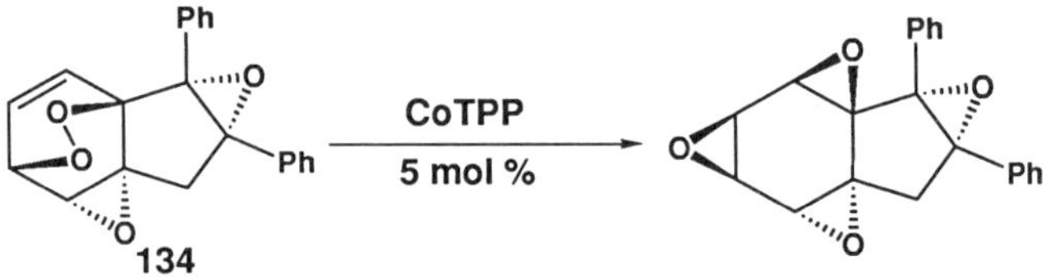

CoTPP 5 mol %

134

CoTPP 5 mol %

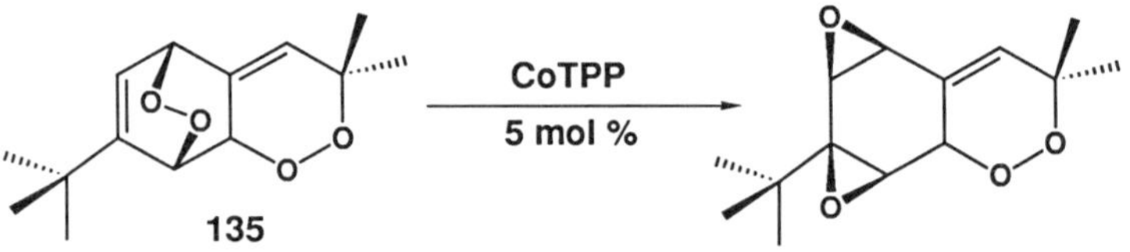

135

The CoTPP-catalyzed rearrangements are stereospecific, giving exclusively the *syn*-bisepoxides. The yields of bisepoxides are much higher in the cobalt-catalyzed processes and the reaction mixtures are not plagued by the formation of the epoxy ketones that is observed on thermolysis of endoperoxides (for an exception, see ref. 129). The reaction of CoTPP with endoperoxide **3**, however, resulted in the formation of a novel dioxacyclooctatriene cyclophane. Careful examination of the reaction mixture at low temperature demonstrated that it was formed by a retro-Diels–Alder reaction of a bisepoxide precursor.[11]

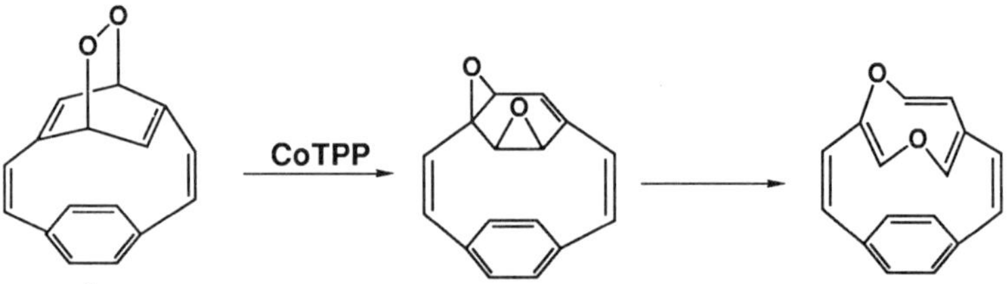

CoTPP

3

The weak reductants zinc *meso*-tetraphenylporphine and *meso*-tetraphenylporphine did not promote the rearrangement to the bisepoxide whereas *N*,*N*,*N*′,*N*′-tetramethyl-*p*-phenylenediamine, a stronger reductant than CoTPP, promoted only a slow rearrangement. The lack of a well defined correlation with the oxidation potential of the catalyst suggests an inner- rather than outer-sphere electron-transfer mechanism. A one-electron radical process is also supported by the observation of β-cleavage during the CoTPP-catalyzed decomposition of endoperoxide **136**.[217] It is unlikely, however, that the β-cleavage produces a free carbon-centered radical, since the radical-stabilizing *gem*-dimethyl group in **137** promotes the formation of the bisepoxide rather than cleavage. These results are most consistent with concerted β-cleavage and epoxide formation (Figure 14).[218] The *gem*-dimethyl group in **137** sterically destabilizes conformation *A*, which is necessary for this concerted process, and exclusive formation of the bisepoxide is observed as a result (Figure 14).

CoTPP

136

CoTPP

137

In 1983, Yamada *et al.*[211] examined the reaction of vitamin D endoperoxide, **138**, with CoTPP. The reaction did not produce the expected bisepoxide, but formed a hemiacetal instead. The authors speculated that the highly regio-

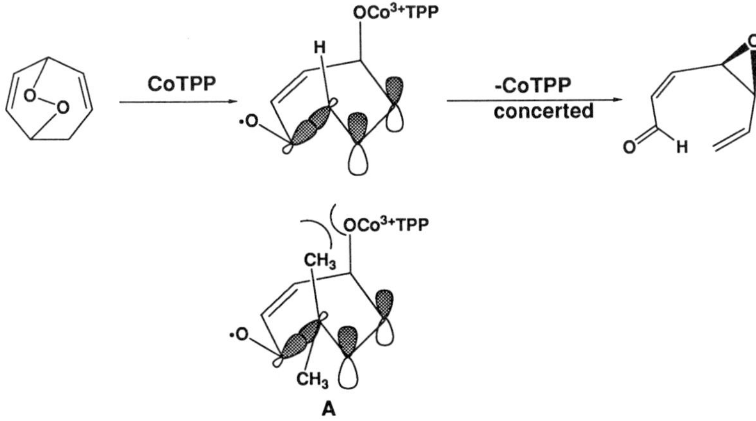

Figure 14. Steric effects in CoTPP-catalyzed rearrangement

selective formation of the hemiacetal reflected the preferential complexation of the cobalt to the least hindered oxygen in the peroxide linkage.

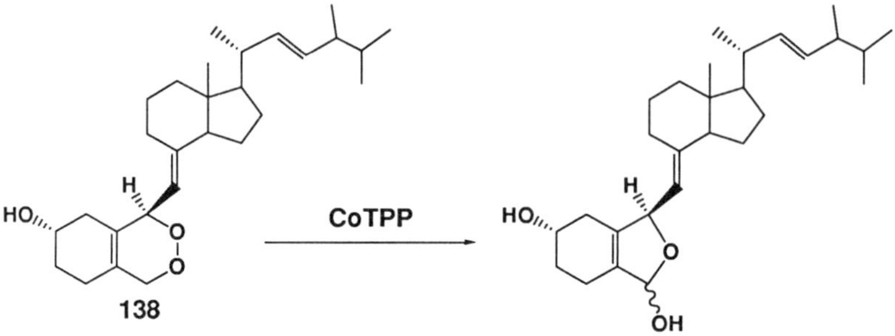

In 1989, O'Shea and Foote examined the Co(II)-catalyzed decomposition of *cis*-3,6-dimethyl-1,2-dioxene, **139**, and 3,6-diphenyl-1,2-dioxene in detail.[219] The four-coordinate low-spin planar complexes CoTPP and cobalt(II)salen were effective catalysts for a remarkably clean conversion to the hemiacetal. Six-coordinate complexes [bis(salicylaldehyde)cobalt(II) dihydrate and cobalt(II) acetylacetonate] did not react with **139**. These results suggest that the mechanism probably involves catalytic cleavage of the peroxide linkage by inner-sphere electron transfer followed by a hydrogen shift, loss of the catalyst and closure to the hemiacetal (Figure 15).

CoTPP-catalyzed decompositions of saturated bicyclic endoperoxides **45**–**47** and **49** demonstrate that the formation of an epoxyaldehyde is characteristic of saturated endoperoxides with a one-carbon bridge. Formation of ethylene, on the other hand, is characteristic of endoperoxides with a two-carbon bridge.[220] These products can be rationalized by invoking β-scission of either a one- or two-carbon bridge in an alkoxy radical formed by inner-sphere electron transfer from CoTPP to the peroxide linkage (Figure 16). Disproportionation of the alkoxy radical intermediates can be invoked to explain the formation of the diols and keto alcohols (Figure 16).

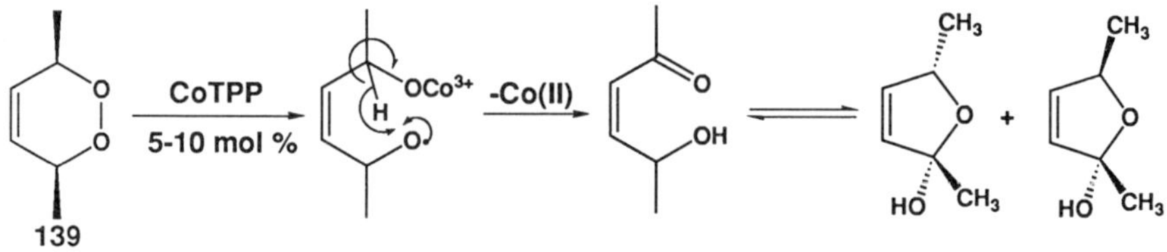

Figure 15. Rearrangement of hexadiene endoperoxide

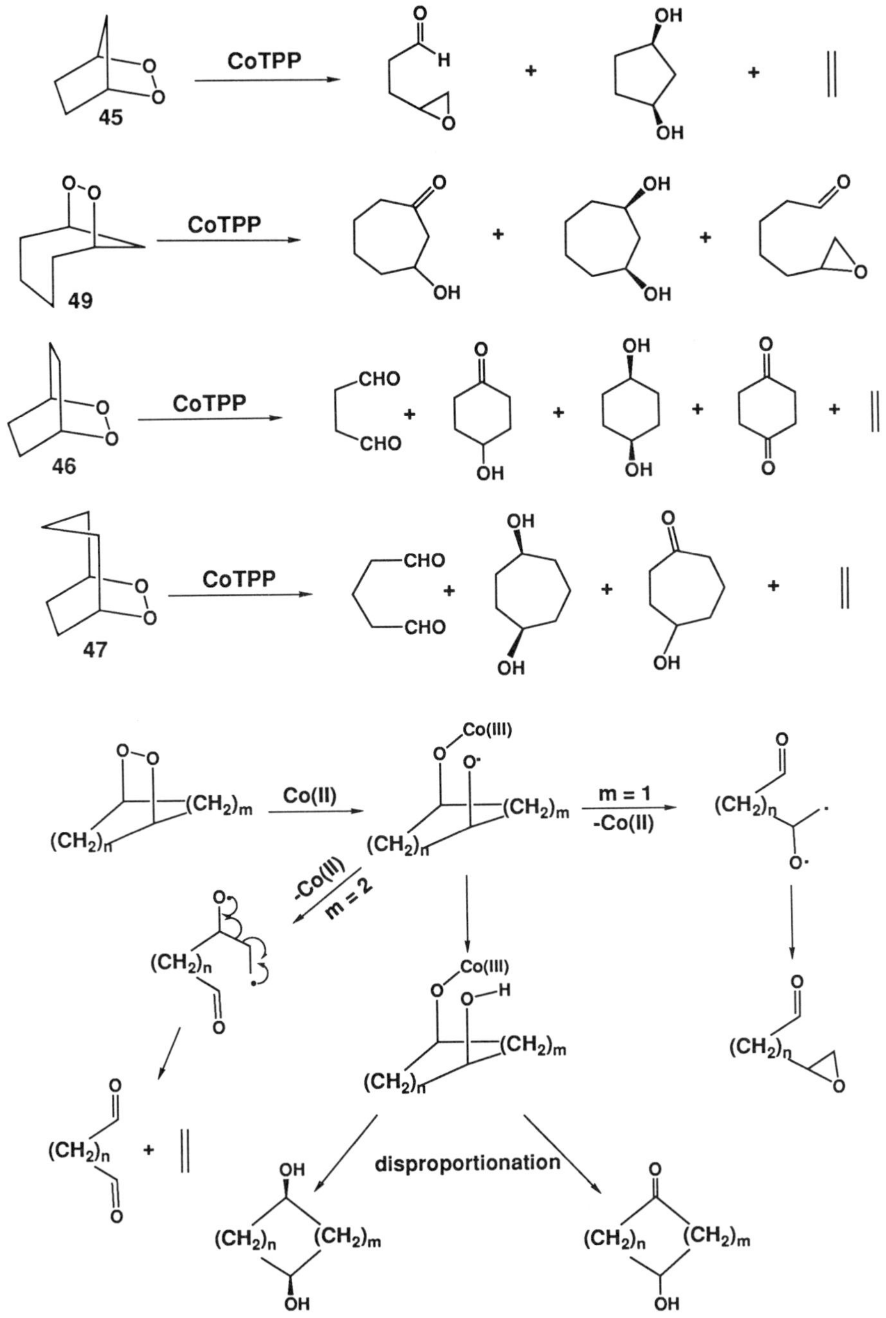

Figure 16. Products of CoTPP-catalyzed rearrangements of endoperoxides

Ruthenium(II) catalyzed decompositions of endoperoxides can also be rationalized by an inner-sphere one-electron transfer process.[221] Reactions of the homogeneous catalyst $RuCl_2(PPh_3)_3$ with unsaturated and saturated endoperoxides give product mixtures similar to those obtained with Co(II).

In contrast to the decompositions induced by Co(II) and Ru(II), an outer-sphere electron-shuttle mechanism (Figure 17) was suggested in order to rationalize the Fe(II)-catalyzed decompositions of the endoperoxides derived from α-phellandrene, **140** and **141**.[222] This suggestion is supported by the observation that $FeSO_4$-catalyzed decomposition of endoperoxide **142** gives the novel remote oxidation products **143** and **144**.[223] In the absence of a double bond, intermediate **145** undergoes hydrogen abstraction. The resulting carbon-centered radical is then oxidized with Fe(III) to form a tertiary carbocation, which collapses to **143** and **144**. The alkoxy radical intermediates, **146**, formed in the reactions of monocyclic endoperoxides, undergo intramolecular hydrogen abstraction to give γ-hydroxyaldehydes, which ultimately condense to form furans.[224] The alkoxy radical formed from the reaction of Fe(II) with dihydroascaridole, **147**, undergoes β-cleavage to 4-hydroxy-4-methylcyclohexanone and isopropyl radical.[225,226]

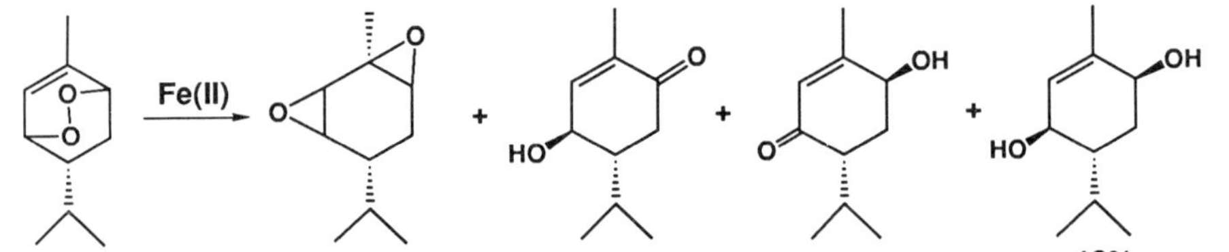

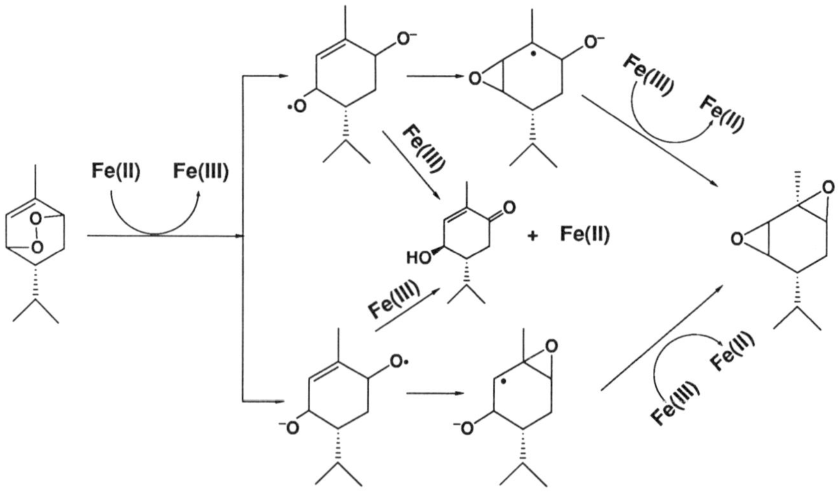

Figure 17. Iron-catalyzed rearrangements

Fe(II)

minor products

141 16% 28%

FeSO$_4$

142 143 144

145

Fe(II) Fe(III) furan

146

147

The pivotal anion radical intermediate in these reactions has also been suggested to be the key intermediate in the biosynthesis of prostaglandin E (PGE), PGI, and thromboxanes from prostaglandin endoperoxides.[227] The viability of this suggestion for thromboxane formation was demonstrated by Kishi and Takahashi,[228] who observed formation of the thromboxane skeleton, **149**, during a study of the Fe(II)-catalyzed decomposition of endoperoxide **148**. The electron-shuttle mechanism results in the formation of carbocation **150**, which ultimately collapses to the thromboxane skeleton.[216] Carbocation **150** is also formed in a much slower Fe(III) Lewis acid-catalyzed route.[229] The viability of PGI_2 formation from alkoxy radical cyclization was demonstrated by Porter and Mebane[230] during a study of the Fe(II)-catalyzed decomposition of endoperoxide **151**.

Copper(I) reacts with endoperoxide **45** via a one-electron transfer to give *cis*-1,3-dihydroxycyclopentane.[231] The reaction does not occur via the one-electron inner- or outer-sphere shuttle mechanism exhibited by Co(II) and Fe(II), respectively, but requires a stoichiometric amount [2Cu(I)/**45**] of copper. Copper(II) also reacts with **45** via a Lewis acid-catalyzed route to give 3-hydroxycyclopentanone.

Palladium(0), which is a two-electron donor, catalytically decomposes both saturated, **47**,[232] and unsaturated, **68**,[233] endoperoxides via competing one- and two-electron inner-sphere processes. The formation of 4-hydroxy ketones can be rationalized by either oxidative insertion of palladium into the O—O bond or S_N2 displacement to generate a zwitterion (Figure 18). On the other hand, the formation of acetone and the decrease in the yield of the 1,4-diketone when the decomposition of **47** was run in the presence of 10 equiv. of isopropanol suggests that hydrogen abstraction by an inner-sphere radical is responsible for diol formation. A similar 1e/2e mechanism can also account for the products in $Rh_2(CO)_4Cl_2$-catalyzed endoperoxide decompositions.[234,235]

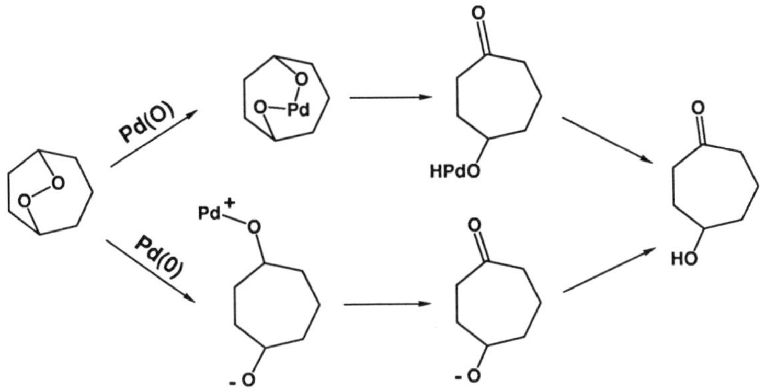

Figure 18. Palladium-catalyzed rearrangements

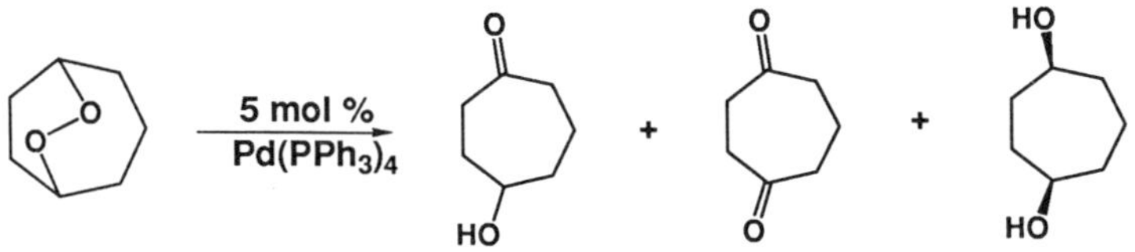

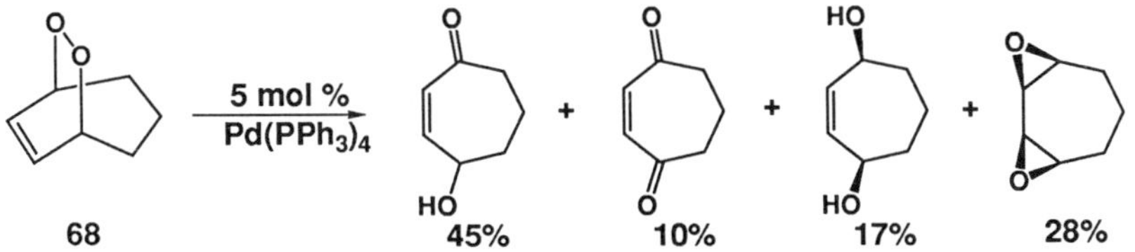

Tin(II) chloride undergoes oxidative addition to the peroxide linkage to give tetracoordinate tin intermediates, **152**, which can decompose by either Sn—O or C—O cleavage.[236] The competition between these reaction modes reflects the similarity in C—O (85 kcal mol^{-1}) and Sn—O (95 kcal mol^{-1}) bond strengths. The synthetic potential of the reaction has been investigated. In particular, the reaction has been utilized to deoxygenate endoperoxides (e.g. **78**) which do not undergo thermal (4 + 2) retrocycloadditions,[236] and the susceptibility of the tetravalent intermediate to nucleophilic attack has been exploited to form C—C bonds (e.g. **153**).[237–239] It has been suggested that low-valent titanium also deoxygenates endoperoxides via a very similar mechanism.[240]

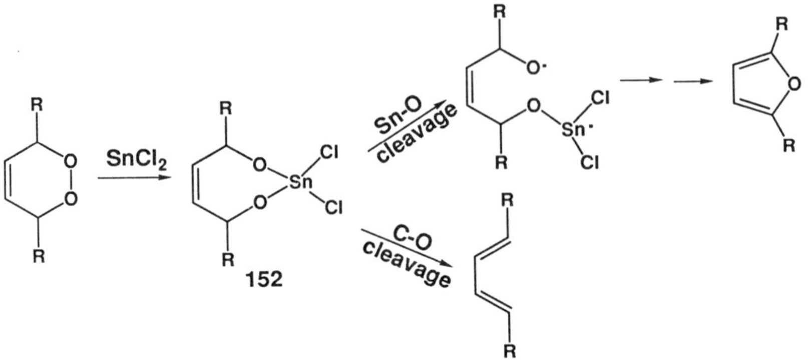

5.2 Reductions

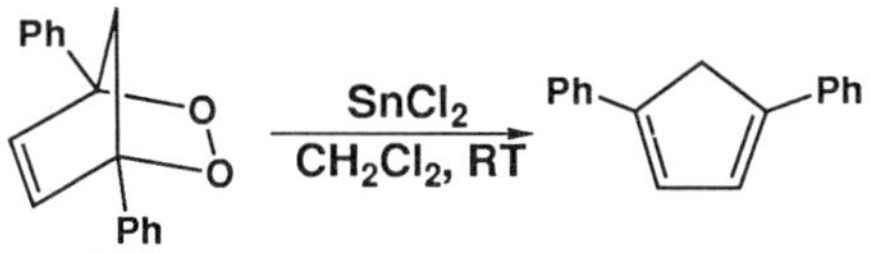

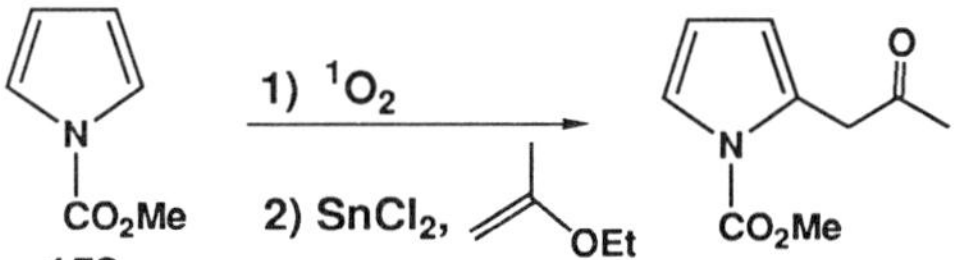

Reducing agents for unsaturated endoperoxides fall into one of three categories: (1) those capable of selective reduction of the O—O bond, (2) those capable of selective reduction of the double bond and (3) unselective reducing agents. Examples of these processes are illustrated by the reductions of endoperoxides **2**, **85**, **138** and **154**–**157**.

In 1988, Corey *et al.*[241] reported that Al(Hg) could reduce the O—O bond in the key intermediate, **154**, in the synthesis of (±)-forskolin without double bond reduction. Thiourea, a mild reducing agent often used to reduce endoperoxides, was used to reduce intermediate **2** in the synthesis of conduritol F.[10] Sodium tetrahydroborate ($NaBH_4$) has been reported to reduce selectively the strained O—O bond in 6,6-dimethylfulvene-1,4-endoperoxides, **85**.[144] Reduction of O—O bonds with $NaBH_4$, however, is restricted to strained ring systems. The strain-free [5.2.1] endoperoxide **158** reacts with $NaBH_4$ to cleave the C—Hg bonds and leaves the O—O bond intact.[65] $LiAlH_4$ is a powerful reducing agent that selectively cleaves the O—O bond in vitamin D_2 endoperoxides, **138**. The formation of byproducts resulting from the ability of

$LiAlH_4$ to act as a base (e.g. **159**)[242] or to reduce other functional groups, however, has been noted and decreases its general utility. Iodide selectively reduces the sterically more accessible peroxide linkage in bisendoperoxide **155**[243] and has the added advantage that its extent of reaction can be monitored colorimetrically. However, iodide is rarely useful for the quantitative reduction of endoperoxides.

154 → Al/Hg, 23°C, 20:1 THF/H_2O

2 → $(NH_2)_2CS$, CH_3OH

85 → $NaBH_4$, CH_3OH

138 → $LiAlH_4$

155 → I^-

Diimide is a generally useful reducing agent which selectively reacts with double bonds in endoperoxides such as **156** without attacking the O—O bond.[244] Bloodworth and Eggelte[245] examined the reaction of diimide with endoperoxides **65**, **68**, **74** and **77** and suggested that both steric and electronic factors dictate the reaction stereochemistry. Heterogeneous catalysts such as

Pd/C, PtO_2 or Ni are usually, but not always,[246] unselective and reduce both the O—O linkage and double bonds (e.g. **157**).[247]

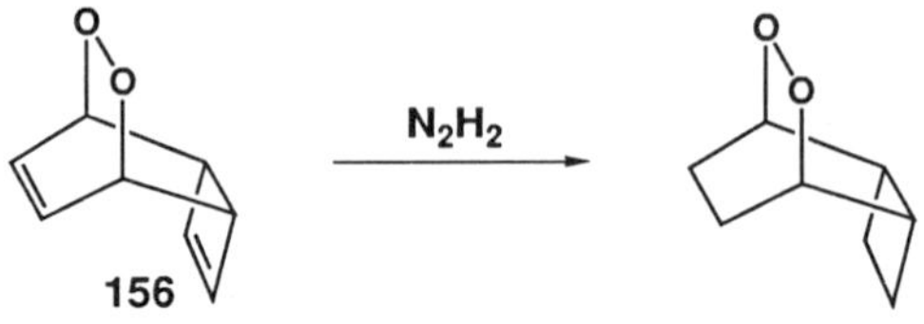

156 $\xrightarrow{N_2H_2}$

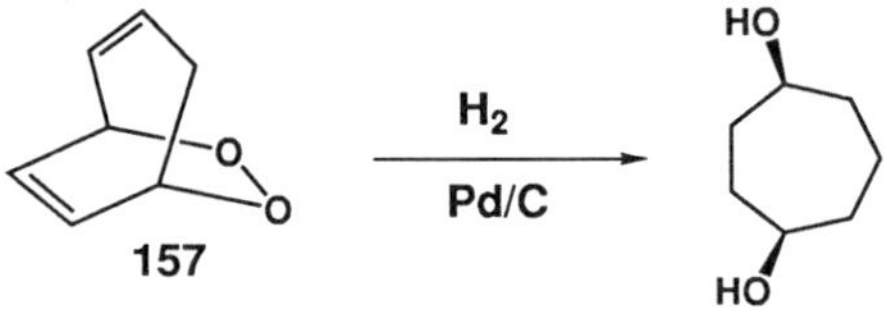

157 $\xrightarrow[Pd/C]{H_2}$

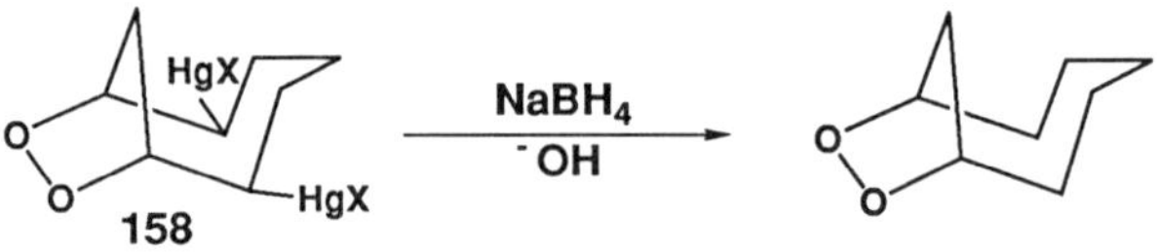

158 $\xrightarrow[^-OH]{NaBH_4}$

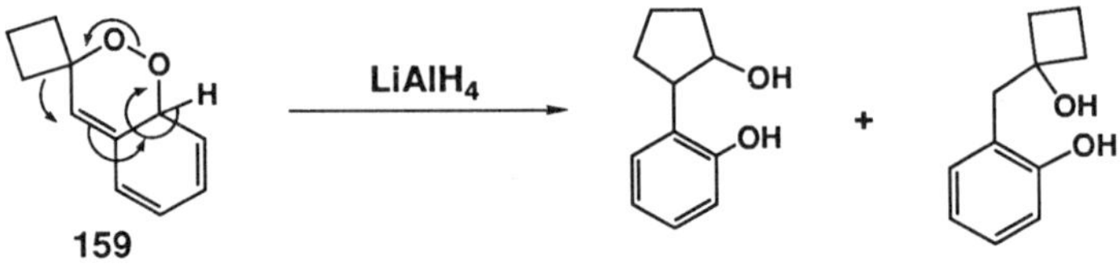

159 $\xrightarrow{LiAlH_4}$ +

$\xrightarrow[CH_3CO_2D]{K^{+}\ {}^{-}O_2CN{=}NCO_2^{-}\ {}^{+}K}$ +

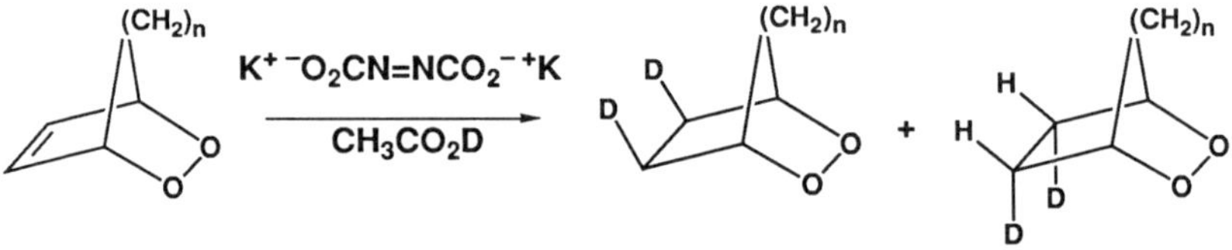

	Anti	*Syn*
77 $n = 1$	100%	0%
65 $n = 2$	34%	66%
68 $n = 3$	0%	100%
74 $n = 4$	0%	100%

5.3 Free Radical Substitutions (the S_H2 Reaction)

An internal S_H2 reaction (S_Hi) is the initiating step in styrene–oxygen copolymer decomposition.[248] Endoperoxides are also susceptible to bimolecular homolytic substitution reactions. Porter and co-workers[249,250] examined the stereochemistry of the S_Hi reaction by systematically varying the relative orientation of the attacking radical and the endoperoxide linkage. The yields of endoperoxides and epoxy alcohol products (Figure 19) formed in the reactions of tri-*n*-butyltin radical with bromoendoperoxides **160**–**165** were measured and the rate ratios $k(S_H i)/k_H$ were derived. From the known value of k_H[251] the values for $k(S_H i)$ were determined (Table 7). From these data, it is clear that a colinear backside attack (dihedral angle $\theta_{1234} = 180°$) is preferred in the $S_H i$ reaction. Backside approach in the five-membered ring endoperoxides **160** and **162** is more difficult to achieve than in the six-membered ring endoperoxide **161** and $k(S_H i)$ is one to two orders of magnitude smaller. The dihedral angle θ_{1234} in the seven-membered ring compound **163** is constrained to less than 70–100° and very little epoxy alcohol is observed. Further verification of the requirement for backside attack is provided by the reactions of closely related endoperoxides **164** and **165**.[250] The dihedral angle in **164** ($\theta_{1234} = 176°$) is close to ideal and $k(S_H i)$ is large. In contrast, the dihedral angle in **165** ($\theta_{1234} = 60°$) is far from the required 180° and no epoxy alcohol was formed.

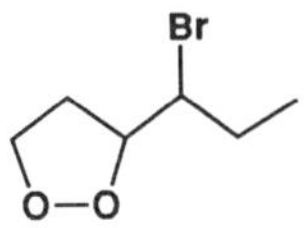

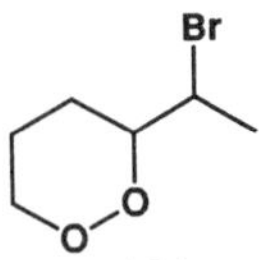

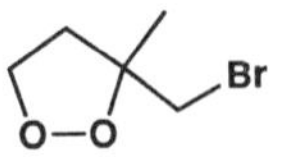

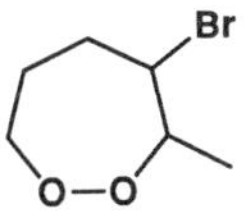

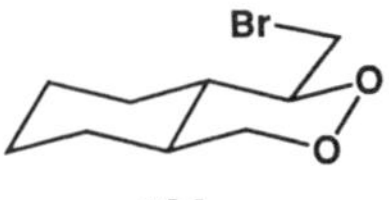

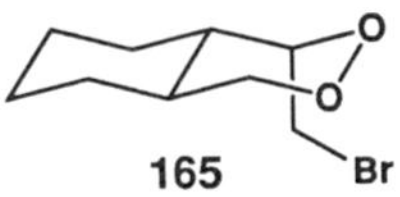

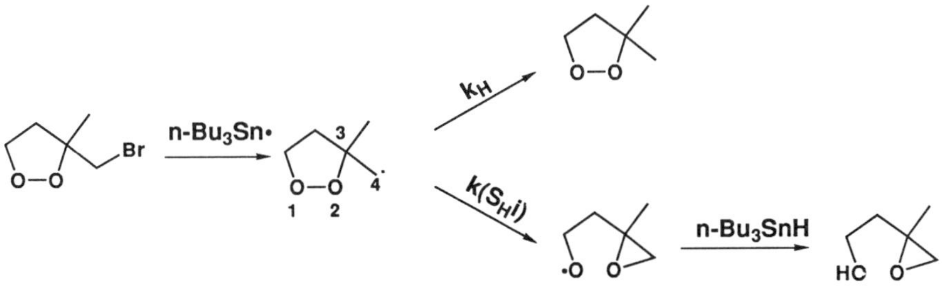

Figure 19. Homolytic substitution

Table 7. Rates of homolytic substitution

Compound	$k\ (S_H i)\ (\mathrm{l\ mol^{-1}\ s^{-1}})$	Compound	$k\ (S_H i)\ (\mathrm{l\ mol^{-1}\ s^{-1}})$
160	7.5×10^4	**163**	< 1
161	8.7×10^5	**164**	1.2×10^4
162	1×10^4	**165**	No epoxide observed

5.4 Reactions of Trivalent Phosphorus Compounds

The reactions of triphenylphosphine (Ph_3P) with unsaturated endoperoxides have frequently been reported.[6] These reactions proceed by initial cleavage of the peroxide linkage followed by S_N2' displacement to give the unsaturated epoxide and triphenylphosphine oxide. The reactions of a number of unsymmetrical endoperoxides with Ph_3P have been examined, but the regiochemistry has not been firmly established.[252,253] However, in the case of **166**, only one of the two possible epoxyalkene products was formed.[125]

CH₂OCOPh O O O 166 → PPh₃ or P(OMe)₃ → ⁻O CH₂OCOPh O ŌPPh₃ → S_N2' → O CH₂OCOPh O

The reactions of Ph_3P with the saturated bicyclic endoperoxides PGG_2 and PGH_2 were reported to give a *cis*-1,3-diol as the product.[254] On the other hand, the reaction of **45** with PPh_3 in moist dichloromethane resulted in the slow decomposition of the peroxide and production of triphenylphosphine oxide, *trans*-1,3-cyclopentane diol and trace amounts of two unidentified products, but no *cis*-1,3-diol.[255] A mechanism proceeding via the formation of a phosphorane, **167**, followed by ionization and displacement of triphenylphosphine oxide by water was suggested in order to explain the exclusive formation of the *trans*-diol.

45 → PPh₃ → 167 (PPh₃) ⇌ O⁻ O—P⁺Ph₃ → H₂O → HO ...OH + Ph₃PO

This mechanism was supported by several experimental observations, including the following: (1) the rate of the reaction was identical in dichloromethane [$(0.99 \pm 0.06) \times 10^{-2}$ l mol^{-1}s^{-1}] and benzene [$(0.93 \pm 0.06) \times 10^{-2}$ l mol^{-1}], as expected for a rate-determining biphilic insertion[256,257] to give **167**; (2) the Hammett ρ^+ values (Table 8) are much smaller than expected for nucleophilic reactions of phosphines and consistent with the contributions of both nucleophilic and electrophilic (biphilic) components to the transition state;[258] and (3) phosphorane **167** was directly observable ($\delta_p{}^{31} = -63.89$ ppm relative to 85% H_3PO_4) under anhydrous conditions. The formation of **167** is visibly exothermic as a result of the relief of a considerable amount of ring strain present in **45** and the formation of two stronger P—O bonds (45 kcal mol^{-1}) at the expense of a weaker O—O bond (35 kcal mol^{-1}). The activation barriers also decrease as expected with increasing strain in the peroxide [**168**[259] < **67** < **45** < **46** (Table 8)].

Table 8. Kinetic data for the reactions of peroxides with triarylphosphines

Compound	$\Delta H\ddagger$ (kcal mol^{-1})	$\Delta S\ddagger$ (cal mol^{-1}K^{-1})	ρ^{+}
45	11.0 ± 0.67	-29.6 ± 2.2	-0.44 ± 0.01
46	14.9 ± 1.2	-20.6 ± 3.9	-0.38 ± 0.01
67	10.4 ± 2.6	-27.0 ± 8.7	
168	9.2	-28	-0.82

Phosphoranes have also been established as intermediates in the reactions of **45** with the phenyl dialkylphosphine **169**,[260] phosphinite **170**, phosphonite **171**, and phosphite,**172**.[261]

168 169 $Ph_2P(OMe)$ 170 $PhP(OMe)_2$ 171 $P(OMe)_3$ 172

5.5 Acid- and Base-catalyzed Reactions

The most prevalent base-catalyzed reaction of an endoperoxide is the Kornblum–De La Mare decomposition.[262] The reaction is initiated by removal of a proton from the carbon adjacent to the peroxy linkage, as illustrated by the reaction of **156**, and results in the formation of a hydroxy ketone.[126,244] An attempt to extend the synthetic utility and to induce asymmetry into the rearrangement using endoperoxides with enantiotopic hydrogens and chiral bases met with only moderate success.[234] The saturated endoperoxide **45** also undergoes Kornblum–De La Mare decomposition in the presence of DABCO.[263] The initially formed keto alkoxide in this case has the option of being protonated or undergoing a retroaldol condensation. A large negative entropy of activation ($\Delta S\ddagger = -30 \pm 3$ e.u.) and kinetic isotope effect [k(**45**)/k(**45-d$_8$**) = 8] is consistent with the rate-determining removal of a bridgehead hydrogen by DABCO and supports the Kornblum–De La Mare mechanism.

156 $\xrightarrow{Et_3N}$

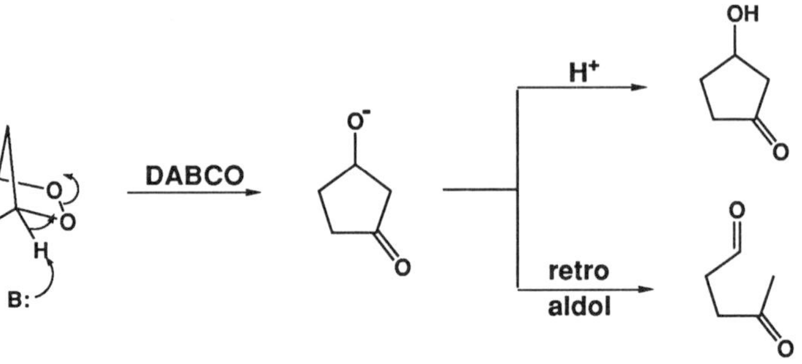

Treatment of allylic or benzylic endoperoxides with either Brønsted or Lewis acids promotes heterolysis of the C—O bond and the formation of a peroxy allylic or benzylic cation. This key intermediate has available several possible reactions, including: (1) collapse at the other end of the allylic unit to form a dioxetane[264] (e.g. **173**),[265] (2) capture by aldehydes or ketones to form 1,2,4-trioxanes (e.g. **78**);[51] or (3) capture by alcohols to give hydroperoxy ethers (e.g. **103**).[266] Treatment of 1,4-dimethyl-1,4-dihydronaphthalene-1,4-endoperoxide, **101**, with hydrochloric, hydrobromic, formic or trifluoroacetic acid results in functionalization of one of the methyl groups.[267] A key step in this novel reaction is loss of a proton from the peroxy cation intermediate.

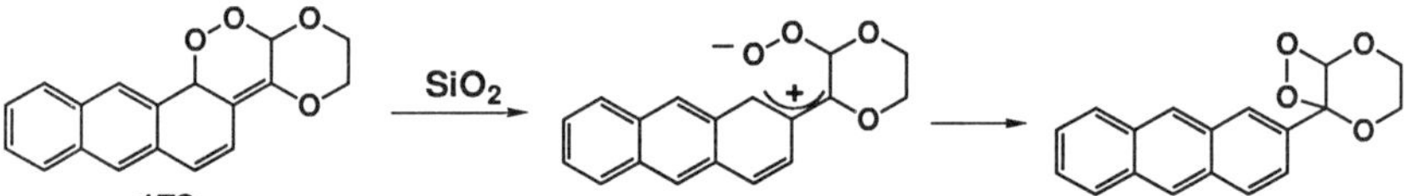

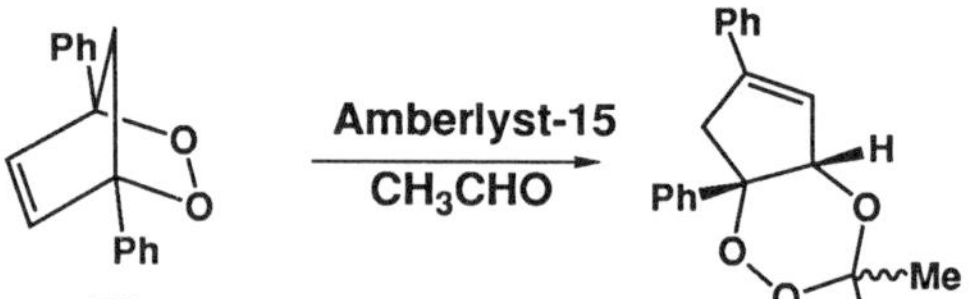

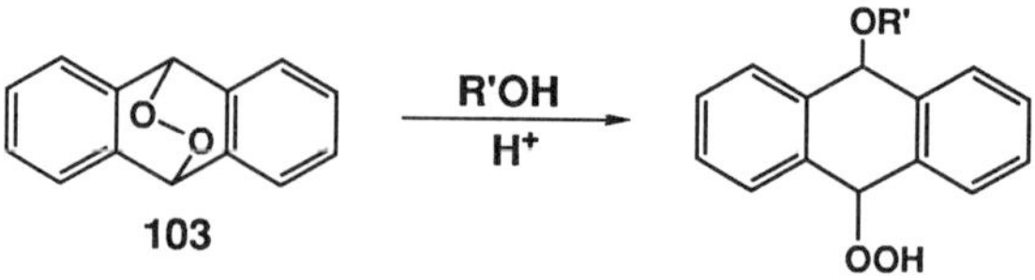

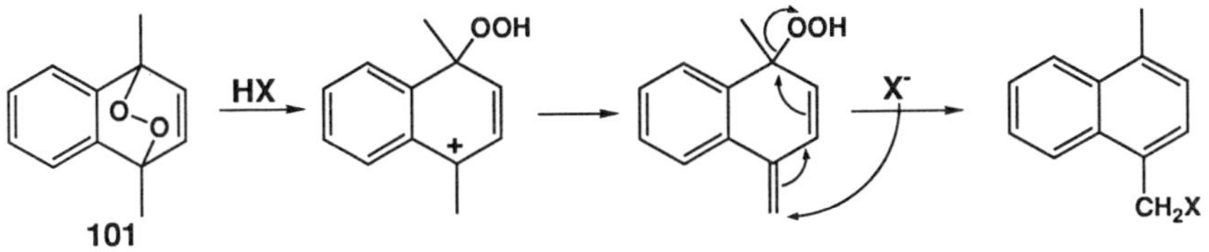

As is evident from this review, the field of endoperoxide chemistry is extremely active. The potential for the synthetic utility of these compounds is only now being recognized, and some of the recent developments described suggest that the full potential of this field has not yet been explored.

6 ACKNOWLEDGEMENTS

This work was supported by the NSF and the Donors of the Petroleum Research Fund, administered by the American Chemical Society (E. L. C.) and NSF Grant No. CHE89-11916 and NIH GM-200819 (C. S. F.). E. L. C. thanks the Department of Chemistry and Biochemistry of the University of California, Los Angeles, for hosting a sabbatical during the early stages of the preparation of this chapter.

7 REFERENCES

1. A. Windaus and J. Brunken, *Justus Liebigs Ann. Chem.*, **460**, 225 (1928).
2. G. O. Schenck and K. Ziegler, *Naturwissenschaften*, **32**, 157 (1944).
3. D. H. Nugteren, R. K. Beerthuis and D. A. Van Dorp, *Recl. Trav. Chim. Pays-Bas*, **85**, 405 (1966).
4. M. Hamberg and B. Samuelsson, *J. Biol. Chem.*, **242**, 5336 (1967).
5. K. Gollnick and G. O. Schenck, in *Oxygen as a Dienophile.* (J. Hamer, Ed.), Academic Press, New York (1967), p. 255.
6. M. Balci, *Chem. Rev.*, **81**, 91 (1981).
7. I. Saito and S. S. Nittala, in *The Chemistry of Peroxides* (S. Patai, Ed.), Wiley, Chichester (1983), p. 311.
8. A. J. Bloodworth and H. J. Eggelte, in *Singlet O_2* (A. A. Frimer, Ed.), CRC Press, Boca Raton, FL (1985), p. 93.
9. N. A. Porter and D. G. Wujek, in *The Autoxidation of Polyunsaturated Lipids* (A. Quinfanilla, Ed.), Plenum Press, New York (1987), p. 55.
10. B. H. Secen, Y. Sutbeyaz and M. Balci, *Tetrahedron Lett.*, **31**, 1323 (1990).
11. I. Erden, P. Golitz, R. Nader and A. de Meijere, *Angew. Chem., Int. Ed. Engl.*, **20**, 583 (1981).
12. M. Stobbe, S. Kirchmeyer, G. Adiwidjaja and A. de Meijere, *Angew. Chem., Int. Ed. Engl.*, **25**, 171 (1986).

13. I. Rosenthal, in *Singlet O_2* (A. A. Frimer, Ed.), CRC Press, Boca Raton, FL (1985), p. 13.
14. A. P. Schapp, A. L. Thayer, E. C. Blossey and D. C. Neckers, *J. Am. Chem. Soc.*, **97**, 3741 (1975).
15. E. J. Corey and W. C. Taylor, *J. Am. Chem. Soc.*, **86**, 3881 (1964).
16. R. W. Murray and H. L. Kaplan, *J. Am. Chem. Soc.*, **91**, 5358 (1969).
17. A. P. Schaap, A. L. Thayer, G. F. Faler, K. Goda and T. Kimura, *J. Am. Chem. Soc.*, **96**, 4025 (1974).
18. A. A. Gorman, I. Hamblett, C. Lambert, B. Spencer and M. C. Standen, *J. Am. Chem. Soc.*, **110**, 8053 (1988).
19. J. R. Hurst and G. B. Schuster, *J. Am. Chem. Soc.*, **105**, 5756 (1983).
20. F. Wilkinson and J. G. Brummer, *J. Phys. Chem. Ref. Data*, **10**, 809 (1981).
21. B. M. Monroe, *J. Am. Chem. Soc.*, **103**, 7252 (1981).
22. E. L. Clennan, *Tetrahedron*, **47**, 1343 (1991).
23. K. Gollnick and A. Griesbeck, *Tetrahedron Lett.*, **24**, 3303 (1983).
24. K. E. O'Shea and C. S. Foote, *J. Am. Chem. Soc.*, **110**, 7167 (1988).
25. Y. Kayama, M. Oda, and Y. Kitahara, *Chem. Lett.*, 345 (1974).
26. W. Adam and I. Erden, *Tetrahedron Lett.*, 2781 (1979).
27. W. Adam and G. Klug, *Tetrahedron Lett.*, **23**, 3155 (1982).
28. D. H. R. Barton, G. Le Clerc, P. D. Magnus, and I. D. Menzies, *J. Chem. Soc., Chem. Commun.*, 447 (1972).
29. D. H. R. Barton, R. K. Haynes, P. D. Magnus and I. D. Menzies, *J. Chem. Soc., Chem. Commun.*, 511 (1974).
30. R. K. Haynes, *Aust. J. Chem.*, **31**, 121 (1978).
31. D. J. Bellville, D. D. Wirth and N. L. Bauld, *J. Am. Chem. Soc.*, **103**, 718 (1981).
32. M. F. Arain, R. K. Haynes, S. C. Vonwiller and T. W. Hambley, *J. Am. Chem. Soc.*, **107**, 4582 (1985).
33. S. F. Nelsen, M. F. Teasley and D. L. Kapp, *J. Am. Chem. Soc.*, **108**, 5503 (1986).
34. E. L. Clennan, W. Simmons and C. W. Almgren, *J. Am. Chem. Soc.*, **103**, 2098 (1981).
35. J. Eriksen, C. S. Foote and T. L. Parker, *J. Am. Chem. Soc.*, **99**, 6455 (1977).
36. K. Gollnick and A. Schnatterer, *Tetrahedron Lett.*, **25**, 185 (1984).
37. K. Mizuno, N. Kamiyama, N. Ichinose and Y. Otsuji, *Tetrahedron*, **41**, 2207 (1985).
38. L. T. Spada and C. S. Foote, *J. Am. Chem. Soc.*, **102**, 391 (1980).
39. A. P. Schaap, K. A. Zaklika, B. Kaskar and W.-M. Fung, *J. Am. Chem. Soc.*, **102**, 389 (1980).
40. L. E. Manring, M. K. Kramer and C. S. Foote, *Tetrahedron Lett.*, **25**, 2523 (1984).
41. L. E. Manring, C. Gu and C. S. Foote, *J. Phys. Chem.*, **87**, 40 (1983).
42. Y. Araki, D. C. Dobrowolski, T. E. Goyne, D. C. Hanson, Z. Q. Jiang, K. J. Lee and C. S. Foote, *J. Am. Chem. Soc.*, **106**, 4570 (1984).
43. W. Adam, S. Grabowski and R. M. Wilson, *Acc. Chem. Res.*, **23**, 165 (1990).
44. W. Adam, H. Platsch, and J. Wirz, *J. Am. Chem. Soc.*, **111**, 6896 (1989).
45. W. Adam, P. Hossel, W. Hummer and H. Platsch, *J. Am. Chem. Soc.*, **109**, 7570 (1987).
46. W. Adam. K. Hannemann, and R. M. Wilson, *J. Am. Chem. Soc.*, **106**, 7646 (1984).
47. R. M. Wilson and F. Geiser, *J. Am. Chem. Soc.*, **100**, 2225 (1978).
48. M. Yoshioka, M. Oka, Y. Ishikawa, H. Tomita and T. Hasegawa, *J. Chem. Soc., Chem. Commun.*, 639 (1986).
49. W. Adam, U. Kliem, T. Mosandl, E.-M. Peters, K. Peters and H. G. von Schnering, *J. Org. Chem.*, **53**, 4986 (1988).
50. W. R. Roth and B. P. Scholz, *Chem. Ber.*, **115**, 1197 (1982).

51. C. W. Jefford, D. Jaggi, J. Boukouvalas and S. Kohmoto, *J. Am. Chem. Soc.*, **105**, 6497 (1983).
52. C. W. Jefford, F. Favarger, S. Ferro, D. Chambaz, A. Bringhen, G. Bernardinelli and J. Boukouvalas, *Helv. Chim. Acta*, **69**, 1778 (1986).
53. C. W. Jefford, E. C. McGoran, J. Boukouvalas, G. Richardson and B. L. Robinson, *Helv. Chim. Acta*, **71**, 1805 (1988).
54. J. A. Kepler, A. Philip, Y. W. Lee, H. A. Musallam and F. I. Carroll, *J. Med. Chem.*, **30**, 1505 (1987).
55. O. R. Idowu, J. L. Maggs, S. A. Ward and G. Edwards, *Tetrahedron* **46**, 1871 (1990).
56. A. J. Bloodworth, T. Melvin and J. C. Mitchell, *J. Org. Chem.*, **53**, 1078 (1988).
57. A. J. Bloodworth, T. Melvin, J. C. Mitchell and T. Melvin, *J. Org. Chem.*, **51**, 2612 (1986).
58. K. Takahashi, M. Shiro and M. Kishi, *J. Org. Chem.*, **53**, 3098 (1988).
59. N. A. Porter and D. W. Gilmore, *J. Am. Chem. Soc.*, **99**, 3503 (1977).
60. N. A. Porter, J. D. Byers, K. M. Holden and D. B. Menzel, *J. Am. Chem. Soc.*, **101**, 4319 (1979).
61. A. J. Bloodworth and H. J. Eggelte, *J. Chem. Soc., Chem. Commun.*, 741 (1979).
62. A. J. Bloodworth, J. A. Khan and M. E. Loveitt, *J. Chem. Soc., Perkin Trans. 1*, 621 (1981).
63. A. J. Bloodworth and M. D. Spencer, *Tetrahedron Lett.*, **31**, 5513 (1990).
64. W. Adam., A. J. Bloodworth, H. J. Eggelte and M. E. Loveitt, *Angew. Chem., Int. Ed. Engl.*, **17**, 209 (1978).
65. A. J. Bloodworth and J. A. Khan, *Tetrahedron Lett.*, 3075 (1978).
66. J. E. Baldwin, *J. Chem. Soc., Chem. Commun.*, 734 (1976).
67. J. R. Nixon, M. A. Cudd and N. A. Porter, *J. Org. Chem.*, **43**, 4048 (1978).
68. N. A. Porter, M. O. Funk, D. Gilmore, R. Isaac and J. Nixon, *J. Am. Chem. Soc.*, **98**, 6000 (1976).
69. C.-H. Lin, D. L. Alexander, C. G. Chidester, R. R. Gorman and R. A. Johnson, *J. Am. Chem. Soc.*, **104**, 1621 (1982).
70. M. F. Salomon and R. G. Salomon, *J. Am. Chem. Soc.*, **99**, 3500 (1977).
71. R. G. Salomon and M. F. Salomon, *J. Am. Chem. Soc.*, **99**, 3501 (1977).
72. J. Yoshida, S. Nakatani, K. Sakaguchi and S. Isoe, *J. Org. Chem.*, **54**, 3383 (1989).
73. J. Yoshida, S. Nakatani and S. Isoe, *Tetrahedron Lett.*, **31**, 2425 (1990).
74. N. A. Porter and M. O. Funk, *J. Org. Chem.*, **40**, 3614 (1975).
75. N. A. Porter, L. S. Lehman, B. A. Weber and K. J. Smith, *J. Am. Chem. Soc.*, **103**, 6447 (1981).
76. D. E. O'Connor, E. D. Mihelich and M. C. Coleman, *J. Am. Chem. Soc.*, **106**, 3577 (1984).
77. H. W.-S. Chan, J. A. Matthew and D. T. Coxon, *J. Chem. Soc., Chem. Commun.*, 235 (1980).
78. A. L. J. Beckwith and R. D. Wagner, *J. Am. Chem. Soc.*, **101**, 7099 (1979).
79. A. L. J. Beckwith and R. D. Wagner, *J. Chem. Soc., Chem. Commun.*, 485 (1980).
80. A. L. J. Beckwith, C. J. Easton and A. K. Serelis, *J. Chem. Soc., Chem. Commun.*, 482 (1980).
81. N. A. Porter and P. J. Zuraw, *J. Org. Chem.*, **49**, 1345 (1984).
82. H. A. J. Carless and R. J. Batten, *Tetrahedron Lett.*, **23**, 4735 (1982).
83. N. A. Porter, A. N. Roe and A. T. McPhail, *J. Am. Chem. Soc.*, **102**, 7574 (1980).
84. A. N. Roe, A. T. McPhail and N. A. Porter, *J. Am. Chem. Soc.*, **105**, 1199 (1983).
85. J. A. Khan and N. A. Porter, *Angew. Chem., Suppl.*, 513 (1982).
86. A. J. Bloodworth, R. J. Curtis and N. Mistry, *J. Chem. Soc., Chem. Commun.*, 954 (1989).

87. A. J. Bloodworth and R. J. Curtis, *J. Chem. Soc., Chem. Commun.*, 173 (1989).
88. D. Cremer, in *The Chemistry of Peroxides* (S. Patai, Ed.), Wiley, Chichester (1983), p. 1.
89. D. R. Kearns, *J. Am. Chem. Soc.*, **91**, 6554 (1969).
90. C. Guidotti, U. Lamanna, M. Maestro and R. Moccia, *Theor. Chim. Acta*, **27**, 55 (1972).
91. R. L. Miller and D. F. Horning, *J. Chem. Phys.*, **34**, 265 (1961).
92. R. S. Brown, *Can. J. Chem.*, **53**, 3439 (1975).
93. S. F. Nelsen, M. F. Teasley, A. J. Bloodworth and H. J. Eggelte, *J. Org. Chem.*, **50**, 3299 (1985).
94. S. F. Nelsen, M. F. Teasley, D. L. Kapp and R. M. Wilson, *J. Org. Chem.*, **49**, 1843 (1984).
95. D. J. Coughlin, R. S. Brown and R. G. Salomon, *J. Am. Chem. Soc.*, **101**, 1533 (1979).
96. T. Kondo, M. Tanimoto, M. Matsumoto, K. Nomoto, Y. Achiba and K. Kimura, *Tetrahedron Lett.*, **21**, 1649 (1980).
97. T. Kondo, M. Matsumoto and M. Tanimoto, *Tetrahedron Lett.*, 3819 (1978).
98. M. L. Kaplan and G. N. Taylor, *Tetrahedron Lett.*, 295 (1973).
99. E. L. Clennan and K. Nagraba, *J. Org. Chem.*, **52**, 294 (1987).
100. M. Albericci, M. Collart-Lempereur, J. C. Braekman, D. Daloze, B. Tursch, J. P. Declerq, G. Germain and M. Van Meerssche, *Tetrahedron Lett.*, 2687 (1979).
101. P. Monaco, M. Parrilli and L. Previtera, *Tetrahedron Lett.*, **28**, 4609 (1987).
102. S. Sakemi, T. Higa, U. Anthoni and C. Christophersen, *Tetrahedron*, **41**, 263 (1987).
103. A. A. L. Gunatilaka, Y. Gopichand, F. J. Schmitz and C. Djerassi, *J. Org. Chem.*, **46**, 3860 (1981).
104. M. Rotem and Y. Kashman, *Tetrahedron Lett.*, 3193 (1979).
105. R. J. Capon and J. K. MacLeod, *Tetrahedron*, **41**, 3391 (1985).
106. M. D. Higgs and D. J. Faulkner, *J. Org. Chem.*, **43**, 3454 (1978).
107. D. Stierle and D. J. Faulkner, *J. Org. Chem.*, **44**, 964 (1979).
108. D. Stierle and D. J. Faulkner, *J. Org. Chem.*, **45**, 3396 (1980).
109. R. J. Wells, *Tetrahedron Lett.*, 2637 (1976).
110. H. Schwarz and H.-M. Schiebel, in *The Chemistry of Peroxides* (S. Patai, Ed.), Wiley, Chichester (1983), p. 105.
111. L. S. Silbert, in *Organic Peroxides*, (D. Swern, Ed.), Wiley–Interscience, New York (1971), p. 637.
112. T. Keough and A. J. DeStefano, *Spectra*, **8**, 7 (1982).
113. H. M. Fales, E. A. Sokolsoki, L. K. Pannell, P. Quan-Long, D. L. Klayman, A. J. Lin, A. Brossi and J. A. Kelley, *Anal. Chem.*, **62**, 2494 (1990).
114. E. L. Clennan and R. Heppner, unpublished results (1984).
115. J.-P. Kintzinger, *Oxygen NMR Characteristic Parameters and Applications*, Springer, New York (1981), p. 1.
116. M. G. Zagorski, D. S. Allan, R. G. Salomon, E. L. Clennan, P. C. Heah and R. P. L'Esperance, *J. Org. Chem.*, **50**, 4484 (1985).
117. T. Sugawara, Y. Kawada, M. Katoh and H. Iwamura, *Bull. Chem. Soc. Jpn.*, **52**, 3391 (1979).
118. J. K. Crandall and M. A. Centeno, *J. Org. Chem.*, **44**, 1183 (1979).
119. Y. Ogata, K. Tomizawa and K. Furuta, in *The Chemistry of Peroxides* (S. Patai, Ed.), Wiley, Chichester (1983), p. 711.
120. K. K. Maheshwari, P. De Mayo and D. Wiegand, *Can. J. Chem.*, **48**, 3265 (1970).
121. A. C. Baldwin, in *The Chemistry of Peroxides* (S. Patai, Ed.), Wiley, Chichester (1983), p. 97.

122. S. W. Benson, *Thermochemical Kinetics*, Wiley, New York (1976).
123. J. Boche and O. Runquist, *J. Org. Chem.*, **33**, 4285 (1968).
124. C. H. Foster and G. A. Berchtold, *J. Org. Chem.*, **40**, 3743 (1975).
125. B. Ganem, G. W. Holbert, L. B. Weiss and K. Ishizumi, *J. Am. Chem. Soc.*, **100**, 6483 (1978).
126. W. Adam, O. Cueto and O. De Lucchi, *J. Org. Chem.*, **45**, 5220 (1980).
127. W. Adam, O. Cueto, O. De Lucchi, K. Peters, E.-M. Peters and H. G. von Schnering, *Angew. Chem., Int. Ed. Engl.*, **20**, 580 (1981).
128. T. Knoechel, W. Pickl and J. Daub. *J. Chem. Soc., Chem. Commun.*, 337 (1982).
129. A. de Meijere, D. Kaufmann and I. Erden, *Tetrahedron*, **42**, 6487 (1986).
130. B. Atasoy and M. Balci, *Tetrahedron*, **42**, 1461 (1986).
131. H. A. J. Carless, R. Atkins and G. K. Fekarurhobo, *Tetrahedron Lett.*, **26**, 803 (1985).
132. D. M. Floyd and C. M. Cimarusti, *Tetrahedron Lett.*, 4129 (1979).
133. W. Adam, *Angew. Chem., Int. Ed. Engl.*, **13**, 619 (1974).
134. K. H. Schulte-Elte, B. Willhalm and G. Ohloff, *Angew. Chem., Int. Ed. Engl.*, **8**, 985 (1969).
135. W. Adam and D. J. Trecker, *Tetrahedron*, **27**, 2631 (1971).
136. H. Takeshita, H. Kanamori and T. Hatsui, *Tetrahedron Lett.*, 3139 (1973).
137. L. A. Paquette, R. V. C. Carr, E. Arnold and J. Clardy *J. Org. Chem.*, **45**, 4907 (1980).
138. S. J. Hathaway and L. A. Paquette, *Tetrahedron*, **41**, 2037 (1985).
139. E. L. Clennan and K. Matusuno, *J. Org. Chem.*, **52**, 3483 (1987).
140. T. Sato, I. Moritani and M. Matsuyama, *Tetrahedron Lett.*, 5113 (1969).
141. Y. Nakadaira, T. Nomura, S. Kanouchi, R. Sato, C. Kabuto and H. Sakurai, *Chem. Lett*, 209 (1983).
142. F. Lin, *Dissertation*, University of California, Los Angeles (1989).
143. N. Harada, S. Suzuki, H. Uda and H. Ueno, *J. Am. Chem. Soc.*, **94**, 1777 (1972).
144. N. Harada, H. Uda, H. Ueno and S. Utsumi, *Chem. Lett.*, 1173 (1973).
145. J.-P. LeRoux and J.-J. Basselier, *C.R. Acad. Sci., Ser. C*, **271**, 461 (1970).
146. W. Skorianetz, K. H. Schulte-Elte and G. Ohloff, *Helv. Chim. Acta*, **54**, 1913 (1971).
147. W. Skorianetz, K. H. Schulte-Elte and G. Ohloff, *Angew. Chem., Int. Ed. Engl.*, **11**, 330 (1972).
148. N. Harada, H. Uda and H. Ueno, *Chem. Lett.*, 663 (1972).
149. N. Harada, S. Suzuki, H. Uda and H. Ueno, *Chem. Lett.*, 803 (1972).
150. W. Adam and J. Sanabia, *J. Am. Chem. Soc.*, **99**, 2735 (1977).
151. E. S. Huyser and R. M. VanScoy, *J. Org. Chem.*, **33**, 3524 (1968).
152. A. J. Bloodworth and D. S. Baker, *J. Chem. Soc., Chem. Commun.*, 547 (1981).
153. D. J. Coughlin and R. G. Salomon, *J. Am. Chem. Soc.*, **101**, 2761 (1979).
154. W. Adam and N. Duran, *J. Am. Chem. Soc.*, **99**, 2729 (1977).
155. R. G. Salomon, M. F. Salomon and D. J. Coughlin, *J. Am. Chem. Soc.*, **100**, 660 (1978).
156. R. M. Wilson and J. W. Rekers, *J. Am. Chem. Soc.*, **103**, 206 (1981).
157. R. M. Wilson, J. S. Goudar and J. E. Sidenstick, *J. Am. Chem. Soc.*, **109**, 6895 (1987).
158. A. J. Bloodworth, D. S. Baker and H. J. Eggelte, *J. Chem. Soc., Chem. Commun.*, 1034 (1982).
159. C. J. M. van den Heuvel and T. J. de Boer, *Recl. Trav. Chim. Pays-Bas*, **96**, 157 (1977).
160. C. J. M. van den Heuvel, A. Hofland, H. Steinberg and T. J. de Boer, *Recl. Trav. Chim. Pays-Bas*, **99**, 275 (1980).

161. M. Schafer-Ridder, U. Brockes and E. Vogel, *Angew. Chem., Int. Ed. Engl.*, **15**, 228 (1976).
162. C. J. M. van den Heuvel, H. Steinberg and T. J. de Boer, *Recl. Trav. Chim. Pays-Bas*, **99**, 109 (1980).
163. P. F. Southern and W. A. Waters, *J. Chem. Soc.*, 4340 (1960).
164. J. Rigaudy, J. Baranne-Lafont, A. Defoin and N. K. Cuong, *Tetrahedron*, **34**, 73 (1978).
165. J. Rigaudy and D. Sparfel, *Tetrahedron*, **34**, 2263 (1978).
166. J. Rigaudy and D. Sparfel, *Tetrahedron*, **34**, 113 (1978).
167. R. Schmidt and H.-D. Brauer, *J. Photochem.*, **34**, 1 (1986).
168. H. H. Wasserman and J. R. Scheffer, *J. Am. Chem. Soc.*, **89**, 3073 (1967).
169. H. H. Wasserman, J. R. Scheffer and J. L. Cooper, *J. Am. Chem. Soc.*, **94**, 4991 (1972).
170. T. Wilson, A. U. Khan and M. M. Mehrotra, *Photochem. Photobiol.*, **43**, 661 (1986).
171. I. Saito, R. Nagata and T. Matsuura, *J. Am. Chem. Soc.*, **107**, 6329 (1985).
172. N. J. Turro, M.-F. Chow and J. Rigaudy, *J. Am. Chem. Soc.*, **101**, 1300 (1979).
173. N. J. Turro, M.-F. Chow and J. Rigaudy, *J. Am. Chem. Soc.*, **103**, 7218 (1981).
174. N. J. Turro and M.-F. Chow, *J. Am. Chem. Soc.*, **101**, 3701 (1979).
175. D. R. Kearns and A. U. Khan, *Photochem. Photobiol.*, **10**, 193 (1969).
176. N. J. Turro, V. Ramamurthy, W. Cherry and W. Farneth, *Chem. Rev.*, **78**, 125 (1978).
177. J. G. Calvert and J. N. Pitts, *Photochemistry*, Wiley, New York (1966), p. 443.
178. J. Rigaudy, C. Breliere and P. Scribe, *Tetrahedron Lett.*, 687 (1978).
179. P.-T. Chou, S. Khan and H. Frei, *Chem. Phys. Lett.*, **129**, 463 (1986).
180. P.-T. Chou and H. Frei, *J. Chem. Phys.*, **87**, 3843 (1987).
181. R. Schmidt, *J. Photochem.*, **23**, 379 (1983).
182. R. Schmidt, K. Schaffner, W. Trost and H.-D. Brauer, *J. Phys. Chem.*, **88**, 956 (1984).
183. W. Drews, R. Schmidt and H.-D. Brauer, *Chem. Phys. Lett.* **70**, 84 (1980).
184. B. Stevens and W. A. Glauser, *Chem. Phys. Lett.*, **129**, 120 (1986).
185. R. Schmidt and H.-D. Brauer, *J. Photochem.*, **15**, 85 (1981).
186. H.-D. Brauer and R. Schmidt, *Photochem. Photobiol.*, **41**, 119 (1985).
187. S. Tokita, S. Suge, M. Toya and H. Nishi, *Nippon Kagaku Kaishi*, 97 (1988).
188. S. Tokita, T. Arai, M. Toya and H. Nishi, *Nippon Kagaku Kaishi*, 814 (1988).
189. S. Tokita, T. Arai, M. Ohaka and H. Nishi, *Nippon Kagaku Kaishi*, 876 (1989).
190. H.-D. Brauer and R. Schmidt, *Photochem. Photobiol.*, **37**, 587 (1983).
191. R. Schmidt, W. Drews and H.-D. Brauer, *J. Am. Chem. Soc.*, **102**, 2791 (1980).
192. H.-D. Brauer, W. Drews and R. Schmidt, *J. Photochem.*, **12**, 293 (1980).
193. S.-Y. Hou, C. G. Dupuy, M. J. McAuliffe, D. A. Hrovat and K. B. Eisenthal, *J. Am. Chem. Soc.*, **103**, 6982 (1981).
194. H.-D. Brauer and R. Schmidt, *J. Photochem.*, **27**, 17 (1984).
195. K. B. Eisenthal, N. J. Turro, C. G. Dupuy, D. A. Hrovat, J. Langan, T. A. Jenny and E. V. Sitzmann, *J. Phys. Chem.*, **90**, 5168 (1986).
196. R. Schmidt, H.-D. Brauer and J. Rigaudy, *J. Photochem.*, **34**, 197 (1986).
197. T. Blumenstock, F. J. Comes, R. Schmidt and H.-D. Brauer, *Chem. Phys. Lett.*, **127**, 452 (1986).
198. T. Blumenstock, K. Jesse, F. J. Comes, R. Schmidt and H.-D. Brauer, *Chem. Phys.*, **130**, 289 (1989).
199. K. Jesse, F. J. Comes, R. Schmidt and II.-D. Brauer, *Chem. Phys. Lett.*, **160**, 8 (1989).
200. N. P. Ernsting, R. Schmidt and H.-D. Brauer, *J. Phys. Chem.*, **94**, 5252 (1990).
201. A. J. Bloodworth and H. J. Eggelte, *Tetrahedron Lett.*, **25**, 1525 (1984).
202. R. Srinivasan, K. H. Brown, J. A. Ors and L. S. White, *J. Am. Chem. Soc.*, **101**, 7424 (1979).

203. R. J. Conca and W. Bergmann, *J. Org. Chem.*, **18**, 1104 (1953).
204. N. Harada, A. Kawamoto, Y. Takuma and H. Uda, *J. Chem. Soc., Perkin Trans. 1*, 1796 (1976).
205. J. Rigaudy, C. Deletang, D. Sparfel and N. K. Cuong, *C.R. Acad. Sci, Ser. C*, **267**, 1714 (1968).
206. J. Rigaudy, P. Scribe and C. Breliere, *Tetrahedron*, **37**, 2585 (1981).
207. J. Rigaudy, N. K. Cuong and J.-Y. Godard, *Bull. Soc. Chim. Fr.*, 78 (1985).
208. J. Rigaudy, J. Baranne-Lafont, A. Ranjon and A. Caspar, *Bull. Soc. Chim. Fr.*, 187 (1984).
209. J. Rigaudy, A. Caspar, J. Baranne-Lafont and C. Chassagnard, *Bull. Soc. Chim. Fr.*, 195 (1984).
210. S. J. Buckland and R. S. Davidson, *Phosphorus Sulfur*, **18**, 225 (1983).
211. S. Yamada, K. Nakayama, H. Takayama, A. Itai and Y. Iitaka, *J. Org. Chem.*, **48**, 3477 (1983).
212. J. D. Boyd, C. S. Foote and D. K. Imagawa, *J. Am. Chem. Soc.*, **102**, 3641 (1980).
213. N. Akbulut, A. Menzek and M. Balci, *Tetrahedron Lett.*, 1689 (1987).
214. M. Balci and Y. Sutbeyaz, *Tetrahedron Lett.*, **24**, 311 (1983).
215. H. A. J. Carless, J. R. Billinge and O. Z. Oak, *Tetrahedron Lett.*, 3133 (1989).
216. K. Takahashi and M. Kishi, *Tetrahedron*, **44**, 4737 (1988).
217. M. Balci and Y. Sutbeyaz, *Tetrahedron Lett.*, **24**, 4135 (1983).
218. Y. Sutbeyaz, H. Secen and M. Balci, *J. Org. Chem.*, **53**, 2312 (1988).
219. K. E. O'Shea and C. S. Foote, *J. Org. Chem.*, **54**, 3475 (1989).
220. M. Balci and N. Akbulut, *Tetrahedron*, **41**, 1315 (1985).
221. M. Suzuki, R. Noyori and N. Hamanaka, *J. Am. Chem. Soc.*, **104**, 2024 (1982).
222. J. A. Turner and W. Herz, *J. Org. Chem.*, **42**, 1895 (1977).
223. W. Herz, R. C. Ligon, J. A. Turner and J. F. Blount, *J. Org. Chem.*, **42**, 1885 (1977).
224. J. A. Turner and W. Herz, *J. Org. Chem.*, **42**, 1900 (1977).
225. D. Brown, B. T. Davis, T. G. Halsall and A. P. Hands, *J. Chem. Soc.*, 4492 (1962).
226. D. Brown, B. T. Davis and T. G. Halsall, *J. Chem. Soc.*, 1095 (1963).
227. J. A. Turner and W. Herz, *Experientia*, **33**, 1133 (1977).
228. K. Takahashi and M. Kishi, *J. Chem. Soc., Chem. Commun.*, 722 (1987).
229. K. Takahashi and M. Kishi, *Tetrahedron Lett.*, **29**, 4595 (1988).
230. N. A. Porter and R. C. Mebane, *Tetrahedron Lett.*, **23**, 2289 (1982).
231. N. A. Porter, J. R. Nixon and D. W. Gilmore, *ACS Symp. Ser.*, **69**, 89 (1978).
232. M. Suzuki, R. Noyori and M. Hamanaka, *J. Am. Chem. Soc.*, **103**, 5606 (1981).
233. M. Suzuki, Y. Oda and R. Noyori, *Tetrahedron Lett.*, **22**, 4413 (1981).
234. J.-P. Hagenbuch and P. Vogel, *J. Chem. Soc., Chem. Commun.*, 1062 (1980).
235. J.-P. Hagenbuch, J.-L. Birbaum, J.-L. Metral and P. Vogel, *Helv. Chim. Acta*, **65**, 887 (1982).
236. S. Kohmot, S. Kasai, M. Yamamoto and K. Yamada, *Chem. Lett.*, 1477 (1988).
237. M. Natsume and H. Muratake, *Tetrahedron Lett.*, 3477 (1979).
238. M. Natsume, Y. Sekine and M. Ogawa, *Tetrahedron Lett.*, 3473 (1979).
239. M. Natsume, I. Utsumomiya, K. Yamaguchi and S. Sakai, *Tetrahedron*, **41**, 2115 (1985).
240. R. Riguera and E. Quinoa, *J. Chem. Soc., Chem. Commun.*, 1120 (1984).
241. E. J. Corey, P. Da Silva Jardine and J. C. Rohloff, *J. Am. Chem. Soc.*, **110**, 3672 (1988).
242. J. Rigaudy, M. Maumy, P. Capdevielle and L. Breton, *Tetrahedron*, **33**, 53 (1977).
243. G. Rio, D. Bricout and L. Lacombe, *Tetrahedron Lett.*, 3583 (1972).
244. W. Adam, G. Klug, E.-M. Peters, K. Peters and H. G. von Schnering, *Tetrahedron*, **41**, 2045 (1985).
245. A. J. Bloodworth and H. J. Eggelte, *J. Chem. Soc., Perkin Trans. 2*, 2069 (1984).

246. P. Bladon, R. B. Clayton, C. W. Greenhalgh, H. B. Henbest, E. R. H. Jones, B. J. Lovell, G. Silverstone, G. W. Wood and G. F. Woods, *J. Chem. Soc.*, 4883 (1952).
247. W. Adam and M. Balci, *J. Am. Chem. Soc.*, **101**, 7537 (1979).
248. F. R. Mayo, *Acc. Chem. Res.*, **1**, 193 (1968).
249. N. A. Porter and J. R. Nixon, *J. Am. Chem. Soc.*, **100**, 7116 (1978).
250. N. A. Porter, M. A. Cudd, R. W. Miller and A. T. McPhail, *J. Am. Chem. Soc.*, **102**, 414 (1980).
251. D. J. Carlsson and K. U. Ingold, *J. Am. Chem. Soc.*, **90**, 7047 (1968).
252. P. A. Burns, C. S. Foote and S. Mazur, *J. Org. Chem.*, **41**, 899 (1976).
253. P. A. Burns and C. S. Foote, *J. Org. Chem.*, **41**, 908 (1976).
254. M. Hamberg and G. Samuelsson, *Proc. Natl. Acad. Sci. USA*, **71**, 3400 (1974).
255. E. L. Clennan and P. C. Heah, *J. Org. Chem.*, **46**, 4105 (1981).
256. D. B. Denney, D. Z. Denney, C. D. Hall and K. L. Marsi, *J. Am. Chem. Soc.*, **94**, 245 (1972).
257. A. L. Baumstark, C. J. McCloskey, T. E. Williams and D. R. Chrisope, *J. Org. Chem.*, **45**, 3593 (1980).
258. E. L. Clennan and P. C. Heah, *J. Org. Chem.*, **49**, 2284 (1984).
259. A. Baumstark, M. Barrett and K. M. Kral, *J. Heterocyc. Chem.*, **19**, 201 (1982).
260. E. L. Clennan and P. C. Heah, *J. Org. Chem.*, **48**, 2621 (1983).
261. E. L. Clennan and P. C. Heah, *J. Org. Chem.*, **47**, 3329 (1982).
262. N. Kornblum and H. E. De La Mare, *J. Am. Chem. Soc.*, **73**, 880 (1951).
263. M. G. Zagorski and R. G. Salomon, *J. Am. Chem. Soc.*, **102**, 2501 (1980).
264. W. Adam and I. Erden, *Tetrahedron Lett.*, 1975 (1979).
265. A. P. Schaap, P. A. Burns and K. A. Zaklika, *J. Am. Chem. Soc.*, **99**, 1270 (1977).
266. A. A. Frimer, *Chem. Rev.*, **79**, 359 (1979).
267. C. W. Jefford, J.-C. Rossier, S. Kohmoto and J. Boukouvalas, *J. Chem. Soc., Chem. Commun.*, 1496 (1984).

7 Diacyl Peroxides

KEN FUJIMORI

Department of Chemistry, University of Tsukuba, Tsukuba, Ibaraki 305, Japan

Organic Peroxides. Edited by W. Ando

This chapter deals with basic diacyl peroxide chemistry and advances therein since Hiatt's review in 1970.[1]

1 PREPARATION, PURIFICATION AND SAFEGUARDS

1.1 Preparation

Symmetrical diacyl peroxides are usually prepared by the reaction between sodium peroxide and acylating agents, e.g. acid halides, acid anhydrides (equation 1).[2–7] A few six-membered cyclic diacyl peroxides have been synthesized in the same manner, but in low yield.[7,8] Sodium peroxide can be replaced with a combination of hydrogen peroxide and base.[9–14] The reaction of sodium superoxide with acylating agents also gives diacyl peroxides in aprotic solvents (equation 2).[15]

$$2\,R{-}\overset{\overset{\displaystyle O}{\|}}{C}{-}X + Na_2O_2\ (H_2O_2/2B{:}) \longrightarrow R{-}\overset{\overset{\displaystyle O}{\|}}{C}{-}O{-}O{-}\overset{\overset{\displaystyle O}{\|}}{C}{-}R + 2\,NaX\ (2BH^+) \quad (1)$$

$$2\,R{-}\overset{\overset{\displaystyle O}{\|}}{C}{-}X + 2\,NaO_2 \longrightarrow R{-}\overset{\overset{\displaystyle O}{\|}}{C}{-}O{-}O{-}\overset{\overset{\displaystyle O}{\|}}{C}{-}R + 2\,NaX + O_2 \quad (2)$$

The condensation reaction of carboxylic acids with hydrogen peroxide in the presence of dicyclohexylcarbodiimide (DCC) is a convenient synthetic method to obtain diacyl peroxides on a laboratory scale. This method has the advantages that carboxylic acids can be used directly without conversion into acid halides or acid anhydrides and the reaction can be carried out in a single organic solvent without any acid or base catalyst and it proceeds cleanly.[16–18] Greene and Kazan[16] improved the yield of phthaloyl peroxide by this method (equation 3).

$$\text{phthalic anhydride} + H_2O_2 + \underset{\text{DCC}}{C_6H_{11}{-}N{=}C{=}N{-}C_6H_{11}} \longrightarrow \text{phthaloyl peroxide} + (C_6H_{11}{-}NH)_2CO \quad (3)$$

Martin and King[19] converted phenylmaleic anhydride to diperoxyphenylmaleic acid, which then cyclized to phenylmaleoyl peroxide. Adam and co-workers[20] prepared disubstituted malonyl peroxides by stirring a mixture of the malonic acid, 98% hydrogen peroxide and methanesulfonic acid (equation 4).

$$R_2C(CO_2H)_2 + H_2O_2 \xrightarrow{CH_3SO_3H} R_2C\!<\!\begin{matrix}C(=O){-}O\\ C(=O){-}O\end{matrix}\!\!\mid + 2\,H_2O \quad (4)$$

Ramirez *et al.*[21] reported another good procedure to obtain cyclic diacyl peroxides by a series of sequential reactions. Interestingly, in this procedure acyclic diketones afford acyclic peroxides in poorer yield than the cyclic compounds,[21] suggesting the intervention of the intermediate shown in equation 5. On the other hand, ozonolysis of 1, 2-diphenylethylenediol carbonate gives benzoyloxycarbonyl benzoyl peroxide (equation 6).[22]

$$\text{(phenanthrenequinone–}P(OCH_3)_3\text{ adduct)} + O_3 \longrightarrow \left[\text{ozonide–}P(OCH_3)_3 \longrightarrow \text{zwitterion–}P(OCH_3)_3\right] \longrightarrow \text{(cyclic diacyl peroxide)} + O{=}P(OCH_3)_3 \qquad (5)$$

$$\text{(4,5-diphenyl-1,3-dioxol-2-one)} + O_3 \longrightarrow Ph{-}\overset{O}{\overset{\|}{C}}{-}O{-}O{-}\overset{O}{\overset{\|}{C}}{-}O{-}\overset{O}{\overset{\|}{C}}{-}Ph \qquad (6)$$

Unsymmetrical diacyl peroxides are generally synthesized from peracids and acylating agents in the presence of base (equation 7).[23–27] A prolonged reaction time should be avoided, since the base catalyzes the acyl transfer reaction to give symmetrical diacyl peroxides. Here again DCC is an effective dehydrating agent to condense a peracid and a carboxylic acid producing an unsymmetrical diacyl peroxide (equation 8).

$$R{-}\overset{O}{\overset{\|}{C}}{-}X + HOO\overset{O}{\overset{\|}{C}}{-}R' + {:}B \longrightarrow R{-}\overset{O}{\overset{\|}{C}}{-}O{-}O{-}\overset{O}{\overset{\|}{C}}{-}R' + HB^+ \qquad (7)$$

$$R{-}\overset{O}{\overset{\|}{C}}{-}OH + HOO\overset{O}{\overset{\|}{C}}{-}R' + DCC \longrightarrow R{-}\overset{O}{\overset{\|}{C}}{-}O{-}O{-}\overset{O}{\overset{\|}{C}}{-}R' + \left(C_6H_{11}{-}NH\right)_2CO \qquad (8)$$

Autoxidation of aldehydes in the presence of acylating agents is a convenient synthetic route to unsymmetrical diacyl peroxides in which peracids formed by the autoxidation react with the acylating agent. If $^{18}O_2$ or $^{17}O_2$ is used, one obtains the unsymmetrical diacyl peroxides in which the peroxidic oxygen atoms are labeled with ^{18}O or ^{17}O (equation 9).[28]

$$C_6H_5{-}\overset{O}{\overset{\|}{C}}H \xrightarrow{^{17}\bullet_2,\ (CD_3CO)_2O} C_6H_5{-}\overset{O}{\overset{\|}{C}}{-}^{17}\bullet{-}^{17}\bullet{-}\overset{O}{\overset{\|}{C}}{-}CD_3 + CD_3CO_2H \qquad (9)$$

1.2 Purification and determination

Following on from the synthetic procedures outlined above, crude diacyl peroxides are obtained as solutions in alkanes, dichloromethane, diethyl ether, etc. When the solution contains an organic base such as pyridine, it should be washed with a cold dilute aqueous acid solution followed by cold aqueous sodium hydrogen-carbonate solution. Evaporation can be achieved below 0°C under vacuum and stopped before dryness to retain sufficient solvent to crystallize the peroxides by cooling in a dry-ice–acetone bath. The mother liquid is sucked out through a small-diameter glass tube with a sintered-glass filter attached to the end. The glass tube is connected to a suction pump via a trap which receives the mother liquid. The residual crystalline or oily peroxide is dissolved into the minimum amount of a chosen solvent (alkanes, methanol, diethyl ether, etc.) at a suitable temperature depending on the stability of the peroxide (usually lower than room temperature) and kept in a freezer to recrystallize the peroxides, if necessary, cooled with a dry-ice–acetone bath. Repeated recrystallization of the peroxides from two different kinds of solvents, e.g. hexane and methanol, gives pure peroxides. In a few cases diacyl peroxides can be purified by distillation.[20b]

Methods for the determination of diacyl peroxides have been summarized by Mair and Hall.[29] The following iodimetric method is widely applicable. A 45-ml volume of freshly distilled propan-2-ol is placed in a 100 ml conical flask. Acetic acid (1 ml), a freshly prepared saturated aqueous potassium iodide solution (1 ml) and a small chip of dry-ice are added to the flask. A solution containing about 0.1 mmol of diacyl peroxide is pipetted into the flask, which is then heated for 5–15 min on a hot-plate up to incipient boiling. The iodine thus liberated in the hot solution is immediately titrated with 0.02 mol l^{-1} sodium thiosulfate solution without any indicator. A control experiment should be done in the same manner without the peroxide. When the peroxide is labile, sampling should be carried out in a cold room.

1.3 Explosive Hazards and Safeguards

Decomposition of organic peroxides can be initiated by heat, mechanical shock, friction and contamination of redox reaction catalysts such as heavy metal ions.[30] The sensitivity of peroxides to various hazardous conditions and the explosive power differ depending on the structure of the peroxides.[31] Mageli and Sheppard[30] reviewed test methods for the safety evaluation of peroxides for industrial uses.

Both the sensitivity and the power of explosion of organic peroxides increase with increasing active oxygen concentration. Since polymeric phthaloyl peroxides, $(OOCC_6H_4COO)_n$, and succinyl peroxides, $(OOCCH_2CH_2COO)_n$, are extremely explosive,[32] handling of the dry peroxides should be avoided. The rates of unimolecular decomposition of monomeric phthaloyl peroxide and 2,3-dimethylsuccinyl peroxide are slow in solution; however, since these six-membered cyclic peroxides are highly explosive and sensitive to mechanical shock, only small amounts of solid peroxides should be handled.[8,33] Perfluoroalkanoyl peroxides, such as $(CF_3CO_2)_2$ and $(C_3F_7CO_2)_2$, are not only very unstable but also vigorously explosive and hence the isolation of neat peroxides should be avoided for safety. Although acetyl peroxide decomposes fairly slowly in solution, crystalline acetyl peroxide is highly explosive because of its high peroxide oxygen concentration. When neat diacyl peroxides for which the activation energies of unimolecular decomposition in solution are fairly small, e.g. cyclohexaneformyl peroxide, 2-methylphenylpropionyl peroxide or cyclopropaneformyl peroxide, are allowed to stand at room temperature, these inherently unstable peroxides explode spontaneously.

Since heavy metal ions induce degradation or explosion of organic peroxides, a metallic spatula cannot be used for handling peroxides. Pure peroxides are much more stable than contaminated compounds; however, the presence of pure water in solid peroxides reduces their explosive characteristics. Diacyl peroxides in pure organic solution are hardly explosive. Aromatic solvents cannot be used to dissolve perfluoroalkanoyl peroxides.[34]

Diacyl peroxides should be stored in a deep freezer. All experiments should be carried out behind a safety shield in a hood.

2 PHYSICAL PROPERTIES

2.1 Infrared Spectra

The carbonyl functions in diacyl peroxides exhibit a very strong characteristic doublet in the $\nu_{c=O}$ region (aliphatic, 1784–1796 and 1811–1820 cm^{-1}; aromatic, 1758–1783 and 1780–1805 cm^{-1}). The corresponding acid anhydrides and acyl alkyl (or aryl) carbonates (the carboxy-inversion products formed from the diacyl peroxides via an ionic reaction path) also show characteristic doublet absorptions in a similar $\nu_{c=O}$ region; however, the splitting of the $\nu_{c=O}$ doublet for diacyl peroxides (25–30 cm^{-1}) differs from that of acyl alkyl carbonates (55–65 cm^{-1}). Some typical examples are listed in Table 1. Since the carbonyl stretching frequencies for both carboxylic acids (aliphatic, 1712 ± 12 cm^{-1}; aromatic, 1690 ± 10 cm^{-1}) and esters (aliphatic, 1740 ± 10 cm^{-1}; aromatic, 1723 ± 7 cm^{-1}) are lower than those for the carboxy-inversion products, $\nu_{c=O}$ absorption bands are useful for the quantitative analysis of the reaction products of the decomposition of diacyl peroxides. For example, thermolysis of benzoyl cyclohexaneformyl peroxide affords benzoyl cyclohexyl carbonate (the carboxy-inversion product: $\nu_{c=O}$ 1750 and 1805 cm^{-1}), cyclohexyl benzoate ($\nu_{c=O}$ 1719 cm^{-1}) and benzoic acid ($\nu_{c=O}$ 1700 cm^{-1}), the yields of which were successfully determined by IR spectrometry analysis.[27]

2.2 Structure

X-ray crystallographic data are available for substituted[35,36] and unsubstituted[37] benzoyl peroxides, acetyl benzoyl peroxide,[28] phthaloyl peroxide[38] and

Table 1. Infrared spectral data for some selected diacyl peroxides (RCO—O—O—COR′) and the corresponding acid anhydrides (RCO—O—COR′) and acyl carbonates (ROCO—O—COR′)[a]

R	R′	Carbonyl stretching vibration (cm^{-1})		
		Diacyl peroxide	Acid anhydride	Acyl carbonate
CH_3	CH_3	1820, 1796	1824, 1748	
CF_3	CF_3	1860, 1840	1884, 1818	
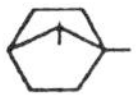		1789, 1766		1800, 1756
$PhCH_2(CH_3)CH$—	$PhCH_2(CH_3)CH$—	1775, 1805		1815, 1750
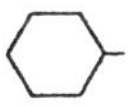	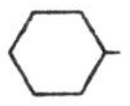	1787, 1767		1790, 1745
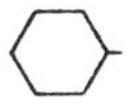	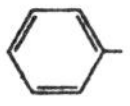	1800, 1770		1805, 1750
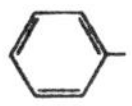		1789, 1767	1789, 1727	1780, 1725
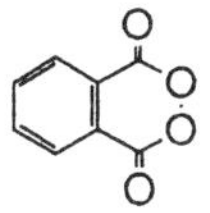		1768, 1750 (s)	1845, 1775	
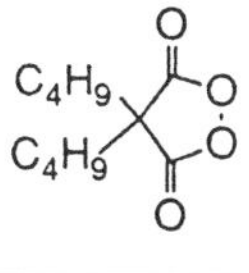 C_4H_9, C_4H_9		1783		

[a]Carboxy conversion product of the corresponding peroxide.

a few aliphatic diacyl peroxides with long alkyl chains.[35,39] Three of them are illustrated in Figures 1–3.

These structural data reported for diacyl peroxides show the following characteristics: (a) The peroxide dihedral angle (CO—O—O—CO) in acyclic diacyl peroxides resembles that of anhydrous crystalline hydrogen peroxide,[38] while the introduction of bulky substituents at the *ortho* positions of the phenyl ring in benzoyl peroxide increases the dihedral angle. MO calculations by Sawada *et al.*[41] revealed that the introduction of a fluorine atom at the α-carbon of aliphatic diacyl peroxides widens the peroxide dihedral angle. (b) Two carbonyl groups in acyclic diacyl peroxides are located facing inwards towards each other. (c) The length of the O—O bond is 1.460 ± 0.015 Å, which is similar to those reported for other organic peroxides.[35] (d) In phthaloyl peroxide, a

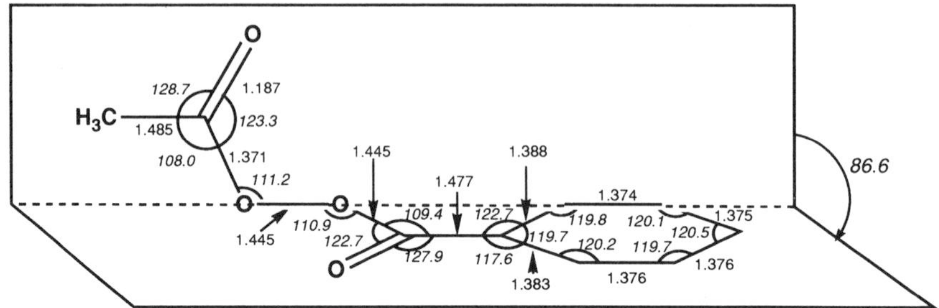

Figure 1. Structure of acetyl benzoyl peroxide.[28] Bond lengths in Ångströms, angles (italics) in degrees

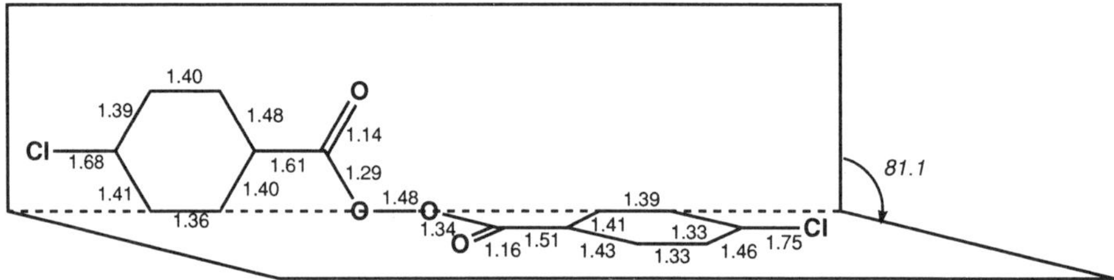

Figure 2. Structure of 4-chlorobenzoyl peroxide.[36] Bond lengths in Ångströms, angles (italics) in degrees

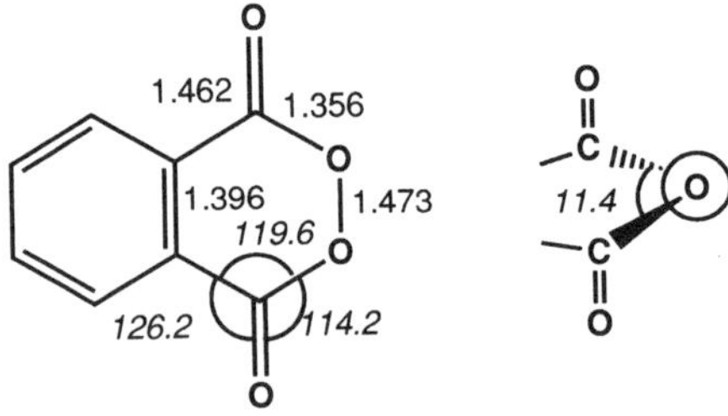

Figure 3. Structure of phthaloyl peroxide. Right: Newman projection along the peroxide bond.[38] Bond lengths in Ångströms; angles (italics) in degrees

six-membered aromatic diacyl peroxide, the peroxide dihedral angle is forced to 11.4° and the O—O bond is stretched to 1.473 Å because of its cyclic structure.

The conformations of acyclic diacyl peroxides determined by the X-ray method are consistent with MO calculation results.[37]

3 UNIMOLECULAR REACTIONS

Numerous kinetic and product analysis data have been accumulated for the thermal decomposition of diacyl peroxides.[1] Generally, the rate equation for the thermal decomposition of diacyl peroxides involves a unimolecular reaction process, a molecular-induced reaction process with a closed shell molecule (D) and a radical-induced reaction path, whose rate constants are denoted by k_d, k_{mi} and k_{ri}, respectively. The experimentally observable apparent first-order rate constant is expressed by k_d^{obs}. The apparent experimentally observable k_d is the sum of first-order rate constants of radical (k_r) and ionic (k_i) decomposition processes.

$$\begin{aligned} -\mathrm{d}[\text{peroxide}]/\mathrm{d}t &= k_d^{obs}[\text{peroxide}] \\ &= k_d[\text{peroxide}] + k_{mi}[\text{peroxide}][\text{D}] + k_{ri}[\text{peroxide}]^n \\ &= (k_r + k_i)[\text{peroxide}] + k_{mi}[\text{peroxide}][\text{D}] + k_{ri}[\text{peroxide}]^n \end{aligned} \tag{10}$$

$$k_d^{obs} = k_d + k_{mi}[\text{D}] + k_{ri}[\text{peroxide}]^{n-1} \tag{11}$$

This section deals with unimolecular reactions of diacyl peroxides which do not contain functional groups that initiate intramolecular induced decompositions. Scheme 1 represents all possible unimolecular decomposition paths of diacyl peroxides other than reactions with the solvent.[27] The bars above radical pairs and ion pairs in Scheme 1 mean that the species beneath the bars exist in a solvent cage. Neither the diffusion process of these radical pairs and ion pairs nor the reactions which occur outside the solvent cage are represented in Scheme 1. The reactions of radicals and ions that have escaped from the solvent cage belong to the category of general radical and ion chemistry, and hence this chapter is focused on the chemical behavior of the radical pairs and the ion pairs generated from diacyl peroxides. The reactivities of diacyl peroxides (RCO_2O_2R', **1**) depend totally on the structures of R and R′ and the polarity of the medium. As shown in Table 2, Fujimori and Oae[6,43] classified diacyl peroxides into three cases from structural points of view. Table 2 also lists the mechanisms of the major decomposition pathways and the fraction of cage return (f_r) of the acyloxy radical intermediate for each case. The thermal instability of diacyl peroxides originates primarily from the weak peroxide bond. Depending on the stability of R^+, R'^+, $R^\bullet$, and $R'^\bullet$, O—O bond scission accompanies simultaneous R—CO and/or R′—CO bond cleavage in either a homolytic or a heterolytic manner. Primary aliphatic radicals and carbocations are unstable. Aryl radicals and aryl carbocations are more unstable. Therefore, thermal decomposition of Case IA, IIA and III peroxides proceeds mainly by a homolytic one-bond scission mechanism, and hence all with similar rates. Since secondary and tertiary alkyl radicals and carbocations are stable, Case IB and IIB peroxides decompose mainly by either heterolytic or

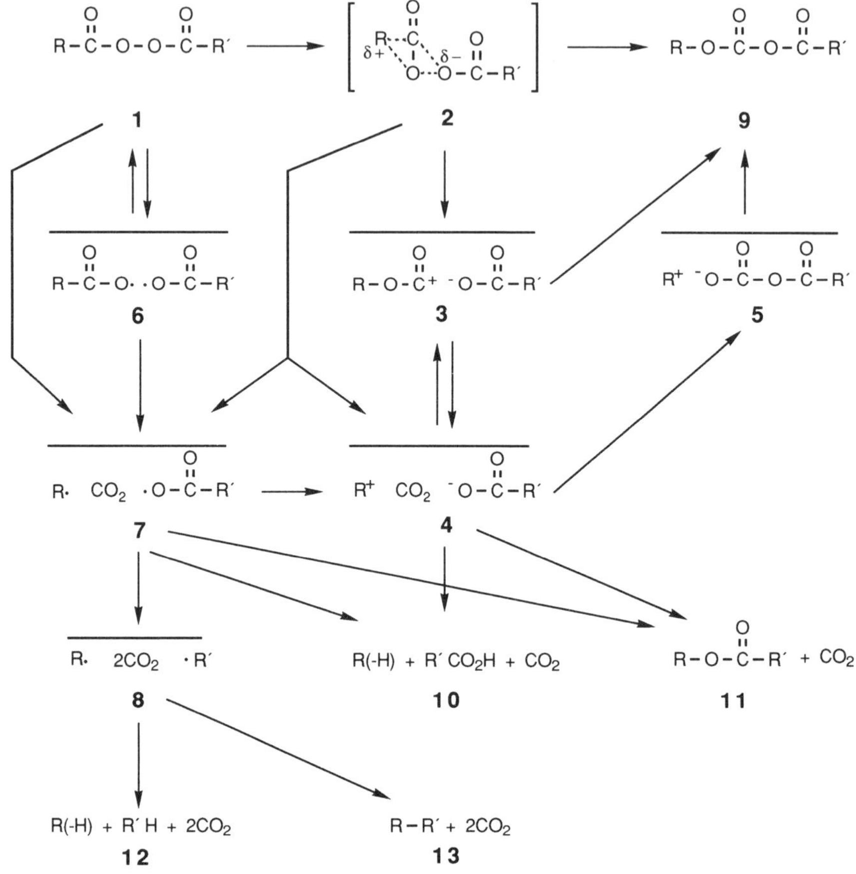

Scheme 1. Decomposition of diacyl peroxide in solution

homolytic two-bond scission mechanisms. Experimental data for the thermal decomposition of some typical diacyl peroxides are given in Table 3.

3.1 Thermal Decomposition by Radical Reaction Mechanisms

3.1.1 Classification of diacyl peroxides—structural effect on fates of acyloxy radical pair

Karsch *et al.*[28] photolyzed a single crystal of acetyl benzoyl peroxide at 77 K and found that the benzoyloxy radical is stable enough to exist at this temperature; however, they could not detect acetoxy radicals but observed methyl radicals. Recent advances in laser flash photolysis have made it possible to

Table 2. Classification of diacyl peroxides (RCO—O—O—COR′)

Case	R	R′	Mode of O—O bond scission[a]	Rate of thermal decomposition	f_r[b]
IA	Primary alkyl	Primary alkyl	Homolytic	Medium	Large
IB	*sec*-Alkyl *tert*-Alkyl	*sec*-Alkyl *tert*-Alkyl[c]	Heterolytic, Homolytic	Fast	Small
IIA	Primary alkyl	Aryl	Homolytic	Slow	Very small
IIB	*sec*-Alky- *tert*-Alkyl[c]	Aryl	Heterolytic	Fast	Medium
III	Aryl	Aryl	Homolytic	Slow	Small

[a]Major reaction mode.
[b]Fraction of cage return of acyloxy radical pair.
[c]Primary alkyl groups, such as cyclopropylmethyl, benzyl, allyl, which stabilize positive charge are also involved in this class.

observe aroyloxy radicals directly,[44–51] but the direct detection of aliphatic acyloxy radicals has never been reported. Jaffe *et al.*[52] estimated that the decarboxylation of benzoyloxy radical is thermally neutral, whereas those of acetoxy, propionyloxy and butyryloxy radicals are 17, 14 and 13 kcal mol^{-1} exothermic, respectively.[52] Figure 4 shows that the relative activation free

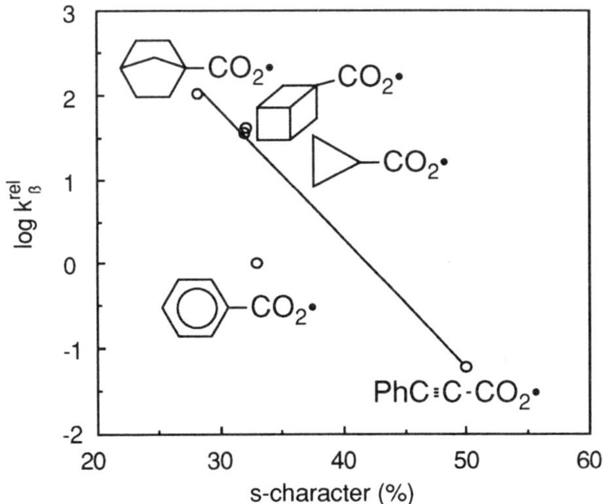

Figure 4. Plot of logarithm of relative rate constant of decarboxylation (k_β^{rel}) of $RCO_2\cdot$ against s-character in R—CO_2 bond[53]

Table 3. First-order rate constants for decomposition (k_d) and ^{18}O scrambling (k_s) and yield of carboxy-inversion product (C_i) in thermolysis of some selected diacyl peroxides (RCO—O—O—COR′)

Case	R	R′	Solvent	T (°C)	$k_d \times 10^6$ (s^{-1})	$k_s \times 10^6$ (s^{-1})	$C_i \times 100$[a]	Ref.
IA	CH_3	CH_3	Isooctane	80	72.0	44.0	0	70
	CH_3	CH_3	Mineral oil	80	52	63.7	0	74a
	CH_3	CH_3	Gas phase	75.5	41.0	28.4	0	77
	CH_3	C_2H_5	Isooctane	80	78.9	16.6	0	74b
	C_2H_5	C_2H_5	Isooctane	80	78.9	8.1	0	74b
IB	$(CH_3)_2CH$	$(CH_3)_2CH$	CCl_4	40	240		30	151
	cyclopropyl-CH_2—	cyclopropyl-CH_2—	CCl_4	55	320	0	37	156
	bicyclo[2.2.1]heptyl	bicyclo[2.2.1]heptyl	CCl_4	80	104	58.2	52	6
	bicyclo[2.2.2]octyl	bicyclo[2.2.2]octyl	CCl_4	30	171		91[b]	151
	$(CH_3)_3C$—	$(CH_3)_3C$—	CCl_4	10	190		85[b]	151
IIA	cyclopropyl	phenyl	CCl_4	80	56.5	5.15		54

IIB	$(CH_3)_2CH$		CCl_4	70	6000		65	157
			CCl_4	70	17.4	*ca* 0.7	51	26
			CCl_4	45	67.2	0	71.4	27
		Cl	CCl_4	45	193	0	74	27
III			Isooctane	80	27	1.30	0	75
			Mineral oil	80	28.9	6.27	0	75
			CCl_4	60	180			96
IIC			CCl_4	80	66.5	5.0	0	84
IIIC			CCl_4	80	0.312	13.0	0	84

[a]Yield of carboxy-inversion product (%).
[b]Total contribution of ionic path.

energy for the decarboxylation of $RCO_2\cdot$($\log k_\beta^{rel}$) decreases linearly with increase in the s-character of the R—CO bond. The point for the benzoyloxy radical is no longer on the line, suggesting that it is highly stabilized.[53] These results show that aliphatic acyloxy radicals are highly unstable in comparison with aroyloxy radicals. Figure 5 illustrates free energy diagrams for the reactions of acyloxy radical pairs which are formed in a solvent cage by the homolysis of O—O bonds based on the striking difference in stability between aromatic and aliphatic acyloxy radicals.[6,43,54]

The extent of the cage return of the acyloxy radical pair to regenerate peroxide (f_r) is controlled by the relative height of the free energy barriers of the dissociation (DIS), decarboxylation (β) and recombination processes (REC) of the geminate acyloxy radical pair. To a first order approximation, the structural effect of R and R′ on the free energy of activation for diffusion of the acyloxy radical pair out of the solvent cage is much less than the structural effect on the free energy barriers for both the recombination and the decarboxylation of the acyloxy radical pair which depend greatly on the stability of the acyloxy radicals. Thus, since Case I peroxide produces a very unstable acyloxy radical pair, the free energy of activation for the recombination is less than those for Case II and III peroxides, and hence f_r is larger than those for the other cases, although the decarboxylation is very fast ($k_{REC} \approx k_\beta > k_{DIS}$). The Case I acyloxy radical pair may be more stable than the two independent acyloxy radicals. A detailed discussion of this is given in Section 3.1.9.3. Since the free

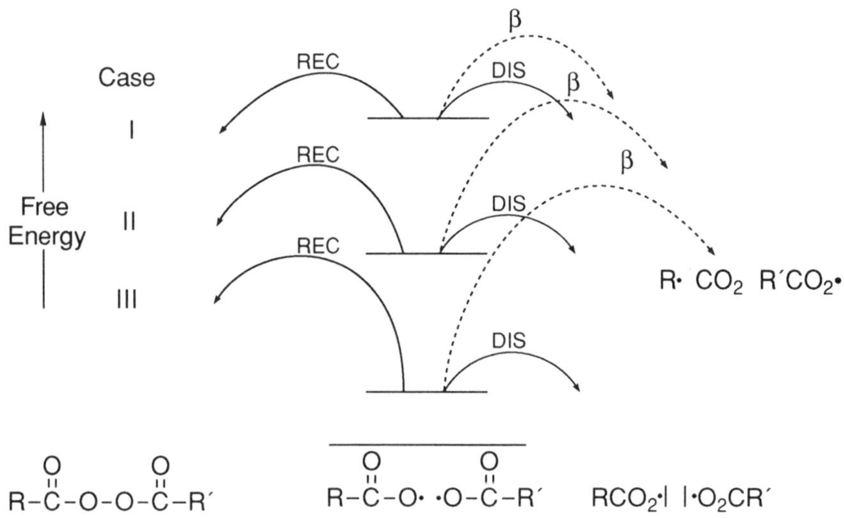

Figure 5. Postulated free energy diagrams for reactions of caged acyloxy radical pairs, i.e. recombination (REC), dissociation (DIS) and decarboxylation (β)

energy barrier of the recombination of the Case II acyloxy radical pair is greater than that of the Case I acyloxy radical pair, f_r for Case II becomes less than that for Case I. In the thermolysis of Case III peroxides, since the recombination of stable aroyloxy radical pairs requires a higher free energy of activation than the Case I acyloxy radical pair, k_{REC} is small and hence the diffusion process from the solvent cage prodominates over both the recombination and the decarboxylation processes in the usual solvents ($k_{DIS} \gg k_{REC} > k_{\beta}$). Therefore, f_r for the Case III acyloxy radical pair is fairly small in the usual solvents, but extremely high in highly viscous solvents ($k_{REC} > k_{\beta} > k_{DIS}$).

3.1.2 Experimental criteria to diagnose the mode of radical decomposition of diacyl peroxides

Depending on the stabilities of R· and R′·, the mode of decomposition of **1** varies from one-bond to multi-bond scission (Schemes 2 and 3). The modes of homolysis have been diagnosed by the following criteria.

3.1.2.1 Kinetic parameters ($\Delta H^{\ddagger}$, $\Delta S^{\ddagger}$)

Large $\Delta H^{\ddagger}$ and positive $\Delta S^{\ddagger}$ values are obtained for one-bond initiators, whereas multi-bond initiators show small $\Delta H^{\ddagger}$ and negative $\Delta S^{\ddagger}$ values.[55,56]

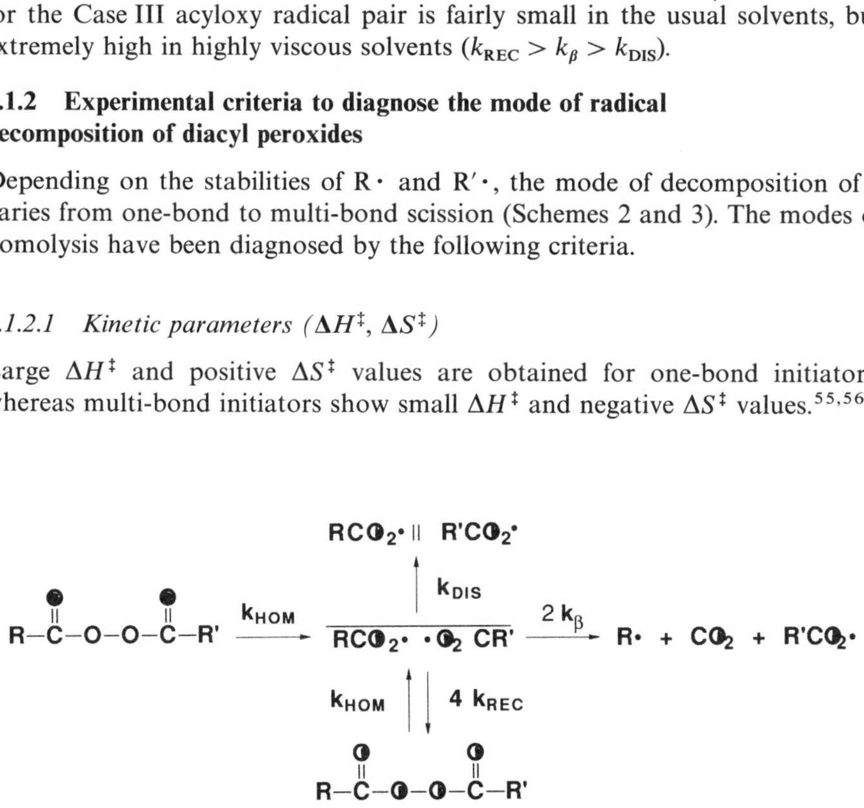

R–C(=O)–O–O–C(=O)–R′ —k_{HOM}→ RCO_2· ·O_2CR' —$2k_{\beta}$→ R· + CO_2 + $R'CO_2$·

RCO_2· ·O_2CR' —k_{DIS}→ RCO_2· ‖ $R'CO_2$·

RCO_2· ·O_2CR' ⇌ (k_{HOM}, $4k_{REC}$) R–C(=O)–O–O–C(=O)–R′

Scheme 2. One-bond scission mechanism (acyloxy radical pair mechanism)

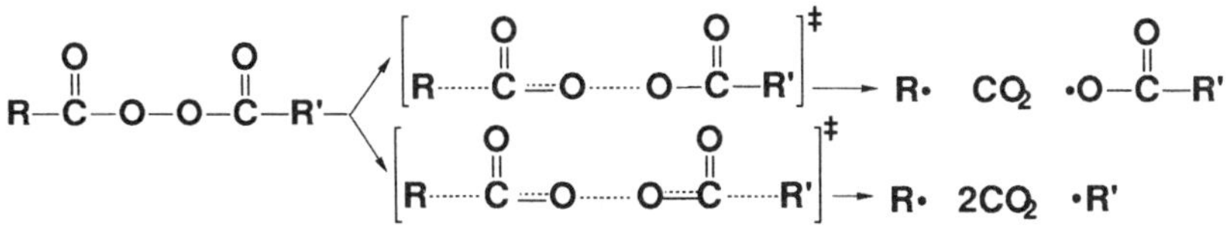

R–C(=O)–O–O–C(=O)–R′ → [R⋯C(=O)⋯O⋯O–C(=O)–R′]‡ → R· CO_2 ·O–C(=O)–R′

R–C(=O)–O–O–C(=O)–R′ → [R⋯C(=O)⋯O⋯O⋯C(=O)⋯R′]‡ → R· $2CO_2$ ·R′

Scheme 3. Two- and three-bond scission mechanisms

3.1.2.2 *Trapping of acyloxy radicals by radical scavengers*

If the peroxide is a one-bond initiator, addition of a radical scavenger diminishes the yield of carbon dioxide. The radical scavenger used for this purpose is required not to induce decomposition of the peroxide tested.[57–59]

3.1.2.3 *Solvent viscosity effect*

According to the solvent cage concept,[60–64] the decomposition rate of a one-bond initiator decreases with increasing viscosity of the medium, since the rate of diffusion of an acyloxy radical pair out of the solvent cage is slowed in a viscous medium. However, the rate of thermal decomposition of multi-bond initiators is insensitive to the viscosity of the medium. For this purpose, a series of alkanes with different chain lengths is often used.[64–66]

The rate constant for the unimolecular thermal decomposition of a one-bond scission initiator (Scheme 2) is expressed by equation 12. Pryor and co-workers[65] correlated the rates of the decomposition of some radical initiators with the viscosity of hydrocarbon solvents of the reaction by equation 13, where $f(\eta)$ is expressed by equation 14, i.e. the rate constant (k_β) of the β-elimination (decarboxylation) of the geminate radical pair was assumed to be negligibly small in comparison with those of the recombination (k_{REC}) and the diffusion out of the solvent cage (k_{DIS}). Later, for the case in which the size of k_β is comparable to k_{REC} and k_{DIS}, Neuman and Lockyer[66] modified the Pryor and co-workers' expression for $f(\eta)$ to equation 15 by taking the k_β value into consideration.

$$k_d = k_{HOM}(2k_\beta + k_{DIS})/(4k_{REC} + 2k_\beta + k_{DIS}) = k_{HOM}(1 - f_r) \quad (12)$$

$$1/k_d = (4k_{REC}/k_{HOM}A_D)f(\eta) + 1/k_{HOM} \quad (13)$$

$$f(\eta) = (\eta/A_V)^{0.7} \quad (14)$$

$$f(\eta) = (\eta/A_V)^{0.64}/[1 + (2k_\beta/A_D)(\eta/A_V)^{0.64}] \quad (15)$$

where

A_D = pre-exponential factor for the diffusion of geminate radical pair;
A_V = pre-exponential factor of the self-diffusive flow of the solvent;
η = viscosity of solvent.

One can obtain k_{HOM} by treating the solvent viscosity-dependent k_d values with equation 13. Then the fraction of cage return of the geminate acyloxy radical pair (f_r) can be calculated from k_{HOM} and k_d by equation 16.

$$f_r = 4k_{REC}/(4k_{REC} + 2k_\beta + k_{DIS}) = 1 - (k_d/k_{HOM}) \quad (16)$$

3.1.2.4 *Activation volume ($\Delta V^{\ddagger}$, pressure effect)*

The rates of decomposition of one-bond scission peroxides decrease with increase in pressure, i.e. the reaction requires a positive activation volume ($\Delta V^{\ddagger}$). Multi-bond scission peroxides decompose faster under higher pressure with negative $\Delta V^{\ddagger}$.[67,68]

3.1.2.5 *Kinetic isotope effects*

Primary kinetic $^{16}O/^{18}O$ and $^{12}C/^{13}C$ isotope effects[69,70] and secondary $^{1}H/^{2}H$ kinetic isotope effects[71–73] have been used.

3.1.2.6 *^{18}O scrambling in diacyl peroxides*

Symmetrical diacyl peroxides involve two kinds of four oxygen atoms. Unsymmetrical diacyl peroxides all have different kinds of four oxygen atoms. Hence, the ^{18}O or ^{17}O tracer experiments provide a lot of useful information for determining the reaction mechanisms.[6,25–27,59,70,74–77] When an [^{18}O]-carbonyl-labeled one-bond scission diacyl peroxide is subjected to thermolysis, the original ^{18}O label scrambles between the carbonyl and the peroxidic oxygen atoms of the remaining peroxide (Scheme 2).[6,26,70,74–77] An ideal multi-bond scission diacyl peroxide may not cause the ^{18}O scrambling in the peroxide. If one assumes that the ^{18}O scrambling occurs solely by the recombination of the acyloxy radical pair as shown in Scheme 2, the first-order rate constant of the ^{18}O scrambling in a diacyl peroxide (k_s) is expressed by equation 17 and the f_r value can be obtained from experimentally observable k_d and k_s values by equation 18.

$$k_s = 4k_{HOM}k_{REC}/(4k_{REC} + 2k_{\beta} + k_{DIS}) = k_{HOM}f_r \qquad (17)$$

$$f_r = 4k_{REC}/(4k_{REC} + 2k_{\beta} + k_{DIS}) = [(k_d/k_s) + 1]^{-1} \qquad (18)$$

3.1.3 Case IA diacyl peroxides

3.1.3.1 *Acetyl peroxide* (**14**), *acetyl propionyl peroxide, propionyl peroxide and butyryl peroxide*

The mechanism of thermal decomposition of acetyl peroxide (**14**), the simplest diacyl peroxide, has been most extensively investigated. The thermal decomposition of **14** in isooctane affords methyl acetate in 12.4% yield at 80°C[70] and 18% at 60°C (equation 19).[64] Since the ester is not formed in the gas phase, the ester is a cage product.[61a] Szwarc and co-workers[78,79] attributed essentially the same $\Delta H^{\ddagger}$ for the thermal decomposition of Case IA peroxides (*ca* 30 kcal mol^{-1}) to the one-bond scission mechanism.

$$\underset{\mathbf{14}}{(CH_3\overset{O}{\overset{\|}{C}}O)_2} \xrightarrow[60\,^{\circ}C\,(80\,^{\circ}C)]{n\text{-}C_8H_{18}} \underset{18\,\%\,(12.4\,\%)}{CH_3\text{-}\overset{O}{\overset{\|}{C}}\text{-}OCH_3} + \underset{5.5\,\%\,(2.9\,\%)}{CH_3\text{-}CH_3} + CH_4 + CO_2 \qquad (19)$$

Based on secondary $^1H/^2H$ kinetic isotope effects on the decompositions of acetyl peroxide and trideuterioacetyl peroxide, Koenig and Cruthoff[71b] concluded that **14** decomposes by a one-bond scission mechanism in which the O—O bond cleavage involves some CH_3—CO bond reorganization. However, radical trapping experiments using I_2–H_2O, galvinoxyl, diphenylpicrylhydrazyl and 9,10-dihydroanthracene failed to detect acetoxy radicals.[58] The thermolysis of **14** in cyclohexene proceeded with a 26% higher rate than in CCl_4,[58] and gave cyclohexyl acetate and 1,2-bisacetoxycyclohexane in which the ^{18}O label scrambled completely between the carbonyl and the ether oxygen atoms when [^{18}O]carbonyl-labeled **14** was used.[59] These results were explained by the one-bond scission mechanism, i.e. the acetoxy radical pair reacts with cyclohexene (Scheme 4) but there is no bimolecular reaction between acetyl peroxide and cyclohexene.[59]

$$CH_3-\overset{\bullet}{\overset{\|}{C}}-O-O-\overset{\bullet}{\overset{\|}{C}}-CH_3 \longrightarrow [\,2\ CH_3CO_2^{\bullet}\,] \rightleftharpoons CH_3-\overset{O}{\overset{\|}{C}}-O-O-\overset{O}{\overset{\|}{C}}-CH_3$$

cyclohexene ↓

1,2-bis(O_2CCH_3)cyclohexane ⟵ [cyclohexene + $\cdot O_2CCH_3$, $CH_3CO_2^{\bullet}$] ⟶ ⟶ 1-(O_2CCH_3)-2-(CH_3)cyclohexane

Scheme 4

Taylor and Martin[70] found that ^{18}O scrambling in **14** took place during the thermal decomposition of [^{18}O]carbonyl-labeled **14** in isooctane. Assuming the one-bond scission mechanism shown in Scheme 2, they calculated that 38% of the acetoxy radical pair recombined back to afford the ^{18}O scrambled **14** at 80°C. An increase in the solvent viscosity retarded the rate of decomposition, but increased the rate of ^{18}O scrambling in **14**, in keeping with the solvent cage concept.[65a,73,74] Table 4 summarizes kinetic data and f_r values determined from k_d and k_s values by equation 16 in comparison with those determined from the solvent viscosity effect on k_d by equation 14. Good agreement between the f_r values obtained by the two different methods supports the one-bond scission mechanism. The thermal decompositions of **14** and propionyl peroxide proceed at almost identical rates, whereas ^{18}O scrambling for **14** proceeds five times faster than that for propionyl peroxide at 80°C in isooctane.[74b] This observation

Table 4. Rate constants and f_r values in the decomposition of acetyl peroxide **14** at 80°C[65a]

Solvent	$k_s \times 10^5$ (s^{-1})	$k_d \times 10^{-5}$ (s^{-1})	f_r (^{18}O)[a]	f_r (viscosity)[b]
Isooctane	4.00	7.28	0.35	0.28
Dodecane	4.68	6.26	0.43	0.39
Octadecane	5.25	5.18	0.50	0.49

[a]Determined by ^{18}O tracer with equation 18.
[b]Determined by viscosity of solvent on k_d with equation 16.

by Martin *et al.* indicates these Case IA diacyl peroxides are one-bond initiators and is in accordance with propionyloxy radical decarboxylating much faster than acetoxy radical. All these observations seem to support the one-bond scission mechanism shown in Scheme 2.

On the other hand, Goldstein *et al.*[69] proposed a multi-bond scission mechanism for the decomposition of **14** based on their kinetic $^{12}C/^{13}C$ and $^{16}O/^{18}O$ isotope effects. They determined two modes of experimentally observable ^{18}O scrambling in [^{18}O]carbonyl-labeled **14** in solution and calculated that a [3,3] sigmatropic shift can account for 63% (in cumene) and 85% (in isooctane) of the ^{18}O scrambling in **14** (Scheme 5).[76] Goldstein and Haiby[77] found that the ^{18}O scrambling of **14** took place even in the gas phase, in which there is no solvent cage effect, and took this observation to be concrete evidence that the ^{18}O scrambling of **14** occurs by [3,3] and [1,3] sigmatropic shifts. A detailed discussion of this discrepancy is given in Section 3.1.6.

Scheme 5. Sigmatropic mechanisms

3.1.3.2 Cyclopropaneformyl peroxide (**15_3**)[54]

The rate of thermal decomposition of cyclopropaneformyl peroxide (**15_3**) decreases with increase in the viscosity of the hydrocarbon solvent; however, the rate of ^{18}O scrambling of **15_3** in Nujol is twice that in hexane. As shown in Figure 6, the solvent viscosity-dependent kinetic data fit better with the Pryor–Neuman equation[66] than Pryor equation,[65] revealing that **15_3** is a

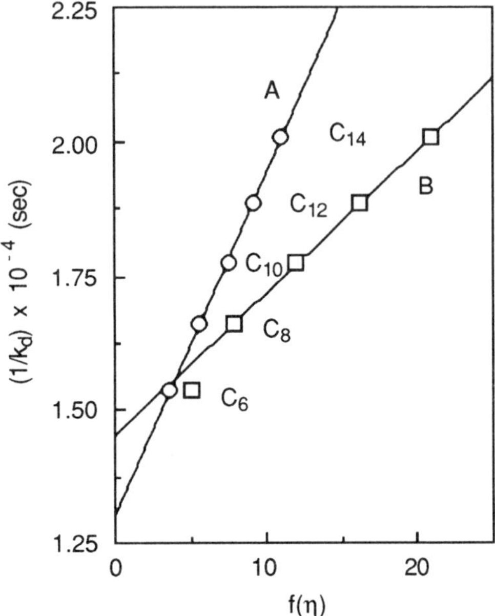

Figure 6. Correlation between k_d of $\mathbf{15_3}$ and viscosity of n-C_nH_{2n+2} solvent at 80°C. (A) Pryor–Neuman equation; (B) Pryor equation

one-bond initiator and the rate of decarboxylation (k_β) of cyclopropaneformyloxy radical is comparable to k_{DIS} and k_{REC}. Values of f_r for the geminate cyclopropaneformyloxy radical pair are calculated to be 0.22 for thermolysis in hexane at 80°C and 0.33 in Nujol from k_{HOM} obtained from the intercept of the straight line in Figure 6 and k_d values by equation 16.

On the other hand, the f_r values were determined to be 0.24 in hexane and 0.35 in Nujol from k_s and k_d at 80°C by equation 18. The good agreement between these f_r values determined by the two different methods indicates that the ^{18}O scrambling in $\mathbf{15_3}$ take place solely through the recombination of the intermediate cyclopropaneformyloxy radical pair. The rate constant for the dissociation of the cyclopropaneformyloxy radical pair out of the solvent cage is estimated to be one order of magnitude greater than those for the recombination and the decarboxylation of the acyloxy radical pair in octane (k_{REC}:k_{DIS};k_β = 1:11.5:1.5). The enthalpy of activation for the decomposition of $\mathbf{15_3}$ is 2 kcal mol^{-1} greater than that for the ^{18}O scrambling of $\mathbf{15_3}$ in CCl_4, which corresponds to the energy barrier for the dissociation of the cyclopropaneformyloxy radical pair in the solvent cage. The entropy for the ^{18}O scrambling is small but positive, in contrast to the large negative values of entropies of activation for the usual concerted sigmatropic shifts. Johnston *et al.*[80] concluded that $\mathbf{15_3}$ is a two-bond scission initiator on the basis of a spin trapping experiment with phenyl-*tert*-butylnitrone. Thus, both the viscosity effect on k_d and the ^{18}O tracer experiment are more sensitive tests than the spin trapping experiments with the nitrone for the detection of acyloxy radical pairs formed in the initial stage of the decomposition of some diacyl peroxides for which k_β is fairly large.[54]

3.1.4 Case IB diacyl peroxides

3.1.4.1 *Isobutyryl peroxide* (**16**)

The activation energy for the decomposition of isobutyryl peroxide (**16**) is 5 kcal mol^{-1} less than those for Case IA diacyl peroxides, which are uniformally 30 kcal mol^{-1}. Szwarc and co-workers[81] explained this fact in terms of **16** being a two-bond initiator (Scheme 3). Since the decarboxylation of acetoxy, propionyloxy and butyryloxy radicals has to overcome a substantial activation energy barrier, unimolecular rupture of the O—O bond proceeds in the thermolysis of Case IA diacyl peroxides. The exothermicity of the decarboxylation of isobutyryloxy radical helps in the O—O bond scission process, i.e. two-bond scission.[81] Later, Lamb *et al.*[82] found that the decomposition of **16** gives a carboxy-inversion product (27% in CCl_4 at 50°C) and is accelerated by polar solvents (Scheme 6). Walling *et al.*[83] reported that both radical and ionic decomposition processes of **16** were accelerated by increasing polarity of the solvent, and proposed that both decomposition processes share a common radical pair–ion pair intermediate. Hence the two-bond scission mechanism for the thermal decomposition of **16** would involve a polar intermediate (or transition state) (Figure 7, path B).

$$\left(\begin{matrix}H_3C\\H_3C\end{matrix}\!\!>CH\text{-}\overset{O}{\overset{\|}{C}}\text{-}O\right)_2 \xrightarrow[\text{cyclohexane (in } CH_3CN)]{40\,^{\circ}C} \left[\begin{matrix}\text{Scavengable radicals}\\ \text{by galvinoxyl}\\ 41.5\,\%\ (10.2\,\%)\end{matrix}\right]$$

16

$$\begin{matrix}H_3C\\H_3C\end{matrix}\!\!>CH\text{-}O\text{-}\overset{O}{\overset{\|}{C}}\text{-}O\text{-}\overset{O}{\overset{\|}{C}}\text{-}CH\!\!<\begin{matrix}CH_3\\CH_3\end{matrix} \;+\; \begin{matrix}H_3C\\H_3C\end{matrix}\!\!>CHCO_2H \;+\; \begin{matrix}H_3C\\H_3C\end{matrix}\!\!>CH\text{-}\overset{H}{N}\text{-}\overset{O}{\overset{\|}{C}}CH_3$$

17.9 % (34.3 %) (22.1 %) (16.6 %)

$$+\; \begin{matrix}H_3C\\H_3C\end{matrix}\!\!>CH\text{-}O\text{-}\overset{O}{\overset{\|}{C}}\text{-}CH\!\!<\begin{matrix}CH_3\\CH_3\end{matrix} \;+\; \left(\begin{matrix}H_3C\\H_3C\end{matrix}\!\!>CH\right)_2 \;+\; CH_3CH_2CH_3 \;+\; CH_2{=}C\text{-}CH_3$$

5.7 % (10.7 %) 25.1% (1.1 %) 3.7 % (0.3 %) 4.1 % (6.6 %)

Scheme 6

3.1.4.2 Cycloalkaneformyl peroxides ($\mathbf{15}_{4-6}$)[84] *(Scheme 7)*

Kinetic data for both the decomposition and the ^{18}O scrambling of cycloalkaneformyl peroxides ($\mathbf{15}_{3-6}$) are listed in Table 5. The peroxide $\mathbf{15}_3$ does not afford a carboxy-inversion product (**9**) and, as mentioned in Section 3.1.3.2, $\mathbf{15}_3$ is a one-bond initiator which decomposes along the path A in Figure 7. On the other hand, the following facts (Table 5) suggest that $\mathbf{15}_6$ is a typical two-bond initiator along path B in Figure 7. (1) The thermal decomposition gives 50% of **9** in CCl_4. (2) Both $\Delta H^{\ddagger}$ and $\Delta S^{\ddagger}$ are much smaller than for $\mathbf{15}_3$. (3) The rate of ^{18}O scrambling is much smaller than that of decomposition, i.e. $k_s/k_d = 0.01$ in CCl_4 at 45°C whereas $k_s/k_d = 0.35$ for $\mathbf{15}_3$ in CCl_4 at 65°C.

$\mathbf{15}_3$ $\mathbf{15}_4$ $\mathbf{15}_5$ $\mathbf{15}_6$

Scheme 7

The ring size of **15** affects not only its ground states but also the transition states of the decompositions. The fact that although the contribution of the ionic path to give **9** (C_i in Table 3) varies widely from 0 to 0.5, $\Delta H^{\ddagger}$ for the decomposition correlates linearly with the s-character of the ring carbon, suggesting that the decomposition path of $\mathbf{15}_{3-6}$ shifts successively from path A to path B with increase in the polar nature of the transition (Figure 7), in keeping with Walling *et al.*'s mechanism.[83]

If the ^{18}O scrambling mechanism for $\mathbf{15}_3$[54] is also true for $\mathbf{15}_{4-6}$, $\Delta H_s^{\ddagger}$ is taken as the potential energy required to form a cycloalkaneformyloxy radical pair

Table 5. Kinetic data for thermal decomposition and ^{18}O scrambling of cycloalkaneformyl peroxides ($\mathbf{15}_{3-6}$) in CCl_4 at 45°C[84]

Reaction	Peroxide	$k_d \times 10^6$ (s^{-1})	$\Delta H_d^{\ddagger}$ (kcal mol^{-1})	$\Delta S_d^{\ddagger}$ (cal mol^{-1} K^{-1})
Decomposition	$\mathbf{15}_3$	0.290	31.5 ± 0.4	10.9 ± 1.1
	$\mathbf{15}_4$	1.32 ± 0.02	27.2 ± 0.1	4.6 ± 0.05
	$\mathbf{15}_5$	4.05 ± 0.06	26.7 ± 0.2	4.6 ± 0.1
	$\mathbf{15}_6$	260	25.8 ± 1.7	5.5 ± 0.5
		$k_s \times 10^7$ (s^{-1})	$\Delta H_s^{\ddagger}$ (kcal mol^{-1})	$\Delta S_s^{\ddagger}$ (cal mol^{-1} K^{-1})
^{18}O scrambling	$\mathbf{15}_3$	1.29	29.5 ± 0.5	2.7 ± 1.4
	$\mathbf{15}_4$	3.93 ± 0.34	27.9 ± 0.5	−0.4 ± 1.5
	$\mathbf{15}_5$	13.2		
	$\mathbf{15}_6$	17.9 ± 0.6	27.1 ± 1.2	0.372 ± 3.8

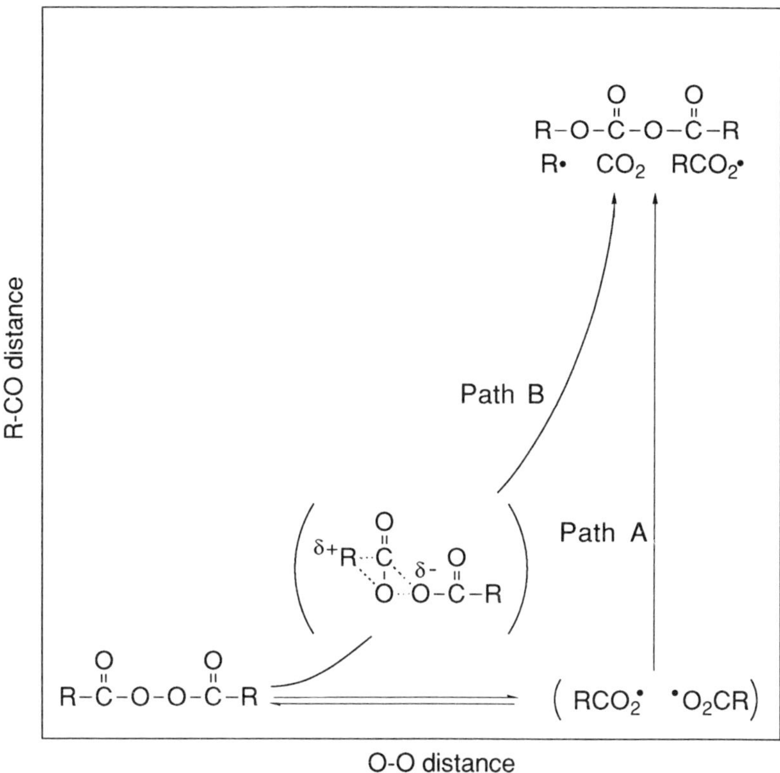

Figure 7. Shift of transition state of decomposition of cycloalkaneformyl peroxides (**15**$_3$–**15**$_6$)

from **15**$_{3-6}$. The positive enthalpy difference between the decomposition and the ^{18}O scrambling ($\Delta H_d^\ddagger - \Delta H_s^\ddagger = 2.4$ kcal mol^{-1}) is regarded as the $\Delta H^\ddagger$ for the dissociation of the cyclopropaneformyloxy radical pair as mentioned in Section 3.1.3.2[54] (see also Section 3.1.9.3). Then the negative values of $\Delta H_d^\ddagger - \Delta H_s^\ddagger$ for **15**$_{4-6}$ are attributed to the enthalpy difference between the acyloxy radical pair, **6**, and the polar intermediate (or transition state), **2**. Hence the decomposition of **15**$_3$ occurs exclusively through path A, whereas **15**$_{4-6}$ decompose mainly by path B (Figure 7). If one assumes that the effect of the ring size on the energy level of cycloalkaneformyloxy radical pairs can be ignored to a first-order approximation, although the energy levels of cycloalkaneformyloxy radical pairs are of course affected by the ring size, as will be discussed in Section 3.1.9.3, the potential energy levels for **15**$_{4-6}$ are lower than the level for the cycloalkaneformyloxy radical pair [$\Delta H_s^\ddagger$(**15**$_3$) – $\Delta H_s^\ddagger$(**15**$_{4-6}$)]. The relative potential energy levels of **15**$_{4-6}$, cycloalkaneformyloxy radical pair and the transition state of each **15** are illustrated in Figure 8.

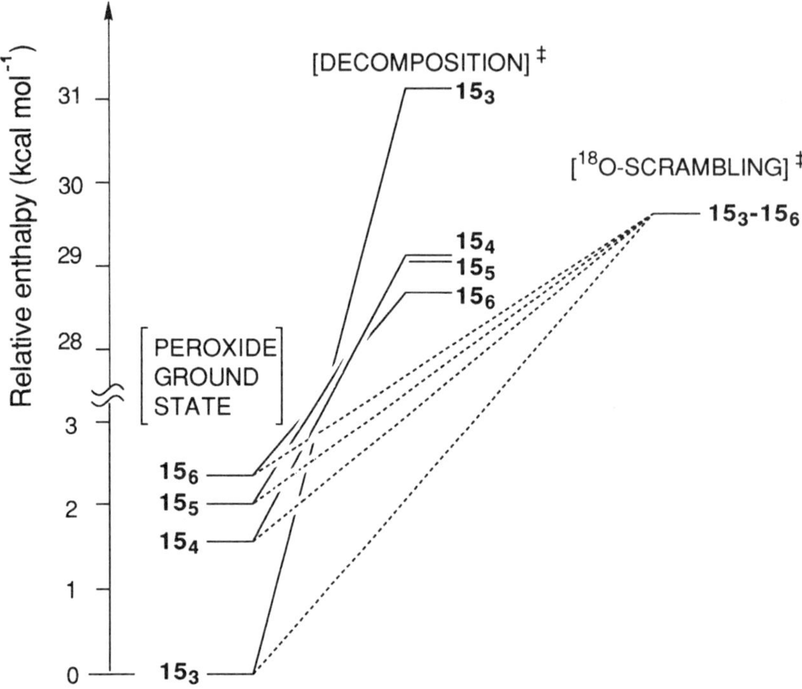

Figure 8. Potential energy diagram for decomposition and ^{18}O scrambling of cycloalkaneformyl peroxides (**15_3–15_6**)

3.1.4.3 1-Apocamphoryl peroxide (**17**)

Since the major path of decomposition for the usual Case IB diacyl peroxides is the two-bond scission process, the relative rate of the ^{18}O scrambling vs the decomposition of these peroxides is very small. 1-Apocamphoryl peroxide (**17**) is the only known exception, i.e. the rate constant of the ^{18}O scrambling (k_s) is comparable to the rate constants of the radical decomposition (k_r) and the ionic (carboxy-inversion) reaction (k_i), e.g. $k_s = 1.85 \times 10^6\ s^{-1}$, $k_r = 1.26 \times 10^6\ s^{-1}$, $k_i = 1.50 \times 10^6\ s^{-1}$ in CCl_4 at 70°C.[6] The yield of 1-apocamphyl 1-apocamphorate is only 2.2% (equation 20).[6]

$$(\text{R-C(=O)-O})_2 \ (\mathbf{17}) \xrightarrow[CCl_4]{70\,°C} \text{R-O-C(=O)-O-C(=O)-R} \ (54.5\%) + \text{R-Cl} \ (36\%) \quad (20)$$

$$+ \ \text{R-C(=O)-O-R} \ (2.2\%) + \text{R-R} \ (9\%) + CO_2$$

3.1.4.4 Perfluoroalkanoyl peroxides

Chengxue *et al.*[85] and Yoshida *et al.*[86] prepared perfluoroalkanoyl peroxides from the acyl halides by treatment with hydrogen peroxide and sodium hydroxide in Freon-113. The first-order rate constants of decomposition of these peroxides are compared with those for the corresponding alkanoyl peroxides in Table 6.[41] These perfluorinated peroxides are very explosive, and the k_d values are 2–4 orders of magnitude greater than those for the corresponding alkanoyl peroxides. Analysis by the MNDO molecular orbital method by Sawada *et al.*[41] suggests that the fluorine atom introduced at the α-carbon atom makes the peroxide bond long and the dihedral angle between the two CO_2 planes wide, and this extreme instability of the perfluorinated peroxides is due to the destabilization of the resonance energy of the O—O bond and not to the electrostatic energy.[41] The major product of the thermal decomposition of bisperfluoroalkanoyl peroxides in Freon-113 is perfluoroalkyl radical dimer (**1** → **6** → **7** → **8** → **13** in Scheme 1) (equation 21). The yield of perfluoroalkyl dimer is markedly higher than that of alkyl dimer in the thermolysis of alkanoyl peroxides,[87] suggesting that the decarboxylation of perfluoroalkanoyloxy radical is extremely fast. The perfluoro-α-isopropoxyethyl radicals undergo substantial β-scission (equation 22).[85]

$$(CF_3CF_2CF_2CO)_2 \xrightarrow[\text{Freon-113}]{\text{r.t.}} \underset{89\%}{(CF_3CF_2CF_2)_2} + CO_2 \tag{21}$$

$$\left((F_3C)_2CFOCF(F_3C)\text{-}\overset{O}{\overset{\|}{C}}\text{-}O \right)_2 \xrightarrow[\text{Freon-113}]{\text{r.t.}} \underset{21\%}{[(CF_3)_2CFOCF(CF_3)]_2} + \underset{11\%}{[(CF_3)_2CF]_2} + \underset{34\%}{(CF_3)_2CFOCF(CF_3)CF(CF_3)_2} + CO_2 \tag{22}$$

Table 6. Comparison of first-order rate constants (k_d) of thermal decomposition of symmetrical perfluoroalkanoyl peroxides and those of the corresponding dialkanoyl peroxides [$(RCO_2)_2$] at 25°C[41]

R	$k_d \times 10^8$ (s^{-1})	R	$k_d \times 10^8$ (s^{-1})
CF_3	100	CH_3	2
C_2F_5	1160	C_2H_5	3
n-C_3F_7	3320	n-C_3H_7	4
n-C_7F_{15}	5860	n-C_7H_{15}	6
CH_3CF_2	96 000	CF_3CH_2	1

3.1.5 Case IIA diacyl peroxides

3.1.5.1 Acetyl benzoyl peroxide (**18**)

Acetyl benzoyl peroxide (**18**) is very susceptible to radical-induced decomposition and decomposes unimolecularly in chlorobenzene at 70°C, $k_{HOM} = 2 \times 10^{-5}\ s^{-1}$.[88] The ^{18}O scrambling in [^{18}O]acetyl-labeled **18** takes place faster than in [^{18}O]benzoyl-labeled **18** during thermal decomposition.[89] The rates of ^{18}O scrambling in **18** are much lower than in **14**, i.e. the corresponding Case I diacyl peroxide.

3.1.5.2 Cyclopropaneformyl benzoyl peroxide (**19**)

[^{18}O]cyclopropaneformyl-labeled cyclopropaneformyl benzoyl peroxide (**19**) decomposes at the same rate as $\mathbf{15_3}$ at 80°C, whereas the k_s of **19** in the cyclopropaneformyloxy side is only 35% of the k_s for $\mathbf{15_3}$.[54]

3.1.6 Case IIB diacyl peroxides

3.1.6.1 1-Apocamphoryl benzoyl peroxide (**20**)

The thermal decompositions of **17** and **20** occur at similar rates (Table 2) and afford almost identical yields of the carboxy-inversion products (equation 23).[6,26] On the other hand, ^{18}O scrambling in **20** is much slower than that in **17**, while the ester yield in the thermal decomposition of **20** is *ca* ten times greater than that of **17**. These results are generally observed and rationalized as follows.[6] The ester is formed by the combination of the radical pair **7** in a solvent cage (Scheme 1). The yield of the radical pair **7** in the thermal decomposition of Case II diacyl peroxides is greater than that for Case I peroxides,[26] and hence the ester yield in the thermal decomposition of Case II peroxides becomes greater than that of Case I peroxides.

20 —(70°C, CCl_4)→ 51% + 20% + 20% —Cl + 17% —Cl + CO_2 + C_2Cl_6 59% (23)

3.1.7 Case III diacyl peroxides

3.1.7.1 Benzoyl peroxide (**21**)

Many studies have been published on the kinetics and product analysis of the thermal decomposition of **21** and substituted **21**. These were reviewed by Hiatt in 1970.[1] DeTar *et al.*[90] carefully analyzed the yields of reaction products in the thermal decomposition of **21** in benzene as a function of reaction time and initial concentration of **21**. Several major products determined by DeTar *et al.* are shown in Scheme 8. Their experimental results could be simulated by reaction pathways constituted by a combination of 101 elemental reactions, and the rate constants for each step were assigned.[91]

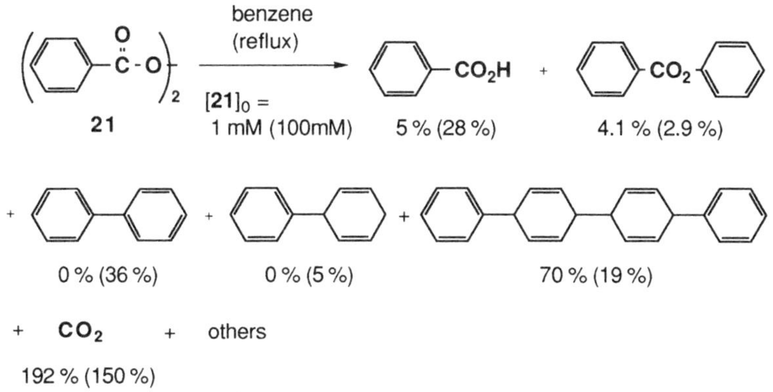

benzene (reflux); $[\mathbf{21}]_0$ = 1 mM (100mM)

21 → $PhCO_2H$ 5 % (28 %) + $PhCO_2Ph$ 4.1 % (2.9 %) + 0 % (36 %) + 0 % (5 %) + 70 % (19 %) + CO_2 192 % (150 %) + others

$[\mathbf{21}]_0$: initial concentration of **21**.

Scheme 8

Nozaki and Bartlett[92] showed that **21** is susceptible to radical-induced decomposition and the observed apparent first-order rate constant (k_d^{obs}) of the thermal decomposition of **21** consists of the unimolecular homolysis of the O—O bond (k_d) and the radical-induced decomposition (k_{ri}) (see Section 4.1). They obtained k_d by analysis of the dependence of the apparent k_d on the initial concentration of **21**. Hammond and Soffer[57] succeeded in trapping benzoyloxy radicals by iodine in 100% yield in the thermal decomposition of **21** in wet CCl_4 containing I_2, clearly showing that **21** is a one-bond initiator. The determination of k_d has also been done by measuring the kinetics of the decrease in a radical scavenger such as galvinoxyl[93a] or α,α-diphenyl-β-picrylhydrazyl[93b] in the thermal decomposition of **21**. Janzen *et al.*[94] have obtained k_{HOM} by measuring the ESR signal intensity of the spin adduct of the benzoyloxy radical to PBN (equation 24). An Ahrrenius plot (30–80°C) of k_d values measured by different workers using various methods in different temperature ranges is a straight line with the equation $\log k_d\,(s^{-1}) = 16.0 - (33.1/2.3RT)(\text{kcal mol}^{-1})$.[94]

$$(PhCO_2)_2 + PhCH{=}\overset{\overset{O}{|}}{N}{-}C(CH_3)_3 \longrightarrow Ph{-}\underset{\underset{OCOPh}{|}}{CH}{-}\overset{\overset{O\cdot}{|}}{N}{-}C(CH_3)_3 + Ph{-}\underset{\underset{Ph}{|}}{CH}{-}\overset{\overset{O\cdot}{|}}{N}{-}C(CH_3)_3 + Ph{-}\underset{\underset{O}{\|}}{C}{-}\overset{\overset{O\cdot}{|}}{N}{-}C(CH_3)_3 \quad (24)$$

21 (PBN)

The effect of polar substituents on k_d was investigated by Swain *et al.*[95] The k_d values of **21** and substituted **21** fit the Hammett equation with $\rho = -0.32$. This was rationalized by the substituents controlling the electron density of the lone pairs on the peroxidic oxygen atoms.

The ^{18}O scrambling in [^{18}O]carbonyl-labeled **21** was tested by Martin and Hargis,[75] who expected a higher f_r value than that of **14**, since the benzoyloxy radical pair is stable enough to have a good chance of recombining to afford ^{18}O scrambled **21**. However, they observed small f_r values (0.046 in isooctane and 0.20 in mineral oil at 80°C) in comparison with **14** (0.38 in isooctane and 0.55 in mineral oil). This observation can be rationalized in terms of $k_{DIS} \gg k_{REC}$, k_β for the benxoyloxy radical pair.[6,43]

3.1.7.2 *Bis(1-naphthoyl) peroxide* (**22**) *and benzoyl 1-naphthoyl peroxide* (**27**)

Leffler and Zepp[96] refined earlier work on the thermal decomposition of bis(1-naphthoyl) peroxide (**22**) by Kharasch and Dannley,[97a] Cooper[9] and Takebayashi *et al.*[97b] Products of the thermal decomposition of **22** in two different solvents are shown in Scheme 9.[96]

The relative magnitudes of k_d for **21:22:27** are 1:11:64 at 60°C in benzene. The amounts of free radical efficiency measured by galvinoxyl are 1.71 mol mol^{-1} of **21**, but only 0.71 mol mol^{-1} of **22**. The low free radical efficiency in the thermal decomposition of **22** may be due to the relatively high yields of cage products (**23, 24**); coupling of the 1-naphthoyloxy radical pair takes place easily to give **26**, which rearranges to **24**. In contrast, cage products are almost absent in the thermal decomposition of **21**.

3.1.7.3 *Case III diacyl peroxides containing heteroaromatic substituents*

Kinetic data for the thermal decomposition of Case III diacyl peroxides containing furan and thiophene rings are available.[98]

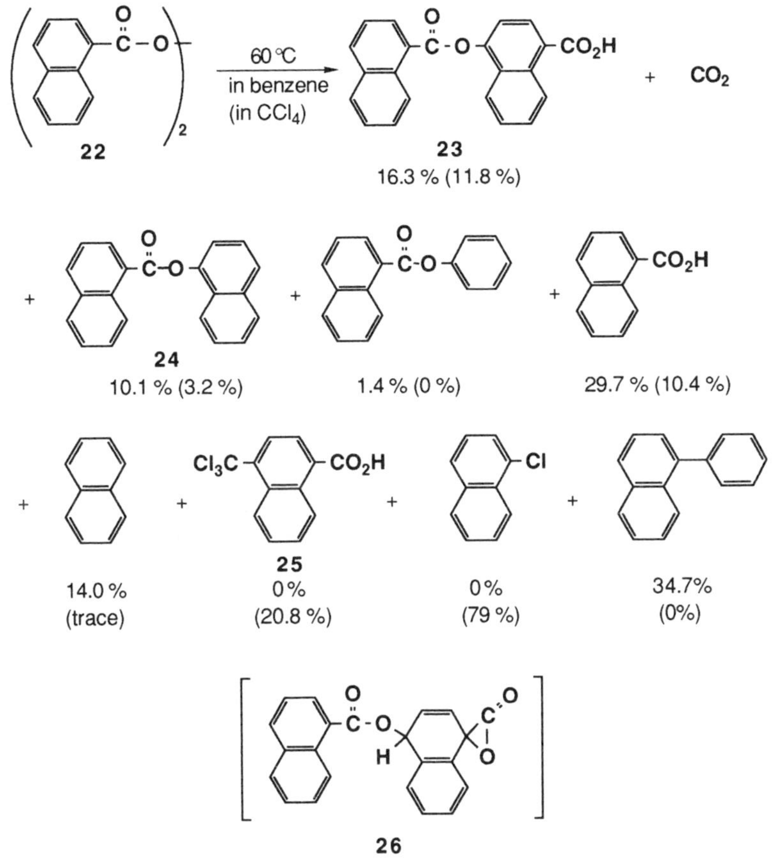

60 °C in benzene (in CCl_4)

22 → 23 16.3 % (11.8 %) + CO_2

+ 24 10.1 % (3.2 %) + 1.4 % (0 %) + 29.7 % (10.4 %)

+ 14.0 % (trace) + 25 0% (20.8 %) + 0% (79 %) + 34.7% (0%)

26

Scheme 9

3.1.8 Cyclic diacyl peroxides

Russell,[7a] Greene[33] and Greene and Rees[99] showed in their pioneering works that k_d of phthaloyl peroxide (**28**), a typical cyclic Case III diacyl peroxide, is small whereas **28** is highly reactive toward electron donors such as alkenes and aromatic hydrocarbons as compared with **21**. Fujimori *et al.*[100] reinvestigated the unimolecular decomposition of **28** by means of ^{18}O tracer in CCl_4 at various temperatures to verify their postulate mentioned in Section 3.1.1.[6,43] According to that postulate, f_r of **28** should be extremely large, because the unescapable phthaloyl radical pair has no dissociation path ($k_{DIS} = 0$) whereas this is the fastest process for the benzoyloxy radical pair.[6] In fact, the f_r value for **28** was found to be 0.98. Qualitative potential energy diagrams for the thermal

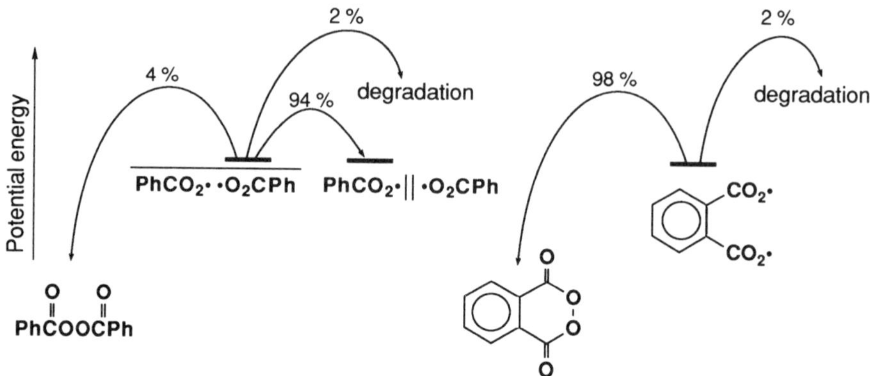

Figure 9. Fates of benzoyloxy radical pair (in isooctane at 80°C) and phthaloyloxy radical (in CCl_4 at 80°C) which are produced by O—O bond homolysis of the parent diacyl peroxides

decomposition of **21** and **28** are illustrated in Figure 9 based on kinetic data for k_d and k_s.[100] The k_{HOM} values were calculated from experimentally observed k_d and k_s values by equation 18 to be $1.33 \times 10^{-5}\ s^{-1}$ for **28** (in CCl_4)[100] and $2.83 \times 10^{-5}\ s^{-1}$ for **21** (in isooctane)[75] at 80°C. The small difference in k_{HOM} for **21** and **28** reveals that the much smaller k_d of **28** than that of **21** (1:87) cannot be attributed to the hindrance of O—O bond stretching by the cyclic structure of **28** but is due to the lack of a dissociation process for the two acyloxy radical groups, which is usually the fastest path in the fates of the benzoyloxy radical pair formed from **21**.

The characteristic feature of the cyclic Case III diacyl peroxide mentioned above also applies to the cyclic Case IB diacyl peroxide. *trans*-Hexahydrophthaloyl peroxide (**29**) decomposes faster than the corresponding acyclic compound, i.e. cyclohexaneformyl peroxide (**15₆**), although the stretching vibration of the peroxide linkage of **29** would be hindered by its cyclic structure. The activation enthalpies for the decomposition of **29** and $\mathbf{15_6}$ are almost identical. For the acyloxy radical pair generated from the usual Case IB diacyl peroxides in a solvent, k_β is greater than k_{REC} and k_{DIS}. Therefore, the f_r value for an acylic Case IB peroxide is considered not to increase on going to the cyclic Case IB peroxide as much as is observed on going to Case III diacyl peroxides. In fact, f_r (0.07) for the unescapable *trans*-hexahydrophthaloyloxy radical is 3.5 times greater than that for the cyclohexaneformyloxy radical pair in CCl_4, but is still small (Scheme 10). Thermal decomposition of **29** gives *cis*-cyclohexene as a volatile product. No evidence for the formation of *trans*-cyclohexene was reported.[101]

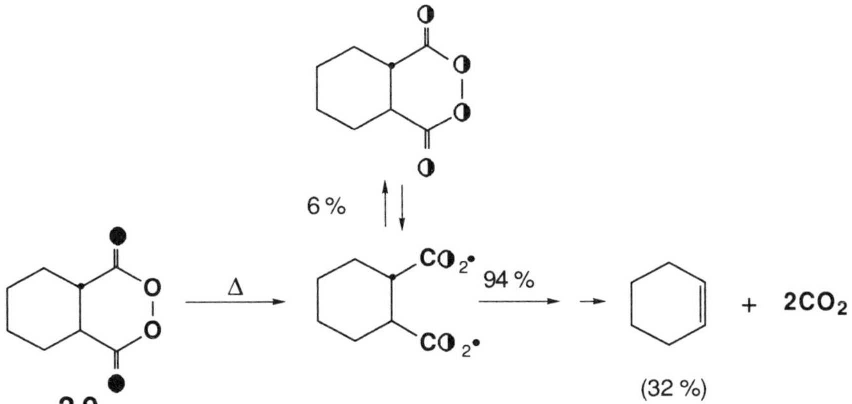

Scheme 10

The thermal decompositions of both *threo*- and *erythro*-2,3-dimethylsuccinyl peroxide afford the same mixture of the four products shown in Scheme 11 in 50–70% overall yield. Scheme 11 illustrates the reaction pathways and relative yields of the four products.[101] Thermolysis of phenylmaleoyl peroxide gives phenylacetylene as the only volatile product (13%).[102]

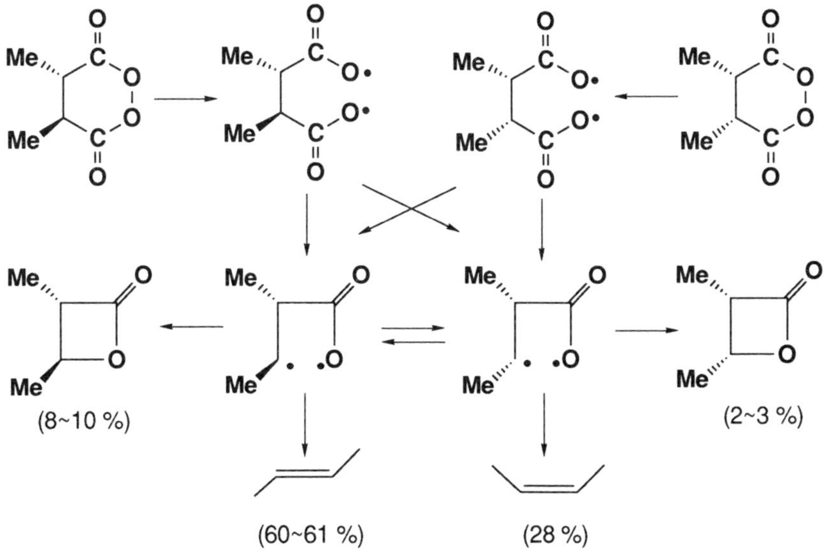

Scheme 11

3.1.9 Initial stage of O—O bond homolysis of diacyl peroxides—nature of acyloxy radical pair

3.1.9.1 *Effect of solvent viscosity on k_d and k_s*

In the thermal decomposition of Case IA diacyl peroxides, an increase in viscosity results in a decrease in k_d, but an increase in k_s.[64,65,84] These observations seem to agree well with the acyloxy radical pair mechanism, i.e. the viscosity of the solvent controls the rate constant of the dissociation of the geminate acyloxy radical pair formed by the homolytic scission of the O—O bond (k_{DIS}). However, the acyloxyradical pair mechanism based on the usual solvent cage concept cannot explain the occurrence of the ^{18}O scrambling of **14** in the gas phase.[77]

3.1.9.2 *Comparison of the ^{18}O scrambling in diacyl peroxides and the Cope rearrangement of hexa-1,5-dienes*

The Cope rearrangement is a kind of intramolecular deuterium scrambling reaction in hexa-1,5-dienes, whereas the ^{18}O scrambling of diacyl peroxides is a kind of intramolecular rearrangement reaction. Since a diacyl peroxide is a 1,3,4,6-tetraoxahexa-1,5-diene (Scheme 12), the ^{18}O scrambling can be regarded as a tetraoxa-Cope rearrangement. Hence the comparison of the Cope rearrangement and ^{18}O scrambling is helpful in understanding the mechanisms

Case I: R=R'=H ; Case II: R=H,R'=Ph ; Case III: R=R'=Ph

Bond forming type

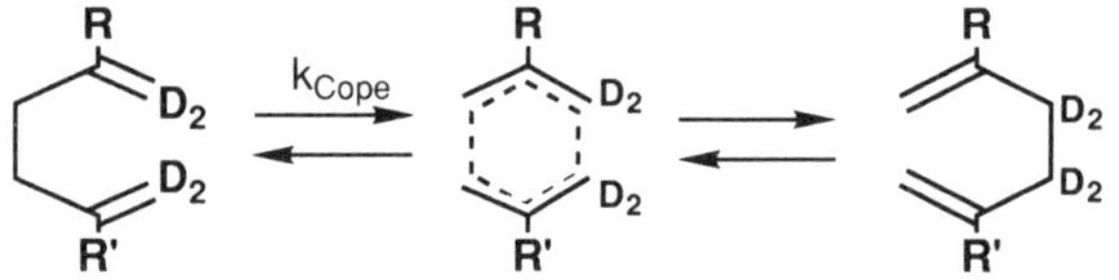

R, R', D_2, D_2, k_{Cope}

k^{rel}_{Cope}: Case I: Case II: Case III = 1: 69 : 4900 (189.8 °C)[103]

Bond breaking type

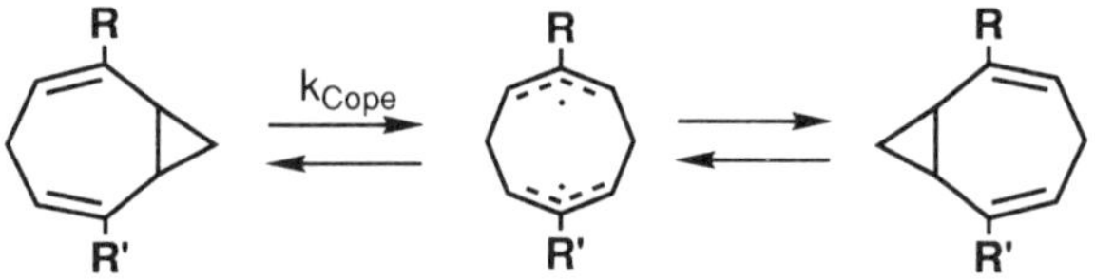

R, R', k_{Cope}

k^{rel}_{Cope}: Case I: Case III = 15:1 (80 °C)[107]

Scheme 12. Cope rearrangements

of these rearrangements. If the [3,3] sigmatropic shift plays a major role in the ^{18}O scrambling,[76,77] the substituent effect on the ^{18}O scrambling may show a similar trend to that in the Cope rearrangement.

The successive replacement of hydrogens at the 2- and 5-positions of hexa-1,5-diene by phenyl groups increases the extent of formation of a new bond between the 1- and 6-positions of the diene, but decreases the extent of cleavage of the old bond between the 3- and 4-positions in the transition state of the Cope rearrangement to resemble a cyclohexane-1,4-diyl structure.[103–106] Thus the rates of the usual Cope rearrangement increase in the order Case I < Case II < Case III (Scheme 12). As a result, the rate of the bond-forming type Cope rearrangement is accelerated by the introduction of each phenyl group.[103,104] In contrast, successive displacement of cyclopropyl groups of **15**$_3$ by phenyl groups reduces the rate of the ^{18}O scrambling [**15**$_3$ (Case I) > **19** (Case II) > **21** (Case III)], as exemplified by the linear correlation between the logarithms of the relative rates of the bond-forming type Cope rearrangement ($\log k^{\mathrm{rel}}_{\mathrm{Cope}}$) and those of the ^{18}O scrambling ($\log k^{\mathrm{rel}}_{\mathrm{s}}$) with a negative slope (Figure 10). This trend is, however, in good agreement with the substitution effect on the initial bond-breaking type Cope rearrangement of barbaralanes and homotropylidenes (Scheme 12).[107] These facts suggest that the tetraoxa-Cope rearrangement is of the bond-breaking type, i.e. the ^{18}O scrambling occurs via an acyloxy radical pair or a state which is just like an acyloxy radical pair.

According to the acyloxy radical pair mechanism in Scheme 2, k_s/k_d can be expressed by equation 25.

$$k_s/k_d = 4k_{\mathrm{REC}}/(k_{\mathrm{DIS}} + 2k_\beta) \tag{25}$$

In the thermal decomposition of **15**$_3$, **19** and **22**, since not only are k_β negligibly smaller than k_{DIS} but also k_{DIS} have similar values, equation 25 can be approximated as $(k_s/k_d)^{\mathrm{rel}} = k_{\mathrm{REC}}{}^{\mathrm{rel}}$ in these peroxides. Thus the $k_{\mathrm{REC}}{}^{\mathrm{rel}}$ values calculated from experimentally obtained k_s and k_d values are **15**$_3$:**19**:**21** =

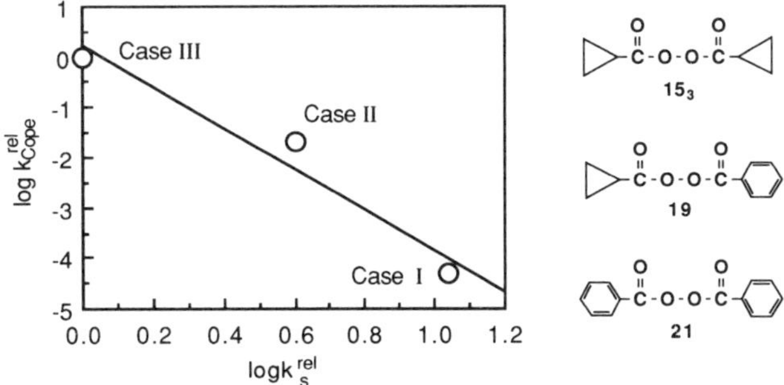

Figure 10. Linear free energy relationship between the ^{18}O scrambling in diacyl peroxides (**15**$_3$, **19**, **21**) and the Cope rearrangement of hexa-1,5-dienes

6.6:1.9:1, in accordance with the earlier postulate of Fujimori and Oae[6] that the k_{REC} values of acyloxy radical pairs decrease in the order Case I > Case II > Case III (see Table 2 and Figure 4).

3.1.9.3 Unification of the acyloxy radical pair mechanism and the sigmatropic mechanism for ^{18}O scrambling in diacyl peroxides

Molecular orbital calculations on acyloxy radicals suggest that the three σ-radicals and one π-radical shown in Scheme 13 have close energy levels.[50,108–114] Scheme 13 shows the singly occupied molecular orbital (SOMO) of the acyloxy radical of the four states. In an acyloxy radical with relatively small OCO angle, the unsymmetrical $^2A'$ σ-radical is the most stable, whereas the $^2B_2$ σ-radical has the lowest energy at C_{2v} symmetry. When the OCO angle is wide, the $^2A_1$ σ-radical is the most stable, although the $^2A_1$ σ-radical has a higher energy than the optimum $^2B_2$ σ-radical. There have been arguments as to which σ-radical is at a lower energy level, unsymmetrical ($^2A'$) or symmetrical ($^2B_2$). Most papers report that the $^2A'$ σ-radical is the most stable. However, the results of MCSCF MO calculations on the formyloxy radical by McLean *et al.*[114] suggest that the optimum structure is the C_{2v} symmetrical $^2B_2$ radical.[114] Figure 11 illustrates the results calculated by Feller *et al.*[113] with the MCSCF MO method. Anyhow, decarboxylation of the $^2B_2$ σ-radical is forbidden, since the process produces an electronically excited state of CO_2.[108] Therefore, the decarboxylation of an acyloxy radical takes place via a $^2A_1$ σ-radical, which gives ground-state CO_2.[108,109,113]

Stretching of the O—O bond of a diacyl peroxide would generate primarily an interacting unsymmetrical $^2A'$ σ-radical pair, $[^2A' \cdot {}^2A']$, which then follows one of three possible pathways shown in Scheme 14, depending on the structure of R as follows.

Case III diacyl peroxides: the ground state of the benzoyloxy radical has been

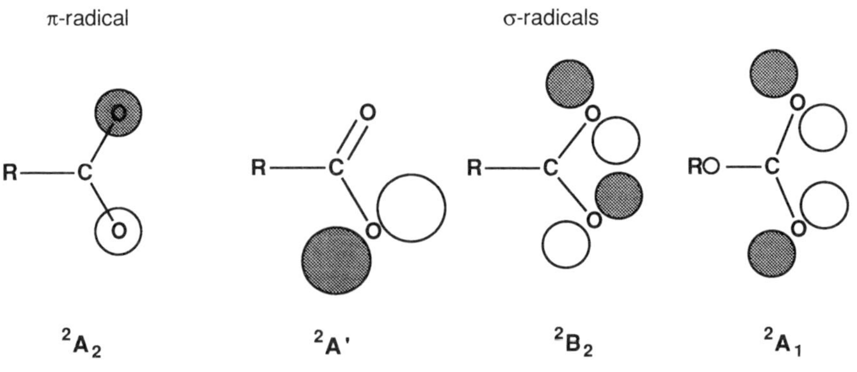

Scheme 13

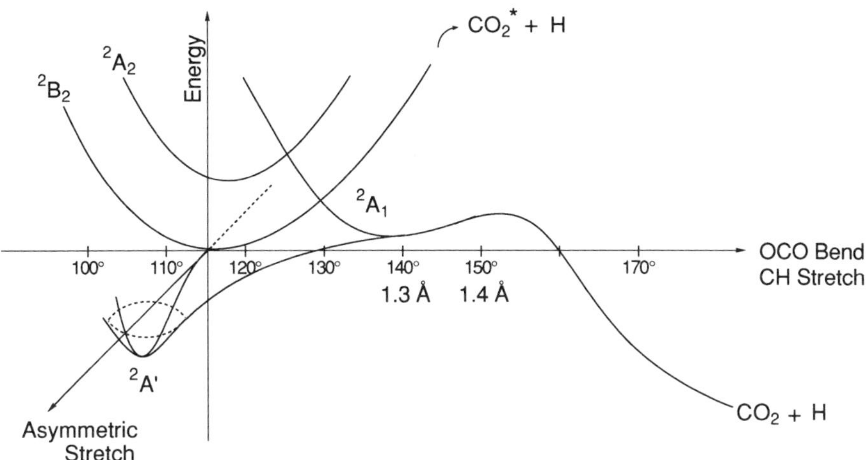

Figure 11. Potential surfaces of formyloxy radical calculated with MCSCF MO method (Feller *et al.*[113])

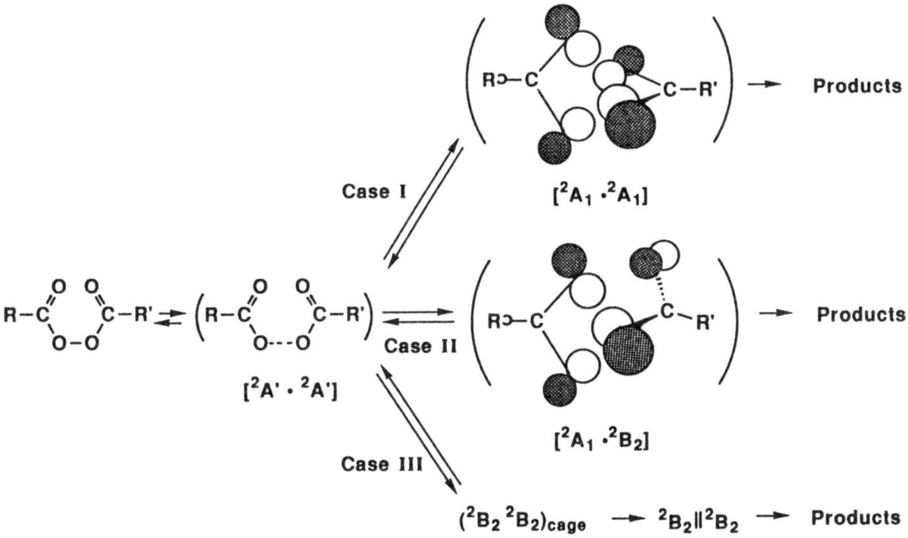

Case I

$[^2A_1 \cdot {}^2A_1]$ → Products

Case II

$[^2A' \cdot {}^2A']$

$[^2A_1 \cdot {}^2B_2]$ → Products

Case III

$(^2B_2\ {}^2B_2)_{cage}$ → $^2B_2 \| {}^2B_2$ → Products

Scheme 14

experimentally proved to be a $^2B_2$ σ-radical[28] and the MO calculations suggest that the benzoyloxy radical favors a $^2B_2$ σ-radical over a $^2A_1$ σ-radical.[113] Since the decarboxylation of a $^2B_2$ σ-radical is forbidden, the benzoyloxy radical derived from Case III peroxides is longer lived than aliphatic radicals derived from Case I peroxides. The energy barrier to recombine such a stabilized $^2B_2$

σ-benzoyloxy radical pair would be substantial, resulting in a small f_r, i.e. the most favorable path of the benzoyloxy radical pair is the dissociation process in a solvent with the usual viscosity.

Case I diacyl peroxides: when the two $^2A_1$ σ-radicals meet in a manner in which the two planes of radicals meet at 90° to each other to avoid an unfavorable two sets of lone pair–lone pair interactions between oxygen atoms, the interaction of the two SOMO (6a1 orbital) may result in a four-center bonding-type MO which accommodates the two electrons just like 1,3-dehydro-5,7-adamantanediyl dication (**30** in Scheme 15).[115] Such a weakly interacting $^2A_1$ σ-radical pair, $[^2A_1 \cdot {}^2A_1]$ in which four oxygen atoms are equivalent may be more stable than the two independent $^2A_1$ σ-radicals. Fujimori and Oae[116] proposed $[^2A_1 \cdot {}^2A_1]$ as a common intermediate of both the ^{18}O scrambling and the decomposition of Case I peroxides. Since the coefficient of the SOMO (6a1 orbital) on R is appreciable in $^2A_1$ state $RCO_2\cdot$, the energy level of the $[^2A_1 \cdot {}^2A_1]$ intermediate would be lowered with an increase in the radical-stabilizing ability of the R group attached to the carboxyl carbon. This may be a rational explanation for the occurrence of ^{18}O scrambling in $\mathbf{15_6}$, although the major decomposition path of $\mathbf{15_6}$ is two-bond scission via polar intermediate **3** (path B in Figure 7).[116]

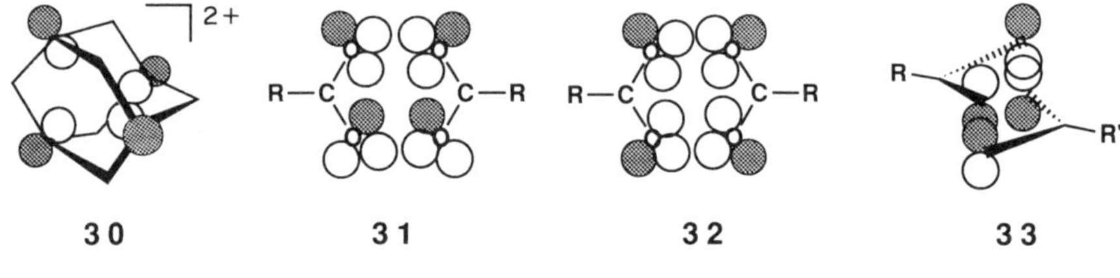

Scheme 15

The ^{18}O scrambling of **14** in the gas phase can be rationalized by the mechanism involving a $[^2A_1 \cdot {}^2A_1]$ intermediate (Case I in Scheme 14). Since four oxygen atoms in $[^2A_1 \cdot {}^2A_1]$ are equivalent, recombination of $[^2A_1 \cdot {}^2A_1]$ which formed from a [^{18}O]carbonyl-labeled diacyl peroxide results in a completely ^{18}O scrambled diacyl peroxide (equation 26).[116] Since $\Delta H^{\ddagger}$ of ^{18}O scrambling of **14** in the gas phase is only 0.7 kcal mol^{-1} less than $\Delta H^{\ddagger}$ of decomposition,[77] the energy gain for pairing two acetoxy radicals in the manner $[^2A_1 \cdot {}^2A_1]$ is small. This tetraoxa-Cope rearrangement via $[^2A_1 \cdot {}^2A_1]$ is peculiar in comparison with the usual Cope rearrangement, where the transition state (or intermediate) can be regarded as the interaction of two π-allyl radicals (**33** in Scheme 15)[117] and is at a 23.3 kcal mol^{-1} lower energy level than two non-interacting allyl radicals.[105]

$$\mathrm{R}\overset{\overset{\bullet}{\|}}{\mathrm{C}}\mathrm{OO}\overset{\overset{\bullet}{\|}}{\mathrm{C}}\mathrm{R} \longrightarrow [^2\mathrm{A}_1\cdot{}^2\mathrm{A}_1] \rightleftharpoons 1/4\ \mathrm{R}\overset{\overset{\bullet}{\|}}{\mathrm{C}}\mathrm{OO}\overset{\overset{\bullet}{\|}}{\mathrm{C}}\mathrm{R} + 2/4\ \mathrm{R}\overset{\overset{\bullet}{\|}}{\mathrm{C}}\mathrm{O}{\bullet}\overset{\overset{\mathrm{O}}{\|}}{\mathrm{C}}\mathrm{R} + 1/4\ \mathrm{R}\overset{\overset{\mathrm{O}}{\|}}{\mathrm{C}}{\bullet}{\bullet}\overset{\overset{\mathrm{O}}{\|}}{\mathrm{C}}\mathrm{R} \quad (26)$$

Both the $^2B_2 \cdot {}^2B_2$ radical pair (**31**) and the $^2A_1 \cdot {}^2A_1$ radical pair (**32**) with a planar conformation would be more unstable than the two non-interacting σ-radicals, since two sets of lone pair–lone pair repulsions of oxygen atoms raise the energy of both **31** and **32**.

Case II diacyl peroxides: a benzoyloxy radical in the $^2B_2$ state resists a change to the $^2A_1$ state, whereas aliphatic $RCO_2\cdot$ in the $^2B_2$ state tends to change to the $^2A_1$ state. The pairing between $^2B_2$ and $^2A_1$ σ-radicals may form a two-electron three-center bond with a small energy gain affording $[^2A_1 \cdot {}^2B_2]$ (Scheme 14). The postulation of weakly interacting $[^2A_1 \cdot {}^2B_2]$ can explain nicely the experimental observations that ^{18}O scrambling in Case II peroxides (**18**, **20**) takes place much faster in the aliphatic than the aromatic counterpart (equation 27).[26,89]

$$\text{R-}\overset{\bullet}{\underset{}{\ddot{C}}}\text{-O-O-}\overset{\bullet}{\ddot{C}}\text{-R}' \rightleftharpoons [\,^2A_1 \cdot {}^2B_2\,] \rightleftharpoons \text{R-}\overset{O}{\ddot{C}}\text{-}\bullet\text{-O-}\overset{\bullet}{\ddot{C}}\text{-R}' \qquad (27)$$

(R = alkyl, R' = aryl)

The ^{18}O scrambling in diacyl peroxides through $[^2A_1 \cdot {}^2A_1]$ or $[^2A_1 \cdot {}^2B_2]$ may take place in the gas phase and may be accelerated in solution by an increase in the viscosity of the solvent, since the returns of $[^2A_1 \cdot {}^2A_1]$ and $[^2A_1 \cdot {}^2B_2]$ to the ^{18}O scrambled diacyl peroxides are forced in a viscous solvent with a high internal pressure, reasonably interpreting the data in Table 4.

3.2 Photochemical Decomposition by Radical Reaction Mechanisms

3.2.1 Photolysis of crystalline diacyl peroxides

Photolysis of a single crystal of diacyl peroxide is a good system for investigating chemical reactions in the solid phase.[28] When the experiments are carried out at low temperature, one can measure ESR, IR, UV and visible spectra to collect information about intermediate free radicals frozen in a single-crystal lattice.

Box *et al.*[118] detected phenyl radical pairs by irradiating crystals of benzoyl peroxide (BPO) at 4.2 K. Barchuk *et al.*[119] observed the phenyl–benzoyloxy radical pair (PB pair) in crystals of BPO photolyzed at 62 K and suggested that photoexcitation with polarized light is localized in half of the peroxide. This was later criticized by Vary and McBride,[120] who investigated carefully the behavior of radical pairs formed during photolysis (300–400 nm light) of a single crystal of BPO at 5–65 K by ESR spectroscopy. They found four different kinds of PB pairs and two kinds of benzoyloxy radical pairs (BB pair). Although one of the PB pairs is observable at 5 K, the BB pair cannot be trapped at the same temperature. The BB pair generated at 10 K is infinitely stable at 10 K. These observations indicate that the BB pair formed at 5 K recombines back to the peroxide rapidly, whereas at 10 K some geometric change takes place to hinder the recombination of the BB pair.[120]

McBride and co-workers[28,110] irradiated a single crystal of acetyl-d_3 benzoyl peroxide labeled with ^{17}O at either the carbonyl or peroxidic oxygen atoms at 64 K. They observed a methyl–benzoyloxy radical pair (MB) but could not detect an acetoxy–benzoyloxy radical pair. Irradiation of the MB pair with visible light changed MB to the methyl–phenyl radical pair (equation 28). From the observed ^{17}O hyperfine splitting in the ESR signal of the MB pair, they suggested that the ground state of the benzoyloxy radical is the $^2B_2$ state.[28]

$$\left[CD_3-\overset{O}{\overset{\|}{C}}-{}^{17}\bullet-{}^{17}\bullet-\overset{O}{\overset{\|}{C}}-C_6H_5\right]_{crystal} \xrightarrow{h\nu\ (64\ K)} \left[CD_3^{\bullet}\ \ CO_2\ \ {}^{17}\bullet-\overset{O}{\overset{\|}{C}}-C_6H_5\right]_{crystal} \tag{28}$$

$$\xrightarrow{\text{visible light}} \left[CD_3^{\bullet}\ \ 2\,CO_2\ \ {}^{\bullet}C_6H_5\right]_{crystal}$$

Hollingsworth and McBride[121] measured the IR spectra of single crystals of long n-alkanoyl peroxides irradiated by UV light at 20 K to investigate the van der Waals CO_2 dimer.

3.2.2 Photolysis in solution with continuous UV light

3.2.2.1 Case I diacyl peroxides

Walker and Wild[122] and Szwarc *et al.*[81] reported that the photochemical decomposition of acetyl peroxide in solution (equation 29) and in the solid state gives ethane, methane, CO_2 and small amounts of CO and O_2. Kashiwagi *et al.*[123] photolyzed [^{18}O]carbonyl-labeled optically active 2-methyl-3-phenylbutyryl peroxide (**34**) with a low-pressure mercury lamp. The alcohol moiety of the ester (30% yield) retained 68% of the original configuration while the carbonyl oxygen retained 85% of the original [^{18}O]carbonyl-label (equation 30). Similar stereochemical results were observed by Feldhues and Schafer[87] in the photolysis of the optically active unsymmetrical peroxide **35** (equation 31).

$$CH_3-\overset{O}{\overset{\|}{C}}-O-O-\overset{O}{\overset{\|}{C}}-CH_3 \xrightarrow{h\nu\ (16\sim18\ ^\circ C)} C_2H_6 + CH_4 + CO_2 + CO + O_2 \tag{29}$$

14

	C_2H_6	CH_4	CO_2	CO	O_2
crystals	25.6 %	5.1 %	67.4 %	1.3 %	0.4 %
in cyclohexane	8.4 %	28.9 %	59.7 %	1.3 %	0.9 %

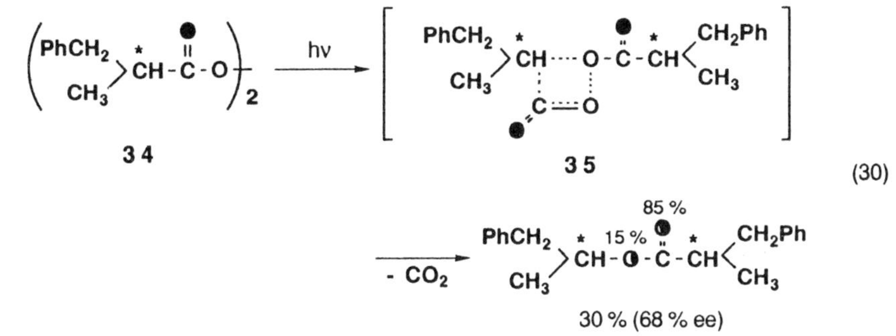

$$\left(\begin{matrix}PhCH_2\\CH_3\end{matrix}\!>\!\overset{*}{C}H-\overset{\bullet}{\overset{\|}{C}}-O\right)_2 \xrightarrow{h\nu} \left[\begin{matrix}PhCH_2\\CH_3\end{matrix}\!>\!\overset{*}{C}H\cdots O-\overset{\bullet}{\overset{\|}{C}}-\overset{*}{C}H\!<\!\begin{matrix}CH_2Ph\\CH_3\end{matrix}\;\; \vdots\; C\!=\!O\right]$$

34 35 (30)

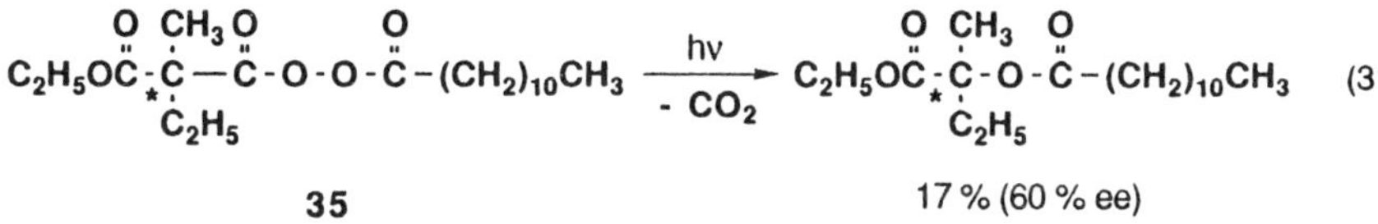

$$\xrightarrow{-CO_2} \begin{matrix}PhCH_2\\CH_3\end{matrix}\!>\!\overset{*}{C}H-O-\overset{\|}{C}-\overset{*}{C}H\!<\!\begin{matrix}CH_2Ph\\CH_3\end{matrix}$$

85 %, 15 %

30 % (68 % ee)

$$C_2H_5O\overset{O}{\overset{\|}{C}}-\overset{*}{\underset{C_2H_5}{\overset{CH_3}{C}}}-\overset{O}{\overset{\|}{C}}-O-O-\overset{O}{\overset{\|}{C}}-(CH_2)_{10}CH_3 \xrightarrow[-CO_2]{h\nu} C_2H_5O\overset{O}{\overset{\|}{C}}-\overset{*}{\underset{C_2H_5}{\overset{CH_3}{C}}}-O-\overset{O}{\overset{\|}{C}}-(CH_2)_{10}CH_3 \quad (31)$$

35 17 % (60 % ee)

The ^{18}O tracer results for the photolysis of [^{18}O]benzoyl-labeled **18** varied depending on the reaction conditions, i.e. the original ^{18}O label was completely scrambled in the two oxygens of methyl benzoate at 273 K in ethanol, but 73% and 62% retained in the carbonyl oxygen of the ester obtained by photolysis in an ethanol matrix at 77 K and in single crystals at 203 K, respectively.[110]

Photolysis of a Case I diacyl peroxide with 250–350 nm light in cyclopropane or in pentane at low temperature (−25 to 105°C) was reported to be a good method for generating alkyl radicals for ESR studies by Kochi and Krusic,[124] who recorded well resolved ESR spectra of 18 primary alkyl radicals by this method. Sheldon and Kochi[125] reported products and ESR spectral data for intermediate carbon radicals formed in the photolysis of a variety of Case I peroxides. One example is shown in Scheme 16. The rearrangement of the ω-

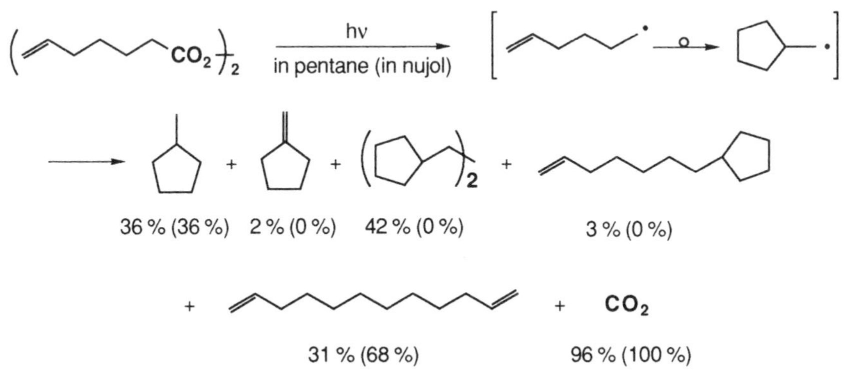

hν, in pentane (in nujol)

36 % (36 %) 2 % (0 %) 42 % (0 %) 3 % (0 %)

+ CO_2

31 % (68 %) 96 % (100 %)

Scheme 16

hexenyl radical to the cyclopentylmethyl radical is often used as a useful clock reaction which occurs at a rate of $10^5\ s^{-1}$ at 25°C according to the equation[126]

$$\log k_{\text{rearrangement}}\ (s^{-1}) = 9.1 - (10.4/2.3RT)(\text{kcal mol}^{-1})$$

Thirty-two unsymmetrical Case I diacyl peroxides were photolyzed at −78°C without solvent for the synthetic purpose of obtaining alkyl coupling products ($R—C_{11}H_{23}$) by Schäfer *et al.* (equation 32).[87] The yield of the alkyl dimer depends on the structure of R (20–76%).[87] The use of a solvent decreased the yield of the alkyl dimers. The same products were obtained by thermolysis and square-pulse electrolysis of the same Case IA diacyl peroxides but in lower yields than by photolysis (equations 32 and 33).[87]

$$R\text{-}\overset{O}{\overset{\|}{C}}\text{-}O\text{-}O\text{-}\overset{O}{\overset{\|}{C}}\text{-}(CH_2)_{10}CH_3 \xrightarrow[\text{or square pulse radiolysis}]{h\nu,\ \text{or}\ \Delta,} R\text{-}(CH_2)_{10}CH_3 + CO_2 \quad (32)$$
(20~76 %)

$$\text{AcO}\ldots\text{(steroid)}\text{-}\underset{O}{\underset{\|}{C}}OO\underset{O}{\underset{\|}{C}}(CH_2)_{10}CH_3 \xrightarrow[-\ CO_2]{h\nu} \text{AcO}\ldots\text{(steroid)}\text{-}(CH_2)_{10}CH_3 \quad (33)$$
OAc OAc

3.2.2.2 *Case III diacyl peroxides*

Photochemical decomposition of benzoyl peroxide (**21**) gives essentially the same products as thermal decomposition. Tokumaru and co-workers[127] carefully determined the yields of the major reaction products of both the photochemical and thermal decomposition of **21** in solution (equation 34).

$$(C_6H_5\text{-}\overset{O}{\overset{\|}{C}}\text{-}O)_2 \xrightarrow{\text{benzene}} C_6H_5\text{-}\overset{O}{\overset{\|}{C}}\text{-}O\text{-}C_6H_5 + C_6H_5\text{-}CO_2H + C_6H_5\text{-}C_6H_5 \quad (34)$$

21				
	hν (LPML)	13 %	45 %	23 %
	Δ (80 °C)	2 %	31 %	26 %

Photochemical decomposition gives phenyl benzoate in a 5–6 times higher yield than thermal decomposition. The ester was established to be a coupling product of the geminate radical pair in a solvent cage by cross-over experiments using unlabeled BPO and BPO-d_6. In the presence of a sufficient amount of styrene, which scavanges benzoyloxy radicals, the yield of CO_2 in thermolysis at 80°C (30%) is much lower than that in photolysis at room temperature (70%).

Both phenyl and benzoyloxy radicals are trapped by *N*-benzylidene-*tert*-butylamine *N*-oxide in the photolysis, whereas only the spin adduct of the benzoyloxy radical is detected in thermolysis (40°C, 8 min). Although decarboxylation of the benzoyloxy radical takes place by irradiation with visible light (750 nm gives the highest efficiency),[28] irradiation with a tungsten–bromine lamp (TBL) does not increase the yield of the ester in either photolysis with a low-pressure mercury lamp (LPML) or thermolysis at 80°C. A change in the intensity of the light does not alter the yield of the ester, obviously suggesting that the ester formed by a one-photon process. Based on these observations, the difference in the ester yields in photolysis and thermolysis was attributed to a simultaneous two-bond scission mechanism or a mechanism in which the benzoyloxy radical pair is electronically or vibrationally excited and undergoes rapid decarboxylation to generate a phenyl–benzoyloxy radical pair, which combines to afford the ester.[127]

Addition–elimination mechanisms for the *ipso* substitution by radicals formed during the photochemical decomposition of BPO and thenoyl peroxide were investigated in detail by Tokumaru and co-workers.[128]

Benzoyl peroxide and decanoyl peroxide are efficient quenchers of various triplet sensitizers. Electronic energy transfer and charge transfer to the peroxides are suggested as being the important factors contributing to the quenching process.[129]

3.2.2.3 Cyclic diacyl peroxides

Adam and Rucktäschel[20a] found that photolysis of α-substituted malonyl peroxides (**35**) in alcohol undergo decarboxylation to afford primarily α-lactones (**36**), which react readily with alcohol to give eventually α-alkoxylmalonic acids in high yield, whereas photolysis in a hydrocarbon solvent yields a polyester (**37**). At 77 K the primary product, **36**, could be characterized by IR spectrometry ($\nu_{C=O}$ 1895 cm^{-1}). Prolonged irradiation of **36** at 77 K causes decarbonylation as shown in Scheme 17. On warming to 180 K, **36** (R = *n*-Bu) polymerizes.[20d] The α-lactone **36** (R = CF_3) prepared in this manner is fairly stable at room temperature (half-life *ca* 8 h in the gas phase).[20c]

On thermolysis of phthaloyl peroxide (**28**) in the gas phase, a benzyne intermediate was trapped by tetraphenylcyclopentadienone to yield 1,2,3,4-tetraphenylnaphthalene.[130] In contrast, photodecarboxylation of **28** in an argon matrix at 8 K affords a β-lactone (**39**) as a primary product, which is further converted to the ketene **40** by irradiation at wavelengths shorter than 315 nm. Irradiation of **40** at wavelengths longer than 340 nm gives **39** at 8 K, as shown in Scheme 18.[131]

Thermal and photochemical transformation of phenylmaleoyl peroxide (**41**) to phenylacetylene was considered to proceed via β-lactone **42** (equation 35).[19]

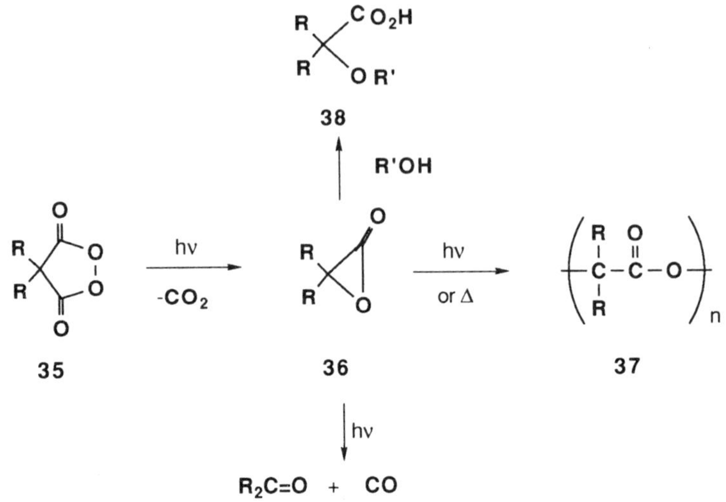

Scheme 17

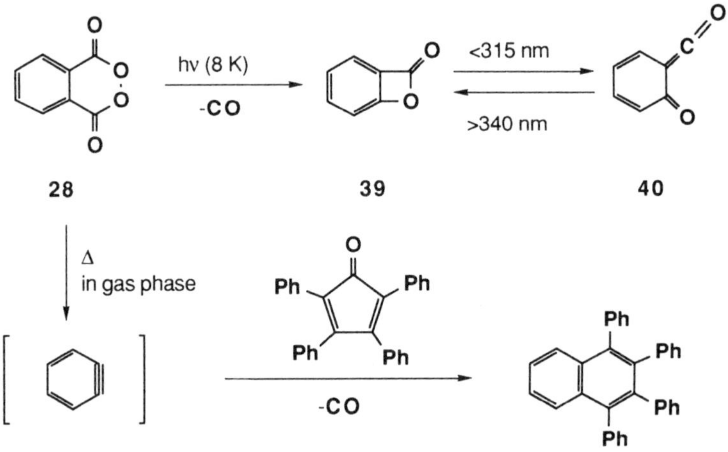

Scheme 18

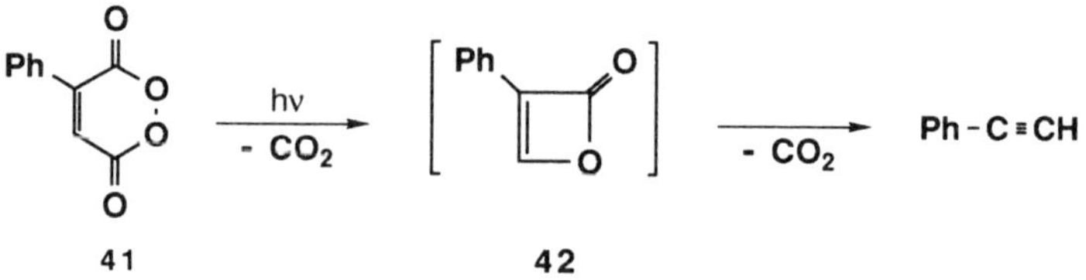

(35)

3.2.3 Direct observations of acyloxy radicals in solution generated by photolysis of diacyl peroxides

Recent advances in laser flash photolysis (LFP) have made it possible to observe acyloxy radicals in solution by spectroscopy. The first direct study of aroyloxy radical kinetics was reported by Yamauchi and co-workers,[44,46] who succeeded in following three acyloxy radicals (**47, 47a, c**) generated by 308 nm LFP of benzoyl peroxide (**21**) and substituted compounds (**21a, c**, X = *p*-MeO, *p*-Cl) in CCl_4 with time-resolved ESR spectroscopy (Scheme 19). These aroyloxy radicals were assigned to be a σ-radical based on the relatively large *g*-values and small hyperfine coupling constants, consistent with the results of McBride *et al.*[28] Ingold and co-workers[47–51] and Tokumaru and co-workers[45,132]

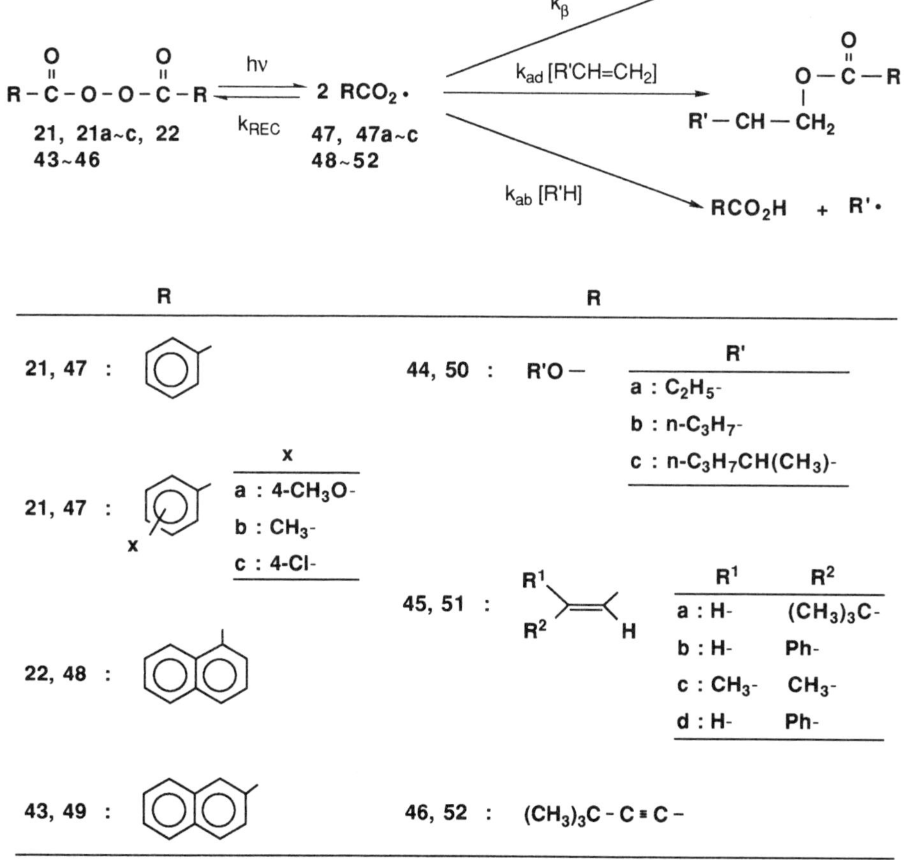

Scheme 19

developed aroyloxy radical kinetics by following the decay of aroyloxy radicals generated by LFP of **21** and **21a–c** on the nanosecond to microsecond time scale with time-resolved UV–visible spectrophotometry. Ingold and co-workers also succeeded in detecting and characterizing alkoxycarbonyloxy radicals ($R'OCO_2\cdot$, **50**),[51] alkenoyloxy radicals ($R^1R^2C{=}CHCO_2\cdot$, **51**)[133] and alkynoyloxy radicals (t-$BuC{\equiv}CCO_2\cdot$, **52**)[133] produced by LFP of alkyl peroxycarbonates (**44**), alkenoyl peroxides (**45**) and alkynoyloxy peroxides (**46**) by the same technique. Data on naphthoyloxy radicals (**48, 49**) were obtained with LFP of naphthoyl peroxides (**22, 43**) by Tokumaru and co-workers.[132]

Recently, values of k_β for some acyloxy radicals (**54**) were estimated by Hilborn and Pincock[134] from the yield of ionic product (**56**) and $k_{ET} = 2.6 \times 10^{10}\ s^{-1}$ in the photolysis of 1-naphthylmethyl alkanoates, i.e. $k_\beta = k_{ET}$ [(100 − yield of **56**)/yield of **56**] assuming that k_{ET} is independent of the structure of $RCO_2\cdot$ (Scheme 20). Falvey and Schuster[135] assigned k_3 of the 9-methyl-9-fluoreneformyloxy radical (**58**) from the time lag of the formation of 9-methyl-9-fluorenyl radical (**59**) in the picosecond time scale LFP of the perester **57** (equation 36).

$$\mathbf{57}\ (\text{9-methylfluorene-9-}C(=O)\text{-}O\text{-}O\text{-}t\text{-}Bu) \xrightarrow[\text{(LFP)}]{h\nu} \left[\mathbf{58}\ (\text{9-methylfluorene-9-}CO_2\cdot)\ \ \cdot O\text{-}t\text{-}Bu\right] \xrightarrow[-CO_2]{k_\beta} \mathbf{59}\ (\text{9-methyl-9-fluorenyl}\cdot)\ \ \cdot O\text{-}t\text{-}Bu \qquad (36)$$

$$\underset{\mathbf{53}}{ArCH_2\text{-}O\text{-}C(=O)\text{-}R} \xrightarrow[CH_3OH]{h\nu} \underset{\mathbf{54}}{ArCH_2^{\cdot}\ \cdot O_2CR} \xrightarrow[(-CO_2)]{k_\beta} ArCH_2^{\cdot}\ \cdot R \longrightarrow \underset{\mathbf{55}}{ArCH_2R}$$

$$\mathbf{54} \xrightarrow{k_{ET}} ArCH_2^{+}\ ^{-}O_2CR \xrightarrow{CH_3OH} \underset{\mathbf{56}}{ArCH_2OCH_3} + HO_2CR$$

Ar	R	
1-naphthyl	a : CH_3-	d : $(CH_3)_3C$-
	b : CH_3CH_2-	e : $PhCH_2$-
	c : $(CH_3)_2CH$-	f : $PhCH_2CH_2$-

Scheme 20

Kinetic data on decarboxylation (k_β), hydrogen abstraction from C—H bonds (k_{abs}) and addition to alkenes (k_{add}) of acyloxy radicals are summarized in Tables 7 and 8.

3.2.3.1 Electronic and ESR spectra of acyloxy radicals

All acyloxy radicals (**47–52**) hitherto measured show very characteristic broad, structureless absorption bands in the visible region of the spectrum extending from about 400 nm to beyond 800 nm, e.g. $\varepsilon_{720\ nm} \approx 290\ l\ mol^{-1}\ cm^{-1}$ for **47a**. Ingold *et al.* attributed this visible absorption to a transition from the $^2B_2$ σ-electronic ground state to the $^2A_1$ excited-state potential energy surface (see Scheme 13 and Figure 11). This assignment is consistent not only with the experimental result that irradiation of benzoyloxy radical with visible light (*ca* 700 nm) accelerates the decarboxylation,[28,48] but also with the theoretical result that the decarboxylation of acyloxy radicals in the $^2B_2$ σ-electronic state is not allowed, whereas the $^2A_1$ σ-radical is allowed to decarboxylate.[108,109] The fact that both **47** and **50** have similar structureless, broad visible absorption bands indicates that the odd electron is located only in the carboxyl group and not in the aromatic ring in **47**, consistent with the assignment of the ground state to be the $^2B_2$ σ-state.

Continuous UV irradiation of solutions of diacyl peroxides **21** and **44–46** at 145–162 K allowed ESR spectra of acyloxy radicals **50–52** to be recorded (Figure 12).[136] The very small hyperfine coupling constants of ^{1}H and ^{13}C suggest that the spin is almost exclusively concentrated on the two carboxyl oxygen atoms, in agreement with ESR experimental results for the benzoyloxy radical in a single crystal[110] and the theoretical value for the $^2B_2$ state acyloxy radical, suggesting that the ground state of these acyloxy radicals is $^2B_2$. However, one cannot eliminate completely the possibility that the experimentally observed ESR spectra are due to fast interconversion of two mirror-image forms of $^2A'$ σ-radicals via the $^2B_2$ σ-radical on the ESR time scale.[112]

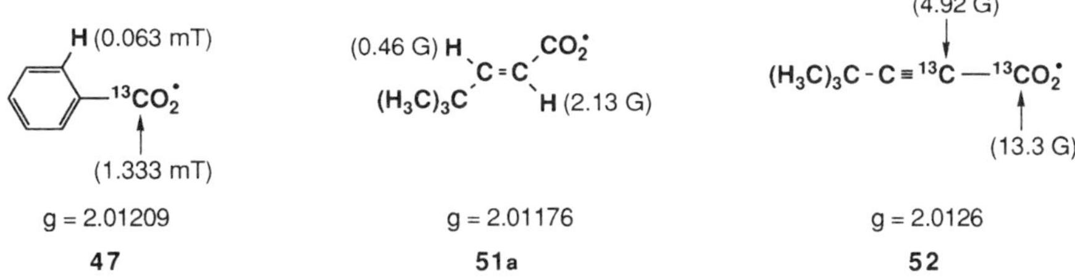

Figure 12. Hyperfine splitting constants and g-values of acyloxy radicals in cyclopropane solution at −125°C[136b]

Table 7. First-order rate constants of decarboxylation (k_β) of $RCO_2\cdot$

$RCO_2\cdot$	Method[a]	T (°C)	Solvent	$k_\beta \times 10^{-6}$ (s^{-1})	Log A (log s^{-1})	E_a (kcal mol^{-1})	Ref.
47	A	24	CCl_4	2.0 ± 1.0	12.6 ± 0.1	8.6 ± 0.3	49
			CH_3CN	5.9 ± 0.2			
47	A	25	CCl_4	4	12.1 ± 0.4	7.3 ± 0.6	45
47	B	20	CCl_4	4.5	10.8	5.8	46
47	C	100–130		10			142
47a	A	25	CCl_4	0.41 ± 0.7	13.6 ± 0.4	11.5 ± 0.5	45
47b	A	24	CCl_4	0.34 ± 0.01	12.3 ± 0.8	9.2 ± 1.0	49
47b	A	25	CH_3CN	1.8	13.1	9.3	132c
47c	A	24	CCl_4	1.4 ± 0.3	12.3 ± 0.4	8.4 ± 0.5	49
47c	A	25	CH_3CN	1.9 ± 0.1	13.2 ± 0.3	9.4 ± 0.4	45
48	A	25	CCl_4	5.1	12.1	7.5	132c
49	A	25	CCl_4	4.0	13.0	8.7	132c
50	A	r.t.	CCl_4	11			51
51a	A	20	CCl_4	1			133
51b	A	20	CCl_4	0.5			133
52	A	20	CCl_4	< 0.1			133

R[b]					
CH_3	D	20	CH_3OH	1,300 ± 200	134
CH_3	E	60	*n*-Alkane	1,600	64
CH_3	C	110	$(CCl_3)_2CO$	2,500	143
CH_3CH_2	D	20	CH_3OH	2,000 ± 300	134
$PhCH_2CH_2$	D	20	CH_3OH	2,300 ± 400	134
$(CH_3)_2CH$	D	20	CH_3OH	6,500 ± 800	134
$(CH_3)_3C$	D	20	CH_3OH	11,000 ± 2,000	134
$PhCH_2$	D	20	CH_3OH	5,000 ± 800	134
H_3C	F	r.t.	CH_3CN–hexane	18,000	135

[a](A)LFP–visible; (B) LFP–ESR; (C) Δ–CIDNP; (D) Product yields and clock reaction (Scheme 20); (E) solvent viscosity effect on k_d; (F) equation 36 (see text).

[b] R in $RCO_2\cdot$.

Table 8. Second-order rate constants for reaction of $RCO_2\cdot$ with selected substrates in CCl_4 at 24°C[a]

	$k \times 10^{-6}$ mol^{-1} s^{-1}					
Substrate	**47a**	**47**	**47c**	**50b**[b]	**51b**[c]	**48**[d]
Cyclohexane	0.53 ± 0.03	1.4 ± 0.05	12 ± 0.4	15.8 ± 0.7	0.73 ± 0.11	
Triethylsilane	4.8 ± 0.1	5.6 ± 0.7	38 ± 3	89 ± 8	6.9 ± 0.9	
Cyclohexene	64 ± 3	12.2	140 ± 30	1720 ± 130	95 ± 9	
Styrene	55 ± 3	51 ± 4	140 ± 30	2030 ± 110	170 ± 10	111
Benzene	2.3 ± 0.2	78 ± 14	220 ± 60	7.4 ± 0.4	2.7 ± 0.3	11.4
Anisole						42
Toluene	8.4 ± 0.9[a]					92
ArOH[f]	1000 ± 100					

[a]Reference 49. [b]Reference 51. [c]Reference 136. [d]Reference 132 (25°C). [e]Reference 48. [f]2,4,6-Tri-*t*-butylphenol.

3.2.3.2 Kinetics of decarboxylation of acyloxy radicals

Values of k_β were directly determined from the first-order decay curve of the visible region absorbance due to acyloxy radicals (**47–52**) generated by LFP of the corresponding diacyl peroxides in CCl_4 or CH_3CN. It is noteworthy that the k_β of **47a** is much less in CH_3CN than that in CCl_4. The decay of **47** in CH_3CN obeys almost second-order kinetics. This result was rationalized by the recombination of the radicals to regenerate 4-methoxybenzoyl peroxide (**21a**) with an activation energy of 10.4 ± 1.0 kcal mol^{-1}.[45] The acetoxy radical is also more stable in acetic acid than in CCl_4.[135] Electron-donating substituents on the phenyl ring reduce the k_β of benzoyloxy radicals (**47**), although the effect is small. Values of k_β for $RCO_2\cdot$ increase with increasing radical-stabilizing ability of R (aryl < primary alkyl < secondary alkyl ≈ benzyl < tertiary alkyl). This trend is the same as for decarbonylation of the acyl radical, which, however, is 10^6 times more sensitive to the structural effect than the decarboxylation of acyloxy radicals. The results reflect the higher extrusibility of CO_2 than that of CO.[137–139]

The discrepancy between k_β values estimated with indirect methods by earlier investigators[140,141] and the newer direct kinetic data has been discussed in detail by Chateauneuf *et al.*[49]

3.2.3.3 Hydrogen atom abstraction and addition to multiple bonds

Acyloxy radicals (**47–52**) have a strong affinity to hydrogen donors such as alkanes. Radical **47a** abstracts a hydrogen atom from the *p*-methoxy group of the parent peroxide in the LFP of **21a** in CCl_4.[44–48] Compared with the *tert*-butoxy radical, the benzoyloxy radical has a similar reactivity in hydrogen atom abstractions but is much more reactive in additions to multiple bonds.[48] For example, the second-order rate constant (k_{obs}) of hydrogen atom abstraction from cyclohexane with **47a** is one third of that for the *tert*-butoxy radical, whereas **47a** is 60 times more reactive toward styrene than the *tert*-butoxy radical. The extremely high reactivity of **47** toward phenol and nitroxide radical was explained by a single electron transfer from these substrates to **47**.[48] The fairly high electronegativities of acyloxy radicals seem to characterize their reactivities.

3.3 Thermal Decomposition by Ionic Reaction Mechanisms

Leffler[144] found that the decomposition of 4-methoxy-4′-nitrobenzoyl peroxide (**60**), a highly polarized Case III peroxide, is converted mainly through a carboxy inversion reaction into 4-methoxyphenyl 4-nitrobenzoyl carbonate (**62**) in a polar solvent (equation 37). Denney[25] proposed a mechanism involving the intimate ion pair **61** as the intermediate in the carboxy inversion based on ^{18}O tracer experimental results, i.e. α-^{18}O labeled **60** gives the carboxy-inversion product **62** in which the a-oxygen does not contain any ^{18}O label, whereas the ^{18}O label in the δ oxygen of **60** is retained 66% in the d-oxygen of **62**.[145]

$$CH_3O-C_6H_4-\overset{\alpha\,(100\%)}{C(=O^*)}-O_\beta-O_\gamma-\overset{\delta\,(100\%)}{C(=O^*)}-C_6H_4-NO_2 \longrightarrow \left[CH_3O-C_6H_4^{(+)}{<}\,CO_2^* \;\; {}^{-}O-C(=O)-C_6H_4-NO_2\right] \longrightarrow CH_3-C_6H_4-O_a-C(=O_b)-O_c-C(=O_d)-C_6H_4-NO_2 \quad (37)$$

60 → **61** → **62** (a: 0%, d: 66%)

The carboxy-inversion process is known to be a major reaction process in the thermal decomposition of Case IB and IIB diacyl peroxides.[6,26,27,82,83,146–159] Denney and Sherman[159] reported that the carboxy-inversion reaction of Case IIB diacyl peroxides is a good synthetic route to convert carboxylic acids into alcohols with one less carbon.

The following observations demonstrate the ionic nature of the carboxy inversion. (a) The carboxy inversion is accelerated by polar media and is catalyzed by both Brønsted and Lewis acids.[144,148–151] (b) The configuration of the migrating alkyl group is maintained throughout the inversion process, as exemplified in equations 38[152] and 39.[153] The high yield of ester formation with retention of configuration in the thermal decomposition of bis(2-norbornylcarbonyl) peroxide[160] can also be explained by the same path as shown in equation 38. (c) The rate of the carboxy inversion increases as the ability of the migrating alkyl group to form the corresponding carbocation increases. In the same way, an increase in the electronegativity of the migrating carboxylate group accelerates the reaction.[27,67,82,148–160] (d) The kinetics of the carboxy-inversion process give a large negative $\Delta V^{\ddagger}$[67,82] and lower $\Delta H^{\ddagger}$ and $\Delta S^{\ddagger}$ values than the radical decomposition process.[26,27,150,156,157] (e) The migrating group rearranges specifically to the adjacent peroxidic oxygen atom.[25–27,149,150,152,156]

100 % 100 %
R-C-O-O-C-R → [δ+ R··C··O δ- O-C-R] → R-O-C-O-C-R
34 63

(38)

→ R-O-C-R 82 % ee 79~89 % 11~21 % R = PhCH₂ CH₃ >CH-
64

C_2H_5-C(S)(CH$_3$)(H)-C(O)-O-O-C(O)-C$_6$H$_4$Cl →(CH_3CN)
65
(66 % ee)

(39)

C_2H_5-C(S)(CH$_3$)(H)-O-C(O)-O-C(O)-C$_6$H$_4$Cl + C_2H_5-C(R)(CH$_3$)(H)-N(COCH$_3$)(CO-C$_6$H$_4$Cl)
66 67
36 % (66 % ee) 40 % (9.6 % ee)

Walling *et al.*[83] proposed the following mechanism for the thermal decomposition of Case IB and IIB diacyl peroxides based on their observations that an increase in the polarity of the solvent results in a large increase in the rate of ionic decomposition (k_i) and a small increase in the rate of radical decomposition (k_r), i.e. the peroxides first give an intimate radical pair–ion pair intermediate (**64**) through a single rate-determining step and then **64** is partitioned into radical and ionic reaction products. Thermal decomposition of (*S*)-(+)-3-chlorobenzoyl 2-methylbutanoyl peroxide (**65**) in CH_3CN gives a mixture in 36% yield of carboxy-inversion product (**66**) in which the 1-methylpropyl group migrates with 100% retention of configuration and *N*-acetyl-*N*-3-chlorobenzoyl-2-butylamine (**67**) with 14.4% net inversion of configuration (equation 39). Walling and Sloan[153] explained the results as a consequence of nucleophilic capture of the intimate ion pair **64** by solvent acetonitrile.[153] The lack of CIDNP signals due to the reaction products during thermal decomposition of **65** was explained by the mechanism that the radical

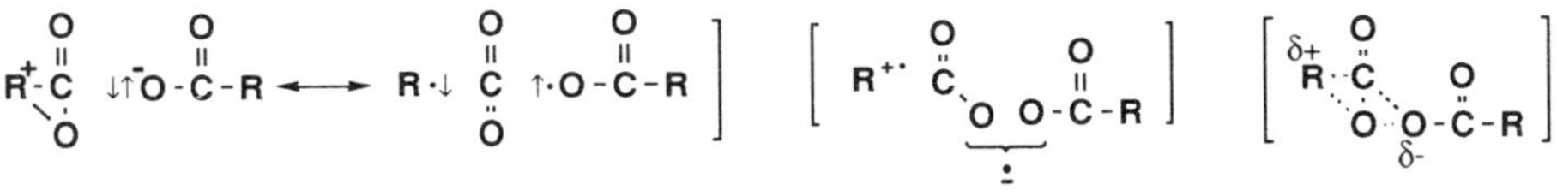

Figure 13. Structures proposed for polar intermediates involved in thermal decomposition of diacyl peroxides

pair–ion pair exchange occurs in tight ion pair **64** too rapidly and with too little separation for polarization to occur (Figure 13).[83]

Decomposition of cyclopropaneacetyl peroxide (**68**) as a member of the family of cycloalkaneacetyl peroxides is anomalously rapid.[160] Oae *et al.*[156] found that thermal decomposition of [^{18}O]carbonyl-labeled **68** gives the carboxy-inversion product (**72**) in 37% yield among other ionic and radical reaction products. The a-oxygen in **69** involves 20% of the original ^{18}O label in **68**.[156] Since ^{18}O scrambling in **68** by the mechanism in Scheme 2 does not take place ($k_s = 0$), equilibration of the ^{18}O label into the three b-, c- and d-oxygen atoms of **69** is due to the ^{18}O scrambling in **72** as shown in equation 40 after the carboxy-inversion reaction. Since the a-oxygen in **69** does not equilibrate with the other remaining b-, c- and d-oxygens under the conditions applied for thermal decomposition of **68**, the incorporation of the ^{18}O label (20%) in the a-oxygen of **69** occurs during the carboxy-inversion process itself.[27] This fact is best explained in terms of the complete heterolysis of the C—C bond in the positively charged half of **2** or **64** and its recombination, i.e. **1(68)** → **2** → **3** → **4** → **5** → **9(69)**.[156] The other possible path, **1(68)**(→ **2**) → **7** → **4** → **5** → **9(69)**, would contribute as a minor path (equation 40).

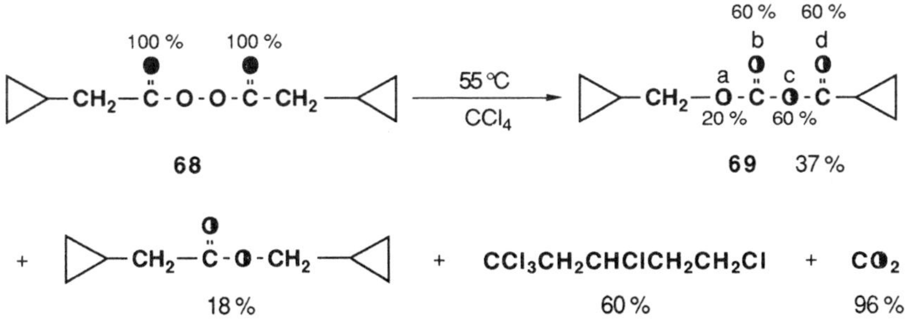

(40)

Taylor *et al.*[158] explored the carboxy-inversion mechanism by a different approach using their structural probes. They decomposed cyclobutaneformyl 3-chlorobenzoyl peroxide (**70**), cyclopropaneacetyl 3-chlorobenzoyl peroxide and 4-pentenoyl 3-chlorobenzoyl peroxide. If the carboxy-inversion products are formed through the intermediate **4** (**72**), Wagner–Meerwein rearrangement in the carboxy-inversion products should occur. This was the case, as shown in

Scheme 21. They obtained a mixture of unrearranged **73** (62%) and Wagner–Meerwein-rearranged **74** (13%) together with unrearranged ester (4%) and Wagner–Meerwein-rearranged ester (16%). The intervention of **4** (**72**) in the carboxy-inversion process was further evidenced by Sigman *et al.*[161] When [^{13}C]cyclobutaneformylcarbonyl-labeled **70** is decomposed in supercritical carbon dioxide, a mixture of **73** (43.6% yield) which retained the ^{13}C label and Wagner–Meerwein-rearranged **74** (4.9% yield) in which ^{13}C-label is 12% exchanged were obtained.[161]

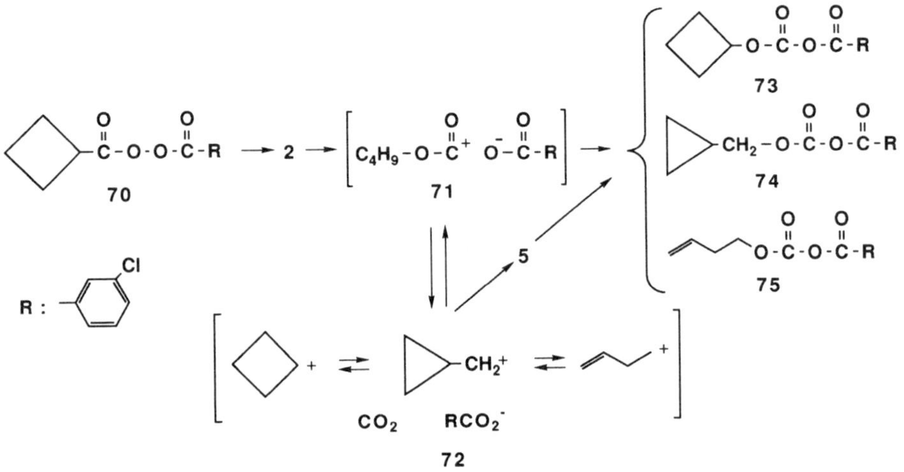

Scheme 21

Signals of ^{13}C CIDNP are detected for cyclobutyl 3-chlorobenzoate and cyclopropylcarbinyl 3-chlorobenzoate during thermal decomposition of **70** in CCl_4. The latter polarization suggests the route **7** → **4** → **11**.[158] A similar CIDNP signal indicating the occurrence of **7** → **4** was observed by Lawler *et al.*[162] in the thermal decomposition of *tert*-butylacetyl 3-chlorobenzoyl peroxide, i.e. Wagner–Meerwein-rearrangement products, 1,1-dimethylcyclopropane and 2-methylbut-1-ene, showed ^{1}H CIDNP signals.

Taylor *et al.*[158] discussed the structure of the primary intermediate in comparison with the structural effect of R on the rate of decomposition of RCO—O—O—COR′ and that on the heterolysis of the O—O bond in $RC(CH_3)_2$—O—O—$COC_6H_4NO_2$ reported by Hedaya and Winstein.[163] They concluded that in the primary intermediate C to O migration of R has not occurred and Leffler and More's unbridged structure (**65**)[151] best accounts for the results, as a transition-state model, since the rates of decomposition of diacyl peroxides are much less sensitive to the structure of R than those for solvolysis of the peresters.[158]

When the ability of R to stabilize positive charge is relatively poor, e.g. the 1-

apocamphyl group, R migrates exclusively to the peroxidic oxygen in the carboxy inversion.[26] Therefore, the bonding electron pair of the R—CO linkage undoubtedly interacts with σ^*_{O-O} to assist heterolysis of the peroxidic linkage. However, since rupture of the O—O bond takes place early in the reaction coordinate, such assistance by R would appear to be small, as shown in Figure 7, path B. In the primary intermediate R bridges between C and O, and only when R^+ is stable can decarboxylation from the positively charged counterpart take place to afford the ion pair **4**. Recombination of **4** leads to the ^{18}O-randomized carboxy-inversion product **9**. It is not yet clear whether **2** is an intermediate or a transition state and further efforts are needed to solve the problem.

^{18}O tracer is useful for determining the quantitative assignment of the contribution of each intermediate involved in the carboxy-inversion process. When one applies the ^{18}O-tracer technique to investigate the mechanism of the carboxy inversion, one must take the ^{18}O scrambling of diacyl peroxide (Scheme 2)[6,27] and the other type of ^{18}O scrambling of the inversion product (equation 41)[26,164,165] into consideration. Fujimori and Oae[27] investigated the mechanism of the thermal decomposition of cyclohexaneformyl benzoyl peroxide (**76**, X = H) and cyclohexaneformyl 3-chlorobenzoyl peroxide (**76**, X = Cl) by means of the ^{18}O-tracer method. Ionic reaction products are the major products (Scheme 22).

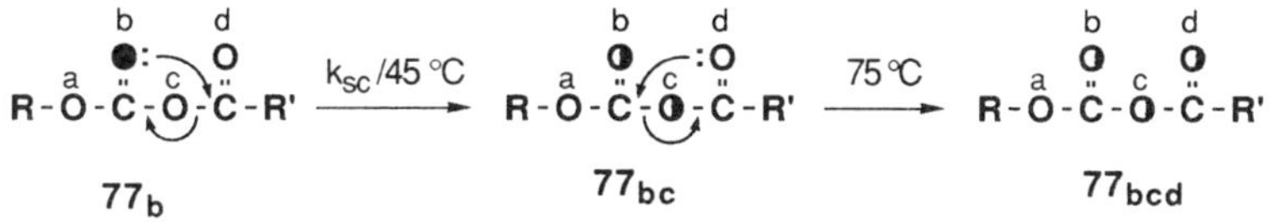

R-O-C(=O)-O-C(=O)-R' (**77b**) —k_{sc}/45 °C→ **77bc** —75 °C→ **77bcd** (41)

R = cyclohexyl, R' = phenyl

k_{sc} : $2.85 \times 10^{-5}\ s^{-1}$ in CCl_4

$3.39 \times 10^{-5}\ s^{-1}$ in 2 M sulfolane-CCl_4

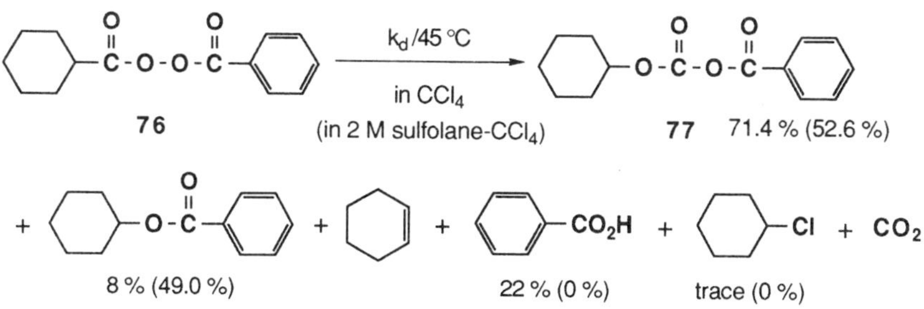

76 —k_d/45 °C, in CCl_4 (in 2 M sulfolane-CCl_4)→ **77** 71.4 % (52.6 %)

\+ cyclohexyl benzoate 8 % (49.0 %) + cyclohexene + $C_6H_5CO_2H$ 22 % (0 %) + cyclohexyl-Cl trace (0 %) + CO_2

k_d : $(6.72 \pm 0.3)\times 10^{-5}\ s^{-1}$ in CCl_4

$(4.23 \pm 0.02)\times 10^{-4}\ s^{-1}$ in 2 M sulfolane-CCl_4

Scheme 22

In these ^{18}O-tracer experiments, the ^{18}O scrambling in **76** did not take place at a measurable rate. The rate of the ^{18}O scrambling of the carboxy-inversion product (**77**)(k_{sc}) (equation 41) was determined in independent experiments using ^{18}O-labeled **77**. Fortunately, the k_{sc} values of **77** were found to be smaller than the k_d values of **76**, although not negligibly small. Then the ^{18}O distributions in the carboxy-inversion products, **77**, formed from α-^{18}O-labeled **76** and δ-^{18}O-labeled **76** were analyzed taking the kinetic data on ^{18}O scrambling of **77** (k_{sc} in equation 41) into consideration. The results are summarized in Scheme 23. The relatively small solvent and substituent effects on the distribution of precursors of the inversion products are remarkable. However, the polarity of solvent affects the distribution of products, i.e. on going from CCl_4 to 2 M sulfolane–CCl_4 the yield of **77** decreases and the ester yield increases markedly, indicating that the rate of recapture of CO_2 by the cyclohexyl cation in **4** decreases in polar solvents. The ion pair **4** gives mainly elimination products in CCl_4 but gives only a coupling product, i.e. ester, in polar solvents.[27]

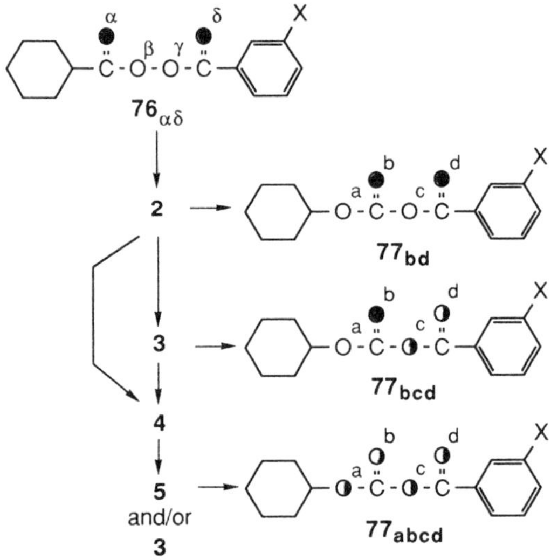

	X=H A	X=H B	X=Cl A	X=Cl B
77_{bd} (from 2)	32%	38%	32%	26%
77_{bcd} (from 3)	62%	54%	62%	66%
77_{abcd} (from 5 and/or 3)	6%	8%	6%	8%

A : in CCl_4
B : in 2 M sulfolane-CCl_4

Scheme 23

Hence the thermal decomposition of Case IIB diacyl peroxides is a good method for the generation of ion pairs in aprotic solvents. Murr and co-workers[166,167] discussed the nature of ion pairs generated by solvolytic decomposition of Case IIB diacyl peroxides and the corresponding carboxy-inversion products in comparison with the ion pairs produced from alkylnitroamides and protonation of diazoalkanes.

4 BIMOLECULAR REACTIONS

4.1 Radical-induced Decomposition

Nozaki and Bartlett[92] found that the kinetics of the thermal decomposition of **21** can be expressed by equation 42. The second term originates from the induced decomposition of the peroxide by solvent radicals. The apparent rate constant of the radical-induced reaction (k_{ri}) is given by equation 46.[92] The size of k_{ri} depends on the nature of the solvent. Hiatt[1] gave values of k_{ri} for **21**. The general solution of the first-order plus χ-order rate expression was proposed by Lamb *et al.*[168]

$$-\mathrm{d[peroxide]}/\mathrm{d}t = k_{\mathrm{d}} + k_{\mathrm{ri}}[\mathrm{peroxide}]^{3/2} \tag{42}$$

$$\text{peroxide} \xrightarrow{k_{\mathrm{HOM}}} 2\mathrm{R}\cdot \tag{43}$$

$$2\mathrm{R}\cdot \xrightarrow{k_{\mathrm{term}}} \mathrm{RR} \tag{44}$$

$$\mathrm{R}\cdot + \text{peroxide} \xrightarrow{k_2} \mathrm{X} + \mathrm{R}\cdot \tag{45}$$

$$k_{\mathrm{ri}} = k_2(k_{\mathrm{d}}/k_{\mathrm{term}})^{1/2} \tag{46}$$

Four types of radical-induced decomposition of diacyl peroxides are known and typical examples are shown in equations 47–50. Nucleophilic radicals attack peroxidic oxygen. Denney and Feig[169] confirmed that the diethyl ether radical attacks the peroxidic oxygen with ^{18}O tracer (equation 47).[170]

$$(C_6H_5-\overset{\bullet}{C}(=O)-O-)_2 \;(100\%) + CH_3-\dot{C}H-O-C_2H_5 \longrightarrow C_6H_5-C(=O)-O-CH(OC_2H_5)(CH_3) \;(70\text{-}80\%,\ 20\text{-}30\%) + C_6H_5-CO_2\cdot \tag{47}$$

The trichloromethyl radical attacks the phenyl ring of **21** (equation 48).[171] Acetyl benzoyl peroxide is susceptible to this type of methyl radical-induced reaction.[88] Attack of trichloromethyl radical on the aromatic ring occurs rapidly in the thermal decomposition of **22** in CCl_4 (Scheme 9). A similar induced decomposition was reported by Hart and Chloupek[172] for an alkenic diacyl peroxide (equation 49).[172]

A single-electron reduction with radical species induces the decomposition of diacyl peroxides (equation 50).[173]

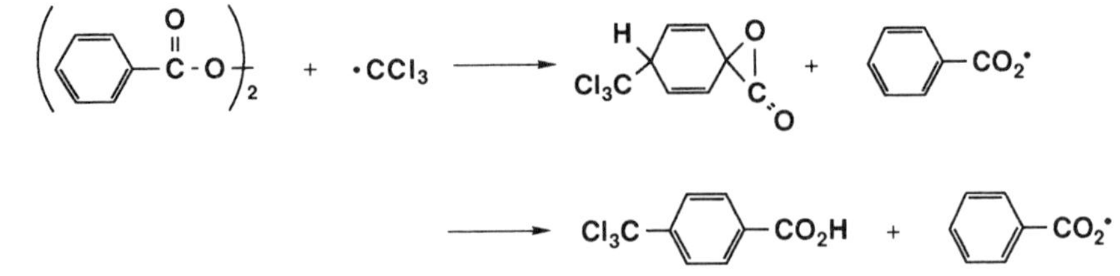

(48)

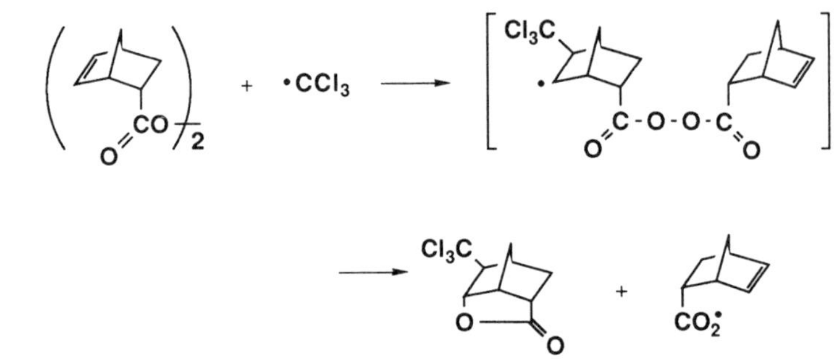

(49)

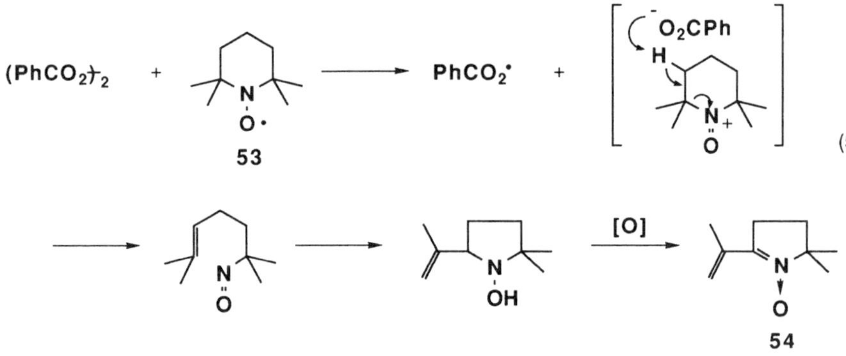

(50)

4.2 Molecular-induced Decomposition

Hard nucleophiles such as alkoxide anion and hydroxide anion attack the carbonyl carbon (charge-controlled reaction) of diacyl peroxides to give peracids. The reaction of **21** with sodium methoxide is a convenient synthetic method for perbenzoic acid (equation 51).[174]

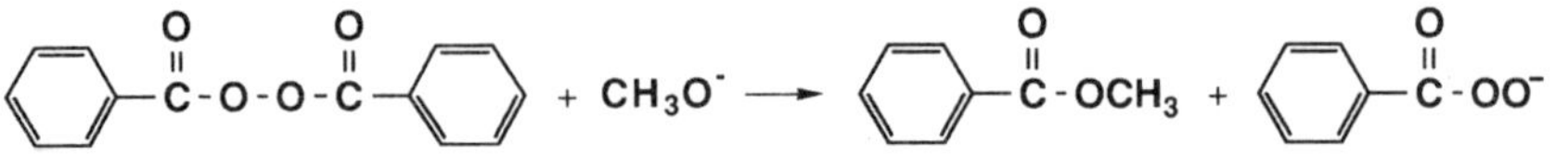

$$PhC(O)OOC(O)Ph + CH_3O^- \longrightarrow PhC(O)OCH_3 + PhC(O)OO^- \quad (51)$$

On the other hand, soft nucleophiles, such as carbanions,[175,176] secondary and tertiary amines,[177–179] divalent sulfur compounds,[180,181] trivalent phosphorus compounds,[182] phenols[183–187] and iodide ion[188] attack the peroxidic oxygen (orbital-controlled reaction). Induced decomposition of diacyl peroxides with closed-shell molecules (D:) diverges into two paths depending on the oxidation potential of D:, i.e. a single-electron transfer reaction (SET) to form the radical pair intermediate **83** and a second-order nucleophilic substitution reaction (S_N2) affording ion pair **84**. The SET path proceeds via the radical pair pathway, whereas the S_N2 path proceeds via the ion pair pathway. Interconversion between **83** and **84** may be possible in some cases (Scheme 24), and hence detection of radicals or formation of radical products cannot be concrete evidence of an initial SET reaction.[179]

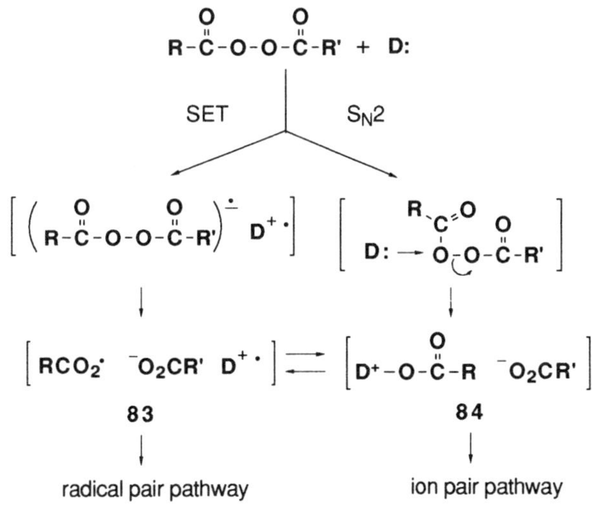

Scheme 24

The following experimental procedures may be useful for distinguishing the two reaction paths in Scheme 24. (a) Product analysis. Srinivas and Taylor's structural probes, **70**, and analogues (Section 3.3), can also be applicable to solve this problem.[179] (b) Effect of oxidation potential of D: on the rate. (c) ^{18}O-tracer experiment. The SET process would afford the ^{18}O scrambling product. (d) Secondary kinetic hydrogen isotope effect. In the reaction of the peroxide with β-deuteriated D:, e.g. $(D_3C)_2S$, $(D_3C)_2NPh$, $(D_3C)_3N$, Ph_2ND, the S_N2 path results in an inverse isotope effect ($k_H/k_D < 1$), whereas the normal isotope effect ($k_H/k_D > 1$) is observed in the SET reaction.[181]

Schank *et al.*[175] reported the acyloxylation of active methylenes with diacyl peroxide as shown in Scheme 25. The neutral diketone **78** gives the monoacyloxylated product **79**. The reaction of the sodium salt **80** with **21** results eventually in **82** via **81**.

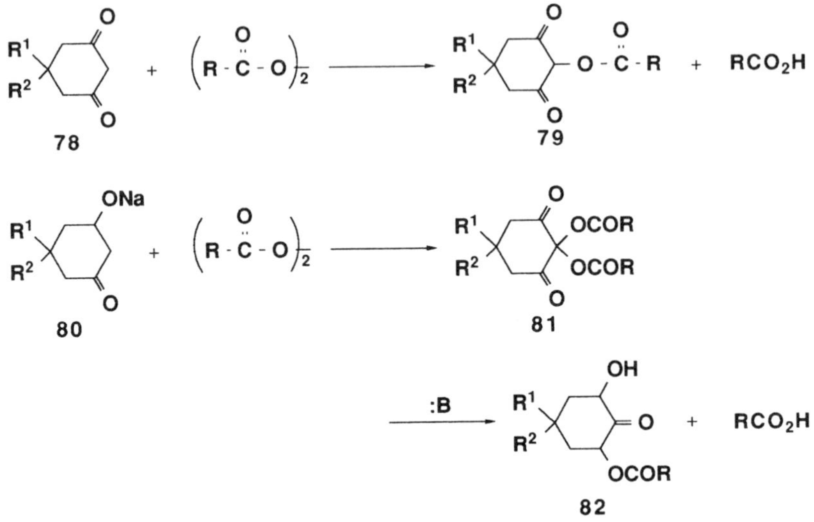

Scheme 25

Reaction of benzoyl peroxide and sulfide **85** is initiated by an S_N2 reaction affording the ion pair **86**, which eventually gives a mixture of sulfoxide **87** and Pummerer rearrangement product **88** (Scheme 26).[180] Pryor and Hendrickson[181] proved the mechanism by their isotope effect criteria.

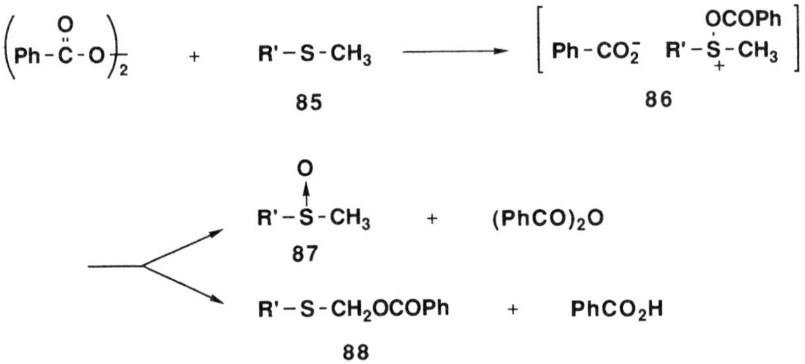

Scheme 26

Bimolecular reaction between **21** and aliphatic tertiary amines (quinuclidine, diazabicyclo[2.2.2]octane, *N,N*-dimethylaniline) are considered to be initiated by S_N2 reactions (Scheme 27).[179]

Highly electron-donating *N,N'*-tetramethyl-*p*-phenylenediamine reacts with **70** by the SET path (Scheme 28).[179]

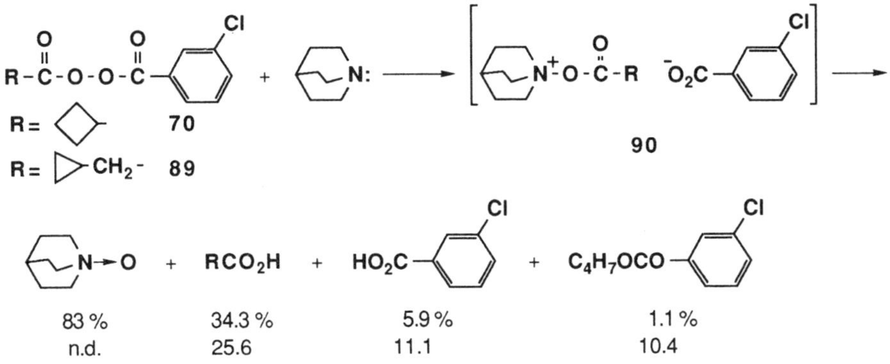

R = cyclobutyl 70
R = cyclopropyl-CH_2- 89

90

	N-oxide	RCO_2H	HO_2C-C_6H_4-Cl	C_4H_7OCO-C_6H_4-Cl
	83 %	34.3 %	5.9 %	1.1 %
	n.d.	25.6	11.1	10.4

Scheme 27

- 55 °C
$CHCl_3$

R = cyclobutyl 70
R = cyclopropyl-CH_2· 89

91

RCO_2H	HO_2C-C_6H_4-Cl	R-O-CO-C_6H_4-Cl	CO_2	RH
2 %	57.9 %	0.6 %		3~7 %
2	47.9	2.3		n.d.

92 a~c
15.4 %
19.9

92		from **70**	from **89**
C_4H_7 : a	cyclobutyl	15.4 %	0 %
b	cyclopropyl-CH_2	0	16.3
c	allyl-ethyl (butenyl)	0	3.6

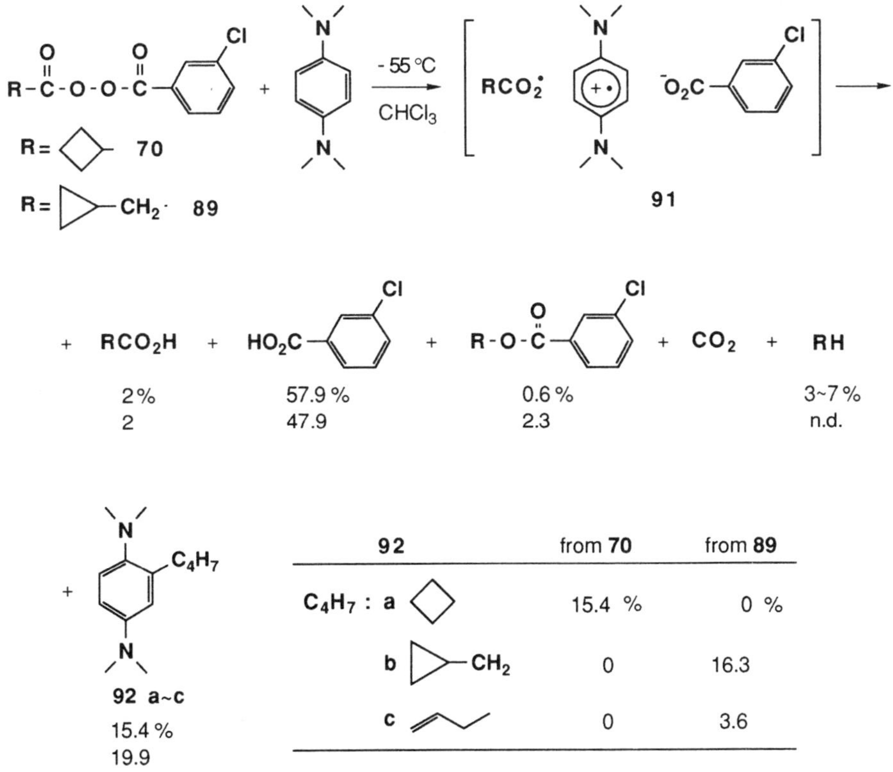

Scheme 28

Alkenes react with diacyl peroxides by an SET process.[189-192] Schuster and co-workers[193,194] found that logarithms of the rates of bimolecular reactions between cyclic diacyl peroxides (**28**, **93**) and aromatic hydrocarbons are linearly correlated with the oxidation potentials of donors and suggested an SET path. The linear correlation between log k_2 and the oxidation potentials of nucleophiles (donors) is usually observed in orbital-controlled S_N2 reactions.[195] Since the S_N2 reaction on the peroxidic oxygen of diacyl peroxides is a typical orbital-controlled reaction, one must be careful in concluding the reaction mechanism from this kind of correlation. Although decomposition of **93** is sufficiently exothermic to produce electronically excited **97**, no evidence was obtained to support this. However, interestingly, the induced decomposition of **93** with aromatics emits light. The phenomenon was explained by the chemically induced electron-exchange luminescence (CIEEL) mechanism shown in Scheme 29.[193]

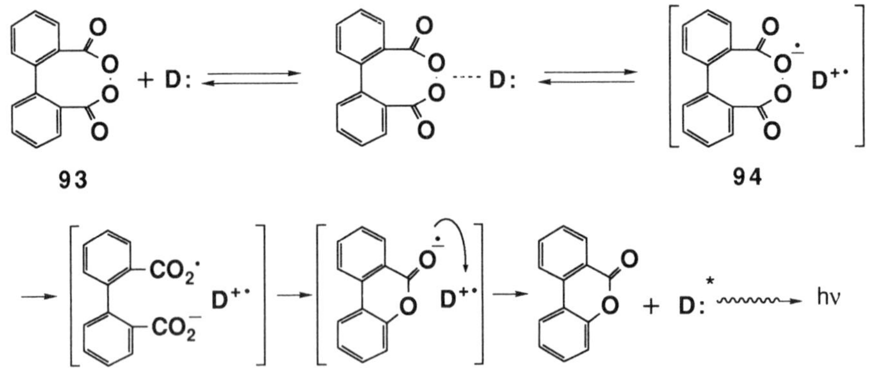

93 + D: ⇌ [93 ···· D:] ⇌ [94 D$^{+\cdot}$] → [95 D$^{+\cdot}$] → [96 D$^{+\cdot}$] → 97 + D:* ⟿ hν

Scheme 29

Highly electron-deficient perfluoroalkanoyl peroxides react with aromatic hydrocarbons (Scheme 30) and aromatic heterocycles (equations 52–54). These reactions were reported to be good synthetic methods for introducing the perfluoroalkyl group (R_F) into these aromatic rings.[196–198]

$$(R_FCO_2)_2 + \left(CH_2-CH(C_6H_5)\right)_n \longrightarrow \left(CH_2-CH(C_6H_4R_F)\right)_x\left(CH_2-CH(C_6H_5)\right)_y \quad (52)$$

$$(C_3F_7CO_2)_2 + \text{thiophene} + C_9F_{19}CO_2^- \ PyH^+ \longrightarrow \text{2-}C_3F_7\text{-thiophene (32\%)} + \text{2-}C_9H_{19}\text{-thiophene (20\%)} \quad (53)$$

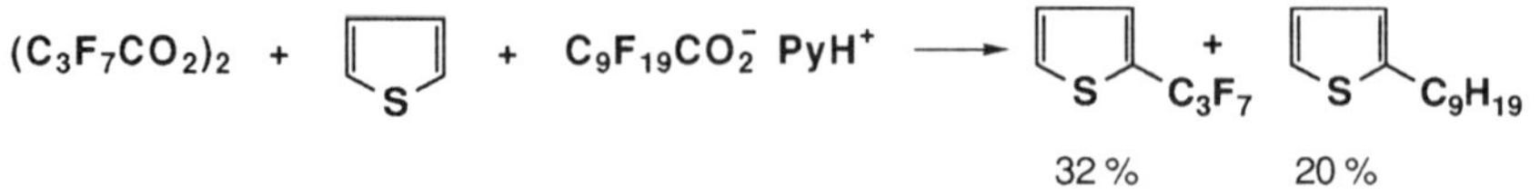

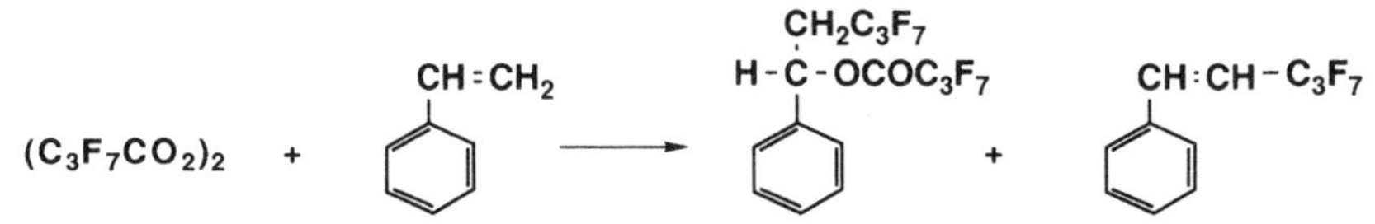

$$(C_3F_7CO_2)_2 + C_6H_5CH{=}CH_2 \longrightarrow C_6H_5\text{-}CH(OCOC_3F_7)CH_2C_3F_7 + C_6H_5CH{=}CH\text{-}C_3F_7 \quad (54)$$

$$(R_FCO_2)_2 + C_6H_5X \longrightarrow (R_FCO_2)_2^{-\bullet} + [C_6H_5X]^{+\bullet}$$

$$\longrightarrow R_F\bullet + R_FCO_2^- + CO_2 + [C_6H_5X]^{+\bullet}$$

$$\longrightarrow [XC_6H_5(H)(R_F)]^+ + R_FCO_2^- + CO_2$$

$$\longrightarrow XC_6H_4R_F + R_FCO_2H + CO_2$$

(45 ~ 97 %)

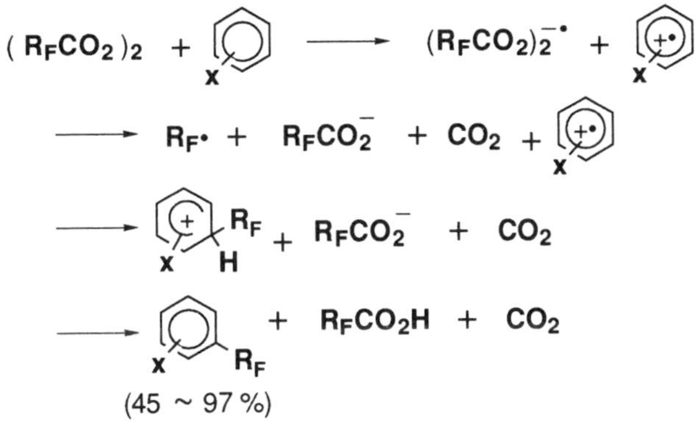

Scheme 30

Eberson[199] tried to treat Schuster and co-workers' kinetic results[193,194] by the Marcus theory. Walling[200] proposed a mechanism in which the radical pair pathway and the ion pair pathway merge around the transition state in which only partial electron transfer is occurring for the endothermic molecular induced decomposition of peroxides.

5 REFERENCES

1. R. Hiatt, in *Organic Peroxides* (D. Swern, Ed.), Wiley–Interscience, New York (1970), p. 749.
2. (a) C. C. Price and E. Krebs, *Org. Synth., Coll. Vol.*, **3**, 649 (1955); (b) C. C. Price and H. Moritz, *J. Am. Chem. Soc.*, **75**, 3686 (1953).
3. J. R. Slagle and H. J. Shine, *J. Org. Chem.*, **24**, 107 (1959).
4. L. S. Silbert, D. Swern and T. Asahara, *J. Org. Chem.*, **33**, 3670 (1968).
5. D. E. van Sickle, *J. Org. Chem.*, **34**, 3446 (1969).
6. K. Fujimori and S. Oae, *Tetrahedron*, **29**, 65 (1973).
7. (a) K. E. Russell, *J. Am. Chem. Soc.*, **77**, 4814 (1955); (b) M. Jones, Jr, and M. R. DeCamp, *J. Org. Chem.*, **36**, 1536 (1971).
8. P. B. Dervan and C. R. Jones, *J. Org. Chem.*, **44**, 2116 (1979).
9. W. Cooper, *J. Chem. Soc.*, 3106 (1951).
10. L. S. Silbert and D. Swern, *J. Am. Chem. Soc.*, **81**, 2364 (1959).
11. J. K. Kochi and H. E. Mains, *J. Org. Chem.*, **30**, 1862 (1965).
12. D. F. DeTar and L. A. Carpino, *J. Am. Chem. Soc.*, **77**, 6370, (1955).
13. R. G. Woolford and R. N. Gedye, *Can. J. Chem.*, **45**, 291 (1967).

14. T. Koenig and R. Cruthoff, *J. Am. Chem. Soc.*, **91**, 2562 (1969).
15. R. A. Johnson, *Tetrahedron Lett.*, 331 (1976).
16. F. D. Greene and J. Kazan, *J. Org. Chem.*, **28**, 2168 (1963).
17. D. F. DeTar and R. Silverstein, *J. Am. Chem. Soc.*, **88**, 1013 (1966).
18. J. K. Kochi, A. Bemis and L. L. Jenkins, *J. Am. Chem. Soc.*, **90**, 4616 (1968).
19. M. M. Martin and J. M. King, *J. Am. Chem. Soc.*, **38**, 1588 (1973).
20. (a) W. Adam and R. Rucktäschel, *J. Am. Chem. Soc.*, **93** 557 (1971); (b) W. Adam and R. Rucktäschel, *J. Org. Chem.*, **37**, 4128 (1972); (c) W. Adam, J. C. Liu and O. Rodriguez, *J. Org. Chem.*, **38**, 2269 (1973); (d) O. L. Chapman, P. W. Wojtkowski, W. Adam, O. Rodriguez and R. Rucktäschel, *J. Am. Chem. Soc.*, **94**, 1365 (1972).
21. F. Ramirez, S. B. Bhatia, R. B. Mitra, Z. Hamlet and N. B. Desai, *J. Am. Chem. Soc.*, **86**, 4394 (1964).
22. J. R. Hurst and G. B. Schuster, *J. Org. Chem.*, **45**, 1053 (1980).
23. H. Wieland, T. Ploetz and K. Indest, *Justus Liebigs Ann. Chem.*, **532**, 179 (1937).
24. D. R. Bryne, F. M. Gruen, D. Priddy and R. D. Schuetz, *J. Heterocycl. Chem.*, **3**, 369 (1966).
25. D. B. Denney, *J. Am. Chem. Soc.*, **78**, 590 (1956).
26. S. Oae, K. Fujimori and S. Kozuka, *Tetrahedron*, **28**, 5327 (1972).
27. K. Fujimori and S. Oae, *J. Chem. Soc., Perkin Trans. 2*, 1335 (1989).
28. J. M. McBride and R. A. Merill, *J. Am. Chem. Soc.*, **102**, 1923 (1980).
29. R. D. Mair and R. T. Hall, in *Organic Peroxides* (D. Swern, Ed.), Wiley-Interscience, New York (1970) p. 535.
30. O. L. Mageli and C. S. Sheppard, in *Organic Peroxides* (D. Swern, Ed.), Wiley-Interscience, New York (1970), p. 90.
31. R. W. van Dolah, *Ind. Eng. Chem.*, **53**, 59A (1961).
32. K. E. Russell, *Can. J. Chem.*, **38**, 1602 (1960).
33. F. D. Greene, *J. Am. Chem. Soc.*, **78**, 2246 (1956).
34. H. Sawada, M. Yoshida, T. Yoshida and N. Kamigata, *J. Fluorine Chem.*, **46**, 423 (1990).
35. (a) J. M. Gougoutas and J. C. Clardy, *Acta Crystallogr., Sect. B*, **26**, 1999 (1970); (b) J. M. Gougoutas and D. G. Nuae, *J. Phys. Chem.*, **82**, 393 (1978); (c) J. M. Gougoutas, *Pure Appl. Chem.*, **27**, 305 (1971); (d) J. M. Gougoutas and K. H. Chang, *Cryst. Struct. Commun.*, **8**, 977 (1979).
36. S. Caticha-Ellis and S. C. Abrahams, *Acta Crystallogr., Sect. B*, **24**, 277 (1968).
37. M. Sax and R. K. McMullan, *Acta Crystallogr.*, **22**, 281 (1967).
38. J. M. Gougoutas, in *The Chemistry of Peroxides* (S. Patai, Ed.), Wiley, Chichester (1983), p. 375.
39. K. D. M. Harris and M. D. Hollingsworth, *Proc. R. Soc. London, Ser. A*, **431**, 245 (1990).
40. (a) S. C. Abrahams, R. L. Collin and W. N. Lipscomb, *Acta Crystallogr.*, **4**, 15 (1951); (b) W. R. Busing and H. A. Levy, *J. Chem. Phys.*, **42**, 3054 (1965).
41. H. Sawada, M. Nakayama, O. Kikuchi and Y. Yokoyama, *J. Fluorine Chem.*, **50**, 393 (1990).
42. O. Kikuchi, A. Hiyama, H. Yoshida and K. Suzuki, *Bull. Chem. Soc. Jpn.*, **51**, 11 (1978).
43. S. Oae and K. Fujimori, in *The Chemistry of Peroxides* (S. Patai, Ed.), Wiley, Chichester (1983), p. 585.
44. S. Yamauchi, N. Hirota, S. Takahara, H. Sakuragi and K. Tokumaru, *J. Am. Chem. Soc.*, **107**, 5021 (1985).

45. H. Misawa, K. Sawabe, S. Takahara, H. Sakuragi and K. Tokumaru, *Chem. Lett.*, 357 (1988).
46. S. Yamauchi, N. Hirota, S. Takahara, H. Misawa, K. Sawabe, H. Sakuragi and K. Tokumaru, *J. Am. Chem. Soc.*, **111**, 4402 (1989).
47. J. Chateauneuf, J. Lusztyk and K. U. Ingold, *J. Am. Chem. Soc.*, **109**, 897 (1987).
48. J. Chateauneuf, J. Lusztyk and K. U. Ingold, *J. Am. Chem. Soc.*, **110**, 2877 (1988).
49. J. Chateauneuf, J. Lusztyk and K. U. Ingold, *J. Am. Chem. Soc.*, **110**, 2886 (1988).
50. H. G. Korth, W. Muller, J. Lusztyk and K. U. Ingold, *Angew. Chem., Int. Ed. Engl.*, **28**, 183 (1989).
51. J. Chateauneuf, J. Lusztyk, B. Maillard and K. U. Ingold, *J. Am. Chem. Soc.*, **110**, 6727 (1988).
52. L. Jaffe, E. J. Prosen and M. Szwarc, *J. Chem. Phys.*, **27**, 416 (1957).
53. J. K. Kochi, in *Free Radicals*, Vol. II (J. K. Kochi, Ed.), Wiley, New York (1973), p. 700.
54. K. Fujimori, Y. Hirose and S. Oae, submitted for publication.
55. P. D. Bartlett and C. Ruchart, *J. Am. Chem. Soc.*, **82**, 1756 (1960).
56. W. A. Pryor and K. Smith, *Intern. J. Chem. Kinet.*, **3**, 387 (1971).
57. G. S. Hammond and L. M. Soffer, *J. Am. Chem. Soc.*, **72**, 4711 (1950).
58. (a) H. J. Shine and J. R. Slagle, *J. Am. Chem. Soc.*, **81**, 6309 (1959); (b) H. J. Shine and D. M. Hoffman, *J. Am. Chem. Soc.*, **83**, 2782 (1961).
59. J. C. Martin, J. W. Taylor and D. H. Drew, *J. Am. Chem. Soc.*, **89**, 129 (1967).
60. (a) J. Franck and E. Rabinowitsch, *Trans. Faraday Soc.*, **30**, 120 (1934); (b) R. M. Noys, *J. Am. Chem. Soc.*, **78**, 54886 (1956).
61. (a) L. Herk, H. Feld and M. Szwarc, *J. Am. Chem. Soc.*, **83**, 2998 (1961); (b) O. Dobis, J. M. Pearson and M. Szwarc, *J. Am. Chem. Soc.*, **90**, 278 (1968); (c) K. Chakravorty, J. M. Pearson and M. Szwarc, *J. Am. Chem. Soc.*, **90**, 283 (1968).
62. (a) S. F. Nelsen and P. D. Bartlett, *J. Am. Chem. Soc.*, **88**, 143 (1966); (b) H. Kiefer and T. G. Traylor, *J. Am. Chem. Soc.*, **89**, 6667 (1967); (c) F. D. Greene, M. A. Berwick and J. C. Stowell, *J. Am. Chem. Soc.*, **92**, 867 (1970); (d) N. Nodelman and J. C. Martin, *J. Am. Chem. Soc.*, **98**, 6597 (1976).
63. (a) T. Koenig and M. Deinzer, *J. Am. Chem. Soc.*, **90**, 7014 (1968); (b) T. Koenig, *J. Am. Chem. Soc.*, **91**, 2558 (1969).
64. W. Braun, L. Rajbenbach and F. R. Eirich, *J. Phys. Chem.*, **66**, 1591 (1962).
65. (a) W. A. Pryor and K. Smith, *J. Am. Chem. Soc.*, **92**, 5403 (1970); (b) W. A. Pryor, E. H. Morkved and H. T. Bickley, *J Org. Chem.*, **37**, 1999 (1972).
66. R. C. Neuman, Jr, and F. D. Lockyer, Jr, *J. Am. Chem. Soc.*, **105**, 3982 (1983).
67. C. Walling, H. N. Moulden, J. H. Waters and R. C. Neuman, Jr, *J. Am. Chem. Soc.*, **87**, 518 (1965).
68. (a) R. C. Neuman, Jr, *Acc. Chem. Res.*, **5**, 381 (1972); (b) R. C. Neuman, Jr, and R. P. Pankratz, *J. Am. Chem. Soc.*, **92**, 4122 (1973).
69. (a) M. J. Goldstein, *Tetrahedron Lett.*, 1601 (1964); (b) M. J. Goldstein, H. A. Judson and M. Yoshida, *J. Am. Chem. Soc.*, **92**, 4122 (1970).
70. J. W. Taylor and J. C. Martin, *J. Am. Chem. Soc.*, **89**, 6904 (1967).
71. (a) T. W. Koenig and W. D. Brewer, *Tetrahedron Lett.*, 2773 (1965); (b) T. Koenig and R. Cruthoff, *J. Am. Chem. Soc.*, **91**, 2562 (1969).
72. T. Koenig and R. Wolf, *J. Am. Chem. Soc.*, **91**, 2569 (1969).
73. (a) T. Koenig and R. Wolf, *J. Am. Chem. Soc.*, **89**, 2948 (1967); (b) T. Koenig and R. Wolf, *J. Am. Chem. Soc.*, **91**, 2574 (1969).
74. (a) J. C. Martin and S. A. Dombchik, in *Oxidation of Organic Compounds, Advances in Organic Chemistry Series*, Vol. 75 (R. F. Gould, Ed.), American Chemical

Society, Washington (1968), p. 269; (b) S. A. Dombchik, *PhD Thesis*, University of Illinois, Urbana, IL (1968).
75. J. C. Martin and J. H. Hargis, *J. Am. Chem. Soc.*, **91**, 5399 (1969).
76. M. J. Goldstein and H. A. Judson, *J. Am. Chem. Soc.*, **92**, 4120 (1970).
77. M. J. Goldstein and W. A. Haiby, *J. Am. Chem. Soc.*, **96**, 7358 (1974).
78. (a) A. Rembaum and M. Szwarc, *J. Am. Chem. Soc.*, **76**, 5975 (1954); (b) M. Levy, M. Steinberg and M. Szwarc, *J. Am. Chem. Soc.*, **76**, 5978 (1954).
79. (a) A. Rembaum and M. Szwarc, *J. Phys. Chem.*, **23**, 909 (1955); (b) J. Smid, A. Menbaum and M. Szwarc, *J. Am. Chem. Soc.*, **78**, 3315 (1956).
80. L. A. Johnston, J. C. Scaiano and K. U. Ingold, *J. Am. Chem. Soc.*, **106**, 4877 (1984).
81. (a) J. Smid and M. Szwarc, *J. Chem. Phys.*, **29**, 432 (1958); (b) M. Szwarc and L. Herk, *J. Chem. Phys.*, **29**, 438 (1958).
82. R. C. Lamb, J. G. Pacifici and P. W. Ayers, *J. Am. Chem. Soc.*, **87**, 3928 (1965).
83. C. Walling, H. P. Waits, J. Milovanovic and C. G. Papianou, *J. Am. Chem. Soc.*, **92**, 4927 (1970).
84. K. Fujimori, Y. Oshibe and S. Oae, in preparation.
85. Z. Chengxue, Z. Penmo, P. Hegi, J. Xiangshan, Q. Yangling, W. Chengjiu and J. Xikui, *J. Org. Chem.*, **47**, 2009 (1982).
86. M. Yoshida, H. Amemiya, M. Kobayashi, H. Sawada, H. Hagii and K. Aoyama, *J. Chem. Soc., Chem. Commun.*, 234 (1985).
87. (a) M. Feldhues and H. J. Schafer, *Tetrahedron*, **41**, 4195 (1985); (b) M. Feldhues and H. J. Schafer, *Tetrahedron*, **41**, 4213 (1985).
88. C. Walling and Z. Cekovic, *J. Am. Chem. Soc.*, **89**, 6681 (1967).
89. M. Kobayashi, H. Minato and R. Hisada, *Bull. Chem. Soc. Jpn.*, **44**, 2271 (1971).
90. D. F. DeTar, R. A. J. Long, J. Rendleman, J. Bradley and P. Duncan, *J. Am. Chem. Soc.*, **89**, 4051 (1967).
91. D. F. DeTar, *J. Am. Chem. Soc.*, **89**, 4058 (1967).
92. (a) K. Nozaki and P. Bartlett, *J. Chem. Soc.*, **68**, 1686 (1946); (b) P. D. Bartlett and K. Nozaki, *J. Am. Chem. Soc.*, **69**, 2299 (1947).
93. (a) G. B. Gill and C. H. Williams, *J. Chem. Soc.*, 995 (1965); (b) C. E. H. Bawn and S. F. Mellish, *Trans. Faraday Soc.*, **47**, 1216 (1951).
94. E. G. Janzen, C. A. Evans and Y. Nishi, *J. Am. Chem. Soc.*, **94**, 8238 (1972).
95. C. G. Swain, W. T. Stockmayer and J. T. Clark, *J. Am. Chem. Soc.*, **72**, 5426 (1950).
96. J. E. Leffler and R. G. Zepp, *J. Am. Chem. Soc.*, **92**, 3713 (1970).
97. (a) M. S. Kharasch and R. L. Dannley, *J Org. Chem.*, **10**, 406 (1945); (b) M. Takebayashi, T. Shinkai and H. Matsui, *Bull. Chem. Soc. Jpn.*, **27**, 371 (1954).
98. D. R. Bryne, S. R. Burton, A. Cin, F. Kabbe, F. M. Gruen and R. D. Schuetz, *J. Chem. Eng. Data*, **17**, 507 (1972).
99. (a) F. D. Greene, *J. Am. Chem. Soc.*, **78**, 2250 (1956); (b) F. D. Greene and W. W. Rees, *J. Am. Chem. Soc.*, **80**, 3432 (1958); (c) F. D. Greene, *J. Am. Chem. Soc.*, **81**, 1503 (1959); (d) F. D. Greene and W. W. Rees, *J. Am. Chem. Soc.*, **82**, 890 (1960); (e) F. D. Greene and W. W. Rees, *J. Am. Chem. Soc.*, **82**, 893 (1960).
100. K. Fujimori, Y. Oshibe and S. Oae, in preparation.
101. P. B. Dervan and C. R. Jones, *J. Org. Chem.*, **44**, 2116 (1979).
102. M. M. Martin and J. M. King, *J. Org. Chem.*, **38**, 1588 (1973).
103. M. J. S. Dewar and L. E. Wade, Jr, *J. Am. Chem. Soc.*, **99**, 4417 (1977).
104. M. J. S. Dewar and C. Jie, *J. Am. Chem. Soc.*, **109**, 5893 (1987).
105. W. von E. Doering, V. G. Toscano and G. H. Beasley, *Tetrahedron*, **18**, 67 (1962).
106. (a) J. J. Gagewski and N. D. Conrad, *J. Am. Chem. Soc.*, **101**, 6693 (1979); (b) J. J. Gagewski, *Acc. Chem. Res.*, **13**, 142 (1980).
107. H. Kessler and W. Otto, *J. Am. Chem. Soc.*, **98**, 5014 (1976).

108. T. Koenig, R. A. Wielesek and J. G. Huntington, *Tetrahedron Lett.*, 2283 (1974).
109. (a) O. Kikuchi, K. Utsumi and K. Suzuki, *Bull. Chem. Soc. Jpn.*, **50**, 1339 (1977); (b) O. Kikuchi, *Tetrahedron Lett.*, 2421 (1977); (c) O. Kikuchi, A. Hiyama, H. Yoshida and K. Suzuki, *Bull. Chem. Soc. Jpn.*, **51**, 11 (1978).
110. N. J. Karsch, E. T. Koh, B. L. Whitsel and J. M. McBride, *J. Am. Chem. Soc.*, **97**, 6729 (1975).
111. S. D. Peyerimhoff, P. S. Skell, D. May and R. J. Buenker, *J. Chem. Soc.*, **104**, 4515 (1982).
112. M. J. S. Dewar, A. H. Pakiari and A. B. Pierini, *J. Am. Chem. Soc.*, **104**, 3242 (1982).
113. D. Feller, E. S. Huyser, W. T. Borden and E. R. Davidson, *J. Am. Chem. Soc.*, **105**, 1459 (1983).
114. A. D. McLean, B. H. Leansfield, III, J. Pacansky and Y. Ellinger, *J. Chem. Phys.*, **83**, 3567 (1985).
115. M. Bremer, P. von R. Schleyer, K. Schotz, M. Kausch and M. Schindler, *Angew. Chem., Int. Ed. Engl.*, **26**, 761 (1987).
116. K. Fujimori and S. Oae, in *Medical, Biochemical and Chemical Aspects of Free Radicals* (O. Hayaishi, E. Niki, M. Kondo and T. Yoshikawa, Eds), Elsevier, Amsterdam (1989), p. 989.
117. R. Hoffmann and R. B. Woodward, *J. Am. Chem. Soc.*, **87**, 4389 (1965).
118. H. C. Box, E. E. Budzinski and H. G. Freund, *J. Am. Chem. Soc.*, **92**, 5305 (1970).
119. V. I. Barchuk, A. A. Dubinsky, O. Y. Grinberg and Y. S. Levedev, *Phys. Lett.*, **34**, 476, (1975).
120. M. W. Vary and J. M. McBride, *Mol. Cryst. Liq. Cryst.*, **52**, 133 (1979).
121. M. D. Hollingsworth and J. M. McBride, *Chem. Phys. Lett.*, **130**, 259 (1986).
122. O. J. Walker and G. L. E. Wild, *J. Chem. Soc.*, 1132 (1937).
123. T Kashiwagi, K. Fujimori, S. Kozuka and S. Oae, *Tetrahedron*, **26**, 3639 (1970).
124. J. K. Kochi and P. J. Krusic, *J. Am. Chem. Soc.*, **91**, 3940 (1969).
125. R. A. Sheldon and J. K. Kochi, *J. Am. Chem. Soc.*, **92**, 4395 (1970).
126. D. Griller and K. U. Ingold, *Acc. Chem. Res.*, **13**, 317 (1980).
127. A. Kitamura, H. Sakuragi, M. Yoshida and K. Tokumaru, *Bull. Chem. Soc. Jpn.*, **53**, 1393 (1980).
128. S. Takahara, T. Urano, A. Kitamura, H. Sakuragi, O. Kikuchi, M. Yoshida and K. Tokumaru, *Bull. Chem. Soc. Jpn.*, **58**, 688 (1985).
129. K. U. Ingold, L. J. Johnston, J. Lusztyk and J. C. Scaiano, *Chem. Phys. Lett.*, 433 (1984).
130. G. Wittig and H. F. Ebel, *Justus Liebigs Ann. Chem.*, **650**, 20 (1961).
131. O. L. Chapman, C. L. McIntosch, J. Pacansky, J. W. Calder and G. Orr, *J. Am. Chem. Soc.*, **95**, 4061 (1973).
132. (a) K. Sawabe, H. Misawa, H. Sakuragi and K. Tokumaru, *Abstracts of Symposium on Photochemistry, Tokyo* (1988) p. 103; (b) T. Tateno, J. Wang, H. Sakuragi and K. Tokumaru, *Abstracts of Symposium on Photochemistry, Kyoto* (1990) p. 411; (c) H. Sakuragi, S. Takahara, H. Misawa, T. Tateno, J. Wang and K. Tokumaru, *Abstracts of 40th Symposium on Basic Organic Chemistry, Tsukuba* (1990), p. 411.
133. H. G. Korth, J. Chateauneuf, J. Lusztyk and K. U. Ingold, *J. Org. Chem.*, **56**, 2405 (1991).
134. J. Hilborn and J. A. Pincock, *J. Am. Chem. Soc.*, **113**, 2683 (1991).
135. D. E. Falvey and G. B. Schuster, *J. Am. Chem. Soc.*, **108**, 7419 (1986).
136. (a) H. G. Korth, J. Chateauneuf, J. Lusztyk and K. U. Ingold, *J. Am. Chem. Soc.*, **110**, 5929 (1988); (b) H. G. Korth, J. Chateauneuf, J. Lusztyk and K. U. Ingold, *J. Chem. Soc., Perkin Trans. 2*, 1997 (1990).
137. M. Levey and M. Szwarc, *J. Am. Chem. Soc.*, **76**, 5981 (1954).

138. H. Fisher and H. Paul, *Acc. Chem. Res.*, **20**, 202 (1987).
139. A. J. Paine and J. Warkentin, *Can. J. Chem.*, **59**, 491 (1981).
140. (a) J. C. Bevington, J. Toole and L. Trossarelli, *Trans. Faraday Soc.*, 863 (1958); (b) J. C. Bevington and J. Toole, *J. Polym. Sci.*, **28**, 413 (1958); (c) J. C. Bevington and T. D. Lewis, *Trans. Faraday Soc.*, 134 (1958).
141. E. G. Janzen and C. A. Evans, *J. Am. Chem. Soc.*, **94**, 205 (1975).
142. (a) J. A. D. Hollander and J. P. M. v. d. Ploeg, *Tetrahedron*, **32**, 2433 (1976); (b) R. E. Scherzel, R. G. Lawler and G. T. Evans, *Chem. Phys. Lett.*, **29**, 109 (1974).
143. (a) R. Kaptein, J. Brokken-Zijp and J. J. Kanter, *J. Am. Chem. Soc.*, **94**, 6280 (1972); (b) R. Kaptein, in *Advances in Free Radical Chemistry*, Vol. 5 (G. H. Smith, Ed.), Academic Press, New York (1975), p. 319.
144. J. E. Leffler, *J. Am. Chem. Soc.*, **72**, 67 (1950).
145. D. B. Denney and D. G. Denney, *J. Am. Chem. Soc.*, **79**, 4806 (1957).
146. P. D. Bartlett and F. D. Greene, *J. Am. Chem. Soc.*, **76**, 1088 (1954).
147. F. D. Greene, H. P. Stein, C. -C. Chu and F. M. Vane, *J. Am. Chem. Soc.*, **86**, 2080 (1964).
148. J. E. Leffler and C. C. Petropoulos, *J. Am. Chem. Soc.*, **79**, 3068 (1957).
149. D. Z. Denney, T. M. Velega and D. B. Denney, *J. Am. Chem. Soc.*, **86**, 46 (1964).
150. T. Kashiwagi and S. Oae, *Tetrahedron*, **26**, 3631 (1970).
151. J. E. Leffler and A. A. More, *J. Am. Chem. Soc.*, **94**, 2483 (1972).
152. T. Kashiwagi, S. Kozuka and S. Oae, *Tetrahedron*, **26**, 3619 (1970).
153. C. Walling and J. P. Sloan, *J. Am. Chem. Soc.*, **101**, 7679 (1979).
154. H. H. Hau and H. Hart, *J. Am. Chem. Soc.*, **81**, 4897 (1959).
155. R. C. Lamb, L. L. Vestal, G. R. Cipau and S. Debnath, *J. Org. Chem.*, **39**, 2096 (1974).
156. S. Oae, K. Fujimori, S. Kozuka and Y. Uchida, *J. Chem. Soc., Perkin Trans. 2*, 1844 (1974).
157. R. C. Lamb and J. R. Sanderson, *J. Am. Chem. Soc.*, **91**, 5034 (1969).
158. K. G. Taylor, C. K. Govindan and M. S. Kaelin, *J. Am. Chem. Soc.*, **101**, 2091 (1979).
159. D. B. Denney and N. Sherman, *J. Org. Chem.*, **30**, 3760 (1965).
160. (a) H. Hart and D. P. Wyman, *J. Am. Chem. Soc.*, **81**, 4891 (1959); (b) H. Hart and R. A. Cioriani, *J. Am. Chem. Soc.*, **84**, 3697 (1962).
161. M. E. Sigman, J. T. Barbas and J. E. Leffler, *J. Org. Chem.*, **52**, 1754 (1987).
162. R. G. Lawler, P. F. Barbara and D. Jacobs, *J. Am. Chem. Soc.*, **100**, 4912 (1978).
163. D. E. Hedaya and S. Winstein, *J. Am. Chem. Soc.*, **89**, 1661 (1967).
164. S. Oae, K. Fujimori and S. Kozuka, *Tetrahedron*, **28**, 5321 (1972).
165. S. Oae, Y. Uchida, K. Fujimori and S. Kozuka, *Bull. Chem. Soc. Jpn.*, **46**, 1741 (1973).
166. R. J. Linhardt and B. L. Murr, *Tetrahedron Lett.*, **20**, 1007 (1979).
167. R. J. Linhardt, B. L. Murr, E. Montgomery, J. Osby and J. Shebine, *J. Org. Chem.*, **47**, 2242 (1982).
168. R. C. Lamb, W. E. McNew, Jr, J. R. Sanderson and D. L. Lunney, *J. Org. Chem.*, **36**, 174 (1971).
169. D. B. Denney and G. Feig, *J. Am. Chem. Soc.*, **81**, 5322 (1959).
170. W. v. E. Doering, K. Okamoto and H. Krauch, *J. Am. Chem. Soc.*, **82**, 3579 (1960).
171. C. Walling and E. S. Savas, *J. Am. Chem. Soc.*, **82**, 1738 (1960).
172. H. Hart and F. J. Chloupek, *J. Am. Chem. Soc.*, **85**, 1155 (1963).
173. G. Moad, E. Rizzardo and D. H. Solomon, *Tetrahedron Lett.*, **22**, 1165 (1981).
174. G. Braun, *Org. Synth., Coll. Vol.*, **1**, 431 (1967).
175. K. Schank, R. Blattner, V. Schmidt and H. Hasenfratz, *Chem. Ber.*, **114**, 1938 (1981).

176. S. -O. Lawesson, C. Frisell, D. Z. Denney and D. B. Denney, *Tetrahedron*, **19**, 1229 (1963).
177. L. Horner and E. Schwenk, *Justus Liebigs Ann. Chem.*, **566**, 69 (1950).
178. D. B. Denney and D. Z. Denney, *J. Am. Chem. Soc.*, **82**, 1389 (1960).
179. S. Srinivas and K. G. Taylor, *J. Org. Chem.*, **55**, 1779 (1990).
180. L. Horner and E. Jurgens, *Justus Liebigs Ann. Chem.*, **602**, 135 (1952).
181. W. A. Pryor and W. H. Hendrickson, Jr, *J. Am. Chem. Soc.*, **97**, 1582 (1975).
182. L. Horner, W. Jurgeleit and K. Klupfel, *Justus Liebigs Ann. Chem.*, **591**, 108 (1955).
183. L. Horner and W. Jurgeleit, *Justus Liebigs Ann. Chem.*, **591**, 138 (1955).
184. C. Walling and R. B. Hodgdon, Jr, *J. Am. Chem. Soc.*, **80**, 228 (1958).
185. D. B. Denney and D. Z. Denney, *J. Am. Chem. Soc.*, **82**, 1389 (1960).
186. M. A. Greenbaum, D. B. Denney and A. K. Hoffmann, *J. Am. Chem. Soc.*, **78**, 2563 (1956).
187. D. B. Denney and M. A. Greenbaum, *J. Am. Chem. Soc.*, **79**, 979 (1957).
188. (a) G. Tsuchihashi, S. Miyajima, T. Otsu and O. Simamura, *Tetrahedron*, **21**, 1039 (1965); (b) G. Tsuchihashi, S. Miyajima, T. Otsu and O. Simamura, *Tetrahedron*, **21**, 1049 (1965).
189. F. D. Greene, W. Adam and J. E. Cantrill, *J. Am. Chem. Soc.*, **83**, 3461 (1961).
190. F. D. Greene and W. Adam, *J. Org. Chem.*, **29**, 136 (1964).
191. R. Hisada, M. Kobayashi and H. Minato, *Bull. Chem. Soc. Jpn.*, **45**, 2035 (1972).
192. M. Yoshida, K. Moriya, H. Sawada and M. Kobayashi, *Chem. Lett*, **755** (1985).
193. (a) J. -y. Koo and G. B. Schuster, *J. Am. Chem. Soc.*, **99**, 6107 (1977); (b) J. -y. Koo and G. B. Schuster, *J. Am. Chem. Soc.*, **100**, 4496 (1978).
194. J. J. Zupancic, K. Horn and G. B. Schuster, *J. Am. Chem. Soc.*, **102**, 5279 (1980).
195. (a) S. Oae, N. Asi and K. Fujimori, *J. Chem. Soc., Perkin Trans. 2*, 571 (1978); (b) S. Oae, N. Asai and K. Fujimori, *J. Chem. Soc., Perkin Trans. 2*, 1124 (1978).
196. (a) M. Yoshida, M. Kobayashi, H. Sawada, H. Hagii and K. Aoshima, *Kagaku Kaishi*, 1958 (1985).
197. (a) H. Sawada, M. Mitani, M. Nakayama, M. Yoshida and N. Kamigata, *Polym. Commun.* **31**, 63 (1990); (b) M. Mitani, H. Sawada, M. Nakayama, M. Yoshida and N. Kamigata, *Polym. J.*, **22**, 623 (1990).
198. M. Yoshida, S. Sasage, N. Kamigata, H. Sawada and M. Nakayama, *Bull. Chem. Soc. Jpn.*, **62**, 2416 (1989).
199. L. Eberson, *Chem. Scr.*, **20**, 29 (1982).
200. C. Walling, *J. Am. Chem. Soc.*, **102**, 6854 (1980).

8 Sulfur and Phosphorus Peroxides

YONG HAE KIM
Department of Chemistry, Korea Advanced Institute of Science and Technology, 373-1 Kusung-dong Yusung-gu, Taejon 305-701, Korea

1 SULFUR PEROXIDES

Numerous sulfur and phosphorus peracids (Figure 1) such as peroxymonosulfuric acid (Caro's acid, **1a**),[1] peroxymonosulfate (Oxone, **1b**),[2] tetra-*n*-butylammonium peroxymonosulfate (tetra-*n*-butylammonium Oxone, **1d**),[3] peroxydisulfate (**2a**),[4] tetra-*n*-butylammonium disulfate (**2b**),[5] symmetrical bissulfonyl peroxide (**3**),[6] unsymmetrical sulfonyl peroxides (**5**),[6] sulfonyl peroxy radical intermediates (**6**),[7] sulfonimidoyl peroxy radical intermediate (**7**),[8] symmetrical bisphosphinyl peroxides (**9**)[5] and symmetrical phosphinyl peroxides (**9**),[5] and phosphinyl peroxy radical intermediates (**10**)[9] have been synthesized, characterized and applied in organic synthesis. These heteroatom peroxides show different oxidizing abilities depending on the substrates and the reaction conditions.

Recently, sulfinyl and sulfonyl peroxy radical intermediates (**6**) have been prepared by the reactions of aryl sulfinyl chloride or aryl sulfonyl chloride with superoxide anion radical, respectively. These peroxy intermediates show strong oxidizing abilities in various oxidation reactions such as selective epoxidation of alkenes, desulfurizations of thioureas and thioamides, oxidations of methylene groups to ketones and sulfoxidations of sulfides under mild reaction conditions. The literature on sulfur and phosphorus peroxides is so vast that the selection of appropriate material for an overview in the limited space available is difficult.

Organic Peroxides. Edited by W. Ando

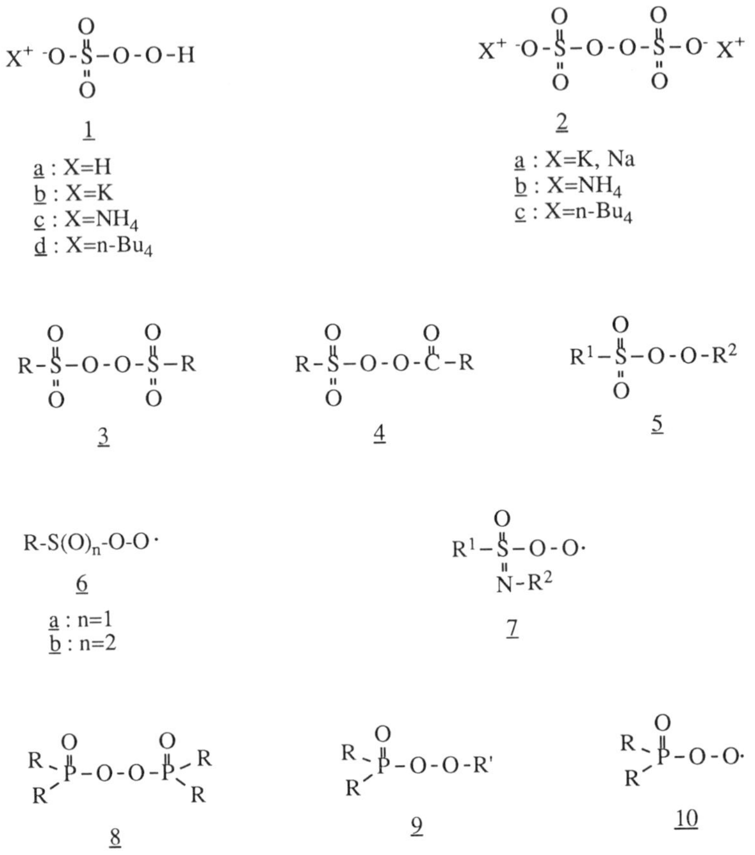

$X^+ \, ^-O-S(=O)(=O)-O-O-H$

1

a : X=H
b : X=K
c : X=NH_4
d : X=n-Bu_4

$X^+ \, ^-O-S(=O)(=O)-O-O-S(=O)(=O)-O^- \, X^+$

2

a : X=K, Na
b : X=NH_4
c : X=n-Bu_4

$R-S(=O)(=O)-O-O-S(=O)(=O)-R$

3

$R-S(=O)(=O)-O-O-C(=O)-R$

4

$R^1-S(=O)(=O)-O-O-R^2$

5

$R-S(O)_n-O-O\cdot$

6

a : n=1
b : n=2

$R^1-S(=O)(=N-R^2)-O-O\cdot$

7

$R_2P(=O)-O-O-P(=O)R_2$

8

$R_2P(=O)-O-O-R'$

9

$R_2P(=O)-O-O\cdot$

10

Figure 1. Structures of sulfur and phosphorus peroxides

An insight into what might be expected of this broad class should be the goal here, rather than a comprehensive review. For a more complete picture, readers should consult the basic papers cited in areas that most interest them. This chapter treats fundamental principles and also presents a selection of newer results.

1.1 Peroxysulfates

Peroxymonosulfuric acid (**1a**), peroxymonosulfates (**1b**, **1c**) and peroxydisulfates (**2a**, **2b**), which are considered to be derivatives of hydrogen peroxides, are stable and commercially available, except **1d** and **2c**. Compound **1a** is also known as Caro's acid, and is prepared *in situ* from either hydrogen peroxide,[1] $K_2S_2O_8$ or ammonium persulfate.[1]

The reagent converts aldehydes into esters in the presence of alcohols[1] and primary aromatic amines into azoxy,[10] nitroso[11] or nitro[12] compounds. Potassium peroxymonosulfate (**1b**) is well known under the trade-name of Oxone. Although the use of the reagent is fairly limited owing to its poor solubility in organic solvents, it is very versatile; good yields are obtained, depending on the substrates. Oxidations of alcohols to ketones and esters,[13] of amines to nitroso compounds[13] and of sulfides to sulfoxides[14] can be effectively carried out. Since Oxone is almost insoluble in organic solvents, a phase-transfer technique using tetrabutylammonium bromide is advantageous (equation 1).[15] The phase-transfer catalyst controls the oxidation of $KHSO_5$ with diaryl sulfides yielding predominantly sulfoxides.

$$R-S-R \xrightarrow[CH_2Cl_2\text{-}H_2O,\ 25\ ^\circ C]{KHSO_5,\ Bu_4N^+Br^-\ (cat)} R-\overset{O}{\overset{\|}{S}}-R \quad + \quad R-\underset{O}{\underset{\|}{\overset{O}{\overset{\|}{S}}}}-R \tag{1}$$

This is in contrast to the reaction in polar solvents, where sulfones are the major products. Oxone (**1b**) converts various tosyl hydrazones into ketones effectively by cleavage of the C=N bond in 50% aqueous organic solvents (e.g. CH_3CN, THF, MeCOMe).[16] The deprotecting reaction for the ketones appears to be initiated via the formation of oxaziridine, followed by fragmentation to the carbonyl compounds as shown in equation 2.

$$R^1R^2C{=}NNHTos \xrightarrow{KOSO_4H} \left[R^1R^2C\overset{O}{\frown}N-NHTos \right] \longrightarrow R^1R^2C{=}O \; + \; N_2 \; + \; TosOH \tag{2}$$

90-98 %

R^1= aryl, R^2= alkyl or R^1=R^2= aryl, R^1=R^2= alkyl

Tetra-*n*-butylammonium Oxone, readily prepared from Oxone, is soluble in organic solvents.[3] The alkylammonium Oxone performs oxidations in anhydrous dichloromethane. Under these conditions, sulfides are selectively oxidized to sulfones in the presence of amines, ketones, esters, carbamates, alkenes and hydroxyl functionalities without reacting with these functional groups.[3] Sulfur-containing amino acid derivatives are oxidized under these conditions. Peroxydisulfate ion is one of the strongest oxidizing agents known in aqueous solution. The standard oxidation–reduction potential for the reaction is estimated to be -2.01 V (equation 3).[17]

$$2\,SO_4^{2-} \longrightarrow S_2O_8^{2-} \; + \; 2\,e^- \tag{3}$$

Kinetic studies using ammonium peroxydisulfate (**2b**) are well documented in a review.[18] Since potassium and sodium peroxydisulfates (**2a**) are insoluble in organic solvents, their utilization is limited. However, a few effective oxidations have been reported. Isatin, a cyclic α-ketoamide, is oxidized with potassium persulfate to 2,3-dioxo-1,4-benzoxazine (equation 4).[19]

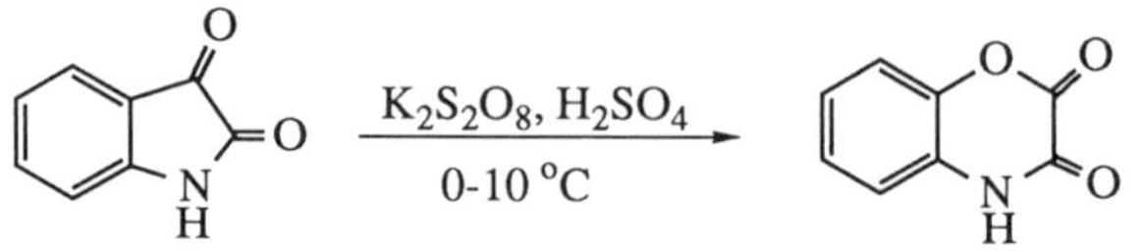

$$\text{Isatin} \xrightarrow[\text{0-10 °C}]{K_2S_2O_8,\ H_2SO_4} \text{2,3-dioxo-1,4-benzoxazine (95 \%)} \quad (4)$$

The regiospecific oxidation of dimethylanisoles to methoxymethylbenzaldehydes is accomplished with potassium persulfate and copper(II) sulfate, which oxidize selectively only the methyls in the *ortho* or *para* position with respect to the methoxy groups (equation 5).[20]

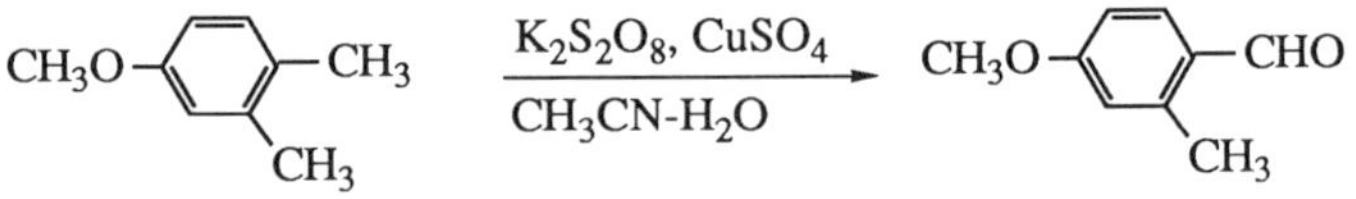

$$CH_3O\text{-}C_6H_3(CH_3)\text{-}CH_3 \xrightarrow[CH_3CN\text{-}H_2O]{K_2S_2O_8,\ CuSO_4} CH_3O\text{-}C_6H_3(CH_3)\text{-}CHO \quad (5)$$

Secondary alcohols react with peroxydisulfate to give the corresponding ketones.[21] The reaction mechanism shown in Figure 2 has been proposed.

The formation and characterization of phenylselenyl cation by the reaction of diphenyl diselenide with ammonium peroxydisulfate have been well established.[22] Various applications of the phenylselenyl cation have been reported for use in organic synthesis. Methoxyselenenylations of several alkenes are effected by oxidation of diphenyl diselenide with ammonium peroxydisulfate in methanol. The reaction mechanism proposed is shown in Figure 3.

Good yields (72–95%) of the methoxyselenylation products are obtained in every case, indicating that this procedure has general applicability. The reactions are regiospecific: the addition products are formed with Markovnikov orientation. Further, the reactions proceed stereospecifically affording the

$$S_2O_8^{2-} \longrightarrow 2\ \cdot OSO_3^-$$

$$\cdot OSO_3^- + R_2CHOH \longrightarrow R_2\dot{C}OH + HSO_4^-$$

$$R_2\dot{C}OH + S_2O_8^{2-} \longrightarrow R_2C{=}O + HSO_4^{2-} + \cdot OSO_3^-$$

or

$$R_2\dot{C}OH + \cdot OSO_3^- \longrightarrow R_2C{=}O + HSO_4^{2-}$$

Figure 2. Mechanism of oxidation of alcohols to ketones. Reprinted with permission from K. B. Wiberg, *J. Am. Chem. Soc.*, **81**, 252 (1959). Copyright (1959) American Chemical Society

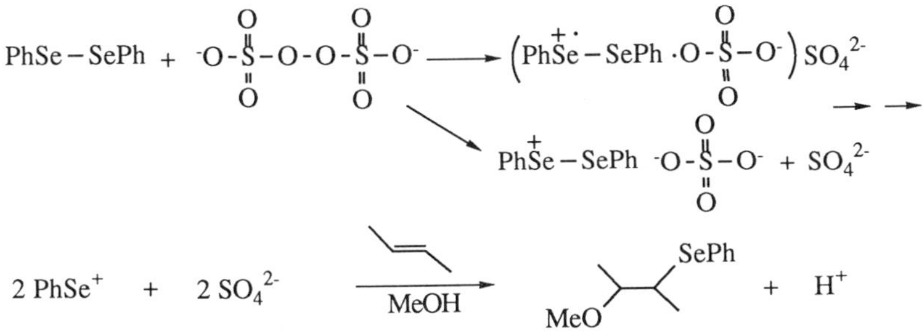

$$PhSe-SePh + {}^{-}O-S(O_2)-O-O-S(O_2)-O^{-} \longrightarrow (Ph\overset{+\cdot}{Se}-SePh \cdot O-S(O_2)-O^{-})\, SO_4^{2-} \longrightarrow \longrightarrow$$

$$\searrow Ph\overset{+}{Se}-SePh \; {}^{-}O-S(O_2)-O^{-} + SO_4^{2-}$$

$$2\,PhSe^{+} + 2\,SO_4^{2-} \xrightarrow[MeOH]{\text{propene}} MeO\text{-}CH(CH_3)\text{-}CH(CH_3)\text{-}SePh + H^{+}$$

Figure 3. Mechanism of methoxyselenylation to alkene. Reprinted with permission from M. Tieco, L. Testaferri, M. Tingoli, D. Chiauelli and D. Bartoli, *Tetrahedron Lett.*, **30**, 1417 (1989). Copyright (1989) Pergamon Press PLC

products of *anti* addition (equation 6).

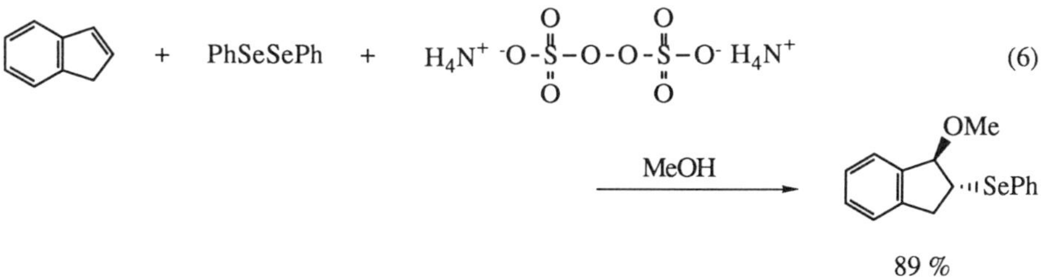

$$\text{indene} + PhSeSePh + H_4N^{+}\; {}^{-}O-S(O_2)-O-O-S(O_2)-O^{-}\; H_4N^{+} \xrightarrow{MeOH} \text{trans-1-OMe-2-SePh-indane} \quad (89\%) \qquad (6)$$

In a similar reaction, selenium-catalyzed conversions of vinyl halides into α-alkoxyacetals have been reported.[23] Vinyl halides reacted with a catalytic amount of diphenyl diselenide in the presence of an excess of ammonium peroxydisulfate in methanol to give the corresponding α-methoxyacetals as the main products (equation 7). The reaction proceeds through an intermediate which then undergoes solvolysis.[23]

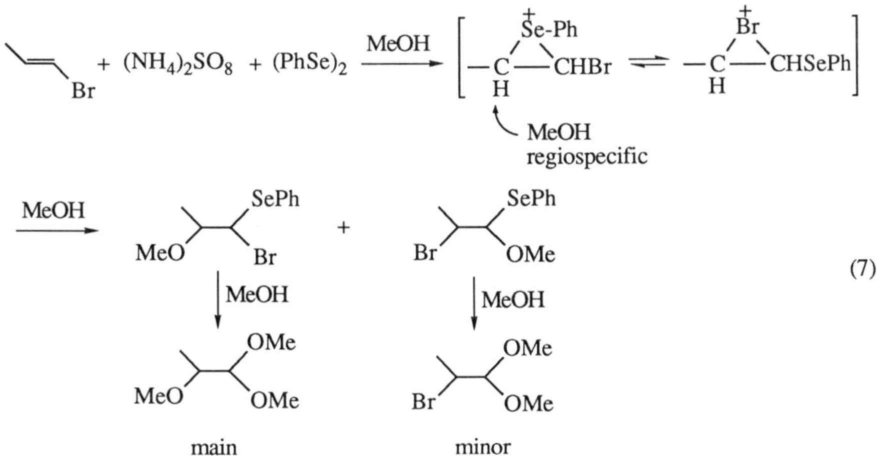

$$CH_3CH{=}CHBr + (NH_4)_2SO_8 + (PhSe)_2 \xrightarrow{MeOH} \left[-\underset{H}{C}\overset{\overset{+}{Se}\text{-Ph}}{\frown}CHBr \rightleftharpoons -\underset{H}{C}\overset{\overset{+}{Br}}{\frown}CHSePh \right]$$

MeOH regiospecific

$$\xrightarrow{MeOH} \text{MeO-CH(CH}_3\text{)-CH(Br)SePh} + \text{Br-CH(CH}_3\text{)-CH(OMe)SePh} \qquad (7)$$

$$\downarrow MeOH \qquad\qquad \downarrow MeOH$$

$$\text{MeO-CH(CH}_3\text{)-CH(OMe)}_2 \;(\text{main}) \qquad \text{Br-CH(CH}_3\text{)-CH(OMe)}_2 \;(\text{minor})$$

Various alkenes react with diphenyl diselenide and ammonium peroxydisulfate in aqueous nitriles as solvent in the presence of trifluoromethanesulfonic acid to afford the amidoselenylation products in good yields (equation 8).[24]

$$R^1CH{=}CHR^2 + (NH_4)_2S_2O_8 + (PhSe)_2 \xrightarrow[CF_3SO_3H]{RCN\text{-}H_2O} \left[\text{R}^1,\text{R}^2\text{-episelenonium } \overset{+}{Se}Ph \right] \longrightarrow R^1CH(SePh)CH(R^2)N{=}C(R)OH \longrightarrow R^1CH(SePh)CH(R^2)NHCOR \quad (8)$$

The intramolecular version of this reaction, using appropriate unsaturated nitriles as the starting materials and dioxane as solvent, takes a different course and affords phenylselenolactones in good yields.[24] This ring-closure reaction proceeds through the initial formation of the hydroxyselenylation products in which the hydrolysis of the cyano group is suggested to be intramolecularly assisted by the hydroxyl group (equation 9).

$$N{\equiv}C(CH_2)_nCH{=}CH_2 \xrightarrow[\text{dioxane},\ CF_3SO_3H,\ H_2O]{(PhSe)_2,\ (NH_4)_2S_2O_8} N{\equiv}C(CH_2)_nCH(OH)CH_2SePh \longrightarrow \text{lactone } O{=}C\!-\!(CH_2)_n\!-\!CH(CH_2SePh)\!-\!O \quad (9)$$

This reaction is employed to effect selenium-induced ring-closure reactions starting from alkenes containing internal nucleophiles. Unsaturated alcohols and amides, β-diketones and β-keto esters give the product of phenylselenoetherification. The same process occurs with dienes and unsaturated ketones when the reaction is carried out in the presence of water or methanol, respectively.[25] The reaction of various methyl ketones with the same reaction system in methanol proceeds smoothly to give α-keto acetals (equation 10).[26]

$$RCOCH_3 \underset{MeOH}{\overset{(PhSe)_2,\ (NH_4)_2S_2O_8}{\rightleftharpoons}} RC(OH){=}CH_2 \rightleftharpoons RC(OH)(OMe)CH(SePh)_2 \longrightarrow RCOCH(OMe)_2 \quad (10)$$

Recently, tetra-n-butylammonium disulfate (**2c**) was successfully prepared by the reaction of tetra-n-butylammoniumsulfonic acid with potassium peroxydisulfate according to equation 11 [white solid, m.p. 119°C (decomp.)].[27]

$$2\ \text{n-Bu}_4\text{NHSO}_4 + \text{K}_2\text{S}_2\text{O}_8 \longrightarrow \text{n-Bu}_4\text{N}^+\ {}^-\text{O}-\overset{\text{O}}{\underset{\text{O}}{\overset{\|}{\underset{\|}{\text{S}}}}}-\text{O}-\text{O}-\overset{\text{O}}{\underset{\text{O}}{\overset{\|}{\underset{\|}{\text{S}}}}}-\text{O}^-\ {}^+\text{NBu}_4\text{-n} \qquad (11)$$

Metal peroxydisulfates (**1a**) and ammonium peroxydisulfate (**1b**) are soluble in water and protic solvents such as methanol. However, in contrast to **1a** and **1b**, tetra-*n*-butylammonium peroxydisulfate (**2c**) is soluble in aprotic solvents such as chloroform, dichloromethane, acetonitrile, dichloroethane and acetone. Thus, it turned out to be an excellent source for the formation of sulfate ion radical ($^{\bullet}OSO_3^-$) under anhydrous conditions.[27] The new oxidizing peroxydisulfate (**2c**) oxidizes various primary and secondary alcohols such as benzylic and allylic alcohols and also alkyl and aryl secondary alcohols to the corresponding aldehydes and ketones, respectively, in almost quantitative yields in aprotic solvents under anhydrous conditions (equation 12).[27]

$$\text{R}^1\text{R}^2\text{CHOH} + (\text{n-Bu}_4\text{NSO}_4)_2 \xrightarrow[\text{or Me}_2\text{CO}]{\text{C}_6\text{H}_6,\ \text{CH}_2\text{Cl}_2} \text{R}^1\text{R}^2\text{C}{=}\text{O} \qquad (12)$$

Under these conditions, primary or secondary alcohols are oxidized to aldehydes or ketones (87–99%) in the presence of alkenes and pyridyl functionalities without reacting with these groups. Under anhydrous conditions, side-reactions due to hydroxyl radicals and an aqueous medium can be avoided and hence a thermodynamically stable allylic or benzylic carbon radical is involved in this reaction (equation 13).

$$\text{R}^1\text{R}^2\text{CHOH} \xrightarrow{{}^-\text{OSO}_3^{\bullet}\ \ \text{SO}_4^{2-}} \text{R}^1\text{R}^2\text{CHO}^{\bullet} \rightleftharpoons \text{R}^1\text{R}^2\dot{\text{C}}\text{OH} \xrightarrow{{}^-\text{OSO}_3^{\bullet}\ \ \text{SO}_4^{2-}} \text{R}^1\text{R}^2\text{C}{=}\text{O} \qquad (13)$$

Both tetrahydrofuranyl ethers and tetrahydropyranyl ethers are utilized for the protection of alcohols. However, it has been a problem to prepare tetrahydrofuranyl ethers by tetrahydrofuranylation of alcohols in mineral acid conditions because the ethers are sensitive and hydrolyzed under acidic conditions.[28] Recently, it was found that various alcohols containing functional moieties are readily tetrahydrofuranylated in excellent yields by the reaction of tetra-*n*-butylammonium disulfate with tetrahydrofuran as a solvent.[27] Some of the results are given in Table 1.

This reaction is carried out, without using an acid catalyst, under mild conditions. The reaction appears to be initiated by the formation of sulfate anion radical, and then of both alkoxy and tetrahydrofuranyl radicals, followed

Table 1. Preparation of tetrahydrofuranyl ethers by the reaction of alcohols with **2c** in tetrahydrofuran

Alcohol	Ether	Yields (%)[a] (isolated)
		97
		95
		90
		96
		85

[a]Isolated yields

by radical coupling (equation 14). A small amount of tetrahydrofuran dimer was also obtained (equation 15). Under these conditions, benzylic, alkyl and allylic alcohols containing alkenes, sulfides and acetals are converted into the

$$\text{n-Bu}_4\text{N}^{+}\,{}^{-}\text{O-S(O}_2\text{)-O-O-S(O}_2\text{)-O}^{-}\ {}^{+}\text{Nn-Bu}_4 \longrightarrow 2\ \text{n-Bu}_4\text{N}^{+}\,{}^{-}\text{O-S(O}_2\text{)-O}^{\cdot}$$

$$\xrightarrow[\text{THF}]{\text{ROH}} \text{RO}^{\cdot} + \text{THF}^{\cdot} + 2\ \text{n-Bu}_4\text{N}^{+}\,{}^{-}\text{O-S(O}_2\text{)-OH}$$

$$\longrightarrow \text{RO-THF} + 2\ \text{n-Bu}_4\text{N}^{+}\,{}^{-}\text{O-S(O}_2\text{)-OH} \quad (14)$$

(15)

tetrahydrofuranyl ethers without fragmentation[29] and without destroying the functional moieties.

Tetra-*n*-butylammonium peroxydisulfate (**2c**) is also utilized for the formation of 1,3-dioxolanyl radical, which is used for C—C bond formation. Novel masked formylation at the β-position of alkenes through 1,4-addition using 1,3-dioxolanyl radical is effectively established. Cyclic α,β-unsaturated ketones, esteric alkenes and sulfonyl alkenes react with **2c** in 1,3-dioxolane as solvent to afford the corresponding 1,3-dioxolanyl adducts in excellent yields.[27] In this reaction, 1,2-addition products are not formed. This reaction appears to proceed through the 1,4-addition of 1,3-dioxolanyl radical to the electron-deficient Δ-carbon adjacent to the carbonyl or sulfonyl moiety (equation 16). The reaction is effective in introducing a formyl function at the β-position of cyclic pentanone, hexanone or heptanone (equation 17).

n-Bu_4N^+ $^-$O-S(O)$_2$-O· + 1,3-dioxolane ⟶ 1,3-dioxolan-2-yl· + n-Bu_4NSO_4H

(16)

R^1=R^3=CO_2Me, R^2=H : R^3=CO_2Me, R^1=R^2=H
R^3=$PhSO_2$, R^1=R^2=H

n=2, 3, 4

+ n-Bu_4N^+ $^-$O-S(O)$_2$-O· ; 1,3-dioxolane, 20 °C, Ar

(17)

It is possible to introduce the formyl group at the α-position of a cyclic hexanone by the reaction of nitrohexene with **2c** in 1,3-dioxolane (equation 18).[27]

NO_2 + n-Bu_4N^+ $^-$O-S(O)$_2$-O· ; 1,3-dioxolane, 20 °C, Ar

(18)

Most known oxidative denitrations have been carried out under strongly basic conditions.[30] However, this reaction proceeds smoothly under neutral and mild reaction conditions, together with 1,3-dioxolanylation at the α-position.

1.2 Symmetrical Sulfonyl Peroxides (3)

Symmetrical sulfonyl peroxides stabilize the peroxide link by inductive and resonance withdrawal of electrons from the peroxide oxygens. The thermal stability is further enhanced by attachment of arenesulfonyl groups, where an aryl group is substituted with an electron-withdrawing group such as 2- or 4-nitrophenyl, 3-nitrophenyl, 3-trifluoromethylphenyl or 3,5-ditrifluoromethylphenyl. These peroxides represent the most stable examples and can be isolated. The pure peroxides decompose exothermically but not violently, and they are easily handled and stored. One of the best routes to the peroxides **3** is a two-phase transfer reaction in which a two-phase mixture of aqueous ethanol containing hydrogen peroxide and potassium carbonate, and chloroform containing the sulfonyl chloride are stirred vigorously (equation 19). The solid product is collected and purified by recrystallization to yield peroxides of high purity.[31]

$$2\ R\text{-}SO_2Cl\ +\ H_2O_2\ \xrightarrow[H_2O\text{-}EtOH\text{-}CHCl_3,\ -20\ ^\circ C]{K_2CO_3}\ R\text{-}S(=O)_2\text{-}O\text{-}O\text{-}S(=O)_2\text{-}R \qquad (19)$$

Only a few dialkyl peroxides are known. Dimethylbissulfonyl peroxide (m.p. 79°C) is stable for 1 week or more at 25°C.[32] On the other hand, ditrifluoromethylbissulfonyl peroxide (liquid) is very unstable and decomposes within a few minutes at 25°C.[33] These are two typical reactions of sulfonyl peroxides. One is a thermal decomposition, which can be divided into three principal models: homolytic, polar and free radical-induced decomposition.[6,34] Only the first two are important in thermal decompositions of disulfonyl peroxides and are usually first-order processes. The other category is the intermolecular reactions of sulfonyl peroxides with electron donors, which are second-order reactions in general and considered to be ionic processes. Bis-2-nitrobenzenesulfonyl peroxide decomposes at near ambient temperature to the sulfite radical, which abstracts hydrogen from the chloroform solvent to form the corresponding sulfonic acid together with the products derived from the trichloromethyl radical (equation 20).

$$(2\text{-}NO_2C_6H_4)\text{-}S(=O)_2\text{-}O\text{-}O\text{-}S(=O)_2\text{-}(C_6H_4\text{-}2\text{-}NO_2)\ \xrightarrow[25\text{-}40\ ^\circ C]{CHCl_3}\ 2\ (2\text{-}NO_2C_6H_4)\text{-}S(=O)_2\text{-}O\cdot\ +\ CHCl_3 \qquad (20)$$

$$\longrightarrow\ (2\text{-}NO_2C_6H_4)\text{-}S(=O)_2\text{-}OH\ +\ \cdot CCl_3$$

These results suggest that the kinetics are first order and radical-induced decomposition is not important, and the activation energy of the O—O bond, E_a, is 24.5 kcal mol^{-1} (1 kcal = 4.184 kJ) (cf. E_a = 30 kcal mol^{-1} for the O—O bond energy of benzoyl peroxides).[34] Recently, Korth *et al.* succeeded in generating methanesulfonyloxyl radical ($CH_3S(O)_2O\cdot$) and 4-trifluoromethylbenzenesulfonyloxyl radical ($4\text{-}CF_3C_6H_4S(O)_2O\cdot$) by 308 nm laser flash photolysis of the parent symmetrical bissulfonyl peroxides. These sulfonyloxyl radicals exhibit a broad structureless absorption similar to that for $SO_4^{-\cdot}$ with λ_{max} 450 nm and are more reactive than any other oxygen-centered radical toward alkane (hydrogen abstraction) and carbon double bonds (addition). Cleavage of the S—C bond of the methanesulfonyloxyl radical does not proceed thermally, however, but takes place photochemically affording a methyl radical and sulfur trioxide.[35] Bissulfonyl peroxides react with electron donors such as π-donors of arenes and alkenes, n-donors of phosphines and amines, and σ-donors of electron-rich carbon—metal σ-bonds to give the corresponding donor–peroxide adducts. The results and a detailed discussion of the literature prior to 1982 are well documented in a review of sulfur peroxides.[6] Recently, Hoffman and co-workers have demonstrated that bisnitrobenzenesulfonyl peroxides react with electron-rich alkenes attached to oxygen or nitrogen to afford the corresponding sulfonoxylation adducts in excellent yields.[6] For example, the addition of 4-nitrobenzenesulfonyl peroxide to 3,4-dihydro-2*H*-pyran in various alcohols gives high yields (77–91%) of 3-{[(4-nitrophenyl)sulfonyl]oxy}-2-alkoxytetrahydropyrans (equation 21).

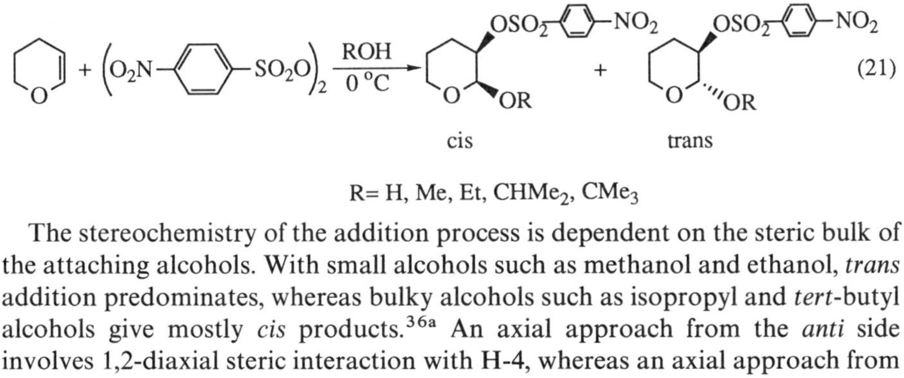

$$\text{3,4-dihydro-2}H\text{-pyran} + (O_2N\text{-}C_6H_4\text{-}SO_2O)_2 \xrightarrow[0\,^{\circ}C]{ROH} \text{cis} + \text{trans} \quad (21)$$

cis trans

R= H, Me, Et, $CHMe_2$, CMe_3

The stereochemistry of the addition process is dependent on the steric bulk of the attaching alcohols. With small alcohols such as methanol and ethanol, *trans* addition predominates, whereas bulky alcohols such as isopropyl and *tert*-butyl alcohols give mostly *cis* products.[36a] An axial approach from the *anti* side involves 1,2-diaxial steric interaction with H-4, whereas an axial approach from the *anti* side involves 1,2-diaxial steric interaction with H-3 (equation 22).

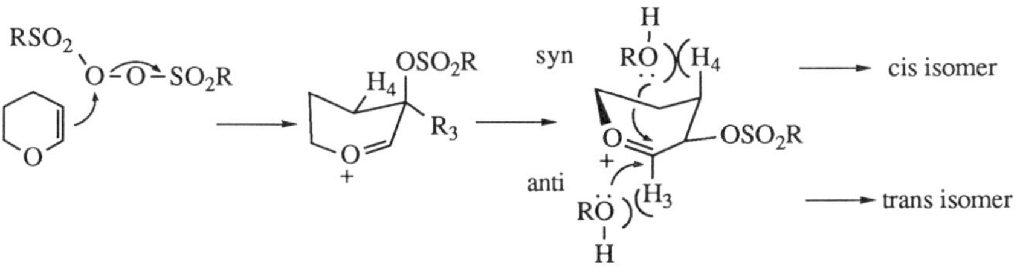

$RSO_2O\text{-}O\text{-}SO_2R$; OSO_2R, H_4, R_3 ; syn: $RO(H)$ ... H_4 → cis isomer ; OSO_2R (22) ; anti: $RO(H)$ ... H_3 → trans isomer

Various cyclic or chain silyl enols are effectively converted into α-arylsulfonoxy ketones by reaction with 4-nitrobenzenesulfonyl peroxide (equation 23).[36b].

$$\text{OSiMe}_3 + (4\text{-NO}_2\text{C}_6\text{H}_4\text{SO}_2\text{O})_2 \xrightarrow[25\,^\circ\text{C}]{\text{EtOAc}} \left[\overset{+}{\text{O}}\text{SiMe}_3,\ \text{OSO}_2\text{C}_6\text{H}_4\text{-NO}_2\text{-4}\right] \xrightarrow{\text{H}_2\text{O}} \text{O},\ \text{OSO}_2\text{C}_6\text{H}_4\text{-NO}_2\text{-4}\ \ (>95\ \%) \tag{23}$$

Enamines are also very common and useful carbonyl derivatives.[37] A series of enamines of morpholine derivatives are satisfactory substrates for the conversion to α-arylsulfonoxy ketones.[36] When a small amount of methanol was added to the reaction solvent, high yields of the products were obtained. Presumably, nucleophilic trapping of the iminium ion by methanol gives a more stable tetrahedral intermediate that provides the ketone on aqueous work-up (equation 24).

$$\text{O, N, R}^1\text{, R}^2\text{, H} \xrightarrow[\text{EtOAc, 2 \% MeOH, -78 °C}]{(4\text{-NO}_2\text{C}_6\text{H}_4\text{SO}_2\text{O})_2} \text{O, }\overset{+}{\text{N}}\text{, OSO}_2\text{C}_6\text{H}_4\text{-NO}_2\text{-4, R}^1\text{, R}^2 \xrightarrow{\text{MeOH}} \text{O, }\overset{+}{\text{N}}\text{H, OSO}_2\text{C}_6\text{H}_4\text{-NO}_2\text{-4, MeO, R}^1\text{, R}^2 \xrightarrow{\text{H}_3\text{O}^+} \text{O, OSO}_2\text{C}_6\text{H}_4\text{-NO}_2\text{-4, R}^1\text{, R}^2 \tag{24}$$

In connection with these reactions of enol acetates, silyl enol ethers and enamines, a series of *O*-trimethylsilyl ketenes and acetals were reacted with $(4\text{-NO}_2\text{C}_6\text{H}_4\text{SO}_2\text{O})_2$ in ethyl acetate and methanol to give the corresponding α-arylsulfonoxy ketones in good yields.[38] These reactions provide a very efficient and regioselective route to the ketones (equation 25).

$$\text{R}^1\text{, R}^2\text{, OSiMe}_3\text{, OR}^3 \xrightarrow[\text{MeONa, 0 °C}]{(\text{RSO}_2\text{O})_2} \text{RSO}_2\text{O, R}^1\text{, R}^2\text{, }\overset{+}{\text{O}}\text{SiMe}_3\text{, OR}^3 \xrightarrow{\text{MeOH}} \text{RSO}_2\text{O, R}^1\text{, R}^2\text{, OSiMe}_3\text{, OMe, OR}^3 \xrightarrow{\text{H}_3\text{O}^+} \text{RSO}_2\text{O, R}^1\text{, R}^2\text{, O, OR}^3\ \ (70\text{-}90\ \%) \tag{25}$$

$\text{R}=4\text{-NO}_2\text{C}_6\text{H}_4$, $\text{R}^1=\text{H}$, $\text{R}^2=\text{Me, Et, Ph}$
$\text{R}^3=\text{Me, Et, t-Bu}$

Little is known of the reaction of sulfonyl peroxides with electron donors such as phosphines,[39] sulfoxides,[40] enamines and amine derivatives. Triphenylphosphine is used to analyze 3-nitrobenzenesulfonyl peroxide, and phosphine was found to attack the peroxidic oxygen exclusively by the use of ^{18}O-labelled peroxide (equation 26).[39]

$$R-S(=^{18}O)_2-O-O-S(=^{18}O)_2-R + :PPh_3 \longrightarrow RS^{18}O_2O^- + Ph_3\overset{+}{P}-O-S^{18}O_2R \qquad (26)$$

$$\longrightarrow R-S(=^{18}O)_2-O-S(=^{18}O)_2-R + Ph_3P{=}O$$

$R = 3\text{-}NO_2C_6H_4$

The reactions of primary and secondary amines with 4-nitrobenzenesulfonyl peroxide at $-78°C$ yield imines which can be converted into carbonyl products in good yields (66–91%) by hydrolytic work-up (equation 27).[41] Probably the

$$R^1CH_2NHR^2 + (RSO_2O)_2 \xrightarrow[-78\,°C]{EtOAc,\ NaOH} [R^1CH_2\text{-}N(R^2)OSO_2R] \xrightarrow{:B} \qquad (27)$$

$$R^1CH{=}NR^2 + RSO_3^- \xrightarrow{H_2O} R^1CHO + R^2NH_2$$

amine attacks the O—O bond nucleophilically to give an *O*-sulfonylhydroxylamine intermediate which forms the imine by elimination.[42] Earlier, it was reported that *N*-substituted amines such as chloroamines[43] and *N*-acylhydroxamines[44] undergo base-promoted elimination to imines. Alternatively, the *O*-sulfonylhydroxylamine could undergo homolysis of the N—O bond and yield the imine by a radical process.[45] Finally, the amine could undergo an electron-transfer reaction with the peroxide to give a nitrogen cation radical and hence the imine (equation 28).[46]

$$R^1CH_2NHR^2 + (RSO_2O)_2 \longrightarrow R^1CH_2\overset{+\cdot}{N}R^2 + \begin{matrix} {}^-OSO_2R \\ \cdot OSO_2R \end{matrix} \longrightarrow R^1CH{=}NR^2 \qquad (28)$$

The stable *O*-sulfonylhydroxylamine intermediate in equation 27 can be produced easily by the reaction of amines and arenesulfonyl peroxides. These intermediate adducts have been very useful for the study of base-promoted, imine-forming eliminations,[47] as sources of nitrenium ions by solvolysis[48] and as aminating agents.[49] Recently, an effective halogenation of anisole by the reaction of anisole with nitrobenzenesulfonyl peroxides in the presence of ammonium halides was reported.[50] Peroxides react in general as electrophiles in the presence of electron donors.[51] Halide ion may react with sulfonyl peroxide, which is a good electrophile, to produce hypohalite.[52] In the sulfonyl hypochlorite, the chlorine should be positively charged owing to the effect of the strong electron-withdrawing nitrobenzenesulfonyl group and may react as an electrophile to be introduced into anisole (equation 29).[50]

$$Et_3NHX + (RSO_2O)_2 \longrightarrow RSO_3X + RSO_3^- N^+Et_3H \xrightarrow{C_6H_5\text{-OMe}} X\text{-}C_6H_4\text{-OMe} + C_6H_4(X)\text{-OMe} + RSO_3H + Et_3N \quad (29)$$

Some results of the halogenations of anisoles are given in Table 2.

Direct halogenations on the methyl carbon of toluene or at the 2-position in thiophene have been examined to afford mixtures of 4- and 2-halogenated benzyl products or solely 2-substituted thiophene in reasonable yields (equation 30). The mechanism is interesting: the formation of benzyl bromide suggests the generation of radical species in the reaction; the unimolecular homolysis of sulfonyl hypobromite may occur competitively with bimolecular electrophilic aromatic substitution.

$$RSO_3X \longrightarrow RSO_3\cdot + Br\cdot \xrightarrow{Ph\text{-}CH_3} RSO_3H + PhCH_2\cdot \xrightarrow{RSO_3X \;\to\; RSO_3\cdot} PhCH_2X \quad (30)$$

R=3- or 4-$NO_2C_6H_4$
X=Cl or Br

Table 2. Halogenation of anisoles with arenesulfonyl peroxides and ammonium halides

Peroxide R	Solvent	Salt	Yield (%)	Isomer ratio, 4- : 2-
4-$NO_2C_6H_4$	MeCN	Et_3NHCl	76	88:12
3-$NO_2C_6H_4$	MeCN	Et_3NHCl	62	87:13
4-$NO_2C_6H_4$	CH_2Cl_2	Et_3NHCl	49	96:4
3-$NO_2C_6H_4$	CH_2Cl_2	Et_3NHCl	58	93:7
4-$NO_2C_6H_4$	MeCN	Et_3NHBr	76	100:0
3-$NO_2C_6H_4$	MeCN	Et_3NHBr	74	96:4

1.3 Unsymmetrical Sulfonyl Peroxides

Unsymmetrical sulfonyl peroxides (**4** and **5**) are generally more unstable than symmetrical sulfonyl peroxides. The unsymmetrical character of these peroxides leads to a greater difference in reaction types than for bissulfonyl peroxides. A few acyl sulfonyl peroxides (**4**) have been prepared by reacting peracid and sulfonyl chloride in the presence of base.[53] Benzoyl 4-toluenesulfonyl peroxide reacts with triphenylphosphine to give triphenylphosphine oxide and benzoic 4-toluenesulfonic anhydride (equation 31). A plausable mechanism derived from ^{18}O labeling experiments has been proposed in which the reaction is initiated by phosphorus attack on the peroxide oxygen attached to the carbonyl function.[53]

$$\mathrm{Ph{-}C({=}O){-}O{-}O{-}S(O_2){-}C_6H_4{-}Me + {:}PPh_3 \longrightarrow Ph{-}C({=}O){-}O{-}\overset{+}{P}Ph_3 + {}^{-}O{-}S(O_2){-}C_6H_4{-}Me \longrightarrow Ph{-}C({=}O){-}O{-}S(O_2){-}C_6H_4{-}Me + Ph_3P{=}O} \quad (31)$$

On the other hand, when 4-toluenesulfonyl peroxide is reacted with 4-tolylmagnesium bromide, benzoic acid and 4-tolyl 4-toluenesulfonate are obtained. This reaction appears to be initiated by attack of the Grignard reagent on the peroxide oxygen bond to the sulfonyl moiety (equation 32). Interaction between magnesium and the carbonyl oxygen of the peroxide probably directs such specificity of attack.

$$\mathrm{Ph{-}C({=}O){-}O{-}O{-}S(O_2){-}C_6H_4{-}Me + Me{-}C_6H_4{-}MgBr \rightleftharpoons [Br{-}\overset{+}{Mg}\cdots O{=}C(Ph){-}O{-}O{-}S(O_2){-}C_6H_4{-}Me,\ {}^{-}C_6H_4{-}Me] \longrightarrow Ph{-}C({=}O){-}OH + Me{-}C_6H_4{-}O{-}S(O_2){-}C_6H_4{-}Me} \quad (32)$$

Hydrazine and sodium methoxide attack the carbonyl carbon of the unsymmetrical peroxide to yield benzoyl hydrazide and methyl benzoate, respectively.[53] Thus, the unsymmetrical sulfonyl peroxides have three active sites, both peroxide oxygens and the carbonyl carbon, which can be attacked by various electron donors. These various modes of attack by nucleophiles have not yet been elucidated. Unsymmetrical aryl sulfonyl peroxides undergo more complicated thermal decompositions in solution. Homolysis of the peroxide bond gives sulfonyl radicals and acyloxy radicals, which undergo the usual decarboxylation, coupling and addition reactions (equation 33).[34,54]

$$\mathrm{R^1SO_2{-}O{-}O{-}C({=}O){-}R^2 \xrightarrow{\Delta} R^1SO_2{-}O\cdot + \cdot O_2CR^2 \longrightarrow R^1SO_2{-}O\cdot + R^1\cdot + CO_2} \quad (33)$$

Activation energies of the order of 25 kcal mol^{-1} are observed in the decompositions of various unsymmetrical peroxides.[55,56] Thus, acetyl cyclohexanesulfonyl peroxide and acetyl *s*-heptanesulfonyl peroxide are commonly used as polymerization initiators because they provide a source of free radicals at relatively low temperatures.[57] Heterolytic cleavage of the O—O bond of acylsulfonyl peroxides is expected as the carboxy inversion rearrangement occurs in unsymmetrical aroyl peroxides.[58,59] The mechanism of the carboxy inversion in acyl sulfonyl peroxides has been elucidated using ^{18}O labeling (equation 34).[60]

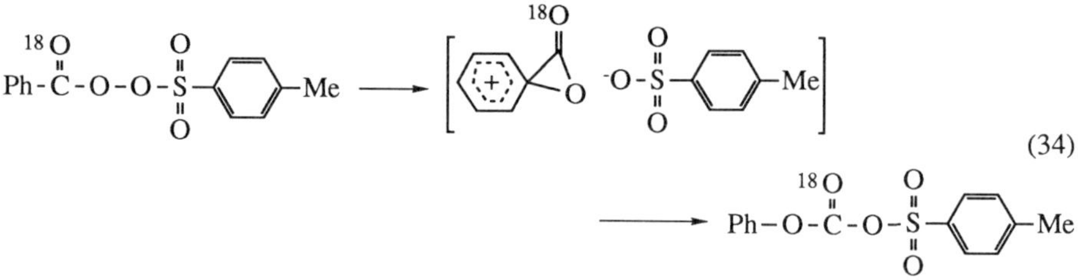

$$Ph-C(={}^{18}O)-O-O-SO_2-C_6H_4-Me \longrightarrow \left[\text{(bridged phenonium)} \; {}^{-}O-SO_2-C_6H_4-Me \right] \longrightarrow Ph-O-C(={}^{18}O)-O-SO_2-C_6H_4-Me \qquad (34)$$

There is no scrambling of the label. The phenyl bridge may be an important feature of the transition state leading to the ion-pair intermediate. Alkyl arenesulfonyl peroxides **5** are known to be unstable, hence little is known of these compounds and of their reactions. *tert*-Butyl sulfonyl peroxide decomposes in alcohol solution without the production of free radicals. In absolute methanol, acetone dimethyl ketal is formed together with the arylsulfonic acid (equation 35).[61] The rearrangement of this reaction is reported to be sensitive to solvent polarity and not subject to acid catalysis.

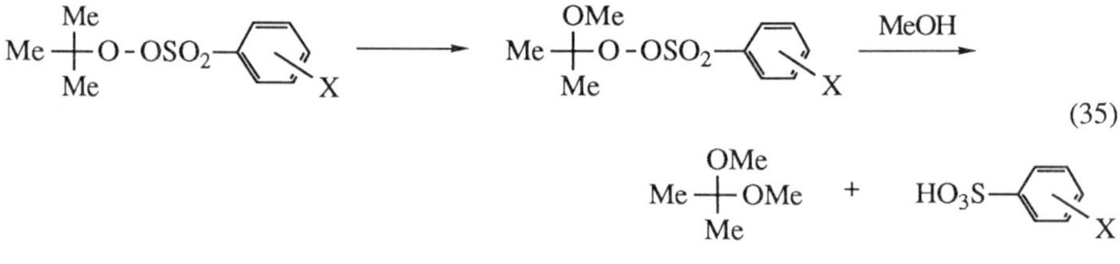

$$Me_3C-O-OSO_2-C_6H_4X \longrightarrow Me_2C(OMe)-O-OSO_2-C_6H_4X \xrightarrow{MeOH} Me_2C(OMe)_2 + HO_3S-C_6H_4X \qquad (35)$$

Since peroxysulfur intermediates such as peroxysulfenate, peroxysulfinate and peroxysulfonate (Figure 4) are unstable, they have neither been isolated nor

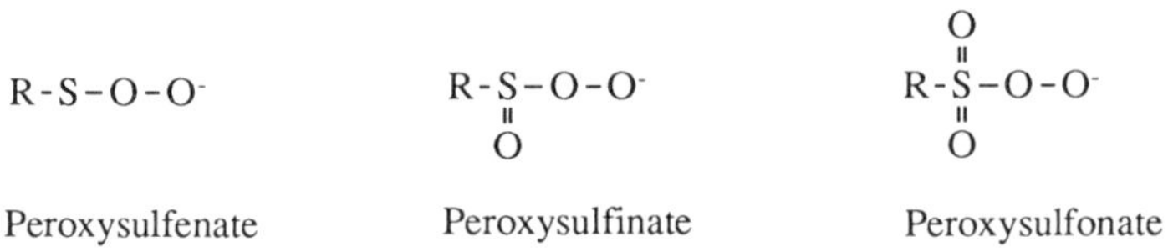

$$R-S-O-O^- \qquad R-S(=O)-O-O^- \qquad R-S(=O)_2-O-O^-$$

Peroxysulfenate Peroxysulfinate Peroxysulfonate

Figure 4. Peroxysulfur intermediates

confirmed. Earlier, it was postulated that these peroxysulfur compounds are formed as intermediates in the strong alkaline autoxidations of thiols to sulfinic and sulfonic acids.[61]

Superoxide radical anion ($O_2^{-\bullet}$) displays four basic modes of action, including deprotonation as a base, H-atom abstraction as a radical, nucleophilic attack as an anion and as an electron-transfer reagent.[62] Superoxide is known to act as a moderate reducing agent but a very weak oxidizing agent.[63] Hence numerous reports dealing with the activation of superoxide through reaction of various organic substrates with superoxide have appeared within the last decade. It has been reported that these peroxysulfur intermediates formed in the reactions of disulfides, thiosulfinates or thiosulfonates with superoxide anion have oxidizing ability.[64]

Most of these peroxy intermediates have been studied in connection with the reactivity of superoxide or biological toxicity, but are not sufficiently strong oxidizing agents for use in organic synthesis, probably owing to their instability. Various sulfonyl chlorides have been found to react with superoxide radical anion to generate sulfonylperoxy intermediates, which have been neither isolated nor confirmed. Although stable sulfonylperoxy intermediates have not been isolated, 2-nitrobenzenesulfonyl peroxy intermediates have been found to be stable enough for use in organic synthesis at low temperatures. Both 2-nitro and 4-nitrobenzenesulfonyl chlorides readily react with $O_2^{-\bullet}$ in acetonitrile under mild conditions to form 2-nitro- and 4-nitrobenzenesulfonyl peroxy intermediates. However, 2-nitrobenzenesulfonyl peroxy intermediates (**I** or **I′**) have been shown to give better yields than 4-nitrobenzenesulfonyl peroxy intermediate in the oxidation of alkenes, as shown in Table 3. A neighboring

Table 3. Effect of substituent of $ArSO_2Cl$ in the epoxidation of alkenes[a] at $-35°C$

$$\text{R-C}_6\text{H}_4\text{-SO}_2\text{Cl} + \text{KO}_2 + \text{>C=C<} \xrightarrow[-35\,^{\circ}\text{C}]{\text{CH}_3\text{CN}} \text{>C(-O-)C<}$$

$ArSO_2$ R	Reaction time (h)	Yield (%) Epoxide	Chalcone[b]
o-NO_2	1.5	84	5
m-NO_2	4.0	60	32
p-NO_2	6.0	75	—

[a]Chalcone was used as the alkene
[b]Recovery of starting material

group effect of NO_2 at the 2-position appears to play an important role in stabilizing the peroxy intermediate (**I** or **I′**) (equation 36).

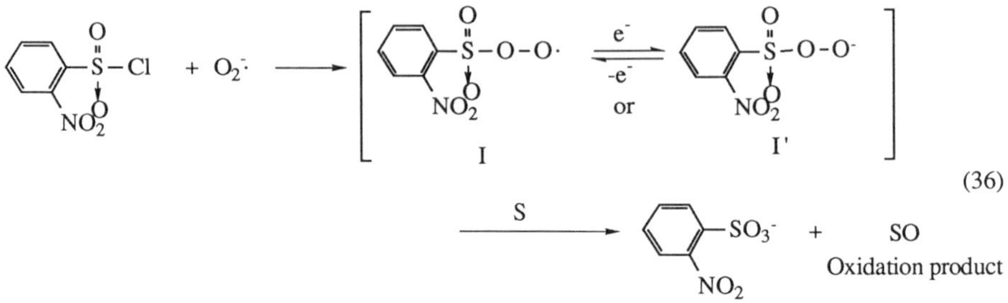

$$2\text{-}NO_2C_6H_4S(O)_2Cl + O_2^{-\bullet} \longrightarrow \left[2\text{-}NO_2C_6H_4S(O)_2\text{-}O\text{-}O\cdot \ (\text{I}) \underset{-e^-}{\overset{e^-}{\rightleftharpoons}} \text{or} \ 2\text{-}NO_2C_6H_4S(O)_2\text{-}O\text{-}O^- \ (\text{I}') \right] \xrightarrow{S} 2\text{-}NO_2C_6H_4SO_3^- + SO \qquad (36)$$

Oxidation product

In addition, 2-nitrobenzenesulfinyl chloride was found to react with $O_2^{-\bullet}$ to form 2-nitrobenzenesulfinyl peroxy intermediate (**II** or **II′**) (equation 37), which shows an excellent oxidizing ability for the oxidation of sulfides to sulfoxides.[65]

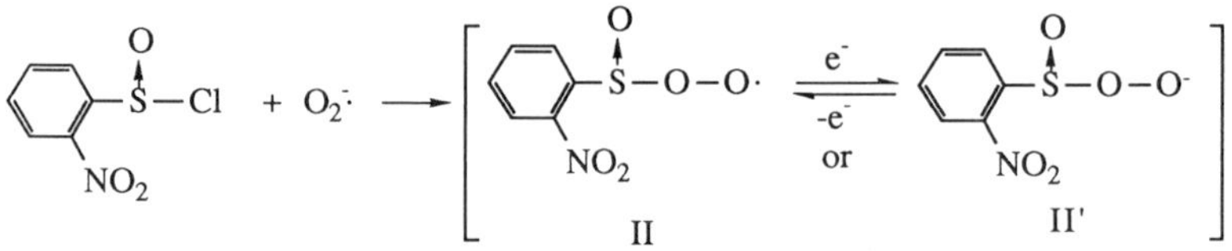

$$2\text{-}NO_2C_6H_4S(O)Cl + O_2^{-\bullet} \longrightarrow \left[2\text{-}NO_2C_6H_4S(O)\text{-}O\text{-}O\cdot \ (\text{II}) \underset{-e^-}{\overset{e^-}{\rightleftharpoons}} \text{or} \ 2\text{-}NO_2C_6H_4S(O)\text{-}O\text{-}O^- \ (\text{II}') \right] \qquad (37)$$

Here the question of whether the active intermediate involves a radical **I** or anion species **I′** which can be formed by one electron transfer from $O_2^{-\bullet}$ to **I** arises. Judging from recent results obtained from the various chemical reactions, an electrophilic radical species (**I**) of a sulfonyl peroxy intermediate is believed to be involved in the oxidation of various substrates.[66–68] Both 2-nitrobenzenesulfonyl peroxy radical (**I**) and 2-nitrobenzenesulfinyl peroxy radical (**II**) have been trapped by spin trapping studies using 5,5-dimethyl-1-pyrroline-1-oxide (DMPO).[69]

The sulfonylperoxy intermediate (**I** or **I′**) and sulfinyl peroxy intermediate (**II** or **II′**) seem to be stabilized by the neighboring group effect of the nitro group as shown in Figure 5, as in the case of methyl 2-nitrophenyl sulfenate,[70] 2-nitrobenzeneseleninic anhydride[71] or 2-nitrobenzeneperoxyseleninic acid.[72]

I II

Figure 5. Neighboring group (NO_2) effects of sulfonyl- and sulfinylperoxy radical

Various alkenes containing one or more double bonds in the molecule were readily oxidized to their epoxides in high regioselectivity and in excellent yields with 2-nitrobenzenesulfonyl peroxy intermediate (**I**) generated *in situ* from 2-nitrobenzenesulfonyl chloride and superoxide at $-35°C$ in acetonitrile[66] (equation 38) (Table 4).

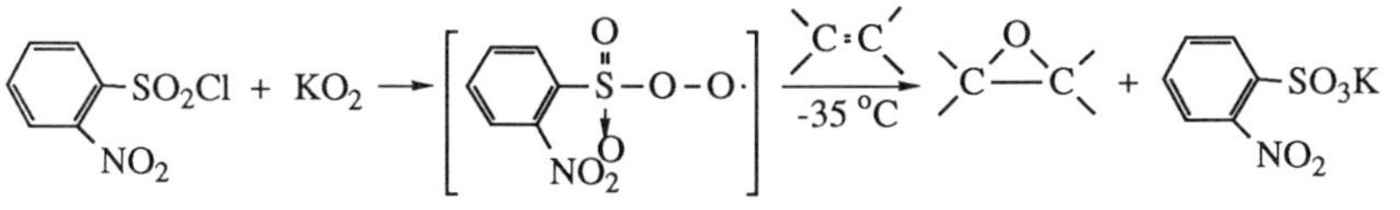

$$2\text{-}NO_2C_6H_4SO_2Cl + KO_2 \longrightarrow \left[2\text{-}NO_2C_6H_4S(O)_2\text{-}O\text{-}O\cdot\right] \xrightarrow[-35\,°C]{>C=C<} >C\overset{O}{—}C< + 2\text{-}NO_2C_6H_4SO_3K \quad (38)$$

Table 4. Epoxidation of alkenes by 2-nitrobenzenesulfonyl peroxy intermediate at *ca.* $-30°C$

Substrate	Time (h)	Product	Yield (%)
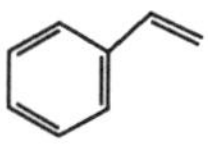 Styrene	4.5	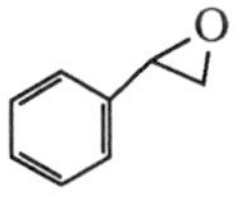 Styrene oxide	92
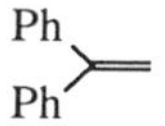 $Ph_2C=CH_2$	5.0	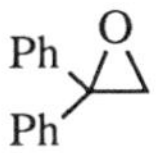 1,1-Diphenyloxirane	82
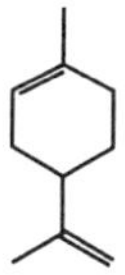 Limonene	4.5	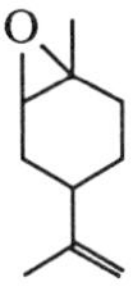 Limonene 1,2-epoxide	87
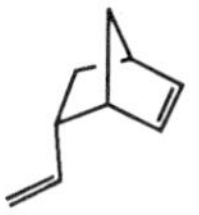 Allyl-substituted norbornene	5.0	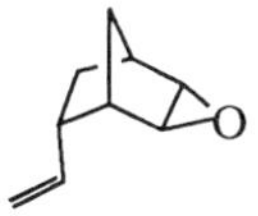 Allyl-substituted norbornene epoxide	78
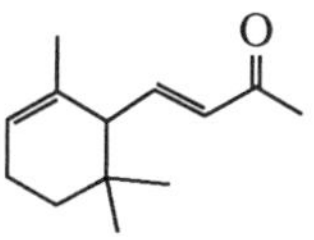 α-Ionone	3.0	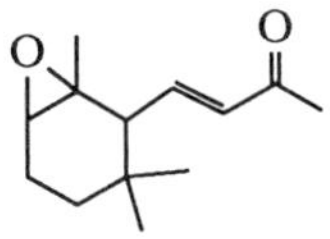 Epoxy-α-ionone	85
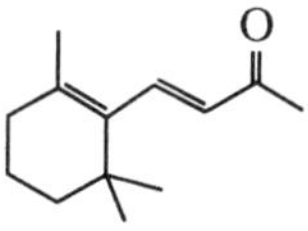 β-Ionone	2.5	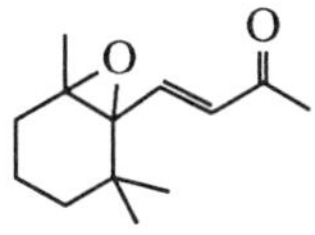 5,6-Epoxy-β-ionone	84
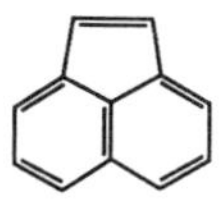 Acenaphthylene	12	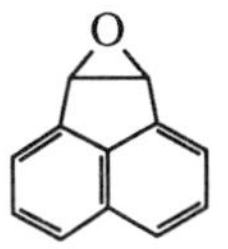 Acenaphthylene oxide	95

Epoxidation of alkenes containing an α,β-unsaturated ketone moiety such as 4-(2,6,6-trimethylcyclohex-2-ene-1-yl)but-3-en-2-one (α-ionone) and 4-(2,6,6-trimethylcyclohex-1-ene-1-yl)but-3-en-2-one (β-ionone) affords the corresponding epoxides in good yields.

Since $O_2^{-\bullet}$ is known to be a fairly strong base,[73] if the peroxy intermediate is an anion (**I′**), epoxidation of double bonds attached to ketone can be expected. However, no epoxidation of the double bonds in the α,β-unsaturated ketone moiety took place. These results suggest that the peroxy intermediate **I** is electrophilic rather than nucleophilic. Further, a regioselective epoxidation on the ring double bonds occurs predominantly.

Highly strained acenaphthylene was smoothly epoxidized in quantitative yields at -30°C in acetonitrile.[74] Acenaphthylene oxide is known to be unstable under acidic conditions, but more stable under basic conditions.[75] Hence, acenaphthylene oxide could be isolated in higher yields than those given by the known methods,[76,77] perhaps owing to the stability of the product under basic conditions. The sulfonylperoxy intermediate is likely to be of the character of a radical, rather than an anion, as in the case of the acylperoxyl radical [ArC(O)OO$^\bullet$][78] and phenyl nitroso oxide radical (PhNOO$^\bullet$),[79] judging from the fact that oxidation of *cis*-stilbene by **I** gave both *trans*- (70%) and *cis*-oxides (30%) under the same conditions (equation 39). If the peroxy intermediate **I** is anionic, stereospecific epoxidation of *cis*-stilbene to *cis*-epoxide may occur.

Ph–CH=CH–Ph (cis) —**I**→ Ph-epoxide-Ph (trans 70 %) + Ph-epoxide-Ph (cis 30 %) (39)

As an example of monooxygenation, polyaromatic compounds such as phenanthrene and pyrene, which are inert to superoxide itself, were readily oxidized to the corresponding K-region arene oxides by treatment with **I**.[74] Anthracene was oxidized to anthraquinone (70%) under the same conditions (equation 40).

Anthracene —**I**, -30 °C, CH_3CN, 10 h→ anthraquinone (40)

It is interesting that various arenes were readily oxidized to their oxides under mild conditions. Such arene oxides are important metabolites *in vivo* formed from processes catalyzed by monooxygenase.[80] Hypoxanthine is known to be metabolized *in vivo* to xanthine by xanthine oxidase, with no evidence to show any intervention by activated species of oxygen such as superoxide, which is distributed widely in living cells.[81] Biomimetic oxidation of triacetylated inosine to triacetylated xanthosine (80%) was examined by intermediate **I** generated *in situ* from 2-nitrobenzenesulfonyl chloride and superoxide under mild conditions as shown in equation 41.[7] The oxidation appears to be initiated via the formation of an oxaziridine, and then convertion into the monooxygenated product by an intramolecular rearrangement.

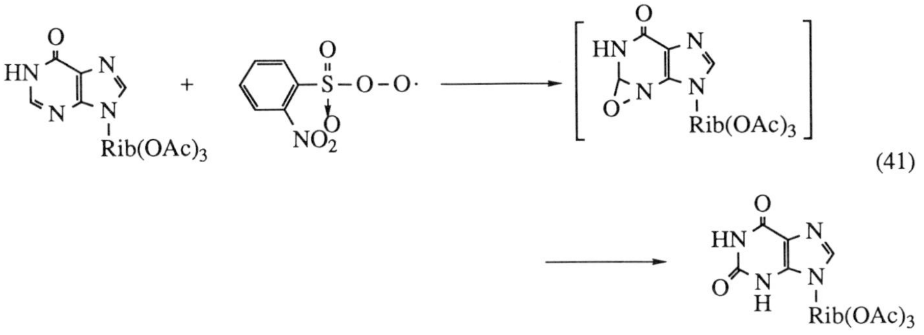

(41)

Metabolic degradation of polyaromatic hydrocarbons is believed to be oxidized with a certain activated oxygen by an enzyme in all aerobic organisms and also soil bacteria.[82] 2-Methylnaphthalene is oxidized to phthalic acid together with other unidentified products by **I** under mild conditions.[5]

The oxidation is considered to be initiated via epoxidation of an aromatic ring, followed by fragmentation to the carbonyl compound (equation 42). It is intriguing that the aromatic ring can be readily destroyed at low temperature because destruction of the aromatic ring usually requires high energy under drastic conditions.

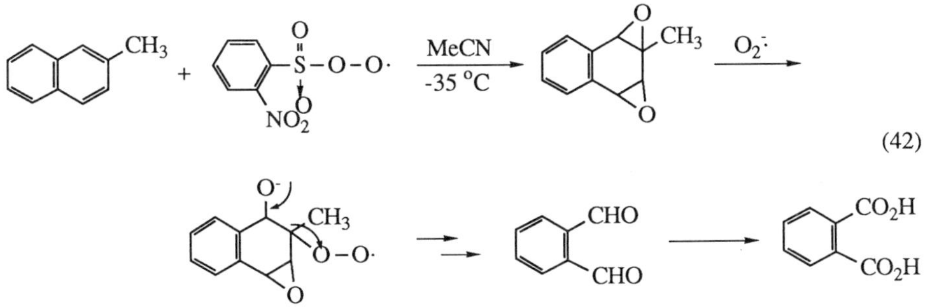

(42)

It is well known that parathion is used as an insecticide, but that it is biologically inert: it depends for its effectiveness on oxidative desulfurization in the insect. Parathion is oxidized to paraoxon through oxidative desulfurization by superoxide *in vivo*.[83] However, superoxide alone is unreactive with parathion in an aprotic solvent such as acetonitrile.

Thionyl chloride is expected to react with superoxide to form the sulfinylperoxy radical **III** or **III′** (equation 43).

$$Cl-\overset{O}{\overset{\|}{S}}-Cl \; + \; O_2^{\cdot -} \longrightarrow \underset{III}{\left[Cl-\overset{O}{\overset{\|}{S}}-O-O\cdot\right]} \;\text{or}\; \underset{III'}{\left[\cdot OO-\overset{O}{\overset{\|}{S}}-OO\cdot\right]} \tag{43}$$

Methyl parathion and other thiophosphates react with thionyl chloride and superoxide to give desulfurized methyl paraoxon and phosphates as shown in equation 44.[5]

$$RO-\overset{S}{\overset{\|}{P}}(OR^1)-OR^2 \; + \; Cl-\overset{O}{\overset{\|}{S}}-Cl \; + \; O_2^{\cdot -} \xrightarrow[-35\,^{\circ}C,\,1\,h]{MeCN} \underset{94\text{-}97\,\%}{RO-\overset{O}{\overset{\|}{P}}(OR^1)-OR^2} \tag{44}$$

$R=R^1=MeO$, $R^2=PhCH_2$, Ph
$R=R^1=EtO$, $R^2=PhCH_2$, Ph
$R=R^1=EtO$, $R^2=4\text{-}NO_2\text{-}C_6H_4$, $4\text{-Br-}C_6H_4$

Recently, important oxidative cleavages of C=X to C=O using **I** have been established (equation 45).

$$>C=X \; + \; 2\text{-}NO_2C_6H_4-\overset{O}{\overset{\|}{S}}(\to O)-O-O\cdot \xrightarrow{CH_3CN} >C=O \tag{45}$$

X=NNHTos, S, H_2

The protection and deprotection of ketones are important reactions. Tosylhydrazones are good crystals and often used for protecting liquid ketones. Various alkyl, aryl and dialkyl tosylhydrazones react readily with **I** to give the corresponding carbonyl compounds in mostly quantitative yields at $-30°C$ under mild conditions (equation 46).[84] In the absence of 2-nitrobenzenesulfonyl chloride, tosylhydrazones were recovered quantitatively under the same conditions; superoxide is inert to tosylhydrazones.

$$R_1R_2C=NNHTos \xrightarrow{I} \left[\begin{matrix} R_1 \\ R_2 \end{matrix} > \overset{O}{C}-N-\overset{H\;\;\curvearrowleft O_2^{\cdot -}}{N}-Tos \right] \longrightarrow \begin{matrix} R_1 \\ R_2 \end{matrix} > C=O \; + \; {}^{-}N=N\text{-}Tos \longrightarrow N_2 \; + \; Tos^{-} \tag{46}$$

The regeneration of carbonyl compounds from tosylhydrazones seems to be initiated via the oxidation of the imino double bond of the tosylhydrazones with **I** to oxaziridine derivatives. The oxaziridine may be immediately converted to the carbonyl compound by a fragmentation reaction induced by superoxide, which is known to be a fairly strong base.[73] The results obtained are shown in Table 5.

In connection with the oxidative cleavage of the C=N bond to C=O, **I** is effectively used for the cleavage of the C=S bond to C=O under the same conditions. Thiocarbonyl derivatives such as thioureas, thioamides and thiocarbamates are readily desulfurized into their corresponding carbonyl compounds at **I** at $-35°C$ in acetonitrile (equation 47).[85]

$$R-\overset{S}{\overset{\|}{C}}-NHR' \xrightarrow[-35\ °C]{I} \left[R-\overset{S=O}{\overset{\|}{C}}-NHR'\right] \longrightarrow R-\overset{O}{\overset{\|}{C}}-NHR' + \underset{88\ \%}{1/8\ S_8} \qquad (47)$$

R=aryl, amine, R'= alkyl, aryl

This method has the advantage that it results in a clean and almost quantitative conversion of thiocarbonyl compounds into their corresponding oxygen analogs (86–97%) without the formation of side or intermediate products at $-35°C$. The desulfurization appears to be initiated via oxidation with sulfonylperoxy intermediate **I**, which may have the ability to oxidize substrates to sulfines. The sulfine intermediates may quickly be converted into the carbonyl products by an intramolecular desulfurization as in the case of the oxidation of thiones to ketones through sulfine intermediates.[86] The reaction mechanism is apparently different from that of the direct desulfurization of thiourea by O_2^- [63,87] in the absence (80% in equation 48) of 2-nitrobenzenesul-

$$Ar-\overset{S}{\overset{\|}{C}}-NHAr' \xrightarrow{I} \left[Ar-\overset{SO_n^-}{\overset{|}{C}}=NHAr'\right]_{n=2,3} \longrightarrow \underset{Ar,\ Ar'=aryl}{Ar-\overset{O}{\overset{\|}{C}}-Ar'} + \underset{80\ \%}{SO_4^{2-}} \qquad (48)$$

Table 5. Oxidative cleavage of tosylhydrazone derivatives of **I** in acetonitrile at *ca.* $-30°C$

R_1	R_2	Reaction time (h)	Yield (%)
p-ClC_6H_4	CH_3	5.3	90
p-$NO_2C_6H_4$	H	5.0	92
$(Ph)_2CH$	CH_3	9.0	97
C_6H_5	CH_3	8.0	93
CH_3CH_2	CH_3	6.0	98
CH_3	CH_3	7.0	93
m-$CH_3OC_6H_4$	H	7.5	83

fonyl chloride, where potassium sulfate was obtained in contrast to elemental sulfur (88% in equation 47) together with ureas and guanidines at 25°C, but desulfurization did not occur at −35°C.

It is noteworthy that although the standard desulfurization of 1,3-diphenyl-1-ethylthiourea gave poor yields of the urea analog, probably owing to the steric hindrance of the bulky groups of diphenyl and ethyl groups, this reaction afforded an excellent yield of the urea (88%) under mild conditions. As a typical radical reaction, various benzylic methylene groups are readily oxidized under mild conditions to the ketones in excellent yields (equation 49).[49]

$$\mathrm{ArCH_2R} \xrightarrow[\text{MeCN, -35 °C}]{\mathbf{I}} \mathrm{Ar\dot{C}H\text{-}R} \longrightarrow \mathrm{Ar\text{-}CHR(O\text{-}O\cdot)} \longrightarrow \mathrm{Ar\text{-}C(=O)\text{-}R} \quad (49)$$

82-98 %

Ar=aryl, R=alkyl, aryl

One of the hydrogen atoms of the benzylic methylene moiety is abstracted by **I**. The benzylic radical may couple with molecular oxygen or $O_2^{-\bullet}$ to form a peroxy radical which is then converted into the carbonyl group. Tetralin is oxidized stepwise to 1,4-naphthoquinone through α-tetralone and 1,4-dihydroxynaphthalene (equation 50).

tetralin —**I**→ α-tetralone → [1,4-dione] ↔ [1,4-dihydroxynaphthalene (OH, OH)] (50)

—**I** or $O_2^{-\bullet}$→ [semiquinone radical (O·, OH)] → 1,4-naphthoquinone 85 %

The question has been raised of whether the active intermediate involves a radical (**I**) or anion species which can be formed by one-electron transfer from $O_2^{-\bullet}$ to **I**. Diaryl, dialkyl and alkylaryl sulfoxides are readily oxidized to sulfones at −30°C in excellent yields (82–98%) without epoxidation on the double bond.[82] In addition, from the competitive oxidations of 4-substituted-phenyl methyl sulfoxides to sulfones, a ρ value of −1.39 is obtained as the reactivity ratio for the 4-substituted-phenyl methyl sulfoxides;[88] hence this reaction appears to involve the electrophilic attack of an electrophilic oxidant.[89] Further, spin trapping studies by ESR have demonstrated that $O_2^{-\bullet}$ efficiently reacts with 2-nitrobenzenesulfinyl chloride to form their peroxy radicals (**6**), some of which are key intermediates for site-selective oxidation of various organic molecules, and it has been argued whether they exist in a radical and/or anion form (equation 51).[69] The radicals were trapped using dimethylpyrrolidine *N*-oxide (DMPO) to form a radical adduct (**11**).

n=1, a_H=10.0 G, a_N=12.8 G (51)

n=2 (I), a_H=10.1 G, a_N=12.8 G

a_H=12.0 G, a_N=14.2 G

11

1.4 Peroxysulfur Peroxides (Miscellaneous)

A few examples of the formation of peroxy sulfenate intermediates have been reported. Desulfurization of thioureas to guanidines by superoxide has been suggested to involve peroxy sulfenate (equation 52).[90]

$$\mathrm{RNH{-}C({=}S){-}NHR} \rightleftharpoons \mathrm{RNH{-}C(SH){=}NHR} \xrightarrow{O_2^{-\cdot}} \mathrm{RNH{-}C(S\cdot){=}NHR} \longrightarrow \left[\mathrm{RNH{-}C(SOO^-){=}NHR}\right] \longrightarrow \longrightarrow \mathrm{RNH{-}C({=}N{-}R){-}NHR} \quad (52)$$

Desulfurization of thioureas by alkaline autoxidation is reported to involve a peroxy sulfenate intermediate (**12** in equation 53).[91]

O_2 – t-BuO⁻

12

(53)

An efficient cyclodesulfurization of *N*-(2-hydroxyphenyl)thioamides by superoxide has been reported.[92] Treatment of thioamides with $O_2^{-\bullet}$ at 25°C in acetonitrile resulted in the formation of 2-substituted benzoxazoles in excellent yields (87–92%) (equation 54). This equation is an example of the utilization of a combination of the neighboring group effect of hydroxylate anion and a good leaving group such as a sulfinate or sulfonate moiety.

n = 1 or 2

(54)

Recently, a sulfonimidoyl peroxy radical intermediate (**7**), which is generated *in situ* from *N*-carbo-(−)-menthoxy-4-tolylsulfonimidoyl chloride and superoxide in the presence of 18-crown-6 at 0°C in acetonitrile, has been reported to have a good oxidizing ability for the selective oxidation of sulfides to sulfoxides and for the epoxidation of alkenes under mild conditions, as shown in equation 55.[93]

(55)

The sulfonimidoyl peroxy intermediate can contain chirality on the sulfur atom. Thus, optically active (+)-(*S*)-*N*-carbo-(−)-menthoxy-4-tolylsulfonimidoyl chloride was prepared[94] and examined in the oxidation of methyl 4-tolyl sulfide to the optically active sulfoxide by **7** or **7′**. However, no distinguishable enantiomeric excess of methyl 4-tolyl sulfoxide was obtained.[93] This result suggests that the peroxy intermediate is likely to be a radical (**7**).

2 PHOSPHORUS PEROXIDES

Peroxides of organophosphorus compounds are unstable in general. Several examples of stable organophosphorus compounds have been reported. Most phosphorus peroxides have been investigated as possible intermediates without isolation. References to organophosphorus compounds containing the peroxide bond prior to 1981 were well reviewed by Konieczny and Sosnovsky.[95] Only a stable bisdiphenylphosphinyl peroxide (**13**) was prepared from metal peroxides (equation 56).[96]

$$Ph_2P(=O)-Cl \quad + \quad M_2O_2 \longrightarrow Ph_2P(=O)-O-O-P(=O)Ph_2 \;(\underline{13}) \tag{56}$$

Since the first synthesis of a phosphorus peroxy ester (equation 57),[97] considerable efforts have been made to characterize the expanding chemistry of organophosphorus peroxy esters (**14**).[98] However, most of the papers on

$$(RO)_2P(=O)-Cl \quad + \quad NaOOR' \longrightarrow Ph_2P(=O)-O-O-R' \;(\underline{14}) \quad + \quad NaCl \tag{57}$$

R=Me, Et, R'=t-Bu

phosphorus peroxides have treated unstable intermediates. As interesting phosphorus peroxy radicals, tetraalkoxy phosphoranyl peroxy radical,[99] tetrachlorophosphoranyl peroxy radical,[100] phenylphosphorus peroxy radical[101] and diphenylphosphinic peroxy radical (**15**)[102] have been reported. This section treats only topics in recent papers and some of the aspects of phosphorus peroxides owing to the limited space available.

2.1 Symmetrical Phosphorus Peroxides

Only one example of a bisdiphenylphosphinyl peroxide (**13**) has been prepared and characterized.[96] It is noteworthy that no peroxides were isolated when substituted arenephosphinyl chlorides were used in the same procedure (equation 57). The peroxide **13** is stable at low temperature but decomposes exothermically on warming to room temperature. The thermal, photochemical, acid-catalyzed and base-catalyzed decompositions of the peroxide have been studied. The thermal decomposition of **13** in several solvents yields diphenylphosphinic acid (**17**) and phenyl hydrogen phenylphosphonate (**18**) after hydrolysis of the reaction mixture. This decomposition appears to proceed via formation of an unsymmetrical anhydride (**16**), as shown in equation 58.[96]

$$Ph_2P(=O)-O-O-P(=O)Ph_2 \;(\underline{13}) \longrightarrow Ph_2P(=O)-O-P(=O)Ph_2 \;(\underline{16}) \xrightarrow{H_2O} Ph_2P(=O)-OH \;(\underline{17}) \quad + \quad Ph(PhO)P(=O)-OH \;(\underline{18}) \tag{58}$$

Thermal decomposition of ^{18}O-labeled **13** results in no scrambling of the label in the rearranged product (**16**), which suggests that the decomposition involves either a concerted process or an ion-pair intermediate (equation 59).

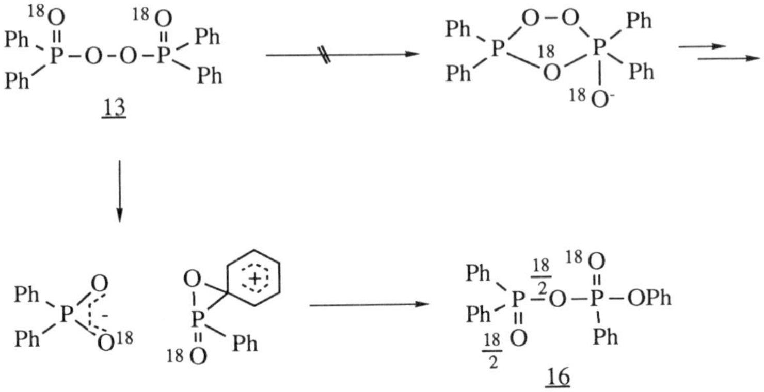

(59)

However, photolysis of the same labeled peroxide affords the same products in the randomization of ^{18}O (equation 60). This decomposition involves complete oxygen scrambling in the free radical intermediate.

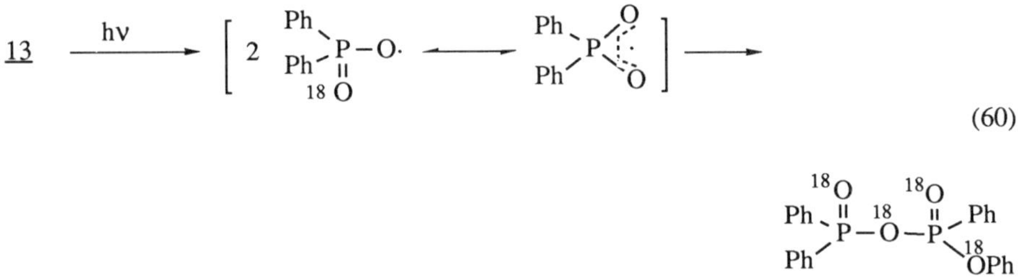

(60)

Nucleophilic oxidation of amines with **13** to give the corresponding *N*,*N*-dialkyl-*O*-diphenylphosphinylhydroxylamines has been reported (equation 61).[100]

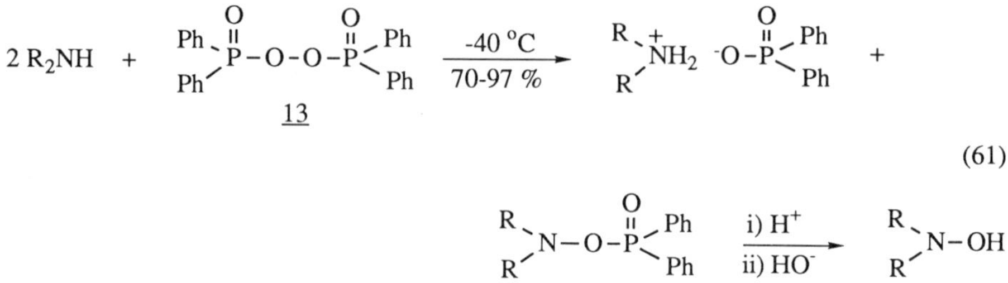

$$2\,R_2NH + Ph_2P(O)-O-O-P(O)Ph_2\ (\mathbf{13}) \xrightarrow[70\text{-}97\,\%]{-40\,^{\circ}C} R_2\overset{+}{N}H_2\ {}^{-}O-P(O)Ph_2 + R_2N-O-P(O)Ph_2 \xrightarrow[\text{ii) }HO^-]{\text{i) }H^+} R_2N-OH \quad (61)$$

Various sulfides react with **13** to give sulfoxides in excellent yield under mild conditions without further oxidation to sulfones (equation 62).[101]

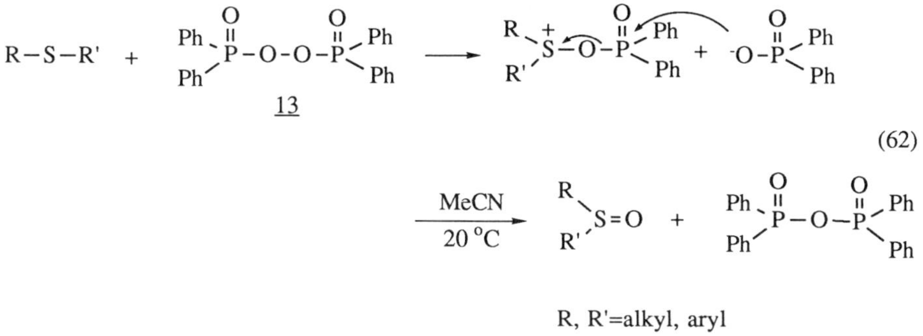

$$R{-}S{-}R' + Ph_2P(O){-}O{-}O{-}P(O)Ph_2\ (\underline{13}) \longrightarrow RR'\overset{+}{S}{-}O{-}P(O)Ph_2 + {}^{-}O{-}P(O)Ph_2 \xrightarrow[20\ ^{\circ}C]{MeCN} RR'S{=}O + Ph_2P(O){-}O{-}P(O)Ph_2 \quad (62)$$

R, R'=alkyl, aryl

Peroxide **13** was also decomposed in chloroform in the presence of an excess of cyclohexene, with sodium hydroxide as catalyst, to give *trans*-2-hydroxycyclohexyl diphenylphosphinate (low yield) (equation 63).

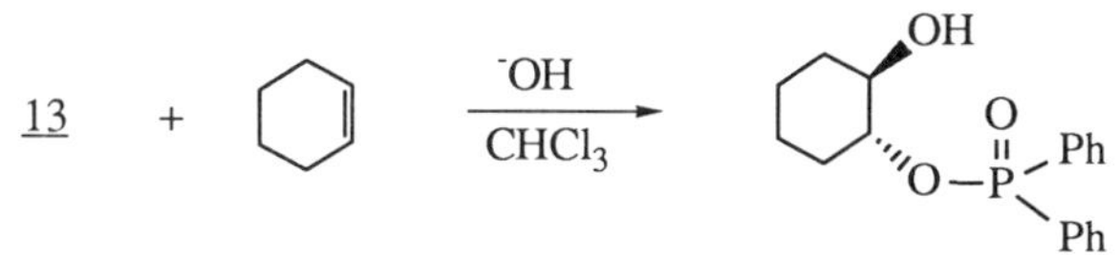

$$\underline{13} + \text{cyclohexene} \xrightarrow[CHCl_3]{{}^{-}OH} \textit{trans}\text{-}2\text{-}HO{-}C_6H_{10}{-}O{-}P(O)Ph_2 \quad (63)$$

2.2 Unsymmetrical Phosphorus Peroxides

As unsymmetrical phosphorous peroxides, many peroxy esters such as peroxyphosphates (**19**),[103] peroxy phosphonates (**20**)[104] and peroxyphosphinates (**21**),[104] have been prepared by the reaction of the chloridates or chloride with a tertiary carboxylic acid in the presence of bases such as of pyridine, sodium hydroxide or sodium hydride, as shown in Figure 6.

Pure peroxy esters **19** were first prepared *in situ* by the reaction of sodium hydride and *tert*-butyl hydroperoxide in diethyl ether, followed by treatment with the phosphorochloridate. The peroxy ester **19** (R = Ph) prepared by this method is stable for several days at 5°C in diethyl ether, presumably because of the absence of trace amounts of diphenyl phosphate. The method was used for

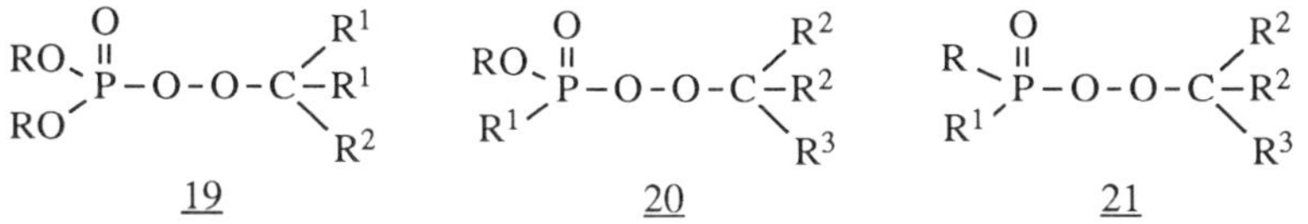

$$(RO)_2P(O){-}O{-}O{-}CR^1_2R^2\ (\underline{19}) \qquad (RO)R^1P(O){-}O{-}O{-}CR^2_2R^3\ (\underline{20}) \qquad RR^1P(O){-}O{-}O{-}CR^2_2R^3\ (\underline{21})$$

Figure 6. Unsymmetrical phosphorus peroxides (**9**)

the preparation of the ring peroxyphosphates **22** and **23** (Figure 7) as a colorless oil which decomposes at 25°C to yield polymeric products and volatile by-products.[105]

Most peroxyphosphates **19** with a group other than *tert*-butylperoxy are unstable even for short periods. For instance, in the preparation of the peroxide **19** ($R = n\text{-}C_4H_9$, $R^1 = Me$, $R^2 = Ph$), it was found that the alleged product was a mixture of products (**24** and **25**) arising from the decomposition of **19** ($R = n\text{-}C_4H_9$, $R^1 = Me$, $R^2 = Ph$) (equation 64).[105]

$$(n\text{-}C_4H_9O)_2\overset{O}{\overset{\|}{P}}\text{-}Cl + NaO_2CMe_2Ph \longrightarrow (n\text{-}C_4H_9O)_2\overset{O}{\overset{\|}{P}}\text{-}OOCMe_2Ph \longrightarrow$$

$$(n\text{-}C_4H_9O)_2\overset{O}{\overset{\|}{P}}\text{-}OCMe_2(OPh) \longrightarrow \underset{\underline{24}}{(n\text{-}C_4H_9O)_2\overset{O}{\overset{\|}{P}}\text{-}OH} + \underset{\underline{25}}{CH_2{=}CMe(OPh)} \qquad (64)$$

The first synthesis of a peroxyphosphate (**2c**), *tert*-butylperoxy alkylphosphonate (**26**) was performed by the condensation of the corresponding alkyl alkylphosphonochloridate with a *tert*-butyl hydroperoxidate in anhydrous diethyl ether (equation 65).[104]

$$\begin{matrix}RO\\R^1\end{matrix}\!\!>\!\overset{O}{\overset{\|}{P}}\text{-}Cl + NaO_2CMe_3 \xrightarrow{Et_2O} \underset{\underline{26}}{\begin{matrix}RO\\R^1\end{matrix}\!\!>\!\overset{O}{\overset{\|}{P}}\text{-}O\text{-}O\text{-}CMe} \qquad (65)$$

R=Me, R'=Me, Et, n-Pr, i-Pr

Peroxyphosphonate can be prepared by a similar reaction in the presence of pyridine[106] instead of the sodium salt of the hydroperoxide. Peroxyphosphonates (**21**) are generally synthesized by the condensation of the corresponding phosphinic chloride with the sodium salt of *tert*-butyl hydroperoxide in a neutral solvent in the presence of sodium sulfate[107] or, alternatively, by the reaction of *tert*-butyl hydroperoxide with phosphinic chloride in the presence of pyridine.[108]

The only fluorine-containing peroxy ester of phosphorus (**27**) was obtained in 87% yield by the condensation of μ-oxobis(phosphonyl difluoride) ($P_2O_3F_4$)

[Structures 22 and 23: cyclic (five- and six-membered ring) $\overset{O}{\overset{\|}{P}}\text{-}O\text{-}O\text{-}CMe_3$ peroxyphosphates]

$\underline{22}$ $\underline{23}$

Figure 7. Ring peroxyphosphates

and trifluoromethyl hydroperoxide (equation 66).[109]

$$P_2O_3F_4 + CF_3OOH \longrightarrow F_2P(O)\text{-}O\text{-}O\text{-}CF_3 \;(\underline{27}) + F_2P(O)OH \qquad (66)$$

The fluoroperester **27** is a liquid and was purified by distillation under vacuum and studied extensively as a synthetic intermediate.[109]

The preparation of the diperoxy esters **28** has been examined and several **28** have been isolated in low yields according to equation 67.[104,110]

$$R\text{-}P(O)Cl_2 \xrightarrow[\text{or } HO_2CMe_3\text{-pyridine}]{NaO_2CMe_3,\ \text{n-hexane}} R\text{-}P(O)(O\text{-}O\text{-}CMe_3)_2 \;(\underline{28}) \qquad (67)$$

R=Me, Et, n-Bu, i-Bu

Dialkyl peroxyphosphates (**19**)[111] or monoperoxy phosphates (**20**)[112] react with phenylmagnesium bromide Grignard reagent to afford the corresponding alkylated aromatic *tert*-butyl phenyl ester in high yields together with dialkylphosphonic acid or alkylphosphonic acid in moderate yields (equation 68).

$$RR^1P(O)\text{-}O\text{-}O\text{-}CMe_3 \;(\underline{19} \text{ or } \underline{20}) + PhMgBr \longrightarrow [RR^1P(=O\cdots MgBr(Ph))\text{-}O\text{-}O\text{-}CMe_3] \longrightarrow PhOCMe_3 + RR^1P(O)\text{-}OMgBr \qquad (68)$$

19 : R=R'= O-alkyl
20 : R= O-alkyl, R'= H

The peroxy phosphates **19** or the peroxy phosphates **20** also react with cyclohexane in the presence of copper(I) ion thermally or photochemically to give the corresponding trialkyl phosphates or phosphonate esters in good yields (equation 69).[113,114] These reactions appear to proceed by a mechanism involving ionic and radical species.

$$\underline{19} \text{ or } \underline{20} + \text{cyclohexene} \xrightarrow[\text{or CuBr, h}\nu,\ 10\ ^\circ C]{CuBr,\ 80\ ^\circ C} \text{cyclohexenyl-O-P(O)RR'} + Me_3COH \qquad (69)$$

R=R'= O-alkyl
R= O-alkyl, R'= H

2.3 Phosphorus Peroxy Radicals

Tetraalkoxyphosphoranylperoxy radical (**29**) is thought to be formed as an intermediate by the reaction of a dialkyl peroxide with a trialkyl phosphite in the presence of oxygen, but without any detailed experimental evidence (equation 70).[99]

$$(RO)_3P \;+\; R'OOR' \xrightarrow{O_2} \underset{\underline{29}}{R'O(RO)_3\,P{-}O{-}O\cdot} \qquad (70)$$

Recently, in connection with sulfonyl peroxy radical intermediates (**6**), various phosphorus peroxy radicals (**10**) such as the diphenylphosphinic peroxy radical **30**, the diphenylphosphoryl peroxy radical **31** and the phenylchlorophosphinic peroxy radical **32** or **33′** have been characterized (Figure 8).

Diphenylphosphinic chloride reacts with the superoxide anion radical ($O_2{}^{-\bullet}$) in acetonitrile under mild conditions to form the diphenylphosphinic peroxy radical intermediate **30**, which shows strong oxidizing abilities for the epoxidation of alkenes, oxidation of sulfides to sulfoxides, desulfurization of thioamides to amides and oxidation of triarylphosphines to phosphine oxides.[115] The anion form (**30′**) can exist by one electron transfer to the radical form (**30**) in equation 71. However, the radical **30** is trapped with dimethylpyrrolidine *N*-oxide (DMPO) to form a radical adduct (**33**).

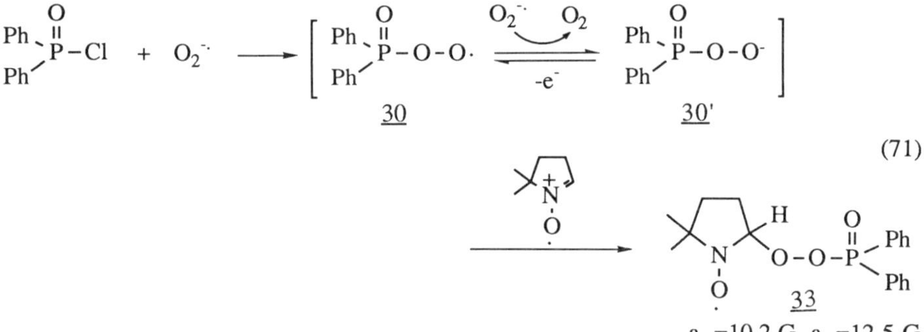

$$Ph_2P(O){-}Cl \;+\; O_2^{-\cdot} \longrightarrow \left[\underset{\underline{30}}{Ph_2P(O){-}O{-}O\cdot} \;\underset{-e^-}{\overset{O_2^{-\cdot}\;\;O_2}{\rightleftharpoons}}\; \underset{\underline{30'}}{Ph_2P(O){-}O{-}O^-}\right] \qquad (71)$$

$$\xrightarrow{\text{DMPO}} \underset{\underline{33}}{\text{DMPO adduct: } \dot{O}{-}N(\text{pyrrolidine}){-}CH{-}O{-}O{-}P(O)Ph_2}$$

$a_H=10.2$ G, $a_N=12.5$ G

$$\underset{\underline{30}}{Ph_2P(O){-}O{-}O\cdot} \qquad \underset{\underline{31}}{(PhO)_2P(O){-}O{-}O\cdot} \qquad \underset{\underline{32}}{Ph{-}P(O)(Cl){-}O{-}O\cdot} \quad \text{or} \quad \underset{\underline{32'}}{Ph{-}P(O)(O{-}O\cdot)_2}$$

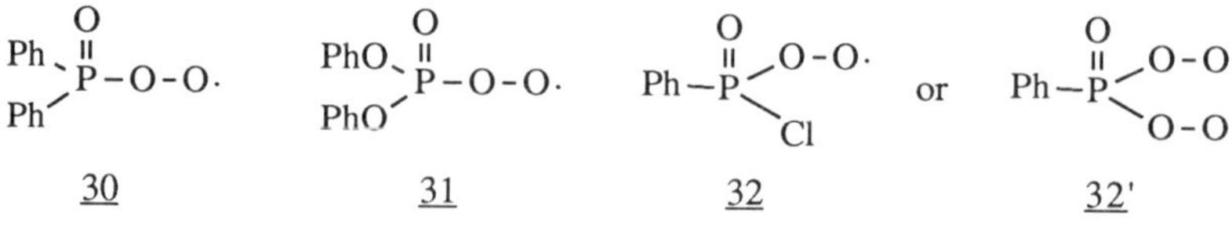

Figure 8. Phosphorus peroxy radicals

In the same way, the diphenylphosphoryl peroxy radical **31** is also trapped using a spin trapping method (equation 72).[69]

$$(PhO)_2P(O)\text{-}Cl + O_2^{-\cdot} \xrightarrow{\text{DMPO}} \text{DMPO-O-O-P(O)(OPh)}_2 \quad (72)$$

a_H=10.3 G, a_N=12.6 G

There have been arguments that the dialkoxy phosphorus peroxy intermediates are radical and/or anionic species.[116] It has also been postulated that the phosphate peroxy intermediate generated from a nucleotide and superoxide may be involved as a radical species in base liberation reactions.[117] The results described above indicate that phosphorus chlorides are good substrates to activated superoxide. A phosphorus peroxy intermediate (**32** or **32'**) generated from phenylphosphinic dichloride and superoxide at −4°C oxidizes thioamides to amides in excellent yields (equation 73).[118]

$$Ph\text{-}P(O)Cl_2 \xrightarrow[-4\,^\circ C,\ MeCN]{2\,O_2^{-\cdot}} [\ \underline{32} \text{ or } \underline{32'}\ ] \xrightarrow{R'\text{-}C(S)\text{-}NR^2R^3} R'\text{-}C(O)\text{-}NR^2R^3 + Ph\text{-}P(O)(O^-)\text{-}O^- \quad (73)$$

R'=aryl, alkyl, R^2=H, R^3=aryl, alkyl

NR^2R^3 : morpholine

In view of the tendency of phosphorus peroxy radicals to be generated from the phosphorus chloride and superoxide, they appear to be stable enough at low temperatures to be used for various oxidation reactions and they may provide fruitful methods for oxidations under mild conditions.

3. REFERENCES

1. (a) A. Nishihara and I. Kubota, *J. Org. Chem.*, **33**, 2525 (1968); (b) G. B. Payne, *Org. Synth., Coll. V.*, **5**, 805 (1973).
2. B. M. Trost and D. P. Curran, *Tetrahedron Lett.*, **22**, 1287 (1981).
3. B. M. Trost and R. Blaslau, *J. Org. Chem.*, **53**, 532 (1988).

4. (a) F. M. Hause and S. R. Ellenberger, *Synthesis*, 723 (1987); (b) C. S. Foote and S. Wexler, *J. Am. Chem. Soc.*, **86**, 3879 (1964).
5. Y. H. Kim and J. C. Jung, unpublished data.
6. R. V. Hoffman, in *Organic Sulfur and Phosphorus Peroxides* (S. Patai, Ed.), Wiley, Chichester (1983), p. 259.
7. Y. H. Kim and K. S. Kim, in *Reviews on Heteroatom Chemistry* (S. Oae, Ed.), MYU, Tokyo (1990), p. 289.
8. Y. H. Kim and D. C. Yoon, *Synth. Commun.*, **19**, 1569 (1987).
9. Y. H. Kim, S. C. Lim and H. S. Chang, *J. Chem. Soc., Chem. Commun.*, 37 (1990).
10. C. M. Atkinson, C. W. Brown, J. McIntyre and J. C. E. Simpson, *J. Chem. Soc.*, 2023 (1954).
11. J. Meisenheimer and E. Hesse, *Chem. Ber.*, **52**, 1161 (1919).
12. R. H. Wiely and J. L. Hartman, *J. Am. Chem. Soc.*, **73**, 494 (1951).
13. R. J. Kennedy and A. M. Stock, *J. Org. Chem.*, **25**, 1901 (1960).
14. I. A. Pearl, *Org. Synth. Coll. Vol.* **4**, 972 (1963).
15. T. L. Evans and M. M. Grade, *Synth. Commun.*, **16**, 1207 (1986).
16. Y. H. Kim, J. C. Jung, and K. S. Kim, unpublished data.
17. W. M. Latimer, *The Oxidation States of the Elements and Their Potentials in Aqueous Solutions*, Prentice Hall, New York (1952), p. 78.
18. D. A. House, *Chem. Rev.*, **62**, 185 (1962).
19. G. Reisenweber and D. Mangold, *Angew. Chem., Int. Ed. Engl.*, **19**, 222 (1980).
20. F. M. Hauser and S. R. Ellenberger, *Synthesis*, 723 (1987).
21. K. B. Wiberg, *J. Am. Chem. Soc.*, **81**, 252 (1959).
22. M. Tiecco, L. Testaferri, M. Tingoli, D. Chinelli and D. Bartoli, *Tetrahedron Lett.*, **30**, 141 (1979).
23. M. Tiecco, L. Testaferri, M. Tingoli, D. Chinelli and D. Bartoli, *Tetrahedron Lett.*, **44**, 2273 (1988).
24. M. Tiecco, L. Testaferri, M. Tingoli and D. Bartoli, *Tetrahedron Lett.*, **45**, 6019 (1989).
25. M. Tiecco, L. Testaferri, M. Tingoli, D. Bartoli and R. Balducci, *J. Org. Chem.*, **55**, 429 (1990).
26. M. Tiecco, L. Testaferri, M. Tingoli and D. Bartoli, *J. Org. Chem.*, **55**, 4523 (1990).
27. Y. H. Kim and J. C. Jung, unpublished data.
28. E. L. Eliel, B. E. Nowark, R. A. Daignalt and V. G. Badding, *J. Org. Chem.*, **30**, 2441 (1965).
29. (a) A. Ledwith, P. J. Russell and L. H. Sutcliffe, *J. Chem. Soc., Perkin Trans. 2*, 630 (1973); (b) T. Carouua, A. Citterio, L. Grossi, F. Minisci and K. Ogawa, *Tetrahedron*, **23**, 2741 (1976).
30. W. E. Notan, *Chem. Rev.*, **55**, 137 (1955).
31. (a) R. L. Dannley, J. E. Gagen and O. J. Stewart, *J. Org. Chem.*, **35**, 3076 (1970); (b) R. L. Dannley and G. E. Corbett, *J. Org. Chem.*, **31**, 153 (1966).
32. (a) R. W. Hazeldine, R. B. Heslop and J. W. Lethbridge, *J. Chem. Soc.*, 4901 (1964); (b) C. J. Myall and D. Pletcher, *J. Chem. Soc., Perkin Trans. 1*, 953 (1975).
33. R. E. Noftle and G. H. Cady, *Inorg. Chem.*, **4**, 1010 (1965).
34. R. Hiatt, in *Organic Peroxides*, Vol. II, (D. Swern, Ed.), Wiley–Interscience, New York, London, p. 812.
35. R. B. Hoffman and G. A. Buntain, *J. Org. Chem.*, **48**, 3308 (1983).
36. R. B. Hoffman, C. S. Carr and B. C. Jankowski, *J. Org. Chem.*, **50**, 5148 (1985).
37. P. W. Hickmotl, *Tetrahedron*, **38**, 1975, 3363 (1982).
38. R. B. Hoffman and H. O. Kim, *J. Org. Chem.*, **53**, 3855 (1988).

39. Y. Yokoyama, H. Wada, M. Kobayashi and H. Minato, *Bull. Chem. Soc. Jpn.*, **44**, 3475 (1971).
40. R. L. Dannley, J. E. Kagen and O. J. Stewart, *J. Org. Chem.*, **35**, 3076 (1970).
41. R. V. Hoffman, *J. Am. Chem. Soc.*, **98**, 6902 (1976).
42. R. V. Hoffman and R. Cadena, *J. Am. Chem. Soc.*, **99**, 8226 (1977).
43. (a) R. A. Bartsch and B. R. Cho, *J. Am. Chem. Soc.*, **101**, 3587 (1979); (b) S. Dayagi and Y. Degani, in *The Chemistry of the Carbon—Nitrogen Double Bond* (S. Patai, Ed.), Wiley, Chichester (1970), p. 117.
44. S. Oae and T. Sakurai, *Bull. Chem. Soc. Jpn.*, **49**, 730 (1976).
45. R. G. Roy and G. A. Swann, *Chem. Commun.*, 427 (1966).
46. (a) L. A. Hull, G. T. Davis, D. H. Rosenblatt and C. K. Mann, *J. Phys. Chem.*, **73**, 2124 (1969); (b) C. A. Audeh and J. R. L. Smith, *J. Chem. Soc.* B, 1741 (1971).
47. R. V. Hoffman and E. L. Belfoure, *J. Am. Chem. Soc.*, **101**, 5687 (1979).
48. (a) P. G. Gassman, *Acc. Chem. Res.*, **3**, 26 (1970); (b) P. G. Gassman and G. D. Hartman, *J. Am. Chem. Soc.*, **95**, 449 (1973).
49. (a) Y. Tamura, J. Maromikava and M. Iheda, *Synthesis*, 1 (1977); (b) G. Boche, N. Mayer, M. Bernheimer and K. Wagner, *Angew. Chem., Int. Ed. Engl.*, **17**, 687 (1978).
50. M. Yoshida, H. Mochizuki and N. Kamigata, *Chem. Lett.*, 2017 (1988).
51. W. A. Pryor and W. H. Hendrickson, Jr, *J. Am. Chem. Soc.*, **105**, 7114 (1983).
52. (a) J. K. Kochi, B. M. Graybill and M. Kurz, *J. Am. Chem. Soc.*, **86**, 5257 (1964); (b) N. J. Bunce and D. P. Tanner, *J. Am. Chem. Soc.*, **91**, 6096 (1969).
53. R. Hisada, M. Kobayashi and H. Minato, *Bull. Chem. Soc. Jpn.*, **45**, 2035 (1972).
54. (a) E. G. E. Hawkins, *Organic Peroxides*, Van Nostrand, Princeton, NJ (1961), p. 311; (b) M. Szware, in *Peroxide Reaction Mechanisms* (J. O. Edwards, Ed.), Interscience, New York (1962), p. 153.
55. R. Hisada, M. Kobayashi and H. Minato, *Bull. Chem. Soc. Jpn.*, **45**, 564 (1972).
56. (a) G. A. Razuvacv, L. M. Terman, V. R. Likhterov and U. S. Etlis, *J. Polym. Sci.*, **52**, 123 (1961); (b) G. A. Razuvacv, U. R. Likterov and U. S. Etlis, *Zh. Obsch. Khim.*, **32**, 2033 (1962).
57. See, e.g., (a) R. L. Friedman and R. N. Leevis, *U.S. Pat.* 4 151 339 (1979); *Chem. Abstr.*, **91**, 21448w (1979); (b) J. A. Manner, *U.S. Pat.* 3 998 888 (1977); *Chem. Abstr.*, **86**, 139498n (1977).
58. J. E. Leffler, *J. Am. Chem. Soc.*, **72**, 67 (1950).
59. D. B. Denney, *J. Am. Chem. Soc.*, **78**, 590 (1956).
60. R. Hisada, M. Kobayashi and H. Minato, *Bull. Chem. Soc. Jpn.*, **45**, 2902 (1972).
61. H. Bager, *Recl. Trav. Chim. Pays-Bas.*, **82**, 773 (1963).
62. (a) E. L. Luff, *Chem. Soc. Rev.*, **6**, 195 (1977); (b) D. T. Sawyer and J. S. Valentine, *Acc. Chem. Res.*, **14**, 393 (1981).
63. D. T. Sawyer, M. J. Gibian, M. M. Morison and E. T. Seo, *J. Am. Chem. Soc.*, **100**, 627 (1978).
64. (a) T. Takata, Y. H. Kim and S. Oae, *Tetrahedron Lett.*, 821 (1979); (b) S. Oae, T. Takata and Y. H. Kim, *Tetrahedron*, **37**, 37 (1981).
65. Y. H. Kim and D. C. Yoon, *Tetrahedron Lett.*, **29**, 6453 (1988).
66. Y. H. Kim and B. C. Jung, *J. Org. Chem.*, **48**, 1562 (1983).
67. H. K. Lee and Y. H. Kim, *Sulfur Lett.*, **7**, 1 (1987).
68. Y. H. Kim, K. S. Kim and H. K. Lee, *Tetrahedron Lett.*, **30**, 6357 (1989).
69. Y. H. Kim, S. C. Lim, M. Hoshino, Y. Ohtsuka and T. Ohishi, *Chem. Lett.*, 167 (1989).
70. W. C. Hamilton and S. J. Laplace, *J. Am. Chem. Soc.*, **86**, 2289 (1964).

71. R. Erikson and H. Hauge, *Acta Chem. Scand.*, **26**, 3152 (1972).
72. L. Syper and J. Mechouski, *Tetrahedron Lett.*, **43**, 207 (1987).
73. A. A. Frimer, P. Gilinsky-Sharon, G. Aljadeff, H. E. Gottlip, J. Hameiri-Buch, V. Marks, R. Philosof and Z. Rosental, *J. Org. Chem.*, **54**, 4853 (1989).
74. H. K. Lee, K. S. Kim, J. C. Kim and Y. H. Kim, *Chem. Lett.*, 561 (1988).
75. K. Ishikawa, H. C. Charles and G. W. Griffin, *Tetrahedron Lett.*, 427 (1977).
76. S. Krishnan, D. G. Kuhn and G. A. Hamilton, *J. Am. Chem. Soc.*, **99**, 8121 (1977).
77. T. H. Kinstle and P. J. Ihig, *J. Org. Chem.*, **35**, 257 (1970).
78. Y. Sawaki and Y. Ogata, *J. Org. Chem.*, **49**, 3344 (1984).
79. Y. Sawaki, S. Ishikawa and H. Iwamura, *J. Am. Chem. Soc.*, **109**, 584 (1987).
80. (a) F. Hirata and O. Hayashi, *J. Biol. Chem.*, **250**, 5960 (1975); (b) T. Taniguchi, F. Hirata and C. Hayashi, *J. Biol. Chem.*, **252**, 2774 (1977).
81. J. M. McCord and I. Fvidovich, *J. Biol. Chem.*, **244**, 6049 (1969).
82. M. H. Rogoff, *J. Bacteriol.*, **83**, 998 (1062).
83. A. N. Davison, *Nature (London)*, **174**, 1056 (1954).
84. Y. H. Kim, H. K. Lee and H. S. Chang, *Tetrahedron Lett.*, **28**, 4285 (1987).
85. Y. H. Kim, B. C. Chung and H. S. Chang, *Tetrahedron Lett.*, **26**, 109 (1985).
86. J. Bland, *J. Chem. Educ.*, **55**, 151 (1978).
87. Y. H. Kim, G. H. Yon and H. J. Kim, *Chem. Lett.*, 312 (1984).
88. Y. H. Kim and H. K. Lee, *Chem. Lett.*, 312 (1984).
89. (a) R. Curci, A. Giovince and G. Modena, *Tetrahedron*, **22**, 1235 (1966); (b) S. Oae and T. Tahata, *Tetrahedron Lett.*, **21**, 3213 (1980).
90. Y. H. Kim, H. J. Kim and G. H. Yon, *J. Chem. Soc., Chem. Commun.*, 715 (1983).
91. Y. H. Kim, H. J. Kim and G. H. Yon, *J. Chem. Soc., Chem. Commun.*, 1064 (1984).
91. Y. H. Kim, H. J. Kim and G. H. Yon, *J. Chem. Soc., Chem. Commun.*, 1064 (1984).
92. Y. H. Kim, Y. I. Kim, H. S. Chang and D. C. Yoon, *Heterocycles*, **29**, 213 (1989).
93. Y. H. Kim and D. C. Yoon, *Synth. Commun.*, **19**, 1569 (1989).
94. M. J. Jones and D. J. Cram, *J. Am. Chem. Soc.*, **96**, 2183 (1974).
95. M. Konieczny and G. Sosnovsky, *Chem. Rev.*, **81**, 49 (1981).
96. R. L. Dannley and K. Kabre, *J. Am. Chem. Soc.*, **87**, 4085 (1965).
97. A. Rieche, G. Hilgetag and G. Schramm, *Angew. Chem.*, **71**, 285 (1959).
98. (a) A. Rieche, G. Hilgetag and G. Schramm, *Chem. Ber.*, **95**, 381 (1962); (b) G. Sosnovaky and J. H. Brown, *Chem. Ber.*, 529 (1966).
99. (a) R. L. Danny, R. L. Waller, R. V. Hoffman and R. F. Hudson, *J. Org. Chem.*, **37**, 418 (1972); (b) R. L. Danny, R. L. Waller, R. V. Hoffman and R. F. Hudson, *J. Chem. Soc., Chem. Commun.*, 1362 (1971).
100. (a) J. J. Yaouanc, G. Masse and G. Sturtz, *Synthesis*, 808 (1985); (b) G. Boche and R. H. Sommerlade, *Tetrahedron*, **42**, 2703 (1986).
101. Y. H. Kim, S. C. Lim and H. C. Choi, unpublished data.
102. S. C. Lim and Y. H. Kim, *Heteroatom. Chem.* **1**, 261 (1990).
103. G. Sosnovsky and E. H. Zaret, *J. Org. Chem.*, **34**, 968 (1969).
104. T. I. Yurzhenko and B. I. Kaspruk, *Zh. Obshch. Khim.*, **41**, 1644 (1971).
105. (a) G. Sosnovsky and E. H. Zaret, *Synthesis*, 202 (1972); (b) G. Sosnovsky and E. H. Zaret, *Z. Naturforsch., Teil. B*, **30**, 732 (1972).
106. V. P. Maslennikov and V. P. Sergeeva, *Zh. Obshch. Khim.*, **40**, 2529 (1970).
107. V. P. Maslennikov, V. P. Sergeeva and N. G. Sukhikh, *Zh. Obshch. Khim.*, **40**, 2019 (1970).
108. E. Avar and W. P. Neumann, *J. Organomet. Chem.*, **131**, 207 (1977).
109. P. A. Bernstein and D. D. Desmarteau, *J. Fluorine Chem.*, **2**, 315 (1972).
110. T. I. Yurzhenko and B. I. Kaspruk, *Dokl. Akad. Nauk SSSR*, 113 (1966).

111. G. Sosnovsky and M. Konieczny, *Synthesis*, 114 (1971).
112. K. D. Berlin, M. E. Perterson, *J. Org. Chem.*, **32**, 125 (1967).
113. G. Sosnovsky and G. Karas, *Z. Naturforsch.*, *Teil B*, **33**, 1165 (1973).
114. G. Sosnovsky and G. Karas, *Z. Naturforsch.*, *Teil B*, **33**, 1177 (1978).
115. S. C. Lim and Y. H. Kim, *Heteroatom. Chem.*, **1**, 42 (1990).
116. M. Miura, M. Nojima and S. Kusabayashi, *J. Chem. Soc., Chem. Commun.*, 1352 (1982).
117. H. Yamane, N. Yada, E. Katori, T. Mashino, T. Nagano and M. Hirobe, *Biochem. Biophys. Res. Commun.*, **142**, 1104 (1987).
118. Y. H. Kim, S. C. Lim and H. S. Chang, *J. Chem. Soc., Chem. Commun.*, 36 (1990).

9 Peroxy Acids and Peroxy Esters

YASUHIKO SAWAKI
Department of Applied Chemistry, Faculty of Engineering, Nagoya University, Chikusa-ku, Nagoya 464-01, Japan

Organic Peroxides. Edited by W. Ando

1 PEROXY ACIDS

The syntheses, properties and reactions of peroxy acids (peroxycarboxylic acids) have been summarized in comprehensive reviews.[1–4] In this chapter are described the fundamental principles in the chemistry of peroxy acids together with recent results.

1.1 Synthesis of Peroxy Acids

1.1.1 Synthesis from carboxylic acids

Peroxy acids (PAs) are prepared by the acid-catalyzed equilibrium between carboxylic acids and hydrogen peroxide (equation 1).

$$RCO_2H + H_2O_2 \underset{}{\overset{H^+}{\rightleftharpoons}} RCO_3H + H_2O \qquad (1)$$

Aliphatic PAs are easily synthesized by using sulfuric acid as the catalyst. The equilibrium constants for R = Me, Et, *i*-Pr and *t*-Bu are approximately the same.[5] Usually, PAs in equilibria are applied to oxidations without isolation and hence are appropriate for large-scale reactions. For the synthesis of aromatic PAs, methanesulfonic acid has conveniently been used both as an acid catalyst and as a solvent.[6] In this case, highly concentrated hydrogen peroxide should be used with caution.

Peroxy acids are also obtained by the reaction of acid anhydrides with hydrogen peroxide (equation 2). For example, concentrated solutions of PAs

$$(RCO)_2O + H_2O_2 \xrightarrow{H^+} RCO_3H + RCO_2H \qquad (2)$$

may be obtained by the acid-catalyzed reaction of acetic anhydride;[7] special care should be taken with regard to contamination by highly explosive acetyl peroxide. Trifluoroperoxyacetic acid could be obtained without an acid catalyst.[8] From phthalic anhydride, monoperoxyphthalic acid is obtained in high yield by reaction with alkaline hydrogen peroxide.[9]

1.1.2 Perhydrolysis of acid chlorides

Peroxy acids are conveniently obtained from the perhydrolysis of acid chlorides with alkaline hydrogen peroxide (equation 3).[10,11] Addition of excess amounts

$$\mathrm{RCOCl + H_2O_2 + NaOH \longrightarrow RCO_3H + H_2O + NaCl} \qquad (3)$$

of hydrogen peroxide is pertinent since the molar ratio of RCOC1: H_2O_2 = 2:1 leads to the preparation of diacyl peroxides. The synthetic method for PAs is based on the high nucleophilicity of hydroperoxide ion (HOO^-) having an α-effect.[12] Pyridine could be used as a base in the reaction in organic solvents.[13]

High yields of PAs are obtained by the perhydrolysis of diacyl peroxides (equation 4),[11,14a] which is more appropriate than the method with methoxide

$$\mathrm{RC(=O)OOC(=O)R + H_2O_2 \xrightarrow{OH^-} 2\ RC(=O)OOH} \qquad (4)$$

ion yielding methyl esters as a by-product.[14b] Peroxycarbonic acids, which are useful as neutral oxidizing agents, are synthesized by this method.[15]

1.1.3 Autoxidation of aldehydes

Peroxy acids are obtained by the radical-initiated autoxidation of aldehydes (equation 5); e.g. the photooxidation of acetaldehyde[16] and the ozone-initiated

$$\mathrm{RCHO \xrightarrow{O_2\ /\ initiator} RCO_3H} \qquad (5)$$

autoxidation of benzaldehydes.[17] Metallic catalysts such as Co or Mn are also applicable, and a large number of patents have been filed for the preparation of PAs by the autoxidation of aldehydes. The detailed synthesis of individual PAs were reviewed comprehensively by Swern.[18]

1.2 Physical Properties and Structure of Peroxy Acids

Peroxy acids with short alkyl chains were sometimes distilled with precaution under reduced pressures, as exemplified in Table 1. However, their distillation is not recommended in common laboratories because of the danger of possible explosion. One of the characteristic physical properties of PAs is that their melting points are lower than those of the corresponding parent acids. The lower melting points and high solubility of PAs in organic solvents are important aspects for their use as oxidants.

Table 1. Properties of peroxy acids

Peroxy acids	M.p. (°C)	B.p. (°C/mmHg)	pK_a [a]	Method of preparation [b] (equation No.)
HCO_3H			7.1[19]	1
$MeCO_3H$	0[20]	31/26[21a]	8.2[19]	1, 2, 5
n-$PrCO_3H$	−10[20]	26–29/12[21b]	8.2[19]	1, 2, 5
n-$C_{11}H_{23}CO_3H$	52[22]			1, 2
CF_3CO_3H			~3.7[23]	1, 2
$PhCO_3H$	41[6]		7.78[24]	1, 4
m-$ClC_6H_4CO_3H$	138[2]		7.29,[24] 7.14[26]	1, 3
m,p-$(O_2N)_2C_6H_3CO_3H$	113[25]			1, 3
o-$(HO_2C)C_6H_4CO_3H$	110[2]		8.08[26]	2

[a] pK_a values of peroxy acids in water at 20–25 °C.
[b] Preparation methods often applied.

The boiling points of PAs are significantly lower than those of their parent acids. For example, the boiling points extrapolated to 760 mmHg of peroxyacetic and peroxypropionic acids are 110 and 120°C, respectively,[20] which are lower than those of the parent acids (118 and 141°C). This is significant, since the molecular weights of PAs are higher than those of the parent acids by 16 u (one oxygen atom). The lower boiling points of PAs are due to the high stability of the monomeric form with intramolecular hydrogen bonding (**1**).

```
        O.
       //  ·.
  R—C       H
     \     /
      O—O
```

(1)

An IR study of RCO_3H revealed that in non-coordinating solvents such as carbon tetrachloride, PAs exist in the H-bonded form (**1**), the O–H stretching being observed at 3280–3300 cm^{-1}, regardless of their concentrations.[27] This is contrasted with the case of the parent acids (RCO_2H), where polymeric forms are more stable and the monomeric form is observable only under highly diluted conditions. The high stability of the intramolecular H-bonding of PAs is apparently due to the five-membered ring (**1**) with no angle strain in comparison with the strained four-membered ring in the parent acids.

Other evidence for the intramolecular H-bonding form (**1**) is provided by the measurement of dipole moments. From the dipole moments of peroxyacetic acid

($\mu = 2.384$ D)[28] and peroxybenzoic acid,[29] the H-bonded forms are shown to have a planar five-membered structure. From a gas-phase dipole moment study and a microwave analysis of peroxyformic acid, its detailed structural features as in **2** were obtained.[30] The data for $d(\text{O—O}) = 1.445$ Å, $d(\text{O} \cdots \text{H}) = 1.861$ Å

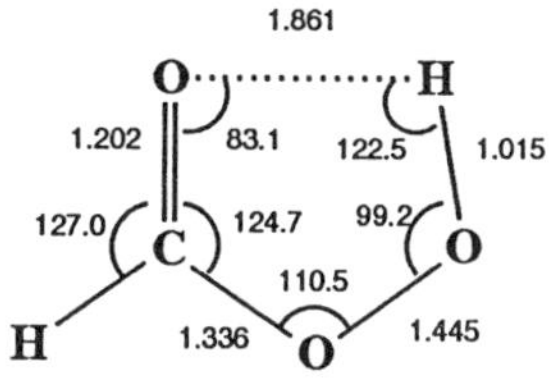

1.861
O······H
1.202 83.1 122.5 1.015
127.0 C 124.7 99.2 O
110.5
H 1.336 O 1.445

(2) (bond length, Å; bond angle, degree)

and $< \text{COO} = 110.5°$ are interesting. The related *ab initio* (4–21G*) theoretical calculation supported the structure of **2**, and the calculated dipole moment of 1.57 D was close to the observed value of 1.398 D.[30] The intramolecular five-membered ring structure of PAs is also supported by other theoretical calculations.[31]

In coordinating solvents containing oxygen or nitrogen atoms, PAs exist as a complex (**3**) with intermolecular H-bonding. By an elegant study with

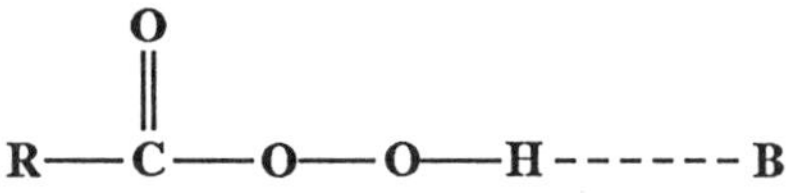

R—C(=O)—O—O—H - - - - - - B

(3)

calorimetric and ^{1}H NMR analyses, Plesničar *et al.*[32] showed that the formation of intermolecular complex (**3**) is proportional to the solvent basicities and the acidities of PAs. It is natural that the H-bonding becomes stronger with increasing basicity of the solvent. Thus, PAs in ethers or amides exist in the intermolecular H-bonding form (**3**).[33]

Although hydrogen peroxide (pK_a = 11.6) is a stronger acid than water, the acidities of PAs are considerably weaker than those of the parent acids. As shown in Table 1, the pk_a values of PAs are 7–8 in water, which are higher than those of the parent acids by 3–4 pk_a units. The large difference is easily understood if it is considered that the inductive effect of acyl groups is significantly decreased by oxygen incorporation in comparison with the parent acids. The acidity of peroxybenzoic acids increases with electron-attracting substituents, affording a Hammett ρ-value of +0.67.[26] The pk_a of peroxytrifluoroacetic acid was estimated to be *ca* 3.7 owing to the strong electron-attracting CF_3 group.[23]

Peroxy acids are, as described in the following sections, useful as oxidants in organic reactions and syntheses. Precautions should be taken not to touch the oxidants, which are irritants to the skin. Some PAs are applied in bleaching in laundries.[34] For example, the magnesium salt of monoperoxyphthalic acid[35] is stable and suitable for oxidations under mild conditions.[36]

1.3 Electrophilic Reactions of Peroxy Acids

The most characteristic reaction of PAs is electrophilic oxygen transfers. The oxygen atoms of PAs are transferred to substrate nucleophiles with lone-pair or π-bonding electrons according to equation 6. The oxygen transfers are

$$\mathrm{S\!:} + \mathrm{HOOC(=O)R} \longrightarrow \left[\mathrm{S\!:}\ \ \mathrm{O(H)\!-\!OC(=O)R} \right] \longrightarrow \mathrm{S{=}O} + \mathrm{HOC(=O)R} \qquad (6)$$

accelerated by electron-donating substituents on the substrates and by electron-attracting groups on the PAs. Some typical polar effects are listed in Table 2.

The epoxidation of alkenes is the most important reaction of PAs and oxygen insertion in C—H bonds is interesting. As described below, electrophilic oxygen transfers to various compounds are known. The enhanced reactivity of PAs, much higher than that of hydroperoxides, is simply based on the fact that the acidities of departing carboxylic acids ($pk_a \approx 4$–5) are different from those of alcohols ($pk_a \approx 15$) as the departing group.[38]

1.3.1 Epoxidation of alkenes

The formation of epoxides by the reaction of alkenes with PAs (the Prilezhaev reaction)[39] has long been known, is important and has been extensively

Table 2. Polar effect in the O-transfers with peroxy acids

Peroxy acids	Nucleophiles	Solvent (temp.)	$k \times 10^3$ ($\mathrm{l\ mol^{-1}\ s^{-1}}$)[a]	ρ[b]	Ref.
$MeCO_3H$	$ArNH_2$	EtOH (30°C)	41.4	−1.86	37a
$PhCO_3H$	Pyridines	50% Dioxane (25°C)	4.8	−2.35	37b
$MeCO_3H$	ArN = 0	47% EtOH (30°C)	0.52	−1.58	37c
$ArCO_3H$	$(p\text{-}ClC_6H_4)_2S$	i-PrOH (−35°C)	0.026	+1.05	37d
$ArCO_3H$	$p\text{-}MeC_6H_4SOMe$	40% Dioxane (25°C)	60.3	+0.75	37e
$ArCO_3^-$	$p\text{-}MeC_6H_4SOMe$	40% Dioxane (25°C)	390	+0.57	37e
$PhCO_3^-$	Ar_2SO	40% Dioxane (25°C)	160	+0.69	37f
$ArCO_3H$	PhCH = CHPh	C_6H_6 (30°C)	0.66	+1.4	37g
$PhCO_3H$	$ArCH = CH_2$	C_6H_6 (30°C)	1.44	−1.30[c]	37h

[a]Second-order rate constants for unsubstituted reagents (i.e. Ar = Ph).
[b]The Hammett ρ-values (vs σ) for substituents on Ar.
[c]Correlated vs the combined value of $(\sigma + \gamma\cdot\Delta\sigma_R{}^+)$ with $\gamma = 0.48$.

reviewed.[4,40–42] The characteristic features of epoxidation with PAs are as follows: the epoxidation is accelerated (a) by increasing the electron density of C=C double bonds, (b) by the electron-attracting groups on PAs and (c) the rates are significantly reduced by coordinating solvents such as ethers, which form intermolecular H-bonds. Thus, the transition state **4**[43] has been long known and is widely accepted (equation 7).

$$RCO_3H + \text{>C=C<} \longrightarrow [\text{transition state } \mathbf{(4)}] \longrightarrow RCO_2H + \text{>C—C<}\ (\text{epoxide, O}) \qquad (7)$$

(4)

Alkylated alkenes are of higher π-electron density and more reactive than the less substituted alkenes. The following order of reactivity[40] is important in predicting the product selectivities:

CH_2=CH_2	<	RCH=CH_2	<	PhCH=CH_2	<	RCH=CHR	<	R_2C=CR_2
1		25		60		500		6000

(R = alkyl group). The figures are the approximate relative rates of epoxidation and suggest that selectivities over 90% are easily attainable in the co-presence of different types of alkenic groups.

Solvent effects in PA epoxidations are significant. For example, the epoxidation rates in diethyl ether or ethyl acetate are approximately one tenth of those in benzene or chloroform. The much slower rate with intermolecular H-bonding (**3**) is suggestive of the cyclic transition state **4** for the epoxidation mechanism (equation 7). No observation of acid catalysis under the usual conditions is another evidence for the epoxidation mechanism involving **4**. When coordinating groups are available in alkenes, the intermolecular hydrogen bonds with PAs lead to site-selective epoxidation (e.g., equation 8).[44,45] Many examples of *syn* selectivity are known.[46,47]

$$\text{(R, OH-substituted cyclohexene)} \xrightarrow{\text{PA}} \text{(R, OH-substituted epoxycyclohexane)} \qquad (8)$$

A detailed mechanistic study of PA epoxidation has been carried out. Kinetic isotope effects suggested that the transition state **4** is unsymmetric.[48] As an alternative epoxidation mechanism, a 1,3-dipolar mechanism was proposed in which PA reacts as a 1,3-dipolar reagent (**5** in equation 9).[49] However, the

$$RCO_3H \rightleftharpoons \underset{(5)}{R-\overset{OH}{\underset{O-O^-}{C^+}}} \xrightarrow{\;>C=C<\;} \underset{(6)}{R-\overset{OH}{C}(-O-O-)\!-\!C\!-\!C-} \longrightarrow RCO_2H + -\overset{|}{C}-\overset{|}{C}- \text{(epoxide, bridged by O)} \qquad (9)$$

relative reactivity of alkenes[50] and the absence of steric retardation in substituted PAs[51] are against the 1,3-dipolar mechanism. It is not reasonable to accept the 1,3-dipolar mechanism, although *ab initio* MO calculations could not deny it.[52] The simple mechanism in equation 7 is acceptable and can explain all the experimental results.

Neutral epoxidations are possible with peroxycarbonic acids (**7**)[53] and peroxycarbamic acids (**8**).[54] These oxidants are appropriate for acid-sensitive

$$\underset{(7)}{RO\overset{O}{\overset{\|}{C}}OOH} \qquad\qquad \underset{(8)}{R^1R^2N\overset{O}{\overset{\|}{C}}OOH}$$

epoxides because there is no formation of carboxylic acids after the epoxidation. In general, epoxides are prone to acid-catalyzed ring openings and hence, to avoid these reactions, bases such as sodium carbonate or phosphate buffer are sometimes added. When PAs are synthesized in the presence of strong acids, preliminary neutralization of the acids is necessary.

An intramolecular epoxidation as pictured in **9**[55] is possible but is not general since the PA group and C=C double bonds should be situated in appropriate positions. Attempted asymmetric epoxidations using optically active PAs (**10**)

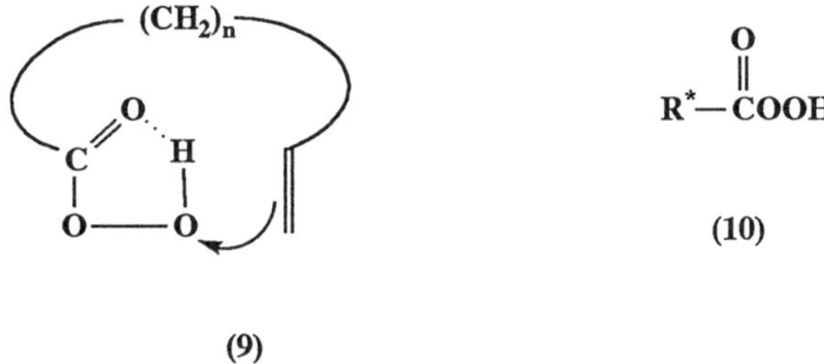

(9) (10)

were unsuccessful (i.e. <4% e.e.).[56] Inspection of the transition state **4** for the epoxidation indicates that high optical yields in the asymmetric induction are not expected because there is no contact interaction between R* and alkenes.

1.3.2 Oxidation of acetylenes

The PA oxidation of acetylenes (alkynes) has not been studied as thoroughly as that of alkenes. This is simply because the oxidations of acetylenes yield complex mixtures of products[57] and are not useful as synthetic methods. For example, the oxidation of cyclodecyne with *m*-chloroperoxybenzoic acid (MCPBA) afforded a mixture of products **11**, **12**, and **13** (equation 10).[58] A characteristic point here is the formation of the *trans*-annular products **11** and **12**.

$(CH_2)_8$ C≡C —MCPBA→ **(11)** + **(12)** + **(13)** (10)

The PA oxidation of phenylacetylenes was accelerated by electron-donating groups on the acetylenes as reflected in the Hammett correlation of $\rho = -1.40$ (σ^+).[59] The solvent effects were just the same with the epoxidation of alkenes.[60] Hence it is reasonable to assume the formation of oxirene intermediates (**14**) (equation 11).[57] MO calculations predicted an unsymmetric transition state for the formation of **14**.[61]

RC≡CR′ —PA→ [R—C(O)=C—R′] **(14)** ⇌ [R—C(=O)—C̈R′] **(15)** —rearrangement→ RR′C=C=O —H_2O→ $RR'CHCO_2H$ (11)

Since oxirenes with four π-electrons will be anti-aromatic, the isomerization of **14** to ketocarbenes (**15**) is expected to be facile.[62] The following rearrangements of ketocarbenes to ketenes are well known and the major products could be explained by the intermediacy of **15**. In the PA oxidation of biphenylacetylene, the migrating group was shown to be a terminal hydrogen atom,[63] which is in agreement with the known migrating aptitude of H > Ar in the rearrangement of ketocarbenes. The above-mentioned formation of *trans*-annular products **11** and **12** (equation 10) is well explained by the intramolecular C—H insertion in ketocarbene intermediates.

The related PA oxidation of allenes affords complex mixtures of products. However, in the particular case of allenes with bulky *tert*-butyl groups, allene oxides (**16**) and spirobioxirane intermediates (**17**) could be isolated (equation 12).[64]

$$R_2C{=}C{=}CR'_2 \xrightarrow{PA} R_2\overset{O}{C\!-\!C}{=}CR'_2 \xrightarrow{PA} R_2\overset{O}{C\!-\!C}\overset{O}{\!-\!C}R'_2 \quad (12)$$

(**16**) (**17**)

1.3.3 Hydroxylation of arenes and alkanes

Aromatic rings are hydroxylated by PAs; for example, the hydroxylation of mesitylene was attained by the uncatalyzed[65] and BF_3-catalyzed reaction[66] of peroxytrifluoroacetic acid. Cationic intermediates (**18**) are probably involved

$$X\!-\!C_6H_5 + HO\!-\!OC(O)R \longrightarrow [X\!-\!C_6H_5(H)(OH)]^+ \longrightarrow X\!-\!C_6H_4\!-\!OH \quad (13)$$

(**18**)

(equation 13) since the NIH shift also occurred.[67] The hydroxylation is faster for arenes with electron-donating substituents, affording *ortho*- and *para*-oriented products. For example, in the preparative hydroxylation of phenol with peroxyacetic acid (60°C in acetic acid), catechol and hydroquinone were obtained in 45 and 35% yields, respectively, based on the PA.[68]

Since the hydroxylated products are more reactive than the starting arenes, the resulting phenols are further oxidized by PAs. For example, the MCPBA oxidation of phenol **19** (3 h in chloroform at room temperature) yielded the

$$\text{(19)} \xrightarrow{MCPBA} \text{hydroquinone (26\%)} + \text{quinone (50\%)} \quad (14)$$

(**19**) 26% 50%

corresponding quinone as the major product as shown in equation 14.[69] The oxidation of *p*-cymene under the same conditions afforded the same quinone in 20% yield.

The selective C—C cleavage of catechol to afford *cis, cis*-muconic acid could be attained by Fe(III)-catalyzed oxidation with peroxyacetic acid.[70] Oxidations of arenes with excess amounts of peroxytrifluoroacetic acid lead to the complete degradation of aromatic rings, converting alkylbenzenes into the corresponding aliphatic carboxylic acids.[71] For example, toluene and pentylbenzene (equation 15) were converted into acetic and hexanoic acids in 95 and 65%

$$\text{n-}C_5H_{11}\text{-}C_6H_5 \xrightarrow{CF_3CO_3H} \underset{65\%}{\text{n-}C_5H_{11}CO_2H} + \underset{10\%}{\text{n-}C_4H_9CO_2H} + \underset{15\%}{CH_3CO_2H} \quad (15)$$

yields, respectively, and acenaphthene yielded 80% of succinic acid.[71b] These degradations are based on the selective attack of the strongly electrophilic PA on the aromatic nuclei.

It is interesting that aliphatic C—H bonds are also hydroxylated by PAs. The electrophilic nature of the hydroxylation has been demonstrated by the fact that the oxidation of methylcyclohexane with peroxytrifluoroacetic acid afforded the tertiary and secondary alcohols in 71 and 29% yields, respectively, the tertiary C—H bond being over 20 times more reactive than the secondary bond.[72] Since then, hydroxylations of aliphatic C—H bonds with PAs such as MCPBA have been thoroughly studied[73] and applied to organic syntheses.[74] The hydroxylation method with PAs is useful because tertiary C—H bonds are selectively hydroxylated according to the reactivity order tertiary ≫ secondary ≫ primary C—H bonds.[73]

Detailed studies by Schneider and Müller[75] revealed that the hydroxylation with PAs proceeds with high regioselectivity and retention of stereochemistry. The electrophilic nature of the hydroxylation was exemplified by the Hammett correlations of $\rho^* = -2.2$ for alkanes and $\rho = +0.63$ for peroxybenzoic acids. These data together with $k_H/k_D = 2.2$ for methylcyclohexane substantiated the electrophilic oxygen insertion into C—H σ-bonds as shown in equation 16. The

$$R\text{—}\overset{|}{\underset{H}{C}} \cdots \text{O(—O—C(=O)R}^1\text{, H}\cdots\text{O)} \longrightarrow R\text{—}\overset{|}{\underset{OH}{C}}\text{—} + R^1CO_2H \quad (16)$$

relative reactivities of tertiary and secondary C—H bonds toward peroxybenzoic acids were as high as 90–500, leading to the highly selective hydroxylation of tertiary C—H bonds. These are true if the radical decomposition of PAs (see Section 1.5.1) is not involved.

1.3.4 Oxidation of carbonyl compounds

Oxygen insertions with PAs in aldehydes and ketones are well known as the Baeyer–Villiger oxidation.[76] As shown in equation 17, the reaction involves the

$$R^1-\underset{\underset{O}{\|}}{C}-R^2 + R^3CO_3H \underset{}{\overset{H^+}{\rightleftharpoons}} R^1-\overset{OH}{\overset{|}{C}}(-R^2)-O-O-\underset{\underset{O}{\|}}{C}R^3 \xrightarrow{-R^3CO_2H} R^2\underset{\underset{O}{\|}}{C}OR^1 \quad (17)$$

(20)

migration of an R^1 group from the carbon to the peroxy oxygen in the carbonyl adduct (**20**). The migratory aptitude of R^1 is as follows:

Ethyl < Neopentyl < Phenyl < Cyclopentyl < Benzyl < i-Propyl < Cyclohexyl < t-Butyl

(0.07) (0.11) (1.0) (1.1) (1.3) (1.9) (3.0) (39)

the values in parentheses are the relative values.[77] Since the rearrangement in **20** is of the Criegee type, the electron-donating substituents on the migrating group accelerate the reaction significantly.

Contrary to the case of epoxidations, the Baeyer–Villiger oxidations are subject to acid catalysis. Both steps, the addition of PA to the C=O bond and the rearrangement in the adduct **20**, are catalyzed by acids. These aspects were clarified by detailed kinetic studies of the oxidation of acetophenones with peroxybenzoic acid.[78] Whereas weak acids such as acetic acid catalyzed only the addition step, the rate-determining step in the presence of sulfuric or trifluoroacetic acid was the rearrangement from C=O adducts (i.e. **21A**), the

$$Ar-\overset{OH}{\overset{|}{C}}-R \text{ with } -O-OCOR\cdots HA \qquad\qquad Ar-\overset{O^-}{\overset{|}{C}}-R \text{ with } -O-OCOR$$

(21A) (21B)

apparent ρ-value being -2.1 (σ^+). In the Baeyer–Villiger oxidation of benzaldehydes (ArCHO), Ar and H groups migrated competitively.[79] The migration of hydride to yield carboxylic acids was independent of the substituents in Ar, the k_H/k_D values being in the range of 1.4–3.0. In contrast, the migration of aryl groups to afford phenols was dependent on the substituents, resulting in large ρ-values of *ca* -4 to -5 (σ^+).

The Baeyer–Villiger reaction was also catalyzed by bases.[79] The addition to C=O groups becomes rapid owing to the high nucleophilicity of PA anions. The migration step is also accelerated by the pushing effect of the α-oxyanion as shown in **21B**.

The PA oxidation of alicyclic ketones yields lactones, and aliphatic aldehydes are oxidized to formates and carboxylic acids.[76] The major products from the PA oxidation of α,β-unsaturated aldehydes are the corresponding vinyl esters, i.e. the product from the migration of vinyl groups.[80] The reaction of 1,2-diketones to yield acid anhydrides[81] has been clarified by an ^{18}O-tracer study.[82]

The use of a Nafion-H resin as an acid catalyst is effective in separating products.[83] It is interesting that the Baeyer-Villiger oxidation proceeds effectively in mixed crystals of ketones and PAs.[84] The selectivity of the Baeyer–Villiger reaction may be altered by the use of micelle surfaces as a reaction site.[85] The regioselectivity of migrating groups, which is the important point in the Baeyer-Villiger reaction, has been shown to be dependent on the kinds of PAs used.[86]

1.3.5 Oxidation of nitrogen compounds

The oxidation of tertiary amines to *N*-oxides proceeds fairly facilely by the nucleophilic attack of nitrogen lone-pair electrons on PAs as shown in equation 18. This is typically shown by the ρ-value of −2.35 in the PA oxidation

$$R_3N: + R^1C(=O)OOH \text{ (cyclic, H}\cdots\text{O=C)} \longrightarrow R_3N{=}O + R^1CO_2H \qquad (18)$$

of substituted pyridine.[37b] Typical synthetic examples are the formation of *N*-oxides from pyridine[87] and pyrimidines.[88]

The PA oxidation of primary amines proceeds stepwise as shown in equation 19, the products being dependent on the kinds of substrates and PAs.

$$RNH_2 \xrightarrow{PA} RNHOH \xrightarrow{PA} RN{=}O \xrightarrow{PA} RNO_2 \qquad (19)$$

Thus, anilines are oxidized to nitrosobenzenes with peroxyacetic acid[89] and to nitrobenzene with peroxymaleic[90] or peroxytrifluoroacetic acid.[91] The oxidation of aliphatic amines to nitro compounds is possible with MCPBA.[92] The oxidations with peroxytrifluoroacetic acid of nitroso compounds[93] and oximes[94] are facile, affording the corresponding nitro compounds in high yields.

An interesting example is the PA oxidation of secondary amines to yield nitroxy radicals (equation 20).[95]

$$\text{2,2,6,6-tetramethyl-4-piperidone (N-H)} \xrightarrow[\text{CHCl}_3,\ 30\%]{\text{MCPBA}} \text{2,2,6,6-tetramethyl-4-oxopiperidine-N-O}^{\bullet} \quad (20)$$

Azo compounds are oxidized to the corresponding *N*-oxides (equation 21).[96] A kinetic study of this reaction showed again the nucleophilic attack of the nitrogen one pair electrons on the peroxy oxygen of PA.[97]

$$\text{ArN=NAr} \xrightarrow{\text{PA}} \text{ArN(\rightarrow O)=NAr} \quad (21)$$

Diazo compounds are easily oxidized by PAs to yield nitrogen and the corresponding carbonyl compounds (equation 22).[98] The substituent effect showed the nucleophilic attack of diazo carbon atoms on PAs.[98a]

$$\text{Ph}_2\text{C}^- - \text{N}^+ \equiv \text{N} + \text{ArCO}_3\text{H} \longrightarrow \text{N}_2 + \text{Ph}_2\text{C=O} + \text{ArCO}_2\text{H} \quad (22)$$

The oxidation of diarylhydrazones with PAs, buffered with sodium hydrogencarbonate, was reported to yield diaryldiazomethanes.[99] The reaction of diaziridines, cyclic isomers of diazo compounds, with PAs was too slow to proceed directly.[100] The PA oxidation of *N*-ylides proceeds via the nucleophilic attack of carbanionic carbons of the ylides[101] just as in the case of diazo compounds (equation 22).

The PA oxidation of imines is the important synthetic route for oxaziridines.[102] Various kinds of optically active oxaziridines have been synthesized and applied to asymmetric oxidations.[103] In general, the oxidations of imines to oxaziridines are very facile, as exemplified in equation 23.[104]

$$\text{PhCH=NBu}^t \xrightarrow[\text{C}_6\text{H}_6/\ 0^\circ\text{C}/\ 99\%]{\text{PhCO}_3\text{H}} \text{PhCH—NBu}^t \text{ (bridged by O, oxaziridine)} \quad (23)$$

Two mechanisms are conceivable for the formation of oxaziridines, viz. the one-step epoxidation mechanism (equation 24) and the two-step mechanism (equation 25).[105]

$$\rangle C{=}N{-} + RCO_3H \longrightarrow -\overset{|}{C}\underset{O}{\diagdown\!\diagup}N{-} + RCO_2H \qquad (24)$$

$$\rangle C{=}N{-} + RCO_3H \longrightarrow -\overset{|}{\underset{OOCOR}{C}}{-}NH{-} \longrightarrow -\overset{|}{C}\underset{O}{\diagdown\!\diagup}N{-} + RCO_2H \qquad (25)$$

(22)

The substituent effect on imines was explained by the one-step epoxidation-type mechanism (equation 24).[106] Detailed kinetic studies revealed that oxaziridines are formed by the Baeyer–Villiger-type mechanism (equation 25).[107] The unexpectedly large accelerating effect of alcohols was explained by the acid-catalyzed addition of PA to the imine double bonds.[107a] In the case of 3,4-dihydroisoquinolines, where the addition to C=N bond is not facile, nitrone formation (equation 26) occurred competitively with the formation of oxaziridines (equation 25).[107b]

$$\rangle C{=}\ddot{N}\langle + \text{H}\cdots\text{O=C(R)–O–O–H (cyclic)} \longrightarrow \rangle C{=}\underset{\downarrow O}{N}{-} + RCO_2H \qquad (26)$$

Ab initio MO calculations supported the two-step mechanism in equation 25.[108] Oxaziridine formation from various types of imines was not always stereospecific and the stereochemistry changed with the solvent.[109] These facts are explicable only by the two-step mechanism. The low-temperature (−78°C) oxidation of imines by monoperoxycamphoric acid [(+)-MPCA] afforded the optically active oxaziridine **23** with 60% e.e. maximum (equation 27).[103,110] The asymmetric induction reflects the relative stability of conformers in adduct **22**.

$$ArCH{=}NR \xrightarrow{(+)MPCA} \underset{H}{\overset{Ar}{}}C\underset{}{\overset{O}{\diagup\!\diagdown}}N\underset{R}{} \qquad (27)$$

(23)

1.3.6 Oxidation of sulfur compounds

Sulfides are easily oxidized by PA to sulfoxides and sulfones (equation 28). Sulfoxides are obtained by use of equimolar PA and sulfones are synthesized with a 1:2 molar ratio at higher temperatures.[111]

$$R_2S \xrightarrow{PA} R_2SO \xrightarrow{PA} R_2SO_2 \quad (28)$$

Since PA oxidations of sulfides are facile, the selective oxidation of sulfide groups is possible in the presence of C=C, C≡C or amino groups.[112] The lone-pair electrons of sulfides attack the peroxidic oxygen of PAs as exemplified in the positive ρ-value of $+1.58$ for substituted peroxybenzoic acids.[37d] Selective formation of sulfoxides results since sulfides are *ca* 1000 times more reactive than sulfoxides.[113] The PA oxidation of sulfoxides to sulfones has been shown also to involve nucleophilic attack of sulfoxide lone-pair electrons on PA from the observed substituent effect of $\rho = +0.75$ for $ArCO_3H$[37e] and $\rho = -1.06$ (σ^+) for Ar_2SO.[114] Although a two-step mechanism was suggested,[115] available kinetic data are well explained by the one-step mechanism.

Attempted asymmetric oxidations of sulfides with optically active PA (e.g. R^*CO_3H) were unsuccessful (i.e. $<10\%$ e.e.).[116] This is because the interaction of R* with sulfides is, as pictured in **24**, too weak to induce asymmetry in sulfoxides.

(24)

The PA oxidation of thiols yields disulfides under mild conditions,[117] and with excess amounts of PA sulfinic[118] and sulfonic acids[119] are obtained (equation 29).

$$RSH \xrightarrow{PA} RSSR \xrightarrow{PA} RSO_2H \xrightarrow{PA} RSO_3H \quad (29)$$

The oxidation of disulfides with equimolar PA yielded thiosulfinates (**25**).[120] The expected products with 2 equiv. of PA are α-disulfoxides (**26**), but the disulfoxides were unstable and readily disproportionated to yield various sulfur compounds.[121] Low-temperature NMR studies indicated the initial formation of α-disulfoxides (**26**), followed by the isomerization to thiolsulfonates (**28**) via thiolsulfinate intermediates (**27**) (equation 30).[122] The homolysis of α-disulfoxides is also likely[121,122d] and the decomposition of **26** seems to be changed by the reaction conditions, e.g. temperature, PA and solvent.

$$\underset{(25)}{RS(=O)SR} \xrightarrow{PA} \left[\underset{(26)}{RS(=O)-S(=O)R}\right] \longrightarrow \left[\underset{(27)}{RSOS(=O)R}\right] \longrightarrow \underset{(28)}{RS(=O)_2SR} \quad (30)$$

Thiocarbonyl compounds are desulfurized via sulfine intermediates (**29**) (equation 31). Substituent effects in thiocarbonyls again suggested the nucleophilic attack of the sulfur lone-pair electrons on PA.[123]

$$Ar_2C{=}S \xrightarrow{PA} \underset{(\mathbf{29})}{Ar_2C{=}S{=}O} \xrightarrow{PA} Ar_2C{=}O + SO_2 + S \qquad (31)$$

1.3.7 Oxidation of other nucleophiles

Trivalent phosphorus compounds are potent deoxygenating agents and reduce peroxides effectively.[124] The PA oxidation of phosphites and phosphines to phosphonates and phosphine oxides, respectively, is facile and rapid. An interesting reaction is the PA oxidation of diphosphenes where the P—P cleavage resulted via a phosphene oxide intermediate (**30**) (equation 32).[125]

$$ArP{=}PAr + RCO_3H \xrightarrow[CH_2Cl_2;\ 0^{\circ}C]{} \underset{(\mathbf{30})}{ArP{=}P(O)Ar} + RCO_2H$$

$$\longrightarrow ArPH{-}P(O)(OCOR)Ar \xrightarrow{H_2O} ArPH_2 + ArP(O)(OH)OCOR \qquad (32)$$

In the PA oxidation of strained cyclic phosphine oxides, a Baeyer–Villiger-type reaction proceeded, resulting in an oxygen insertion in P—C bonds.[126,127] For example, the PA oxidation of 7-phosphanorbornene (**31**) afforded, as shown in equation 33, oxaphosphorane (**33**) via the formation of adduct (**32**), which

$$(\mathbf{31}) \xrightarrow{ArCO_3H} (\mathbf{32}) \xrightarrow{-\ ArCO_2H} (\mathbf{33}) \qquad (33)$$

was much faster than the epoxidation of the C=C double bond.[127] Interestingly, the migrating group was the less hindered (i.e. less substituted) group, as exemplified in equation 34.[127]

$$\text{(4-membered cyclic phosphine oxide: Me, Me, Me; P(=O)Ph)} \xrightarrow{\text{MCPBA}} \text{(5-membered O–P(=O)Ph ring, Me, Me, Me)} + \text{(5-membered O–P(=O)Ph ring, Me, Me, Me)} \quad (34)$$

9 : 1

Selenides (R_2Se) are oxidized by PAs to yield, as with sulfides, monooxides and dioxides.[128] An interesting reaction is a Pummerer-type rearrangement for selenides with an electron-attracting group (equation 35).[129]

$$\text{PhSeCH}_2\text{R} \xrightarrow[\text{R = CN, CO}_2\text{Me}]{\text{R}'\text{CO}_3\text{H}} \underset{(34)}{\text{PhSeCHR(OCOR}')} \quad (35)$$

Substitutions of seleno and telluro groups occurred when the PA oxidations of selenides or tellurides were conducted in alcohols (equation 36).[130] This reaction is due to the enhanced departing ability by the oxidation of M. A similar substitution has been reported in the PA oxidation of thiol esters.[131]

$$\underset{(\text{M = Se, Te})}{\text{RMPh}} \xrightarrow[\text{R}'\text{OH}]{\text{PA}} \text{ROR}' \quad (36)$$

The PA oxidation of benzeneselenols (**35**) yields selenic acids (**36**) and then seleninic acids (**37**) (equation 37).[132] The PA oxidation is again the nucleophilic attack of lone-pair electrons of Se. The relative reactivity of selenenyl compounds decreases in the order ArSeH > ArSeOH > ArSeOSeAr > ArSeOEt > ArSeSeAr.[132b]

$$\underset{(35)}{\text{ArSeH}} \xrightarrow{\text{PA}} \underset{(36)}{\text{ArSeOH}} \xrightarrow{\text{PA}} \underset{(37)}{\text{ArSe(=O)OH}} \quad (37)$$

The PA oxidation of organosilicon compounds has been studied extensively. Trialkylsilanes are oxidized to silanols (equation 38).[133]

$$\text{R}_3\text{SiH} \xrightarrow{\text{PA}} \text{R}_3\text{SiOH} \quad (38)$$

The PA oxidation of disilanes is also facile, affording disiloxanes by the oxygen insertion in Si—Si bonds (equation 39).[134] The substituent effect on

$$R_3SiSiR_3 \xrightarrow{\text{PA}} R_3SiOSiR_3 \qquad (39)$$

$ArMe_2SiSiMe_2Ar$ was complex, the ρ-value being -0.3 (σ^+) for electron-donating and $+0.30$ (σ^+) for electron-attracting groups. The complex effect was explained by a 1,3-dipolar mechanism like equation 9,[134] but more detailed studies are needed for more conclusive evidence. Many kinetic studies on the PA oxidation of polysilanes have shown that the reactivity of Si—Si bonds is enhanced by substituents with electron-donating conjugation.[135] The relative reactivities of Si—Si and C=C bonds are of approximately the same order, whereas PA oxidation of allylsilanes resulted in selective epoxidation.[136]

Organotrialkoxy silanes (**38**, X = OR)[137] and trifluorosilanes (**38**, X = F)[138] are easily oxidized by PAs to yield substituted products (equation 40); thus, the SiX_3 groups are equivalent to OH group. The PA oxidation of silyl iodides (R_3SiI) afforded similarly silanols.[139]

$$\underset{(38)}{RSiX_3} \xrightarrow{\text{PA}} ROH \qquad (40)$$

1.3.8 Oxidation of halogen compounds

The reaction between PAs and halide ions (equations 41 and 42) has long been known, the formation of hypohalides being rate determining.[140] The reaction of PAs with iodide ion is especially fast and is well known in iodimetric titrations.

$$X^- + RCO_3H \longrightarrow HOX + RCO_2^- \qquad (41)$$

$$HOX + X^- \longrightarrow X_2 + HO^- \qquad (42)$$

It is interesting that halogens (X_2) are also oxidized further by PAs. For example, the aromatic iodination proceeds effectively with iodine and peroxyacetic acid according to equation 43.[141] The intermediacy of acetyl

$$ArH + I_2 + MeCO_3H \longrightarrow ArI + MeCO_2H \qquad (43)$$

hypoiodide (AcOI) has been proposed for the aromatic iodination. The bromination of arenes was similarly achieved with bromine and PA.[142] The oxidation of alkenes with iodine and PAs resulted in an effective haloacyloxylation (equation 44).[7b,143] The rate-determining step was shown to be the PA oxidation of the charge-transfer complex between alkenes and iodine.

$$2 \; \gt C{=}C\lt \; + I_2 + MeCO_3H \xrightarrow[MeCO_2H]{} 2 \; -\overset{|}{\underset{I}{\underset{|}{C}}}-\overset{|}{\underset{OCOMe}{\underset{|}{C}}}- \tag{44}$$

The use of primary alkyl iodides in place of iodine resulted in a similar acyloxylation.[144] With secondary and tertiary alkyl iodides, aromatic iodination was not observed and the iodides were oxidized to iodoacetates.[145] The final products were, as shown in equation 45, α-acetoxylated iodides via the oxidative elimination of hypoiodide.

$$\text{(cyclohexyl iodide)} + MeCO_3H \longrightarrow \text{(cyclohexene)} + HOI + MeCO_2H \longrightarrow \text{(2-iodocyclohexyl acetate; I, OCOMe)} \tag{45}$$

Since then, the oxidative elimination of alkyl iodides has been extensively studied.[146–148] The PA oxidation of primary alkyl iodides to yield alcohols was suggested to involve the rearrangement of iodoso compounds (**39**) to afford hypoiodites (**41**), as shown in equation 46.[146] The involvement of a cationic

$$RCH_2CH_2I \xrightarrow[CH_2Cl_2]{PA} \underset{(\mathbf{39})}{RCH_2CH_2I{=}O} \longrightarrow \underset{(\mathbf{40})}{RCH_2CH_2^+ \; IO^-} \longrightarrow \underset{(\mathbf{41})}{RCH_2CH_2OI} \longrightarrow \longrightarrow RCH_2CH_2OH \tag{46}$$

intermediate (**40**) was supported by the observed scrambling of deuterium atoms in the oxidative elimination of 2-phenylethyl-1,1-d_2 iodide. Similar Wagner–Meerwein-type rearrangements have been observed in many cases.[147,148] At present, it is not possible to conclude that the precursor for the carbenium ions is the iodoso compound **39** or diacetoxy iodides.[149]

The reaction of PAs with potassium bromide in the presence of crown ethers is another method for aromatic brominations[150] and the haloacyloxylation of alkenes (equation 47).[151] The KBr–PA–crown ether system could be an efficient means for the production of acyl bromides.

$$\mathrm{ArCO_3H + KBr \xrightarrow{\text{18-crown-6}} ArC(=O)Br \xrightarrow{>C=C<} -C(Br)-C(OCOAr)-} \quad (47)$$

1.4 Nucleophilic Reactions of Peroxy Acids

1.4.1 α-Nucleophiles

Peroxy anions (ROO^-) are α-nucleophiles[12] and their nucleophilic reactivities are much higher than those expected from their pk_a values.[12,152] The efficient syntheses of PAs and diacyl peroxides under alkaline conditions (equation 48) are based on the high nucleophilicity of HOO^- and RCO_3^-.

$$\mathrm{RC(=O)Cl \xrightarrow{HOO^-} RC(=O)OO^- \xrightarrow{RCOCl} RC(=O)OOC(=O)R} \quad (48)$$

The acyl transfer reaction of peroxybenzoate ions with *p*-nitrophenyl acetate (equation 49) was facile with second-order rate constants (k_2) of

$$\mathrm{ArCO_3^- + MeC(=O)O\text{-}C_6H_4\text{-}NO_2 \xrightarrow{k_2} ArC(=O)OOC(=O)Me + {}^-O\text{-}C_6H_4\text{-}NO_2} \quad (49)$$

40–90 l mol^{-1} s^{-1} (25°C, pH 10).[26] The correlation of log k_2 vs the pK_a value of the PAs was linear, affording a Brønsted slope of 0.38 including hydroperoxide ion (HOO^-, pK_a = 11.6). Thus, the α-effect of $ArCO_3^-$ was shown to be 10^2–10^3.[26] A micellar effect was studied in the nucleophilic reaction of MCPBA anion with *p*-nitrophenyl phosphate.[153]

1.4.2 Alkaline epoxidation and oxidation

The reaction of hydroperoxide ions with electron-deficient alkenes (**42**, Y = COR, CN, etc.) is known as alkaline or nucleophilic epoxidation.[154] In the case of alkaline hydrogen peroxide (X = OH), the equilibria between the Michael adducts **43a** and **43b** are established, and *trans*-epoxides are obtained stereoselectively via the intramolecular cyclization of the more stable conformer (**43b**) (equations 50a and 50b).[154]

$$R^1(H)C{=}C(R^2)Y \ (\mathbf{42a}) \xrightarrow{XO^-} R^1(H)(XO)C{-}C(R^2)(Y)^- \ (\mathbf{43a}) \xrightarrow{-X^-} R^1(H)\overset{O}{C{-}C}(R^2)Y \qquad (50a)$$

$$\mathbf{43a} \rightleftarrows \mathbf{43b}$$

$$H(R^1)C{=}C(R^2)Y \ (\mathbf{42b}) \xrightarrow{XO^-} H(R^1)(XO)C{-}C(R^2)(Y)^- \ (\mathbf{43b}) \xrightarrow{-X^-} H(R^1)\overset{O}{C{-}C}(R^2)Y \qquad (50b)$$

Stereospecific epoxidations should result if the cyclization of **43** is faster than the isomerization between **43a** and **43b**. Such cases have been found when hypochloride (X = Cl)[155] and PA anion (X = RCO)[156] were applied, the stereospecificity being higher for the former. A detailed theoretical study of these nucleophilic epoxidation has been reported.[157] The results were such that the stereospecificity becomes higher (a) with increasing nucleofugacity (i.e. leaving group ability) or X, (b) with increasing hyperconjugation ability of COX bonds, and (c) with decreasing electron-attracting ability of Y.

Whereas sulfides do not react with PA anions, sulfoxides are oxidized facilely. The oxidation of sulfoxides with PA anions is much faster than that with the corresponding PAs.[37e] The positive ρ-value of $+0.69$ for substituted sulfoxides[37f] suggests the nucleophilic attack of PA anion on sulfoxides, followed by the oxidative decomposition of the adduct (**44**) (equation 51). The positive ρ-value of $+0.57$ for substituted peroxybenzoic acids[37e] demonstrated clearly that the decomposition of **44** is surely rate determining.

$$ArCO_3^- + R_2S{=}O \rightleftarrows ArC(=O){-}O{-}O{-}S(R)_2{-}O^- \ (\mathbf{44}) \xrightarrow{\text{slow}} ArCO_2^- + R_2SO_2 \qquad (51)$$

1.4.3 Alkaline decomposition

The alkaline decomposition of PAs has long been known to be facile.[158] A detailed study of the decomposition of peroxybenzoic acids revealed that the decomposition is fastest at their half-dissociation, yielding oxygen and benzoic acid (over 95%) together with a trace amount of benzoyl peroxide (equation 52).[24]

$$RCO_3H + RCO_3^- \longrightarrow RCO_2H + RCO_2^- + O_2 \tag{52}$$

$$v = k_2[RCO_3H][RCO_3^-] \tag{53}$$

For the case of aliphatic PAs, the decomposition likewise had the maximum rate at the half-dissociations satisfying the rate equation 53. Two competitive mechanisms for the PA decomposition have been proposed from ^{18}O-tracer studies.[159] One is the retention mechanism via a C=O adduct (**45**) to yield $^*O—^*O$, and the other is the nucleophilic attack on the peroxy oxygen atom (**46**) resulting in the formation of scrambled $O—O^*$. The ^{18}O-tracer studies indicated that retention pathways by **45** were 83%, 26% and 24% for R = Me, $C_6H_4(o\text{-}CO_2^-)$ and *t*-Bu, respectively.[159]

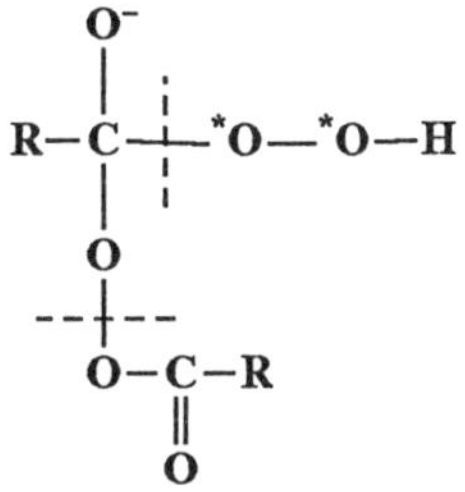

O⁻
R—C—*O—*O—H
O
O—C—R
O

(45)

O
C
R *O—*O H
⁻O—O—C—R
O

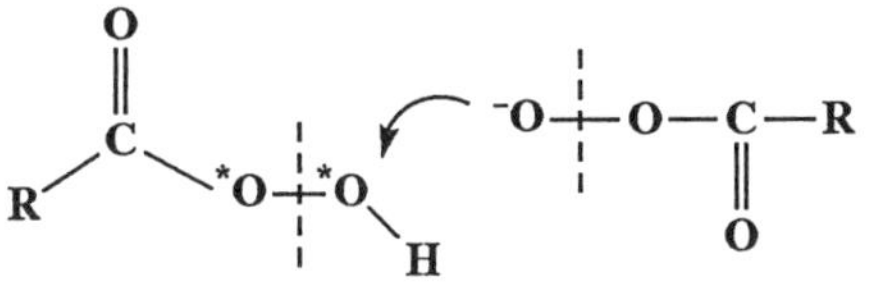

(46)

Another retentive pathway also seems to be possible since the reaction between hydrogen peroxide and PA yields oxygen gas with a retention mechanism.[160] Hydrogen peroxide may be produced by the alkaline hydrolysis of PAs. At any rate, the mechanism for the alkaline decomposition of PAs is complex. One difficulty in these studies is due to the accompanying decomposition induced by contaminated metallic ions.

1.5 Radical and Catalyzed Reactions of Peroxy Acids

1.5.1 Radical reactions

The O—O bonds in peroxides are weak and easily cleaved by heat and light. The gas-phase decomposition kinetics (130–240°C) of aliphatic PAs, RCO_3H (R = Me, Et, and Bu), afford an O—O bond energy of 138 kJ mol^{-1} (33 kcal mol^{-1}).[161] This value is approximately the same as those of diacyl peroxides.

Thus, any heating of PAs above 80°C starts the O—O homolysis to yield radicals according to equation 54.

$$\text{RC(=O)OOH} \xrightarrow{\Delta \text{ or } h\nu} \text{RC(=O)O}\cdot + \cdot\text{OH} \tag{54}$$

The photolysis of peroxides is highly effective in yielding radicals at any temperatures. For example, the irradiation of peroxyacetic acid in cyclohexane affords cyclohexanol in 90% yield (equation 55).[162] The formation of methane

$$CH_3CO_3H \xrightarrow[\text{cyclohexane}]{h\nu} CH_4 + CO_2 + \text{cyclohexyl-OH} \tag{55}$$

strongly suggests hydrogen atom abstraction by methyl radicals. Photolysis in toluene or xylenes yielded complex mixtures of products resulting from ring methylation and side-chain hydroxylation.[163,164] An interesting phenomenon has been reported that the irradiation of mixed crystals of peroxydicarboxylic acids yields highly stable carbon radicals at room temperature.[165]

Leffort *et al.*[166] systematically studied the radical decomposition of PAs. Various radicals were produced by the thermolysis of several PAs and their reactivities with PAs were examined. The reaction scheme is outlined in equations 56a–e.

$$RCO_3H \longrightarrow RCO_2\cdot + \cdot OH \tag{56a}$$

$$RCO_2\cdot \longrightarrow R\cdot + CO_2 \tag{56b}$$

$$R\cdot \text{ (or } S\cdot) + \text{H–O–OCOR} \longrightarrow \text{ROH (or SOH)} + RCO_2\cdot \tag{56c}$$

$$R\cdot + H\text{-}S \longrightarrow R\text{-}H + S\cdot \tag{56d}$$

$$X\cdot + H\text{–}OOCOR \longrightarrow X\text{-}H + RCO_3\cdot \tag{56e}$$

The decarboxylation of aliphatic carboxy radicals (equation 56b) is very fast. Most important here is the radical substitution on the peroxy oxygen of PAs (i.e. an S_H2 reaction on PAs) as shown in equation 56c. It is interesting that more nucleophilic radicals react faster with the peroxy oxygen of PAs, i.e. the relative reactivity for equation 56c being in the decreasing order tertiary > secondary > primary carbon radicals. For example, relative reactivities are shown in equation 57 ($R = n\text{-}C_{10}H_{21}$).[166] These relative reactivities of radicals with PAs are correlated with their ionization potentials and are explained by the important polar effect pictured in **47**.

$$\underset{100}{R\dot{C}Me_2} > \underset{11}{R\dot{C}HMe} > \underset{0.4}{R\dot{C}HCl} > \underset{0.016}{R\dot{C}H_2} \qquad (57)$$

$$\left[\overset{\delta^+}{R\cdot}\cdots\cdots O(H)\cdots\cdots \overset{\delta^-}{O}-C(=O)-R'\right]$$

(47)

On the other hand, the predominant reaction of electrophilic radicals ($HO\cdot$, $ROO\cdot$, $CH_3\cdot$, etc.) is hydrogen atom abstraction from substrates (HS in equation 56d) or from a PA (equation 56e). The consequence of these electrophilic radicals is hydroxylation of solvents according to equation 56c.[167,168] The reaction of various types of radicals with PAs is rationalized by the schematic equations 58a and 58b. This rationalization is

$$RC(=O)O{-}O{-}H + \cdot R_E \longrightarrow R_EH + RCO_3\cdot \qquad (58a)$$

$$RC(=O)O{-}O{-}H + \cdot R_{Nu} \longrightarrow R_{Nu}OH + RCO_2\cdot \qquad (58b)$$

reasonable and applicable as a practical method for the hydroxylation of C—H bonds. It should be noted that the radical hydroxylation of C—H bonds affords, in contrast to the previously mentioned case of non-radical hydroxylation (Section 1.3.3), racemized alcohols.[169]

1.5.2 Catalyzed reactions

The O—O bonds in PAs are easily cleaved by one-electron reductions. For example, the Ti(III) reduction of aromatic PAs $ArCO_3H$ yielded $ArCO_2\cdot$ radicals as studied by ESR.[170] In the Fenton-type reaction of PAs with Fe(II), isopropyl alcohol was oxidized to acetone.[171]

Some catalytic oxidations utilizing the facile PA oxidation of Co(II) and Cr(IV) to Co(III) and Cr(VI), respectively, have been reported. Thus, the $Co(OAc)_3$-catalyzed PA oxidation of alcohols is effective for the conversion of secondary alcohols into ketones.[172] An interesting approach is the PA oxidation of secondary alcohols to ketones catalyzed by a Cr(VI) ester (**48**) (equation 59).[173]

(48) + R_2CHOH → Cr(=O)(OH)($OCHR_2$) ester → Cr(OH)$_2$ ester + $R_2C{=}O$; $MeCO_3H$ regenerates (48) (59)

As model reactions for cytochrome P-450 monooxygenases, catalyzed oxidation with metalloporphyrins and PAs have been extensively studied.[174] These reactions are reviewed in Chapter 11.1.

2 PEROXY ESTERS

The synthesis and reactions of peroxy esters have been reviewed extensively.[1,4,175] In the following, important aspects of the chemistry of peroxy esters and recent results are summarized.

2.1 Synthesis of Peroxy Esters

The most important synthetic method for peroxy esters is the reaction of acyl chlorides with tertiary hydroperoxides in the presence of pyridine (equation 60).

$$RCOCl + R'OOH \xrightarrow{\text{pyridine}} RC(=O)OOR' \quad (60)$$

The reaction is conducted in solvents such as diethyl ether or dichloromethane at low temperatures by the final addition of hydroperoxides[176] or pyridine.[177]

The reaction of acid chlorides with sodium *tert*-butyl peroxide at lower temperatures was applied for the preparation of unstable peroxy esters.[178] The use of aqueous alkaline solutions is suitable for large-scale preparations.[179] Peroxyformates could be synthesized by the uncatalyzed reaction.[180]

Peroxycarbonates (**49**) are prepared similarly from the base-catalyzed reactions of chlorocarbonates (equation 61).[179] Peroxy carbamates (**50**) are

$$\mathrm{RO\overset{\overset{O}{\|}}{C}Cl + R'OOH \xrightarrow{\text{base}} RO\overset{\overset{O}{\|}}{C}OOR'} \quad (61)$$

(49)

conveniently obtained from isocyanates under neutral conditions (equation 62).[181] A similar reaction of ketenes is an alternative for the preparation of peroxy esters under neutral conditions.[182]

$$\mathrm{RN{=}C{=}O + R'OH \longrightarrow RNH\overset{\overset{O}{\|}}{C}OOR'} \quad (62)$$

(50)

2.2 Physical Properties of Peroxy Esters

Peroxy esters are relatively stable and easier to treat in comparison with peroxy acids and diacyl peroxides. In Table 3 are listed the boiling and melting points of some peroxy esters. Comprehensive tabulations are available elsewhere.[185]

The boiling points of peroxy esters are slightly higher, owing to their higher molecular weights, than those of parent esters. As indicated in Table 3, lower

Table 3. Properties of peroxy esters

RCO_3R'					
R	R′	Starting materials	M.p. (°C)	B.p. (°C/mmHg)	Ref.
H	*t*-Bu	RCO_2H + *t*-BuOOH		42/24	180
Me	*t*-Bu	RCOCl + *t*-BuOOH		22/1	176
CF_3	*t*-Bu	RCO_2COR + *t*-BuOOH		21/22	176
Ph	*t*-Bu	RCOCl + *t*-BuOOH	8	75–77/22	183
n-$C_{11}H_{23}$	*t*-Bu	RCOCl + *t*-BuOOH	8.0–8.6		177
Ph_2CH	*t*-Bu	RCOCl + *t*-BuOOH	58–60		184a
Ph_3C	*t*-Bu	RCOCl + *t*-BuOONa	61–64		184b
PhNH	*t*-Bu	RNCO + *t*-BuOOH	83		181
EtO	*t*-Bu	RCOCl + *t*-BuOOH		30/0.8	179

peroxy esters were purified by distillation under reduced pressures; such distillations should be done with appropriate precautions behind safety shields. Solid peroxy esters were purified by recrystallization. The purity of peroxy esters may be checked by observing the characteristic C=O stretching at 1780–1850 cm^{-1} in the IR spectra. Their purity is finally determined by iodimetric titration.

Theoretical calculations on the structures of peroxy esters are relatively few in comparison with those of peroxy acids. EHMO calculations predicted that the *cis* conformer **51a** is more stable than **51b**.[186] The predominance of the *cis*

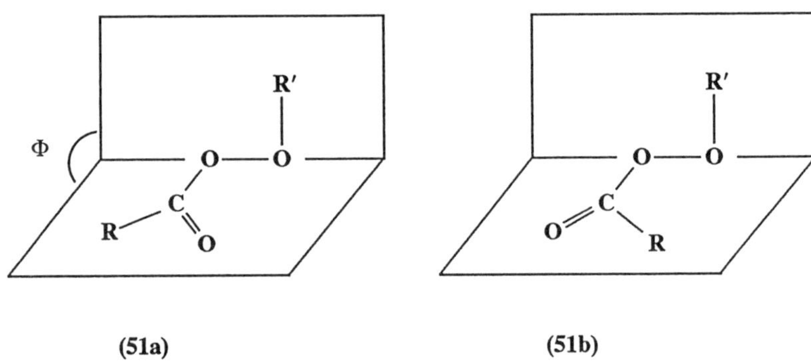

(51a) **(51b)**

structure was supported by measurement of the dipole moments of *tert*-butyl peroxy esters.[187] From the detailed dipole moment analysis of peroxy esters (R′ = Me, *t*-Bu) in benzene, the dihedral angles (ϕ) were estimated to be 110–150° and 160–180° for R′ = Me and *t*-Bu, respectively.[188] In the crystal state, the dihedral angles of *tert*-butyl peroxynitrobenzoates were 129° and 120° for *m*- and *p*-nitro isomers, respectively.[189]

The polalographic reduction data ($E_{1/2}$) of peroxy esters are in the range -0.8 to -1.0 V vs SCE,[190] which are significantly lower than those of peroxy acids ($E_{1/2} \approx 0.0$ V). The lower values for peroxy esters suggest a lower reactivity as oxidants, which is exemplified by the slow reaction rates with iodide ion in iodometric titrations. The substituent effect on the ^{13}C NMR shifts of peroxy esters has been reported.[191]

2.3 Reactions with Nucleophiles

Excellent reviews have been published on the decomposition[175,192] and the catalyzed reactions of peroxy esters.[193] In the following sections several types of reactions of peroxy esters are reviewed.

The electrophilic reactivity of peroxy esters is low and hence they can react only with strong nucleophiles. For example, phosphines deoxygenate *tert*-butyl peroxybenzoates to yield phosphine oxides and the corresponding esters.[194]

Trivalent phosphines attack nucleophilically since the oxygen transfer was accelerated by electron-attracting groups on the peroxy esters ($\rho = +1.24$).[195] An ^{18}O tracer study indicated attack on the peroxy ether oxygen, as shown in equation 63.

$$ArC(=O^*)OOBu^t + :PPh_3 \longrightarrow ArC(=O^*)OBu^t + Ph_3P{=}O \qquad (63)$$

tert-Butyl ethers can be conveniently prepared by the reaction of Grignard reagents with *tert*-butyl peroxybenzoates (equation 64), e.g. R = Ph (80%), Et (77%) and cyclohexyl (74%).[196]

$$PhC(=O)OOBu^t + RMgX \xrightarrow[\text{ether}]{} ROBu^t + PhCO_2MgX \qquad (64)$$

The efficient alkoxylation of the carbanionic reagents may be pictured as in **52**. The use of aryllithium gave similar results.

R, C, O, O—Bu^t, O------Mg——R, X

(52)

The alkoxylation of amines is also possible (equation 65), with lower yields of 20–40%.[197] The accompanying formation of amides (PhCONHR) suggests the

$$PhC(=O)OOBu^t + RNHLi \longrightarrow PhNHOBu^t \qquad (65)$$

nucleophilic attack of amide anion on the acyl carbon as a competitive reaction. In contrast, hydroxide ion attacks only the acyl carbon, resulting in hydrolysis of the peroxy ester via a tetrahedral intermediate (**53**) (equation 66).[198]

$$PhC(=O)OOBu^t + OH^- \rightleftharpoons Ph{-}C(O^-)(OH){-}OOBu^t \longrightarrow PhCO_2^- + t\text{-}BuOOH \qquad (66)$$

(53)

In summary, the reactions of nucleophiles with peroxy esters occur either on the peroxy oxygen (B in **54**) or on the acyl carbon (A). Carbanionic reagents react on B, amide ions on A and B and hydroxide ion solely on A That is, soft nucleophiles attack on B and hard nucleophiles on A. Reactions on C or D are not known, in contrast to the case with peroxy acids (R′ = H), which react mostly on C.

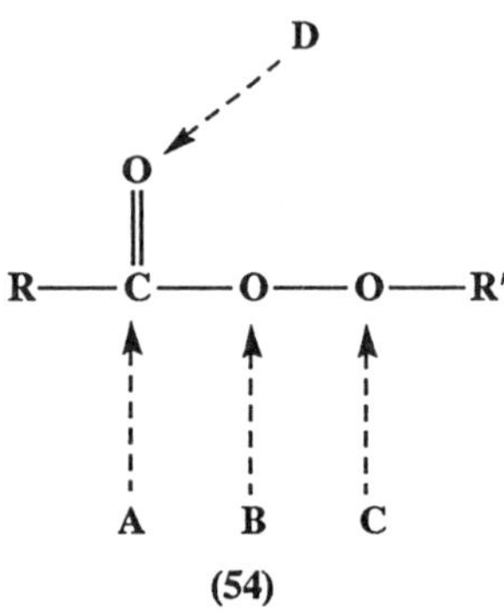

(54)

2.4 Ionic Rearrangements of Peroxy Esters

trans-9-Decalyl peroxybenzoate (**55**) yields **56** thermally by carbon to oxygen migration (equation 67), which is termed the Criegee rearrangement.[199] This type of rearrangement proceeds via a concerted ionic mechanism as deduced from (a) the accelerating effect of electron-attracting groups on Ar,[200a] (b) the faster reaction in polar solvents,[200b] and (c) the lack of scrambling of carbonyl carbons in the products.[200c]

OOCOAr (55) → OCOAr (56) —H_2O→ OH, O (67)

More detailed studies have been carried out. The reaction (equation 68) of cumyl peroxybenzoates (**57**, R = phenyl) was accelerated by electron-attracting

$Me_2C(R)$—O—O—C(=O)Ar (57) → [Me, Me, C, R, O, δ^+, δ^- OCOAr] (58) → Me_2C(OR)(OCOAr) (59) (68)

groups on Ar ($\rho = +1.21$)[201a] and by electron-donating groups on cumyls ($\rho = -5.1$).[201b] Hence the rearrangement is pictured as proceeding via a polar transition state (**58**). The migrating aptitudes of the R groups are shown in equation 69.[202] It is apparent that the nucleophilic 1,2-rearrangement becomes

$$\begin{array}{ccccccccccc} \text{Me} & < & \text{Et} & < & \text{PhCH}_2 & < & \text{i-Pr} & < & \text{Ph} & < & \text{t-Bu} \\ 1 & & 45 & & 1630 & & 2940 & & 11800 & & 22800 \end{array} \quad (69)$$

faster with increasing stability of carbenium ions. The 2-norbornyl group shifted more slowly than *i*-Pr.[203] Interestingly, stable carbonium ions may be formed as a product, as exemplified in equation 70.[204]

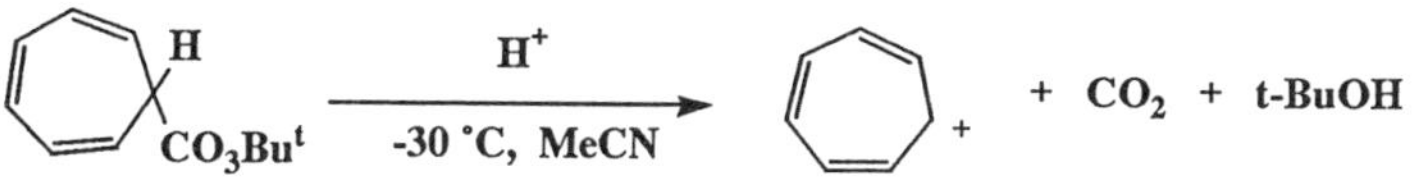

$$\text{C}_7\text{H}_7(\text{H})\text{CO}_3\text{Bu}^t \xrightarrow[\text{-30 °C, MeCN}]{\text{H}^+} \text{C}_7\text{H}_7^+ \ (70\%) + \text{CO}_2 + \text{t-BuOH} \quad (70)$$

A base-catalyzed concerted decomposition has been reported with peroxyformates.[205] The reaction was accelerated by increasing strength of the base and the k_H/k_D for the formate hydrogen was 4.1 (pyridine-catalyzed, 90°C). The reaction scheme shown in equation 71 was reasonably proposed.

$$\text{B:} + \text{H–C(=O)–O–OBu}^t \longrightarrow \text{B:} + \text{CO}_2 + \text{HOBu}^t \quad (71)$$

In contrast, the use of a strong base such as *t*-BuOK resulted in another type of reaction. Thus, α-alkoxyacetic acids (**61**) were obtained via α-carbanion intermediates (**60**), the yields being 86% with $R^1 = R^2 = H$ and 39% with $R^1 = H$, $R^2 = Me$ (equation 72).[206]

$$\text{R}^1\text{R}^2\text{CHC(=O)OOBu}^t \xrightarrow{\text{t-BuO}^-} \underset{(60)}{\text{R}^1\text{R}^2\overset{-}{\text{C}}\text{C(=O)OOBu}^t} \longrightarrow \underset{(61)}{\text{R}^1\text{R}^2\text{C(OBu}^t)\text{CO}_2\text{H}} \quad (72)$$

In the acid-catalyzed reaction of peroxyacetates (**62**), a ring expansion was observed, leading to 4-oxa-pent-2-enones (equation 73).[207]

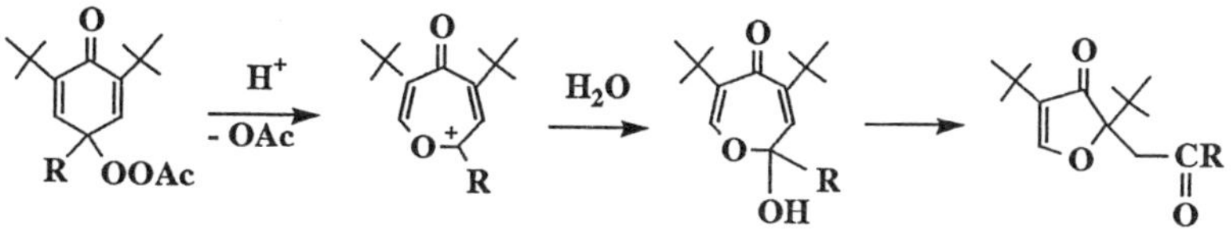

$$\underset{(62)}{\text{R, OOAc}} \xrightarrow[\text{- OAc}]{\text{H}^+} \text{O}^+\text{–R} \xrightarrow{\text{H}_2\text{O}} \text{O–C(OH)R} \longrightarrow \text{O–CH}_2\text{C(=O)R} \quad (73)$$

2.5 Radical Decomposition of Peroxy Esters

2.5.1 One- and two-bond homolysis

Peroxy esters are important and useful as radical initiators in polymerizations, and hence numerous studies and patents on their homolyses have been published. Radical-forming decompositions of peroxy esters may be classified in several types, as shown in equations 74a–d, that is, (a) one-bond O—O homolysis, (b) two-bond concerted homolysis, (c) radical-induced decompositions and (d) metal-catalyzed decompositions. The last is discussed in Section 2.6.

$$RC(=O)OOR' \longrightarrow RC(=O)O\cdot + \cdot OR' \qquad (74a)$$

$$RC(=O)OOR' \longrightarrow R\cdot + CO_2 + \cdot OR' \qquad (74b)$$

$$RC(=O)OOR' \xrightarrow{S\cdot} RC(=O)OS + \cdot OR' \qquad (74c)$$

$$RC(=O)OOR' \xrightarrow{M^{n+}} M^{(n+1)+} + RCO_2^- + \cdot OR' \qquad (74d)$$

Table 4 lists the half-lives and activation parameters for the thermolysis of various peroxy esters. It is apparent that the homolysis rates change with the structures of R in RCO_3Bu^t. Bartlett and Hiatt[176] noted that peroxy esters more stable than benzoyl peroxide decompose via one-bond homolysis according to

Table 4. Kinetic data for the thermolysis of *tert*-butyl peroxy esters

R in RCO_3Bu^t	Solvent	Half-life (60°C) (min)	$\Delta H^\ddagger$ kJ mol^{-1}	$\Delta H^\ddagger$ kcal mol^{-1}	Ref.
Me	PhCl	5×10^5	159	38.0	176
Ph	MeC_6H_4Cl	3×10^4	142	34.0	176
(Benzoyl peroxide)	PhH	6000	137	32.7	176
$PhCH_2$	PhCl	1700	120	28.7	176
t-Bu	PhCl	300	128	30.6	176
$PhOCH_2$	PhH	80	109	26.0	208
Ph_2CH	PhCl	26	102	24.4	176
CO_3Bu^t	*i*-PrPh	6.8	107	25.6	209
Ph_3C	*i*-PrPh	1	101	24.1	184b

equation 75a. The other peroxy esters which decompose faster undergo two-bond concerted homolysis (equation 75b).

$$RCOOBu^t \longrightarrow RC(=O){-}O\cdots\cdots OBu^t \ (\mathbf{63}) \longrightarrow RCO\cdot + t\text{-}BuO\cdot \qquad (75a)$$

$$RCOOBu^t \longrightarrow R\cdots\cdots C(=O){-}O\cdots\cdots OBu^t \ (\mathbf{64}) \longrightarrow R\cdot + CO_2 + t\text{-}BuO\cdot \qquad (75b)$$

The thermolysis of *tert*-butyl peroxybenzoates was accelerated by electron-donating groups, as shown by the Hammett ρ-values of -0.7 (σ)[183b] and -0.7 ($\gamma = 0.3$).[210] Since the peroxy esters decompose by one-bond fission (**63**), the negative ρ-values suggest that O—O bonds become weaker by substituting electron-donating groups. The γ-value of 0.3 in the modified Hammett correlation was explained by the resonance stabilization of benzoyloxy radicals.[210]

When R = $PhCH_2$ or *t*-Bu, the thermolyses of RCO_3Bu^t were facile and the resulting $\Delta H^{\ddagger}$ values were substantially lowered (Table 4). These facile thermolyses reflect the importance of the stability of the radicals R· in the concerted O—O and C—C homolyses as shown in **64**. The large variation in the relative rates for Me (1), *i*-Pr (50), Cl_3C (515) and *t*-Bu (1660) is well understood as a consequence of relative stabilities of radicals (R·). The observed ρ-value of -1.04 (σ^+, 100°C) for $ArCH_2CO_3Bu^t$ indicates clearly the importance of the polar effect (**65b**) in the transition state.[211]

$$\overset{\bullet}{Ar}CH_2\cdots\cdots C(=O)\cdots O\cdots\overset{\bullet}{O}Bu^t \ (\mathbf{65a}) \longleftrightarrow Ar\overset{+}{C}H_2\cdots\cdots C(=O)\cdots O\cdots\cdots {}^{-}OBu^t \ (\mathbf{65b})$$

The large acceleration by an α-alkoxy group (e.g. R = $R'OCH_2$) is well explained by a transition state like **65**, i.e. the large stabilizing effect of radicals (**65a**) and cations (**65b**) by α-oxygen atoms.[208,212] Steric acceleration by bulky groups could not be ascertained.[213] In the thermolysis of twelve bridgehead peroxy esters, the important factor was the polar rather than the steric acceleration.[214] The mechanism of peroxy ester homolysis was analyzed by the least-motion model.[215] The thermolysis of *N*-phenyl peroxycarbamates (R = PhNH) resulted in a similar substituent effect of $\rho = -2.2$ (σ^+),[216] suggesting a transition state like **65** (i.e. ArNH in place of $ArCH_2$).[69]

The thermolysis of peroxyoxalates (**66**) has been studied extensively. Two-bond homolysis (**67**) was concluded from the low activation enthalpies ($\Delta H^{\ddagger}$ = 110–120 kJ mol^{-1} or 26–28 kcal mol^{-1}) and the relative rates of *t*-Bu > *t*-BuO > Et > $PhCH_2$ for R in **66**.[209,217] The formation of dialkyl carbonates was additional evidence for the concerted homolysis as in **67**. The slower rates for **67** with R = *t*-BuO precluded the expected three-bond cleavage of O—O, C—C and C—O bonds.[218]

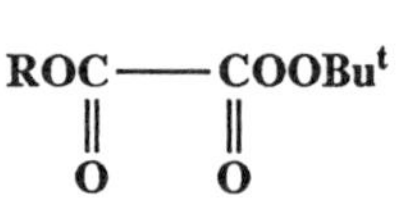

ROC(=O)—C(=O)OOBu^t (**66**) ROC(=O) ┆ C(=O)—O ┆ OBu^t (**67**) ROC(=O)OOBu^t (**68**)

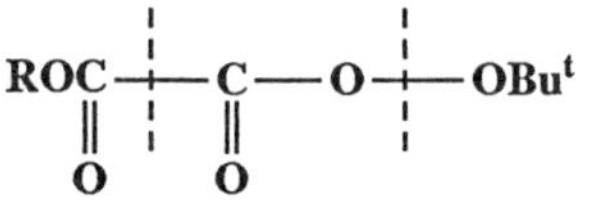

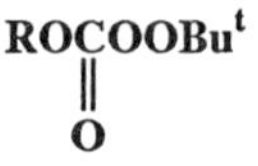

Peroxycarbonates (**68**) with R = *t*-Bu[219] and 2,3-epoxypropyl[220] decomposed slowly via one-bond O—O homolysis. Di-*tert*-butyl diperoxycarbonate (**68**, R = *t*-BuO) decomposed similarly by one-bond cleavage.[221]

2.5.2 Photolysis

Irradiation of peroxides is another routine method leading to their O—O homolysis, as previously reviewed.[222] The photolysis is certainly of the one-bond fission type for any kind of peroxy esters. The initially formed acyloxy radicals are decarboxylated very readily, affording high yields of alkyl radicals and carbon dioxide, as shown in equation 76.[223] For example, ethers were

$$\mathrm{RC(=O)OOBu^t} \xrightarrow{h\nu} \left[\mathrm{RC(=O)O\cdot} + \cdot\mathrm{OBu^t}\right] \longrightarrow \left[\mathrm{R\cdot} + \mathrm{CO_2} + \cdot\mathrm{OBu^t}\right] \longrightarrow \longrightarrow \mathrm{CO_2} + \mathrm{ROBu^t} + \mathrm{RH} + t\text{-}\mathrm{BuOH} \quad (76)$$

obtained as cage products in the photolysis of cyclopropyl peroxyacetate (**69**), as shown in equation 77.[223] The photolysis of related peroxy esters was suitable for examining the isomerization of cyclopropyl carbinyl radicals.

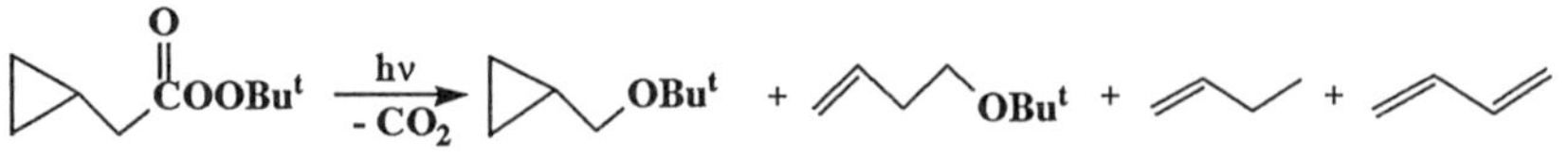

$$\text{c-C}_3\text{H}_5\text{CH}_2\text{C(=O)OOBu}^t\ (\mathbf{69}) \xrightarrow[-\mathrm{CO_2}]{h\nu} \text{c-C}_3\text{H}_5\text{CH}_2\text{OBu}^t\ (33\%) + \mathrm{CH_2{=}CHCH_2CH_2OBu^t}\ (45\%) + \mathrm{CH_2{=}CHCH_2CH_3}\ (20\%) + \mathrm{CH_2{=}CHCH{=}CH_2}\ (\text{trace}) \quad (77)$$

It is interesting that the photolysis of *tert*-butyl peroxyformate yields hydrogen radical (H$\cdot$).[224] Another intriguing reaction is the photochemical decomposition of β-peroxylactones.[225]

The absorbance of aromatic peroxy esters is much stronger than that of aliphatic peroxy esters. The peroxy esters with benzoyl groups absorb light strongly and are photolyzed efficiently,[226] e.g. as in equation 78.

$$ArC(=O)-C_6H_4-C(=O)OOBu^t \xrightarrow{h\nu} ArC(=O^{*3})-C_6H_4-C(=O)OOBu^t$$

$$\longrightarrow ArC(=O)-C_6H_4-C(=O)O\cdot + t\text{-}BuO\cdot \quad (78)$$

Since the photolysis of peroxides proceeds efficiently at any temperature, their application in polymerizations is highly promising.

2.5.3 Anchimerically assisted homolysis

Ortho substituents such as alkylthio or iodo groups in *tert*-butyl peroxybenzoates accelerated the homolysis rates dramatically. For example, the rate order for several *ortho* substituents is shown in equation 79, the figures in parentheses being relative rates.[227]

$$\begin{array}{ccccccccccc} MeSO_2 & < & H & < & t\text{-}Bu & < & I & < & Ph_2C{=}CH & < & PhS \\ (0.4) & & (0.1) & & (2.4) & & (81) & & (150) & & (65000) \end{array} \quad (79)$$

It is apparent that steric acceleration is not operative since the rates of bulky $MeSO_2$ or *t*-Bu groups are low. As shown, the *o*-thiophenoxy group accelerated the homolysis more than 60 000 times and lowered the $\Delta H^{\ddagger}$ value to 23 kcal mol^{-1}.[227] The homolysis of peroxybenzoates with *o*-thiophenoxyls was accelerated by electron-donating groups [$\rho = -1.3$ (σ)] and by polar solvents.[228] These results seem to indicate an ionic decomposition, but the major reaction was shown, by using radical scavengers, to be the radical-forming homolysis. Thus, the anchimerically induced homolysis could be explained by the polar effect as pictured as **70b** or **70c** in equation 80.

$$o\text{-}X\text{-}C_6H_4\text{-}C(=O)O\text{-}OR \longrightarrow \left[\underset{(70a)}{\dot{X}\cdots O \ \cdot OR} \longleftrightarrow \underset{(70b)}{X^{+}\cdots O \ \ \bar{O}R} \longleftrightarrow \underset{(70c)}{X^{+\cdot} \ \ O^{-} \ \cdot OR} \right] \quad (80)$$

The sulfuranyl radicals **70a** (X = RS) have, in fact, been detected in photolysis at low temperature (−85°C).[229] The radicals were shown by ESR analysis to be

9-S-3 (nine-electron, three coordinated) sulfur radicals. Based on a CIDNP study, the reaction of the singlet radical pair **71** has been proposed to proceed as shown in equation 81.[230]

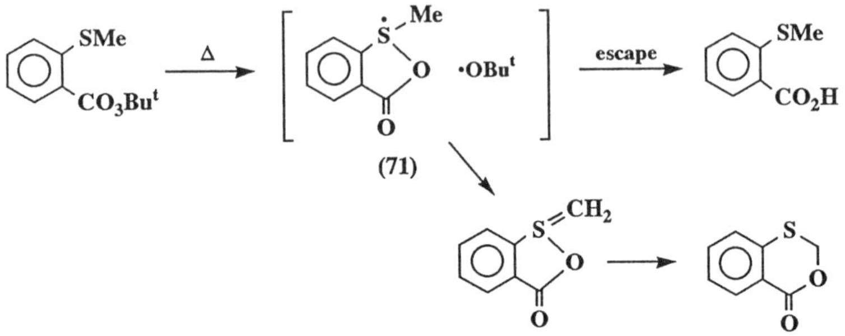

(81)

2.5.4 Molecule-induced homolysis

The homolysis of peroxy esters is also induced intermolecularly by sulfides. For example, in the reaction of peroxybenzoates with sulfides, the molecule-induced homolysis as in equation 82 was proposed from the rates of radical formation and the substituent effect on sulfides [$\rho = +1.34$ (σ) for thioanisoles].[231] A pertinent point here is that the peroxy ester decomposition is induced by the one-electron transfer (ET) from sulfides, suggesting the importance of **72** in equation 82. It is interesting to recall that the one-electron reduction of peroxides leads to a facile O—O cleavage (Section 2.2).

$$PhCO_3Bu^t + Me_2S \xrightarrow{ET} [Me_2S^{+\cdot}\ PhCO_2^-\ \cdot OBu^t] \longrightarrow$$

(72)

$$[Me_2\dot{S}OCOPh\ \ t\text{-}BuO\cdot] \rightarrow \text{Free radicals}; \quad \rightarrow MeSCH_2OCOPh + t\text{-}BuOH \qquad (82)$$

Chemiluminescence was observed when secondary alkyl peroxy esters were decomposed in the presence of fluorescers (Fl).[232] The key process here is again the electron transfer from the fluorescent electron donors to the peroxy esters as shown in equation 83. The chemiluminescence intensities were higher with decreasing oxidation potential of the fluorescer. This interesting chemiluminescence is called 'chemically initiated electron-exchange luminescence' (CIEEL).[233] For the particular case of the secondary peroxy ester **73**, a weak chemiluminescence was observed even in the absence of a fluorescer and was explained by a six-membered cyclic transition state.[232b]

$$\underset{\textbf{(73)}}{PhCH(Me)OOC(=O)Me} \xrightarrow{Fl} \underset{\textbf{(74)}}{\left[\begin{matrix} Me \quad H \qquad O^{-\bullet} \\ C \qquad\quad C \\ Ph \quad O{-}O \quad Me \end{matrix} \quad Fl^{+\bullet} \right]} \xrightarrow{-AcOH}$$

$$\begin{matrix} Me \\ C{=}O^{-\bullet} \\ Ph \end{matrix} + Fl^{+\bullet} \longrightarrow PhCOMe + Fl^{*} \longrightarrow \text{light} \qquad (83)$$

2.5.5 Radical-induced decompositions

The decomposition of peroxy esters becomes much faster in hydrogen-donating solvents such as ethers or alcohols. For example, the homolysis of *tert*-butyl peroxyacetate in *sec*-butyl and isopropyl alcohols was about 100 times faster than that in benzene.[234] The solvent alcohols were oxidized, as shown in equation 84, to aldehydes or ketones in 9–140% yields. The yields of carbon

$$RCO_3Bu^t + R^1R^2CHOH \longrightarrow$$

$$\underset{60\text{-}100\%}{t\text{-}BuOH} + \underset{5\text{-}33\%}{Me_2C{=}O} + \underset{5\text{-}16\%}{RH} + \underset{4\text{-}12\%}{CO_2} + \underset{9\text{-}140\%}{R^1R^2C{=}O} \qquad (84)$$

dioxide were low whereas those of *tert*-butyl alcohols were fairly high. The major reaction involves hydrogen atom abstraction by the *tert*-butoxy radical (equation 85a), followed by radical substitution on the peroxy oxygen (equation 85b). The induced decomposition in peroxy esters proceeds by the chain reactions 85a and 85b. In the S_H2 reaction 85b, the carbon radicals attack the acyl-side peroxy oxygen because of steric hindrance by *t*-Bu.

$$t\text{-}BuO\cdot + R^1R^2CHOH \longrightarrow t\text{-}BuOH + R^1R^2\dot{C}OH \qquad (85a)$$

$$R^1R^2\dot{C}OH + RC(=O)OOBu^t \longrightarrow R^1R^2C(OH)(OCOR) + t\text{-}BuO\cdot \qquad (85b)$$

$$R^1R^2C(OH)(OCOR) \longrightarrow R^1R^2C{=}O + RCO_2H \qquad (85c)$$

Radical-induced decompositions of peroxyoxalates are also known,[209] but are not as significant as for peroxy acids or diacyl peroxides. An induced decomposition involving a triple bond is known (equation 86).[235]

$$\mathrm{PhC{\equiv}CCO_3Bu^t + PhCH_2\cdot \longrightarrow PhC{\equiv}CCH_2Ph + CO_2 + t\text{-}BuO\cdot} \qquad (86)$$

Radical-induced decompositions are also started by the abstraction of α-hydrogen in peroxy esters as shown in equation 87.[236] The intramolecular radical cyclization of **75** yields an α-lactone (**76**), leading to polyesters as final products.

$$\mathrm{RR'CHCO_3Bu^t \xrightarrow{t\text{-}BuO\cdot} RR'\dot{C}{-}C({=}O){-}OOBu^t \ (75) \xrightarrow{-\,t\text{-}Bu\cdot} RR'C\underset{O}{\diagdown\diagup}C{=}O \ (76) \longrightarrow Polyester} \qquad (87)$$

2.5.6 Stereochemistry of radicals

The homolysis of peroxy esters has sometimes been studied to clarify the structure and reactivity of radicals. Important aspects are briefly described in the following.

It is well known that carbenium ions are of planar trigonal structure and carbanions possess an sp^3 pyramidal structure. However, the stable structure of carbon radicals has not always been clarified adequately.[237] When 9-decalyl peroxycarboxylate was thermolyzed under oxygen, the stereochemistry of the resulting 9-decalyl hydroperoxide was the same (i.e. *cis*: *trans* = 1:9), regardless of the starting isomers.[238] However, when the oxygen pressure was raised, the proportion of *cis*-hydroperoxide increased. This interesting result indicates that the radicals are non-planar and possess a different *cis* and *trans* stereochemistry.

The relative stabilities of bridgehead radicals, which could not be planar, have been estimated as in equation 88 from the thermolysis rates.[239] The differences were not as large as for the corresponding carbenium ions. This indicates that the demand for coplanarity is not as important for radicals.

$$\mathrm{Me_3C\cdot} \;(1.0) > \text{1-adamantyl}\cdot \;(0.4) > \text{1-bicyclo[2.2.2]octyl}\cdot \;(0.05) > \text{1-norbornyl}\cdot \;(4 \times 10^{-4}) \qquad (88)$$

The stereochemistry of vinyl radicals has been studied extensively. The thermolysis of peroxycinnamates (**77**) afforded styrenes in 40–70% yields (equation 89). When R = Me, *cis*- and *trans*-**77** afforded the same reaction mixture.[240a,b] However, the stereochemistry of the resulting alkenes was significantly different when R = Br or Cl, i.e. most of the starting stereochemistry was retained in the alkenes produced.[240c] This means that the *cis–trans* isomerization between **78a** and **78b** is slower when R = Br and Cl.

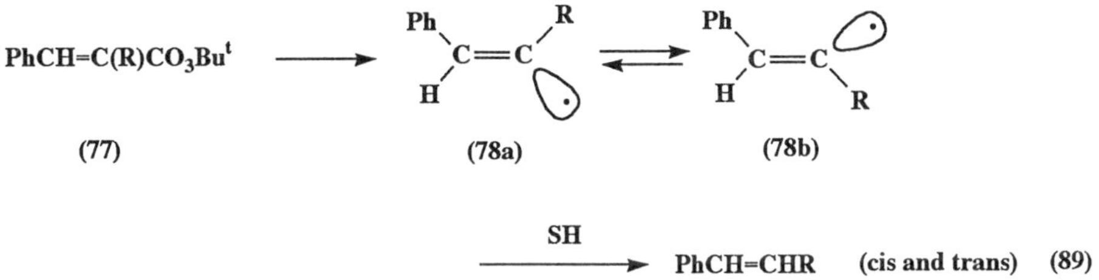

$$\mathrm{PhCH{=}C(R)CO_3Bu^t}\ (\mathbf{77}) \longrightarrow \mathrm{Ph(H)C{=}\dot{C}R}\ (\mathbf{78a}) \rightleftharpoons (\mathbf{78b}) \xrightarrow{\mathrm{SH}} \mathrm{PhCH{=}CHR}\ \text{(cis and trans)} \qquad (89)$$

A stereoelectronic effect has been observed in the disproportionation of cyclohexyl radicals produced from peroxy esters.[241]

2.5.7 Some utilization of radical reactions

Homolyses of peroxy esters have been applied to some synthetic reactions. One example is the photolysis of *tert*-butyl peroxyacetate as a source of methyl radicals. Thus, 8-methylation of caffeine (**79**)[242] and 2-methylation of adenine (**80**) under acidic conditions[243] could be achieved. The methylation is based on nucleophilic radical substitution by methyl radical.[244]

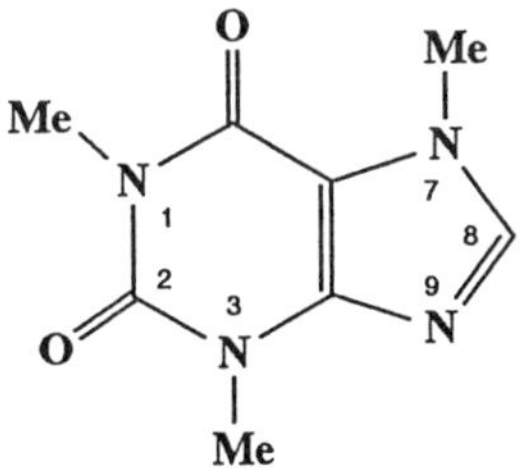

(**79**) 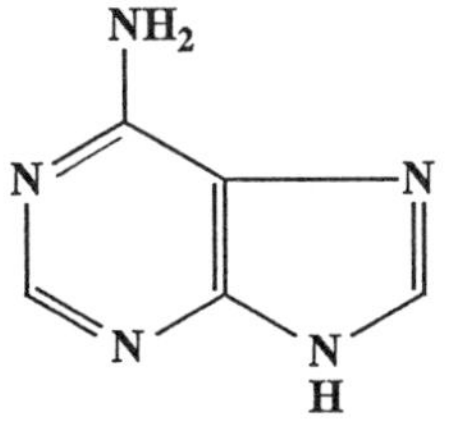 (**80**)

Another interesting application is the intramolecular cyano transfer by the photolysis of α-cyano peroxy esters (**81**). The cyano transfer is based on the intramolecular δ-H abstraction in the alkoxy radical **82** to yield **83** (equation 90).[245] The cyano group transfer is attained by the intramolecular addition of radical **83** to the C≡N group followed by ring opening. The final product is the γ-cyano ketone **84**. This transformation is interesting because nitriles are easily converted, by base, oxygen and AcCl, into α-cyano peroxy esters and then reiterative functionalization becomes possible by the repeated use of the cyano group.

(81) $\xrightarrow{h\nu}$ (82) $\longrightarrow$ (83) $\longrightarrow$ $\longrightarrow$ (84) (90)

Conversion of alcohols to halides is possible by the homolysis of peroxyoxalates according to equation 91.[246] The yields of chlorides were 40–60% from secondary alcohols and 30% from primary alcohols. The method seems to be useful when the alcohols are prone to cationic rearrangements in conventional methods.

$$\text{ROH} \longrightarrow \text{ROC(=O)}\text{—C(=O)Cl} \xrightarrow{\text{t-BuOH}} \text{ROC(=O)}\text{—C(=O)OOBu}^t \xrightarrow{\Delta} \text{R}\cdot \xrightarrow{\text{CCl}_4} \text{RCl} \quad (91)$$

Radical alkylations of ethers are possible by applying the induced decomposition of peroxy esters. One example is the alkylation of tetrahydrofuran with the alkene peroxy ester **85** (equation 92).[247] The

$$\underset{(85)}{CH_2{=}CHCH_2CH_2C(=O)OOBu^t} \xrightarrow[\text{THF}]{\Delta} \underset{(86)}{\text{THF-2-yl}{-}CH_2{-}\text{(pentanolide)}} \quad (92)$$

pentanolide **86** was produced via intramolecular cyclization of the adduct radical between **85** and THF radical. The same type of reaction has been applied to the functionalization of crown ethers.[248]

2.6 Metal-catalyzed Reactions of Peroxy Esters

The acyloxylation of C—H bonds with peroxy esters and copper catalysts is useful in organic synthesis and has been reviewed extensively.[193,249–251] The acyloxylation involves a chain reaction of peroxy esters, as shown in equations 93–95.[249]

$$RCO_3Bu^t + Cu(I) \longrightarrow t\text{-}BuO\cdot + Cu(II)(OCOR) \quad (93)$$

$$t\text{-}BuO\cdot + R'H \longrightarrow t\text{-}BuOH + R'\cdot \quad (94)$$

$$R'\cdot + Cu(II)(OCOR) \longrightarrow R'OCOR + Cu(I) \quad (95)$$

The peroxy esters are decomposed by reductive homolysis by Cu(I), yielding *tert*-butoxy radical (equation 93). The butoxy radical then abstracts hydrogen from active C—H bonds, usually allylic or α-hydrogen of ethers (equation 94). The final step is oxidative ligand transfer (equation 95). Whereas common carbon radicals are rapidly oxidized by the Cu(II) ion,[252] radicals with electron-attracting groups such as α-cyano or α-carboxy are not easy to oxidize, resulting in a shorter chain length.[253]

The acyloxylations are often conducted at 70–90°C in the presence of a catalytic amount of CuBr or CuOAc, but are also operative at room temperature when irradiated.[254] Sometimes the reactions are inhibited by oxygen, which is caused by the oxidation of Cu(I) by oxygen.[255]

2.6.1 Acyloxylation of allylic and related hydrogens

Allylic C—H bonds are effectively acyloxylated by the copper-catalyzed reactions of peroxy esters. Since the acyloxylation involves allylic radical intermediates, the isomerization of double bonds is inevitable. For example, the major products from terminal alkenes are 3-acetoxyalk-1-enes, as shown in equation 96.[255,256]

$$CH_3(CH_2)_nCH_2-CH=CH_2 + RCO_3Bu^t \xrightarrow[\text{AcOH, 80-90\%}]{\text{CuX}} CH_3(CH_2)_nCH(OAc)-CH=CH_2 + CH_3(CH_2)_nCH=CH-CH_2OAc \quad (96)$$

85 : 15

Cyclohexene afforded solely 3-acetoxycyclohexene in good yield.[257] The acyloxylation of bicyclo[3.2.1.]oct-2-ene (**87**) resulted in the selective formation of *exo*-**89** (equation 97).[258] The optically active **87** yielded the racemized **89**,

$$(87) \xrightarrow[\text{CuBr/ PhH}]{PhCO_3Bu^t} (88) \longrightarrow (89) \quad (97)$$

(**87**) (**88**) (**89**)

suggesting the involvement of free allyl radical **88**. A similar *exo* selectivity has been observed in the acyloxylation of β-pinene.[259]

Propargylic C—H bonds are likewise acyloxylated with copper(I) salts and peroxy esters (equation 98).[260]

$$RC\equiv CCHR^1R^2 \xrightarrow{PhCO_3Bu^t/CuCl} RC\equiv CCR^1R^2(OCOPh) \quad (98)$$

Rearrangement of the alkyl groups during acyloxylation is good evidence for the one-electron oxidation of radicals **90** by Cu(II) ions (equation 99).[255]

$$\text{5,5-dimethyl-1,3-cyclohexadiene} \xrightarrow{\text{t-BuO}\cdot} \mathbf{(90)} \xrightarrow{\text{Cu(II)}} \text{cation} \xrightarrow{-H^+} \text{o-xylene} \quad (99)$$

(90)

2.6.2 Mechanism of allylic acyloxylation

The first step in allylic acyloxylations is abstraction of allylic hydrogens by *tert*-butoxy radical (equation 100). The kinetic isotope effect of k_H/k_D was 2.90

$$RCH_2-CH=CH_2 \xrightarrow[k_H]{\text{t-BuO}\cdot} R\dot{C}H-CH=CH_2 \longleftrightarrow RCH=CH-\dot{C}H_2 \quad (100)$$

(91a) (91b)

$$\mathbf{(91)} \xrightarrow[k_X]{\text{Cu(II)X}} RCH(X)-CH=CH_2 + RCH=CH-CH_2X \quad (101)$$

(92a) (92b)

for allylbenzene [PhCH(D)CH=CH_2] with no temperature effect.[261] The moderate primary isotope effect was analyzed to suggest the cyclic transition state **93** rather than a linear hydrogen abstraction (**94**).

(93) (94)

The subsequent reaction (equation 101) is the product-determining step. It is not easy to conclude whether the reaction of allyl radicals **91** with Cu(II)X proceeds via a ligand transfer (**95b**) or a one-electron transfer (**95a**). The situation may be changed by the nature of X. For example, the **92a**:**92b** ratio for the acyloxylation of butenes in acetic acid was 9:1 with CuOAc but 3:7 with CuCl.[256a] The latter ratio was explained as reflecting an importance of ligand transfer (**95b**) when X = Cl.

R–CH···CH···CH$_2$ (+) Cu(I)X ⟷ R–CH(X)–CH–CH$_2$–Cu(I)

(95a) (95b)

According to a detailed study by Beckwith and Zavitsas,[262] but-1- and -2-enes afforded the same product ratio (90:10) of 3-acetoxybut-1-ene and *trans*-1-acetoxybut-2-ene, whereas *cis*- and *trans*-pent-2-enes yielded solely *trans*-4-acetoxypent-2-ene. These results led them to suggest a cyclic transition state involving an allyl copper complex (**96**) as shown in equation 102.

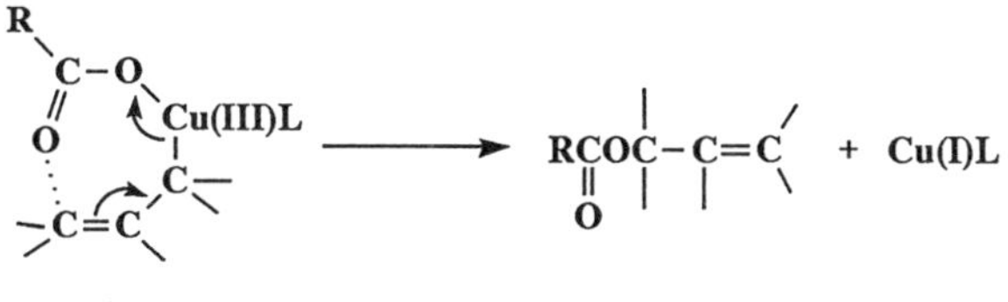

$$(96) \longrightarrow \mathrm{RCOC{-}C{=}C} + \mathrm{Cu(I)L} \qquad (102)$$

(96)

The predominant formation of thermodynamically unstable 3-acetoxy alkenes could be reasonably explained by the intramolecular acyl transfer in the copper complex **96**.

2.6.3 α-Acyloxylation of ethers and thioethers

α-Acyloxylation of ethers has been extensively studied.[193,250] Typical examples are the α-acyloxylation of tetrahydrofuran (equation 103)[254] and benzyl ethers.[263] The α-acyloxylation is based on the fact that the *tert*-butoxy radical attacks selectively the α-C—H bonds activated by ether oxygen.

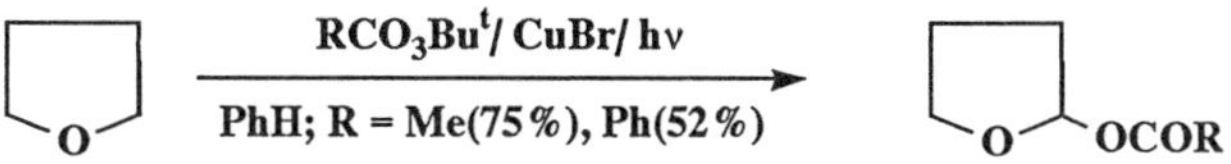

$$\text{THF} \xrightarrow[\text{PhH; R = Me(75\%), Ph(52\%)}]{\mathrm{RCO_3Bu^t/\,CuBr/\,h\nu}} \text{2-(OCOR)-THF} \qquad (103)$$

Thioethers are likewise functionalized at α-C—H bonds, e.g. equation 104.[264] The C—H bonds α to thioethers are more reactive than those α to the oxygens, as indicated by equation 105.[263]

$$(n\text{-}C_3H_7)S \xrightarrow[78\%]{PhCO_3Bu^t/\,CuBr} n\text{-}C_3H_7S\text{-}\underset{\displaystyle OCOPh}{\underset{|}{C}}HCH_2CH_3 \quad (104)$$

$$\text{1,4-oxathiane} \xrightarrow{PhCO_3Bu^t} \text{2-(benzoyloxy)-1,4-oxathiane (OCOPh at C}\alpha\text{ to S)} \quad (105)$$

2.6.4 Miscellaneous acyloxylations

The acyloxylation of benzylic C—H bonds as in cumene[265] and tetralin[266] is facile. Benzyl esters (equation 106)[266] and aldehydes (equation 107)[266] are likewise acyloxylated, although in lower yields.

$$PhCH_2OCOMe \xrightarrow[35\%]{PhCO_3Bu^t/\,CuBr} Ph\underset{\displaystyle OCOPh}{\underset{|}{C}}HOCOMe \quad (106)$$

$$PhCHO \xrightarrow{PhCO_3Bu^t/\,CuBr} R\underset{O}{\underset{\|}{C}}O\underset{O}{\underset{\|}{C}}Ph \quad (107)$$

The α-acyloxylation of alcohols yields hemiacetals, which are readily converted into carbonyl compounds (equation 108).[266,267] Functionalization with peroxy esters and copper salts may be applied to the β-lactam rings as shown in equation 109.[268] α-Amidoxylations can be achieved by using peroxyamidates as shown in equation 110.[269]

$$R_1R_2CHOH \xrightarrow{RCO_2Bu^t/\,Cu(I)} R^1R^2C\begin{matrix} OH \\ OCOR \end{matrix} \longrightarrow R^1R^2C{=}O \quad (108)$$

$$\text{penam (R, H, S, N, O, }CO_2Me) \xrightarrow[R\,=\,\text{phthalimido; }41\%]{PhCO_3Bu^t/CuCl} \text{penam with OCOPh in place of H} \quad (109)$$

$$\text{cyclohexene} + Me_2N\underset{O}{\underset{\|}{C}}OOBu^t \xrightarrow[20\%]{CuX} \text{3-(}OCONMe_2\text{)cyclohexene} \quad (110)$$

3 ACKNOWLEDGEMENT

The author thanks Mr Morio Inaishi for helpful assistance with the manuscript.

4 REFERENCES

1. A. G. Davies, *Organic Peroxides*, Butterworths, London (1961), pp. 55-63.
2. D. Swern, *Organic Peroxides*, Vol. I, Wiley-Interscience, New York (1970), pp. 313-516.
3. L. S. Silbert, in *Organic Peroxides*, Vol. II (D. Swern Ed.), Wiley-Interscience, New York (1970), pp. 637-798.
4. G. Bouillon, C. Lick and K. Schank, in *The Chemistry of Peroxides* (S. Patai, Ed.), Wiley, Chichester (1983), pp. 287-303.
5. (a) Y. Sawaki and Y. Ogata, *Bull. Chem. Soc. Jpn.*, **38**, 2103 (1965); (b) Y. Sawaki and Y. Ogata, *Tetrahedron*, **21**, 3381 (1965).
6. L. S. Silbert, E. Siegel and D. Swern, *Org. Synth. Coll. Vol.*, **5**, 904 (1973).
7. (a) D. Swern, *Org. React.*, **7**, 395 (1953); (b) Y. Ogata and K. Aoki, *J. Org. Chem.*, **31**, 4181 (1966).
8. (a) W. D. Emmons and G. B. Lucas, *J. Am. Chem. Soc.*, **77**, 2287 (1955); (b) K. B. Wiberg and S. T. Waddell, *J. Am. Chem. Soc.*, **112**, 2194 (1990).
9. G. B. Payne, *Org. Synth. Coll. Vol.*, **5**, 805 (1973).
10. A. Kergomard and J. Bigou, *Bull. Soc. Chim. Fr.*, 486 (1956).
11. Y. Ogata and Y. Sawaki, *Tetrahedron*, **23**, 3327 (1967).
12. J. O. Edwards and R. G. Pearson, *J. Am. Chem. Soc.*, **84**, 4607 (1962).
13. J. Y. Nedelec, J. Sorba and D. Lefort, *Synthesis*, 821 (1976).
14. (a) R. M. McDonald, R. N. Steppel and J. E. Dorsey, *Org. Synth.*, **50**, 15 (1970). (b) G. Braun, *Org. Synth., Coll. Vol.*, **1**, 431 (1941).
15. R. M. Coates and J. W. Williams, *J. Org. Chem.*, **39**, 3054 (1974).
16. B. Phillips, F. C. Frostick, Jr, and P. S. Starcher, *J. Am. Chem. Soc.*, **79**, 5982 (1957).
17. C. R. Dick and R. F. Hanna, *J. Org. Chem.*, **29**, 1218 (1964).
18. D. Swern, *Organic Peroxides*, Vol. I, Wiley-Interscience, New York (1970), pp. 475-516.
19. A. J. Everett and G. J. Minkoff, *Trans. Faraday Soc.*, **49**, 410 (1953).
20. A. C. Egerton, W. Emte and G. J. Minkoff, *Discuss. Faraday Soc.*, 278 (1951).
21. (a) E. Koubek and J. O. Edwards, *J. Org. Chem.*, **28**, 2157 (1963); (b) H. Krimm, *US Pat.*, 2, 813, 896 (1957).
22. W. E. Parker, C. Ricciuti, C. L. Ogg and D. Swern, *J. Am. Chem. Soc.*, **77**, 4037 (1955).
23. W. D. Emmons and A. S. Pagano, *J. Am. Chem. Soc.*, **77**, 89 (1955).
24. J. F. Goodman, P. Robson and E. R. Wilson, *Trans. Faraday Soc.*, **58**, 1846 (1962).
25. W. H. Rastetter, T. J. Richard and M. D. Lewis, *J. Org. Chem.*, **43**, 3163 (1978).
26. D. M. Davies and P. Jones, *J. Org. Chem.*, **43**, 769 (1978).
27. D. Swern, L. P. Witnauer, C. R. Eddy and W. E. Parker, *J. Am. Chem. Soc.*, **77**, 5537 (1955).
28. J. A. Cugley, W. Bossert, A. Bauder and H. H. Guenthard, *Chem. Phys.*, **16**, 229 (1976).
29. O. Exner and B. Plesnicar, *Collect, Czech. Chem. Commun.*, **43**, 3079 (1978).
30. M. Oldani, T.-K. Ha and A. Baulder, *J. Am. Chem. Soc.*, **105**, 360 (1983).

31. (a) L. M. Hjelmiland and G. H. Loew, *Tetrahedron*, **33**, 1029 (1977); (b) C. W. Bock, M. Trachtman and P. Gorge, *J. Mol. Struct.*, **71**, 327 (1981).
32. (a) J. Skerjanc, A. Regent and B. Plesničar, *J. Chem. Soc., Chem. Commun.*, 1007 (1980); (b) B. Plesnicar, V. Menart, M. Hodoscek, J. Koller, F. Kovac and J. Skerjanc, *J. Chem. Soc., Perkin Trans. 2*, 1397 (1986).
33. (a) R. Curci, R. A. DiPrete, J. O. Edwards and G. Modena, *J. Org. Chem.*, **35**, 740 (1970); (b) R. Kavcic and B. Plesnicar, *J. Org. Chem.*, **35**, 2033 (1970).
34. T. A. B. M. Bolsman, R. Kok and A. D. Vreugdenhil, *J. Am. Oil Chem. Soc.*, **65**, 1211 (1988).
35. W. P. Griffith, A. C. Skapski and A. P. West, *Inorg. Chim. Acta*, **65L**, 249 (1982).
36. P. Brougham, M. S. Cooper, D. A. Cummerson, H. Heaney and N. Thompson, *Synthesis*, 1015 (1987).
37. (a) K. M. Ibne-Rasa, and J. O. Edwards, *J. Am. Chem. Soc.*, **84**, 763 (1962); (b) A. Dondoni, G. Modena and P. E. Todesco, *Gazz. Chim. Ital.*, **91**, 613 (1961); (c) K. M. Ibne-Rasa, C. G. Lauro and J. O. Edwards, *J. Am. Chem. Soc.*, **85**, 1165 (1963); (d) C. G. Overberger and R. W. Cummins, *J. Am. Chem. Soc.*, **75**, 4250 (1953); (e) R. Curci, A. Giovine and G. Modena, *Tetrahedron*, **22**, 1235 (1966); (f) R. Curci and G. Modena, *Tetrahedron Lett.*, 1749 (1963); (g) B. M. Lynch and K. H. Pausacker, *J. Chem. Soc.*, 1525 (1955); (h) Y. Ishii and Y. Inamoto, *Kogyo Kagaku Zasshi*, **63**, 765 (1960).
38. (a) T. C. Bruice, *J. Chem. Soc., Chem. Commun.*, 14 (1983); (b) T. C. Bruice, J. B. Noar, S. S. Ball and U. V. Venkataram, *J. Am. Chem. Soc.*, **105**, 2452 (1983).
39. N. Prileschajew, *Chem. Ber.*, **42**, 4811 (1909).
40. D. Swern, in *Organic Peroxides*, Vol. II (D. Swern, Ed.), Wiley-Interscience, New York (1971), pp. 355–533.
41. V. G. Dryuk, *Tetrahedron*, **32**, 2855 (1976).
42. B. Plesnicar, in *The Chemistry of Peroxides* (S. Patai, Ed.), Wiley, Chichester (1983), pp. 521–584.
43. P. D. Bartlett, *Rec. Chem. Prog.*, **11**, 47 (1950).
44. (a) T. Itoh, K. Jitsukawa, K. Kaneda and S. Teranishi, *J. Am. Chem. Soc.*, **101**, 159 (1979); (b) R. B. Dehnel and G. H. Whitham, *J. Chem. Soc., Perkin Trans. 1*, 953 (1979).
45. B. A. McKittrick and B. Ganem, *Tetrahedron Lett.*, **26**, 4895 (1985).
46. G. Berti, *Top. Stereochem.*, **7**, 93 (1973).
47. K. B. Sharpless and T. R. Verhoeven, *Aldrichim. Acta*, **12**, 63 (1979).
48. R. P. Hanzlik and G. O. Schearer, *J. Am. Chem. Soc.*, **97**, 5231 (1975).
49. H. Kwart and D. H. Hoffman, *J. Org. Chem.*, **31**, 419 (1966).
50. K. D. Bingham, G. D. Meakins and G. H. Witham, *Chem. Commun.*, 445 (1966).
51. H. -J. Schneider, N. Becker and K. Philippi, *Chem. Ber.*, **114**, 1562 (1981).
52. (a) A. Azman, B. Borstnik and B. Plesničar, *J. Org. Chem.*, **34**, 971 (1969); (b) B. Plesnicar, M. Tasevski and A. Azman, *J. Am. Chem. Soc.*, **100**, 743 (1978).
53. R. D. Bach, M. W. Klein, R. A. Ryntz and J. W. Holubka, *J. Org. Chem.*, **44**, 2569 (1979).
54. (a) N. Matsumura, N. Sonoda and S. Tsutsumi, *Tetrahedron Lett.*, 2029 (1970); (b) J. Rebek, Jr, R. McCready, S. Wolf and A. Mossman, *J. Org. Chem.*, **44**, 1485 (1979).
55. (a) E. J. Corey, H. Niwa and J. R. Falk, *J. Am. Chem. Soc.*, **101**, 1586 (1979); (b) J. Rebeck, Jr, and R. McCready, *J. Am. Chem. Soc.*, **102**, 5602 (1980).
56. J. D. Morrison and H. S. Mosher, *Asymmetric Organic Reactions*, Prentice-Hall, Englewood Cliffs, NJ (1971), p. 258.
57. (a) R. N. McDonald and P. A. Schwab, *J. Am. Chem. Soc.*, **86**, 4866 (1964); (b) J. K. Stille and D. D. Whitehurst, *J. Am. Chem. Soc.*, **86**, 4871 (1964).

58. J. Ciabattoni, R. A. Campbell, C. A. Renner and P. W. Concannon, *J. Am. Chem. Soc.*, **92**, 3826 (1970).
59. Y. Ogata, Y. Sawaki and H. Inoue, *J. Org. Chem.*, **38**, 1044 (1973).
60. K. M. Ibne-Rasa, R. H. Pater, J. Ciabattoni and J. O. Edwards, *J. Am. Chem. Soc.*, **95**, 7894 (1973).
61. J. Koller and B. Plesničar, *J. Chem. Soc., Perkin Trans. 2*, 1361 (1982).
62. K. Tanaka and M. Yoshimine, *J. Am. Chem. Soc.*, **102**, 7655 (1980).
63. P. R. Ortiz de Montellano and K. L. Kunze, *J. Am. Chem. Soc.*, **102**, 7373 (1980).
64. (a) R. L. Camp and F. D. Greene, *J. Am. Chem. Soc.*, **90**, 7349 (1968); (b) J. K. Crandall, W. W. Conover, J. B. Komin and W. H. Machleder, *J. Org. Chem.*, **39**, 1723 (1974).
65. R. D. Chambers, P. Goggin and W. K. R. Musgrave, *J. Chem. Soc.*, 1804 (1959).
66. H. Hart, *Acc. Chem. Res.*, **4**, 337 (1971).
67. D. Jerina, J. Daly, W. Landis, B. Witkop and S. Undenfriend, *J. Am. Chem. Soc.*, **89**, 3347 (1967).
68. T. Tsuchiya, M. Andoh and J. Imamura, *Nippon Kagaku Kaishi*, 370 (1979).
69. Y. Asakawa, R. Matsuda, M. Tori and M. Sono, *J. Org. Chem.*, **53**, 5453 (1988).
70. A. J. Pandell, *J. Org. Chem.*, **48**, 3908 (1983).
71. (a) N. C. Deno, B. A. Greigger, L. A. Messner, M. D. Meyer and S. G. Stroud, *Tetrahedron Lett.*, 1703 (1977); (b) R. Liotta and W. S. Hoff, *J. Org. Chem.*, **45**, 2887 (1980).
72. U. Frommer and V. Ullrich, *Z. Naturforsch., Teil B*, **26**, 322 (1971).
73. (a) W. Müller and H. -J. Schneider, *Angew. Chem., Int. Ed. Engl.*, **18**, 407 (1979); (b) H. -J. Schneider, N. Becker and K. Phillipi, *Chem. Ber.*, **114**, 1562 (1981).
74. (a) M. Tori, M. Sano and Y. Asakawa, *Bull. Chem. Soc. Jpn.*, **58**, 2669 (1985); (b) M. Tori, R. Matsuda and Y. Asakawa, *Tetrahedron*, **42**, 1275 (1986).
75. H. J. Schneider and W. Müller, *J. Org. Chem.*, **50**, 4609 (1985).
76. (a) C. H. Hassall, *Org. React.*, **9**, 73 (1957); (b) P. A. S. Smith, in *Molecular Rearrangements*, Vol. I (P. de Mayo, Ed.), Interscience, New York (1963), p. 457.
77. M. F. Hawthorne, W. D. Emmons and K. S. McCallum, *J. Am. Chem. Soc.*, **80**, 6393 (1958).
78. Y. Ogata and Y. Sawaki, *J. Org. Chem.*, **37**, 2953 (1972).
79. Y. Ogata and Y. Sawaki, *J. Am. Chem. Soc.*, **94**, 4189 (1972).
80. L. Syper, *Tetrahedron*, **43**, 2853 (1987).
81. G. Speier and Z. Tyeklar, *Chem. Ber.*, **112**, 389 (1979).
82. P. M. Cullis, J. R. P. Arnold, M. Clarke, R. Howell, M. DeMira, M. Naylor and D. Nicholls, *J. Chem. Soc., Chem. Commun.*, 1088 (1987).
83. G. A. Olah, T. Yamato, P. S. Iyer, N. J. Trivedi, B. P. Singh and G. K. S. Prakash, *Mater. Chem. Phys.*, **17**, 21 (1987).
84. F. Toda, M. Yagi and K. Kiyoshige, *J. Chem. Soc., Chem. Commun.*, 958 (1988).
85. Y. Fujise, K. Fujiwara and Y. Ito, *Chem. Lett.*, 1475 (1988).
86. (a) Z. Grudzinski, S. M. Roberts, C. Howard and R. F. Newton, *J. Chem. Soc., Perkin Trans, 1*, 1182 (1978); (b) G. R. Krow, C. A. Johnson, J. P. Guare, D. Kubrak, K. J. Henz, D. A. Shaw, S. W. Szczepanski and J. T. Carey, *J. Org. Chem.*, **47**, 5239 (1982).
87. H. S. Mosher, L. Turner and A. Carlsmith, *Org. Synth. Coll. Vol.*, **4**, 828 (1963).
88. M. V. Jovanovic, *Can. J. Chem.*, **62**, 1176 (1984).
89. R. R. Holmes and R. P. Bayer, *J. Am. Chem. Soc.*, **82**, 3454 (1960).
90. R. W. White and W. D. Emmons, *Tetrahedron*, **17**, 31 (1962).
91. A. S. Pagano and W. D. Emmons, *Org. Synth. Coll. Vol.*, **5**, 367 (1973).
92. P. E. Eaton and G. E. Wicks, *J. Org. Chem.*, **53**, 5353 (1988).
93. W. D. Emmons and A. F. Ferris, *J. Am. Chem. Soc.*, **75**. 4623 (1953).

94. (a) W. D. Emmons and A. S. Pagano, *J. Am. Chem. Soc.*, **77**, 4557 (1955); (b) R. J. Sundberg and P. A. Bukowick, *J. Org. Chem.*, **33**, 4098 (1968).
95. (a) T. Toda, E. Mori and K. Murayama, *Bull. Chem. Soc. Jpn.*, **45**, 1904 (1972); (b) T. Yoshioka, S. Higashida and K. Murayama, *Bull. Chem. Soc. Jpn.*, **45**, 636 (1972).
96. G. M. Badger, R. G. Buttery and G. E. Lewis, *J. Chem. Soc.*, 2143 (1953).
97. T. Mitsuhashi, O. Simamura and Y. Tezuka, *Chem. Commun.*, 1300 (1970).
98. (a) R. Curci, F. DiFuria and F. Marcuzzi, *J. Org. Chem.*, **36**, 3774 (1971); (b) R. Curci, F. DiFuria, J. Ciabattoni and P. W. Concannon, *J. Org. Chem.*, **39**, 3295 (1974).
99. K. L. Handoo and S. K. Handoo, *Indian J. Chem.*, **20B**, 699 (1981); *Chem. Abstr.*, **95**, 2198 (1981).
100. M. T. H. Liu and I. Yamamoto, *Can. J. Chem.*, **57**, 1299 (1979).
101. M. Schulz, N. Grossmann and W. Schauer, *J. Prakt. Chem.*, **318**, 586 (1976).
102. L. A. Paquette, *Principles of Modern Heterocyclic Chemistry*, Benjamin, New York (1968), p. 63.
103. F. A. Davis and R. H. Jenkins, Jr, in *Asymmetric Synthesis*, Vol. IV (J. D. Morrison and J. W. Scott, Eds), Academic Press, New York (1984), p. 322.
104. W. D. Emmons and A. S. Pagano, *Org. Synth. Coll. Vol.*, **5**, 191 (1973).
105. E. Schmitz, R. Ohme and D. Murawski, *Chem. Ber.*, **98**, 2516 (1965).
106. (a) V. Madan and L. B. Clapp, *J. Am. Chem. Soc.*, **91**, 6078 (1969); (b) Y. Madan and L. B. Clapp, *J. Am. Chem. Soc.*, **92**, 4902 (1970).
107. (a) Y. Ogata and Y. Sawaki, *J. Am. Chem. Soc.*, **95**, 4687 (1973); (b) Y. Ogata and Y. Sawaki, *J. Am. Chem. Soc.*, **95**, 4692 (1973).
108. A. Azman, J. Koller and B. Plesničar, *J. Am. Chem. Soc.*, **101**, 1107 (1979).
109. D. R. Boyd, D. C. Neill, C. G. Watson and W. B. Jennings, *J. Chem. Soc., Perkin Trans. 2*, 1813 (1975).
110. W. H. Pirke and P. L. Rinaldi, *J. Org. Chem.*, **43**, 4475 (1978).
111. C. R. Johnson, H. Difenbach, J. E. Keiser and J. C. Scharp, *Tetrahedron*, **25**, 5649 (1969).
112. (a) G. A. Russell and L. A. Ochrymowycz, *J. Org. Chem.*, **35**, 2106 (1970); (b) J. F. Carson and L. E. Boggs, *J. Org. Chem.*, **36**, 611 (1971).
113. Y. Sawaki and Y. Ogata, *J. Org. Chem.*, **49**, 3344 (1984).
114. Y. Sawaki and Y. Ogata, *J. Am. Chem. Soc.*, **103**, 3832 (1981).
115. R. Curci, R. A. DiPrete, J. O. Edwards and G. Modena, *J. Org. Chem.*, **35**, 740 (1970).
116. (a) A. Maccioni, F. Montanari, M. Secci and M. Montanari, *Tetrahedron Lett.*, 607 (1961); (b) U. Folli, D. Iarossi, F. Montanari and G. Torre, *J. Chem. Soc. C*, 1317 (1968).
117. C. C. Price and G. W. Stacy, *Org. Synth. Coll. Vol.*, **3**, 86 (1955).
118. W. G. Filby, K. Günther and R. D. Penhorn, *J. Org. Chem.*, **38**, 4070 (1973).
119. C. J. Cavallito and D. M. Fruchauf, *J. Am. Chem. Soc.*, **71**, 2248 (1949).
120. L. D. Small, J. H. Baily and C. J. Cavallito, *J. Am. Chem. Soc.*, **69**, 1710 (1947).
121. M. M. Chau and J. L. Kice, *J. Am. Chem. Soc.*, **98**, 7711 (1976).
122. (a) F. Freeman and C. N. Angeletakis, *J. Am. Chem. Soc.*, **104**, 5766 (1982); (b) F. Freeman and C. N. Angeletakis, *J. Org. Chem.*, **47**, 3403 (1982); (c) F. Freeman and C. N. Angeletakis, *J. Am. Chem. Soc.*, **105**, 4039 (1983); (d) F. Freeman and C. N. Angeletakis, *J. Org. Chem.*, **50**, 793 (1985).
123. (a) A. Battaglia, A. Dondoni, P. Giorgianni, G. Maccagnani and G. Mazzanti, *J. Chem. Soc. B*, 1547 (1971); (b) A. Battaglia, A. Dondoni, G. Maccagnani and G. Mazzanti, *J. Chem. Soc., Perkin Trans. 2*, 610 (1974).

124. L. F. Fieser and M. Fieser, *Reagents Org. Synth.*, **1**, 1212, 1238 (1967).
125. M. Yoshifuji, K. Ando, K. Toyota, I. Shima and N. Inamoto, *J. Chem. Soc., Chem. Commun.*, 419 (1983).
126. Y. Kashman and O. Awerbouch, *Tetrahedron*, **31**, 53 (1975).
127. L. D. Quin, J. C. Kisalus and K. A. Mesh, *J. Org. Chem.*, **48**, 4466 (1983).
128. (a) B. Lindgren, *Acta Chem. Scand.*, **26**, 2560 (1972); (b) M. Tiecco, L. Testaferri, M. Tingoli, D. Chianelli and D. Bartoli, *Gazz. Chim. Ital.*, **117**, 423 (1987).
129. G. Galambos and V. Simonidesz, *Tetrahedron Lett.*, 4371 (1982).
130. S. Uemura and S. Fukuzawa, *J. Chem. Soc., Perkin Trans. 1*, 471 (1985).
131. H. Minato, H. Kodama, T. Miura and M. Kobayashi, *Chem. Lett.*, 413 (1977).
132. (a) J. L. Kice, S. Chiou and L. Weclas, *J. Org. Chem.*, **50**, 2508 (1985); (b) J. L. Kice and S. Chiou, *J. Org. Chem.*, **51**, 290 (1986).
133. (a) Y. Nagai, K. Honda and T. Migita, *J. Organomet. Chem.*, **8**, 372 (1967); (b) L. H. Sommer, L. A. Ulland and G. A. Parker, *J. Am. Chem. Soc.*, **94**, 3469 (1972).
134. H. Sakurai, T. Imoto, N. Hayashi and M. Kumada, *J. Am. Chem. Soc.*, **87**, 4001 (1965).
135. (a) G. A. Razuvaev, T. N. Brevnova, A. N. Kornev, M. A. Lopatin and A. N. Egorochkin, *Zh. Obshch. Khim.*, **57**, 375 (1987); (b) V. V. Semenov, A. N. Kornev, T. N. Brevnova and G. A. Razuvaev, *Zh. Obshch. Khim.*, **59**, 151 (1989).
136. T. A. Dixon, K. P. Steele and W. P. Weber, *J. Organomet. Chem.*, **231**, 299 (1982).
137. A. Hosomi, S. Iijima and H. Sakurai, *Chem. Lett.*, 243 (1981).
138. (a) K. Tamao, T. Kakui, M. Akita, T. Iwahara, R. Kanatani, J. Yoshida and M. Kumada, *Tetrahedron*, **39**, 983 (1983); (b) K. Tamao, M. Akita and M. Kumada, *J. Organomet. Chem.*, **254**, 13 (1983).
139. A. I. Al-Wassil, C. Eaborn and A. K. Saxena, *J. Chem. Soc., Chem. Commun.*, 974 (1983).
140. J. O. Edward, *Peroxide Reaction Mechanisms*, Interscience, New York (1962), p. 73.
141. (a) Y. Ogata and G. Nakajima, *Tetrahedron*, **20**, 43, 2751 (1964); (b) Y. Ogata and K. Aoki, *J. Am. Chem. Soc.*, **90**, 6187 (1968).
142. Y. Ogata, Y. Furuya and K. Okano, *Bull. Chem. Soc. Jpn.*, **37**, 960 (1964).
143. (a) Y. Ogata, K. Aoki and Y. Furuya, *Chem. Ind. (London)*, 304 (1965); (b) Y. Ogata, and K. Aoki, *J. Org. Chem.*, **31**, 1625 (1966).
144. Y. Ogata and K. Aoki, *J. Org. Chem.*, **34**, 3974 (1969).
145. Y. Ogata and K. Aoki, *J. Org. Chem.*, **34**, 3978 (1969).
146. H. J. Reich and S. L. Peake, *J. Am. Chem. Soc.*, **100**, 4888 (1978).
147. T. L. MacDonald, N. Narasimhan and L. T. Burka, *J. Am. Chem. Soc.*, **102**, 7760 (1980).
148. R. I. Davidson and P. J. Kropp, *J. Org. Chem.*, **47**, 1904 (1982).
149. D. G. Morris and A. G. Shepherd, *J. Chem. Soc., Chem. Commun.*, 1250 (1981).
150. M. Srebnik, R. Mechoulam and I. Yona, *J. Chem. Soc., Perkin Trans. 1*, 1423 (1987).
151. M. Srebnik, *Synth. Commun.*, **19**, 197 (1989).
152. (a) J. Gerstein and W. P. Jencks, *J. Am. Chem. Soc.*, **86**, 4655 (1964); (b) J. E. Dixon and T. C. Bruice, *J. Am. Chem. Soc.*, **93**, 6592 (1971).
153. C. A. Bunton, M. M. Mhala, J. R. Moffatt, D. Monarres and G. Savelli, *J. Org. Chem.*, **49**, 426 (1984).
154. S. Patai and Z. Rappoport, in *The Chemistry of Alkenes* (S. Patai, Ed.), Wiley, Chichester (1964), p. 512.
155. J. R. Doherty, D. D. Keane, K. G. Karathe, W. I. O'Sullivan, E. M. Philbin, R. M. Simons and P. C. Teague, *Tetrahedron Lett.*, 441 (1968).
156. R. Curci and F. DiFuria, *Tetrahedron Lett.*, 4085 (1974).

157. Y. Apeloig, M. Karni and Z. Rappoport, *J. Am. Chem. Soc.*, **105**, 2784 (1983).
158. W. Kirmse and L. Horner, *Chem. Ber.*, **89**, 836 (1956).
159. (a) E. Koubek, M. L. Haggett, C. J. Battaglia, K. M. Ibne-Rasa, H. Y. Pyun and J. O. Edwards, *J. Am. Chem. Soc.*, **85**, 2263 (1963); (b) R. E. Ball, J. O. Edwards, M. L. Haggett and P. Jones, *J. Am. Chem. Soc.*, **89**, 2331 (1967); (c) E. Koubek and J. E. Welsch, *J. Org. Chem.*, **33**, 445 (1968).
160. K. Akiba and O. Simamura, *Tetrahedron*, **26**, 2519, 2527 (1970).
161. S. S. Levush, Z. P. Prisyazhnyuk and A. M. Kovalskaya, *Kinet. Katal.*, **24**, 1294 (1983).
162. D. L. Heywood, B. Phillips and H. A. Stansburg, Jr, *J. Org. Chem.*, **26**, 281 (1961).
163. Y. Ogata and K. Tomizawa, *J. Org. Chem.*, **43**, 261 (1978).
164. Y. Ogata and K. Tomizawa, *J. Org. Chem.*, **45**, 785 (1980).
165. G. I. Nikishin, A. T. Koritzky, S. V. Lindeman, R. G. Gerr, Y. T. Struchkov and E. K. Starostin, *J. Am. Chem. Soc.*, **111**, 9214 (1989).
166. D. Lefort, J. Fossey, M. Gruselle, J. Y. Nedelec and J. Sorba, *Tetrahedron*, **41**, 4237 (1985).
167. J. Fossey and D. Lefort, *Tetrahedron*, **36**, 1023 (1980).
168. J. Sorba, J. Fossey, J. Y. Nedelec and D. Lefort, *Tetrahedron*, **35**, 1509 (1979).
169. J. Fossey, D. Lefort, M. Massoudi, J. Y. Nedelec and J. Sorba, *Can. J. Chem.*, **63**, 678 (1985).
170. B. Ashworth, B. C. Gilbert, R. G. G. Holms and R. O. C. Norman, *J. Chem. Soc., Perkin Trans. 2*, 951 (1978).
171. E. S. Huyser and G. W. Howkins, *J. Org. Chem.*, **48**, 1705 (1983).
172. T. Morimoto, M. Hirano, M. Wachi and T. Murakami, *J. Chem. Soc., Perkin Trans. 2*, 1949 (1984).
173. E. J. Corey, E. P. Barrette and P. A. Magriotis, *Tetrahedron Lett.*, **26**, 5855 (1985).
174. E.g. (a) J. T. Groves and Y. Watanabe, *J. Am. Chem. Soc.*, **110**, 8443 (1988); (b) P. N. Balasubramanian and T. C. Bruice, *J. Am. Chem. Soc.*, **108**, 5495 (1986); (c) T. G. Traylor and F. Xu, *J. Am. Chem. Soc.*, **112**, 178 (1990).
175. L. A. Singer, in *Organic Peroxides*, Vol. I (D. Swern, Ed.), Wiley–Interscience, New York (1970), pp. 265–312.
176. P. D. Bartlett and R. R. Hiatt, *J. Am. Chem. Soc.*, **80**, 1398 (1958).
177. L. S. Silbert and D. Swern, *J. Am. Chem. Soc.*, **81**, 2364 (1959).
178. P. D. Bartlett and R. E. Pincock, *J. Am. Chem. Soc.*, **84**, 2445 (1962).
179. A. G. Davies and H. J. Hunter, *J. Chem. Soc.*, 1808 (1953).
180. C. Rüchardt and R. Hecht, *Chem. Ber.*, **97**, 2716 (1964).
181. E. L. O'Brien, F. M. Beringer and R. B. Mesrobian, *J. Am. Chem. Soc.*, **79**, 6238 (1957).
182. R. Naylor, *J. Chem. Soc.*, 244 (1945).
183. (a) N. A. Milas and D. M. Sargenor, *J. Am. Chem. Soc.*, **68**, 642 (1946); (b) A. T. Blomquist and I. A. Bernstein, *J. Am. Chem. Soc.*, **73**, 5546 (1951).
184. (a) P. D. Bartlett and L. B. Gortler, *J. Am. Chem. Soc.*, **85**, 1864 (1963); (b) J. P. Lorand and P. D. Bartlett, *J. Am. Chem. Soc.*, **88**, 3294 (1966).
185. O. L. Magelli and C. S. Sheppard, in *Organic Peroxides*, Vol. I (D. Swern, Ed.), Wiley–Interscience, New York (1970), p. 73.
186. T. Yonezawa, H. Kato and O. Yamamoto, *Bull. Chem. Soc. Jpn.*, **40**, 307 (1967).
187. F. D. Verderame and J. C. Miller, *J. Phys. Chem.*, **66**, 2185 (1962).
188. B. Plesničar and O. Exner, *Collect. Czech. Chem. Commun.*, **46**, 490 (1981).
189. L. Golic and I. Leban, *Acta Crystallogr., Sect. C*, **40**, 447 (1984).
190. (a) L. S. Silbert, L. P. Witnauer, D. Swern and C. Ricciuti, *J. Am. Chem. Soc.*, **81**, 3244 (1959); (b) M. Schulz and K. H. Schwarz, *Z. Chem.*, **7**, 176 (1967).

191. G. Goermar, R. Radeglia, E. Scharn and M. Schulz, *J. Prakt. Chem.*, **325**, 333 (1983).
192. C. Rüchardt, *Fortschr. Chem. Forsch.*, **6**, 251 (1966).
193. G. Sosnowsky and S. -O. Lawesson, *Angew. Chem.*, **76**, 218 (1964).
194. L. Horner and W. Jurgeleit, *Justus Liebigs Ann. Chem.*, **591**, 138 (1955).
195. D. B. Denney, W. F. Goodyear and B. Goldstein, *J. Am. Chem. Soc.*, **83**, 1726 (1961).
196. (a) S. -O. Lawesson and N. C. Yang, *J. Am. Chem. Soc.*, **81**, 4231 (1959); (b) C. Friesell and S. -O. Lawesson, *Org. Synth., Coll. Vol.*, **5**, 924 (1973); (c) F. Marcuzzi and G. Melloni, *Synthesis*, 451 (1976).
197. A. C. M. Meesters and M. H. Benn, *Synthesis*, 679 (1980).
198. V. L. Antonovskii, M. M. Buzlanova and Z. S. Frolova, *Kinet. Katal.*, **8**, 671 (1967).
199. R. Criegee, *Chem. Ber.*, **77**, 722 (1944).
200. (a) P. D. Bartlett and J. L. Kice, *J. Am. Chem. Soc.*, **75**, 5591 (1953); (b) H. L. Goering and A. C. Olson, *J. Am. Chem. Soc.*, **75**, 5853 (1953); (c) D. B. Denney, *J. Am. Chem. Soc.*, **77**, 1706 (1955).
201. (a) N. Y. Yablokava, V. A. Yablokov and V. A. Shushunov, *Kinet. Katal.*, **7**, 165 (1966); (b) S. Winstein and G. C. Robinson, *J. Am. Chem. Soc.*, **80**, 169 (1958).
202. E. Hedaya and S. Winstein, *Tetrahedron Lett.*, 563 (1962); *J. Am. Chem. Soc.*, **89**, 1661 (1967).
203. H. Langhals, G. Range, E. Wistuba and C. Rüchardt, *Chem. Ber.*, **114**, 3813 (1981).
204. C. Rüchardt and H. Schwarzer, *Angew. Chem.*, **74**, 251 (1962); *Chem. Ber.*, **99**, 1878 (1966).
205. R. E. Pincock, *J. Am. Chem. Soc.*, **86**, 1820 (1964); **87**, 1274 (1965).
206. A. Nishinaga, K. Nakamura and T. Matsuura, *J. Org. Chem.*, **48**, 3696, 3700 (1983).
207. A. Nishinaga, K. Nakamura and T. Matsuura, *J. Org. Chem.*, **47**, 1431 (1982).
208. D. R. Dixon and A. Rajaczknowski, *Chem. Commun.*, 337 (1966).
209. P. D. Bartlett, E. P. Benzing and R. E. Pincock, *J. Am. Chem. Soc.*, **82**, 1762 (1960).
210. Y. Nakashio, M. Hasegawa, O. Harada and T. Yamamoto, *Nippon Kagaku Kaishi*, 1863 (1988).
211. P. D. Bartlett and C. Rüchardt, *J. Am. Chem. Soc.*, **82**, 1756 (1960).
212. C. Rüchardt, H. Böck and I. Ruthardt, *Angew. Chem.*, **78**, 268 (1966).
213. W. Duismann and C. Rüchardt, *Justus Liebigs. Ann. Chem.*, 1834 (1976).
214. C. Rüchardt, V. Golzke and G. Range, *Chem. Ber.*, **114**, 2769 (1981).
215. R. A. Firestone, *J. Org. Chem.*, **45**, 3604 (1980).
216. (a) C. J. Pederson, *J. Org. Chem.*, **23**, 255 (1958); (b) E. L. O'Brien, F. M. Beringer and R. B. Mesrobian, *J. Am. Chem. Soc.*, **81**, 1506 (1959).
217. P. D. Bartlett and R. E. Pincock, *J. Am. Chem. Soc.*, **82**, 1769 (1960).
218. P. D. Bartlett, B. A. Contarev and H. Sakurai, *J. Am. Chem. Soc.*, **84**, 3101 (1962).
219. P. D. Bartlett and H. Sakurai, *J. Am. Chem. Soc.*, **84**, 3269 (1962).
220. V. S. Etlis, N. N. Trofimov and G. A. Razuvaev, *Zh. Org. Khim.*, **2**, 973 (1966).
221. M. M. Martin, *J. Am. Chem. Soc.*, **83**, 2869 (1961).
222. Y. Ogata, K. Tomizawa and K. Furuta, in *The Chemistry of Peroxides* (S. Patai, Ed.), Wiley, Chichester (1983), pp. 711–775.
223. R. A. Sheldon and J. K. Kochi, *J. Am. Chem. Soc.*, **92**, 5175 (1970).
224. W. A. Pryor and R. W. Henderson, *J. Am. Chem. Soc.*, **92**, 7234 (1970).
225. (a) W. Adam, O. Cueto, L. N. Guedes and L. O. Rodriguez, *J. Org. Chem.*, **43**, 1466 (1978); (b) W. Adam, O. Cueto and L. N. Guedes, *J. Am. Chem. Soc.*, **102**, 2106 (1980).
226. (a) L. Thijs, S. N. Gupta and D. C. Neckers, *J. Org. Chem.*, **44**, 4123 (1979); (b) I. I. Abu-Abdoun, L. Thijs and D. C. Neckers, *Macromolecules*, **17**, 282 (1984).

227. T. T. H. Fisher and J. C. Martin, *J. Am. Chem. Soc.*, **88**, 3382 (1966).
228. D. L. Tuleen, W. G. Bentrude and J. C. Martin, *J. Am. Chem. Soc.*, **85**, 1938 (1963).
229. C. W. Perkins, J. C. Martin, A. J. Arduengo, W. Lau, A. Alegria and J. K. Kochi, *J. Am. Chem. Soc.*, **102**, 7753 (1980).
230. W. Nakanishi, Y. Kusuyama, Y. Ikeda and H. Iwamura, *Bull. Chem. Soc. Jpn.*, **56**, 3123 (1983).
231. W. A Pryor and W. H. Hendrickson, Jr, *J. Am. Chem. Soc.*, **105**, 7114 (1983).
232. (a) B. G. Dixon and G. B. Schuster, *J. Am. Chem. Soc.*, **101**, 3116 (1979); (b) B. G. Dixon and G. B. Schuster, *J. Am. Chem. Soc.*, **103**, 3068 (1981).
233. G. B. Schuster, *Acc. Chem. Res.*, **12**, 366 (1979).
234. C. Walling and J. C. Azar, *J. Org. Chem.*, **33**, 3888 (1968).
235. N. Muramoto, T. Ochiai, O. Simamura and M. Yoshida, *Chem. Commun.*, 717 (1968).
236. (a) P. D. Bartlett and L. B. Gortler, *J. Am. Chem. Soc.*, **85**, 1864 (1963); (b) P. D. Bartlett and J. M. McBride, *J. Am. Chem. Soc.*, **87**, 1727 (1965).
237. E.g. W. A. Pryor, *Free Radicals*, McGraw-Hill, New York (1966), p. 30.
238. P. D. Bartlett, R. E. Pincock, J. H. Rolston, W. G. Schindel and L. A. Singer, *J. Am. Chem. Soc.*, **87**, 2590 (1965).
239. R. C. Fort, Jr, and R. E. Franklin, *J. Am. Chem. Soc.*, **90**, 5267 (1968).
240. (a) J. A. Kampmeier and R. M. Fantazier, *J. Am. Chem. Soc.*, **88**, 1959 (1966); (b) L. A. Singer and N. P. Kong, *J. Am. Chem. Soc.*, **88**, 5213 (1966); (c) L. A. Singer and N. P. Kong, *J. Am. Chem. Soc.*, **89**, 5251 (1967).
241. (a) A. L. Beckwith and C. Easton, *J. Am. Chem. Soc.*, **100**, 2913 (1978); (b) A. L. Beckwith and C. J. Easton, *J. Chem. Soc., Perkin Trans. 2*, 661 (1983).
242. (a) M. F. Zady and J. L. Wong, *J. Am. Chem. Soc.*, **99**, 5096 (1977); (b) M. F. Zady and J. L. Wong, *Nucleic Acid Chem.*, **1**, 29 (1978).
243. M. F. Zady and J. L. Wong, *J. Org. Chem.*, **44**, 1450 (1979).
244. F. Minisci, R. Mondelli, G. P. Gardini and O. Porta, *Tetrahedron*, **28**, 2403 (1972).
245. D. S. Watt, *J. Am. Chem. Soc.*, **98**, 271 (1976).
246. F. R. Jensen and T. I. Moder, *J. Am. Chem. Soc.*, **97**, 2281 (1975).
247. A. Kharrat, C. Gardrat and B. Maillard, *Can. J. Chem.*, **62**, 2385 (1984).
248. B. Maillard, M. J. Bourgeois, R. Lalande and E. Montaudon, *NATO ASI, Ser. C*, **260**, 283 (1989).
249. J. K. Kochi, *Tetrahedron*, **18**, 483 (1962).
250. D. J. Rawlington and G. Sosnovsky, *Synthesis*, 1 (1972).
251. R. A. Sheldon and J. K. Kochi, *Metal-Catalyzed Oxidation of Organic Compounds*, Academic Press, New York (1981), pp. 42, 289.
252. J. K. Kochi and R. D. Gilliom, *J. Am. Chem. Soc.*, **86**, 5251 (1964).
253. J. K. Kochi and D. M. Mog, *J. Am. Chem. Soc.*, **87**, 522 (1965).
254. G. Sosnowsky, *Tetrahedron*, **21**, 871 (1965).
255. C. Walling and A. A. Zavitsas, *J. Am. Chem. Soc.*, **85**, 2084 (1963).
256. (a) J. K. Kochi, *J. Am. Chem. Soc.*, **84**, 774 (1962); (b) J. K. Kochi and H. E. Mains, *J. Org. Chem.*, **30**, 1862 (1965).
257. K. Pedersen, P. Jacobsen and S. -O. Lawesson, *Org. Synth., Coll. Vol.*, **5**, 70 (1973).
258. H. L. Goering and U. Mayer, *J. Am. Chem. Soc.*, **86**, 3753 (1964).
259. J. J. Villenave, H. Francois and R. Lalande, *Bull. Soc. Chim. Fr.*, 599 (1970).
260. H. Kropf, R. Schröeder and R. Fölsing, *Synthesis*, 894 (1977).
261. (a) H. Kwart, D. A. Benko and M. E. Bromberg, *J. Am. Chem. Soc.*, **100**, 7093, (1978); (b) H. Kwart, M. Brechbiel, W. Miles and L. D. Kwart, *J. Org. Chem.*, **47**, 4524 (1982).
262. A. L. J. Beckwith and A. A. Zavitsas, *J. Am. Chem. Soc.*, **108**, 8230 (1986).

263. G. Berglund and S. -O. Lawesson, *Ark. Kemi*, **20**, 225 (1963).
264. G. Sosnowsky, *Tetrahedron*, **18**, 15 (1962).
265. M. S. Kharash and A. Fono, *J. Org. Chem.*, **23**, 324 (1958).
266. G. Sosnowsky and N. C. Yang, *J. Org. Chem.*, **25**, 899 (1960).
267. S. -O. Lawesson and C. Berglund, *Ark. Kemi*, **17**, 485 (1961).
268. H. Matsumura, T. Yano, M. Ueyama, K. Tori and W. Nagata, *J. Chem. Soc., Chem. Commun.*, 485 (1979).
269. D. J. Rawlinson, M. Konieczny and G. Sosnowsky, *Z. Naturforsch, Teil B*, **34**, 76 (1979).

10 Polyoxides

BOŽO PLESNIČAR
Department of Chemistry, University of Ljubljana, PO Box 537, 61001 Ljubljana, Slovenia

Organic Peroxides. Edited by W. Ando

1 INTRODUCTION

This review deals with the formation, characterization, decomposition and reactivity of polyoxides, i.e. compounds of the general formula $R—O_n—R$, where R is hydrogen or other atoms or groups and $n \geqslant 3$. These species, which may be regarded as higher homologs of hydrogen peroxide, alkyl hydroperoxides, dialkyl peroxides and diacyl peroxides:

$H—O_n—H$	hydrogen polyoxides
$R—O_n—H$	alkyl hydropolyoxides
$R—O_n—R$	dialkyl polyoxides
$R—C(=O)—O_n—C(=O)—R$	diacyl polyoxides

are key intermediates in low-temperature oxidations, atmospheric chemistry, the chemistry of combustion and flames and biochemical oxidations. Since the older literature on organic polyoxides has already been reviewed,[1] this chapter is not exhaustive in coverage but rather concentrates on more recent advances in the field. However, for the sake of completeness, the most relevant older work is also discussed.

2 HYDROGEN POLYOXIDES

2.1 Condensed-phase Experiments

The hydrogen polyoxides, HO_nH ($n = 3$, 4) were proposed as possible intermediates in the decomposition of hydrogen peroxide long ago. However, in the earlier studies the evidence for their existence was only indirect, i.e. it was based mainly on either evolution of oxygen or evolution of heat on warming the samples. The early history of the discovery of hydrogen polyoxides has already been reviewed,[2–6] and will therefore not be repeated here.

It was only relatively recently that direct spectroscopic evidence for the existence of these polyoxides became available. In the 1970s, Giguère and co-workers, in a series of papers,[7–10] reported for the first time the fundamental skeletal vibrations of H_2O_3 and H_2O_4 and the deuteriated analogs by studying infrared and Raman spectra of the products from electrically dissociated water and related hydrogen-oxygen-containing systems (H_2O_2; H_2O and O_2) at cryogenic temperatures. All the observed frequencies for these polyoxides were assigned on the basis of assumed linear structures of C_2 symmetry and are consistent with the temperature behavior, ^{18}O isotope shifts, and depolarization of various bonds. The fundamental skeletal vibrations for H_2O_3 and H_2O_4 are summarized in Table 1.[9] It should be pointed out that these values are for hydrogen-bonded species.

Yagodovskaya *et al.*[11] studied the low-temperature (−198°C) reaction of ozone with atomic hydrogen in liquid ozone and the dissociation of water and hydrogen peroxide vapor with subsequent freezing of the products into the matrix. The IR spectra of the reaction products showed bands which they assigned to H_2O_3 and H_2O_4.

Bielski and Schwartz[12] provided UV spectroscopic evidence for the existence of H_2O_3 in the pulse radiolysis of air-saturated perchloric acid solutions. A UV absorption spectrum very similar to, but more intense than, that of H_2O_2 was reported.

However, Lee and co-workers[13,14] by studying the Fourier transform IR spectra of the products obtained by the photolysis of formaldehyde or glyoxal ($H_2C_2O_2$) in solid O_2 at 13–18 K, found no absorptions that could be attributed to the inner and outer O—O stretch modes of H_2O_3 and H_2O_4, reported by Giguère and co-workers. They concluded that these polyoxides did not form in the photolysis system under investigation or, if they formed, they decomposed under conditions investigated. Since Lee and co-workers did not study the IR

Table 1. Vibrational frequency assignments for the observed IR and Raman (R) absorption bands of hydrogen polyoxides[9]

Frequency (cm^{-1})		Activity	Assignment
HOOOOH	DOOOOD		
855	857	R	O—O sym. stretch
764	768	R	O—O stretch (med.)
450	440	R	O—O—O sym. bend
98	98	R	Torsion (O—O med.)
823	828	R, IR	O—O asym. stretch
435	430	IR	O—O—O asym. bend
HOOOH	DOOOD		
855	857	R, IR	O—O sym. stretch
500	497	R	O—O—O bend
755	760	R, IR	O—O asym. stretch

spectra at longer wavelengths, it is not clear whether the two groups were observing the same species.

Sugimoto, Sawyer and Kanofsky[15] reported that electrolytic reduction of protons in the presence of oxygen in acetonitrile produced $^1O_2$ at a platinum electrode surface. The involvement of HOOOOH in this reaction, which subsequently undergoes a concerted homolytic fragmentation to yield H_2O_2 and $^1O_2$, was suggested on the basis of the results of experiments performed with a 50:50 mixture of $^{36}O_2$ and $^{32}O_2$. HOO· radicals, which are formed during electrolysis, are adsorbed at the electrode surface in such a way (with the H-ends oriented to the surface) that their ends are in a favorable position for radical–radical recombination to produce HOOOOH (reaction 1).

$$\begin{aligned} H^+ &\longrightarrow (Pt)H\cdot \xrightarrow{O_2} (Pt)HOO\cdot \\ &\longrightarrow HOOOOH \longrightarrow 0.5\,H_2O_2 + 0.5\,{}^1O_2 \end{aligned} \tag{1}$$

Giguère and Herman[6,7] reported that H_2O_4 begins to decompose at -100°C, whereas H_2O_3 decomposes at about -55°C. An absorption at 760 cm^{-1}, assigned to DOOOD, disappeared with a half-life of approximately 5 h, corresponding to an apparent first-order rate constant of about 3.8×10^{-5} s^{-1} at -65°C.[7]

Recently, ^1H NMR spectroscopic evidence for the involvement of a transient intermediate, tentatively assigned to H_2O_3, in the decomposition of silyl hydrotrioxides was given[16] (see below). This polyoxide begins to decompose appreciably at about -60°C (δOOOH = 13.3 ppm, acetone-d_6), in agreement with the results of Giguère and Herman.[7]

An oxygen-rich intermediate, tentatively assigned to H_2O_3, has recently been generated independently by UV irradiation of H_2O_2 in methyl acetate at -78°C.[16] This polyoxide decomposes in the temperature range -60 to -30°C to form singlet oxygen ($^1O_2$). The presence of small amounts of water in H_2O_2 appears to assist in the formation and/or stabilization of this polyoxide, sharpens the OOOH absorption (δOOOH = 14.2 ± 1.0 ppm) and shifts its position downfield. The following sequence of reactions is probably involved in the formation of H_2O_3:

$$H_2O_2 + h\nu \rightarrow HO\cdot + HO\cdot \tag{2}$$

$$HO\cdot + H_2O_2 \rightarrow H_2O + HOO\cdot \tag{3}$$

$$HO\cdot + HOO\cdot \rightarrow HOOOH \tag{4}$$

An oxygen-rich intermediate, tentatively assigned to HOOOH, is also formed in the low-temperature (-78°C) ozonation of anhydrous H_2O_2 in methyl acetate.[17] Again, singlet oxygen is formed in its decomposition (isolation of 9,10-dimethylanthracene endoperoxide).

2.2 Gas-phase Experiments

2.2.1 Hydrogen tetraoxide, HO_4H

2.2.1.1 Kinetic data

Experimental evidence has accumulated in recent years that the self-reaction of hydroperoxyl radicals, HOO·, in the gas phase involves an intermediate, H_2O_4, which decomposes to H_2O_2 and O_2:[18–32]

$$HOO\cdot + HOO\cdot \rightarrow [H_2O_4] \rightarrow H_2O_2 + O_2 \tag{5}$$

Hence the original hypothesis that the end products are formed by direct hydrogen abstraction of one HOO radical from the other is inconsistent with positive pressure and negative temperature dependences (a negative activation energy) of the rate constant of reaction 5.

The rate constant for reaction 5 is linearly dependent on pressure, ranging from $(1.91 \pm 0.29) \times 10^{-12}$ cm^3 molecule^{-1} s^{-1} at 100 Torr to $(2.97 \pm 0.45) \times 10^{-12}$ cm^3 molecule^{-1} s^{-1} at 700 Torr in nitrogen (1 Torr = 133.3 Pa). A pressure-independent contribution to the reaction with a rate constant of 1.75×10^{-12} cm^3 molecule^{-1} s^{-1} was calculated (298 K).[21,22] It has also been reported that the rate constant for the self-reaction of DOO radicals:

$$DOO\cdot + DOO\cdot \rightarrow D_2O_2 + O_2 \tag{6}$$

is less dependent on pressure than the corresponding reaction involving HOO radicals. Several other low-pressure (< 10 Torr) studies have given comparable rate constants for reaction 5 [k(298 K) = $(1.6 \pm 0.2) \times 10^{-12}$ cm^{-3} molecule^{-1} s^{-1}].[19,20,27–32]

The self-reaction of HOO radicals shows a surprisingly strong negative temperature dependence of approximately $\exp(600/T)$ or $T^{-1.8}$[20,22] with the pressure-dependent part having an even stronger negative T dependence [$\exp(980/T)$ or T^{-3}].[22] The activation parameters for a pressure-independent channel of this reaction are $E_a = -1.2 \pm 0.4$ kcal mol^{-1} and $A = (2.3 \pm 0.6) \times 10^{-13}$ cm^3 molecule^{-1} s^{-1}.

The mechanism involving the formation of H_2O_2 and O_2 by a molecular elimination from an H_2O_4 intermediate was suggested (reaction 7):

$$\begin{array}{c} HOO\cdot + HOO\cdot \rightleftharpoons H_2O_4{}^* \rightarrow H_2O_2 + O_2 \\ \downharpoonleft\upharpoonright \text{ M} \\ H_2O_4 \end{array} \tag{7}$$

According to this mechanism, H_2O_4 is formed in a highly vibrationally excited state. The pressure-dependent contribution arises from the stabilization of

H_2O_4*, while the zero-pressure contribution originates from the vibrationally excited H_2O_4 molecules which decompose to give H_2O_2 and O_2.

The catalytic effect of the water vapor (2–50 Torr), and also NH_3, on the self-reaction of HOO• radicals has been reported.[27,28] It has been suggested that the increase in the rate constant results from the formation of the complexed hydrogen-bonded HOO radicals which react faster with the HOO than does the uncomplexed HOO radical. Still another possibility is that the larger number of available vibrational modes in the complexed H_2O_4 results in a stabilization of this entity with respect to dissociation to reactants compared with the uncomplexed form.[19,22]

The nature of the intermediate and its decomposition mechanism are still not settled. The observation of a relatively high deuterium isotope effect of k_H/k_D = 2.8,[26] and the low *A* factor for the reaction, suggest that the decomposition reaction proceeds via a cyclic four-center transition state in which hydrogen is transferred intramolecularly, as shown below:

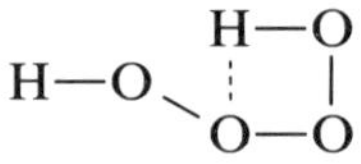

```
          H—O
H—O   ⋮   |
    ╲ O—O
```

The proposed mechanism is also consistent with the observation that disproportionation of HOO radicals proceeds much more slowly in aqueous solution ($k_1 = 1.26 \times 10^{-15}$ cm^3 molecule^{-1} s^{-1}).[33] The engagement of the terminal hydrogens in hydrogen bonds, thus preventing the formation of a cyclic transition state, is believed to be the cause of this phenomenon. In general, the self-reaction of HOO radicals proceeds much faster in hydrocarbon solvents than in polar media.[34]

The other suggested mechanisms for the decomposition of H_2O_4 involving cyclic transition states:

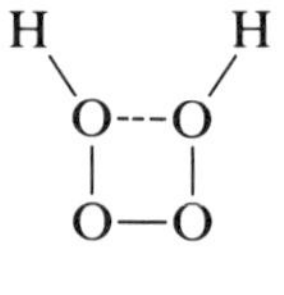

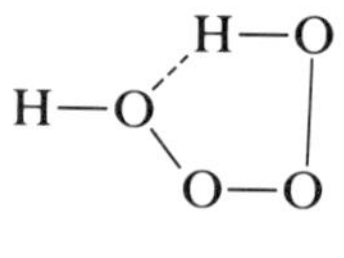

```
H       H                   H—O
 ╲     ╱                  ⋰   |
  O---O             H—O       |
  |   |               ╲       |
  O—O                  O—O
```

were ruled out on the basis of ^{18}O isotope studies by Niki *et al.*,[25] which showed that no isotope mixing in the reaction of $H^{16}O_2$ and $H^{18}O_2$ radicals, thus ruling out the five-centered transition state. At the same time, virtually no ozone was detected in the reaction mixture. However, these studies could not distinguish between a linear HOOOOH molecule leading to a cyclic transition state and a hydrogen-bonded cyclic dimer, $(HOO)_2$. Such an intermediate has been suggested to be involved in the disproportionation of HOO radicals in the gas phase (reaction 8).[35]

$$HOO\cdot + \cdot OOH \rightleftharpoons \left[\begin{array}{c} O \cdots H - O \\ | \qquad\quad | \\ O - H \cdots O \end{array} \right] \rightarrow H_2O_2 + O_2 \tag{8}$$

The HOO radical has a relatively large dipole moment (2.1 D), so that hydrogen-bonded entities of either the head-to-tail type (OOH ··· OOH) or a doubly hydrogen-bonded cyclic dimer are reasonable structures. Tso and Lee[14] argued in favor of the formation of cyclic dimers even in the solid state by studying the formation of HOO radicals in an O_2 matrix between 12 and 20 K.

The results of RRKM calculations using either a linear HOOOOH intermediate or a cyclic, doubly hydrogen-bonded intermediate are consistent with the observed pronounced temperature and pressure dependences and the reported isotope effect for the self-reaction of HOO radicals.[30] Quantum-mechanical tunneling, a very likely possibility in this type of reaction (particularly at low temperatures), was not considered in these calculations.

Further work is needed to clarify the details of the mechanism of this reaction. A direct spectroscopic identification of H_2O_4 in solution, and the study of its decomposition, would be particularly helpful in this respect.

2.2.1.2 Structure and thermochemical data

The results of *ab initio* studies on the structure of H_2O_4, by using the double zeta plus polarization (DZ + P) basis set [O(9s5p1d/4s2p1d), H(4s1p/2s1p)] revealed that the straight-chain structure of H_2O_4 is slightly more stable (11 kcal mol^{-1}, DZ + P CI) with respect to two HOO radicals than the cyclic two hydrogen-bonded entity (7 kcal mol^{-1}).[36] An *ab initio* BAC-MP4 calculation gives an even greater stability, i.e. 15 kcal mol^{-1}, for HOOOOH species.[37]

The calculated SCF equilibrium geometries for the monomeric and dimeric hydroperoxyl radical and for the open-chain HOOOOH are shown in Figure 1. The chain structure, i.e. a minimum on the potential energy surface, possesses no elements of symmetry (C_1). Hydrogen atoms are directed toward the opposite HO_2 group, thus reducing the distances from the hydrogen atom to the nearest oxygen of the other HO_2 group to 2.51 and 2.54 Å (DZ + P). The presence of intramolecular hydrogen bonds might account for the asymmetry of this species.[36] The possibility of the existence of several conformational forms of H_2O_4 of similar energy, including some with intramolecular hydrogen bonds, has been demonstrated previously in a theoretical study at a lower level of sophistication.[38]

At present, the lowest estimate for ΔH_f° (H_2O_4) is $-(6.2 \pm 2.1)$ kcal mol^{-1}

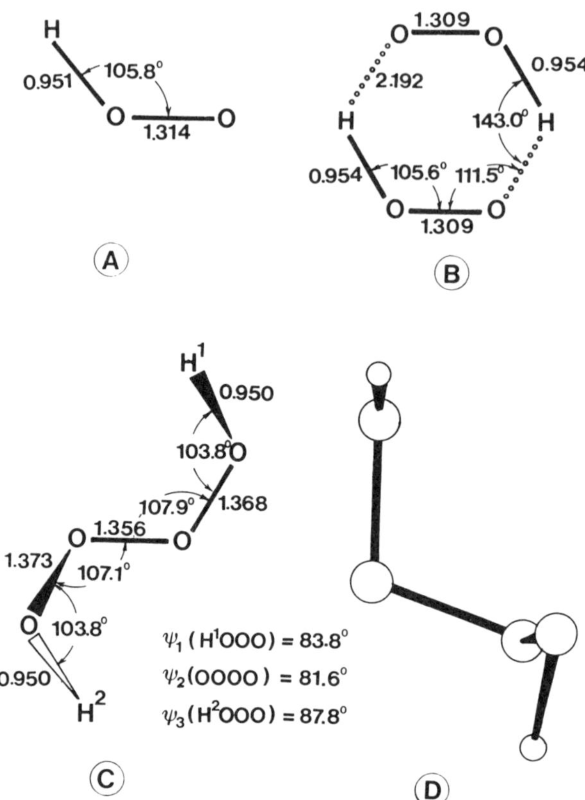

Figure 1. Equilibrium geometries calculated by using the double zeta plus polarization basis set [O(9s5p1d/4s2p1d), H(4s1p/2s1p)]:(A) hydroperoxyl radical; (B) hydroperoxyl radical dimer; (C) hydrogen tetraoxide, H_2O_4; (D) H_2O_4 as viewed within the plane of one HOO group. The angle between the planes of the OOH groups is 24.5°. The bond lengths are in ångströms and the bond angles in degrees[36]

and the bond dissociation energy is D(HOO—OOH) = (12.1 ± 2.1) kcal mol^{-1}.[39] The latter value is in good agreement with the prediction of an *ab initio* DZ + P CI study, which gives D(HOO—OOH) = 11 kcal mol^{-1}.[36] The self-reaction of HOO radicals to produce H_2O_2 and O_2 is exothermic, i.e. $\Delta H_r^\circ = -17$ kcal mol^{-1} ($^1O_2$) and -39 kcal mol^{-1} ($^3O_2$).[40]

2.2.2 Hydrogen trioxide, HO_3H

There is much less consensus of opinion about the existence of a linear HOOOH as an intermediate in the gas-phase reaction 9.

$$HO\cdot + HOO\cdot \rightarrow H_2O + O_2 \tag{9}$$

2.2.2.1 Kinetic data

This reaction has received a great deal of attention mainly because of its importance in atmospheric chemistry.[26,41–49] The reaction is much faster [k(298 K) = (6.2–8.0) × 10^{-11} cm^3 molecule^{-1} s^{-1} at pressures < 10 Torr[41a,43,44,46]] than the self-reaction of HOO radicals. A low-pressure temperature dependence study gave an exp(420/T) or $T^{-1.3}$ dependence.[45] The corresponding activation parameters are $E_a = -0.8 \pm 0.4$ kcal mol^{-1} and $A = 1.7 \times 10^{-11}$ cm^3 molecule^{-1} s^{-1}. Higher values for the rate constants were obtained at atmospheric pressure [He, Ar, N_2, k(298 K) = (1.0–1.2) × 10^{-10} cm^3 molecule^{-1} s^{-1}].[42]

Any attempt to explain the mechanism of this reaction must therefore account for the pronounced negative temperature and positive pressure dependences of the rate of the reaction. It has been suggested that a vibrationally excited intermediate of the structure HOOOH is involved.[41b,43,46,48] This could decompose either back to reactants or in a four-center process to eliminate H_2O (the zero-pressure contribution), or be collisionally stabilized by removing enough energy from the intermediate so that it can no longer dissociate to reactants (the pressure-dependent contribution):

$$\mathrm{HO\cdot} + \mathrm{HOO\cdot} \rightleftharpoons \mathrm{HOOOH^*} \longrightarrow \begin{cases} \longrightarrow \mathrm{H_2O + O_2} \\ \xrightarrow{\mathrm{M}} \mathrm{HOOOH} \end{cases} \tag{10}$$

However, isotopic labeling experiments,[41] i.e. the lack of production of appreciable amounts of labeled O_2 from labeled OH reactant, seem to rule out such a possibility as a predominant reaction pathway. Rather, it was suggested on the basis of the results of *ab initio* calculations[50] that an internally excited weakly hydrogen-bonded intermediate, which stabilizes the tight 'early' transition state for the abstraction of the hydrogen atom to such an extent that it is no longer the rate-determining one, is formed first (reaction 11).

$$\mathrm{HO\cdot} + \mathrm{HOO\cdot} \rightleftharpoons [\mathrm{HO \cdots HOO}] \rightarrow \mathrm{H_2O + O_2} \tag{11}$$

In contrast to the self-reaction of HOO radicals, which deviate significantly from the bond energy–bond order model, the reaction of $HO^{\bullet}$ with $HO_2^{\bullet}$ follows this model very well.[49] The negative temperature dependence was explained as being due to the preferential decomposition of this adduct to reform reactants at higher temperatures. However, the hydrogen-bonded intermediate is probably too weakly bound to produce a significant pressure dependence. Hence it is plausible that two reaction pathways are available for this reaction, i.e. the first involving the hydrogen atom abstraction and the second the formation of HOOOH.

2.2.2.2 Structure and thermochemical data

Several *ab initio* molecular orbital studies, employing minimal[51–53] and extended basis sets,[16,54,55] indicate a zig-zag skew-chain structure for H_2O_3. The geometry parameters for the hydrogen bonded adduct HO···HOO [DZ + CI(SDQ)][55] and the monomeric form of H_2O_3 (MP2/6-31G*)[54] are shown in Figure 2.

Very similar geometry parameters for H_2O_3 were reported recently by using the 6-31G basis set.[16] Since it is known that the dihedral angle in peroxides is not well reproduced unless the highest level calculations including extensive corrections for electron correlation are used, it was stated that this is due to the fact that in H_2O_3 the rotational potential is much steeper than that in H_2O_2. Therefore, the 6-31G basis set was also used to study the cyclic dimeric H_2O_3 (Figure 2). The calculated relatively strong binding energy for the intermolecularly hydrogen-bonded cyclic dimer, $BE = 7.7$ kcal mol^{-1} (6-31G**//6-31G), indicates that self-association might be the characteristic structural feature of this species.

The implications of the structures on the decomposition of H_2O_3 are obvious. Either the proton is transferred intramolecularly to the terminal oxygen atom via a four-center transition state as shown in reaction 12:

$$\left[\begin{array}{c} \text{O}{=}{=}\text{O} \\ /\quad\quad\backslash \\ \text{H}\text{-------}\text{O—H} \end{array}\right]^{\ddagger} \rightarrow H_2O + {}^1O_2 \qquad (12)$$

or a two-proton transfer involving a cyclic dimeric structure is operative in the decomposition of this polyoxide (reaction 13).

$$\begin{array}{ccc} & \text{O—O} & \\ & / \quad\quad \backslash & \\ \text{H—O} & & \text{H} \\ \vdots & & \vdots \\ \text{H} & & \text{O—H} \\ & \backslash \quad\quad / & \\ & \text{O—O} & \end{array} \rightarrow 2\,H_2O + 2\,{}^1O_2 \qquad (13)$$

At least in solutions at low temperatures, these 'polar' pathways[16a] are preferable to the radical ones suggested previously for the decomposition of H_2O_3,[40] and unless the intramolecular proton transfer is assisted in some way (for example in HOOOH-oxygen base complexes),[16a] we prefer the intermolecular proton transfer shown in reaction 13. Namely, our *ab initio* calculations on reaction 12 showed that the activation energy of 55.2 kcal mol^{-1} at the MP4/6-31G**//6-31G level of sophistication (51.5 kcal mol^{-1}, 6-31G//6-31G) is too high for this reaction to proceed on the singlet potential energy surface. The reaction was calculated to be exothermic by 2.7 kcal mol^{-1} (2.5 kcal mol^{-1}, 6-31G//6-31G).[16b]

The enthalpy change (ΔH_r°) for the reaction of HO with HOO radicals to produce H_2O and O_2 is -48 kcal mol^{-1} ($^1O_2$) and -70 kcal mol^{-1} ($^3O_2$). The reaction is, therefore, exothermic enough to open channels to the production of singlet oxygen ($^1\Delta_g$ and $^1\Sigma_g$).

The covalently-bonded HOOOH was found to be 26.4 kcal mol^{-1} more stable than the HO and HOO radicals.[55] The heat of formation of the H_2O_3 open-chain form was estimated on thermochemical grounds to be $\Delta H_f^\circ = -19.4 \pm 1.4$ kcal mol^{-1} and D(HOO—OH) $= 31.7 \pm 1.4$ kcal mol^{-1}.[39] The binding energy in the adduct HO···HOO was calculated to be 4.7 kcal mol^{-1}.[55]

The Mulliken population analysis for HOOOH shows a considerable accumulation of charge on both terminal oxygen atoms with only little charge accumulated on the central oxygen (O_1, O_3, -0.31 e; O_2, -0.07 e; H, -0.35 e (6-31G**/CI).[55] The calculated dipole moment of HOOOH is 1.26 D (6-31G**/CI)[55] and 1.13 D (6-31G**//6-31G).[16]

3 DIALKYL POLYOXIDES

The known role of dialkyl polyoxides, RO_nR, goes back to the kinetic analysis of autoxidation, which shows that kinetic chains are terminated by a bimolecular reaction between two alkylperoxyl radicals[56–63] to form short-lived intermediates, i.e., dialkyl tetraoxides.[57]

3.1 Tetraoxides

3.1.1 Tertiary tetraoxides

Bartlett and Guaraldi[64] were the first to demonstrate unambiguously the involvement of a dialkyl tetraoxide in the low-temperature self-reaction of *tert*-butylperoxyl radicals generated by irradiation of di-*tert*-butyl peroxycarbonate in frozen dichloromethane at -196°C with 254.0 nm radiation, or by the oxidation of *tert*-butyl hydroperoxide with lead tetraacetate at -90°C. They observed a reversible change in their concentration below -85°C and concluded that below this temperature the *tert*-butylperoxyl radicals were in equilibrium with di-*tert*-butyl tetraoxide (reaction 14). This polyoxide decomposes at higher temperatures (> -80°C) to produce two *tert*-butoxyl radicals (presumably in a concerted process) and oxygen. The alkoxyls may recombine in the solvent cage to form di-*tert*-butyl peroxide or they may escape from the cage (reaction 15).

$$ROO\cdot + ROO\cdot \rightleftharpoons ROOOOR \qquad (14)$$

$$ROOOOR \rightarrow [RO\cdot + O_2 + \cdot OR] \longrightarrow \begin{cases} ROOR + O_2 \\ 2RO\cdot + O_2 \end{cases} \qquad (15)$$

Milas and Plesničar[65] used the criterion of oxygen evolution to demonstrate the involvement of di-*tert*-butyl tetraoxide in the low-temperature reaction of iodosobenzene or (diacetyloxy)iodobenzene with *tert*-butyl hydroperoxide in dichloromethane below $-80°C$. On warming to *ca* $-75°C$, strong evolution of oxygen was observed.

Di-*tert*-butyl tetraoxide has also been prepared by photolysis of azoisobutane, azoisobutyronitrile and azocyclohexylnitrile in the presence of oxygen.[66]

The existence of equilibrium between tetraoxides and *tert*-alkylperoxyl radicals at low temperatures was later confirmed independently by two groups. Bennett and co-workers[67–69] studied this equilibrium by photolysing di-*tert*-butyl peroxide in oxygenated alkanes, while Ingold, Howard and Adamic[70–74] used photolysis of azo compounds in oxygen-saturated dichlorodifluoromethane (Freon 12) and photolysis of hydroperoxides as a source of the peroxyl radicals.

Thermodynamic parameters for the equilibrium between tertiary alkylperoxyl radicals and the corresponding tetraoxides can be described as $\Delta H° = -(8.8 \pm 1.0)$ kcal mol^{-1} and $\Delta S° = -(34 \pm 6)$ cal K^{-1} mol^{-1} (the standard state being 1 M).[68,71] These thermodynamic parameters can, together with the rate constants for termination of tertiary peroxyl radicals, which range from $10^{2.5}$ to 10^5 l mol^{-1} s^{-1} (at ambient temperature),[67–78] be used to calculate activation parameters for irreversible tetraoxide decomposition (Table 2).[68,71,73]

Isotope studies[79] of the self-reactions of t-Bu$^{18}O^{18}O$ and t-Bu$^{16}O^{16}O$ radicals showed that there is head-to-head and not head-to-tail collision as suggested earlier.[80] The relative amounts of $^{32}O_2$, $^{34}O_2$ and $^{36}O_2$ were found to be those expected from statistical scrambling.

Table 2. Arrhenius parameters for the irreversible decomposition of some tetraoxides (ROOOOR)

R	E_a (kcal mol^{-1})	Log A	Ref.
tert-Butyl	17.5	16.6	71
Cumyl	16.5	17.1	71
2,2,3-Trimethylbut-3-yl	16.2 ± 2.0	17.5 ± 2.4	68
2-Methylpent-2-yl	18.2 ± 1.7	19.6 ± 2.2	68

The suggestion that only one bond in the tetraoxide is cleaved in the rate-determining step (reaction 16):[71]

$$ROOOOR \rightarrow ROOO\cdot + RO\cdot \rightarrow RO\cdot + O_2 + RO\cdot \qquad (16)$$

was rejected on thermochemical grounds.[78] The formation of $ROOO^{\bullet}$ from $RO^{\bullet}$ and O_2 is endothermic by *ca* 12 kcal mol^{-1} ($R = t$-Bu). Also, all attempts to detect *t*-BuOOO radicals by ESR at $-196°C$ failed.[71] However, the question of the mechanism of decomposition of di-*tert*-alkyl tetraoxides (concerted or two-step process) deserves further consideration.

3.1.2 Primary and secondary tetraoxides

In 1956, Russell[57] suggested that the self-reaction of secondary and primary peroxyl radicals involves tetraoxides. These polyoxides are believed to be formed rapidly and reversibly with subsequent decomposition to molecular products. The formation of ketone, alcohol and oxygen together with the observations of a deuterium isotope effect ($k_H/k_D = 1.37 \pm 0.14$, at 30°C)[81] suggest a cyclic transition state for the decomposition of the tetraoxides (reaction 17).

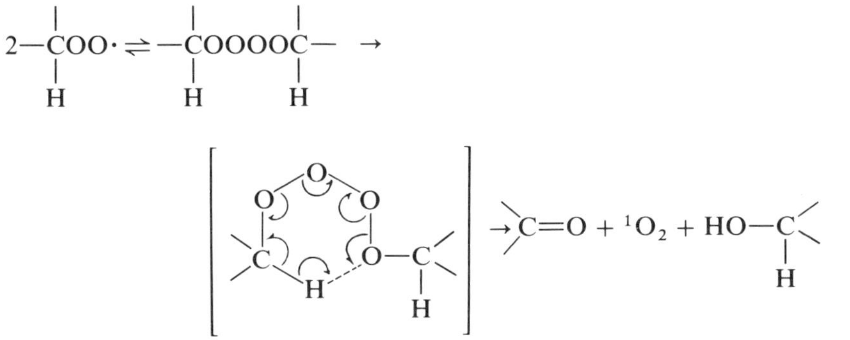

$$2\,\text{—}\overset{|}{\underset{\text{H}}{\text{C}}}\text{OO}\cdot \rightleftharpoons \text{—}\overset{|}{\underset{\text{H}}{\text{C}}}\text{OOOO}\overset{|}{\underset{\text{H}}{\text{C}}}\text{—} \rightarrow$$

$$\left[\text{cyclic transition state: } >C(\text{–O–O–O–O–}C(H)<)\cdots H\right] \rightarrow \;>C{=}O + {}^{1}O_2 + HO\text{—}\overset{|}{\underset{\text{H}}{\text{C}}}< \qquad (17)$$

This mechanism is consistent with the observed low activation energies for the termination of primary and secondary peroxy radicals ranging from 1 to 4 kcal mol^{-1} with an *A* factor of 10^8–10^9 l mol^{-1} s^{-1}.[68,82] The reported termination second-order rate constants, $2k_t$ (the overall termination rate constants), are in the range 10^7–10^8 l mol^{-1} s^{-1} (30°C, determined from the Arrhenius plot at temperatures $> -80°C$).[68,76,82] However, the kinetic evidence indicates that below $-80°C$ the self-reaction of secondary (and probably primary peroxyl radicals also) is not a simple bimolecular reaction involving a cyclic transition state but is most likely a complex radical reaction.[82]

The isotopic distribution of the oxygen evolved from the self-reaction of a number of primary and secondary alkylperoxyl radicals in an inert solvent indicates a head-to-head reaction.[79]

A concerted termination scheme for secondary peroxyl radicals (the Russell mechanism) requires that oxygen evolved should be formed in an excited singlet state. Howard and Ingold[84] were the first to confirm this prediction by trapping $^1O_2$ with 9,10-diphenylanthracene. Both singlet states of oxygen have also been detected by spectroscopic methods.[85,86]

These reactions are sufficiently exothermic to produce electronically excited ketones. The spin conservation rule requires that an electronically excited triplet ketone be produced if ground-state triplet oxygen is formed. Lee and Mendenhall[87] measured the yields of excited ketones from self-reactions of various alkylperoxyl radicals by measuring the chemiluminescence emission from solutions of $R^1R^2CHO_2H$, or oxygen-saturated solutions of $R^1R^2CH_2$, in the presence of a radical initiator *trans*-RON=NOR (R = *t*-Bu or R^1R^2CH) in paired experiments. They reported an activation energy of about 9 kcal mol^{-1} higher for the production of a ketone in the triplet state than in the ground state.

Structural effects on the yields of singlet oxygen from alkylperoxyl radical recombination were studied recently.[88] The peroxyl radicals were generated from radical-initiated autoxidation of various hydrocarbons. The yields of singlet oxygen ranged from 3.4 to 6% for primary ($PhCMe_3$, PhMe) and from 3.9 to 14% for secondary (n-$C_{12}H_{26}$, PhEt) alkylperoxyl terminations. Although several other mechanistic possibilities for the generation of $^1O_2$ in these reactions were considered (reactions 18b–d), it was concluded that a concerted

$$R^1R^2C(H)\text{—O—O—O—OCHR}^1R^2 \rightarrow R^1R^2CHOH + R^1R^2CO(S_0) + {}^1O_2 \quad (18a)$$

$$\rightarrow R^1R^2CO + R^1R^2CHOOOH \rightarrow R^1R^2CHOH + {}^1O_2 \quad (18b)$$

$$\rightarrow 2R^1R^2CHO + {}^3O_2 \rightarrow R^1R^2CO + R^1R^2CHOH \quad (18c)$$

$$\rightarrow \overline{R^1R^2CO(T_0) + {}^3O_2} \rightarrow R^1R^2CO(S_0) + {}^1O_2 \quad (18d)$$

termination scheme (reaction 18a) is the predominant reaction pathway. It is possible that this reaction yields predominantly $^1O_2$ ($^1\Sigma_g$), which partitions between $^1O_2$ ($^1\Delta_g$) and $^3O_2$ ($^3\Sigma_g$). This presumption is supported by the observed relative insensitivity of $^1O_2$ ($^1\Delta_g$) to alkyl structure.[88]

While the Russell mechanism remains an adequate explanation for the generation of singlet oxygen by primary and secondary peroxyl radicals, a recent demonstration of the production of singlet oxygen from the bimolecular reactions of tertiary peroxy radicals (*tert*-butylperoxyl, cumylperoxyl)[89] appears to suggest that for *tert*-butyl peroxyl radicals, the decomposition of the tetroxide (reaction 19), is the main reaction pathway. However, careful product

$$ROOOOR \rightarrow ROOR + {}^1O_2 \quad (19)$$

studies on these or similar systems are necessary to substantiate this claim.

It has been suggested recently, that the $^1O_2$ emission from the oxygenation of various halocarbons[90] and hexafluorobenzene[91] with superoxide anion results from the dimerization of the primarily formed peroxyl radicals to form the tetraoxide intermediate (reactions 20 and 21).

$$RX + O_2^- \rightarrow ROO\cdot + X^- \tag{20}$$

$$2ROO\cdot \rightarrow ROOOOR \rightarrow {}^1O_2({}^1\Delta_g) + \text{products} \tag{21}$$

The observation that secondary peroxyl radicals behave similarly to tertiary peroxyl radicals in forming tetraoxides reversibly below −100°C (ESR),[82] together with the difficulties in accommodating the kinetic and/or thermochemical data with the concerted mechanism (the Russell mechanism) discussed above, led Nangia and Benson[92] to suggest the involvement of the Criegee zwitterion in the initial step of the self-reaction of primary and secondary peroxyl radicals (reaction 22).

$$2RCH_2OO\cdot \rightarrow RCH_2OOH + \overline{RCHOO} \tag{22}$$

$$\overline{RCHOO} + RCH_2OO\cdot \rightarrow (RCHO + RCH_2OOO\cdot) \tag{23}$$

$$(RCHO + RCH_2OOO\cdot) \rightarrow RCHO + RCH_2O\cdot + {}^1O_2 \tag{24}$$

$$RCH_2O\cdot + RCH_2OO\cdot \rightarrow RCH_2OH + \overline{RCHOO} \tag{25}$$

Although this is an attractive mechanistic possibility, recent experiments by Niki *et al.*[93] using the Fourier transform IR method failed to provide evidence for the involvement of a carbonyl oxide (trapping technique) in the gas-phase self-reaction of CH_3OO radicals in O_2–N_2 mixtures at 700 Torr at ambient temperature. Similar observations were also made by Kan *et al.*[94] The self-reaction of methylperoxyl radicals is believed to proceed mainly by three reaction paths, all involving dimethyl tetraoxide as the primary intermediate (reactions 26a–26c).[95–100]

$$2CH_3OO\cdot \rightarrow CH_3OOOOCH_3 \rightarrow 2CH_3O\cdot + O_2 \tag{26a}$$

$$\rightarrow CH_2O + CH_3OH + O_2 \tag{26b}$$

$$\rightarrow CH_3OOCH_3 + O_2 \tag{26c}$$

$$\rightarrow CH_3OOH + \overline{CH_2OO} \tag{26d}$$

The relative rate constants were determined to be $k_{26a}/k_{26c} = 1.32 \pm 0.16$, $k_{26b}/k_{26c} > 7$ and $k_{26d} \ll k_{26b}$. The reported apparent rate constants for the second-order disappearance of $CH_3OO^\cdot$ ($2CH_3OO^\cdot \rightarrow$ products) was found to be $(3.7 \pm 0.6) \times 10^{-13}$ cm^3 molecule^{-1} s^{-1} (25°C).[94–99] However, the actual value for k is about 15% lower than that derived from a purely second-order analysis owing to the secondary removal of CH_3OO radicals by $HOO^\cdot$, formed by the reaction of methoxyl radicals with O_2 (reaction 27).

$$CH_3O\cdot + O_2 \rightarrow CH_2O + HOO\cdot \tag{27}$$

Table 3. Vibrational frequency assignments for the observed IR absorption bands of some isotopically labeled dimethyl tetraoxide species[101]

Frequency (cm^{-1})				
$CH_3{}^{16}O_4CH_3$	$CD_3{}^{16}O_4CD_3$	$CH_3{}^{18}O_4CH_3$	$CD_3{}^{18}O_4CD_3$	Assignment
978	902	950 ± 5	896	C—O stretch
960	893	945 ± 5	872	C—O stretch
775	—	736 ± 5	—	O—O stretch
580	570	548	541	O—O—O bend
457	416	442	400	C—O—O bend
296		283		O—O—O bend

This leads to a revised Arrhenius expression for k over the temperature range 228–380 K:[100]

$$k = (1.4 \pm 0.4) \times 10^{-13} \exp(220 \pm 70)/T \text{ cm}^3 \text{ molecule}^{-1} \text{ s}^{-1}$$

All the above evidence, i.e. the small A factor ($< 10^{-12}$ cm^3 $molecule^{-1}$ s^{-1}) and strong negative temperature dependence (negative activation energy), suggest a decomposition of a stable intermediate which involves a sterically constrained cyclic transition state.

Dimethyl tetraoxide, $CH_3OOOOCH_3$, has recently been identified by an IR matrix technique.[101] It was generated by allowing methyl radicals to interact with matrices of Ar plus 10% O_2 (isotopically labeled forms). Vibrational frequency assignments for this polyoxide are summarized in Table 3.

3.2 Trioxides

3.2.1 Di-*tert*-butyl and dicumyl trioxides

The oxygen evolution temperature was used independently by Bartlett and Günther[102] and Milas and Plesničar[65] to characterize di-*tert*-butyl trioxide in the low-temperature reaction of *tert*-butyl hydroperoxide with lead tetraacetate[102] and (diacetyloxy)iodobenzene and iodosobenzene,[65] respectively.

When present in large excess, *tert*-butyl hydroperoxide is decomposed by *tert*-butoxyl radicals present after the decomposition of di-*tert*-butyl tetraoxide (< -80°C) (reactions 28 and 29). Some *tert*-butylperoxyl radicals formed in this process are trapped by *tert*-butoxyl radicals below -30°C to form the corresponding trioxide (reaction 30). Thus, a high yield of the trioxide is expected when *tert*-butoxyl and *tert*-butylperoxyl radicals are generated in close vicinity of each other. Bartlett and Lahav[103] prepared di-*tert*-butyl trioxide by photolyzing solid di-*tert*-butyl diperoxymonocarbonate at -50°C (reaction 31), or by the reaction of ozone with either dissolved *tert*-butyl hydroperoxide or suspended hydrated sodium salt of the hydroperoxide in Freon 12 (CF_2Cl_2) or methyl chloride.

The ozone–sodium salt method was also used for the preparation of dicumyl trioxide.[102] By using a special low-temperature procedure for freeing the trioxides from accompanying alcohol and other impurities, they obtained crystalline di-*tert*-butyl and dicumyl trioxides.

$$\text{ROOOOR} \rightarrow 2\text{RO}\cdot + \text{O}_2 \quad (\text{and } \rightarrow \text{ROOR} + \text{O}_2) \tag{28}$$

$$\text{ROOH} + \text{RO}\cdot \rightarrow \text{ROO}\cdot + \text{ROH} \tag{29}$$

$$\text{ROO}\cdot + \text{RO}\cdot \rightarrow \text{ROOOR} \tag{30}$$

$$\text{ROOC(=O)OOR} \rightarrow \text{ROOOR} + \text{CO}_2 \tag{31}$$

Both trioxides are unstable at temperatures above −30°C. NMR characteristics and kinetic and activation parameters for the decomposition of di-*tert*-butyl and dicumyl trioxide are given in Tables 4 and 5, respectively.

Table 4. ^{1}H NMR spectroscopic data for di-*tert*-butyl and dicumyl trioxides and the corresponding homologs in $CFCl_3$ at −70°C

Compound	Cumyl[a], δCH_3 (ppm)	*tert*-Butyl[b], δ (ppm)
ROOOR	1.67	1.25
ROOR	1.53	1.16
ROOH	1.55	1.19
ROH	1.58	1.21

[a]Ref. 103.
[b]Ref. 66.

Table 5. Kinetic and activation parameters for the decomposition of dicumyl and di-*tert*-butyl trioxide

Trioxide	Solvent	T (°C)	$k \times 10^4$ (s^{-1})	E_a (kcal mol^{-1})	Log A
Dicumyl	$CFCl_3$[a]	−25.0	0.93	24.1	16.9
		−17.5	2.1		
		−15.0	6.2		
	CH_2Cl_2[a]	−24.8	2.1		
Di-*tert*-butyl	CH_2Cl_2[b]	−33.3	0.72	23.8	17.6
		−24.8	4.0 ± 0.6		

[a]Ref. 103.
[b]Ref. 102.

Bennett *et al.*[104] reported ESR evidence for the production of methyl *tert*-butyl trioxide in the low-temperature ($< -130°C$) photolysis of a solution of di-*tert*-butyl peroxide in oxygenated cyclopropane or Freon 12. This trioxide is unstable even at $-130°C$. The most plausible explanation for its formation is the sequence of reactions 32–35. The trioxide can decompose by two pathways (reaction 36).

$$t\text{-BuOOBu-}t \rightarrow 2t\text{-BuO}\cdot \tag{32}$$

$$t\text{-BuO}\cdot \rightarrow \text{Me}\cdot + \text{Me}_2\text{C}{=}\text{O} \tag{33}$$

$$\text{Me}\cdot + \text{O}_2 \rightarrow \text{MeOO}\cdot \tag{34}$$

$$\text{MeOO}\cdot + t\text{-BuO}\cdot \rightarrow \text{MeOOOBu-}t \tag{35}$$

$$\text{MeOOOBu-}t \rightarrow \text{MeOO}\cdot + t\text{-BuO}\cdot \tag{36a}$$

$$\text{MeOOOBu-}t \rightarrow \text{MeO}\cdot + t\text{-BuOO}\cdot \tag{36b}$$

The absence of an ESR signal of $\text{MeOO}^{\bullet}$ in the decomposition is not surprising since the cleavage to *tert*-butylperoxyl and methoxyl radicals is energetically more favorable than that producing *tert*-butoxyl and methylperoxyl radicals [$D(t\text{-BuOO—OMe}) = 30$ kcal mol^{-1}, $D(t\text{-BuO—OOMe}) = 21$ kcal mol^{-1})]. At the same time, the self-reaction of MeOO radicals is several orders of magnitude faster than that for the *t*-BuOO radicals.[104]

Several other unsymmetrically substituted trioxides have also been prepared by photolyzing di-*tert*-butyl peroxide in oxygenated hydrocarbons.[104]

3.2.2 Bis(trifluoromethyl) trioxide and other fluoro-substituted trioxides

A surprisingly stable bis(trifluoromethyl) trioxide was isolated by Anderson and Fox[105] in yields of up to 84% from the reaction of F_2O with COF_2 over CsF catalyst. This trioxide has also been prepared independently by the direct fluorination of salts of trifluoroacetic acid.[107] The isolation of three other perfluorotrioxides, i.e. $CF_3OOOC_2F_5$, $C_2F_5OOOC_2F_5$ and $CF_3OOOCF_2OOCF_3$ were reported.[106,107] All these polyoxides were characterized by mass, IR[107b] (CF_3OOOCF_3, νO—O—O, 875 (sym. stretch), 773 (asym. stretch), 288 cm^{-1} (bend)) and ^{19}F NMR[106] spectra (^{19}F NMR, CF_3OOOCF_3, δ 68.7; CF_3OOCF_3, δ 69.0; CF_3OCF_3, δ 58.3 ppm downfield from $CFCl_3$ as internal standard and solvent, 25°C).

Bis(trifluoromethyl) trioxide decomposes to give mainly bis(trifluoromethyl) peroxide and oxygen. The half-life for this reaction is about 65 weeks at 25°C.[106] Hohorst *et al.*[107a] found that the decomposition is first order with an activation energy about 30 kcal mol^{-1}. The estimated value[39] for the bond dissociation energy $D(CF_3OO\text{—}OCF_3)$ is 29 ± 4 kcal mol^{-1} (the experimental value is

30.3 kcal mol^{-1}).[108] The mechanism of decomposition is probably similar to that already proposed for other trioxides.

$$CF_3OOOCF_3 \rightarrow CF_3OO\cdot + CF_3O\cdot \tag{37}$$

$$2CF_3OO\cdot \rightarrow CF_3OOOOCF_3 \rightarrow [CF_3O\cdot O_2\cdot OCF_3] \begin{cases} \rightarrow CF_3OOCF_3 + O_2 \\ \rightarrow 2CF_3O\cdot + O_2 \end{cases} \tag{38}$$

It is interesting that oxygen addition to trifluoromethoxyl radical (reaction 39), although endothermic, must occur even at $-196°C$.

$$CF_3O\cdot + O_2 \rightarrow CF_3OOO\cdot \tag{39}$$

Fessenden[109] reported low-temperature ESR evidence for trifluoromethyltrioxyl radicals in the low-temperature photolysis of bis-(trifluoromethyl)peroxide in the presence of oxygen. This appears to be the only trioxyl radical detected so far.

3.2.3 Bis(pentafluorosulfur) trioxide

SF_5OOOSF_5, another relatively stable trioxide, was first reported by Czarnowski and Schumacher.[110] They found that thermal decomposition between 5 and 25°C in the presence of a sufficiently high pressure of oxygen yields SF_5OOSF_5 and O_2 as the only products. In the absence of oxygen and at higher temperatures, the formation of SF_5OSF_5 and O_2 is favored. In the presence of over 100 Torr of oxygen, the homogeneous decomposition of the trioxide is strictly first order with respect to the trioxide pressure and independent of the total pressure, inert gases and the reaction products. The activation parameters ($E_a = 26.0$ kcal mol^{-1}, $\log A = 16.1$) indicate a mechanism of decomposition similar to that already proposed for simple dialkyl trioxides (equations 40–44).

$$SF_5OOOSF_5 \rightarrow SF_5OOSF_5 + \tfrac{1}{2}O_2 \tag{40}$$

$$SF_5OOOSF_5 \rightarrow SF_5OSF_5 + O_2 \tag{41}$$

$$SF_5OOOSF_5 \rightarrow SF_5OO\cdot + SF_5O\cdot \tag{42}$$

$$2SF_5OO\cdot \rightarrow 2SF_5O\cdot + O_2 \tag{43}$$

$$2SF_5O\cdot \rightarrow SF_5OOSF_5 \tag{44}$$

3.2.4 Miscellaneous trioxides

There is some indication in the literature that oxidative cleavage of single C—C bonds by ozone at low temperatures ($-45°C$) to yield substantial amounts of ketones involves trioxides as reactive intermediates. For example, Keinan and co-workers[111] reported the isolation of 8% acetone as the only product of the ozonation of neopentane and explained it by the direct insertion of ozone into the C—C bond. The transition state for the insertion is believed to be analogous to that already proposed by Yoneda and Olah[112] for the ozonation of alkanes in superacid media (equation 45). The oxidative cleavage of the C—C bond in bicyclo [*n*.1.0] alkanes[113] and the formation of ketones from 3,7-dimethyloctyl acetate[114] have been explained in the same way.

$$-\overset{|}{\underset{|}{C}}-\overset{|}{\underset{|}{C}}- + O_3 \rightarrow \left[\begin{array}{c} O \\ \overset{+}{O} \quad O^- \\ -\overset{|}{\underset{|}{C}}\cdots \\ C \\ /|\backslash \end{array} \right]^{\ddagger} \rightarrow -\overset{|}{\underset{|}{C}}OOO\overset{|}{\underset{|}{C}}- \rightarrow 2\,\rangle C{=}O \qquad (45)$$

The proposed mechanism is supported by the observed similar yields of ketones by the cleavage of either primary, secondary or tertiary alkyl groups. It is also interesting that the relative yields of ketones from the cleavage of alkyl groups in acyclic hydrocarbons are higher than those from alkyl groups in cyclic hydrocarbons.[111]

Trioxides have also been proposed as reactive intermediates in the ozonation of organometallic compounds of the type R_4M and R_3MMR_3 (M = Si, Ge, Sn, Pb). These reactions have been extensively studied by Alexandrov and Tarunin.[115] For example, tetraethylsilane reacts with ozone presumably forming the trioxide (reaction 46), which rearranges with migration of hydrogen to the middle oxygen atom to form the corresponding hydroperoxide and acetaldehyde (reaction 47).

$$(C_2H_5)_3Si{-}CH_2CH_3 \rightarrow (C_2H_5)_3SiOOOC_2H_5 \qquad (46)$$

$$^{-}:\ddot{O}{-}\overset{+}{O}{=}\ddot{O}:$$

$$\begin{array}{c} H{-}CH{-}CH_3 \\ | \\ (C_2H_5)_3Si{-}O{-}O{-}O \end{array} \rightarrow (C_2H_5)_3SiOOH + CH_3CHO \qquad (47)$$

Hexaethyldistannane is also believed to react at low temperatures (below −100°C) to form the trioxide (reaction 48), which at higher temperatures decomposes to hexaethylstannoxane and oxygen (reaction 49).[116]

$$(C_2H_5)_3SnSn(C_2H_5)_3 + O_3 \rightarrow (C_2H_5)_3SnOOOSn(C_2H_5)_3 \qquad (48)$$

$$(C_2H_5)_3SnOOOSn(C_2H_5)_3 \rightarrow (C_2H_5)_3SnOSn(C_2H_5)_3 + O_2 \qquad (49)$$

In the ozonation of organometallic compounds of the type R_4M, where the alkyl group has no α-hydrogen atom, the initially formed trioxide may decompose in either a heterolytic (reaction 50) or a homolytic process (reaction 51), depending on the substituent at the atom M and on the nature of the heteroatom.[117] However, there has so far been no direct spectroscopic evidence for this type of trioxides.

$$\begin{array}{c} CH_3 \\ | \\ CH_3{-}C{-}CH_3 \\ | \\ -\overset{|}{\underset{|}{M}}{-}O{-}O{-}O \end{array} \rightarrow -\overset{|}{\underset{|}{M}}{-}O{-}O{-}CH_3 + (CH_3)_2C{=}O \qquad (50)$$

$$-\overset{|}{\underset{|}{M}}{-}O{-}O{-}O{-}C(CH_3)_3 \rightarrow -\overset{|}{\underset{|}{M}}{-}O{-}O\cdot + \cdot OC(CH_3)_3 \qquad (51)$$

3.3 Structure and Thermochemical Data

Little of a definitive nature is known about the structure of dialkyl polyoxides. It seems safe to predict that dialkyl tetraoxides and trioxides are structurally similar to hydrogen tetraoxide and hydrogen trioxide, respectively. An *ab initio* MO SCF study on dimethyl trioxide by using the minimal basis set (STO-3G) revealed a staggered zig-zag structure with a dihedral angle, ψ, of 83°.[118]

The heats of formation, $\Delta H_f°$, of $CH_3O_4CH_3$ and $CH_3O_3CH_3$ were estimated to be -3.8 ± 2.1 and -16.9 ± 1.4 kcal mol^{-1}, respectively. From these values the bond dissociation energies of the weakest bond in $CH_3OO{-}OOCH_3$ (9.2 ± 2.4 kcal mol^{-1}) and in $CH_3O{-}OOCH_3$ (23.7 ± 1.2 kcal mol^{-1}) were derived.[39]

4 DIACYL TETRAOXIDES

Although no direct spectroscopic evidence has been reported so far for the existence of acyl tetraoxides, there is strong evidence that such polyoxides are formed in the autoxidation of various aliphatic and aromatic aldehydes.[119–123]

The results of product studies,[119] carbon dioxide evolution[120] and labeling experiments[121] on the free-radical-initiated autoxidation of acetaldehyde by Traylor and co-workers indicate that aldehyde termination is preceded by the formation of acetyl tetraoxide, which decomposes in a concerted process to methyl radicals, carbon dioxide and oxygen (reaction 52). The absence of acetyl peroxide among the products of these reactions appears to rule out a concerted cyclic process (reaction 53a) or appreciable cage collapse (reaction 53b).

$$2R{-}\underset{\underset{O}{\|}}{C}{-}O{-}O^{\cdot} \rightarrow R{-}\underset{\underset{O}{\|}}{C}{-}O{-}O{-}O{-}O{-}\underset{\underset{O}{\|}}{C}{-}R \rightarrow 2R\cdot + 2CO_2 + O_2 \quad (52)$$

$$2R{-}\underset{\underset{O}{\|}}{C}{-}O{-}O\cdot \not\rightarrow \left[\text{cyclic: } R{-}C(=O)\text{-}O{-}O{-}O{\cdots}O{-}C(=O){-}R\right] \rightarrow R{-}\overset{\overset{O}{\|}}{C}OO\overset{\overset{O}{\|}}{C}{-}R + O_2 \quad (53a)$$

$$2R{-}\underset{\underset{O}{\|}}{C}{-}O{-}O\cdot \not\rightarrow \left[R{-}C(=O)O\cdot \quad O_2 \quad \cdot OC(=O){-}R\right] \rightarrow (\text{products})_{cage} \quad (53b)$$

The results of the radical-induced decomposition of peroxyacetic acid in acetic acid in the presence of oxygen also confirm the involvement of acetyl tetraoxide in this reaction.[122] The actual termination in the autoxidation of acetaldehyde involves the cross-reaction of methylperoxyl radicals either with acetylperoxyl radicals or with themselves to form the corresponding tetraoxides, which decompose by either the Russell mechanism (reaction 56), or an alternative acyclic mechanism giving the same products.[122]

$$CH_3\cdot + O_2 \rightarrow CH_3OO\cdot \quad (54)$$

$$CH_3OO\cdot + \cdot OOCH_3 \rightarrow (CH_3OOOOCH_3) \rightarrow CH_2O + CH_3OH + O_2 \quad (55)$$

$$CH_3\underset{\underset{O}{\|}}{C}OO\cdot + \cdot OOCH_3 \rightarrow CH_3\underset{\underset{O}{\|}}{C}{-}O{-}O{-}O{-}\underset{\underset{H-CH_2}{|}}{O} \rightarrow CH_3COOH + CH_2O + O_2 \quad (56)$$

It is more difficult to explain the very fast termination (close to the diffusion-controlled limit) of benzaldehyde autoxidation. Ingold and coworkers[123] suggested that benzoyl tetraoxide, $PhC(O)O_4C(O)Ph$, is formed irreversibly in these reactions and that, if this tetraoxide decomposes to give benzoyloxy radicals, most of these radicals recombine in the cage. Still another plausible explanation for the slower termination of acetylperoxyl radicals is the much faster rate of decarboxylation of acetyloxy radicals in comparison with benzoyloxyl radicals or the reversible formation of the acetyl tetraoxide, which decomposes more slowly than benzoyl tetraoxide.

5 ALKYL HYDROTRIOXIDES

Alkyl hydrotrioxides, ROOOH, can be regarded as higher homologs of alcohols and alkyl hydroperoxides. These polyoxides have been proposed as unstable intermediates in the low-temperature ozonation of various saturated organic compounds,[124] e.g. hydrocarbons,[125–128] alcohols,[125] ethers,[129–131] aldehydes,[132,133] acetals,[134–141] amines,[142–144] diazo compounds[145] and silanes,[146–150] but the spectroscopic evidence for their existence became available only recently.[151–162]

5.1 Mechanism of Hydrotrioxide Formation

Several different mechanisms for the oxidation of the C—H bond of saturated organic compounds to form hydrotrioxides have been proposed. A concerted 1,3-dipolar insertion mechanism (reaction 59) was proposed for the ozonolysis of saturated hydrocarbons,[111,124,126] ethers,[129–131] acetals,[137,138] aldehydes[132,133] and silanes.[146,147] A mechanism involving hydrogen atom abstraction (reaction 58) has been proposed in the reaction of ozone with hydrocarbons,[127,128] alcohols,[125] ethers[130] and acetals.[138] A hydride ion transfer to form a carbonium ion and hydrotrioxide ion (reaction 57) was suggested as a mechanistic possibility for hydrocarbons[127,128] and acetals.[134,135,138] The hydride abstraction by ozone has recently been proposed by Nangia and Benson,[163] based mainly on thermochemical grounds, to be the mechanistic feature of the oxidation of the C—H bonds.

$$\geq\text{C—H} + O_3 \rightarrow \left[\geq\text{C}^+ \quad HO_3^-\right] \rightarrow \geq\text{COOOH} \tag{57}$$

$$\geq\text{C—H} + O_3 \rightarrow \left[\geq\text{C}\cdot \quad HO_3\cdot\right] \rightarrow \geq\text{COOOH} \tag{58}$$

$$\geq\text{C—H} + O_3 \rightarrow \left[\geq\text{C}\overset{\cdots\text{O}\diagdown}{\underset{\diagdown\text{H}\cdots\text{O}\diagup}{}}\text{O}\right]^{\ddagger} \rightarrow \geq\text{COOOH} \tag{59}$$

Although considerable efforts have been made in recent years to elucidate the mechanism of these important reactions, an unambiguous substantiation of either of the proposed mechanisms is still lacking. Let us briefly review the experimental facts concerning the mechanism of these reactions.

5.1.1 Stereochemistry

The reactions of ozone with hydrocarbons to yield the corresponding alcohols and ketones proceeds with considerable retention of configuration, with the percentage retention being higher in more viscous solvents.[127,128,158] Retention of configuration was also reported in the ozonation of silanes.[146]

5.1.2 Relative reactivities

The relative reactivity of primary, secondary and tertiary hydrogens (0.003:1:87)[164,165] is greater than that for typical radical reactions ($RO^{\bullet}$, 0.08:1.0:33.7; $ROO^{\bullet}$, 0.024:1.0: 10.0).[166] Some steric requirements of the reaction are indicated by the fact that equatorial tertiary hydrogens react several times more readily than axial hydrogens.[127,128]

The following benzylic reactivity was reported: cumene > ethylbenzene > toluene and triphenylmethane > diphenylmethane > toluene.[167] However, the benzylic reactivity of fluorene is only three times that of diphenylmethane and one fifteenth that of 9,10-dihydroanthracene-type compounds. It should be pointed out that hydrogen abstraction from fluorene by benzyl radical proceeds eight times more readily than from diphenylmethane and derivatives of fluorene form carbocations less readily than do the corresponding derivatives of diphenylmethane.[167]

The relative reactivity order of tertiary > secondary > primary has also been reported for ethers.[127,130,164]

5.1.3 Kinetics and substituent effects

The stoichiometry of the reaction of ozone with acetals is 1:1 in each reactant and the reaction is first order in acetal and first order in ozone.[137] Electron-donating substituents accelerate the reaction with ρ values ranging from -1.10 to -1.58 for acetals[137,140,141] (vs Hammett or Taft substituent constants) and -2.07 for the ozonation of substituted toluenes. The latter ρ value is more negative than those for hydrogen abstraction from the same substrate by alkylperoxyl radicals (-0.76) and alkoxyl radicals (-0.34),[168] suggesting that at least for the ozonation of alkanes the transition state for the ozonation is more polar than that for the radical abstraction of the H atom.

The polarity of the solvent has been reported to have little effect on the rate of ozonation of acetals, which would seem to rule out the involvement of ionic intermediates unless tight ion pairs are involved.[137] However, the actual yield of acetal hydrotrioxides, as determined by ^1H NMR spectroscopy, appears to be strongly dependent on the solvent (polarity?). Thus, much higher yields of the hydrotrioxides were obtained in acetone and diethyl ether than in dichloromethane and pentane.[153,154]

A comparison of log *A* values for the ozonation of ethers shows that these range from 5.5 (di-2-chloroethyl ether) to 7.9 (di-isopropyl ether). The log *A* values for 2,3-dimethylbutane and cyclohexane are 7.3 and 7.7, respectively.[164] All these values are lower than those expected for radical abstractions (8.7 for ethers and aldehydes and 9.4 for tertiary alkanes[169]) but higher than those of typical 1,3-dipolar insertions.[170] Nevertheless, a highly oriented transition state for these reactions is indicated. Nangia and Benson[163] suggested a log *A* value of about 6.5 for hydride abstraction from a hydrocarbon by ozone. The activation energies are in the range 6–8 kcal mol^{-1}.[137,164]

Considerable C—H bond breaking in the transition state for the reaction is indicated by a relatively large kinetic isotope effect ($k_H/k_D = 4.0–4.5$ for toluene, saturated hydrocarbons[130] and benzyl *tert*-butyl ether;[130] $k_H/k_D = 6.5$ for acetals[171]).

5.1.4 Stereoelectronic effects

Ozonation of acetals is controlled by stereoelectronic factors. It was demonstrated by Deslongchamps *et al.*[135,138] that in order for ozonation to proceed, one non-bonded electron pair on each oxygen atom has to lie antiperiplanar to the C—H bond of the acetal function (equation 60).

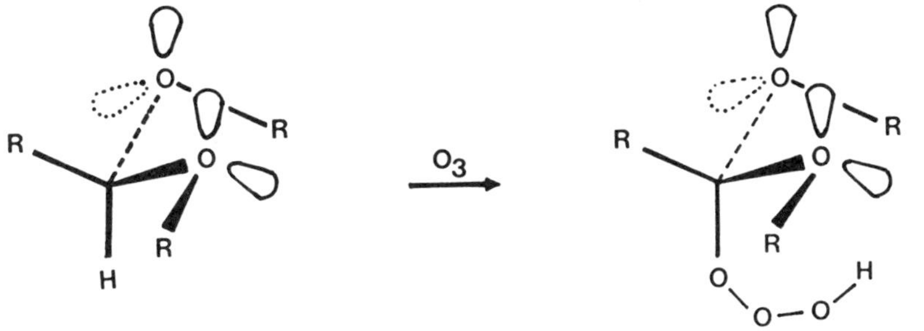

(60)

Ab initio calculations support the presumption that the antiperiplanar electron pair should increase the electron density of the C—H bond, which would then be more susceptible to attack by ozone as an electrophile.[172] At the same time, this effect should also stabilize the incipient dialkoxycarbonium ion and/or the corresponding radical (H-atom abstraction), if either of these two intermediates is formed in the ozonation.[138] It is interesting in this connection that an isokinetic relationship was found for the ozonation of acyclic acetals of heptaldehyde with the isokinetic temperature below the experimental temperature range, i.e. in a domain of temperatures where entropy factors control the reactivity.[137] In cyclic acetals of the same aldehyde the isokinetic temperature is above the experimental temperatures, i.e. where the reactivity depends mainly on enthalpy factors. Taillefer and co-workers[137] interpreted these results in terms of a conformational change which must take place before ozonation in the case of acyclic acetals.

Most of the above-mentioned experimental evidence indicates that either dipolar insertion or hydride ion transfer is the mechanistic feature of the formation of hydrotrioxides. In order to distinguish between these two mechanisms, Giamalva *et al.*[173] investigated the reaction of ozone with norbornane, adamantane and bicyclo[2.2.2]octane. They found that norbornane behaves like a secondary hydrocarbon, reacting with ozone at a rate only slightly higher than cyclopentane [norbornane, $E_a = 13.6$ kcal mol^{-1}, log $A = 7.6$ (per H)]. If the reaction of the C—H bond were to proceed via a hydride abstraction as the rate-determining step, norbornane should react at a rate comparable to that of a tertiary hydrocarbon. These results were interpreted as supporting a 1,3-dipolar insertion mechanism for the reaction of ozone with C—H bonds with possible transition states represented as resonance structures including as extremes those having a high degree of radical and carbocation character. The central transition structure represents a concerted insertion with the O—H bond formation preceding the C—O formation (Scheme 1). It has already been suggested that hydride abstraction and hydrogen-atom abstraction might be two extreme resonance forms of intermediates involved in a common single mechanistic pathway.[128,135,138]

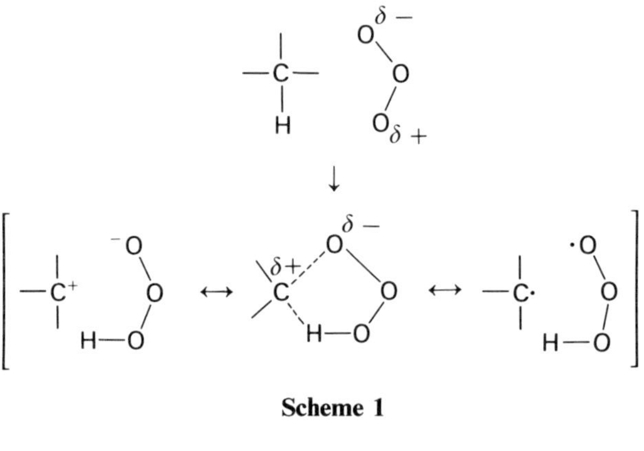

Scheme 1

Still another mechanism, depicted in reaction 61 and involving attack of ozone at the oxygen atom, was proposed for the ozonation of ethers.[131] The initial attack of ozone at the heteroatom has also been proposed to be the first step of the ozonation of amines and sulfides.

Although this mechanism cannot be completely ruled out, most of the available experimental evidence favors a common single mechanism as discussed above for the ozonation of saturated hydrocarbons, ethers and acetals.

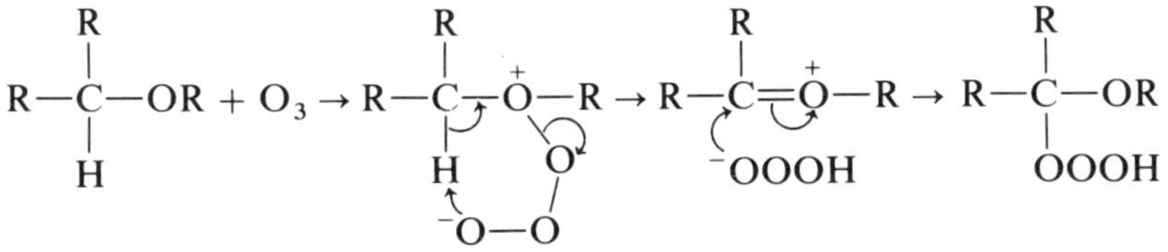

$$\mathrm{R{-}\underset{H}{\overset{R}{C}}{-}OR + O_3 \rightarrow R{-}\underset{\overset{|}{H}\quad O{-}O^{-}\;O}{\overset{R}{C}}{-}\overset{+}{O}{-}R \rightarrow R{-}\underset{^{-}OOOH}{\overset{R}{C}}{=}\overset{+}{O}{-}R \rightarrow R{-}\underset{OOOH}{\overset{R}{C}}{-}OR} \qquad (61)$$

The relatively greater reactivity of acetals and ethers compared with hydrocarbons can be explained by resonance stabilization of a cation or radical intermediate by the alkoxy group and/or by the weakening of the C—H bond α to oxygen.

Olah *et al.*[174] suggested that ozonation of alkanes in superacid (F_3SO_3H–SbF_5–SO_2ClF) solution at −78°C involves electrophilic insertion by protonated ozone, HO_3^+, into a C—H bond through two-electron, three-center-bonded pentacoordinated carbonium ions. These could yield, after the loss of proton, either the neutral hydrotrioxide or cleave in other ways as shown in Scheme 2.

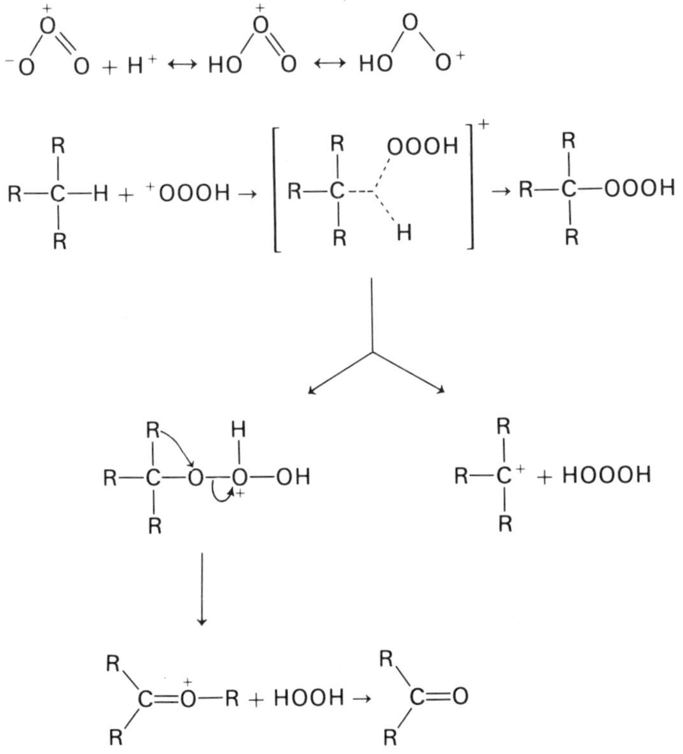

$$\mathrm{^{-}O{-}\overset{+}{O}{=}O + H^+ \leftrightarrow HO{-}\overset{+}{O}{=}O \leftrightarrow HO{-}O{-}O^+}$$

$$\mathrm{R{-}\underset{R}{\overset{R}{C}}{-}H + {}^{+}OOOH \rightarrow \left[R{-}\underset{R}{\overset{R}{C}}\cdots\begin{matrix}OOOH\\ H\end{matrix} \right]^+ \rightarrow R{-}\underset{R}{\overset{R}{C}}{-}OOOH}$$

$$\mathrm{R{-}\underset{R}{\overset{R}{C}}{-}O{-}\overset{H}{\overset{|}{\overset{+}{O}}}{-}OH \qquad\qquad R{-}\underset{R}{\overset{R}{C^+}} + HOOOH}$$

$$\downarrow$$

$$\mathrm{\begin{matrix}R\\ \\ R\end{matrix}\!\!>C{=}\overset{+}{O}{-}R + HOOH \rightarrow \begin{matrix}R\\ \\ R\end{matrix}\!\!>C{=}O}$$

Scheme 2

The relative order of reactivity of bonds in alkanes with $^{+}O_3H$ was found to be tertiary C—H > secondary C—H, primary C—H > C—C. All attempts to detect HO_3^{+} directly by ^{1}H NMR spectroscopy failed. Recent theoretical studies on this species showed that a proton is attached to open chain ozone in a *trans* arrangement. The corresponding *cis* isomer is 3.6 kcal mol^{-1} higher in energy. The protonation energy of ozone was calculated to be 148 kcal mol^{-1}.[175]

Best *et al.*[176] have recently provided indirect evidence for the formation of protonated alkyl hydrotrioxides in a heterolytic reaction involving nucleophilic attack of hydroperoxide on *gem*-dialkylperoxonium ions (Scheme 3).

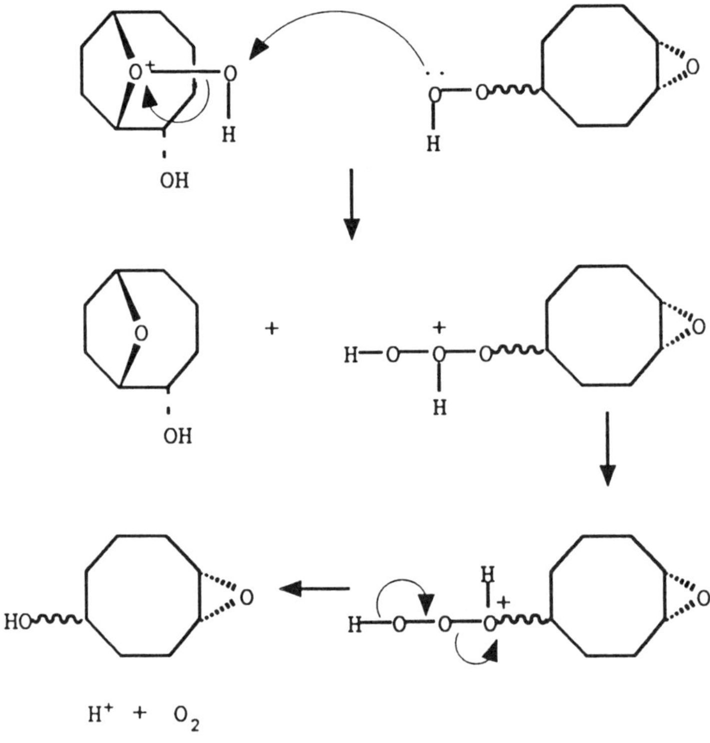

Scheme 3

5.2 Characterization and Structure of Hydrotrioxides

5.2.1 NMR spectroscopic studies

Murray and co-workers[151,152] provided the first spectroscopic evidence for the existence of hydrotrioxides by studying the ozonolysis of ethers at low temperatures. The ^{1}H NMR spectral analysis of ozonized solutions of diisopropyl ether, 2-methyltetrahydrofuran, and methyl isopropyl ether showed the presence of a greatly deshielded absorption at *ca* $\delta 13 \pm 1$ ppm downfield from Me_4Si. The intermediacy of hydrotrioxides was later verified in the ozonation of acetals,[153–155] cumene[156,157] and other saturated hydrocarbons,[158] benzylic ethers and alcohols[159,160] and most recently silanes.[160,161,162] Selected NMR chemical shifts of hydrotrioxides are given in Table 6.

Since the position of the OOOH absorption did not change much with dilution (<0.15 ppm), it was tentatively assigned to the OOOH absorptions of the intramolecularly hydrogen-bonded six-membered ring of the hydrotrioxides of ethers,[152,169] aldehydes[152] and acetals.[154]

The appearance of two OOOH absorptions in the ^{1}H NMR spectra of the hydrotrioxides of 2-methyltetrahydrofuran and acetals was originally attributed to the presence of two forms of the hydrotrioxide with OOOH bonded intramolecularly to either of the lone pairs of the tetrahedral oxygen atom[152] or to two conformations of the six-membered ring with R groups on the oxygen atom axial or equatorial. Two distinct forms of the hydrotrioxide of norcarane, presumably resulting from the attack of ozone at the *endo*- and *exo*-C—H bond, have also been reported.[158]

A more detailed study of the NMR spectra of hydrotrioxides of alkyl α-methylbenzyl ethers showed that oxygen bases are able to disrupt the chelate ring with subsequent formation of intermolecularly hydrogen-bonded associates, and/or favor the formation of such associates in the process of formation of hydrotrioxides (Scheme 4).

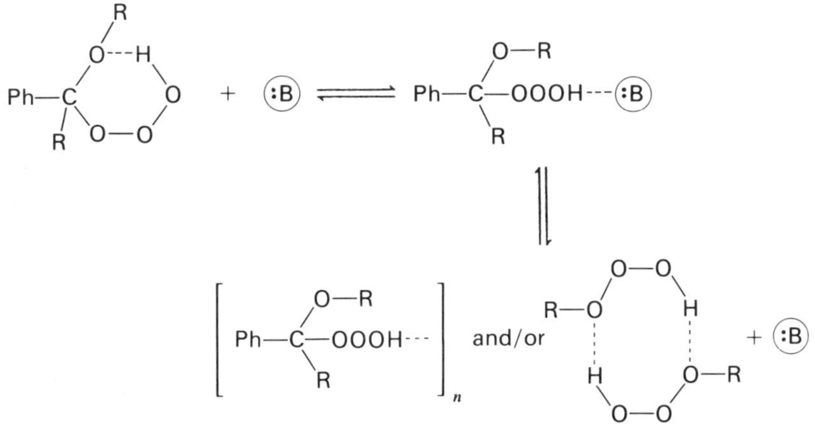

Ph—C(R)(O—R)—OOOH ring (O---H intramolecular) + (:B) ⇌ Ph—C(O—R)(R)—OOOH---(:B)

⇅

[Ph—C(O—R)(R)—OOOH---]$_n$ and/or R—O---H—O—O / O—O—H---O—R (dimer) + (:B)

Scheme 4

Table 6. Selected NMR chemical shifts for some representative hydrotrioxides and their analogs

Compound	Y	Solvent	^{1}H NMR			^{13}C NMR				^{29}Si NMR:
			δOOOH	δCH_3	δOCH_3	δC(1)	δC(2)	δC(3)	δC(4)	δSi
Ph^{2}–$C^{1}(^{3}CH_3)(CH_3)$–Y	OOOH[a]	$(CD_3)_2CO$	13.67	1.64		104.3	134.5	—	—	—
	OOOH		10.95	1.52		84.0	145.5	26.6		
	OH		5.18	1.50		72.1	150.7	32.3		
	H			1.15, 1.19 (d)		34.6	148.8	24.3		
Ph^{2}–$C^{1}(^{3}CH_3)(O^{4}CH_3)$–Y	OOOH[b]	$(CD_3)_2CO$	13.19, 13.54 (d)	1.60	3.25	108.1	134.3			
		$MeCO_2Et$	12.90, 13.15 (d)							
	OOH	$(CDCl_3)$	8.85	1.60	3.35	106.7	140.5	49.9	24.6	
		$(CD_3)_2CO$	10.6			106.4	142.5	49.7	25.4	
	H	$(CDCl_3)$	4.12 (q)	1.35	3.05	79.7	143.9	56.1	24.0	
Ph^{1}–$Si(^{3}CH_3)(CH_3)$–Y	OOOH[c]	$(CD_3)_2CO$	13.96	0.55		138.5	2.7			17.31
	OOH		11.2	0.43		139.6	2.8			14.31
	OH		4.5–5.5	0.32		143.5	3.1			3.30
	H		4.40	0.30		140.6	−1.1			−16.95

[a]Ref. 157.
[b]Ref. 160a.
[c]Ref. 16a.

The observed change in the relative intensities for OOOH absorptions of hydrotrioxides generated by ozonizing enantiomeric starting materials in ethyl acetate and acetone-d_6 seems to support the belief that the NMR doublets are due to the formation of two distinct cyclic dimeric forms of hydrotrioxides, i.e. '*cis*' and/or '*trans*' homochiral (*R,R, S,S*) and heterochiral (*R,S, S,R*) ensembles which are probably solvated with the oxygen base.[160a] The possibility that the OOOH doublets are due to the presence of two pairs of diastereoisomeric forms of the intramolecularly hydrogen-bonded forms with a chemical shift of the sensor nuclei being different in both forms also cannot be completely ruled out.[160a]

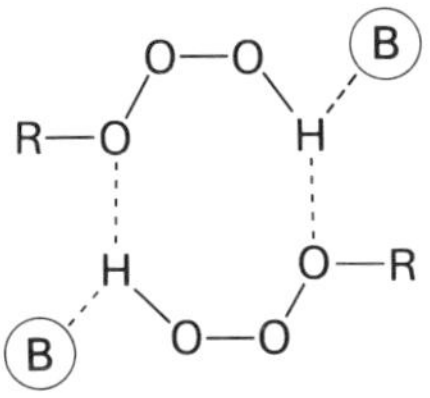

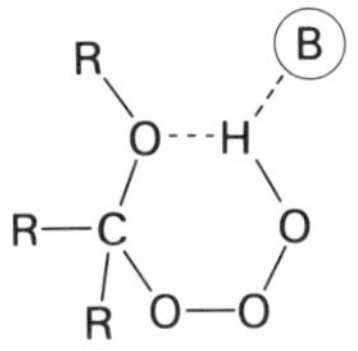

The experimental evidence strongly indicates that the structure of these hydrotrioxides is critically dependent on the reaction conditions during their formation.[160]

5.2.2 Theoretical studies

Recently, *ab initio* quantum mechanical calculations have been carried out on monomeric and dimeric methyl hydrotrioxide (CH_3OOOH) and silyl hydrotrioxide ($H_3SiOOOH$), and also the corresponding hydroperoxides.[177] The calculated relatively strong binding energies for the intermolecularly hydrogen-bonded cyclic dimers of the hydrotrioxides (8.0–8.6 kcal mol^{-1}; hydroperoxides 7.2–7.9 kcal mol^{-1}) support the belief that self-association is, analogously to hydroperoxides, the structural feature of these polyoxides. The equilibrium geometries of the monomeric and dimeric forms of HOOOH and alkyl hydrotrioxides are shown in Figures 2 and 3 respectively.

The Mulliken population analysis with the use of the 6–31G**//6–31G basis set shows a considerable accumulation of charge on both terminal oxygen atoms. The calculated net atomic charges on oxygen hydrogen atoms are similar to those calculated previously by Jackels and Phillips[55] for HOOOH at a still higher level of sophistication (6–31G**/CI, O_1, O_3, -0.31 e; O_2, -0.07 e; H, 0.35 e). The calculated electric dipole moments of CH_3OOOH and $H_3SiOOOH$ are 1.38 and 1.16 D, respectively.

It has been shown on thermochemical grounds that the split into RO and OOH radicals in hydrotrioxides is the lowest energy and the fastest radical decomposition pathway available (BDE: CH_3O—OOH, 24 ± 1 kcal mol^{-1}; CH_3OO—OH, 29 ± 1 kcal mol^{-1}).[39,163] The calculated bond orders of the

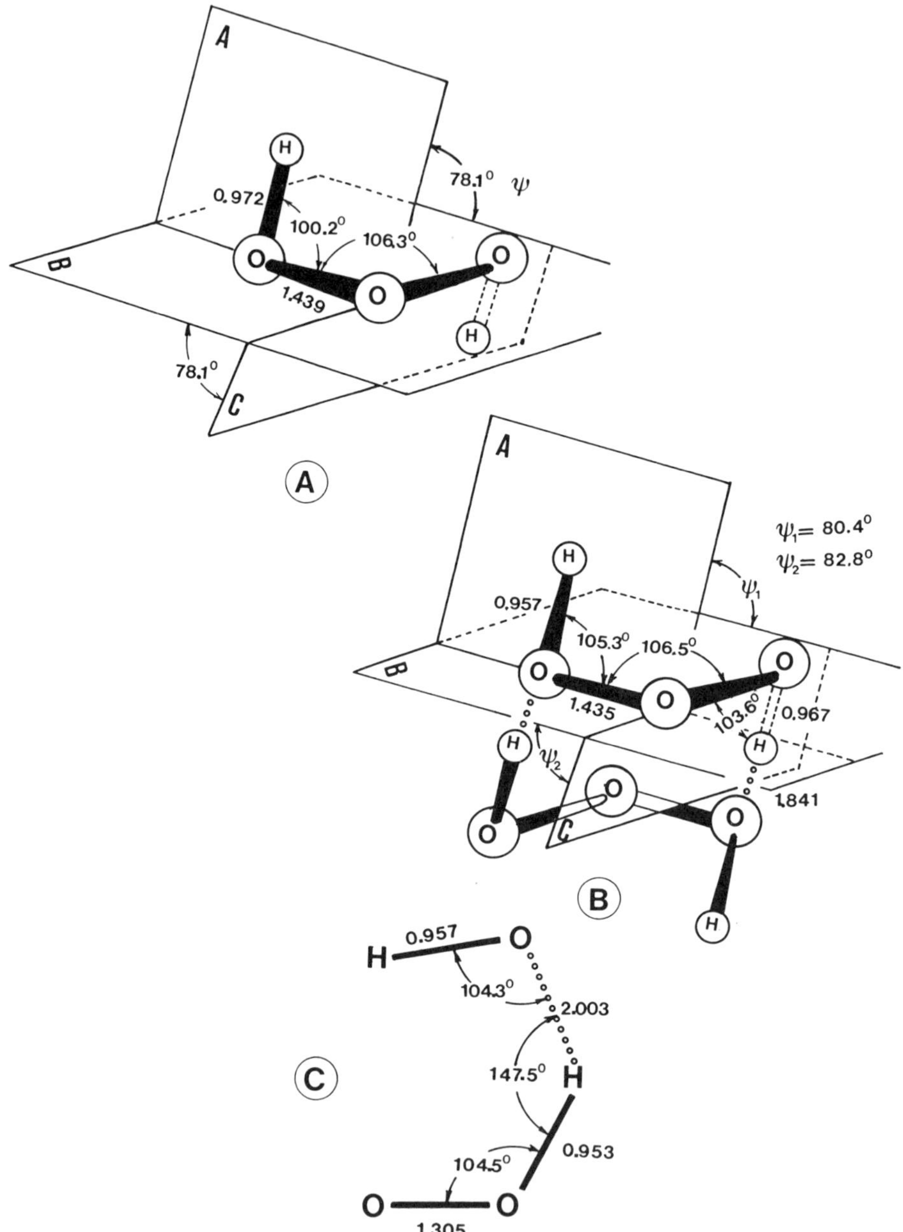

Figure 2. Calculated equilibrium geometries of (A) monomeric hydrogen trioxide, H_2O_3 (MP2/6-31G* basis set),[54,55] (B) cyclic dimeric H_2O_3 (6-31G),[16] and (C) hydrogen-bonded HO ··· HOO complex (UHF SCF level).[55] The bond lengths are in ångströms and the bond angles in degrees

respective O—O bonds in both hydrotrioxides (CH_3O—OOH, 1.036; CH_3OO—OH, 1.045; H_3SiO—OOH, 1.039; H_3SiOO—OH, 1.055) unambiguously confirm these predictions.

The order of theoretical acidities, estimated as the energy difference between the energy minimum for the neutral molecules and that of the corresponding anion, ΔE_{eq}(kcal mol^{-1}), calculated at the RHF 6–31 + + G//6–31 + + G level of sophistication is: HOOOH (352.3) > $H_3SiOOOH$ (353.3) > CH_3OOOH (358.3) > H_3SiOOH (365.3) > CH_3OOH (371.4). This order is in accordance with the expectation that an increased number of oxygen atoms in the members of the homologous series should increase their acidity. It appears that the main reason for the relatively high acidity of HOOOH, in comparison with other hydrotrioxides, is the stabilization of the hydroperoxide anion by intramolecular hydrogen bonding.

5.3 Decomposition of Hydrotrioxides

The observation of evolution of oxygen when various saturated organic substrates were ozonized at −78°C and then allowed to warm up to ambient temperature has been used as a criterion for the involvement of hydrotrioxides in these reactions. More recent systematic investigations (including kinetic studies) of the decomposition of hydrotrioxides formed in the low-temperature ozonation of ethers, aldehydes, acetals, hydrocarbons and silanes allow some generalizations to be made regarding the mechanisms of decomposition of this type of polyoxides.

5.3.1 Ether, alcohol and acetal hydrotrioxides

Basically, two mechanisms can be envisaged for the decomposition of hydrotrioxides capable, at least theoretically, of existing in the intramolecularly hydrogen-bonded form. Murray and co-workers[151] were the first to propose, on the bases of relatively low log *A* values for the decomposition of the hydrotrioxides of methyl isopropyl ether, 2-methyltetrahydrofuran and benzaldehyde and the formation of singlet oxygen ($^1O_2$) among the decomposition products, the concerted process (shown in reactions 63 and 64) with a cyclic six-membered ring transition state.[151,152]

$$\text{cyclic } R^1R^3C(OR^2\cdots H)OOO \rightarrow [\text{cyclic transition state}]^{\ddagger} \rightarrow R^1\!-\!C(=O)\!-\!R^3 + R^2OH + {}^1O_2 \qquad (63)$$

$$R-C\overset{O\cdots H}{\underset{O-O}{\langle\ \ \rangle}}O \rightarrow \left[R-C\overset{O\cdots H}{\underset{O\cdots O}{\langle\ \ \rangle}}O \right]^{\ddagger} \rightarrow R-\overset{O}{\overset{\|}{C}}-OH + {}^1O_2 \qquad (64)$$

It is known from previous studies of the pyrolysis of halides, esters and cyclobutanes that an intramolecular six-membered transition state, of the type shown above, can participate with both heterolytic ('pericyclic' process) and homolytic character. The overall efficiency of ${}^1O_2$ formation in the above-mentioned reactions (80% of the absorbed ozone was able to react with 1,3-diphenylisobenzofuran in the case of 2-methyltetrahydrofuran and 30% in the case of methyl isopropyl ether) already suggested that other reactions also occur.

The importance of radical processes in the decomposition of ether hydrotrioxides was recently revealed in studies of the products and kinetics of decomposition of hydrotrioxides of benzylic ethers and alcohols.[160] A detailed study of the decomposition products of methyl α-methylbenzyl ether hydrotrioxide in acetone or ethyl acetate (0.15 M) showed acetophenone (50 ± 7% yield per mole of the hydrotrioxide), methanol (14 ± 2%), methyl benzoate (20 ± 3%), benzoic acid (5 ± 1%), hydrogen peroxide and singlet oxygen (45 ± 7%, 1,3-diphenylisobenzofuran; 25 ± 4%, tetraphenylcyclopentadienone). Acetophenone (65 ± 9%), water, benzoic acid (5 ± 1%), hydrogen peroxide and singlet oxygen (60 ± 9%, 1,3-diphenylisobenzofuran; 25 ± 4%, tetraphenylcyclopentadienone) were found to be the major products in the decomposition of α-methylbenzyl alcohol hydrotrioxide. Evidence for the presence of small amounts of a metastable hydroperoxide, most likely α-hydroxy-α-methylbenzyl hydroperoxide, in the decomposition mixtures from both hydrotrioxides, was also presented.[160a]

Kinetic and activation parameters for the first-order decay of hydrotrioxides of benzylic ethers and alcohols are summarized in Table 7. Solvent effects on the decomposition are relatively small. Electron-withdrawing groups accelerate decomposition, whereas electron-donating groups retard it. The Hammett ρ values for the decomposition of substituted methyl α-methylbenzyl ether hydrotrioxides are 0.7 ± 0.1 (ethyl acetate), 1.0 ± 0.2 (diethyl ether) and 1.4 ± 0.2 (acetone).

Although a non-radical ('pericyclic') decomposition mechanism cannot be completely ruled out, the above-mentioned experimental evidence strongly suggests the involvement of radical processes in these decompositions. A split into RO and OOH radicals is, as described above, the lowest energy and the fastest radical path available for the decomposition of hydrotrioxides. The formation of a cage radical pair in the decomposition of these polyoxides was first proposed by Nangia and Benson.[163] The radicals can either undergo cage

Table 7. Kinetic and activation parameters for the decomposition of some ether and alcohol hydrotrioxides

Hydrotrioxide	Solvent	k (s^{-1})	T (°C)	E_a (kcal mol^{-1})	Log A	Ref.
$Me_2C(OMe)OOOH$	Neat	7.32×10^{-4}	−25.5	16.6	11.5	152
2-methyltetrahydrofuran-2-yl-OOOH	Et_2O	1.88×10^{-4}	−26.5	17.4	11.7	152
Ph—C(OR)(Me)—OOOH						
R = Me	Me_2CO	2.5×10^{-2}	−25	19.9	16.0	160
	$MeCO_2Et$	2.17×10^{-2}	−25	18.0	14.2	160
	Et_2O			15.1	12.9	160
R = H	Me_2CO	2.63×10^{-2}	−25	16.9	13.4	160
	$MeCO_2Et$	3.68×10^{-2}	−25	17.0	13.6	160
	Et_2O	2.76×10^{-2}	−25	16.8	13.3	160

disproportionation to form the tetrahedral intermediate and singlet oxygen or diffuse out of the cage. The tetrahedral intermediate can react with hydrogen peroxide to form the corresponding α-hydroxy hydroperoxide, which is probably the source of benzoic acid in the decomposition mixture.

It should be pointed out that, in addition to the 'pericyclic' pathway, the abstraction of hydrogen from the hydrotrioxide to produce trioxyl radicals can also be a source of singlet oxygen (reactions 65 and 66).

$$ROOO\cdot \rightarrow RO\cdot + O_2\ (^1O_2 \text{ and/or } ^3O_2) \qquad (65)$$

$$2ROOO\cdot \rightarrow 2ROO\cdot + O_2\ (^1O_2 \text{ and/or } ^3O_2) \qquad (66)$$

A relatively small effect of an added radical inhibitor, 2,6-di-*tert*-butyl-4-methylphenol, i.e. slightly reduced rates of decomposition and increased activation energies, implies that decomposition proceeds predominantly by the 'pericyclic' mechanism, or involves homolysis of the RO—OOH bond with subsequent cage disproportionation of the radical pair, as suggested by Benson.[163] Weaker O—H bonds in hydrotrioxides as compared with the corresponding hydroperoxides (bond order: CH_3OOO—H, 1.680; CH_3OO—H, 1.703) render these polyoxides even more susceptible to induced decomposition. It is possible that this reactivity is strongly reduced owing to the engagement of the O—H bonds of the hydrotrioxides in relatively strong hydrogen bonds.

Both mechanistic possibilities described above are also available for the decomposition of the dimeric hydrotrioxides, as shown in reactions 67 and 68.

$$\left[\begin{array}{ccc} & O{=}O & \\ R{-}\overset{\delta^-}{O} & & H^{\delta^+} \\ {}^{\delta^+}H & & \underset{\delta^-}{O}{-}R \\ & O{=}O & \end{array}\right]^{\ddagger} \rightarrow 2ROH + 2\,{}^1O_2 \qquad (67)$$

$$\left[\begin{array}{ccc} & {}^{\delta\cdot}O{-}O & \\ R{-}\overset{\delta\cdot}{O} & & H \\ H & & \overset{\delta\cdot}{O}{-}R \\ & O{-}O_{\delta\cdot} & \end{array}\right]^{\ddagger} \rightarrow 2RO\cdot + 2HOO\cdot \qquad (68)$$

The observed substituent effect (positive Hammett ρ values) suggests that either there is some build-up of negative charge in the transition state, or radicals derived from benzylic ethers remaining after ozonation attack the 'electrophilic' middle oxygen atom in the hydrotrioxide in an S_H2 process (reaction 69), thus causing chain decomposition.

$$ROOOH + R\cdot \rightarrow RO\cdot + ROOH \qquad (69)$$

In 1971, Deslongchamps and Moreau[134] first demonstrated that ozone reacts with the acetal function from an aldehyde to produce the corresponding ester, alcohol and oxygen. As already mentioned, the reaction is controlled by stereoelectronic factors. It was assumed that hydrotrioxides are involved in these reactions.

The intermediacy of hydrotrioxides in these reactions has been verified in recent years (see Table 6). Their decomposition was investigated by following the decay of OOOH absorptions in several solvents and was found to obey first-order kinetics. The kinetic and activation parameters for the decomposition of various acetal hydrotrioxides are given in Table 8.[154,155]

As in the case of hydrotrioxides of benzylic ethers and alcohols, comparable rate constants for decomposition of acetal hydrotrioxides were obtained in various solvents. Electron-withdrawing groups in the hydrotrioxide derivatives of benzaldehyde dimethyl acetals accelerate decomposition whereas electron-donating groups retard it. A Hammett ρ value of 1.2 ± 0.2 was obtained in diethyl ether.

Table 8. Kinetic and activation parameters for the decomposition of some representative acetal hydrotrioxides[154]

Hydrotrioxide	Solvent	k (s^{-1})	T (°C)	E_a (kcal mol^{-1})	Log A
$R^1C(OR^2)_2OOOH$					
R^1 = Me, R^2 = Me	Et_2O[a]	2.9×10^{-2}	−10	16.1	11.8
	Neat	2.7×10^{-2}	−10	20.7	15.6
R^1 = Me, R^2 = Et	Et_2O[a]	7.6×10^{-2}	−10	13.2	9.8
	Neat	1.1×10^{-1}	−10	22.0	18.2[b]
Ph—C(—O_3H)(—O—CH_2—CH_2—O—) (cyclic)	Et_2O[a]	6.6×10^{-2}	−30	15.9	13.1
	CH_2Cl_2[a]	6.4×10^{-2}	−30	20.5	16.6
$PhC(OR)_2O_3H$					
R = Me	Et_2O[a]	6.6×10^{-2}	−30	19.0	15.9
R = Et	Et_2O[a]	7.6×10^{-2}	−30	19.8	16.7

[a]0.1–0.3 M.
[b]Ref. 155.

The decomposition of the hydrotrioxide of acetaldehyde diethyl acetal yields ethyl acetate (0.80–0.85 mol per mole of the hydrotrioxide), ethanol (0.75–0.80 mol), water, acetaldehyde, acetic acid, ethyl formate, diethyl carbonate (0.06–0.12 mol) and a mixture of gases. α-Hydroperoxydiethyl ether, α,α-diethoxydiethyl peroxide and hydrogen peroxide were also detected. The same types of peroxides were also found in the decomposition mixture of aromatic acyclic acetal hydrotrioxides. Relatively high yields of singlet oxygen, ranging from 55% (aromatic acyclic acetals) to 85% (aromatic cyclic acetals, ethyl acetate), were obtained in the decomposition of all the acetal hydrotrioxides investigated.

The proposed mechanism for the decomposition of acetal hydrotrioxides is similar to that for the decomposition of hydrotrioxides of benzylic ethers and alcohols.[154,155] Whereas ethyl acetate, ethanol and singlet oxygen could result from a non-radical ('pericyclic' mechanism) process, the presence of the other decomposition products indicates the involvement of radical processes (the Benson mechanism). The activation parameters (particularly the relatively large log A values), which were deduced from the observed rate of decomposition which is probably proceeding by several simultaneous first-order pathways, are also in accordance with these presumptions.

5.3.2 Hydrocarbon hydrotrioxides

Although several groups of workers have provided evidence for the involvement of hydrotrioxides in the reaction of alkanes with ozone, it was only recently that the NMR spectroscopic evidence for their existence became available.[156–158]

Pryor, Ohto and Church[156] were the first to detect by NMR spectroscopy a hydrotrioxide derivative in the low-temperature ozonation of cumene.[156,157] They provided evidence that cumyl hydrotrioxide is formed in a series of steps, which initially involve the formation of a charge-transfer complex (with absorption at 360 nm in acetone) that is photolysed (at $-78°C$) or thermolyzed ($> -40°C$) to the hydrotrioxide via a hydride ion shift, or an electron transfer followed by a hydrogen atom shift (reactions 70–72)

$$RH + O_3 \rightarrow \underset{\text{CT complex}}{[RH, O_3]} \tag{70}$$

$$[RH, O_3] \rightarrow [R^+ \ ^-OOOH] \tag{71}$$

$$[RH, O_3] \rightarrow [RH^{\dot{+}}OOO^{\dot{-}}] \rightarrow [R^+ \ ^-OOOH] \rightarrow ROOOH \tag{72}$$

The decomposition of cumyl hydrotrioxide, which was monitored by following the decay of either the CH_3 or the OOOH absorptions, was first order and gave the activation parameters $E_a = 16.0 \pm 0.1$ kcal mol^{-1} and $\log A = 10.4 \pm 0.1$. These activation parameters are comparable to those reported for the decomposition of hydrotrioxides of acetals, benzaldehyde and benzylic ethers and alcohols. However, in the presence of 2,6-di-*tert*-butyl-4-methylbenzene (BHT), the decomposition of the hydrotrioxide was retarded and the activation parameters became $E_a = 23.9 \pm 0.1$ kcal mol^{-1} and $\log A = 16.4 \pm 0, 1$, indicating that the low activation parameters obtained for the decomposition in the absence of radical inhibitor were due to a chain decomposition process.[157] The Arrhenius parameters obtained are also in excellent agreement with thermochemical predictions for the homolytic scission of the RO—OOR bond.[39,163]

The observation that BHT and the hydrotrioxide decomposed at the same rate indicates that radicals are formed quantitatively. In the presence of BHT, the hydrotrioxide decomposed to form cumyl alcohol as the only organic product (reactions 73–75).[157]

$$ROOOH \rightarrow RO\cdot + \cdot OOH \tag{73}$$

$$RO\cdot(\cdot OOH) + BHT \rightarrow ROH\,(HOOH) + ArO\cdot \tag{74}$$

$$RO\cdot(\cdot OOH) + ArO\cdot \rightarrow \text{non-radical products} \tag{75}$$

In the absence of BHT, small amounts of acetophenone and cumyl hydroperoxide were also formed. It was suggested that acetophenone probably resulted from the cumyltrioxyl radical produced in a chain decomposition and its exothermic fragmentation to the cumyloxyl radical. The scission of the latter then yields acetophenone (reactions 76–78).[157]

$$RO\cdot(\cdot OOH) + ROOOH \rightarrow ROH\ (HOOH) + ROOO\cdot \quad (76)$$

$$ROOO\cdot \rightarrow RO\cdot + {}^1O_2 \quad (77)$$

$$RO\cdot \rightarrow PhCOCH_3 + CH_3\cdot \quad (78)$$

$$RO\cdot + RH \rightarrow ROH + R\cdot \quad (79)$$

$$R\cdot + O_2 \rightarrow ROO\cdot + RH \rightarrow ROOH + R\cdot \quad (80)$$

The cumyl hydroperoxide is believed to be produced by hydrogen abstraction from cumene by any oxy radical present in the system, and subsequent series of reactions (reactions 79 and 80). One further piece of evidence for a homolytic scission of the hydrotrioxide is the observation of the ESR spectrum of the cumylperoxyl radical, which could be detected in the ozonized solutions of cumene at low temperatures.[157]

5.3.3 Silane hydrotrioxides

Silane hydrotrioxides were proposed as transient intermediates in the reaction of ozone with various silanes about 20 years ago.[146,147] However, it was only recently that their chemistry attracted attention.

Recently, NMR spectroscopic evidence for the existence of silyl hydrotrioxides became available (see Table 6).[16a, 161, 162] As discussed above, these polyoxides exist in 'inert' solvents (i.e. solvents of low basicity) probably in the dimeric (or polymeric) form, whereas stronger bases disrupt the intermolecular hydrogen bonds in the dimers to form the opened hydrotrioxide–oxygen base adducts.

A study of the decomposition products of dimethylphenylsilyl hydrotrioxide in acetone-d_6 revealed dimethylphenylsilanol (yield $90 \pm 3\%$ per mole of the hydrotrioxide), dimethylphenyldisiloxane ($8 \pm 2\%$), hydrogen peroxide ($9 \pm 2\%$) and singlet oxygen ($65 \pm 5\%$, 1,3-diphenylisobenzofuran; $50 \pm 5\%$, 9,10-diphenylanthracene; $25 \pm 5\%$, tetraphenylisobenzofuran). Dimethylphenylsilyl hydroperoxide ($5 \pm 2\%$) was also detected by NMR spectroscopy in the decomposition mixture below -45°C. This hydroperoxide, which is otherwise reasonably stable at ambient temperature, decomposes fairly rapidly in the decomposition mixture above -45°C.[16a]

In addition to the decomposition products mentioned, the ^1H NMR spectra revealed that still another transient polyoxide intermediate was formed during the decomposition of dimethylphenylsilyl hydrotrioxide. The disappearance of

OOOH and CH_3 absorptions was accompanied by the simultaneous appearance of another low-field absorption at $\delta 13.2 \pm 0.1$ ppm. Since this OOOH absorption evidently belonged to a species with exchangeable protons, it was tentatively assigned, in the absence of any other resonance that could be attributed to an organic compound, to hydrogen trioxide, HOOOH. Lower concentrations of this intermediate were detected in methyl acetate and dimethyl ether as solvents. However, when methanol was added to solutions of the hydrotrioxide in methyl acetate, significantly higher yields of this intermediate were formed.[16]

The kinetic and activation parameters for the decomposition of dimethylphenylsilyl hydrotrioxide in various solvents are given, together with those for the decay of the OOOH absorption tentatively assigned to HOOOH, in Tables 9 and 10.[16]

An Arrhenius plot of log k vs $1/T$ was found to be non-linear ('concave upward') for the decomposition of dimethylphenylsilyl hydrotrioxide in acetone-d_6; the activation parameters were thus obtained from the slopes of approximately linear portions of this plot. It was suggested that this curvature might be due to the fact that the second hydrotrioxide, tentatively assigned to HOOOH, begins to decompose appreciably at about -60°C (the formation of water), i.e. approximately the same temperature as that reported by Giguère *et al.* for the decomposition of HOOOH.[7]

The kinetic and activation parameters for the decomposition of dimethylphenylsilyl hydrotrioxide, i.e. large negative activation entropies (small log A values), a significant substituent effect on the decomposition in ethyl acetate (Hammett ρ value of 1.2 ± 0.1), the observed dependence of the rate of decomposition on solvent polarity (acetone-d_6 > methyl acetate > dimethyl ether), and the fact that a radical inhibitor, 2,6-di-*tert*-butyl-4-methylphenol, had no measurable effect on the rate of decomposition of dimethylphenylsilyl hydrotrioxide, indicate the importance of polar decomposition pathways. Some of the mechanistic possibilities involving solvated dimeric and/or polymeric hydrogen-bonded forms of the hydrotrioxide are shown in Scheme 5.

The possibility of the involvement of the intermediates and/or transition states with pentacoordinated silicon (reactions 81–83) also appears plausible.[16]

Whereas the existence of dialkyl trioxides (ROOOR, R = alkyl) is well documented (see above), there have been no reports of spectroscopic evidence for the silicon analogs. Such species, if formed in these reactions, are expected to break down either in a heterolytic (reaction 82a) or homolytic process (reaction 82b). An analogous heterolytic pathway is also available for the monomeric silyl hydrotrioxide, complexed with the oxygen base (reaction 83). The formation of structures with pentacoordinated silicon, which is in accordance with the *ab initio* calculations showing a substantial accumulation of negative charge on the terminal oxygen atoms in silyl hydrotrioxide,[177] appear to explain both the observed substituent effect on the decomposition of dimethylphenylsilyl hydrotrioxides and the formation of dimethylphenylmethoxysilane and the observed higher yields of HOOOH in the presence of methanol (reaction 84).

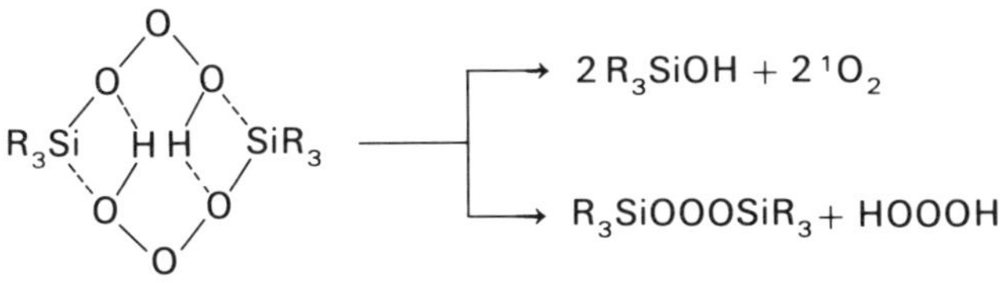

$$R_3Si(\cdots HOOO\cdots)_2SiR_3 \rightarrow 2\,R_3SiOH + 2\,{}^1O_2 \quad (81a)$$

$$\rightarrow R_3SiOOOSiR_3 + HOOOH \quad (81b)$$

$$R_3SiOOOSiR_3 \rightarrow \left[R_3Si\cdots O{=}O\cdots OSiR_3\right] \rightarrow R_3SiOSiR_3 + {}^1O_2 \quad (82a)$$

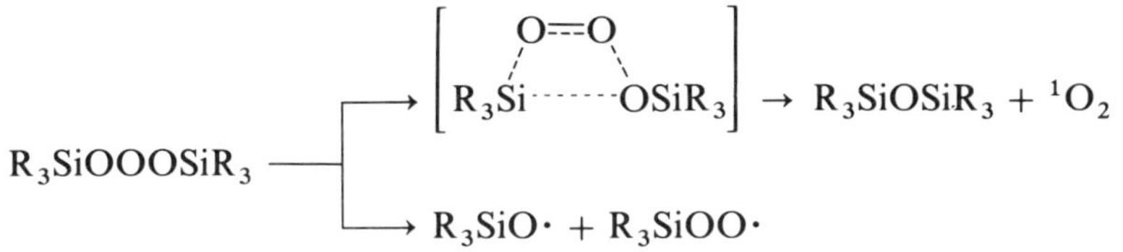

$$\rightarrow R_3SiO\cdot + R_3SiOO\cdot \quad (82b)$$

$$R_3SiOOOH\cdots(B) \rightarrow \left[R_3Si\cdots O{=}O\cdots O{-}H\cdots(B)\right] \rightarrow R_3SiOH\cdots(B) + {}^1O_2 \quad (83)$$

$$R_3Si(O{-}O{-}O{-}H)\cdots O(H)CH_3 \rightarrow R_3SiOCH_3 + HOOOH \quad (84)$$

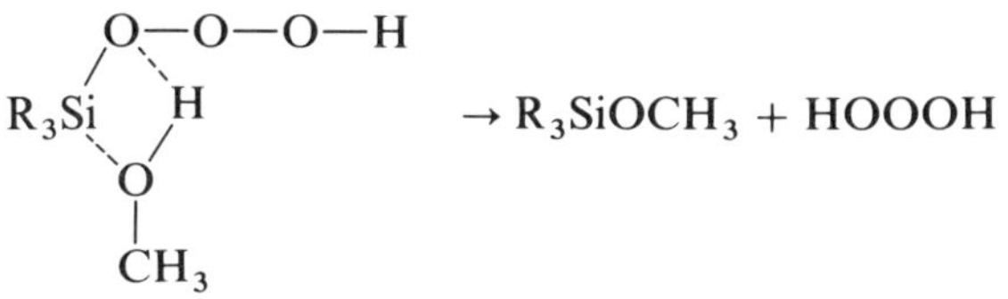

$$\left[R_3Si{-}O{-}O{-}O{-}H\cdots O(SiR_3){-}O{-}O{-}H\cdots\right](B) \rightarrow 2\,R_3SiOH + 2\,{}^1O_2$$

$$\rightarrow R_3SiOOO^- + HOOOH + R_3Si^+ + {}^-OOOH$$

$$R_3SiOOO^- \rightarrow R_3SiO^- + {}^1O_2; \quad {}^-OOOH \rightarrow HO^- + {}^1O_2$$

$$\left[R_3SiO(\cdots HO)OO{-}O(\cdots HO)(\cdots HO)\right](B) \rightarrow R_3SiOOO^- + HOOOH + R_3Si^+ + {}^-OOOH$$

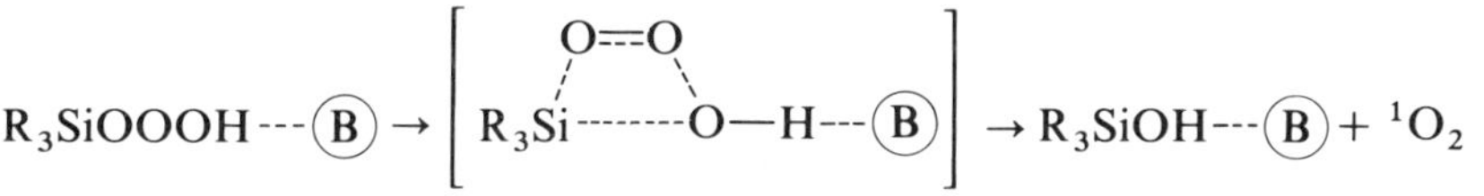

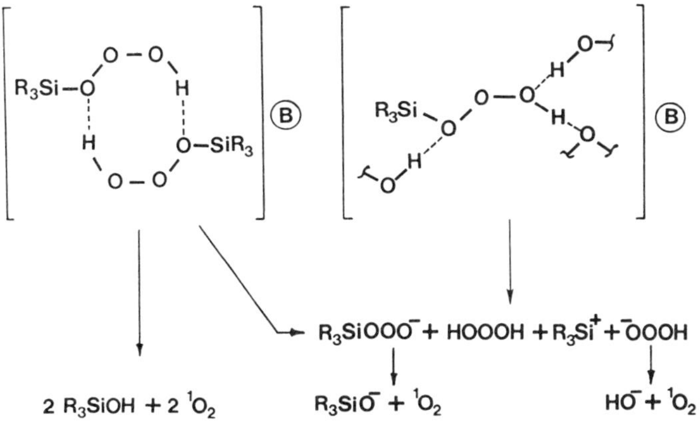

Scheme 5

Table 9. Kinetic and activation parameters for the decomposition of dimethylphenylsilyl hydrotrioxide[16a]

ROOOH	Solvent	T (°C)	$k_1 \times 10^4$ (s^{-1})^a	E_a (kcal mol^{-1})	Log A	ΔS (e.u.)
$PhSi(Me)_2$—OOOH	$(CD_3)_2CO$	−80	0.55	4.7	1.0	−58
		−75	0.60			
		−70	0.70			
		−65	1.20			
		−60	1.31	9.5	5.5	−35
		−50	1.91			
		−45	2.52			
		−40	3.70			
		−30	10.7^b			
		−20	17.1			
	$MeCO_2Me$	−50	1.1	10.0	5.9	−33
	Me_2O	−50	0.8	12.6	8.2	−23

^{a}Relative deviation 8%.

^{b}$Bu_3SiOOOH$ [Bu_3SiH as solvent, δOOOH = 14 ppm (−70°C)], k (−50°C) = 1×10^{-1} s^{-1}, E_a = 6.0 kcal mol^{-1}, log A = 6.8.[162a] $Et_3SiOOOH$ [$(CD_3)_2CO$, δOOOH = 13.7 ppm (−78°C)], k (−50°C) = 1.25×10^{-4} s^{-1}, $\Delta H\ddagger$ = 17.2 kcal mol^{-1}, $\Delta S\ddagger$ = 1.6 e.u. (temperature range: −55 to −40°C).[162b]

Table 10. Kinetic and activation parameters for the decay of the OOOH absorption tentatively assigned to hydrogen trioxide: temperature dependence of the OOOH absorption (acetone-d_6)[16a]

T (°C)	δOOOH (ppm)	$10^4\ k_1$ (s^{-1})[a]	E_a (kcal mol^{-1})	Log A	ΔS (e.u.)
−60	13.37				
−50	13.29	0.20	11.0	6.1	−32
−40	13.16	0.56			
−30	13.08				
	(12.73)[b]	1.6 (1.3)[b]			
−20		4.0			
−10		9.3			

[a]Relative standard deviations 10%
[b]In dimethyl ether (relative standard deviation 15%).

It was not possible on the basis of kinetic results and experiments with an added radical inhibitor to exclude completely the possibility of some radical contributions to the overall decomposition mechanism of dimethylphenylsilyl hydrotrioxide, particularly at temperatures above −50°C (fast warm-up procedure) (reactions 85–88).

$$(R_3SiOOOH)_n \rightarrow [R_3SiO\cdot\ \cdot OOH] \rightarrow \begin{cases} R_3SiOH + {}^1O_2 \\ \xrightarrow{H\cdot} R_3SiOH + HOOH \end{cases} \tag{85}$$

$$R_3SiH + Z\cdot \rightarrow R_3Si\cdot + HZ \xrightarrow{O_2} R_3SiOO\cdot \rightarrow R_3SiOOH \tag{86}$$

$$R_3SiOOOH + R_3Si\cdot \rightarrow R_3SiOOH + R_3SiO\cdot \tag{87}$$

$$R_3SiOOOH + Z\cdot \rightarrow R_3SiOOO\cdot \rightarrow R_3SiO\cdot + {}^1O_2 + HZ \tag{88}$$

The possibility still exists that the oxygenated products, i.e. dimethylphenylsilyl hydrotrioxide, dimethylphenylsilyl hydroperoxide, dimethylphenylsilanol or oxygen, are better traps for silyl and/or silyloxyl radicals than the phenol. However, it appears that these processes are only minor (<10%) under the conditions investigated.

The activation parameters for the decomposition of hydrogen trioxide, HOOOH, suggest a predominantly polar decomposition mechanism, as already described (see Section 2.2.2).

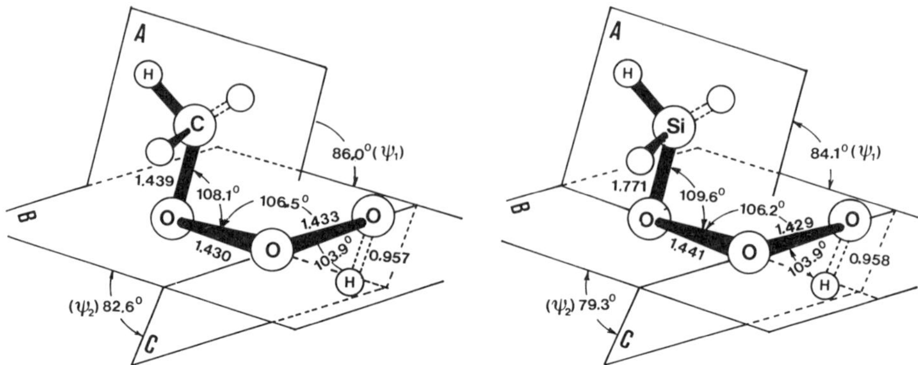

Figure 3. Calculated equilibrium geometries of methyl hydrotrioxide and silyl hydrotrioxide (6–31G). The bond lengths are in ångströms and the bond angles in degrees[177]

5.4 Reactivity of Hydrotrioxides

5.4.1 Generators of singlet oxygen

Corey *et al.*[148] have shown that triethylsilyl hydrotrioxide, produced by the low-temperature (−78°C) ozonation of triethylsilane, decomposes at higher temperatures to give singlet oxygen.[148] A number of preparatively very attractive characteristic transformations involving singlet oxygen have been reported (Table 11).

The formation of singlet oxygen, $^1O_2$, was also confirmed by the near-IR spectroscopy. The decomposition of the trioxide in dichloromethane and carbon disulfide at −60°C was measured by following the decay in emission at 1278 nm with time ($^1O_2 \rightarrow {}^3O_2$). The half-life, $\tau_{1/2}$, was found to be 150 and 10 s, corresponding to first-order rate constants of 4.6×10^{-3} and 6.9×10^{-2} s^{-1} in CH_2Cl_2 and CS_2, respectively. Since the lifetime of $^1O_2$ in CS_2 is known to be considerably longer than that in CH_2Cl_2, it is obvious that the hydrotrioxide is more stable in CH_2Cl_2 than in CS_2. It was suggested that CH_2Cl_2 stabilizes triethylsilyl hydrotrioxide, presumably by intermolecular hydrogen bonding.

5.4.2 Dioxetane-forming reagents

Posner *et al.*[150] reported that triethylsilyl hydrotrioxide is an efficient reagent for direct conversion of electron-rich alkenes to 1,2-dioxetanes (e.g. reaction 89).[150a]

Unactivated alkenes are oxidatively cleaved by triethylsilyl hydrotrioxide. For example, this oxidant reacts with methyl oleate followed by aluminum lithium hydride to produce nonan-1-ol and nonane-1,9-diol in 64 and 74%

Table 11. Oxygenation of various substrates using singlet oxygen from the decomposition of triethylsilyl hydrotrioxide[148]

Substrate	Product	Yield (%)
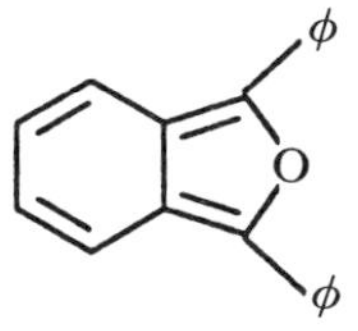 ϕ, O, ϕ	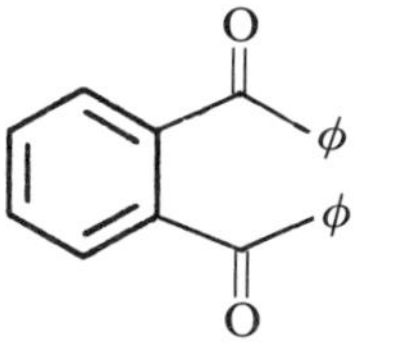 O, ϕ, ϕ, O	91
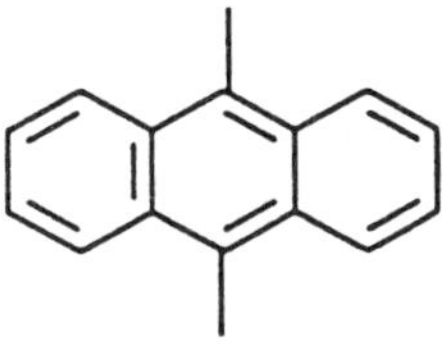	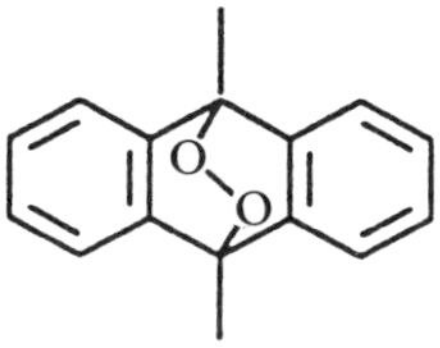 O, O	92
	 O, O	46
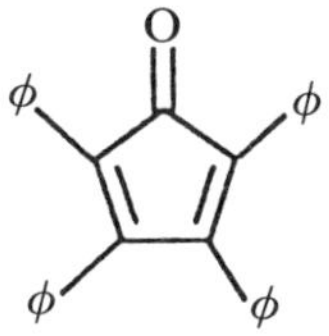 O, ϕ, ϕ, ϕ, ϕ	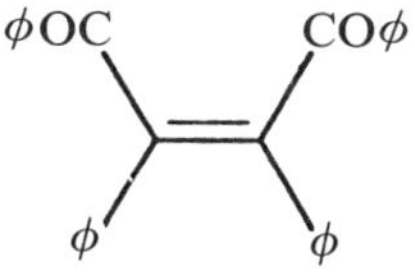 ϕOC, COϕ, ϕ, ϕ	62
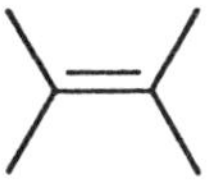	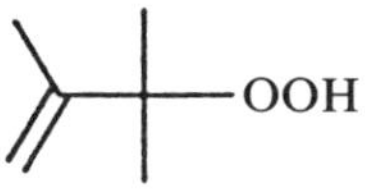 OOH	40
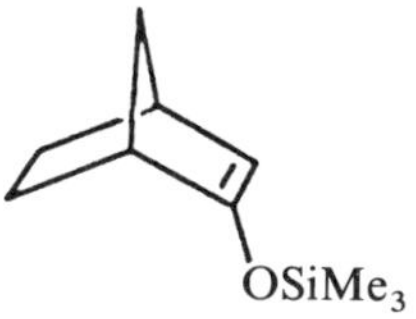 $OSiMe_3$	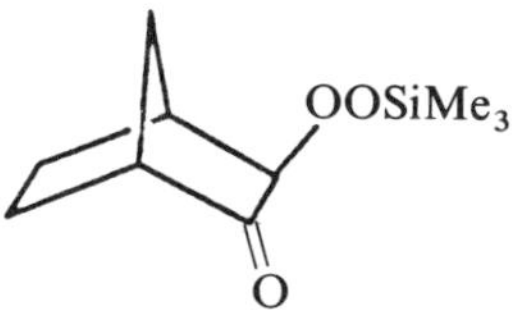 $OOSiMe_3$, O	50
	 O, O	90

$$\xrightarrow[\text{CH}_2\text{Cl}_2,\ -78\,^\circ\text{C}]{\text{Et}_3\text{SiO}_3\text{H}} \qquad (89)$$

$R^1 = OMe$; $R^2 = m\text{-}HCO_6H_4$ 46%
$R^1 = H$; $R^2 = C_6H_5$ 0%

yields, respectively, and with 2-phenylhex-1-ene to yield valerophenone. In neither of these cases could evidence for the involvement of a dioxetane intermediate be found.

The mechanism shown in Scheme 6 was proposed to account for triethylsilyl hydrotrioxide reacting with electron-rich alkenes as a dioxetane-forming reagent (pathway a) and with unactivated alkenes as a non-dioxetane carbonyl-forming reagent (pathway b).[150a]

By using triethylsilyl hydrotrioxide, a very short and efficient synthesis of new, simple, and potent antimalarial 1,2,4-trioxanes has recently been accomplished (reactions 90 and 91).[150b]

1) $Et_3SiOOOH$
2) t-$BuMe_2SiOTf$

OMe → H, OMe 42 % (90)

1) $Et_3SiOOOH$
2) t-$BuMe_2SiOTf$

Ph O → H, Ph O 48 % (91)

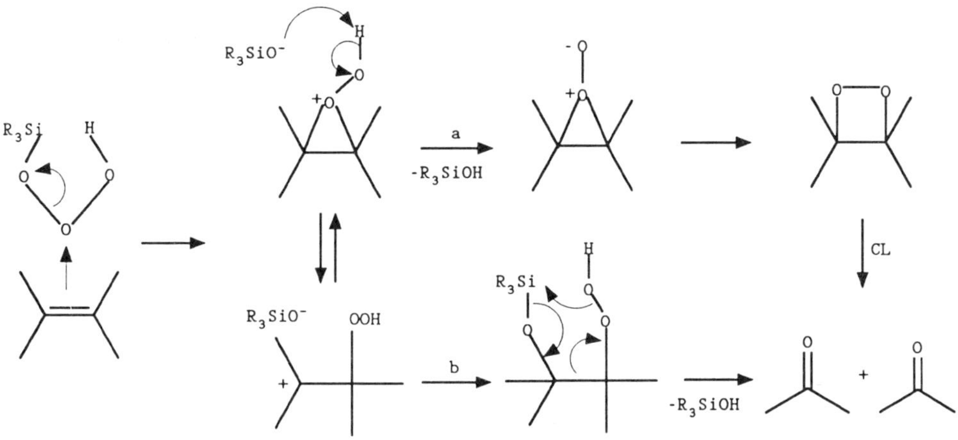

Scheme 6

5.4.3 Oxidation of sulfides and sulfoxides

Soon after the discovery of hydrotrioxides, it was demonstrated[160,161] that hydrotrioxides of acetals, alcohols and ethers oxidize simple dialkyl sulfides (e.g. dimethyl and diethyl sulfides) to the corresponding sulfoxides, and the latter to sulfones in high yields.

Recently, it was reported that dimethylphenylsilyl hydrotrioxide oxidizes phenyl methyl sulfides to sulfoxides in various solvents at $-78°C$ in yields ranging from 70 to 80%.[16a] These oxidations, which were run in the presence of excess sulfide, produced virtually no sulfone. When phenyl methyl sulfoxides were allowed to react with the hydrotrioxide, relatively low yields of the sulfones were obtained ($5 \pm 1\%$). A series of competitive kinetic experiments with *para*-substituted phenyl methyl sulfides gave Hammett ρ values ranging from -2.0 (acetone) to -1.6 (pentane). The oxidation of the corresponding sulfoxides gave the ρ value of -1.1 (acetone).

The negative Hammett ρ values and the observed reactivity order PhSMe $\gg$ PhSOMe indicate an electrophilic oxygen transfer from the hydrotrioxide to the sulfide and sulfoxide. A mechanism, similar to that already proposed by Posner *et al.*,[150] involving the rate-determining nucleophilic attack of the sulfide/sulfoxide on the middle oxygen atom in the hydrotrioxide was proposed (Scheme 7).

The results of *ab initio* studies on $H_3SiOOOH$ show unambiguously that the middle oxygen atom is (as in CH_3OOOH) most probably the 'electrophilic' center in the hydrotrioxide.[177] However, some alternative explanations of the mechanism of these oxidations, for example single-electron transfer processes, cannot be ruled out.

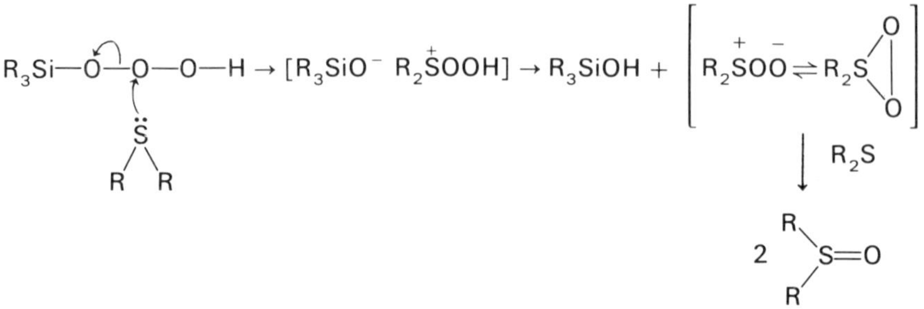

$$R_3Si{-}O{-}O{-}O{-}H \;(+\; R_2\ddot{S}) \rightarrow [R_3SiO^- \; R_2\overset{+}{S}OOH] \rightarrow R_3SiOH + \left[R_2\overset{+}{S}O\overset{-}{O} \rightleftharpoons R_2S\begin{smallmatrix}O\\|\\O\end{smallmatrix} \right] \xrightarrow{R_2S} 2\, R_2S{=}O$$

Scheme 7

6 ALKYL HYDROTETRAOXIDES

Relatively little of a definitive nature is known about alkyl hydrotetraoxides, ROOOOH. These species have not so far been detected spectroscopically. However, the strong negative temperature dependence of the rate constant and the low A factor for the gas-phase reaction of $CH_3O_2\cdot$ with $HO_2\cdot$[100a,178–181] [$k = (4.4 \pm 0.7) \times 10^{-13} \exp(780 \pm 55)/T$ cm^3 molecule^{-1} s^{-1}, measured between 248 and 573 K, and independent of pressure between 210 and 760 Torr][180] suggest the involvement of methyl hydrotetraoxide as an intermediate in this reaction (reaction 92).

$$CH_3OO\cdot + HOO\cdot \rightleftharpoons CH_3OOOOH \rightarrow CH_3OOH + O_2 \qquad (92)$$

Methyl hydrotetraoxide may either redissociate to reactants or decompose to products. The four-membered cyclic transition state shown, analogous to that already proposed for the decomposition of HOOOOH (see discussion above), has been suggested to be involved in the decomposition of this species.[180] It is

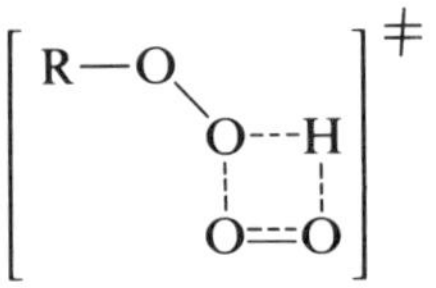

$$\left[\begin{matrix} R{-}O & \\ & O{\cdots}H \\ & \vdots \quad\; \vdots \\ & O{=}O \end{matrix} \right]^{\ddagger}$$

possible that the equilibrium conformations of these polyoxides correspond to intramolecularly hydrogen bonded structures. However, the self-association of these species, e.g. dimerization, also cannot be ruled out. The proton transfer would be particularly favorable in such assemblies.

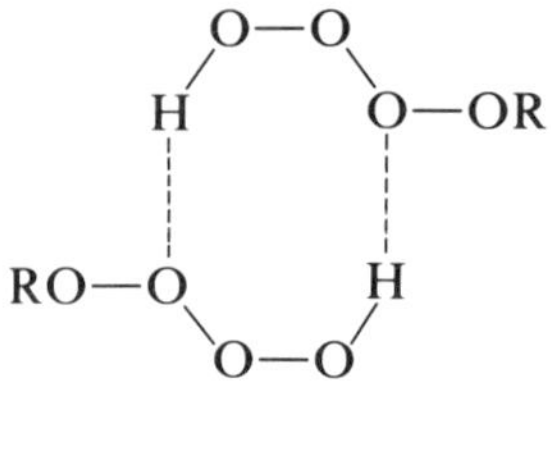

$$\begin{matrix} & O{-}O & \\ H & & O{-}OR \\ \vdots & & \vdots \\ RO{-}O & & H \\ & O{-}O & \end{matrix}$$

The observation of significant amounts of HDO in the reaction of $CD_3O_2^{\bullet}$ with $HO_2^{\bullet}$ suggests, at least in the gas-phase reaction at higher temperatures, still another decomposition channel (reaction 93):

$$CH_3O_2\cdot + HO_2\cdot \rightarrow \text{[cyclic } H_2C(H)\text{–O–O–O–O(H) adduct]} \rightarrow H_2O + HCHO + O_2 \tag{93}$$

A radical decomposition of the alkyl hydrotetraoxide to ROO and OOH radicals, with subsequent in-cage abstraction of the hydrogen atom (BDE, $CH_3OO—OOH$, 9 ± 1 kcal mol^{-1}), or direct decomposition to CH_3O and OH radicals (reactions 94a and b), could also be important decomposition pathways (condensed-phase reactions).

$$CH_3OOOOH \rightarrow [CH_3OO\cdot \cdot OOH] \rightarrow CH_3OOH + O_2 \tag{94a}$$

$$CH_3OOOOH \rightarrow CH_3O\cdot + O_2 + \cdot OH \tag{94b}$$

A relatively large negative temperature dependence and a low *A* factor for reaction 92 appear to be not in accordance with a process that involves the formation of intermolecularly hydrogen-bonded adduct $CH_3OO \cdots HOO$ and subsequent transfer of the hydrogen atom (reaction 95).

$$CH_3O_2^{\bullet} + HO_2^{\bullet} \rightarrow CH_3OO\cdot \cdots HOO\cdot \rightarrow [CH_3OO \cdots H \cdots O_2]^{\ddagger} \rightarrow CH_3OOH + O_2 \tag{95}$$

A MINDO/3 method has recently been used to study the geometrical structure and vibrational frequencies of CH_3O_4H.[182] However, high-level *ab initio* calculations on several aspects of the structure and decomposition of alkyl hydrotrioxides and alkyl hydrotetraoxides are still lacking.

7 ACKNOWLEDGEMENTS

I am grateful to my colleagues, especially F. Kovač, J. Koller and J. Cerkovnik, for their contributions to the research results from our group which are presented in this review. The studies from our laboratory were supported by the Ministry of Science and Technology of the Republic of Slovenia and the Yugoslav-United States Joint Fund for Scientific and Technological Cooperation (US National Science Foundation, JFP 538 and JF 834). I am also grateful to Professors G. A. Russell and J. C. Martin for their encouragement and advice.

8 REFERENCES

1. B. Plesničar, in *The Chemistry of Peroxides* (S. Patai, Ed.), Wiley, Chichester (1983), Chap. 16.
2. M. Venugopalan and R. G. Jones, *The Chemistry of Dissociated Water Vapour and Related Systems*, Wiley, New York (1968), Chap. 7.
3. E. A. V. Ebsworth, J. A. Connor and J. J. Turner, in *Comprehensive Inorganic Chemistry* (J. C. Bailar, Jr, H. J. Emeleus, R. Nyholm and A. F. Trotman-Dickenson, Eds), Pergamon Press, Oxford (1975), Chap. 22.
4. P. A. Giguère, *Trans. N.Y. Acad. Sci.*, **34**, 334 (1972).
5. P. A. Giguère, *Peroxyde d'Hydrogène et Polyoxyde d'Hydrogène, Vol. 4, Compléments au Nouveau Traité de Chimie Minérale*, Manon, Paris, 1975.
6. P. A. Giguère, *Chemistry (Washington)*, **48**, 20 (1975).
7. P. A. Giguère and K. Herman, *Can. J. Chem.*, **48**, 3473 (1970).
8. X. Deglise and P. A. Giguère, *Can. J. Chem.*, **49**, 2242 (1971).
9. J. L. Arnau and P. A. Giguère, *J. Chem. Phys.*, **60**, 270 (1974).
10. P. A. Giguère and T. K. K. Srinivasan, *Chem. Phys. Lett.*, **33**, 479 (1975).
11. T. V. Yagodovskaya, D. Zhogin and L. I. Nekrasov, *Zh. Fiz. Khim.*, **50**, 2736 (1976); T. V. Yagodovskaya and L. I. Nekrasov, *Zh. Fiz. Khim.*, **51**, 2434 (1977); **60**, 922 (1986).
12. B. H. J. Bielski and H. A. Schwartz, *J. Phys. Chem.*, **72**, 3836 (1968).
13. M. Diem, T.-L. Tso and E. K. C. Lee, *J. Chem. Phys.*, **76**, 6452 (1982).
14. T.-L. Tso and E. K. C. Lee, *J. Phys. Chem.*, **89**, 1618 (1985).
15. H. Sugimoto, D. T. Sawyer and J. R. Kanofsky, *J. Am. Chem. Soc.*, **110**, 8707 (1988).
16. (a) B. Plesničar, J. Cerkovnik, J. Koller and F. Kovač, *J. Am. Chem. Soc.*, **113**, 4946 (1991); (b) J. Koller and B. Plesničar, in preparation. See also: C. Gonzales, J. Theisen, L. Zhu, H. B. Schlegel, W. L. Hase and E. W. Kaiser, *J. Phys. Chem.*, **95**, 6784 (1991).
17. J. Cerkovnik and B. Plesničar, in preparation.
18. M. Kaufman and S. Sherwell, *Prog. React. Kinet.*, **12**, 1 (1983).
19. R. A. Cox and J. P. Burrows, *J. Phys. Chem.*, **83**, 2560 (1979).
20. B. A. Trush and J. P. T. Wilkinson, *Chem. Phys. Lett.*, **66**, 441 (1979).
21. S. P. Sander, M. Peterson, R. T. Watson and R. Patrick, *J. Phys. Chem.*, **86**, 1236 (1982).
22. C. C. Kircher and S. P. Sander, *J. Phys. Chem.*, **88**, 2082 (1984).
23. W. B. DeMore, J. J. Margitan, M. J. Molina, R. T. Watson, D. M. Golden, R. F. Hampson, M. J. Kurylo, C. J. Howard and A. R. Ravishankara, *Chemical Kinetics and Photochemical Data for Use in Stratospheric Modeling: Evaluation Number 7*, JPL Publishing 85–37, Jet Propulsion Laboratory, Pasadena (1985).
24. G. A. Takacs and C. J. Howard, *J. Phys. Chem.*, **90**, 687 (1986).
25. H. Niki, P. D. Maker, C. M. Savage and L. P. Breitenbach, *Chem. Phys. Lett.*, **73**, 43 (1980).
26. C. J. Hochanadel, T. J. Sworski and P. J. Ogren, *J. Phys. Chem.*, **84**, 3274 (1980).
27. E. J. Hamilton, Jr and R. R. Lii, *Int. J. Chem. Kinet.*, **9**, 875 (1977), and references cited therein.
28. R. R. Lii, R. A. Gorse, Jr, M. C. Sauer, Jr, and S. Gordon, *J. Phys. Chem.*, **84**, 813 (1980).
29. R. Patrick and M. J. Pilling, *Chem. Phys. Lett.*, **91**, 343 (1982).
30. R. Patrick, J. R. Barker and D. M. Golden, *J. Phys. Chem.*, **88**, 128 (1984).
31. F. Kaufman, *J. Phys. Chem.*, **88**, 4909 (1984).
32. (a) V. B. Rozenshtein, Yu, M. Gershenzon, S. D. Ilin and O. P. Kishkovitch, *Chem.*

Phys. Lett., **112**, 473 (1984); (b) P. D. Lightfoot, B. Veyret and R. Lesclaux, *Chem. Phys. Lett.*, **150**, 120 (1988).
33. B. H. J. Bielski and A. D. Allen, *J. Phys. Chem.*, **81**, 1046 (1977).
34. J. A. Howard and K. U. Ingold, *Can. J. Chem.*, **45**, 785 (1967).
35. P. A. Giguère, *J. Phys. Chem.*, **85**, 3733 (1981).
36. G. Fitzgerald, T. J. Lee and H. F. Schaefer, III and R. J. Bartlett, *J. Chem. Phys.*, **83**, 6275 (1985).
37. C. F. Melius, J. S. Binkley and M. Frisch, unpublished results, cited in Ref. 14.
38. B. Plesničar, S. Kaiser and A. Ažman, *J. Am. Chem. Soc.*, **95**, 5476 (1973).
39. J. S. Francisco and I. H. Williams, *Int. J. Chem. Kinet.*, **20**, 455 (1988).
40. L. G. S. Shum and S. W. Benson, *J. Phys. Chem.*, **87**, 3479 (1983); P. S. Nangia and S. W. Benson, *J. Phys. Chem.*, **83**, 1138 (1979).
41. (a) M. J. Kurylo, O. Klais and A. H. Laufer, *J. Phys. Chem.*, **85**, 3674 (1981); (b) P. Dransfeld and H. Gg. Wagner, *Z. Naturforsch.*, **42a**, 471 (1987).
42. L. F. Keyser, *J. Phys. Chem.*, **85**, 3667 (1981); **92**, 1193 (1988).
43. W. B. DeMore, *J. Phys. Chem.*, **86**, 121 (1982).
44. J. J. Schwab, W. H. Brune and J. G. Anderson, *J. Phys. Chem.*, **93**, 1030 (1989).
45. U. C. Sridharan, L. X. Qui and F. Kaufman, *J. Phys. Chem.*, **86**, 4569 (1982).
46. U. C. Sridharan, L. X. Qiu, F. Kaufman, *J. Phys. Chem.*, **88**, 1281 (1984).
47. F. Temps and H. Gg. Wagner, *Ber. Bunsenges. Phys. Chem.*, **86**, 119 (1982).
48. M. Mozurkewich, *J. Phys. Chem.*, **90**, 2216 (1986).
49. D. W. Toohey, Wm. H. Brune and J. G. Anderson, *J. Phys. Chem.*, **91**, 1215 (1987).
50. D. W. Toohey and J. G. Anderson, *J. Phys. Chem.*, **93**, 1049 (1989).
51. L. Radom, W. J. Hehre and J. A. Pople, *J. Am. Chem. Soc.*, **93**, 289 (1971).
52. R. J. Blint and M. D. Newton, *J. Chem. Phys.*, **59**, 6220 (1973).
53. B. Plesničar, D. Kocjan, S. Murovec and A. Ažman, *J. Am. Chem. Soc.*, **98**, 3143 (1976).
54. D. Cremer, *J. Chem. Phys.*, **69**, 4456 (1978).
55. C. F. Jackels and D. H. Phillips, *J. Chem. Phys.*, **84**, 5013 (1986).
56. H. S. Blanchard, *J. Am. Chem. Soc.*, **81**, 4548 (1959).
57. G. A. Russell, *Chem. Ind. (London)*, 1483 (1956).
58. G. A. Russell, *J. Am. Chem. Soc.*, **79**, 3871 (1957).
59. T. G. Traylor and P. D. Bartlett, *Tetrahedron Lett.*, 30 (1960).
60. P. D. Bartlett and T. G. Traylor, *J. Am. Chem. Soc.*, **85**, 2407 (1963).
61. T. G. Traylor, *J. Am. Chem. Soc.*, **85**, 2411 (1963).
62. R. Hiatt and T. G. Traylor, *J. Am. Chem. Soc.*, **87**, 3766 (1965).
63. J. R. Thomas, *J. Am. Chem. Soc.*, **89**, 4872 (1963).
64. P. D. Bartlett and G. Guaraldi, *J. Am. Chem. Soc.*, **89**, 4799 (1967).
65. N. A. Milas and B. Plesničar, *J. Am. Chem. Soc.*, **90**, 4450 (1968).
66. T. Mill and R. S. Stringham, *J. Am. Chem. Soc.*, **90**, 1062 (1968).
67. J. E. Bennett, D. M. Brown and B. Mile, *Chem. Commun.*, 504 (1969).
68. J. E. Bennett, D. M. Brown and B. Mile, *Trans. Faraday Soc.*, **66**, 397 (1970).
69. J. E. Bennett, B. Mile and B. Ward, *Adv. Phys. Org. Chem.*, **8**, 1 (1970).
70. K. Adamic, J. A. Howard and K. U. Ingold, *Chem. Commun.*, 504 (1969).
71. K. Adamic, J. A. Howard and K. U. Ingold, *Can. J. Chem.*, **47**, 3803 (1969).
72. K. U. Ingold, *Acc. Chem. Res.*, **2**, 1 (1969).
73. J. A. Howard, *Adv. Free-Radical Chem.*, **4**, 49 (1972).
74. J. A. Howard, *Am. Chem. Soc. Symp. Ser.*, **69**, 413 (1978).
75. S. Fukuzumi and Y. Ono, *J. Chem. Soc., Perkin Trans. 2*, 622 (1977).
76. S. Fukuzumi and Y. Ono, *J. Phys. Chem.*, **81**, 1895 (1977).
77. S. Korcek, J. H. B. Chenier, J. A. Howard and K. U. Ingold, *Can. J. Chem.*, **50**, 2285 (1972).

78. P. S. Nangia and S. W. Benson, *Int. J. Chem. Kinet.*, **12**, 29 (1980).
79. J. E. Bennett and J. A. Howard, *J. Am. Chem. Soc.*, **95**, 4008 (1973).
80. J. R. Thomas, *J. Am. Chem. Soc.*, **87**, 3935 (1965).
81. J. A. Howard and K. U. Ingold, *J. Am. Chem. Soc.*, **90**, 1058 (1968).
82. J. A. Howard and J. E. Bennett, *Can. J. Chem.*, **50**, 2375 (1972).
83. D. F. Bowman, T. Gillan and K. U. Ingold, *J. Am. Chem. Soc.*, **93**, 6555 (1971).
84. J. A. Howard and K. U. Ingold, *J. Am. Chem. Soc.*, **90**, 1056 (1968).
85. M. Nakano, K. Takayama, Y. Shimizu, Y. Tsuji, H. Inaba and T. Migita, *J. Am. Chem. Soc.*, **98**, 1974 (1976).
86. H. Inaba, Y. Shimizu and A. Yamagishi, *Photochem. Photobiol.*, **30**, 169 (1979).
87. S.-H. Lee and G. D. Mendenhall, *J. Am. Chem. Soc.*, **110**, 4318 (1988).
88. Q. Niu and G. D. Mendenhall, *J. Am. Chem. Soc.*, **112**, 1656 (1990).
89. J. R. Kanofsky, *J. Org. Chem.*, **51**, 3386 (1986).
90. J. R. Kanofsky, H. Sugimoto and D. T. Sawyer, *J. Am. Chem. Soc.*, **110**, 3698 (1988).
91. H. Sugimoto, S. Matsumoto, D. T. Sawyer, J. R. Kanofsky, A. K. Chowdhury and C. L. Wilkins, *J. Am. Chem. Soc.*, **110**, 5193 (1988).
92. P. S. Nangia and S. W. Benson, *Int. J. Chem. Kinet.*, **12**, 43 (1980).
93. H. Niki, P. D. Maker, C. M. Savage and L. P. Breitenbach, *J. Phys. Chem.*, **85**, 877 (1981).
94. C. S. Kan, J. G. Calvert and J. H. Shaw, *J. Phys. Chem.*, **84**, 3411 (1980).
95. C. J. Hochanadel, J. A. Ghormley, J. W. Boyle and P. J. Ogren, *J. Phys. Chem.*, **81**, 3 (1977).
96. D. A. Parkes, *Int. J. Chem. Kinet.*, **9**, 451 (1977); C. Anastasi, J. W. M. Smith and D. A. Parkes, *J. Chem. Soc., Faraday Trans. 1*, 1693 (1978).
97. E. Sanhueza, R. Simonaitis and J. Heicklen, *Int. J. Chem. Kinet.*, **11**, 907 (1979).
98. C. S. Kan, R. D. McQuigg, M. R. Whitbeck and J. G. Calvert. *Int. J. Chem. Kinet.*, **11**, 921 (1979).
99. S. P. Sander and R. T. Watson, *J. Phys. Chem.*, **85**, 2960 (1981).
100. (a) P. Dagaut, T. J. Wallington and M. J. Kurylo, *J. Phys. Chem.*, **92**, 3833, 3836 (1988); (b) P. D. Lightfoot, R. Lesclaux and B. Veyret, *J. Phys. Chem.*, **94**, 700 (1990).
101. P. Ase, W. Bock and A. Snelson, *J. Phys. Chem.*, **90**, 2099 (1986).
102. P. D. Bartlett and P. Günther, *J. Am. Chem. Soc.*, **88**, 3288 (1966).
103. P. D. Bartlett and M. Lahav, *Is. J. Chem.*, **10**, 101 (1972).
104. J. E. Bennett, G. Brunton and R. Summers, *J. Chem. Soc., Perkin Trans. 2*, 981 (1980).
105. L. R. Anderson and W. B. Fox, *J. Am. Chem. Soc.*, **89**, 4313 (1967).
106. P. G. Thompson, *J. Am. Chem. Soc.*, **89**, 4316 (1967).
107. (a) F. A. Hohorst, D. D. DesMarteau, L. R. Anderson, D. E. Gould and W. B. Fox, *J. Am. Chem. Soc.*, **95**, 3866 (1973); (b) J. D. Witt, J. R. Durig, D. D. DesMarteau and R. M. Hammaker, *Inorg. Chem.*, **12**, 807 (1973).
108. J. Czarnowski and H. J. Schumacher, *Int. J. Chem. Kinet.*, **13**, 639 (1981).
109. R. W. Fessenden, *J. Chem. Phys.*, **48**, 3725 (1968).
110. J. Czarnowski and H. J. Schumacher, *J. Fluorine Chem.*, **12**, 497 (1978); J. Czarnowski and H. J. Schumacher, *Int. J. Chem. Kinet.*, **11**, 613 (1979).
111. D. Tal, E. Keinan and Y. Mazur, *J. Am. Chem. Soc.*, **101**, 502 (1979); E. Keinan and H. T. Varkony, in *The Chemistry of Peroxides* (S. Patai, Ed.), Wiley, Chichester (1983), Chap. 19.
112. N. Yoneda and G. A. Olah, *J. Am. Chem. Soc.*, **99**, 3113 (1969).
113. T. Preuss, E. Proksch and A. deMeijere, *Tetrahedron Lett.*, 833 (1978).
114. A. L. J. Beckwith, C. L. Bodkin and T. Duong, *Aust. J. Chem.*, **30**, 2177 (1977).

115. Yu. A. Alexandrov and B. I. Tarunin, *J. Organomet. Chem.*, **238**, 125 (1982).
116. Yu. A. Alexandrov, N. G. Scheyanov and V. A. Shushunov, *Dokl. Akad, Nauk SSSR*, **192**, 91 (1970); L. C. Willemsens and G. J. Van der Kerk, *J. Organomet. Chem.*, **15**, 117 (1968).
117. Yu. A. Alexandrov and N. G. Sheyanov, *Zh. Obshch. Khim.*, **40**, 1664 (1970).
118. B. Plesničar, D. Kocjan, S. Murovec and A. Ažman, *J. Am. Chem. Soc.*, **98**, 3143 (1976).
119. N. A. Clinton, R. A. Kenley and T. G. Traylor, *J. Am. Chem. Soc.*, **97**, 3746 (1975).
120. N. A. Clinton, R. A. Kenley and T. G. Traylor, *J. Am. Chem. Soc.*, **97**, 3752 (1975).
121. N. A. Clinton, R. A. Kenley and T. G. Traylor, *J. Am. Chem. Soc.*, **97**, 3757 (1975).
122. R. A. Kenley and T. G. Traylor, *J. Am. Chem. Soc.*, **97**, 4700 (1975).
123. G. E. Zaikov, J. A. Howard and K. U. Ingold, *Can. J. Chem.*, **47**, 3017 (1969).
124. P. S. Bailey, *Ozonation in Organic Chemistry*, Vol. 2, Academic Press, New York (1982).
125. M. C. Whiting, A. J. N. Bolt and J. H. Parish, *Adv. Chem. Ser.*, **77**, 4 (1968).
126. J. E. Batterbee and P. S. Bailey, *J. Org. Chem.*, **32**, 899 (1967).
127. G. A. Hamilton, B. S. Ribner and T. M. Hellman, *Adv. Chem. Ser.*, **77**, 15 (1968).
128. T. M. Hellman and G. A. Hamilton, *J. Am. Chem. Soc.*, **96**, 1530 (1974).
129. C. C. Price and A. L. Tumolo, *J. Am. Chem. Soc.*, **86**, 4691 (1964).
130. R. E. Erickson, R. T. Hansen and J. Harkins, *J. Am. Chem. Soc.*, **90**, 6777 (1968).
131. P. S. Bailey and D. A. Lerdal, *J. Am. Chem. Soc.*, **100**, 5820 (1978).
132. H. M. White and P. S. Bailey, *J. Org. Chem.*, **30**, 3037 (1965).
132. R. E. Erickson, D. Bakalik, C. Richards, M. Scanion and D. Huddleston, *J. Org. Chem.*, **31**, 461 (1966).
134. P. Deslongchamps and C. Moreau, *Can. J. Chem.*, **49**, 2465 (1971).
135. P. Deslongchamps, P. Atlani, D. Frehel and C. Moreau, *Can. J. Chem.*, **52**, 3651 (1974).
136. P. Deslongchamps, C. Moreau, D. Frehel and R. Chenevert, *Can. J. Chem.*, **53**, 1204 (1975).
137. (a) R. J. Teillefer, S. E. Thomas, Y. Nadeau and H. Beierbeck, *Can. J. Chem.*, **57**, 3041 (1979); (b) R. J. Teillefer, S. E. Thomas, Y. Nadeau, S. Fliszar and H. Henry, *Can. J. Chem.*, **58**, 1138 (1980).
138. P. Deslongchamps, *Stereoelectronic Effects in Organic Chemistry*, Pergamon Press, Oxford (1983).
139. P. Deslongchamps, A. Belanger, D. J. F. Berney, H.-J. Borschberg, R. Brousseau, A. Doutheau, R. Durand, H. Katayama, R. Lapalme, D. M. Leturc, C.-C. Liao, F. N. MacLachlan, J.-P. Maffrand, F. Marazza, R. Martino, C. Moreau, L. Ruest, L. Saint-Laurent, R. Saintonge and P. Soucy, *Can. J. Chem.*, **68**, 153 (1990).
140. B. M. Brudnik, S. S. Zlotskii, U. B. Imashev and D. L. Rakhmankulov, *Dokl. Akad, Nauk SSSR*, **241**, 129 (1978).
141. (a) D. L. Rakhmankulov, E. M. Kuramshin and S. S. Zlotskii, *Usp. Khim.*, **54**, 923 (1985); (b) L. G. Kulak, E. M. Kuramshin, D. L. Rakmankulov and S. S. Zlotskii, *Izv. Vyssh. Uchebn. Zaved. Khim. Khim. Tekhnol.*, **31**, 3 (1988).
142. P. S. Bailey, D. A. Mitchard and A.-I. Khashab, *J. Org. Chem.*, **33**, 2675 (1968).
143. P. S. Bailey and J. E. Keller, *J. Org. Chem.*, **33**, 2680 (1968).
144. P. S. Bailey, T. P. Carter, Jr, and L. M. Southwick, *J. Org. Chem.*, **37**, 2997 (1972).
145. P. S. Bailey, A. M. Reader, P. Kolssaker, H. M. White and J. C. Barborak, *J. Org. Chem.*, **30**, 3042 (1965).
146. L. Spialter, L. Pazdernik, S. Bernstein, W. A. Swansiger, G. R. Buell and M. E. Freeburger, *Adv. Chem. Ser.*, **112**, 65 (1972).
147. R. J. Ouellette and D. L. Marks, *J. Organomet. Chem.*, **11**, 407 (1968).

148. E. J. Corey, M. M. Mehrotra and A. U. Khan, *J. Am. Chem. Soc.*, **108**, 2472 (1986).
149. G. H. Posner, M. Weitzberg, W. M. Nelson, B. L. Murr and H. H. Seliger, *J. Am. Chem. Soc.*, **109**, 278 (1987).
150. G. H. Posner, K. S. Webb, W. M. Nelson, T. Kishimoto and H. H. Seliger, *J. Org. Chem.*, **54**, 3252 (1989); G. H. Posner, C. H. Oh and W. K. Milhous, *Tetrahedron Lett.*, **32**, 4235 (1991).
151. R. W. Murray, W. C. Lumma, Jr, and W.-P. Lin, *J. Am. Chem. Soc.*, **92**, 3205 (1970).
152. F. E. Stary, D. E. Emge and R. W. Murray, *J. Am. Chem. Soc.*, **98**, 1880 (1976).
153. F. Kovač and B. Plesničar, *J. Chem. Soc. Chem. Commun.*, 122 (1978).
154. F. Kovač and B. Plesničar, *J. Am. Chem. Soc.*, **101**, 2677 (1979).
155. N. Ya. Shafikov, R. A. Sadikov, V. V. Shereshovets, A. A. Panasenko and V. D. Komissarov, *Izv. Akad. Nauk SSSR, Ser. Khim.*, 1923 (1981).
156. W. A. Pryor, N. Ohto and D. F. Church, *J. Am. Chem. Soc.*, **104**, 5813 (1982).
157. W. A. Pryor, N. Ohto and D. F. Church, *J. Am. Chem. Soc.*, **105**, 3614 (1983).
158. M. Zarth and A. de Meijere, *Chem. Ber.*, **118**, 2429 (1985).
159. B. Plesničar, F. Kovač, M. Hodoščček and J. Koller, *J. Chem. Soc., Chem. Commun.*, 515 (1985).
160. (a) B. Plesničar, F. Kovač and M. Schara, *J. Am. Chem. Soc.*, **110**, 214 (1988); (b) V. V. Shereshovets, V. D. Komissarov and F. A. Galieva, *Bull. Acad. Sci. USSR, Div. Chem. Sci.*, **37**, 230 (1988); (c) V. V. Shereshovets, F. A. Galieva, R. A. Sadykov and V. D. Komissarov, *Bull. Acad. Sci, USSR Div. Chem. Sci.*, **38**, 2025 (1989).
161. B. Plesničar, F. Kovač and J. Cerkovnik, International Symposium on Activation of Dioxygen and Homogeneous Catalytic Oxidations, Tsukuba, Japan, July 1987, Abstract M-27.
162. (a) B. I. Tarunin, V. N. Tarunina and Yu. A. Kurskii, *Zh. Obsch. Khim.*, **58**, 1060 (1988); (b) M. Koenig, J. Barrau and N. B. Hamida, *J. Organomet. Chem.*, **356**, 133 (1988).
163. P. S. Nangia and S. W. Benson, *J. Am. Chem. Soc.*, **102**, 3105 (1980).
164. D. H. Giamalva, D. F. Church and W. A. Pryor, *J. Am. Chem. Soc.*, **108**, 7678 (1986).
165. D. G. Williamson and R. J. Cvetanovic, *J. Am. Chem. Soc.*, **92**, 2949 (1970).
166. K. U. Ingold, in *Free Radicals*, Vol. 2 (J. K. Kochi, Ed.), Wiley, New York (1973), Chap. 2.
167. W. A. Pryor, G. J. Gleicher and D. F. Church, *J. Org. Chem.*, **49**, 2574 (1984).
168. W. A. Pryor, T. H. Lin, J. P. Stanley and R. W. Henderson, *J. Am. Chem. Soc.*, **95**, 6993 (1973).
169. D. G. Hendry, T. Mill, L. Piszkiewicz, J. A. Howard and H. K. Eigenmann, *J. Phys. Chem. Ref. Data*, **3**, 937 (1974).
170. R. Huisgen, *Angew. Chem., Int. Ed. Engl.*, **2**, 633 (1963).
171. R. J. Teillefer and S. E. Thomas, unpublished results, cited in Ref. 138.
172. J.-M. Lehn, G. Wipff and H.-B. Bürgi, *Helv. Chim. Acta*, **57**, 493 (1974).
173. D. H. Giamalva, D. F. Church and W. A. Pryor, *J. Org. Chem.*, **53**, 3429 (1988).
174. G. A. Olah, N. Yoneda and D. G. Parker, *J. Am. Chem. Soc.*, **98**, 5261 (1976); G. A. Olah, O. Farooq and G. K. S. Prakash, in *Activation and Functionalization of Alkanes* (C. L. Hill, Ed.), Wiley-Interscience, New York (1989), Chap. 2, p. 27.
175. C. Meredith, G. E. Quelch and H. F. Schaefer, III, *J. Am. Chem. Soc.*, **113**, 1186 (1991).
176. S. P. Best, A. J. Bloodworth and M. D. Spencer, *J. Chem. Soc., Chem. Commun.*, 416 (1990).
177. J. Koller, M. Hodoščček and B. Plesničar, *J. Am. Chem. Soc.*, **112**, 2124 (1990).
178. R. A. Cox and G. S. Tyndall, *J. Chem. Soc., Faraday Trans. 2*, **76**, 153 (1980).

179. (a) M. J. Kurylo, P. Dagaut, T. J. Wallington and D. M. Neuman, *Chem. Phys. Lett.*, **139**, 513 (1987); (b) P. Dagaut, T. J. Wallington and M. J. Kurylo, *J. Phys. Chem.*, **92**, 3836 (1988).
180. (a) K. McAdam, B. Veyret and R. Lesclaux, *Chem. Phys. Lett.*, **133**, 39 (1987); (b) P. D. Lightfoot, B. Veyret and R. Lesclaux, *J. Phys. Chem.*, **94**, 708 (1990).
181. M. E. Jenkin, R. A. Cox, G. Hayman and L. J. Whyte, *J. Chem. Soc., Faraday Trans. 2*, **84**, 913 (1988).
182. A. F. Dmitruk, V. V. Lobanov and L. I. Kholoimova, *Teor. Eksp. Khim.*, **22**, 363 (1986).

11.1 Transition Metal Peroxides and Related Compounds

NOBUMASA KITAJIMA, MUNETAKA AKITA and YOSHIHIKO MORO-OKA
Research Laboratory of Resources Utilization, Tokyo Institute of Technology, 4259 Nagatsuta, Midori-ku, Yokohama 227, Japan

1 INTRODUCTION

Transition metal species containing a dioxygen ligand have been assumed to be key intermediates in various oxidative transformation of organic substrates and biological oxygenation processes. For a better understanding of the functions of the active species, it is essential to elucidate the modes of interaction of a dioxygen species with a transition metal center(s) in isolated complexes and to characterize its chemical reactivity toward various substrates.[1]

The chemistry of transition metal dioxygen complexes has been studied from three different points of view. (1) Development of selective catalytic oxidative transformations of organic compounds:[2] this field stems from the autoxidation of organic compounds, and related alkylperoxo metal species have attracted much attention as an active species in the catalyzed epoxidation of alkenes such as the Halcon process[3] and the Sharpless oxidation.[4] (2) Oxidative addition reaction to low-valent transition metal complexes:[5] this reaction has been studied as one of the basic reactions in the field of organotransition metal

Organic Peroxides. Edited by W. Ando

chemistry together with reductive elimination, insertion reactions, etc. (3) Oxygen carrier and oxygenation reactions in biological systems: Co and Fe complexes including porphyrin iron complexes are the best characterized family of dioxygen complexes. Several modes of interaction with polynuclear systems have been proposed and corresponding model complexes have been prepared recently.

This chapter briefly summarizes preparative methods, structural features and basic reaction aspects of transition metal peroxides, which are divided into two categories, transition metal peroxides and those relevant to biological systems. Some detailed reaction mechanisms of particular systems are discussed.

2 TRANSITION METAL PEROXIDES

2.1 Preparative Methods

Transition metal dioxygen complexes are usually prepared by either oxidative addition of dioxygen (equation 1) or ligand-exchange reactions with hydrogen peroxide or its derivatives (equation 2). The former method is applicable to coordinatively unsaturated or labile low-valent transition metal complexes, and accompanies an increase in the formal oxidation number of the central metal by one or two to form a superoxide (O_2^-) or peroxide ligand (O_2^{2-}), respectively. The O_2 adduct of porphyrin iron complexes[6] and Vaska's complex[7] have been prepared by this method, and they are recognized as Fe(III) superoxide complexes and an Ir(III) peroxide complex, respectively. The latter method may work well for non-hydrolyzable and less oxidizable higher valent metal halides or hydroxides. Related anionic dioxygen species such as superoxide (KO_2) also serve as the dioxygen source, as shown in equation 3.[8] Nucleophilic attack by O_2^- at the coordinatively unsaturated metal center followed by electron transfer may give the Pd–O_2 complex with evolution of gaseous O_2.

$$M^n + O_2 \longrightarrow M^{n+1}(O_2^-) \quad \text{or} \quad M^{n+2}(O_2^{2-}) \tag{1}$$

$$M^nX_2 + H_2O_2 \longrightarrow M^n(O_2^{2-}) \tag{2}$$

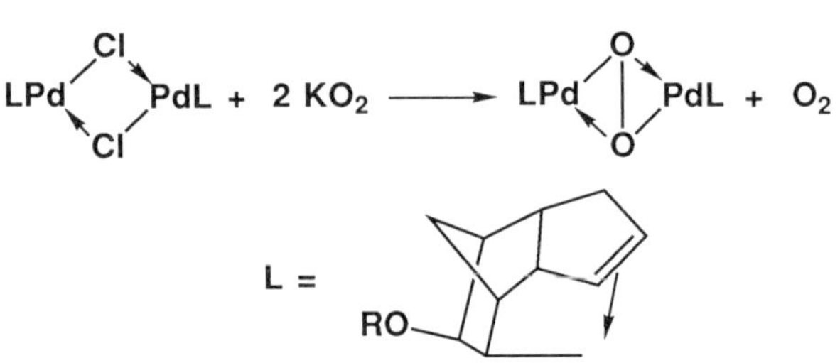

$$LPd(\mu\text{-}Cl)_2PdL + 2\,KO_2 \longrightarrow LPd(\mu\text{-}O_2)PdL + O_2 \tag{3}$$

L =

RO

2.2 Structure

To date six coordination modes of the dioxygen ligand (I–VI) have been structurally characterized[1] (Scheme 1). A wide variety of central metals (Ti–Cu) and oxidation states [Pt(II) (d^8)–W(IV) (d^0)] may be adopted, and various kinds of ancillary ligands involving Group VB donors (amine, imine, phosphine, arsine), Group VIB donors (oxo, alkoxy), their combinations (porphyrin, Schiff base) and even organometallic ligands (cyclopentadienyl, alkyl, alkene) appear to be compatible with the O_2 ligand.

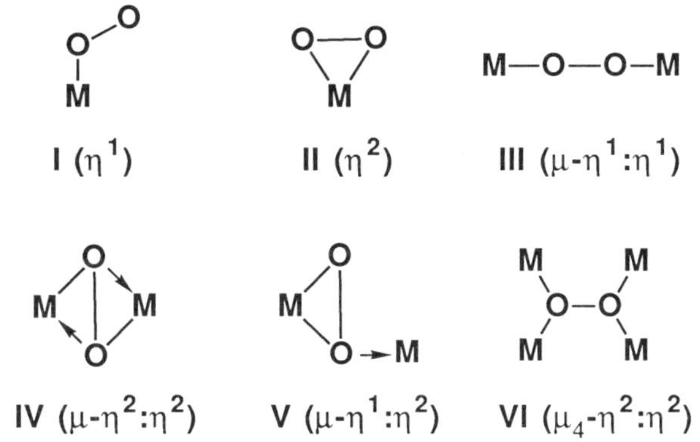

M—O—O—M

I (η^1) II (η^2) III (μ-η^1:η^1)

IV (μ-η^2:η^2) V (μ-η^1:η^2) VI (μ_4-η^2:η^2)

Scheme 1

The η^1-coordination mode (I) is found in a class of paramagnetic bioinorganic model compounds, e.g. cobalt Schiff base complexes[9] and iron porphyrins.[6] In some cases, the nucleophilic O_2 part interacts with another metal center to give a dinuclear μ-η^1:η^1-complex (III).[10] A cyclic version with a metal—metal bond was prepared recently by oxygenation of an iridium A-frame complex[11] (equation 4).

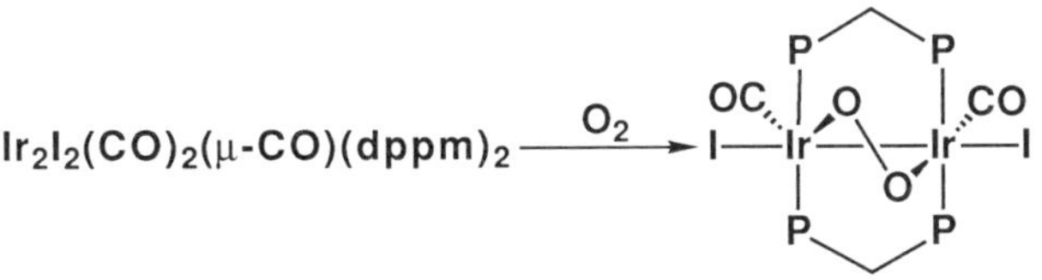

$$Ir_2I_2(CO)_2(\mu\text{-}CO)(dppm)_2 \xrightarrow{O_2} \quad (4)$$

On the other hand, most diamagnetic organometallic peroxides[12] and some metal porphyrin complexes[13] contain a cyclic η^2-structure (II). In this case also the O_2 ligand may coordinate to another metal center to give a dimeric complex with a μ-η^1:η^2-structure (V). Most Rh complexes tend to dimerize through this mode. For example, the O_2 adduct of the Wilkinson complex dimerizes via dissociation of one of the phosphine ligands (Scheme 2),[14] and $[Cp^*RhCl_2]_2(O_2)$ is found to possess a similar structure.[15]

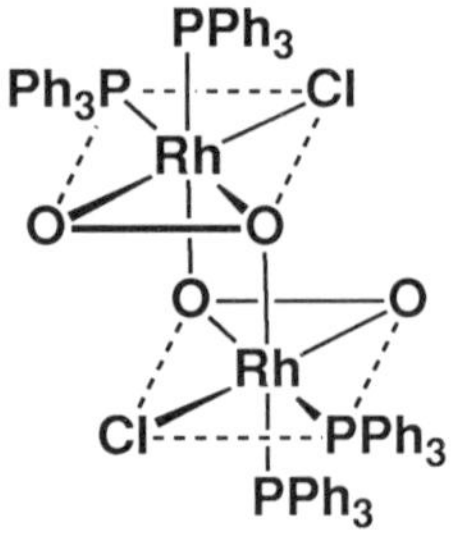

Scheme 2

Multi-ligating modes of the O_2 ligand have also been reported. The first example of a μ-η^2:η^2-complex (IV) was prepared via the ligand-exchange reaction of dinuclear μ-halide complexes with KO_2[8] (equation 3). A dicopper analog, a model complex for the hemocyanine center, was prepared by either oxygenation of a Cu(I) complex or ligand-exchange reaction of a Cu(II) hydroxide with hydrogen peroxide, and was structurally characterized[16] (equation 5). A more complicated tetrairon complex with a μ_4-η^2:η^2-coordination mode was prepared as a model complex relevant to the oxygen evolving center in photosystem II.[17]

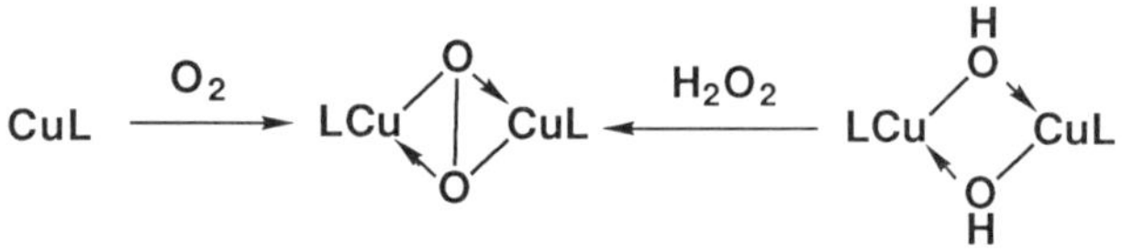

(5)

L = HB(3,5-dialkylpyrazolyl)$_3$

It may be difficult to distinguish one structure from the others. Although X-ray crystallography is a definitive method for determining the coordination mode of the O_2 ligand, this method cannot always be applicable to complexes of poor crystallinity or thermal instability. In such cases vibrational spectra (IR and Raman spectra) may be of much help.[18] The $\nu(O_2)$ of O_2 adducts has been observed in the range 1300–700 cm^{-1} depending on the extent of donation of electrons from the metal center to the O_2 ligand, which has been typically classified into two groups: superoxide $\nu(O_2) = 1200$–1070 cm^{-1} and peroxide $\nu(O_2) = 930$–750 cm^{-1}. Because the $\nu(O_2)$ region may overlap with other absorptions, an assignment should be confirmed by isotopic labeling. In addition, introduction of a mixture of isotopes may permit discrimination of the symmetrical structures II, III, IV and VI from the unsymmetrical structures I and V. For example, when a mixture of $^{16}O_2$, $^{16}O^{18}O$ and $^{18}O_2$ is introduced, complexes with the symmetrical structure II may exhibit three O—O stretching vibrations [ν(M—^{16}O—^{16}O), ν(M—^{16}O—^{18}O) and ν(M—^{18}O—^{18}O)] and, on the other hand, those with the unsymmetrical structure I may exhibit four O—O stretching vibrations [$\nu(M—^{16}O—^{16}O)$, $\nu(M—^{16}O—^{18}O)$, $\nu(M—^{18}O—^{16}O)$ and $\nu(M—^{18}O—^{18}O)$]. In principle, a structure could be distinguished by a combination of information on ν (O_2) values, splitting pattern on introduction of a mixture of O_2 isotopes, magnetic properties and analysis of nuclearity, e.g. by molecular weight analysis.

2.3 Typical Reactivities

Although few reaction patterns of the η^1-complexes (type I) have been documented, a representative one is dimerization leading to type **III** complexes as mentioned above (equation 6). Similarly, a proton may attack the distal oxygen atom to give a hydroperoxy intermediate. It is generally accepted that the active species of the P-450 system, a highly reactive cationic metal oxo species, may be formed by double protonation followed by elimination of water[19] (equation 6).

$$\text{M-O-O} + \text{M or E} \longrightarrow \begin{cases} \text{M-O-O-M} \\ \text{M-O-O-E} \longrightarrow \text{M}^{+}\text{=O} + {}^{-}\text{OE} \end{cases} \quad (6)$$

The reactivities of the O_2 adducts of Group VIII metals with the cyclic structure **II** were studied intensively by Tatsuno and Otsuka.[20] The O_2 part is also so nucleophilic as to react with proton, alkylating and acylating reagents and polar multiple bonds (equation 7). The first three reagents afford η^1-hydro-, alkyl- and acylperoxy complexes, respectively.[20] Organic substrates with a polar multiple bond such as ketone, CO, TCNE (tetracyanoethylene) and SO_2 insert into the M—O bond to give a five-membered peroxymetallacycle, which may occasionally isomerize to a four-numbered dioxametallacycle when X = O.[21] A similar species is assumed as an intermediate for the Rh-catalyzed Wacker-type transformation of an alk-1-ene to a methyl ketone[22] (equation 8). In some cases the O_2 part behaves as a Brønsted base to abstract proton from an acidic substrate such as acacH as shown in equation 9.[23] The Group VIII dioxygen complexes can oxidize nucleophilic substrates such as PR_3, CNR and SR_2 to give the corresponding oxides OPR_3, OCNR and OSR_2, respectively.[24]

$$\text{M}(\text{O}_2) \begin{cases} + \text{E} \longrightarrow \text{M-O-O-E} \\ + \text{X=Y} \longrightarrow \text{M(O-O-Y-X)} \longrightarrow \text{M(O)(O)Y=X} \quad \text{(when X = O)} \end{cases} \quad (7)$$

$$R\text{-}CH{=}CH_2 \xrightarrow[Rh^{III}\,/\,Cu^{II}\,cat.]{O_2} R\text{-}CO\text{-}CH_3 \tag{8}$$

$$\left[Rh(O_2) \xrightarrow{R\text{-}CH=CH_2} \text{Rh–O–O–C(R)(H)–CH}_2 \right]$$

$$RhCl(O_2)(PPh_3)_3 \xrightarrow{acac\text{-}H} RhCl(OOH)(acac)(PPh_3)_2 \tag{9}$$

The Group V and VI metal complexes of type II studied extensively by Chaumette *et al.*[25] exhibit higher reactivity toward alkenes to give oxygenated products such as epoxide, methyl ketone and carbonyl compounds derived from C=C bond cleavage. A peroxymetallacyclic species, which has never been isolated, is postulated as an intermediate for the formation of these products.

The reactivity of the bridging O_2 ligand in polynuclear complexes has been explored less extensively. The O—O bond fission in dinuclear μ-η^1:η^1-O_2 iron porphyrin complexes (type III) to give μ-oxo dimers may be quoted as one of the established reactions (equation 10).

$$M\text{-}O\text{-}O\text{-}M \longrightarrow 2\ M\text{-}O\cdot \longrightarrow M\text{-}O\text{-}M \tag{10}$$

A series of Group VIII dinuclear μ-η^2:η^2-O_2 complexes (type IV) were prepared and their reactivities were examined in our laboratory. The Pd complexes were the first examples of the dioxygen complexes in which a dioxygen ligand and a potential reactant, i.e. an alkenic substrate, coexist in the same coordination sphere[8] (equation 3). The O_2 part was so basic that the Pd complexes reacted with acidic organic compounds (HX) to give H_2O_2 and $(\mu\text{-}X)_2$-bridged dinuclear complexes[26] (equation 11). Conversion to the μ-η^1:η^1-form was achieved by addition of a donor such as acetonitrile. A series of Rh complexes containing both O_2 and an alkene ligand were prepared in a

$$LPd(\mu\text{-}O_2)PdL \xrightarrow{HX} LPd(\mu\text{-}X)_2PdL \quad (X = {}^-OR,\ {}^-SR,\ {}^-NR_2.)$$

$$LPd(\mu\text{-}O_2)PdL \xrightarrow{L'} L'LPd(\mu\text{-}O\text{-}O)PdLL' \quad (L' = MeCN) \tag{11}$$

similar manner[27] (equation 12). They were also basic enough to abstract proton from CpH and ketone. In the latter case the ketone dimerized via base-catalyzed aldol condensation to coordinate to the Rh center as an acac-type ligand. Thermolysis of the cyclooctadiene (COD) complex at 150°C in the presence of PPh_3 produced $OPPh_3$ in 97% yield, but attempts to oxidize the coordinated alkene ligands were unsuccessful.

LRh(μ-Cl)₂RhL —KO₂→ LRh(μ-O₂)RhL; then acacH → LRh(acac); CpH → LRhCp; Δ → $OPPh_3$ (12)

L = 1,5-cyclooctadiene (COD)
norbornadiene
dicyclopentadiene

Later, we succeeded in isolating Rh complexes of another type with the composition of $[(diene)Rh(O_2)]_2$ [diene = cycloocta-1,5-diene (COD), norbornadiene] by the reaction of cationic diene complexes with KO_2, but their structure is still ambiguous[28] (equation 13).

$[(COD)_2Rh^+]$ —KO_2→ $[(COD)_2Rh(O_2)]_2$; then PPh_3, Δ → $OPPh_3$; $^{18}O_2$, Δ → [^{16}O]cyclooct-4-enone + [^{16}O]cyclooctanone; cyclohexene, Δ → cyclooctanone (13)

In a manner similar to the above-mentioned μ-η^2:η^2-complexes, added PPh_3 was oxidized on thermolysis to give $OPPh_3$ in 97% yield. In addition, thermolysis in benzene without any additives resulted in the formation of oxygenated products of the COD ligand, i.e. cyclooct-4-enone (20%), cyclooctanone (30%) and small amounts of unidentified C_8 carbonyl compounds. An intramolecular process was verified by the pyrolysis of the $^{16}O_2$ complex under an $^{18}O_2$ atmosphere giving [^{16}O]cyclooct-4-enone of 90% isotopic purity. While thermolysis in the presence of cyclohexene led to the exclusive formation of cyclooctanone as the C_8 product, no oxygenated product derived from cyclohexene was detected but benzene and cyclohexane were formed via a hydrogen transfer reaction.

In conclusion, dioxygen complexes of any coordination mode are not generally as reactive toward organic substrates as expected, and in order to activate the O_2 complex as an oxidant it appears to be essential to forcibly cleave the O—O bond in a heterolytic manner to generate a highly reactive cationic metal oxo species as reported for cytochrome P-450 model systems (equation 6).

2.4 Transition Metal Alkyl- and Hydroperoxides

Transition metal alkyl- and hydroperoxides have also been postulated as candidates for intermediates in the metal-catalyzed oxidation of organic compounds.

Although MOOR species have been implicated as intermediates in the autoxidation of hydrocarbons, such species have seldom been isolated from a reaction mixture. Nishinaga *et al.*[29] succeeded in the isolation and structural characterization of a peroxyquinolato-cobalt(III) complex, which was obtained during oxygenation of phenol derivatives by Co(II) Schiff base complexes (equation 14). Similarly, a Pt(IV) alkylperoxide was isolated by oxidative addition of alkyl halide to a Pt(II) complex in the presence of dioxygen[30] (equation 15). As another route to an MOOR complex, a ligand-exchange reaction with ROOH or ROOM (M = Li, K) has been employed.

LCo + [2,6-Bu$^t_2$-4-R-phenol] $\xrightarrow{O_2}$ LCo—O—O—[4-R-2,6-Bu$^t_2$-cyclohexadienone] (14)

L = Schiff base

(N N)Pd(Me)$_2$ + R-X $\xrightarrow{O_2}$ (N N)Pd(Me)$_2$(O-O-R)(X) (15)

N N = phenanthroline, bipy.

Heterolytic O—O bond cleavage appears to be a major reaction pathway except for the Co(III) complexes. As a typical example, thermal decomposition of a d^0 complex $Cp^*_2Hf(R)(OOBu^t)$ proceeds via initial hapticity change to an η^2-coordination followed by simultaneous migration of the R and OBu^t groups to give $Cp^*_2Hf(OR)(OBu^t)$[31] (equation 16). For the Co(III) complexes homolytic cleavage of M—O and O—O bonds has been observed and the resulting radical species abstract hydrogen atoms from hydrocarbons to give alcohol, ketone and dialkylperoxide as final products.[32]

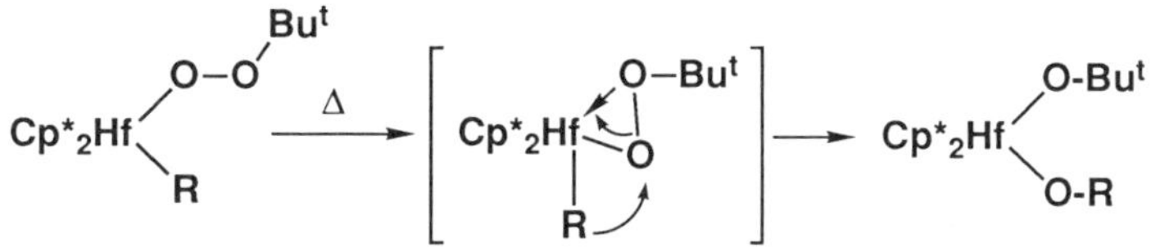

(16)

The MOOR species is also attributed to the active species of the catalyzed epoxidation of alkenes such as in the Halcon process and the Sharpless oxidation. Whereas an alkylperoxy metal species is proposed as an intermediate, some other reaction mechanisms involving an intact alkyl hydroperoxide or a peroxymetallacyclic species have also been postulated, and the reaction mechanism still remains controversial. According to Mimoun and co-workers,[33] the catalytic conversion of an alk-1-ene into epoxide or methyl ketone is interpreted in terms of a *β-tert*-butylperoxyethyl intermediate formed by insertion of alkene into an M—OOBu^t bond (Scheme 3). The decomposition pathway depends on the central metal. When M is an electrophilic Group V and VI d^0 metal, electron transfer to the α-carbon results in the formation of an epoxide. On the other hand, when M is a less electrophilic late transition metal, electron transfer to the β-carbon atom with simultaneous H transfer affords a methyl ketone.

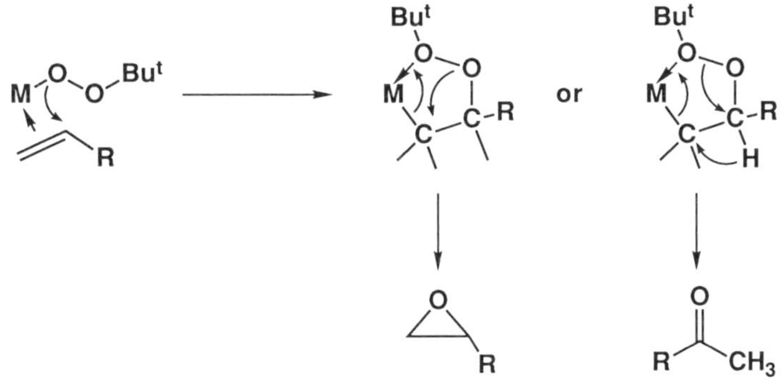

Scheme 3

The interaction of O_2^- and *t*-BuOO$^-$ with Group VIII metal–alkene complexes was also examined. As a result, it was found that reactions of coordinatively unsaturated 16e complexes occurred at the metal center to give an alkylperoxo complex (Scheme 4). Treatment of $[(\eta^3\text{-}C_3H_5)Pd^+(PPh_3)_2]PF_6$ and $[(COD)_2Rh]BF_4$ with KOOBu^t afforded $(\eta^3\text{-}C_3H_5)Pd(PPh_3)(OOBu^t)$ and (COD)Rh(OOBu^t)·KBF_4, respectively[34] (equations 17 and 18). The latter complex showed a reactivity similar to the Rh–O_2 complex. Thermolysis gave a mixture of cyclooctenone and cyclooctanone, and co-thermolysis with PPh_3

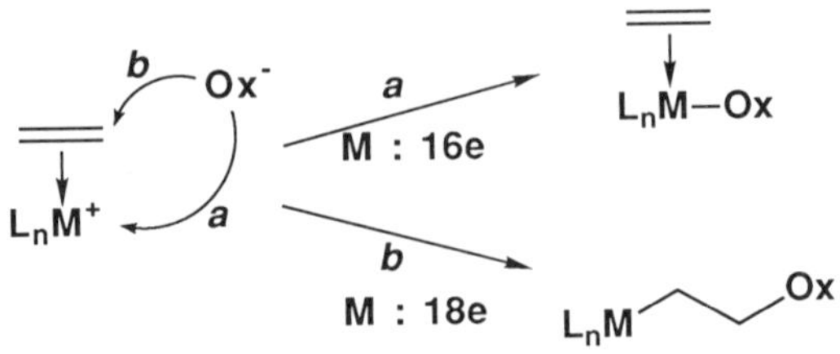

Scheme 4

gave $OPPh_3$. On the other hand, reaction with coordinatively saturated 18e alkene complexes resulted in nucleophilic addition to the alkenic ligand from the *exo* side. As shown in equation 19, $[(COD)Pd^+Cp]BF_4$ reacted with *t*-BuOOLi to give an adduct in which the *t*-BuOO group is bonded to the COD group[34] (equation 19). Similarly, reactions of a series of coordinatively saturated $[(\eta^6\text{-benzene})Ru^+(\text{cyclic-}\eta^5\text{-dienyl})]$ complexes with KO_2 afforded $(\eta^6\text{-benzene})Ru(\text{cyclic-}\eta^4\text{-dienone})$ complexes, probably via initial nucleophilic attack at the terminal position of the dienyl moiety[35] (equation 20).

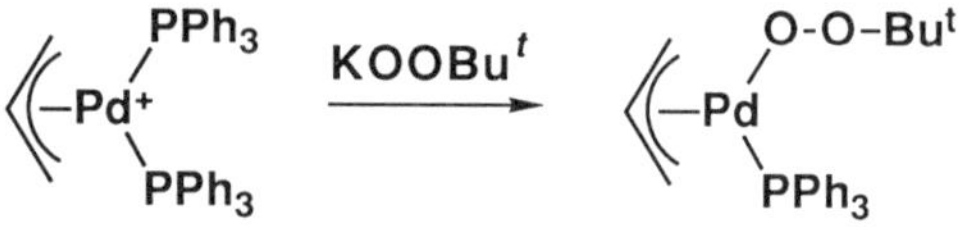

(17)

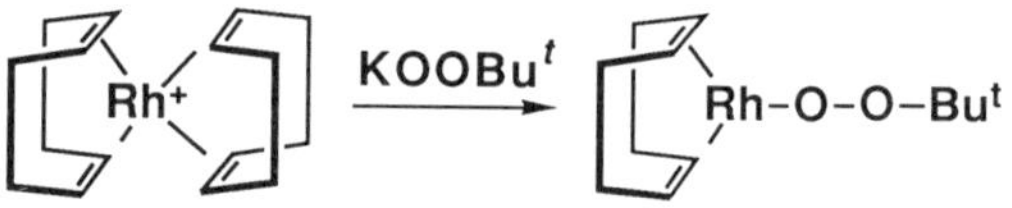

(18)

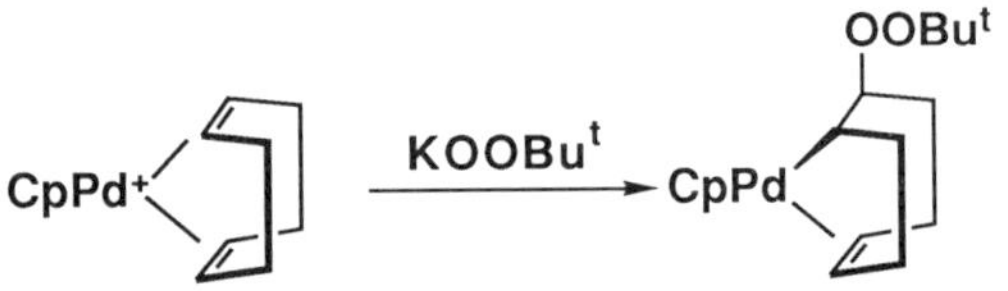

(19)

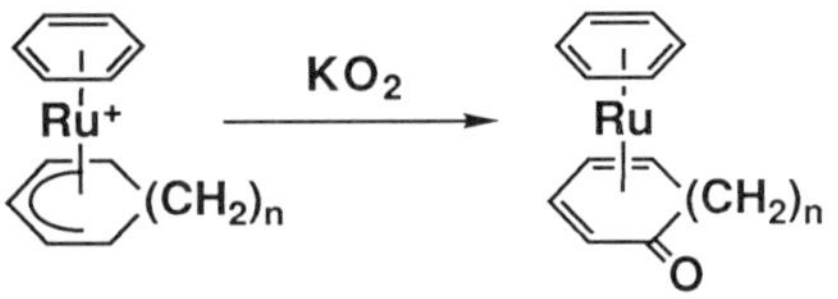

(20)

The chemistry of transition metal hydroperoxy complexes is far less well explored than that of dioxygen and alkylperoxy complexes, presumably owing to the limited number of examples, whereas MOOH species are likely precursors for high-valent metal oxo species (M=O), which is assumed to be an active species in biological oxygenation systems such as P-450 and bleomycin (equation 6; E = H^+). Only three isolated examples have been reported so far, an Rh(III) complex and Pd(II) and Pt(II) phosphine complexes with an electron-withdrawing alkyl group. The former complex was prepared according to equation 9.[23] Thermal decomposition of the Rh complex resulted in oxidation of the coordinated PPh_3 ligand to give $OPPh_3$, and an intramolecular mechanism was confirmed by isotopic labeling with ^{18}O. However, oxidation of externally added alkenes was not observed. The Pd(II) and Pt(II) complexes were prepared by a ligand-exchange reaction of the hydroxide complex with H_2O_2[36] (equation 21). The Pt(II) hydroperoxy complex appears to be more reactive than the corresponding alkylperoxy complex. The oxidation of PPh_3 to $OPPh_3$ by the former complex proceeds even at 25°C, whereas the latter complex requires heating at 100°C. In addition, whereas the Pt(II)–aqueous H_2O_2 system is reported to catalyze the epoxidation of alkenes, the intermediate does not appear to be a PtOOH species.[37] Epoxide may be formed by decomposition of β-hydroperoxyethyl species, which should be formed, in a manner similar to path b in Scheme 4, by addition of $^-$OOH to an alkene ligand bound to the Pt center ionized in a protic medium.

$$(R_3P)_2Pt(CF_3)(OH) \xrightarrow{H_2O_2} (R_3P)_2Pt(CF_3)(OOH) \qquad (21)$$

3 TRANSITION METAL PEROXO COMPLEXES RELEVANT TO BIOLOGICAL SYSTEMS

3.1 Dioxygen Binding and Activation in Proteins

Many metalloproteins have been recognized to participate in biological redox reactions of dioxygen. Biological systems utilize dioxygen as an electron acceptor. The fundamental mechanism is ascribed to the initial one- or two-electron reduction of the metal center, followed by dioxygen binding and highly efficient electron transfer from the metal to the dioxygen, resulting in the formation of a superoxo or peroxo intermediate. Heme iron-containing oxygen carriers, hemoglobin and myoglobin, form a stable superoxo adduct, while hemerythin and hemocyanin, of which the active center consists of a diiron or dicopper site, bind dioxygen as a peroxide ion. In addition to oxygen carriers, whose function is simple reversible dioxygen binding, a number of metalloproteins containing iron or copper are known to activate dioxygen, catalyzing the O—O bond cleavage and subsequent oxygen atom incorporation into the substrates. These include monooxygenases and some types of dioxygenases and oxidases. On the other hand, catalase, superoxide dismutase and dioxygen evolution center (OEC) in photosynthetic system II catalyze dioxygen formation from hydrogen peroxide, superoxide ion or water; hence the involvement of a peroxo species as a key reaction intermediate is reasonably suggested. Except for oxygen carriers, the dioxygen adducts of the metalloproteins are generally not stable enough to be characterized by spectroscopic means, so information on the structure and activation mechanism of the coordinated dioxygen is seriously lacking.

By using a suitable ligand which mimics the coordination environment of the metal center in the proteins, it is possible to synthesize dioxygen complexes modeling the dioxygen adduct of the proteins. This synthetic bioinorganic approach provides the structural and mechanistic basis for understanding the structure and function of the proteins, which are not available from the protein chemistry. In this section, recent progress in the synthesis and characterization of dioxygen complexes of iron, copper and manganese relevant to biological systems is briefly reviewed.

3.2 Peroxo Iron Complexes with a Porphyrin Ligand

The synthesis and structural characterization of a superoxoiron(III) complex with a hindered porphyrin as a model for hemoglobin and myoglobin was achieved by Collman *et al.*[38] more than 15 years ago. This success is recognized as one of the most important contributions in bioinorganic chemistry, since the X-ray structure and physicochemical comparison of the complex and proteins proved the precise coordination structure of the dioxygen and how the proteins control the function as an oxygen carrier. The chemistry of the superoxoiron(III) complexes has been reviewed in detail previously,[39] and no further explanation is added here.

Whereas the chemistry of heme iron oxygen carriers was developed in the 1970s, modelling studies relevant to the catalysis of the related protein, cytochrome P-450, were extensively explored in the 1980s. Cytochrome P-450 is a family of monooxygenases which catalyzes oxygen incorporation reactions of a wide variety of organic substrates including alkenes and alkanes.[19,40] The active site consists of a heme ion which is similar to hemoglobin and myoglobin, but the distinct feature is that cytochrome P-450 has cysteine as an axial ligand, whereas histidine is the ligand for hemoglobin and myoglobin.

In cytochrome P-450, dioxygen initially binds to the reduced Fe(II) form to give a superoxoiron(III) intermediate, as in hemoglobin and myoglobin. Then a second reduction, originating from NAD(P)H, occurs to give a peroxoiron(III) species (equation 22). The superoxo intermediate is reasonably stable at low temperatures and can be characterized by spectroscopic methods such as resonance Raman spectroscopy.[41] However, the peroxo species is not detectable

$$\mathrm{Fe^{II}} \xrightarrow{O_2} \mathrm{Fe^{III}{-}O{-}O^-} \xrightarrow{e^-} \mathrm{Fe^{III}(O{-}O)^{2-}} \qquad (22)$$

owing to its instability. Thus, the modeling approach with a synthetic porphyrin complex was expected to offer an opportunity to clarify the chemistry. Valentine *et al.*[42] first succeeded in preparing such a peroxo complex by treating TPPFe(Cl) with KO_2. Based on the physicochemical characterization, the peroxide ion was suggested to coordinate to the iron in a side-on η^2 structure. A peroxoiron(III) complex was also prepared by dioxygen addition to a super-reduced porphyrin complex of Fe(I).[43] Although neither of the peroxo complexes was crystallized, the X-ray structure of the related peroxo-manganese(III) complex has since been reported.[44] As predicted for the iron complex, the side-on coordination of the peroxide ion was definitely established as shown in Figure 1.

In cytochrome P-450 catalysis, the peroxo intermediate is assumed to be protonated and the subsequent O—O bond heterolysis of the hydroperoxo intermedia formed gives an oxoiron(IV) porphyrin π-cation radical, generally accepted as the species responsible for the oxo-transfer reactions (equation 23).[19] Oxygen incorporation reactions catalyzed by cytochrome P-450 occur even under anaerobic reaction conditions, in the presence of an appropriate oxidant such as mCPBA (*m*-chloroperbenzoic acid) or PhIO. This

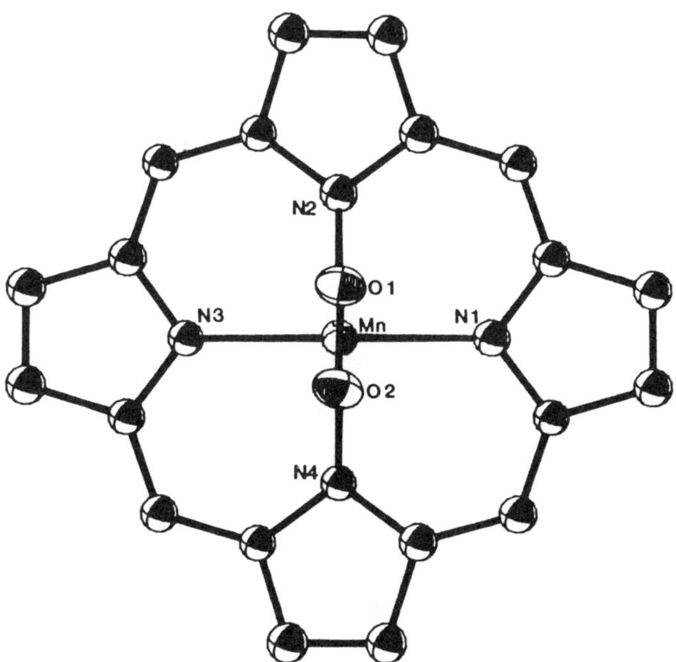

Figure 1. Molecular view of a peroxomanganese(III) complex, $[TPPMn(O_2)]^-$. The phenyl rings of the porphyrin ligand were omitted for clarity

anaerobic oxidation is called a shunt pathway. A number of studies have been accomplished to mimic this type of oxo-transfer reaction, and it is now well known that such a reaction is catalyzed by a synthetic porphyrin complex. The reaction mechanism of the biomimetic system has been extensively investigated, and it is generally accepted that the reaction proceeds via an oxoiron(IV) porphyrin π-cation radical, supporting the proposed mechanism for cytochrome P-450 involving also such an oxoiron(IV) π-cation radical intermediate as an active species. The oxoiron(IV) porphyrin π-cation radical complex can be observed by spectroscopic means in the model system, and its electronic structure has been extensively characterized.[45–48]

$$\mathrm{Fe^{III}(O{-}O^{2-})} \xrightarrow{H^+} \mathrm{Fe^{III}{-}O{-}OH} \xrightarrow[-H_2O]{H^+} \mathrm{O{=}Fe^{IV}}^{+\cdot} \qquad (23)$$

Although there is no example of a structurally well defined hydroperoxoiron(III) porphyrin complex, the related acyl- and alkylperoxo complexes have been prepared and their reaction chemistry provides the basis for an understanding of the O—O bond activation in cytochrome P-450. The acyl- or alkylperoxo complex is also a good model for the intermediate involved in the shunt pathway. With a hindered porphyrin, a hydroxoiron(III) complex is isolable. The reaction of this complex with mCPBA gave an acylperoxo complex which was identified by IR spectrometry.[49] In the presence of a trace amount of proton, the acylperoxo complex undergoes heterolytic cleavage of the O—O bond, resulting in the formation of the reactive oxoiron(IV) porphyrin π-cation radical. This result supports the mechanism of cytochrome P-450 shown in equation 23. An alkylperoxo complex was prepared by reaction of alkyliron(III) complex[50] with dioxygen at low temperature.[51] The reaction chemistry of the species may be essentially identical with that of the acylperoxo complex, but this aspect was not explored in detail.

Dioxygen addition to an iron(II) porphyrin generally causes irreversible decomposition to a μ-oxo diiron(III, III) complex. Although its biological relevance is not straightforward, the mechanism has been studied, and it is now well established that the reaction proceeds via initial formation of a superoxoiron(III) intermediate followed by its coupling with an iron(II) complex, to yield a μ-peroxo intermediate. The μ-peroxo intermediate is not stable enough to be isolable, but its magnetic properties and reactivity were explored, mainly by ^{1}H NMR spectroscopy. The antiferromagnetic dinuclear complex undergoes homolysis of the O—O bond which is promoted by the addition of a base such as imidazole, affording an oxoiron(IV) porphyrin complex.[52] Kinetic experiments indicated that the rate-determining step for the formation of the μ-oxo complex is homolysis.[53] This implies that the oxoiron(IV) complex formed reacts immediately with the iron(II) complex to give the μ-oxo complex. The reactivity of the oxoiron(IV) complex toward oxo-transfer reactions is apparently much lower than that of the oxoiron(IV) porphyrin π-cation radical.

3.3 Peroxo Complexes Relevant to Non-heme Iron Proteins

In addition to heme iron proteins described above, a number of non-heme iron proteins participating in dioxygen metabolism are now known. The best characterized example in this family is hemerythrin, an oxygen carrier for some marine worms.[54] The structure and dioxygen binding mechanism of this protein have been established in detail. It has an asymmetric diiron center with bis(carboxylato)hydroxo bridging ligands in its reduced state. The iron on one side is six-coordinate whereas the iron on the other side is five-coordinate. Dioxygen binds to the five-coordinate iron, and following two-electron transfer from the two iron(II) ions and proton transfer from the hydroxo ligand, results in the construction of an unusual terminal hydroperoxo structure with an oxo bridging ligand (equation 24). It is remarkable that the hydroperoxo intermediate is stable so that hemerythrinn can function as an oxygen carrier. This is contrary to the fact that with cytochrome P-450 the hydroperoxo intermediate is very unstable and is readily converted irreversibly into a reactive oxo iron intermediate. The factors that control the striking difference between these two hydroperoxo species are not clear.

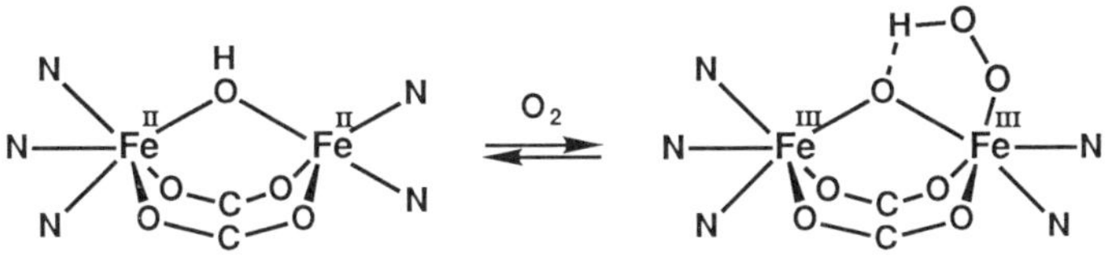

(24)

Because of these unusual structural features, the synthesis of the dinuclear iron peroxo (or hydroperoxo) complex modeling hemerythrin has been a topic of current interest in bioinorganic chemistry.[55] Despite many efforts, however, no success has yet been achieved. Serious synthetic difficulties come from the asymmetric structure. Recently, Lippard[55] reported the synthesis of an asymmetric diiron(II, II) complex in which the iron on one side is five-coordinate as in hemerythrin. Dioxygen reacts rapidly with the complex irreversibly to give a symmetric μ-oxo diiron(III, III) complex. Unfortunately, no evidence for the formation of a peroxo intermediate was obtained in this reaction.

Kitajima *et al.*[56] and Menage *et al.*[57] independently succeeded in preparing the first examples of well characterized μ-peroxo diiron(III, III) complexes with a non-porphyrin ligand. A mononuclear five-coordinate iron(II) benzoato complex was synthesized, whose structure is reasonably similar to that of the five-coordinate iron in hemerythrin, and it was found that the complex binds dioxygen reversibly at $-20°C$.[56] Resonance Raman experiments together with the stoichiometry of dioxygen consumption established that the species formed is a μ-peroxo diiron complex (equation 25). By using a dinucleating ligand, Menage *et al.*[57] prepared a diiron(II, II) complex in which each iron is five-coordinate. The two iron ions are bridged by one phenoxo and one carboxylate group. The complex reacts with dioxygen irreversibly to afford a μ-peroxo complex, which was characterized by resonance Raman spectroscopy (equation 26). Both of the peroxo complexes have a symmetric structure, and hence they cannot serve as accurate synthetic models for hemerythrin. However, these studies clearly indicate that coordinatively unsaturated iron(II) complexes evenwith non-porphyrin ligands can bind dioxygen under appropriate reaction conditions.

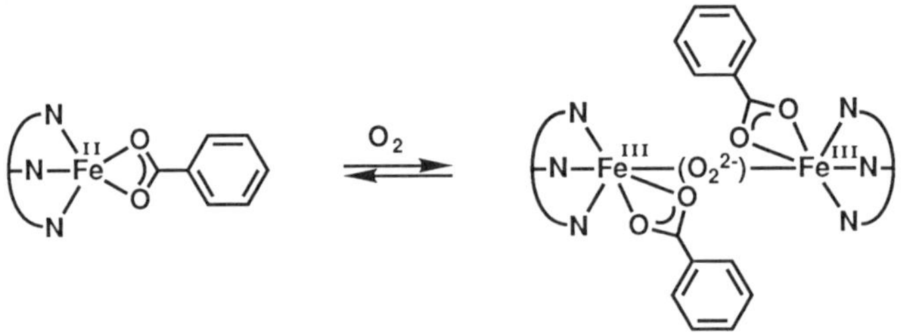

(25)

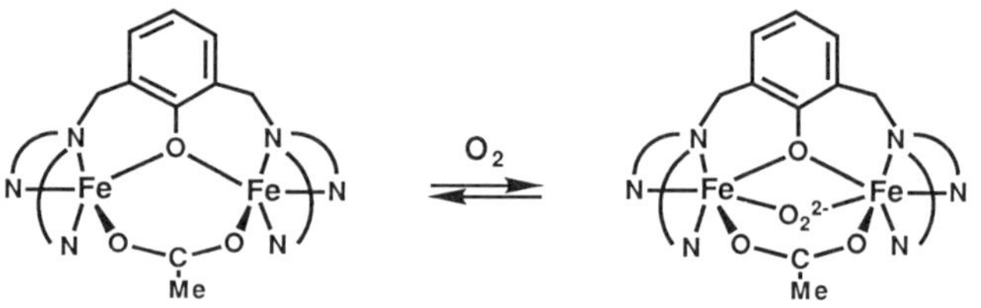

(26)

Recently, several other iron proteins have emerged as non-heme iron proteins containing a dinuclear iron site in addition to hemerythrin.[58] These include a remarkable bio-oxidation catalyst, methane monooxygenase.[59] Further investigations, especially on the reaction aspects of dinuclear iron dioxygen complexes, may shed light on the structure and catalysis of these fascinating proteins.

It is noteworthy that a considerable number of non-heme iron proteins which catalyze oxygenations possess a mononuclear iron site. The structural and mechanistic details of these proteins are still almost completely unknown. In this regard, the chemistry of mononuclear peroxo iron complexes should be explored. The well known mononuclear peroxoiron(III) complex with EDTA may be a good synthetic model for these proteins.[60] The peroxo structure of the complex was recently ascertained by resonance Raman spectroscopy.[61]

3.4 Peroxo Copper Complexes

Hemocyanin is a ubiquitous oxygen carrier for invertebrates.[62] The reduced state of hemocyanin has been structurally characterized by X-ray crystallography.[63] Hemocyanin contains a dicopper site (Cu—Cu separation *ca* 3.6 Å) in which each copper ion is coordinated by three histidyl nitrogen atoms. The structure of the dioxygen adduct has not been elucidated, although dioxygen is known to bind the dicopper site in a symmetric fashion.[64] The dicopper–dioxygen moiety in the dioxygen adduct is thus denoted $Cu^{2+}(O_2{}^{2-})Cu^{2+}$. This chromophore exhibits very characteristic properties: (1) it is diamagnetic; (2) it shows an unusually low O—O stretching frequency, *ca* 750 cm^{-1}; and (3) there are strong absorption bands at 350 and 580 nm. It is generally accepted that an unidentified endogenous bridging ligand other than peroxide ion sits between the two copper ions and mediates the strong magnetic interaction responsible for the diamagnetism. However, no experimental evidence for the existence of such an endogenous ligand has been provided.

The striking characteristics of hemocyanin have attracted much attention from inorganic chemists, and considerable efforts have been made to synthesize μ-peroxo dicopper complexes which closely mimic the characteristic properties of the dioxygen binding state of hemocyanin.[65] Recently, several μ-peroxo dicopper complexes were successfully prepared. The structures were identified by resonance Raman spectroscopy or EXAFS.[66–68] Further, Jacobson *et al.*[69] reported the first X-ray structure of such a complex; the coordination mode of the peroxide ion is *trans*-μ-1,2, as shown in Figure 2. Unfortunately, none of these examples can closely mimic the physicochemical characteristics of hemocyanin, and thus provided negative evidence that hemocyanin does not possess such coordination structures.

By using a hindered tris(pyrazolyl)borate as a ligand, the preparation and structure determination of the first μ-η^2:η^2 peroxo dicopper complex was successfully achieved (Figure 3).[16,70–72] The copper ions are coordinated by three pyrazole nitrogen atoms and the Cu—Cu separation is *ca* 3.6 Å. These structural features are reasonably similar to those known for hemocyanin. Now, this μ-η^2:η^2 peroxo complex shows remarkable similarities in its physicochemical characteristics to the dioxygen adduct of hemocyanin. Therefore, it is highly likely that hemocyanin also binds dioxygen as a peroxide ion in the unusual side-on mode, μ-η^2:η^2, without the existence of any other bridging ligand.

A monooxygenase, tyrosinase, is known to possess a dicopper site, the structure of which is very similar to that in hemocyanin, so that the reaction chemistry of the μ-η^2:η^2 peroxo complex may give mechanistic insight into the catalysis of tyrosinase. The reactivity and mechanism of μ-η^2:η^2 were investigated in order to address this aspect, and it was concluded that the complex does not undergo heterolysis to give a highly reactive oxo-transfer

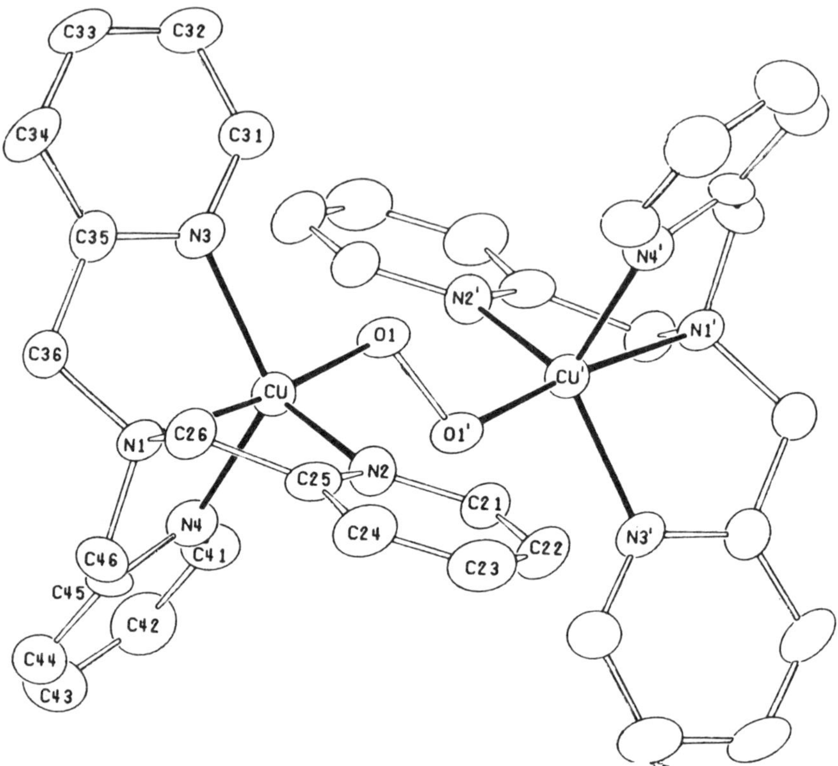

Figure 2. ORTEP drawing of a *trans*-μ1,2 peroxo dicopper complex

intermediate such as oxocopper(IV).[73] Rather, the reactions were explicable in terms of homolysis of the O—O bond. The radical species formed, possibly oxocopper(III) or copper(II)-oxygen anion radical, shows no oxo-transfer reactivity, but works as a weak hydrogen acceptor. Therefore, it seems likely that the oxidation mechanism catalyzed by tyrosinase is distinct from that of cytochrome P-450. The reaction presumably proceeds via radical character intermediates. On the other hand, a number of hydroxylation reactions of the ligand aromatic rings have been found when the corresponding dicopper(I, I) complexes were treated with dioxygen.[74–80] The oxidations were claimed to proceed via μ-peroxo dicopper(II, II) intermediates which are effective for oxo-transfer reactions. However, all these proposed mechanisms are very speculative, and no experimental evidence to support the hypothetical mechanism has been provided.

In addition to tyrosinase, several copper-containing monooxygenases are known. All these proteins contain a mononuclear copper site. As in cytochrome

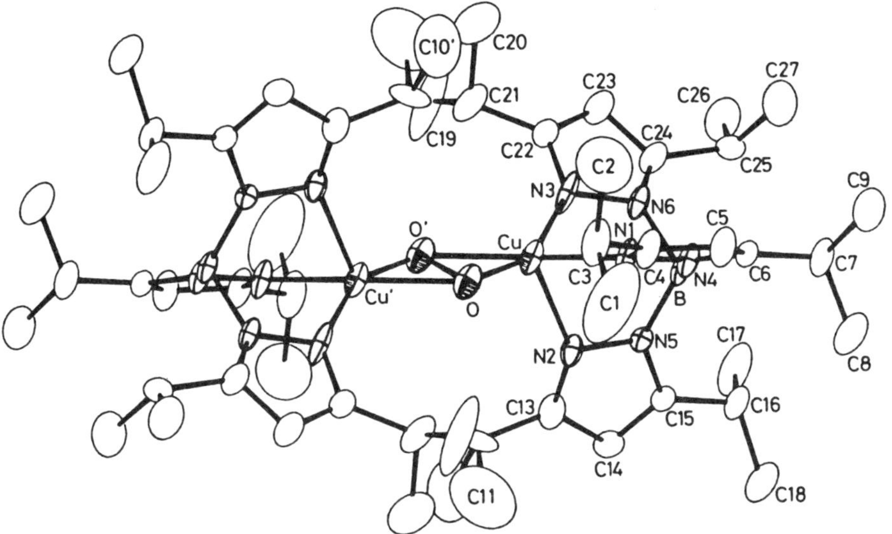

Figure 3. ORTEP drawing of a μ-η^2:η^2 peroxo dicopper complex

P-450, hydroperoxo or acyl- or alkylperoxo intermediates may have an important role in these enzymatic reactions. Karlin and co-workers[81,82] reported the preparation of dicopper hydroperoxo and acylperoxo complexes. The complexes were obtained by protonation or acylation of a μ-peroxo dicopper complex. The structure of the acylperoxo complex was determined by X-ray crystallography. Both of these complexes showed very low oxo-transfer reactivity; they are effective for oxygenation of PPh_3 or sulfide, but not for cyclohexene. On the other hand, a mononuclear acyl- or alkylperoxo copper(II) complex was successfully prepared for the first time.[83,84] The complexes were not active for epoxidation of cyclohexene either. It is therefore inferred that the peroxo copper(II) complexes are essentially ineffective for oxo-transfer reactions. This is consistent with the fact that copper monooxygenases merely catalyze effortless oxidations as compared with cytochrome P-450.

3.5 Peroxo Manganese Complexes

Although there is no example of a manganese protein in which the dioxygen binding state has been identified, dioxygen–manganese complexes have received increasing interest from a bioinorganic point of view.[85] The research inspiration stems mainly from its potential relevance to physiologically the most important oxidation, the oxidation of water to dioxygen in photosynthetic system II.[86] A manganese cluster in OEC is known to be responsible for this reaction, although

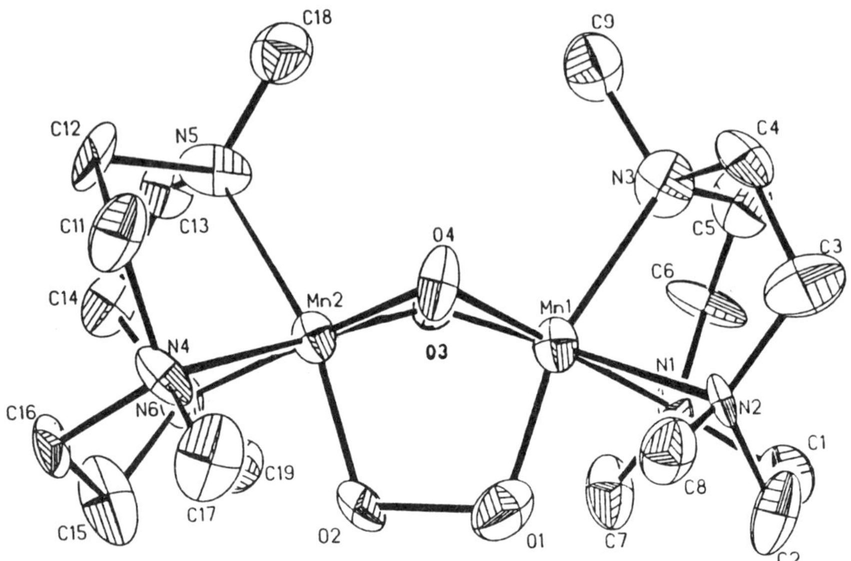

Figure 4. Molecular structure of a peroxo dimanganese(IV, IV) complex

there is no consensus about the precise arrangement of the manganese ions or the coordination structure. This means that one should keep in mind that there are no criteria at present to indicate which complex can serve as a synthetic model for this particular protein.

Except for the peroxomanganase(III) porphyrin complex already described, no isolable peroxo manganese complex is known. Bhula *et al.*[87] successfully prepared a trinuclear peroxo manganese complex. Bossek *et al.*[88] reported a remarkable dinuclear manganese(IV, IV) complex which contain a di-μ-oxo backbone. The structure of the complex is shown in Figure 4. Since it is most likely that OEC contains at least one di-μ-oxo manganese unit, based on EXAFS data, this complex may be a good synthetic model for the final reaction intermediate of OEC. In fact, the complex decomposes at room temperature in solution, releasing dioxygen. The peroxo complex may also be relevant to pseudocatalase[89] and ribonucleotide reductase,[90] which were recently identified as possessing a dimanganese site.

4 REFERENCES

1. (a) J. A. Lyons, *Aspects Homog. Catal.*, **3**, 1 (1977); (b) M. H. Gubelmann and A. F. Williams, *Struct. Bonding*, **55**, 1 (1983).
2. R. A. Sheldon and J. K. Kochi, *Metal-Catalyzed Oxidations of Organic Compounds*, Academic Press, New York (1981).

3. R. Landau, G. A. Sullivan and D. Brown, *Chemtech.*, 602 (1979).
4. (a) K. B. Sharpless and T. R. Verhoeven, *Aldrichim. Acta*, **12**, 63 (1979); (b) M. G. Finn and K. B. Sharpless, in *Asymmetric Synthesis*, Vol. 5 (J. D. Morrison, Ed.), Academic Press, New York (1985), Chapter 8, p. 247.
5. J. P. Collman, L. S. Hegedus, J. R. Norton and R. G. Finke, *Principles and Applications of Organotransition Metal Chemistry*, 2nd ed., University Science Books, Mill Valley, CA (1987).
6. J. P. Collman, R. R Gagne, C. A. Reed, T. Halbert, G. Lang and W. T. Robinson, *J. Am. Chem. Soc.*, **97**, 1427 (1975).
7. (a) L. Vaska, *Science*, **140**, 809 (1963); (b) J. Ibers and S. LaPlaca, *Science*, **145**, 920 (1964); (c) S. LaPlaca and J. Ibers, *J. Am. Chem. Soc.*, **87**, 2581 (1965).
8. H. Suzuki, K. Mizutani, Y. Moro-oka and T. Ikawa, *J. Am. Chem. Soc.*, **101**, 748 (1979), P. J. Chung, H. Suzuki, Y. Moro-oka and T. Ikawa, *Chem. Lett.*, 63 (1980).
9. See, e.g., R. S. Gall, J. F. Rogers, W. P. Shaefer and G. C. Christoph, *J. Am. Chem. Soc.*, **98**, 3135 (1976).
10. Many examples of cobalt complexes have been reported; see, e.g., (super oxide complex) F. R. Fronczek, W. P. Schaefer and R. E. Marsch, *Inorg. Chem.*, **14**, 611 (1975); (peroxide complex) F. R. Fronczek and W. P. Schaefer, *Inorg. Chim. Acta*, **9**, 143 (1974).
11. B. A. Vaarstra, J. Xiao and M. Cowie, *J. Am. Chem. Soc.*, **112**, 9425 (1990).
12. See, e.g. T. Yoshida, K. Tatsumi, M. Matsumoto, K. Nakatsu, A. Nakamura, T. Fueno and S. Otsuka, *Nouv. J. Chim.*, **3**, 761 (1979).
13. (a) R. Guilard, J. M. Latour, C. Le Comte, J.-C. Marchon, J. Protas and D. Ripoll, *Inorg. Chem.*, **17**, 1228 (1978); (b) R. B. Van Atta, C. E Strouse, L. K. Hanson and J. S. Valentine, *J. Am. Chem. Soc.*, **109**, 1425 (1987).
14. M. J. Bennett and P. B. Donaldson, *J. Am. Chem. Soc.*, **93**, 3307 (1971).
15. P. R. Sharp, D. W. Hoard and C. L. Barns, *J. Am. Chem. Soc.*, **112**, 2024 (1990).
16. N. Kitajima, K. Fujisawa, K. Toriumi and Y. Moro-oka, *J. Am. Chem. Soc.*, **111**, 8975 (1989).
17. W. Michlitz, S. G. Bott, J. G. Bentsen and S. J. Lippard, *J. Am. Chem. Soc.*, **111**, 372 (1989).
18. N.-T. Yu and E. A. Kepp, in *Biological Applications of Raman Spectroscopy*, Vol. 3 (T. G. Spiro, Ed.), Wiley-Interscience, New York (1987), p. 39.
19. P. R. Ortiz de Montellano (Ed.), *Cytochrome P-450*, Plenum Press, New York (1986).
20. Y. Tatsuno and S. Otsuka, *J. Am. Chem. Soc.*, **103**, 5832 (1981).
21. R. Ugo, F. Conti, S. Conti, S. Cenini, R. Mason and G. Robertson, *J. Chem. Soc., Chem. Commun.*, 1498 (1968).
22. H. Mimoun, M. M. P. Machirant and I. S. de Roch, *J. Am. Chem. Soc.*, **100**, 5437 (1978).
23. H. Suzuki, S. Matsuura, Y. Moro-oka and T. Ikawa, *Chem. Lett.*, 1011 (1982); *J. Organomet. Chem.*, **286**, 247 (1985).
24. (a) J. Birk, J. Halpern and A. Pickard, *J. Am. Chem. Soc.*, **90**, 4491 (1968); (b) J. Halpern and A. Pickard, *Inorg. Chem.*, **9**, 2798 (1970).
25. P. Chaumette, H. Mimoun, L. Saussune, J. Fischer and A. Mitscher, *J. Organomet. Chem.*, **250**, 291 (1983).
26. R. Sugimoto, H. Eikawa, H. Suzuki, Y. Moro-oka and T. Ikawa, *Bull. Chem. Soc. Jpn.*, **54**, 2897 (1981).
27. F. Sakurai, H. Suzuki, Y. Moro-oka and T. Ikawa, *J. Am. Chem. Soc.*, **102**, 1749 (1980).
28. R. Sugimoto, H. Suzuki, Y. Moro-oka and T. Ikawa, *Chem. Lett.*, 1863 (1982).

29. A. Nishinaga, H. Tomita, K. Nishizsawa and T. Matsuura, *J. Chem. Soc., Dalton Trans.*, 1504 (1981).
30. G. Ferguson, P. K. Monghan, M. Parvez and R. J. Puddephatt, *Organometallics*, **4**, 1669 (1985).
31. A. van Asselt, B. D. Santansiero and J. E. Bercaw, *J. Am. Chem. Soc.*, **108**, 8291 (1986).
32. L. Saussine, E. Brazi, A. Robine, H. Mimoun, J. Fischer and R. Weiss, *J. Am. Chem. Soc.*, **107**, 3534 (1985).
33. H. Mimoun, R. Charpentier, A. Mitchler, J. Fischer and R. Weiss, *J. Am. Chem. Soc.*, **102**, 1047 (1980); H. Mimoun, M. Mignard, P. Brechot and L. Saussine, *J. Am. Chem. Soc.*, **108**, 3711 (1986).
34. N. Oshima, Y. Hamatani, H. Fukui, H. Suzuki and Y. Moro-oka, *J. Organomet. Chem.*, **303**, C21 (1986).
35. N. Oshima, H. Suzuki and Y. Moro-oka, *Inorg. Chem.*, **25**, 3407 (1986).
36. G. Strukul, R. Ros, and R. A. Michelin *Inorg. Chem.*, **21**, 495 (1982).
37. G. Strukul and R. A. Michelin, *J. Am. Chem. Soc.*, **107**, 7563 (1985).
38. J. P. Collman, R. R. Gagne, C. A. Reed, W. T. Robinson and G. A. Rodley, *Proc. Natl. Acad. Sci. USA*, **71**, 1325 (1974).
39. J. P. Collman, *Acc. Chem. Res.*, **10**, 265 (1977).
40. F. P. Guengerich and T. L. Macdonald, *Acc. Chem. Res.*, **17**, 9 (1984).
41. O. Bangcharoenpaurpong, A. K. Rizos, P. M. Champion, D. Jollie and S. G. Sligar, *J. Biol. Chem.*, **261**, 8089 (1986).
42. E. McCandlish, A. R. Mikszal, M. Nappa, A. Q. Sprenger, J. S. Valentine, J. D. Strong and T. G. Spiro, *J. Am. Chem. Soc.*, **102**, 4268 (1980).
43. C. H. Welborn, D. Dolphin and B. R. James, *J. Am. Chem. Soc.*, **103**, 2869 (1981).
44. R. B. VanAtta, C. E. Strouse, L. K. Hanson and J. S. Valentine, *J. Am. Chem. Soc.*, **109**, 1425 (1987).
45. J. T. Groves, R. C. Haushalter, M. Nakamura, T. E. Nemo and B. J. Evans, *J. Am. Chem. Soc.*, **103**, 2884 (1981).
46. D. H. Chin, A. L. Balch and G. N. La Mar, *J. Am. Chem. Soc.*, **102**, 1446 (1980).
47. J. T. Groves, R. Quinn, T. J. McMurry, M. Nakamura, G. Lang and B. Boso, *J. Am. Chem. Soc.*, **107**, 357 (1980).
48. J. E. Penner-Hahn, T. J. McMurry, M. Renner, L. Latos-Grazynsky, K. S. Eble, I. M. Davis, A. L. Balch, J. T. Groves, J. H. Dawson and K. O. Hodgson, *J. Biol. Chem.*, **258**, 12761 (1983).
49. (a) J. T. Groves and Y. Watanabe, *J. Am. Chem. Soc.*, **108**, 154 (1986); (b) J. T. Groves and Y. Watanabe, *Inorg. Chem.*, **26**, 785 (1987).
50. A. L. Balch and N. W. Renner, *Inorg. Chem.*, **25**, 303 (1986).
51. R. D. Arasasingham, A. L. Balch and L. Latos-Grazynski, *J. Am. Chem. Soc.*, **109**, 5846 (1987).
52. D. H. Chin, J. D. Gaudis, G. N. La Mar and A. L. Balch, *J. Am. Chem. Soc.*, **99**, 5486 (1977).
53. D. H. Chin, G. N. La Mar and A. L. Balch, *J. Am. Chem. Soc.*, **102**, 4344 (1980).
54. (a) I. M. Klotz and D. M. Kurtz, *Acc. Chem. Res.*, **17**, 16 (1984); (b) P. C. Wilkins and P. G. Wilkins, *Coord. Chem. Rev.*, **79**, 195 (1987).
55. (a) S. J. Lippard, *Angew. Chem., Int. Ed. Engl.*, **27**, 344 (1988); (b) D. M. Kurtz, *Chem. Rev.*, **90**, 585 (1990).
56. N. Kitajima, H. Fukui, Y. Moro-oka, Y. Mizutani and T. Kitagawa, *J. Am. Chem. Soc.*, **112**, 6402 (1990).
57. S. Menage, B. A. Brennan, C. Juarez-Garcia, E. Munck and L. Que, Jr, *J. Am. Chem. Soc.*, **112**, 6423 (1990).
58. L. Que, Jr, and A. E. True, *Prog. Inorg. Chem.*, **38**, 97 (1990).

59. (a) M. P. Woodland, D. S. Patil, P. Cammack and H. Dalton, *Biochim. Biophys. Acta*, **873**, 237 (1986); (b) A. Ericson, B. Hedman, K. O. Hodgson, J. Green, H. Dalton, J. G. Bentsen, R. H. Beer and S. J. Lippard, *J. Am. Chem. Soc.*, **110**, 2330 (1988).
60. H. J. Shugar, A. T. Hubbard, F. C. Anson and H. B. Gray, *J. Am. Chem. Soc.*, **91**, 71 (1969).
61. S. Ahmad, J. D. McCallum, A. K. Shiemke, E. H. Appleman, T. M. Loehr and J. Sanders-Loehr, *Inorg. Chem.*, **27**, 2230 (1988).
62. E. I. Solomon, in *Metal Clusters in Proteins* (L. Que, Jr, Ed), ACS Symposium Series, No. 372, American Chemical Society, Washington, DC (1988), p. 116.
63. (a) W. P. J. Gaykema, W. G. J. Hol, J. M. Vereijken, N. M. Soeter, H. J. Bak and J. J. Beintema, *Nature* (London), **309**, 23 (1984); (b) A. Volbeda and W. G. J. Hol, *J. Mol. Biol.*, **209**, 249 (1989).
64. T. J. Thammann, J. S. Loehr and T. M. Loehr, *J. Am. Chem. Soc.*, **99**, 4187 (1977).
65. (a) T. N. Sorrell, *Tetrahedron*, **45**, 3 (1989); (b) Z. Tyeklar and K. D. Karlin, *Acc. Chem. Res.*, **22**, 241 (1989).
66. K. D. Karlin, R. W. Cruse, Y. Gultneh, J. C. Hayes and J. Zubieta, *J. Am. Chem. Soc.*, **106**, 3372 (1988).
67. J. S. Thomson, *J. Am. Chem. Soc.*, **106**, 8308 (1984).
68. N. J. Blackburn, R. W. Strange, A. Farooq, M. S. Haka and K. D. Karlin, *J. Am. Chem. Soc.*, **110**, 4263 (1988).
69. T. R. Jacobson, Z. Tyeklar, A. Farooq, K. D. Karlin, S. Liu and J. Zubieta, *J. Am. Chem. Soc.*, **110**, 3690 (1988).
70. N. Kitajima, T. Koda, S. Hashimoto, T. Kitagawa and Y. Moro-oka, *J. Chem. Soc., Chem. Commun.*, 151 (1988).
71. N. Kitajima, T. Koda, S. Hashimoto, T. Kitagawa and Y. Moro-oka, *J. Am. Chem. Soc.*, **113**, 5664 (1991).
72. N. Kitajima, K. Fujisawa, C. Fujimoto, Y. Moro-oka, S. Hashimoto, T. Kitagawa, K. Toriumi, K. Tatsumi and A. Nakamura, *J. Am. Chem. Soc.*, **114**, 1277 (1992).
73. N. Kitajima, T. Koda, Y. Iwata, and Y. Moro-oka, *J. Am. Chem. Soc.*, **112**, 8833 (1990).
74. K. D. Karlin, J. C. Hayes, Y. Gultneh, R. W. Cruse, J. W. McKown, J. P. Hutchinson and J. Zubieta, *J. Am. Chem. Soc.*, **106**, 2121 (1984).
75. T. N. Sorrell, M. R. Malachowski and D. L. Jameson, *Inorg. Chem.*, **21**, 3250 (1982).
76. L. Casella, M. Gullotti, G. Pallanza and L. Rigoni, *J. Am. Chem. Soc.*, **110**, 4221 (1988).
77. K. D. Karlin, B. I. Cohen, R. R. Jacobson and J. Zubieta, *J. Am. Chem. Soc.*, **109**, 6194 (1987).
78. O. J. Gelling, F. van Bolhuis, A. Meetsma and B. L. Feringa, *J. Chem. Soc., Chem. Commun.*, 552 (1988).
79. M. Reglier, M. Amadei, R. Tadayoni and B. Waegell, *J. Chem. Soc., Chem. Commun.*, 447 (1989).
80. R. Menif and A. E. Martell, *J. Chem. Soc., Chem. Commun.*, 1521 (1989).
81. K D. Karlin, P. Ghosh, R. W. Cruse, A. Farooq, Y. Geltneh, R. R. Jacobson, N. J. Blackburn, R. W. Strange and Z. Zubieta, *J. Am. Chem. Soc.*, **110**, 6769 (1988).
82. P. Ghosh, Z. Tyeklar, K. D. Karlin, R. R. Jacobson and J. Zubieta, *J. Am. Chem. Soc.*, **109**, 6889 (1987).
83. N. Kitajima, K. Fujisawa and Y. Moro-oka, *Inorg. Chem.*, **29**, 357 (1990).
84. N. Kitajima, T. Katayama, K. Fujisawa, Y. Iwata and Y. Moro-oka, to be published.
85. K. Wieghardt, *Angew. Chem., Int. Ed. Engl.*, **28**, 1153 (1989).

86. (a) G. Christou, *Acc. Chem. Res.*, **22**, 328 (1989); (b) J. B. Vincent and G. Christou, *Adv. Inorg. Chem.*, **33**, 197 (1989).
87. R. Bhula, G. J. Gaeme, J. Gainsford and D. C. Weatherburn, *J. Am. Chem. Soc.*, **110**, 7550 (1988).
88. U. Bossek, T. Weyhermuller, K. Wieghardt, B. Nuber and J. Weiss, *J. Am. Chem. Soc.*, **112**, 6387 (1990).
89. R. M. Fronco, J. E. Penner-Hahn and C. J. Bender, *J. Am. Chem. Soc.*, **110**, 7554 (1988).
90. A. Willing, H. Follman and G. Auling, *Eur. J. Biochem.*, **170**, 603 (1988).

11.2 Transition Metal Catalyzed Oxidation. The Role of Peroxometal Complexes

VALERIA CONTE, FULVIO DI FURIA and GIORGIO MODENA

Centro di Studio sui Meccanismi Reazioni Organiche-CNR, Dipartimento Chimica Organica dell'Universitá, via Marzolo 1, I-35131 Padova, Italy

1 INTRODUCTION

The synthetic scope of peroxides as oxidants of organic substrates has been investigated for almost a century.[1–3] Commonly used compounds have been inorganic and organic peracids, hydrogen peroxide and alkyl hydroperoxides, mostly in polar reactions[3] which formally amount to an oxygen transfer from the peroxide to the nucleophilic substrate.

For a long time it was considered that only organic peracids were of real synthetic interest because of their remarkable reactivity, usually much greater than that of other peroxidic species.[3,4] In fact, by using organic peracids, the smooth oxidation of all the substrates reported in Scheme 1 can be achieved under relatively mild conditions and with good yields.[1–4] However, organic peracids do have a series of drawbacks, the most serious one, apart from the relatively high cost, being a safety problem both in their preparation and in their storage. Therefore, the possibility of replacing organic peracids with hydrogen peroxide or alkyl hydroperoxides has been carefully investigated.[4] In fact,

Organic Peroxides. Edited by W. Ando

$$XOOH + Sub \longrightarrow XOH + SubO$$

X = H, Alkyl, Acyl
Sub = Phosphines, tertiary amines, thioethers, sulfoxides, alkenes, allylic alcohols

Scheme 1

because of the low reactivity of these species, their utilization in synthetically attractive oxidation reactions requires that an appropriate catalyst is present.

In the mid-1930s Milas and co-workers[5] discovered that some metal ions, in particular V, Mo and Cr derivatives, were excellent catalysts in oxidations by hydrogen peroxide. Later, it was found that the same metals, usually employed as soluble complexes, were also able to catalyze oxidations by alkyl hydroperoxides.[6,7] The resulting oxidizing systems are remarkably effective, their reactivity sometimes approaching that of organic peracids.

As a result there are today industrial processes based on the use of alkyl hydroperoxides or hydrogen peroxide. Typical examples are the Halcon process[8] employing a hydroperoxide together with an Mo(VI) derivative, the Shell process[9] where a hydroperoxide together with silica-supported titanium are used and the H_2O_2–Ti heterogenous system where the titanium catalyst is supported on zeolites,[10] e.g. titanium silicalite.

The mechanism of the catalytic effect of such metal ions was obscure for a long time. Such a mechanism could be assigned to the general class of acid-catalyzed processes where the protonated peroxide is a better oxidant than the neutral peroxide because of the increased fugacity of the leaving group, XO^- vs XOH (see Scheme 2) and of the increased electrophilicity of the peroxide.[11] However, the rate enhancements observed on addition of even minute amounts of some derivatives of the transition metal ions mentioned above are very large, suggesting that other factors are playing a role. Even

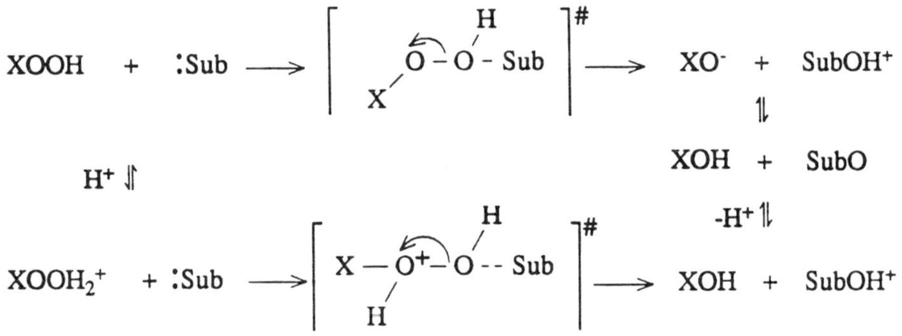

$$XOOH + :Sub \longrightarrow \left[X{-}O{-}O(H){-}Sub \right]^{\#} \longrightarrow XO^- + SubOH^+ \rightleftharpoons XOH + SubO$$

$$H^+ \rightleftharpoons \qquad\qquad -H^+ \rightleftharpoons$$

$$XOOH_2^+ + :Sub \longrightarrow \left[X{-}O^+(H){-}O(H){\cdot\cdot}Sub \right]^{\#} \longrightarrow XOH + SubOH^+$$

Scheme 2

though Brønsted acid catalysis,[12] in addition to catalysis by early-row Lewis acids,[13] may also be very effective, high acid concentrations under anhydrous conditions are required.[13] As a consequence, such systems have to be treated with extreme care since the risk of explosive decomposition of peroxides is high.

As will be discussed in detail in this chapter, we now know that the catalytic effect is due to the formation, in the reaction mixtures, of peroxometal species[14–17] whose reactivity is far higher than that of the peroxide, either hydrogen peroxide or an alkyl hydroperoxide, from which they originated. This conclusion, which was reached some years ago, establishes a close link between the chemistry of metal-catalyzed oxidations and that of peroxometal complexes.[14–17] These are, in fact, well defined species which in many instances may be isolated and fully characterized.

The first family of peroxometal complexes displaying remarkable oxidizing ability, as measured by their capability to epoxidize alkenes, was synthesized by Mimoun *et al.* in 1969[18] (Figure 1). These are oxo–diperoxo molybdenum(VI) complexes isolated by addition of hydrogen peroxide to a Mo(VI) precursor.

Later it was realized[19] that from a general point of view such complexes belong to the same family of peroxometal species which may be obtained by addition of dioxygen to suitable metal derivatives in a low oxidation state,[20] whose parent compound is the Vaska complex synthesized in 1963[21] (Figure 2). In fact, independently of their method of formation and of the nature of the

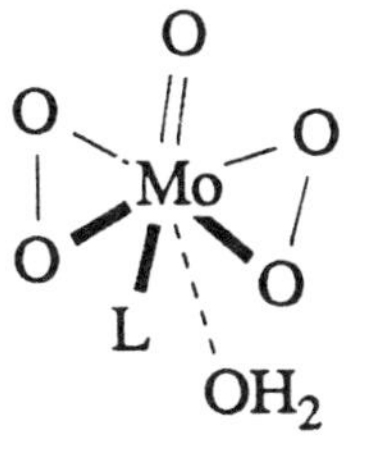

L = HMPT, Py, DMF etc.

Figure 1

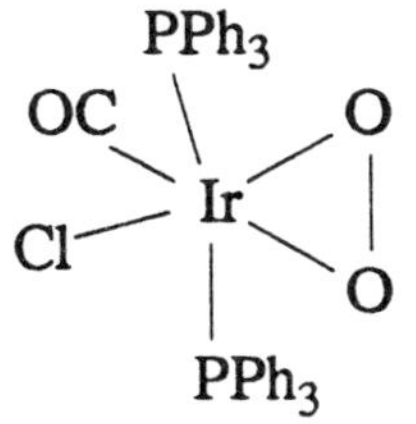

Figure 2

metal, peroxo complexes containing at least one η^2-bonded peroxo moiety have similar structural and spectroscopic features.[14,22] Some of these will be discussed later. On the other hand, their chemical behavior and, in particular, their oxidizing ability are vastly different. As an example, the Vaska complex is a very poor oxygen-transfer reagent, being able almost only to oxidize sulfur dioxide to sulfate.[19,23] It may therefore be concluded that any attempt to correlate the chemical behavior of peroxometal complexes with some of their general physical properties is dangerous and often misleading.

The availability of stable peroxometal species, which may be used as stoichiometric oxidants of various nucleophiles, has stimulated a host of mechanistic studies focused on the crucial point of the oxidation reaction, i.e. the oxygen-transfer step.[15,24–28] These studies, carried out under controlled conditions which undoubtedly are simpler than those of the catalytic reactions, allowed, as we shall see, a much better understanding of the metal-catalyzed oxidations by peroxides. Moreover, the mechanistic results turned out to be highly beneficial in improving the selectivity and hence the synthetic applicability of these reactions.[22–29]

2 LIMITATIONS AND SCOPE

The field of transition metal-catalyzed oxidations by peroxides has several facets. These include biochemical,[30] industrial,[8–10,17] synthetic[15,29] and mechanistic[15,24,25,28] aspects. To review all these topics would be outside the scope of this chapter, also because some of them are dealt with in other parts of this book. We shall focus our attention on the mechanistic aspects of the catalytic reactions and consequently on the chemistry of the peroxo complexes of those transition metal ions which are efficient catalysts and, in fact, the actual oxidizing species in these reactions.

This choice originates from our firm belief that mechanistic proposals must be subjected to periodic re-evaluation in the light of new experimental facts which become available in the literature. In this respect, the information on metal-catalyzed oxidations is increasing at such a remarkable speed that the time has come for a 'mise au point' of the matter. Our main aim has been to accommodate the behavior of peroxometal complexes within the general reactivity of peroxides.[1,3] In order to do this, we have reviewed what is known on the general properties of peroxometal complexes so that a comparison with the chemistry of classical peroxides could be made.

Of the many modes of reaction of peroxometal complexes, we have discussed only the one which is generally classified as 'electrophilic oxidation[11] where an oxygen atom is transferred from the peroxide to the nucleophilic substrate, whereas we have merely mentioned other important reactions of those complexes, such as the oxidation of alcohols, where, at least in some cases, radicals are involved.[31]

Even though we have emphasized the mechanistic problems, we are aware of the impact that catalyzed oxidations have on synthetic organic chemistry. Therefore, we decided to include in this chapter some of the most straightforward examples of the synthetic achievements obtained.

The coverage of the relavant literature is up to the end of 1990, but it was outside the scope of this chapter to make an exhaustive review of the very many and often important reports that have appeared in the last two decades.

3 STRUCTURE AND PROPERTIES OF PEROXOMETAL COMPLEXES

The family of peroxometal complexes is now large. However, the number of species which have been found to act as effective oxygen-transfer agents toward synthetically significant substrates, i.e. those species whose chemistry is directly connected with that of metal-catalyzed oxidations, is more limited.[15] Here we shall focus our attention on Ti(IV), V(V), Mo(V) and W(VI) derivatives. Moreover, we shall concentrate on those characteristics of the complexes, either in the solid state or in solution, which might be related to their chemical behavior as oxidants.

It has already been pointed out that the solid-state structure and, in particular, the bond lengths and the bond angles measured for several peroxo complexes, appear to have only a loose connection with their reactivity.[14b,22] In fact, whereas the pentagonal bipyramidal structure appears (Figure 3) to be almost ubiquitous for a large variety of species, their reactivity toward the same substrates under identical conditions is vastly different.

In Table 1 some isolated peroxo complexes are listed together with their metal—peroxide oxygen and oxygen—oxygen bond lengths. Recent data on the ^{17}O NMR chemical shifts of the oxo and peroxo oxygens are also reported. Further, the type of reactivity is indicated. It is evident from the data that no significant correlation can be found between structural parameters and reactivity.

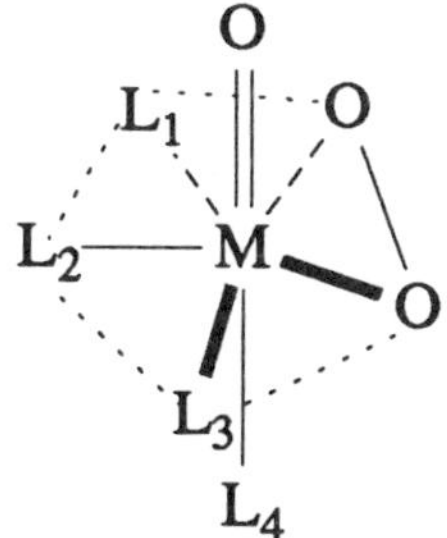

Figure 3. See Table 1 for examples of various metals and ligands

Table 1. Selected physicochemical properties of representative metal peroxo complexes

Complex	Bond lengths (Å)			^{17}O-NMR chem. shifts		Typical substrates	Ref.
	M=O	M—O	O—O	M=O	M(O—O)	oxidized	
(Dipic)Ti(O_2)·$2H_2O$	—	1.85	1.46	—	585	—	32, 48
(Pic)VO(O_2)·$2H_2O$	1.58	1.87	1.43	1221	641	Aromatics	33, 46–48
(Pico)VO(O_2)·$4H_2O$	1.59	1.88	1.43	1270	730	Thioethers	34, 47
(Dipic)VO(OOtBu)·H_2O	1.57	1.87	1.44	—	—	Aromatics	35
$K_3[VO(O_2)_2(Ox)]\cdot H_2O$	1.62	1.89	1.45	—	—	—	36
$Mo(O_2)_2$HMPT,H_2O	1.66	1.94	1.49	840	468	Alkenes	21, 37, 45–47
$MoO(O_2)_2$·(S)-DML	1.67	1.93	1.45	—	—	Alkenes (asym.)	38, 47
$[MoO(O_2)_2(Pic)]^-Me_4N^+$	1.68	1.95	1.46	844	431	Alcohols	39, 47
$[MoO(O_2)_2(Pico)]^-Bu_4N^+$	1.67	1.94	1.46	834	455	Alcohols	39, 47
(Dipic)MoO(O_2)·H_2O	1.67	1.90	1.45	—	—	Ketones	40
$K_2[MoO(O_2)_2(Ox)]$	1.68	1.95	1.47	—	—	—	41
Mo(O_2)Cl(Pic)HMPT	1.66	2.00	1.41	—	—	Epoxides	42
$WO(O_2)_2$HMPT·H_2O	1.62	1.96	1.52	692	400	Alkenes	43, 48
$CrO(O_2)_2$HMPT	—	—	—	1324	770	Thioethers	44, 48
$CrO(O_2)_2$Py	1.57	1.81	1.40	—	—	Alcohols	45

It is worth noting that, at least in the solid state, the triangular (side-on) arrangement of the peroxide moiety seems to be the preferred one also when the peroxo species is formed from an alkyl hydroperoxide (Figure 4).[35]

On the other hand, it is reasonable to assume that the reactivity of a given reagent should be better correlated with its structure in solution. Indeed, the structure in solution can be substantially different from that observed in the solid state. Therefore, physico-chemical studies of such complexes in solution are becoming increasingly important. Among the properties investigated, acid–base equilibria appear to have particular relevance since peroxo species may be fairly strong Lewis acids.[49]

As far as the structure of peroxo complexes in solution is concerned, let us consider, as an example, the reactions leading to the formation of peroxovanadium species from the interaction of hydrogen peroxide[50] or alkyl hydroperoxides[51] and soluble vanadium compounds in protic solvents such as methanol. When vanadium(IV) derivatives are used, a rapid oxidation of the metal to its highest oxidation state, i.e. V(V) takes place but, at least at low metal ion concentrations, no typical Fenton chemistry is observed. At the same time, the acac ligand is removed and the peroxometal complex is formed. The pertinent reactions are given in Scheme 3.

Equation 3 takes into account the possibility of an equilibrium between the peracid-like and the cyclic side-on structure of the V(V) peroxo complex. By contrast, in the peroxo complex formed from an alkyl hydroperoxide, only one alkoxo ligand is shown to be displaced to give the end-on peroxo compound (equation 2′). However, it cannot be ruled out that even with alkyl hydroperoxides the distal oxygen may coordinate to the metal, displaying a weakly bonded neutral molecule of the solvent. Such a coordination in solution does not appear to be energetically very important (see below), but its role in the oxygen-transfer step is certainly critical. As a further comment, it should be mentioned that the studies in solution do not provide much information on the

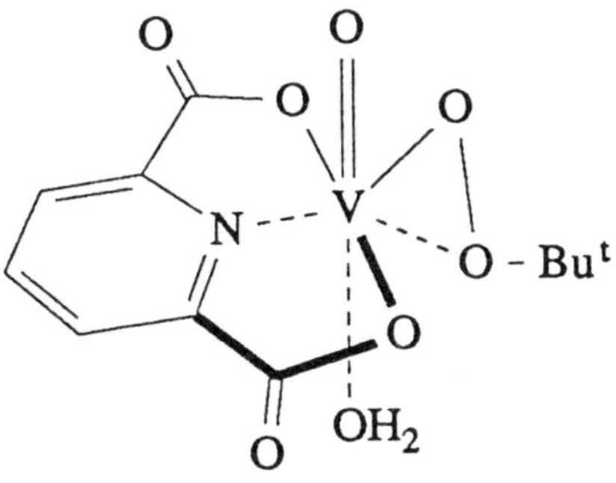

Figure 4

$$2\,VO(acac)_2 + XO_2H + 6\,ROH \longrightarrow 2\,VO(OR)_3 + 4\,acacH + XOH + H_2O \quad (1)$$

$X = H, R'$

$$VO(OR)_3 + H_2O_2 \rightleftharpoons VO(OR)_2O_2H + ROH \quad (2)$$

$$VO(OR)_3 + R'O_2H \rightleftharpoons VO(O_2R')(OR)_2 + ROH \quad (2')$$

$$VO(OR)_2O_2H \rightleftharpoons VO(O_2)OR + ROH \quad (3)$$

$$VO(O_2)OR + H_2O_2 \overset{ROH}{\rightleftharpoons} VO(O_2)_2^- + ROH_2^+ \quad (4)$$

Scheme 3

role of other molecules of the solvent, which, in the crystal structure, are present in the coordination sphere of the complexes.

In addition to the information stemming from crystallographic data, there is definite evidence from studies in solution that the peroxo complexes derived from hydrogen peroxide and d^0 transition metal ions such as those of Ti(IV), V(V), Mo(VI) and W(VI) have a cyclic η^2 structure whereas the coordination of alkyl hydroperoxides to the same metal ions is significantly different. This is based on the finding that there are substantial differences between the association constants of the H_2O_2 and that of the ROOH to the metal cited above, whereas associations to first-row electrophiles are similar.[24,52] Pertinent data are given in Table 2.

A key feature revealed by the data in Table 2 is that the association constants to the carbonyl group and to B(III) are almost the same for the two kinds of peroxides. This is consistent with their nucleophilicity as measured by the pK_a values, which are similar,[59] 11.6 for H_2O_2 and 12.8 for *t*-BuOOH. Therefore, in the association to the metal derivatives, there must be other factors which play a relevant role. Such factors have to be related to the nature of the nucleophile, either hydrogen peroxide or alkyl hydroperoxides, and with the nature of the electrophilic centers.

Hydrogen peroxide is, in fact, a bidentate ligand and if it behaves as such it might have association constants much higher than those for similar but monodentate ligands because of entropic factors.[60] However, this is not enough to explain the results in Table 2. For example, the hydrogen peroxide adduct to carbonyl groups has no tendency to evolve towards dioxiranes (equation 5). Instead these very reactive oxidants are only formed by high-energy intermediates via peroxidic oxygen—oxygen bond breaking[61] (equation 6). Such a reluctance of hydrogen peroxide to form three-membered rings with carbon or boron, which contrasts with the easy cyclization of diols to such centers, as for example in the formation of 1,3-dioxolane or 1,3-dioxane[62] (equation 7), may be related to the strong p–p electron repulsion in planar

Table 2. Comparison of K_{ass} (mol l^{-1}) to various electrophilic centers for H_2O_2 and t-BuO_2H

Reagent	$K_{ass}(H_2O_2)/K_{ass}(t\text{-}BuO_2H)^a$	Ref.
CH_3CHO	2	53
$B(OH)_4^-$	2	52, 54
$VO(OEt)_3$	> 1000	51, 52, 55
$MoO_2(acac)_2$	v.l.	56
$Ti(acac)_2$	v.l.	57
$W(CO)_6$	v.l.	58

^{a}v.l. = Very large as indicated by kinetic data.

H_2O_2, particularly in its cisoid conformation[63,64] exerted by the formally unshared electrons of the oxygens.

$$\text{(OH)C–O–O–H} \longrightarrow \text{C(O–O)} + H_2O \qquad (5)$$

$$\text{(}^{18}\text{OH)C–O–O–EWG} \longrightarrow \text{C(}^{18}\text{O–O)} + \text{EWG-OH} \qquad (6)$$

$$\text{(OH)C–O–CH}_2\text{CH}_2\text{–OH} \longrightarrow \text{C(O–CH}_2\text{CH}_2\text{–O)} + H_2O \qquad (7)$$

The transition metal ions, particularly those cited above, may release such a repulsion by accepting electron density from the filled antibonding orbitals to the empty d orbitals of appropriate symmetry. The high stability of the already discussed triangular arrangement of peroxometal complexes may be related to this interaction.

The crystallographic[35] and kinetic[65] evidence that the alkyl hydro-peroxides may adopt similar structures with both oxygens coordinated to the metal offers further support to the importance of this fact. Of course, such a bonding cannot have a strength comparable to that of hydrogen peroxide, and hence comparable association constants, since the neutral oxygen, which should become more and more positively charged with the extent of bonding, is a much weaker base (or nucleophile) than the negatively charged terminal oxygen. It is this fundamental difference that makes the truly dicoordinated H_2O_2 such a better ligand for the transition metal ions than alkyl hydroperoxides, as the association constants in Table 2 show.

The example discussed above, concerning peroxovanadium complexes, underlines the conclusion that the design of an appropriate coordination sphere of the metal precursor capable of determining a desired kind of reactivity or selectivity is hampered by the occurrence of solvolytic equilibria involving the ligands in solution. This is particularly unfortunate when chiral ligands are placed in such a coordination sphere in order to obtain asymmetric oxidations. Representative of this situation are complexes of Ti(IV), V(V) or Mo(VI) in catalytic reactions where the oxidant is either hydrogen peroxide or an alkyl hydroperoxide.

Early attempts to obtain enantioselective epoxidation of allylic alcohols by alkyl hydroperoxides and chiral V(V) complexes with ligands of the family of hydroxamic acids[27,66] and of alkylephedrine derivatives,[67] very strong bidentate ligands, gave enantiomeric excesses (e.e.) from fair to good (up to

80%), but the reaction proceeds only with the very reactive allylic alcohols and simple alkenes were not epoxidized. It was also observed that the catalyst prepared *in situ* with an excess of ligand (five-fold) gave the best optical yields.[66]

A similar approach was proposed for the enantioselective oxidation of thioethers[68] and also alkenes[69,70] under catalytic conditions by using large excesses of commercial chiral alcohols with again low efficiency (e.e. up to 10%).

The latter system proved to be totally ineffective in the enantioselective epoxidation of allylic alcohols because under the catalytic conditions the allylic alcohol in great excess completely displaced the chiral alcohol from the coordination sphere of the metals.[49] It was later observed that by drastically reducing the concentration of the allylic alcohol, enantioselective epoxidation may be obtained, albeit only with low e.e.s (*ca* 15%).[71]

It was also observed that a change of the oxidant from alkyl hydroperoxides to hydrogen peroxide caused a large decrease in the enantioselectivity in the oxidation of sulfides[72] (from *ca* 10% to 0.5–1% e.e.; see Scheme 4). It is evident that the much better bidentate ligand H_2O_2 leaves only one chiral alcohol molecule bonded to the metal (see also Scheme 3), as compared with the two left in the case of alkyl hydroperoxides, thus making the chiral complex a less efficient chiral reagent.

$$VO(OR^*)_3 + \text{t-BuO}_2H \rightleftharpoons VO(OR^*)_2(O_2Bu\text{-t}) + R^*OH$$

$$VO(OR^*)_2(O_2Bu\text{-t}) + \text{p-ClPhSCH}_3 \longrightarrow \underset{\text{e.e.} \sim 10\%}{\text{p-ClPhS}^*(O)CH_3}$$

$$VO(OR^*)_3 + HOOH \rightleftharpoons VO(O_2)(OR^*) + 2R^*OH$$

$$VO(O_2)(OR^*) + \text{p-ClPhSCH}_3 \longrightarrow \underset{\text{e.e.} < 1\%}{\text{p-ClPhS}^*(O)CH_3}$$

R^*OH = (-)Menthol

Scheme 4

These early attempts show how complex are the equilibria among ligands, peroxides and substrates in their coordination to the metal center, eventually opening the route to better enantioselective oxidants. Indeed, the titanium–tartrate catalyst developed by Sharpless for the epoxidation of allylic alcohols[73] is, perhaps, the best example of an enantioselective reagent so far reported. Similar enantioselective catalysts,[74,75] still based on titanium, but with different metal:tartrate:water ratios (see Scheme 5), are now available also for the enantioselective oxidation of thioethers with, often,[74–79] very high e.e. values.

$$\text{p-CH}_3\text{Ph-S-CH}_3 \xrightarrow[\text{chiral ligand; -20°C}]{\text{ROOH\textbackslash cat}} \text{p-CH}_3\text{Ph-S}^*\text{(O)-CH}_3$$

Ti^{IV} (eq)	H_2O (eq)	(+)-DET (eq)	t-BuOOH (eq)	Subst. (eq)	solvent	e.e.%	Ref.
1	-	1	2	1	CH_2Cl_2	-	73
1	-	4	2	1	DCE	88	74
1	1	2	1	1	CH_2Cl_2	91	75

Scheme 5

Depending on the nature of the metal and on the experimental conditions, it has been observed, again on the basis of kinetic data, that peroxo complexes derived from hydrogen peroxide and the metals cited above can coordinate nucleophilic molecules.[42,80,81] Moreover, taking advantage of the fact that stable diperoxo complexes of Mo(VI) and W(VI) are available,[18] spectroscopic measurements have been carried out.

^{1}H NMR data in alcoholic solvents show that when complexes of the general formula MO_5HMPT (M = Mo or W) are used as catalysts in reactions where H_2O_2 is the oxidant, displacement of the phosphoramido ligand takes place.[82] The equilibria are sufficiently slow on the NMR time scale to allow the identification of the free and the bound ligand and the dissociation constants are easily measured (Scheme 6).

$$\text{MoO(O}_2)_2\text{HMPT} + \text{ROH} \rightleftharpoons \text{MoO(O}_2)_2\text{(ROH)} + \text{HMPT}$$

$$K_{diss} = \frac{[\text{MoO(O}_2)_2\text{(ROH)}]\,[\text{HMPT}]}{[\text{MoO(O}_2)_2\text{HMPT}]}$$

Scheme 6

In contrast, ^{1}H NMR and ^{31}P NMR, showed that in non-coordinating solvents the HMPT ligand is tightly bound to the molybdenum peroxo species[83,84] mentioned previously. In addition, no direct evidence has been obtained of the coordination of a second molecule of HMPT in the apical position of the Mo peroxo species even though such a coordination is suggested by kinetic data (equation 8).[56]

$$MoO(O_2)_2HMPT \quad + \quad HMPT \rightleftharpoons MoO(O_2)_2(HMPT)_2 \qquad (8)$$

Detailed spectroscopic, kinetic and potentiometric studies have also been carried out in order to elucidate the acidic properties of metal peroxo species formed in protic solvents on addition of hydrogen peroxide to V(V), Mo(VI) and W(VI) precursors.[50,58,82,85] A typical example[50] of such an acidity is represented by the peroxo species formed in alcoholic solution from H_2O_2 and $VO(OEt)_3$. It has been shown that, depending on the excess of peroxide over the metal, both a monoperoxo and a diperoxo complex may be subsequently formed, as illustrated in Scheme 7.

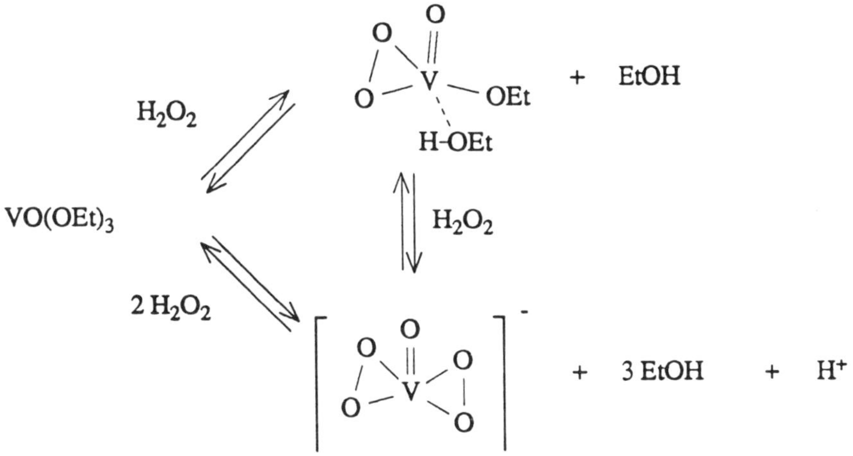

$$VO(OEt)_3 \xrightleftharpoons{H_2O_2} VO(O_2)(OEt)(HOEt) + EtOH$$

$$VO(OEt)_3 \xrightleftharpoons{2\,H_2O_2} [VO(O_2)_2]^- + 3\,EtOH + H^+$$

$$VO(O_2)(OEt)(HOEt) \xrightleftharpoons{H_2O_2} [VO(O_2)_2]^-$$

Scheme 7

The acidities of both species have been compared with those of other acids of known pK_a, using potentiometry.[50] The results revealed that the monoperoxo complex has an acidity between those of acetic acid and of methanesulfonic acid, while the diperoxidic species is even stronger than methanesulfonic acid (see Figure 5). The difference in the acidities of the complexes is reflected in their capability to act as electrophilic oxidants. In fact, the diperoxo complex is a less efficient oxidant of thioethers than the monoperoxo complex.[50,85] This behavior is consistent with the presence on the complex of a negative charge that reduces its electrophilic character.

More recently, reports have appeared on the acidity of Mo and W peroxo complexes in alcoholic solutions.[82] The addition of hydrogen peroxide to Mo(VI) and W(VI) derivatives, as already mentioned, leads to the formation of

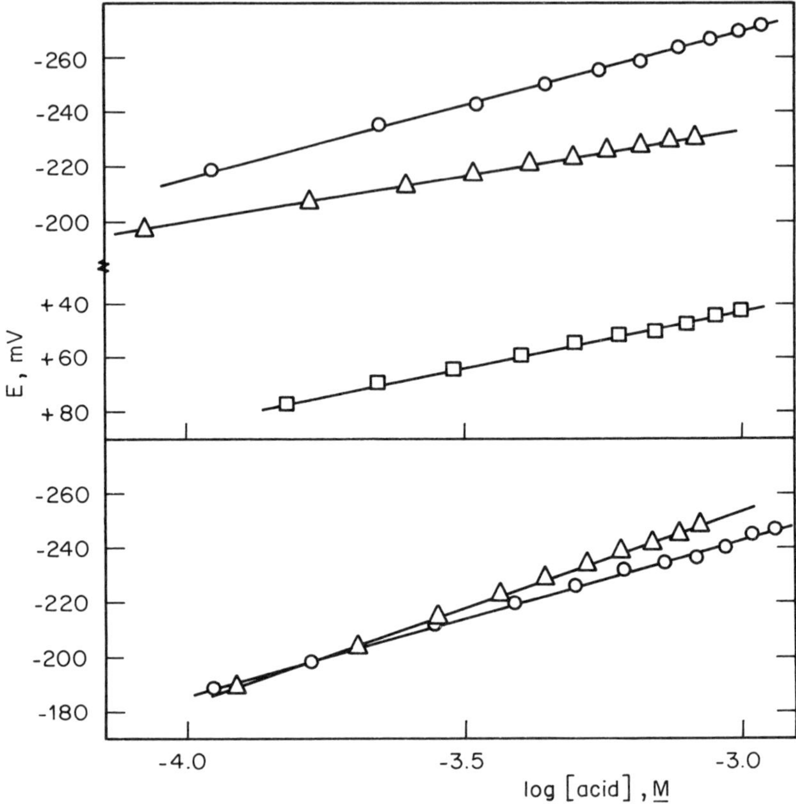

Figure 5. Variation of e.m.f. of 3.0×10^{-3} mol l^{-1} (top) and of 0.15 mol l^{-1} (bottom) H_2O_2 in ethanol–0.45 mol l^{-1} water as a function of added methanesulfonic acid (○), $VO(OiPr)_3$ (△) and acetic acid (□). Reproduced from ref. 50a by permission of Elsevier Sequoia

diperoxo complexes whose acidities have been directly measured by potentiometric titration[82] (see Figure 6).

In particular, although the structure of the diperoxotungsten[43] complex is very similar to that of the diperoxomolybdenum complex,[37] its acidity is much higher. As a consequence, in protic solvents the diperoxotungsten is more dissociated than the corresponding molybdenum species. Again, evidence for this situation can be found in the oxidation of nucleophilic substrates, where it has been confirmed that the less dissociated compound is the better oxidant.[82] However, if the correct comparison of the oxidative abilities of the two complexes is carried out by adjusting the acidity of the reaction medium, so that both complexes are undissociated, the diperoxotungsten complexes are far more efficient electrophilic oxidants than diperoxomolybdenum species.[82] Such a

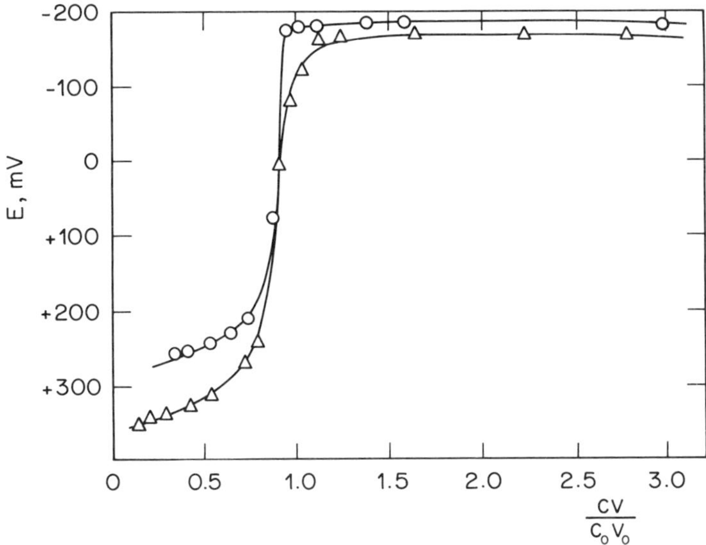

Figure 6. Potentiometric titration of solutions (25 ml, 1.3 mol l^{-1}) of $WoO_5HMPT \cdot H_2O$ (△) and of $MoO_5HMPT \cdot H_2O$ (○) with NaOEt (0.11 mol l^{-1}) in ethanol–0.5 mol l^{-1} water at 25°C. Reproduced from ref. 82 by permission of Elsevier Sequoia

behavior has also been confirmed by comparing the oxidative abilities of the isolated complexes MO_5HMPT ($M = Mo$ or W) toward nucleophilic substrates in aprotic solvents where the acid–base equilibria discussed above cannot take place, confirming that the tungsten derivative is more efficient.[43]

A schematic representation of the acid–base equilibria of MO_5L species in alcoholic media is shown in equation 9.

$$\mathrm{O{=}M(O_2)_2(O(H)R)} \overset{\mathrm{ROH}}{\rightleftharpoons} \left[\mathrm{O{=}M(O_2)_2(OR)}\right]^- + \mathrm{ROH_2^+} \qquad (9)$$

The information so far available does not allow a decision as to where the negative charge of the anionic peroxo complexes is localized. However, two likely structures are represented in Fig. 7.

Two pieces of evidence favoring the localization of the negative charge on the metal oxo group structure may be mentioned. The peroxoanion-like structure **1** would suggest that in highly basic solution dianions should be formed as a result of the opening of the second peroxo bridge, whereas there is no evidence for the

1 **2**

Figure 7

existence of such dianions. In fact (see Figure 6), the potentiometric titration, in alcoholic solution, of the two peroxo complexes MO_5HMPT (M = Mo or W) indicates that they behave as monoprotic acids.[82] Moreover, these anionic peroxo complexes do not oxidize electrophilic substrates, such as ketones or electron-poor alkenes, at variance with the behavior generally observed for peroxo anions.[62]

The second information available comes from recent ^{17}O NMR measurements, which indicate that in the absence of a ligand capable of accommodating the negative charge, e.g. the anion of picolinic acid (see, e.g., the complex MoO_5Pic in Table 1), the oxo signal of anionic peroxo complexes shifts about 50 ppm upfield with respect to the signal of neutral species.[47]

4 KINETIC ASPECTS OF METAL-CATALYZED OXIDATIONS

In a transition metal-catalyzed reaction where the oxidant is a peroxide, either hydrogen peroxide or an alkyl hydroperoxide, there is the *in situ* formation of a peroxometal complex which is the real oxidant in solution. The overall process is shown in Scheme 8.

In order to explain the catalytic effect observed, it must be assumed that the oxidative ability of the peroxometal species formed on interaction of the peroxide with the catalyst is greater than that of the peroxide itself.[6,7,86–88] In fact, dramatic enhancements of the oxidation rates are observed when even minute amounts of the metal catalyst are added to reaction mixtures containing the substrate and the peroxide so that, in some instances, the catalyzed (k'_2) rate constant is orders of magnitude larger than the uncatalyzed (k_2) value.[86–88]

The occurrence of a catalytic process requires also that the metal species formed after the transfer of the peroxidic oxygen to the substrate is still able to

$$XOOH + L_mM \underset{k_{-1}}{\overset{k_1}{\rightleftharpoons}} L_nM(O_2X)_y + y\,LH \quad (10)$$

$$XOOH + Sub \xrightarrow{k_2} Prod + XOH \quad (11)$$

$$L_nM(O_2X)_y + Sub \xrightarrow{k_2'} Prod + L_nM(OX)_y \quad (12)$$

$$L_nM(OX)_y + y\,XOOH \underset{k_{-3}}{\overset{k_3}{\rightleftharpoons}} L_nM(O_2X)_y + y\,XOH \quad (13)$$

X = H, Alkyl
M = Ti^{IV}, V^{V}, Mo^{VI}, W^{VI}
y = 1, 2

Scheme 8

add the peroxide (equation 13), thus restoring the peroxometal complex.[15,16,24,25]

Although it is possible to accommodate the catalyzed oxidations by hydrogen peroxide and by an alkyl hydroperoxide within the same general scheme, there are several differences between the two systems that affect their oxidative behavior. The first and most obvious difference is in the nature of the peroxometal complex formed. The elucidation of this aspect has received a great help from research work carried out in the past 20 years aimed at the isolation and characterization of peroxometal compounds.[14b] Although this topic has been considered above, it may be recalled here that, at least in the solid state, the triangular arrangement of the peroxidic moiety is the preferred one even in the case of hydroperoxides[35,37] (see Figures 1 and 2). There is, in fact, evidence that the peroxometal complexes derived from hydrogen peroxide maintain the triangular structure in solution even of highly solvating solvents (see the previous section). For those derived from alkyl hydroperoxides such a triangular arrangement should occur at least in the transition state.[26,27,89,90]

A second difference between the two kinds of peroxides may be found in the kinetic behavior of the oxidation reactions that they carry out. Generally, when hydrogen peroxide is employed, the rate law of the oxidation of a nucleophile (Nu) is overall second order, first order in the substrate and first order in the catalyst.[16,24,25,50,91] The zero-order dependence of the rates on hydrogen peroxide concentration is rationalized as resulting from a very high value of the association constant of the peroxide to the metal in the process leading to the formation of the peroxometal complex.[52] Such a ‘saturation’ of the catalyst, which implies that an increase in hydrogen peroxide concentration would not enhance the oxidation rates, is further evidence that the contribution of the uncatalyzed oxidation to the overall rate is negligible.

In contrast, the rate law of the catalyzed reactions where the oxidant is an alkyl hydroperoxide are, generally, third order, first order each in substrate, hydroperoxide and catalyst.[51,65] Such behavior not only indicates that the association constant of hydroperoxides to transition metal derivatives should be smaller than that of hydrogen peroxide but also that the absolute value of such a constant should be very small. Indeed, in a kinetic scheme which involves an equilibrium process followed by irreversible oxygen transfer to the substrate, one would expect a kinetic order of the peroxide less than one on the basis of the equations in Scheme 9, which lead to the final calculated equation 17.

$$M(OR')_n + ROOH \underset{k_{-1}}{\overset{k_1}{\rightleftharpoons}} (R'O)_{n-1}M\text{-}OOR + R'OH \tag{14}$$

$$(R'O)_{n-1}MO_2R + Nu \xrightarrow{k_2} NuO + (R'O)_{n-1}MOR \tag{15}$$

$$(R'O)_{n-1}MOR + ROOH \underset{k_{-3}}{\overset{k_3}{\rightleftharpoons}} (R'O)_{n-1}MOOR + ROH \tag{16}$$

$k_1/k_{-1} = K$

Scheme 9

$$R = \frac{k_2[\mathrm{Nu}]_0[\mathrm{ROOH}]_0[\mathrm{catalyst}]_0}{1 + K[\mathrm{ROOH}]_0} \tag{17}$$

Therefore, the occurrence of the first order of the oxidant, i.e. that $1 \gg K[\mathrm{ROOH}]_0$, confirms that K is in fact very small.

The information provided by kinetic studies has been subjected to a direct test in order to confirm that hydrogen peroxide indeed adds to transition metal derivatives much better than alkyl hydroperoxides[51,55] (see above).

A further consequence of the small value of the association constant of hydroperoxides to the metal catalysts is the autoinhibitory kinetic behavior which is often observed.[7] This is due to the competition between the hydroperoxide and the alcohol co-produced in the oxidation step for the addition to the metal ion (see equation 16 in Scheme 9).

Such an effect may in some instances be fairly severe so that the oxidation reaction may become inconveniently slow. Aside from practical considerations such as the low yields of oxidized products obtained, the occurrence of an autoinhibition also has a negative effect on the kinetic analysis of the reactions. In fact, most of the early studies were carried out by using the initial rates of oxidation[7,65,92] instead of the more accurate kinetic rate constants obtained from integrated plots.[49] A way to circumvent such an inconvenience is to study the oxidation reactions in an alcoholic solvent. This implies, of course, that the reaction rates are usually much slower than those measured in non-protic solvents.[49,51,52]

On the other hand, the autoinhibition is 'masked' and clean first-order plots of log[Ox] vs time are obtained which allow the rate constants to be calculated with a high degree of precision. Usually, these studies have been carried out by employing very reactive nucleophiles in order to speed up the oxidation rates. Typical substrates were organic sulfides whose reactivity with peroxides closely resembles that of alkenes (see the following section).[49]

In the case of catalyzed oxidations by hydrogen peroxide, the water obtained as a by-product hardly competes with hydrogen peroxide so that, from a kinetic point of view, the complications encountered with alkyl hydroperoxides are eliminated.[50,52,56] On the other hand, dilute hydrogen peroxide is soluble only in water-miscible solvents, such as alcohols or cyclic ethers, where the oxidation reactions are usually slower than in non-polar solvents, probably because of an effective stabilization of the peroxometal complex through solvation. This has stimulated the development of two-phase procedures[93–100] where the peroxometal species is formed in an aqueous phase and then transferred, by an appropriate phase-transfer agent, into an organic phase where the oxidation takes place. In particular, depending on the nature of the peroxo complex in aqueous solution, which, as we have seen before,[50,82] can be present either as a neutral or an anionic species, different phase-transfer agents have been used.[93,95,96] Thus W(VI) and Mo(VI) neutral peroxo complexes are extracted by neutral lipophilic ligands[94,95] belonging to the class of phosphoric amides, whereas anionic peroxo species, which are also good oxidants of the alcoholic function,[39,101] are transferred into the organic phase by classical cationic phase-transfer agents[96,98] such as tetraalkylammonium salts. Further details on such systems can be found in recent publications.[93–100]

5 MECHANISM OF OXYGEN TRANSFER REACTIONS

The understanding of the mechanism of oxygen transfer reactions from peroxometal complexes to nucleophilic substrates, i.e. the detailed description of the transition state of such a process, is a central point in the chemistry of catalytic oxidations. As an example, on the basis of such information one may be able to predict not only the nature and the regiochemistry of the products but also the stereochemical behavior of the reaction when asymmetric oxidations are carried out.[28,89,102] This, in turn, may eventually allow the development of peroxometal species able to oxidize a wide variety of prochiral substrates with a large degree of enantioselectivity.

It must be pointed out, however, that oxygen-transfer reactions from peroxides may follow different mechanistic pathways depending on the nature of both the substrate and the oxidant.[11,24] In fact, this is a fairly common feature that such a process shares with several other chemical reactions that are hardly accommodated within a single mechanistic scheme in spite of the many efforts made toward providing unified mechanisms.[25,103]

The first dichotomy encountered in oxygen-transfer reactions from peroxides is whether or not an electron-transfer process proceedes the bond formation with the oxygen delivered to the substrate. In addition, if a clean polar reaction not involving radical species is taking place, it may be difficult to ascertain whether the peroxo species is acting as an electrophile or as a nucleophile (Scheme 10).

The occurrence of path 1 (equation 18), which is representative of several kinds of reactions, is easily observed both by product studies and by kinetic analysis if a radical chain takes place,[104,105] whereas it may be difficult to detect if very short chains or simple bimolecular reactions occur.[31] Often these reactions are more complex than simple oxygen transfers, the peroxide or the peroxometal only being involved in the initiation step whereas the transformation of the reagent into the product is carried out by dioxygen.[15] Such reactions are reviewed in other chapters and we shall not discuss them at any length in this section.

The nucleophilic reactivity of peroxo species, depicted in path 3 (equation 20), is a well defined mechanistic pathway even though it has been unambiguously identified only in a limited number of reactions. A typical case is that of the oxidation of sulfoxides to sulfones by the anion of perbenzoic acid[106] (Figure 8),

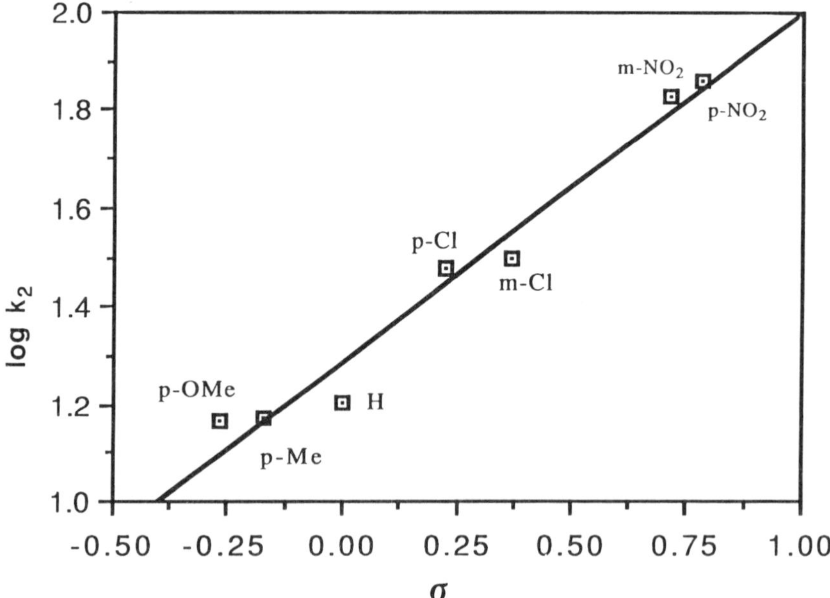

Figure 8. Substituent effect on the rate of oxidation of substituted diphenyl sulfoxides by perbenzoate anion at pH 11.7 in dioxane–water (40:60) at 25°C. Taken from ref. 102

$$\text{H-O-O-H} + {:}\text{Sub} \longrightarrow [\text{HO}^{\bullet}, \text{Sub}^{+\bullet}] + \text{HO}^- \longrightarrow \text{HO-Sub}^+ + \text{HO}^- \longrightarrow \text{SubO} + H_2O \qquad (18)$$

$$\text{H-O-O-H} + {:}\text{Sub} \longrightarrow \text{H-O-Sub}^+ + \text{HO}^- \longrightarrow \text{SubO} + H_2O \qquad (19)$$

$$\text{H-O-O-H} + \text{B} \rightleftharpoons \text{HOO}^- + \text{BH}^+ \xrightarrow{{:}\text{Sub}} [\text{SubOOH}]^- \underset{}{\overset{\text{BH}^+}{\rightleftharpoons}} \text{SubO} + H_2O + \text{B} \qquad (20)$$

Scheme 10

where a positive ρ value is observed, indicating the nucleophilic character of the oxidant in the rate-determining step.

Anionic peroxospecies, such as HOO^-, $ArCOO^-$ and t-$BuOO^-$, are also able to epoxidize 'electrophilic alkenes' containing an electron-withdrawing group (EWG)[107–109] (Scheme 11).

$$R\text{-}O\text{-}O^- + \text{alkene-EWG} \rightleftarrows \text{C(OOR)-C}^-\text{(EWG)} \rightarrow \text{epoxide-EWG} + RO^-$$

R = H, Alkyl, Acyl
EWG = $-NO_2$, -CO-, $-SO_2$- etc.

Scheme 11

Also in this case the reaction begins with the nucleophilic attack of the anionic oxygen at the most positive carbon of the substrate to give a carbanion which in the subsequent step attacks the peroxidic oxygen to give the epoxide. The reaction differs in a number of features from the normal alkene epoxidation by percarboxylic acids. An important difference is the stereochemical course. In fact, whereas the latter process is completely stereospecific, providing (Z)-epoxides from (Z)-alkenes, the reaction in Scheme 11 is a stereoselective oxidation which provides the thermodynamically more stable (E)-epoxide, as a result of the possibility of C—C rotation in the intermediate. This mechanistic scheme is further confirmed by the finding that if an oxidant such as ClO^- is used, the lifetime of the intermediate is greatly diminished, owing to the effectiveness of Cl^- as a leaving group as compared with RO^-, and the reaction becomes stereospecific again.[107–109]

Much more abundant are the examples concerning the reactivity of peroxo compounds, including peroxometal species, with nucleophiles. In fact, oxygen transfers to nucleophilic substrates are significant synthetic procedures in organic chemistry and they are also encountered in industrial processes.[8–10] However, in spite of the large number of reactions investigated, the mechanism is still a matter of debate.[24,25] Two different descriptions of the oxygen-transfer step have been offered: (a) a nucleophilic attack of the substrate to the peroxide oxygen along the O—O bond,[110–113] leading in the case of alkene epoxidations to Bartlett's[110] transition state, or (b) a 1,3-dipolar cycloaddition of the peroxo compound to the nucleophile followed by the cleavage of the intermediate thus formed.[25,114] (see Scheme 12).

Further studies[115] showed that the suggested intermediate in the latter mechanism could not be along the reaction coordinate, particularly because it was found to be a fairly stable compound. Therefore, such a mechanism was modified by suggesting that the intermediate was never reached and the reaction was assumed to evolve directly from reagents to products.

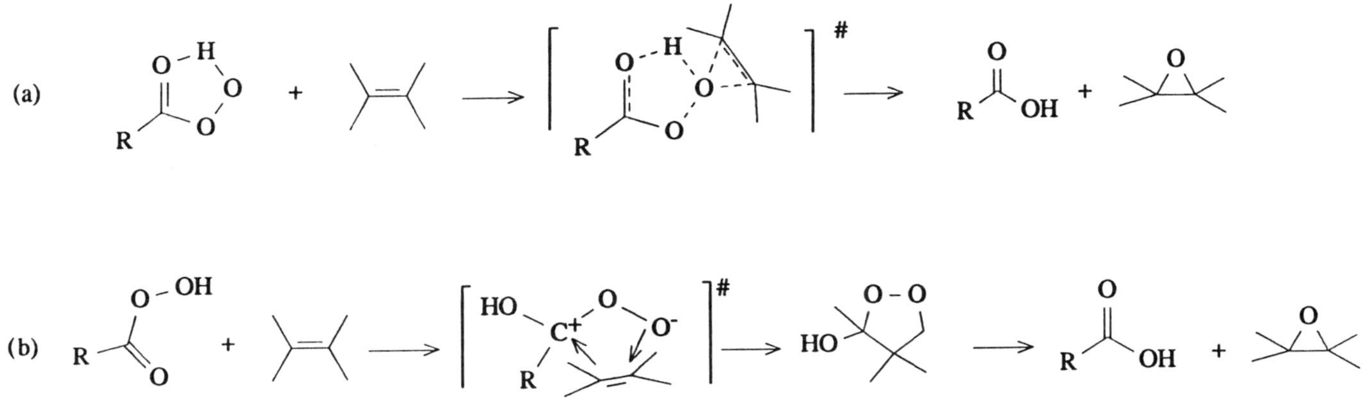

Scheme 12

The extension of these mechanistic hypotheses to the oxygen-transfer processes from peroxometal compounds was straightforward for the former, as shown in route (a) in Scheme 13. Indeed, the transfer of the oxygen atom is assumed to occur via a mechanism strictly reminiscent of that of Bartlett.[24,26,27] The latter mechanistic proposal was also extended to peroxometal complexes in the so-called peroxometallacycle mechanism,[25,80,103] which has been largely based on the studies of molybdenum peroxo complexes, as shown in route (b) in Scheme 13.

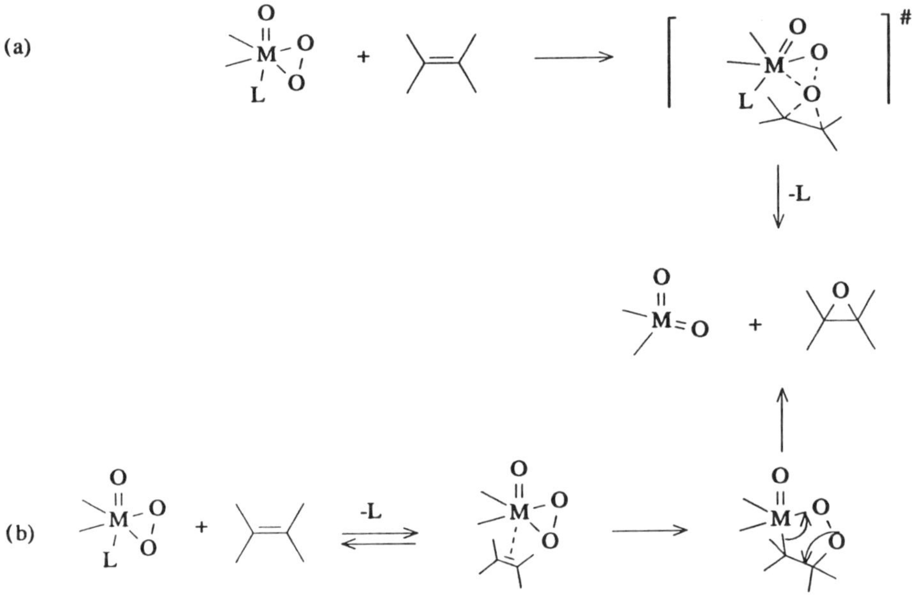

(a)

-L

(b)

-L

Scheme 13

In the light of such controversy, it is useful to analyze in some detail the features of both mechanistic pathways.

5.1 Electrophilic Mechanism

The most striking example of such a mechanism is the oxidation of n-nucleophiles such as phosphines, thioethers and tertiary amines by hydrogen peroxide or alkyl hydroperoxide (equation 21). In fact, it is difficult to envisage mechanistic routes different from that shown, where the nucleophile attacks the antibonding σ-orbital of the peroxidic group. Such a mechanism bears a close resemblance to the classical reaction of chlorine with the hydroxy anion (or a water molecule) (equation 22) and in a more general way with the classical S_N2 mechanism.[62]

On the basis of such a very simple mechanistic scheme, several features of the oxidation of nucleophiles by peroxo compounds are easily rationalized, as follows.

i. The direct dependence of the reaction rates on the nucleophilicity of the substrate and on the nucleofugacity of the leaving group. Therefore, phosphines are usually more easily oxidized than thioethers because the phosphorus atom is more nucleophilic than the sulfur atom. In turn, protonated hydrogen peroxide is a better oxidant than hydrogen peroxide itself because H_2O is a better leaving group than HO^- (equation 23).[11,24,116]

ii. The formation of charged species, as required by the stoichiometry of the reaction, should imply an energetically unfavorable situation of the transition state owing to a considerable charge separation. This is avoided by the system by using different alternatives. The simplest occurs when an acid is present by the intervention of either specific or a general acid catalysis, i.e. by protonation of the leaving group either in a prior step or in the rate-determining step. More complex is the process to avoid charge separation in protic solvents where one or more solvent molecules transfer, via a sequence of hydrogen bond shifts, the terminal proton of the hydroperoxide to the oxygen proximal to the alkyl group (equation 24).[117]

   In aprotic solvents the hydroperoxide itself may intervene in the proton-transfer process. In such a case the reaction rate becomes quadratic in the oxidant.[117] In the case of organic peracids, the charge separation is avoided by an internal hydrogen bonding.[118–121] As a result, the rates of oxidation are faster in aprotic than in protic media, where the internally hydrogen-bonded structure is destabilized with respect to the open structure because of the solvation.[121]

iii. The steric factors in such reactions are usually modest.[3,122] This is consistent with the fact that the transferred oxygen is directly bonded to the very small hydrogen and to another oxygen, also a small atom. It follows that any interaction between the nucleophile and the oxidant is with an atom or a group which is distant from the reaction center. The intrinsic weakness of the interactions, unless particularly rigid systems are involved, makes the enantioselective oxidation inefficient.[123]

The oxygen transfers from a peroxide to an n-nucleophile, i.e. a sulfide, or to a π-nucleophile, i.e. an alkene, are strictly related processes and show very similar features. For example, it was found that the susceptibility to solvent change of *S*-oxidation and epoxidation are linearly correlated[116,121] (see Figure 9).

$$R_2S + R'O_2H \longrightarrow \left[R'{-}O{-}O(H) \leftarrow :SR_2 \right]^{\#} \longrightarrow R'O^- + R_2SO^+H \rightleftharpoons R_2SO + R'OH \qquad (21)$$

$$Cl{-}Cl + {}^{\cdot}OH \longrightarrow \left[Cl{-}{-}Cl{-}{-}O{-}H \right]^{\#} \longrightarrow Cl^- + ClOH \qquad (22)$$

$$R_2S + R'O_2H_2^+ \longrightarrow \left[R'(H)O^+{-}O(H) \leftarrow :SR_2 \right]^{\#} \longrightarrow R'OH + R_2SO^+H \rightleftharpoons R'OH_2^+ + R_2SO \qquad (23)$$

$$Nu: + R'OOH \xrightarrow{ROH} \left[Nu: \rightarrow O(H){-}O(R')\cdots H{-}O{-}R \right]^{\#} \longrightarrow NuO + R'OH + ROH \qquad (24)$$

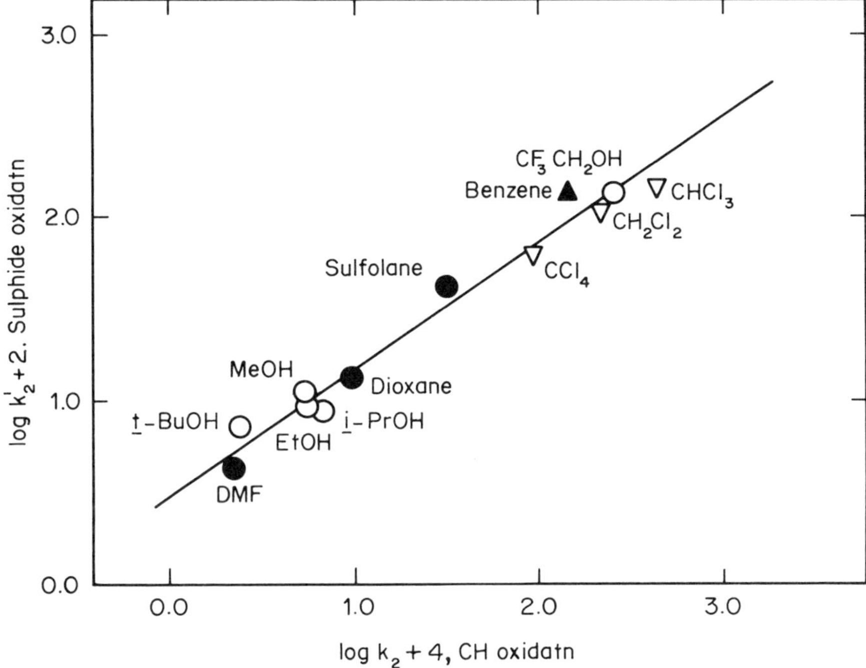

Figure 9. Linear trend in the plot of the logarithms of *p*-nitrodiphenyl thioether oxidation rate constants vs the logarithms of rate constants for cyclohexene epoxidation by perbenzoic acid in various solvents. Reprinted with permission from R. Curci, R. A. Di Prete, J. O. Edwards and G. Modena, *J. Org. Chem.*, **35**, 740 (1970). Copyright 1970, American Chemical Society

There are therefore no major difficulties or logic breaks to extend the considerations made for n-nucleophiles to π-nucleophiles. The classic Bartlett mechanism[110] shows, as indicated by the stereospecificity of alkene epoxidations, that the π-electron pairs play the same role as that of the n-electrons of the n-nucleophiles (see Scheme 12).

The simple mechanistic scheme formulated for typical peroxidic reagents such as hydrogen peroxide, alkyl hydroperoxides and percarboxylic acids may be extended to the reactions with H_2O_2 or ROOH either catalyzed by typical Lewis acids, such as BF_3 and $AlCl_3$, or by transition metal ions in their highest oxidation state. Indeed, typical Lewis acids may act as the proton in increasing the nucleofugacity of the leaving group, making the oxygen-transfer process more facile. Also, the peroxo complexes derived from d^0 transition metal ions discussed in the previous section represent a form of electrophilic activation of the peroxide as the 'leaving' oxygen is incorporated in a strong oxo acid (see Scheme 13).

In this respect, the finding that the peroxo complexes have a triangular or quasi-triangular structure in the solid state is fully consistent with the mechanism depicted because the oxygen, which is part of the leaving group, is already bonded to the metal.

This mode of thought tends to keep and develop the simple, obliged scheme of the elementary reaction such as that of equations 21 and 23 and of Scheme 12, which are also supported by quantomechanical calculations,[124] to more complex reagents and substrates, leaving the search for mechanistic changes only to those systems where they are required by the particular results obtained. In this context it is typical the case of those reactions in which clean evidence of radical chain processes are involved.[33,104,125]

5.2 The Peroxometallacycle Approach

The alternative proposal of the peroxometallacycle intermediate[25] can be related to the well documented metallacycle formed in methatesis reactions where two molecules of alkenes are converted into a different pair of alkenes (Scheme 14).[126,127] A similar metallacycle is formed by reacting peroxometals such as those derived from platinum or palladium with electrophilic alkenes (equation 25).[128,129] In these cases the peroxometallacycles are stable species and when they are forced to react they usually fragment, yielding carbonyl compounds (equation 26).[128,129]

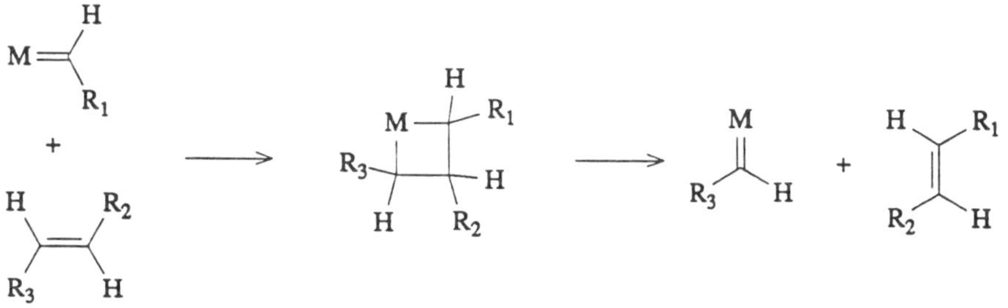

M=CH(R1) + (H)(R3)C=C(R2)(H) ⟶ M–C(H)(R1)–C(H)(R2)–C(R3)(H)–M (metallacycle) ⟶ M=C(R3)(H) + (H)(R1)C=C(R2)(H)

Scheme 14

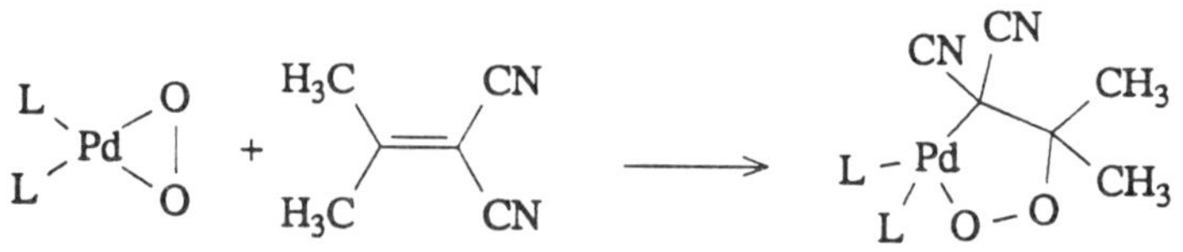

$$L_2Pd(O_2) + (H_3C)_2C{=}C(CN)_2 \longrightarrow L_2Pd\text{–}C(CN)_2\text{–}C(CH_3)_2\text{–}O\text{–}O \quad (25)$$

$L = Ph_3P$

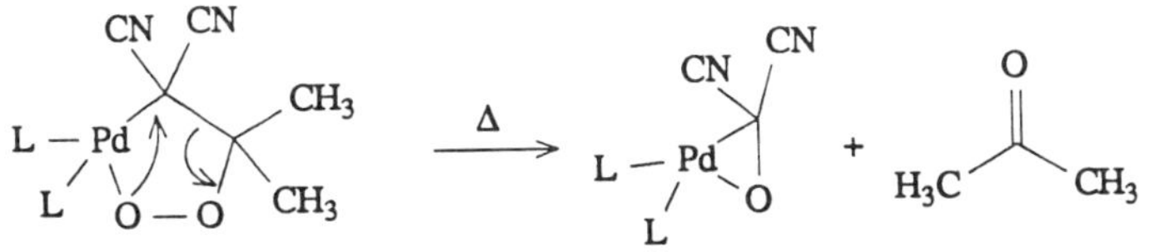

(26)

$L = Ph_3P$

Another system where the occurrence of a peroxometallacycle intermediate is likely is that of Rh(III) peroxo compounds which oxidize terminal alkenes to methyl ketones (Scheme 15).[130,131] It has been shown that the product is formed by β-heteroelimination, as might have been expected.[131] Indeed, the instability of primary and secondary alkylhydroperoxides, because of the ease of this type of reaction, is well documented (equation 27).[132]

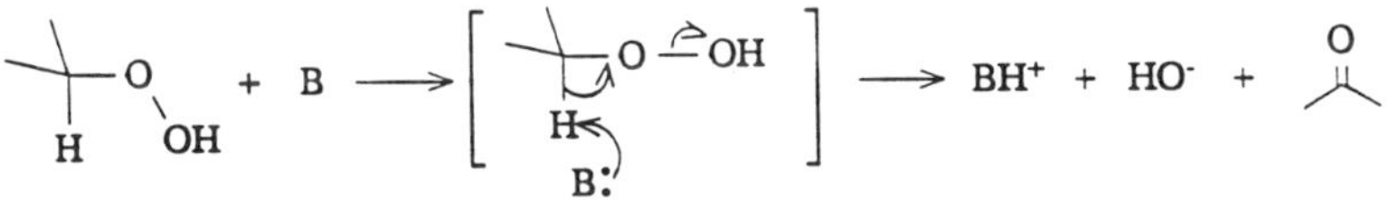

(27)

As far as derivatives of d^0 metals, such as Ti(IV), V(V), Mo(VI) and W(VI), are concerned, direct evidence of peroxometallacyclic intermediates is not available.[24,56,92] Their involvement has been proposed mainly on the basis of kinetic results which indicate that some sort of intermediate is formed in the epoxidation of alkenes by V(V) and Mo(VI) peroxo complexes.[83,133] In particular, it has been observed that saturation of the oxidant occurs when large excesses of substrates are employed.[133] This kind of evidence does not give any indication of whether the intermediate is along the reaction coordinate or on a parallel path simply leading to the same final product at a lower rate by subtracting the more efficient reagent from the main path.[24,56]

Indeed, in the case of the much studied MoO_5HMPT oxidation of alkenes, it was shown that the saturation kinetics are probably due to the formation of substrate peroxo compound less reactive than the original reagent.[56] Further, the basic hypothesis that the alkene has to substitute HMPT in the complex (equation 28) requires also that the reaction be autoinhibited because of the build-up of HMPT concentration with the progress of the reaction.

$$MoO(O_2)_2HMPT + Alkene \rightleftharpoons MoO(O_2)_2(Alkene) + HMPT \qquad (28)$$

This phenomenon was not observed. The inhibition effect of initially added HMPT, which has been well documented[56,83] may be simply related to the build-up of the complex $MO_5(HMPT)_2$, which is less reactive than its precursor. This has been demonstrated to be the case in one example.[56] Another requirement would be that optically active monodentate ligands should not be effective in asymmetric oxidation, whereas it was observed that they do yield

Scheme 15

optically active epoxides[134] although the e.e. are low (*ca* 8%). Bidentate ligands, however, are better in inducing enantioselectivity[38b,c] (up to 35%), as expected.

The problem of the intermediate structure is even more difficult to assess. Theoretical calculations would preclude the intervention of a peroxometallacycle intermediate at least as far as Mo(VI) compounds are concerned.[135] Indeed, in this case the cyclic structure of a substrate–oxidant complex formed with Mo(VI)-oxo-diperoxo complexes seems to be at much higher energy than that of species resulting from the coordination of the nucleophilic alkene to the electrophilic metal center in a sort of acid–base interaction. Moreover, the chemical behavior of known peroxometallacycles indicates that their preferred mode of decomposition is, as already mentioned, the formation of carbonyl compounds.

In the much less studied case of V(V) peroxo complexes, it has more recently been shown[133] that the coordination to the metal peroxo complex gives rise to a 'competitive' intermediate which should therefore be along the main reaction path. Also in this case, however, no specific evidence on the structure of the intermediate has been obtained.

In conclusion, even though in some special cases an intermediate may be involved in the oxidation with transition metal peroxo complexes, it is not necessarily always along the reaction coordinate and in even fewer cases, if any, they are of the peroxometallacycle type. In particular, the oxidation of n-nucleophiles would suggest the absence of such an intermediate or at least any major role for it.

In fact, in the oxidation of thioethers, the peroxometallacycle, which has to be formulated as arising from the insertion of a sulfur atom in the metal peroxo oxygen bond, is improbable since it should be a sulfurane-like derivative. In the oxidation of tertiary amines no intermediates other than those resulting from an acid–base equilibrium may be envisaged because of the lack of electrons available for the formation of bonds. Finally, in the oxidation of allylic alcohols, which are known to coordinate to the metal through the hydroxo group, a concomitant coordination of the double bond is geometrically difficult.

We shall close this section as we opened it. There is no physical justification to force every reaction within the same rigid mechanism for the very obvious reason that the reaction coordinate is, by definition, the lowest energy path connecting reagents to products. Hence there is enough evidence that palladium, platinum and rhodium peroxo complexes give rise with appropriate substrates to peroxometallacycle intermediates before yielding oxidized products, and that there are no alternatives to a simple nucleophilic attack on the oxygen–oxygen bond of the peroxo to rationalize the oxidation of n-nucleophiles with peroxides in catalyzed or uncatalyzed reaction. Moreover, many of the oxygen transfer reactions to π-nucleophiles, i.e. the epoxidation reaction, can be easily accommodated in an analogous scheme. Whether there is a cross-over of the mechanisms with particular transition metal peroxo complexes and, perhaps, with specific ligands is difficult at this stage to state, even if it cannot be ruled out.

In conclusion, whereas a series of problems are encountered in accommodating the experimental results in a mechanistic pathway which involves a two-step reaction where the intermediate is a peroxometallacycle, no major objections can be found for the simple, bimolecular electrophilic mechanism.

6 SOME SELECTED SYNTHETIC APPLICATIONS

The epoxidation of simple alkenes has been widely used as a test reaction for establishing the effectiveness of peroxometal complexes either in stoichiometric or in catalytic processes. However, from a synthetic point of view, the alkene epoxidation is only one of the several reactions which can be carried out by these oxidants.

Processes employing either hydrogen peroxide or alkyl hydroperoxides together with a catalyst for alkene epoxidation are of particular interest from an industrial point of view, since the previously mentioned drawbacks of peracids become dramatic because of the large amounts of material used. In contrast, in the laboratory preparations, where usually small amounts of reagents are used, such drawbacks are less important. However, in the field of organic synthesis, the utilization of peroxides in metal-catalyzed reactions may be considered in competition with, and often is to be considered superior to, that of peracids when selectivity problems must be solved.[15] In fact, by acting on the coordination sphere of the metal, the reactivity of peroxometal complexes is much more easily tunable than that of peracids.[31,105,136,137] This is immediately clear when one looks at the epoxidation of allylic alcohols.[29,73]

In Table 3, the results of the epoxidation of typical allylic alcohols, i.e. geraniol and cyclohexen-3-ol, with MCPBA, V(V)–ROOH and Mo(VI)–ROOH are shown, together with the relative rates.[138]

These data illustrate not only that different regioselectivities can be achieved by the metal-catalyzed systems compared with peracids but also indicate that the choice of the appropriate catalyst is crucial. In fact, the rate acceleration observed with the vanadium system is very large and may be rationalized as arising from the formation of alkoxovanadium complexes by addition of the substrate to the metal.[27,139]

The ability of vanadium derivatives to form such complexes has already been mentioned. Such an ability is less important in the case of molybdenum. As a consequence, the conclusion may be drawn that in the epoxidation of simple alkenes molybdenum catalysts are superior reagents,[15,70,140] whereas in the epoxidation of allylic alcohols the vanadium catalysts are to be preferred because of their higher reactivity.[28]

Table 3. Epoxidation of typical allylic alcohols with various oxidants

Substrate	Oxidant	Relative rate[a]	Product(s)
[structure: allylic alcohol, OH]	MCPBA	0.5	6,7-Epoxide
As above	Mo(VI)-TBHP	45	2,3-Epoxide
As above	V(V)-TBHP	100	2,3-Epoxide
[structure: cyclohex-2-enol, OH]	MCPBA	0.55	*syn*:*anti* Epoxide 92:8
As above	Mo(VI)-TBHP	4.5	*syn*:*anti* Epoxide 98:2
As above	V(V)-TBHP	> 200	*syn*:*anti* Epoxide 98:2

[a]Relative rates with respect to the epoxidation of cyclohexene by MCPBA; see ref. 138.

A comparison between peracids and metal-catalyzed systems can also be made also in the oxidation of thioethers. Here, no major reactivity problems are present because of the large nucleophilicity of the sulfur atom, which allows its smooth oxidation by several electrophilic reagents.[14a,141–143] However, a selectivity problem exists.

As shown in Scheme 16, the oxidation of thioethers may proceed further, leading eventually to sulfones. Moreover, the same reagents which carry out the first step of the oxidation are also able to oxidize sulfoxides to sulfones.[14a,141–143] However, by taking advantage of the fact that, in general, thioethers are more reactive than the corresponding sulfoxides, high yields of sulfoxides can be obtained without using an excess of the substrate over the oxidant if such an oxidant is a peroxometal complex whose selectivity appears to be higher than that of peracids.

$$\text{R-S-R}' + \text{Oxidant} \longrightarrow \text{R-S(O)-R}'$$

$$\text{R-S(O)-R}' + \text{Oxidant} \longrightarrow \text{R-S(O)}_2\text{-R}'$$

Oxidant = peracids, XO_2H/metals

Scheme 16

Also in stereoselective oxidations, metal-catalyzed systems appear to be more effective than peracids. This is illustrated by the well known results of the epoxidation of allylic and homoallylic alcohols (see Table 4).

Table 4. Stereochemistry of epoxidation of allylic and homoallylic alcohols

Substrate	Oxidant	*erythro*-Epoxide	*threo*-Epoxide	Ref.
		[structure: CH_3, OH, Bu^n, H, O]	[structure: CH_3, OH, H, Bu^n, O]	139
[structure: CH_3, Bu^n, OH]	MCPBA	41	59	
	Mo(VI)-TBHP	16	84	
	V(V)-TBHP	2	98	
		[structure: OH, O]	[structure: OH, O]	144
[structure: HO]	MCPBA	55	45	
	V(V)-TBHP	66	33	

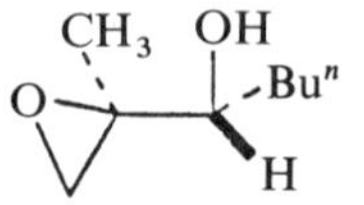

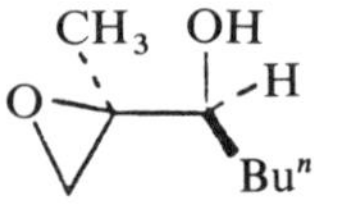

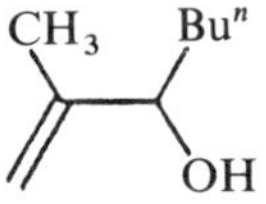

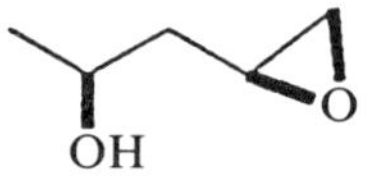

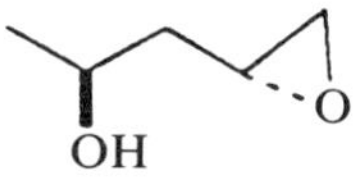

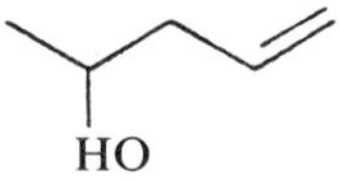

There is no doubt that the superiority of metal-catalyzed oxidations by peroxides manifests itself in the most straightforward way in asymmetric oxidations. In this field spectacular results have been obtained by using the titanium-hydroperoxide reagent developed by Sharpless in 1980 for allylic alcohol asymmetric epoxidations,[73] where e.e. > 98% are routinely obtained with a large variety of substrates.[145] Later, modifications to the original Sharpless reagent capable of oxidizing prochiral thioethers with good enantioselectivity were obtained[74,75] (see Scheme 5 for details of the composition of the oxidizing systems). Some representative examples[76–79] of such oxidations are reported in Table 5.

The nature of the oxidizing species in solution is still the subject of careful investigations.[78,148,149] Even though such an aspect is outside the scope of this chapter, a general consideration is in order. In particular, it is observed that the three systems display different behaviors toward the same substrates.[78]

In this field a general understanding of the origin of the asymmetric induction is not yet fully available.

7 CONCLUSIONS

The aim of this chapter was to give a general overview of the mechanism(s) of the transition metal-catalyzed epoxidation and heteroatom oxidation, i.e. those

Table 5. Titanium-catalyzed asymmetric oxidations of thioethers

Thioether	Diastereomeric ratio	e.e. (%)	Ref.
Tolyl methyl thioether	—	>85	74, 75
Methyl *n*-octyl thioether	—	94	78
Methyl 2-naphthyl thioether	—	90	76
Phenyl cyclopropyl thioether	—	95	146
2-Ph-2-CH_3-1,3-dithiane (S, S; Ph, CH_3)	100:0	78	77
2-H-2-COOBz-1,3-dithiolane (S, S; H, COOBz)	92:8	>98[a]	147
1,1′-Binaphthalene 2,2′-Dimethyl thioether	57:43	>98	79

[a]After crystallization

reactions where one oxygen atom is transferred from the peroxide to the substrate. This, of course, involved a description and discussion of the chemistry of metal peroxides, which under catalytic conditions are the true oxidants in these systems.

In the limited space available, which did not allow an exhaustive examination of all the reactions studied and of all the mechanisms proposed, we have made an attempt to review critically the evidence available for the two most widely accepted mechanisms of the critical step of the reaction, i.e. the oxygen-transfer process.

We reached the conclusion that the simple description of a nucleophilic attack of the substrates on the electrophilic peroxide, probably along the oxygen—oxygen bond, is sufficient to explain most of the results and that this mechanism is preferred to more complex ones unless there is specific evidence for their occurrence. Nature does not necessarily choose the simplest route, but often it does.

8 ABBREVIATIONS

acac	Acetylacetonate
DEC	Dichloroethane
DET	(+)-Diethyl tartrate
Dipic	Pyridine-2,6-dicarboxylate
(*S*)-DML	(*S*)-*N*,*N*-Dimethyllactamide
HMPT	Hexamethylphosphoric triamide
MCPBA	*meta*-Chloroperoxybenzoic acid
Ox	Oxalate
Pic	Pyridine-2-carboxylate
Pico	Picolinate-*N*-oxido anion
Py	Pyridine
TBHP	*tert*-Butyl hydroperoxide

9 REFERENCES

1. S. Patai (Ed.), *The Chemistry of Peroxides*, Wiley–Interscience, Chichester (1983).
2. A. G. Davies, *Organic Peroxides*, Butterworths, London (1961).
3. D. Swern, in *Organic Peroxides*, Vol. II (D. Swern, Ed.), Wiley–Interscience, New York (1971), p. 355.
4. R. A. Sheldon, *J. Mol. Catal.*, **7**, 107 (1980).
5. (a) N. A. Milas and S. Saussman, *J. Am. Chem. Soc.*, **58**, 1302 (1936); (b) N. A. Milas and S. Saussman, *J. Am. Chem. Soc.*, **59**, 342 (1937); (c) N. A. Milas, S. Saussman and H. S. Mason, *J. Am. Chem. Soc.*, **61**, 1844 (1939); (d) N. A. Milas and L. S. Maloney, *J. Am. Chem. Soc.*, **62**, 1841 (1940).
6. (a) W. F. Brill and N. Indictor, *J. Org. Chem.*, **29**, 710 (1964); (b) N. Indictor and W. F. Brill, *J. Org. Chem.*, **30**, 2074 (1965).
7. E. S. Gould, R. R. Hiatt and K. C. Irwin, *J. Am. Chem. Soc.*, **90**, 4573 (1968).
8. J. Kollar, to Halcon International, *US Pat.*, 3 350 422 (1967); 3 351 635 (1967); 3 507 809 (1970); 3 625 981 (1971).
9. (a) Shell Oil, *Br. Pat.*, 1 249 079 (1971); (b) H. P. Wulff, to Shell Oil, *US Pat.*, 3 923 843 (1975).
10. U. Romano, A. Esposito, F. Maspero, C. Neri and M. G. Clerici, *Chim. Ind. (Milan)*, **72**, 610 (1990), and references cited therein.
11. R. Curci and J. O. Edwards, in *Organic Peroxides*, Vol. I (D. Swern, Ed.), Wiley–Interscience, New York (1970), Chap. 4.
12. G. A. Olah, *Acc. Chem. Res.*, **20**, 422 (1987).
13. M. Pralus, J. C. Lecoq and J. P. Schirmann, in *Fundamental Research in Homogenous Catalysis*, Vol 3 (M. Tsutsui, Ed.), Plenum Press, New York (1979) p. 327.
14. (a) R. A. Sheldon, in *The Chemistry of Peroxides* (S. Patai, Ed.), Wiley–Interscience, Chichester (1983), Chap. 6; (b) H. Mimoun, in *The Chemistry of Peroxides* (S. Patai, Ed.), Wiley, Chichester (1983), Chap. 15.
15. R. A. Sheldon and J. K. Kochi, *Metal Catalyzed Oxidations of Organic Compounds*, Academic Press, New York (1981).
16. F. Di Furia and G. Modena, *Rev. Chem. Intermed.*, **6**, 51 (1985).

17. R. A. Sheldon, *Bull. Soc. Chim. Belg.*, **94**, 651 (1985).
18. H. Mimoun, I. Sérée de Roch and L. Sajus, *Bull. Soc. Chim. Fr.*, 1481 (1969).
19. L. Vaska, *Acc. Chem. Res.* **9**, 175 (1976).
20. B. Bosnich, W. G. Jackson, S. T. D. Lo and J. W. McLaren, *Inorg. Chem.*, **13**, 2605 (1974).
21. L. Vaska, *Science*, **140**, 809 (1963).
22. See, e.g., A. F. Ghiron and R. C. Thompson, *Inorg. Chem.*, **29**, 4457 (1990).
23. J. Valentine, D. Valentine and J. P. Collman, *Inorg. Chem.*, **10**, 219 (1971).
24. F. Di Furia and G. Modena, *Pure Appl. Chem.*, **54**, 1853 (1982).
25. H. Mimoun, *Angew. Chem., Int. Ed. Engl.*, **21**, 734 (1982).
26. A. O. Chong and K. B. Sharpless, *J. Org. Chem.*, **42**, 1587 (1977).
27. K. B. Sharpless and T. R. Verhoeven, *Aldrichim. Acta.*, **12**, 63 (1979).
28. M. G. Finn and K. B. Sharpless, in *Asymmetric Synthesis*, Vol. 5 (J. D. Morrison, Ed.), Academic Press, New York (1985), Chap. 8.
29. K. B. Sharpless, C. H. Behrens, T. Katsuki, A. W. M. Lee, V. S. Martin, M. Takatani, S. M. Viti, F. J. Walker and S. S. Woodard, *Pure Appl. Chem.*, **55**, 589 (1983).
30. See, e.g., P. R. Ortiz de Montellano, *Cytocrome P-450, Structure, Mechanism and Biochemistry*, Plenum Press, New York (1986).
31. S. Campestrini, F. Di Furia, G. Modena and F. Novello, in *Proceedings of the 4th International Symposium on Dioxygen Activation and Homogeneous Catalytic Oxidations*, (L. I. Símandi, Ed.), Elsevier, Amsterdam (1991), p. 375.
32. D. Schwarzenbach, *Helv. Chim. Acta*, **55**, 2990 (1972).
33. H. Mimoun, L. Saussine, E. Daire, M. Postel, J. Fischer and R. Weiss, *J. Am. Chem. Soc.*, **105**, 3101 (1983).
34. M. Bonchio, V. Conte, F. Di Furia, G. Modena and G. Valle, unpublished results.
35. H. Mimoun, P. Chaumette, M. Mignard, L. Saussine, J. Fischer and R. Weiss, *Nouv. J. Chim.*, **7**, 467 (1983).
36. D. Begin, F. W. B. Einstein and J. Field, *Inorg. Chem.*, **14**, 1785 (1975).
37. J. M. Le Carpentier, R. Schlupp and R. Weiss, *Acta Crystallogr., Sect. B*, **28**, 1278 (1972).
38. (a) W. Winter, C. Mark and V. Schurig, *Inorg. Chem.*, **19**, 2045 (1980); (b) H. B. Kagan, H. Mimoun, C. Mark and V. Schurig, *Angew. Chem., Int. Ed. Engl.*, **19**, 485 (1979); (c) V. Schurig, K. Hintzer, U. Leyrer, C. Mark, P. Pitchen and H. B. Kagan, *J. Organomet. Chem.*, **370**, 81 (1989).
39. O. Bortolini, S. Campestrini, F. Di Furia, G. Modena and G. Valle, *J. Org. Chem.*, **52**, 5467 (1987).
40. S. E. Jacobson, R. Tang and F. Mares, *Inorg. Chem.*, **17**, 3055 (1978).
41. C. Djordjevic, K. J. Covert and E. Sinn, *Inorg. Chim. Acta*, **101**, L37 (1985).
42. P. Chaumette, H. Mimoun, L. Saussine, J. Fischer and A. Mitschler, *J. Organomet. Chem.*, **250**, 291 (1983).
43. G. Amato, A. Arcoria, F. P. Ballisteri, G. A. Tomaselli, O. Bortolini, V. Conte, F. Di Furia, G. Modena and G. Valle, *J. Mol. Catal.*, **37**, 165 (1986).
44. R. Curci, G. Fusco, O. Sciacovelli and L. Troisi, *J. Mol. Catal.*, **32**, 251 (1985).
45. (a) R. Stomberg, *Ark. Kemi*, **22**, 2178 (1964); (b) G. W. J. Fleet and W. Little, *Tetrahedron Lett.*, **42**, 3749 (1977).
46. M. Postel, C. Brevard, H. Arzoumanian and J. G. Riess, *J. Am. Chem. Soc.*, **105**, 4922 (1983).
47. V. Conte, F. Di Furia, G. Modena and O. Bortolini, *J. Org. Chem.*, **53**, 4581 (1988).
48. M. Camporeale, L. Cassidei, R. Mello, O. Sciacovelli, L. Troisi and R. Curci, in *The Role of Oxygen in Chemistry and Biochemistry (Studies in Organic Chemistry, Vol. 33)* (W. Ando and Y. Moro-Oka, Eds.), Elsevier, Amsterdam (1988), p. 201.

49. R. Curci, F. Di Furia and G. Modena, in *Fundamental Research in Homogenous Catalysis*, Vol. 2 (Y. Ishii and M. Tsutsui, Eds), Plenum Press, New York (1978), p. 255.
50. (a) O. Bortolini, F. Di Furia, P. Scrimin and G. Modena, *J. Mol. Catal.*, **7**, 59 (1980); (b) O. Bortolini, F. Di Furia, P. Scrimin and G. Modena, *J. Mol. Catal.*, **9**, 323 (1980).
51. R. Curci, F. Di Furia, R. Testi and G. Modena, *J. Chem. Soc., Perkin Trans. 2*, 752 (1974).
52. F. Di Furia, G. Modena, R. Curci, S. J. Bachofer, J. O. Edwards and M. Pomerantz, *J. Mol. Catal.*, **14**, 219 (1982).
53. (a) P. L. Kooijman and W. L. Ghijsen, *Recl. Trav. Chim. Pays-Bas*, **66**, 205 (1947); (b) M. C. V. Sauer and J. O. Edwards, *J. Phys. Chem.*, **75**, 3377 (1971).
54. (a) H. Menzel, *Z. Phys. Chem.*, **105**, 401 (1923); (b) J. O. Edwards, *J. Am. Chem. Soc.*, **75**, 6154 (1953); (c) D. M. Kern, *J. Am. Chem. Soc.*, **77**, 5458 (1955); (d) P. J. Antikainen, *Acta Chem. Scand.*, **10**, 756 (1956).
55. S. Cenci, F. Di Furia, G. Modena, R. Curci and J. O. Edwards, *J. Chem. Soc., Perkin Trans. 2*, 979 (1978).
56. O. Bortolini, V Conte, F. Di Furia and G. Modena, *J. Mol. Catal.*, **19**, 331 (1983).
57. O. Bortolini, F. Di Furia and G. Modena, *J. Mol. Catal.*, **16**, 69 (1982).
58. A. Arcoria, F. P. Ballistreri, G. A. Tomaselli, F. Di Furia and G. Modena, *J. Mol. Catal.*, **18**, 177 (1983).
59. A. J. Everett and G. J. Minkoff, *Trans. Faraday Soc.*, **49**, 410 (1953).
60. F. A. Cotton and G. Wilkinson, *Advanced Inorganic Chemistry*, Wiley, New York (1988), p. 45.
61. W. Adam, R. Curci and J. O. Edwards, *Acc. Chem. Res.*, **22**, 205 (1989), and references cited therein.
62. J. March, *Advanced Organic Chemistry*, Wiley-Interscience, New York (1985).
63. D. Cremer, in *The Chemistry of Peroxides* (S. Patai, Ed.), Wiley, Chichester (1983), Chap. 1.
64. B. M. Gimarc, *Acc. Chem. Res.*, **7**, 384 (1974).
65. S. S. Woodard, M. G. Finn and K. B. Sharpless, *J. Am. Chem. Soc.*, **113**, 106 (1991).
66. R. C. Michaelson, R. E. Palermo and K. B. Sharpless, *J. Am. Chem. Soc.*, **99**, 1990 (1977).
67. S.-I. Yamada, T. Mashiko and S. Terashima, *J. Am. Chem. Soc.*, **99**, 1988 (1977).
68. F. Di Furia, G. Modena and R. Curci, *Tetrahedron Lett.*, **50**, 4637 (1976).
69. R. Curci, F. Di Furia, J. O. Edwards and G. Modena, *Chim. Ind. (Milan)*, **60**, 41 (1978).
70. K. Tani, M. Hanafusa and S. Otsuka, *Tetrahedron Lett.*, **32**, 3017 (1979).
71. O. Bortolini, F. Di Furia and G. Modena, unpublished results.
72. F. Di Furia and G. Modena, in *Fundamental Research in Homogeneous Catalysis*, Vol. 3 (M. Tsutsui, Ed.), Plenum Press, New York (1979), p. 433.
73. T. Katsuki and K. B. Sharpless, *J. Am. Chem. Soc.*, **102**, 5974 (1980).
74. F. Di Furia, G. Modena and R. Seraglia, *Synthesis*, 325 (1984).
75. P. Pitchen and H. B. Kagan, *Tetrahedron Lett.*, **25**, 1049 (1984).
76. P. Pitchen, E. Duñach, M. N. Deshmukh and H. B. Kagan, *J. Am. Chem. Soc.*, **106**, 8188 (1984).
77. O. Samuel, B. Ronan and H. B. Kagan, *J. Organomet. Chem.*, **370**, 43 (1989).
78. V. Conte, F. Di Furia, G. Licini, G. Modena and G. Sbampato, in *Proceedings of the 4th International Symposium on Dioxygen Activation and Homogeneous Catalytic Oxidations* (L. I. Símandi, Ed.), Elsevier, Amsterdam (1991), p. 385.
79. F. Di Furia, G. Licini, G. Modena and G. Valle, *Bull. Soc. Chim. Fr.*, **127**, 734 (1990).

80. H. Mimoun, *J. Mol. Catal.*, **7**, 1 (1980).
81. O. Bortolini, C. Campello, F. Di Furia and G. Modena, *J. Mol. Catal.*, **14**, 63 (1982).
82. A. Arcoria, F. P. Ballisteri, G. A. Tomaselli, F. Di Furia and G. Modena, *J. Mol. Catal.*, **24**, 189 (1984).
83. H. Mimoun, I. Sèrée de Roch and L. Sajus, *Tetrahedron*, **26**, 37 (1970).
84. O. Bortolini, V. Conte, F. Di Furia, G. Modena and R. Seraglia, *Atual. Fis. Quim. Org.*, 47 (1984).
85. F. Di Furia and G. Modena, *Recl. Trav. Chim. Pays-Bas*, **98**, 181 (1979).
86. J. Sobczak and J. J. Ziolkowski, *Inorg. Chim. Acta*, **19**, 15 (1976).
87. M. N. Sheng and J. G. Zajecek, *J. Org. Chem.*, **35**, 1839 (1970).
88. J. Sobczak and J. J. Ziolkowski, *J. Mol. Catal*, **13**, 11 (1981).
89. K. A. Jørgensen, *Chem. Rev.*, **89**, 431 (1989).
90. R. D. Bach, G. J. Wolber and B. A. Coddens, *J. Am. Chem. Soc.*, **106**, 6098 (1984).
91. O. Bortolini, F. Di Furia, G. Modena, C. Scardellato and P. Scrimin, *J. Mol. Catal.*, **11**, 107 (1981).
92. R. A. Sheldon, *Recl. Trav. Chim. Pays-Bas*, **92**, 253 (1973).
93. O. Bortolini, V. Conte, F. Di Furia and G. Modena, in *The Role of Oxygen in Chemistry and Biochemistry (Studies in Organic Chemistry, Vol. 33)* (W. Ando and Y. Moro-Oka, Eds), Elsevier, Amsterdam (1988), p. 301.
94. O. Bortolini, L. Bragante, F. Di Furia and G. Modena, *Can. J. Chem.*, **64**, 1189 (1986).
95. O. Bortolini, F. Di Furia, G. Modena and R. Seraglia, *J. Org. Chem.*, **50**, 2688 (1985).
96. O. Bortolini, V. Conte, F. Di Furia and G. Modena, *J. Org. Chem.*, **51**, 2661 (1986).
97. C. Venturello, E. Alneri and M. Ricci, *J. Org. Chem.*, **48**, 3831 (1983).
98. C. Venturello and R. D'Aloisio, *J. Org. Chem.*, **53**, 1553 (1988).
99. F. P. Ballisteri, S. Failla and G. A. Tomaselli, *J. Org. Chem.*, **53**, 830 (1988).
100. Y. Ishii and Y. Sakata, *J. Org. Chem.*, **55**, 5545 (1990).
101. S. Campestrini, F. Di Furia, G. Modena and O. Bortolini, *J. Org. Chem.*, **55**, 3658 (1990).
102. K. A. Jørgensen, R. A. Wheeler and R. Hoffmann, *J. Am. Chem. Soc.*, **109**, 3240 (1987).
103. H. Himoun, *Isr. J. Chem.*, **23**, 451 (1983).
104. V. Conte, F. Di Furia and G. Modena, *J. Org. Chem.*, **53**, 1665 (1988).
105. M. Bonchio, V. Conte, F. Di Furia, G. Modena, C. Padovani and M. Sivak, *Res. Chem. Intermed.*, **12**, 111 (1989).
106. R. Curci and G. Modena, *Gazz. Chim. Ital.*, **94**, 1257 (1964).
107. R. Curci and F. Di Furia, *Tetrahedron Lett.*, 4085 (1974).
108. R. Curci and F. Di Furia, *Int. J. Chem. Kinet.*, **7**, 341 (1975).
109. R. Curci, F. Di Furia and M. Meneghin, *Gazz. Chim. Ital.*, **108**, 123 (1978).
110. P. D. Bartlett, *Rec. Chem. Prog.*, **11**, 47 (1950).
111. R. G. Pews, *J. Am. Chem. Soc.*, **89**, 5605 (1967).
112. D. R. Campbell, J. O. Edwards, J. Maclachlan and K. Polgar, *J. Am. Chem. Soc.*, **80**, 5308 (1958).
113. K. D. Bingham, G. D. Meakins and G. H. Whitham, *Chem. Commun.*, 445 (1966).
114. H. Kwart and D. M. Hoffman, *J. Org. Chem.*, **31**, 419 (1966).
115. H. Kwart, P. S. Starcher and S. W. Tinsley, *Chem. Commun.*, 335 (1966).
116. R. Curci, R. Di Prete, J. O. Edwards and G. Modena, in *Hydrogen-Bonded Solvent Systems* (A. K. Covington and P. Jones, Eds), Taylor and Francis, London (1968), p. 303.
117. M. A. P. Dankleff, R. Curci, J. O. Edwards and H.-Y. Pyun, *J. Am. Chem. Soc.*, **90**, 3209 (1968).

118. P. A. Giguère and A. W. Olmos, *Can. J. Chem.*, **30**, 821 (1952).
119. C. G. Overberger and R. W. Cummins, *J. Am. Chem. Soc.*, **75**, 4783 (1953).
120. D. Swern, L. P. Witnauer, C. R. Eddy and W. E. Parker, *J. Am. Chem. Soc.*, **77**, 5537 (1955).
121. R. Curci, R. A. Di Prete, J. O. Edwards and G. Modena, *J. Org. Chem.*, **35**, 740 (1970), and references cited therein.
122. (a) G. Modena, *Gazz. Chim. Ital.*, **89**, 834 (1959); (b) A. Cerniani, G. Modena and P. E. Todesco, *Gazz. Chim. Ital.*, **90**, 3 (1960); (c) G. M. Gasperini, G. Modena and P. E. Todesco, *Gazz. Chim. Ital.*, **90**, 12 (1960).
123. (a) A. Mayr, F. Montanari and F. Tramontini, *Gazz. Chim. Ital.*, **90**, 739 (1960); U. Folli, D. Iarossi, F. Montanari and G. Torre, *J. Chem. Soc. C*, 1317 (1968).
124. R. D. Bach, A. L. Owensby, C. Gonzales and H. B. Schlegel, *J. Am. Chem. Soc.*, **113**, 2338 (1991).
125. M. Bonchio, V. Conte, F. Di Furia and G. Modena, *J. Org. Chem.*, **54**, 4368 (1989).
126. J. P. Collman, L. S. Hegedus, J. R. Norton and R. G. Finke, *Principles and Application of Organotransition Metal Chemistry*, University Science, Mill Valley (1987), p. 477.
127. R. H. Grubbs and A. Miyashita, in *Fundamental Research in Homogeneous Catalysis*, Vol. 2 (Y. Ishii and M. Tsutsui, Eds), Plenum Press, New York (1978), p. 207.
128. R. Ugo, *Engelhard Ind. Tech. Bull.*, **11**, 45 (1971).
129. R. A. Sheldon and J. A. Van Doorn, *J. Organomet. Chem.*, **94**, 115 (1975).
130. H. Mimoun, M. M. Perez Machirant and I. Sèrée de Roch, *J. Am. Chem. Soc.*, **100**, 5437 (1978).
131. O. Bortolini, F. Di Furia, G. Modena and R. Seraglia, *J. Mol. Catal.*, **22**, 313 (1984).
132. See, e.g., V. A. Yablokov, *Russ. Chem. Rev.*, **49**, 833 (1980).
133. H. Mimoun, M. Mignard, P. Brechot and L. Saussine, *J. Am. Chem. Soc.*, **108**, 3711 (1986).
134. O. Bortolini, F. Di Furia, G. Modena and A. Schionato, *J. Mol. Catal.*, **35**, 47 (1986).
135. K. A. Jørgensen and R. Hoffman, *Acta Chem. Scand, Ser. B.*, **40**, 411 (1986).
136. S. E. Jacobson, D. A. Mucigrosso and F. Mares, *J. Org. Chem.*, **44**, 921 (1979).
137. M. Bonchio, V. Conte, F. Coppa, F. Di Furia and G. Modena, in *Proceedings of the 4th International Symposium on Dioxygen Activation and Homogeneous Catalytic Oxidations* (L. I. Símandi, Ed.), Elsevier, Amsterdam (1991), p. 497.
138. K. B. Sharpless and R. C. Michaelson, *J. Am. Chem. Soc.*, **95**, 6136 (1973).
139. B. E. Rossiter, T. R. Verhoeven and K. B. Sharpless, *Tetrahedron Lett.*, 4733 (1979).
140. O. Bortolini, F. Di Furia and G. Modena, *J. Mol. Catal.*, **19**, 319 (1983).
141. G. A. Tolstikov, V. P. Yur'ev and U. M. Dzhemilev, *Russ. Chem. Rev.*, **44**, 319 (1975).
142. M. Madesclaire, *Tetrahedron*, **42**, 5459 (1986).
143. S. Patai, Z. Rappoport and C. Stirling (Eds), *The Chemistry of Sulphones and Sulphoxides*, Wiley-Interscience, Chichester (1988).
144. P. A. Bartlett and K. K. Jernsted, *J. Am. Chem. Soc.*, **99**, 4829 (1977).
145. M. G. Finn and K. B. Sharpless, in *Asymmetric Synthesis*, Vol. 5, (J. D. Morrison, Ed.), Academic Press, New York (1985), Chap. 7.
146. E. Duñach and H. B. Kagan, *Nouv. J. Chim.*, **89**, 1 (1985).
147. F. Di Furia, G. Licini and G. Modena, *Gazz. Chim. Ital.*, **120**, 165 (1990).
148. M. G. Finn and K. B. Sharpless, *J. Am. Chem. Soc.*, **113**, 113 (1991).
149. E. J. Corey, *J. Org. Chem.*, **55**, 1693 (1990).

12 Peroxides from Photosensitized Oxidation of Heteroatom Compounds

TAKESHI AKASAKA and WATARU ANDO
Department of Chemistry, University of Tsukuba, Tsukuba, Ibaraki 305, Japan

In the last 20 years, increasing interest has been shown in peroxides in which the oxygen atom bonds to the heteroatom in connection with chemical and biological actions.[1–4] Photooxidation of heteroatom compounds in living systems, which plays an important role, may produce peroxides by the action of

Organic Peroxides. Edited by W. Ando

active oxygen species, resulting in significant deactivation of several enzymes, for example phosphoglucomutase and chymotrypsin.[4,5] Such photodegradation of substrates by oxygen, light, and pigments such as flavins and porphyrins, in biological systems is called 'photodynamic action,' i.e. dye-sensitized photoinduced damage to living species.[1–4]

Since some of the naturally occurring sulfur compounds so far isolated have S—O bonds, these S-oxidation products can play important roles in biochemical reactions.[6,7] These S—O compounds include sulfoxides such as sulforaphene,[8] biotin 1-oxide,[9] and *S*-methylcysteine sulfoxide,[10] as well as thiosulfinates[6,7] such as β-lipoic acid[11] and allicin.[6] For instance, methionine undergoes photosensitized oxidation to give the sulfoxide.[12–17] The degradation of methionine is responsible for the loss of activity of several important enzymes which are damaged in photodynamic action.[4]

Much attention has also been focused on the structures and reactivities of peroxidic intermediates derived from the photosensitized oxidation of other heteroatom compounds such as organoselenium,[18] organonitrogen[4] and organophosphorus compounds[19] which also have some significance in connection with biological action.

The oxidation of organosilicon compounds such as disilanes and polysilanes by molecular oxygen has been the subject of continued interest.[20] When some disilanes and polysilanes are heated or photo-irradiated in air, rapid oxidation takes place to give insoluble products containing Si—O—Si bonds, indicative of loss of utility of polysilanes. The role of the oxygen molecule in the oxidation and oxidative degradation of polysilanes, which have potential technological usefulness, is significant.

Photosensitized oxidation can be simply described by equation 1, in which substrate (R) gives an initial oxygen adduct (RO_2).[5] Several kinds of dyes, such

$$R + O_2 \xrightarrow{h\nu/\text{sens}} RO_2 \tag{1}$$

as methylene blue, Rose Bengal, and so forth, are commonly used as sensitizers having relatively low triplet energies (*ca* 35–40 kcal mol^{-1}). The two major fates of a triplet dye sensitizer ($^3D^*$) in a photosensitized oxidation (Types I and II processes)[21] are summarized in equations 2 and 3. In the Type I process, a substrate is activated either directly, by a sensitizer triplet, or indirectly to give a free radical species which reacts with molecular oxygen. The Type II process involves singlet oxygen ($^1O_2$) which is formed by energy transfer from the triplet sensitizer to the ground state oxygen. Recently, however, it has been suggested that photosensitized oxidations via a non-$^1O_2$ mechanism involve an electron-transfer process producing the substrate cation radical and/or superoxide anion radical.[22–29] In this chapter, except where stated, photosensitized oxidation simply means oxidation by the Type II process in which $^1O_2$ is the active oxidizing species.

$$D \xrightarrow{h\nu} {}^1D^* \longrightarrow {}^3D^*$$

$${}^3D^* \xrightarrow[(1)]{{}^3O_2} {}^1O_2 + D \quad ; \quad {}^3D^* \xrightarrow[(1)]{{}^3O_2} O_2^{\cdot -} + D^{\cdot +}$$

$${}^1O_2 \xrightarrow{R} RO_2 \quad ; \quad {}^1O_2 \longrightarrow {}^3O_2 \quad ; \quad {}^1O_2 \xrightarrow{R} R^{\cdot +} + O_2^{\cdot -}$$

$${}^3D^* \xrightarrow[(2)]{R} D^{+} + R^{\cdot -} \quad ; \quad R^{\cdot -} \xrightarrow{{}^3O_2} O_2^{\cdot -} \qquad (2)$$

$${}^3D^* \xrightarrow[(2)]{R} D^{\cdot -} + R^{\cdot +} \quad ; \quad D^{\cdot -} \xrightarrow{{}^3O_2} O_2^{\cdot -} \quad ; \quad R^{\cdot +} + O_2^{\cdot -} \longrightarrow RO_2 \text{ (or } {}^1O_2\text{)}$$

$${}^3D^* \xrightarrow[(3)]{D} D^{\cdot +} + D^{\cdot -}$$

$$\begin{aligned} D^{\cdot +} + R &\longrightarrow D + R^{\cdot +} \\ D^{\cdot -} + {}^3O_2 &\longrightarrow D + O_2^{\cdot -} \end{aligned} \qquad (3)$$

This chapter describes the photosensitized oxidation of heteroatom-containing compounds such as organosulfur, organoselenium, organonitrogen and organophosphorus compound in which the reaction center involves the heteroatom. This chapter also briefly surveys photooxidation of organosilicon and organogermanium compounds.

1 PEROXIDIC INTERMEDIATES IN PHOTOSENSITIZED OXIDATION OF ORGANOSULFUR AND ORGANOSELENIUM COMPOUNDS

Photosensitized oxidation, induced by singlet oxygen, is of particular interest because methionine is one of the amino acids attacked most rapidly in photodynamic action (the destructive action of dye sensitizers, light, and oxygen on organisms.[12,16,17] The formation of hydrogen peroxide and dehydromethionine in the photosensitized oxidation of methionine has been rationalized in terms of respectively inter- and intramolecular displacement at sulfur in the intermediate persulfoxide.[17] The photosensitized oxidation of methionine (**1**) depends on the pH value, with one mole of oxygen consumed per mole of methionine at high pH; at pH of 5 or lower, one mole of oxygen oxidizes two moles of methionine (equation 4).[30] Thus, the photosensitized oxidation of sulfur-containing amino acids[30,31] i.e. methionine, glutathione, cysteine, and so on, is important.

$$H_2NCH(COO^-)CH_2CH_2SCH_3 \ (\underline{1}) \xrightarrow{h\nu/sens/O_2} H_2NCH(COO^-)CH_2CH_2\overset{+}{S}(CH_3)O\text{-}O^- \quad (4)$$

(zwitterion: $\xrightarrow[H_2O]{OH^-}$ $H_2NCH(COO^-)CH_2CH_2S(\rightarrow O)CH_3$ ($\underline{2}$) $+ H_2O_2$; $\xrightleftharpoons{H^+}$ $H_2NCH(COO^-)CH_2CH_2\overset{+}{S}(CH_3)O\text{-}OH \longrightarrow$ cyclic sulfonium (HN, S^+–CH_3, COO$^-$) $+ H_2O_2$; $\xrightarrow[\text{(at low pH)}]{\underline{1}}$ 2 x $\underline{2}$)

The photosensitized oxidation of organosulfur compounds continues to yield fascinating results.[32] Much attention has focused on the structures and reactivities of initially formed peroxidic intermediates.[33]

This section describes the photosensitized oxidation (singlet oxygenation) of organosulfur, organoselenium and organotellurium compounds and also briefly surveys photooxidation occurring via non-singlet oxygen mechanisms.[34]

1.1 Sulfides

The photosensitized oxidation of sulfides, first reported by Schenck and Krauch,[35,36] continues to yield important results. They reported that dialkyl sulfides undergo photosensitized oxidation to give 2 mol of sulfoxide per mole of absorbed oxygen.[37] This stoichiometry suggests that oxygen first reacts with the sulfide to give a highly reactive peroxidic intermediate, which immediately reacts with another molecule of sulfide to afford the stable sulfoxide molecule (equation 5).[21]

$$RSR' + O_2 \xrightarrow{h\nu/sens} [\text{peroxidic intermediate}] \xrightarrow{RSR'} 2\ RS(\rightarrow O)R' \quad (5)$$

Experimental evidence suggests that the photoreaction often involves molecular oxygen in the excited singlet $^1\Delta_g$ state, as originally proposed by Kautsky[38,39] and subsequently by Foote and Wexler.[40]

Since $^1O_2$ is a weak electrophile, a nucleophilic sulfide molecule could be easily oxidized.[41,42] In the photo-oxidation of sulfides, much attention has been devoted to the structures and the reactivities of initially formed reactive intermediates,[33] for which persulfoxide (**3**)[16,43] diradical (**4**),[37] thiadioxirane (**5**),[44] tight ion pair (**6**),[45] hydrogen-bonded persulfoxide (**7**),[46,47] or sulfurane (**8**)[48,49] structures have been suggested.

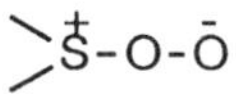

$>\overset{+}{S}\text{-O-}\overset{-}{O}$ $>\dot{S}\text{-O-}\dot{O}$ $>S\langle{}^{O}_{O}$ $>\overset{+}{S}\cdot O_2^{\cdot -}$

3 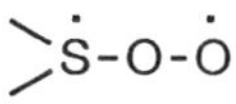   4 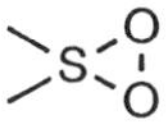   5 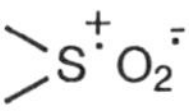   6

$>\overset{+}{S}\text{-O-}\overset{-}{O}\cdots\text{H—OMe}$ OOH–S(–OMe)

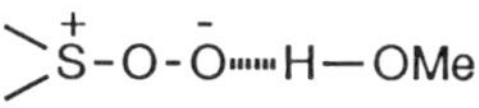

7 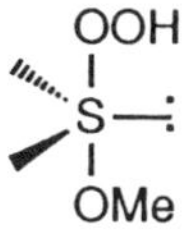   8

In a quenching study of $^1O_2$, Foote *et al.*[50] suggested that diethyl sulfide is oxidized by way of an unstable intermediate formed by addition of $^1O_2$; this intermediate would then oxidize a second sulfide molecule. Foote and Peters[16,43,51] reported experimental verification of an unstable reactive intermediate by detailed reaction kinetic studies and trapping of the intermediate. Although Krauch *et al.*[37] suggested a diradical intermediate (**4**) as the unstable oxidizing species formed initially in the reaction between alkyl sulfide and $^1O_2$, Foote and Peters tentatively assigned a zwitterionic persulfoxide structure (**3**) and others have proposed a cyclic thiadioxirane (**5**).

Further evidence for a zwitterionic persulfoxide comes from various studies involving co-oxidation of mixtures of sulfides, oxidation of certain bisulfides and studies of solvent effects.

Foote and Peters[16,43] demonstrated that diphenyl sulfide reacts 170 times more slowly than diethyl sulfide with $^1O_2$ and co-oxidation of the two sulfides with $^1O_2$ gives similar proportions of diphenyl and diethyl sulfoxides. It is suggested that the oxidation of the normally unreactive diphenyl sulfide involves the diethyl persulfoxide or thiadioxirane intermediate as oxidizing agent (equation 6).

In some cases the persulfoxide is postulated to react intramolecularly in the case of bisulfides or to afford sulfone, as shown in equation 7.[16,43]

In compounds containing two sulfide groups, the second sulfide group can serve as an intramolecular trap for the intermediate formed by the addition of $^1O_2$ to the first; the major product is the corresponding disulfoxide together with some monosulfoxide. With 1,4-dithianes, both an inter- and an intramolecular reaction take place. Exclusive formation of the *cis*-disulfoxide, however, clearly indicates that it results from the intramolecular reaction. The process producing sulfone may involve a cyclic persulfoxide, thiadioxirane

$$\text{Sens} \xrightarrow{h\nu} {}^1\text{Sens}^* \longrightarrow {}^3\text{Sens}^* \xrightarrow{{}^3O_2} {}^1O_2 \xrightarrow{Et_2S} [\,Et_2\overset{+}{S}\text{-O-}\overset{-}{O}\,] \xrightarrow{Et_2S} 2\ Et_2SO \tag{6}$$

(${}^1O_2 \rightarrow$ Sens; ${}^1O_2 \rightarrow {}^3O_2$; $[Et_2\overset{+}{S}\text{-O-}\overset{-}{O}] \xrightarrow{Ph_2S} Et_2SO + Ph_2SO$)

$$CH_3S(CH_2)_nSCH_3 \xrightarrow{{}^1O_2} CH_3S(O)(CH_2)_nS(O)CH_3 + CH_3S(O)(CH_2)_nSCH_3 \tag{7}$$

n = 2	major	minor
3	minor	major

$$Et_2S \xrightarrow[-78°C]{{}^1O_2} Et_2SO + Et_2SO_2$$
$$1 : 2.3$$

intermediate. It has been shown that sulfoxides are not oxidized to sulfones under photosensitized oxidation conditions, so that the direct formation of sulfone at low conversion of sulfide is thought to provide further evidence for the presence of an intermediate containing two oxygens.

Cauzzo *et al.*[52] studied the oxidation of sulfoxides by the persulfoxide intermediate formed during the chrysene-sensitized photooxidation of n-Bu_2S to n-Bu_2SO. Based on the results of kinetic experiments, they proposed a thiadioxirane (**5**) rather than a zwitterion (**3**) as the peroxidic intermediate.

Ando *et al.*[53] reported that when diaryl sulfides are used as trapping agents on methanol, they act as nucleophiles, i.e. the intermediate is an electrophile. Sawaki and Ogata[46] independently reported that sulfoxides function as electrophiles toward the intermediate in benzene, i.e. the intermediate is a nucleophile under their conditions. The lowered nucleophilicity of a persulfoxide intermediate in methanol is attributed to hydrogen bonding of the alcohol proton with the terminal oxygen of the persulfoxide (**7**). Another possibility is that methanol actually adds to the intermediate to give a peroxysulfurane (**8**) similar to those (**9**) characterized by Martin and Martin[54] (equation 8). This intermediate would be expected to react with sulfoxides,

$$\xrightarrow{-\text{RfOH}} \qquad (8)$$

RfO OOH; H_3C; F_3C CF_3; O-O$^-$; F_3C CF_3; H_3C; F_3C CF_3

9

Rf = $C(CF_3)_2Ph$

sulfides and alkenes, similarly to the case with (**9**).[54] Since alkenes are inert to the intermediate formed in photooxidation of sulfide in methanol, however, a peroxysulfurane (**8**) might be ruled out. These two observations and the previous kinetic results strongly suggested that there might be more than one intermediate involved in the diethyl sulfide–$^1O_2$ system.

In the photosensitized oxidation of alkyl sulfides, it was concluded that no quenching occurs in methanol as solvent but that in benzene over 95% of the reactions of $^1O_2$ lead to quenching and only a few lead to sulfoxide.[16,43,51]

The photosensitized oxidations of sulfides in methanol are very different reactions to those observed in aprotic solvents. Physical quenching does not compete with chemical reaction even at room temperature and, as a result, the formation of oxidized products is more efficient in methanol than in benzene. In addition, in protic solvents it is unnecessary to involve a second intermediate to describe the kinetic behavior in sulfoxide-trapping experiments.[46,47]

In protic solvents the persulfoxide intermediate is stabilized either by hydrogen bonding as **7** or by rapid conversion to a sulfurane (**8**). The function of the alcohol in **7** was interpreted as decreasing the negative charge density on the persulfoxide, thus promoting nucleophilic attack by a second sulfide.[38,39]

Foote and co-workers[47,55,56] reinvestigated the photosensitized oxidation of sulfides and reported that in competitive trapping of the intermediate formed in the photosensitized oxidation of Et_2S the relative ratio of the trapping is 1:18:51:6 for Ph_2S, Et_2S, Ph_2SO and Me_2SO in benzene and 1:37 for Ph_2S and Et_2S in acetonitrile. However, the ratio of Ph_2S to Ph_2SO changes to 1:5 in favor of Ph_2SO in methanol. These results confirm that those are two intermediates, in which Et_2S and Ph_2S compete for an electrophilic intermediate, but Ph_2SO reacts at an earlier stage.

As a result of substantial research, a general mechanism, summarized in equation 9, has emerged for the photosensitized oxidation of a sulfide.

Foote and co-workers[56] proposed that there are two intermediates in the $^1O_2$ reaction of sulfide in aprotic solvents, in which an initial nucleophilic persulfoxide intermediate (**3**) reacts with an electrophile such as diphenyl

$$[R_2\overset{+}{S}\text{-O-}\overset{-}{O}\cdots H\text{-OMe}]\ (\mathbf{7}) \xrightarrow{R_2S} 2\ R_2SO$$

$$R_2S + {}^1O_2 \longrightarrow [R_2\overset{+}{S}\text{-O-}\overset{-}{O}]\ (\mathbf{3}) \longrightarrow [R_2S\langle O_2\rangle]\ (\mathbf{5}) \longrightarrow R_2SO_2 \quad (9)$$

$$\mathbf{3} \xrightarrow{MeOH} \mathbf{7};\quad \mathbf{3} \longrightarrow R_2S + {}^3O_2;\quad \mathbf{3} \xrightarrow{Ph_2S} R_2SO + Ph_2SO;\quad \mathbf{5} \xrightarrow{R_2S} 2\ R_2SO$$

sulfoxide, loses triplet oxygen or collapses to an electrophilic thiadioxirane intermediate (**5**) that reacts with a nucleophile such as sulfide.

Although it is clear that two intermediates actually exist, their structures are not certain. The structures described are hypothetical simply to explain that the first intermediate is nucleophilic and the second is electrophilic.[56]

Unusually stable persulfoxides are thought to be formed from the reaction of $^1O_2$ with allyl sulfides.[57] Reaction of $^1O_2$ with deuterated allyl phenyl sulfide (**11**) causes the isomerization to **12** (equation 10). It is suggested for acceleration of the thio-allylic rearrangement that $^1O_2$ forms a reversible donor-acceptor complex with allyl phenyl sulfide. Matsumoto and Kuroda[58] also found that allyl phenyl sulfide is easily oxidized by $^1O_2$ to the corresponding sulfoxides, despite Kwart *et al.*'s report[57] that no sulfoxide is formed.

Proposed active oxidizing intermediates, **3**, **4** and **5**, have been known hitherto to oxidize phosphite[59] and α-C—H bonds,[24,32,60–62] to cause oxidative decarboxylation of α-keto carboxylic acids[63] and desulfurization of thiiranes giving alkenes[64] and to trap with alcohol affording peroxysulfenic acid[65] and hydroperoxymethoxysulfurane,[48,49] in addition to oxidation of sulfide and sulfoxide.

$$\mathbf{10} \xrightarrow{140^\circ C} \text{rubrene} + {}^1O_2$$

$$\mathbf{10} + PhSCH_2CH{=}CD_2\ (\mathbf{11}) \xrightarrow{160^\circ C} PhSCD_2CH{=}CH_2\ (\mathbf{12}) + \text{rubrene} \quad (10)$$

$$\mathbf{11} \longrightarrow [\,\text{persulfoxide (PhS-O-O}^-\text{), D}_2\text{ terminus}\ \overset{\varphi}{\rightleftharpoons}\ \text{persulfoxide, D}_2\text{ at S-carbon}\,] \longrightarrow \mathbf{12}$$

Trapping experiments of the intermediate formed in the singlet oxygenation of sulfide were also carried out with trimethyl phosphite.[59] The phosphite is relatively inert toward $^1O_2$ but very efficient in trapping the intermediate in the photosensitized oxidation of Et_2S (equation 11). Although the mechanism has not yet been clarified, the phosphite could act as a persulfoxide trapping agent.

$$Et_2S \xrightarrow{^1O_2} [Et_2\overset{+}{S}\text{-O-}\overset{-}{O}] \xrightarrow{(CH_3O)_3P} Et_2SO + (CH_3O)_3PO \qquad (11)$$

Meanwhile, it has been reported that the $^1O_2$ reaction of sulfides bearing active α-C—H bonds, such as benzyl sulfide (equation 12)[60] and 9-fluorenyl

$$\text{R-S-CH}_2\text{Ph} \xrightarrow{^1O_2} \text{R-S(}\rightarrow\text{O)-CH}_2\text{Ph} + \text{R-S(}\rightarrow\text{O)}_2\text{-CH}_2\text{Ph} + \text{PhCHO} + [\text{RSOH}]$$

$$\left[\text{R-}\overset{+}{S}(\text{O-O}^-)\text{-CH}_2\text{Ph} \longrightarrow \text{R-}\overset{+}{S}(\text{OOH})\text{-}\overset{-}{C}\text{HPh} \longrightarrow \text{R-S(}\rightarrow\text{O)-CH(OH)Ph}\ \underline{13}\right] \qquad (12)$$

ethyl sulfide (equation 13),[24] affords fragmentation products such as benzaldehyde and thiolsulfinate by C—S bond cleavage together with the usual

9-(ethylthio)fluorene $\xrightarrow{^1O_2}$ 9-fluorenyl-S(→O)Et + 9-fluorenyl-S(→O)₂Et + 1-hydroxyfluorenone + fluorenone + $(EtS)_2$

[9-fluorenyl-$\overset{+}{S}$(OO$^-$)Et ⟶ $^-$C9-fluorenyl-$\overset{+}{S}$(OOH)Et ⟶ 9-HOO-9-SEt-fluorene $\underline{14}$] (13)

S-oxidized products, sulfoxides and sulfones. The primary peroxidic intermediate in both reactions is thought to be a persulfoxide followed by intramolecular α-hydrogen abstraction to give the secondary intermediate. In both cases, the persulfoxide is postulated to act as a base. It was proposed that an α-hydroxy sulfoxide (**13**) is a key intermediate derived from the persulfoxide by rearrangement. It was pointed out, however, that the α-hydroperoxy sulfide **14** is a reasonable intermediate on the basis of detailed product analysis.

Recently, it has been found that $^1O_2$ oxidation of the thiazolidine derivatives **15a** and subsequent reaction selectively affords the corresponding α-hydroxy sulfides (**16a**) in high yields (equation 14).[61,62] Abstraction of an α-proton in the persulfoxide intermediate **17a** leading to the α-hydroperoxy thiazolidine **18a** via Pummerer-type rearrangement competes with *S*-oxidation. This hydroperoxide (**18a**) is stable at 0°C and is capable of oxidizing sulfide and phosphine to give the sulfoxide and phosphine oxide. More interestingly, co-oxidation of alkenes to epoxides has been reported in singlet oxygenation of thiazolidine derivatives and alkyl benzyl sulfides in the presence of alkenes (equations 14 and 15).[32] Intramolecular α-proton abstraction in the initially

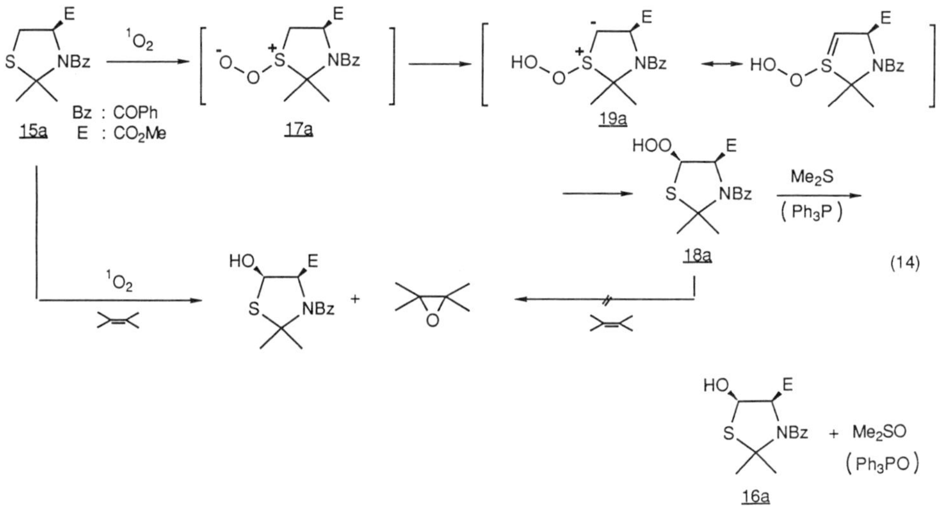

(14)

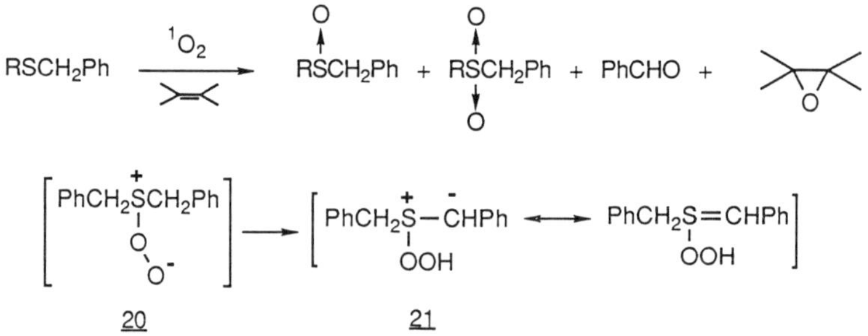

(15)

formed persulfoxide intermediates (**17a** and **20**) gives *S*-hydroperoxysulfonium ylide intermediates (**19a** and **21**, respectively), which are capable of epoxidizing alkenes, themselves inert toward $^1O_2$. An intramolecular hydrogen bonding may stabilize the intermediate **22** and would activate the electrophilic oxygen, so that it could react with an alkene, similarly to the case of the peracid **23**.

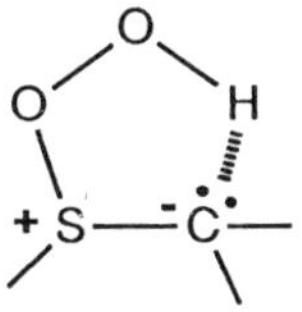

22 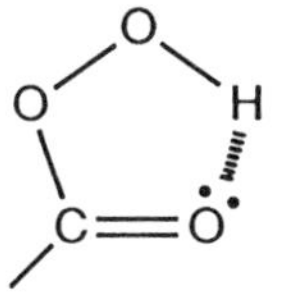 23

Recently, Clennan and Chen[48] contended that the diastereoselective oxidations of hydroperoxy sulfides can be attributed to the selective formation of cyclic sulfuranes (**24**) (equation 16). In the photosensitized oxidation of γ-hydroxy sulfides, experimental evidence for the formation of a sulfurane intermediate (**25**) was provided by detailed product analysis and ^{17}O labeling experiments.[49]

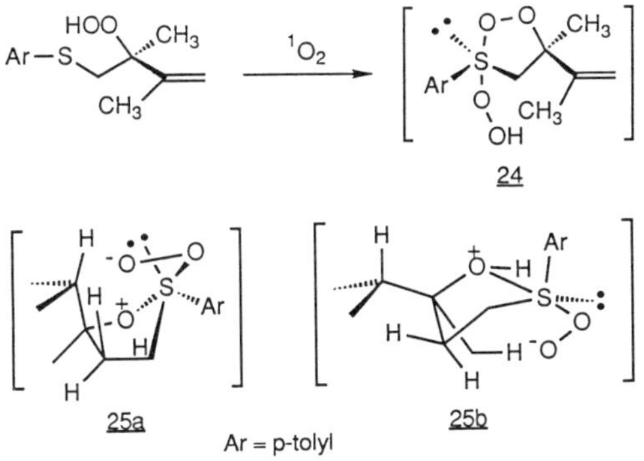

24

25a 25b

Ar = p-tolyl

(16)

A persulfoxide is used as a monooxygenase model in the decarboxylation of α-keto carboxylic acids. It has been found that this decarboxylation is caused by a persulfoxide generated *in situ* during the photosensitized oxidation of Et_2S, affording the corresponding carboxylic acid and gaseous CO_2 (equation 17).[63]

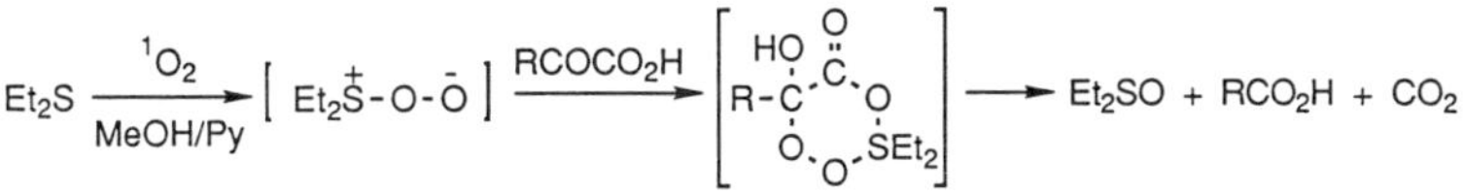

$$Et_2S \xrightarrow[MeOH/Py]{^1O_2} [Et_2\overset{+}{S}\text{-}O\text{-}\overset{-}{O}] \xrightarrow{RCOCO_2H} [\ \ ] \longrightarrow Et_2SO + RCO_2H + CO_2 \qquad (17)$$

Decarboxylations were conveniently performed in aprotic solvents in which the nucleophilic nature of the persulfoxide is manifested. However, the nucleophilicity of the persulfoxide is substantially lowered in methanol, and the reaction occurs only under basic conditions. Since the C—C bond of an α-keto carboxylic acid is elongated by hydrogen bonding in the presence of pyridine, the nucleophilic attack of the persulfoxide on carbonyl carbon and subsequent cyclization should be facilitated in acetonitrile, which enhances the charge separation in α-keto carboxylic acids.

Photosensitized oxidation of biadamantylidene thiirane (**26**) afforded, in addition to the corresponding thiirane oxide, the desulfurization products, i.e. biadamantylidene and its oxidized products (equation 18).[64] A spirothiadioxirane intermediate (**27**) derived from a peroxythiirane oxide (**28**) is thought to be formed.

$^1O_2$

Ad=Ad (18)

26

Ad = 2-adamantyl

Ad=Ad + SO_2

28 27

Recently, the reaction of $^1O_2$ with the thiirane **29** in methanol has been investigated.[65] The trapping of a persulfoxide intermediate (**30**) of thiirane **29** by methanol affords the peroxysulfenic acid (**31**), which is able to oxidize alkenes to epoxides and sulfides to sulfoxides. The peroxysulfenic acid intermediate might be derived from ring opening of a thiirane peroxide intermediate by nucleophilic attack of methanol, similarly to the case of the acid-catalyzed ring opening of thiirane oxides (equation 19).[66–68] This is the first example of the efficient epoxidation of alkenes by the active oxidizing species generated in the photooxidation of sulfides.

$^1O_2$ MeOH

SOOH OMe

29 30 31

SOH OMe (19)

Jensen and Foote[69] independently investigated the reaction of $^1O_2$ with thiirane and proposed a hydroperoxymethoxysulfurane (**32**) as an alternative intermediate in the reaction with methanol (equation 20). The primary products in alcoholic solvents are thiirane oxides and sulfinic esters (**33**). In aprotic solvents, they are thiirane oxides. These products can be rationalized by assuming that a reactive intermediate, a peroxythiirane oxide, is formed from

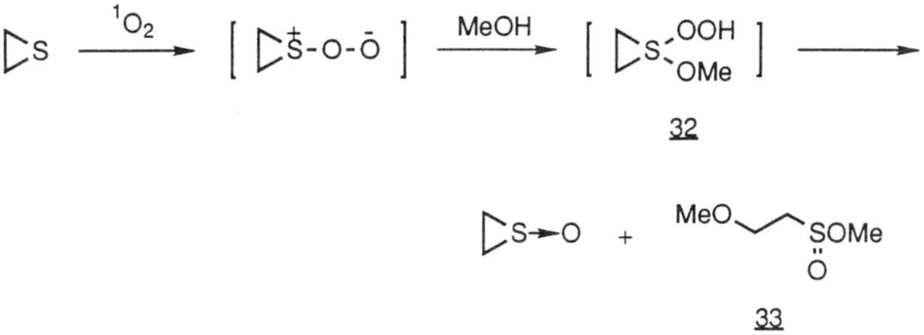

(20)

$^1O_2$ and thiirane. This intermediate can undergo trapping by the thiirane itself or by nucleophilic solvents. The structure of the peroxythiirane oxide has been investigated by *ab initio* calculations.[67] Based on frontier molecular orbital theory and *ab initio* calculations, it is suggested that the formation of the observed sulfinic esters might go through a hydroperoxymethoxysulfurane.

The persulfoxide intermediate can transfer an oxygen atom to iron-porphyrin affording the oxene species, similarly to the case with iodosobenzene and *m*-chloroperbenzoic acid.[71] The oxene oxidizes alkenes to epoxides. An oxenoid mechanism might be proposed for the epoxidation reactions (equation 21).

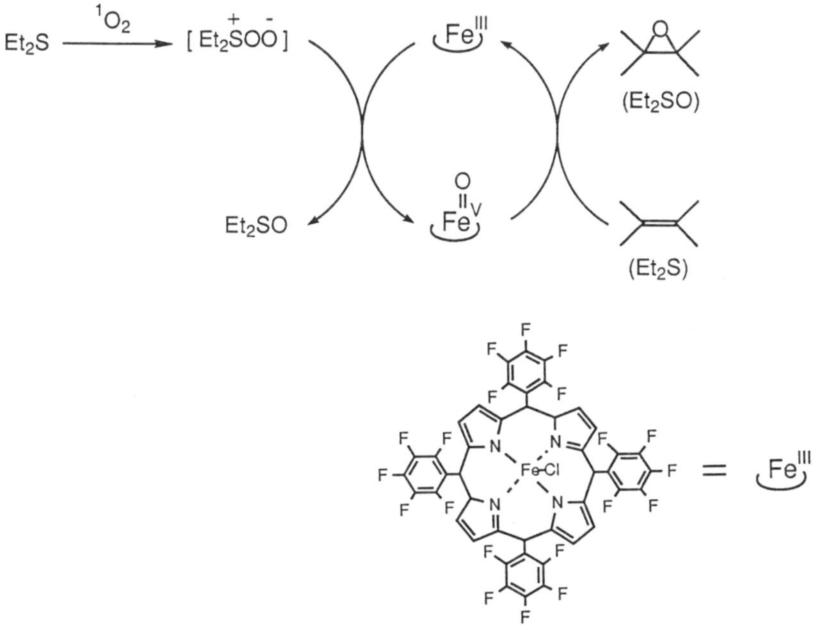

(21)

A recent theoretical study of the additions of $^1O_2$ to H_2S, Me_2S and thiirane (CH_2SCH_2) was unable to locate a minimum on the potential energy surface corresponding to the thiadioxirane (**5**).[67] This is a surprising outcome in view of the fact that the dioxirane **34**[72,73] and isolable dimethyldioxirane **35**[74–76] are well characterized species.

34 **35**

As an example of a reaction of a persulfoxide intermediate formed in the singlet oxygenation of sulfide, the oxidation of sulfoxide with the persulfoxide was carried out (equation 22).[77] Evidence for a linear sulfurane intermediate

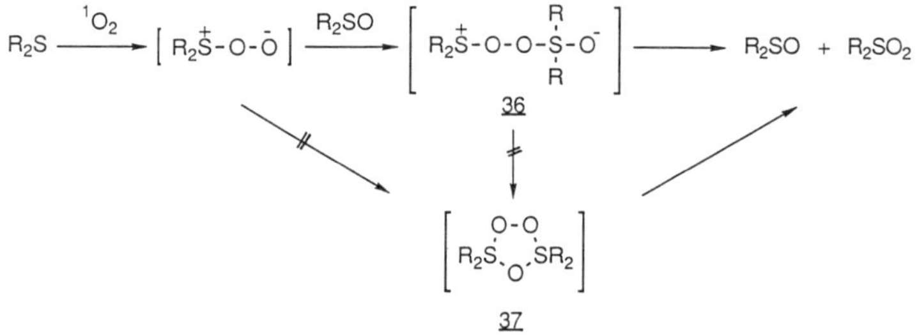

$$R_2S \xrightarrow{^1O_2} [R_2\overset{+}{S}\text{-O-}\overset{-}{O}] \xrightarrow{R_2SO} [R_2\overset{+}{S}\text{-O-O-}SR_2\text{-}O^-]\ (\mathbf{36}) \longrightarrow R_2SO + R_2SO_2 \quad (22)$$

36 ↛ **37** $[R_2S(\text{O-O})(\text{O})SR_2]$

(**36**), but not a cyclic one (**37**), was reported in experiments using isotopically labeled compounds in the two systems: $(CH_3)_2S$-$^{18}O_2$-$(CD_3)_2SO$ and $(CH_3)_2$-$^{16}O_2$-$(CD_3)_2S^{18}O$.

Differentiating the nucleophilic character of the active oxidizing species in their oxygen transfer, Adam *et al.*[78] developed thianthrene 5-oxide (SSO) as a convenient mechanistic probe, containing both a nucleophilic sulfide and an electrophilic sulfoxide site. Indeed, the oxygen adduct in the photosensitized oxidation of diethyl sulfide produced proportionally more sulfone SSO_2 than disulfoxide SOSO (equation 23). These results reveal a dominant nucleophilic character, causing mainly oxidation of the sulfoxide sulfur.

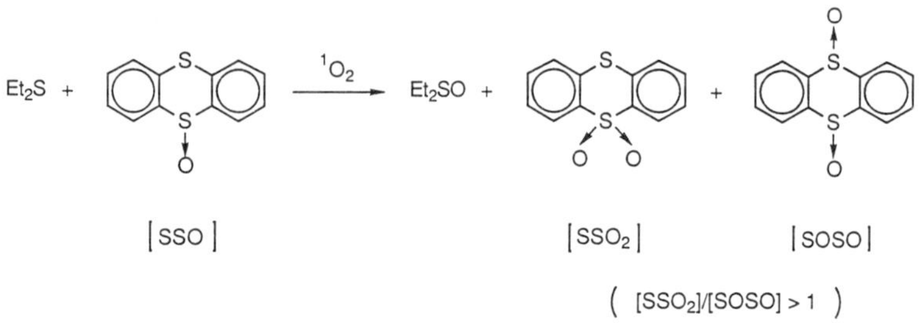

$$Et_2S + [SSO] \xrightarrow{^1O_2} Et_2SO + [SSO_2] + [SOSO] \quad (23)$$

$([SSO_2]/[SOSO] > 1)$

The assumption of a possible intermediate such as a persulfoxide and a thiadioxirane in the reaction of $^1O_2$ with sulfide led to the idea that a related species might be stable enough to allow direct observation or isolation.

Martin and Martin[54] carried out the oxidation of sulfides and alkenes by the oxidizing species generated in the reaction of the sulfurane **38a** with hydrogen peroxide (equation 24). The intermediate affords a small amount of sulfide and oxygen together with sulfone and sulfoxide on warming to room temperature. This oxidation is rather electrophilic and a ρ-value of -0.86 was obtained in the oxidation of (p-X-$C_6H_4)_2S$ by the sulfurane–H_2O_2 system. It is likely, therefore, that the reactive intermediate is not persulfoxide **39a**, which is expected to be nucleophilic, but rather hydroperoxysulfurane **40a**. The intermediate did not give any epoxide on reaction with cyclohexene.

Bartlett *et al.*[79] also invoked the participation of a persulfoxide intermediate (**41**) in the reaction of sulfurane **38a** with alkyl hydroperoxides (equation 24).

The oxidation of sulfoxide with caroate affords the corresponding sulfone.[80] However, careful stoichiometric experiments revealed that some oxygen is obtained during the sulfoxide oxidation. Loss of ^{18}O label in the recovered sulfoxide occurs. A difference in the two oxygen atoms coordinated to sulfur atom in the thiadioxirane intermediate (**42**) may contribute to the result, since the apical and equatorial oxygens have different oxidizing abilities (equation 25).

A number of electron-rich compounds, such as enamines,[27] vinyl ethers[81] and phenols,[82] react with $^1O_2$ in a manner different from less electron-rich compounds. The reactions may go by way of an electron transfer from the electron-rich molecule (R) to $^1O_2$.[27] It is suggested that quenching of $^1O_2$ by amines and sulfides occurs via partial or complete electron transfer to $^1O_2$.[83] The ion pair formed on electron transfer could react either to give product (RO_2) or revert to starting material (R) and $^3O_2$, the net result of which is quenching of $^1O_2$.

Evidence for the involvement of ion radicals in the photosensitized oxidation of sulfides comes from the observation of the cleavage of certain sulfides with the concomitant formation of unoxidized disulfides.[34,84] On oxidation with several photosensitizers, di-*tert*-butyl sulfide (**43**) gives not only the corresponding sulfoxide and sulfone but also di-*tert*-butyl disulfide (**44**) (equation 26). No disulfide was produced with sensitizer in the absence of oxygen. Oxygen plays an essential part in generating the cation radicals, even though direct attack of $^1O_2$ on the sulfide is not involved in this process.

Photo-oxidation of chlorpromazine in water and ethanol ultimately affords the corresponding sulfoxide (equation 27).[85] The electron-transfer process was considered to generate the cation radical which reacts with $^3O_2$ to give a persulfoxide-type intermediate (**45**).

The photosensitized oxidation of sulfides in the micelle in the presence of phenothiazine derivatives (MPT) resulted in formation of sulfoxides.[86] $^1O_2$ was implicated as the primary oxidizing agent and is thought to arise from the intramicellar recombination of the transient $MPT^{+\bullet}$–$O_2^{-\bullet}$ ion pair.

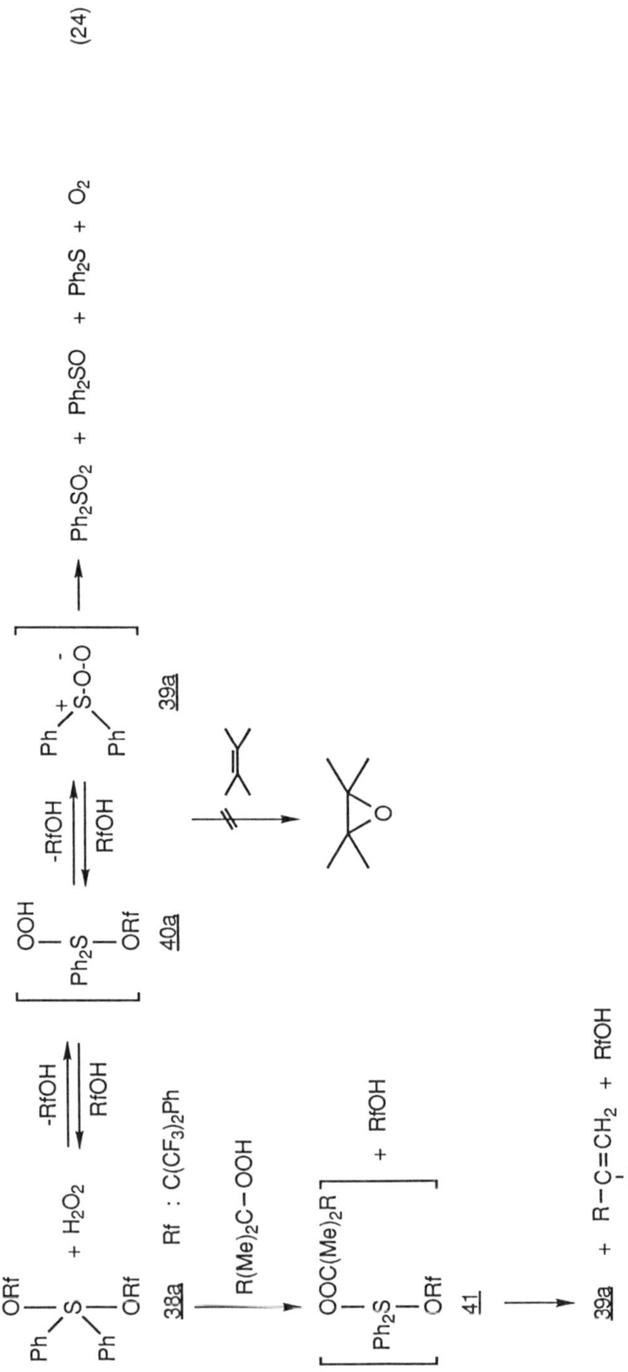

$$\underset{\underline{38a}}{Ph_2S(ORf)_2} + H_2O_2 \underset{RfOH}{\overset{-RfOH}{\rightleftharpoons}} \left[\underset{\underline{40a}}{Ph_2S(OOH)(ORf)} \underset{RfOH}{\overset{-RfOH}{\rightleftharpoons}} \underset{\underline{39a}}{Ph_2\overset{+}{S}\text{-O-}\overset{-}{O}} \right] \longrightarrow Ph_2SO_2 + Ph_2SO + Ph_2S + O_2 \qquad (24)$$

Rf : $C(CF_3)_2Ph$

$\underline{39a}$ + $Me_2C{=}CMe_2$ $\not\longrightarrow$ tetramethyloxirane

$$\underline{38a} \xrightarrow{R(Me)_2C-OOH} \left[\underset{\underline{41}}{Ph_2S(OOC(Me)_2R)(ORf)} \right] + RfOH$$

$$\underline{41} \longrightarrow \underline{39a} + R-C(Me){=}CH_2 + RfOH$$

$R{-}S(\rightarrow{}^{18}O){-}R + ZOO^- \longrightarrow {}^-({}^{18}O){-}SR_2{-}OOZ \xrightarrow{-ZO^-}$

● = ^{18}O

Z ; SO_3^-

$\downarrow -ZO^-$

$R_2S(\rightarrow O)(\rightarrow {}^{18}O)$ (25)

42

$\xrightarrow{ZOO^-}$ R_2S● + OO + ZO^-

$\xrightarrow{ZOO^-}$ R_2SO + ●O + ZO^-

$$Me_3CSCMe_3 \;(\underline{43}) \xrightarrow{^1O_2} Me_3C\overset{+\cdot}{S}CMe_3 + O_2^{\cdot-} \longrightarrow Me_3C^+ + Me_3CS\cdot \longrightarrow 1/2\ Me_3CSSCMe_3 \;(\underline{44}) \quad (26)$$

$[\text{chlorpromazine}]^{+\cdot}$ + $^3O_2$ ⟶ $[\underline{45}]^{+\cdot}$ ⟶ ⟶ (27)

Cl, N, S, $(CH_2)_3NHMe_2$

45

O ← S; Cl; N; $(CH_2)_3NHMe_2$

Electron-transfer from sulfides[87] and amines[88] to $^1O_2$ has actually been observed in aqueous solution. Noteworthy are these reports by Saito and co-workers who have shown that superoxide anion is formed from the reaction of amines or sulfides with $^1O_2$ using a water-soluble $^1O_2$ source (equation 28). They detected the short-lived superoxide anion radical with the aid of an enzymic assay utilizing superoxide dismutase.

$^1O_2$ + $CH_3S-C_6H_4-OH$ (46) →

[$^1O_2^{\delta-}\cdots CH_3S^{\delta+}-C_6H_4-OH$] $\xrightarrow{\text{aq.solvent}}$ $(O_2^{\cdot-})_{solv}$ + $(CH_3\overset{+\cdot}{S}-C_6H_4-OH)_{solv}$ → O_2 + $CH_3S-C_6H_4-OH$

→ $CH_3\overset{+}{S}(OO^-)-C_6H_4-OH$ $\xrightarrow{46}$ 2 $CH_3S(\rightarrow O)-C_6H_4-OH$ (28)

Foote and co-workers[23] have reported 9,10-dicyanoanthracene (DCA)-sensitized photo-oxidation involving a donor radical cation and DCA radical anion, which subsequently reduces $^3O_2$ to superoxide anion radical. For example, Foote and co-workers[23] observed that in DCA-sensitized photo-oxidation, diphenyl sulfide is three times more reactive than diethyl sulfide (in striking contrast to the order of reactivity seen with other sensitizers) and that the photo-oxidation is not quenched by β-carotene as an effective $^1O_2$ quencher. A mechanism involving cation and anion radicals intermediate in this oxidation has been proposed (equation 29).

$$\text{DCA (9,10-dicyanoanthracene)} \xrightarrow{h\nu} {}^1\text{DCA}^*$$

$${}^1\text{DCA}^* + Ph_2S \longrightarrow \text{DCA}^{\cdot-} + Ph_2S^{\cdot+} \qquad (29)$$

$$\text{DCA}^{\cdot-} + {}^3O_2 \longrightarrow \text{DCA} + O_2^{\cdot-}$$

$$Ph_2S^{\cdot+} + O_2^{\cdot-} \longrightarrow Ph_2\overset{+}{S}\text{-O-}\overset{-}{O} \xrightarrow{Ph_2S} 2\,Ph_2SO$$

Dithiocine sulfoxide (**47**) was obtained by both the singlet oxygenation of dithiocine (**48**) and the reaction of cation radical $\mathbf{48}^{+\cdot}$ and superoxide anion radical in which the direct coupling reaction affords a persulfoxide (**49**) (equation 30).[89] Thus, the reaction of the sulfide cation radical and superoxide anion radical may show similarity in possible reaction modes with singlet oxygenation.

$$\mathbf{48} \xrightarrow{^1O_2} \mathbf{49} \xrightarrow{\mathbf{48}} 2\ \mathbf{47} \qquad (30)$$

$$\mathbf{48}^{\cdot+} + O_2^{\cdot-} \longrightarrow \mathbf{48} + O_2 \quad (\text{also} \rightarrow \mathbf{49})$$

Of particular interest is the ability of semiconductors to sensitize photo-oxidation reactions. Titanium oxide has been shown to sensitize the oxidation of sulfides to give sulfoxides and sulfones.[90] Evidence was presented against the involvement of singlet oxygen and in favor of the sulfide cation radicals as key intermediates. In the case of diphenyl sulfide, formation of biphenyl suggests that C—S bond cleavage in the sulfide cation radical intermediate is involved (equation 31).

$$TiO_2 \xrightarrow{h\nu} TiO_2(h^+) + e^-$$

$$TiO_2(h^+) + R_2S \longrightarrow TiO_2 + R_2S^{\cdot+}$$

$$O_2 + e^- \longrightarrow O_2^{\cdot-}$$

$$R_2S^{\cdot+} + O_2^{\cdot-} \longrightarrow [\,R_2\overset{+}{S}O\overset{-}{O}\,] \longrightarrow \text{oxidation products}$$

$$R_2S^{\cdot+} \longrightarrow R^+ + RS^{\cdot}$$

$$R_2S^{\cdot+} \longrightarrow R^{\cdot} + RS^+ \quad (R = Ph)$$

$$2\,RS^{\cdot} \longrightarrow RSSR$$

$$2\,R^{\cdot} \longrightarrow R\text{–}R \qquad (31)$$

Photo-oxidation of substituted diphenyl sulfides on TiO_2 powder in oxygen-saturated acetonitrile affords the corresponding sulfoxides, which are converted into the sulfone on continued photo-oxygenation. Detailed kinetic experiments and product analysis provided evidence for the involvement of the photo-generated sulfide cation radical, which gave the initial oxygen adduct **50** (equation 32).[91]

$$Ph-\overset{+\cdot}{S}-Ph \xrightarrow{^{3}O_2} Ph-\underset{+}{\overset{O-O^{\cdot}}{S}}-Ph \;(\underline{50}) \longrightarrow \longrightarrow Ph-\underset{\downarrow O}{S}-Ph \qquad (32)$$

Direct irradiation of the sulfide-oxygen charge-transfer band (λ_{max} = 300–350 nm) in the absence of sensitizers can also lead to sulfoxide formation, possibly by a process involving a sulfide cation radical and superoxide anion radical (equation 33).[92] Such unsensitized photo-oxidation of sulfides can even occur in the solid state.[93]

$$R_2S + {}^{3}O_2 \longrightarrow [R_2S \cdots O_2] \xrightarrow{h\nu} [R_2S \cdots O_2]^{*} \longrightarrow [R_2S^{+\cdot}\, O_2^{-\cdot}] \longrightarrow \longrightarrow R_2SO \qquad (33)$$

λ_{max} = 300~350 nm

R = Et,t-Bu,Ph,etc

In the photo-oxidation of sulfides with molecular oxygen, much attention has been devoted to the reactivities and structures of the initially formed peroxidic intermediates, for which persulfoxide (**1**), diradical (**2**) or thiadioxirane (**3**) structures have been suggested.[33] Since no direct observation of the intermediates has been achieved so far, however, the structures are still controversial. On irradiation exciting contact charge-transfer bands of sulfides in an oxygen matrix at 13 K with UV radiation, the first observation of the matrix-isolated persulfoxides was achieved by Fourier transform IR spectroscopy.[94] The IR spectra suggest that these species (**51**) at least formulated as the persulfoxide zwitterions. Theoretical studies supported these experimental results.[70]

$Ph\overset{*}{S}Me(O-\overset{*}{O})$ <u>51a</u> $\quad$ $Ph\overset{*}{S}Ph(O-\overset{*}{O})$ <u>51b</u> $\quad$ $\overset{*}{S}-O-\overset{*}{O}$ (bicyclic) <u>51c</u> $\quad$ $*$ = +/- or $\cdot$

1.2 Thiols

The photo-oxidation of thiols is of particular relevance to biological systems.[95–98] Complex products are usually afforded. Additional factors complicating the photo-oxidation are the facility with which autoxidation[99] and hydrolysis in aqueous solution occur. Thiol-containing amino acids, such as cysteine and glutathione, are photo-oxidized to the corresponding disulfides both with and without a sensitizer.[99] The mechanism of the singlet oxygenation of thiols is not generally described in the literature. Thiols are also known to act as $^1O_2$ quenchers.[100,101]

Alkanethiols were photo-oxygenated in the presence and absence of sensitizers, such as eosin in benzene.[96] UV irradiation of methanethiol in the absence of a sensitizer and in the presence of oxygen yields dimethyl disulfide as the main product, accompanied by H_2O_2 and SO_2 (equation 34).[102] The

$$\mathrm{RSH} \xrightarrow{h\nu/\mathrm{sens}/O_2} \mathrm{RSSR} \qquad (34)$$

reaction of the thiyl radical intermediate with oxygen is presumably very slow. Bently[103] obtained several *S*-oxygenated products, including methanesulfinic acid and methanesulfonic acid in addition to dimethyl disulfide.

Without sensitization, both cysteine and cystine undergo photo-oxidation and degradation to afford cysteic acid, cysteinesulfinic acid, alanine, ammonia and related products.[96]

Although cysteic acid and H_2O_2 are formed in the crystal violet-sensitized photo-oxidation of cysteine, cystine was the only product obtained with $^1O_2$ (equation 35).[104]

$$\mathrm{HO_2C\text{-}CH(NH_2)\text{-}CH_2SH} + H_2O + 2O_2 \xrightarrow{h\nu/\text{crystal violet}} \mathrm{HO_2C\text{-}CH(NH_2)\text{-}CH_2SO_3H} + H_2O_2$$

$$\mathrm{HO_2C\text{-}CH(NH_2)\text{-}CH_2SH} \xrightarrow{^1O_2} \left(\mathrm{HO_2C\text{-}CH(NH_2)\text{-}CH_2S}\right)_2 \qquad (35)$$

In addition to autoxidation, one of the important reactions of thiols involving molecular oxygen is co-oxidation.[99,105] Thiols and alkenes are co-oxidized by molecular oxygen giving a 2-hydroxyethyl sulfoxide (**52**) as the final product (equation 36). The hydroperoxide **53** has been isolated as an intermediate in the thiophenol–indene system. The co-oxidation reaction has been applied to the synthesis of the prostanoid precursor 1,2-dioxolane (equation 37).[106]

The photo-oxidation of cysteine to cystine is sensitized by aqueous suspensions of cadmium sulfide with good efficiency.[107] Other bioorganic thiols (including glutathione, penicillamine, cysteamine and dithiothreitol) are also oxidized. Quenchers show that $^1O_2$ is not involved in the photo-oxidation

RSH ⟶ RS· $\xrightarrow{CH=CHR'}$ $RSCH_2\dot{C}HR'$ $\xrightarrow{O_2}$ $RSCH_2CHR'(O{-}O\cdot)$ $\xrightarrow{RSH,\ -RS\cdot}$ $RSCH_2CHR'(O{-}OH)$ **53** ⟶ $RS(\to O)CH_2CHR'(OH)$ **52** (36)

(37)

process. The partial involvement of superoxide in the reaction mechanism, however, is suggested, since superoxide dismutase inhibits the yield of cysteine by photo-oxidation.

The photo-oxidation of thiophenolates sensitized by tetraphenylporphin leads to the corresponding benzenesulfonates (equation 38).[108] A proposed mechanism is as follows. The reaction proceeds via attack of $^1O_2$ on the

$^1O_2$ + ArS^- ⟶ $[ArS{-}O{-}O^-]$ **54** $\xrightarrow{ArS^-}$ $[ArSO^-]$ **55** $\xrightarrow{^1O_2}$ $[ArS(\to O){-}O{-}O^-]$ **56** ⟶ $ArSO_3^-$

54 $\xrightarrow{-O_2^{\bar{\cdot}}}$ [ArS·] ⟶ ArSSAr (38)

thiophenolate anion to give a peroxysulfenate intermediate (**54**), which can react with thiophenolate resulting in the formation of a sulfenate (**55**). The latter is oxidized by $^1O_2$ to afford the peroxysulfinate (**56**), followed by rearrangement giving sulfonate.

1.3 Disulfides

A number of biologically important molecules containing disulfide bonds are known to be susceptible to photodynamic action.

Weil *et al.*[12] originally reported that cystine is photo-oxidized in the presence of methylene blue (MB). Subsequently, however, they noted that cystine reacted only sluggishly in photosensitized oxidation.[109] The possibility of photodynamic action in biological substrates containing the cystine residue has been examined with regard to photosensitized oxidation.[14,15] Murakami[110] reported that in the MB-sensitized oxygenation of cystine in phosphate buffer at pH 8.4, oxygen is absorbed at a rate much slower than those observed for histidine, tyrosine, tryptophan and methionine. It was suggested that the product was cysteic acid. Organic disulfides, with the exception of amino acids, have been known to undergo *S*-oxidation without S—S bond cleavage. Calvin *et al.*[111] found that the dithiolane **57**, a derivative of α-lipoic acid, can be readily photo-oxidized to the corresponding monooxide **58** in the presence of a sensitizer (equation 39). Murray and Jindal[112] also studied the MB-sensitized

$$2 \times \underset{\underline{57}}{\text{S–S (ring)}} + O_2 \xrightarrow{h\nu/\text{ZnTPP}} \underset{\underline{58}}{\text{S(→O)–S (ring)}} \qquad (39)$$

ZnTPP ; zinc tetraphenylporphine

oxygenation of cystine and its derivatives. They speculated that the products are the corresponding thiolsulfinates [RS(O)SR] in all cases, although the products were not actually identified. Involvement of $^1O_2$ was considered as an active oxygen species. Murray and Jindal reported $^1O_2$ oxidation of simple dialkyl disulfides to afford the corresponding thiolsulfinates and thiolsulfonates (equation 40).[112–116] The thiolsulfinate itself is not oxidized further under the reaction conditions. Thiolsulfonates may come from the disproportionation of the thiolsulfinate. Alternatively, thiolsulfonate can be formed by the intramolecular rearrangement of a peroxy intermediate (**59**) (equation 41).[114]

$$\text{RSSR} \xrightarrow{^1O_2} \text{RS(→O)SR} + \text{RS(→O)(→O)SR} \qquad (40)$$

R = Me,Et

$$\mathrm{RSSR} + {}^1\mathrm{O_2} \longrightarrow \left[\mathrm{R\overset{+}{S}(O{-}O^-)SR}\right]\ (\mathbf{59}) \xrightarrow{\mathrm{RSSR}} 2\,\mathrm{RS(O)SR} + \mathrm{RS(O_2)SR} \qquad \left(\mathrm{R{-}S(O{-}O){-}S{-}R}\right)\ (\mathbf{60}) \tag{41}$$

Such a process would be analogous to that described by Foote and Peters.[16,43] They demonstrated that a sulfone can be formed from the intramolecular reaction of a persulfoxide intermediate. It is noteworthy that in the photosensitized oxidation of a mixture of *t*-BuSSBu-*t* and *i*-PrSSPr-*i* no mixed thiolsulfinate, i.e. *t*-BuS(O)SPr-*i* and *t*-BuSS(O)Pr-*i*, was obtained, indicative of no S—S bond cleavage during the course of the photo-oxidation.[114] As shown in equation 41, the stoichiometry of the reaction indicates that 1 mol of oxygen is absorbed per 2 mol of disulfide. This is consistent with initial formation of a zwitterionic peroxy intermediate **59**, with some alternatives such as **60**. The peroxy intermediate **59**, like that proposed in the photosensitized oxidation of sulfide, can oxidize a second disulfide molecule to yield 2 mol of thiolsulfinate. Intermediate **59** or its alternatives can also oxidize diphenyl disulfide, itself inert to ${}^1O_2$, to give EtS(O)SEt and PhS(O)SPh when the mixture of EtSSEt and PhSSPh is photo-oxygenated.[114] The intermediate peroxy species, therefore, has a strong oxidizing ability and is much less selective than ${}^1O_2$ (equation 42).

$$\mathrm{PhS(O)SPh} \xleftarrow{{}^1\mathrm{O_2}}\!\!\!\!/\;\; \left.\begin{matrix}\mathrm{EtSSEt}\\+\\\mathrm{PhSSPh}\end{matrix}\right\} \xrightarrow{{}^1\mathrm{O_2}} \left[\mathrm{Et\overset{+}{S}(O{-}O^-)SEt}\right] \begin{cases} \xrightarrow{\mathrm{EtSSEt}} 2\,\mathrm{EtS(O)SEt} \\ \xrightarrow[\mathrm{PhSSPh}]{} \mathrm{EtS(O)SEt} + \mathrm{PhS(O)SPh} \end{cases} \tag{42}$$

Stary *et al.*[115] investigated the MB-sensitized oxygenation of the methyl ester of α-lipoic acid (**61**), affording four diastereomeric thiolsulfinates (**62**–**65**) as the major products, in addition to two regioisomeric thiolsulfonates (**66** and **67**) (equation 43). As in the case of linear disulfides, the reaction proceeds via an initial peroxy intermediate. The reaction of **61** with ${}^1O_2$ can generate two regioisomeric peroxy intermediates (**68**), which can either rearrange intramolecularly to the two thiolsulfonates **66** and **67** or oxidize a second disulfide molecule affording four thiolsulfinates (**62**–**65**).

Recently, it has been found that in ${}^1O_2$ oxidation of dialkyl disulfides, co-oxidation of alkene to epoxide occurs in methanol (equation 44).[117] Isotopic labeling experiments suggested that no S—S bond cleavage occurs during the course of the photo-oxidation as shown in equation 45. The hydroperoxy intermediate **69** has been proposed to have an ability to oxidize alkenes, in contrast to the photosensitized oxidation of sulfide in methanol in which the

61 → ($^1O_2$, MeOH) 68a + 68b; Intermolecular → 62 (11%) + 63 (22%) + 64 (19%) + 65 (24%); Intramolecular → 66 (2%) + 67 (11%)

R = $-(CH_2)_4-CO_2Me$ (43)

RSSR → ($^1O_2$, MeOH/norbornene) RS(O)SR + RS(O)S(O)R + epoxide (44)

solvated intermediate **8** was suggested. In aprotic solvents epoxidation also takes place.[117] The intermediate **70**, which may come from the insertion of molecular oxygen into an S—S bond, has been speculated, similar to that (**71**) in the reaction of disulfide with triazolinedione.

[RS-O-O-SR]

70

71

1.4 Thioketones and Related Compounds

The photosensitized oxidation of thiocarbonyl compounds to give the corresponding carbonyl compounds has been known for a long time. Many photosensitized oxidations of thioketones are sensitized by dyes known to generate $^1O_2$,[118,119] and the effect of $^1O_2$ quenchers confirms the involvement

$R_HSSR_D \xrightarrow{^1O_2} \left[R_H\text{-}\overset{+}{S}(\text{-}O\text{-}O^-)\text{-}S\text{-}R_D + R_H\text{-}S\text{-}\overset{+}{S}(\text{-}O\text{-}O^-)\text{-}R_D \right] \xrightarrow{MeOH} \left[R_H\text{-}S(OMe)(OOH)\text{-}S\text{-}R_D + R_H\text{-}S\text{-}S(MeO)(HOO)\text{-}R_D \right] \xrightarrow[-MeOH]{Me_2C{=}CMe_2} R_H\text{-}S(\rightarrow O)\text{-}S\text{-}R_D + R_H\text{-}S\text{-}S(\rightarrow O)\text{-}R_D + \text{tetramethyloxirane}$

$R_H = CH_2CH_3$
$R_D = CD_2CH_3$

69

$\left[R_H\text{-}\overset{+}{S}(\text{-}O\text{-}O^-)\text{-}S\text{-}R_D + R_H\text{-}S\text{-}\overset{+}{S}(\text{-}O\text{-}O^-)\text{-}R_D \right] \overset{MeOH}{\not\rightarrow} \left\{ \begin{matrix} R_H\text{-}SOOH + R_D\text{-}SOMe \\ R_H\text{-}SOMe + R_D\text{-}SOOH \end{matrix} \right\} \xrightarrow{Me_2C{=}CMe_2 \; / \; -\text{epoxide}} \left\{ \begin{matrix} R_H\text{-}SOH + R_D\text{-}SOMe \\ R_H\text{-}SOMe + R_D\text{-}SOH \end{matrix} \right\} \longrightarrow \begin{matrix} R_H\text{-}S(\rightarrow O)\text{-}S\text{-}R_D + R_H\text{-}S\text{-}S(\rightarrow O)\text{-}R_D \\ R_H\text{-}S(\rightarrow O)\text{-}S\text{-}R_D + R_H\text{-}S\text{-}S(\rightarrow O)\text{-}R_D \end{matrix}$

(45)

of $^1O_2$ as an active oxidizing species (equation 46).[120] In the absence of an added sensitizer, thioketones are also known to be susceptible to light and oxygen.[121–124] Nevertheless, this photochemical reaction of thioketones has been found to involve $^1O_2$.[120,125] $^1O_2$ is produced by energy transfer from a photoactivated thioketone to $^3O_2$.

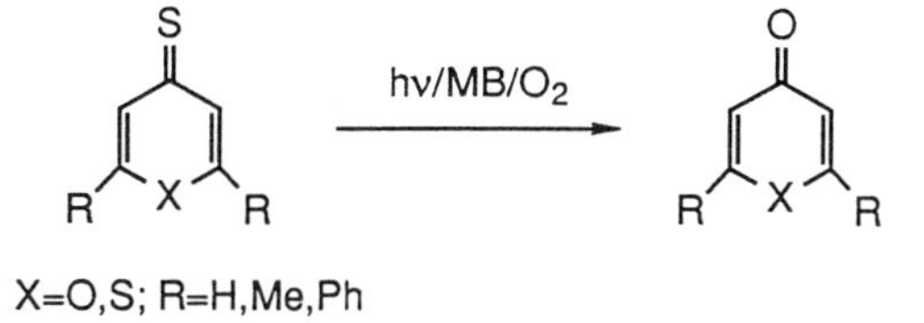

hv/MB/O_2 (46)

X=O,S; R=H,Me,Ph

Studies with xanthene-9-thione and related compounds showed that a weak chemiluminescence accompanies the oxidation, which can be enhanced by adding a fluorescer, and this is taken as evidence for a 1,2,3-dioxathietane intermediate (**72**) (equation 47).[126] The coproducts of most oxidations are sulfur

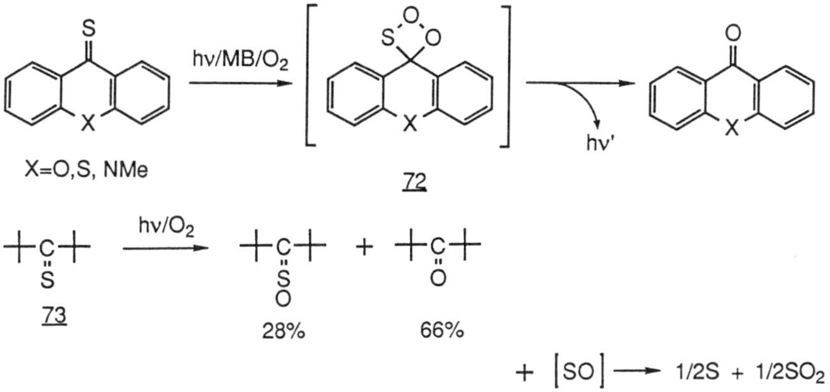

hv/MB/O_2 hv'

X=O,S, NMe

72

hv/O_2

73

28% 66%

+ [SO] ⟶ 1/2S + 1/2SO_2 (47)

and sulfur dioxide, which may arise by disproportionation of the sulfur monooxide that is eliminated from a dioxathietane. However, the results are more complex with some aliphatic thioketones. Di-*tert*-butyl thioketone (**73**) gives a sulfine in addition to the ketone (equation 48).[127,128] The ratio of the

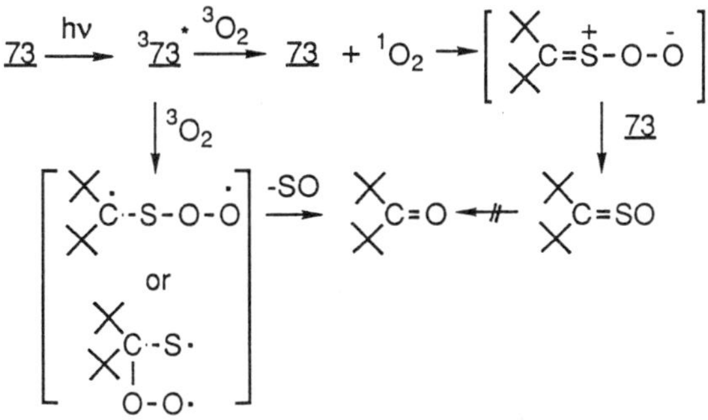

73 $\xrightarrow{hv}$ ^{3}73* $\xrightarrow{^3O_2}$ 73 + $^1O_2$ ⟶ [C=S⁺-O-O⁻]

$^3O_2$ 73

-SO C=O ⟵/⟶ C=SO (48)

or

two products is affected by the solvent polarity and by the presence of or absence of an added sensitizer. Tamagaki *et al.*[127] reported that the photosensitized oxidation of **73** gave ketone and sulfine in a 2:1 ratio. They suggested that $^1O_2$ oxidation is competitive with $^3O_2$. Oxidation with $^1O_2$ yielded predominantly the sulfine. The sulfine does not react with either $^1O_2$ or $^3O_2$. Two reaction mechanisms were postulated in which triplet thioketone interacts with $^3O_2$ to afford the ketone via a biradical intermediate, and the thioketone can also interact with $^1O_2$ generated *in situ* to give the sulfine through a zwitterionic intermediate.

Steric effects play an important role in determining the ratio of ketone to sulfine. Under conditions where di-*tert*-butyl thioketone gives mainly sulfine, 2,2,4,4-tetramethylcyclobutanethione affords largely ketone.[129] Similarly, norbornane-2-thione (**74**) leads to ketone, whereas thiocamphor (**75**) gives both ketone and sulfine (equation 49).[130] These observations are interpreted in terms

(49)

	hv/O2 → ketone (=O)	+ sulfine (=SO)
74; R=H	62%	0%
75; R=Me	35%	14%

of a mechanism in which a dipolar acyclic intermediate (**76**) can cyclize to a dioxathietane (**77**), which in turn breaks down to give a ketone, or it can eliminate an oxygen atom to give the sulfine directly or via an oxathiirane (**78**) (equation 50). The possibility of two different acyclic intermediates is raised by

$$R_2C{=}S \xrightarrow{^1O_2} [R_2C{=}\overset{+}{S}{-}O{-}\overset{-}{O}]\ (\mathbf{76}) \longrightarrow [R_2C(S)(O{-}O)]\ (\mathbf{77}) \longrightarrow R_2C{=}O + [SO]$$

$$\mathbf{76} \xrightarrow{-[O]} [R_2C(S)(O)]\ (\mathbf{78}) \longrightarrow R_2C{=}SO; \quad \mathbf{76} \xrightarrow{-[O]} R_2C{=}SO \qquad (50)$$

consideration of different $^1O_2$ interactions with a non-bonding orbital on sulfur or with a π-orbital.[131] π-Orbital interaction, which predominates with diaryl thioketones, leads to ketone only, but *n*-orbital interaction, which is the major process for dialkyl or alkyl aryl thioketones, leads mainly to sulfine with some ketone formation. In a study of a wide range of alkyl thioketones, the quenching of $^1O_2$ was found to parallel the ionization energies of the *n*- rather than the π-electrons.[130]

Carlsen[132] carried out CNDO/B calculations on the reaction of thioformaldehyde (**79**) with $^1O_2$. He suggested the initial formation of an oxathiirane *O*-oxide intermediate (**80**) via a cheletropic reaction followed by rearrangement into a 1,2,3-dioxathietane (**81**), which decomposes via a diradical intermediate to give the ketone (equation 51).

$$H_2C{=}S \xrightarrow{^1O_2} \underset{\underline{80}}{\text{perepoxide-like adduct}} \longrightarrow \underset{\underline{81}}{\text{1,2,3-dioxathietane}} \longrightarrow \text{diradical} \qquad (51)$$

Thiocarbonyl compounds that can be photo-oxidized to the corresponding carbonyl compounds or sulfines include a cyclopropenethione,[133] *O*-alkyl thioesters,[134] 1,2-benzodithiole-3-thione,[135] di-*tert*-butyl thioketene[136,137] and sulfines.[138] Photosensitized oxidation of diphenylcyclopropenethione (**82**) yields diphenylcyclopropenone, in which a 1,2,3-dioxathietane intermediate (**83**) has been postulated (equation 52).[133]

$$\underset{\underline{82}}{Ph_2C_3{=}S} \xrightarrow{^1O_2} \left[\underset{\underline{83}}{\text{dioxathietane}}\right] \longrightarrow Ph_2C_3{=}O \ (+\ SO) \qquad (52)$$

O-Ethyl thioacetate is oxidized by light and oxygen to yield the corresponding ester in which a dioxathietane (**84**) is postulated as an intermediate (equation 53).[134].

$$CH_3C(S)OEt \xrightarrow{h\nu/O_2} \left[\underset{\underline{84}}{CH_3{-}C(OEt)(SOO)}\right] \longrightarrow CH_3C(O)OEt \ (+SO) \qquad (53)$$

Tamagaki and Hotta[135] reported the isolation of the sulfine in addition to the ketone in polymer-bound Rose Bengal-sensitized oxygenation of 1,2-benzodithio-3-thione (**85**) (equation 54).

$$\underset{\underline{85}}{\text{1,2-benzodithiole-3-thione}} \xrightarrow{h\nu/O_2/\text{(P)-RB}} \underset{5\%}{\text{C=SO}} + \underset{3\%}{\text{C=O}} \qquad (54)$$

(P)-RB; polymer-bound rose bengal

In the $^1O_2$ oxidation of the thioketene **86**, the thioketene *S*-oxide **87** was obtained in addition to other products (**88** and **89**) (equation 55).[136,137] This oxidation resembles that of thioketone.

$$\text{(}t\text{-Bu)}_2\text{C=C=S}\ (\underline{86}) \xrightarrow{^1O_2} \text{(}t\text{-Bu)}_2\text{C=C=SO}\ (\underline{87}\ 20\%) + \text{(}t\text{-Bu)}_2\text{C=SO}\ (\underline{88}\ 10\%) + t\text{-BuCOCH=CHCO}t\text{-Bu}\ (\underline{89}\ 30\%) \quad (55)$$

Zwannenberg *et al.*[138] reported that MB-sensitized oxygenation of sulfines afforded the corresponding ketones, although some sterically hindered sulfines are inert toward $^1O_2$. They proposed that $^1O_2$ gives a 1,2-cycloaddition reaction at the C—S bond followed by fragmentation to the ketone and SO_2 (equation 56).

$$\text{>C=SO} \xrightarrow{^1O_2} \left[\begin{array}{c} \text{>C—SO} \\ | \quad | \\ \text{O—O} \end{array} \right] \longrightarrow \text{>C=O} + SO_2 \quad (56)$$

Some nitrogen-containing derivatives behave differently. Thiourea has long been known to undergo chlorophyll-sensitized oxidation to cyanamide and sulfuric acid in a process that has been used in actinometry (equation 57).[139]

$$H_2N\text{–C(=S)–}NH_2 \rightleftarrows H_2N\text{–C(SH)=}NH_2 \xrightarrow[\text{chlorophyll}]{h\nu/O_2} \left[H_2N\text{–C(SOOH)=}NH_2 \right] (\underline{91}) \longrightarrow H_2N\text{–C(SO}_2\text{H)=}NH_2\ (\underline{90}) \longrightarrow\longrightarrow H_2NCN + H_2SO_4 \quad (57)$$

The sulfinic acid **90** can be isolated as an intermediate, which might come from a peroxysulfenic acid intermediate (**91**).[140,141] A similar product is formed from benzaldehyde thiosemicarbazide (equation 58).[140]

$$\text{PhCH=N–NH–C(=S)–}NH_2 \xrightarrow[\text{eosin}]{h\nu/O_2} \text{PhCH=N–N=C(SO}_2\text{H)–}NH_2 \quad (58)$$

1.5 Sulfur Ylides

Photosensitized oxidation of carbonyl-stabilized sulfur ylides has been investigated. Ando and co-workers[142,143] reported that ylides are relatively reactive toward $^1O_2$. Dimethylsulfonium and dimethyloxosulfonium biscarbomethoxymethylides (**92** and **93**) are photo-oxidized to give dimethyl sulfoxide, dimethyl sulfone and dimethyl mesoxalate (equation 59). Control

$$Me_2\overset{+}{S}-\overset{-}{C}(CO_2Me)_2 \xrightarrow{^1O_2} Me_2SO + Me_2SO_2 + O=C(CO_2Me)_2 \quad (59)$$

92 → 60 %, 38 %, 82 %

$$Me_2\overset{+}{S}(\rightarrow O)-\overset{-}{C}(CO_2Me)_2 \xrightarrow{^1O_2} Me_2SO + Me_2SO_2 + O=C(CO_2Me)_2$$

93 → 66 %, 16 %, 58 %

experiments indicate that these ylides do not react with $^3O_2$.[143] The elimination of dimethyl sulfoxide from the ylides (**92** and **93**) and the low yield of dimethyl sulfone from **93** suggest that the reaction involves not only the 1,2-dioxetane intermediate **94** but also the corresponding carbonyl oxide **95** (equation 60).[143]

$$Me_2\overset{+}{S}(\rightarrow O)-\overset{-}{C}< \xrightarrow{^1O_2} [Me_2\overset{+}{S}(\rightarrow O)-C(-O-O^-)<] \longrightarrow$$

$$\downarrow$$

$$[Me_2S(\rightarrow O)(-O-O-)C<] \longrightarrow Me_2SO_2 + >C=O \quad (60)$$

94

$$Me_2SO + [>\overset{+}{C}=O-\overset{-}{O}] \begin{cases} \xrightarrow{R_2SO} >C=O + R_2SO_2 \\ \xrightarrow{Ph_2SO} >C=O + Ph_2SO \end{cases}$$

95

The photo-oxidation of the ylides in the presence of diphenyl sulfide affords diphenyl sulfoxide as an oxygen-transfer product, presumably obtained from the reaction of the carbonyl oxide and diphenyl sulfide. Diphenyl sulfide is unreactive toward $^1O_2$.

In the photosensitized oxidation of dimethyloxosulfonium phenylphenacyl ylide (**96**), benzoic acid is obtained from the decomposition of a 2,3-dioxetane-like intermediate (**97**) (equation 61).[143] The cleavage of the C—S bond by elimination of dimethyl sulfoxide is observed as the main reaction process. Dimethyl sulfone is also produced either by the decomposition of the 1,2-

(61)

dioxetane-like intermediate **98** or by oxygen transfer from the carbonyl oxide **99** to dimethyl sulfoxide formed during the reaction. The corresponding reaction in methanol affords methyl benzoate, which is probably produced by attack of methanol on the intermediate **100**. This methanol attack on the ozonide **100** predominates over both the formation of the dioxetane **98** and the carbonyl oxide **99**.

1.6 Selenides and Tellurides

Photosensitized oxidation of dialkyl and alkyl aryl selenides affords the corresponding selenoxides, similar to the case with sulfides (equation 62).[144] The

$$R_1\text{-Se-}R_2 \xrightarrow[\text{MeOH}]{^1O_2} R_1\text{-Se}(\rightarrow O)\text{-}R_2$$

R_1	R_2
Ph	Me
Ph	CH_2Ph
Ph	C_3H_7–CHOH–$CH(C_3H_7)$–
Me	C_9H_{19}–$CH(CH_3)$–
Me	C_3H_7–CHOH–$CH(C_3H_7)$–
p-Tolyl	p-Tolyl

(62)

yields of selenoxides are very high and no selenone formation was observed. Photosensitized oxidation of dibenzyl selenide, however, gives benzaldehyde and diselenide instead of the *Se*-oxygenated product.[145] In methanol, diaryl selenides are also photo-oxygenated, where the initial formation of a perselenoxide intermediate (**101a**) or its alternatives (**101b** and **101c**) may be conceivable.[145]

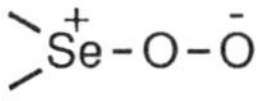

Recently, heavy atom quenching of $^1O_2$ by members of the $PhXCH_3$ series (X = Te, Se, S, O) has been reported.[18,146] A high value of the quenching rate constant was found for $PhTeCH_3$, $3.8 \times 10^9\ l\ mol^{-1}\ s^{-1}$.

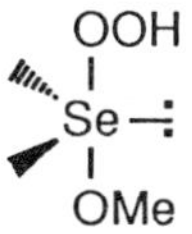

The mechanism of the oxidation of tellurapyrylium dyes (**102a–d**) by $^1O_2$ can be compared with the well studied oxidation of sulfides to sulfoxides, in which the formation of a persulfoxide or thiadioxirane intermediate is considered (equation 63).[147] Similar intermediates can be proposed for the reaction of the tellurapyrylium dyes with $^1O_2$ followed by the initial oxidized intermediate reacting with unoxidized tellurapyrylium dye. The tellurapyrylium dye **102d**

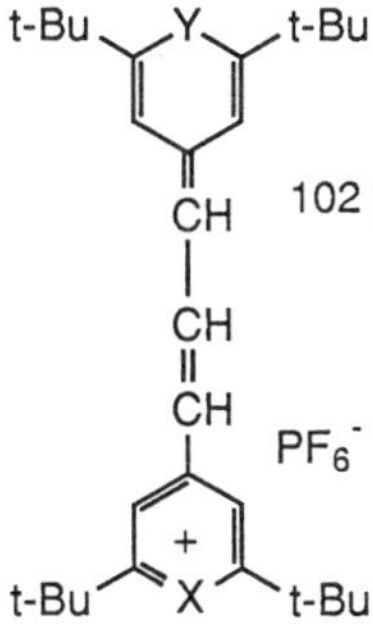

	X	Y
102 a	Te	O
b	Te	S
c	Te	Se
d	Te	Te

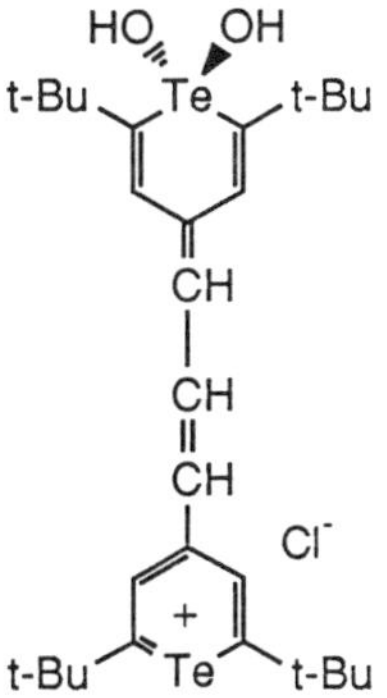

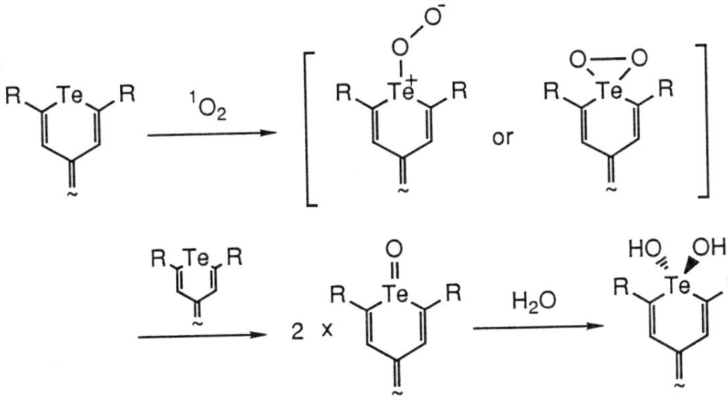

(63)

catalyzes the conversion of $^1O_2$ and water to hydrogen peroxide.[148] The catalytic reactions of **102d** utilize tellurium(IV) species (**103**) as an intermediate. The formation of **103** mediates the reduction of $^1O_2$ to hydrogen peroxide in the presence of water (equation 64).

$$\begin{aligned}&\underline{102d} + O_2 \xrightarrow{h\nu} \underline{102d} + {}^1O_2\\&2 \times \underline{102d} + 2\,H_2O + {}^1O_2 \longrightarrow 2 \times \underline{103}\\&\underline{103} \rightleftharpoons \underline{102d} + H_2O_2\end{aligned} \quad (64)$$

2 PEROXIDIC INTERMEDIATES IN PHOTOSENSITIZED OXIDATION OF ORGANONITROGEN COMPOUNDS

2.1 Amines

The photosensitized oxidation of aliphatic and aromatic amines has been extensively investigated for many years. Amines are well known as $^1O_2$ quenchers, with the quenching efficiency decreasing in the order tertiary > secondary > primary amines, for which Foote[83] and Bellus[149] have critically reviewed the data obtained up to 1978. Based on a correlation between quenching rates and ionization potentials, Ogryzlo and Tang[150] first proposed that the quenching of $^1O_2$ by alkyl amines may proceed via a charge-transfer intermediate. Gollnick and Lindner[151] observed that the photo-oxygenation rate of amines is also a function of their ionization potentials and proposed that both quenching and reaction of $^1O_2$ with amines pass through common intermediates. Young *et al.*[152] demonstrated that $^1O_2$ suffers physical quenching by a series of substituted *N*,*N*-dimethylanilines through partial charge-transfer intermediates in methanol.

Although it is well established that aliphatic amines may quench $^1O_2$ via the physical process (k_{phy}),[83,149,150,152–154] certain amines did react with $^1O_2$ to yield the oxygenated products via the chemical process (k_{chem}) (equations 65 and 66).[151,154–156]

$$^1O_2 + NR_3 \rightleftharpoons [\,^1O_2^{-} \cdots\cdots \overset{+}{N}R_3\,] \begin{cases} \xrightarrow{k_{phys}} {}^3O_2 + NR_3 \\ \xrightarrow[k_{chem}]{} \text{products} \end{cases} \quad (65)$$

$$-\overset{|}{\underset{H}{C}}-\overset{|}{N}- + {}^1O_2 \xrightarrow{H_2O} \rangle C{=}O + H{-}N\langle + H_2O_2 \qquad (66)$$

1) primary and secondary amines

$$-\overset{|}{\underset{H}{C}}-\overset{|}{N}-H + {}^1O_2 \longrightarrow -\overset{|}{\underset{H}{C}}-\overset{|+}{\underset{O-O^-}{N}}-H \rightleftharpoons -\overset{|}{\underset{H}{C}}-\overset{|}{\underset{OOH}{N}} \longrightarrow \rangle C{=}N- + H_2O_2$$

$$-\overset{|}{\underset{H}{C}}-\overset{|+}{\underset{O-O^-}{N}}-H \longrightarrow -\overset{|}{\underset{H}{C}}-\overset{|}{N}-H + {}^3O_2$$

$$\rangle C{=}N- \xrightarrow{H_2O} \rangle C{=}O + -NH_2$$

2) tertiary amines

$$-\overset{|}{\underset{H}{C}}-\overset{|}{N}- + {}^1O_2 \longrightarrow -\overset{|}{\underset{H}{C}}-\overset{|+}{\underset{O-O^-}{N}}- \longrightarrow \rangle C{=}\overset{+}{N}\langle + HOO^-$$

$$-\overset{|}{\underset{H}{C}}-\overset{|+}{\underset{O-O^-}{N}}- \longrightarrow -\overset{|}{\underset{H}{C}}-\overset{|}{N}- + {}^3O_2$$

$$\rangle C{=}\overset{+}{N}\langle \rightleftharpoons -\overset{|}{\underset{OOH}{C}}-N\langle \rightleftharpoons -\overset{|}{\underset{^-OO}{C}}-\overset{|+}{\underset{H}{N}}- \xrightarrow{H_2O} \rangle C{=}O + \rangle NH + H_2O_2$$

There have been several reports of the photosensitized oxidation of amines,[157] in which α-oxidation,[156,158] β-oxidation,[159] dehydrogenation[158,159] and dealkylation[155,159] have been observed. These reactions may take place through either or both of two well recognized mechanisms, Type I and Type II($^1O_2$) mechanisms.[21] In Type I photo-oxygenation the excited sensitizer interacts with the substrate to give a substrate radical, which reacts with oxygen to give the oxygenated products.

Schenck[157] first reported the photosensitized oxidation of amines, in which hydroperoxyamines (**104** and **105**) were the initial products of the oxygenation (equation 67).

Schaefer and Zimmerman[159] reported the first definitive product study of the photosensitized oxidation of primary, secondary and tertiary alkylamines (equation 68). Complex mixtures of products resulted from further reactions of the initially formed imine. The predominant processes were (a) hydrolysis to alkylamines and aldehydes, (b) β-oxidation of *N*-alkylidene groups to give hydroperoxides that led to formamides or α-keto aldehydes and (c) addition of hydroperoxide to imine followed by base-catalyzed rearrangement to amides.

The predominant products of the aerobic oxidation of alkylarylamines sensitized by benzophenone were carbonyl compounds and arylamines resulting from *N*-dealkylation (equation 69).[160]

$$Me_2CH{-}CH_2{-}NH_2 \xrightarrow{h\nu/sens/O_2} Me_2CH{-}CH(OOH){-}NH_2 \; (\underline{104})$$

$$\underline{104} \xrightarrow{-H_2O} Me_2CH{-}C(=O){-}NH_2$$

$$\underline{104} \xrightarrow{Red.} Me_2CH{-}CH(OH){-}NH_2 \longrightarrow Me_2CH{-}CH{=}O + NH_3$$

$$\text{pyrrolidine } (H_2C{-}CH_2{-}CH_2{-}NH{-}CH_2) \xrightarrow{h\nu/sens/O_2} \text{2,5-bis(hydroperoxy)pyrrolidine } (\underline{105}) \xrightarrow{Red.} \left[\text{2,5-dihydroxypyrrolidine}\right] \longrightarrow \text{pyrrole} \quad (67)$$

$$Bu_2NH \xrightarrow{h\nu/sens/O_2} PrCH{=}NBu \xrightarrow{H_2O_2} PrCH(OOH)NHBu \longrightarrow PrCONHBu$$

$$PrCH{=}NBu \xrightarrow{h\nu/sens/O_2} EtCH(OOH)CH{=}NBu \longrightarrow EtCHO + OHCNHBu \quad (68)$$

$$^3Ph_2C{=}O^* + (PhCH_2)_3N \longrightarrow Ph_2\dot{C}{-}OH + (PhCH_2)_2N\dot{C}HPh$$

$$Ph_2\dot{C}{-}OH + O_2 \longrightarrow Ph_2CO + HO_2\cdot$$

$$HO_2\cdot + (PhCH_2)_3N \longrightarrow H_2O_2 + (PhCH_2)_2N\dot{C}HPh$$

$$(PhCH_2)_2N\dot{C}HPh + O_2 \longrightarrow (PhCH_2)_2NCH(OO\cdot)Ph \longrightarrow (PhCH_2)N(\dot{C}HPh){-}CH(OOH)Ph$$

$$\longrightarrow PhCH_2N{=}CHPh + PhCHO + H_2O \quad (69)$$

Mechanisms of photosensitized oxidations of amines have been controversial. A Type I mechanism involving hydrogen transfer to give amino radical **106** via a triplet dye–amine charge-transfer complex was first suggested (equation 70).[160] Kinetic studies were interpreted in terms of the reaction of $^1O_2$ with amine in a charge-transfer process to give the same radical **106** (Type II) (equation 71).[154]

$$^3Dye^* + (RCH_2)_3N \longrightarrow Dye^{\dot{-}}(RCH_2)_3N^{\dot{+}} \longrightarrow DyeH^{\bullet} + (RCH_2)_2N\dot{C}HR \quad \underset{\textbf{106}}{} \qquad (70)$$

$$^1O_2 + (RCH_2)_3N \longrightarrow (RCH_2)_2\overset{+\bullet}{N}CH_2R + O_2^{\dot{-}} \longrightarrow (RCH_2)_2N\dot{C}HR + HO_2^{\bullet} \quad \underset{\textbf{106}}{} \qquad (71)$$

On the basis of kinetic experiments, Davidson and Trethewey[161] have shown that in the case of triethylamine the two processes, Type I and II, are operative simultaneously (equation 72), but no primary peroxidic intermediate has been proposed for an amine–$^1O_2$ adduct.

$$^3Dye^* \begin{cases} \xrightarrow{^3O_2} {}^1O_2 \xrightarrow{R_3N} \\ \xrightarrow[R_3N]{} \text{radicals} \xrightarrow[^3O_2]{} \end{cases} \longrightarrow \text{Oxidation products} \qquad (72)$$

Photosensitized oxidations of tropinone (**107a**) and pseudopelletierine (**107b**)[156] and related polycyclic *N*-methylamines and steroidal *N,N*-dimethylamines[14] gave mixtures of demethylation products and *N*-formyl derivatives (equation 73).

$$(CH_2)_n\ NMe \text{ bicyclic ketone } (=O) \xrightarrow{^1O_2} (CH_2)_n\ NCHO \text{ bicyclic ketone } (=O) \qquad (73)$$

107a; n=1
107b; n=0

Photosensitized oxidation of cyclohexylamine and dicyclohexylamine in organic media gave the respective hydroperoxides (**108** and **109**) as primary products,[163] in contrast to the charge-transfer photo-oxygenation, which afforded *N*-cyclohexylidenecyclohexylamine and cyclohexanone oxime (equation 74).[164] The initiation of the product formation could be reviewed according to an electron-transfer reaction, probably followed by the acid–base reaction (equation 75).

CT/O_2 hν/sens/O_2

NH_2 OOH 108 (74)

CT/O_2 hν/sens/O_2

NH OOH 109

$NH_2 \cdots O_2 \xrightarrow{h\nu} \dot{N}^{+}H_2 + O_2^{\cdot-}$ (75)

$\dot{N}^{+}H_2 + O_2^{\cdot-} \longrightarrow \dot{N}H + HO_2\cdot$

Photosensitized oxidation of triethylamine produced acetaldehyde and diethylamine.[165] The radical intermediates produced by the excitation of O_2–triethylamine in the region of the charge-transfer band were studied by the ESR method at various temperatures from 77 to 300 K, and $CH_3\cdot$, $CH_3CH_2\cdot$ and possibly $Et_2N(CH_3)CH\cdot$, $HOO\cdot$ and $Et_2NO\cdot$ were observed. The processes for photosensitized oxidation of triethylamine shown in equation 76 are conceivable.

1) $Et_3N \cdots O_2 \xrightarrow{h\nu} Et_3N^{+} \cdots O_2^{-}$

$Et_2N^{+}\text{-}CH(H)\text{-}CH_3 \cdots O\text{-}O^{-} \longrightarrow Et_2N\text{-}\dot{C}H\text{-}CH_3 \cdots \cdot O\text{-}OH$ (76)

2) $^3[Et_3N^{+} \cdots O_2^{-}] \longrightarrow {}^1[Et_3N^{+} \cdots O_2^{-}]$ or $^3[Et_3N] \cdots O_2 \xrightarrow{h\nu'} \dot{C}H_3, \dot{C}_2H_5, Et_2\dot{N}, Et_2N\dot{C}HCH_3$, etc.

$Et_2N\text{-}\dot{C}H\text{-}CH_3 + O_2 \longrightarrow Et_2N\text{-}CH(OO\cdot)\text{-}CH_3 \longrightarrow Et_2NO\cdot + CH_3CHO$

Saito *et al.*[166] reported the photo-oxidation of 4,5-bis(*N*,*N*-dimethylamino)-*o*-xylene (**110**) in which both Type I and II processes are operative. Of particular interest is that the two processes gave rise to different products. The formation of *N*-formyl derivative **111** depended on the type of sensitizer but was unaffected by known $^1O_2$ quenchers, and that of epoxyenone **112** was independent of the sensitizer type but was inhibited by $^1O_2$ quenchers (equation 77). The Type II process was believed to proceed via successive rearrangement of a Diels–Alder adduct (**113**).

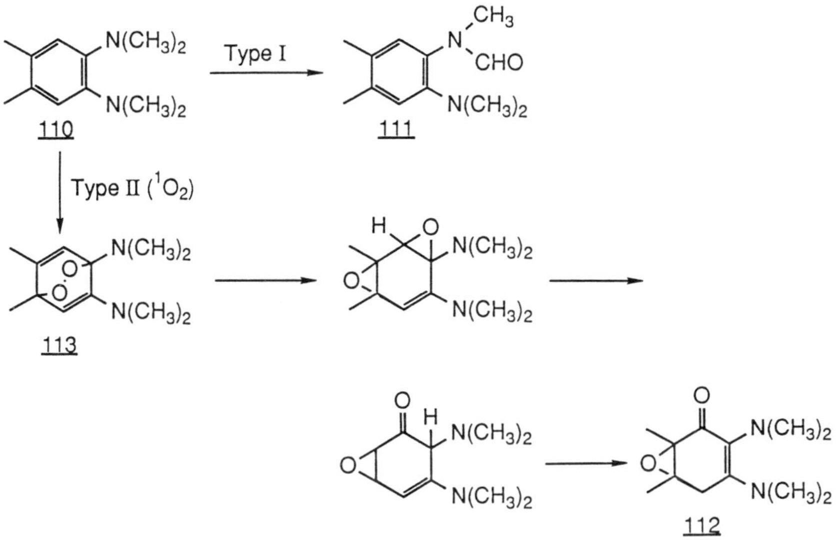

$$\mathbf{110} \xrightarrow{\text{Type I}} \mathbf{111}; \quad \mathbf{110} \xrightarrow{\text{Type II } (^1O_2)} \mathbf{113} \longrightarrow \longrightarrow \longrightarrow \mathbf{112} \tag{77}$$

$^1O_2$ has been reported to be involved in photosensitized oxidations resulting in the formation of stable nitroxyl radicals.[167]

In the $^1O_2$ reaction of diethylhydroxylamine, a high yield of hydrogen peroxide was obtained.[168] A chemical path such as that in equation 78 is dominant in the interaction between $^1O_2$ and the hydroxylamine.

$$Et_2NOH + {}^1O_2 \longrightarrow [Et_2NO \cdots H \cdots O_2]$$

$$\longrightarrow Et_2NO\cdot + HO_2\cdot \xrightarrow{Et_2NOH} Et_2NO\cdot + H_2O_2$$

$$\longrightarrow Et\overset{+}{N}(=CHCH_3)O^- + H_2O_2 \tag{78}$$

Aziridines have been shown to undergo photosensitized oxidation to a variety of products depending on the nature of the ring substituents (equation 79).[169]

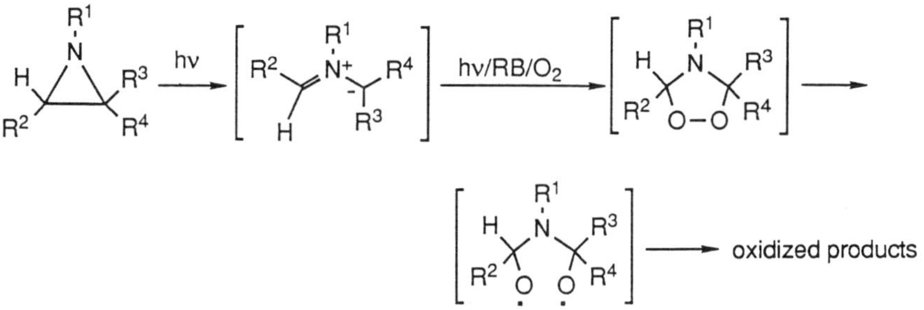

$$\xrightarrow{h\nu} \xrightarrow{h\nu/RB/O_2} \longrightarrow \longrightarrow \text{oxidized products} \tag{79}$$

Recently, Clennan *et al.*[170] have reported that hydrazines are among the most potent physical quenchers of $^1O_2$ known. The quenching mechanism is best characterized as a contact charge-transfer spin–orbit coupling-induced process.

A variety of organic amines are among substrates shown to react with $^1O_2$ to generate superoxide and donor cation radical. Manring and Foote[171] reported the detection of a cation radical in the reaction between photochemically generated $^1O_2$ and tetramethyl-*p*-phenylenediamine in aqueous media.

Saito *et al.*[172] reported chemical evidence for the formation of $O_2^{-\bullet}$ via one-electron transfer in the reaction between $^1O_2$ and *N,N*-dimethyl-*p*-anisidine in aqueous media (equation 80).

$$^1O_2 + \text{MeO}-C_6H_4-\text{NMe}_2 \longrightarrow O_2^{\overline{\bullet}} + \text{MeO}-[C_6H_4]^{+\bullet}-\text{NMe}_2 \quad (80)$$

Haugen and Whitten[173] recently reported a novel reaction for the photo-oxidative fragmentation of amino alcohols in which both $^1O_2$ and superoxide ion are sequentially involved as active intermediates (equation 81).

$$^1O_2 + H_2O + R^1\text{-CH(OH)-CH(NR}^3{}_2)R^2 \xrightarrow[\text{RB}]{h\nu} R^1CHO + R^2CHO + R^3{}_2NH + H_2O_2 \qquad \text{RB=rose bengal}$$

$$^1O_2 + R^1\text{-CH(OH)-CH(NR}^3{}_2)R^2 \longrightarrow O_2^{\overline{\bullet}}\ ,\ R^1\text{-CH(OH)-CH(}\overset{+\bullet}{N}R^3{}_2)R^2$$

$$O_2^{\overline{\bullet}}\ \ R^1\text{-CH(O-H)-CH(}\overset{+\bullet}{N}R^3{}_2)R^2 \longrightarrow HO_2^{\bullet},\ R^1CHO,\ R^3{}_2N\dot{C}HR^2 \qquad (81)$$

$$HO_2^{\bullet},\ R^1CHO,\ R^3{}_2N\dot{C}HR^2 \longrightarrow HO_2^- + R^1CHO + R^3{}_2N{=}CHR^2$$

Photo-oxidation of tertiary amines sensitized by electron acceptors such as 9,10-dicyanoanthracene has been applied as a useful *N*-dealkylation method (equation 82).[174,175]

$$R^1R^2N\text{-}CH_3 \xrightarrow{h\nu/\text{DCA}/O_2} R^1R^2NH \qquad (82)$$

Although amines have often been used as single-electron donors in many photoinduced redox studies, little is known of the chemical fate of the oxidized species so generated. Cadmium sulfide has been shown to sensitize the oxidative dealkylation of *N*-methylated aromatic amines[176] and titanium dioxide to sensitize the oxidation of lactams (equation 83), *N*-alkylamines[177] and aliphatic amines.[178]

hv/TiO2/O2

Me O

Me O

(83)

hv/TiO2/O2

O

Me

Me

Fox and co-workers[178,179] reported the identity of products formed by photocatalytic oxidation of a primary and a secondary aliphatic amine on irradiated TiO_2 powder suspended in oxygen-saturated acetonitrile. Two pathways are observed with a primary amine in which the reactivity is apparently controlled by the reaction conditions (e.g. initial amine concentration, nature of the sensitizer) (equation 84). These experiments represent the first characterizations of solution-phase aliphatic amine photo-oxidations sensitized by a heterogeneous metal oxide catalyst.

$$RCH_2NH_2 \xrightarrow{h\nu/TiO_2/O_2} RCHO + RCH_2NHCHO \qquad R{=}Ph(CH_2)_3$$

$$RCH_2\overset{+\bullet}{N}H_2 + O_2^{\bar{\bullet}} \xrightarrow{-H^+} R\dot{C}HNH_2 \xrightarrow{O_2} RCH(OO^\bullet)NH_2 \longrightarrow RCH(OOH)NH_2 \xrightarrow{-NH_2OH} RCHO \qquad (84)$$

$$R\dot{C}HNH_2 \underset{+e^-}{\overset{-e^-}{\rightleftarrows}} RCH{=}\overset{+}{N}H_2 \longrightarrow RCH{=}NH{-}CH_2R \xrightarrow{-H^+} R'CH{=}CH{-}NHCH_2R$$

$$R'CH{=}CH{-}NHCH_2R \xrightarrow{h\nu/TiO_2/O_2} R'\underset{O-O}{CHCH}NHCH_2R \longrightarrow R'CHO + RCH_2NHCHO \qquad R' = Ph(CH_2)_2$$

2.2 Azides

It is known that arylnitrenes react with oxygen to yield nitroarenes. Abramovitch and Challand[180] studied the photoreaction of phenyl azide with oxygen in acetonitrile and found nitrobenzene and tars as major photoproducts, together with trace amounts of azobenzene and aniline.

The photo-oxidation of phenyl azide (**114**) in oxygen-saturated acetonitrile solutions was investigated by Go and Waddell.[181] A reaction sequence is presented to account for the photo-oxidation of **114** (equation 85). Irradiation of **114** results in the loss of nitrogen and formation of phenylnitrene. The latter reacts quantitatively with oxygen to form nitrosobenzene, the only stable primary photoproduct. Nitrosobenzene then is either photo-oxidized to nitrobenzene or affords azoxybenzene.

$$PhN_3 \xrightarrow{h\nu} PhN \xrightarrow{O_2} PhNOO \xrightarrow{h\nu} PhNO_2$$

114

PhNOO ⇌ (hν, hν) azoxybenzene (PhN(O)=NPh); (Z)-PhN=NPh ⇌ (hν, hν) azoxybenzene ⇌ (hν, hν) (E)-PhN=NPh; (Z)-PhN=NPh ⇌ (hν, hν) PhN=NPh (85)

From trapping and tracer studies the reaction reveals the formation of phenyl nitroso oxides (**115**) as an electrophilic peroxy radical followed by isomerization unimolecularly to nitrobenzene.[182] This feature is described by structure **115b** rather than **115a** and cyclic **116**.

$Ph{-}N{=}\overset{+}{O}{-}O^-$ (115a) $Ph{-}\dot{N}{-}O{-}O\cdot$ (115b) $Ph{-}N\langle O{-}O \rangle$ (116)

The photolysis of 4-nitrophenyl azide gives triplet 4-nitrophenyl nitrene, which reacts with oxygen affording (4-nitrophenyl)nitroso oxide (**117**).[183] The latter may exist either as zwitterions or biradicals. In either case, dimerization of **117** to the six-membered ring diperoxide (**118**) is expected to occur rapidly. Two fragmentation paths from **118** are possible: one gives nitrosobenzene **119** and O_2 and the other forms dinitrobenzene (equation 86).

An ESR study in low-temperature matrices revealed some structural features of intermediate nitroso oxides.[184] Photolysis of *p*-diazidobenzene (**120**) in hydrocarbon solvents at 77 K results initially in the formation of the corresponding nitrene (**121**). This nitrene undergoes a slow chemical reaction with molecular oxygen to form two different oxygen adducts. One of these adducts is a ground-state triplet which is assigned as the *p*-azidophenyliminodioxy diradical structure **122a**. The second species is diamagnetic and has been assigned the dipolar nitroso oxide structure **122b**. This assignment is supported by their facile phototransformation into *p*-nitrophenyl azide **123**. The latter presumably proceeds via the dioxazirane **124a** or the dimeric species **124b** (equation 87).

Ar–N=N–Ar
Dimerization
$Ar-N_3 \xrightarrow{h\nu} {}^1[Ar-N_3]^* \xrightarrow[2.\ (ISC)]{1.\ (-N_2)} {}^3[Ar-N] \xrightarrow{RNO} Ar-N(\rightarrow O)=NR$
R=Ph, $(CH_3)_3C$
ISC
$^3[Ar-N_3]^*$
1. $(+O_2)$ 2. $(-N_2O)$
O_2
[Ar–N-O-O] 117
Dimerization
Ar–N=O 119 ← $-O_2$ [Ar–N(O-O)(O-O)N–Ar] 118 → $Ar-NO_2$ (86)
Ar=p-$NO_2C_6H_4$

120 → hν → [121] → O_2, 77K → [122a], [122b] → [124a], [124b] → 123 (87)

Photolysis of phenyl azide in an oxygen matrix at 15 K reveals the formation of the corresponding didehydroazepine (**125**).[185] Photochemical and thermochemical addition of molecular oxygen to **125** results in the formation of the isocyanate, which may come from a dioxetane intermediate (**126**) as shown in equation 88.

$PhN_3 \xrightarrow{h\nu}$ 125 $\xrightarrow[O_2]{\Delta \text{ or } h\nu}$ [126] → N=C=O / =O (88)

3 PEROXIDIC INTERMEDIATES IN PHOTOSENSITIZED OXIDATION OF ORGANOPHOSPHORUS COMPOUNDS

3.1 Phosphines and Phosphites

Only a few photosensitized oxidations of phosphorus compounds are known. It was shown that irradiation of trialkyl phosphite and benzyl diethyl phosphite in dry air or oxygen readily gives the corresponding phosphates in quantitative yield.[186] Triphenyl phosphite was only partially oxidized under the standard conditions of reaction. The process probably involves homolysis of otherwise stable unidentified impurities followed by a radical chain reaction (equation 89).

$$R\cdot + O_2 \longrightarrow ROO\cdot$$

$$ROO\cdot + (EtO)_3P \longrightarrow (EtO)_3\dot{P}OOR \longrightarrow (EtO)_3P{=}O + RO\cdot \qquad (89)$$

$$RO\cdot + (EtO)_3P \longrightarrow (EtO)_3\dot{P}OR \longrightarrow (EtO)_3P{=}O + R\cdot$$

The photosensitized oxidation of several trialkyl phosphites gave the corresponding phosphates (equation 90).[19]

$$(RO)_3P \xrightarrow{^1O_2} [\,(RO)_3P\cdot O_2\,] \xrightarrow{(RO)_3P} 2\,(RO)_3PO \qquad (90)$$

The photo-oxidation of triphenylphosphine in micelles in the presence of phenothiazine (MPT) resulted in the formation of triphenylphosphine oxide.[86] $^1O_2$ was suggested to be generated by the intramicellar recombination of $MPT^{+\cdot}$ and $O^{-\cdot}$ radical ion pairs.

Foote and co-workers reported that trimethyl phosphite is relatively inert toward $^1O_2$ but very efficient in trapping the intermediates in the photo-oxidation of diethyl sulfide[59] and biadamantylidene (equation 91).[187]

$$Et_2S \xrightarrow{^1O_2} [\,Et_2\overset{+}{S}{-}O{-}\overset{-}{O}\,] \xrightarrow{(MeO)_3P} Et_2SO + (MeO)_3PO$$

$$[\,Et_2\overset{+}{S}{-}O{-}\overset{-}{O}\,] \downarrow \quad [\,Et_2S\langle{}^{O}_{O}\,] \xrightarrow{Et_2S} 2\,Et_2SO$$

$$Ad{=}Ad \xrightarrow{^1O_2} \left[\,\text{Ad–Ad perepoxide } (O^+{-}O^-)\,\right] \longrightarrow \text{Ad–Ad dioxetane } (O{-}O)$$

$$\left[\,\text{perepoxide}\,\right] \xrightarrow{(PhO)_3P} \text{Ad–Ad epoxide} + (PhO)_3P{=}O \qquad \text{Ad = 2-adamantyl} \qquad (91)$$

Recently, it has been found that in $^1O_2$ oxidation of **127** co-oxidation of alkene to epoxide occurred in dichloromethane (equation 92).[188]

$$127 \xrightarrow[\text{norbornene}]{h\nu/MB/O_2} \text{P}{=}\text{O} + \text{epoxide} \qquad (92)$$

The photosensitized oxidation of coordinated phosphine in dithiolato cobalt complexes (**128**) proceeds via oxidative elimination of the coordinated phosphine by $^1O_2$ (equation 93).[189]

$$\text{CpCo(S}_2\text{C}_2\text{(CN)}_2) + Bu_3P \longrightarrow 128 \xrightarrow{^1O_2} \text{CpCo(S}_2\text{C}_2\text{(CN)}_2) + Bu_3P{=}O \qquad (93)$$

3.2 Phosphonium Ylides

The action of oxygen on phosphonium ylides affords a carbonyl compound and a phosphine oxide.[190] The carbonyl compound can react with excess of the starting ylide to form a symmetrical alkene (equation 94).

$$R^1R^2C{=}PPh_3 + O_2 \longrightarrow R^1R^2C{=}O + O{=}PPh_3$$

$$R^1R^2C{=}O + R^1R^2C{=}PPh_3 \longrightarrow R^1R^2C{=}CR^1R^2 \qquad (94)$$

The photo-oxygenation of the phosphonium ylide **129** gives triphenylphosphine oxide and diphenyl dithiocarbonate via a dioxetane intermediate (**130**).[191] The reaction is rapid and is self-sensitized by the yellow ylide (equation 95).

$$\underset{129}{Ph_3P{=}C(SPh)_2} \xrightarrow{h\nu/O_2} \underset{130}{\left[\begin{array}{c}O{-}O\\ |\quad\ |\\ Ph_3P{-}C(SPh)_2\end{array}\right]} \longrightarrow Ph_3P{=}O + O{=}C(SPh)_2 \qquad (95)$$

The reaction of methoxycarbonylalkylidenephosphoranes (**131**) with $^1O_2$ gave α-ketocarboxylic esters.[192] The mechanism is assumed to involve an electrophilic attack by $^1O_2$ on the carbanionic center of **131** to give the zwitterionic peroxide **132**, which then closes to the phosphetane **133** followed by cleavage to the products (equation 96).

$$Ph_3P{=}CRCO_2Me + {}^1O_2 \longrightarrow \left[Ph_3P^+{-}C(R)(CO_2Me){-}O{-}O^- \right] \; \underline{132}$$

131

$$\longrightarrow \left[Ph_3P{-}C(R)(CO_2Me){-}O{-}O \text{ (ring)} \right] \; \underline{133} \longrightarrow Ph_3P{=}O + O{=}C(R)CO_2Me \qquad (96)$$

The direct observation of phospha-1,2-dioxetanes (**134** and **135**), which could be derived from photo-oxygenation of the corresponding phosphonium ylide (**136**) and phosphazine (**137**), has been achieved by ^{31}P NMR studies at low temperature (equation 97).[193]

$$Ph_3P{=}C(Ph)CO_2Me \; (\underline{136}) \xrightarrow{{}^1O_2} Ph_3P{-}C(Ph)(CO_2Me){-}O{-}O \text{ (ring)} \; (\underline{134})$$

$$\underline{137}\; (\text{indanylidene}){=}N{-}N{=}PPh_3 \xrightarrow[-N_2]{{}^1O_2} \underline{135} \qquad (97)$$

4 PEROXIDES FROM PHOTOSENSITIZED OXIDATION OF ORGANOSILICON AND RELATED COMPOUNDS

4.1 Oxygenation of Disilenes and Digermenes

Disilenes (**138**) readily react with oxygen.[194–198] Some crystalline disilenes lose their characteristic color when they are exposed to air, yielding the oxygen adduct. The adduct was at first believed to have a 1,2-dioxadisiletane structure (**139**),[194] but was later shown to be a 1,3-isomer (**140**) (equation 98).[199] In solution the oxidation initially affords two compounds, a major amount of a 1,2-dioxetane (**139**) and a minor amount of an oxadisilirane (**141**).[195,196] Compound **139** slowly rearranges to **140**; **141** undergoes further oxidation to a 1,3-cyclodisiloxane (**140**). When (*Z*)- and (*E*)-**142** were oxidized, (*Z*)- and (*E*)-

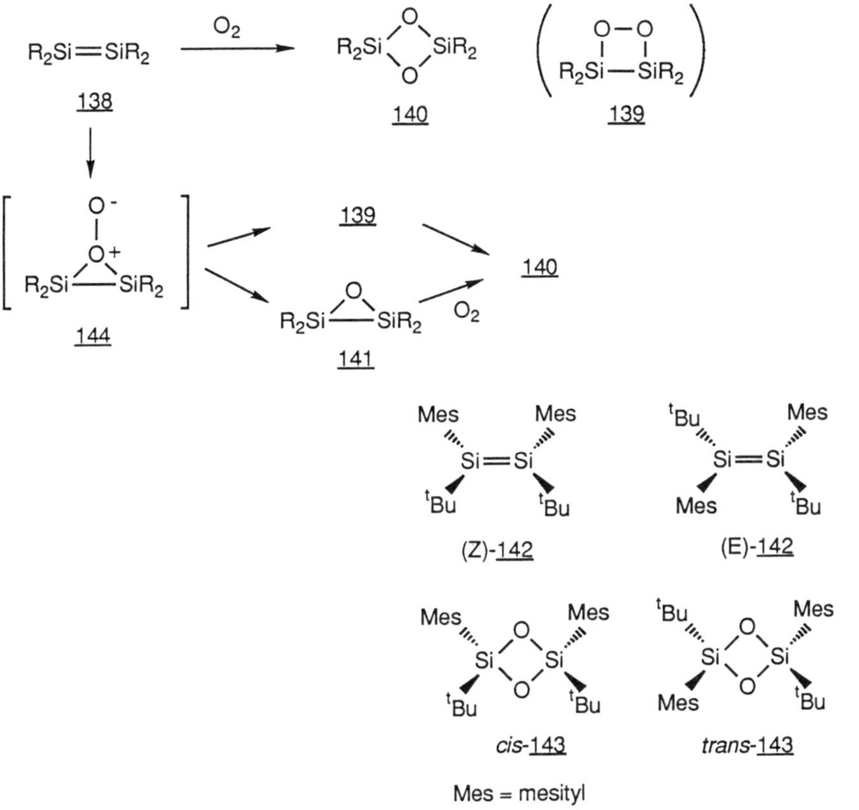

(98)

Mes = mesityl

1,3-cyclodisiloxane (**143**) were produced, respectively.[200] The results reveal that both oxygenation pathways through **139** and **141** undergo stereospecifically.

Although singlet oxygenation of alkenes gives 1,2-dioxetanes,[201,202] the silicon analog **139** arises from reaction of disilenes with triplet oxygen. Spin-pairing must occur at some point in the latter case. Moreover, 1,2-dioxetanes decompose to give the carbonyl compounds with accompanying chemiluminescence. Conversion of **139** to **140** does not, however, go through the silanone ($R_2Si{=}O$) species as an intermediate.[199,203] Since both **139** and **141** are formed simultaneously in the initial stage, it seems likely that they arise from a common intermediate such as the perepoxide **144**. Watanabe *et al.*[198] reported that the oxygenation of tetrakis(2,4,6-triisopropylphenyl)disilene (**145**) gives a 1-oxa-2-silacyclopent-3-ene derivative (**146**), which comes from the intermediate **147** via homolytic cleavage of the O—O bond followed by abstraction of α-hydrogen from *o*-isopropyl groups in aryl substituents (equation 99).

Exposure of a digermene (**148**) to oxygen leads to a 1,2-digermadioxetane (**149**), which on heating is isomerized to **150**, as seen in its silicon analog (equation 100).[204] Interestingly, photolysis of **149** yields **151**.

$$Ar_2Si{=}SiAr_2 \ (\underline{145}) \xrightarrow{O_2} \left[\begin{matrix} O-O \\ | \quad | \\ Ar_2Si-SiAr_2 \end{matrix} \ (\underline{147}) \right] \longrightarrow \underline{146} \quad (99)$$

Ar = 2,4,6-triisopropylphenyl

$$R_2Ge{=}GeR_2 \xrightarrow{O_2} \begin{matrix} O-O \\ | \quad | \\ R_2Ge-GeR_2 \end{matrix} \ (\underline{149}) \xrightarrow{h\nu} R_2Ge(\mu\text{-}O)_2GeR_2 \ (\underline{151}); \quad \underline{149} \xrightarrow{\Delta} \underline{150} \quad (100)$$

148a; R = 2,6-diethylphenyl
148b; R = 2,6-diisopropylphenyl

150a; R' = H, R = 2,6-diethylphenyl
150b; R' = Me, R = 2,6-diisopropylphenyl

Recently, the first cyclodistannoxane (**152**), an orange crystalline compound, was obtained by treatment of Sn_2R_4 [$R = CH(SiMe)_2$] with Me_3NO in n-C_6H_{14} as shown in equation 101.[205] Treatment of [SnR_2] with oxygen in the presence of [$Mo_2(OAc)_2(MeCN)_6$][BF_4] afforded an unexpected product, 1,2,4,3,5-trioxadistanolane (**153**), in a low yield (equation 102).[206] The mode of formation of **153** is not understood.

$$R_2Sn{=}SnR_2 \xrightarrow{Me_3NO} R_2Sn(\mu\text{-}O)_2SnR_2 \ (\underline{152}) \quad (101)$$

R = CH(SiMe$_3$)$_2$

$$2\,[R_2Sn] \xrightarrow[{[Mo_2(OAc)_2(MeCN)_6][BF_4]}]{O_2} \underline{153} \quad (102)$$

R = CH(SiMe$_3$)$_2$

4.2 Silenes

The reaction of silenes with oxygen is exemplified by the pyrolysis of dimethylsilacyclobutane (**154**) in the presence of oxygen,[207–210] which produced several oligomers of dimethylsilanone. The reaction presumably proceeds via cycloaddition of molecular oxygen to the transient silene followed by rapid decomposition of the siladioxetane to formaldehyde and dimethylsilanone (equation 103). There is no evidence for its being chemiluminescent.

$$\underset{\underline{154}}{\text{(silacyclobutane, } SiMe_2)} \xrightarrow{\Delta} H_2C{=}CH_2 + [H_2C{=}SiMe_2] \xrightarrow{O_2} [\text{siladioxetane: } SiMe_2\text{–}CH_2\text{, O–O}] \longrightarrow O{=}CH_2 + [O{=}SiMe_2] \rightarrow \text{oligomer} \qquad (103)$$

A solid silene (**155**), which is stable under an argon atmosphere, has been isolated from the photolysis of the acylsilane **156**.[208] When it was exposed to dry oxygen, however, vigorous oxidation took place. The major products are the cyclic trimer of bis(trimethylsilyl)silanone (**157**) and adamantanecarboxylic acid trimethylsilyl ester (**158**), which are presumably formed from breakdown of an intermediate siladioxetane (**159**) (equation 104).

$$\underset{\underline{156}}{(Me_3Si)_3Si\text{–}C(=O)\text{–}R} \xrightarrow{h\nu} \underset{\underline{155}}{(Me_3Si)_2Si{=}C(OSiMe_3)R} \xrightarrow{O_2} \underset{\underline{159}}{[(Me_3Si)_2Si\text{–}C(OSiMe_3)R\text{, O–O}]}$$

$$\longrightarrow \underset{\underline{157}}{[(Me_3Si)_2Si{=}O]} + \underset{\underline{158}}{Me_3SiO\text{–}C(=O)\text{–}R} \qquad (104)$$

$$\underline{157} \longrightarrow \text{cyclo-}[(SiMe_3)_2SiO]_3$$

R = 1-adamantyl

4.3 Oxygenation of Silicon–Silicon σ-Bonds

It is known that strained silicon–silicon σ-bonds are exothermically oxidized by molecular oxygen to afford disiloxanes as monooxygenated products.[20]

Carberry and West[211] reported that the cyclopentasilane **160** was gradually oxygenated to produce the corresponding oxidized compound **161** (equation 105). Kumada and co-workers[212,213] also investigated the oxidation of (*Z*)- and (*E*)-1,2-difluoro-1,2-dimethyl-1,2-disilacyclohexane (**162**) with molecular oxygen to afford stereospecifically (*Z*)- and (*E*)-2,7-dimethyl-2,7-difluoro-1-oxa-2,7-disilacycloheptane (**163**) respectively (equation 106).

160 $\xrightarrow{O_2}$ 161 (105)

(Z)-162 $\xrightarrow{O_2}$ (Z)-163

(E)-162 $\xrightarrow{O_2}$ (E)-163 (106)

Recently, Ando and co-workers reported results on the photosensitized oxidation of disiliranes **164**[214] and oxadisiliranes **165**[215] to give cyclic peroxides, 1,2,3,5-dioxadisilolanes **166** and 1,2,4,3,5-trioxadisiloranes **167** demonstrating the first example of dioxygen insertion into a silicon–silicon σ-bond. The results of trapping experiments and theoretical studies of the initially formed peroxidic intermediate revealed the intermediacy of peroxonium ions **168** and **169** (equation 107).

Ar_2Si–X–$SiAr_2$ $\xrightarrow{^1O_2}$ [168, 169] $\longrightarrow$ 166, 167 (107)

164a; X=CH_2, Ar=mesityl
164b; X=CH_2, Ar=H
165a; X–O, Ar=mesityl
165b; X=O, Ar=2,6-dimethylphenyl
165c; X=O, Ar=2,6-diethylphenyl
165d; X=O, Ar=2,6-diisopropylphenyl

X-ray crystal analysis of **167a** shows that the molecule possesses a twofold rotation axis which passes through the O-1 atom and bisects the O-2—O-2′ bond, and the central pentagonal core adopts a C_2 half-chair conformation as shown in Figure 1.[215] The aryl groups occupy a helical arrangement about each silicon atom. The structure may be regarded roughly as a 'similar figure' of that of secondary ozonides in the carbon series.[216] Compounds **166a**[214] and **167a**[215] smoothly rearranged to **170a** and **171a**, respectively. Both rearrangements presumably proceed via a Criegee-type mechanism. Reduction of **166a**[214] and **167a**[215] with triphenylphosphine afforded **172a** and **173a**, respectively (equations 108 and 109).

170a ←(silica gel)— 166a —(Ph_3P)→ 172a + Ph_3PO (108)

Ar=mesityl

171a ←(100°C)— 167a —(Ph_3P)→ 173a + Ph_3PO (109)

Ar=mesityl

Photosensitized oxidation of **164**[214] and **165**[215] in the presence of a sulfoxide as a nucleophilic oxygen-atom acceptor gave **172** and **173** with accompanying **166** and **167** and the corresponding sulfones, indicative of the intermediacy of peroxonium ions **168** and **169**, respectively (equation 110).

Theoretical studies of the products of the reaction of **164b** with $^1O_2$ offer a basis for the possible structure of the intermediate **168b**.[214] The optimized

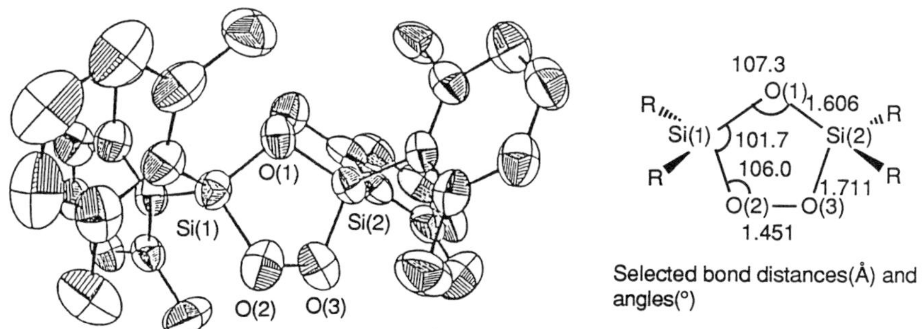

Figure 1. ORTEP diagram of **167a**

$$\left[\ Ar_2Si(\text{X})(\text{O}^+\text{–O}^-)SiAr_2\ \right] \xrightarrow{RSOR'} Ar_2Si(\text{X})(\text{O})SiAr_2 + RSO_2R' \qquad (110)$$

168; X=CH_2
169; X=O

172
173

structure of **168b** at the HF/6–31G* level is shown in Figure 2. The most stable conformation has C_s symmetry with pendant oxygen bending out of the molecular plane.

The UV absorption spectrum of disilirane (**164a**) and oxadisilirane (**165**) in an oxygen-saturated acetonitrile solution reveals a weak, broad contact charge-transfer (CT) band with a maximum at 300 nm.[217] The CT photo-oxygenation of disilirane (**164a**) and oxadisiliranes (**165c** and **165d**) gave **166a**, and **167c** and **167d**, respectively. The reaction intermediate was studied by Fourier transform IR spectroscopy and a disilirane–oxygen adduct (**174a**) was observed from UV irradiation of the contact CT band of **164a** in an oxygen matrix at 16 K.[217] A theoretical study of the intermediate also revealed the adduct **174b** (Figure 3).[217]

The photosensitized oxidation of digermiranes **175** also afforded the corresponding 1,2,3,5-dioxadigermolane (**176**) (equation 111).[218] The molecular structure of **176a** is shown in Figure 4.

$$Ar_2Ge(\text{X})GeAr_2 \xrightarrow{^1O_2} Ar_2Ge(\text{X})(\text{O–O})GeAr_2 \qquad (111)$$

175a; X=CH_2, Ar=2,6-diethylphenyl
175b; X=NPh, Ar=mesityl

176

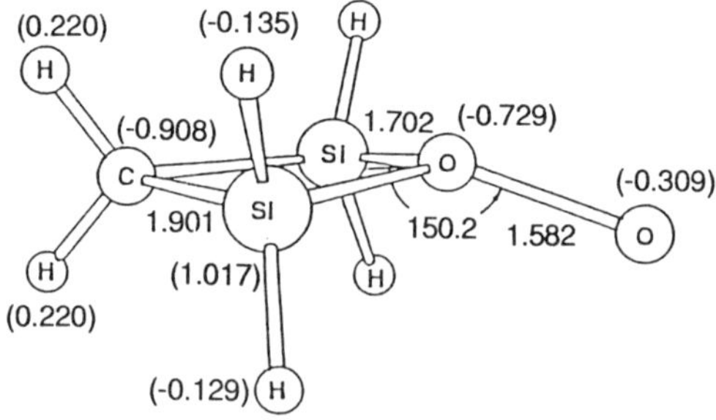

Selected bond distances(Å) and angles(°) and net charges in parentheses

Figure 2. HF/6–31G* optimized geometry of **168b**

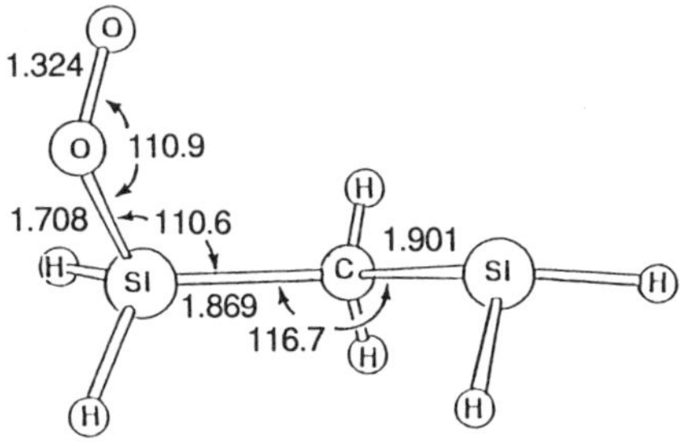

Selected bond distances(Å) and angles(°)

Figure 3. HF/6-31G* optimized geometry of **174b**

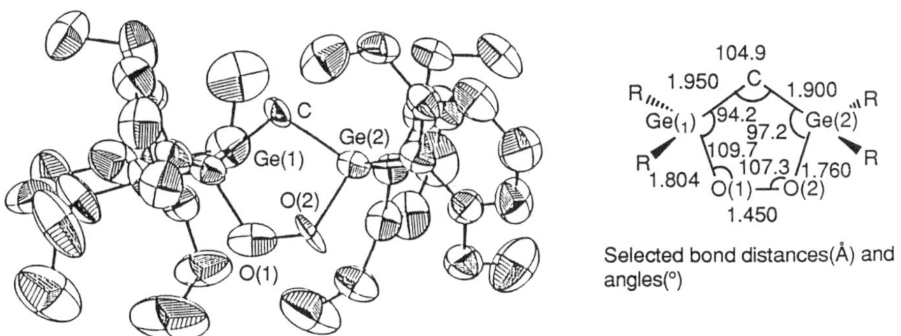

Figure 4. ORTEP diagram of **176a**

H_2
C
Ar_2Si $SiAr_2$ *
O O *
Ar=mesityl

175a

Photoinduced electron-transfer oxygenation of 1,2-disiletanes (**177**) afforded the corresponding 1,2,3,6-dioxadisilins (**178**) as a dioxygen insertion product into a silicon–silicon σ-bond together with 1,2,5-oxadisilolenes (**179**) (equation 112).[219] The results are reasonably accounted for by an electron transfer between disiletanes **177** and the sensitizer followed by addition of oxygen.

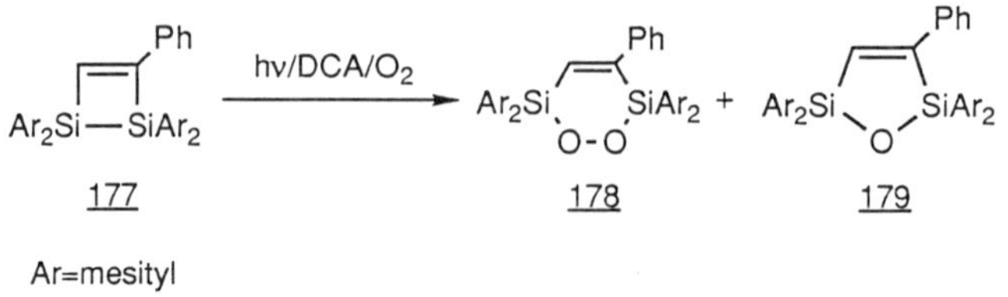

(112)

Ar=mesityl

4.4 Silylenes

The reaction of carbenes with molecular oxygen yields carbonyl *O*-oxides (**180**), dioxiranes (**181**) or carbonyl compounds (equation 113).[220,221] These species

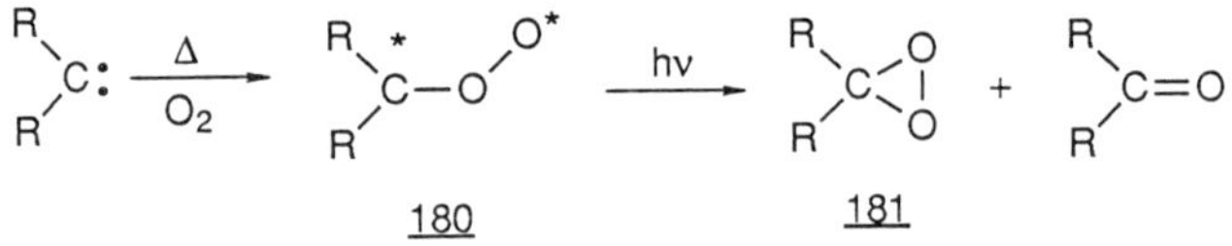

(113)

are powerful oxygen-atom transfer reagents; dioxiranes (**181**) especially have been proved to be useful in synthetic organic chemistry.[75,76] Kinetic experiments on the reaction of silylenes with molecular oxygen have been reported.[222] On the basis of the observation of an IR absorption at 16 K, the matrix-isolated dimesitylsilylene **182a**–O_2 adduct was assigned as the structure of dimesitylsilanone *O*-oxide (**183a**) (equation 114).[223] In contrast to **182a**, the

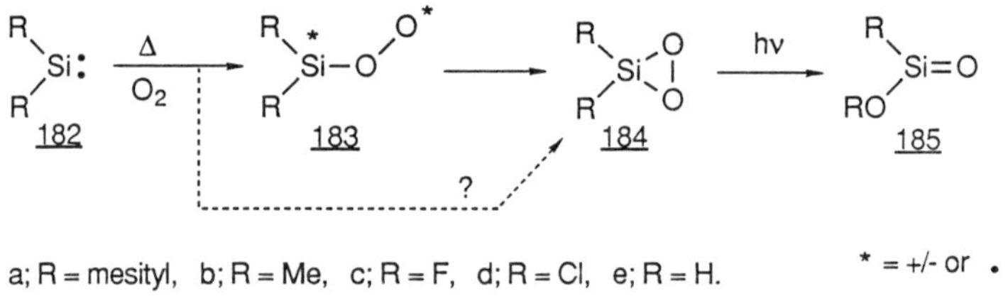

(114)

a; R = mesityl, b; R = Me, c; R = F, d; R = Cl, e; R = H. * = +/- or •

thermal reaction of dimethylsilylene (**182b**) with dioxygen at 35–42 K in an oxygen-doped argon matrix leads to the formation of dimethyldioxasilirane (**184b**) as the sole product, which on irradiation rearranges to methoxymethylsilanone (**185b**).[224,225] According to *ab initio* calculations (MP2/6-31G*) performed by Cremer *et al.*,[226] the rearrangement of **180** to **181** (R = H) shows an activation barrier of 22.8 kcal mol^{-1} and is exothermic by 34.5 kcal mol^{-1} (Figure 5). Therefore, the thermal rearrangement does not occur in solution or in the matrix. In contrast, Nagase *et al.*[227] and Patyk *et al.*[225] reported independently that the rearrangement of silanone *O*-oxide (**183e**) to dioxiranes (**184e**) exhibits an activation barrier of only 4–6 kcal mol^{-1}, which is sufficiently small for the isomerization to proceed rapidly even at low temperatures. However, as the high exothermicity of the isomerization (*ca*

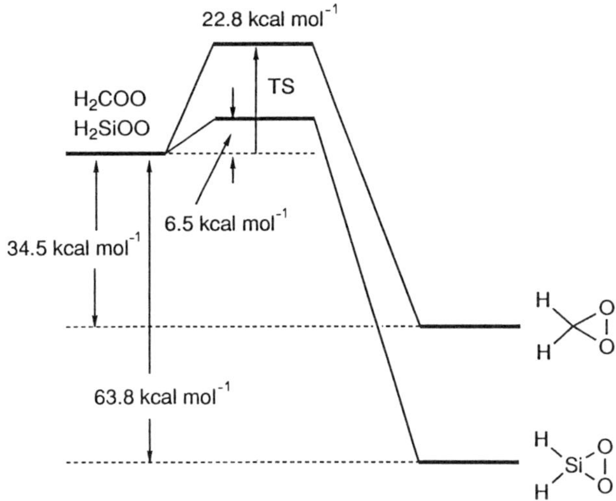

Figure 5. MP2/6-31G*-calculated activation barrier and reaction energies for the rearrangements of **180** and **182e**

60 kcal mol^{-1}) suggests, the resultant siladioxiranes are 'hot' with considerable excess energy and are subject to complex decomposition. The discrepancy between Ando's and Sander's results might result from the reaction conditions such as the substrates, oxygen concentration and temperature.

5 ACKNOWLEDGEMENTS

It is a pleasure to acknowledge the experimental efforts of our co-workers whose name are contained in the references. Special thanks are due to Professor S. Nagase (Yokohama National University) and Dr A. Yabe (National Chemical Laboratory for Industry) for their contributions. Financial support from the Ministry of Education, Science and Culture and University of Tsukuba is also greatly appreciated.

6 REFERENCES

1. C. S. Foote, in *Free Radicals in Biology*, Vol. 2 (W. A. Pryor, Ed.), Academic Press, New York (1976), Chap. 3.
2. N. I. Krinsky, in *Singlet Oxygen* (H. H. Wasserman and R. W. Murray, Eds.), Academic Press, New York (1979), Chap. 12.
3. J. D. Spikes, in *Oxygen and Oxy-Radicals in Chemistry and Biology*, (M. A. J. Rodgers and E. L. Powers, Eds), Academic Press, New York (1981), pp. 421–488.

4. D. C. Straight and J. D. Spikes, in *Singlet O_2*, Vol. IV (A. A. Frimer, Ed.), CRC Press, Boca Raton, FL (1984), Chap. 2.
5. M. Koizumi, S. Kato, N. Mataga, T. Matsuura and Y. Usui, *Photosensitized Reactions*, Kagakudojin Publishing, Kyoto (1978), Chap. 12.
6. S. Oae, in *Organic Chemistry of Sulfur* (S. Oae, Ed.), Plenum Press, New York (1977), Chap. 8.
7. E. Block, *Reactions of Organosulfur Compounds*, Academic Press, New York (1978), Chap. 1.
8. H. Schmidt and P. Karrer, *Helv. Chim. Acta*, **31**, 1017 (1948).
9. L. D. Wright, E. L. Cresson, D. E. Valliant, D. E. Wolf and K. Folkers, *J. Am. Chem. Soc.*, **76**, 4163 (1954).
10. C. J. Morris and J. F. Thompson, *J. Am. Chem. Soc.*, **78**, 1605 (1956).
11. L. J. Leed, B. G. DeBusk, C. S. Hormberger, Jr, and I. C. Gunsalus, *J. Am. Chem. Soc.*, **75**, 1271 (1953).
12. L. Weil, W. G. Gordon and A. R. Buchert, *Arch. Biochem. Biophys.*, **33**, 90 (1951).
13. W. J. Ray, Jr, H. G. Latham, Jr, M. Katsonlis and D. E. Koshland, Jr, *J. Am. Chem. Soc.*, **82**, 4743 (1960).
14. C. A. Benassi, E. Scoffone, G. Galiazzo and G. Jori, *Photochem. Photobiol.*, **6**, 857 (1967).
15. G. Jori, G. Galiazzo, A. Marzotto and E. Scoffone, *Biochim. Biophys. Acta*, **154**, 1 (1968).
16. C. S. Foote and J. W. Peters, *Int. Congr. Pure Appl. Chem., Spec. Lect., 23rd*, **4**, 129 (1971).
17. P. K. Sysak, C. S. Foote and T.-Y. Chang, *Photochem. Photobiol.*, **26**, 19 (1977).
18. R. Scurlock, M., Rougee, R. V., Bensasson, M. Evers and N. Derev, *Photochem. Photobiol.*, **40**, 733 (1991).
19. P. R. Bolduc and G. L. Goe, *J. Org Chem.*, **39**, 3178 (1974).
20. R. West, in *The Chemistry of Organic Silicon Compounds*, S. Patai and Z. Rappoport (Eds), Vol. 2, Wiley, Chichester (1989), Chap. 19.
21. K. Gollnick, *Adv. Photochem.*, **6**, 1 (1968).
22. C. W. Jefford and A. F. Boschung, *Helv. Chim. Acta*, **60**, 2673 (1977).
23. J. Ericksen, C. S. Foote and T. L. Packer, *J. Am. Chem. Soc.*, **99**, 6455 (1977).
24. W. Ando, T. Nagashima, K. Saito and S. Kohmoto, *J. Chem. Soc., Chem. Commun.*, 154 (1979).
25. I. Saito, K. Tamoto and T. Matsuura, *Tetrahedron Lett.*, 2899 (1979).
26. N. Berenjian, P. de Mayo, F. H. Phoenix and A. C. Weeden, *Tetrahedron Lett.*, 4197 (1979).
27. C. S. Foote, A. A. Dzakpasu and J. W.-P. Lin, *Tetrahedron Lett.*, 1247 (1975).
28. R. Tang, H. J. Hue, J. F. Wolf and F. Mares, *J. Am. Chem. Soc.*, **100**, 5248 (1978).
29. D. H. R. Barton, R. K. Haynes, G. Lecherc, P. D. Magnus and I. D. Menzies, *J. Chem. Soc., Perkin Trans.* **1**, 2055 (1975).
30. J. D. Spikes and M. L. Macknight, *Ann. N.Y. Acad. Sci.*, **171**, 149 (1970).
31. J. D. Spikes, *Ann. N.Y. Acad. Sci.*, **244**, 496 (1975).
32. T. Akasaka, A. Sakurai and W. Ando, *J. Am. Chem. Soc.*, **113**, 2696 (1991), and references cited therein.
33. W. Ando, T. Takata, in *Singlet O_2*, Vol. III (A. A. Frimer, Ed.), CRC Press, Boca Raton, FL (1985), Chap. 1.
34. P. D. Bartlett, in *Organic Free Radicals* (W. A. Pryor, Ed.), American Chemical Society, Washington, DC (1978), pp. 15–32.
35. G. O. Schenck and C. H. Krauch, *Angew. Chem.*, **74**, 510 (1962).
36. G. O. Schenck and C. H. Krauch, *Chem. Ber.*, **96**, 517 (1963).

37. C. H. Krauch, D. Hess and G. O. Schenck, unpublished results, cited in ref. 2.
38. H. Kautsky, *Biochem. Z.*, **291**, 271 (1937).
39. H. Kautsky, *Trans. Faraday Soc.*, **35**, 216 (1939).
40. C. S. Foote and S. Wexler, *J. Am. Chem. Soc.*, **86**, 3880 (1964).
41. B. M. Monroe, *Photochem. Photobiol.*, **29**, 761 (1979).
42. M. L. Kacher and C. S. Foote, *Photochem. Photobiol.*, **29**, 765 (1979).
43. C. S. Foote and J. W. Peters, *J. Am. Chem. Soc.*, **93**, 3795 (1971).
44. C. S. Foote and F. Jensen, *J. Am. Chem. Soc.*, **110**, 2368 (1988).
45. K. Inoue, T. Matsuura and I. Saito, *Tetrahedron*, **41**, 2177 (1985).
46. Y. Sawaki and Y. Ogata, *J. Am. Chem. Soc.*, **103**, 5947 (1981).
47. C.-L. Gu, C. S. Foote and M. L. Kacher, *J. Am. Chem. Soc.*, **103**, 5949 (1981).
48. E. L. Clennan and X. Chen, *J. Am. Chem. Soc.*, **111**, 5787, 8212 (1989).
49. E. L. Clennan and K. Yang, *J. Am. Chem. Soc.*, **112**, 4043 (1990).
50. C. S. Foote, R. W. Denny, L. Weaver, Y. Chang and J. W. Peters, *Ann. N.Y. Acad. Sci.*, **171**, 139 (1970).
51. C. S. Foote, *Pure Appl. Chem.*, **27**, 635 (1971).
52. G. Cauzzo, G. Gennari, F. Da Re and R. Curci, *Gazz. Chem. Ital.*, **109**, 541 (1979).
53. W. Ando, Y. Kabe and H. Miyazaki, *Photochem. Photobiol.* **31**, 191 (1980).
54. L. D. Martin and J. C. Martin, *J. Am. Chem. Soc.*, **99**, 3511 (1977).
55. C.-L. Gu, C. S. Foote and M. L. Kacher, *J. Am. Chem. Soc.*, **104**, 6060 (1982).
56. J.-J. Liang, C.-L. Gu, M. L. Kacher and C. S. Foote, *J. Am. Chem. Soc.*, **105**, 4717 (1983).
57. H. Kwart, N. A. Johnson, T. Eggerichs and T. J. Geroge, *J. Org. Chem.*, **42**, 172 (1977).
58. M. Matsumoto and K. Kuroda, *Abstracts of Symposium on Photochemistry, Sendai, Japan* (1977), p. 4.
59. K. Nahm and C. S. Foote, *J. Am. Chem. Soc.*, **111**, 1909 (1989).
60. E. J. Corey and C. Quannes, *Tetrahedron Lett.*, 4263 (1976).
61. T. Takata, Y. Tamura and W. Ando, *Tetrahedron*, **41**, 2133 (1985).
62. W. Ando, T. Akasaka and T. Takata, *J. Synth. Org. Chem. Jpn.*, **44**, 97 (1986), and references cited therein.
63. W. Ando, H. Miyazaki and T. Akasaka, *Tetrahedron Lett.*, **23**, 2655 (1982).
64. W. Ando, H. Sonobe and T. Akasaka, *Tetrahedron Lett.*, **27**, 4473 (1986).
65. T. Akasaka, M. Kako, H. Sonobe and W. Ando, *J. Am. Chem. Soc.*, **110**, 494 (1988).
66. G. E. Manser, A. D. Mesure and J. G. Tilett, *Tetrahedron Lett.*, 3153 (1969).
67. K. Kondo, A. Negishi and G. Tsuchihashi, *Tetrahedron Lett.*, 3173 (1969).
68. K. Kondo, A. Negishi and I. Ojima, *J. Am. Chem. Soc.*, **94**, 5786 (1971).
69. F. Jensen and C. S. Foote, *J. Am. Chem. Soc.*, **109**, 1478 (1987).
70. C. S. Foote and F. Jensen, *J. Am. Chem. Soc.*, **110**, 2368 (1988).
71. T. Akasaka, M. Haranaka and W. Ando, *J. Am. Chem. Soc.*, **113**, 9898 (1991).
72. F. J. Lovas and R. D. Suenram, *Chem. Phys. Lett.*, **51**, 453 (1977).
73. R. D. Suenram and F. J. Lovas, *J. Am. Chem. Soc.*, **100**, 5117 (1978).
74. R. W. Murray and R. Jeyaraman, *J. Org. Chem.*, **50**, 2847 (1985).
75. R. W. Murray, *Chem. Rev.*, **89**, 1187 (1989).
76. W. Adam, R. Curci and J. O. Edwards, *Acc. Chem. Res.*, **22**, 205 (1989).
77. T. Akasaka and W. Ando, *J. Chem. Soc., Chem. Commun.*, 1203 (1983).
78. W. Adam, W. Haas and G. Sieker, *J. Am. Chem. Soc.*, **106**, 5020 (1984).
79. P. D. Bartlett, T. Aida, H.-K. Chu and T.-S. Fang, *J. Am. Chem. Soc.*, **102**, 3515 (1980).
80. J. O. Edwards, R. H. Pater, R. Curci and F. DiFuria, *Photochem. Photobiol.*, **30**, 63 (1979).

81. P. D. Bartlett, G. D. Mendenhall and A. P. Schaap, *Ann. N.Y. Acad, Sci.*, **171**, 79 (1970).
82. C. S. Foote, M. Thomas and T.-Y. Ching, *J. Photochem.*, **5**, 172 (1976).
83. C. S. Foote, in *Singlet Oxygen* (H. H. Wasserman and R. W. Murray, Eds), Academic Press, New York (1979), Chap. 5, p. 114.
84. M. J. Shapiro and M. E. Landis, unpublished results, cited in ref. 22.
85. T. Iwaoka and M. Kondo, *Bull. Chem. Soc. Jpn.*, **47**, 980 (1974).
86. M. C. Hovey, *J. Am. Chem. Soc.*, **104**, 4196 (1982).
87. K. Inoue, T. Matsuura and I. Saito, *Tetrahedron*, **41**, 2177 (1985).
88. I. Saito, T. Matsuura and K. Inoue, *J. Am. Chem. Soc.*, **103**, 188 (1981); **105**, 3200 (1983).
89. T. Akasaka and W. Ando, *Tetrahedron Lett.*, **26**, 5049 (1985).
90. R. S. Davidson and J. E. Pratt, *Tetrahedron Lett.*, **24**, 5903 (1983).
91. M. A. Fox and A. A. Abdel-Wahab, *Tetrahedron Lett.*, **31**, 4533 (1990).
92. T. Tezuka, H. Miyazaki and H. Suzuki, *Tetrahedron Lett.*, 1959 (1978).
93. D. Sinnreich, H. Lind and H. Batzer, *Tetrahedron Lett.*, 3541 (1976).
94. T. Akasaka, A. Yabe and W. Ando, *J. Am. Chem. Soc.*, **109**, 8085 (1987).
95. E. Collinson and A. J. Swallow, *Chem. Rev.*, **56**, 471 (1956).
96. W. E. Savige and J. A. MacLaren, in *Organic Sulfur Compounds*, Vol. 2. (N. Kharasch, Ed.), Pergamon Press, Elmsford, NY (1966), Chap. 15.
97. M. Friedman, *The Chemistry and Biochemistry of the Sulfhydryl Group in Amino Acids, Peptides and Proteins*, Pergamon Press, Elmsford, NY (1973), Chap. 3.
98. G. Capozzi and G. Modena, in *The Chemistry of the Thiol Group, Part 2* (S. Patai, Ed.), Wiley, Chichester (1974), Chap. 17.
99. A. A. Oswald and T. J. Wallace, in *Organic Sulfur Compounds*, Vol. 2 (N. Kharasch, Ed.), Pergamon Press, Elmsford, NY (1966), Chap. 8.
100. M. Rougee, R. V. Bensasson, E. J. Land and R. Pariente, *Photochem. Photobiol.*, **47**, 485 (1988).
101. P. diMascio, T. P. A. Devasagayam, S. Kaiser and H. Sies, *Biochem. Soc. Trans.*, **18**, 1054 (1990).
102. S. F. Nelsen, P. J. Kinlen and D. H. Evans, *J. Am. Chem. Soc.*, **103**, 7045 (1981).
103. M. D. Bently, *US NITS PB Rep.*, PB-243599 (1975), p. 17; *Chem. Abstr.*, **84**, 168903e (1975).
104. G. Gennari, G. Cauzzo and G. Jori, *Photochem. Photobiol.*, **20**, 497 (1974).
105. M. S. Kharasch, W. Nudenberg and G. J. Mantell, *J. Org. Chem.*, **16**, 524 (1951).
106. A. J. Beckwith and R. D. Wagner, *J. Am. Chem. Soc.*, **101**, 7099 (1979).
107. J. D. Spikes, *Photochem. Photobiol.*, **34**, 549 (1981).
108. S. J. Jongsma and J. Cornelisse, *Tetrahedron Lett.*, **22**, 2919 (1981).
109. L. Weil, *Arch. Biochem. Biophys.*, **110**, 57 (1965).
110. E. Murakami, *Seikagaku*, **36**, 84, 88, 126, 797 (1964).
111. M. Calvin, H. Griseback and R. C. Fuller, *J. Am. Chem. Soc.*, **77**, 2659 (1959).
112. R. W. Murray, S. L. Jindal, *Photochem. Photobiol.*, **16**, 147 (1972).
113. R. W. Murray, R. D. Smetana and E. Block, *Tetrahedron Lett.*, 299 (1971).
114. R. W. Murray, S. L. Jindal, *J. Org. Chem.*, **37**, 3516 (1972).
115. F. E. Stary, S. L. Jindal and R. W. Murray, *J. Org. Chem.*, **40**, 58 (1975).
116. E. Block and J. O'Connor, *J. Am. Chem. Soc.*, **96**, 3921 (1974).
117. T. Akasaka, A. Sakurai and W. Ando, unpublished results.
118. N. Ishibe, M. Odani and M. Sunami, *J. Chem. Soc. B*, 1837 (1971).
119. N. Ishibe, M. Odani and K. Teramura, *J. Chem. Soc., Chem. Commun.*, 371 (1970).
120. R. Rajee and V. Ramamurthy, *Tetrahedron Lett.*, 5127 (1978).
121. L. Gattermann and H. Schultz, *Chem. Ber.*, **29**, 2944 (1896).
122. H. Staudinger and H. Freudenberger, *Chem. Ber.*, **61**, 1836 (1928).

123. A. Schönberg, O. Schultz and S. Nickel, *Chem. Ber.*, **61**, 2175 (1928).
124. A. Schönberg and A. Mostafa, *J. Chem. Soc.*, 275 (1943).
125. J. J. Worman, M. Schen and P. C. Nichols, *Can. J. Chem.*, **50**, 3923 (1972).
126. N. Suzuki, K. Sano, S. Wakatsuki, N. Tani and Y. Izawa, *Bull. Chem. Soc. Jpn.*, **55**, 3351 (1982).
127. S. Tamagaki, R. Akatsuka, M. Nakamura and S. Kozuka, *Tetrahedron Lett.*, 3665 (1979).
128. V. J. Rao and V. Ramamurthy, *Indian J. Chem.*, **19B**, 143 (1980).
129. V. J. Rao, K. Muthuramu and V. Ramamurthy, *J. Org. Chem.*, **47**, 127 (1982).
130. N. Ramnath, V. Ramesh and V. Ramamurthy, *J. Org. Chem.*, **48**, 214 (1983).
131. N. Ramnath, V. Ramesh and V. Ramamurthy, *J. Chem. Soc., Chem. Commun.*, 112 (1981).
132. L. Carlsen, *J. Chem. Soc., Perkin Trans. 2*, 188 (1980).
133. S. Singh, M. M. Bhadbhade, K. Venkatesan and V. Ramamurthy, *J. Org. Chem.*, **47**, 3550 (1982).
134. J. Gano and S. Atik, *Tetrahedron Lett.*, 4635 (1979).
135. S. Tamagaki and K. Hotta, *J. Chem. Soc., Chem. Commun.*, 598 (1980).
136. V. J. Rao and V. Ramamurthy, *J. Chem. Soc., Chem. Commun.*, 638 (1981).
137. V. J. Rao, V. Ramamurthy, E. Schassmann and H. Nimmesgern, *J. Org. Chem.*, **49**, 615 (1984).
138. B. Zwannenberg, A. Wagenaar and J. Strating, *Tetrahedron Lett.*, 4683 (1970).
139. A. Warburg and S. Schocken, *Arch. Biochem.*, **21**, 363 (1949).
140. G. O. Schenck and H. Wirth, *Naturwissenschaften*, **40**, 141 (1953).
141. G. O. Schenck, *Angew. Chem.*, **69**, 579 (1957).
142. W. Ando, S. Kohmoto and K. Nishizawa, *J. Chem. Soc., Chem. Commun.*, 894 (1978).
143. W. Ando, S. Kohmoto, K. Nishizawa and H. Tsumaki, *Photochem. Photobiol.*, **30**, 81 (1979).
144. L. Hevesi and A. Krief, *Angew. Chem., Int. Ed. Engl.*, **15**, 381 (1976).
145. T. Akasaka, M. Abe and W. Ando, unpublished results.
146. E. J. Corey, A. U. Khan and D.-C. Ha, *Tetrahedron Lett.*, **31**, 1389 (1990).
147. M. R. Detty and P. B. Merkel, *J. Am. Chem. Soc.*, **112**, 3845 (1990).
148. M. R. Detty and S. L. Gibson, *J. Am. Chem. Soc.*, **112**, 4086 (1990).
149. D. Bellus, *Adv. Photochem.*, **11**, 105 (1979).
150. E. A. Ogryzlo and C. W. Tang, *J. Am. Chem. Soc.*, **92**, 5034 (1970).
151. K. Gollnick and J. H. E. Lindner, *Tetrahedron Lett.*, 1903 (1973).
152. R. H. Young, R. L. Martin, D. Feriozi, D. Brewer and R. Kayser, *Photochem. Photobiol.*, **17**, 233 (1973).
153. C. Quannes and T. Wilson, *J. Am. Chem. Soc.*, **90**, 6527 (1968).
154. W. F. Smith, Jr, *J. Am. Chem. Soc.*, **94**, 186 (1972).
155. J. H. E. Lindner, H. J. Kuhn and K. Gollnick, *Tetrahedron Lett.*, 1705 (1972).
156. M. H. Fisch, J. C. Gramain and J. A. Oleson, *J. Chem. Soc., Chem. Commun.*, 13 (1970); 663 (1971).
157. G. O. Schenck, *Angew. Chem.*, **69**, 579 (1957).
158. R. F. Bartholomew and R. S. Davidson, *J. Chem. Soc., Chem. Commun.*, 1174 (1970).
159. F. C. Schaefer and W. D. Zimmerman, *J. Org. Chem.*, **35**, 2165 (1970).
160. R. F. Bartholomew and R. S. Davidson, *J. Chem. Soc. C*, 2342, 2347 (1971).
161. R. S. Davidson and K. R. Trethewey, *J. Chem. Soc., Perkin Trans. 2*, 173, 178 (1977).
162. D. Herlem, Y. Hubert-Brierre, F. Khuong-Huu and R. Goutarel, *Tetrahedron*, **29**, 2195 (1973).
163. E. G. E. Hawkins, *J. Chem. Soc., Perkin Trans. 1*, 13 (1972).

164. N. Kulevsky, C. Niu and V. I. Stenberg, *J. Org. Chem.*, **38**, 1154 (1973).
165. H. Tsubomura, T. Yagishita and H. Toi, *Bull. Chem. Soc. Jpn.*, **46**, 3051 (1973).
166. I. Saito, S. Abe, Y. Takahashi and T. Matsuura, *Tetrahedron Lett.*, 4001 (1974).
167. V. B. Ivanov, V. Y. Shlyapintokh, O. M. Khvostach, A. B. Shapiro and E. G. Rozantsev, *J. Photochem.*, **4**, 313 (1975).
168. M. V. Encinas, E. Lemp and E. A. Lissi, *J. Chem. Soc., Perkin Trans 2*, 1125 (1987).
169. V. Bhat and M. V. Geroge, *Tetrahedron Lett.*, 4133 (1977).
170. E. L. Clennan, L. J. Noe, E. Szneler and T. Wen, *J. Am. Chem. Soc.*, **112**, 5080 (1990).
171. L. E. Manring and C. S. Foote, *J. Phys. Chem.*, **86**, 1257 (1982).
172. I. Saito, T. Matsuura and K. Inoue, *J. Am. Chem. Soc.*, **103**, 188 (1981); **105**, 3200 (1983).
173. C. M. Haugen and D. G. Whitten, *J. Am. Chem. Soc.*, **111**, 7281 (1989).
174. J. Santamaria, R. Ouchabane and J. Rigaudy, *Tetrahedron Lett.*, **30**, 2927, 3977 (1989).
175. J. Santamaria, M. T. Kaddachi and J. Rigaudy, *Tetrahedron Lett.*, **31**, 4735 (1990).
176. T. Watanabe, T. Takizawa and K. Honda, *J. Phys. Chem.*, **81**, 1845 (1977).
177. J. W. Pavlike and S. Tantayanon, *J. Am. Chem. Soc.*, **103**, 6755 (1981).
178. M. A. Fox and M.-J. Chen. *J. Am. Chem. Soc.*, **105**, 4497 (1983).
179. M. A. Fox and J. N. Younathan, *Tetrahedron*, **22**, 6285 (1986).
180. R. A. Abramovitch and S. R. Challand, *J. Chem. Soc., Chem. Commun.*, 964 (1972).
181. C. L. Go and W. H. Waddell, *J. Org. Chem.*, **48**, 2897 (1983).
182. Y. Sawaki, S. Ishikawa and H. Iwamura, *J. Am. Chem. Soc.*, **109**, 584 (1987).
183. T.-Y. Liang and G. B. Schuster, *J. Am. Chem. Soc.*, **109**, 7803 (1987).
184. J. S. Brinen and B. Singh, *J. Am. Chem. Soc.*, **93**, 6623 (1971).
185. T. Akasaka, A. Yabe and W. Ando, unpublished results.
186. J. I. G. Cadogan, M. Cameron-Wood and W. R. Foster, *J. Chem. Soc.*, 2549 (1963).
187. M. Stratakis, M. Orfanopoulos and C. S. Foote, *Tetrahedron Lett.*, **32**, 863 (1991).
188. T. Akasaka, I. Kita and W. Ando, unpublished results.
189. H. Hatano, M. Kajitani, T. Akiyama, Y. Sakaguchi, J. Nakamura, H. Hayashi and A. Sugimori, *Chem. Lett.*, 1089 (1990).
190. H. J. Bestmann, L. Kisielowski and W. Distler, *Angew. Chem., Int. Ed. Engl.*, **15**, 298 (1976).
191. W. Adam and J.-C. Lin, *J. Am. Chem. Soc.*, **94**, 1205 (1972).
192. C. W. Jefford and G. Barchietto, *Tetrahedron Lett.*, 4531 (1977).
193. T. Akasaka, R. Sato and W. Ando, *J. Am. Chem. Soc.*, **107**, 5539 (1985).
194. R. West, M. J. Fink and J. Michl, *Science (Washington, DC)*, **214** 1343 (1981).
195. M. J. Michalczyk, R. West and J. Michl, *J. Chem. Soc., Chem. Commun.*, 1525 (1984).
196. H. B. Yokelson, A. J. Millevolte, G. R. Gillette and R. West, *J. Am. Chem. Soc.*, **109**, 6865 (1987).
197. S. Masamune, Y. Eriyama and T. Kawase, *Angew. Chem., Int. Ed. Engl.*, **26**, 584 (1987).
198. H. Watanabe, K. Takeuchi, K. Nakajima, Y. Nagai and M. Goto, *Chem. Lett.*, 1343 (1988).
199. M. J. Fink, D. J. DeYoung, R. West and J. Michl, *J. Am. Chem. Soc.*, **105**, 1070 (1983).
200. R. West, *Angew. Chem., Int. Ed. Engl.*, **26**, 1201 (1987).
201. P. D. Bartlett and M. E. Landis, in *Singlet Oxygen* (H. H. Wasserman and R. W. Murray, Eds), Academic Press, New York (1979), Chap. 7.
202. A. L. Baumstark, in *Singlet O_2*, Vol. II (A. A. Frimer, Ed.), CRC Press, Boca Raton, FL (1985), Chap. 1.

203. I. M. T. Davidson and A. Fenton, *Organometallics*, **4**, 2060 (1985).
204. S. Masamune, S. A. Batcheller, J. Park, W. M. Davis, O. Yamashita, Y. Ohta and Y. Kabe, *J. Am. Chem. Soc.*, **111**, 1888 (1989).
205. M. A. Edelman, P. B. Hitchcock and M. F. Lappert, *J. Chem. Soc., Chem. Commun.*, 1116 (1990).
206. C. J. Cardin, D. J. Cardin, M. M. Devereux and M. A. Convery, *J. Chem. Soc., Chem. Commun.*, 1461 (1990).
207. I. M. T. Davidson, C. E. Dean and F. T. Lawrence, *J. Chem. Soc., Chem. Commun.*, 52 (1981).
208. A. G. Brook, S. C. Nyburg, F. Abdesaken, B. Gutekunst, G. Gutekunst, R. K. M. R. Kallury, Y. C. Poon, Y.-M. Chang and W. Wong-Ng, *J. Am. Chem. Soc.*, **104**, 5667 (1982).
209. T. J. Barton, S. A. Burns, I. M. T. Davidson, S. Ijadi-Maghsoodi and I. T. Wood, *J. Am. Chem. Soc.*, **106**, 6367 (1984).
210. N. J. Wiberg, *J. Organomet. Chem.*, **273**, 141 (1984).
211. E. Carberry and R. West, *J. Organomet. Chem.*, **6**, 582 (1966).
212. K. Tamao, M. Kumada and M. Ishikawa, *J. Organomet. Chem.*, **31**, 17 (1971).
213. K. Tamao and M. Kumada, *J. Organomet. Chem.*, **31**, 35 (1971).
214. W. Ando, M. Kako, T. Akasaka, S. Nagase, T. Kawai, Y. Nagai and T. Sato, *Tetrahedron Lett.*, **30**, 6705 (1989).
215. W. Ando, M. Kako, T. Akasaka and Y. Kabe, *Tetrahedron Lett.*, **31**, 4177 (1990).
216. J. G. Gougoutas, in *The Chemistry of Peroxides*, (S. Patai, Ed.), Wiley, Chichester (1983), Chap. 12.
217. T. Akasaka, M. Kako, S. Nagase, A. Yabe and W. Ando, *J. Am. Chem. Soc.*, **112**, 7804 (1990).
218. M. Kako, T. Akasaka and W. Ando, *J. Chem. Soc., Chem. Commun.*, 457 (1992).
219. T. Akasaka, K. Sato, M. Kako and W. Ando, *Tetrahedron Lett.*, **32**, 6605 (1992); *Tetrahedron*, **48**, 3283 (1992).
220. W. Sander, *J. Org. Chem.*, **54**, 333 (1989), and references cited therein.
221. W. Sander, *Angew. Chem., Int. Ed. Engl.*, **29**, 344 (1990).
222. P. P. Gasper, D. Holton and S. Konieczny, *Acc. Chem. Res.*, **20**, 329 (1987), and references cited therein.
223. T. Akasaka, S. Nagase, A. Yabe and W. Ando, *J. Am. Chem. Soc.*, **110**, 6270 (1988).
224. A. Patyk, W. Sander, J. Gauss and D. Cremer, *Angew. Chem., Int. Ed. Engl.*, **28**, 898 (1989).
225. A. Patyk, W. Sander, J. Gauss and D. Cremer, *Chem. Ber.*, **123**, 89 (1990).
226. D. Cremer, T. Schmidt, J. Gauss and T. P. Radhakrishnan, *Angew. Chem., Int. Ed. Engl.*, **28**, 427 (1988).
227. S. Nagase, T. Kudo, T. Akasaka and W. Ando, *Chem. Phys. Lett.*, **163**, 23 (1989).

Addendum

A recent study of the photosensitized oxidation of sulfide shows that a stable structure for thiadioxirane **5** is calculated: Y. Watanabe, N. Kuriki, K. Ishiguro and Y. Sawaki, *J. Am. Chem. Soc.*, **113**, 2677 (1991).

13 Peroxides from Ozonization

KEVIN J. McCULLOUGH
Department of Chemistry, Heriot-Watt University, Edinburgh EH14 4AS, UK

and

MASATOMO NOJIMA
Department of Process Engineering, Faculty of Engineering, Osaka University, Yamadaoka, Suita, Osaka 565, Japan

1 INTERMEDIATE PEROXIDES IN OZONIZATION

The reaction of ozone with organic compounds, discovered in the middle of the 19th century, results in the formation of a variety of organic peroxides as either intermediates or stable products. Thus, ozone has been shown to be a versatile,

Organic Peroxides. Edited by W. Ando

alternative reagent to hydrogen peroxide and dioxygen (singlet or triplet) for the synthesis of organic peroxides.

The ozonization of organic compounds has already been comprehensively reviewed by Bailey in two volumes published in 1978 and 1982.[1] In this chapter recent developments in this topical area of chemistry, particularly over the past decade, are discussed with particular emphasis on the nature of the peroxidic products isolated from ozonization reactions.

Section 1 deals with the intermediate peroxides generated in the ozonization of alkenes, alkynes and saturated compounds. Although this subject has been amply reviewed,[1–4] recent spectroscopic studies have provided a more thorough understanding of the nature of reactive peroxidic species.

In Section 2, recent examples of carbonyl oxide cycloaddition reactions are described. Although the intermediate carbonyl oxides possess a high degree of 1,3-dipolar character,[5] they were generally considered to participate in cycloaddition reactions solely with carbonyl compounds, especially aldehydes, to produce ozonides (1,2,4-trioxolanes). Recently, however, it has been discovered that formaldehyde *O*-oxide can undergo cycloaddition to ethyl vinyl ether forming the corresponding 1,2-dioxolane.[6] In addition, several other new cycloaddition reactions of carbonyl oxides leading to novel cyclic peroxides have been subsequently reported.

Recent studies of the capture of ozonolysis intermediates by 'participating' (protic, nucleophilic) solvents are described in Section 3. Such reactions are reported to be useful methods for preparing various hydroperoxides. In Section 4, some transformations of the ozonization products into novel cyclic peroxides are summarized. Section 5 includes a number of miscellaneous peroxidic products isolated from ozonization reactions.

1.1 Alkenes

The classical Criegee mechanism, outlined in Scheme 1, describes the major features of the transformation of an alkene by ozonolysis into a secondary (or final) ozonide involving the formation of a primary ozonide **2**, decomposition of **2** to give a 'carbonyl oxide' (**3**) and a carbonyl compound (**4**), followed by recombination of the two fragments **3** and **4** to yield the final ozonide (**7**). Several recent studies have focused on the detailed chemistry of the primary ozonide **2** and the Criegee intermediate **3**, and the possibility of interconversion of **3** and the corresponding 1,2-dioxirane **6**.[7]

1.1.1 Ozone–alkene complexes

It has been speculated that ozone and an alkene substrate could initially form a weak π-complex prior to the formation of the primary ozonide.[8,9] A weakly bound charge-transfer (CT) complex between ozone and ethylene in the gas

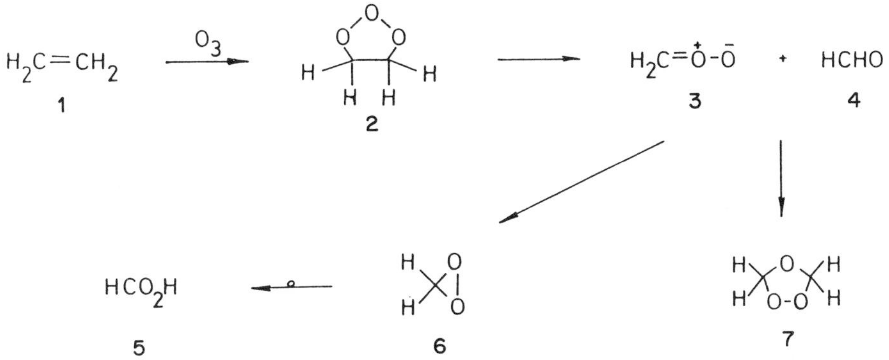

$H_2C{=}CH_2$ (1) $\xrightarrow{O_3}$ 2 $\longrightarrow$ $H_2C{=}\overset{+}{O}{-}\overset{-}{O}$ (3) + HCHO (4)

3 $\longrightarrow$ 6 $\longrightarrow$ HCO_2H (5)

3 + 4 $\longrightarrow$ 7

Scheme 1

phase has been observed and its structure investigated by pulsed Fourier transform microwave spectroscopy (Figure 1).[10] In the complex, the perpendicular separation of the two parallel molecular planes and the electronic dipole moment were estimated to be 3.30 Å and 0.461 D (cf. 0.532 D for ozone), respectively. A 1:1 complex between ethylene and sulfur dioxide was also found to exhibit similar structural features.[11]

The results of *ab initio* calculations, although in general agreement with a parallel plane approach of the two components, were consistent with transition states in which the ozone and the alkene are more tightly bound (C—O distance 2.4 and 2.0 Å, respectively) than in the observed π-complex between ethylene

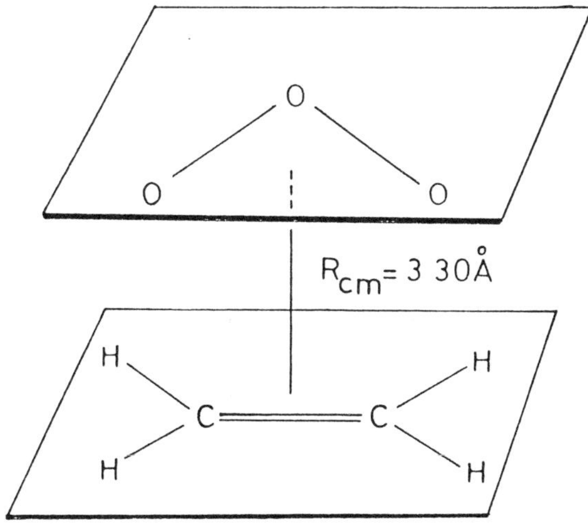

Figure 1. A charge-transfer complex between ethylene and ozone

and ozone.[12,13] One of the theoretical studies at the MP2(SDTQ) level also predicted that ozone and ethylene should form a weakly bound π-complex (*ca* 0.8 kcal mol^{-1} more stable than the reactants) prior to the cycloaddition transition state ($E_a \approx 1.1$ kcal mol^{-1}).[12]

Ozone–alkene mixtures, deposited in argon or krypton matrices at 12 K, produced broad, intense UV–visible absorptions which were assignable to ozone–alkene CT complexes.[14] With tetramethylethylene, trimethylethylene, (*Z*)- and (*E*)-but-2-ene, isobutene and propene, UV–visible absorption maxima were readily correlated with the ionization potentials of the corresponding alkenes. With (Z)-but-2-ene, for example, as the temperature of the matrix was raised the intensity of the absorption of the π-complex at 395 nm in the UV–visible spectrum was observed to decrease with concomitant appearance of an infrared absorption band at 1383.4 cm^{-1}, attributable to the corresponding primary ozonide. Estimates of the activation energy from Arrhenius plots for the transformation of the CT complex to the primary ozonide were in reasonable agreement with the calculated value for ozone and ethylene.[12]

1.1.2 Primary ozonides

There is considerable evidence to support the existence of a primary ozonide (PO) as a distinct chemical species formed irreversibly from the cycloaddition of ozone to the alkene. In earlier studies,[15–18] the intermediacy of a PO was deduced by the isolation of a diol from (*E*)- and terminal alkenes in reaction sequences similar to that outlined in Scheme 2. Since diols could not be isolated from (*Z*)-alkenes under similar conditions, it was concluded that (*Z*)-POs are less stable than the corresponding (*E*)-stereoisomers.

8 $\xrightarrow{O_3}$ 9 $\xrightarrow{(CH_3)_2CHMgBr}$ 10

Scheme 2

IR spectra of 1,2,3-trioxolanes, derived from the reaction of ozone with a series of simple ethylene derivatives in either carbon disulfide or ethane at −178°C, were found to contain three characteristic bands at 690, 970 and 1100 cm^{-1}. At higher temperatures, the 1,2,3-trioxolane decomposed with the formation of 1,2,4-trioxolanes and peroxy compounds.[19] The IR spectrum of the primary ozonide of ethylene (**2**) in a solid xenon matrix at 80–100 K has been recorded.[20] On using ozone enriched with 50% oxygen-18, the observed sextet splitting of the antisymmetric O—O—O stretching band at 647 cm^{-1} was consistent with the primary ozonide structure and directly characterized the weak O—O—O single bonds.

^1H NMR studies of ozonolysis reaction mixtures prepared by the rapid injection of tetramethylethylene into solutions of ozone in dichloromethane-d_2 at $-85°C$ indicated that the alkene was rapidly consumed and, of the two readily discernible peaks in the spectrum, the minor one at δ 2.20 ppm corresponded to acetone and the major peak at δ1.36 ppm was assigned to the primary ozonide.[21] The latter peak disappeared with time, as the peaks of acetone and the complex mixture of peroxide products (δ 1.40–1.63 ppm) increased. No signals that could be assigned to acetone *O*-oxide were observed. The decomposition of the primary ozonide in CD_2Cl_2 had $\Delta H^{\ddagger} = 39$ kJ mol^{-1} and $\Delta S^{\ddagger} = 28$ J mol^{-1} K^{-1}. Although the decomposition rate of the primary ozonide was unaffected by the presence of an eight fold excess of tetracyanoethylene, the yield of peroxide products was reduced by 80%. These results suggest that tetracyanoethylene did not react with the primary ozonide, but with another species formed from the primary ozonide, probably acetone *O*-oxide.

In similar studies of the ozonolysis of (*Z*)-but-2-ene between -90 and $-100°C$, the transient signals observed at δ 1.30 (CH_3) and δ 4.97 ppm (CH) were assigned to the primary ozonide 4,5-dimethyl-1,2,3-trioxolane. Consistent with the earlier observations, the primary ozonide from (*Z*)-but-2-ene was much less stable than that from tetramethylethylene, the half-lives being 0.06 and 0.74 s, respectively.

As determined by microwave spectroscopy, the lowest energy conformation adopted by the 1,2,3-trioxolane **2** in the gas phase has been shown to be an oxygen envelope (Figure 2).[22] The estimated O_2—$O_{1,3}$ moment in **2** (0.33 D) is similar in magnitude but opposite in sign to the O_2—$O_{1,3}$ moment of ozone (0.51 D). Thus, the central oxygen in **2** is more electronegative than the neighboring oxygens bonded to the ring carbons, whereas in ozone the central oxygen is electropositive with respect to the terminal oxygens. Ethylene PO **2** is

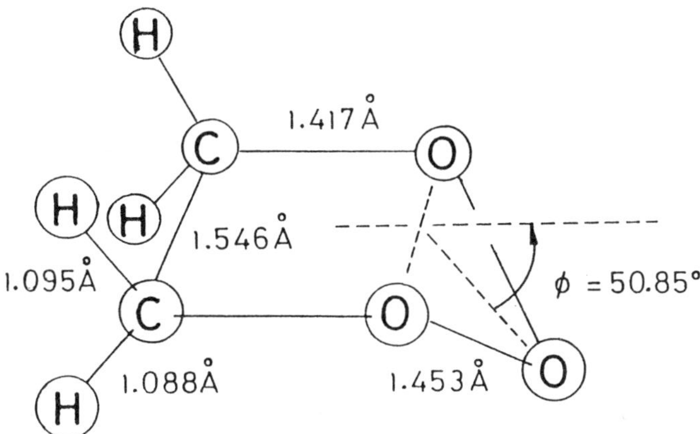

Figure 2. Structure of 1,2,3-trioxolane

thermally stable up to *ca* −90°C. On thermal decomposition it produced the secondary ozonide **7** and the dioxirane **6**.

A 1:1 adduct formed between ozone and 9-*tert*-butyl-10-methylanthracene has been isolated and characterized by X-ray crystallographic analysis.[23] Similar 1:1 adducts with ozone are obtained from 9-alkoxyanthracenes also.[24]

1.1.3 Carbonyl oxides

Carbonyl oxides are the key intermediates in the Criegee ozonolysis mechanism. The finite existence of such an intermediate has been demonstrated by the generation of **12** from the deuteriated cyclopentenylacetone derivative **11** and subsequent isolation of a 1:1 mixture of the ozonides, **13** and **14** (Scheme 3).[25] Direct observation of Criegee intermediates from the ozonolysis of alkenes by spectroscopic techniques has been precluded thus far because in-cage recombination of the carbonyl oxide with the co-produced carbonyl compound appears to be a rapid and efficient process. To investigate the nature of this important chemical species, some alternative methods of *in situ* generation of carbonyl oxides in the absence of carbonyl compounds have been developed, e.g. the photolysis of diazoalkanes in the presence of molecular oxygen.[26–28]

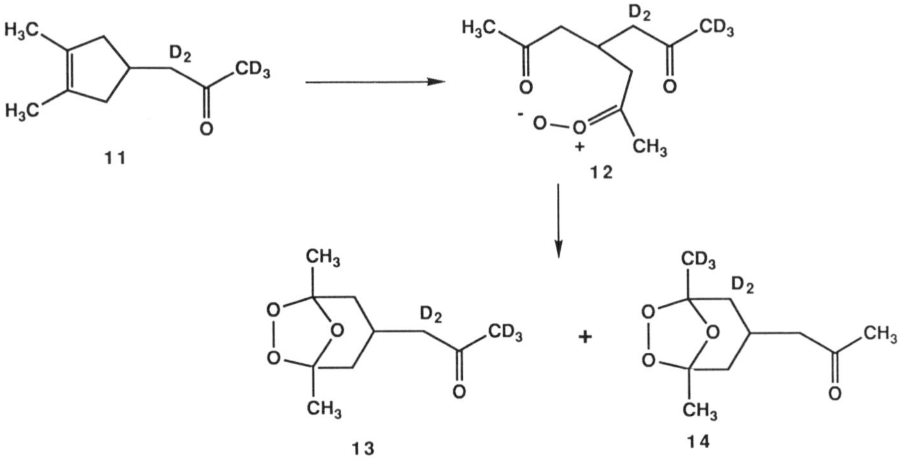

H_3C H_3C D_2 CD_3 O **11** → H_3C D_2 CD_3 O O $^-$O—O$^+$ CH_3 **12** → CH_3 O O O D_2 CD_3 O CH_3 **13** + CD_3 D_2 O O O CH_3 O CH_3 **14**

Scheme 3

The carbonyl oxide **16**, generated by the dye-sensitized photo-oxygenation of diphenyldiazomethane (**15**), behaved normally; [3 + 2] cycloaddition with added benzaldehyde yielded the ozonide **17** and capture by the solvent methanol provided the α-methoxy hydroperoxide **18** (Scheme 4).[29,30] Similarly, carbenes generated by direct photolysis of diazomethanes can also trap triplet oxygen to yield the corresponding carbonyl oxides. The rates of decay of the carbonyl oxides and their rates of reaction with added aldehydes have been determined.[26,31]

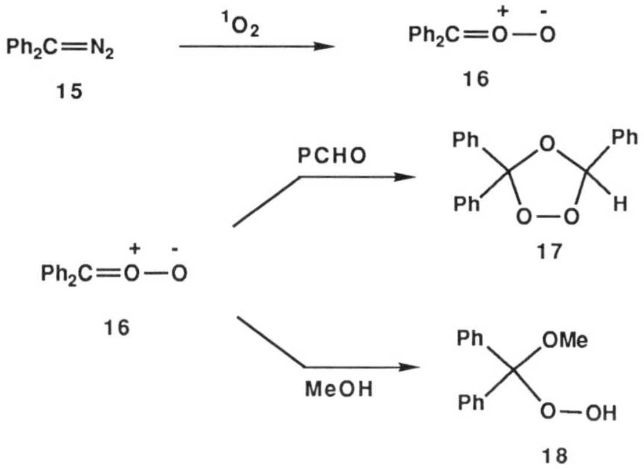

Scheme 4

The thermal and photochemical reactions of carbenes in oxygen-doped matrices have also been investigated.[32–35] The primary thermal adducts of free carbenes and molecular oxygen are carbonyl *O*-oxides, which were characterized spectroscopically by intense O—O vibrational stretches at $\nu \approx 900\ \text{cm}^{-1}$ and π–π^* electronic transitions at $\lambda_{max} \approx 400$ nm. On irradiation with long-wavelength radiation (500–630 nm), a carbonyl *O*-oxide such as **16** was found either to rearrange to the dioxirane **22** or to split off an oxygen atom. In contrast, dioxiranes were much more photostable than carbon oxides, requiring irradiation with blue or UV radiation before undergoing photorearrangement to the ester **20** (Scheme 5).

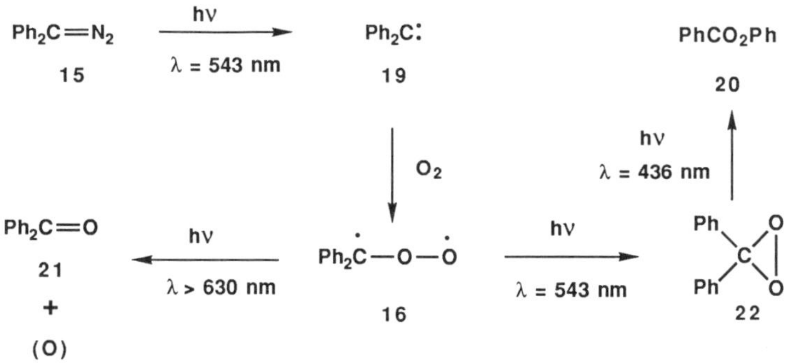

Scheme 5

Carbonyl oxides and/or related oxenoid species have been postulated as intermediates in a number of other reaction sequences. The reaction of 2,5-dimethylfuran with singlet oxygen produced an adduct which exhibited chemical properties more like a dioxirane than a carbonyl oxide.[36] Similarly, tetraphenylporphyrin-sensitized photo-oxygenation of 2-methyl-5-trimethylsilylfuran afforded quantitatively trimethylsilyl 2-oxopent-4-enoate.[37] Conversely, the dye-sensitized photo-oxygenation of 2-methoxy-5-phenylfurans (**23**) seems to proceed by the corresponding carbonyl oxides **24**.[38] Thus, the reaction of **23a**–**c** in anhydrous methanol at $-40°C$ gave quantitatively the α-methoxyalkyl hydroperoxides **25**, whereas the reaction of **23a** and **b** in acetone resulted in the formation of the 1,2,4-trioxolanes **26a** and **b** (Scheme 6).

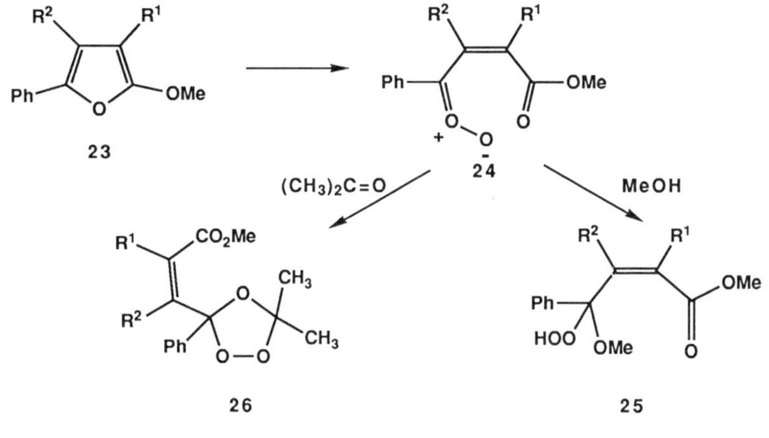

a: R^1 = CO_2Me, R^2 = H b: R^1 = COMe, R^2 = H c: R^1 = R^2 = H

Scheme 6

On heating the furan endoperoxide **27**, however, it rearranged into the enol ester **31**, whereas the corresponding saturated ozonide **28** afforded instead the isomeric ozonide **32** (Scheme 7).[39] The above results have suggested that the energy barrier between the carbonyl oxide and dioxirane valence isomers may not be high enough to prevent their interconversion (see below).

The reaction of the azomethine imine **33** with singlet oxygen provided cyclohexanone *O*-oxide (**34**), which reacted normally in the presence of methanol or acetaldehyde to produce either the corresponding α-methoxy hydroperoxide (**35**) or the 1,2,4-trioxolane (**36**) as appropriate (Scheme 8).[40] Dimethyl oxosulfonium and pyridinium bis(methoxycarbonyl)methylides[41] and adamantanone azine[42] were found to act as precursors of the corresponding carbonyl oxides and/or the related oxenoid species under photo-oxygenation conditions.

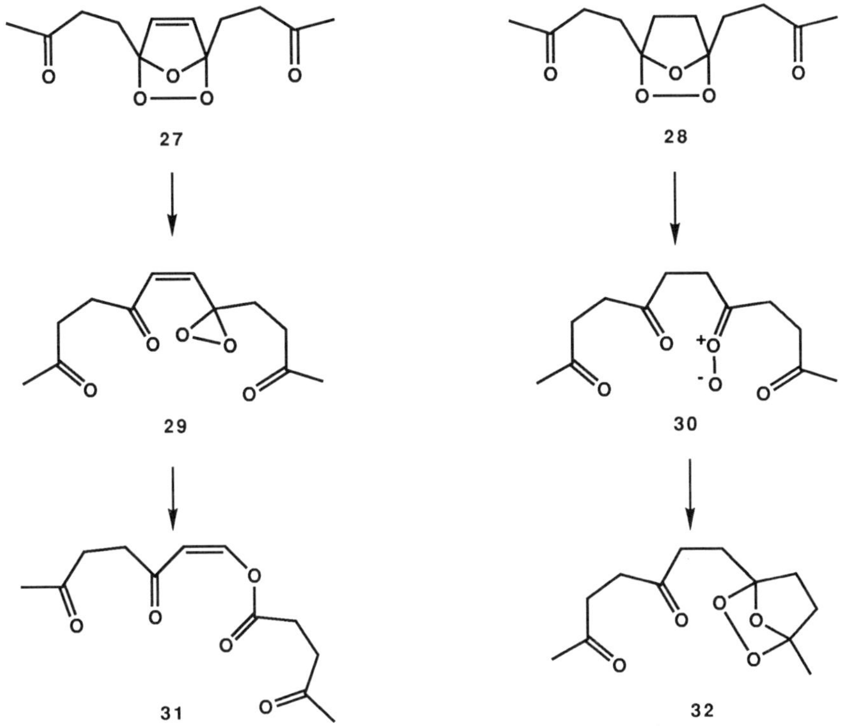

27 28

29 30

31 32

Scheme 7

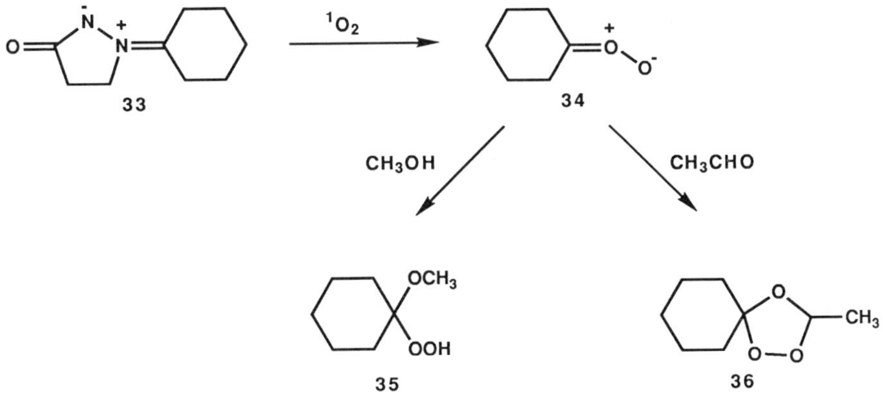

$^1O_2$

33 34

CH_3OH CH_3CHO

OCH_3 OOH CH_3

35 36

Scheme 8

Based on the nature of the isolated non-peroxidic products **40** and **41**, methylene blue-sensitized photo-oxygenation of adamantylidenecyclopropane (**37**) has been postulated to proceed via the cyclobutanone *O*-oxide **39** obtained by ring expansion of the initially formed zwitterionic intermediate **38** (Scheme 9).[43]

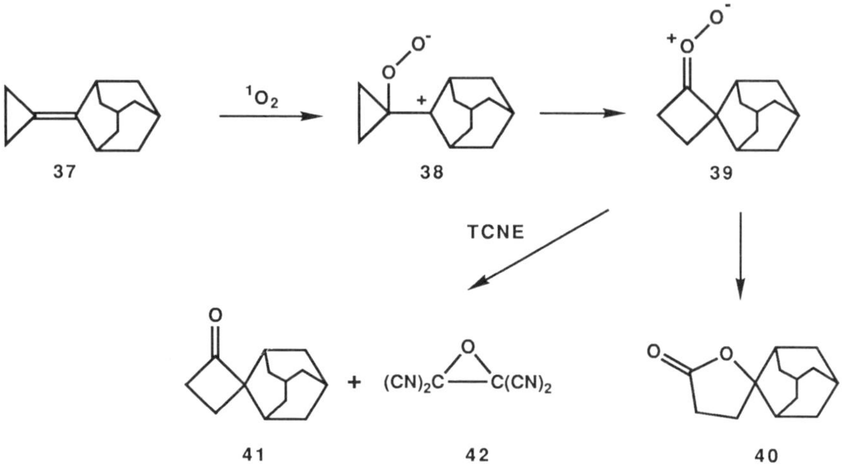

Scheme 9

During the oxidation of alkenes by (*meso*-tetraphenylporphyrinato) manganese(III) chloride–sodium hypochlorite, as a model of cytochrome P-450, it was found that the oxidation can be catalyzed by the co-existing phenylacetaldehyde.[44] This result has been interpreted in terms of an oxygen-atom transfer from the manganese $Mn^{V}{=}O$ species to the aldehyde yielding phenylacetaldehyde *O*-oxide or the isomeric dioxirane, which in turn efficiently epoxidized alkenes.

The ozonolysis of 1,4-dibromo-2,3-dimethyl-2-butene (**43**) in acetone has been found to yield significant amounts of the dimer (**48**) and trimer (**49**) formally derived from acetone *O*-oxide (**47**). Since **47** cannot be obtained directly by ozonolysis of either alkene (**43**) or the solvent acetone (**46**), there must be oxygen-atom transfer from the intermediate bromoacetone *O*-oxide (**44**) (or its dioxirane equivalent **45**, as preferred by the authors) to **46** to yield acetone *O*-oxide (**47**) (Scheme 10).[45]

The best known reactions of carbonyl oxides are [3 + 2] cycloadditions with carbonyl compounds and their capture by protic, nucleophilic solvents as mentioned previously (see also Section 2). More recently it has been reported that, under certain circumstances, carbonyl oxides may also behave as oxygen-atom transfer agents. The range of substrates reported to date includes alkenes,[45–50], aromatic compounds,[46,51–53] saturated hydrocarbons,[46,54]

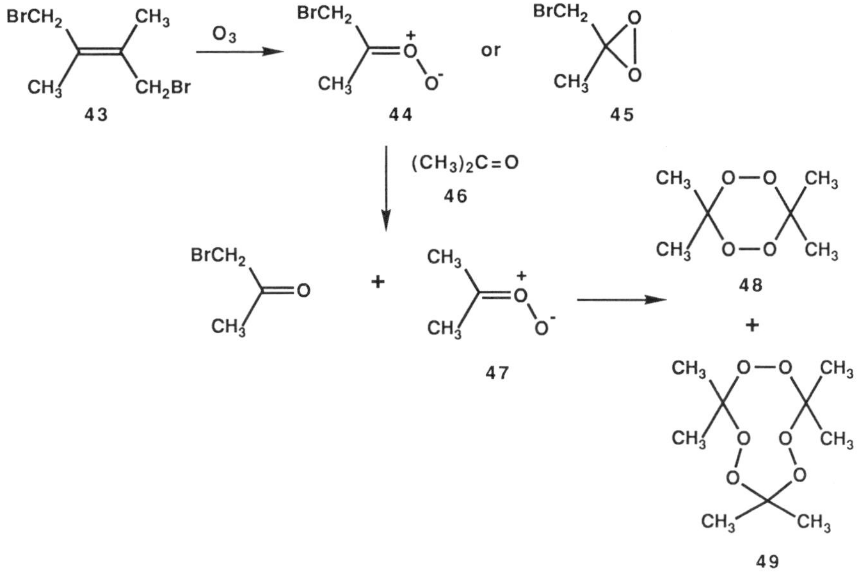

Scheme 10

sulfoxides,[46,55,56], sulfides,[46,55–57] ketenes[58] and sulfur dioxide[59] (Scheme 11), in addition to carbonyl compounds.[45] On the basis of the relative reactivities of these substrates ($Ph_2SO > Ph_2S >$ C=C > benzene) and the substituent

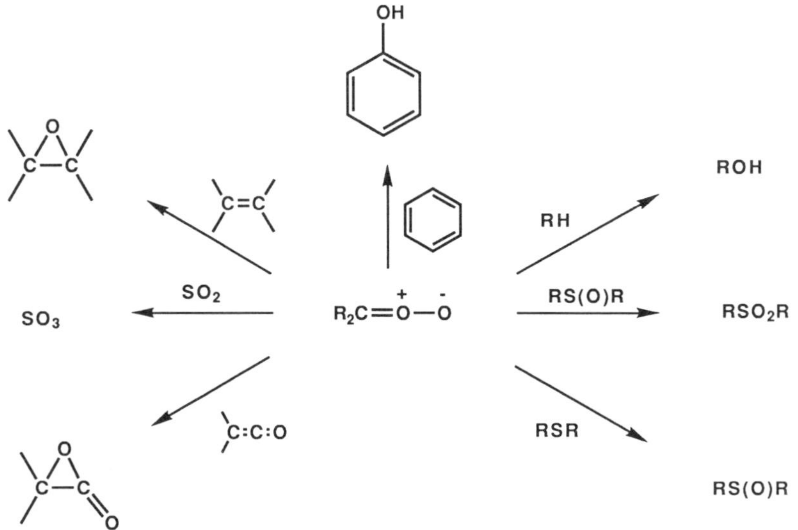

Scheme 11

electronic effects in the oxidation of *para*-substituted diphenyl sulfoxides, it has been claimed that the conventional alkyl- and aryl-substituted carbonyl oxides are nucleophilic oxygen-atom transfer agents although electron-withdrawing substituents could modify the nature of carbonyl oxides significantly. Thus, the electron-deficient α,α,α-trifluoroacetophenone *O*-oxide, generated from phenyltrifluorodiazomethane, oxidized both sulfides and sulfoxides electrophilically.[60] A remarkable difference in behaviour of nucleophilic dibenzotropone *O*-oxide and electrophilic 1,4-naphthoquinone *O*-oxide in the oxidation of thianthrene 5-oxide (**50**) has been noted (Scheme 12)[55] (see also refs. 52 and 57).

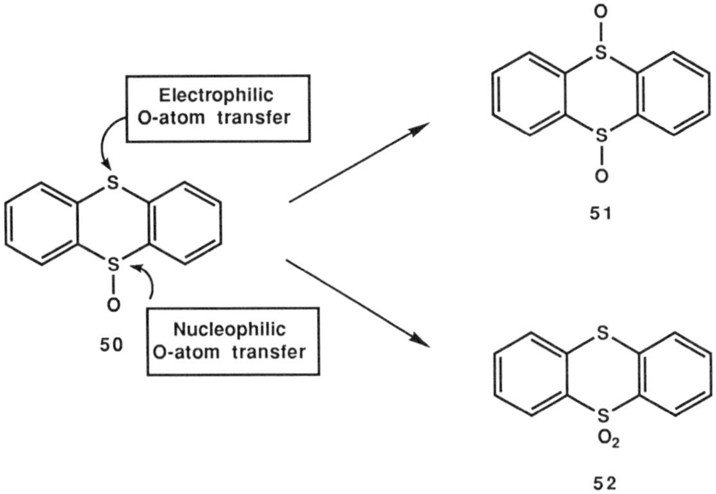

Scheme 12

The ozonolysis of 1,2-difluoroethylene yielded a small amount of 1,2,3-trifluorocyclopropane together with the ozonide and epoxide.[61] The unexpected formation of the cyclopropane implies that formyl fluoride *O*-oxide must have eliminated a molecule of oxygen to generate fluorocarbene, which in turn cycloadded to 1,2-difluoroethylene. In the ozonolysis of tetrafluoroethylene, a strong emission from difluoromethylene was observed, indicating that elimination of an oxygen molecule had also occurred in the case of F_2COO.[62]

Theoretical calculations have provided valuable information on the electronic effects of substituents on the nature of carbonyl oxides. The results of semi-empirical (MINDO/3-UHF) calculations carried out on a variety of substituted carbonyl oxide systems indicate that strong π-donors should enhance the zwitterionic character of carbonyl oxides whereas weak π-donors and π-acceptors do not significantly perturb the biradical character of the parent formaldehyde *O*-oxide **3**.[63] From the calculated charges on the terminal oxygen atoms of the carbonyl oxides and correlations between the latter and measured

X_{SO} values,[55] it has been deduced that π-donors will increase the nucleophilicity of carbonyl oxides with the converse being true for π-acceptors. CNDO/S calculations on the electronic transitions of carbonyl oxides were in excellent agreement with the spectroscopic data available from matrix isolation experiments and laser flash photolysis studies.[63]

1.1.4 1,2-Dioxiranes

1,2-Dioxirane (**6**) has been rigorously established as an intermediate in the low-temperature, gas-phase ozonolysis of ethylene.[64–66] The molecular structure of **6**, as determined by microwave spectroscopy, is depicted in Figure 3. Of particular structural significance is the O—O bond length, which is the longest observed to date (1.52 Å; cf. $O—O_{av}$ = 1.46 Å).[4] The electronic dipole moment was estimated to be 2.479 D.

Consistent with the observation of **6**, formaldehyde *O*-oxide (**3**), generated by ozonolysis of ethylene in the gas phase at 298 K and 1.1 kPa (8 Torr), decomposed primarily via 'hot' formic acid (**53**)[67] to give a product mixture composed of carbon dioxide, water, formaldehyde, formic acid, methanol and an unidentified product of mass 43. Using computer simulation, the mechanism illustrated in Scheme 13 was proposed as the primary decomposition mode of the 'hot' formic acid (**53**).

The chemical behavior of acetone *O*-oxide (**47**) was found, however, to differ substantially from that of formaldehyde *O*-oxide (**3**).[68–70] Investigation of the gas-phase reaction of ozone with tetramethylethylene at 294 K and 530 Pa (4 Torr) in a stopped-flow reactor coupled to a photoionization mass spectrometer[69] showed that the major products were $(CH_3)_2C{=}O$, HCHO,

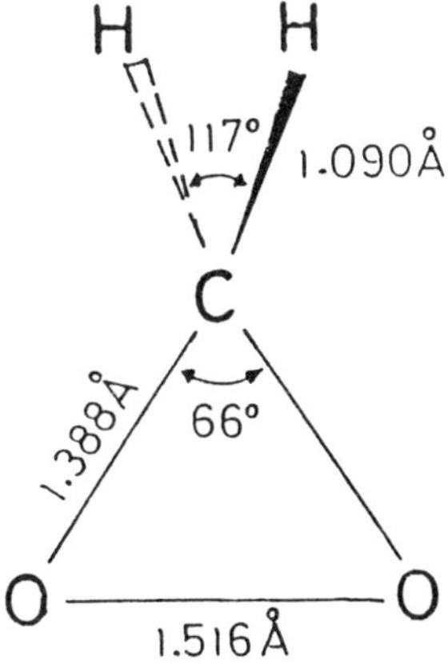

Figure 3. Structure of 1,2-dioxirane

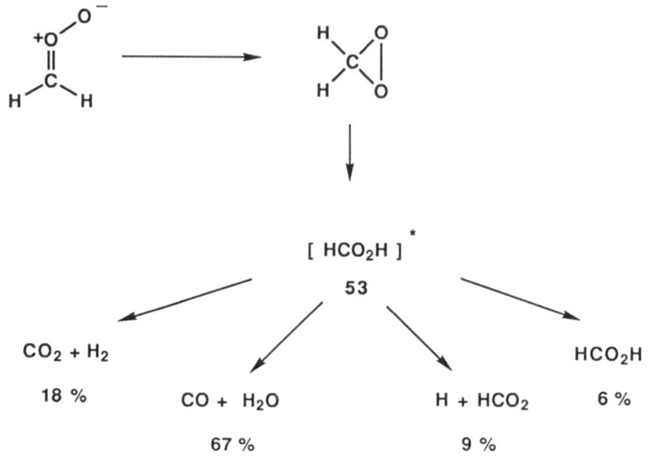

$[HCO_2H]^*$

53

$CO_2 + H_2$ 18 %

$CO + H_2O$ 67 %

$H + HCO_2$ 9 %

HCO_2H 6 %

Scheme 13

CH_3COCH_2OH and CH_3COCHO. The formation of 'hot' ester molecules ($R'R''COO \rightarrow R'COOR''^* \rightarrow$ products), by analogy with H_2COO (**3**) generated by ozonolysis of terminal alkenes $R'R''C{=}CH_2$ (Scheme 13), was not observed for alkyl-substituted carbonyl oxides $R'R''COO$ (for a general discussion on gas-phase ozonolysis see the review by Atkinson and Carter).[71]

In solution-phase ozonolysis reactions, no conclusive evidence has been obtained for the intermediacy of 1,2-dioxiranes. Although there was insufficient evidence to demonstrate that carbonyl oxides, generated from the photo-oxygenation of diazoalkanes in solution had rearranged to the corresponding esters via the isomeric dioxiranes,[72] it has been proposed that the novel ozonide **59** isolated from the ozonolysis of hexamethylbenzene (**54**) on silica gel or Florisil was likely to have resulted from the rearrangement of the carbonyl oxide intermediate **56** to the dioxirane **57** (Scheme 14).[73] As mentioned previously, the thermal behavior of furan endoperoxides[36–38] and lactone formation from the reaction of adamantylidenecyclopropane with singlet oxygen[43] may also provide circumstantial support for similar rearrangement processes.

Detailed procedures have been published for the preparation of stable solutions of 3,3-dialkyldioxiranes in acetone from the reaction of ketones and caroate.[74–76] Since the chemistry of dioxiranes will be described in detail elsewhere, the discussion here is focused on major differences in chemical behavior between dioxiranes and the isomeric carbonyl oxides.

Of particular interest is the thermal stability of dioxiranes; for example, a solution of 3,3-dimethyl-1,2-dioxirane in acetone, when kept at -20°C, is stable for several days. Dioxiranes can efficiently oxidize a variety of electron-rich substrates, including hydrocarbons. Aldehydes are slowly oxidized to carboxylic acids, instead of forming ozonides. Consistent with this, α,β-unsaturated ketones

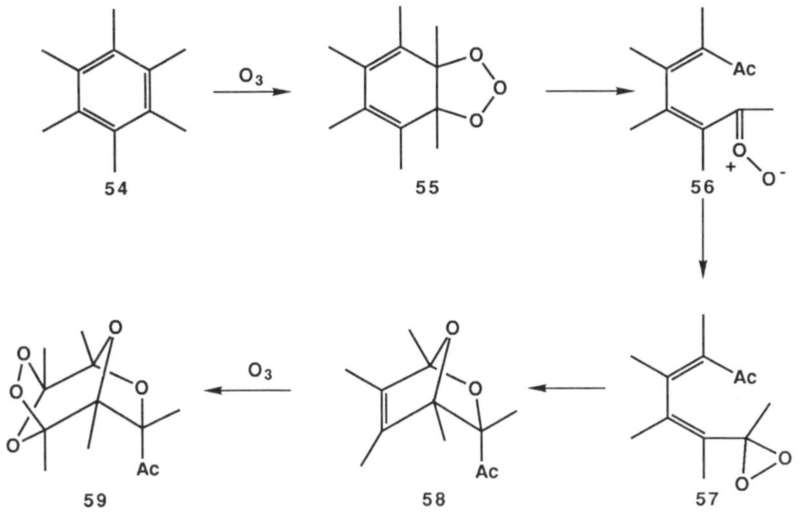

Scheme 14

were conveniently epoxidized by 3,3-dimethyl-1,2-dioxirane.[77] Epoxidation of alkenes with dioxiranes was found to be stereospecific,[74–76,78] whereas with carbonyl oxides, although stereoselective, some loss of alkene stereochemistry was observed.[47] It is also interesting that dioxiranes appeared to be stable in 'participating' solvents such as methanol.[74–76,79] Thus, the reaction of allyl alcohols with dioxiranes resulted in the formation of the corresponding α-hydroxy epoxides.[80]

1.1.5 Peroxy epoxide and the related intermediates

Epoxidation and other 'partial cleavage' reactions have long been known to compete with ozonolysis during the ozonization of alkenes and related unsaturated systems such as allenes, ketenes and ketenimines.[1] In such cases, the product isolated may be either the epoxide itself or an epoxide-derived secondary product. Yields of 'partial cleavage' products vary widely from *ca* 10% to nearly 100%.

Although oxygen-atom transfer from a carbonyl oxide to an alkene, as discussed above, provided a 'partial cleavage' product (Scheme 11), two alternative pathways involving direct oxygen-atom transfer from ozone to the C—C double bond with elimination of a molecule of oxygen have also been reported (Scheme 15). With sterically hindered alkenes, the proportion of 'partial cleavage' increased directly with increasing bulk of the substituents around the double bond. Thus, (*E*)- and (*Z*)-1-(1-naphthyl)-1-phenylpropene[81] and (*E*)- and (*Z*)-1-mesityl-1-phenylpropene[82] afforded the corresponding epoxides in yields of around 70% and 90%, respectively, probably via a π-complex analogous to **60** (Scheme 15). It has been suggested, however, that σ-complexes like **61** may be intermediates in ozonizations of the sterically unhindered alkenes, possessing electron-donating substituents, which produce epoxides or rearrangement products. Intermediates like **61** should be stabilized by the mesomeric effects of electron-donating substituents.

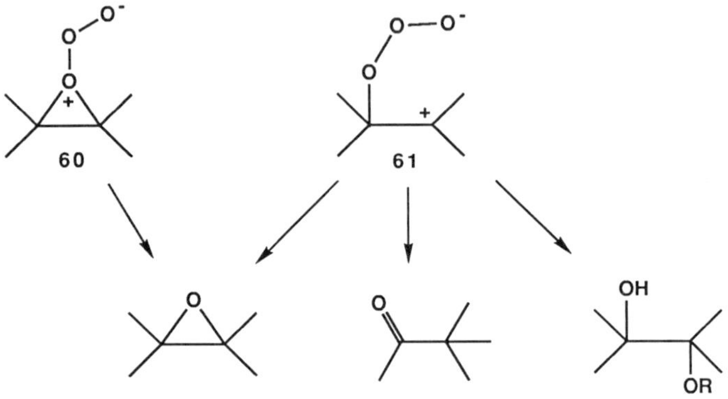

Scheme 15

For a series of hetero-substituted ethenes, the ratio of partial cleavage vs complete cleavage of the double bond was found to increase in the order $(CH_3)_3CCH{=}CH_2 < (CH_3)_3CC(OR){=}CH_2 < (CH_3)_3CC(NR_2){=}CH_2 < (CH_3)_3CC(SR){=}CH_2$.[83] A variety of vinyl sulfides yielded the corresponding 'partial cleavage' products in high yields.[84–87] Although vinyl ethers showed a preference for complete cleavage of the double bond, the sterically congested tetrahydrochromans, however, produced significant amounts of the partial cleavage products.[88,89] Similar trends were noted for a series of silyl vinyl ethers.[90,91]

Ozonization of some halo-substituted ethenes also yielded the corresponding epoxides.[61,92,93] Since the reactions were not stereospecific, the products might be produced via the corresponding σ-complexes (cf. **61**, Scheme 15) although alternative mechanisms involving oxygen-atom transfer from the nucleophilic carbonyl oxide intermediates to the electron-deficient vinyl halides could also be possible.

Reactions of ozone with tetramethoxyethylene (**62**) produced dimethyl carbonate (20–40%), methyl trimethoxyacetate (35–60%) and a dioxetane (**65**) (20–35%), the respective yields of which varied substantially with initial concentration of **62**, temperature and solvent (Scheme 16).[94] To account for the overall stoichiometry of the reaction, it has been proposed that the initial step involves an electron transfer from the alkene **62** to ozone, producing a radical pair which can either combine via the acyclic adduct **63** (cf. **61**, Scheme 15) to give the peroxy epoxide **64**, or initiate a radical chain oxidation reaction (Scheme 16). Singlet oxygen, produced on cleavage of **64**, subsequently enters into a [2 + 2] cycloaddition with **62**. The reactions of (*Z*)-1,2-dimethoxystilbene with ozone are considered to proceed in a similar manner.

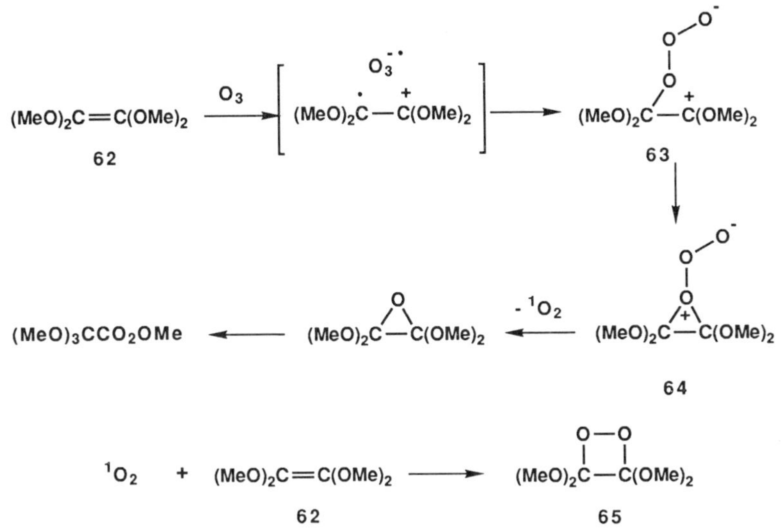

Scheme 16

1.2 Alkynes

Reactions between ozone and alkynes (**66**) are apparently much more complex than those with alkenes. A simplified mechanism highlighting a number of putative intermediates and rearrangement products is illustrated in Scheme 17.[1]

The formation of the corresponding alkoxyalkyl hydroperoxides (**68**) from the ozonization of phenylacetylene and phenylpropyne in methanol or ethanol provides support for the intermediacy of α-keto carbonyl oxides (**69**) (Scheme 17).[95] In addition, the observation of a strong carbonyl band at 1740 cm^{-1} during *in situ* IR spectroscopic studies of the ozonation of several alkynes in liquid carbon dioxide was indicative of the presence of some stabilized form of the carbonyl oxide intermediate **69**, possibly a dimer.[96]

Ozonization of but-2-ync in dichloromethane at -70°C afforded three peroxy species, each of which was capable of epoxidizing cyclohexene and other alkenes added after the ozonation was complete. From the analysis of their different roles as epoxidizing agents, the species were tentatively identified as the trioxolene **67** (R = Me), the dioxirane **70** (R = Me) and peroxyacetic acid.[97]

Based on the observation of chemiluminescence during the decomposition of a relatively stable ozonolysis intermediate in the presence of fluorescers such as

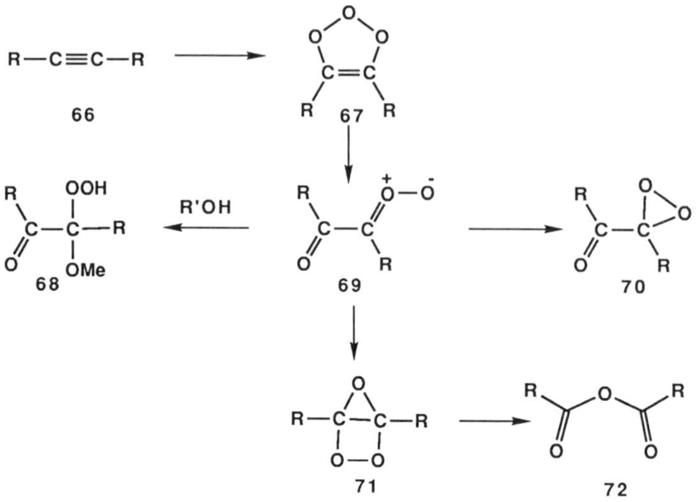

Scheme 17

9,10-diphenylanthracene, it has been suggested that the rearrangement of carbonyl oxides **69** to anhydrides **72** may proceed by an intermediate having an ozonide structure like **71**.[98,99] Photosensitized oxygenation of 2-diazobutan-3-one in dichloromethane at $-78°C$ yielded an adduct which, from low-temperature ^{13}C NMR spectroscopic data, was identified as the ozonide intermediate 1,4-dimethyltrioxabicyclo[2.1.0]pentane (**71**; R = Me). On addition of diphenyl sulfide followed by warming the resulting reaction mixture to room temperature, diphenyl sulfoxide was obtained, albeit in low yield.[100]

By the use of ESR spin trapping techniques, it has been shown that the intermediates from the ozonolysis of dimethylacetylene (DMA) decompose in part via free-radical pathways.[101] Thus, treatment of a DMA–ozone reaction mixture at $-70°C$ with 5,5-dimethyl-1-pyrroline *N*-oxide (DMPO) gave an ESR spectrum assignable to the acetoxyl adduct of DMPO. As the temperature was raised towards $-45°C$, the spectrum began to change and at $-30°C$ the major signals were due to the acetyl radical spin adduct of DMPO. No spin adducts were observed when the ozonolysis was carried out in the presence of methanol. When methanol was added at $-78°C$ after ozonolysis, but before the spin trap, both acetoxyl and acetyl adducts were obtained. Since spin trap adducts were observed even at $-70°C$, it has been proposed that the spin traps assist the decompositions of intermediates **71** (R = Me) and **70** (R = Me) by electron transfer or 1,3-dipolar additions to produce the acetoxyl and acetyl adducts, respectively (Scheme 17).

1.3 Saturated Compounds

Ozone generally reacts with saturated compounds either by insertion in a C—H bond or by formation of 1:1 complexes.[1] Hydrotrioxides, having the —OOOH functionality, have been postulated as unstable intermediates in the low-temperature ozonization of various saturated compounds including ethers, aldehydes, hydrocarbons, alcohols, silanes, amines, diazo compounds and acetals. In certain cases, direct spectroscopic evidence has now become available to substantiate the existence of such intermediates.[102] Conversely, phosphites, sulfides, phosphines, sulfoxides, tertiary amines, and certain highly hindered alkenes preferentially form 1:1 adducts with ozone.[103]

1.3.1 Charge-transfer complexes

It has been reported that ammonia and phosphines, in a similar manner to alkenes and aromatic compounds,[104] form 1:1 CT complexes with ozone.[105] Infrared spectra of solid matrices, prepared by co-deposition of phosphine and ozone in argon at 12–18 K, contained sharp satellite absorption bands at 1037.3, 988.5, 986.3 and 705.2 cm^{-1}, which have been associated with the formation of symmetrical PH_3-O_3 complexes,[105] analogous in structure to the corresponding ammonia–ozone[106] and trimethylamine–sulfur dioxide[107] complexes. The ready photodissociation of the PH_3-O_3 complex, by irradiation with red visible light, is considered to proceed initially by a CT mechanism producing the PH_3^+-O_3^- ion pair, which easily transfers O^- to give the photolysis product. Molecular complexes, formed between phosphorus(III) halides (PF_3, PCl_3 and PBr_3) and ozone in solid argon matrices, were transformed into the corresponding phosphoryl halides, as primary photoproducts, on irradiation with visible light.[108] On the other hand, IR spectral data obtained from matrix-isolated ozone and iodine monochloride were consistent with the formation of an acyclic O—O—O—I—Cl complex.[109]

1.3.2 Zwitterionic or cyclic intermediates

Trialkyl and triaryl phosphites form stable 1:1 complexes with ozone.[110] Thus, treatment of a dilute solution of triphenyl phosphite in dichloromethane with ozone at −75°C afforded a 1:1 ozone–phosphite adduct which was stable at temperatures below −15°C. ^{31}P NMR studies indicated that the adduct existed primarily in the cyclic rather than the open-chain form. A variety of stable phosphite–ozone adducts (Figure 4), including that derived from triphenyl phosphite,[111,112] decompose under mild conditions with concomitant generation of singlet oxygen.[113–116] Two mechanisms have been proposed for the oxidation of alkenes by phosphite–ozone adducts: at room temperature

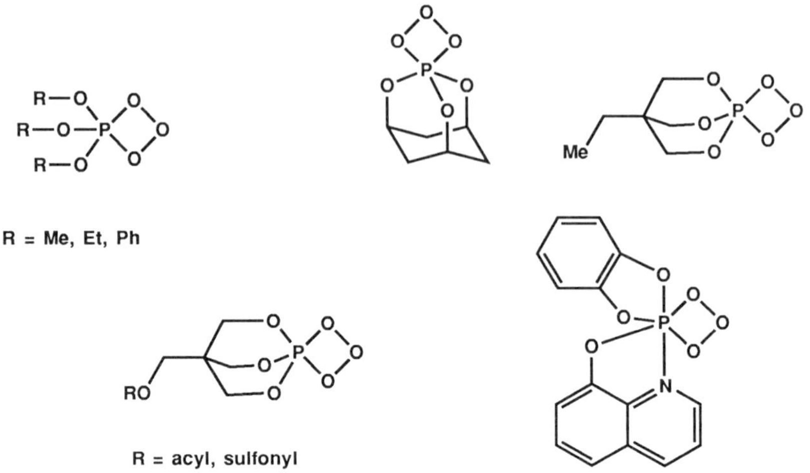

Figure 4. Stable phosphite–ozone adducts

singlet oxygen is involved, whereas at low temperatures the substrate is oxidized directly by the adduct.[117]

Although the reaction of a sulfide with ozone could yield a cyclic intermediate as described above, no chemical species other than the corresponding sulfoxide has been isolated or detected.[1] Similarly, ozonization of various phosphines produced phosphine oxides.[1]

In the gas-phase ozonization of either hydrogen sulfide or methanethiol, chemiluminescence, attributed to emissions by excited-state sulfur dioxide, was observed.[118] The only possible reactions that could yield $SO_2{}^*$ in a single collision are

$$O_3 + H_2S \rightarrow SO_2 + H_2O$$

$$O_3 + CH_3SH \rightarrow SO_2 + CH_3OH$$

A mechanism to account for the complex rearrangement process is outlined in Scheme 18.

Thioketones and arenedithiocarboxylates afford two types of ozonolysis products, ketones or thioesters on the one hand and sulfines on the other, depending on the degree of substitution of the substrates.[119] It has been suggested that ketones are obtained from comparatively unhindered thiocarbonyl compounds via [3 + 2] cycloaddition processes whereas with the more hindered substrates ozone attack is more likely to occur at the terminal sulfur atoms to yield the linear adducts, which on elimination of molecular oxygen provide sulfines.[120,121] Thioketenes are reported to accept an oxygen atom from ozone yielding the corresponding sulfines, probably via the zwitterionic intermediates.[122]

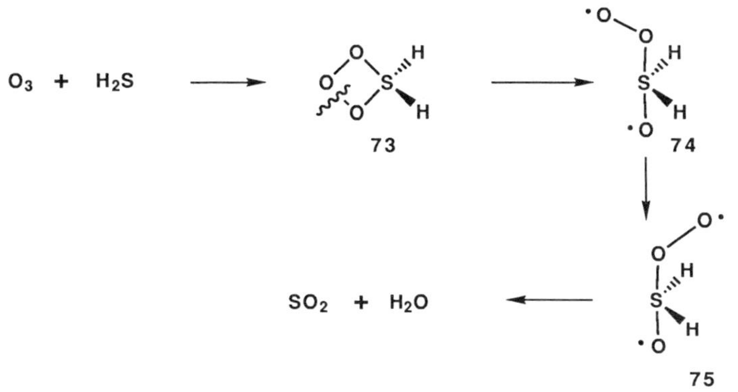

$O_3 + H_2S \longrightarrow$ **73** $\longrightarrow$ **74** $\longrightarrow$ **75** $\longrightarrow SO_2 + H_2O$

Scheme 18

The reaction of ozone with an amine may give rise to a range of oxidation products (Scheme 19).[1] The initial interaction between electrophilic ozone and a nucleophilic amine would be expected to produce an adduct similar to **76**. The subsequent loss of molecular oxygen from intermediate **76** to yield an amine oxide has been found to be an important process for primary and tertiary amines. In accordance with this, singlet oxygen has been detected from gas-phase reactions of ozone with triethylamine.[123] For amines with primary alkyl groups, side-chain oxidation often occurs, providing α-hydroxy amines. Under ozonization conditions, secondary amines are known to produce nitroxide radicals.

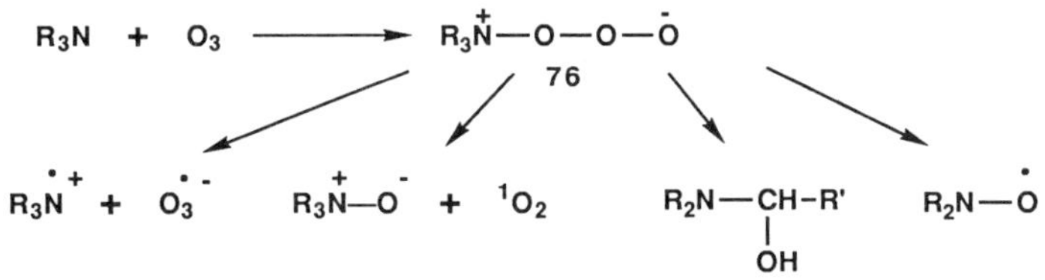

$R_3N + O_3 \longrightarrow R_3\overset{+}{N}{-}O{-}O{-}\overset{-}{O}$ (**76**)

$R_3N^{\bullet +} + O_3^{\bullet -}$ $\quad R_3\overset{+}{N}{-}O^- + {}^1O_2$ $\quad R_2N{-}CH(OH){-}R'$ $\quad R_2N{-}O^{\bullet}$

Scheme 19

Since ozone has a high electron affinity (+2.1 eV), adducts such as **76** can dissociate into radical ion pairs.[124] Thus, during ozonizations of phenothiazine (**77**) and phenoxazine in dichloromethane, the respective radical cations and the ozonate radical anion were detected.[125] Since they have higher ionization potentials, phenoxathiin and thianthrene did not produce analogous radical species on reaction with ozone. The products from phenothiazine (**77**) were 2-chlorophenothiazine (**78**) (in chlorinated solvents), 3*H*-phenothiazin-3-one (**79**) and phenothiazine 5-oxide (**80**), as illustrated in Scheme 20.

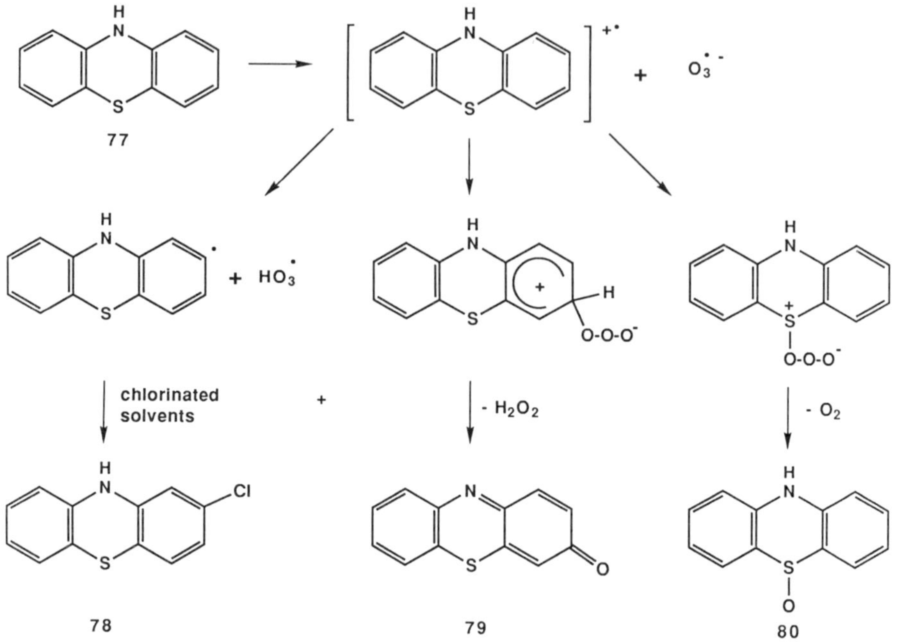

Scheme 20

1.3.3 Hydrotrioxides

The synthesis, characterization and reactions of hydrotrioxides, which has been one of the most extensively studied topics in ozonization chemistry during the past decade, will be discussed in detail elsewhere (see Chapter 10). Several hydrotrioxide derivatives, as depicted in Figure 5, have been characterized to date in solution by ^{1}H NMR spectroscopy. The chemical shift of the hydrotrioxide proton has been consistently observed at around δ13 ppm.[126–131]

2 CYCLOADDITION REACTIONS OF CARBONYL OXIDES

The nature of the isolable peroxidic products derived from the ozonization of alkenes and related compounds containing carbon—carbon double bonds depends on the reactions and reactivity of the discrete, but transient, carbonyl oxide intermediate **81**, the so-called Criegee intermediate, generated on decomposition of the primary ozonide (see Section 1.1). A summary of the reactions of a typical carbonyl oxide is depicted in Scheme 21.

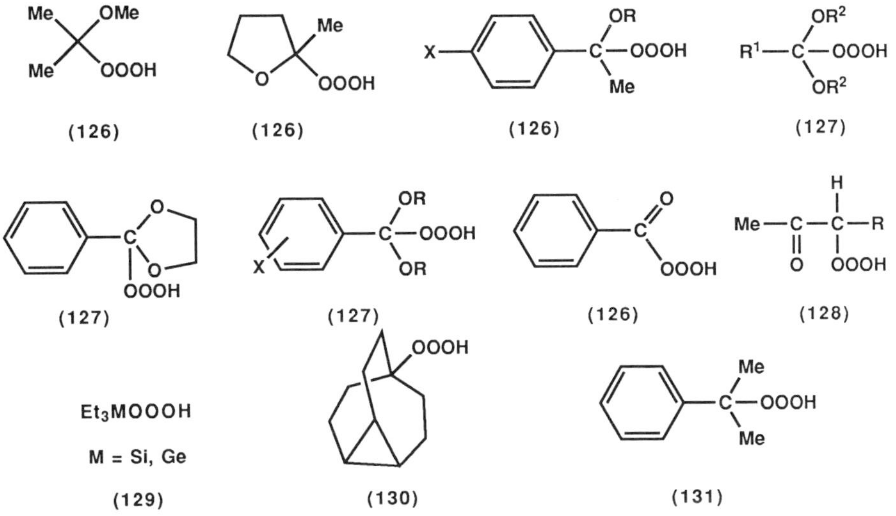
(126)
(126)
(126)
(127)
(127)
(127)
(126)
(128)
Et3MOOOH
M = Si, Ge
(129)
(130)
(131)

Figure 5. The hydrotriperoxides determined by ^{1}H NMR spectroscopy (literature references in parentheses)

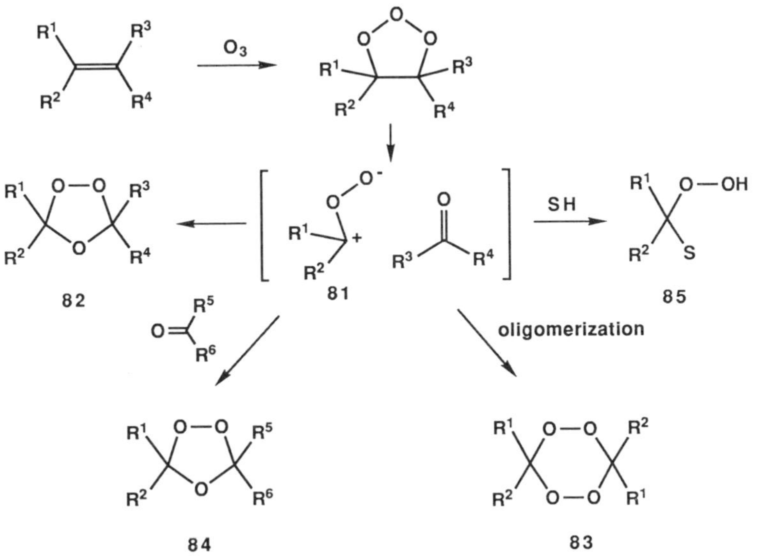
O3
SH
oligomerization
82
81
85
84
83

Scheme 21

As a consequence of solvent interactions, the carbonyl oxide is considered to possess a high degree of 1,3-dipolar character in the solution phase. Thus, a concerted recombination of the carbonyl oxide with the carbonyl fragment, co-produced from decomposition of the primary ozonide, generally affords the corresponding secondary ozonide (1,2,4-trioxolane) (**82**). Alternatively, the carbonyl oxide may undergo self-addition to give cyclic dimers (1,2,4,5-tetroxanes) (**83**) or higher oligomeric peroxides. In the presence of added carbonyl compounds or protic solvents such as alcohols, cross-ozonides (**84**) or hydroperoxides (**85**), respectively, may also be obtained.

2.1 [3 + 2] Cycloaddition Reactions

On the basis of the three-step Criegee mechanism, secondary ozonides (1,2,4-trioxolanes, more usually termed simply 'ozonides') would be the isolable peroxidic products from reactions between ozone and simple alkenes in the absence of participating solvents such as alcohols. The efficiency of ozonide formation from the [3 + 2] cycloaddition of carbonyl oxides with carbonyl compounds is highly dependent on the dipolarophilicity of the carbonyl component. The experimentally established order of reactivity was aldehydes > ketones > esters. Many of the major factors affecting this process have already been reviewed extensively.[1,2,132] Several recent studies have been directed towards providing further insight into the mechanism of the ozonization process.

The extent of cross-ozonide formation from the ozonization of mixtures of ethylene-d_0 and ethylene-d_4 was found to vary in the range 10–90% as a function of solvent polarity. The product variations were consistent with increased stabilization of the intermediate carbonyl oxides with increasing solvent polarity.[133] Detailed analysis of the product ratios from the reactions of ethylene-1,1-d_2 in the presence of acetaldehyde indicated that there was a significant inverse kinetic secondary isotope effect associated with the recombination reactions of the intermediate formaldehyde *O*-oxide. This provided evidence that the recombination step was concerted, at least in ozonolysis reactions involving simple alkenes.[134]

The isomer ratios of the ozonides and cross-ozonides derived from appropriately substituted alkenes have been extensively studied in order to gain information on the ozonolysis mechanism. Interpretation of such data is complicated because it is not fully known how uncertainties in (a) the direction of cleavage of the primary ozonide and (b) the configuration of the resulting carbonyl oxide(s) (*syn–anti* isomerization) affect the stereoselectivity in the recombination process.[1,2]

Ozonization of propene at −78°C in non-participating solvents afforded a mixture of propene ozonide in about 80% yield, accompanied by ethylene and but-2-ene (*E* and *Z*) cross-ozonides, respectively. The *E*/*Z* ratio of the but-2-ene ozonide was around 2:1, consistent with a preponderance of the *anti* isomer of the acetaldehyde *O*-oxide.[135]

The isomeric ratios of ozonides obtained from the ozonolysis of *E*,*Z* pairs of either styrene-β-d_1 or hexene-1-d_1 in pentane or ester solvents, or in the presence of added aldehyde, were considered to be consistent with the intermediate formaldehyde *O*-oxide-d_1 (HDCOO) being generated from the corresponding primary ozonides in at least a 3:1 mixture of non-interconvertible *syn–anti* isomers, although it was uncertain as to which isomer actually predominated. Stereodirecting effects of the dipolarophile in the recombination reaction appeared to follow an order ester > acetaldehyde > benzaldehyde.[136]

The generally observed failure of tetraalkyl-substituted alkenes to yield the corresponding ozonides has usually been attributed to the low dipolarophilicity of the ketone. Ozonolysis of the dibromoalkene **43** in acetone did, however, give significant amounts of the cross-ozonide **89** together with the cyclic acetone peroxides **48** and **49** (see Scheme 10).[45] Similarly, ozonolysis of tetramethylethylene (**86**) in bromoacetone also produced ozonide **89** in comparable yield (Scheme 22). It follows that the failure of tetrasubstituted alkenes to produce the corresponding ozonides cannot be simply due to a lack of reactivity between the respective ozonolysis fragments. To account for the formation of the cyclic peroxides **48** and **49** from **43**, it has been proposed that either the adduct **87** decomposed to give dimethyldioxirane, or there has been oxygen transfer from the carbonyl oxide **44** to acetone to give the carbonyl oxide **47**; either intermediate could in turn dimerize or trimerize as appropriate.

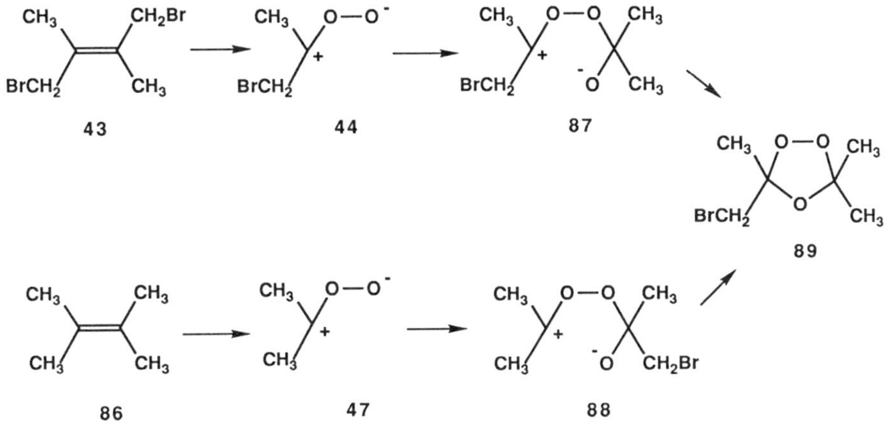

43 44 87 89 86 47 88

Scheme 22

Treatment of 2,3-dimethyl- (**90a**) and 2,3-diphenylbuta-1,3-diene (**90b**) in pentane with 0.9 equiv. of ozone afforded the corresponding mono-ozonides **91a** and **91b** in 65% and 26% yield, respectively.[137,138] Under similar conditions, the unsymmetrically substituted buta-1,3-dienes **92** gave mixtures of both possible ozonides **93** and **94** (Scheme 23).[138] Consistent with its high degree of substitution, solution-phase ozonolysis of 2,3,4,5-tetramethylhexa-2,4-diene did not result in the formation of any mono-ozonide.

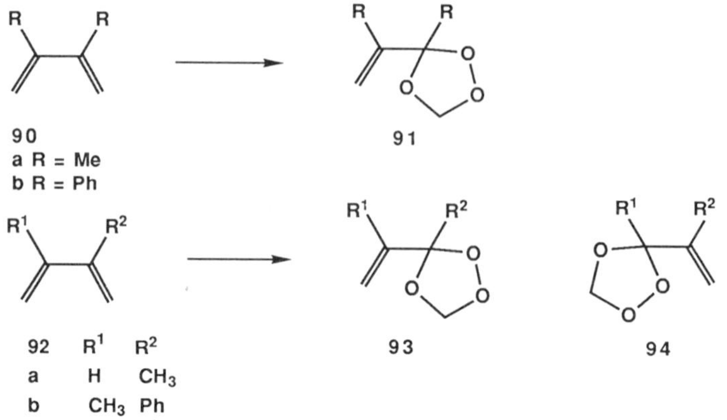

90
a R = Me
b R = Ph

91

92	R^1	R^2
a	H	CH_3
b	CH_3	Ph

93

94

Scheme 23

The general features of the Criegee mechanism would be expected to apply to the ozonization of cyclic alkenes, although modification is required since the carbonyl oxide and carbonyl groups are necessarily located within the same molecule. For cyclopentenes and related compounds, intramolecular [3 + 2] cyclization between the carbonyl oxide and the carbonyl groups was found to be kinetically favored whereas for larger ring sizes intermolecular oligomerization reactions often predominate.[1]

Reactions of substituted indenes (**95**) with ozone in non-participating solvents generally gave rise to the corresponding bicyclic ozonides in high yield as a consequence of intramolecular recombination of either or both possible carbonyl oxide–carbonyl pairs (Scheme 24).[139–144] With 1-substituted indenes

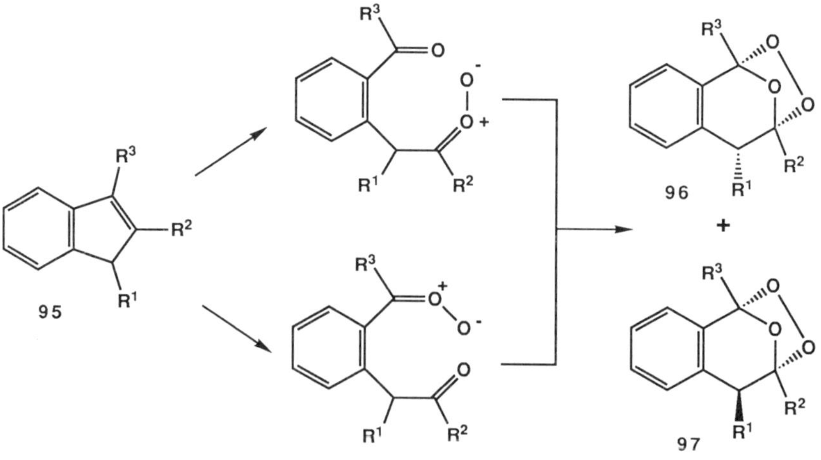

95

96

97

Scheme 24

(**95**; $R^1 \neq H$), there is the additional possibility of forming *endo* and *exo* stereoisomeric ozonides, **96** and **97**, with the latter being the thermodynamically favored.[139,140] The ozonide isomer ratios were found to vary markedly with the degree of substitution and the steric requirements of the 1-substituent.[140] For indenes **95** (R^1 = *t*-Bu), the corresponding *exo* isomer was formed exclusively. Although the observed ozonide isomer ratios provided little specific information on the overall ozonolysis mechanism, in the sterically congested systems such as the 1,2- and 1,3-di- and 1,2,3-trisubstituted indenes, the recombination process was unlikely to have been entirely concerted as proposed for simple acyclic systems.

Ozonolysis of the trioxane derivative **98** in dichloromethane at −78°C afforded the polycyclic ozonide **99** as a single isomer, whose structure was determined by X-ray crystallography, together with an intramolecular oxygen-transfer product (**100**) (Scheme 25).[53] The mechanism of formation of the ozonide **99** would be expected to be similar to that proposed for indenes **95**.

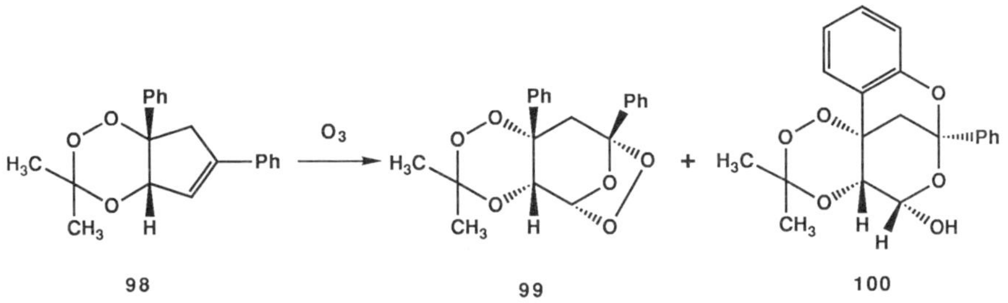

98 99 100

Scheme 25

Although both cholesterol (**101a**) and its acetate derivative (**101b**) were found to give complex product mixtures on ozonization in non-participating solvents, the corresponding isomeric ozonides could be isolated and separated by column chromatography[145] (Scheme 26).

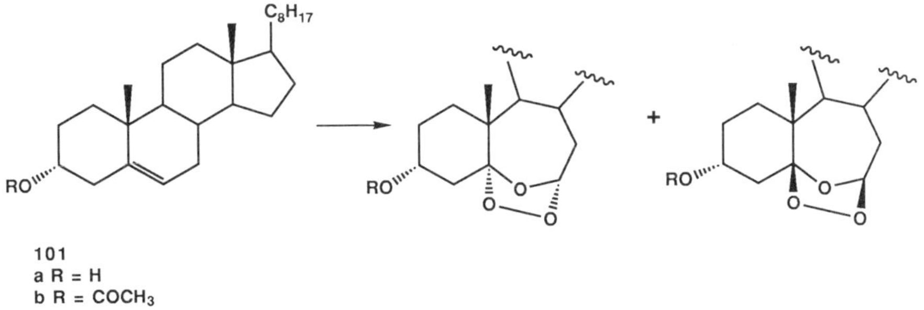

101
a R = H
b R = $COCH_3$

Scheme 26

Since esters are generally recognized as being significantly less reactive than ketones towards carbonyl oxides, they have been commonly used as solvents in ozonolysis reactions. Ozonization of styrene in a series of simple ester solvents afforded, in addition to styrene ozonide (*ca* 75%), the first reported examples of alkoxy ozonides (**102**) in low yield (5–10%), resulting specifically from the [3 + 2] cycloaddition of formaldehyde *O*-oxide to the ester (Scheme 27).[146] The yields of **102** were higher with formate esters than acetates. Following the above observations, 3-methoxy-1,2,4-trioxolane (**103**) was also isolated from reactions of ozone with simple methyl vinyl ether, although the major product was found to be 3-methoxy-1,2-dioxolane (**104**), resulting from a regioselective [3 + 2] cycloaddition of formaldehyde *O*-oxide to unreacted methyl vinyl ether. Increasing the concentration of methyl vinyl ether increased the yield of **104** in a predictable fashion[6,147,148] (Scheme 28). Similar results were obtained for the ozonolysis of 1-ethyl vinyl ether[148] and vinyl acetate.[149]

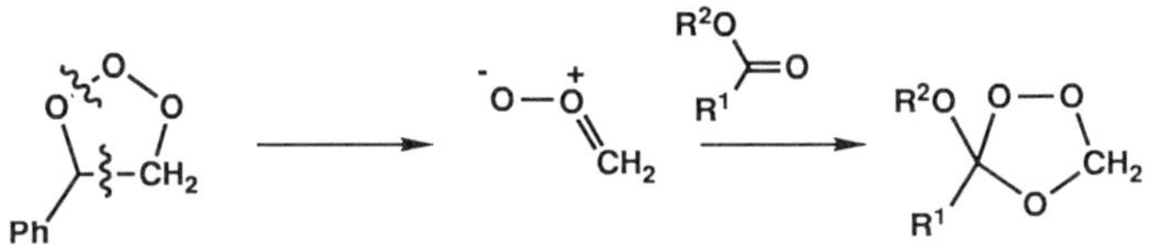

102
R^1 = H, CH_3
R^2 = CH_3, CH_2CH_3

Scheme 27

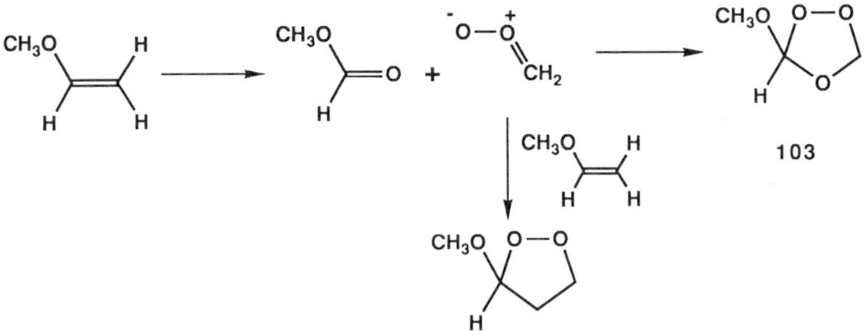

103

104

Scheme 28

In addition to the synthesis of new trioxolane and dioxolane derivatives, the above studies[6,136,146–149] have shown that (a) the primary ozonides derived from simple enol ethers undergo a highly selective cleavage to give the carbonyl oxide and an ester with no evidence for products derived from the alternative alkoxy-substituted carbonyl oxide, (b) the cycloaddition to the enol ethers was a highly regio- and stereoselective process and (c) the relative order of dipolarophilicities towards carbonyl oxides was aldehydes > enol ethers > esters ≈ ketones.

The products obtained from the ozonolysis of vinyl ethers were sensitive to the nature of the substrate. Thus, ozonolysis of the vinyl ethers **105** in non-participating solvents, whose primary ozonides were shown to decompose in an analogous fashion to those described above, did not produce any of the corresponding ozonides or 1,2-dioxolanes; the major peroxidic products were the corresponding tetroxanes (**106**) (Scheme 29).[79,150]

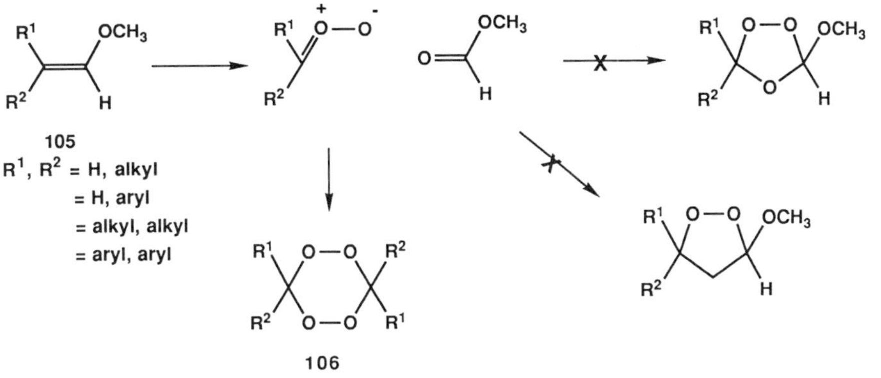

105
R^1, R^2 = H, alkyl
= H, aryl
= alkyl, alkyl
= aryl, aryl

106

Scheme 29

Observed variations in the dioxolane stereoisomer ratios from ozonolysis of (*E*)- and (*Z*)-ethoxypropene have suggested that the substrate configuration may influence the *syn*/*anti* ratios of the intermediate carbonyl oxides, which in turn affects the cycloaddition process.[151]

The isomeric keto enol ethers **103** and **108**, which could be regarded as acyclic equivalents of the indene **109**, gave distinctive ozonolysis reactions. The (*Z*)-isomer **108** and the indene **109** in carbon tetrachloride at 0°C afforded the ozonide **110** in high yield. Under similar conditions, **107** gave mainly polymeric material (Scheme 30).[144] These differences in behavior between **107** and **108** have been attributed to influences of the substrate on the configuration of the respective intermediate carbonyl oxides.

The reaction of the diene **111** with 1 equiv. of ozone in carbon tetrachloride produced only the keto ester **112**, arising from an intramolecular oxygen transfer from the carbonyl oxide to the methoxyvinyl group followed by a 1,2-hydride shift (Scheme 31).[50,79]

No alkoxy ozonides were obtained from ozonization of either 1,1- or 1,2-dimethoxyethylene.[151] Although the primary ozonide from the former cleaves selectively in the predicted fashion, the low dipolarophilicity of dimethyl carbonate resulted in a preferential reaction of the formaldehyde *O*-oxide with

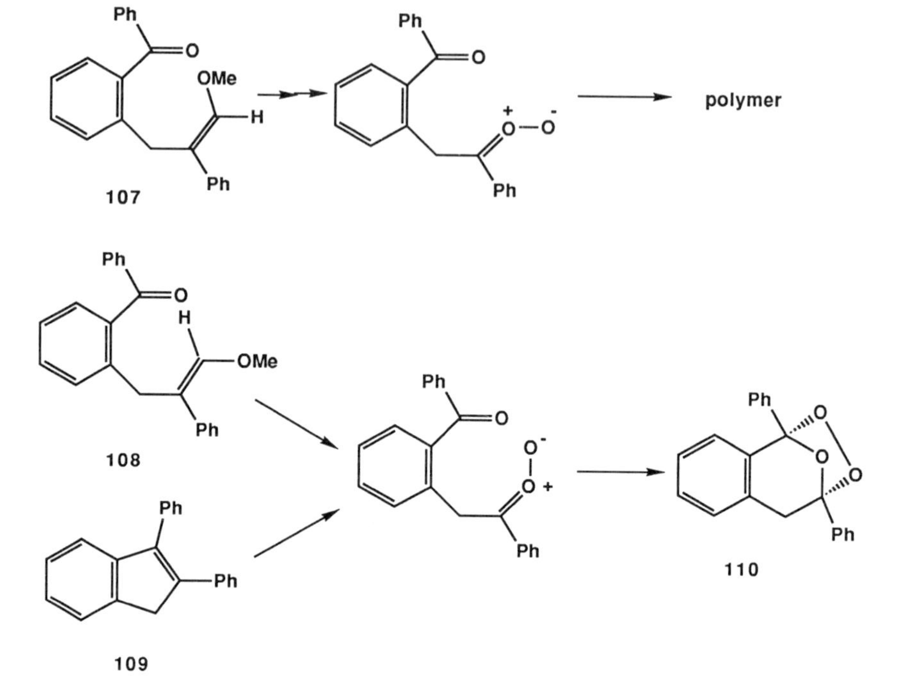

Scheme 30

unreacted substrate to give 3,3-dimethoxy-1,2-dioxolane (**113**) in 68% yield. 1,2-Dimethoxyethene afforded the corresponding dimethoxytetroxane (**114**) (*ca* 15% yield as a mixture of isomers), providing evidence for the intermediacy of the methoxy substituted carbonyl oxide, together with polymeric material. The intermediate methoxy carbonyl oxide was found to be reactive to methanol, aldehydes and, to a lesser extent, ketones, but not to esters (Scheme 32).

Ozonolysis of the methoxymethylene Meldrum's acid derivatives **115** provided the corresponding isolable, 'normal' ozonides in moderate yield (Scheme 33). The mechanism of this reaction has not been investigated.[152]

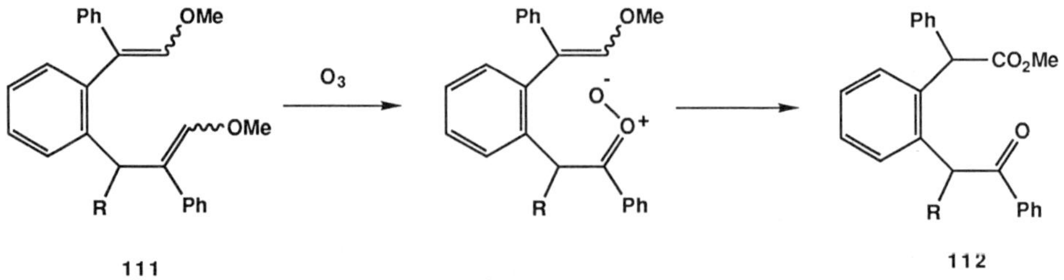

Scheme 31

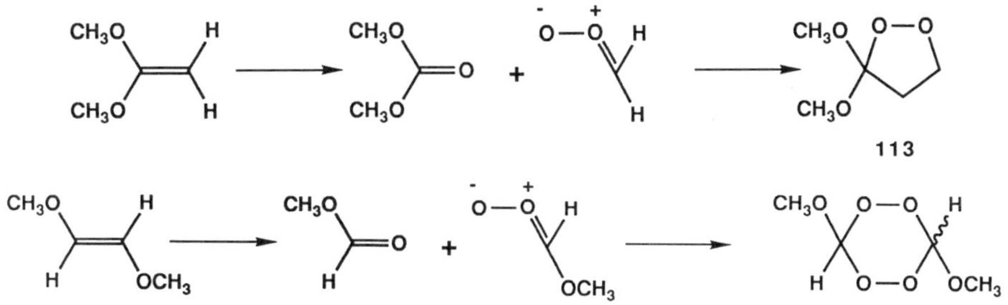

Scheme 32

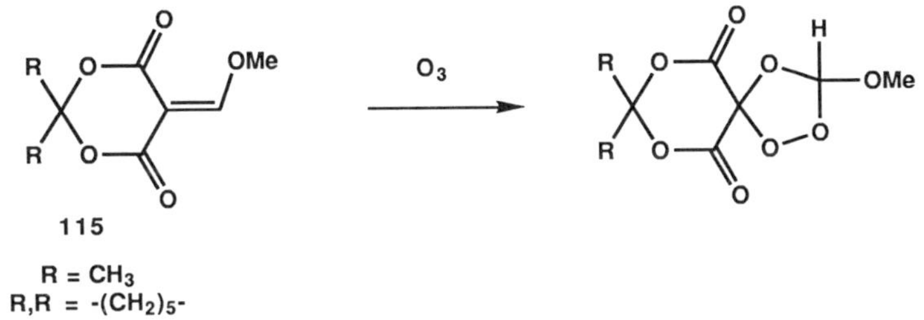

Scheme 33

Since the only peroxidic product formed from the reaction of tetramethoxyethylene (**62**) with ozone is the 1,2-dioxetane **65**, it has been proposed that the initial step involves an electron transfer from the electron-rich alkene to the ozone to give a radical ion pair which collapses to give the peroxy epoxide **64**. The dioxetane **65** is ultimately derived from a [2 + 2] cycloaddition between the alkene and singlet oxygen which results from fragmentation of **64**. A tentative mechanism is outlined in Scheme 16.[94]

Carbonyl oxides, derived from ozonolysis of homoallylic esters (**116**) in dichloromethane, were found to undergo a kinetically favored intramolecular cyclization with the adjacent ester group to yield the bicyclic ozonides **117** (Scheme 34).[153] The formation of the ozonides **117** is, however, governed by a combination of steric and electronic factors. When $R^1 = t$-Bu, the intermediate appears to adopt a conformation favorable to cyclization. The ester group requires to be activated by having an electron-withdrawing group, e.g. $R^2 = p\text{-}O_2NC_6H_4$, attached to the carbonyl carbon.

Ozonolysis of the symmetrical bicyclic diester **118** afforded, in addition to polymeric peroxides, the methoxy ozonide **119** as a minor reaction product (*ca* 15%). Since the CHO group, co-produced in the ozonolysis process, would be spatially too distant for intramolecular reaction, the carbonyl oxide moiety undergoes a cycloaddition with the nearest methoxy carbonyl group[154] (Scheme 35).

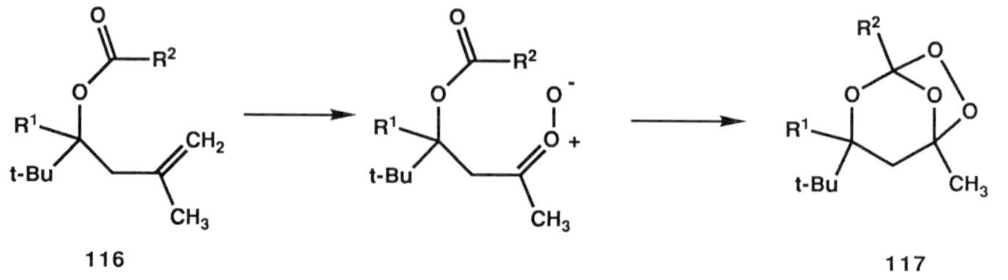

116 117

Scheme 34

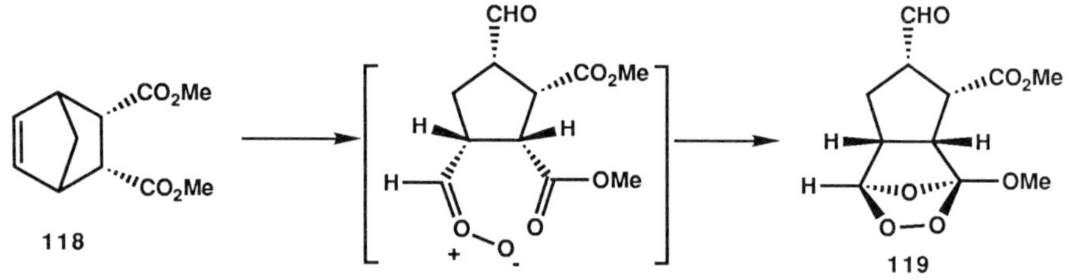

118 119

Scheme 35

As would be expected for five-membered ring systems, ozonolysis of the enol esters **120** and **121** in a range of solvents afforded the corresponding ozonides **122** and **123**, respectively. Ozonide **123** can be isolated in high yield from **121** even with methanol as solvent, indicating a strong driving force for intramolecular cyclization exists in the acenaphthene system (Scheme 36).[155]

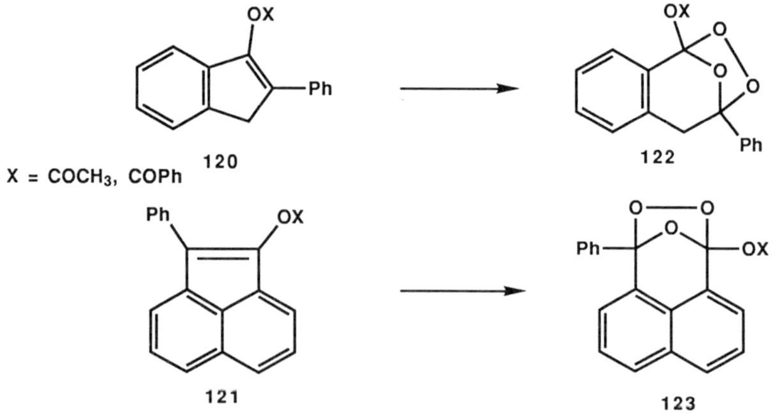

X = $COCH_3$, COPh

120 122

121 123

Scheme 36

By analogy with **120** and **121**, ozonolysis of dihydrofurans and benzofurans would be expected to give rise to bicyclic ozonides as a result of intramolecular cycloaddition processes. Although the cyclic vinyl ethers **124** and **125** do give the expected ozonides **126** and **127** in moderate to good yields, the corresponding acetyl derivatives **128** and **129** afforded, under similar ozonolysis conditions, the anomalous keto ozonides **130** and **131**, respectively, which have incorporated the original acetyl carbonyl group.[156] ^{17}O-labeling studies indicated unambiguously that the primary ozonide from **128** selectively cleaves to give an ester oxide **132**, rather than the expected carbonyl oxide, which preferentially undergoes a cycloaddition with the more remote carbonyl group to give **130** (Scheme 37).[157]

Consistent with the electrophilic nature of ozone, alkenes generally exhibit decreased reactivity towards ozone with increasing halogen substitution. The chemistry of chloro- and bromoalkenes has already been extensively reviewed.[1] More recent studies have been directed towards the reactions of ozone with low molecular weight fluoroalkenes.

Ozonolysis of fluoroethylene, neat or in solution, affords the volatile 3-fluoro-1,2,3-trioxolane together with small amounts of the cross-ozonides. Microwave studies showed that the ring adopted a twisted half-chair conformation in which the fluorine atom occupied an axial position.[158] Similarly, 1,1-difluoroethylene gave the corresponding secondary ozonide (*ca* 20% yield) as the major isolable peroxidic product.[159]

The secondary ozonides derived from (*E*)- and (*Z*)-1,2-difluoroethylene were obtained as a mixture of isomers in which the (*E*) isomer consistently predominated by around 10:1, irrespective of the configuration of the alkene.[61] To rationalize the observed stereoselectivity, it has been proposed that (a) the carbonyl oxide HFCOO is formed primarily as the *syn*-isomer and (b) the two components in the cycloaddition react in a trans fashion in order to minimize F $\cdots$ F repulsion.[160]

Ozonolysis of trifluoroethylene gave the secondary ozonide via both possible recombination pathways and a small amount of the 3,5-difluoro-1,2,4-trioxolane cross-ozonide.[159] Although tetrafluoroethylene does not give the corresponding ozonide, it does yield trifluoroethylene ozonide when the ozonization is carried out in the presence of formyl fluoride, thus providing evidence for the intermediacy of F_2COO.[159]

The formation of ozonides together with appropriate carbonyl compounds from the ozonolysis reactions of most fluoroalkenes has indicated that the Criegee mechanism was operating to some extent although epoxides and, in some cases, cyclopropanes were also often obtained in significant amounts. More detailed discussion of the formation of peroxidic and non-peroxidic products from fluoroalkenes may be found elsewhere.[161]

Following the isolation of 1,2-dioxolanes from ozonolysis of vinyl ethers (see above), the scope of [3 + 2] cycloaddition reactions involving carbonyl oxides

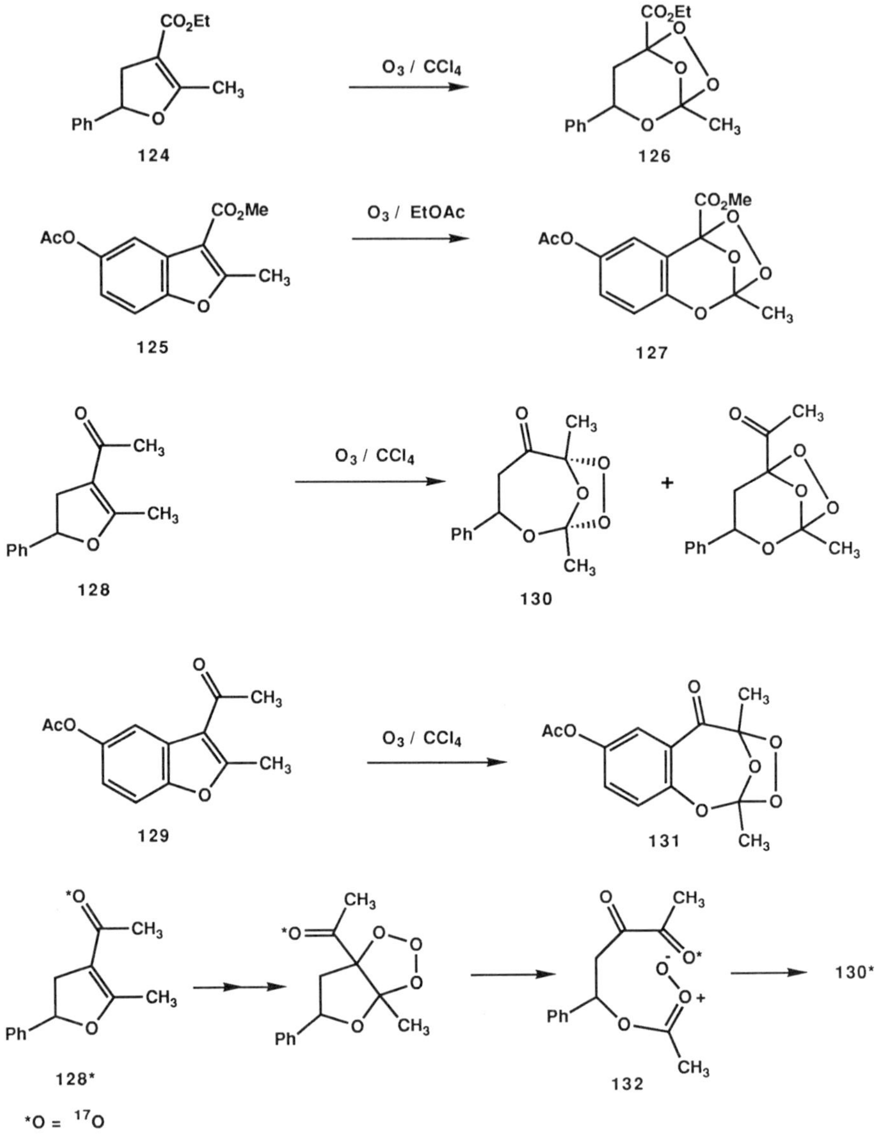

Scheme 37

has been extended further. Thus, carbonyl oxides, derived from enol ethers, react readily with a range of imine derivatives to provide a new, direct synthesis of 1,2,4-dioxazolidines (**133**) (Scheme 38).[162] In competition experiments, the imines were found to be more dipolarophilic than similarly substituted ketones.

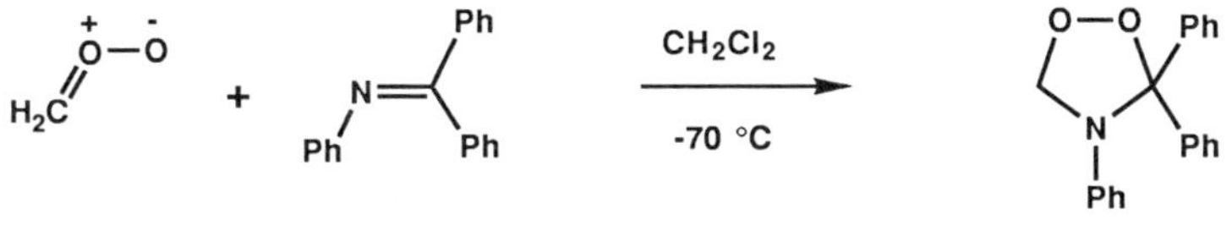

Scheme 38

Surprisingly, the carbonyl oxides **134** formed the corresponding thioozonides **135** by [3 + 2] cycladdition with thioadamantanone, even in the presence of equimolar amounts of adamantanone. With thiobenzophenone, however, the carbonyl oxides behaved as oxygen-transfer agents resulting in the formation of sulfines (**136**) (Scheme 39).[163]

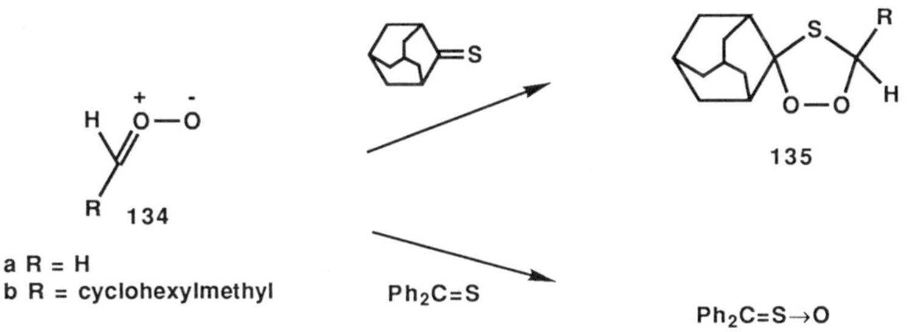

Scheme 39

The direct synthesis of a pure ozonide by ozonolysis of the corresponding alkene in solution is often precluded because of inefficient recombination of the carbonyl oxide and carbonyl components, and/or the predominence of side-reactions of the carbonyl oxide. To circumvent such problems, the technique of 'dry ozonization,' whereby the alkenic substrate is pre-absorbed on either silica gel[164,165] or polyethylene,[150,166–171] and treated with ozone at low temperatures, has been developed. It is presumed that the restricted mobility of the ozonolysis intermediates and the absence of solvent should favor the formation of ozonide.

Dry ozonization on silica gel was found to give improved yields of pure ozonides from alkenes known to produce ozonides in the solution, although the results may vary depending on the water content of the silica.[164,165] Conversely, dry ozonization on polyethylene has been used to prepare a series of mono- and diozonides which could not be prepared by solution-phase reactions. The scope of this method includes the successful syntheses of (a) the elusive ozonide **137** from tetramethylethylene,[166,169] (b) the ozonide **138** and epoxy ozonide **139** from 2,3-di-*tert*-butylbuta-1,3-diene,[168] (c) the diozonides **141a** and **b** from isoprene (**140a**) and hexamethylbuta-1,3-diene (**140b**),[167,169] (d) bicyclic ozonides **143a** and **b** from cyclooctene (**142a**) and cyclodecene (**142b**),[169] (e) acetoxy-substituted ozonides **144** from vinyl esters,[170] (f) alkoxy-substituted ozonides **145** from enol ethers[150] and (g) dichloro-substituted ozonides **146** from dichloroalkenes[171] (Scheme 40).

2.2 [3 + 3] Cycloaddition Reactions

The dimerization of carbonyl oxide intermediates to give 1,2,4,5-tetroxanes as minor products from the ozonization of alkenes and related compounds in solution has been known for some time.[1,172] Since a concerted [3 + 3] cycloaddition between two carbonyl oxides is thermally forbidden, the formation of the tetroxanes is most likely to be stepwise, giving rise to equilibrium mixtures of *cis* and *trans* isomers with unsymmetrically substituted carbonyl oxides.

Ozonolysis of 1-ethoxypropene afforded 3,6-dimethyl-1,2,4,5-tetroxane as a mixture of isomers in addition to ozonides and dioxolanes.[151] In contrast, the 2,2-disubstituted vinyl ethers gave only the corresponding tetroxanes **106** as the sole, isolable ozonolysis products in non-participating solvent (Scheme 29).[79,150] In each case, however, the tetroxanes were derived from carbonyl oxides arising from the predicted cleavage mode of the respective primary ozonides.

(*Z*)-1,2-Dimethoxyethylene, which on ozonolysis must produce the methoxy-substituted carbonyl oxide, yielded the dimethoxy-substituted tetroxane **114** (*ca* 15% yield, mixture of (*E*) and (*Z*) isomers) instead of the corresponding ozonide.[149,173]

Reaction of 1,2,4-tri-*tert*-butylnaphthalene with 1 equiv. of ozone resulted in cleavage of the least sterically hindered double bond to give the intermediate carbonyl oxide–carbonyl pair which dimerized to the tetroxane **147** (Scheme 41).[174] Intramolecular cyclization processes were most likely to have been disfavored on steric grounds.

In principle, carbonyl oxides should be able to undergo [3 + 3] cycloaddition reactions with other 1,3-dipoles. Recently it was found that carbonyl oxides, derived from enol ethers, reacted with nitrones to yield the corresponding 1,2,4,5-trioxazinanes (**148a** and **b**), the first examples of a new class of cyclic peroxide (Scheme 42).[175] Competition experiments indicated that the carbonyl oxides react preferentially with nitrones in the presence of carbonyl compounds such as benzaldehyde or benzophenone. Other 1,3-dipoles such as 2,4,6-trimethylbenzonitrile oxide, *N*-(phenanthridin-5-io)benzimidate or azoxybenzene failed to produce 1:1 adducts with carbonyl oxides generated under similar conditions.

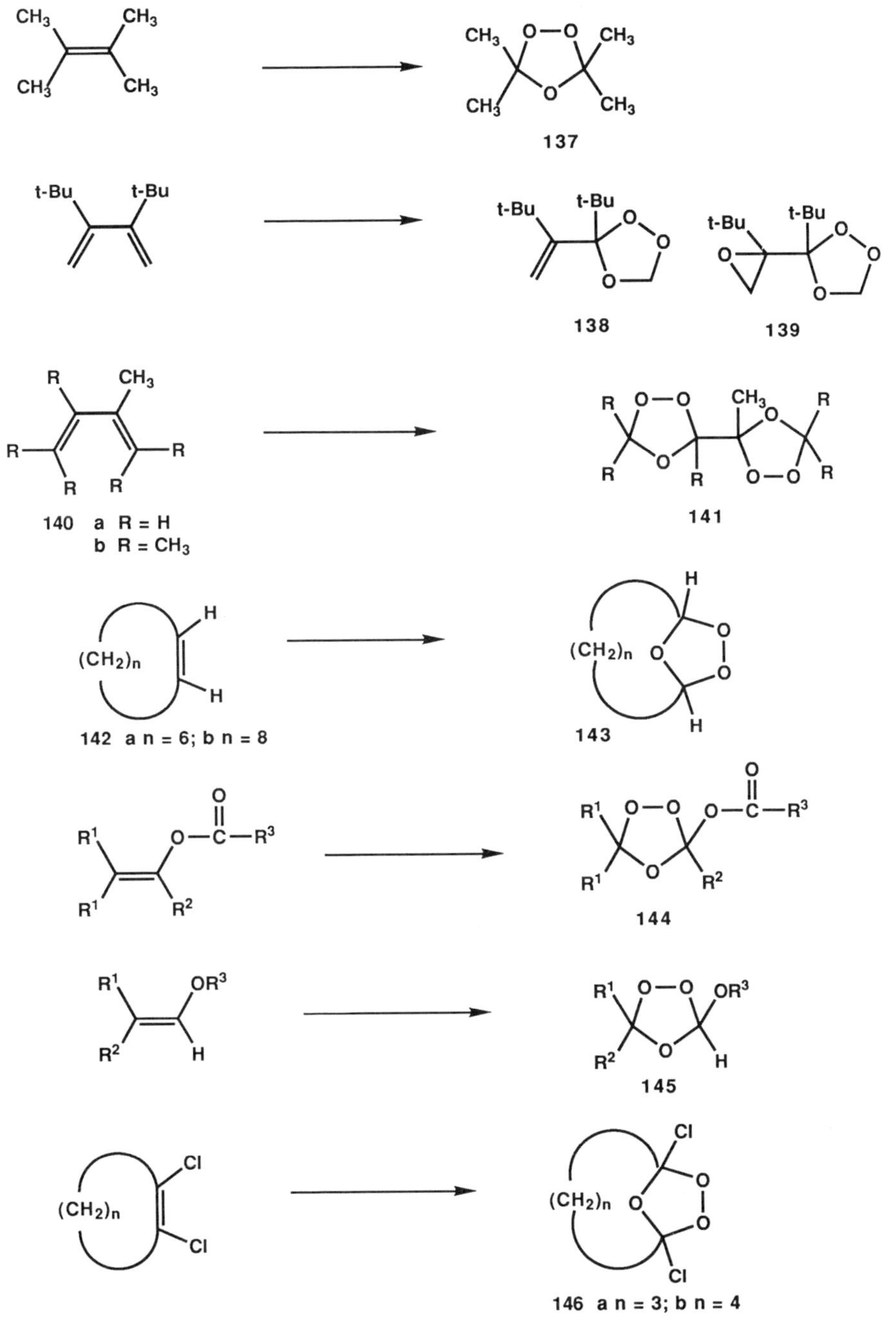

137

138 139

140 a R = H
b R = CH_3

141

142 a n = 6; b n = 8

143

144

145

146 a n = 3; b n = 4

Scheme 40

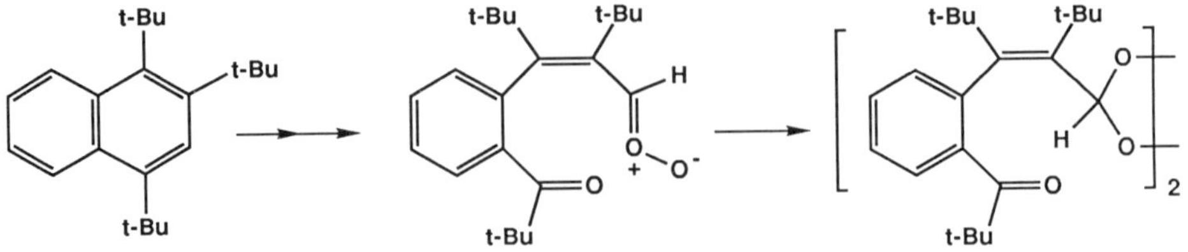

Scheme 41

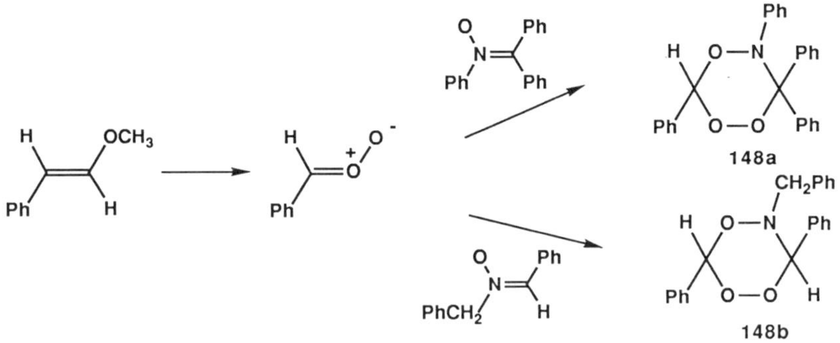

Scheme 42

2.3 Other Cycloaddition Reactions

Since carbonyl oxides undergo self-dimerization and give adducts with nitrones as described above, non-concerted [3 + 4] cycloaddition processes with dienes should be possible in theory.

Mono-ozonization of 1,2,4,5-tetraphenylcyclopenta-1,3-diene afforded a mixture (7:3) of two separable, isomeric bicyclic peroxides (**149** and **150**). The major product **149** was the normal ozonide resulting from intramolecular [3 + 2] cycloaddition of either possible carbonyl oxide **151** intermediate, whereas **150** was derived exclusively from intermediate **151b** by a formal [3 + 4] cycloaddition between the carbonyl oxide and conjugated enone moieties (Scheme 43).[176]

More extended modes of cycloaddition have also been observed for the intermolecular reactions of formaldehyde *O*-oxide with certain keto aldehydes. Thus, ozonolysis of ethyl vinyl ether in the presence of 1-benzoyl-8-formylnaphthalene in dichloromethane at -70°C yielded, in addition to the ozonide **152**, two isomeric 1:1 adducts (**153** and **154**). The formation of the latter adducts can be most readily rationalized on the basis of two stepwise [3 + 2 + 2] cycloaddition processes, which compete with ozonide formation (Scheme

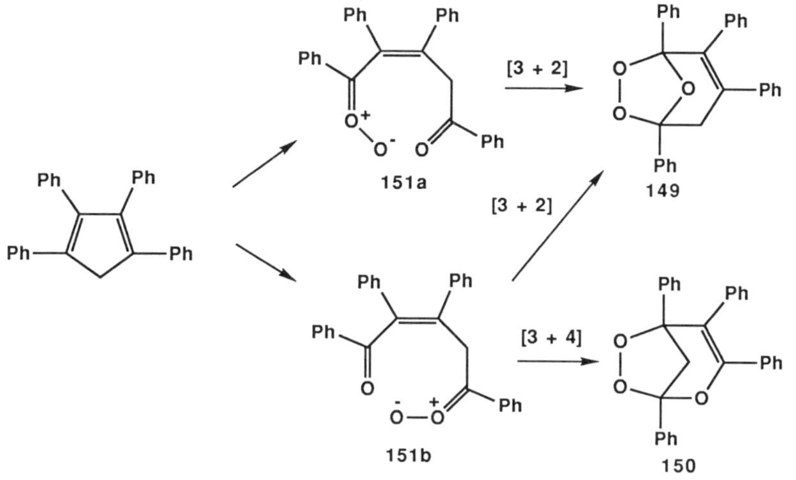

Scheme 43

44).[177] On treatment with trifluoroacetic acid, ozonide **152** was partially isomerized into **153** (25%) and **154** (16%).

Under similar reaction conditions, the keto aldehydes **155** produced only the [3 + 2 + 2] adducts **156**, which were structurally analogous to **153**.

3 CAPTURE OF CARBONYL OXIDES BY PARTICIPATING SOLVENTS AND RELATED REACTIONS

Carbonyl oxides react readily with protic solvents, particularly alcohols, to form the corresponding hydroperoxy adducts **157** (Scheme 45).[1] With unsymmetrically substituted alkenes, quantitative analysis of the two adducts gives an indication of the relative importance of the two possible scission pathways of the primary ozonide. Recent rate studies have established a relative order reactivity of protic solvents towards carbonyl oxides of methanol > ethanol > propanol > isopropanol > water > *tert*-butanol > acetic acid.[178]

Following the reported observation of unusually stable primary ozonides from the ozonization of (*E*)- or (*Z*)-1,4-dichlorobut-2-ene in methanol,[179] reinvestigation of the reaction has shown that the major product was the expected methoxy hydroperoxide **158**, which can also form a hemiperacetal by subsequent reaction with chloroacetaldehyde (Scheme 46).[180]

Cleavage of the primary ozonide from vinyl chloride, in predictable fashion, gave formaldehyde *O*-oxide and formyl chloride, which were efficiently trapped

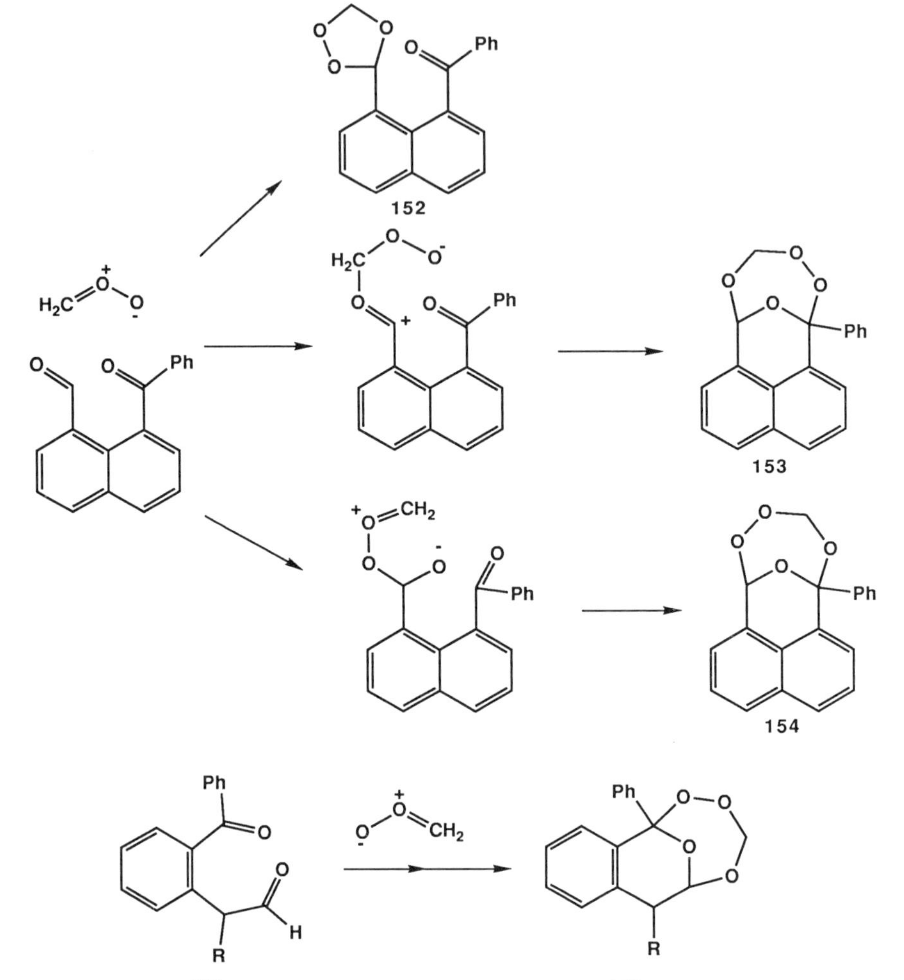

Scheme 44

by methanol to produce methoxymethyl hydroperoxide and methyl formate, respectively.[181]

Although ozonolyses of 1,2-dichloro-substituted ethylenes (**159**) in methanol were expected to follow the normal Criegee mechanism, the resulting α-chloro-α-methoxyalkyl hydroperoxides (**160**), which were too unstable to be isolated, decomposed to give the corresponding methyl esters and hypochlorous acid.[182] Capture of the intermediate chloro-substituted carbonyl oxides by hydrogen chloride afforded the isolable dichloroalkyl hydroperoxides (**161**), which were

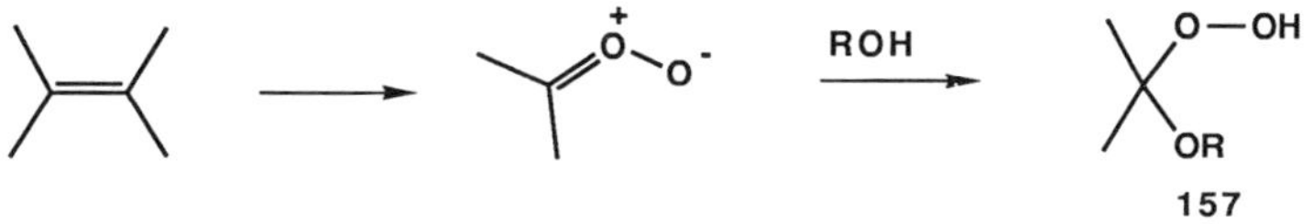

Scheme 45

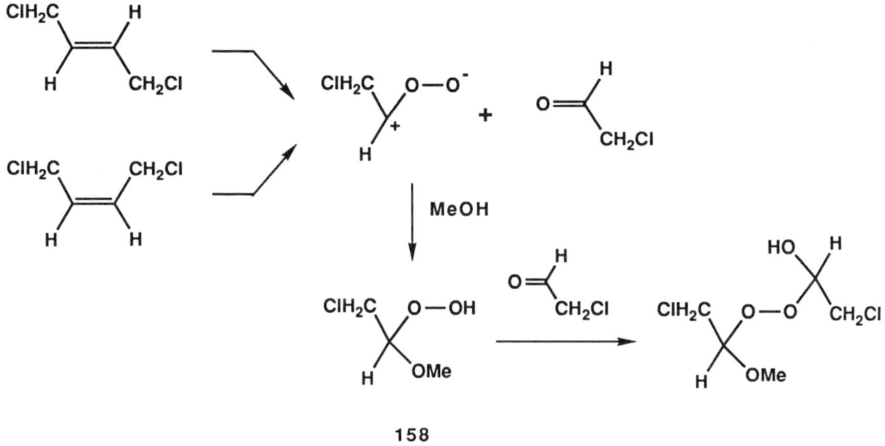

Scheme 46

found to react slowly with acid chlorides yielding peresters (**162**).[183] Peresters **162** could be produced directly by ozonlysis of **159** in the presence of tetrabutylammonium chloride (Scheme 47).[183,184]

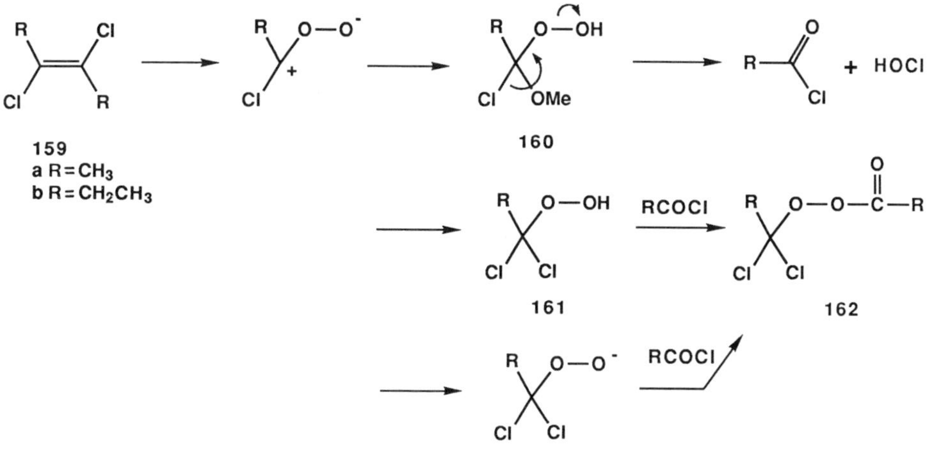

Scheme 47

Treatment of a solution of 2,3-dimethylbuta-1,3-diene (**90a**) in methanol at −78°C with 0.8 equiv. of ozone gave the expected mono-unsaturated α-methoxy hydroperoxide **163** together with a small quantity of the peracetal **164**.[137] On further ozonization, the primary ozonide derived from **163** underwent a highly selective cleavage (> 80%), directed by the electron-withdrawing effects of the methoxy and hydroperoxy groups, which resulted in the formation of the thermally labile keto hydroperoxide **165** and methoxymethyl hydroperoxide. Compound **165** subsequently fragmented to give a series of non-peroxidic products (Scheme 48).[185]

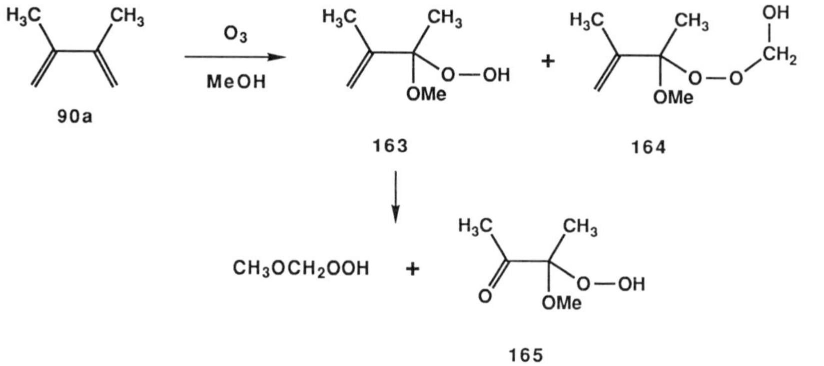

H_3C CH_3 **90a** —O_3, MeOH→ **163** (H_3C, CH_3, O—OH, OMe) + **164** (H_3C, CH_3, O—O—CH_2—OH, OMe)

163 → CH_3OCH_2OOH + **165** (H_3C, CH_3, O, O—OH, OMe)

Scheme 48

The reactions of unsymmetrically substituted butadienes (**92**) in methanol with ozone showed similar trends to **90**a. With 1 equiv. of ozone, both possible unsaturated methoxy hydroperoxides **166** and **167** are formed as the major products.[138] Further treatment with ozone in methanol afforded the corresponding keto hydroperoxides, which fragmented as above (Scheme 49).[186]

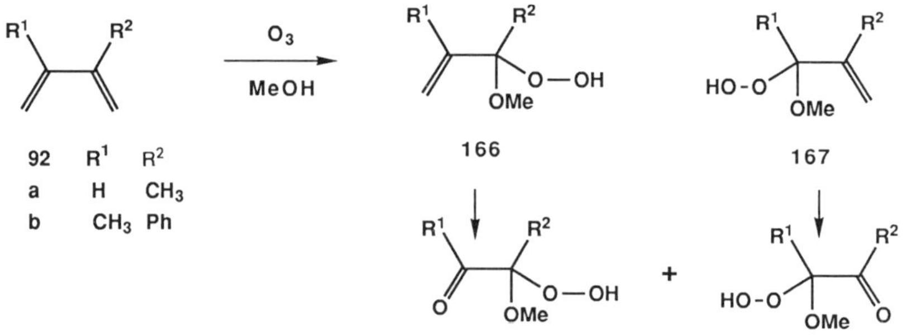

92	R^1	R^2
a	H	CH_3
b	CH_3	Ph

92 —O_3, MeOH→ **166** (R^1, R^2, O—OH, OMe) + **167** (R^1, R^2, HO-O, OMe)

166 → (R^1, R^2, O, O—OH, OMe) + **167** → (R^1, R^2, HO-O, OMe, O)

Scheme 49

Mono-ozonolysis of diene **90b** in methanol gave the methoxy hydroperoxide **168** (21%). On treatment with 2 equiv. of ozone, the epoxy hydroperoxide **169** could be isolated in low yield.[186] The only isolable peroxidic product from the reaction of 2,3,4,5-tetramethylhex-2,4-diene in methanol with 1 equiv. of ozone was the hydroperoxide **170** (47%) (Scheme 50).[138]

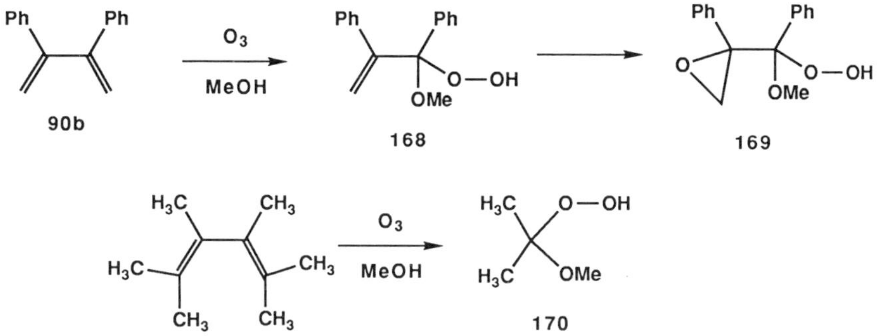

Scheme 50

Diozonolyses of the 1-chloro-substituted dienes **171** in methanol proceeded in a stepwise fashion, similar to those described above. Ozone was shown to attack preferentially at the non-chlorinated double bond to give mainly the hydroperoxide **172** from **171a** (93%) and a mixture of **172** (28%) and the β-chloroenone **173** (59%) from **171b** (Scheme 51). Subsequent ozonolyses of **172** and **173** were considered to proceed according to predicted pathways.[187] The reactions of 2-chloro-substituted butadienes **174** in methanol with ozone were also strongly influenced by the directing effects of the chloro substituent. Thus, the first equivalent of ozone reacted at the non-chlorinated double bond yielding predominantly the corresponding α,β-unsatuated carbonyl compounds which reacted with methanol to give the unsaturated ketals **175** (>90%) and methoxymethyl hydroperoxide. The primary ozonides subsequently derived from **175** fragmented selectively to give acid chlorides which reacted in turn with methanol producing methyl esters.[188]

Ozonolysis of the non-conjugated diene **176** in methanol afforded, after acid work-up, the crystalline cyclic peroxide **177** in 55–60% yield (Scheme 52). The stereochemistry of **177**, as determined by X-ray crystallography, suggests that there must be a high degree of stereocontrol in the trapping of the carbonyl oxide intermediate and in the cyclization of the resulting methoxy hydroperoxide.[189]

Treatment of the manool derivatives **178** with ozone, even in methanol, yielded the hydroperoxy acetals **179**, which were almost certainly formed by intramolecular capture of the respective carbonyl oxide intermediates by the hydroxy group of the adjacent sidechain.[190] In a similar fashion, the carbonyl

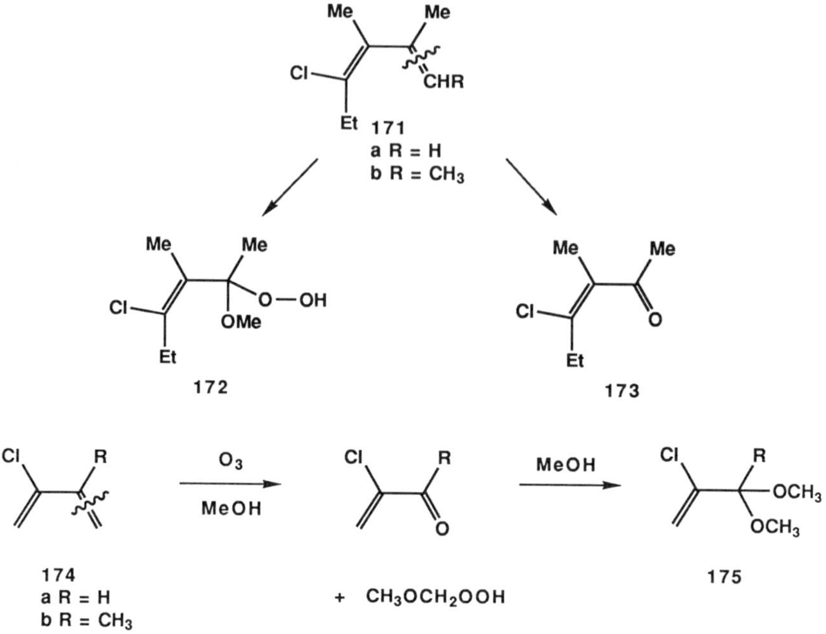

Me Me
Cl
CHR
Et 171
a R = H
b R = CH_3

Me Me
Cl O—OH
OMe
Et
172

Me Me
Cl
O
Et
173

Cl R
O_3
MeOH
Cl R
O
MeOH
Cl R
OCH_3
OCH_3

174
a R = H
b R = CH_3

+ CH_3OCH_2OOH

175

Scheme 51

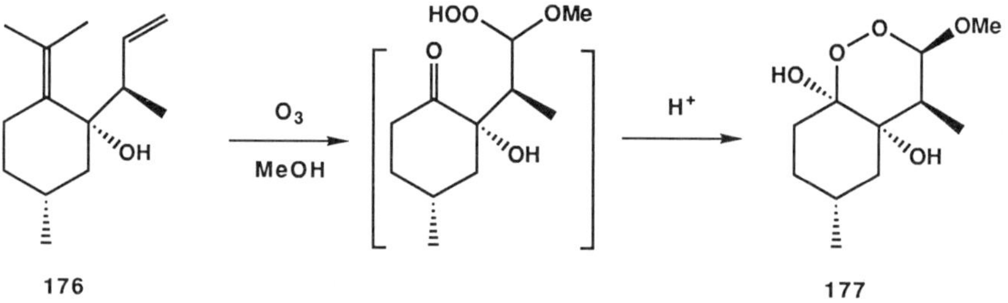

O_3
MeOH
HOO OMe
O
OH
OH
H^+
HO
O
O
OMe
OH

176

177

Scheme 52

oxide **180**, derived selectively from the corresponding unsaturated bicyclic alkene, afforded the hydroperoxide **181** in high yield.[191] The hydroperoxy acetals **179** and **181** were readily transformed into macrocyclic lactones (Scheme 53).

Ozonization of unsubstituted cycloalkenes in methanol resulted in the formation of the solvent-derived cleavage products **182**, which, in the presence of acid catalysts, subsequently formed acyclic oligomeric peroxides.[192] Ozonolyses of 1-methylcyclopentene (**183a**) and 1-methylcyclobutene (**183b**) in methanol were found to proceed mainly through the more highly substituted carbonyl

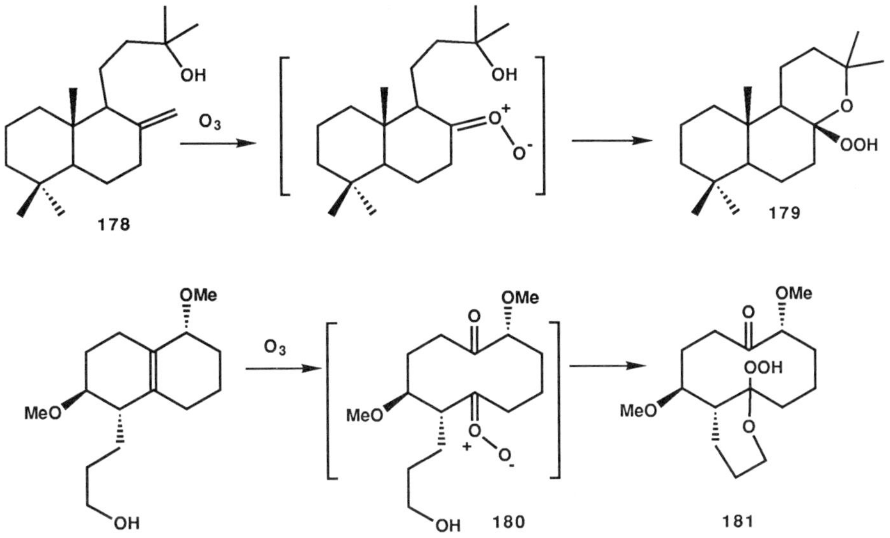

Scheme 53

oxide intermediate **184** in each case (Scheme 54).[193] Thus, capture of **184a** by methanol gave the methoxy **186a**, which on spontaneous intramolecular cyclization formed the hemiperacetal **187** (90%) as a mixture of isomers. The methoxy hydroperoxide **188a**, derived from capture of the alternative carbonyl oxide **185a**, was also observed (*ca* 10%) as an equilibrium mixture (4:1) of the acyclic and cyclic (**189a**) forms. Under similar conditions, **183b** afforded a more complex product mixture consisting of the cyclic hemiperacetals **187b** (mixture of isomers, 43%) and **189b** [*cis* (2%); *trans* (7%)], the hydroperoxy tetrahydrofurans **190** [*cis* (13%); *trans* (3%)] and **191** [*cis* (6%); *trans* (5%)] and the bicyclic ozonide **192** (20%). The formation of the tetrahydrofuran derivatives **190** and **191** clearly indicates that intramolecular partial capture of the carbonyl oxide moiety in **184** and **185** by the carbonyl group oxygen to give the corresponding cyclic zwitterionic intermediates **193** and **194**, respectively, must be competitive with the expected intermolecular solvent-capture process. It also follows that the ozonide **192** was more likely to have been formed by stepwise processes involving intermediates such as **193** and **194** than by a concerted intramolecular [3 + 2] cycloaddition of **184** or **185**.

Previous studies have shown that, even in reactive protic solvents such as methanol, indenes, on reaction with ozone, often give rise to the corresponding ozonides in high yield.[1] The overall reactions of di- and trisubstituted indenes in methanol with ozone are influenced by the nature of the substituents and temperature. For example, yields of the isomeric ozonides derived from 1,2-dimethyl-3-aryl- and 1-methyl-2,3-diarylindenes were found to vary signifi-

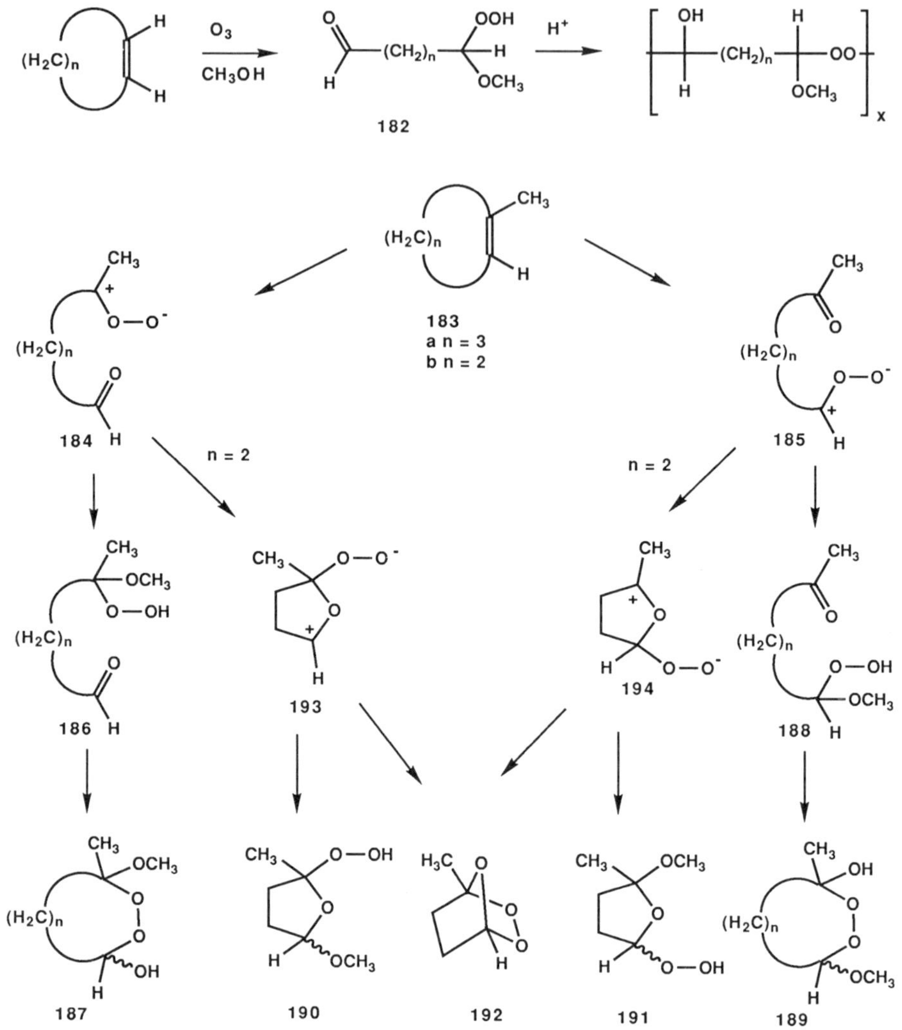

Scheme 54

cantly with reaction temperature; at 20°C the yields were 60–70% whereas at −70°C they dropped to *ca* 30%.[141]

Ozonization of 1-methyl-3-phenylindene (**195**) in methanol at either 20 or −70°C produced the corresponding ozonide **196** (*ca* 40%) as the major isolable product together with a small amount of the solvent-participated product **197** (6%), which does not appear to cyclize to the hemiperacetal.[141] In contrast, ozonolysis of 1-methyl,[141,194] 1-phenyl-[143] and 1-chloro-2,3-diphenylindene[195] (**198**) in methanol at low temperature produced, in high yield (>70%), the novel isochroman derivatives **199** as single isomers in which the hydroperoxy and methoxy groups are *syn*-related. Since the corresponding ozonides were stable in methanol, the solvent-participated products **199** are considered to have arisen from a sequence which involved an unexpectedly selective scission of the primary ozonide, intramolecular partial capture of the carbonyl oxide moiety by the oxygen of the carbonyl group followed by directed incorporation of methanol (Scheme 55).

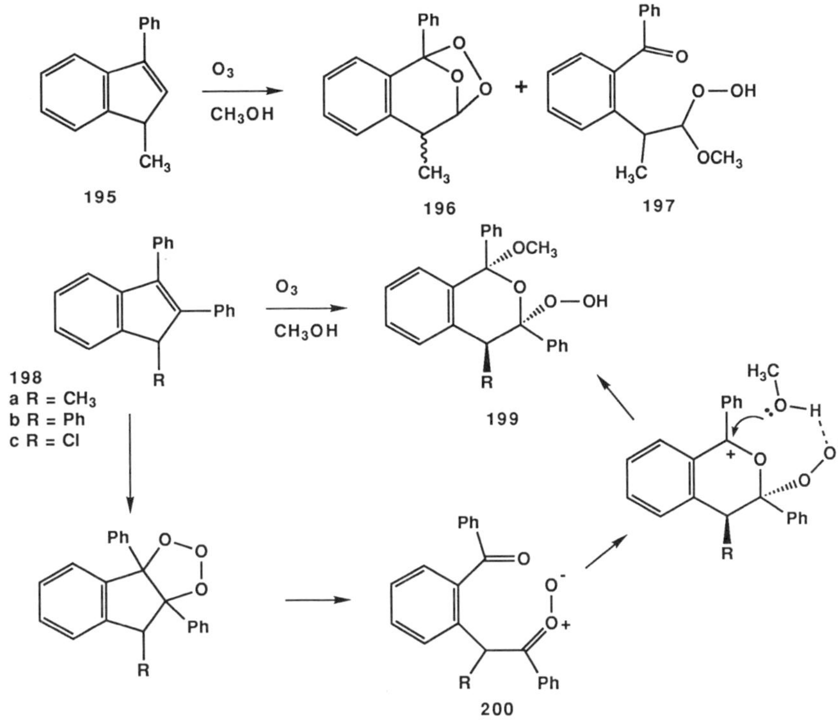

Scheme 55

The structure of the solvent-participated products **199** was consistent with a highly selective decomposition of the primary ozonides derived from **198** and a significant solvent effect on the subsequent intra- and intermolecular reactions of the carbonyl oxide–carbonyl pairs **200**. Thus, ozonization of the enol ethers **201a** and **b**, which should formally be precursors of the same carbonyl oxides **200a** and **b**, respectively, in methanol at low temperature yielded the isomeric solvent-participated products **202** (Scheme 56).[142,143]

Although the major isolable product was the *endo*-ozonide **204**, the 1-acetoxyindene derivative **203**, under similar reaction conditions, afforded the

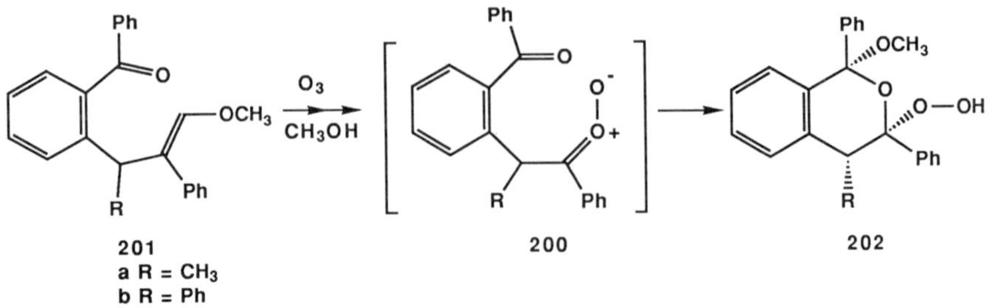

Scheme 56

cyclic hemiperacetal **205** (24%) from the cleavage-cyclization mechanism outlined in Scheme 57.[195] An alternative sequence for the formation of **205**, in which the carbonyl oxide intermediate is captured by methanol followed by cyclization of the resulting methoxy hydroperoxide, seems less likely because, by analogy with the solvent-derived products from 1-methylcyclopentene and cyclobutene,[193] such cyclic hemiperacetals would probably be obtained as mixtures of isomers.

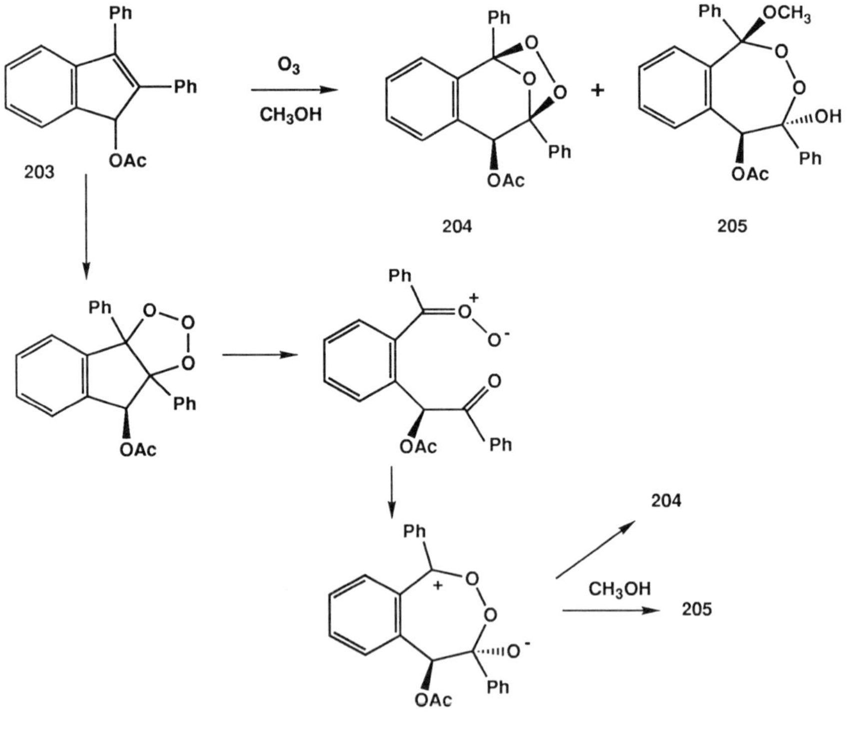

Scheme 57

Reactions of the isomeric enol ethers **107** and **108**, which could be regarded as acyclic equivalents of 2,3-diphenylindene (**109**), with ozone methanol at -70°C gave different ozonolysis products. Whereas **108** and **109** yielded mainly the bicyclic ozonide **110** (49% and 65%, respectively) and the cyclic hemiperacetal **206** (*ca* 15%), **107** produced the isochroman **207** in high yield. To rationalize these differences, particularly for **107** and **108**, it has been proposed that the reactions proceed through non-interconvertible *syn* and *anti* forms of the common carbonyl oxide, which have different chemical properties (Scheme 58).[144]

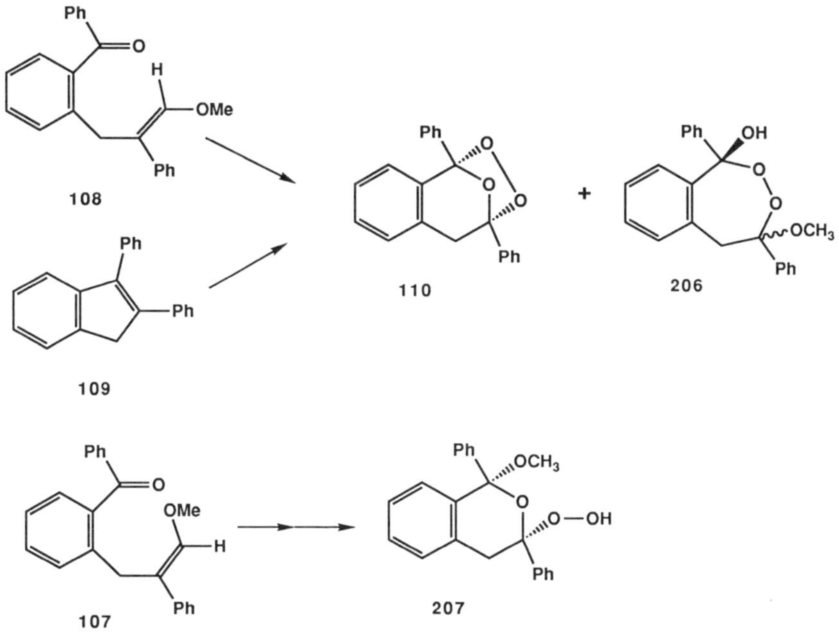

Scheme 58

Like indenes, acenaphthylenes exhibit a strong tendency to form ozonides as the major ozonolysis products, even in the presence of methanol at low temperature. Thus the ozonolysis of the acenaphthylenes **208** in methanol at -70°C afforded the ozonides **209** together with the solvent-participated products **210**. The structure of **210** was consistent with the reaction having proceeded through the more highly substituted carbonyl oxide intermediate. Independent generation of the alternative carbonyl oxide intermediate from the ketoalkene **211** resulted in formation of the methoxy hydroperoxide **212**, the regioisomer of **210a** (Scheme 59).[196]

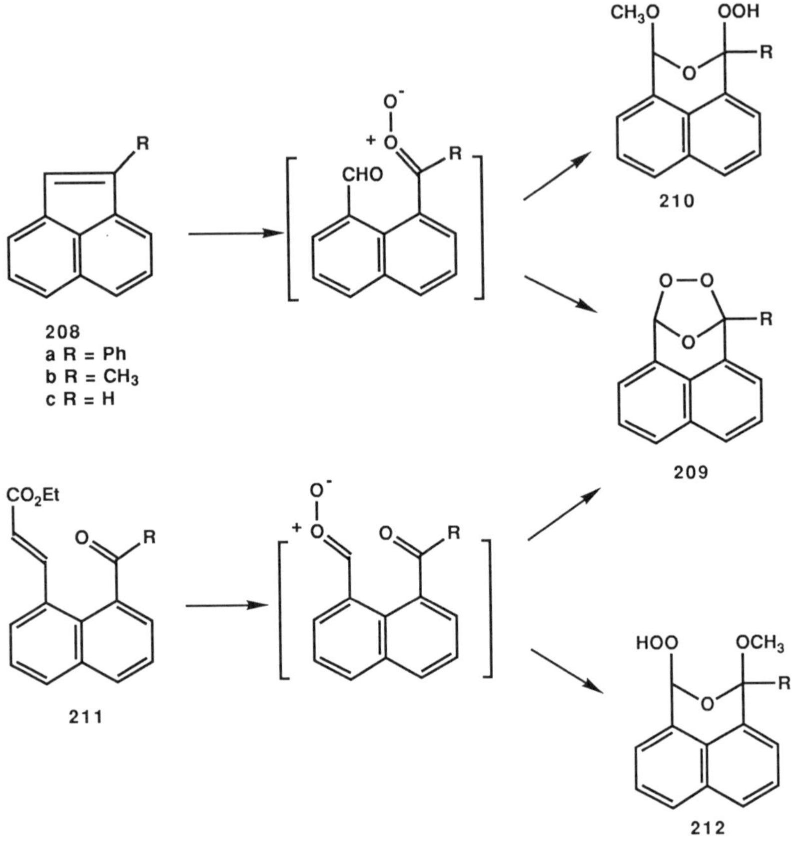

208
a R = Ph
b R = CH_3
c R = H

210

209

211

212

Scheme 59

In addition to the expected ozonide, ozonolysis of the acenaphthylene derivative **213** in 2,2,2-trifluoroethanol at 0°C yielded the debenzoylated hydroperoxide **214** in 42% yield (Scheme 60).[155]

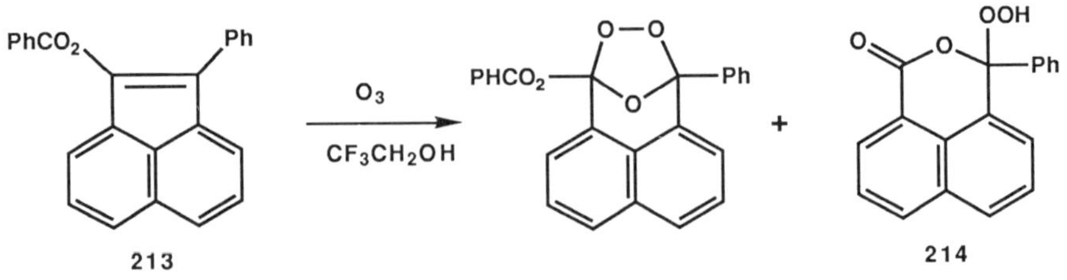

O_3
CF_3CH_2OH

213

214

Scheme 60

Ozonolysis of pyrene in methanol at $-70°C$ yielded a solvent-participated product (**215**; 2:1 mixture of stereoisomers) which is similar in nature to those derived from indenes and acenaphthylenes (Scheme 61).[196]

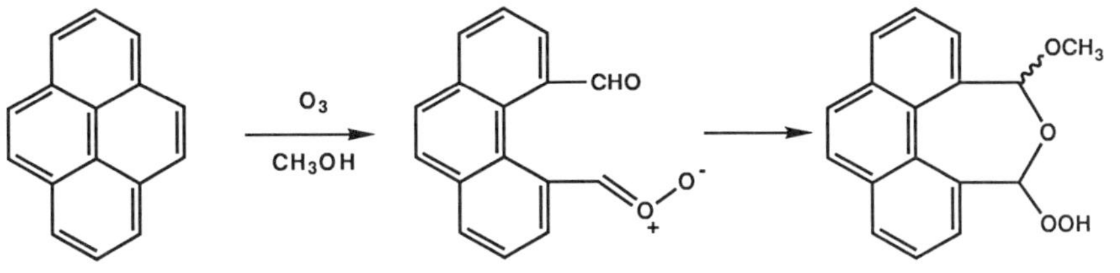

Scheme 61

Based on spectroscopic evidence, the structure of the product obtained quantitatively from the ozonation of cholesterol acetate in methanol has been reassigned as the methoxy hydroperoxide **216**[197] rather than the eight-membered cyclic hemiperacetal **217** previously claimed (Scheme 62).[198]

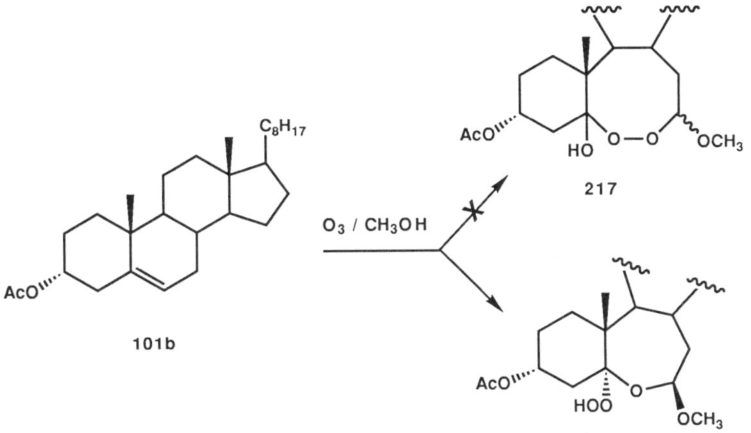

Scheme 62

Treatment of the conjugated cyclohexadienes **218** in methanol at low temperatures with 2 equiv. of ozone resulted in sequential cleavage of the two double bonds to produce mainly the solvent-participated product **219** plus small amounts of the 1,2-dioxanes **220**–**222**. On warming the reaction mixture to room temperature, **219** decomposed with concomitant formation of **220**–**222** (Scheme 63).[199]

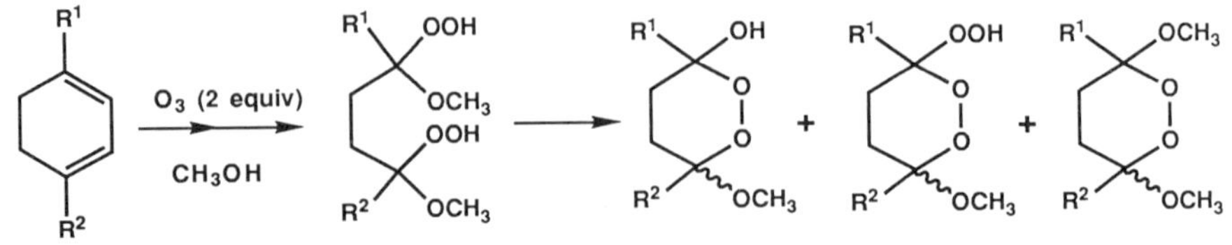

Scheme 63

4 SECONDARY PEROXIDES FROM OZONIDES

As described in Sections 2 and 3, the nature of the peroxidic products isolated from ozonolysis reactions is dependent not only on the substrate but also on the reaction solvent and temperature. Ozonides (1,2,4-trioxolanes) arising from recombination of the carbonyl oxide with the carbonyl fragment are usually produced in non-participating solvents such as hydrocarbons, whereas in the presence of protic solvents the carbonyl oxide intermediates may be trapped, e.g. by alcohols to give the corresponding α-alkoxy hydroperoxides. Under favorable circumstances, the ozonolysis intermediates may also form cyclic adducts with electron-rich alkenes, imines or nitrones.

Although often referred to as 'final' peroxidic ozonolysis products, ozonides can undergo rearrangement in the presence of catalytic amounts of antimony pentachloride or chlorosulphonic acid without cleavage of the peroxide bond. Thus, the monocyclic ozonide **223** was transformed into the tetroxane **224** by acid-catalyzed cleavage of one of the C—O bonds to the peroxide linkage followed by elimination of a molecule of benzaldehyde to generate a protonated or complexed carbonyl oxide species which can subsequently induce the ring opening of a second ozonide molecule (Scheme 64).[200,201] Under similar conditions, the unsymmetrically substituted ozonide from 1-phenylhexene afforded a mixture containing the three possible 1,4-disubstituted tetroxanes.

In addition to tetroxane derivatives, acid-catalyzed rearrangement of the bicyclic ozonides **225** gave the pentaoxabicyclo[5.3.1]undecanes **226** (Scheme 65).[200–202] An extensive range of analogous bicyclic peroxides have been obtained from acid-catalyzed reactions of bicyclic ozonides such as **227** and **228** with (a) carbonyl compounds in hydrogen peroxide (equation 1),[202,203] (b) dihydroxy peroxides (equation 2),[204,205] and (c) other ozonides (equation 3)[204,205] (Scheme 66).

The isolation of peroxidic products from the acid-catalyzed reactions described above requires that protonation or complexation should occur preferentially at one of the oxygen atoms of the peroxide bridge. In contrast, however, an *ab initio* theoretical study of the parent 1,2,3-trioxolane suggests

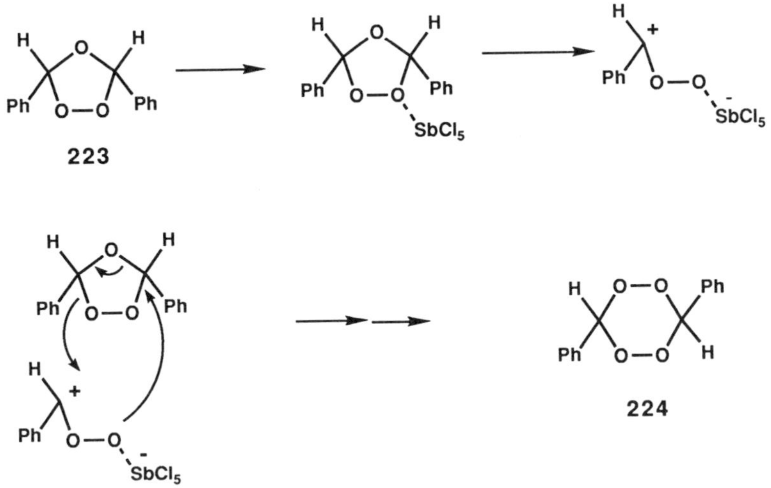

223

224

Scheme 64

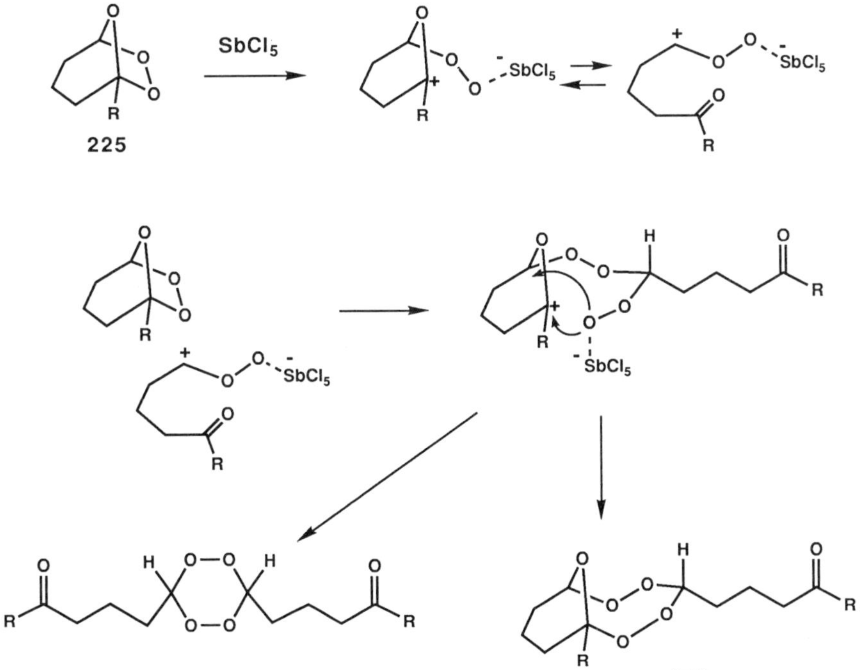

SbCl5

225

226

Scheme 65

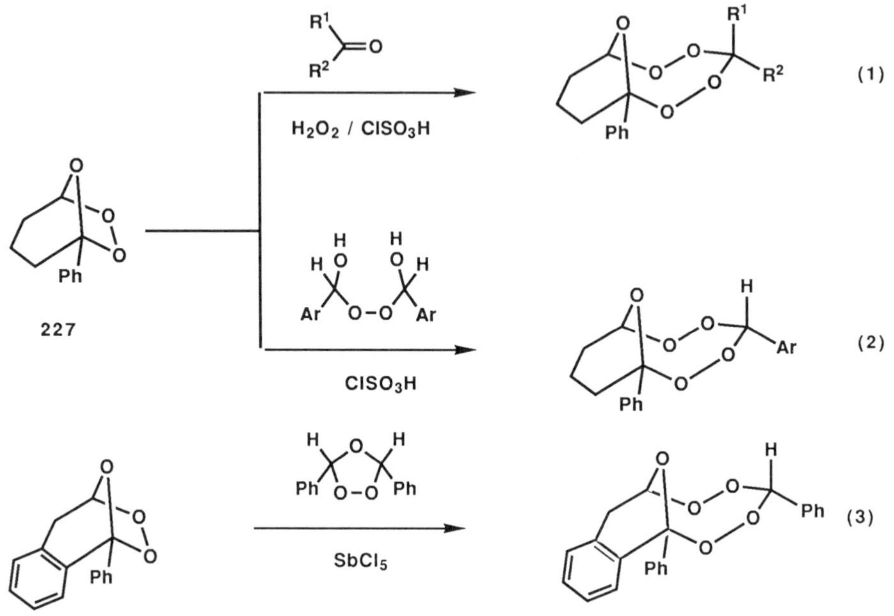

Scheme 66

that protonation should be thermodynamically more favored at the ether oxygen.[206]

Acid-catalyzed methanolysis of the bicyclic ozonides **227** and **228** resulted in the formation of the corresponding dimethoxy cyclic peracetals **229** and **230** (Scheme 67).[207] Owing to relief of ring strain, bicyclic ozonides were generally found to ring open significantly faster than analogous monocyclic systems. Cyclic hemiperacetals, isolated from the ozonolysis of indenes or their acyclic equivalents in methanol, were also transformed into the corresponding dimethoxy compounds on treatment with trifluoroacetic acid in methanol.[144]

1,2-Dioxolanes have been isolated directly from the ozonolysis of certain simple enol ethers.[6,148,151] Boron trifluoride-catalyzed reaction of ozonides **231** with 1,1-disubstituted alkenes provided the trisubstituted 1,2-dioxolanes **232** in variable yield as outlined in Scheme 68.[208,209]

Ozonolysis of the acenaphthodioxin derivative **233** afforded a stable crystalline solid which has been shown by X-ray crystallography to be the cyclic perester **234** rather than the expected ozonide.[210] *A priori*, it was tentatively proposed that **234** had been formed by rearrangement of the intermediate ozonide derived from **233** (Scheme 69). A perester of similar structure (**235**) has also been isolated from the ozonolysis of 1,2-dichloroacenaphthylene.[184]

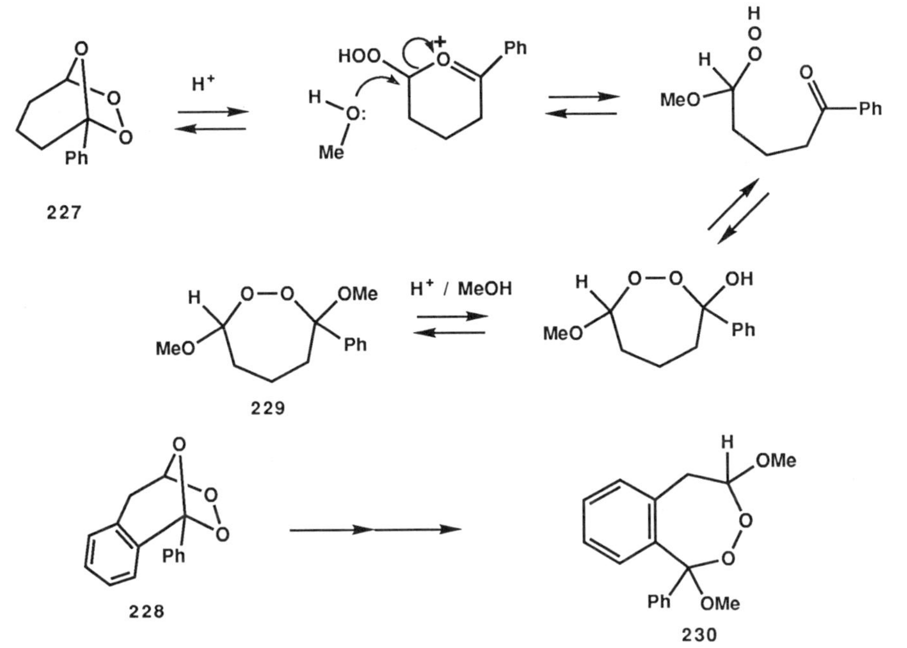

Scheme 67

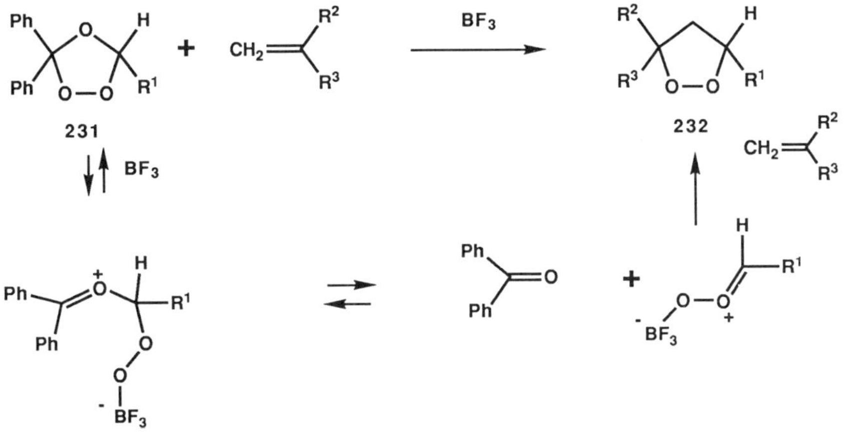

Scheme 68

There have been comparatively few reports of chemical transformations performed on ozonides which leave the central trioxolane ring intact, but some recent examples can be given. Treatment of the chloro vinyl ozonide **236** in

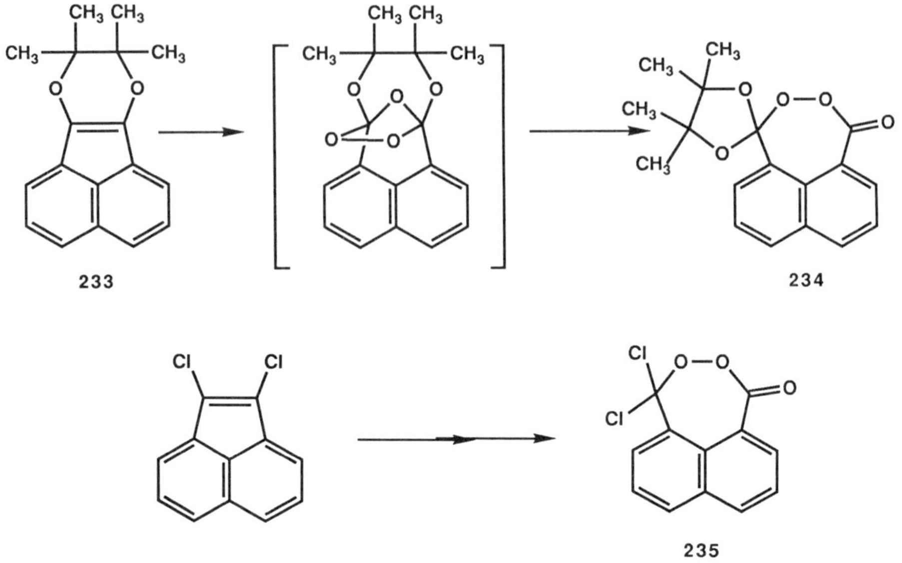

233

234

235

Scheme 69

methanol afforded the ester ozonide **237**, via the expected acid chloride.[211] Reaction of the unsaturated ozonide **238** gave the corresponding epoxide, which on $AgBF_4$-catalyzed epoxide ring opening provided the keto ozonides **239** and **240**.[211] Nucleophilic substitution of the trioxolane ring has been successfully achieved by reaction of the ozonides **241** and **141b** with alcohols or phenol in the presence of sodium hydrogen carbonate to produce a series of new ether ozonides (**242**)[212] (Scheme 70).

5 MISCELLANEOUS PEROXIDES FROM OZONIZATION

In addition to conventional ozonides and solvent-participated products, a variety of unusual peroxides has been detected as intermediates in, or isolated from, ozonization reactions. Studies of the reactions of hex-3-ene derivatives with ozone have indicated that the solvent affects the proportion and nature of the oligomeric peroxides obtained.[213,214]

Ozonolysis of cyclopropylidenecycloalkanes (**243**) afforded the 4-oxo-1,2-dioxane (**245**), which is thought to have been produced by homolytic cleavage of one O—O bond in the primary ozonide followed by ring opening of the adjacent cyclopropylidene ring and intramolecular coupling of the resulting 1,6-diradical (**244**) (Scheme 71).[215] The formation of the cyclic keto peroxide **247**

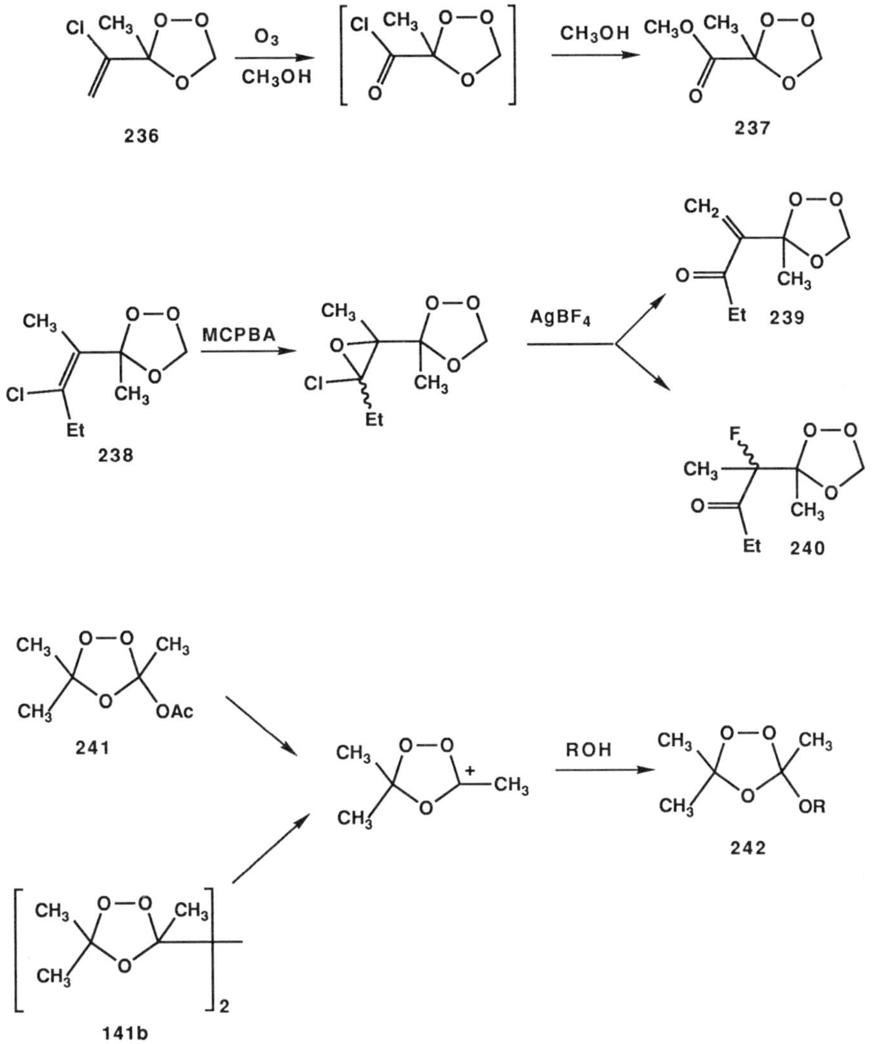

Scheme 70

from the methylenecyclobutane **246** has been rationalized by an analogous double ring fission process.[216]

It has been proposed that cyclobutane-1,2-diones (**251**), obtained from ozonolyses of alkenylidenecyclopropanes (**248**), were derived from the corresponding carbonyl oxides (**249**) by a rearrangement sequence which involved isomerization, ring expansion to the dioxetene **250** and ring opening as depicted in Scheme 72.[217]

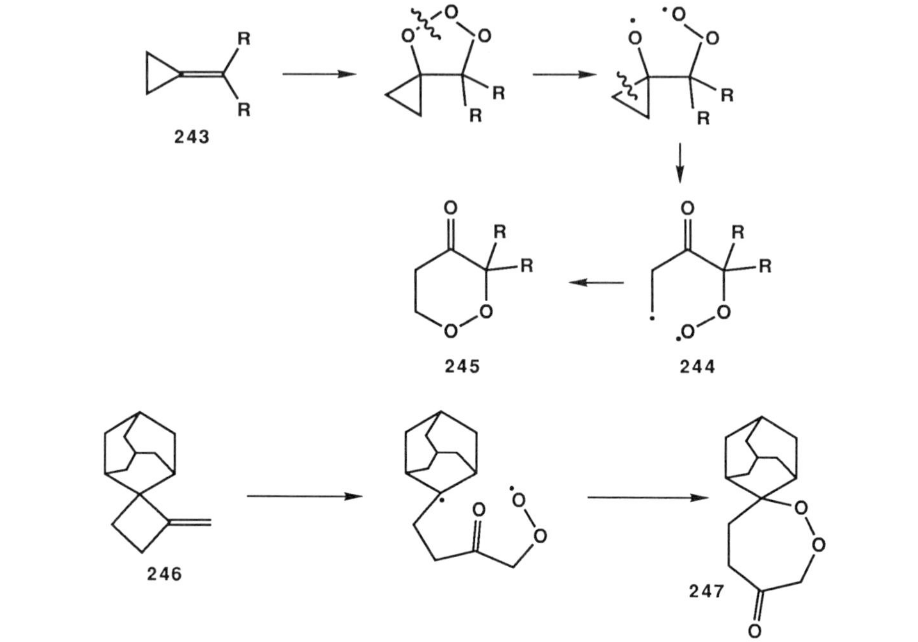

243

245

244

246

247

Scheme 71

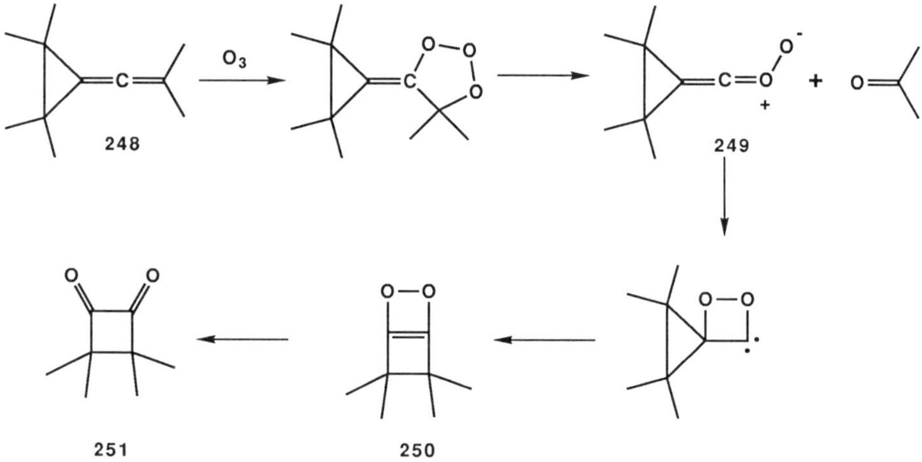

O_3

248

249

251

250

Scheme 72

Treatment of the vinylsilane **252** in dichloromethane at -75°C with ozone afforded a mixture of the α-peroxy ketone **254**, which had resulted from partial cleavage of the O—O bond of the primary ozonide **253** followed by migration of the trimethylsilyl group, and the dioxetane **255** (Scheme 73).[218] Silylperoxy ketones similar to **254** have also been prepared from the corresponding unsaturated silyl peroxides **256** by successive reaction with ozone in methanol followed by dimethyl sulfide.[219]

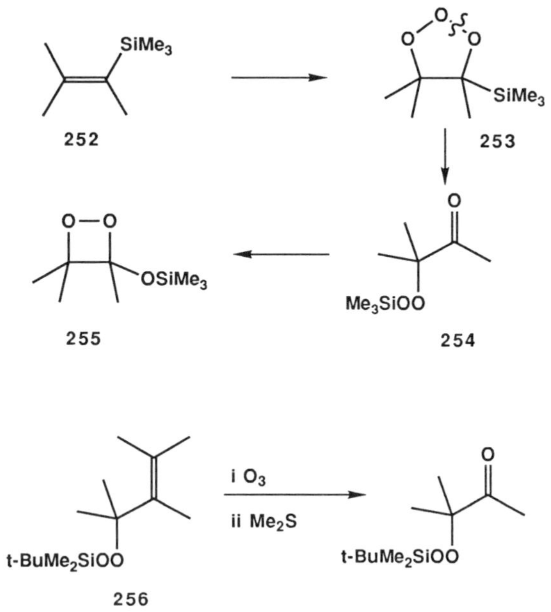
SiMe3
252
SiMe3
253
O
Me3SiOO
254
O—O
OSiMe3
255
i O3
ii Me2S
t-BuMe2SiOO
256
O
t-BuMe2SiOO

Scheme 73

The above strategy has been elegantly applied to the synthesis of a C,D-ring fragment of artemisinin.[220,221] Thus, treatment of **257** with ozone in methanol at −78°C and subsequent removal of the solvent at low temperature (*ca* 0°C) yielded the dioxetane **258** as a single diastereoisomer. Thermal rearrangement of **258** proceeded, even at room temperature, to give **259** as the major product in 57% yield (Scheme 74).

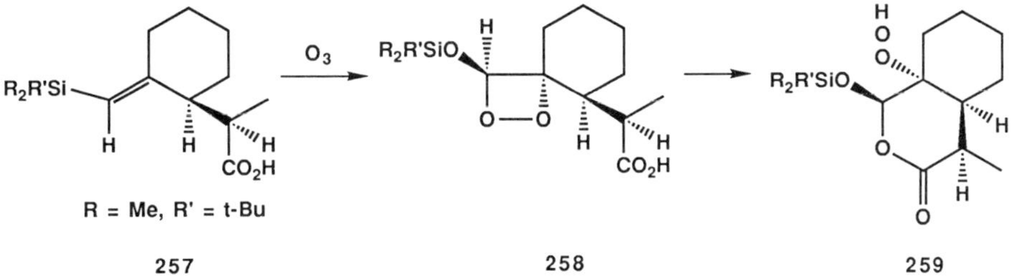
R2R'Si
H
H
H
CO2H
R = Me, R' = t-Bu
257
O3
R2R'SiO
H
O—O
H
H
CO2H
258
H
O
O
R2R'SiO
H
O
O
H
259

Scheme 74

It has been found that the ozonolysis of diphenylvinylene carbonate (**260**) yielded dibenzoyl monoperoxycarbonate (**261**), the first example of a diacyl monoperoxycarbonate (Scheme 75).[222]

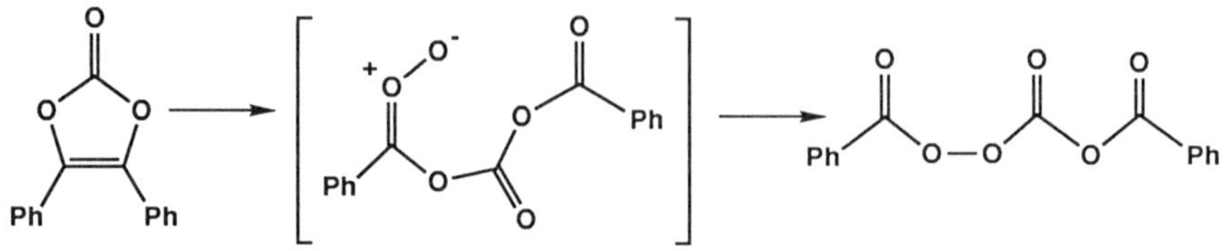

260

261

Scheme 75

Although ozonides are more usually obtained from the reactions of ozone with alkenes, ozonization of the saturated cholestane derivatives **262**, either supported on silica gel ('dry ozonization')[223] or in solution,[224] resulted in cleavage of the C-14—C-15 bond with concomitant formation of the ozonides **263** (Scheme 76).

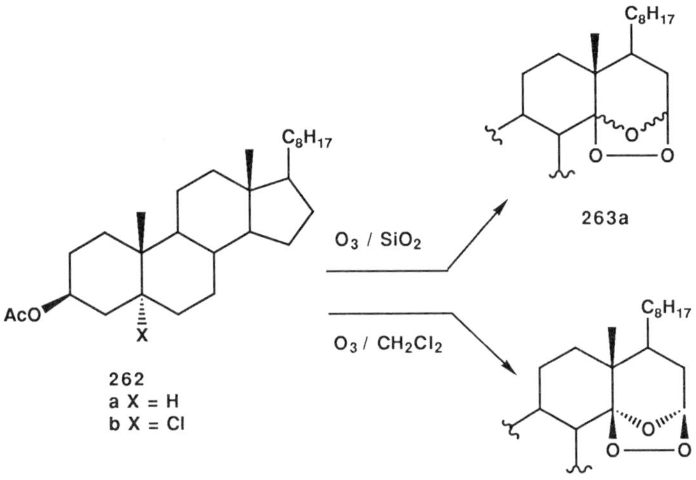

O_3 / SiO_2

O_3 / CH_2Cl_2

262
a X = H
b X = Cl

263a

263b

Scheme 76

Reaction of the diphosphene **264** with 2 equiv. of ozone in toluene at $-80°C$ gave a relatively stable cyclic diperoxide (**265**) (Scheme 77).[225]

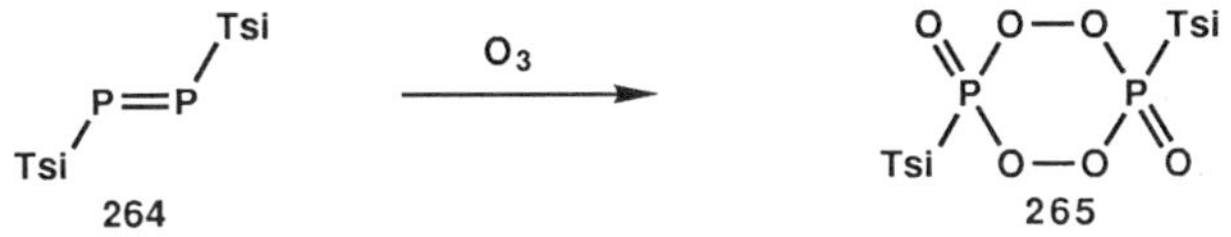

Tsi—P=P—Tsi (264) $\xrightarrow{O_3}$ 265

Tsi = $(Me_3Si)_3C$

Scheme 77

6 CONCLUSION

Over the past decade, increased insight into the mechanisms of the reactions between ozone and a variety of organic substrates has been facilitated by advances in spectroscopic, theoretical, structural and synthetic techniques.

Carbonyl oxides, now well established as key intermediates in the ozonolysis process, have been shown to be more versatile 1,3-dipoles than expected by forming adducts not only with aldehydes and ketones, but also with esters, electron-rich alkenes, imines and nitrones.

A range of new ozonides have been prepared by using enol ethers as substrates rather than alkenes and/or the recently developed procedures such as 'dry ozonization.' In addition, the chemistry of ozonides has been extended further, hence providing entry to a range of new mono- and polycyclic peroxides.

Although considerable progress has been made in this important area of chemistry, there is still much more to be learned regarding the reaction between ozone and unsaturated substrates. The pursuit of a better understanding of the factors which affect the initial formation and decomposition of primary ozonides and the chemistry of the resulting carbonyl oxide intermediates should continue to present challenges for future investigation.

7 ACKNOWLEDGEMENTS

Our own contributions in this area have been realized and advanced by the efforts of our capable research co-workers, who are acknowledged as co-authors in the references cited. Our special thanks are due to Professors S. Kusabayashi (University of Osaka), J. G. Buchanan (Heriot-Watt University), K. Griesbaum and S. Nagase and Dr Miura for their support and encouragement.

Financial support from the Ministry of Education, Japan, and the SERC, UK, and support for traveling by DAAD, Germany (to M. N.) and the British Council and the Carnegie Trust for the Universities of Scotland (to K. J. McC.) are also gratefully acknowledged.

8 REFERENCES

1. P. S. Bailey, *Ozonation in Organic Chemistry*, Academic Press, New York, Vol. 1 (1978); Vol. 2 (1982).
2. R. L. Kuczkowski, in A. Padwa (Ed.), *1,3-Dipolar Cycladdition Chemistry*, Vol. 2, Wiley–Interscience, New York (1984) Chap. 11.
3. W. Sander, *Angew. Chem., Int. Ed. Engl.*, **29**, 344 (1990).
4. S. Patai (Ed.), *The Chemistry of Functional Groups, Peroxides*, Wiley, Chichester, (1983).
5. I. Fleming, *Frontier Orbitals and Organic Chemical Reactions*, Wiley–Interscience, London (1976).
6. H. Keul and R. L. Kuczkowski, *J. Am. Chem. Soc.*, **106**, 5370 (1984).
7. R. Criegee, *Angew. Chem., Int. Ed. Engl.*, **14**, 745 (1975).
8. W. G. Alcock and B. Mile, *J. Chem. Soc., Chem. Commun.*, 5 (1976).
9. B. Nelander and L. Nord, *J. Am. Chem. Soc.*, **101**, 3769 (1979).
10. J. Z. Gillies, C. W. Gillies, R. D. Suenram, F. J. Lovas and W. Stahl, *J. Am. Chem. Soc.*, **111**, 3073 (1989).
11. M. S. LaBarge, K. W. Hillig II, and R. L. Kuczkowski, *Angew. Chem., Int. Ed. Engl.*, **27**, 1356 (1988).
12. M. L. McKee and C. M. Rohlfing, *J. Am. Chem. Soc.*, **111**, 2497 (1989).
13. G. Leroy and M. Sana, *Tetrahedron*, **32**, 1379 (1976).
14. K. A. Singmaster and G. C. Pimentel, *J. Phys. Chem.*, **94**, 5226 (1990).
15. R. Criegee and G. Schroder, *Chem. Ber.*, **93**, 689 (1960).
16. F. L. Greenwood and S. Cohen, *J. Org. Chem.*, **28**, 1159 (1963).
17. F. L. Greenwood, *J. Org. Chem.*, **29**, 1321 (1964).
18. F. L. Greenwood, *J. Org. Chem.*, **30**, 3108 (1965).
19. B. Mile, G. W. Morris and W. G. Alcock, *J. Chem. Soc., Perkin Trans. 1*, 1644 (1979).
20. C. K. Kohlmiller and L. Andrews, *J. Am. Chem. Soc.*, **103**, 2578 (1981).
21. J. F. McGarrity and J. Prodolliet, *J. Org. Chem.*, **49**, 4465 (1984).
22. J. Z. Gillies, C. W. Gillies, R. D. Suenram and F. J. Lovas, *J. Am. Chem. Soc.*, **110**, 7991 (1988).
23. Y. Ito, A. Matsuura, R. Otani, T. Matsuura, K. Fukuyama and Y. Katsube, *J. Am. Chem. Soc.*, **105**, 5699 (1983).
24. J. Rigaudy, G. Chelu and N. K. Cuong, *J. Org. Chem.*, **50**, 4474 (1985).
25. R. Criegee, A. Banciu and H. Keul, *Chem. Ber.*, **108**, 1642 (1975).
26. J. C. Scaiano, W. G. McGimpsey and H. L. Casal, *J. Org. Chem.*, **54**, 1612 (1989).
27. Y. Fujiwara, Y. Tanimoto, M. Itoh, K. Hirai and H. Tomioka, *J. Am. Chem. Soc.*, **109**, 1942 (1987).
28. T. Sugawara, H. Iwamura, H. Hayashi, A. Sekiguchi, W. Ando and M. T. H. Liu, *Chem. Lett.*, 1261 (1983).
29. R. W. Murray and D. P. Higley, *J. Am. Chem. Soc.*, **95**, 7886 (1973).
30. D. P. Higley and R. W. Murray, *J. Am. Chem. Soc.*, **96**, 3330 (1974).
31. M. Girard and D. Griller, *J. Phys. Chem.*, **90**, 6801 (1986).
32. W. W. Sander, *J. Org. Chem.*, **54**, 333 (1989).
33. O. L. Chapman and T. C. Hess, *J. Am. Chem. Soc.*, **106**, 1842 (1984).
34. I. R. Dunkin and C. J. Shields, *J. Chem. Soc., Chem. Commun.*, 154 (1986).
35. G. A. Ganzer, R. S. Sheridan and M. T. H. Liu, *J. Am. Chem. Soc.*, **108**, 1517 (1986).
36. W. Adam and A. Rodriguez, *J. Am. Chem. Soc.*, **102**, 404 (1980).
37. W. Adam and A. Rodriguez, *Tetrahedron Lett.*, **22**, 3505 (1981).

38. M. L. Graziano, M. R. Iesce, G. Cimminiello and R. Scarpati, *J. Chem. Soc., Perkin Trans. 1*, 1699 (1988).
39. W. Adam and A. Rodriguez, *Tetrahedron Lett.*, **22**, 3509 (1981).
40. M. Schulz and N. Grossmann, *J. Prakt. Chem.*, **318**, 575 (1976).
41. W. Ando, S. Kohmoto, and K. Nishizawa, *J. Chem. Soc., Chem. Commun.*, 894 (1978).
42. R. Sato, H. Sonobe, T. Akasaka and W. Ando, *Tetrahedron*, **42**, 5273 (1986).
43. C. J. M. van den Heuvel, H. Steinberg and T. J. deBoer, *Recl. Trav. Chim. Pays-Bas*, **104**, 145 (1985).
44. A. W. van der Made, M. J. P. van Gerwen, W. Drenth and R. J. M. Nolte, *J. Chem. Soc., Chem. Commun.*, 888 (1987).
45. R. W. Murray and S. K. Agarwal, *J. Org. Chem.*, **50**, 4698 (1985).
46. Y. Sawaki, H. Kato and Y. Ogata, *J. Am. Chem. Soc.*, **103**, 3832 (1981).
47. T. A. Hinrichs, V. Ramachandran and R. W. Murray, *J. Am. Chem. Soc.*, **101**, 1282 (1979).
48. W. A. Pryor and C. K. Govindan, *J. Am. Chem. Soc.*, **103**, 7681 (1981).
49. D. Cremer and C. W. Bock, *J. Am. Chem. Soc.*, **108**, 3375 (1986).
50. N. Nakamura, M. Nojima and S. Kusabayashi, *J. Am. Chem. Soc.*, **108**, 4671 (1986).
51. S. Kumar and R. W. Murray, *J. Am. Chem. Soc.*, **106**, 1040 (1984).
52. S. K. Agarwal and R. W. Murray, *Photochem. Photobiol.*, **35**, 31 (1982).
53. C. W. Jefford, J. Boukouvalas and S. Kohmoto, *Helv. Chim. Acta*, **69**, 941 (1986).
54. G. A. Hamilton and J. R. Giacin, *J. Am. Chem. Soc.*, **88**, 1584 (1966).
55. W. Adam, H. Durr, W. Haas and B. Lohray, *Angew. Chem., Int. Ed. Engl.*, **25**, 101 (1986).
56. W. Ando, H. Miyazaki and S. Kohmoto, *Tetrahedron Lett.*, 1317 (1979).
57. W. Ando, Y. Kabe and H. Miyazaki, *Photochem. Photobiol.*, **31**, 191 (1980).
58. R. M. Moriarty, K. B. White and A. Chin, *J. Am. Chem. Soc.*, **100**, 5582 (1978).
59. S. Hatakeyama, H. Kobayashi and H. Akimoto, *J. Phys. Chem.*, **88**, 4736 (1984).
60. K. Ishiguro, Y. Hirano and Y. Sawaki, *Tetrahedron Lett.*, **28**, 6201 (1987).
61. J. W. Agopovich and C. W. Gillies, *J. Am. Chem. Soc.*, **104**, 813 (1982).
62. S. Toby and F. S. Toby, *J. Phys. Chem.*, **84**, 206 (1980).
63. D. Cremer, T. Schmidt, W. Sander and P. Bischof, *J. Org. Chem.*, **54**, 2515 (1989).
64. R. D. Suenram and F. J. Lovas, *J. Am. Chem. Soc.*, **100**, 5117 (1978).
65. F. J. Lovas and R. D. Suenram, *Chem. Phys. Lett.*, **51**, 453 (1977).
66. R. I. Martinez, R. E. Huie and J. T. Herron, *Chem. Phys. Lett.*, **51**, 457 (1977).
67. J. T. Herron and R. E. Huie, *J. Am. Chem. Soc.*, **99**, 5430 (1977).
68. R. I. Martinez and J. T. Herron, *J. Phys. Chem.*, **92**, 4644 (1988).
69. R. I. Martinez and J. T. Herron, *J. Phys. Chem.*, **91**, 946 (1987).
70. H. Niki, P. D. Maker, C. M. Savage, L. P. Breitenbach and M. D. Hurley, *J. Phys. Chem.*, **91**, 941 (1987).
71. R. Atkinson and W. P. L. Carter, *Chem. Rev.*, **84**, 437 (1984).
72. K. Ishiguro, Y. Hirano and Y. Sawaki, *J. Org. Chem.*, **53**, 5397 (1988).
73. E. Zadok, S. Rubinraut, F. Frolow and Y. Mazur, *J. Org. Chem.* **50**, 2647 (1985).
74. W. Adam, R. Curci and J. O. Edwards, *Acc. Chem. Res.*, **22**, 205 (1989).
75. R. W. Murray, *Chem. Rev.*, **89** 1187 (1989).
76. R. Curci, in *Advances in Oxygenated Processes*, (A. L. Baumstark, Ed.), Vol. 2 JAI Press, Greenwich (1990), Chap. 1.
77. W. Adam, L. Hadjiarapoglou and B. Nestler, *Tetrahedron Lett.*, **31**, 331 (1990).
78. M. Rahman, M. L. McKee, P. B. Shevlin and R. Sztyrbicka, *J. Am. Chem. Soc.*, **110**, 4002 (1988).
79. N. Nakamura, M. Nojima and S. Kusabayashi, *J. Am. Chem. Soc.*, **109**, 4969 (1987).

80. G. Cicala, R. Curci, M. Fiorentino and O. Laricchiuta, *J. Org. Chem.*, **47**, 2670 (1982).
81. R. W. Murray and A. Suzui, *J. Am. Chem. Soc.*, **95**, 3343 (1973).
82. P. S. Bailey, H. H. Hwang and C. Chiang, *J. Org. Chem.*, **50**, 231 (1985).
83. M. Strobel, L. Morin and D. Paquer, *Tetrahedron Lett.*, **21**, 523 (1980).
84. D. Barillier and M. Vazeux, *J. Org. Chem.*, **51**, 2276 (1986).
85. L. Morin, D. Barillier, M. Strobel and D. Paquer, *Tetrahedron Lett.*, **22**, 2267 (1981).
86. R. Chaussin, P. Leriverend and D. Paquer, *J. Chem. Soc., Chem. Commun.*, 1032 (1978).
87. D. Barillier, M. P. Strobel, L. Morin and D. Paquer, *Nouv. J. Chim.*, **6**, 201 (1982); *Chem. Abstr.*, **97**, 144430r (1982).
88. I. J. Borowitz and R. D. Rapp, *J. Org. Chem.*, **34**, 1370 (1969).
89. P. G. Gassman and X. Creary, *Tetrahedron Lett.*, 4407, 4411 (1972).
90. R. D. Clark and C. H. Heathcock, *J. Org. Chem.*, **41**, 1396 (1976).
91. R. P. C. Cousins and N. S. Simpkins, *Tetrahedron Lett.*, **30**, 7241 (1989).
92. W. J. Gensler, S. Chan and S. Ruchirawat, *J. Org. Chem.*, **46**, 1750 (1981).
93. I. L. Ismoilov, D. P. Kiyukhin, I. M. Barkalov and S. T. Teshabaev, *Khim. Fiz.*, **8**, 98 (1989); *Chem. Abstr.*, **111**, 56789s (1989).
94. K. R. Kopecky, J. Molina and R. Rico, *Can. J. Chem.*, **66**, 2234 (1988).
95. P. S. Bailey, Y. G. Chang and W. W. L. Kwie, *J. Org. Chem.*, **27**, 1198 (1962).
96. W. B. DeMore and C. Lin, *J. Org. Chem.*, **38**, 9851 (1973).
97. R. E. Keay and G. A. Hamilton, *J. Am. Chem. Soc.*, **98**, 6578 (1976).
98. S. Jackson and L. A. Hull, *J. Org. Chem.*, **41**, 3340 (1976).
99. N. C. Yang and J. Libman, *J. Org. Chem.*, **39**, 1782 (1974).
100. W. Ando, H. Miyazaki, K. Ito and D. Auchi, *Tetrahedron Lett.*, **23**, 555 (1982).
101. W. A. Pryor, C. K. Govindan and D. F. Church, *J. Am. Chem. Soc.*, **104**, 7563 (1982).
102. B. Plesničar, in *The Chemistry of Functional Groups, Peroxides* (S. Patai, Ed.), Wiley, Chichester (1983), Chap. 16.
103. R. W. Murray, in *Singlet Oxygen* (H. H. Wasserman and and R. W. Murray, Eds), Academic Press, New York, (1979) Chap. 3.
104. P. S. Bailey, J. W. Ward, T. P. Carter, Jr, E. Nieh, C. M. Fischer and A. Y. Khashab, *J. Am. Chem. Soc.*, **96**, 6136 (1974).
105. R. Withnall, M. Hawkins and L. Andrews, *J. Phys. Chem.*, **90**, 575 (1986).
106. R. R. Lucchese, K. Haber and H. F. Schaeffer III, *J. Am. Chem. Soc.*, **98**, 7617 (1976).
107. D. van der Helm, J. D. Childs and S. D. Christian, *J. Chem. Soc., Chem. Commun.*, 887 (1969).
108. B. W. Moores and L. Andrews, *J. Phys. Chem.*, **93**, 1902 (1989).
109. M. Hawkins, L. Andrews, A. J. Downs and D. J. Drury, *J. Am. Chem. Soc.*, **106**, 3076 (1984).
110. Q. E. Thompson, *J. Am. Chem. Soc.*, **83**, 845 (1961).
111. R. W. Murray and M. L. Kaplan, *J. Am. Chem. Soc.*, **91**, 5358 (1969).
112. G. D. Mendenhall, in *Advances in Oxygenated Processes*, (A. L. Baumstark, Ed.), Vol. 2, JAI Press, Greenwich (1990), Chap. 6.
113. M. Koenig, F. El Khatib, A. Munoz and R. Wolf, *Tetrahedron Lett.*, **23**, 421 (1982).
114. G. D. Mendenhall, R. F. Kessick and M. Dobrzelewski, *J. Photochem.*, **25**, 227 (1984).
115. V. V. Shereshovets, N. M. Korotaeva, L. V. Spirikhin, V. D. Komissarov and G. A. Tolstikov, *Izv. Akad. Nauk SSSR, Ser. Khim.*, 1381 (1988); *Chem. Abstr.*, **110**, 212934x (1989).

116. S. M. Ramos, J. C. Owrutsky and P. M. Keehn, *Tetrahedron Lett.*, **26**, 5895 (1985).
117. A. P. Schaap and P. D. Bartlett, *J. Am. Chem. Soc.*, **92**, 6055 (1970).
118. R. J. Glinski, J. A. Sedarski and D. A. Dixon, *J. Am. Chem. Soc.*, **104**, 1126 (1982).
119. B. Zwanenburg and W. A. J. Janssen, *Synthesis*, 617 (1973).
120. S. Tamagaki, R. Akatsuka, M. Nakamura and S. Kozuka, *Tetrahedron Lett.*, 3665 (1979).
121. L. Carlsen, *Tetrahedron Lett.*, 4103 (1977).
122. V. Jayathirtha Rao and V. Ramamurthy, *J. Chem. Soc., Chem. Commun.*, 638 (1981).
123. W. C. Eisenberg, K. Taylor and R. W. Murray, *J. Am. Chem. Soc.*, **107**, 8299 (1985).
124. J. M. Sichel, *Can. J. Chem.*, **51**, 2124 (1973).
125. M. Matsui, Y. Miyamoto, K. Shibata and Y. Takase, *Bull. Chem. Soc. Jpn.*, **57**, 2526 (1984).
126. F. E. Stary, D. E. Emge and R. W. Murray, *J. Am. Chem. Soc.*, **98**, 1880 (1976); B. Plesničar, F. Kovac and M. Schara, *J. Am. Chem. Soc.*, **110**, 214 (1988).
127. E. M. Kuramshin, L. G. Kulak, S. S. Zlotskii and D. L. Rakhmankulov, *Izv. Khim.*, **18**, 236 (1985); *Chem. Abstr.*, **104**, 148813d (1986); F. Kovac and B. Plesničar, *J. Am. Chem. Soc.*, **101**, 2677 (1979); E. M. Kuramshin, L. G. Kulak, S. S. Zlotskii and D. L. Rakhmankulov, *Zh. Org. Khim.*, **22**, 1986 (1986); *Chem. Abstr.*, **106**, 213803h (1986).
128. V. V. Shereshovets, N. N. Kabal'nova, V. D. Komissarov and G. A. Tolstikov, *Izv. Akad, Nauk. SSSR, Ser. Khim.*, 1660 (1985); *Chem. Abstr.*, **103**, 214517t (1985).
129. M. Koenig, J. Barrau and N. Ben Hamida, *J. Organomet. Chem.*, **356**, 133 (1988).
130. M. Zarth and A. de Meijere, *Chem. Ber.*, **118**, 2429 (1985).
131. W. A. Pryor, N. Ohto and D. F. Church, *J. Am. Chem. Soc.*, **104**, 5813 (1982).
132. R. L. Kuczkowski, *Acc. Chem. Res.*, **16**, 1159 (1983).
133. G. D. Fong and R. L. Kuczkowski, *J. Am. Chem. Soc.*, **102**, 4763 (1980).
134. J.-I. Choe and R. L. Kuczkowski, *J. Am. Chem. Soc.*, **105**, 4839, (1983); J.-I. Choe, M. K. Painter and R. L. Kuczkowski, *J. Am. Chem. Soc.*, **106**, 2891, (1984).
135. J.-I. Choe, M. Scrinivasan and R. L. Kuczkowski, *J. Am. Chem. Soc.*, **105**, 4703, (1983).
136. H. Keul and R. L. Kuczkowski, *J. Org. Chem.*, **50**, 3371 (1985).
137. K. Griesbaum, H. Keul, S. Agarwal and G. Zwick, *Chem. Ber.*, **116**, 409 (1983).
138. K. Griesbaum and G. Zwick, *Chem. Ber.*, **118**, 3041 (1985).
139. M. Miura, A. Ikegami, M. Nojima, S. Kusabayashi, K. J. McCullough and S. Nagase, *J. Am. Chem. Soc.*, **105**, 2414 (1983).
140. M. Miura, M. Nojima, S. Kusabayashi and K. J. McCullough, *J. Am. Chem. Soc.*, **106**, 2932 (1984).
141. M. Miura, T. Fujisaka, M. Nojima, S. Kusabayashi and K. J. McCullough, *J. Org. Chem.*, **50**, 1504 (1985).
142. N. Nakamura, M. Nojima and S. Kusabayashi, *The Role of Oxygen in Chemistry and Biochemistry (Studies in Organic Chemistry, Vol. 33)*, Elsevier, Amsterdam (1988) pp. 129–132.
143. N. Nakamura, T. Fujisaka, M. Nojima, S. Kusabayashi and K. J. McCullough, *J. Am. Chem. Soc.*, **111**, 1799 (1989).
144. K. J. McCullough, N. Nakamura, T. Fujisaka, M. Nojima and S. Kusabayashi, *J. Am. Chem. Soc.*, **113**, 1786 (1991).
145. K. Jaworski and L. L. Smith, *J. Org. Chem.*, **53**, 845 (1988).
146. H. Keul and R. L. Kuczkowski, *J. Am. Chem. Soc.*, **106**, 3383 (1984).
147. M. S. LaBarge, H. Keul, R. L. Kuczkowski, M. Wallasch and D. Cremer, *J. Am. Chem. Soc.*, **110**, 2081 (1988).

148. H. Keul, H.-S. Choi and R. L. Kuczkowski, *J. Org. Chem.*, **50**, 3365 (1985).
149. B. J. Wojciechowski, C.-Y. Chiang and R. L. Kuczkowski, *J. Org. Chem.*, **55**, 1120 (1990).
150. K. Griesbaum, W.-S. Kim, N. Nakamura, M. Mori, M. Nojima and S. Kusabayashi, *J. Org. Chem.*, **55**, 6153 (1990).
151. B. J. Wojciechowski, W. H. Pearson and R. L. Kuczkowski, *J. Org. Chem.*, **54**, 115 (1989).
152. K. Schank and Ch. Schuhknecht, *Chem. Ber.*, **115**, 2000 (1982).
153. W. H. Bunnelle and E. O. Schlemper, *J. Am. Chem. Soc.*, **109**, 613 (1987).
154. V. N. Odinokov, O. S. Kukovinet, L. M. Khalilov, G. A. Tolstikov, A. Yu. Kosnikov, S. V. Lindeman and Y. T. Strychkov, *Tetrahedron Lett.* **26**, 5843 (1985).
155. T. Sugimoto, N. Nakamura, M. Nojima and S. Kusabayashi, *J. Org. Chem.*, **56**, 1672 (1991).
156. W. H. Bunnelle, L. A. Meyer and E. O. Schlemper, *J. Am. Chem. Soc.*, **111**, 7612 (1989).
157. W. H. Bunnelle, *J. Am. Chem. Soc.*, **111**, 7613 (1989).
158. K. W. Hillig, R. P. Lattimer and R. L. Kuckowski, *J. Am. Chem. Soc.*, **104**, 988 (1982).
159. J. W. Agopovich and C. W. Gillies, *J. Am. Chem. Soc.*, **105**, 5047 (1983).
160. D. Cremer, *J. Am. Chem. Soc.*, **103**, 3633 (1981).
161. C. W. Gillies and R. L. Kuczkowski, *Isr. J. Chem.*, **23**, 446 (1983).
162. M. Mori, M. Nojima, S. Kusabasyashi and K. J. McCullough, *J. Chem. Soc., Chem. Commun.*, 1550 (1988).
163. T. Tabuchi, M. Nojima and S. Kusabayashi, *J. Chem. Soc., Chem. Commun.*, 625 (1990).
164. I. E. Den Besten and T. H. Kinstle, *J. Am. Chem. Soc.*, **102**, 5969 (1980).
165. C. Aronovitch, D. Tal and Y. Mazur, *Tetrahedron Lett.* **23**, 3623 (1982).
166. K. Griesbaum, W. Volpp and R. Greinert, *J. Am. Chem. Soc.*, **107**, 5309 (1985).
167. K. Griesbaum and W. Volpp, *Angew. Chem., Int. Ed. Engl.*, **25**, 81 (1986).
168. K. Griesbaum and W. Volpp, *Chem. Ber.*, **121**, 1795 (1988).
169. K. Griesbaum, W. Volpp, R. Greinert, H.-J. Greunig, J. Schmid and H. Henke, *J. Org. Chem.*, **54**, 383 (1989).
170. K. Griesbaum, W. Volpp, T.-S. Huh and I. C. Jung, *Chem. Ber.*, **122**, 941 (1989).
171. K. Griesbaum and R. Greinert, *Chem. Ber.*, **123**, 391 (1990).
172. R. W. Murray, J. W. P. Lin and D. A. Grumke, *Adv. Chem. Ser.*, **112**, 9 (1972).
173. C.-Y. Chiang, W. Butler and R. L. Kuczkowski, *J. Chem. Soc., Chem. Commun.*, 465 (1988).
174. Y. Ito, A. Matsuura and T. Matsuura, *J. Org. Chem.*, **49**, 4300 (1984).
175. M. Mori, T. Sugiyama, M. Nojima, S. Kusabayashi and K. J. McCullough, *J. Am. Chem. Soc.*, **111**, 6884 (1989).
176. M. Mori, M. Nojima and S. Kusabayashi, *J. Am. Chem. Soc.*, **109**, 4407 (1987).
177. T. Sugimoto, M. Nojima, S. Kusabayashi and K. J. McCullough, *J. Am. Chem. Soc.*, **112**, 3690 (1990).
178. Y. Yamamoto, E. Niki and Y. Kamiya, *Bull. Chem. Soc. Jpn.*, **55**, 2677 (1982).
179. E. Tempesti, M. Fornaroli, L. Giuffre, E. Montoneri and G. Airoldi, *J. Chem. Soc., Perkin Trans 1*, 1319 (1983).
180. K. Griesbaum and M. Meister, *Chem. Ber.*, **118**, 845 (1985).
181. M. Meister, G. Zwick and K. Griesbaum, *Can. J. Chem.*, **61**, 2385 (1983).
182. T.-S. Huh, J. Neumeister and K. Griesbaum, *Can. J. Chem.*, **59**, 3188 (1981).
183. S. Gacbe and W. V. Turner, *J. Org. Chem.*, **49**, 2711 (1984).
184. S. Gaebe, W. V. Turner, E. Hellpointer and F. Korte, *Chem. Ber.*, **118**, 2571 (1985).
185. K. Griesbaum, G. Zwick, S. Agarwal, H. Keul, B. Pfeffer and R. W. Murray, *J. Org. Chem.*, **50**, 4194 (1985).

186. K. Griesbaum and G. Zeick, *Chem. Ber.*, **119**, 229 (1986).
187. K. Griesbaum and A. R. Bandyopadhyay, *Can. J. Chem.*, **65**, 487 (1987).
188. K. Griesbaum and M. Meister, *Chem. Ber.*, **120**, 1573 (1987).
189. M. Peirrot, M. El Idrissi and M. Santelli, *Tetrahedron Lett.* **30**, 461 (1989).
190. P. K. Grant, C. K. Lai, J. S. Prasad and T. M. Yap, *Aust. J. Chem.*, **41**, 711 (1988).
191. S. L. Schreiber and W.-F. Liew, *J. Am. Chem. Soc.*, **107**, 2980 (1985).
192. K. Griesbaum and J. Neumeister, *Chem. Ber.*, **115**, 2697 (1982).
193. K. Griesbaum and G. Kiesel, *Chem. Ber.*, **122**, 145 (1989).
194. K. J. McCullough, M. Nojima, M. Miura, T. Fujisaka and S. Kusabayashi, *J. Chem. Soc., Chem. Commun.*, 35 (1984).
195. K. J. McCullough, T. Fujisaka, M. Nojima and S. Kusabayashi, *Tetrahedron Lett.* **29**, 3375 (1988).
196. T. Sugimoto, M. Nojima and S. Kusabayashi, *J. Org. Chem.*, **55**, 3816 (1990).
197. Z. Paryzek, J. Martynow and W. Swoboda, *J. Chem. Soc., Perkin Trans. 1*, 1222 (1990).
198. J. Gamulka and L. L. Smith, *J. Am. Chem. Soc.*, **105**, 1972 (1983); K. Jaworski and L. L. Smith, *Magn. Reson. Chem.*, **216**, 104 (1988).
199. K. Griesbaum, H. Mertens and I. C. Jung, *Can. J. Chem.*, **69**, 1369 (1990).
200. M. Miura and M. Nojima, *J. Chem. Soc., Chem. Commun.*, 467 (1979).
201. M. Miura and M. Nojima, *J. Am. Chem. Soc.*, **102**, 288 (1980).
202. M. Miura, A. Ikegami and M. Nojima, *J. Chem. Soc., Chem. Commun.*, 1279 (1980).
203. K. J. McCullough, M. D. Walkinshaw and M. Nojima, *J. Chem. Res.*, (*S*), 369; (*M*), 4357–4370 (1981).
204. M. Miura, M. Nojima, S. Kusabayashi and S. Nagase, *J. Am. Chem. Soc.*, **103**, 1789 (1981).
205. M. Miura, A. Ikegami, M. Nojima, S. Kusabayashi, K. J. McCullough and M. D. Walkinshaw, *J. Chem. Soc., Perkin Trans, 1*, 1657 (1983).
206. M. Miura, S. Nagase, M. Nojima and S. Kusabayashi, *J. Org. Chem.*, **48**, 2366 (1983).
207. M. Miura, M. Nojima and S. Kusabayashi, *J. Chem. Soc., Perkin Trans. 1*, 2909 (1980).
208. M. Miura, M. Yoshida, M. Nojima and S. Kusabayashi, *J. Chem. Soc., Chem. Commun.*, 397 (1982).
209. M. Yoshida, M. Miura, M. Nojima and S. Kusabayashi, *J. Am. Chem. Soc.*, **105**, 6279 (1983).
210. T.-S. Fang and W.-P. Mei, *Tetrahedron Lett.* **28**, 329 (1987).
211. K. Griesbaum, A. R. Bandyopadhyay and M. Meister, *Can. J. Chem.*, **64**, 1553 (1986).
212. K. Griesbaum, W. Volpp and T.-S. Huh, *Tetrahedron Lett.*, **30**, 1511 (1989).
213. R. W. Murray and V. Ramachandran, *J. Org. Chem.*, **48**, 813 (1983).
214. R. W. Murray and J.-S. Su, *J. Org. Chem.*, **48**, 817 (1983).
215. C. J. M. van den Heuvel, A. Hofland, J. C. van Velzen, H. Steinberg and Th. J. de Boer, *Recl. Trav. Chim. Pays-Bas*, **103**, 233 (1984).
216. J. C. van Velzen, W. C. J. van Tunen and Th. J. de Boer, *Recl. Trav. Chim. Pays-Bas*, **105**, 225 (1986).
217. J. K. Crandall and T. Schuster, *J. Org. Chem.*, **55**, 1973 (1990).
218. G. Buchi and H. Wuest, *J. Am. Chem. Soc.*, **100**, 294 (1978).
219. W. Adam, C. R. Saha-Moeller and B. Will, *Synthesis*, **121** (1989).
220. M. A. Avery, C. Jennings-White and W. K. M. Chong, *J. Org. Chem.*, **54**, 1789 (1989).
221. M. A. Avery, W. K. M. Chong and G. Detre, *Tetrahedron Lett.*, **31**, 1799 (1990).
222. J. R. Hurst and G. B. Schuster, *J. Org. Chem.*, **45**, 1053 (1980).

223. R. L. Wife, D. Kyle, L. J. Mulheirn and H. C. Volger, *J. Chem. Soc., Chem. Commun.*, 306 (1982).
224. D. Miljkovic, D. Canadi, J. Petrovic, S. Stankovic, B. Ribar, G. Argay and A. Kalman, *Tetrahedron Lett.* **25**, 1403 (1984).
225. A. Caminade, C. Couret, J. Escudie and M. Koenig, *J. Chem. Soc., Chem. Commun.*, 1622 (1984).

14 Formation of Organoperoxides from Superoxide

TETSUO NAGANO, HIROMASA TAKIZAWA and MASAAKI HIROBE
Faculty of Pharmaceutical Sciences, University of Tokyo, Hongo, Bunkyo-ku, Tokyo 113, Japan

Organic Peroxides. Edited by W. Ando

1 INTRODUCTION

Superoxide ($O_2^{-\bullet}$), an active oxygen species, plays important roles in various diseases[1–10] caused by oxygen toxicity such as ischemia, carcinogenesis, inflammation, diabetes and aging. Superoxide dismutases (SODs) act as one of the defense systems against oxygen toxicity in living cells by catalyzing the dismutation reaction of $O_2^{-\bullet}$ to hydrogen peroxide (H_2O_2) and dioxygen (O_2).[11] However, the compounds inducing the formation of $O_2^{-\bullet}$ intracellularly, such as paraquat (PQ^{2+}),[12–19] cause serious damage to the cells including lipid peroxidation, protein denaturation and DNA damage in spite of the induction of the specific activity of SOD. Hence $O_2^{-\bullet}$ has been considered as a potent toxic species by many biochemists. Also, many chemists have focused considerable interest on the chemistry of $O_2^{-\bullet}$.[20–26] The reactivity of $O_2^{-\bullet}$ is summarized in Scheme 1.

What is most important is the difference in the reactivity of $O_2^{-\bullet}$ in aprotic solvents and in aqueous media. In aqueous media, $O_2^{-\bullet}$ is readily dismutated to H_2O_2 and O_2, and any reactions caused by $O_2^{-\bullet}$ must compete with this rapid disproportionation. From this point, nucleophilic reactions, radical-induced reactions involving hydrogen abstraction and base-catalyzed reactions by $O_2^{-\bullet}$ would be almost ruled out in aqueous media. However, $O_2^{-\bullet}$ appears to act not only as a one-electron reductant but also as a one-electron oxidant in aqueous media, because the redox potential of O_2-$O_2^{-\bullet}$ is sensitive to the nature of the solvent used for the measurement and the E'_0 value in water (−0.33 V vs NHE) is higher than those observed in aprotic solvents.

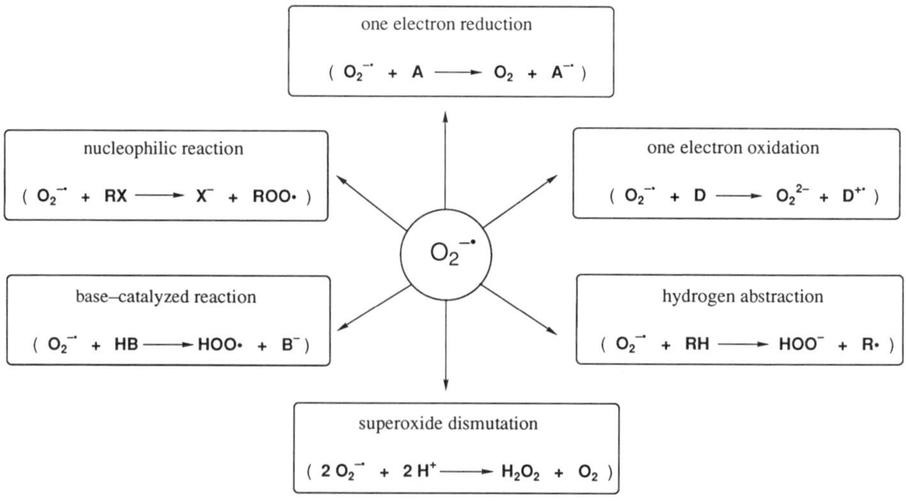

Scheme 1. Reactivity of superoxide ($O_2^{-\bullet}$)

On the other hand, in aprotic solvents $O_2^{-\bullet}$ acts as an effective nucleophile, or causes hydrogen abstraction as a free radical. Moreover, $O_2^{-\bullet}$ can act as a Brønsted base or a one-electron reductant in aprotic solvents. There are many reports of $O_2^{-\bullet}$ being very toxic to living cells,[27] but the reactivity of $O_2^{-\bullet}$ alone is well known to be less vigorous than that of other active oxygen species such as $\cdot$OH $^1O_2$.[28]

How can this be explained? One explanation may be that $O_2^{-\bullet}$ reacts with certain compounds to form reactive peroxides which can oxidize many organic compounds (see Section 2).

The term 'peroxide' is taken to encompass simple peroxides, hydroperoxides, peresters and also unstable peroxy intermediates which cannot be isolated at the peroxide stage.

Here we describe four methods for the generation of $O_2^{-\bullet}$ before introducing various reactions by $O_2^{-\bullet}$. The first is electrolytic reduction of O_2 in dimethylformamide (DMF), acetonitrile (CH_3CN), dichloromethane (CH_2Cl_2), pyridine and dimethyl sulfoxide (DMSO). We can obtain pure $O_2^{-\bullet}$ but only at low concentrations by this method, and it is not useful for synthetic procedures. The second method is to use commercially available potassium superoxide (KO_2) in the presence of 18-crown-6 in an aprotic solvent. This method is useful for product chemistry, although KO_2 contains small amounts of K_2O_2 and KOH as contaminants and so the method is unsuitable for the elucidation of reaction mechanisms. The third is to use tetramethylammonium superoxide prepared from tetramethylammonium hydroxide and KO_2. However, the preparation procedure is fairly difficult and sometimes dangerous. The last method is to use enzyme systems such as xanthine–xanthine oxidase in aqueous media. This method, of course, cannot be used for synthetic procedures. The KO_2–18-crown-6 system has often been used in many of the reactions described in this chapter, because the system can be readily prepared in aprotic solvents.

Oxygen-18-labeled KO_2 is useful for mechanistic studies on the mode of $O_2^{-\bullet}$ attack. $K^{18}O_2$ can be prepared by bubbling $^{18}O_2$ into benzene with dissolved benzhydrol and potassium *tert*-butoxide according to the method of Rosenthal.[29] The $K^{18}O_2$ content of the prepared KO_2 powder can be determined by Raman spectroscopy. The Raman lines at 1088 and at 1154 cm^{-1} are assigned to $^{18}O—^{18}O$ and $^{16}O—^{16}O$ stretching vibrations, respectively. The incorporation into the product is readily determined by mass spectrometry.

The formation of organoperoxides from $O_2^{-\bullet}$ is described in Sections 2–4.

2 PEROXY INTERMEDIATES AS POTENT OXIDANTS

2.1 Acyl Halides

Hirobe and co-workers[30–33] found that $O_2^{-\bullet}$ reacts with substrates such as alkenes, sulfides and compounds bearing labile hydrogens in the presence of acyl halides to give the corresponding oxidized products. Diacyl peroxides are formed as stable products in the reaction of $O_2^{-\bullet}$ with acyl chlorides[34] (see Section 3.2) and peroxy compounds [RC(O)OO·, RC(O)OO$^-$] may be involved as intermediates in this reaction.[35,36] Thus, the KO_2-acyl halide system causes co-oxidation of various substrates. The co-oxidation yield depends on the $O_2^{-\bullet}$/acyl halide ratio. In the co-oxidation of alkenes during the oxidation of acyl halides by $O_2^{-\bullet}$, the yield is extremely low when the KO_2/C_6H_5COCl ratio is below 1. However, yield increases considerably at a ratio $KO_2/C_6H_5COCl = 2$. Above 2 no further increase in yield is observed. The reaction mechanism is proposed as follows. $O_2^{-\bullet}$ makes a nucleophilic attack on C_6H_5COCl to give $C_6H_5C(O)OO\cdot$, which is rapidly reduced by one-electron transfer from $O_2^{-\bullet}$ to produce $C_6H_5C(O)OO^-$. $C_6H_5C(O)OO^-$ oxidizes various alkenes to give the corresponding oxides. On the other hand, C_6H_5COCl and O_2^{2-} readily afford $(C_6H_5COO)_2O_2$ by nucleophilic substitution at the carbonyl carbon by O_2^{2-} without co-oxidation even if some alkene such as the substrate is present in the reaction solution. C_6H_5COCl reacts more slowly with O_2^{2-} than $O_2^{-\bullet}$ and the resulting intermediate $C_6H_5C(O)OO^-$ reacts not with the alkene but with C_6H_5COCl remaining in the reaction solution to give $(C_6H_5COO)_2O_2$, which does not act as an oxidant in the co-oxidation of the alkene. In the reaction with $O_2^{-\bullet}$, it is proposed that since two equimolar amounts of $O_2^{-\bullet}$ react more rapidly with C_6H_5COCl to give $C_6H_5C(O)OO^-$, C_6H_5COCl does not survive in the reaction solution, and hence $C_6H_5C(O)OO^-$ reacts not with C_6H_5COCl but with the alkene.

Phosgene dimer [$(COCl_2)_2$] is most effective for the formation of reactive peroxy intermediates in the reaction with $O_2^{-\bullet}$.[31,32] Polycyclic aromatic compounds are co-oxidized to their epoxides during the oxidation of $(COCl_2)_2$ by $O_2^{-\bullet}$. The KO_2 and $(COCl_2)_2$ system can cooxidize *trans*-stilbene (89%), *p*-nitro-*trans*-stilbene (79%), *p*-chloro-*trans*-stilbene (99%), *p*-methyl-*trans*-stilbene (70%), α-methylstyrene (52%), norbornylene (99%), cyclohexene (80%), phenanthrene (38%), pyrene (12%) and benzpyrene (5%) to the corresponding oxides. Compounds bearing labile hydrogens such as tetralin, fluorene, xanthene and 9,10-dihydroanthracene are also oxidized to the corresponding ketones (34–88%). Also, the $KO_2/(COCl_2)_2$ ratio has been examined in detail. When the ratio is below 4, the yield is extremely low because diacyl peroxide [ClC(O)OO(O)CCl] is formed by the reaction of acyl peroxyanion [ClC(O)OO$^-$] with $COCl_2$. However, the yield increases markedly at a ratio of 8. Therefore, 8 mmol of KO_2 are necessary for 1 mmol of $(COCl_2)_2$.

Oxygen-18-labeled KO_2 is used for mechanistic studies on the mode of $O_2^{-\bullet}$ attack.[33] In the $K^{18}O_2$ and PhCOCl system, ^{18}O incorporation into substrates was examined in four oxidation reactions. In the co-oxidations of styrene to styrene oxide and of ethyl methyl sulfoxide to the sulfone, ^{18}O is incorporated into these substrates almost independently of the concentration of $^{16}O_2$ dissolved in the reaction solution. On the other hand, in the co-oxidations of tetralin to α-tetralone and of ethyl methyl sulfide to the sulfoxide, the percentage of incorporation of ^{18}O decreases with an increase in the dissolved $^{16}O_2$ concentration.

The mechanism shown in Scheme 2 is proposed for these co-oxidations. $^{18}O_2^{-\bullet}$ first reacts by nucleophilic substitution with PhCOCl to form the benzoylperoxy radical [PhC(O)$^{18}O^{18}O\cdot$], which abstracts one electron from tetralin or sulfides. Next, $^{16}O_2$ dissolved in the reaction solution (in part, $^{18}O_2$ derived from $^{18}O_2^{-\bullet}$) attacks the cation radical formed from these substrates, and in so doing the co-oxidation reaction proceeds with formation of the oxidized products containing ^{16}O. On the other hand, in the absence of tetralin or sulfides, PhC(O)$^{18}O^{18}O\cdot$ can be readily reduced by another $O_2^{-\bullet}$ to form the benzoyl peroxy anion PhC(O)$^{18}O^{18}O^-$, which attacks alkenes and sulfoxides directly to cause oxidation reactions, followed by the formation of the products containing ^{18}O.

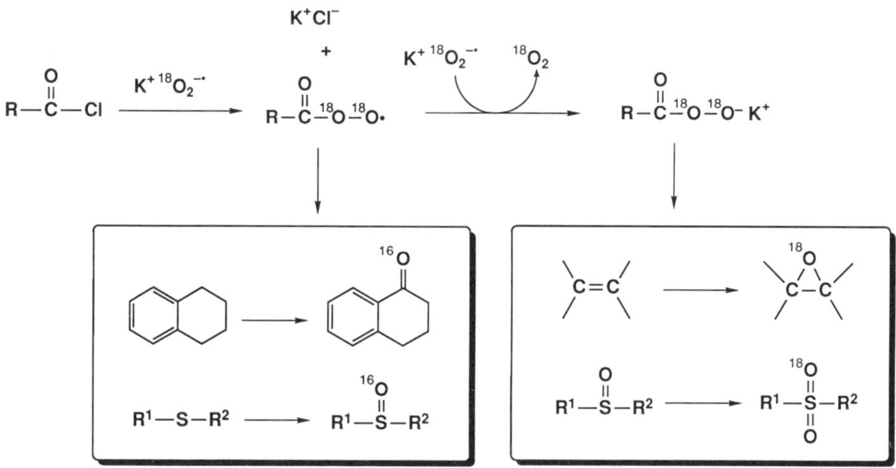

Scheme 2. Co-oxidation by the KO_2 and C_6H_5COCl system

2.2 Sulfides, Sulfones and Sulfonyl Halides

Sulfoxides, added to the reaction system of disulfide, thiosulfinic *S*-ester, thiosulfonic *S*-ester or sulfonyl chloride with $O_2^{-\bullet}$, are oxidized to the sulfones with peroxysulfur intermediates formed *in situ* in the system. Phosphines, added to the reaction system of disulfide or sodium thiolate with $O_2^{-\bullet}$, are also oxidized to the phosphine oxides. Not only stilbene and acenaphthylene but also chalcone and its derivatives, incorporated in the reactions of sulfonyl chloride, sulfonyl chloride and thiosulfonic *S*-ester with $O_2^{-\bullet}$, are found to be oxidized to the corresponding epoxides.[32,37–42]

Formation of peroxysulfenate (**1**), peroxysulfinate (**2**) and peroxysulfonate (**3**) as intermediates (Scheme 3) is proposed as a result of stripping the peroxy oxygen with three kinds of trapping agents such as sulfoxides, phosphines and alkenes, added to the reaction systems of various organic sulfur compounds with $O_2^{-\bullet}$. These sulfoxides, phosphines and alkenes are inert to $O_2^{-\bullet}$ alone.

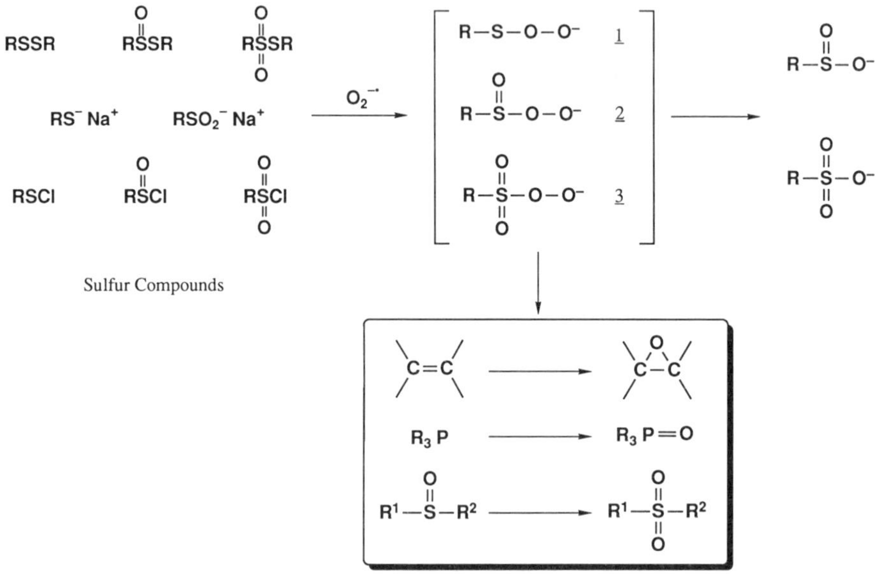

RSSR, RS(O)SR, RS(O)₂SR, $RS^- Na^+$, $RSO_2^- Na^+$, RSCl, RS(O)Cl, RS(O)₂Cl (Sulfur Compounds) $\xrightarrow{O_2^{-\bullet}}$ [R—S—O—O⁻ **1**; R—S(O)—O—O⁻ **2**; R—S(O)₂—O—O⁻ **3**] ⟶ R—S(O)—O⁻; R—S(O)₂—O⁻

Trapping: C=C ⟶ epoxide; R_3P ⟶ $R_3P{=}O$; R^1—S(O)—R^2 ⟶ R^1—S(O)₂—R^2

Scheme 3. Co-oxidation by the KO_2 and sulfur compound system

Especially, a peroxysulfur intermediate such as **3** is an excellent oxidizing agent for the facile and regioselective epoxidation of alkenes.[43] Although the conceivable peroxysulfur intermediate has neither been isolated nor detected, it is proposed to be an oxidant in the oxidation of disulfide-related compounds and sulfonyl chloride. Thus, if the peroxysulfur intermediate is stable enough at low temperature, it may be broadly useful for the synthesis of various epoxides.

Regioselective epoxidation of alkenes containing an α,β-unsaturated ketone moiety such as carvone (equation 1) affords their corresponding epoxides quantitatively. The double bond in the α,β-unsaturated ketone moiety is not oxidized.

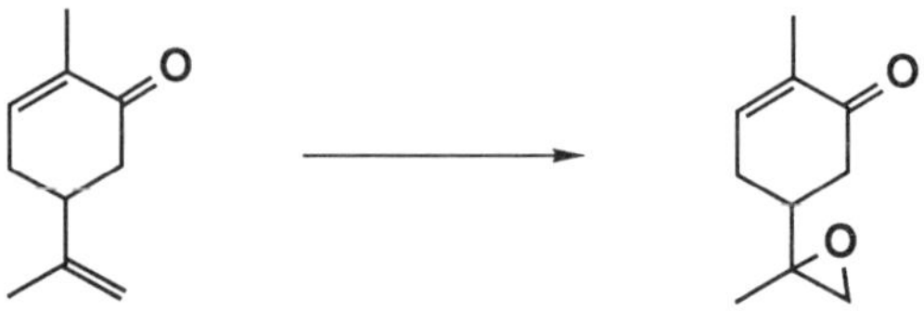

(1)

A peroxysulfur intermediate generated from the reaction of *o*-nitrobenzene sulfonyl chloride with $O_2^{-\bullet}$ gives a better yield of the epoxide in a shorter time than other peroxysulfur intermediates.

Phenanthrene and pyrene are readily oxidized to the corresponding arene oxide by a 2-nitrobenzene sulfonylperoxyl radical in 95%, 50% and 15% yields, respectively, in a polar aprotic solvent such as CH_3CN, CH_3NO_2 or DMF.[42]

2.3 Carbon Dioxide

$O_2^{-\bullet}$ can react with carbon dioxide (CO_2) to yield peroxy intermediates ($CO_4^{-\bullet}$ and $C_2O_6^{-\bullet}$.[33,45,46] Sulfide compounds such as ethyl methyl sulfide (84%), benzyl methyl sulfide (78%) and dibenzyl sulfide (39%) are co-oxidized to the sulfoxide with the yields indicated during the oxidation of CO_2 with $O_2^{-\bullet}$. These species are more reactive than $O_2^{-\bullet}$ alone. Dry-ice can be used as the CO_2 source. In this system, electron-rich alkenes such as α- and β-methylstyrenes are oxidized in better yields than electron-poor alkenes such as *p*-chlorostyrene. In this system, no sulfone is obtained from sulfoxide. The active intermediate is suggested to have an electrophilic character. In the absence of CO_2, these reactions do not proceed and the starting materials are mainly recovered. The peroxy intermediates [$^{-}OC(O)OOH$ and $^{-}OC(O)OC(O)OOH$] generated from $O_2^{-\bullet}$ and CO_2 are essential for oxidation of the substrates (Scheme 4).

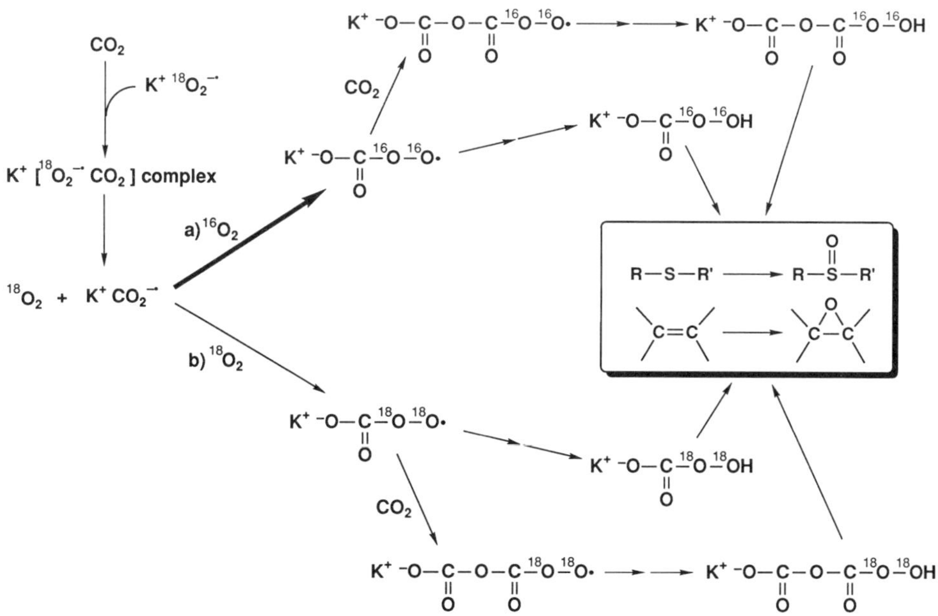

Scheme 4. Co-oxidation by the KO_2 and CO_2 system

The incorporation of ^{18}O into substrates by the $K^{18}O_2$ and CO_2 system has been examined in order to clarify the oxidation mechanisms. The percentage incorporation of ^{18}O decreases with increase in the dissolved $^{16}O_2$ concentration. The initial step is the formation of the $[O_2^{-\bullet}CO_2]$ complex by interaction between $O_2^{-\bullet}$ and CO_2. Under aerobic conditions, $CO_2^{-\bullet}$ is attacked by $^{18}O_2$ or $^{16}O_2$ to form $CO_4^{-\bullet}$, followed by the formation of $C_2O_6^{-\bullet}$ by addition of another molecule of CO_2 (path a or b in Scheme 4). Under an argon atmosphere, $CO_4^{-\bullet}$ and $C_2O_6^{-\bullet}$ are formed only via path b.

2.4 Polyhalides

Sawyer and co-workers[47,48] reported that in aprotic solvents, $O_2^{-\bullet}$ can oxygenate CCl_4 to yield $HOC(O)O^-$ as an overall product via formation of $Cl_3COO\cdot$ radical and Cl_3COO^- anion. The initial step is believed to be an interaction between $O_2^{-\bullet}$ and CCl_4. The resulting activated complex can then dissociate to give chloride ion, O_2 and $\cdot CCl_3$ radical, followed by coupling of the $\cdot CCl_3$ radical and O_2 to form the $Cl_3COO\cdot$ radical.

The radical can oxidize alkenes to the corresponding epoxides (56–83%).[33,36,49] The high reactivity might be due to the electron-withdrawing nature of the trichloromethyl group. Other chlorine compounds are also effective for oxidizing alkenes to epoxides under mild conditions [yield of *trans*-stilbene oxide from *trans*-stilbene: 63% (Cl_3CCCl_3), 53% (F_3CCCl_3), 72% ($BrCCl_3$), 50% ($Cl_2C{=}CCl_2$), 43% ($Cl_2C{=}CHCl$), 42% ($F_3CCBrClH$), 27% ($CHCl_3$), 26% ($PhCCl_3$), 2% (*p,p′*-DDT)]. The oxidation yields in the presence of the polyhalides correlate well with their order of hepatotoxicity: $CBrCl_3 > CCl_4 > CHCl_3 > CH_2Cl_2$. The relative reactivities of these polyhalides with $O_2^{-\bullet}$ also agree with the order of the results obtained using cyclic voltammogram for the $O_2^{-\bullet}$–polyhalide reaction. Cyclic alkenes such as cyclohexane and 1-methylcyclohexene also afford their oxides in 74% and 88% yields, respectively. Polycyclic aromatic compounds are co-oxidized in one step by this system. Tetralin and benzylamine, which bear labile hydrogens, and substrates such as sulfides, ethers and amide compounds are also oxidized under similar conditions.

The reaction of $O_2^{-\bullet}$ with CCl_4 is initiated by a single electron transfer (SET) from $O_2^{-\bullet}$ to CCl_4. Under aerobic conditions, $^{18}O_2$ derived from $^{18}O_2^{-\bullet}$ or $^{16}O_2$ dissolved in the reaction solution attacks $\cdot CCl_3$ to form $CCl_3OO\cdot$ (path a or path b in Scheme 5), which can become CCl_3OOH as an ultimate oxidant. Under an argon atmosphere, CCl_3OOH can be produced only via path b, which would result in 100% incorporation of ^{18}O into the substrate.

2.5 Phosphates

The reaction of nucleotides with $O_2^{-\bullet}$ results in the cleavage of *N*-glycosidic bonds and the release of nucleobases.[50] The yields of the cleavage reaction are

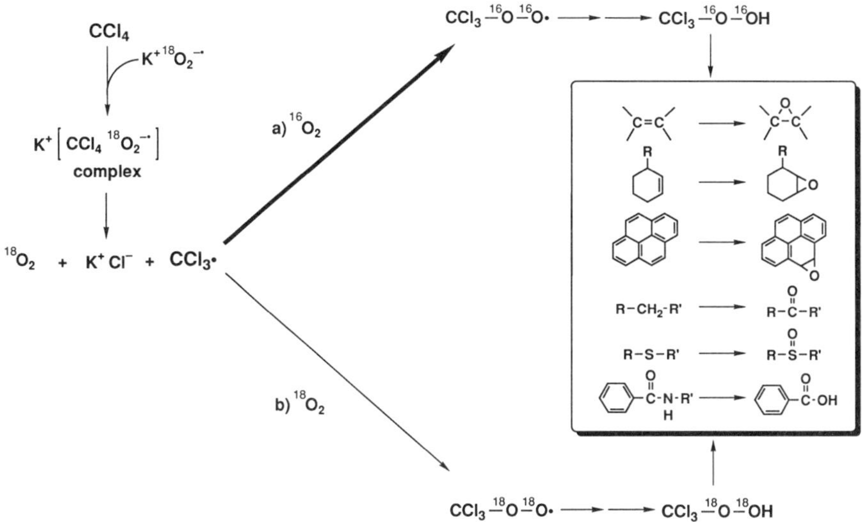

Scheme 5. Co-oxidation by the KO_2 and CCl_4 system

fairly high for nucleotides, whereas those for nucleosides are low. Hence the phosphate moiety undoubtedly enhances the free base formation by $O_2^{-\bullet}$. Addition of some phosphate compounds also enhances the yield of the reaction of $O_2^{-\bullet}$ with the nucleoside. The phosphate peroxy radical or anion, generated *in situ* from $O_2^{-\bullet}$ and the phosphate functional group, may be the species responsible for the degradation. The peroxy intermediate(s) from 5′-GMP and $O_2^{-\bullet}$ can oxidize ethyl methyl sulfoxide to ethyl methyl sulfone in 98% yield. α,β-Unsaturated ketones such as benzoacetophenone and cyclohex-2-enone are also oxidized to the corresponding oxides in 52% and 77% yield, respectively.

^{18}O incorporation into the product by the $K^{18}O_2$ and 5′-GMP system has been examined in the oxidations of sulfide and sulfoxide.[33] The percentage incorporation decreases with increase in the dissolved $^{16}O_2$ concentration in this system. The peroxy anion which has nucleophilic reactivity is an ultimate species in the oxidation of sulfides (Scheme 6).

Miura *et al.*[51] reported that the reaction of $O_2^{-\bullet}$ with diethyl chlorophosphate in the presence of 18-crown-6 gives at least two oxidizing agents, one of which is electrophilic and used in the oxidation of alkenes (equation 2), whereas the other, which is nucleophilic, is important in the oxidation of sulfoxides (equation 3). These two intermediates exist in comparable amounts in the system.

$$(EtO)_2-\overset{\overset{\large O}{\parallel}}{P}-R \xrightarrow{O_2^{-\bullet}} (EtO)_2-\overset{\overset{\large O}{\parallel}}{P}-OO\bullet \tag{2}$$

$$(EtO)_2-\overset{\displaystyle O}{\overset{\|}{P}}-R \xrightarrow{O_2^{-\cdot}} \xrightarrow{O_2^{-\cdot}} (EtO)_2-\overset{\displaystyle O}{\overset{\|}{P}}-OO^- \quad (3)$$

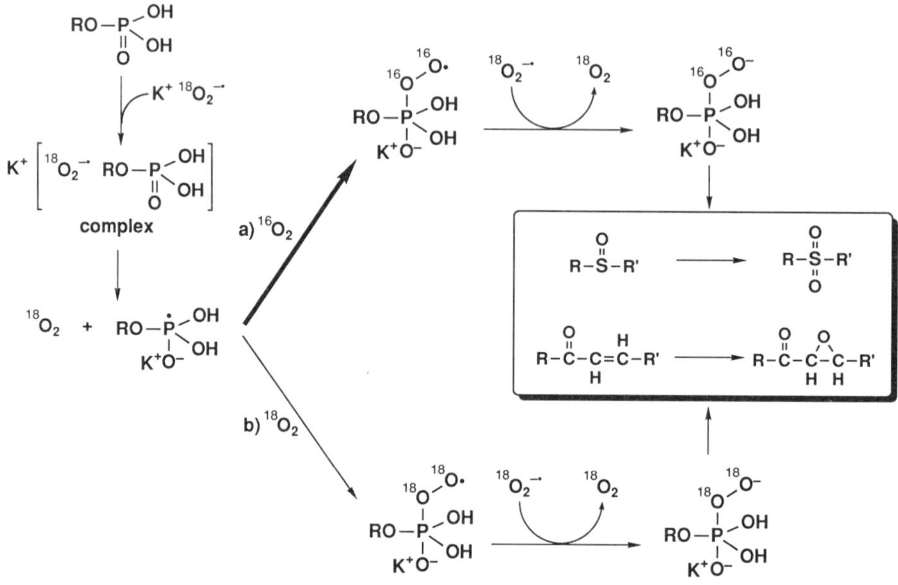

Scheme 6. Co-oxidation by the KO_2 and phosphate system. Reprinted with permission from T. Nagano, H. Yamamoto and M. Hirobe, *J. Am. Chem. Soc.*, **112**, 3529 (1990). Copyright 1990, American Chemical Society

2.6 Carbon Acids[52]

The epoxides of cyclohexenone, 4,4-dimethylcyclohexenone, mesityl oxide and benzylacetophenone are obtained in excellent yields (84–90%) by treating the enones with *in situ*-electrogenerated $O_2^{-\cdot}$ in the presence of an auxiliary carbon acid such as diphenylacetonitrile or, better, diethyl methylmalonate.[53] In the absence of the carbon acid, poor or zero yields are obtained. Sugawara and Baizer[53] concluded that for successful epoxidation a carbon acid should be capable of forming hydroperoxides from $O_2^{-\cdot}$ and O_2 should be present in the reaction mixture (equation 4).

$$H_3CCH(COOEt)_2 \xrightarrow{O_2^{-\cdot}} \left[H_3C\bar{C}(COOEt)_2 \right] \xrightarrow{O_2} \underset{\displaystyle \underset{\displaystyle \downarrow \atop \displaystyle \text{epoxidation}}{OO^-}}{H_3C\underset{|}{C}(COOEt)_2} \quad (4)$$

It is more advantageous to use an auxiliary carbon acid of $pK_a < 14$ as a precursor of some required hydroperoxides.

2.7 Amide Compounds

N-Benzylamides react with $O_2^{-\bullet}$ in benzene to give *o*- and *p*-hydroxylated products.[54] The mechanism is proposed as follows (equation 5). (i) Nucleophilic

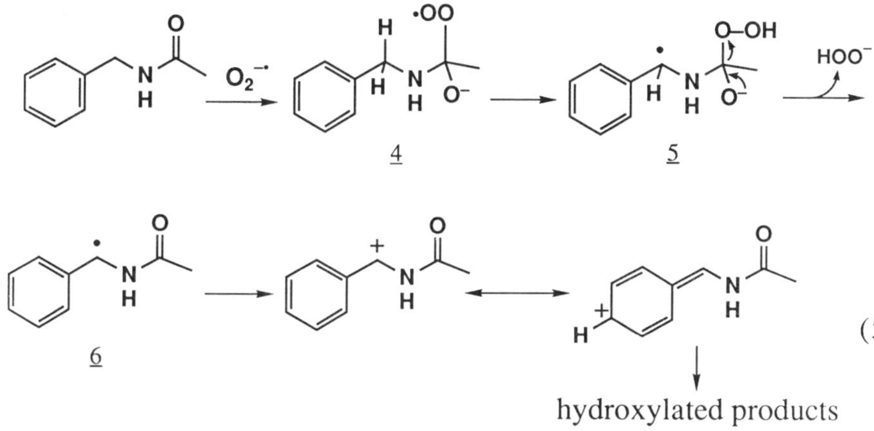

(5)

attack of $O_2^{-\bullet}$ at the CO group of the amide yields a tetrahedral intermediate (**4**), such as proposed in the initial step of ester hydrolysis by $O_2^{-\bullet}$. (ii) The oxygen-centered radical abstracts an H atom attached to the C-α to nitrogen to yield a carbon radical (**5**). This step is favored by a six-membered cyclic transition site. (iii) A peroxide anion is eliminated from the rearranged tetrahedral intermediate. (iv) The resulting carbon radical (**6**) is oxidized by a peroxide species. (v) The cation binds an OH ion. Successive rearomatization of the adduct yields the hydroxylated product. The distribution of *o*- and *p*-hydroxylated isomers rules out any concerted migration of an O atom within any $O_2^{-\bullet}$-substrate adduct.

These results may be significant for studies on $O_2^{-\bullet}$ reactivity because the hydroxylation is based on the formation of the intramolecular peroxide in the reaction of $O_2^{-\bullet}$ with an amide functional group.

3 PEROXY COMPOUNDS AS STABLE PRODUCTS

3.1 Esters

The reaction of $O_2^{-\bullet}$ with the dimesylate **7** yields the peroxide **9**, based on a mechanism[55] in which $O_2^{-\bullet}$ displacement and subsequent electron transfer generate a peroxy mesylate (**8**), which can cyclize to the peroxide **9** (equation 6).

$$\underset{\underline{7}}{H_3C\underset{OMes}{C}HCH_2CH_2\underset{OMes}{C}HCH_2CH_2C_6H_5} \xrightarrow{O_2^{-\bullet}} \left[\underset{\underline{8}}{H_3C\underset{OO^-}{C}HCH_2CH_2\underset{OMes}{C}HCH_2CH_2C_6H_5} \right]$$

$$\xrightarrow{O_2^{-\bullet}} \underset{\underline{9}}{\text{CH}_3\text{(H)-cyclic-O-O-(H)CH}_2\text{CH}_2\text{C}_6\text{H}_5} \qquad (6)$$

Such a process is of value in the synthesis of biologically important prostaglandin endoperoxides such as PGH_2. This is the first reaction in which $O_2^{-\bullet}$ was used as a synthetically useful oxygen nucleophile. $O_2^{-\bullet}$ may be a useful reagent for the S_N2 delivery of nucleophilic oxygen which is superior to conventional reactants such as ^-OH and $RCOO^-$.

The reaction of $O_2^{-\bullet}$ with a representative series of tosylates in DMSO or benzene yields a dialkyl peroxide by the same mechanism as in the reaction of $O_2^{-\bullet}$ with alkyl halides (see Section 3.3) (equation 7).[56]

$$\text{ROTs} + O_2^{-\bullet} \longrightarrow \underset{72\%}{(\text{R-O})_2} + \underset{25\%}{\text{ROH}} \qquad (7)$$

3.2 Acyl Halides

$O_2^{-\bullet}$ reacts as a nucleophile with acyl halides, giving diacyl peroxides in synthetically useful quantities.[34] The reaction proceeds readily without the use of a crown ether, even when benzene is used as the reaction medium, in contrast to the conditions used for dialkyl peroxide synthesis. This reaction provides an alternative method for the preparation of diacyl peroxides under anhydrous conditions (equation 8). As shown in Section 2.1, the ratio of $O_2^{-\bullet}$ to acyl halide is the important factor for reaction yield.

$$2\ \text{R-}\overset{\text{O}}{\overset{\|}{\text{C}}}\text{-Cl} + 2\,KO_2 \longrightarrow \text{R-}\overset{\text{O}}{\overset{\|}{\text{C}}}\text{-OO-}\overset{\text{O}}{\overset{\|}{\text{C}}}\text{-R} + 2KCl + O_2 \qquad (8)$$

R = C_6H_5 (61%), p-CH_3O-C_6H_4 (62%), p-Cl-C_6H_4 (69%), o-Cl-C_6H_4 (57%), o-CH_3-C_6H_4 (53%), $CH_3(CH_2)_{12}$ (74%), $CH_3(CH_2)_{14}$ (50%)

3.3 Halides

The reaction of $O_2^{-\bullet}$ with alkyl halides in DMSO and benzene was reported to proceed by a pathway involving an initial S_N2 displacement leading to alkylperoxy radicals which, in a subsequent one-electron reduction, are converted to peroxy anions.[57] In most instances, the initially isolable principal product is the dialkyl peroxide.[58,59]

The protonated form of $O_2^{-\bullet}$ ($HO_2\bullet$) has a pK_a value in water of 4.69, comparable to that of acetic acid. However, it is well known that anion basicity generally increases dramatically in an aprotic solvent such as DMSO. Thus, it is reasonable to expect that $O_2^{-\bullet}$ is substantially more basic in DMSO than in water. Direct electron transfer from $O_2^{-\bullet}$ to an alkyl bromide is thermodynamically disallowed.

The main peroxidic product obtained from the reaction of KO_2 with primary RX in benzene, DMF or DMSO and from reaction with secondary RX in benzene is the dialkyl peroxide, ROOR.[60] The reaction of KO_2 with secondary RX in DMF gives both hydroperoxide (ROOH) and dialkyl peroxide. The results are consistent with the mechanism outlined in equations 9–12 where in the initial reaction is a nucleophilic displacement of X by $O_2^{-\bullet}$ to give an intermediate alkylperoxy radical. Reaction of this peroxy anion with RX, with H_2O (during the reaction quench) or with DMSO can lead to dialkyl peroxide, alkyl hydroperoxide or alcohol, respectively. Generally, high stereoselectivity is observed in the reaction.

$$\text{R-X} + O_2^{-\bullet} \longrightarrow \text{R-OO}\bullet + X^- \quad (9)$$

$$\text{R-OO}\bullet + e\,(O_2^{-\bullet}) \rightleftharpoons \text{R-OO}^- + (O_2) \quad (10)$$

$$\text{R-OO}^- + \text{R-X} \longrightarrow \text{R-OO-R} + X^- \quad (11)$$

$$\text{R-OO}^- + H^+ \longrightarrow \text{R-OOH} \quad (12)$$

1,2-Dioxenes (equations 13–15) are obtained from the reactions of KO_2 with 1,2-bishalomethylarenes.[61] Rearrangement products are also formed in competition with the cyclization reaction. Formation of cyclic peroxides is presumed to follow a similar mechanism as in the case of acyclic peroxides, namely double nucleophilic substitution of $O_2^{-\bullet}$ with the intervention of a peroxy radical.

The reaction of $O_2^{-\bullet}$ with 9,10-dichloro-9,10-diphenyldihydroanthracene does not give the corresponding endoperoxide.[62] A variety of products formed by nucleophilic substitution and reductive dehalogenation are isolated.

trans-1,2-Dibromocyclohexane reacts with $O_2^{-\bullet}$ in diethyl ether to form cyclohex-2-enol in quantitative yield (equation 16), but not the cyclic peroxide.[55]

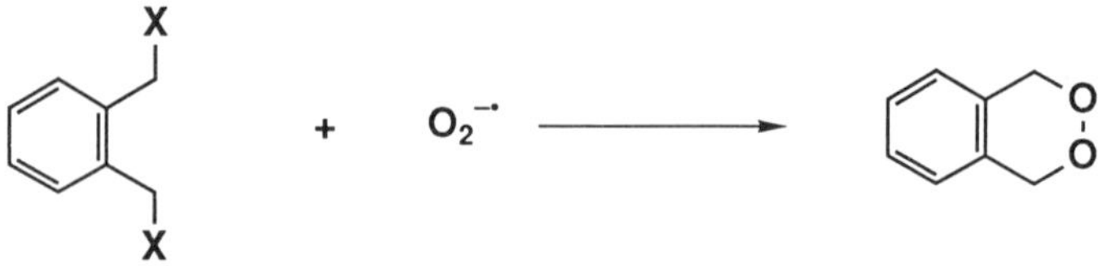

$+ \; O_2^{-\bullet} \longrightarrow$ (13)

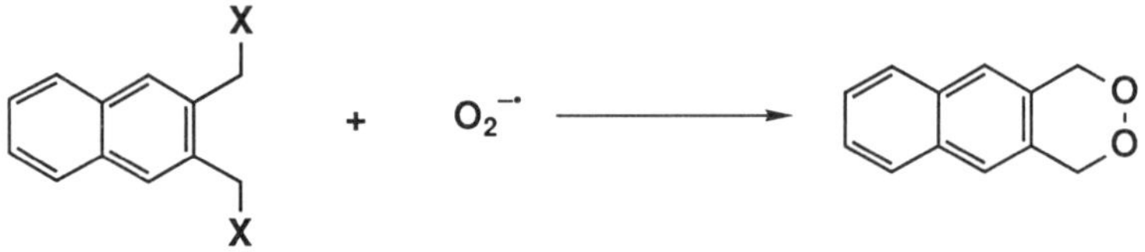

$+ \; O_2^{-\bullet} \longrightarrow$ (14)

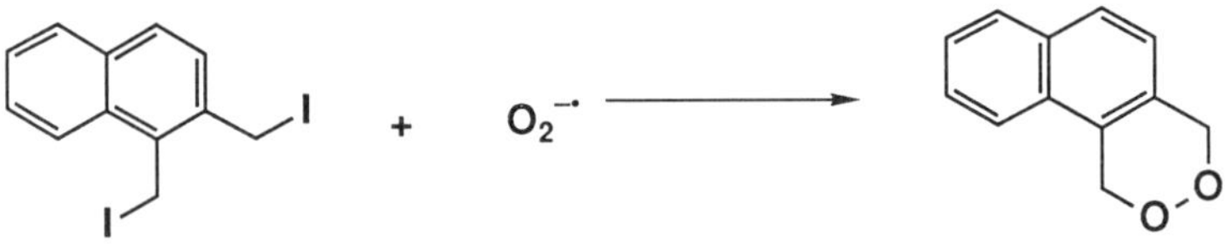

$+ \; O_2^{-\bullet} \longrightarrow$ (15)

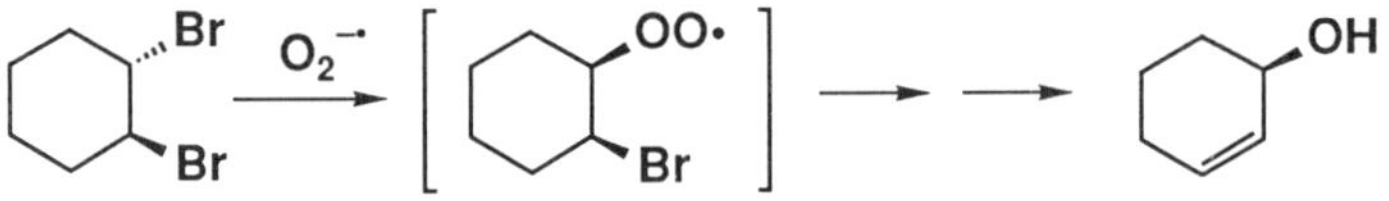

$\xrightarrow{O_2^{-\bullet}}$ [OO• Br] $\longrightarrow \longrightarrow$ OH (16)

4 PEROXY COMPOUNDS AS UNSTABLE INTERMEDIATES

Non-isolable peroxy intermediates formed in the reaction of $O_2^{-\bullet}$ are reviewed in this section.

4.1 Alcohols and Phenols

Generally, aliphatic alcohols disproportionate $O_2^{-\bullet}$ into H_2O_2 and O_2 (equation 17).[63–65] Chin *et al.*[64] reported the proton-induced disproportionation of $O_2^{-\bullet}$ in DMF and in CH_3CH, which was studied by stopped-flow spectrophotometry, cyclic voltammetry and controlled-potential voltammetry at a rotating-disc electrode. For strongly acidic substrates, the disproportionation is too rapid to measure accurately, $k > 10^7\,\mathrm{l\,mol^{-1}\,s^{-1}}$.

$$\text{R-OH} + KO_2 \longrightarrow \text{R-O}^-\text{K}^+ + \boxed{HO_2\bullet} \longrightarrow 1/2\,O_2 + 1/2\,H_2O_2 \qquad (17)$$

With less protic substrates (alcohols, water, catechols, phenols and ascorbic acid) the rate of disproportionation is first order with respect to both $O_2^{-\bullet}$ and substrate. In DMF the second-order rate constants range from $k = 2 \times 10^4\,l\,mol^{-1}\,s^{-1}$ for ascorbic acid to $1 \times 10^{-3}\,l\,mol^{-1}\,s^{-1}$ for butan-1-ol. $HOO\cdot$ as a peroxy intermediate and H_2O_2 as a peroxy product are formed in the reaction.

It has been reported that *o*- and *p*-dihydroxy arene compounds react with $O_2^{-\bullet}$ to produce the corresponding semiquinones, followed by further oxidation to diacylcarboxylic acids; the semiquinones are also obtained by the reaction of quinones with $O_2^{-\bullet}$.[66]

The oxidation of 1,2 and 1,3-dihydroxynaphthalenes with $O_2^{-\bullet}$ leads to 2-hydroxy-1,4-naphthoquinone (**11**) in good yields via the intermediate **10** (equation 18).[67,68] The oxidations are heterogeneous reactions at the solid–liquid interface and the initial step is proposed to be hydrogen atom abstraction by $O_2^{-\bullet}$, in contrast to the reaction with alcohols.

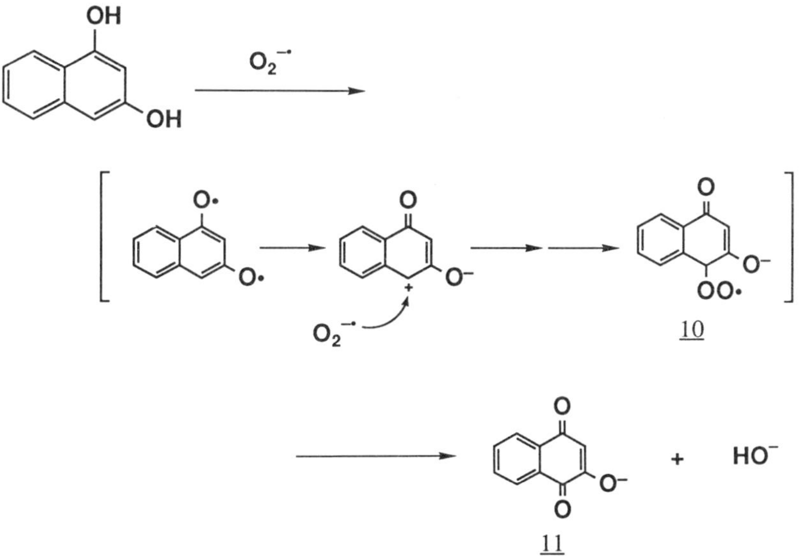

3,5-Di-*tert*-butylcatechol also reacts with $O_2^{-\bullet}$ to form 3,5-di-*tert*-butyl-5-(carboxymethyl)furan-2-one (**12**) and 3,5-di-*tert*-butyl-5-(carboxyhydroxymethyl)furan-2-one (**13**) via hydrogen atom abstraction by $O_2^{-\bullet}$. Some peroxy intermediates have been suggested for oxidative cleavage at the 1,2- and 1,6-positions of the catechol ring (equation 19).[69–71]

The reaction of $O_2^{-\bullet}$ with α-tocopherol yields substrate oxidation product and H_2O_2. However, tocopherol is not oxidized by a direct one-electron transfer to $O_2^{-\bullet}$ or a hydrogen atom abstraction by $O_2^{-\bullet}$. Rather, the reaction

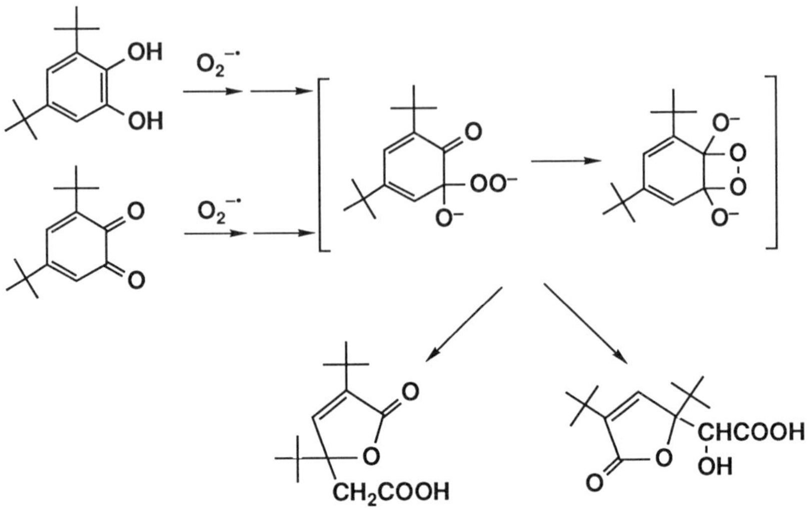

(19)

12 13

mechanism involves abstraction of a proton from the substrate by $O_2^{-\bullet}$ to give substrate anion and the dismutation products of $O_2^{-\bullet}$, H_2O_2 and O_2.[72–75] Nishikimi and co-workers[73,74] reported that an α-tocopherol model compound is oxidized by $O_2^{-\bullet}$ generated from the xanthine–xanthine oxidase system. Matsumoto and co-workers[76–79] determined the structure of the product oxidized at the 5-position, which is probably formed via a hydroperoxide. Deprotonation of $O_2^{-\bullet}$ is assumed to be the initial step of the reaction in the same manner as in a *t*-BuOK-catalyzed oxidation. An α-tocopherol model compound, 2,2,5,7,8-pentamethylchroman-6-ol, was oxidized in the presence of 18-crown-6 with $O_2^{-\bullet}$ to give a diepoxide as the main product, the structure of which was determined by X-ray crystallography (equation 20). The reaction proceeds only slightly under anaerobic conditions. Compound **16** is a unique spiro compound containing two five-membered rings. The reaction is suggested to be $O_2^{-\bullet}$-catalyzed oxygenation, being characteristic of epoxidations and recyclizations, in which hydroperoxide **14** is a key intermediate.

4.2 Esters

Esters are hydrolyzed via an initial reversible nucleophilic addition of $O_2^{-\bullet}$ to the carbonyl carbon, followed by loss of alkoxide from the tetrahedral intermediate. The intermediate forms the corresponding diacyl peroxide in the reaction with another ester and then produces the carbonic acid (equation 21).[80,81]

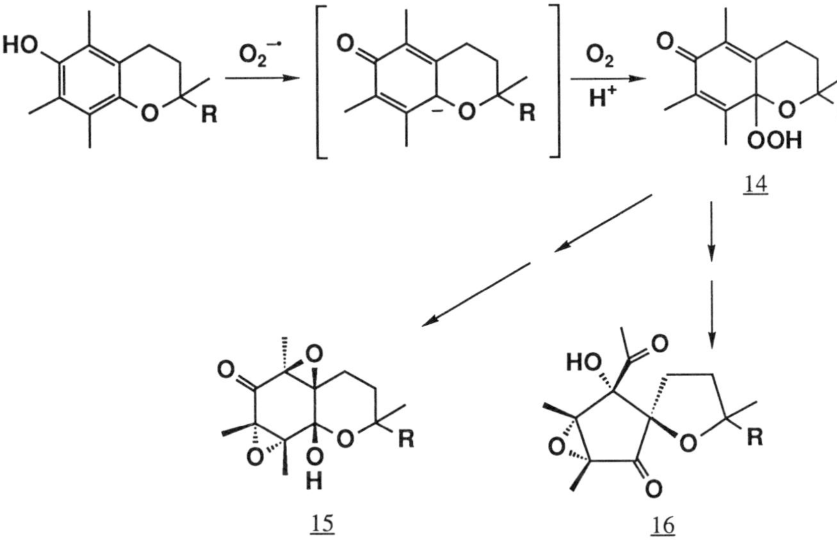

(20)

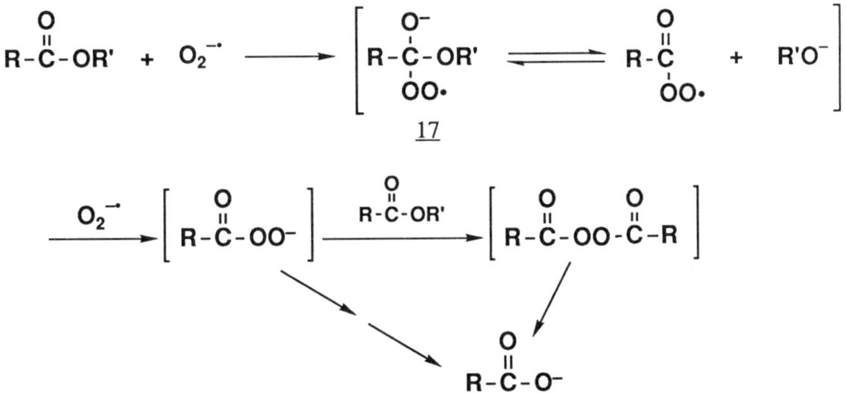

(21)

Forrester and Purushotham[82,83] also investigated the mechanism of the reaction of $O_2^{-\bullet}$ with carboxylic esters. They reported that the peroxide **17** is not considered to be an intermediate from examination of the products obtained when *o*-phenylbenzoyl peroxide was treated with $O_2^{-\bullet}$.

4.3 Peroxides

Conversion of acyl peroxides into carboxylic acids by $O_2^{-\bullet}$ involves either electron transfer to (equation 22) or an S_N2 reaction on the peroxidic group (equation 23).[82] The feasibility of equation 23 was examined by generating the acyltrioxyl radical **18** from an independent route by the reaction of potassium

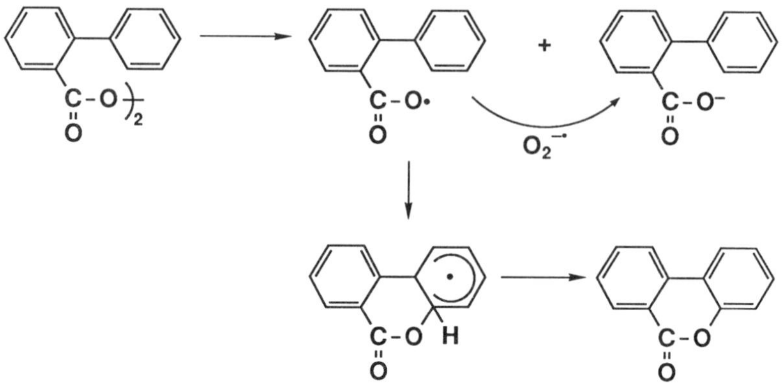

(22)

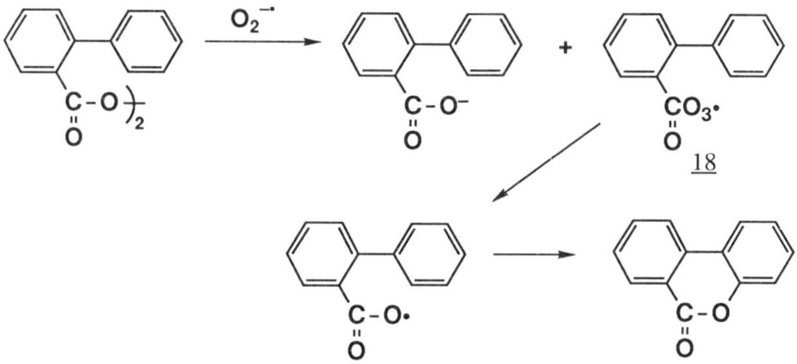

18

(23)

ozonate and *o*-phenylbenzoyl chloride. The intermediate, the acyltrioxyl radical **18**, appears to be of particular interest as a novel oxidant.

The reactions of $O_2^{-\cdot}$ with diacyl peroxides are complex, but this reaction system produces intermediates capable of epoxidation of a number of alkenes.[84] Danen and Arudi[85] reported the efficient production of singlet oxygen ($^1O_2$) in the reaction of $O_2^{-\cdot}$ with benzoyl peroxide and lauroyl peroxide (equation 24).

$$2\,O_2^{-\cdot} + R\text{-}\overset{\text{O}}{\overset{\|}{\text{C}}}\text{-OO-}\overset{\text{O}}{\overset{\|}{\text{C}}}\text{-}R \longrightarrow 2\,{}^1O_2 + 2\,RCO_2^- \qquad (24)$$

Whereas *t*-BuOOH acts solely as a proton source toward $O_2^{-\cdot}$ in toluene and pyridine, an attack on the solvent by the peroxide anion occurs in CH_3CN, leading ultimately to products of peroxide decomposition.[86–88] The results (rapid O_2 evolution and reversible *t*-BuOOH disappearance) indicate removal of a proton from *t*-BuOOH (equations 25 and 26), followed by the well known disproportionation reaction of $O_2^{-\cdot}$ with $\cdot$OOH. Neither dialkyl peroxides nor alkyl hydroperoxides act as electron acceptors from $O_2^{-\cdot}$.

$$\text{tert-BuOOH} + O_2^{-\bullet} \rightleftharpoons \text{tert-BuOO}^- + HO_2^\bullet \quad (25)$$

$$HO_2^\bullet + O_2^{-\bullet} \longrightarrow HO_2^- + O_2 \quad (26)$$

4.4 Ketones

$O_2^{-\bullet}$ induces the oxidative cleavage of ketone compounds.[86–90] The mechanism involves base-catalyzed autoxidation (equation 27). It was reported that α- and

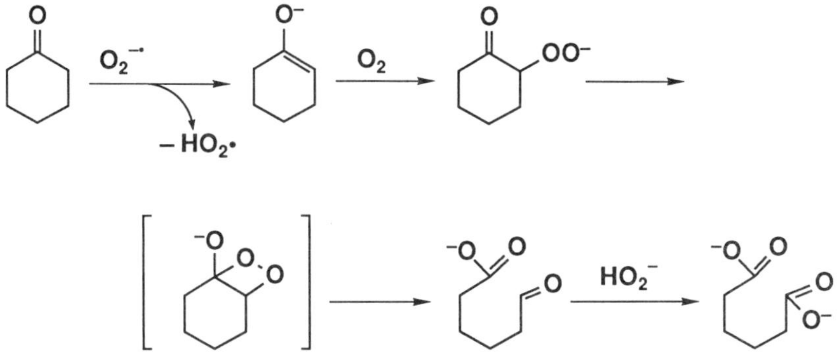

(27)

β-tetralones are oxidized in THF by means of $O_2^{-\bullet}$ with 18-crown-6 to give 2-hydroxy-1,4 naphthoquinone.[91] 3-Keto steroids are also oxidized regioselectively by $O_2^{-\bullet}$.[92–94] In all cases, the reaction products are directly related to the enolization process of the substrates. The reaction mechanism is proposed to proceed via the peroxy intermediate. The mechanism of the $O_2^{-\bullet}$-mediated process entails initial proton (not hydrogen atom) abstraction, which is followed by the various steps typical of base-catalyzed autoxidation.

Tropone (**21**) reacts with $O_2^{-\bullet}$ in DMSO to give salicylaldehyde, but not with peroxide ion (equation 29), whereas tropylium ion (**19**) reacts with both $O_2^{-\bullet}$ and peroxide to give benzene, benzaldehyde, cycloheptatriene and carbon monoxide (equation 28).[95] These oxidations by $O_2^{-\bullet}$ are interpreted as follows. $O_2^{-\bullet}$ reacts as a nucleophile in the reaction with tropylium ion to give the peroxy radical, which can be easily converted into the peroxy anion by electron transfer from $O_2^{-\bullet}$. Ditropyl peroxide (**20**), formed from the reaction of peroxy anion with tropylium ion, equilibrates with the norcaradiene isomer, which may be oxidized by another tropylium ion to benzene and benzaldehyde as shown in equation 28. In contrast, $O_2^{-\bullet}$ probably behaves as an electron donor in the reaction with tropone to give a tropone radical anion from which a peroxy radical may arise. The peroxy radical may collapse to give the anion by electron transfer from $O_2^{-\bullet}$ (equation 29).

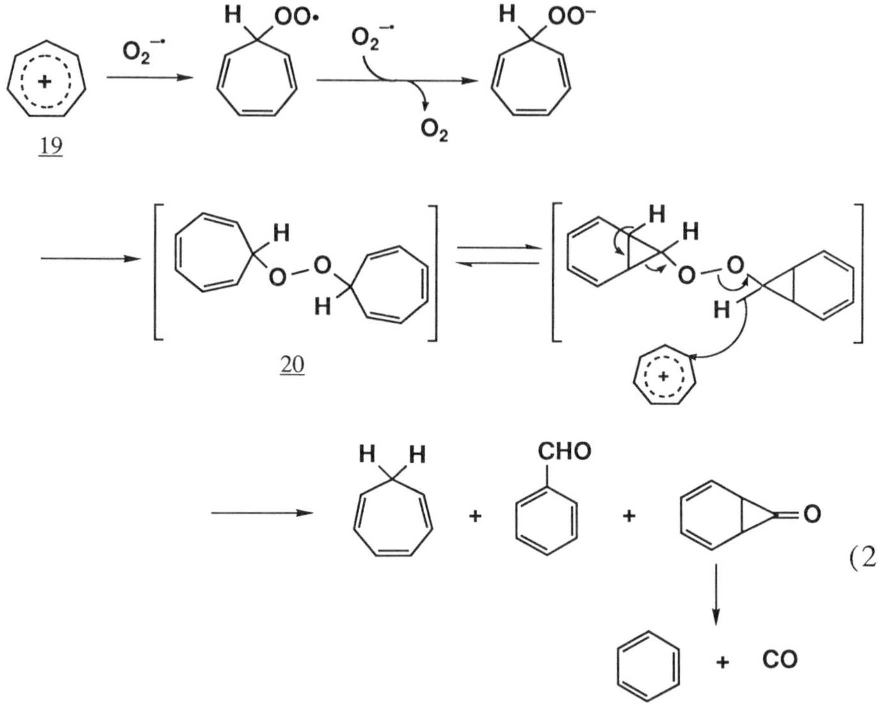

19

20

(28)

21

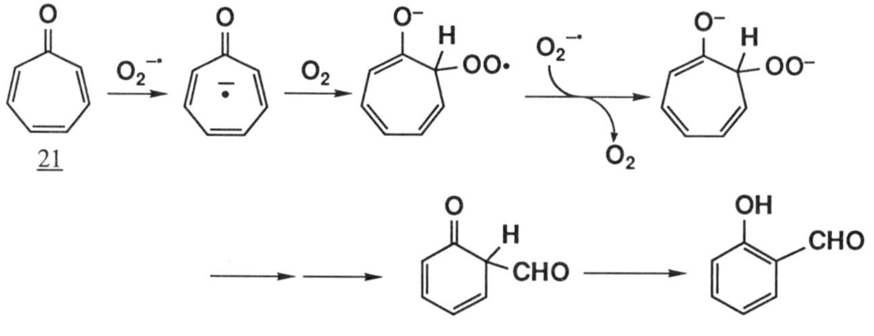

(29)

The reactions of α-keto-, α-hydroxy- and α-haloketones with $O_2^{-\bullet}$ in benzene result in the oxidative cleavage of these compounds to carboxylic acids in a reaction which, in several respects, is reminiscent of the behavior of certain dioxygenases.[96] The dominant net reaction of $O_2^{-\bullet}$ with α1-dicarbonyls such as butane-2,3-dione is also proton abstraction from the enol tautomer.[97,98] For the reaction of benzil, which cannot enolize, the process appears to involve an initial nucleophilic addition by $O_2^{-\bullet}$ to carbonyl carbon, followed by a dioxetane closure on the other carbonyl carbon (equation 30).

Neckers and Hauck[99] and Rosenthal and Frimer[100] reported the reactions of diphenylcyclopropenone and tetracyclones with $O_2^{-\bullet}$.

$$PhC(=O)C(=O)Ph + O_2^{-\bullet} \longrightarrow \left[PhC(OO^{\bullet})(O^-)C(=O)Ph \longrightarrow PhC(O^-)(O\!-\!O)C(O^{\bullet})Ph \right] \longrightarrow 2\,PhC(=O)O^- + O_2 \quad (30)$$

4.5 Enones and Quinones[101]

Whereas $O_2^{-\bullet}$ does not normally react with alkenes, it does react with activated double bonds. Thus, $O_2^{-\bullet}$ oxidatively cleaves certain α,β-unsaturated ketones[102] such as chalcones and tetracyclones. EPR and oxygen labeling ($K^{18}O_2$) experiments confirmed an electron-transfer mechanism.[103]

Frimer and co-workers[104,105] found that chalcones are effectively cleaved by $O_2^{-\bullet}$ to yield carboxylic acids. From the results obtained using ^{18}O-enriched KO_2, the oxygen is not incorporated in the product by direct nucleophilic attack of $O_2^{-\bullet}$. It is probable that the reaction proceeds by a preliminary electron transfer from $O_2^{-\bullet}$ to the enone system of the substrate. The resulting radical anion reacts in turn with the surrounding molecular dioxygen to form the peroxide. In the reaction of $O_2^{-\bullet}$ with coumarins, abstraction of the acidic enolic hydrogens by $O_2^{-\bullet}$ functioning as a base is the most facile process, followed by nucleophilic attack of $O_2^{-\bullet}$ on the lactone moiety, affording saponification products.[105]

Frimer and co-workers examined in detail the reaction of $O_2^{-\bullet}$ with cyclohexenones. The 4,4,6,6-tetrasubstituted cyclohexenones are totally inert, whereas those cyclohexenones possessing available acidic α'- or γ-hydrogens undergo $O_2^{-\bullet}$-mediated base-catalyzed autoxidation (BCA) generating various products depending on the nature and location of the substituents. A detailed explanation is given in Section 4.7.

The reaction of $O_2^{-\bullet}$ with vitamin K_1 and related compounds gives the corresponding 2,3-oxide via a peroxy intermediate (equation 31) and phthalic acid.[106]

4.6 Alkenes

Electron-poor alkenes such as tetrachloroethylene and nitroalkenes react with $O_2^{-\bullet}$ to form the corresponding carboxylic acid.[107,108] A reasonable mechanism for these oxidations is an initial nucleophilic addition of $O_2^{-\bullet}$ to the

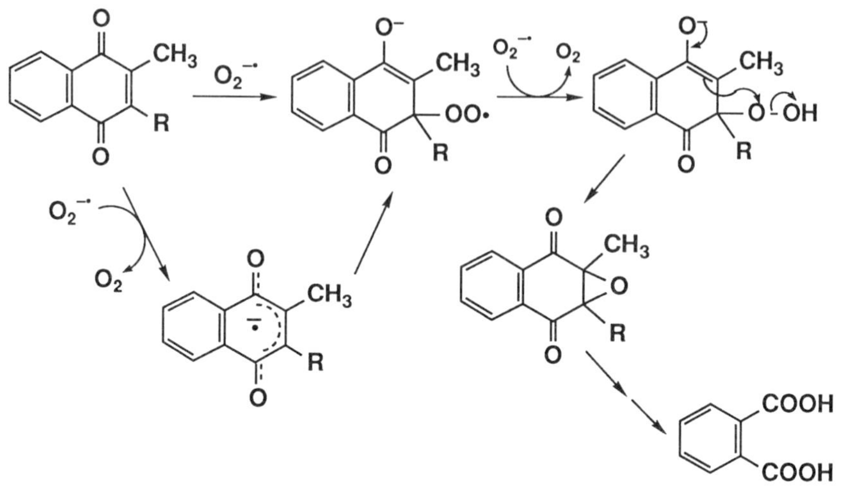

(31)

chloroalkenes. Subsequent loss of chloride ion would give a vinyl peroxy radical, which can cyclize and decompose to a chloroacyl radical and phosgene (equation 32).

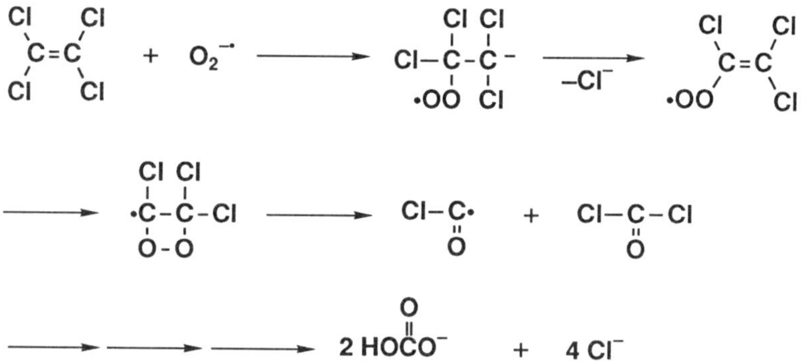

(32)

4.7 Hydrocarbons

Interestingly, the pK_a value (obtained from aqueous pulse radiolysis studies) of HOO·, the conjugated base of $O_2^{-\bullet}$, is 4.69, suggesting that $O_2^{-\bullet}$ should be about as basic as acetate.[109–111] However, $O_2^{-\bullet}$ is reported to mediate efficiently the oxidations of carbon acids having pK_a values in DMSO as high as 35, and to effect the ionization of ketones despite a difference of 20 pK_a units. Further, the basicity of $O_2^{-\bullet}$ should be all the more pronounced in aprotic media in which this hard anion is essentially 'naked.' In the light of this, a general consensus has developed that C—H linkages with low pK_a values react with $O_2^{-\bullet}$ via initial proton transfer. Indeed, Sawyer[108] determined a higher pK_a of 12 for HOO· in DMF, based on electrochemical studies. For kinetic reasons, $O_2^{-\bullet}$-mediated oxidations are base-catalyzed autoxidative (BCA) processes involving deprotonation but not hydrogen atom abstraction as the initial step.

It would seem certain, then, that C—H linkages with low pK_a values react with $O_2^{-\bullet}$ via initial proton transfer. Indeed, cyclopentadiene (p$K_a = 16$), diethyl malonate (p$K_a = 13$), dibenzoylmethane and cyclohexane-1,3-dione are rapidly deprotonated by $O_2^{-\bullet}$, although the resulting anions in the case of the last three are stable to O_2 and $O_2^{-\bullet}$. Similarly, the oxidations of benzoylacetonitrile, malononitrile (p$K_a = 11$), benzyl cyanide, diphenylacetonitrile, alkyl and aryl malonate esters and nitroalkanes (p$K_a = 10$) at the α-position should be considered BCA processes. $O_2^{-\bullet}$ also acts as a base catalyst in autoxidation of 9,10-dihydroanthracene.[112]

In α,β-unsaturated carbonyl systems, two acidic protons are generally present, positioned at the α'- and γ-carbons. Of the two, H-α' is more acidic for inductive reasons; nevertheless, abstraction of H-γ is thermodynamically preferred since the completely conjugated dienolate ion formed is more stable than its cross-conjugated isomer (equation 33). Equation 34 shows the mechanism for the formation of enols.

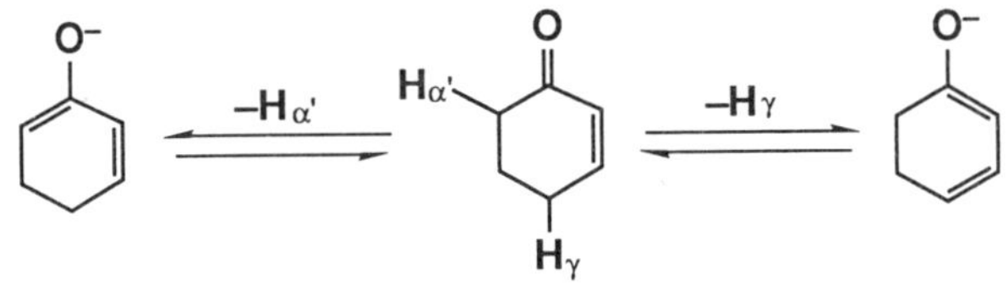

O⁻ –Hα' Hα' O –Hγ O⁻ Hγ (33)

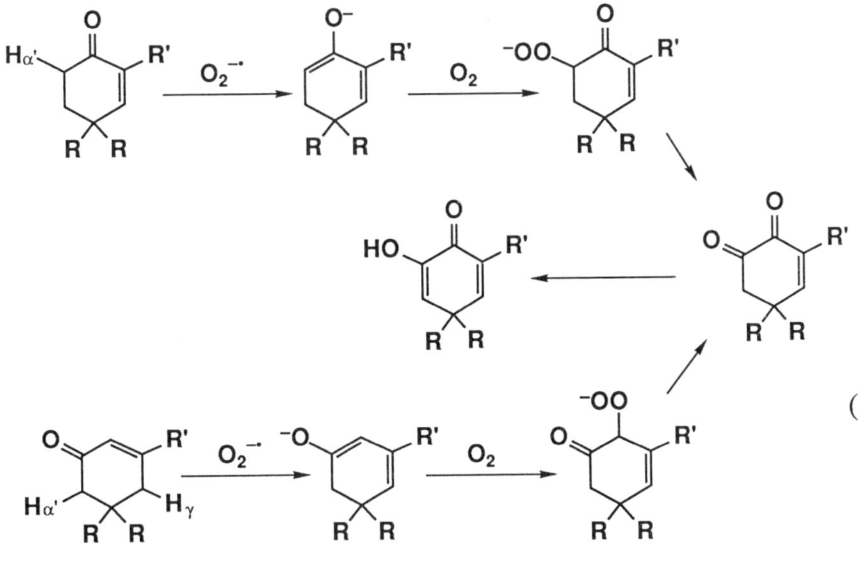

Hα' O R' R R $O_2^{-\bullet}$ O⁻ R' R R O_2 ⁻OO O R' R R

HO O R' R R O O R' R R

(34)

O R' Hα' Hγ R R $O_2^{-\bullet}$ ⁻O R' R R O_2 ⁻OO O R' R R

$O_2^{-\bullet}$ mediates oxidation of diarylmethanes.[113,114] The $O_2^{-\bullet}$-mediated oxidation of diphenylmethane, at least in benzene, proceeds via an initial rate-determining deprotonation by $O_2^{-\bullet}$ generating a benzylic carbanion. Tricarbonylchromium enhances an attack toward the benzylic position by $O_2^{-\bullet}$.[115] The benzylic position of arylmethylamines is readily oxidized to ketone by $O_2^{-\bullet}$.[116]

9,10-Dihydroanthracene reacts instantaneously with $O_2^{-\bullet}$ and gives mainly the dehydrogenated product, anthracene.[117,118] The yield of anthraquinone is very low under a nitrogen stream but increases under an oxygen atmosphere. The reaction seems to follow a mechanism involving a peroxy intermediate.

4.8 Halides

In DMF, $O_2^{-\bullet}$ oxygenates compounds with a trichloromethyl group: CCl_4, $HCCl_3$ and $FCCl_3$ yield hydrogencarbonate ion; $PhCCl_3$ yields a mixture of $PhC(O)OO^-$ and $PhC(O)O^-$; CF_3CCl_3 and $HOCH_2CCl_3$ give their carboxylate anions, $RC(O)O^-$.[47,48,119] Alkyltrichloromethyl compounds are unreactive with $O_2^{-\bullet}$ within a 10-min reaction time at millimolar concentrations. On the basis of the relationship between the relative reaction rates and the electrophilic character of the substrates, as measured by the peak reduction potentials (E_p), the initial step is an electron transfer from the nucleophile to the electrophilic trichloromethyl group. The relative rates of reaction for the primary rate-limiting step follow the order $CCl_4 > CHCl_3 > CH_3Cl > CH_2Cl_2$. The observed substrate reactivity is benzyl > primary > secondary > tertiary > aryl and I > Br > Cl.[120]

$O_2^{-\bullet}$ reacts readily with aliphatic halides to produce the corresponding alcohols.[121,122] The transformations occur with essentially complete inversion of configuration typical of an S_N2 Walden inversion mechanism. The main course of reaction of $O_2^{-\bullet}$ with RX involves an initial rate-limiting displacement of X by $O_2^{-\bullet}$ to generate an intermediate peroxy radical ROO·, and this species is rapidly reduced by a second $O_2^{-\bullet}$. The differences in reduction potentials between $O_2^{-\bullet}$ and aliphatic halides are apparently too great to allow a transfer from $O_2^{-\bullet}$ to the halides.

In DMF $O_2^{-\bullet}$ reacts with PhCHBrCHBrPh, $CH_3CHBrCHBrCH_3$ and CH_2BrCH_2Br via 2:1 stoichiometry to produce PhCHO, CH_3CHO and HCHO, respectively.[123] A similar oxygenation occurs when $O_2^{-\bullet}$ and $CH_2BrCHBrCH_2Cl$ are combined in 5:1 stoichiometry to yield two HCHO and one $HOC(O)O^-$ per substrate. The relative rates of reaction decrease in the order CH_2BrCH_2Br > PhCHBrCHBrPh > $CH_3CHBrCHBrCH_3$. On the basis of the reaction rates and the products, the initial step for each substrate is believed to be a nucleophilic attack on a bromo carbon (R_2CHBr) with displacement of Br^- and formation of R_2CHOO·. Subsequent steps involve reduction by a second $O_2^{-\bullet}$ and a dioxetane closing on an adjacent CX group (equation 35).

$$\mathrm{PhCHBrCHBrPh} + O_2^{-\bullet} \xrightarrow{-Br^-} \mathrm{H{-}C(Ph)(OO\bullet){-}C(Br)(H){-}Ph} \xrightarrow{O_2^{-\bullet} \;\to\; O_2}$$

$$\mathrm{H{-}C(Ph)(O{-}O^-){-}C(Br)(H){-}Ph} \xrightarrow{-Br^-} \left[\mathrm{H{-}C(Ph){-}C(Ph){-}H},\ \text{O—O ring}\right] \longrightarrow 2\,\mathrm{PhCH(O)} \quad (35)$$

Polychloroaromatic hydrocarbons are oxidized by $O_2^{-\bullet}$ in aprotic media.[124,125] The more positive the potential, the greater is the reactivity with $O_2^{-\bullet}$, which is in accord with the oxygenation of alkyl chlorohydrocarbons. A reasonable initial step for these oxygenations is nucleophilic addition of $O_2^{-\bullet}$ to the polyhalobenzene. Subsequent loss of chlorine ion will give a benzoperoxy radical, which will be reduced by a second $O_2^{-\bullet}$ to become a peroxo nucleophile that can attack the adjacent carbochloro center with displacement of its chlorine and apparent formation of an orthoquinone. The C—F bond of fluorocarbons is inert to $O_2^{-\bullet}$ because only the chlorine atoms of F_3CCCl_3 are displaced. However, C_6F_6 and $F_5C_6CF_3$ undergo facile attack by $O_2^{-\bullet}$ in DMF with the displacement of one fluorine and formation of an arylperoxy radical (ArOO·) in the primary step,[126] in addition to C_6Cl_6.[124]

4.9 Nitro Compounds

$O_2^{-\bullet}$ can react with substituted nitrobenzenes to yield the corresponding nitrophenols.[103,107] In the case of *o*-, *m*- and *p*-dinitrobenzenes, the nitro group functions as a leaving group to yield the corresponding nitrophenol. The mechanism is proposed to involve electron transfer from $O_2^{-\bullet}$ to the substituted benzene, yielding an anion radical which is subsequently scavenged by O_2 (equation 36).

$$X\text{-}C_6H_4\text{-}NO_2 + O_2^{-\bullet} \longrightarrow [X\text{-}C_6H_4\text{-}NO_2]^{-\bullet} + O_2 \longrightarrow [\bullet OO(X)C_6H_4NO_2]^{-}$$

$$\longrightarrow \longrightarrow C_6H_5O^- + XO^- \quad (36)$$

Secondary nitroalkanes are deprotonated by electrochemically generated $O_2^{-\bullet}$ and the anions are then oxidized with O_2, thereby being converted to ketones.[128]

4.10 Amines[129]

$O_2^{-\bullet}$ is a selective oxidant for aromatic compounds with the order of susceptibility of substituent groups being SH > NH_2 > OH;[130] *o*- and *p*-disubstituted amines and naphthylamines are readily oxidized.[131–133] Crank *et al.* and Chern and Filippo[135] reported the reaction of various hydrazines.

The oxidation of NH_3 is initiated by $\cdot$OH generated from $O_2^{-\bullet}$ plus H_2O_2 by the metal-catalyzed Haber–Weiss reaction.[136] The xanthine oxidase reaction causes a co-oxidation of NH_3 to NO_2^-, which is inhibitable by superoxide dismutase, catalase, hydroxyl radical scavengers or the chelating agents desferrioxamine or diethylenetriaminepentaacetic acid. The reaction proceeds via a peroxy intermediate (equation 37).

$$\left.\begin{array}{l} H_3N + HO\bullet \longrightarrow H_2N\bullet + H_2O \\ H_2N\bullet + O_2^{-\bullet} + H^+ \longrightarrow H_2N\text{–}OOH \\ H_2N\text{–}OOH + H_2O \longrightarrow H_2N\text{–}OH + H_2O_2 \\ H_2N\text{–}OH + \text{Metal}^{n} \longrightarrow HO\dot{N}H + \text{Metal}^{n-1} + H^+ \\ HO\dot{N}H + O_2^{-\bullet} \longrightarrow HO\text{–}NH\text{–}OO^- \longrightarrow H_2O + NO_2^- \end{array}\right\} \quad (37)$$

Although $O_2^{-\bullet}$ often acts as a one-electron reductant or less frequently as an oxidant, it rarely undergoes covalent bond formation with simple organic or inorganic compounds in water, perhaps owing to its poor nucleophilicity. However, $O_2^{-\bullet}$ can react with nitric oxide (NO) to form the peroxonitrite anion in deaerated aqueous solutions at pH 12–13 (equation 38).[137] This reaction represents one of the few examples of a radical–radical coupling of $O_2^{-\bullet}$ with another odd-electron species to form a diamagnetic product.

$$O_2^{-\bullet} + NO \longrightarrow {}^{-}OONO \quad (38)$$

Takizawa *et al.*[138] found that $O_2^{-\bullet}$ reacts with compounds which have an R_2NH group (e.g. imidazole, adenine, theophylline) and can be protonated, followed by the formation of H_2O_2 and O_2 in the dismutation. Thus, the KO_2

and adenine system causes the oxidation of enones in high yields. Shibata *et al.*[139] reported the reaction of Schiff bases with $O_2^{-\bullet}$ in CH_3CN.

4.11 Amide Compounds

Nitriles are converted into the corresponding amides by $O_2^{-\bullet}$,[140] whereas both benzamide and *N*-(α-phenylethyl)acetamide are inert toward $O_2^{-\bullet}$. However, *N,N*-disubstituted amides, depending on structure, yield α-hydroxylated and cleavage products.[141] The presumed hydroperoxide anion formed by the action of $O_2^{-\bullet}$ and O_2 on the compounds is shown in equation 39.

$$Ph_2CHCONEt_2 \xrightarrow{O_2^{-\bullet}} Ph_2\bar{C}CONEt_2 \xrightarrow{O_2} Ph_2C(OO^-)\text{-}CONEt_2 \longrightarrow {}^-NEt_2 \xrightarrow{H^+} HNEt_2 \quad + \quad Ph_2C(\text{-O-O-})C{=}O \xrightarrow{-CO_2} Ph_2C{=}O \tag{39}$$

N-Acylglycine derivatives are cleaved by the action of $O_2^{-\bullet}$ in diethyl ether or pyridine to give the corresponding carboxamides.[142]

4.12 Cyano Compounds

The reaction of $O_2^{-\bullet}$ with CH_3CN appears to be analogous to that of HOO^-.[143] The decomposition of $O_2^{-\bullet}$ is fast in the presence of CH_3CN in benzene, affording acetamide. Sugawara and Baizer[144] reported the conversion of ethyl cyanoacetate to ethyl glyoxylate by electrogenerated $O_2^{-\bullet}$.

4.13 Sulfur Compounds

Treatment of 1,3-diarylthioureas with $O_2^{-\bullet}$ in THF or CH_3CN results in the formation of 1,2,3-triarylguanidines in excellent yields.[145,146] The thiolate anion of **22** can be converted into the thionyl radical by one-electron transfer to $O_2^{-\bullet}$. The radical **23** then couples with $O_2^{-\bullet}$ to form **24**. The formation of peroxysulfenate or its functional equivalent (peroxysulfinate or peroxysulfonate) is indicated by the oxidation of dimethyl sulfoxide to the sulfone (80%) when the former is used as a trapping agent for the activated oxygen of **24** (equation 40).

The possibility of the formation of a tetrahedral intermediate formed by a direct nucleophilic attack of $O_2^{-\bullet}$ on the thiocarbonyl carbon is ruled out

$$\mathrm{RNH{-}\overset{S}{\overset{\|}{C}}{-}NHR} \rightleftharpoons \underset{\underline{22}}{\mathrm{RNH{-}\overset{SH}{\overset{|}{C}}{=}NR}} \xrightarrow{O_2^{-\cdot}} \underset{\underline{23}}{\mathrm{RNH{-}\overset{S\cdot}{\overset{|}{C}}{=}NR}}$$

$$\xrightarrow{O_2^{-\cdot}} \underset{\underline{24}}{\mathrm{RNH{-}\overset{SOO^-}{\overset{|}{C}}{=}NR}} \quad (40)$$

because tetraphenylthiourea, whose more electrophilic thiocarbonyl carbon is expected to be more readily attacked by $O_2^{-\cdot}$ than 1,3-diphenylthiourea, does not react with $O_2^{-\cdot}$ under the same conditions.[147] Thus, the oxidation reaction of thiourea derivatives requires at least one hydrogen which is necessary for the tautomeric change from thioureas to the thiol forms. $O_2^{-\cdot}$ also converts thiourea and monosubstituted thioureas into cyanamides in an aprotic solvent.[148]

However, Paez *et al.*[149] reported that the first step in the reaction of $O_2^{-\cdot}$ with thioamides [thiobenzamide, thionicotinamide, thioisonicotinamide, 2-ethyl-4-pyridinethiocarboxamide (ethioamide), thioacetamide and thioacetanilide] in DMSO may be nucleophilic addition (equation 41). Subsequent

$$\mathrm{Ph{-}\overset{S}{\overset{\|}{C}}{-}NH_2} \xrightarrow{O_2^{-\cdot}} \mathrm{Ph{-}\underset{OO\cdot}{\underset{|}{\overset{S^-}{\overset{|}{C}}}}{-}NH_2} \xrightarrow{O_2^{-\cdot}} \mathrm{Ph{-}\underset{OO^-}{\underset{|}{\overset{S^-}{\overset{|}{C}}}}{-}NH_2}$$

$$\xrightarrow{\mathrm{Ph{-}\overset{S}{\overset{\|}{C}}{-}NH_2}} 2\ \mathrm{Ph{-}C(S^-){=}NH} + \mathrm{HOOH} \quad (41)$$

reaction of the initially formed peroxythiolate anion with a second $O_2^{-\cdot}$ yields O_2 and the peroxythiolate dianion; the latter undergoes nucleophilic addition with a second thioamide to give, after cleavage, $2RC(S^-){=}NH$ and HOOH. The overall stoichiometry is one $O_2^{-\cdot}$ per thioamide. The formed thioamide anion and O_2 slowly react to form the corresponding nitrile ($RC{\equiv}N$), polysulfides and a second HOOH.

Katori *et al.*[150] also reported that several types of thioamides and thioureas, including thiouracils, are readily desulfurized to the carbonyl compounds by $O_2^{-\cdot}$, generated by KO_2 and 18-crown-6, or electrolysis of O_2 in an aprotic solvent. Such desulfurization may provide a model of metabolic reactions catalyzed by oxygenases. Akasaka and Ando[151] described the reaction of sulfides and sulfoxides with $O_2^{-\cdot}$ in the presence and absence[152,153] of dithiocin. Disulfide bonds are cleaved by $O_2^{-\cdot}$, which suggests a cystine moiety as the damage site in protein denaturation.[154,155]

4.14 Heterocycles

Indolamines react with $O_2^{-\bullet}$ to form products with oxidative cleavage of the five-membered ring via a peroxy intermediate.[156] *N*-Alkylthiazolium halides including thiamine can be oxidized by $O_2^{-\bullet}$,[157] the mechanism of which is shown in equation 42. Although route (a) is supported by conversion of **25** into the products **26** and **27** in the same ratio under an argon atmosphere as on exposure to air, route (b) cannot be ruled out because the molecular oxygen required in this sequence may be provided by one of the steps of route (a).

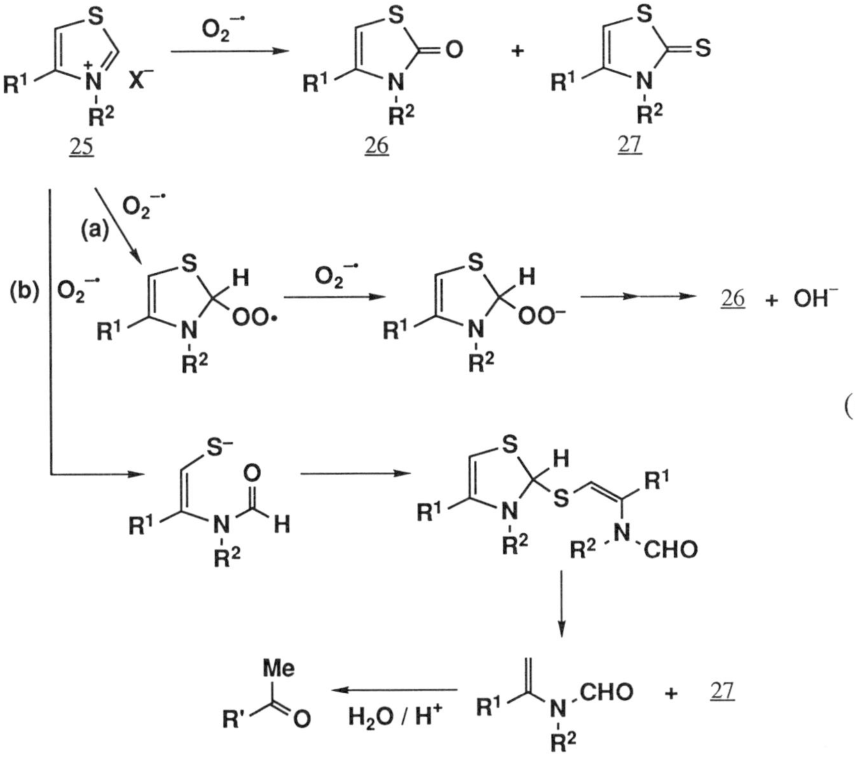

(42)

The reaction of 2,4,6-trisubstituted pyrylium tetrafluoroborates with $O_2^{-\bullet}$ gives three main products, 3,5-disubstituted-2(3*H*)-furanone (**28**), 3-acyl-3,5-disubstituted-2(3*H*)-furanone (**29**) and 2-acyl-3,5-disubstituted furan (**30**).[158] In the initial stage of this reaction, two routes can be considered, a direct nucleophilic addition to the α-position of the pyrylium ring and one-electron transfer from $O_2^{-\bullet}$ to the electron-deficient pyrylium ring. The mechanism shown in equation 43 presumably rationalizes the formation of the products.

$O_2^{-\bullet}$ is considered to be a major factor in the toxicity of paraquat (PQ^{2+}). PQ cation radical ($PQ^{+\bullet}$) reacts with O_2 rapidly ($k_2 = 7.7 \times 10^8\ \mathrm{l\,mol^{-1}\,s^{-1}}$) to

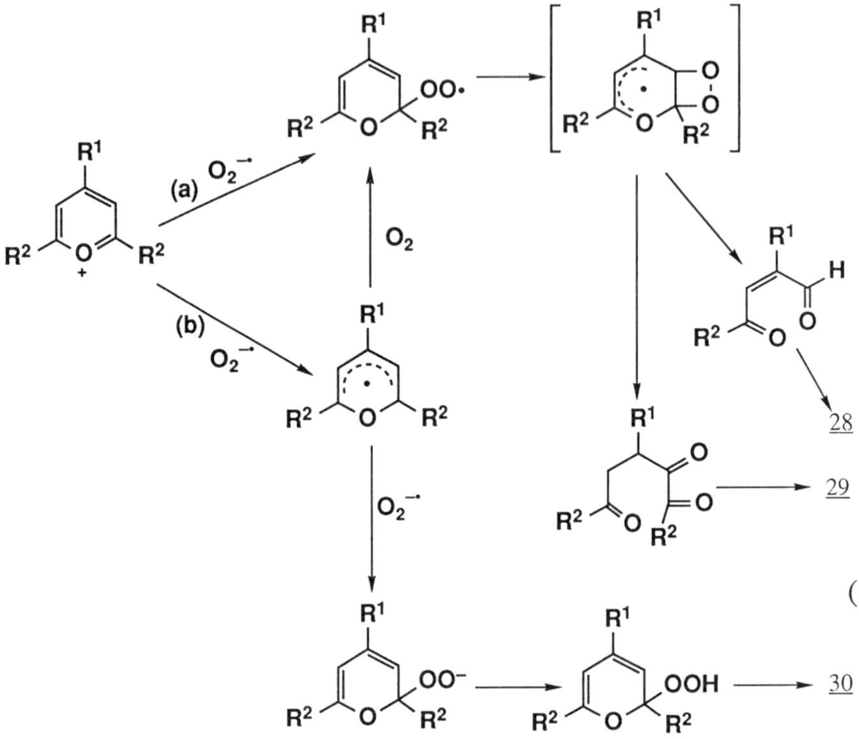

(43)

generate $O_2^{-\bullet}$. PQ^{2+} augments the production of $O_2^{-\bullet}$ by chloroplasts, lung microsomes and homogenates of lung, liver and kidney, and O_2 augments the toxicity of PQ^{2+} in plants, rats and *Escherichia coli.* It is found that 1 equiv. of $O_2^{-\bullet}$ is combined with 1 equiv. of $PQ^{+\bullet}$, and a diamagnetic adduct is the initial product, which is consistent with a primary radical–radical coupling mechanism.[159] In equation 44, $[PQ^{+\bullet}O_2^{-\bullet}]$ rearranges to an unstable dioxetane intermediate which cleaves to form a polymer.

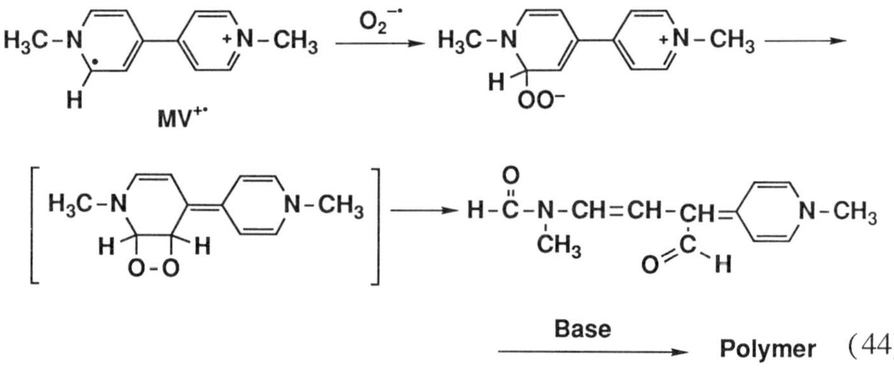

(44)

Alkylated thymine and thymidine derivatives are transformed by $O_2^{-\bullet}$ into the corresponding ring-contracted imidazolone derivatives.[160] A plausible mechanism is shown in equation 45. The 2-methyl-1,2,3-triazinium iodide **31** reacts with $O_2^{-\bullet}$ to give the 5-oxotriazine **33** via a peroxy intermediate (**32**) (equation 46).[161]

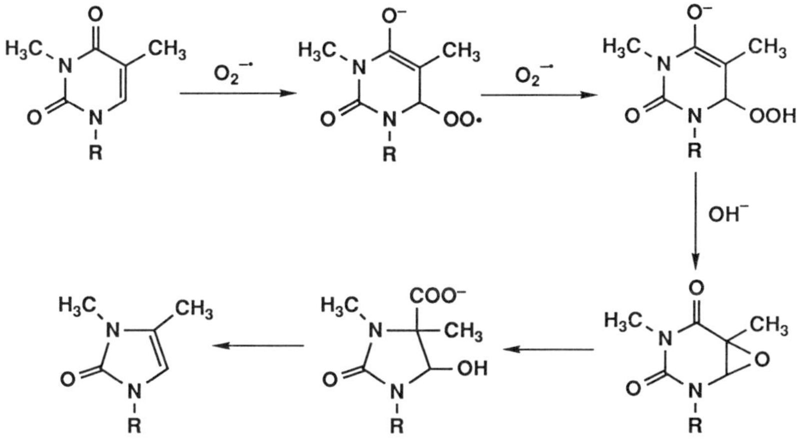

(45)

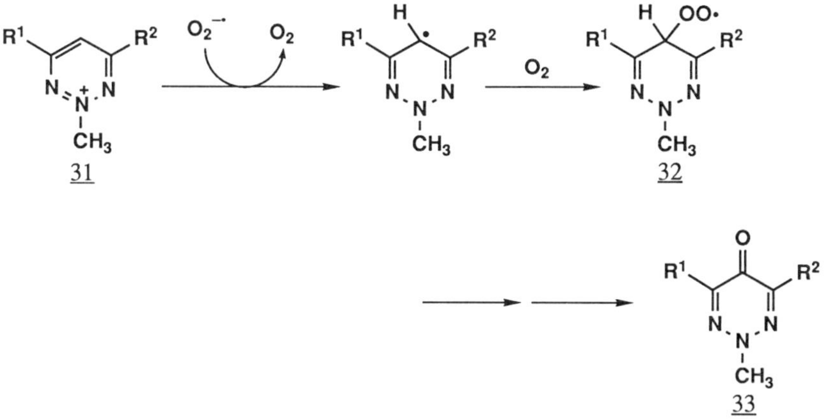

31 32 33

(46)

3-Nitro-2-phenyl-2*H*-1-benzopyrans on treatment with $O_2^{-\bullet}$ in DMSO are degraded mainly to the corresponding salicyclic acids and benzoic acids.[162] $O_2^{-\bullet}$ causes a nucleophilic attack at the carbonyl group of oxazolinones to form ring-opened products.[163]

Coordinated saturated cationic ruthenium(II) complexes are attacked at the terminal position of the dienyl moiety by $O_2^{-\bullet}$ to yield ruthenium(0) complexes containing a cyclic dienone ligand.[164]

5 CONCLUSION

$O_2^{-\bullet}$, which is an anion radical species, has various reactivities, and it has been examined whether $O_2^{-\bullet}$ is useful as a synthetic oxidant. As shown in this chapter, $O_2^{-\bullet}$ reacts with many compounds to form peroxy compounds, including unstable intermediates, which can act as novel and potent oxidants in unique oxidations. In this respect, $O_2^{-\bullet}$ has the possibility of becoming a significant and noteworthy species in organic synthesis.

6 REFERENCES

1. J. M. McCord, *Science*, **185**, 529 (1974).
2. H. C. Birnboim and M. Kanabus-Kaminska, *Proc. Natl. Acad. Sci. USA*, **81**, 6820 (1985).
3. B. N. Ames, *Science*, **221**, 1256 (1983).
4. R. Zimmerman and P. Cerutti, *Proc. Natl. Acad. Sci. USA*, **81** 2085 (1984).
5. L. Fucci, C. N. Oliver, M. J. Coon and E. R. Stadman, *Proc. Natl. Acad. Sci. USA*, **80**, 1521 (1983).
6. B. N. Ames, R. Cathcart, E. Schwiers and P. Hochstein, *Proc. Natl. Acad. Sci. USA*, **78**, 6858 (1981).
7. I. Semsei, G. Rao and A. Richardson, *Biochem. Biophys. Res. Commun.*, **164**, 620 (1989).
8. D. R. Ambruso, B. G. J. M. Bolscher, P. M. Stokman, A. J. Verhoeven and D. Roos, *J. Biol. Chem.*, **265**, 924 (1990).
9. M. Fontecave, A. Graslund and P. Reichard, *J. Biol. Chem.*, **262**, 12332 (1989).
10. K. Kobayashi, K. Hayashi and M. Sono, *J. Biol. Chem.*, **264**, 15280 (1989).
11. J. M. McCord and I. Fridovich, *J. Biol. Chem.*, **244**, 6049 (1969).
12. H. M. Hassan and I. Fridovich, *J. Biol. Chem.*, **254**, 10846 (1979).
13. H. M. Hassan and I. Fridovich, *J. Biol. Chem.*, **253**, 8143 (1978).
14. J. A. Farrington, M. Ebert, E. J. Land and K. Fletcher, *Biochim. Biophys. Acta*, **314**, 372 (1973).
15. R. W. Miller and F. D. H. MacDowall, *Biochim. Biophys. Acta*, **387**, 176 (1975).
16. J. J. Harbour and J. R. Bolton, *Biochem. Biophys. Res. Commun.*, **64**, 803 (1975).
17. M. R. Montgomery, *Res. Commun. Chem. Pathol. Pharmacol.*, **16**, 155 (1977).
18. R. C. Baldwin, A. Pasi, J. T. MacGregor and C. H. Hine, *Toxicol. Appl. Pharmacol.*, **32**, 298 (1975).
19. H. K. Fisher, in *Biochemical Mechanisms of Paraquat Toxicity* (A. P. Autor, Ed.), Academic Press, New York (1977), pp. 57–65.
20. S. Kim, R. DiCosmio and J. S. Filippo, Jr, *Anal. Chem.*, **51**, 679 (1979).
21. A. A. Frimer, in *The Chemistry of Functional Groups, Peroxides* (S. Patai, Ed.), Wiley, Chichester (1983), pp. 429–461.
22. E. Lee-Ruff, *Chem. Soc. Rev.*, **6** 195 (1977).
23. B. H. J. Bielski, D. E. Cabelli, R. L. Arudi and A. B. Ross. *J. Phys. Chem. Ref. Data*, **14** 1041 (1985).
24. J. A. Fee and J. S. Valentine, in *Superoxide and Superoxide Dismutases* (A. M. Michelson, J. M. McCord and I. Fridovich, Eds), Academic Press, New York, 1977, pp. 19–60.

25. A. A. Frimer, in *Superoxide Dismutase*, Vol. II (L. W. Oberley, Ed.), CRC Press, Boca Raton, FL (1982), pp. 83–125.
26. D. T. Sawyer and M. J. Gibian, *Tetrahedron Lett.*, 989 (1977).
27. I. Fridovich, *Acc. Chem. Res.*, **15**, 200 (1982).
28. D. T. Sawyer and J. S. Valentine, *Acc. Chem. Res.*, **14** 393 (1981).
29. I. Rosenthal, *J. Labelled Compd. Radiopharm.*, **12**, 317 (1978).
30. T. Nagano, K. Arakane and M. Hirobe, *Chem. Pharm. Bull.*, **28**, 3719 (1980).
31. T. Nagano, K. Yokoohji and M. Hirobe, *Tetrahedron Lett.*, **25**, 965 (1984).
32. T. Nagano, K. Yokoohji and M. Hirobe, *Tetrahedron Lett.*, **24**, 3481 (1983).
33. T. Nagano, H. Yamamoto and M. Hirobe, *J. Am. Chem. Soc.*, **112**, 3529 (1990).
34. R. A. Johnson, *Tetrahedron Lett.*, 331 (1976).
35. H. Yamamoto, M. Miura, M. Nojima and S. Kusabayashi, *J. Chem. Soc., Perkin Trans. 1*, 173 (1986).
36. K. Akutagawa, N. Furukawa and S. Oae, *Bull. Chem. Soc. Jpn.*, **57**, 1104 (1984).
37. S. Oae, T. Takata and Y. H. Kim, *Bull. Chem. Soc. Jpn.*, **54**, 2712 (1981).
38. Y. H. Kim, K. S. Kim and H. K. Lee, *Tetrahedron Lett.*, **30**, 6357 (1989).
39. Y. H. Kim and D. C. Yoon, *Tetrahedron Lett.*, **29**, 6453 (1988).
40. Y. H. Kim and H. K. Lee, *Chem. Lett.*, 1499 (1987).
41. Y. H. Kim, B. C. Chung and H. S. Chang, *Tetrahedron Lett.*, **26**, 1079 (1985).
42. S. Oae and T. Takata, *Tetrahedron Lett.*, **21**, 3689 (1980).
43. Y. H. Kim and B. C. Chung, *J. Org. Chem.*, **48**, 1562 (1983).
44. H. K. Lee, K. S. Kim, J. S. Kim and Y. H. Kim, *Chem. Lett.*, 561 (1988).
45. H. Yamamoto, T. Mashino, T. Nagano and M. Hirobe, *Tetrahedron Lett.*, **30**, 4133 (1981).
46. J. L. Roberts, Jr, T. S. Calderwood and D. T. Sawyer, *J. Am. Chem. Soc.*, **106**, 4667 (1984).
47. J. L. Roberts, Jr and D. T. Sawyer, *J. Am. Chem. Soc.*, **103**, 712 (1981).
48. J. L. Roberts, Jr, T. S. Calderwood and D. T. Sawyer, *J. Am. Chem. Soc.*, **105**, 7691 (1983).
49. H. Yamamoto, T. Mashino, T. Nagano and M. Hirobe, *J. Am. Chem. Soc.*, **108**, 539 (1986).
50. H. Yamane, N. Yada, E. Katori, T. Mashino, T. Nagano and M. Hirobe, *Biochem. Biophys. Res. Commun.*, **142**, 1104 (1987).
51. M. Miura, M. Nojima and S. Kusabayashi, *J. Chem. Soc., Chem. Commun.*, 1352 (1982).
52. P. M. Allen, U. Hess, C. S. Foote and M. M. Baizer, *Synth. Commun.*, **12**, 123 (1982).
53. M. Sugawara and M. M. Baizer, *J. Org. Chem.*, **48**, 4931 (1983).
54. G. Galliani and B. Rindone, *Tetrahedron*, **37**, 2313 (1981).
55. E. J. Corey, K. C. Nicolaou, M. Shibasaki, Y. Machida and C. S. Shiner, *Tetrahedron Lett.*, 3183 (1975).
56. A. A. Frimer and P. Gilinsky, *Tetrahedron Lett.*, 4331 (1975).
57. C. Chern, R. DiCosmio, R. De Jesus and J. S. Filippo, Jr, *J. Am. Chem. Soc.*, **100**, 7317 (1978).
58. R. A. Johnson and E. G. Nidy, *J. Org. Chem.*, **40**, 1680 (1975).
59. M. V. Merritt and D. T. Sawyer, *J. Org. Chem.*, **35**, 2157 (1970).
60. R. A. Johnson, E. G. Nidy and M. V. Merritt, *J. Am. Chem. Soc.*, **100**, 7960 (1978).
61. L. H. Dao, A. C. Hopkinson, E. Lee-Ruff and J. Rigaudy, *Can. J. Chem.*, **55**, 3791 (1977).
62. A. R. Forrester and V. Purushotham, *Tetrahedron Lett.*, **28**, 3279 (1987).
63. A. Le Berre and Y. Berguer, *Bull. Soc. Chim. Fr.*, 2363 (1966).
64. D. Chin, G. Chiericato, Jr, E. J. Nanni, Jr, and D. T. Sawyer, *J. Am. Chem. Soc.*, **104**, 1296 (1982).

65. C. L. Greenstock and R. W. Miller, *Biochim. Biophys. Acta*, **396**, 11 (1975).
66. E. Lee-Ruff, A. B. P. Lever and J. Rigaudy, *Can. J. Chem.*, **54**, 1837 (1976).
67. D. Vidril-Robert, M. Maurette, E. Oliveros, M. Hocquaux and B. Jacquet, *Tetrahedron Lett.*, **25**, 529 (1986).
68. M. De Min, M. T. Maurette, E. Oliveros, M. Hocquaux and B. Jacquet, *Tetrahedron*, **42**, 4953 (1986).
69. Y. Moro-oka and C. S. Foote, *J. Am. Chem. Soc.*, **98**, 1510 (1976).
70. D. T. Sawyer, T. S. Calderwood, C. L. Johlman and C. L. Wilkins, *J. Org. Chem.*, **50**, 1409 (1985).
71. D. T. Sawyer, M. J. Gibian, M. M. Morrison and E. T. Seo, *J. Am. Chem. Soc.*, **100**, 627 (1978).
72. E. J. Nanni, Jr, M. D. Stallings and D. T. Sawyer, *J. Am. Chem. Soc.*, **102**, 4481 (1975).
73. M. Nishikimi and L. J. Machlin, *Arch. Biochem. Biophys.*, **170**, 684 (1975).
74. K. Yagi and M. Nishikimi, *Taisha*, **15**, 1287 (1978).
75. T. Ozawa and A. Hanaki, *Biochem. Biophys. Res. Commun.*, **129**, 461 (1985).
76. S. Matsumoto and M. Matsuo, *Tetrahedron Lett.*, 1999 (1977).
77. M. Matsuo, S. Matsumoto, Y. Iitaka, A. Hanaki and T. Ozawa, *J. Chem. Soc., Chem. Commun.*, 105 (1979).
78. M. Matsuo, S. Matsumoto and Y. Iitaka, *J. Org. Chem.*, **52**, 3514 (1987).
79. S. Matsumoto, M. Matsuo and Y. Iitaka, *J. Org. Chem.*, **51**, 1435 (1986).
80. J. S. Filippo, Jr, L. S. Romano, C. Chern and J. S. Valentine, *J. Org. Chem.*, **41**, 586 (1976).
81. M. J. Gibian, D. T. Sawyer, T. Ungerman, R. Tangpoonpholvivat and M. M. Morrison, *J. Am. Chem. Soc.*, **101**, 640 (1979).
82. A. R. Forrester and V. Purushotham, *J. Chem. Soc., Perkin Trans. 1*, 945 (1987).
83. A. R. Forrester and V. Purushotham, *J. Chem. Soc., Chem. Commun.*, 1505 (1984).
84. J. P. Stanley, *J. Org. Chem.*, **45**, 1413 (1980).
85. W. C. Danen and R. L. Arudi, *J. Am. Chem. Soc.*, **100**, 3944 (1978).
86. M. Lissel, E. V. Dehmlow, *Tetrahedron Lett.*, 3689 (1978).
87. M. J. Gibian and T. Ungermann, *J. Am. Chem. Soc.*, **101**, 1291 (1979).
88. J. W. Peters and C. S. Foote, *J. Am. Chem. Soc.*, **98**, 873 (1976).
89. M. Lissel, *Tetrahedron Lett.*, **25**, 2213 (1984).
90. M. Lissel, *Z. Naturforsch., Teil B*, **42**, 12 (1987).
91. M. Hocquaux, B. Jacquet, D. Vidril-Robert, M. Maurette and E. Oliveros, *Tetrahedron Lett.*, **25**, 533 (1984).
92. E. Alvarez, C. G. Francisco, R. Freire, R. Hernandez, J. A. Salazar, E. Saurez and C. Betancor, *J. Chem. Soc., Perkin Trans. 1*, 1523 (1986).
93. E. Alvarez, C. Betancor, R. Freire, A. Martin and E. Suarez, *Tetrahedron Lett.*, **22**, 4335 (1981).
94. A. A. Frimer, P. Gilinski-Sharon, J. Hameiri and G. Aljadeff, *J. Org. Chem.*, **47**, 2818 (1982).
95. S. Kobayashi, T. Tezuka and W. Ando, *J. Chem. Soc., Chem. Commun.*, 508 (1979).
96. J. S. Filippo, Jr, C. Chern and J. S. Valentine, *J. Org. Chem.*, **41**, 1077 (1976).
97. D. T. Sawyer, J. J. Stamp and K. A. Menton, *J. Org. Chem.*, **48**, 3733 (1983).
98. K. Boujlel and J. Simonet, *Tetrahedron Lett.*, 1063 (1979).
99. D. C. Neckers and G. Hauck, *J. Org. Chem.*, **48**, 4691 (1983).
100. I. Rosenthal and A. A. Frimer, *Tetrahedron Lett.*, 3731 (1975).
101. R. Poupko and I. Rosenthal, *J. Phys. Chem.*, **77**, 1722 (1973).
102. M. J. Gibian and S. Russo, *J. Org. Chem.*, **49**, 4304 (1984).
103. A. A. Frimer and I. Rosenthal, *Photochem. Photobiol.*, **28**, 711 (1978).
104. I. Rosenthal and A. A. Frimer, *Tetrahedron Lett.*, **23**, 2805 (1976).

105. A. A. Frimer, G. Aljadeff and P. Gilinski-Sharon, *Isr. J. Chem.*, **27**, 39 (1986).
106. I. Saito, T. Otsuki and T. Matsuura, *Tetrahedron Lett.*, 1693 (1979).
107. A. A. Frimer, I. Rosenthal and S. Hoz, *Tetrahedron Lett.*, 4631 (1977).
108. T. S. Calderwood, R. C. Neuman, Jr, and D. T. Sawyer, *J. Am. Chem. Soc.*, **105**, 3337–3339 (1983).
109. A. A. Frimer, P. Gilinsky-Sharon, G. Aljadeff, H. J. E. Gottlieb, J. Hameiri-Buch, V. Marks, R. Philosof and Z. Rosenthal, *J. Org. Chem.*, **54**, 4853 (1989).
110. A. A. Frimer, P. Gilinsky-Sharon, G. Aljadeff, V. Marks and Z. Rosenthal, *J. Org. Chem.*, **54**, 4866 (1989).
111. A. A. Frimer, P. Gilinsky-Sharon and G. Aljadeff, *Tetrahedron Lett.*, **23**, 1301 (1982).
112. T. Osa, M. Ohkatsu and M. Tezuka, *Chem. Lett.*, 99 (1973).
113. E. Lee-Ruff and N. Timms, *Can. J. Chem.*, **58**, 2138 (1980).
114. A. A. Frimer, T. Farkash-Solomon and G. Aljadeff, *J. Org. Chem.*, **51**, 2093 (1986).
115. S. Top, G. Jaouen and M. McGlinchey, *J. Chem. Soc., Chem. Commun.*, 643 (1980).
116. A. A. Frimer, G. Aljadeff and J. Ziv, *J. Org. Chem.*, **48**, 1700 (1983).
117. Y. Moro-oka, P. J. Chung, H. Arakawa and T. Ikawa, *Chem. Lett.*, 1293 (1976).
118. M. Tezuka, Y. Ohkatsu and T. Osa, *Bull. Chem. Soc. Jpn.*, **48**, 1471 (1975).
119. S. T. Purrington and G. B. Kenion, *J. Chem. Soc., Chem. Commun.*, 731 (1982).
120. J. S. Filippo, Jr, C. Chern and J. S. Valentine, *J. Org. Chem.*, **40**, 1678 (1975).
121. W. C. Danen and R. J. Warner, *Tetrahedron Lett.*, 989 (1977).
122. R. Dietz, A. E. J. Forno, B. E. Larcombe and M. E. Peover, *J. Chem. Soc. B*, 816 (1970).
123. T. S. Calderwood and D. T. Sawyer, *J. Am. Chem. Soc.*, **106**, 7185 (1984).
124. H. Sugimoto, S. Matsumoto and D. T. Sawyer, *J. Am. Chem. Soc.*, **109**, 8081 (1987).
125. P. F. Levonovich, H. P. Tannenbaum and R. C. Dougherty, *J. Chem. Soc., Chem. Commun.*, 597 (1975).
126. H. Sugimoto, S. Matsumoto and D. T. Sawyer, *J. Am. Chem. Soc.*, **110**, 5193 (1988).
127. I. Rosenthal and A. A. Frimer, *Tetrahedron Lett.*, 2809 (1976).
128. W. T. Monte, M. M. Baizer and R. D. Little, *J. Org. Chem.*, **48**, 803 (1983).
129. D. J. Stuehr and M. A. Marletta, *J. Org. Chem.*, **50**, 694 (1985).
130. G. Crank and M. I. H. Makin, *Aust. J. Chem.*, **37**, 2331 (1984).
131. G. Crank and M. I. H. Makin, *Tetrahedron Lett.*, 2169 (1979).
132. G. Crank and M. I. H. Makin, *Aust. J. Chem.*, **37**, 845 (1984).
133. G. Crank and M. I. H. Makin, *Tetrahedron Lett.*, **24**, 3159 (1983).
134. G. Crank, G. E. Gately and M. I. H. Makin, *Aust. J. Chem.*, **37**, 2499 (1977).
135. C. Chern and J. S. Filippo, Jr, *J. Org. Chem.*, **42**, 178 (1977).
136. T. Nagano and I. Fridovich, *Arch. Biochem. Biophys.*, **241**, 596 (1985).
137. N. V. Blough and O. C. Zafiriou, *Inorg. Chem.*, **24**, 3502 (1985).
138. H. Takizawa, T. Nagano and M. Hirobe, *Chem. Pharm. Bull.*, **39**, 1894 (1991).
139. K. Shibata, Y. Saito, K. Urano and M. Matsui, *Bull. Chem. Soc. Jpn.*, **59**, 3323 (1986).
140. N. Kornblum and S. Singaram, *J. Org. Chem.*, **44**, 4727 (1979).
141. M. Sugawara, M. M. Baizer, W. T. Monte, R. D. Little and U. Hess, *Acta Chem. Scand., Ser. B*, **37**, 509 (1983).
142. M. Lissel and A. Gau, *Z. Naturforsch., Teil B*, **41**, 367 (1986).
143. Y. Sawaki and Y. Ogata, *Bull. Chem. Soc. Jpn.*, **54**, 793 (1981).
144. M. Sugawara and M. M. Baizer, *Tetrahedron Lett.*, **24**, 2223 (1983).
145. H. S. Chang, G. H. Yon and Y. H. Kim, *Chem. Lett.*, 1291 (1986).
146. Y. H. Kim and G. H. Yon, *J. Chem. Soc., Chem, Commun.*, 715 (1983).
147. Y. H. Kim, G. H. Yon and H. J. Kim, *Chem. Lett.*, 309 (1984).
148. G. Crank and M. I. H. Makin, *J. Chem. Soc., Chem. Commun.*, 53 (1984).

149. O. A. Paez, C. M. Valdez and J. Tor, *J. Org. Chem.*, **53**, 2166 (1988).
150. E. Katori, T. Nagano, T. Kunieda and M. Hirobe, *Chem. Pharm. Bull.*, **29**, 3075 (1981).
151. T. Akasaka and W. Ando, *Tetrahedron Lett.*, **26** 5049 (1985).
152. S. Oae, T. Takata and Y. H. Kim, *Tetrahedron*, **37**, 37 (1981).
153. T. Takata, Y. H. Kim and S. Oae, *Tetrahedron Lett.*, 821 (1979).
154. H. Inoue, T. Nagano and M. Hirobe, *Tetrahedron Lett.*, **25**, 317 (1984).
155. T. Nagano, K. Arakane and M. Hirobe, *Tetrahedron Lett.*, **21**, 5021 (1980).
156. E. Balogh-Hergovich and G. Speier, *Tetrahedron Lett.*, **23**, 4473 (1982).
157. A. Dondoni, G. Galliani, A. Mastellani and A. Medici, *Tetrahedron Lett.*, **26**, 2917 (1985).
158. S. Kobayashi and W. Ando, *Chem. Lett.*, 1159 (1978).
159. E. J. Nanni, Jr, C. T. Angelis, J. Dickson and D. T. Sawyer, *J. Am. Chem. Soc.*, **103**, 4268 (1981).
160. T. Harayama, K. Mori, R. Yanada, K. Iio, Y. Fujita and F. Yoneda, *J. Chem. Soc., Chem. Commun.*, 1171 (1988).
161. T. Itoh, K. Nagata, M. Okada and A. Osawa, *Tetrahedron Lett.*, **31**, 7193 (1990).
162. T. S. Rao and G. K. Trivedi, *Heterocycles*, **26**, 2117 (1987).
163. C. A. Chuaqui, S. Delaney and J. Merritt, *Tetrahedron*, **39**, 2947 (1983).
164. N. Oshima, H. Suzuki and Y. Moro-oka, *Inorg. Chem.*, **25**, 3407 (1986).

15 Peroxides in Biological Systems

ETSUO NIKI
Research Center for Advanced Science and Technology, University of Tokyo, 4–6–1 Komaba, Meguro-ku, Tokyo 153, Japan

1 INTRODUCTION

There is now increasing evidence which suggests the involvement of lipid peroxidation in a variety of pathological events, cancer and aging.[1–8] Active oxygens and free radicals attack lipids, proteins, sugars, enzymes and DNA to induce oxidative damage, which eventually leads to various diseases, cancer and aging (Figure 1). Lipids are important targets of active oxygens and free radicals and the peroxidation of lipids is accepted as being one of the important primary events in oxidative damage in biological systems.

Although it has been generally accepted that lipid peroxidations are toxic and induce cell damage and disease, some lipid peroxidation may be associated with normal biochemical processes. As with prostaglandins, some peroxides function as important, active biofactors *in vivo* and the active oxygens and free radicals are essential in, for example, phagocytosis. Hence oxidation and peroxides are, like the oxygen molecule, a 'double-edged sword.'

Biological lipids which are important targets for oxidation are generally free fatty acids and their esters, such as phospholipids and cholesterol esters (Figure 2). Their oxidizabilities and oxidation products are determined primarily by the composition of the fatty acid moieties, especially the number of double bonds, which are dependent on the type of food ingested. In general, the oxidizabilities increase as the degree of unsaturation increases.

Organic Peroxides. Edited by W. Ando

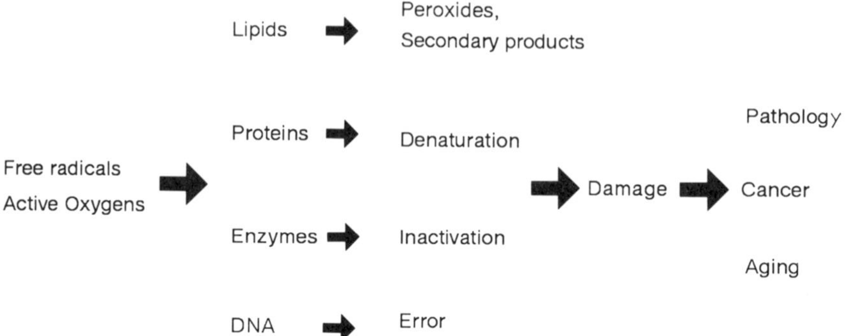

Figure 1. Oxidative damage induced by active oxygens and free radicals

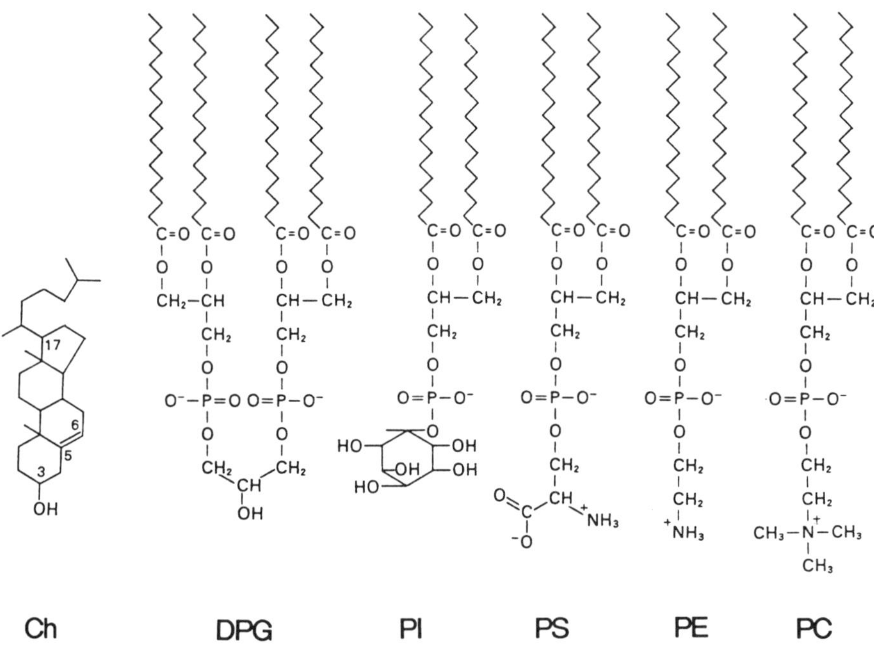

Figure 2. Biological lipids. Ch = cholesterol; DPG = diphosphatidylglycerol or cardiolipin (CL); PI = phosphatidylinositol; PS = phosphatidylserine; PE = phosphatidylethanolamine; PC = phosphatidylcholine

The oxidation of lipids *in vivo* proceeds by both enzymatic and non-enzymatic routes. Enzymatic oxidations sometimes involve radicals, but probably they are not really free. Non-enzymatic oxidation proceeds by both a free-radical chain mechanism and a non-radical mechanism. As in the oxidative deterioration of oil, plastics, rubber and foods, the free-radical chain oxidation must amplify the oxidative damage, since a single hit of the initiating radical induces the chain oxidation of many molecules in sequence.

In this chapter, the oxidations of lipids in membranes and lipoproteins are briefly overviewed, emphasizing the rate and products of free-radical chain oxidation, and the inhibition of oxidations is also discussed.

2 RATE, MECHANISM AND PRODUCTS OF OXIDATION OF LIPIDS IN MEMBRANES

The rate, mechanism and products of the oxidation of lipids in homogeneous solution have been studied extensively and are now well understood.[9–13] The oxidations of lipids in membranes are apparently more complicated, but they proceed by substantially the same mechanism as in homogeneous solution, namely the formation of a lipid radical in the chain initiation followed by chain propagation, where the following four competing reactions determine the rate, route and products of the oxidation: (1) the addition of an oxygen molecule to a lipid radical; (2) the release of oxygen from a lipid peroxyl radical; (3) intermolecular hydrogen atom abstraction from the lipid by a lipid peroxyl radical to give lipid hydroperoxide and a new lipid radical; and (4) intramolecular addition of a lipid peroxyl radical to a double bond to give a cyclic peroxide (Figure 3). The oxidation chain is terminated by the bimolecular interaction of radicals and/or by scavenging of radicals by antioxidant.

The lipid peroxyl radical which serves as a chain carrier is relatively unreactive and it attacks selectively the doubly allylic hydrogen atom of lipids which has the lowest bond dissociation energy and highest reactivity. Therefore, polyunsaturated fatty acids (PUFA), which have two or more double bonds, and their esters are oxidized exclusively and fatty acids which have no or only one double bond are not oxidized. The PUFA which has two double bonds gives conjugated diene hydroperoxides quantitatively, whereas those which have three or more double bonds give cyclic peroxides in addition to conjugated diene hydroperoxides.[11–13] The PUFAs in biological systems have exclusively *cis* double bonds, but their oxidations give *cis,trans*- and *trans,trans*-conjugated diene hydroperoxides (Figure 3). This conformational change produces perturbation in the membranes. It is also known from a spin labeling study that the oxidation of the membrane lipids induces a decrease in fluidity.[14]

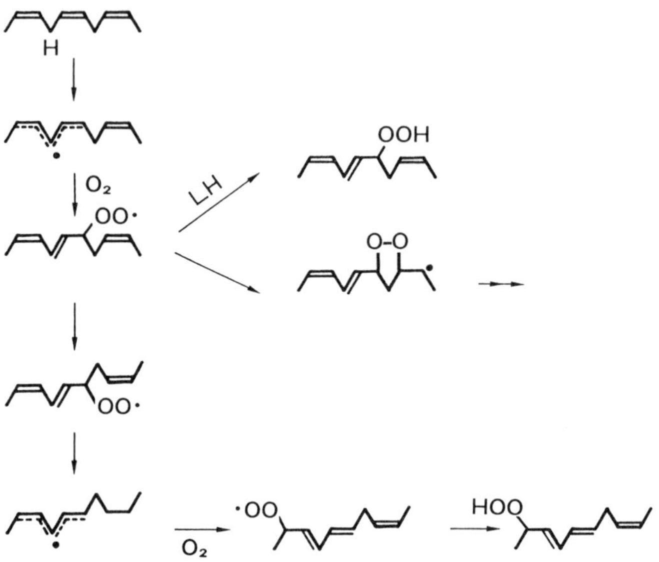

Figure 3. Chain propagation reactions in the oxidation of polyunsaturated lipids

The composition of fatty acids of the phospholipids, and hence the susceptibility to oxidation, depend on the sources. Table 1 gives examples of the fatty acid compositions of a few phosphatidylcholines (PC).[15] In general, the phospholipids from vegetables are rich in linoleic acid, whereas those from fish are rich in higher PUFAs such as pentenoate and hexenoate. It can be seen from Table 1 that the contents of active, doubly allylic hydrogens differ extensively between the lipids.

Table 2 shows the products of fatty acids and their esters in different reaction media.[15] As mentioned above, the oxidation products are dependent on the number of double bonds: when the lipid has only two double bonds (methyl

Table 1. Fatty acid composition of phosphatidylcholines (PC)

PC	Fatty acid (mol%)[a]								Active H[b]
	16:0	18:0	18:1	18:2	18:3	20:4	22:5	22:6	
Egg PC	34.0	12.2	33.2	15.8		4.0			1.11
Rat-liver PC	33.1	13.2	27.6	16.2		6.2	1.3	2.4	2.08
Soybean PC	12.6	3.0	11.2	67.3	5.9				3.16
Dilinoleoyl-PC				100					4.00

[a] The left-hand number indicates the carbon number and the right-hand number the number of double bonds.
[b] Number of bisallylic hydrogens in one molecule.

Table 2. Oxidation of polyunsaturated fatty acid methyl esters and PC at 50°C[a]

Parameter Substrate LH:	18:2	18:2	18:2 PC	18:2 PC	18:3	Egg PC	Egg PC	Rat-liver PC	
Medium:	CH_3CN	Micelle[c]	C_6H_6	Liposome	CH_3CN	CCl_4	Liposome	*t*-BuOH	Liposome
Conversion[b] (%)	3	11	75	64	11	13	16	21	33
Δ O_2	67	157	89	49	226	69	86	55	87
Δ LH	68	135			180	20	42	33	36
Δ Peroxide	70	141	61	53	129	23	39	10	34
Δ (C=C)	74	141	50	38	33	7.2	15	3.9	16
Δ C(=O) − $(C{=}C)_2$	0	0	4.6	2	0	5.6	3.4	3.0	4.1
Δ TBARS	0	0	0	0	33	5.1	6.4	2.3	6.3

[a] Values in μmol.
[b] Δ O_2/initial LH ($\times 10^2$).
[c] In 10 mM Triton X-100 aqueous dispersion. *t*-BuOH, *tert*-butyl alcohol; TBARS, 2-thiobarbituric acid-reactive substrates.

linoleate and dilinoleoyl-PC), the amounts of oxygen uptake, substrate reacted and hydroperoxide and conjugated diene formed agree well and thiobarbituric acid-reactive substrates (TBARS, which is a measure for some hydroperoxides and aldehydes formed) are negligible, suggesting that the oxidation gives conjugated diene hydroperoxides quantitatively. It has been shown that the oxidation of linoleic acid gives four conjugated diene hydroperoxides quantitatively.[11–13] On the other hand, for those which have three or more double bonds, the following correlation holds: oxygen uptake > substrate disappearance > peroxide > conjugated diene > TBARS. These results can be interpreted by the contribution of intramolecular addition of the peroxyl radical to the adjacent double bond, which is possible when the lipid has three or more double bonds (reaction 4). Further, Table 2 shows that this product distribution is independent of the reaction medium, that is, it is determined primarily by the number of double bonds independent of the reaction medium.

Figure 4 shows a typical example of the oxidation of PC liposomal membranes in aqueous dispersions, which are a model of biological membranes. Soybean PC contains about 70% linoleic acid and 5% linoleic acid residues as oxidizable PUFAs, and hence its oxidation can be followed quantitatively by measuring the formation of conjugated diene hydroperoxides. The rates of chain

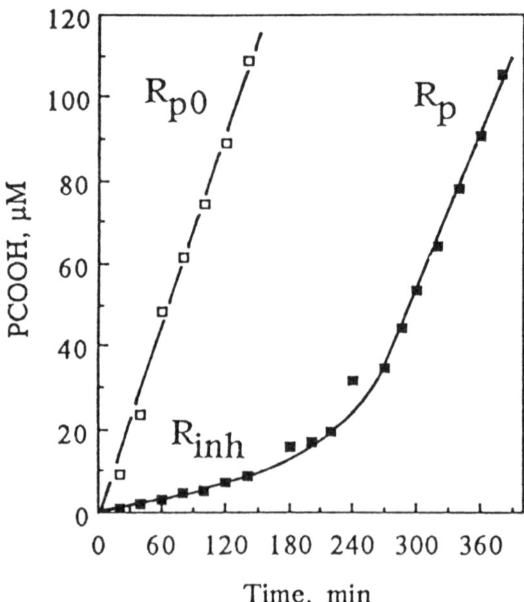

Figure 4. Oxidation of soybean phosphatidylcholine liposomal membranes at 37°C in air. [Soybean PC] = 6.4 mM; [azobis(2,4-dimethylvaleronitrile)] = 0.30 mM; [α-tocopherol] = (□) 0 and (■) 1.0 μM

initiation and oxidation were obtained as 1.47×10^{-10} and 1.35×10^{-8} mol l^{-1} s^{-1}, respectively, and the kinetic chain length was calculated as 92, suggesting that the oxidation proceeds by a free-radical chain mechanism with a long kinetic chain length.[16] It has been also found that erythrocyte membranes[17] and human low-density lipoproteins[18] are also oxidized by a free-radical chain mechanism.

The kinetics of the oxidation of phospholipids in membranes have been studied and it was found that the rate follows the classical rate law,[10,19–21] namely the rate is proportional to the first power of the substrate concentration and the half-power of the rate of chain initiation:

$$R_p = \frac{k_p}{(2k_t)^{1/2}} [\mathrm{LH}] R_i^{1/2} \tag{1}$$

where LH is the substrate, R_p the rate of oxidation, R_i the rate of chain initiation, k_p the rate constant for chain propagation and $2k_t$ the rate constant for chain termination. The ratio of the rate constants $k_p/(2k_t)^{1/2}$ gives the oxidizability of the substrate for a specific reaction medium and conditions.

Table 3 summarizes the rate constants for chain propagation and chain termination. The oxidizabilities are also included.

Of particular interest for biological systems is that the free fatty acids and their derivatives take various forms in aqueous media, such as micelles, liposomes, membranes and also complexes with proteins. For example, linoleic

Table 3. Rate constants for chain propagation and termination and oxidizability of linoleic acid and its esters

Substrate	Medium[a]	Temperature (°C)	k_p l mol^{-1} s^{-1}	k_t l mol^{-1} s^{-1}	$k_p/(2k_t)^{1/2}$ $[(mol\ l^{-1}\ s)^{-1/2}]$	Ref.
Methyl linoleate	In PhCl	30	31	2.7×10^6	2.18×10^{-2}	22
Linoleic acid	Micelles in 0.50 M SDS	30	36.2	3.52×10^5	4.42×10^{-2}	20
Linoleic acid	In 14:0 PC MLV	30			2.76×10^{-2}	22
Linoleic acid	Micelles in 0.50 M SDS	37	39.1	4.17×10^5		19
Dilinoleoyl-PC	MLV	37	36.1	1.32×10^5		19
Dilinoleoyl-PC	SUV	45			0.232	21
Dilinoleoyl-PC	MLV	45			0.116	21
Methyl linoleate	In 14:0 PC MLV	30			1.79×10^{-2}	22
Methyl linoleate	In PhCl	30	6.4	7.6×10^6		23

[a] SDS = sodium dodecyl sulfate; MLV = multilamellar large vesicles; SUV = small unilamellar vesciles.

acid aggregates to give, depending on its concentration, monomeric, oligomeric, spherical and rod-shaped micelles, and it was found that higher structural organization leads to a further increase in chain propagation length.[24]

The oxidation of red blood cells has been studied extensively. They are susceptible to oxidation since they are rich in polyunsaturated fatty acid residues and exposed to high concentrations of oxygen and iron.[25] Their oxidations are induced by various oxidants such as hydrogen peroxide, organic hydroperoxides and superoxide. The free radicals can also induce the oxidation of red blood cells. The oxidation of red blood cells, and their ghost membranes induced by aqueous peroxyl radicals generated from a water-soluble azo compound, 2,2′-azobis(2-amidinopropane) dihydrochloride (AAPH),[26] proceeds by a free-radical chain mechanism.[17,27–29] Both lipids and proteins are oxidized: the oxidation of lipids gives conjugated hydroperoxides and TBARS, whereas that of proteins brings about cross-linking and cleavage. As the oxidation proceeds, leakages of potassium and calcium ions, LDH (lactate dehydrogenase), GOT (glutamic-oxaloacetic transaminase) and hemoglobin are observed and eventually hemolysis takes place. As shown in Figure 5, the rate of hemolysis increased with increasing concentration of AAPH. Interestingly, the extent of hemolysis was proportional to the amount of free radicals formed.[28]

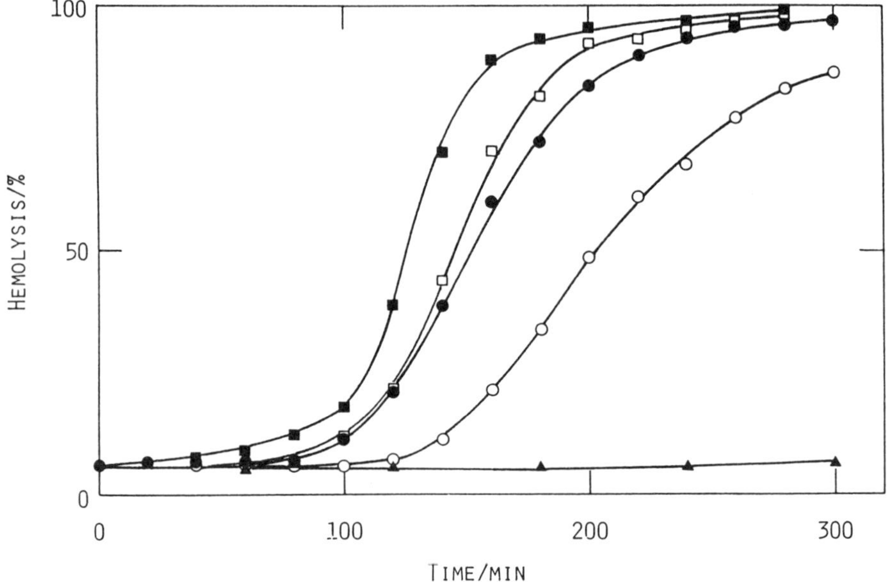

Figure 5. Effect of AAPH concentration on the hemolysis of rabbit red blood cells at 37°C in air. [AAPH] = (▲) 0; (○) 25; (●) 50; (□) 75; (■) 100 mM

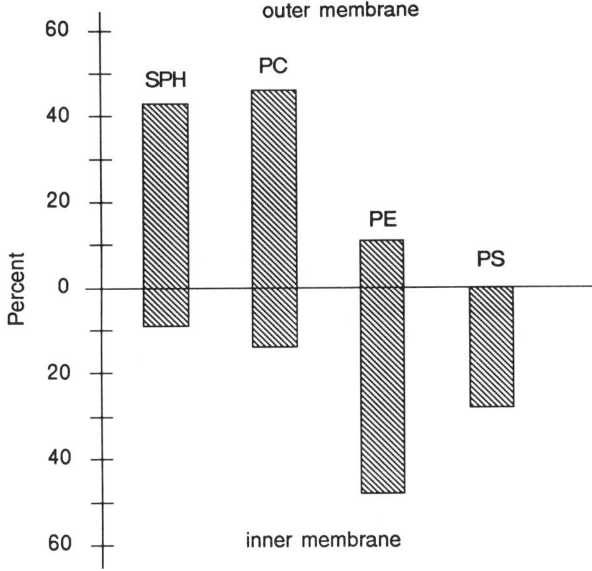

Figure 6. Distribution of phospholipids in human red blood cell bilayer membrane. SPH = sphingomyelin; PC = phosphatidylcholine; PE = phosphatidylethanolamine; PS = phosphatidylserine

Red blood cell membranes are composed of phospholipids, cholesterol and proteins. The distribution of phospholipids in the outer and inner membranes is heterogeneous. As shown in Figure 6, PC and sphingomyelin (SPH) are located predominantly in the outer membrane, whereas phosphatidylethanolamine (PE) and phosphatidylserine (PS) are located exclusively in the inner membrane. When the radicals attack from outside the membranes as in the oxidation induced by AAPH dissolved in the aqueous phase, PC and SPH are oxidized more readily than PE and PS. On the other hand, when the oxidation of red blood cells is induced by hypoxanthine (HX) and xanthine oxidase (XOD), PE is oxidized predominantly, since superoxide formed from HX and XOD gives hydrogen peroxide, which penetrates red blood cell membranes and interacts with hemoglobin to give hydroxyl radical, and the latter attacks lipids from the inside of red blood cells (Figure 7).[30]

Cholesterol is also an important lipid present in biological membranes. Cholesterol is not readily oxidized by itself, but it is co-oxidized with phospholipids to give cholesterol epoxides, cholestanetriol and other products.[31]

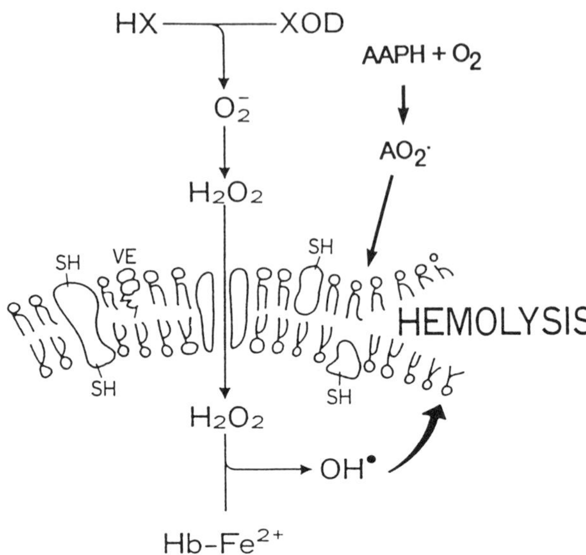

Figure 7. Attack of oxygen radicals on red blood cells from outside and inside the red blood cells

3 OXIDATION OF LIPOPROTEINS

Lipoproteins are the aggregates of lipids and proteins. The oxidation of lipoproteins, especially low-density lipoproteins (LDL), has received much attention recently. LDL is an important target of free radicals in the blood and the oxidation of LDL is considered to be an important event in atherogenesis, since oxidatively modified LDL shows a diminished affinity to the normal LDL receptor and increased affinity to the macrophage scavenger receptor.[32–34] The macrophages rapidly become loaded with lipids and are converted into lipid-laden foam cells, the hallmark of the early atherosclerotic lesion.

LDL has a characteristic structure: PC and free cholesterol are present in the surface membrane, whereas cholesterol esters and triglycerides are located in the inner core. Further, it has a glycoprotein, apo-B. The fatty acid compositions in PC, cholesterol esters and triglycerides depend on the type of food ingested, but usually linoleic acid and arachidonic acid are important PUFAs in the human LDL.[35] It has been found that copper induces the oxidation of LDL to give conjugated diene hydroperoxides as primary products,[35] apparently by the mechanism described above. As the oxidation proceeds, the PUFAs such as linoleic acid and arachidonic acid residues are oxidized and the formation of conjugated dienes, peroxides, and TBARS is observed.[35] Both phosphatidylcholine hydroperoxide and cholesterol ester hydroperoxides are observed, but the formation of triglyceride hydroperoxide is low.[18,36a]

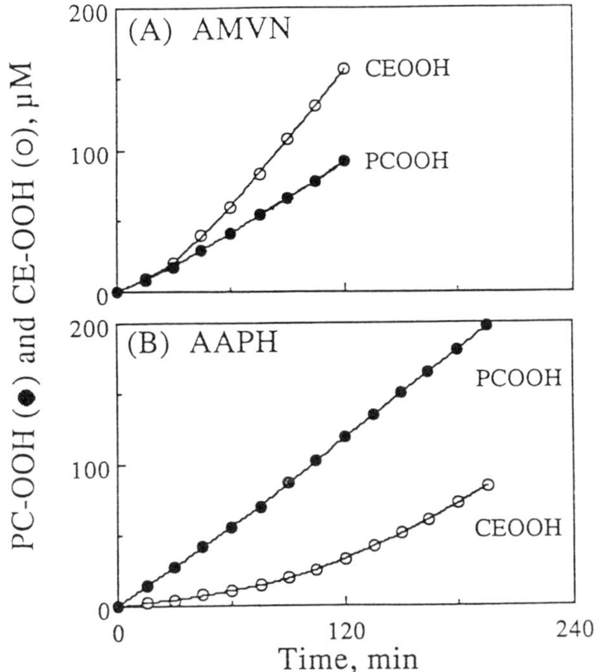

Figure 8. Formation of PC hydroperoxide (PC—OOH) and cholesterol ester hydroperoxide (CE—OOH) in the oxidation of lipid microspheres initiated with (A) 1 mM AMVN or (B) 1 mM AAPH at 37°C in air. [Soybean PC] = 3.2 mM; [cholesteryl linoleate] = 8.6 mM; [ethyl palmitate] = 19.3 mM

The lipid microsphere composed of PC, fatty acid ester and cholesterol ester is a simple model of LDL and its oxidation has also been studied.[36b] The amphipathic PC forms an outer membrane, while hydrophobic fatty acid ester and cholesterol ester are located in the inner core of the lipid microsphere. When the oxidation is initiated with aqueous radicals, the rate of formation of PC hydroperoxide is faster than that of cholesterol ester hydroperoxide, whereas in the lipophilic radical-induced oxidation, cholesterol ester is oxidized faster than PC (Figure 8).

4 REACTIONS OF LIPID HYDROPEROXIDES

Lipid hydroperoxides are not stable *in vivo* and they undergo several secondary reactions,[12,37] the important reactions being enzyme-catalyzed and metal-catalyzed decomposition. The lipid hydroperoxides are toxic themselves and, by giving active oxygen radicals, several enzymes such as glutathione peroxidase, glutathione-S-transferase and peroxidase, decompose hydroperoxides to the corresponding alcohols (equations 2 and 3). Phospholipids are decomposed either as they are or after hydrolysis by phospholipase to give free fatty acid hydroperoxides, which serve as the substrates of the peroxide-decomposing enzymes. The action of the membrane-bound phospholipid hydroperoxide glutathione peroxidase has also been reported.[38]

$$LOOH + 2GSH \rightarrow LOH + GSSG + H_2O \quad (2)$$

$$LOOH + AH_2 \rightarrow LOH + H_2O + A \quad (3)$$

The metal-catalyzed decomposition of hydroperoxides is another important pathway. Iron and copper ions act as redox catalysts in this Haber–Weiss-type reaction (equations 4 and 5).

$$LOOH + M^{n+} \rightarrow LO\cdot + {}^{-}OH + M^{(n+1)+} \quad (4)$$

$$LOOH + M^{(n+1)+} \rightarrow LO_2\cdot + H^+ + M^{n+} \quad (5)$$

The rates of decomposition by Fe^{3+} or Cu^{2+} ions are in general smaller than those by Fe^{2+} and Cu^+ ions, and these higher valence metal ions are reduced by physiological reducing agents such as superoxide and glutathione rather than by reaction 5. The decomposition by Fe^{2+} ion or Cu^+ ion gives a lipid alkoxyl radical, which may abstract a hydrogen atom to give alcohol or cleave to give an aldehyde and an alkyl radical. The alkyl radical thus formed may either react with oxygen to give a peroxyl radical which induces a chain reaction, or abstract hydrogen to give an alkane (equations 6–9). In fact, it has been observed that the oxidative stress increases the concentrations of ethane and pentane in the exhaled gas.[39] Aldehydes, especially unsaturated aldehydes such as 4-hydroxynonenal and malondialdehyde, are toxic for membranes and lipoproteins.[33]

HOO ... COOH

·O ... COOH —LH→ HO ... COOH (6)

→ $CH_2\cdot$ + OHC ... COOH (7)

—O_2→ $CH_2O_2\cdot$ chain initiation (8)

—LH→ (alkane) (9)

The intramolecular addition of an alkoxyl radical to the double bond to give an epoxide has been observed.[40,41] It has also been reported[42] that cholesterol α- and β-epoxides are formed by the oxidation of cholesterol with hepatic microsomal phospholipid hydroperoxides in the presence of Fe^{2+} ion (equations 10 and 11).

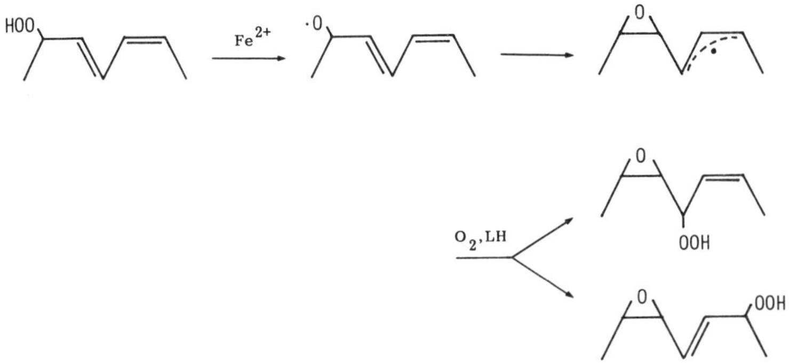

HOO ... $\xrightarrow{Fe^{2+}}$ ·O ... ⟶ O ...

O_2,LH ⟶ O ... OOH (10)

⟶ O ... OOH (11)

Similarly, hydroperoxy epidioxides can be formed by a serial cyclization of lipid peroxyl radicals (equation 12).[43]

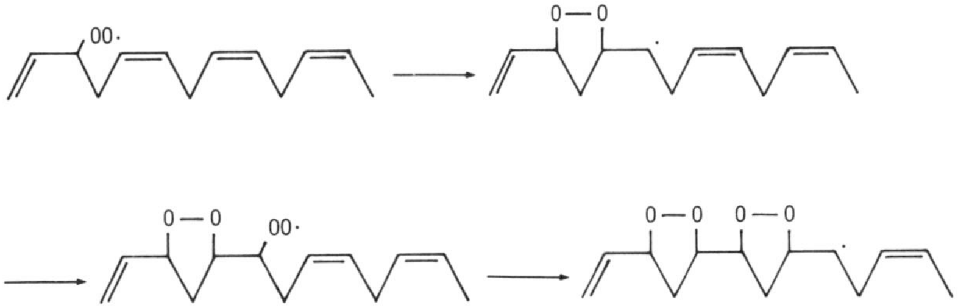

OO· ... ⟶ O—O ...

⟶ O—O OO· ... ⟶ O—O O—O ... (12)

5 PEROXIDES FROM DNA

Many chemical carcinogens are not biologically active themselves, but can be activated through metabolism by the biological host and converted into highly reactive species including radicals as proximate or ultimate carcinogens.[44] Free radicals such as hydroxyl,[45,46] peroxyl[47] and superoxide[48] generated in the vicinity of cellular DNA induce DNA base modification and strand breakage. Especially DNA damage caused by ionizing radiation has received much attention.[49] The metal ions such as iron and copper which are bound to the phosphates of the DNA backbone and to certain amino acids of various proteins also catalyze the site-specific formation of hydroxyl radical. Among the bases, the thymine moiety is the most susceptible to hydroxyl radical attack. Guanine is also a reactive substrate.[51]

It has been found that the irradiation of DNA in the presence of oxygen gives several hydroperoxides, such as *cis*-6-hydroperoxy-5-hydroxy-

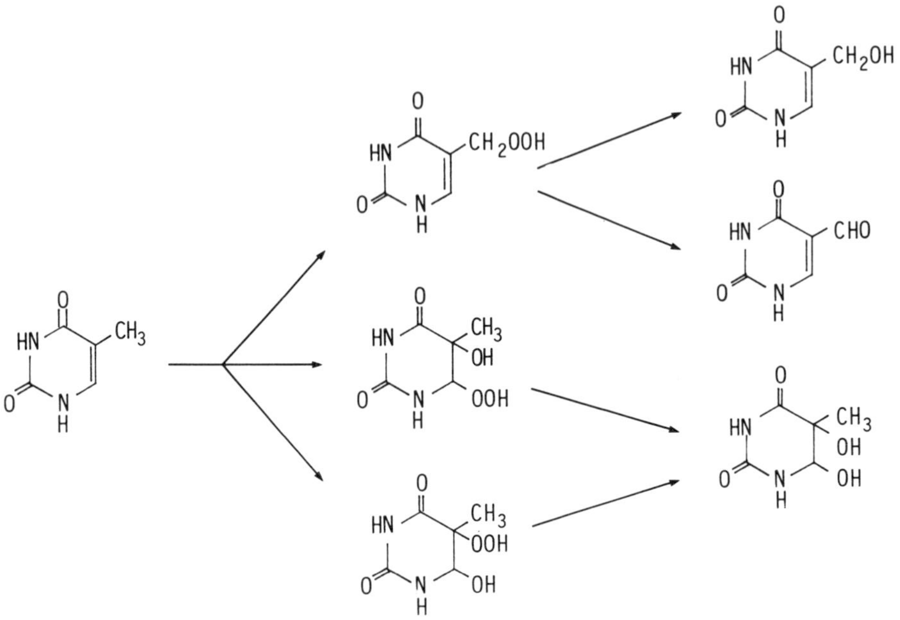

Figure 9. Formation and decomposition of thymine hydroperoxides

5,6-dihydrothymine, *cis*-5-hydroperoxy-6-hydroxy-5,6-dihydrothymine and 5-hydroperoxymethyluracil.[46] Thymine hydroperoxides are mutagenic. These hydroperoxides gradually undergo secondary reactions as shown, as an example, in Figure 9.[52]

6 ENZYMATIC OXIDATION OF LIPIDS

Enzymatic oxidations play a very important role *in vivo*. The enzymes which catalyze oxidation by the oxygen molecule are called oxygenases.[53] There are both monooxygenases and dioxygenases, which insert oxygen atom and dioxygen, respectively, into the substrate.[54] Among others, lipoxygenase and cyclooxygenase are important dioxygenases, especially in arachidonic acid metabolism.[55] Lipoxygenase contains non-heme iron, whereas cyclooxygenase is a heme-iron-containing enzyme. The lipoxygenase enzymes are able to oxygenate arachidonic acid stereospecifically to afford hydroperoxy-eicosatetraenoic acids, which, following enzymatically controlled dehydration, lead to an important class of bioregulators known as leukotrienes.[56]

The lipoxygenase-catalyzed oxidation and autoxidation of PUFA give similar products. For example, soybean lipoxygenase oxidizes linoleic acid and arachidonic acid to give 13-hydroperoxyoctadecadienoate and 15-hydroperoxyeicosatetraenoate, respectively. The overall conversion of PUFA involves the abstraction of bisallylic hydrogen and subsequent addition of an oxygen molecule to the carbon radical after rearrangement, followed by intermolecular abstraction of a hydrogen atom by the resulting peroxyl radical. However, the major difference is that, although the autoxidation is uncontrolled, the enzyme-catalyzed oxidations are stereoselective and regioselective, as shown in Figure 10 and Table 4. It is known that the addition of oxygen is *anti* to hydrogen atom removal and the resultant diene stereochemistry is *trans,cis*.

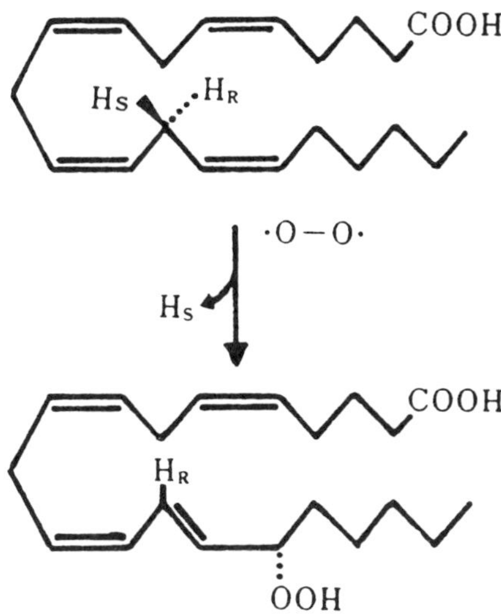

Figure 10. Oxidation of arachidonic acid by lipoxygenase enzymes

Table 4. Stereospecific oxidation by lipoxygenase[a]

Enzyme	Substrate	H abstraction	O insertion	Products
5-Lipoxygenase	AA	7-pro-(*S*)	5*S*	5-HPETE
5-Lipoxygenase	5-HPETE	10-pro-(*R*)	—	5,6-LTA4
12-Lipoxygenase	AA	10-pro-(*R*)	12*S*	12-HPETE
	15-HPETE	10-pro-(*R*)	—	14,15-LTA4
15-Lipoxygenase	AA	13-pro-(*S*)	15*S*	15-HPETE
Cyclooxygenase	AA	PG G_2	9α, 11α, 15*S*	13-pro-(*S*)

[a] AA, arachidonic acid; HPETE, hydroperoxyeicosatetraenoic acid; LT, leukotriene; PG, prostaglandin.

Cyclooxygenase is an enzyme which catalyzes cyclization and oxygenation and it converts arachidonic acid into bicyclic endoperoxide–hydroperoxide PGG_2 stereospecifically.[57]

Figure 11 shows the action of various lipoxygenases and cyclooxygenases in the arachidonic acid cascade which produces various prostaglandins, leukotrienes, thromboxanes and prostacyclins. It is also worth noting that attempts have been made to synthesize chemically these interesting bioactive compounds arising from the lipoxygenase metabolism of arachidonic acid.[56,58] Hydroperoxides and endoperoxides formed by lipoxygenase and cyclooxygenase are readily converted into a variety of biologically important products by other enzymatic conversions.[59,60]

7 DEFENSE SYSTEMS *IN VIVO* AGAINST LIPID PEROXIDATION AND LIPID PEROXIDES

As described above, lipid peroxidation *in vivo* is often a deleterious process and lipid peroxides also exert toxic effects in biological systems. Aerobic organisms are protected from this oxidative damage by an array of defense systems.[61–63] The compounds which are responsible and function in these defense systems are called antioxidants. Some of them are enzymes of high molecular weight and others are small molecules. According to their functions, they are divided into three categories: preventive antioxidants, radical-scavenging antioxidants, and repair and *de novo* compounds (Table 5). These antioxidants with various functions act cooperatively with each other and protect biological systems.

The primary function of the preventive antioxidants is to reduce the rate of chain initiation by suppressing free-radical generation. Lipid hydroperoxides

Table 5. Defense systems *in vivo* against oxidative damage

1. *Preventive antioxidants* reduce formation of free radicals and active oxygens by:
   (a) decomposition of hydroperoxide and hydrogen peroxide:
   glutathione peroxidase, glutathion-s-transferase, peroxidase, catalase
   (b) sequestration of metal ions:
   apoferritin, transferrin, lactoferrin, ceruloplasmin
   (c) quenching and dismutation of active oxygens:
   carotenoids, superoxide dismutase (SOD)
2. *Radical-scavenging antioxidants* inhibit chain initiation and break chain propagation: vitamin C, uric acid, albumin, bilirubin, vitamin E, ubiquinol, carotenoids
3. *Repair* and *de novo* mechanisms:
   phospholipase, protease, transferase
4. *Adaptation* mechanisms

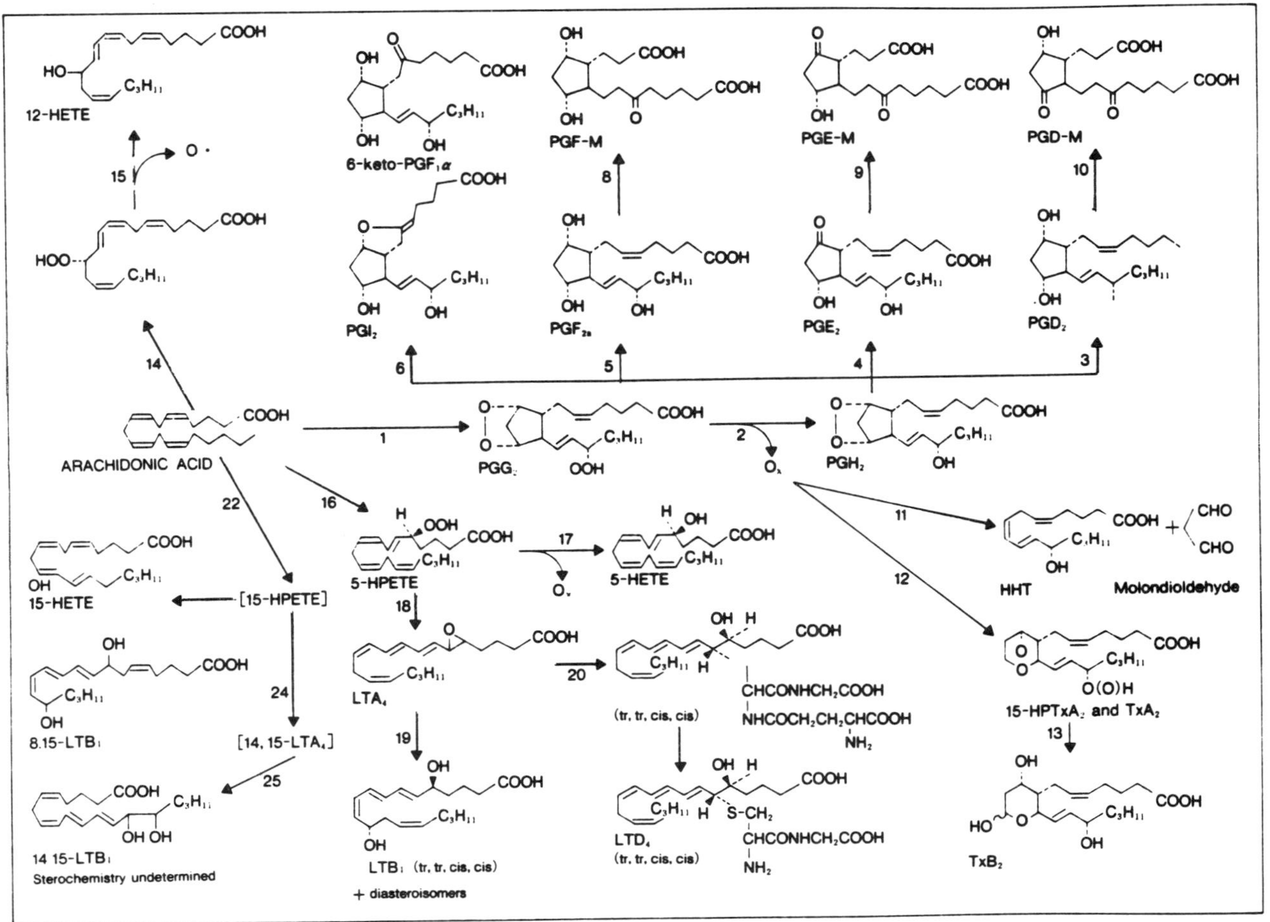

Figure 11. Arachidonic acid cascade

and hydrogen peroxide are important precursors of oxygen radicals. Enzymes such as glutathione peroxidase, glutathione-S-transferase, peroxidase and catalase decompose lipid hydroperoxides and hydrogen peroxide without generating free radicals.[64] Glutathione peroxidase is a selenoprotein and reduces lipid hydroperoxide and hydrogen peroxide to the corresponding alcohol and water, respectively, with the reduced form of glutathione, which in turn is converted into the oxidized form of glutathione (equations 13 and 14).[64]

$$\mathrm{LOOH + 2GSH \rightarrow LOH + H_2O + GSSG} \tag{13}$$

$$\mathrm{HOOH + 2GSH \rightarrow 2H_2O + GSSG} \tag{14}$$

Glutathione-S-transferase reduces lipid hydroperoxide similarly. Catalase is a heme protein, containing one non-covalently bound protoporphyrin iron group in each of its four identical subunits.[65] It catalyzes the reduction of hydrogen peroxide to water and oxygen (equations 15 and 16). Its catalytic turnover rate is one of the highest known, with 44 000 molecules of hydrogen peroxide being decomposed per second, and is diffusion controlled.

$$\mathrm{H_2O_2 + catalase \rightarrow compound\ I + H_2O} \tag{15}$$

$$\mathrm{compound\ I + H_2O_2 \rightarrow catalase + O_2 + H_2O} \tag{16}$$

Peroxidases catalyze the decompositions of hydroperoxides and hydrogen peroxide (equations 17 and 18).

$$\mathrm{LOOH + AH_2 \rightarrow LOH + H_2O + A} \tag{17}$$

$$\mathrm{HOOH + AH_2 \rightarrow 2H_2O + A} \tag{18}$$

Metal ions also play an important role in the generation of free radicals. Proteins which sequester metal ions act as preventive antioxidants. Apoferritin, lactoferrin and transferrin chelate iron and ceruloplasmin chelates copper ion.

Carotenoids[66] and superoxide dismutase (SOD),[67] which deactivate singlet oxygen and superoxide anion radical, respectively, are also considered to be preventive antioxidants.

Thus, the generation of free radicals is suppressed *in vivo*. Nevertheless, some radicals are formed endogenously or may be taken exogenously. The radical-scavenging antioxidants trap these radicals rapidly and inhibit the chain initiation and/or break the chain propagation (equation 19).

$$\mathrm{X\cdot + IH \rightarrow XH + I\cdot} \tag{19}$$

where $X\cdot$ and IH are active radical and antioxidant, respectively, and $I\cdot$ is the newly formed radical from the antioxidant. To be a potent antioxidant, it must scavenge radicals quickly and the antioxidant-derived radicals should be stable enough not to attack the substrate to induce another chain oxidation. The faster the antioxidant scavenges active radicals and the more stable the antioxidant-derived radicals, the stronger its activity as antioxidant.

There are both water-soluble and lipid-soluble radical-scavenging compounds and they function at specific sites. Vitamin C, uric acid, bilirubin, albumin and thiols act as water-soluble antioxidants and scavenge aqueous radicals, but they cannot scavenge lipophilic radicals within the membranes and lipoproteins.

Vitamin E is the most important lipophilic radical-scavenging antioxidant *in vivo*. It scavenges not only lipophilic radicals within the lipid compartment but also aqueous radicals attacking the membrane from outside. The function of ubiquinol, a reduced form of coenzyme Q, as an antioxidant has received much attention recently.[68]

The structures of several important radical-scavenging antioxidants are shown in Figure 12. It is noteworthy that, since the biological systems are heterogeneous, the activities of radical-scavenging antioxidants are determined not only by their chemical reactivities toward radicals but also by their local concentrations and mobilities in the microenvironment at specific sites.[63,69]

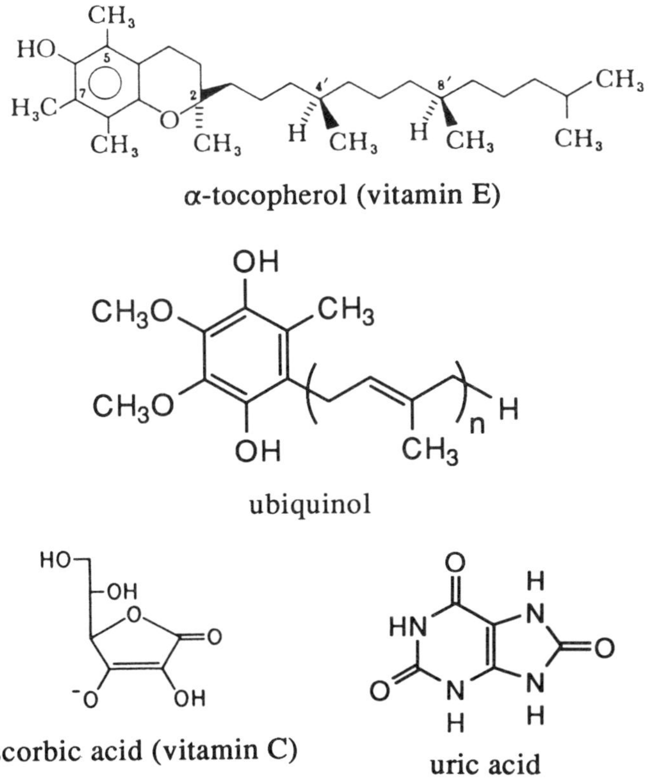

α-tocopherol (vitamin E)

ubiquinol

ascorbic acid (vitamin C)

uric acid

Figure 12. Radical-scavenging antioxidants

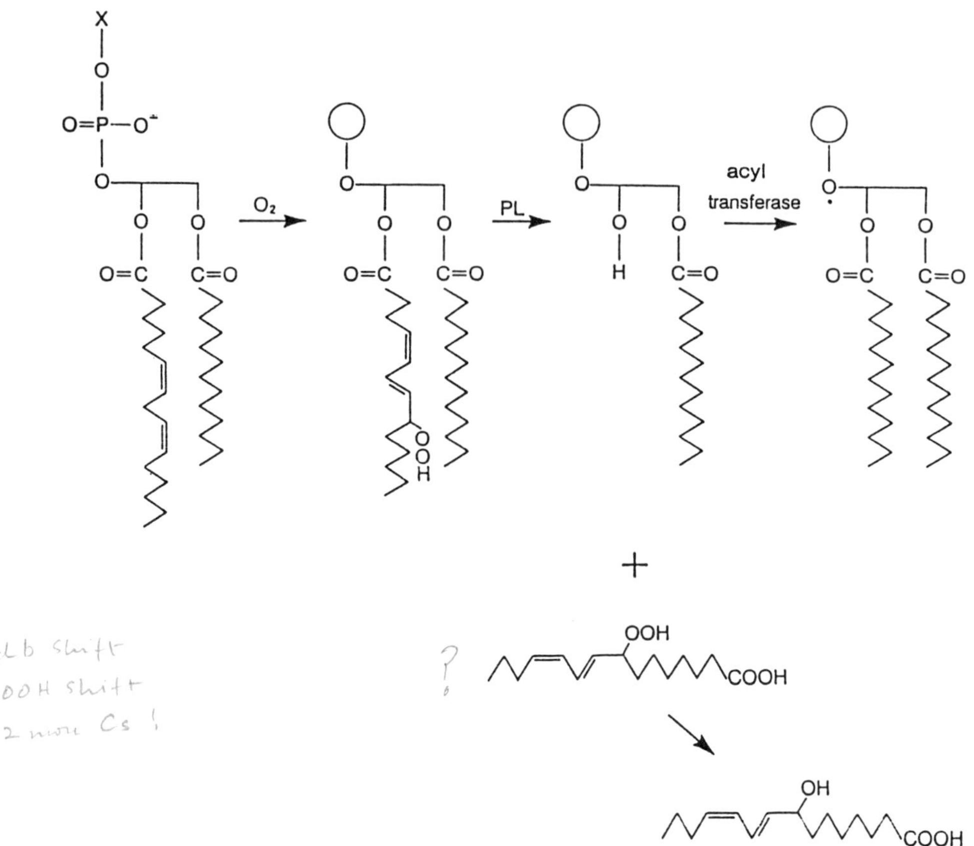

Figure 13. Repair of oxidized phospholipid by phospholipase and transferase

Several enzymes are known to repair the damage and reconstitute the membrane. As shown in Figure 13, phospholipase A_2 acts on the oxidized phospholipids and gives hydrolyzed lipid and free fatty acid hydroperoxide. The former is converted back to phospholipid by acyl transferase and the latter is reduced to alcohol by peroxidases. In this way, the membranes are repaired and reconstituted.

Proteases are known to function as repair enzymes. Further, oxidatively modified DNA is also repaired by specific enzymes.

Another important defense system is adaptation, that is, the oxidative stress induces the formation of a specific antioxidant and its transportation to the target site.

8 REFERENCES

1. W. A. Pryor (Ed.), *Free Radicals in Biology*, Vols I–VI, Academic Press, New York (1976–84).
2. K. Yagi (Ed.), *Lipid Peroxides in Biology and Medicine*, Academic Press, New York (1982).
3. D. C. H. McBrien and T. F. Slater (Eds), *Free Radicals, Lipid Peroxidation and Cancer*, Academic Press, London (1982).
4. B. Halliwell and J. M. C. Gutteridge, *Free Radicals in Biology and Medicine*, Clarendon Press, Oxford (1984).
5. H. Sies (Ed.), *Oxidative Stress*, Academic Press, London (1985).
6. A. Sevanian (Ed.), *Lipid Peroxidation in Biological Systems*, American Oil Chemists' Society, Champaign, IL (1988).
7. M. G. Simic, K. A. Taylor, J. F. Ward and C. von Sonntag (Eds), *Oxygen Radicals in Biology and Medicine*, Plenum Press, New York (1988).
8. O. Hayaishi, E. Niki, M. Kondo and T. Yoshikawa (Eds), *Medical, Biochemical and Chemical Aspects of Free Radicals*, Elsevier, Amsterdam (1990).
9. H. W. S. Chan, G. Levett and J. A. Matthew, *Chem. Phys. Lipids*, **24**, 245 (1970).
10. L. R. C. Barclay and K. U. Ingold, *J. Am. Chem. Soc.*, **103**, 6478 (1981).
11. (a) N. A. Porter, L. S. Lehman, B. A. Weber and K. J. Smith, *J. Am. Chem. Soc.*, **103**, 6447 (1981); (b) N. A. Porter, *Acc. Chem. Res.*, **19**, 262 (1986).
12. E. N. Frankel, *J. Am. Oil Chem. Soc.*, **19**, 262 (1986).
13. H. W. Gardner, *Free Rad. Biol. Med.*, **7**, 65 (1989).
14. M. Takahashi, J. Tsuchiya, E. Niki and S. Urano, *J. Nutr. Sci. Vitaminol.*, **34**, 25 (1988).
15. Y. Yamamoto, E. Niki, Y. Kamiya and H. Shimasaki, *Biochim. Biophys. Acta*, **795**, 332 (1984).
16. E. Niki, Y. Yamamoto, E. Komuro and K. Sato, *Am. J. Clin. Nutr.*, **53**, 2015 (1991).
17. Y. Yamamoto, E. Niki, J. Eguchi, Y. Kamiya and H. Shimasaki, *Biochim. Biophys. Acta*, **819**, 29 (1985).
18. K. Sato, E. Niki and H. Shimasaki, *Arch. Biochem. Biophys.*, **279**, 402 (1990).
19. L. R. C. Barclay, K. A. Baskin, S. J. Locke and M. R. Vingrist, *Can. J. Chem.*, **67**, 1366 (1989).
20. L. R. C. Barclay, K. A. Baskin, S. Jeffrey, S. J. Locke and T. D. Schaefer, *Can. J. Chem.*, **65**, 2529 (1987).
21. L. R. C. Barclay, K. A. Baskin, D. Kon and S. J. Locke, *Can. J. Chem.*, **65**, 2541 (1987).
22. L. R. C. Barclay, S. J. Locke, J. M. MacNeil and J. Vankessel, *Can. J. Chem.*, **63**, 2633 (1985).
23. J. A. Howard and K. U. Ingold, *Can. J. Chem.*, **44**, 1113 (1966).
24. M. Al Sheikhly and M. G. Simic, *J. Phys. Chem.*, **93**, 3103 (1989).
25. D. Chiu, B. Lubin and S. B. Shohet, in W. A. Pryor (Ed.), *Free Radicals in Biology*, Vol. V, Academic Press, New York (1982), pp. 115–169.
26. E. Niki, *Methods Enzymol.*, **186**, 100 (1990).
27. Y. Yamamoto, E. Niki, Y. Kamiya, M. Miki, H. Tamai and M. Mino, *J. Nutr. Sci. Vitaminol.*, **32**, 475 (1986).
28. M. Miki, H. Tamai, M. Mino, Y. Yamamoto and E. Niki, *Arch. Biochem. Biophys.*, **258**, 373 (1987).
29. E. Niki, E. Komuro, M. Takahashi, S. Urano, E. Ito and K. Terao, *J. Biol. Chem.*, **263**, 19809 (1989).
30. H. Tamai, M. Miki and M. Mino, *J. Free Rad. Biol. Med.*, **2**, 49 (1986).

31. L. L. Smith, in C. Vigo-Pelfrey (Ed.), *Membrane Lipid Oxidation*, Vol. I, CRC Press, Boca Raton, FL (1990), pp. 129–154.
32. T. Henriksen, E. M. Nahoney and D. Steinberg, *Proc. Natl. Acad. Sci. USA*, **78**, 6499 (1989); (b) D. Steinberg, S. Parthasarathy, T. E. Carew, J. C. Khoo and J. L. Witztum, *N. Engl. J. Med.*, **320**, 915 (1989).
33. G. Jurgens, H. F. Hoff, G. M. Ghisolm, III and H. Esterbauer, *Chem. Phys. Lipids*, **45**, 315 (1987).
34. (a) J. W. Heinecki, *Free Rad. Biol. Med.*, **3**, 65 (1987); (b) U. P. Steinbrecher, H. Zhang and M. Loughheed, *Free Rad. Biol. Med.*, **9**, 155 (1990).
35. H. Esterbauer, *Ann. N. Y. Acad. Sci.*, **570**, 254 (1989).
36. (a) Y. Yamamoto, M. Kawamura, K. Tasuno, E. Niki and C. Naito, *Free Rad. Biol. Med.*, **9**, Suppl. 1, 66 (1990); (b) K. Tatsuno, Y. Yamamoto, and E. Niki, *Free Rad. Biol. Med.*, **9**, Suppl. 1, 72 (1990).
37. J. Terao, in C. Vigo-Pelfrey (Ed.), *Membrane Lipid Oxidation*, Vol. I, CRC Press, Boca Raton, FL (1990), pp. 219– 238.
38. F. Ursini, M. Maiorino, M. Coassin and A. Roveli, in O. Hayaishi, E. Niki, M. Kondo and T. Yoshikawa (Eds), *Medical, Biochemical and Chemical Aspects of Free Radicals*, Elsevier, Amsterdam (1989), pp. 417–420.
39. A. L. Tappel, *Free Rad. Biol. Med.*, **7**, 193 (1989).
40. H. W. Gardner, D. Weisleder and R. Kleiman, *Lipids*, **13**, 246 (1978).
41. T. A. Kix, R. Fontana, A. Panthani and L. J. Marnett, *J. Biol. Chem.*, **260**, 5358 (1985).
42. T. Watabe, A. Tsubaki, M. Isobe, N. Ozawa and A. Hiratsuka, *Biochim. Biophys. Acta*, **795**, 60 (1984).
43. W. E. Neff, E. N. Frankel and D. Weisleder, *Lipids*, **17**, 780 (1982).
44. P. O. P. Ts'o, W. J. Caspalry and R. L. Lorentzen, in W. A. Pryor (Ed.), *Free Radicals in Biology*, Vol. III, Academic Press, New York (1977), pp. 251–303.
45. R. A. Floyd, J. J. Watson, P. K. Wong, D. H. Altmiller and R. C. Rickard, *Free Rad. Res. Commun.*, **1**, 163 (1986).
46. M. G. Simic, L. Grossman and A. C. Upton (Eds), *Mechanisms of DNA Damage and Repair*, Plenum Press, New York (1986).
47. L. J. Marnett, *Carcinogenesis*, **8**, 1365 (1987).
48. O. I. Aruoma, B. Halliwell and M. Dizdarogln, *J. Biol. Chem.*, **264**, 13024 (1989).
49. (a) P. A. Cerutti, in *Photochemical and Photobiological Nucleic Acids*, Vol. 2 (S. Y. Wand, Ed.), Academic Press, New York (1976), pp. 375–401; (b) P. A. Cerutti, *Science*, **227**, 375 (1985).
50. J. Cadet and R. Teoule, *Photochem. Photobiol.*, **28**, 661 (1978).
51. H. Kasai and S. Nishimura, *Nucleic Acids Res.*, **12**, 2137 (1984).
52. S. Tofigh and K. Fankel, *Free Rad. Biol. Med.*, **7**, 131 (1989).
53. O. Hayaishi, *Oxygenases*, Academic Press, New York (1962).
54. O. Hayaishi, *Molecular Mechanisms of Oxygen Activation*, Academic Press, New York (1974).
55. B. Samuelsson and R. Paoletti (Eds), *Advances in Prostaglandin and Thromboxane Research*, Raven Press, New York (1976).
56. J. Rokach and J. Adams, *Acc. Chem. Res.*, **18**, 87 (1985).
57. S. Yamamoto, in C. Pace-Asciak and E. Grantistrom (Eds), *Prostaglandins and Related Substances*, Elsevier, Amsterdam (1983), pp. 171–202.
58. E. J. Corey, *Experientia*, **38**, 1259 (1982).
59. B. Samuelsson, *Pure Appl. Chem.*, **53**, 1203 (1981).
60. N. A. Porter, in C. Vigo-Pelfrey (Ed.), *Membrane Lipid Oxidation*, Vol. I, CRC Press, Boca Raton, FL (1990), pp. 33–62.

61. G. W. Burton, D. O. Foster, B. Perly, T. F. Slater, I. C. P. Smith and K. U. Ingold, *Philos. Trans. R. Soc. London, Ser. B*, **311**, 565 (1985).
62. B. Halliwell, *Free Rad. Res. Commun.*, **9**, 1 (1990).
63. E. Niki, in K. J. A. Davies (Ed.), *Oxidative Damage and Repair*, Pergamon Press, Oxford (1992).
64. L. Flohe, in W. A. Pryor (Ed.), *Free Radicals in Biology*, Vol. V., Academic Press, New York (1982), pp. 223–254.
65. B. Chance, D. S. Greenstein and F. J. W. Roughton, *Arch. Biochem. Biophys.*, **37**, 301 (1952).
66. N. Krinsky, *Free Rad. Biol. Med.*, **7**, 617 (1989).
67. I. Fridovich, *Adv. Enzymol.*, **58**, 61 (1986).
68. R. E. Beyer and L. Ernster, in G. Lenaz, A. Bernabei, A. Rabbi and M. Battino (Eds), *Highlights in Ubiquinone Research*, Taylor and Francis, London (1990), pp. 191–213.
69. M. Takahashi, J. Tsuchiya and E. Niki, *J. Am. Chem. Soc.*, **111**, 6350 (1989).

Author Index

Note—Numbers in brackets are references quoted in the text on the page immediately following the reference number(s). References not quoted in the text are followed immediately by the page number on which the reference appears in full.

Index compiled by G. Jones

Subject Index

Index compiled by G. Jones